CONCEPTUAL INTEGRATED SCIENCE

PAUL G. HEWITT
City College of San Francisco

SUZANNE LYONS
California State University, Sacramento

JOHN SUCHOCKI
Saint Michael's College

JENNIFER YEH
University of California, San Francisco

San Francisco Boston New York
Cape Town Hong Kong London Madrid Mexico City
Montreal Munich Paris Singapore Sydney Tokyo Toronto

Editor-in-Chief: Adam R.S. Black, Ph.D.
Senior Acquisitions Editor: Lothlórien Homet
Development Editor: Catherine Murphy
Editorial Assistant: Ashley Taylor Anderson
Marketing Director: Christy Lawrence
Managing Editor: Corinne Benson
Production Supervisor: Lori Newman
Manufacturing Buyer: Pam Augspurger
Design Managers: Mark Ong, Marilyn Perry
Cover Designer: Yvo Riezebos Design
Cover Photo Credit: Photo Researchers, Inc.
Text Designer: tani hasegawa
Logo Designer: Mark Ong
Director, Image Resource Center: Melinda Patelli
Manager, Rights and Permissions: Zina Arabia
Photo Researcher: Laura Murray
Project Manager: Crystal Clifton, Progressive Publishing Alternatives
Composition: Progressive Information Technologies
Illustrations: Paul G. Hewitt and Dartmouth Publishing, Inc.
Cover Printer: Phoenix Color Corporation
Text Printer: VonHoffman Press

Library of Congress Cataloging-in-Publication Data

Conceptual integrated science/Paul G. Hewitt ... [et al.].—1st ed.
p.cm.
Includes index.
ISBN 0-8053-9038-3
1. Interdisciplinary approach to knowledge. 2. Science—Philosophy. 3. Science—History. I. Hewitt, Paul G.
Q175.32.K45C66 2006
500—dc22 2006015582

0-8053-9038-3 [Student]
0-8053-9041-3 [Instructor]
0-13-243-285-4 [High School]

4 5 6 7 8 9 10 - VHP - 10 09
www.aw-bc.com

We dedicate Conceptual Integrated Science *to the many scientists and thinkers whose work we explore in this book. To Galileo Galilei, Isaac Newton, Albert Einstein, Marie Curie, Dimitri Mendeleev, Charles Darwin, Alfred Wegener, Edwin Hubble, and all the other scientists whose brilliance and perseverance have elevated the human condition . . . we thank you!*

Contents in Brief

Special Features

MATH CONNECTION

SCIENCE AND SOCIETY

HISTORY OF SCIENCE

TECHNOLOGY

Contents in Detail

The *Conceptual Integrated Science* Photo Album

For scientists and nonscientists alike, discovering the scientific principles underlying everyday things makes science fun. In this book, the authors proudly share some pictures of family and friends enjoying the science of everyday life.

The Part Openers feature our children. Part 1 (Physics) opens on page 15 with Emily Abrams, granddaughter of author Paul Hewitt—and daughter of Leslie and Bob Abrams. Part 2 (Chemistry) on page 215 is Lillian Hewitt's uncle Robert Chew's grandson, Darren Yee. Jennifer Yeh's children Io and Pico introduce Part 3 (Biology) on page 317. Paul's grandchildren Megan and Emily Abrams launch Part 4 (Earth Science). Suzanne Lyons' kids, Tristan and Simone, ponder the cosmos on page 625, introducing the astronomy portion of this book—Part 5. Let the spirit of wonder these children represent forever be nourished!

Author Paul, with wife Lillian on page 48, demonstrates Newton's Third Law—that one cannot touch without being touched. Fire-walking John is shown on page 111 in his capacity as scientist, not shaman.

Leslie Hewitt, earth scientist and important contributor to this book, is shown on page 188 in a colorized rendition of a photo taken of her at age 16. This photo has been shown in all of her dad's books since then. Leslie's brother Paul illustrates the cooling of expanding air on page 112. Their mom and Paul's former wife Millie, recently deceased, is shown on page 112. Author Paul's brother and his wife, David and Barbara Hewitt, are seen pumping water on page 188.

Marjorie Hewitt Suchocki (pronounced Su-hock-ee, with a silent c), theologian and author of several books, John's mom (Paul's sister) illuminates the concept of reflection on page 150. John's wife, Tracy Suchocki, assesses the physical changes in son Evan on page 228. Sam, the Suchocki family dog, cools himself by evaporation on page 224. John's daughter, Maitreya, contemplates her favorite dessert on page 294.

Suzanne's husband, Pete Lang, employs a bit of thermodynamics on page 103. Their daughter Simone muses on page 153, and son Tristan illustrates finger force on page 28. Tristan and Simone demonstrate closeness on page 172.

Jennifer's husband, Nils Gilman, shoots a basket in Golden Gate Park on page 462.

Friends of the authors include Burl Grey who, as discussed in Chapter 2, instilled in Paul a love of science so many years ago. Tenny Lim puts energy into her bow on page 66. Will Maynez shows momentum changes with his air track on page 63. On page C-2 is a caricature of Paul's cartoonist mentor, Ernie Brown, who designed the title logo on the covers of all of Paul's books.

The inclusion of these people who are so dear to the authors makes this book all the more our labor of love.

Paul G. Hewitt

Paul pioneered the conceptual approach to teaching at City College of San Francisco, with guest spots at the University of California at both the Berkeley and Santa Cruz campuses, and the University of Hawaii at both the Manoa and Hilo campuses. He also taught his physics course for 20 years at the Exploratorium in San Francisco, which honored him with its Lifetime Achievement Award in 2000. His books are translated in 12 languages and used worldwide.

Suzanne Lyons

Suzanne received her B.A. in physics from the University of California, Berkeley and her master's degree in education with a focus on science pedagogy and instructional methods at Stanford University. She has been the editor of *Conceptual Physics* and other books in the *Conceptual* series for 16 years and has authored 7 books on physics, hands-on science activities, and other topics in science and education. She has taught physics, physical science, and general science to students of diverse age and ability levels, from elementary school through college. She presently lectures at California State University, Sacramento.

John Suchocki

John is author of *Conceptual Chemistry* as well as coauthor of *Conceptual Physical Science,* and *Conceptual Physical Science Explorations.* John obtained his Ph.D. in organic chemistry from Virginia Commonwealth University. He taught chemistry at the University of Hawaii at Manoa and then at Leeward Community College. Besides his work authoring textbooks, John is currently an adjunct faculty member at Saint Michael's College in Colchester, Vermont. He also produces science education multimedia through his company, Conceptual Productions, **www.cpro.cc.**

Jennifer Yeh

Jennifer earned a Ph.D. in integrative biology from the University of Texas, Austin, for her work on frog skeleton evolution. She obtained her B.A. in physics and astronomy from Harvard University. Following her graduate work, Jennifer was a postdoctoral fellow at the University of California, San Francisco, where she studied the genetics of breast cancer. Jennifer teaches courses in physics, cell biology, human embryology, vertebrate anatomy, and ecology and evolution. She is the author of numerous scientific papers and the book *Endangered Species: Must They Disappear?* (Thomson/Gale, © 2002, 2004).

To the Student

Welcome to *Conceptual Integrated Science*. The science you'll learn here is INTEGRATED. That means we'll explore the individual science disciplines of physics, chemistry, biology, earth science, and astronomy PLUS the areas where these disciplines overlap. Most of the scientific questions you're curious about, or need to know about, involve not just one discipline, but several of them in an overlapping way. How did the universe originate? That's astronomy + physics. How are our bodies altered by the foods we eat, the medicines we take, and the way we exercise? That's chemistry + biology. What's the greenhouse effect? Will it trigger irreversible global warming, threatening life on our planet? Physics, chemistry, biology, and earth science are all needed to understand the answers.

We're convinced that the CONCEPTUAL orientation of this book is the way in which students best learn science. That means that we emphasize concepts *before* computation. Although much of science is mathematical, a firm qualitative grasp of concepts is important too. Too much emphasis on mathematical problem solving early in your science studies can actually distract you from the concepts and prevent you from fully comprehending them. If you continue in science, you may follow up with classes requiring advanced mathematical methods. Whether you do or you don't, we think you'll be glad you learned the concepts first with just enough math to make them clearer.

This course provides plenty of resources beyond the text as well. For example, the interactive figures, interactive tutorials, and demonstration videos on **www.ConceptualSciencePlace.com** will help you visualize science concepts, particularly processes that vary over time such as the velocity of an object in free fall, the phases of the Moon, or the formation of chemical bonds. The activities in the *Laboratory Manual* will build your gut-level feeling for concepts and your analytical skills. Ponder the puzzlers in the *Conceptual Integrated Science Practice Book* and work through the simple review questions—all of this will increase your confidence and mastery of science.

As with all things, what you get out of this class depends on what you put into it. So study hard, ask all the questions you need to, and most of all enjoy your scientific tour of the amazing natural world!

To the Instructor

Conceptual Integrated Science provides an introduction to physics, chemistry, biology, earth science, and astronomy—the full gamut of the natural sciences. How can introductory-level students master such a wide range of material? In this book, we use the **conceptual approach**, which makes integrated science accessible. The conceptual approach:

- relates science to everyday life.
- is personal and direct.
- deemphasizes jargon and vocabulary.
- emphasizes central ideas rather than the details, to avoid information overload.
- puts concepts ahead of computation. Equations are used to clarify concepts rather than as a chance to practice mathematical problem solving.

The conceptual approach was defined over 30 years ago by Paul Hewitt in his groundbreaking text *Conceptual Physics*. Educators around the world today—and their students—testify to its effectiveness.

Integrating the Sciences

While the conceptual approach does much to present a wealth of material in an enjoyable and accessible way, we have employed other pedagogical tactics to specifically target subject integration. First and foremost, the text is written around **unifying concepts**—the relatively small number of essential concepts that underlie various branches of science and tie them together. For example, The Second Law of Thermodynamics, a unifying concept, pops up repeatedly across the disciplines: it underlies the direction of heat flow (physics) and the loss of energy between trophic levels (biology), just to name a few instances.

By emphasizing a relatively small number of unifying concepts, we organize the wealth of material that is integrated science. The unifying concepts provide a cognitive framework to which facts and information can be added. But even more importantly than this, when text material is organized around unifying concepts, the intrinsic order in the sciences becomes apparent. The simplicity and order that science finds in the universe is revealed.

Thus, just as a pile of bricks does not make a house, individual bits of science information do not make a coherent science curriculum. A house rests on a foundation. It has an underlying structure of beams to keep it strong, and walls to give it shape. For *Conceptual Integrated Science*, the conceptual approach is the foundation, unifying concepts provide the underlying structure, and science content from atoms to zoology defines the final shape.

Key Features of the Text

Throughout this book, you will find specific pedagogical elements supporting the conceptual, integrated approach. There are numerous **Integrated Science** sections that profile topics at the crossroads of scientific disciplines. For example, "What Forces Drive the Plates?" discusses the physics principles underlying movement of tectonic plates, and "How Radioactivity Causes Genetic Mutations" details the process by which radiation disrupts biomolecules and cellular function. Importantly, the *Integrated Science* sections run continuously with the text. They are not set aside in boxes, as this might suggest to students that these features are optional reading. The *Integrated Science* features are essential to this course because they focus on substantive topics; they are not merely interesting asides. All *Integrated Science* sections therefore include questions in the text as well as in the end-of-chapter review material, providing a means for you to check student comprehension.

INTEGRATED SCIENCE **EARTH SCIENCE**

Acid Rain

Rainwater is naturally acidic. One source of this acidity is carbon dioxide, the same gas that gives fizz to soda drinks. There are about 810 billion tons of CO_2 in the atmosphere, most of it (about 675 billion tons) from natural sources such as volcanoes and decaying organic matter but a growing amount (about 135 billion tons) from human activities such as the burning of fossil fuels.

Water in the atmosphere reacts with carbon dioxide to form *carbonic acid*:

$$CO_2(g) + H_2O(\ell) \rightarrow H_2CO_3(aq)$$

Carbonic acid, as its name implies, behaves as an acid and lowers the pH of water. The CO_2 in the atmosphere brings the pH of rainwater to about 5.6—noticeably below the neutral pH value of 7. Because of local fluctuations, the normal pH of rainwater varies between 5 and 7. This natural acidity of rainwater may accelerate the erosion of land and, under the right circumstances, can lead to the formation of underground caves, such as those found at Carlsbad Caverns, New Mexico, or Mammoth Cave, Kentucky.

By convention, *acid rain* is a term used for rain having a pH lower than 5. Acid rain is created when airborne pollutants such as sulfur dioxide are absorbed by atmospheric moisture. Sulfur dioxide is readily converted to sulfur trioxide, which reacts with water to form *sulfuric acid*:

$$2\,SO_2(g) + O_2(g) \rightarrow 2\,SO_3(g)$$
$$SO_3(g) + H_2O(\ell) \rightarrow H_2SO_4(aq)$$

When we burn fossil fuels, the reactants that produce sulfuric acid are emitted into the atmosphere. Each year, about 20 million tons of SO_2 are released

The **unifying concept icon**, shown here to the left, highlights places in the text where these concepts come up. Unifying concept icons are dual-purpose. They remind students of the essential, cross-curricular ideas. They also serve as a cross-reference. The icons tell where each unifying concept is introduced so that students can flip back to that section when they need to review. Further, unifying concept icons are all listed in a chart on the cover flap—a chart that can serve as a priority list of essential concepts.

TECHNOLOGY

Direct Measurement of Continental Drift

Today, continental drift is not just deduced from evidence—it can be directly measured. The *Very Long Baseline Interferometry System* (VLBI), for example, was the first system to directly measure the relative motion of Earth's tectonic plates and continents. The VLBI used radio telescopes to detect and record radio signals emitted from quasars. Quasars are so far from Earth (billions of light-years away) that they are virtually pointlike. Their radio emissions, therefore, can be used like a surveyor's beam from a stationary source. The same signal from a quasar arrives at slightly different times at different measuring sites. So, when the VLBI tracked changes in the arrival times of radio signals over a period of years, it showed the rate of movement of the sites relative to each other.

The Global Positioning System (GPS) is currently used to measure the relative motion of different points on Earth. Because GPS results agree with the VLBI results, they provide a cross-check. The GPS system consists of twenty or so satellites that orbit the Earth at an altitude of 20,000 km. These satellites transmit signals back to Earth continuously. Scientists at ground stations around the world use the signals to pinpoint their position in terms of latitude, longitude, and altitude. Scientists repeatedly measure locations of ground stations, monitor change in their relative positions, and thus track co movem

Th of conti one anc predicti show th northw Japan a per year away fr 1.7 cent

SCIENCE AND SOCIETY

Nuclear Technology

When household electricity made its way across the country more than a century ago, it represented a new technology with great potential. But it was not without its hazards. While electric grids could provide citi with a quantity and quality of energy previously unheard of, they co also kill people in ways previously unheard of. Many people oppose the adoption of household electricity because of the inherent danger But safeguards were engineered, and society determined that the ben fits of electricity outweighed the risks. Similar debates continue toda not over electricity, but over nuclear energy and radioactivity.

Without nuclear technology, we would not have X rays, radia- tio nuclear power as

HISTORY OF SCIENCE

Galileo Galilei (1564–1642)

Galileo was born in Pisa, Italy, in the same year in which Shakespeare was born and Michelangelo died. He studied medicine at the University of Pisa and then changed to mathematics. He developed an early interest in motion and was soon in opposition with others around him, who held to Aristotelian ideas about falling bodies. He left Pisa to teach at the University of Padua, where he became an advocate of the new theory of the solar system advanced by the Polish astronomer Copernicus. Galileo was one of the first to build a telescope, and he was the first to direct it to the nighttime sky and discover mountains on our Moon and on the moons of Jupiter. Because he published his findings in Italian instead of Latin, which was pected of so reputable a scholar, and because of the recent inve of the printing press, his ideas reached many people. He soon r afoul of the Church and was warned not to teach, and not to h to, Copernican views. He restrained himself publicly for nearly 15 years. Then he defiantly published his observations and conclusions, which opposed Church doctrine. The outcome was a trial in which he was found guilty, and he was forced to renounce his discoveries. By then an old man and broken in health and spirit, he was sentenced to perpetual house arrest.

MATH CONNECTION

Applying the Work–Energy Theorem

Determine the work done on an object even though you don't know the forces or distances involved.

Problems

1. Calculate the change in kinetic energy when a 50-kg shopping cart moving at 2 m/s is pushed to a speed of 6 m/s.
2. How much work is required to make this change in kinetic energy?

Solutions

1. $\Delta KE = \frac{1}{2} m (v_f^2 - v_o^2)$

$= \frac{1}{2} (50\text{ kg})[(6\text{ m/s})^2 - (2\text{ m/s})^2]$

$= 800\text{ J}$

2. $W = \Delta KE = 800$ J, because the change in KE equals the work done on the shopping cart.

Throughout *Conceptual Integrated Science*, you'll find boxed elements that focus on supplemental topics. **Technology** and **Science and Society** are high-interest features that illustrate the relevancy of science to everyday life. Many of these can be used to prompt class discussion. **History of Science** boxes emphasize that science is personal—a human endeavor. Also, in studying the history of science through these features (as well as in the text) students learn about the process of science, the manner in which scientific discoveries are made. The **Math Connection** boxes are especially helpful to nonscience students; they teach basic problem-solving skills. And for all students, the *Math Connection* boxes are an opportunity to experience the complementary relationship between mathematics and science.

Each chapter begins with an attention-grabbing paragraph. Introductory paragraphs contain questions that students will be able to answer by reading the chapter material. For example in Chapter 8: "Waves," we ask: "How do sound waves differ from light waves? Does the speed of sound differ in various materials? Can it travel in a vacuum? Can one sound wave cancel another—so that two loud noises combine to make zero noise? Are there technological applications for this idea?" Such questions motivate students to

CHECK YOURSELF

A ball rolling along a level surface slowly comes to a stop. How would Aristotle explain this behavior? How would Galileo explain it? How would you explain it? (Think about the Check-Yourself questions throughout this book **before** reading the answers. When you first formulate your answers, you'll find yourself learning more—much more.)

CHECK YOUR ANSWER

Aristotle would probably say that the ball stops because it seeks its natural state of rest. Galileo would probably say that friction overcomes the ball's natural tendency to keep rolling—that friction overcomes the ball's inertia and brings it to a stop. Only you can answer the last question.

Interactive tutorial icon

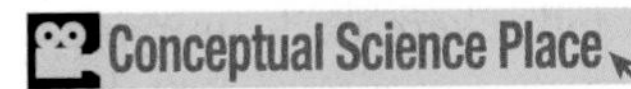

Video demonstration icon

Interactive **Figure**

Interactive figure icon

read the chapter to find the answers. Besides serving as a motivational tool, the interest-piquing initial questions also serve as "advance organizers," to use the parlance of science education. In other words, the introductory paragraphs give students a brief overview of chapter content. And, speaking of advance organizers, the chapter outlines serve this function as well.

Other helpful pedagogical elements in this book include the **Check Yourself** questions. These are short sets of questions sprinkled throughout each chapter to allow students to monitor their own understanding. Answers are provided below the questions to give students immediate feedback. The **Insights** are short margin features providing background information, interesting facts, study tips, and more for added depth and breadth.

Conceptual Integrated Science is the only text in the field with a vast and highly integrated library of **purpose-built media** designed to illustrate the concepts with which students struggle the most. Icons alert students to relevant media they can turn to—**interactive figures, interactive tutorials, and demonstration videos**—for alternative topic presentations. Media features are available to students via the associated website **www.ConceptualSciencePlace.com**. They are also available to instructors for use in presentations via the Media Manager CD-ROM set.

The **art program** in *Conceptual Integrated Science* is distinctive. Of course, there are plenty of engaging Hewitt cartoons. The cartoons keep the mood light and alleviate boredom—a flaw that can render an otherwise worthy text unreadable to introductory students. The cartoons, however, are not just for fun. They reinforce central or challenging concepts in a student-friendly manner. Besides the cartoons, there is generous use of photographs and instructional figures. Diagrams are simple and uncluttered and photographs have been chosen to underscore concepts in an often striking and intriguing way.

At the end of each chapter you will find an extensive **Chapter Review**. Great attention has been paid to the preparation of this material. Students know that the Chapter Review material is what they will be held most accountable for, so it is imperative that the Chapter Review be effective, and for maximum efficiency, we provide an average of 80 questions to choose from and a wide array of question types.

The *Conceptual Integrated Science* Chapter Review consists of a **Summary of Terms** with definitions, **Review Questions, Exercises, Problems, Multiple-Choice Questions**, a list of **online resources**, and **Active Explorations**, which are hands-on activities that require only simple, readily available materials. The important ideas from each chapter are framed in the relatively easy-to-answer Review Questions, which are grouped by chapter sections. These are, as the name implies, a review of chapter material. Their purpose is to provide a structured way to review the chapter. They are not meant to challenge the student's intellect because the answers can be looked up in the chapter. The Exercises, on the other hand, play a different role. Some are designed to prompt the application of physical science to everyday situations, while others are more sophisticated and call for considerable critical thinking. The Problems are mainly simple computations that aid in learning concepts. In order to discourage students from focusing on number crunching at the expense of conceptual understanding, there are fewer Problems than Exercises. Although building confidence in math is a worthy goal, it is not the focus of this book. Multiple-Choice Questions, the last type of assessment questions, require deeper reflection from students and can also function as sample quizzes. Students can find the answers to the odd-numbered Integrated Science Concepts, Exercises and Problems in the back part of the **Practice Book**. Complete answers to all end-of-chapter questions are included in the **Instructor Manual**.

Units of measurement are not emphasized in this text. When used, they are almost exclusively expressed in SI units. (The few exceptions include units in calories, grams per centimeter cubed, and light-years). *Appendix A* details systems of measurement. *Appendix B* provides extended coverage of linear and rotational motion. *Appendix C* provides practice with vector components, and *Appendix D* develops the concepts of exponential growth and decay. *Appendix E* details the physics of fluids, and *Appendix F* provides added coverage of chemical equilibrium.

Organization of the Text

This text is organized into five main parts corresponding to the major subject disciplines—physics, chemistry, biology, earth science, and astronomy. These parts should not be seen as independent silos of topic-specific information, but rather as a sequencing of chapters based on increasing complexity and scale from physics to astronomy, integrated with each other through a focus on unifying concepts.

We begin with physics, the most fundamental science. The physics chapters are adapted from *Conceptual Physics*, but crafted to meet the needs of an integrated science course. The treatment of physics begins with statics so students can start with forces and equilibrium rather than velocity and acceleration. After success with simple forces, students move into the more complex study of kinematics. Newton's laws pick up the pace in Chapter 3, followed by a study of energy and momentum that emphasizes conservation principles in Chapter 4. In Chapter 5, the universal nature of the gravitational force is the backdrop against which interesting Earth-bound examples of gravity are explored. The physics part of *Conceptual Integrated Science* proceeds through the other components of the physics curriculum that are presented in a traditional sequence, from heat, to electricity and magnetism, to wave phenomena (especially sound and light), the atom, and finally nuclear physics.

Part Two, "Chemistry," builds upon the foundation of ideas developed in the physics unit. The overall goal is to enable students to view the macroscopic world in terms of its submicroscopic constituents—atoms and molecules. Our chemistry chapters, in the vein of all books in the *Conceptual* series, emphasize concepts over computation and science in everyday life. Relating chemistry to students' familiar world—the fluorine in their toothpaste, the Teflon on their frying pans, and the flavors produced by various organic molecules—keeps chemistry fun and relevant. *Conceptual Integrated Science* chemistry also connects to social and environmental issues. Students examine such subjects as the causes and effects of acid rain, fuel cells, and the role of polymers in winning World War II; each of these topics helps students see chemistry in its broad context. Chapter 11, the initial chemistry chapter, provides an overview of concepts related to atoms and molecules. Chapter 12 uses the shell model of the atom to explain the chemical bond. We move on to a discussion of chemical reactions, exploring acid–base and redox reactions in considerable depth. Our introduction to chemistry culminates in Chapter 14, which is devoted to organic chemistry.

Part Three builds upon physics and chemistry, taking students to the much more complex realm of biology, the study of life and living things. We begin by asking—what constitutes life? We then focus on three key attributes of living things—cells, genes, and evolution. Chapter 15 addresses the cell, the basic structural and functional unit of living organisms. Chapter 16 covers classical and molecular genetics, the study of heredity, or how traits are passed from parents to offspring. Chapter 17 focuses on evolution, which explains how the diversity of Earth's life is the result of heritable changes in living organisms. Chapter 18 provides an overview of this biological diversity, examining living organisms from bacteria to plants, fungi, and animals. Chapter 19 and Chapter 20 focus on human biology and the functioning of the human body, examining how the body senses the world, moves, thinks, and maintains itself.

Finally, Chapter 21 looks at ecology, the study of how living organisms interact with the living and nonliving components of their environments.

Part Four, "Earth Science," builds on physics, chemistry, and biology in presenting essential concepts from the geosciences. Care is taken to avoid advanced terminology and to keep the discussion simple enough to captivate students at the introductory level. We provide tools to understand topics in the news, from the weather to earthquakes. We address students' innate curiosity about natural events from the seasons to volcanic eruptions. To set the stage for earth science, we begin Part Four with a study of plate tectonics. The overarching concept—that Earth is a dynamic physical system subject to ongoing change as heat is transferred from interior to surface—helps students put the subsequent chapter on rocks and minerals in context. Chapter 24 investigates the surface environment—landforms and bodies of water—to acquaint students with the essential features and physical processes that shape the surface environment. Chapter 25 discusses the individual elements of weather first, then profiles the ways in which weather elements work together. The part ends with Chapter 26: "Earth's History," a highly integrated chapter that ties mass extinctions and other biological phenomena to geological forces.

In Part Five, students expand their focus to astronomy and make use of important ideas learned in physics, chemistry, the earth sciences, and even biology. Chapter 27 details the origin of the solar system according to the nebular theory. Factual information on the planets and other heavenly bodies is tied to this central concept. The final chapter of this book, Chapter 28, provides an in-depth look at the stars with an introduction to the concepts of cosmology, including the Big Bang, Hubble's law, and the search for extraterrestrial life.

A list of recommended websites and a detailed glossary and index complete the book.

Conceptual Integrated Science is a wide-ranging text, offering students tools to understand the natural world and the many scientific issues they will grapple with as citizens of the twenty-first century. May this be one of the most interesting, amazing, and worthwhile courses your students will ever take!

Ancillaries

Conceptual Integrated Science contains more than enough material for a one-year integrated science course. There is also a full suite of ancillary materials in a variety of media developed for this course. Because there are so many resources and because most of the chapters in the text are self-contained, you have great flexibility in tailoring the *Conceptual Integrated Science* program to suit your taste and the needs of your students.

Resources for Instructors

The *Instructor Manual for Conceptual Integrated Science* provides specific suggestions about how to individualize your integrated science course. We have made recommendations about how to structure topics for a semester-length course, for example. Also in the *Instructor Manual* is an abundance of lecture and demonstration ideas and teaching tips. For many chapters, we provide step-by-step lesson plans. The *Instructor Manual* features complete answers to all the Review Questions, Exercises, and Problems in the text. Also look to the *Instructor Manual* for common misconceptions that we have noticed hold students back from acquiring new concepts.

The *Conceptual Integrated Science Media Manager CD-ROM* is a treasure of instructor resources. This cross-platform CD-ROM contains materials for you to use in class as you present concepts. You'll find the largest library available of purpose-built, in-class presentation materials, including interactive tools (applets) and simulations, interactive figures, interactive photos, high-resolution figures from the book and other well-known sources, sample lectures in PowerPoint® slides, PRS-enabled Clicker Questions, and *Next-Time Questions*. The *Next-Time Questions* are a classic ancillary product developed by Paul Hewitt. *Next-Time Questions* are puzzlers designed to provoke lively discussion. It's especially effective to post them at the beginning of class to focus students' attention, or at the end of class to give your students an interesting problem or puzzle to mull over until "next time." As an additional visual aid for your in-class presentations, we offer transparency acetates of more than 100 key figures in the text.

When you need to assess students' mastery of concepts, turn to the *Test Bank* book for more than 2000 multiple-choice questions, plus hundreds of short-answer and essay questions. The questions are categorized by level of difficulty. The test bank material is also available on a dual-platform CD-ROM. This *Test-Gen-EQ Computerized Testing Software* contains everything in the Test Bank and allows you to edit and change the order of the questions, add new questions, and print different versions of a test. For additional assessment and course management resources, Blackboard and CourseCompass™ online course management suites are also available.

Resources for Students

Student resources include the *Laboratory Manual for Conceptual Integrated Science*. The Laboratory Manual features in-depth experiments that develop laboratory skills as well as quick activities requiring only readily available materials. Each laboratory exercise is labeled as *Exploration* or *Application*. Exploration activities can be used to introduce concepts before text reading is assigned. Application activities, on the other hand, provide follow-up practice for students who have studied the relevant concepts in the book. Logistical information on labs plus answers to laboratory questions are available to you, the instructor, in the *Instructor Manual*.

The *Practice Book for Conceptual Integrated Science* is another resource for students. It features a creative blend of cartoons and humor, computational exercises, misconception-busting questions, analogies, and intriguing puzzlers, along with straightforward practice questions and problems, all with a user-friendly tone. The *Conceptual Integrated Science Practice Book* also contains answers to the odd-numbered Exercises and Problems from the main text.

Last, but by no means least, students have *The Conceptual Science Place* (**www.ConceptualSciencePlace.com**), which provides fun and rewarding additional intruction on challenging concepts. This is the website to which students are directed by the media icons in the text, where they will find interactive figures, interactive tutorials, and video demonstrations. In addition, this comprehensive website includes chapter overviews, multiple-choice quizzes, relevant weblinks, flashcards, a glossary, and more. A complementary, one-year access code for the site is provided with every new book.

Acknowledgments

The authors wish to express their sincere appreciation to the many talented and generous people who helped make this book happen. To the professors and teachers who reviewed manuscript, giving so generously of their time and expertise, we express our heartfelt appreciation.

We thank all the contributors to *Conceptual Physics*, *Conceptual Chemistry*, *Conceptual Physical Science* and *Conceptual Physical Science—Explorations*. For helping to shape the physics content over the years for previous editions and supplements, we thank: Dean Baird, Tsing Bardin, Howie Brand, Alexi Cogan, Paul Doherty, Marshall Ellenstein, Ken Ford, Jim Hicks, David Housden, John Hubisz, Dan Johnson, Lillian Lee Hewitt, Tenny Lim, Chelcie Liu, Cedric Linder, Will Maynez, Fred Myers, Ron Perkins, Diane Reindeau, Pablo Robinson, Kenn Sherey, David Williamson, Larry Weinstein, Phil Wolf, and Dean Zollman.

For the chemistry material, we thank: Hilair Chism, Mark Johnson, Kevin Johnson, Frank Lambert, Robley Light, David Lygre, Irene Matusz, Irene Nunes, Frazier Nyasulu, Michael Reese, Mike Stekoll, Joseph Tausta, Margaret Tolbert, and Bob Widing.

Our sincere thanks go to Leslie Hewitt Abrams and Bob Abrams for important input and many contributions to the earth science material in this book. Leslie and Bob have given of their time to ensure the accuracy of the earth science chapters and to share insights gained from their work authoring and teaching earth science.

For advice, expertise, and wide-ranging contributions to the biology section, the authors would like to thank Pamela Yeh, Sarah Ying, Nina Shapley, Nils Gilman, Todd Schlenke, Brian West, Robert Dudley, Vivianne Ding, Mike Fried, W. Bryan Jennings, and Rachel Zierzow.

Our colleagues at Addison Wesley have been our partners in this project and given us much support and guidance. Adam Black, publishing guru, created the concept for this book. Development editors Susan Teahan and Cathy Murphy pulled up their sleeves and worked with us in an integral way to hammer out the manuscript, line by line. Lothlórien Homet, Senior Editor, managed the development of the book, providing sound judgments and creative contributions. Ashley Taylor Anderson, the central administrative hub, has been a crucial player. Her deft and intelligent handling of all things related to work flow and communications is greatly appreciated.

The *Conceptual Integrated Science* authors are fortunate to have helpful, loving, and tolerant (!) spouses who have supported us through the long hours. Thanks go to Lillian Lee Hewitt, Pete Lang (Suzanne's husband), Tracy Suchocki, and Nils Gilman (Jennifer's husband). And to our young children we send our love and gratitude: Tristan and Simone Lyons Lang; Ian, Evan, and Maitreya Suchocki; and Io and Pico Yeh Gilman.

Acknowledgements

Reviewers

Leila Amiri, University of South Florida
Leanne Avery, Indiana University of Pennsylvania
Bambi Bailey, Midwestern State University
Dirk Baron, California State University, Bakersfield
Daniel Berger, Bluffton University
Reginald Blake, City Tech University of New York
Derrick Boucher, King's College
Martin Brock, Eastern Kentucky University
Linda Brown, Gainsville College
Mary Brown, Lansing Community College
Steven Burns, St. Thomas Aquinas College
Erik Burtis, Northern Valley Community College, Woodbridge
Gerry Clarkson, Howard Payne University
Anne Coleman, Cabrini College
Gary Courts, University of Dayton
Red Chasteen, Sam Houston State University
Randy Criss, St. Leo University
Jason Dahl, Bemidji State University
Terry Derting, Murray State University
David DiMattio, St. Bonaventure University
Gary Neil Douglas, Berea College
S. Keith Dunn, Centre College
George Econ, Jackson Community College
Michael S. Epstein, Mount St. Mary's University
Charles Figura, Wartburg College
Lori K. Garrett, Danville Area Community College
David Goldsmith, Westminster College
Brian Goodman, Lakeland College
Nydia R. Hannah, Georgia State University
Carole Hillman, Elmhurst University
James Houpis, California State University, Chico
Thomas Hunt, Jackson Community College
David T. King, Jr., Auburn University
Jeremiah K. Jarrett, Central Connecticut State University
Peter Jeffers, State University of New York, Cortland
Charles Johnson, South Georgia College
Richard Jones, Texas Women's University
Carl Klook, California State University, Bakersfield
Kenneth Laser, Edison Community College
Jeffrey Laub, Rogers State University
Holly Lawson, State University of New York, Fredonia
David Lee, Biola University
Steven Losh, State University of New York, Cortland
Ntungwa Maasha, Coastal Georgia Community College
Kingshuk Majumdar, Berea College
Lynette McGregor, Wartburg College
Preston Miles, Centre College
Frank L. Misiti, Bloomburg University
Matthew Nehring, Adams State College
Marlene Morales, Miami Dade College
Douglas Nelson, Coastal Carolina University
Jan Oliver, Troy State University
Treva Pamer, New Jersey City University
Todd Pedlar, Luther College
Denice Robertson, Northern Kentucky University
Judy Rosovsky, Johnson State College
Steven Salaris, All Saints Christian School
Terry Shank, Marshall University
Sedonia Sipes, Southern Illinois University, Carbondale
Ran Sivron, Baker University
Priscilla Skalac, Olivet Nazarene University
Stanley Sobolewski, Indiana University of Pennsylvania
John Snyder, Lansing Community College
Stuart Snyder, Montana State University
Anne Marie Sokol, Buffalo State College

Pamela Stephens, Midwestern State University
Laura Stumpe, Holy Cross College, Notre Dame
Karen Swanson, William Patterson University
Timothy Swindle, University of Arizona
Rachel Teasdale, California State University, Chico
Diana Treahy, Anderson University
Christos Valiotis, Antelope Valley College
Robin van Tine, St. Leo University
Daniel Vaughn, Southern Illinois University
Stephen Webb, Brescia College
Karen Wehn, Buffalo State University
Adam Wenz, Montana State University, Great Falls
William Wickun, Montana State University, Billings
Bonnie S. Wood, University of Maine, Presque Isle
Robert Zdor, Andrews University

CHAPTER 1

About Science

Two Mars Exploration Rovers, each identical to the one illustrated here, landed on the surface of Mars in January 2004. The buggy-like Rovers were equipped with microscopes, abrasion tools, and spectrometers for analyzing the chemical composition of rock samples.

CHAPTER OUTLINE

Modern civilization is built on science. Nearly all forms of technology—from medicine to space travel—are applications of science. But what exactly *is* science? Where did it originate? How should science be used? What would everyday life be like without it?

Science is an organized body of knowledge about nature. It is the product of observations, common sense, rational thinking, and (sometimes) brilliant insights. People usually do science with other people—it is very much a communal human endeavor. It has been built up over thousands of years and gathered from all parts of the world. Science is an enormous gift to us today, the legacy of countless thinkers and experimenters of the past.

Yet science is more than a body of knowledge. It is also a *method*, a way of exploring nature and discovering the order within it. While some people have a natural aptitude for scientific work, doing science is a skill that must be learned. Importantly, science is also a tool for solving physical problems.

The beginnings of science go back before recorded history, when people first discovered repeating patterns in nature. They noted star patterns in the night sky, patterns in the weather, and patterns in animal migration. From these patterns, people learned to make predictions that gave them some control over their surroundings.

1.1 A Brief History of Advances in Science

When a light goes out in your room, you ask, "How did that happen?" You might check to see if the lamp is plugged in, or if the bulb is burned out, or you might look at homes in your neighborhood to see if there has been a power outage. When you think and act like this, you are searching for *cause-and-effect* relationships—trying to find out what events cause what results. This type of thinking is *rational thinking*, applied to the physical world. It is basic to science.

Today, we use rational thinking so much that it's hard to imagine other ways of interpreting our experiences. But it wasn't always this way. In other times and places, people relied heavily on superstition and magic to interpret the world around them. They were unable to analyze the *physical* world in terms of *physical* causes and effects.

Figure 1.1 A view of the Acropolis, or "high city," in ancient Greece. The buildings making up the Acropolis were built as monuments to the achievements of the residents of the area.

The ancient Greeks used logic and rational thought in a systematic way to investigate the world around them and make many scientific discoveries. They learned that Earth is round and determined its circumference. They discovered why things float and learned that the apparent motion of the stars throughout the night is due to the rotation of Earth. The ancient Greeks founded the science of botany—the systematic study and classification of plants—and even proposed an early version of the principle of natural selection. Such scientific breakthroughs, when applied as technology, greatly enhanced the quality of life in ancient Greece. For example, engineers applied principles articulated by Archimedes and others to construct an elaborate public waterworks, which brought fresh water into the towns and carried sewage away in a sanitary manner.

When the Romans conquered ancient Greece, they adopted much of Greek culture, including the scientific mode of inquiry, and spread it throughout the Roman Empire. When the Roman Empire fell in the fifth century A.D., advancements in science came to a halt in Europe. Nomadic tribes destroyed much in their paths as they conquered Europe and brought in the Dark Ages. But during this time, science continued to advance in other parts of the world.

The Chinese and Polynesians were charting the stars and the planets. Arab nations developed mathematics and learned to make glass, paper, metals, and certain chemicals. Finally, during the tenth through twelfth centuries, Islamic people brought the spirit of scientific inquiry back into Europe when they entered Spain. Then the university emerged. When the printing press was invented by Johann Gutenberg in the fifteenth century, science made a great leap forward. People were suddenly able to communicate easily with one another across distance. The printing press did much to advance scientific thought, just as computers and the Internet are doing today.

Up until the sixteenth century, most people thought Earth was the center of the universe. They thought that the Sun circled the stationary Earth. This thinking was challenged when the Polish astronomer Nicolaus Copernicus quietly published a book proposing that the Sun is stationary and Earth revolves around it. These ideas conflicted with the powerful institution of the Church and were banned for 200 years.

Modern science began in the sixteenth century, when the Italian physicist Galileo Galilei revived the Copernican view. Galileo Galilei used experiments, rather than speculation, to study nature's behavior (we'll say more about Galileo in chapters to follow). Galileo was arrested for popularizing the Copernican theory and for his other contributions to scientific thought. But a century later, his ideas and those of Copernicus were accepted by most educated people.

Scientific discoveries are often opposed, especially if they conflict with what people want to believe. In the early 1800s, geologists were condemned because their findings differed with religious accounts of creation. Later in the same century, geology was accepted, but theories of evolution were condemned. Every age has had its intellectual rebels who were persecuted, vilified, condemned, or suppressed but then later regarded as harmless and even essential to the advancement of civilization and the elevation of the human condition. "At every crossway on the road that leads to the future, each progressive spirit is opposed by a thousand men appointed to guard the past."*

1.2 Mathematics and Conceptual Integrated Science

Pure mathematics is different from science. Math studies relationships among numbers. When it is used as a tool of science, the results can be astounding. Measurements and calculations are essential parts of the powerful science we

* Quote from Maurice Maeterlinck's "Our Social Duty."

MATH CONNECTION

Equations as Guides to Thinking

In *Conceptual Integrated Science*, we recognize the value of equations as guides to thinking. What we mean by this is that simple equations tell you immediately how one quantity is related to another. In Chapter 5, when you study gravity, you will learn about the inverse-square relationship—a mathematical form that comes up over and over again in science. In Appendix D, you can study exponential relations in general. But to start off, consider two basic mathematical relationships.

The direct proportion The more you study for this course, the better you'll do. That's a direct proportion. Similarly, the more coffee you drink, the more nervous you'll feel. The longer time you drive at a constant speed, the farther you travel. If you're paid by the hour, the longer you work, the more money you make. All these examples show relationships between two quantities, and in each case, the relationship is a direct proportion. A direct proportion has the mathematical form $a \sim b$. Direct proportions have graphs of the form shown here.

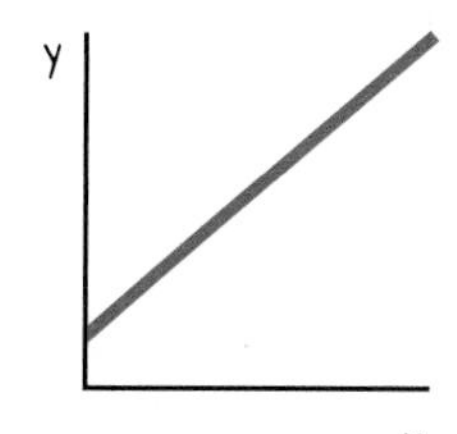

(a) The direct proportion.

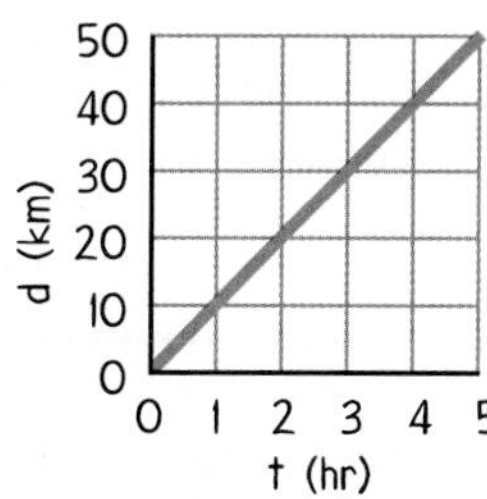

(b) A car travels at a constant speed. The more time it travels, the farther it goes. Distance is directly proportional to time.

The inverse proportion Some quantities are related to one another so that as one increases, the other decreases. The *more* time you spend playing video games, the *less* time you have for homework. The *more* money you spend, the *less* you have in the bank. The *more massive* a grocery cart, the *less it accelerates* when you push it. These quantities are related through the inverse proportion, which has the mathematical form $a \sim \frac{1}{b}$. Inverse proportions have graphs of the form shown here.

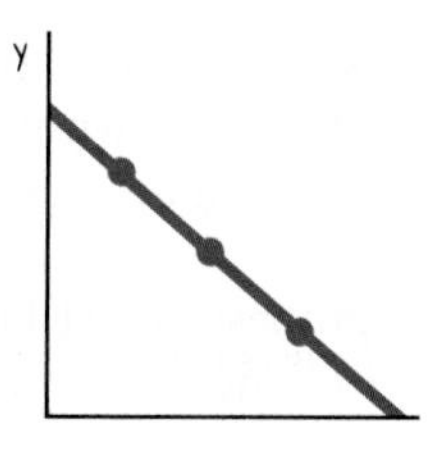

Graph of an inverse proportion.

Note that these mathematical relations have been stated as proportional relations, rather than as *equations*. For a proportional relation to be stated as an equation, the numbers and units on both sides must be the same. We can state a direct or indirect proportion as an exact equation by inserting a *proportionality constant*, k, into the relation. Proportionality constants make the numbers and units on both sides of an equation match up.

For example, consider Hooke's Law. Hooke's Law tells us about springs and other stretchy, elastic objects. Imagine a spring, such as a Slinky. According to Hooke's Law, the more a Slinky is stretched, the harder it is to stretch it further. Stated as a direct proportion, Hooke's Law is written:

$$F \sim x$$

where F is your pulling force and x is the distance the spring is stretched beyond its resting length. But F, a force, has units of newtons (N) and x, a distance, has units such as centimeters (cm). We convert Hooke's Law to an equation by inserting k into the relation. The value of k in this case depends on the shape and material of the spring. For a common Slinky-type metal coil, the proportionality k is about 2.5 N/cm. Now we can state Hooke's Law as an exact equation:

$$F = kx.$$

(a) An unstretched spring. (b) The spring is stretched past its resting length. (c) Stretching the coil farther takes more force. This is Hooke's Law, $F = kx$.

Problems

1. Newton's Law of Cooling tells us that the rate of cooling of an object (R) is approximately equal to the temperature difference between the object and its surroundings (ΔT). This law is expressed like this: $R \sim \Delta T$. What type of mathematical relation is this?
2. A certain spring has a spring constant of 3 N/cm. What size force is needed to stretch this spring 4 cm past its resting length?

Solutions

1. Newton's Law of Cooling is of the form: $a \sim b$, a direct proportion.
2. $F = kx = (3\frac{\text{N}}{\text{cm}}) \times (4\text{ cm}) = 12\text{N}$.

practice today. For example, it would not be possible to send missions to Mars if we were unable to measure the positions of spacecraft or to calculate their trajectories.

You will use some math in this course, especially when you make measurements in lab. In this book, we don't make a big deal about math. Our focus is on understanding concepts in everyday language. We use equations as guides to thinking rather than as recipes for "plug-and-chug" math work. We believe that focusing on math too early, especially on math-based problem solving, is a poor substitute for learning the concepts. That's why the emphasis in this book is on building concepts. Only when concepts are understood does solving problems make sense.

1.3 The Scientific Method—A Classic Tool

The practice of science usually encompasses keen observations, rational analysis, and experimentation. In the sixteenth century, Galileo and the English philosopher Francis Bacon were the first to formalize a particular method for doing science. What they outlined has come to be known as the classic scientific method. This method is essentially as follows:

1. **Observe** Closely observe the physical world around you.
2. **Question** Recognize a question or a problem.
3. **Hypothesize** Make an educated guess—a *hypothesis*—to answer the question.
4. **Predict** Predict consequences that can be observed if the hypothesis is correct. The consequences should be *absent* if the hypothesis is not correct.
5. **Test predictions** Do experiments to see if the consequences you predicted are present.
6. **Draw a conclusion** Formulate the simplest general rule that organizes the hypothesis, predicted effects, and experimental findings.

Although the scientific method is powerful, good science is often done differently, in a less systematic way. In the Integrated Science feature, "An Investigation of Sea Butterflies," you will see a recent application of the classic scientific method. However, many scientific advances involve trial and error, experimenting without guessing, or just plain accidental discovery. More important than a particular method, the success of science has to do with an attitude common to scientists. This attitude is one of inquiry, experimentation, and humility before the facts.

1.4 The Scientific Hypothesis

A scientific **hypothesis** is an educated guess that tentatively answers a question or solves a problem in regard to the physical world. Typically, experiments are done to test hypotheses.

The cardinal rule in science is that all hypotheses must be testable—in other words, they must, at least in principle, be capable of being shown wrong. In science, it is more important that there be a means of proving an idea wrong than that there be a means of proving it right. This is a major factor that distinguishes science from nonscience. The idea that scientific hypotheses must be capable of being proven wrong is a pillar of the philosophy of science and it is stated formally as the **principle of falsifiability**:

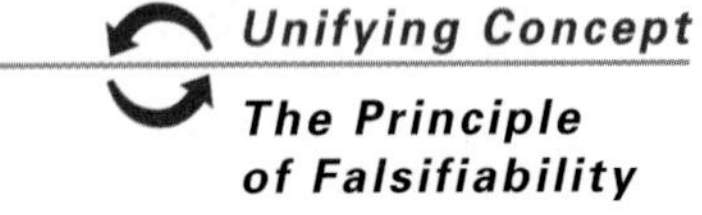

> **For a hypothesis to be considered scientific it must be testable—it must, in principle, be capable of being proven wrong.**

At first this may seem strange, for when we wonder about most things, we concern ourselves with ways of finding out whether they are true. Scientific hypotheses are different. In fact, if you want to distinguish whether a hypothesis is scientific or not, look to see if there is a test for proving it wrong. If there is no test for possible wrongness, then the hypothesis is not scientific. Albert Einstein put it well when he stated, "No number of experiments can prove me right; a single experiment can prove me wrong."

For example, Einstein hypothesized that light is bent by gravity. This might be proven wrong if starlight that grazed the Sun and could be seen during an eclipse were not deflected from a normal path. But starlight *is* found to bend as it passes close to the Sun, just as Einstein's hypothesis would have predicted. If and when a hypothesis or scientific claim is confirmed, it is regarded as useful and a stepping-stone to additional knowledge.

Consider another hypothesis, "The alignment of planets in the sky determines the best time for making decisions." Many people believe it, but this hypothesis is not scientific. It cannot be proven wrong, nor can it be proven right. It is speculation. Likewise, the hypothesis "Intelligent life exists on planets somewhere in the universe besides Earth" is not scientific.* Although it can be proven correct by the verification of a single instance of life existing elsewhere in the universe, there is no way to prove it wrong if no life is ever found. If we searched the far reaches of the universe for eons and found no life, we would not prove that it doesn't exist "around the next corner." A hypothesis that is capable of being proved right but not capable of being proved wrong is not a scientific hypothesis. Many such statements are quite reasonable and useful, but they lie outside the domain of science.

CHECK YOURSELF

Which statements are *scientific* hypotheses?

a. Better stock market decisions are made when the planets Venus, Earth, and Mars are aligned.
b. Atoms are the smallest particles of matter that exist.
c. The Moon is made of Swiss cheese.
d. Outer space contains a kind of matter whose existence can't be detected or tested.
e. Albert Einstein was the greatest physicist of the twentieth century.

CHECK YOUR ANSWERS

All statements are hypotheses, but only statements *a*, *b*, and *c* are scientific hypotheses, because they are testable. Statement *a* can be tested (and proven wrong) by researching the performance of the stock market during times when these planets were aligned. Not only can statement *b* be tested, it has been tested. Although the statement has been found to be untrue (many particles smaller than atoms have been discovered), the statement is nevertheless a scientific one. Likewise for statement *c*, where visits to the Moon have proven that the statement is wrong. Statement *d*, on the other hand, is easily seen to be unscientific, since it can't be tested. Lastly, statement *e* is an assertion that has no test. What possible test, beyond collective opinion, could prove Einstein was the greatest physicist? How could we know? Greatness is a quality that cannot be measured in an objective way.

1.5 The Scientific Experiment

A well-known scientific hypothesis that turned out to be incorrect was that of the greatly respected Greek philosopher Aristotle (384–322 B.C.), who claimed that heavy objects naturally fall faster than light objects. This hypothesis was considered

* The search for intelligent life in the universe is, however, ongoing. This search is based on the *question*: Might there be intelligent life somewhere besides on Earth? This question is the starting point for scientific observations of the physical world, but strictly speaking it is not a scientific hypothesis. A hypothesis is a sharper scientific tool than a question—a better, more finely honed instrument for separating scientific fact from fiction.

to be true for nearly 2000 years—mainly because nearly everyone who knew of Aristotle's conclusions had such great respect for him as a thinker that they simply assumed he couldn't be wrong. Also, in Aristotle's time, air resistance was not recognized as an influence on how quickly an object falls. We've all seen that stones fall faster than leaves fluttering in the air. Without investigating further, it is easy to accept false ideas.

Galileo very carefully examined Aristotle's hypothesis. Then he did something that caught on and changed science forever. He *experimented.* Galileo showed the falseness of Aristotle's claim with a single experiment—dropping heavy and light objects from the Leaning Tower of Pisa. Legend tells us that they fell at equal speeds. In the scientific spirit, one experiment that can be reproduced outweighs any authority, regardless of reputation or the number of advocates.

Scientists must accept their experimental findings even when they would like them to be different. They must strive to distinguish between the results they see and those they wish to see. This is not easy. Scientists, like most people, are very capable of fooling themselves. People have always tended to adopt general rules, beliefs, creeds, ideas, and hypotheses without thoroughly questioning their validity. And sometimes we retain these ideas long after they have been shown to be meaningless, false, or at least questionable. The most widespread assumptions are often the least questioned. Too often, when an idea is adopted, great attention is given to the instances that support it. Contrary evidence is often distorted, belittled, or ignored.

INSIGHT Experiment, not philosophical discussion, decides what is correct in science.

The fact that scientific statements will be thoroughly tested before being believed helps to keep science honest. Sooner or later, mistakes (or deceptions) are found out. A scientist exposed for cheating doesn't get a second chance in the community of scientists. Honesty, so important to the progress of science, thus becomes a matter of self-interest to scientists. There is relatively little bluffing in a game where all bets are called.

INSIGHT We each need a *knowledge filter* to tell the difference between what is true and what only pretends to be true. The best knowledge filter ever invented for explaining the physical world is science.

1.6 Facts, Theories, and Laws

When a scientific hypothesis has been tested over and over again and has not been contradicted, it may become known as a **law** or *principle.* A scientific **fact**, on the other hand, is generally something that competent observers can observe and agree to be true. For example, it is a fact that an amputated limb of a salamander can grow back. Anyone can watch it happen. It is not a fact—yet—that a severed limb of a human can grow back.

Scientists use the word *theory* in a way that differs from everyday speech. In everyday speech, a theory is the same as a hypothesis—a statement that hasn't been tested. But scientifically speaking, a **theory** is a synthesis of facts and well-tested hypotheses. Physicists, as we will learn, use quantum theory to explain the behavior of light. Chemists have theories about how atoms bond to form molecules. The theory of evolution is key to the life sciences. Earth scientists use the theory of plate tectonics to explain why the continents move, and astronomers speak of the theory of the Big Bang to account for the observation that galaxies are moving away from one another.

Facts are revisable data about the world.
Theories interpret facts.

Theories are a foundation of science, but they are not fixed. Rather, they evolve. They pass through stages of refinement. For example, since the theory of the atom was proposed 200 years ago, it has been refined many times in light of new evidence. Those who know only a little about science may argue that scientific theories can't be taken seriously because they are always changing. Those who understand science, however, see it differently. Theories grow stronger and more precise as they evolve to include new information.

SCIENCE AND SOCIETY

Pseudoscience

For a claim to qualify as "scientific" it must meet certain standards. For example, the claim must be reproducible by others who have no stake in whether the claim is true or false. The data and subsequent interpretations are open to scrutiny in a social environment where it's okay to have made an honest mistake, but not okay to have been dishonest or deceiving. Claims that are presented as scientific but do not meet these standards are what we call **pseudoscience**, which literally means "fake science." In the realm of pseudoscience, skepticism and tests for possible wrongness are downplayed or flatly ignored.

Examples of pseudoscience abound. Astrology is an ancient belief system that supposes a person's future is determined by the positions and movements of planets and other celestial bodies. Astrology mimics science in that astrological predictions are based on careful astronomical observations. Yet astrology is not a science because there is no validity to the claim that the positions of celestial objects influence the events of a person's life. After all, the gravitational force exerted by celestial bodies on a person is smaller than the gravitational force exerted by objects making up the earthly environment: trees, chairs, other people, bars of soap, and so on. Further, the predictions of astrology are not borne out; there just is no evidence that astrology works.

For more examples of pseudoscience, turn on the television. You can find advertisements for a plethora of pseudoscientific products. Watch out for remedies to ailments from baldness to obesity to cancer, for air-purifying mechanisms, and for "germ-fighting" cleaning products in particular. While many such products do operate on solid science, others are pure pseudoscience. Buyer beware!

Humans are very good at denial, which may explain why pseudoscience is such a thriving enterprise. Many pseudoscientists themselves do not recognize their efforts as pseudoscience. A practitioner of "absent healing," for example, may truly believe in her ability to cure people she will never meet except through email and credit card exchanges. She may even find anecdotal evidence to support her contentions. The placebo effect, as discussed in Section 20.7, can mask the ineffectiveness of various healing modalities. In terms of the human body, what people believe *will* happen often *can* happen, because of the physical connection between the mind and the body.

That said, consider the enormous downside of pseudoscientific practices. Today there are more than 20,000 practicing astrologers in the United States. Do people listen to these astrologers just for the fun of it? Or do they base important decisions on astrology? You might lose money by listening to pseudoscientific entrepreneurs; or worse, you could become ill. Delusional thinking, in general, carries risk.

Meanwhile, the results of science literacy tests given to the general public show that most Americans lack a basic understanding of basic concepts of science. Some 63 percent of American adults are unaware that the mass extinction of the dinosaurs occurred long before the first human evolved; 75 percent do not know that antibiotics kill bacteria but not viruses; 57 percent do not know that electrons are smaller than atoms. What we find is a rift—a growing divide—between those who have a realistic sense of the capabilities of science and those who do not understand the nature of science and its core concepts or, worse, feel that scientific knowledge is too complex for them to understand. Science is a powerful method for understanding the physical world, and a whole lot more reliable than pseudoscience as a means for bettering the human condition.

Those who can make you believe absurdities can make you commit atrocities—*Voltaire.*

INSIGHT

Science is a way to teach how something gets to be known, what is not known, to what extent things are known (for nothing is known absolutely), how to handle doubt and uncertainty, what the rules of evidence are, how to think about things so that judgments can be made, how to distinguish truth from fraud, and from show.—*Richard Feynman*

1.7 Science Has Limitations

Science deals only with hypotheses that are testable. Its domain is therefore restricted to the observable natural world. While scientific methods can be used to debunk various paranormal claims, they have no way of accounting for testimonies involving the supernatural. The term *supernatural* literally means "above nature." Science works within nature, not above it. Likewise, science is unable to answer philosophical questions, such as "What is the purpose of life?" or religious questions, such as "What is the nature of the human spirit?" Though these questions are valid and may have great importance to us, they rely on subjective personal experience and do not lead to testable hypotheses. They lie outside the realm of science.

1.8 Science, Art, and Religion

The search for a deeper understanding of the world around us has taken different forms, including science, art, and religion. Science is a system by which we discover and record physical phenomena and think about possible explanations for such phenomena. The arts are concerned with personal interpretation and creative expression. Religion addresses the source, purpose, and meaning of it all. Simply put, science asks *how*, art asks *who*, and religion asks *why*.

Science and the arts have certain things in common. In the art of literature, we find out about what is possible in human experience. We can learn about emotions from rage to love, even if we haven't yet experienced them. The arts describe these experiences and suggest what may be possible for us. Similarly, a knowledge of science tells us what is possible in nature. Scientific knowledge helps us to predict possibilities in nature even before they have been experienced. It provides us with a way of connecting things, of seeing relationships between and among them, and of making sense of the great variety of natural events around us. While art broadens our understanding of ourselves, science broadens our understanding of our environment.

Science and religion have similarities also. For example, both are motivated by curiosity about the natural world. Both have great impact on society. Science, for example, leads to useful technological innovations, while religion provides a foothold for many social services. Science and religion, however, are basically different. Science is concerned with understanding the physical universe, whereas religion is concerned with spiritual matters, such as belief and faith. Scientific truth is a matter of public scrutiny; religion is a deeply personal matter. In these respects, science and religion are as different as apples and oranges and do not contradict each other.

Ultimately, in learning more about science, art, and religion, we find that they are not mutually exclusive. Rather, they run parallel to each other like strings on a guitar, each resonating at its own frequency. When played together, the chord they produce can be a chord of profound richness. Science, art, and religion can work very well together, which is why we should never feel forced into choosing one over the other.

That science and religion can work very well together deserves special emphasis. When we study the nature of light later in this book, we will treat light first as a wave and then as a particle. At first, waves and particles may appear contradictory. You might believe that light can be only one or the other, and that you must choose between them. What scientists have discovered, however, is that light waves and light particles *complement* each other, and that, when these two ideas are taken together, they provide a deeper understanding of light. In a similar way, it is mainly people who are either uninformed or misinformed about the deeper natures of both science and religion who feel that they must choose between believing in religion and believing in science. Unless one has a shallow understanding of either or both, there is no contradiction in being religious in one's belief system and being scientific in one's understanding of the natural world.* What your religious beliefs are, and whether you have any religion at all, are of course private matters for you to decide. The tangling up of science and religion has led to many unfortunate arguments over the course of human history.

INSIGHT Art is about cosmic beauty. Science is about cosmic order. Religion is about cosmic purpose.

CHECK YOURSELF

Which of the following activities involves the utmost human expression of passion, talent, and intelligence?

a. painting and sculpture
b. literature
c. music
d. religion
e. science

CHECK YOUR ANSWER

All of them. In this book, we focus on science, which is an enchanting human activity shared by a wide variety of people. With present-day tools and know-how, scientists are reaching further and finding out more about themselves and their environment than people in the past were ever able to do. The more you know about science, the more passionate you feel toward your surroundings. There is science in everything you see, hear, smell, taste, and touch!

* Of course, this does not apply to certain religious extremists who steadfastly assert that one cannot embrace both their brand of religion and science.

1.9 Technology—The Practical Use of Science

Science and technology are also different from each other. Science is concerned with gathering knowledge and organizing it. Technology lets humans use that knowledge for practical purposes, and it provides the instruments scientists need to conduct their investigations.

Technology is a double-edged sword. It can be both helpful and harmful. We have the technology, for example, to extract fossil fuels from the ground and then burn the fossil fuels to produce energy. Energy production from fossil fuels has benefited society in countless ways. On the flip side, the burning of fossil fuels damages the environment. It is tempting to blame technology itself for such problems as pollution, resource depletion, and even overpopulation. These problems, however, are not the fault of technology any more than a stabbing is the fault of the knife. It is humans who use the technology, and humans who are responsible for how it is used.

Remarkably, we already possess the technology to solve many environmental problems. This twenty-first century will likely see a switch from fossil fuels to more sustainable energy sources. We recycle waste products in new and better ways. In some parts of the world, progress is being made toward limiting human population growth, a serious threat that worsens almost every problem faced by humans today. Difficulty solving today's problems results more from social inertia than from failing technology. Technology is our tool. What we do with this tool is up to us. The promise of technology is a cleaner and healthier world. Wise applications of technology *can* improve conditions on planet Earth.

INSIGHT There are many paths scientists can follow in doing science. Scientists who explore the ocean floor or who chart new galaxies, for example, are focused only on making and recording new observations.

1.10 The Natural Sciences: Physics, Chemistry, Biology, Earth Science, and Astronomy

Science is the present-day equivalent of what used to be called *natural philosophy*. Natural philosophy was the study of unanswered questions about nature. As the answers were found, they became part of what is now called science. The study of science today branches into the study of living things and nonliving things: the life sciences and the physical sciences. The life sciences branch into such areas as molecular biology, microbiology, and ecology. The *physical sciences* branch into such areas as physics, chemistry, the earth sciences, and astronomy. In this book, we address the life sciences and physical sciences and the ways in which they overlap—or *integrate*. This gives you a foundation for more specialized study in the future and a framework for understanding science in everyday life and in the news, from the greenhouse effect to tsunamis to genetic engineering.

A few words of explanation about each of the major divisions of science: Physics is the study of such concepts as motion, force, energy, matter, heat, sound, light, and the components of atoms. Chemistry builds on physics by telling us how matter is put together, how atoms combine to form molecules, and how the molecules combine to make the materials around us. Physics and chemistry, applied to Earth and its processes, make up earth science—geology, meteorology, and oceanography. When we apply physics, chemistry, and geology to other planets and to the stars, we are speaking about astronomy. Biology is more complex than physical science, for it involves matter that is alive. Underlying biology is chemistry, and underlying chemistry is physics. So physics is basic to both physical science and life science. That is why we begin with physics, then follow with chemistry and biology, then investigate earth science and conclude with astronomy. All are treated conceptually, with the twin goals of enjoyment and understanding.

1.11 Integrated Science

Just as you can't enjoy a ball game, computer game, or party game until you know its rules, so it is with nature. Because science helps us learn the rules of nature, it also helps us appreciate nature. You may see beauty in a tree, but you'll see more beauty in that tree when you understand how trees and other plants trap solar energy and convert it to the chemical energy that sustains nearly all life on Earth. Similarly, when you look at the stars, your sense of their beauty is enhanced if you know how stars are born from mere clouds of gas and dust—with a little help from the laws of physics, of course. And how much richer it is, when you look at the myriad objects in your environment, to know that they are all composed of atoms—amazing, ancient, invisible systems of particles regulated by an eminently knowable set of laws.

Understanding the physical world—to appreciate it more deeply or to have the power to alter it—requires concepts from different branches of science. For example, the process by which a tree transforms solar energy to chemical energy—photosynthesis—involves the ideas of radiant energy (physics), bonds in molecules (chemistry), gases in the atmosphere (earth science), the Sun (astronomy), and the nature of life (biology). Thus, for a complete understanding of photosynthesis and its importance, concepts beyond biology are required. And so it is for most of the real-world phenomena we are interested in. Put another way, the physical world integrates science, so to understand the world we need to look at science in an integrated way.

If the complexity of science intimidates you, bear this in mind: All the branches of science rest upon a relatively small number of basic ideas. There is a list of some of the most important unifying concepts at the back of this book, and in the page margins where they come up. Learn these underlying ideas, and you will have a toolkit to bring to any phenomenon you wish to understand.

Go to it—we live in a time of rapid and fascinating scientific discovery!

INTEGRATED SCIENCE CHEMISTRY AND BIOLOGY

An Investigation of Sea Butterflies

Let's consider an example of a recent scientific research project that shows how the scientific method can be put to work. Along the way, we'll get a taste of how biology and chemistry are integrated with one another in the physical world.

The Antarctic research team headed by James McClintock, professor of biology at the University of Alabama at Birmingham, and Bill Baker, professor of chemistry at the University of South Florida, was studying the toxic chemicals Antarctic marine organisms secrete to defend themselves against predators (Figure 1.2). McClintock and Baker observed an unusual relationship between two animal species, a sea butterfly and an amphipod—a relationship that led to a question, a scientific hypothesis, a prediction, tests concerning the chemicals involved in the relationship, and finally, a conclusion. The research generally proceeded according to the steps of the classic scientific method.

1. **Observe** The sea butterfly Clione Antarctica is a brightly colored shell-less snail with winglike extensions used in swimming (Figure 1.3a), and the amphipod Hyperiella dilatata resembles a small shrimp. McClintock and Baker observed a large percentage of amphipods carrying sea butterflies on their backs, with the sea butterflies held tightly by the legs of the amphipods (Figure 1.3b). Any amphipod that lost its sea butterfly would quickly seek another—the amphipods were actively abducting the sea butterflies!
2. **Question** McClintock and Baker noted that amphipods carrying butterflies were slowed considerably, making the amphipods more vulnerable to

Figure 1.2 The Chemical Ecology of Antarctic Marine Organisms Research Project was initiated in 1988 by James McClintock, shown here (fifth from left) with his team of colleagues and research assistants. In 1992, he was joined by Bill Baker (second from right). Baker is shown in the inset dressing for a dive into the icy Antarctic water. Like many other science projects, this one was interdisciplinary, involving the efforts of scientists from a wide variety of backgrounds.

predators and less adept at catching prey. Why then did the amphipods abduct the sea butterflies?

3. **Hypothesize** Given their experience with the chemical defense systems of various sea organisms, the research team hypothesized that amphipods carry sea butterflies to produce a chemical that deters a predator of the amphipod.
4. **Predict** Based on their hypotheses, they predicted (a) that they would be able to isolate this chemical and (b) that an amphipod predator would be deterred by it.
5. **Test predictions** To test their hypothesis and prediction, the researchers captured several predator fish species and conducted the test shown in Figure 1.4. The fish were presented with solitary sea butterflies, which they took into their mouths but promptly spat back out. The fish readily ate uncoupled amphipods but spit out any amphipod coupled with a sea butterfly. These are the results expected if the sea butterfly was secreting some sort of chemical deterrent. The same results would be obtained, however, if a predator fish simply didn't like the feel of a sea butterfly in its mouth. The results of this simple test were therefore ambiguous. A conclusion could not yet be drawn.

All scientific tests need to minimize the number of possible conclusions. Often this is done by running an experimental test along with a **control**. Ideally, the experimental test and the control should differ by only one variable.

(a)

(b)

Figure 1.3 (a) The graceful Antarctic sea butterfly is a species of snail that does not have a shell. (b) The shrimplike amphipod attaches a sea butterfly to its back even though doing so limits the amphipod's mobility.

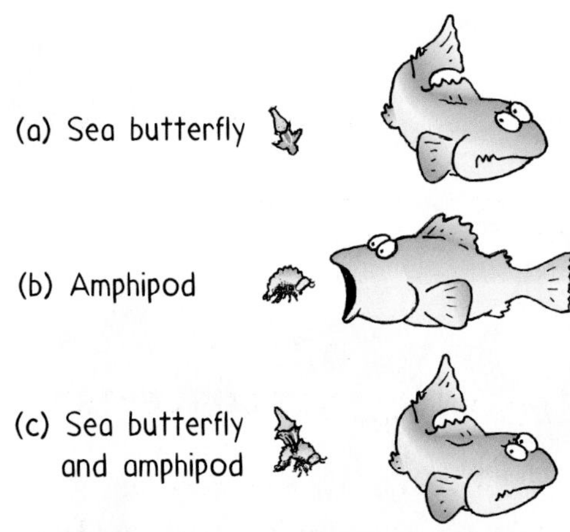

Figure 1.4 In McClintock and Baker's initial experiment, a predatory fish (a) rejected the sea butterfly, (b) ate the free-swimming amphipod, and (c) rejected the amphipod coupled with a sea butterfly.

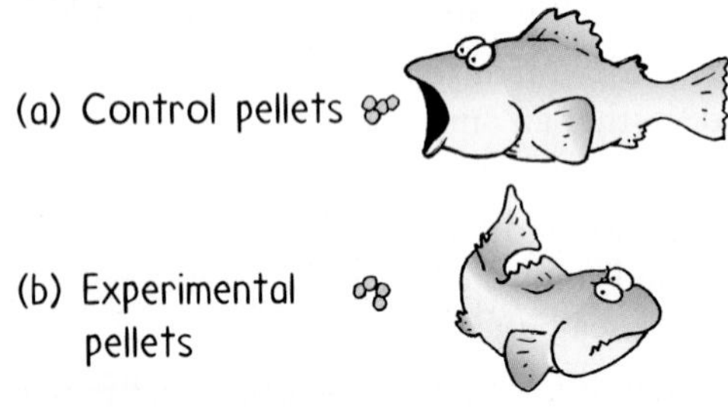

Figure 1.5 The predator fish (a) ate the control pellets but (b) rejected the experimental pellets, which contained sea butterfly extract.

Figure 1.6 Pteroenone is a molecule produced by sea butterflies as a chemical deterrent against predators. Its name is derived from *ptero-*, which means "winged" (for the sea butterfly) and *-enone*, which describes information about the chemical structure. The black spheres represent carbon atoms, the white spheres show hydrogen atoms, and the red spheres represent oxygen atoms.

Any differences in results can then be attributed to how the experimental test differed from the control.

To confirm that the deterrent was chemical and not physical, the researchers made one set of food pellets containing both fish meal and sea butterfly extract (the experimental pellets). For their control test, they made a physically identical set containing only fish meal (the control pellets). As shown in Figure 1.5, the predator fish readily ate the control pellets but not the experimental ones. These results strongly supported the chemical hypothesis.

Further processing of the sea butterfly extract yielded five major chemical compounds, only one of which deterred predator fish from eating the pellets. Chemical analysis of this compound revealed it to be the previously unknown molecule shown in Figure 1.6, which they named pteroenone.

6. **Draw a conclusion** In addition to running control tests, scientists confirm experimental results by repeated testing. In this case, the Antarctic researchers made many food pellets, both experimental and control, so that each test could be repeated many times. Only after obtaining consistent results, can a scientist draw a conclusion. McClintock and Baker were thus able to conclude that amphipods abduct sea butterflies in order to use the sea butterflies' secretion of pteroenone as a defense against predator fish.

Yet, this conclusion would still be regarded with skepticism in the scientific community. Why? There is a great potential for unseen error in any experiment. A laboratory may have faulty equipment that leads to consistently wrong results for example. Because of the potential for unseen error from any particular research group, experimental results must be *reproducible* to be considered valid. This means that other scientists must be able to reproduce the same experimental findings in separate experiments. Thus you can see that it is a long road from bright idea to accepted scientific finding! The plodding, painstaking nature of this process is beneficial though—it is the reason that scientific knowledge is highly trustworthy.

As frequently happens in science, McClintock and Baker's results led to new questions. What are the properties of pteroenone? Does this substance have applications—for example, can it be used as a pest repellent? Could it be useful for treating human disease? In fact, a majority of the chemicals we use were originally discovered in natural sources. This illustrates that there is an important reason for preserving marine habitats, tropical rainforests, and the other diverse natural environments on Earth—they are storehouses of countless yet-to-be-discovered substances.

CHECK YOURSELF

1. Explain how this particular research effort shows the integration of science disciplines.
2. What variable did the experimental fish pellets contain that was not found in the control pellets?
3. If the fish had eaten the experimental pellets, what conclusion could the scientists have drawn?
4. Why must experimental findings be reproducible to be considered valid?

CHECK YOUR ANSWERS

1. The original observation and question were of a biological nature—why do amphipods abduct sea butterflies?
2. Sea butterfly extract.
3. The scientists would have had to conclude that the predator fish were not deterred by the sea butterfly secretions and thus the amphipods did not capture the sea butterflies for this reason.
4. Reproducibility of results is essential because every research project may contain unseen errors.

CHAPTER 1 Review

Summary of Terms

Control A test that excludes the variable being investigated in a scientific experiment.

Fact A phenomenon about which competent observers can agree.

Hypothesis An educated guess or a reasonable explanation. When the hypothesis can be tested by experiment, it qualifies as a *scientific hypothesis*.

Law A general hypothesis or statement about the relationship of natural quantities that has been tested over and over again and has not been contradicted. Also known as a *principle*.

Principle of falsifiability For a hypothesis to be considered scientific it must be testable—it must, in principle, be capable of being proven wrong.

Pseudoscience A theory or practice that is considered to be without scientific foundation but purports to use the methods of science.

Science The collective findings of humans about nature, and a process of gathering and organizing knowledge about nature.

Scientific method An orderly method for gaining, organizing, and applying new knowledge.

Technology The means of solving practical problems by applying the findings of science.

Theory A synthesis of a large body of information that encompasses well-tested hypotheses about certain aspects of the natural world.

Review Questions

1.1 A Brief History of Advances in Science

1. What launched the era of modern science in the sixteenth century?

1.2 Mathematics and Conceptual Integrated Science

2. Why do we believe that focusing on math too early is a mistake in an introductory science course?

1.3 The Scientific Method—A Classic Tool

3. Specifically, what do we mean when we say that a scientific hypothesis must be testable?

1.4 The Scientific Hypothesis

4. Is any hypothesis that is not scientific necessarily unreasonable? Explain.

1.5 The Scientific Experiment

5. How did Galileo disprove Aristotle's idea that heavy objects fall faster than light objects?

1.6 Facts, Theories, and Laws

6. Distinguish among a scientific fact, a hypothesis, a law, and a theory.
7. How does the definition of the word *theory* differ in science versus everyday life?

1.7 Science Has Limitations

8. Your friend says scientific theories cannot be believed because they are always changing. What can you say to counter this argument?

1.8 Science, Art, and Religion

9. What is meant by the term *supernatural*, and why does science not deal with it?
10. Why do religious questions such as "What is the nature of the human spirit?" lie outside of the domain of science?
11. How are science, art, and religion like strings on a guitar?

1.9 Technology—The Practical Use of Science

12. Clearly distinguish between science and technology.

1.10 The Natural Sciences: Physics, Chemistry, Biology, Earth Science, and Astronomy

13. In what sense does chemistry underlie physics? In what sense is biology more complex than physical science?

1.11 Integrated Science

14. Why study integrated science?

INTEGRATED SCIENCE CONCEPTS

Chemistry and Biology—An Investigation of Sea Butterflies

1. What two scientific disciplines were needed to understand the curious behavior of the Antarctic amphipods?
2. When was a control used in the investigation of the amphipods and sea butterflies? Why was a control necessary?
3. What was McClintock and Baker's hypothesis? Was it a scientific hypothesis? Why do you think so?

Active Exploration

Use the scientific method: (1) Based on your observations of your environment, (2) develop a question, (3) hypothesize the answer, (4) predict consequences if your hypothesis is correct, (5) test your predictions, and (6) draw a conclusion. On a sheet of paper, describe in detail how you performed each step of the method from (1) to (6).

Exercises

1. Are the various branches of science separate or do they overlap? Give several examples to support your answer.
2. What do science, art, and religion have in common? How are they different?
3. Can a person's religious beliefs be proven wrong? Can a person's understanding of a particular scientific concept be proven wrong?
4. In what sense is science grand and breathtaking? In what sense is it dull and painstaking?
5. How is the printing press like the Internet in terms of the history of science?

Problems

1. The more candy bars you add to your diet per day, the more weight you gain (all other factors, such as the amount of exercise you get, being equal). Is this an example of a direct proportion or an inverse proportion?
2. State the above relation in mathematical form. (Hint: Don't forget to include a proportionality constant with appropriate units.)
3. What is an example of two quantities that are related as an inverse proportion which you have observed in your daily life? Express this relation in mathematical form.
4. Cite a different example of an inverse proportion that you have observed in your daily life. Show that it is an inverse proportion by (a) expressing it in mathematical form, and (b) representing the relation with a graph. (To do so, you may use estimated rather than real data.)

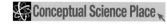

Conceptual Science Place

CHAPTER 1 ONLINE RESOURCES

Quiz • Flashcards • Links

PART ONE

PHYSICS

What I enjoy most is discovering that I understand things. Like learning that our Earth is round for the same reason the Moon and Sun are round—gravity. Every bit of mass inside them pulls on every other bit of mass, all pulling inward and making them ball-shaped. If Earth had corners, gravity would pull them in too. Gravity produces ocean tides, the curved paths of baseballs, and the motion of satellites—which orbit because they continually fall around Earth. Physics tells me that electricity and magnetism connect to become light. Since physics is everywhere, it gives me a foundation for integrating chemistry, biology, Earth science, and astronomy. Best of all, learning science conceptually, starting with physics, is phun!

Figure 2.1 Does a force keep the cannonball moving after it leaves the cannon?

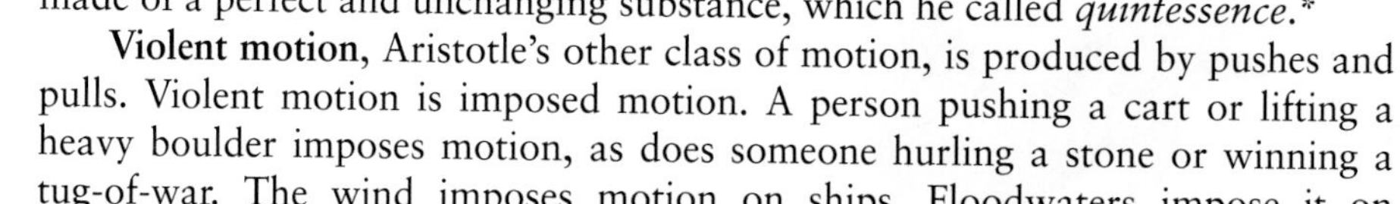

different rules apply in the heavens and that celestial bodies are perfect spheres made of a perfect and unchanging substance, which he called *quintessence.**

Violent motion, Aristotle's other class of motion, is produced by pushes and pulls. Violent motion is imposed motion. A person pushing a cart or lifting a heavy boulder imposes motion, as does someone hurling a stone or winning a tug-of-war. The wind imposes motion on ships. Floodwaters impose it on boulders and tree trunks. Violent motion is externally caused and is imparted to objects, which move not of themselves, not by their nature, but because of impressed *forces*—pushes or pulls.

2.2 Galileo's Concept of Inertia

Unifying Concept

The Principle of Falsifiability

Section 1.4

Aristotle's ideas were accepted as fact for nearly 2,000 years. Then, in the early 1500s, the Italian scientist Galileo demolished Aristotle's belief that heavy objects fall faster than light ones. According to legend, Galileo dropped both heavy and light objects from the Leaning Tower of Pisa (Figure 2.2). He showed that, except for the effects of air friction, objects of different weights fell to the ground in the same amount of time.

Galileo made another discovery. He showed that Aristotle was wrong about forces being necessary to keep objects in motion. Although a force is needed to start an object moving, Galileo showed that, once it is moving, no force is needed to keep it moving—except for the force needed to overcome friction. (We will learn more about friction in Section 2.8.) When friction is absent, a moving object needs no force to keep it moving. It will remain in motion all by itself.

Figure 2.2 Galileo's famous demonstration.

Rather than philosophizing about ideas, Galileo did something that was quite remarkable at the time. Galileo tested his revolutionary idea by *experiment*. This was the beginning of modern science. He rolled balls down inclined planes and observed and recorded the gain in speed as rolling continued (Figure 2.3). On downward-sloping planes, the force of gravity increases a ball's speed. On an upward slope, the force of gravity decreases a ball's speed. What about a ball rolling on a level surface? While rolling on a level surface, the ball neither rolls with nor against the vertical force of gravity—it neither speeds up nor slows down. The rolling ball maintains a constant speed. Galileo reasoned

HISTORY OF SCIENCE

Aristotle (384–322 BC)

Aristotle was the foremost philosopher, scientist, and educator of his time. Born in Greece, he was the son of a physician who personally served the king of Macedonia. At the age of 17, he entered the Academy of Plato, where he worked and studied for 20 years until Plato's death. He then became the tutor of young Alexander the Great. Eight years later, he formed his own school. Aristotle's aim was to systematize existing knowledge, just as Euclid had systematized geometry. Aristotle made critical observations, collected specimens, and gathered, summarized, and classified almost all of the existing knowledge of the physical world. His systematic approach became the method from which Western science later arose. After his death, his voluminous notebooks were preserved in caves near his home and were later sold to the library at Alexandria.

Scholarly activity ceased in most of Europe through the Dark Ages, and the works of Aristotle were forgotten and lost in the scholarship that continued in the Byzantine and Islamic empires. Various texts were reintroduced to Europe during the eleventh and twelfth centuries and were translated into Latin. The Church, the dominant political and cultural force in Western Europe, at first prohibited the works of Aristotle and then accepted and incorporated them into Christian doctrine.

*Quintessence is the fifth essence, the other four being earth, water, air, and fire.

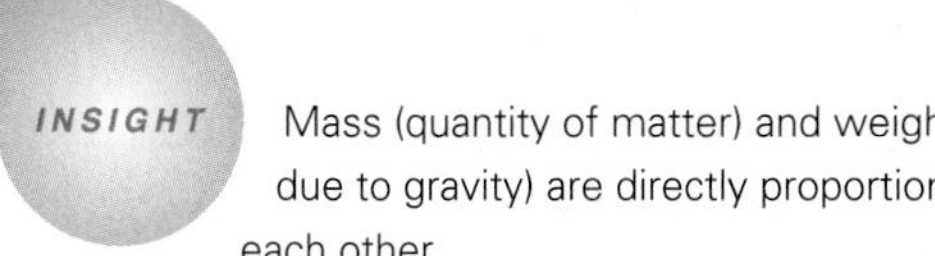

INSIGHT Mass (quantity of matter) and weight (force due to gravity) are directly proportional to each other.

2. Does a 2-kg iron block have twice as much *inertia* as a 1-kg bunch of bananas? Twice as much *mass*? Twice as much *volume*? Twice as much *weight* when weighed in the same location?
3. How does the mass of a gold bar vary with location?

CHECK YOUR ANSWERS

1. The answer is yes to all questions. A 2-kg block of iron has twice as many iron atoms, and therefore twice the amount of inertia, mass, and weight. The blocks consist of the same material, so the 2-kg block also has twice the volume.
2. Two kg of anything has twice the inertia and twice the mass of 1 kg of anything else. Because mass and weight are proportional at the same location, 2 kg of anything will weigh twice as much as 1 kg of anything. Except for volume, the answer to all questions is *yes*. Volume and mass are proportional only when the materials are identical—when they have the same *density*. Iron is much more dense than bananas, so 2 kg of iron must occupy less volume than 1 kg of bananas.
3. It does not vary at all. It consists of the same number of atoms no matter what its location. Although its weight may vary with location, it has the same mass everywhere. This is why mass is preferred to weight in scientific studies.

One Kilogram Weighs 9.8 Newtons

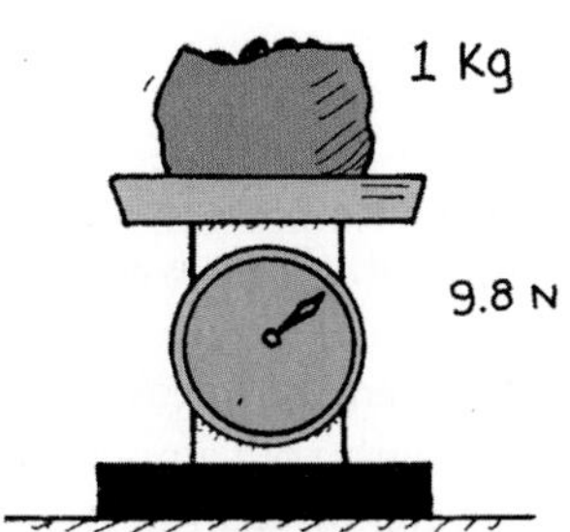
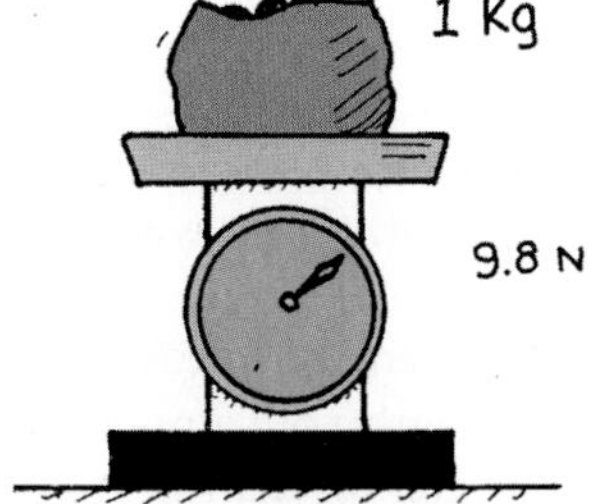

Figure 2.8 One kg of nails weighs 9.8 N, which is equal to 2.2 lb.

The standard unit of mass is the kilogram, abbreviated kg. The standard unit of force is the newton, abbreviated N. The abbreviation is written with a capital letter because the unit is named after a person. A 1-kg bag of any material at the Earth's surface has a weight of 9.8 N in standard units. Away from the Earth's surface, where the force of gravity is less (on the Moon, for example), the bag would weigh less. Except in cases where precision is needed, we will round off 9.8 and call it 10. So 1 kg of something weighs about 10 N. If you know the mass in kilograms and want the weight in newtons, multiply the number of kilograms by 10. Or, if you know the weight in newtons, divide by 10 and you'll have the mass in kilograms. Weight and mass are directly proportional to each other.

The relationship between kilograms and pounds is that 1 kg weighs 2.2 lb at the Earth's surface. (That means that 1 lb is as heavy as 4.45 N.)

Density Is Mass Divided by Volume

An important property of a material, whether solid, liquid, or gaseous, is the measure of its compactness: **density**. Density is a measure of how much mass occupies a given space; it is the amount of matter per unit volume:

$$\text{Density} = \frac{\text{mass}}{\text{volume}}$$

Like mass and weight, density has to do with the "lightness" or "heaviness" of materials. But the distinction is that density also involves the volume of an object, the space it occupies. For example, a kilogram of lead has the same mass as a kilogram of feathers, and, at the surface of the Earth, both of them have the same weight—2.2 lb. But their densities are very different. A kilogram of lead is very dense and would fit into a tennis ball, while the same mass of feathers has a very low density and could adequately stuff the shell of a down-filled sleeping bag. Volume is often measured in cubic centimeters (cm^3) or cubic meters (m^3), so density is most typically expressed in units of g/cm^3 or kg/m^3. Water has a density of 1 g/cm^3. Mercury's density of 13.6 g/cm^3 means that it has 13.6 times as much mass as an equal volume of water. Although different masses of a given material have different volumes, they have the same density.

INSIGHT A pillow is bigger than an auto battery, but which has more matter? Which has more *inertia*? Which has more *mass*?

CHECK YOURSELF

1. Would 1 kg of gold have the same density on the Moon as on Earth?
2. Which has the greater density—an entire candy bar or half a candy bar?

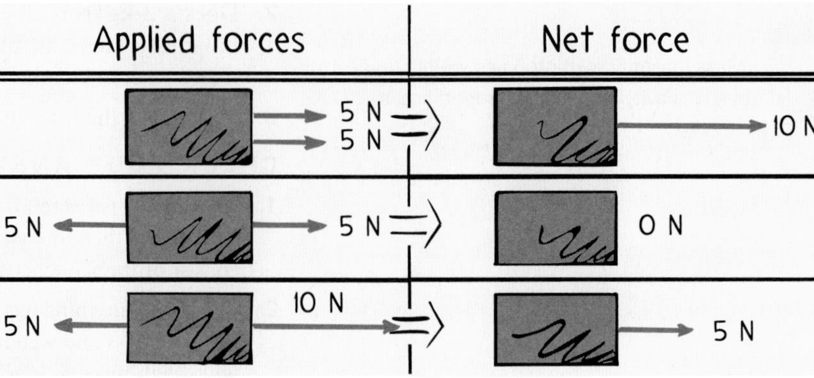

Figure 2.9 Net force.

CHECK YOUR ANSWERS

1. Yes. Since mass and volume remain the same despite gravitational variations, the ratio of mass to volume remains constant as well.
2. Both the half and the entire candy bar have the same density.

2.4 Net Force

In simplest terms, a force is a push or a pull. An object doesn't speed up, slow down, or change direction unless a force acts upon it. When we say "force," we imply the total force, or **net force**, acting on the object. Often more than one force may be acting on an object. For example, when you throw a baseball, the force of gravity and the pushing force you apply with your muscles both act on the ball. When the ball is sailing through the air, the force of gravity and air resistance both act on it. The net force on the ball is the combination of forces. It is the net force that changes an object's state of motion.

For example, suppose you pull on a shoebox with a force of 5 N (slightly more than 1 lb). If your friend also pulls with 5 N in the same direction, the net force on the box is 10 N. If your friend pulls on the box with the same force as you but in the opposite direction, the net force on the box is zero. Now if you increase your pull to 10 N and your friend pulls oppositely with a force of 5 N, the net force is 5 N in the direction of your pull. We see this in Figure 2.9.

The forces in Figure 2.9 are shown by arrows. Forces are vector quantities. A **vector quantity** has both magnitude (how much) and direction (which way). When an arrow represents a vector quantity, the arrow's length represents magnitude and its direction shows the direction of the quantity. Such an arrow is called a *vector*. (You will find more on vectors in the next chapter, in Appendix C, and in the *Practice Book for Conceptual Integrated Science*.)

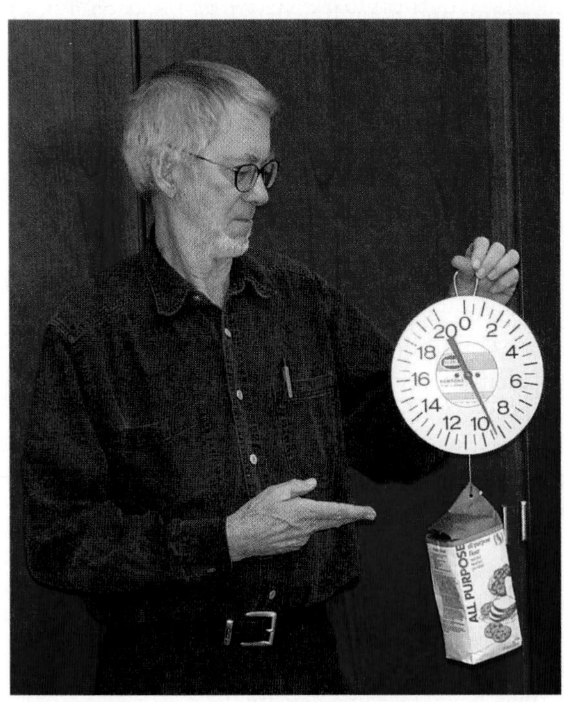

Figure 2.10 Burl Grey, who taught the author about tension forces, suspends a 1-kg bag of flour from a spring scale, showing the weight and tension in the string of nearly 10 N.

2.5 The Equilibrium Rule

If you tie a string around a 2 lb bag of flour and suspend it on a weighing scale (Figure 2.10), a spring in the scale stretches until the scale reads 2 lb. The stretched spring is under a "stretching force" called *tension*. A scale in a science lab is likely calibrated to read the same force as 9 N. Both pounds and newtons are units of weight, which, in turn, are units of force. The bag of flour is attracted to the Earth with a gravitational force of 2 lb—or, equivalently, 9 N. Suspend twice as much flour from the scale and the reading will be 18 N.

There are two forces acting on a bag of flour—tension force acting upward and weight acting downward. The two forces on the bag are equal and opposite, and they cancel to zero. Hence, the bag remains at rest.

SCIENCE AND SOCIETY

Paul Hewitt and the Origin of Conceptual Integrated Science

Paul Hewitt, the founding author of this book, wrote the physics text *Conceptual Physics* when he was a young instructor at City College of San Francisco. *Conceptual Physics* has been the leading physics book for nonscience majors in America for over 30 years, and it's had a major impact on how science is taught—concepts first, math and technical details are brought in later. The following is Paul's personal story about how he discovered the fascination of science by observing physics principles at work in everyday life.

When I was in high school, my counselor advised me not to enroll in science and math classes but instead to focus on my interest in art. I took this advice. For a while, my major interests were drawing comic strips and boxing, but neither of these earned me much success. After a stint in the army, I tried my luck at sign painting, and the cold Boston winters drove me south to Miami, Florida. There, at age 26, I got a job painting billboards and met a new friend who became a great intellectual influence on me, Burl Grey. Like me, Burl had never studied physics in high school. But he was passionate about science in general, and he shared his passion with many questions as we painted together.

I remember Burl asking me about the tensions in the ropes that held up the scaffold we were standing on. The scaffold was simply a heavy horizontal plank suspended by a pair of ropes. Burt twanged the rope nearest his end of the scaffold and asked me to do the same with mine. He was comparing the tensions in both ropes—to determine which was greater. Like a more tightly stretched guitar string, the rope with greater tension twangs at a higher pitch. The finding that Burl's rope had a higher pitch seemed reasonable because his rope supported more of the load.

When I walked toward Burl to borrow one of his brushes, he asked if the tensions in the ropes had changed. Did the tension in his rope change as I moved closer? We agreed that it should have, because even more of the load was supported by Burl's rope. How about my rope? Would its tension decrease? We agreed that it would, for it would be supporting less of the total load. I was unaware at the time that I was discussing physics.

Burl and I used exaggeration to bolster our reasoning (just as physicists do.) If we both stood at an extreme end of the scaffold and leaned outward, it was easy to imagine the opposite of the scaffold rising like the end of a seesaw, with the opposite rope going limp. Then there would be no tension in that rope. We then reasoned that the tension in my rope would gradually decrease as I walked toward Burl. It was fun posing such questions and seeing if we could answer them.

A question we couldn't answer was whether or not the decrease in tension in my rope when I walked away from it would be exactly compensated by a tension increase in Burl's rope. For example, if my rope underwent a decrease of 50 N, would Burl's rope gain 50 N? (We talked pounds back then, but here we use the scientific unit of force, the *newton*—abbreviated N.) Would the gain be exactly 50 N? And, if so, would this be a grand coincidence? I didn't know the answer until more than a year later, when Burl's stimulation resulted in my leaving full-time painting and going to college to learn more about science.

There I learned that any object at rest, such as the sign-painting scaffold I worked on with Burl, is said to be in equilibrium. That is, all the forces that act on it balance to zero ($\Sigma F = 0$). So the sum of the upward forces supplied by the supporting ropes indeed do add up to our weights plus the weight of the scaffold. A 50-N loss in one would be accompanied by a 50-N gain in the other.

I tell this story to make the point that one's thinking is very different when there is a rule to guide it. Now, when I look at any motionless object, I know immediately that all the forces acting on it cancel out. We see nature differently when we know its rules. It makes nature simpler and easier to understand. Without the rules of physics, we tend to be superstitious and to see magic where there is none. Quite wonderfully, everything is beautifully connected to everything else by a surprisingly small number of rules. Physics is a study of nature's rules.

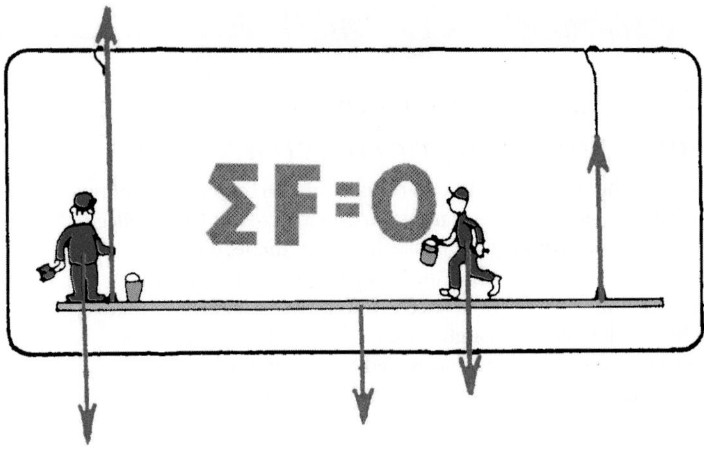

Figure 2.11 The sum of the upward vectors equals the sum of the downward vectors. $\Sigma F = 0$, and the scaffold is in equilibrium.

INSIGHT Everything that isn't undergoing a change in motion is in mechanical equilibrium. That's because $\Sigma F = 0$.

When the net force on something is zero, we say that something is in mechanical equilibrium.* Anything in mechanical equilibrium obeys an interesting rule: In mathematical notation, the **equilibrium rule** is

$$\Sigma F = 0$$

The symbol Σ is the capital Greek letter "sigma." It stands for "the vector sum of" and F stands for "forces." For a suspended body at rest, like the bag of flour, the rule states that the forces acting upward on the body must be balanced by other forces acting downward to make the vector sum equal zero. (Vector quantities take direction into account, so, if upward forces are positive, downward ones are negative, and when summed they equal zero. See Figure 2.11.)

CHECK YOURSELF

Consider the gymnast hanging from the rings.

1. If she hangs with her weight evenly divided between the two rings, how would scale readings in both supporting ropes compare with her weight?
2. Suppose she hangs with slightly more of her weight supported by the left ring. What would a scale on the right read?

CHECK YOUR ANSWERS

1. The reading on each scale will be *half her weight*. The sum of readings on both scales then equals her weight.
2. When more of her weight is supported by the left ring, the reading on the right is *less than half her weight*. No matter how she hangs, the sum of the scale readings equals her weight. For example, if one scale reads two-thirds her weight, the other scale will read one-third her weight. Get it?

2.6 The Support Force

Consider a book lying at rest on a table. It is in equilibrium. What forces act on the book? One is the force due to gravity—the weight of the book. Since the book is in equilibrium, there must be another force acting on it to produce a net force of zero—an upward force opposite to the force of gravity. The table exerts this upward force, called the **support force**. This upward support force, often

*We'll see in Appendix B that another condition for equilibrium is that the net *torque* is zero.

MATH CONNECTION

Applying the Equilibrium Rule

Use the physics you've learned so far to solve these practice problems.

Problems

1. When Burl stands alone in the exact middle of his scaffold, the left side reads 500 N. Fill in the reading on the right scale. The total weight of Burl and the scaffold must be _____ N.

2. Burl stands farther from the left. Fill in the reading on the right scale.

3. In a silly mood, Burl dangles from the right end. Fill in the reading on the right scale.

Solutions Do your answers illustrate the equilibrium rule?

1. The total weight is 1000 N. The right rope must be under 500 N of tension because Burl is in the middle, and both ropes support his weight equally. Since the sum of the tensions is 1000 N, the total weight of Burl and the scaffold must be 1000 N. Let's call the upward tension forces +1000 N. Then the downward weights are −1000 N. What happens when you add +1000 and −1000? The answer is that they equal zero. So we see that $\Sigma F = 0$.
2. Did you get the correct answer of 830 N? Reasoning: We know from Question 1 that the sum of the rope tensions equals 1000 N, and since the left rope has a tension of 170 N, the other rope must make up the difference—that 1000 N − 170 N = 830 N. Get it? If so, great. If not, discuss it with your friends until you do. Then read further.
3. The answer is 1000 N. Do you see that this illustrates $\Sigma F = 0$?

Can you see evidence of $\Sigma F = 0$ in bridges and other structures around you?

called the *normal force*, must equal the weight of the book.* If we designate the upward force as positive, then the downward force (weight) is negative, and the sum of the two is zero. The net force on the book is zero. Stated another way, $\Sigma F = 0$.

To understand better that the table pushes up on the book, compare the case of compressing a spring (Figure 2.12). If you push the spring down, you can feel the spring pushing up on your hand. Similarly, the book lying on the table compresses the atoms in the table, which behave like microscopic springs. The weight of the book squeezes downward on the atoms, and they squeeze upward on the book. In this way, the compressed atoms produce the support force.

When you step on a bathroom scale, two forces act on the scale. One is the downward pull of gravity (your weight) and the other is the upward support force of the floor. These forces compress a spring that is calibrated to show your weight (Figure 2.13). In effect, the scale shows the support force. When you weigh yourself on a bathroom scale at rest, the support force and your weight have the same magnitude.

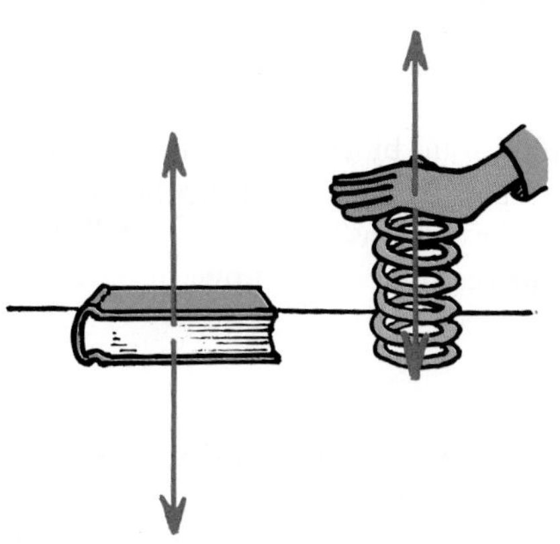

Figure 2.12 The table pushes up on the book with as much force as the downward force of gravity on the book. The spring pushes up on your hand with as much force as you exert to push down on the spring.

CHECK YOURSELF

1. What is the net force on a bathroom scale when a 150-lb person stands on it?
2. Suppose you stand on two bathroom scales with your weight evenly distributed between the two scales. What is the reading on each of the scales?

*This force acts at right angles to the surface. When we say "normal to," we are saying "at right angles to," which is why this force is called a normal force.

Figure 2.13 The upward support force is as much as your weight.

What happens when you stand with more of your weight on one foot than on the other?

CHECK YOUR ANSWERS

1. Zero, because the scale remains at rest. The scale reads the support force (which has the same magnitude as weight), not the net force.
2. The reading on each scale is half your weight, because the sum of the scale readings must balance your weight, so the net force on you will be zero. If you lean more on one scale than the other, more than half your weight will be read on that scale, but less on the other, so they will still add up to your weight. Like the example of the gymnast hanging by the rings, if one scale reads two-thirds of your weight, the other scale will read one-third of your weight.

2.7 Equilibrium of Moving Things

Figure 2.14 When the push on the crate is as great as the force of friction between the crate and the floor, the net force on the crate is zero, and it slides at an unchanging speed.

Equilibrium is a state of no change. Rest is only one form of equilibrium. An object moving at a constant speed in a straight-line path is also in equilibrium. A bowling ball rolling at a constant speed in a straight line is also in equilibrium—until it hits the pins. Whether at rest or steadily rolling in a straight-line path, $\Sigma F = 0$.

An object under the influence of only one force cannot be in equilibrium. The net force couldn't be zero. Only when two or more forces act on it can the object be in equilibrium. We can test whether or not something is in equilibrium by noting whether or not it undergoes changes in its state of motion.

Consider a crate being pushed across a factory floor. If it moves at a steady speed in a straight-line path, it is in equilibrium. This indicates that more than one force is acting on the crate. Another force exists—likely the force of friction between the crate and the floor. The fact that the net force on the crate equals zero tells us that the force of friction must be equal to, and opposite to, the pushing force (Figure 2.14).

2.8 The Force of Friction

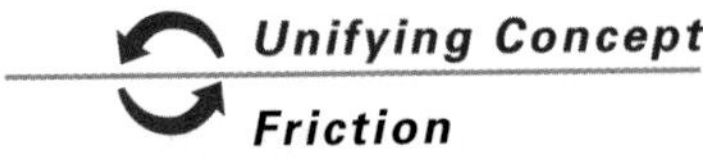

Friction occurs when one object rubs against something else.* Friction occurs for solids, liquids, and gases. An important rule of friction is that it always acts in a direction to oppose motion. If you pull a solid block along a floor to the left, the force of friction on the block will be to the right. A boat propelled to the east by its motor experiences water friction to the west. When an object falls downward through the air, the force of friction, **air resistance**, acts upward. Friction always acts in a direction to oppose motion.

CHECK YOURSELF

An airplane flies through the air at a constant velocity. In other words, it is in equilibrium. Two horizontal forces act on the plane. One is the thrust of the propeller that pushes it forward. The other is the force of air resistance that acts in the opposite direction. Which force is greater?

*Even though it may not seem so yet, most of the concepts in physics are not really complicated. But friction is different. Unlike most concepts in physics, it is a very complicated phenomenon. The findings are empirical (gained from a wide range of experiments) and the predictions are approximate (also based on experiment).

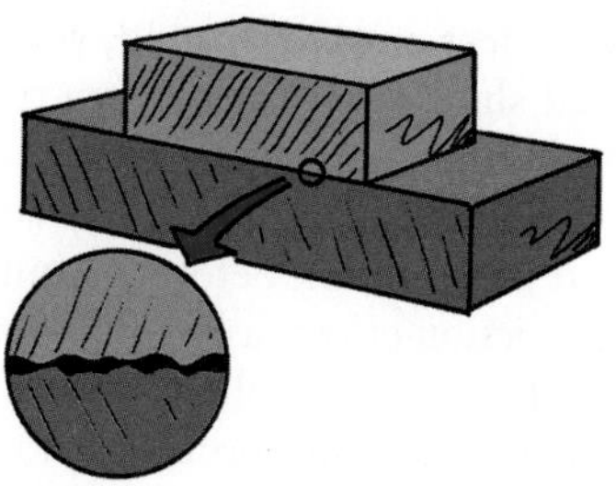

Figure 2.15 Friction results from the mutual contact of irregularities in the surfaces of sliding objects. Even surfaces that appear to be smooth have irregular surfaces when viewed at the microscopic level.

CHECK YOUR ANSWER

Both horizontal forces have the same magnitude. If you call the forward force exerted by the propeller positive, then the air resistance is negative. Since the plane is in equilibrium, can you see that the two forces combine to zero?

The amount of friction between two surfaces depends on the kinds of material and how much they are pressed together. Friction is due to surface bumps and also to the "stickiness" of the atoms on the surfaces of the two materials (Figure 2.15). Friction between a crate and a smooth wooden floor is less than that between the same crate and a rough floor. And, if the crate is full, the friction is more than it would be if it were empty because the crate presses down harder on the floor when it weighs more.

When you push horizontally on a crate and it slides across a factory floor, both your force and the opposite force of friction affect the crate's motion. When you push hard enough on the crate to match the friction, the net force on the crate is zero, and it slides at a constant velocity. Notice that we are talking about what we recently learned—that no change in motion occurs when $\Sigma F = 0$.

Friction

CHECK YOURSELF

1. Suppose you exert a 100-N horizontal force on a heavy crate resting motionless on a factory floor. The fact that it remains at rest indicates that 100 N isn't great enough to make it slide. How does the force of friction between the crate and the floor compare with your push?
2. You push harder—say, 110 N—and the crate still doesn't slide. How much friction acts on the crate?
3. You push still harder, and the crate moves. Once it is in motion, you push with 115 N, which is just sufficient to keep it sliding at a constant velocity. How much friction acts on the crate?
4. What net force does a sliding crate experience when you exert a force of 125 N and the friction between the crate and floor is 115 N?

CHECK YOUR ANSWERS

1. 100 N in the opposite direction. Friction opposes the motion that would occur otherwise. The fact that the crate is at rest is evidence that $\Sigma F = 0$.
2. Friction increases to 110 N, again $\Sigma F = 0$.
3. 115 N, because when it is moving at a constant velocity, $\Sigma F = 0$.
4. 10 N, because $\Sigma F = 125\text{ N} - 115\text{ N}$. In this case, the crate *accelerates*.

INTEGRATED SCIENCE BIOLOGY, ASTRONOMY, CHEMISTRY, AND EARTH SCIENCE

Friction Is Universal

Friction is the opponent of all motion. If an object moves to the left, friction acts on it to the right. If an object moves upward, friction pushes it downward. Whenever two objects are in contact, friction acts in such a way as to prevent or to slow their relative motion.

Your body is well adapted to a friction-filled environment. The fingerprint ridges in your palms and fingers increase surface roughness and so enhance friction between your hands and the things they touch. When your hands are wet, and water partially fills in the troughs between the ridges, friction is reduced—and it's easy to drop a glass or a plate. Your toes and the soles of your feet are similarly patterned with grooves and ridges to help you grip the surface of the

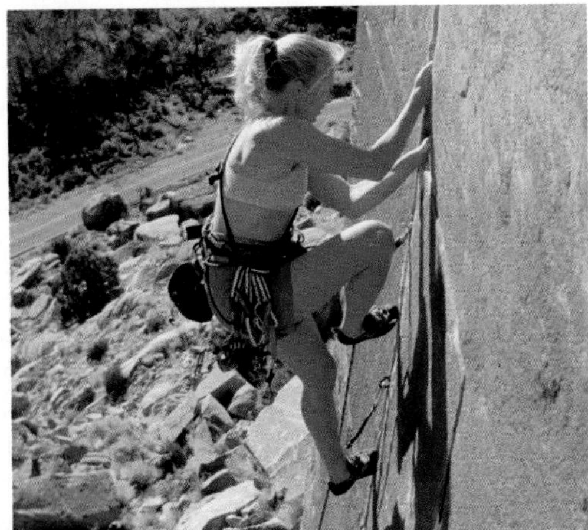

Figure 2.16 The rock climber grips the sheer rock face with her hands. Why is it important that her hands remain dry?

ground. If not for the friction between your feet and the ground, your feet would slip out from under you like smooth-soled shoes on ice when you tried to walk. It's friction between her hands and the rock that holds the climber to the nearly vertical mountain face (Figure 2.16). Can you see the reason why rock climbers often chalk up their hands to absorb hand perspiration before a climb?

Astronomy has its share of interesting friction effects as well. Shooting stars or *meteors* are bits of material falling through Earth's atmosphere. They are heated to incandescence by friction with the gas particles making up the atmosphere. Tiny dust grains, *micrometeoroids,* are so small that, as they fall, air resistance is enough to slow them sufficiently that they do not burn up. Instead, they fall gently to Earth and accumulate, adding hundreds or thousands of tons to Earth's mass every day!

Though friction is often useful, there are many situations in which it reduces efficiency, and minimizing it would save energy—for example, within most machines. So industry employs chemists to develop lubricants that minimize friction. Lubricants reduce friction by separating two contacting surfaces with an intermediate layer of more slippery material. Instead of rubbing against each other, the surfaces rub against the lubricant. Most lubricants are oils or greases. Currently, chemists are trying to develop lubricants that won't evaporate at higher and higher temperatures or freeze at low temperatures; won't lock when called upon to carry heavier loads; or won't leak through gaskets and seals when spun at high speeds.

Earthquakes are an Earth-science phenomenon that depends in an obvious way on friction. Earthquakes happen when adjoining, massive blocks of rock are pushed or pulled in different directions. The blocks of rock, locked together by friction, resist motion until the stress becomes too great. At that point, friction is overcome, the blocks of rock let go, and they slip into new positions, releasing energy that vibrates the Earth.

CHECK YOURSELF

1. Tristan is holding a brick, as shown, by pressing his hands inward on the brick to supply the contact force. In what direction does gravity act? In what direction does friction act? What is the net force on the brick? What does the texture of Tristan's hands have to do with the amount of friction acting on the brick?
2. Why do some lubricants sometimes fail when supporting a heavy load?

CHECK YOUR ANSWERS

1. Gravity pulls the brick downward. Friction opposes gravity, so it acts upward. The net force on the brick is zero, as evidenced by no change in its state of motion. Tristan's hands are covered with grooves and ridges that, like treads on a tire, increase friction and improve his grip.
2. Lubricants fail when they too readily evaporate at high temperatures, or freeze at low temperatures, or leak through gaskets and seals when spun at high speeds.

2.9 Speed and Velocity

Speed

Before the time of Galileo, people described moving things as "slow" or "fast." Such descriptions were vague. Galileo was the first to measure speed by comparing the distance covered with the time that it takes to move that distance. He defined speed as the distance covered per amount of travel time.

$$\text{Speed} = \frac{\text{distance covered}}{\text{travel time}}$$

Figure 2.17 A cheetah can maintain a very high speed, but only for a short time.

For example, if a bicyclist covers 20 km in 1 h, her speed is 20 km/h. Or, if she runs 6 m in 1 s, her speed is 6 m/s.

Any combination of units for distance and time can be used for speed—kilometers per hour (km/h), centimeters per day (the speed of a sick snail), or whatever is useful and convenient. The slash symbol (/) is read as "per," and means "divided

Figure 2.18 A common automobile speedometer. Note that speed is shown both in km/h and in mi/h.

INSIGHT If you get a traffic ticket for speeding, is it because of your *instantaneous speed* or your *average speed?*

TABLE 2.1 Approximate Speeds in Different Units

12 mi/h = 20 km/h = 6 m/s (bowling ball)
25 mi/h = 40 km/h = 11 m/s (very good sprinter)
37 mi/h = 60 km/h = 17 m/s (sprinting rabbit)
50 mi/h = 80 km/h = 22 m/s (tsunami)
62 mi/h = 100 km/h = 28 m/s (sprinting cheetah)
75 mi/h = 120 km/h = 33 m/s (batted softball)
100 mi/h = 160 km/h = 44 m/s (batted baseball)

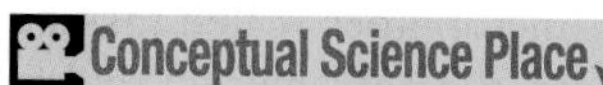

Definition of Speed; Average Speed

by." In science, the preferred unit of speed is meters per second (m/s). Table 2.1 compares some speeds in different units.

Instantaneous Speed

Moving things often have variations in speed. A car, for example, may travel along a street at 50 km/h, slow to 0 km/h at a red light, and speed up to only 30 km/h because of traffic. At any instant, you can tell the speed of the car by looking at its speedometer. The speed at any instant is *instantaneous speed.*

Average Speed

In planning a trip by car, the driver wants to know the travel time. The driver is concerned with the *average speed* for the trip. How is average speed defined?

$$\textbf{Average speed} = \frac{\textbf{total distance covered}}{\textbf{travel time}}$$

Average speed can be calculated rather easily. For example, if you drive a distance of 80 km in 1 h, your average speed is 80 km/h. Likewise, if you travel 320 km in 4 h,

$$\text{Average speed} = \frac{\text{total distance covered}}{\text{travel time}} = \frac{320\text{ km}}{4\text{ h}} = 80\text{ km/h}.$$

Note that, when a distance in kilometers (km) is divided by a time in hours (h), the answer is in kilometers per hour (km/h).

Since average speed is the entire distance covered divided by the total time of travel, it doesn't indicate the various instantaneous speeds that may have occurred along the way. On most trips, the instantaneous speed is often quite different from the average speed.

If we know average speed and travel time, the distance traveled is easy to find. A simple rearrangement of the definition above gives

$$\textbf{Total distance covered} = \textbf{average speed} \times \textbf{travel time}$$

For example, if your average speed on a 4-h trip is 80 km/h, then you cover a total distance of 320 km.

CHECK YOURSELF

1. What is the average speed of a cheetah that sprints 100 m in 4 s? How about if it sprints 50 m in 2 s?
2. If a car travels at an average speed of 60 km/h for 1 h, it will cover a distance of 60 km. (a) How far would it travel if it moved at this rate for 4 h? (b) For 10 h?
3. In addition to the speedometer on the dashboard of every car, there is an odometer, which records the distance traveled. If the initial reading is set at zero at the beginning of a trip and the reading is 40 km after 0.5 h, what was the average speed?
4. Would it be possible to attain this average speed and never go faster than 80 km/h?

CHECK YOUR ANSWERS

(Are you reading this before you have reasoned answers in your mind? As mentioned earlier, *think* before you read the answers. You'll not only learn more, you'll enjoy learning more.)

1. In both cases, the answer is 25 m/s:

$$\text{Average speed} = \frac{\text{total distance covered}}{\text{travel time}} = \frac{100\text{ m}}{4\text{ s}} = \frac{50\text{ m}}{2\text{ s}} = 25\text{ m/s}$$

2. The distance traveled is the average speed × time of travel, so:

(a) Distance = 60 km/h × 4 h = 240 km
(b) Distance = 60 km/h × 10 h = 600 km

3. Average speed $= \dfrac{\text{total distance covered}}{\text{travel time}} = \dfrac{40\text{ km}}{0.5\text{ h}} = 80\text{ km/h}.$

Figure 2.19 Although the car can maintain a constant speed along a circular track, it cannot maintain a constant velocity. Why?

4. No, not if the trip starts from rest and ends at rest. During the trip, there are times when the instantaneous speeds are less than 80 km/h, so the driver must at some time drive faster than 80 km/h in order to average 80 km/h. In practice, average speeds are usually much less than high instantaneous speeds.

Velocity

When we know both the speed and direction of an object, we know its **velocity.** For example, if a vehicle travels at 60 km/h, we know its speed. But, if we say how fast it moves at 60 km/h to the north, we specify its *velocity*. Speed is a description of how fast; velocity is a description of how fast and in what direction. Velocity is a vector quantity.

Constant speed means steady speed, neither speeding up nor slowing down. Constant velocity, on the other hand, means both constant speed and constant direction. Constant direction is a straight line—the object's path doesn't curve. So constant velocity means motion in a straight line at a constant speed—motion with no acceleration.

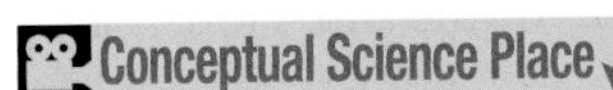

Velocity; Changing Velocity; Definition of Acceleration; Force Causes Acceleration

CHECK YOURSELF

"She moves at a constant speed in a constant direction." Say the same in a few words.

CHECK YOUR ANSWER

"She moves at a constant velocity."

Motion Is Relative

Everything is always moving. Even when you think you're standing still, you're actually speeding through space. You're moving relative to the Sun and stars, although you are at rest relative to Earth. At this moment, your speed relative to the Sun is about 100,000 km/h, and it is even faster relative to the center of our galaxy.

When we discuss the speed or velocity of something, we mean speed or velocity relative to something else. For example, when we say that a space shuttle travels at 30,000 km/h, we mean relative to Earth. Unless stated otherwise, all speeds discussed in this book are relative to the surface of Earth. Motion is relative (Figure 2.20).

Figure 2.20 When sitting on a chair, your speed is zero relative to Earth but 30 km/s relative to the Sun.

2.10 Acceleration

Most moving things usually experience variations in their motion. We say they undergo **acceleration**. The first to formulate the concept of acceleration was Galileo, who developed the concept in his experiments with inclined planes. He found that balls rolling down inclined planes rolled faster and faster. Their velocities changed as they rolled. Further, the balls gained the same amount of speed in equal time intervals (Figure 2.21).

INSIGHT Can you see that a car has three controls that change velocity—the gas pedal (accelerator), the brakes, and the steering wheel?

Galileo defined the rate of change of velocity acceleration.*

$$\text{Acceleration} = \frac{\text{change of velocity}}{\text{time interval}}$$

Figure 2.21 A ball gains the same amount of speed in equal intervals of time. It undergoes acceleration.

*The capital Greek letter Δ (delta) is often used as a symbol for "change in" or "difference in." In "delta" notation, $a = \Delta v/\Delta t$, where Δv is the change in velocity and Δt is the change in time (the time interval). From this we can see that $\Delta v = a\Delta t$. See further development of linear motion in Appendix B.

Figure 2.22 We say that a body undergoes acceleration when there is a change in its state of motion.

Acceleration is experienced when you're in a moving bus. When the driver steps on the gas pedal, the bus gains speed. We say that the bus accelerates. We can see why the gas pedal is called the "accelerator." When the brakes are applied, the vehicle slows. This is also acceleration, because the velocity of the vehicle is changing. When something slows down, we often call this *deceleration*.

Consider driving a car that steadily increases in speed. Suppose that, in 1 s, you steadily increase your velocity from 30 km/h to 35 km/h. In the next second, you steadily go from 35 km/h to 40 km/h, and so on. You change your velocity by 5 km/h each second. Thus we can see that

$$\text{Acceleration} = \frac{\text{change in velocity}}{\text{time interval}} = \frac{5\ \text{km/h}}{1\ \text{s}} = 5\ \text{km/h}\cdot\text{s}$$

In this example, the acceleration is "5 kilometers per hour-second" (abbreviated as 5 km/h·s). Note that a unit for time enters twice: once for the unit of velocity and again for the interval of time in which the velocity is changing. Also note that acceleration is not just the change in velocity; it is the change in velocity per second. If either speed or direction changes, or if both change, then velocity changes.

When a car makes a turn, even if its speed does not change, it is accelerating. Can you see why? Acceleration often occurs because the car's direction is changing. Acceleration refers to a change in velocity. So acceleration involves a change in speed, a change in direction, or change in both speed and direction. Figure 2.22 illustrates this.

CHECK YOURSELF

In 2.0 s, a car increases its speed from 60 km/h to 65 km/h while a bicycle goes from rest to 5 km/h. Which has the greater acceleration?

CHECK YOUR ANSWER

Both have the same acceleration, since both gain the same amount of speed in the same time. Both accelerate at 2.5 km/h·s.

Hold a stone at a height above your head and drop it. It accelerates during its fall. When the only force that acts on a falling object is that due to gravity, when air resistance doesn't affect its motion, we say the object is in **free fall**. All freely falling objects in the same vicinity undergo the same acceleration. At the Earth's surface, a freely falling object gains speed at the rate of 10 m/s each second, as shown in Table 2.2.

$$\text{Acceleration} = \frac{\text{change in speed}}{\text{time interval}} = \frac{10\ \text{m/s}}{1\ \text{s}} = 10\ \text{m/s}$$

We read the acceleration of free fall as "10 meters per second squared." (More precisely, 9.8 m/s^2.) This is the same as saying that acceleration is "10 meters per second per second." Note again that the unit of time, the second, appears twice. It appears once for a unit of velocity and the time during which the velocity changes.

In Figure 2.23, we imagine a freely falling boulder with a speedometer attached. As the boulder falls, the speedometer shows that the boulder goes 10 m/s faster each second. This 10 m/s gain each second is the boulder's acceleration. (The acceleration of free fall is further developed in Appendix B.)

Up-and-down motion is shown in Figure 2.24. The ball leaves the thrower's hand at 30 m/s. Call this the initial velocity. The figure uses the convention of up as positive and down as indicated by a minus sign (−). Notice that the 1-second interval positions correspond to changes in velocity of 10 m/s.

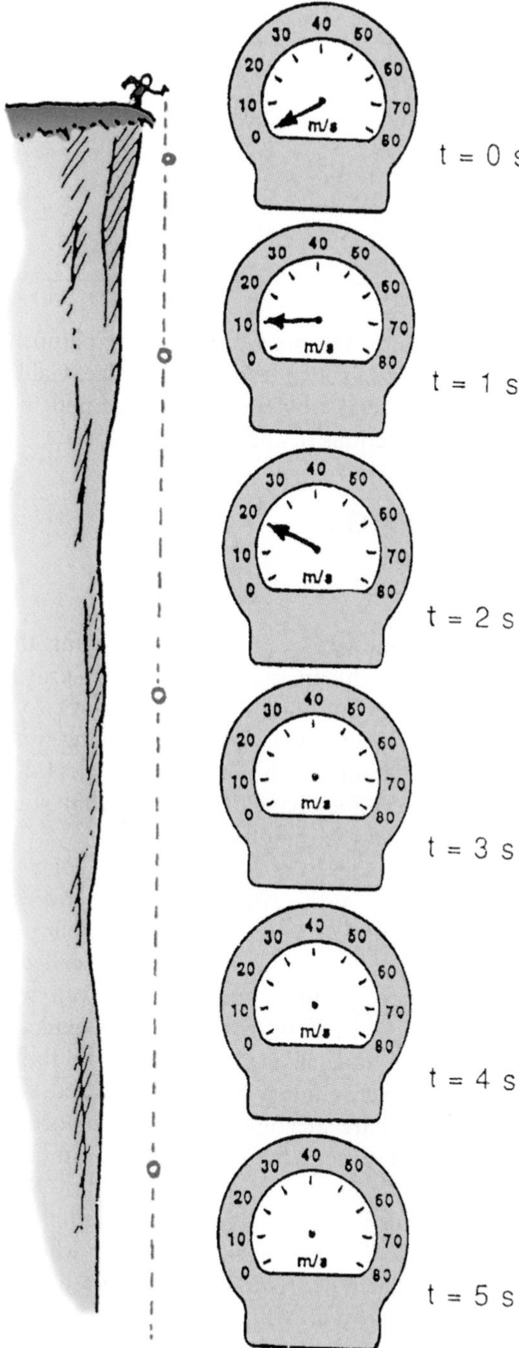

Figure 2.23 Imagine that a falling boulder is equipped with a speedometer. In each succeeding second of fall, you'd find the boulder's speed increasing by the same amount: 10 m/s. Sketch in the missing speedometer needle at $t = 3$ s, $t = 4$ s, and $t = 5$ s.

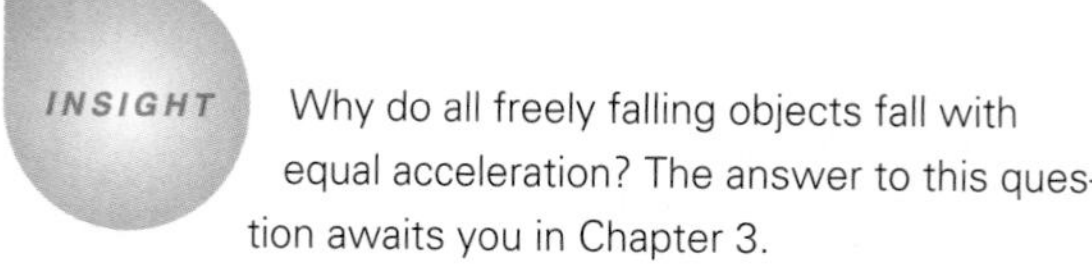

Why do all freely falling objects fall with equal acceleration? The answer to this question awaits you in Chapter 3.

TABLE 2.2 Free Fall

Time of Fall (s)	Speed of Fall (m/s)	Distance of Fall (m)
0	0	0
1	10	5
2	20	20
3	30	45
4	40	80
5	50	125
.	.	.
.	.	.
.	.	.
t	$10t$	$\frac{1}{2}\,10t^2$

Aristotle used logic to establish his ideas of motion. Galileo used experiment. Galileo showed that experiments are superior to logic in testing knowledge. Galileo was concerned with how things move rather than why they move. The path was paved for Isaac Newton to make further connections of concepts in motion.

INTEGRATED SCIENCE BIOLOGY

Hang Time

Some athletes and dancers have great jumping ability. Leaping straight up, they seem to defy gravity, hanging in the air for what feels like at least two or three seconds. In reality, however, the "hang time" of even the best jumpers is almost always less than 1 s. What determines hang time? Just what you'd expect—how high you jump.

Just as we tend to overestimate how long the best jumpers stay in the air, we can easily overestimate how high people jump. You can test your own jumping ability by performing what's called a standing vertical jump: Stand facing a wall with your feet flat on the floor and your arms extended upward. Make a mark on the wall at the top of your reach. Then jump, arms outstretched, and make another mark at your peak. The distance between the two marks measures your vertical jump. If it's more than 0.6 m (2 ft), you're exceptional.

Now, the world record for a standing vertical jump is held by basketball star Spud Webb, who jumped 1.25 m in 1986. What was his hang time when he made that jump? At the top of a jump, right when you stop going up and are about to start coming down, your speed is zero and you are at rest. As we show in Appendix B, the relationship between time down and vertical height for a uniformly accelerating object starting from rest is

$$d = \frac{1}{2}gt^2$$

We can re-arrange this equation to calculate time:

$$t = \sqrt{\frac{2d}{g}}$$

For Spud's jump, we use 1.25 m for d and 10 m/s^2 for g and see that:

$$t = \sqrt{\frac{2d}{g}} = \sqrt{\frac{2(1.25\text{ m})}{10\text{ m/s}^2}} = 0.50\text{ s}$$

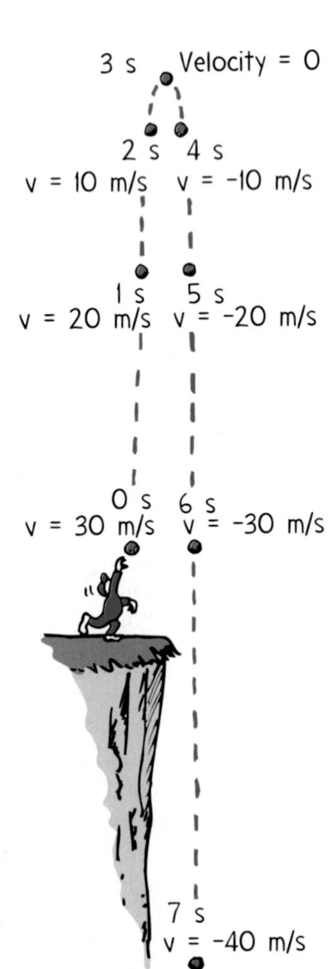

Figure 2.24 Interactive Figure

The rate at which velocity changes each second is the same.

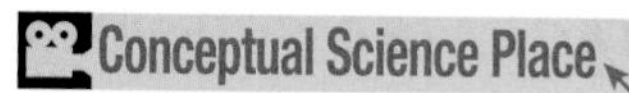

Free Fall: How Far?

The hang time is actually double this, since this is the time for only one way of an up-and-down round trip—so Spud's total hang time is 1 s. And that, remember, is a world record.

What determines jumping ability? Your jumping ability increases with the length of your legs and with the strength of your leg muscles. If you look at the bodies of animals that specialize in jumping—frogs, kangaroos, and rabbits, for example—you'll see that they all have elongated, and very muscular, hind legs.

CHECK YOURSELF

The red kangaroo has a vertical jump height of about 1.8 m (6 ft). What is the kangaroo's hang time?

CHECK YOUR ANSWER

The time it takes for the kangaroo to reach the ground from a height of 1.8 m is:

$$t = \sqrt{\frac{2d}{g}} = \sqrt{\frac{2(1.8 \text{ m})}{10 \text{ m/s}^2}} = 0.60 \text{ s}$$

The hang time is double this, or 1.2 s.

CHAPTER 2 Review

Summary of Terms

Acceleration The rate at which velocity changes with time; the change in velocity may be in magnitude, or in direction, or in both. It is usually measured in m/s^2.

Air resistance The force of friction acting on an object due to its motion through air.

Density A measure of mass per volume for a substance.

Equilibrium rule The vector sum of forces acting on a nonaccelerating object equals zero: $\Sigma F = 0$.

Force Simply stated, a push or a pull.

Free fall Motion under the influence of gravitational pull only.

Friction The resistive force that opposes the motion or attempted motion of an object through a fluid or past another object with which it is in contact.

Inertia The property of things to resist changes in motion.

Kilogram The unit of mass. One kilogram (symbol kg) is the mass of 1 liter (symbol L) of water at 4°C.

Mass The quantity of matter in an object. More specifically, it is a measure of the inertia or sluggishness that an object exhibits in response to any effort made to start it, stop it, deflect it, or change its state of motion in any way.

Net force The combination of all forces that act on an object.

Newton The scientific unit of force.

Speed The distance traveled per unit of time.

Support force The force that supports an object against gravity, often called the normal force.

Vector quantity A quantity that specifies direction as well as magnitude.

Velocity The speed of an object with specification of its direction of motion.

Weight Simply stated, the force of gravity on an object. More specifically, the gravitational force with which a body presses against a supporting surface.

Review Questions

2.1 Aristotle on Motion

1. What were the two main classifications of motion in Aristotle's view of nature?
2. Did Aristotle believe that forces are necessary to keep moving objects moving, or did he believe that, once moving, they'd move by themselves?

2.2 Galileo's Concept of Motion

3. What two main ideas of Aristotle did Galileo discredit?
4. Which dominated Galileo's way of extending knowledge—philosophical discussion or experiment?
5. What is the name of the property of objects to maintain their states of motion?

2.3 Mass—A Measure of Inertia

6. Which depends on gravity—weight or mass?
7. Where would your weight be greater—on Earth or on the Moon? Where would your mass be greater?
8. What are the units of measurement for weight and mass?
9. One kg weighs 9.8 N on Earth. Would it weigh more or less on the Moon?
10. Which has the greater density—1 kg of water or 10 kg of water?

2.4 Net Force

11. What is the net force on a box that is being pushed to the left with a force of 50 N while it is also being pushed to the right with a force of 60 N?
12. What two quantities are necessary to determine a vector quantity?

2.5 The Equilibrium Rule

13. What is the name given to a force that occurs in a rope when both ends are pulled in opposite directions?
14. How much tension is there in a rope that holds a 20-N bag of apples at rest?
15. What does $\Sigma F = 0$ mean?

2.6 The Support Force

16. Why is the support force on an object often called the normal force?
17. When you weigh yourself, how does the support force of the scale acting on you compare with the gravitational force between you and Earth?

2.7 Equilibrium of Moving Things

18. What test tells us whether or not a moving object is in equilibrium?
19. If we push a crate at constant velocity, how does friction acting on the crate compare with our pushing force?

2.8 The Force of Friction

20. How does the direction of a friction force compare with the direction of motion of a sliding object?
21. If you push on a heavy crate to the right and it slides, what is the direction of friction on the crate?
22. Suppose you push on a heavy crate, but not hard enough to make it slide. Does a friction force act on the crate?

2.9 Speed and Velocity

23. What equation shows the relationship between speed, distance, and time.
24. Why do we say that velocity is a vector and speed is not?
25. Does the speedometer on a vehicle show average speed or instantaneous speed?
26. How can you be at rest and also moving at 100,000 km/h at the same time?

2.10 Acceleration

27. What equation shows the relationship between velocity, time, and acceleration?
28. What is the acceleration of an object in free fall at Earth's surface?
29. Why does the unit of time appear twice in the definition of acceleration?
30. When you toss a ball upward, by how much does its upward speed decrease each second?

Multiple Choice Questions

Choose the BEST answer to the following.

1. To be in mechanical equilibrium, an object must be
   (a) at rest.
   (b) moving at constant velocity.
   (c) either a or b.
   (d) neither a nor b.
2. If an object moves along a straight-line path at constant speed, then it must be
   (a) accelerating.
   (b) acted on by a force.
   (c) both of these.
   (d) none of these.
3. What is the net force on a box that is being pushed to the left with a force of 40 N while it is also being pushed to the right with a force of 50 N?
   (a) 10 N to the left
   (b) 10 N to the right
   (c) 90 N to the left
   (d) 90 N to the right
4. Neglecting air resistance, when you toss a rock upward, by about how much does its upward speed decrease each second?
   (a) 10 m/s
   (b) 10 m/s^2
   (c) depends on the initial speed
   (d) none of these
5. In 2.0 seconds, a car increases its speed from 30 km/h to 35 km/h while a bicycle goes from rest to 4 km/h. Which has the greater acceleration?
   (a) The car.
   (b) The bicycle.
   (c) The accelerations are equal.
   (d) Impossible to know from the information provided.

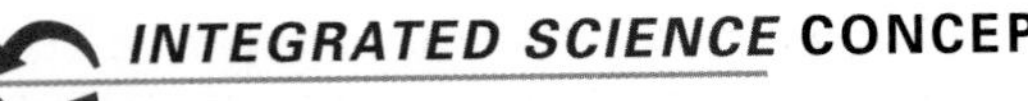

INTEGRATED SCIENCE CONCEPTS

Biology—Friction Is Universal

1. *Joints* are places where bones meet. Many of them, such as the ball-and-socket joints in your shoulders and hips, are bathed with *synovial fluid*, a viscous substance resembling the white of an egg. Speculate about what the purpose of the synovial fluid might be.
2. Describe one phenomenon from each of the major natural sciences—physics, chemistry, biology, Earth science, and astronomy—in which friction plays a major and interesting role.
3. Is it more correct to say that friction prevents earthquakes or that friction causes earthquakes? Justify your answer.

Biology—Hang Time

1. When during a jump is your speed zero?
2. What are some of the anatomical features that affect jumping ability?

Active Explorations

1. Roll cans of different masses across the floor at equal initial speeds and notice the effect of inertia and friction on how far each can rolls.
2. By any method you choose, determine both your walking speed and your running speed.

Exercises

1. A bowling ball rolling along a lane gradually slows as it rolls. How would Aristotle interpret this observation? How would Galileo interpret it?
2. What Aristotelian idea did Galileo discredit in his fabled experiment at the Leaning Tower of Pisa? With his experiments with inclined planes?
3. What physical quantity is a measure of how much inertia an object has?
4. Does a dieting person accurately lose mass or lose weight?
5. One gram of lead has a mass of 11.3 g. What is its density? Two grams of aluminum has a mass of 5.4 g. What is the density of aluminum?
6. Which has the greater density—5 kg of lead or 10 kg of aluminum?
7. Consider a pair of forces, one with a magnitude of 25 N and the other with a magnitude of 15 N. What maximum net force is possible for these two forces? What is the minimum net force possible?
8. The sketch shows a painter's scaffold in mechanical equilibrium. The person in the middle weighs 250 N, and the tensions in each rope are 200 N. What is the weight of the scaffold?

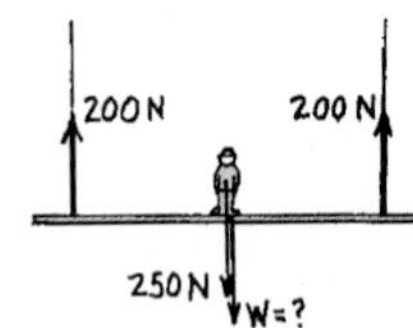

9. A different scaffold that weighs 300 N supports two painters, one weighing 250 N and one weighing 300 N. The reading on the left scale is 400 N. What should the reading on the right scale be?

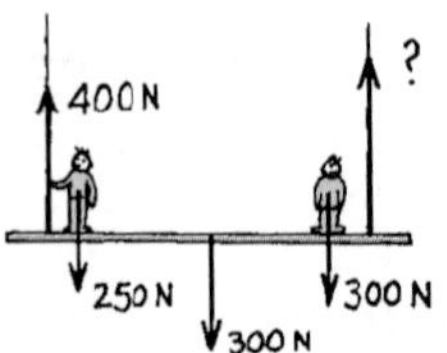

10. Can an object be in mechanical equilibrium when only a single force acts on it? Explain.

11. Nellie Newton hangs at rest from the ends of the rope as shown. How does the reading on the scale compare with her weight?

12. Harry the painter swings year after year from his bosun's chair. His weight is 500 N, and the rope, unknown to him, has a breaking point of 300 N. Why doesn't the rope break when he is supported as shown on the left below? One day, Harry was painting near a flagpole, and, for a change, he tied the free end of the rope to the flagpole instead of to his chair, as shown on the right. Why did Harry end up taking his vacation early?

13. Consider the two forces acting on the person standing still—namely, the downward pull of gravity and the upward support force of the floor. Are these forces equal and opposite?
14. Can we accurately say that, if something moves at constant velocity, there are no forces acting on it? Explain.
15. At the moment an object that has been tossed upward into the air reaches its highest point, is it in equilibrium? Defend your answer.
16. If you push horizontally on a crate and it slides across the floor, slightly gaining speed, how does the friction acting on the crate compare with your push?
17. What is the impact speed when a car moving at 100 km/h bumps into the rear of another car traveling in the same direction at 98 km/h?
18. Harry Hotshot can paddle a canoe in still water at 8 km/h. How successful will he be in canoeing upstream in a river that flows 8 km/h?
19. A destination 120 mi away is posted on a highway sign, and the speed limit is 60 mi/h. If you drive at the posted speed, will you reach the destination in 2 h? Or in more than 2 h?
20. Suppose that a freely falling object were somehow equipped with a speedometer. By how much would its speed reading increase with each second of fall?
21. Suppose that the freely falling object in the previous exercise were also equipped with an odometer. Would the readings of distance fallen each second indicate equal or unequal distances of fall for successive seconds? Explain.
22. When a ballplayer throws a ball straight up, by how much does the speed of the ball decrease each second while it is ascending? In the absence of air resistance, by how much does its speed increase each second while it is descending? How much time elapses for its ascent compared with its descent?
23. Someone standing on the edge of a cliff (as in Figure 2.24) throws a ball straight up at a certain speed and another ball straight down with the same initial speed. If air resistance is negligible, which ball has the greater speed when it strikes the ground below?
24. For a freely falling object dropped from rest, what is its acceleration at the end of the fifth second of fall? At the end of the tenth second? Defend your answer (and distinguish between velocity and acceleration).
25. Two balls, A and B, are released simultaneously from rest at the left end of the equal-length tracks A and B, as shown. Which ball will reach the end of its track first?

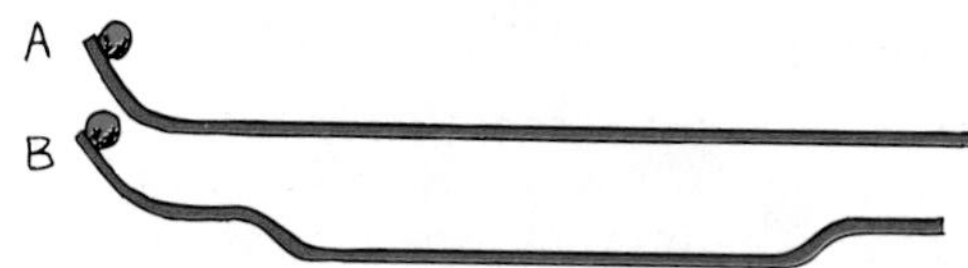

26. Refer to the tracks above.
    (a) Does ball B roll faster along the lower part of track B than ball A rolls along track A?
    (b) Is the speed gained by ball B going down the extra dip the same as the speed it loses going up near the right-hand end? And doesn't this mean that the speed of balls A and B will be the same at the ends of both tracks?
    (c) On track B, won't the average speed dipping down and up be greater than the average speed of ball A during the same time?
    (d) So, overall, does ball A or ball B have the greater average speed? (Do you wish to change your answer to Exercise 25?)

Problems

1. Suppose that a 30-N force and a 20-N force act on an object.
    (a) Show that when both forces act in the same direction the resultant force is 50 N.
    (b) Show that when both forces act in opposite directions the resultant force is 10 N. Is the resultant in the direction of the 30-N or the 20-N force?
2. A horizontal force of 100 N is required to push a box across a floor at a constant velocity.
    (a) Show that the net force acting on the box is zero.
    (b) Show that the friction force that acts on the box is 100 N.
3. A firefighter with a mass of 100 kg slides down a vertical pole at a constant speed. Show that the force of friction provided by the pole is about 1000 N.
4. The ocean's level is currently rising at about 1.5 mm per year. Show that at this rate the sea level will be 3 meters higher in 2000 years.
5. A vehicle changes its velocity from 90 km/h to a dead stop in 10 s. Show that its acceleration in doing so is $-2.5\ m/s^2$.
6. A ball is thrown straight up with an initial speed of 40 m/s.
    (a) Show that its time in the air is about 8 seconds.
    (b) Show that its maximum height, neglecting air resistance, is about 80 m.
7. Extend Table 2.2 (which gives values from 0 to 5 s) from 6 to 10 s, assuming no air resistance.
8. A ball is thrown straight up with enough speed so that it is in the air for several seconds.
    (a) What is the velocity of the ball when it reaches its highest point?
    (b) What is its velocity 1 s before it reaches its highest point?
    (c) What is the change in its velocity during this 1-s interval?
    (d) What is its velocity 1 s after it reaches its highest point?
    (e) What is the change in velocity during this 1-s interval?
    (f) What is the change in velocity during the 2-s interval from 1 s before the highest point to 1 s after the highest point? (Caution: We are asking for velocity, not speed.)
    (g) What is the acceleration of the ball during any of these time intervals and at the moment the ball has zero velocity at the top of its path?

Conceptual Science Place

CHAPTER 2 ONLINE RESOURCES

Interactive Figures 2.24 • ***Videos*** Friction, Definition of Speed, Average Speed, Velocity, Changing Velocity, Definition of Acceleration, Force Causes Acceleration, Free Fall: How Far? • ***Quiz*** • ***Flashcards*** • ***Links***

CHAPTER 3

Newton's Laws of Motion

The parachutist accelerates under the influence of gravity until he reaches his terminal velocity; he then falls with constant velocity to Earth.

CHAPTER OUTLINE

A heavy parachutist falls faster than a lighter one and, therefore, has a rougher landing—but why? Have you tried the party trick where you pull a tablecloth out from under place settings and the dishes stay put? How does this "trick" work, and what law of motion does it demonstrate? Have you heard the expression "You can't touch without being touched"? Does this statement about the objective world of physics have a corollary in the world of human emotions? How did Newton's laws get us to the Moon? How do birds fly, rockets take off, and people walk? How do Newton's laws of motion interface with modern discoveries about motion gained from relativity and quantum mechanics? You'll learn all this and much more in this chapter.

3.1 Newton's First Law of Motion

Galileo's work set the stage for Isaac Newton, who was born shortly after Galileo's death in 1642. By the time Newton was 23, he had developed his famous three laws of motion, which completed the overthrow of Aristotelian ideas about motion. These three laws first appeared in one of the most famous books of all time, Newton's *Philosophiae Naturalis Principia Mathematica*,* often simply known as the *Principia*. The first law is a restatement of Galileo's concept of inertia; the second law relates acceleration to its cause—force; and the third is the law of action and reaction. **Newton's first law** is:

> **Every object continues in its state of rest, or a uniform speed in a straight line, unless acted on by a nonzero force.**

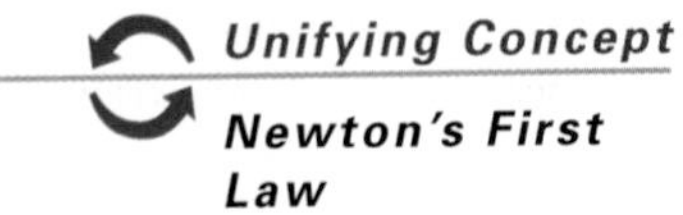

The key word in this law is *continues*; an object continues to do whatever it happens to be doing unless a force is exerted upon it. If the object is at rest, it continues in a state of rest. This is nicely demonstrated when a tablecloth is skillfully whipped from beneath dishes sitting on a tabletop, leaving the dishes in their initial state of rest (Figure 3.1).† On the other hand, if an object is moving, it continues to move without changing its speed or direction, as evidenced by

*The Latin title means "mathematical principles of natural philosophy."

†Close inspection reveals that brief friction between the dishes and the fast-moving tablecloth starts the dishes moving, but then friction between the dishes and the table stops the dishes before they slide very far. If you try this, use unbreakable dishes!

Figure 3.1 Inertia in action.

Figure 3.2 Rapid deceleration is sensed by the driver, who lurches forward—inertia in action!

Conceptual Science Place

Newton's Law of Inertia

space probes that continually move in outer space. This property of objects to resist changes in motion is called **inertia** (Figures 3.1 and 3.2).

CHECK YOURSELF

When a space shuttle travels in a nearly circular orbit around the Earth, is a force required to maintain its high speed? If the force of gravity were suddenly cut off, what type of path would the shuttle follow?

CHECK YOUR ANSWER

There is no force in the direction of the shuttle's motion, which is why it coasts at a constant speed by its own inertia. The only force acting on it is the force of gravity, which acts at right angles to its motion (toward the Earth's center). We'll see later that this right-angled force holds the shuttle in a circular path. If it were cut off, the shuttle would fly off in a straight line at a constant velocity.

3.2 Newton's Second Law of Motion

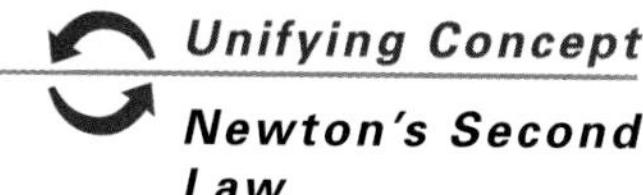

Unifying Concept

Newton's Second Law

Isaac Newton was the first to recognize the connection between force and mass in producing acceleration, which is one of the most central rules of nature, as expressed in his second law of motion. **Newton's second law** states:

> **The acceleration produced by a net force on an object is directly proportional to the net force, is in the same direction as the net force, and is inversely proportional to the mass of the object.**

Or, in shorter notation,

$$\text{Acceleration} \sim \frac{\text{net force}}{\text{mass}}$$

By using consistent units such as newtons (N) for force, kilograms (kg) for mass, and meters per second squared (m/s^2) for acceleration, we produce the exact equation:

$$\text{Acceleration} = \frac{\text{net force}}{\text{mass}}$$

In its briefest form, where a is acceleration, F is net force, and m is mass:

$$a = \frac{F}{m}$$

MATH CONNECTION

Equations as Guides to Thinking: $a = \frac{F}{m}$

Newton's second law is not only simple in form but also widely applicable, so it's a highly useful tool in many problem-solving situations—it is well worth your while to become adept at using it.

For example, if we know the mass of an object in kilograms (kg) and its acceleration in meters per second (m/s^2), then the force will be expressed in newtons (N). One newton is the force needed to give a mass of one kilogram an acceleration of one meter per second squared. We can rearrange Newton's law to read

$$\text{Force} = \text{mass} \times \text{acceleration}$$
$$1\,\text{N} = (1\,\text{kg}) \times (1\,\text{m/s}^2)$$

We can see that

$$1\,\text{N} = 1\,\text{kg}\cdot\text{m/s}^2$$

The centered dot between "1 kg" and "m/s^2" means that one is multiplied by the other.

If we know two of the quantities in Newton's second law, we can calculate the third. For example, how much force, or thrust, must a 20,000-kg jet plane develop to achieve an acceleration of 1.5 m/s^2? Using the equation, we can calculate

$$\begin{aligned} F &= ma \\ &= (20{,}000\,\text{kg}) \times (1.5\,\text{m/s}^2) \\ &= 30{,}000\,\text{kg}\cdot\text{m/s}^2 \\ &= 30{,}000\,\text{N} \end{aligned}$$

Suppose that we know the force and the mass and that we want to find the acceleration. For example, what acceleration is produced by a force of 2000 N applied to a 1000-kg car? Using Newton's second law, we find that

$$\begin{aligned} a &= \frac{F}{m} = \frac{2000\,\text{N}}{1000\,\text{kg}} \\ &= \frac{2000\,\text{kg}\cdot\text{m/s}^2}{1000\,\text{kg}} = 2\,\text{m/s}^2 \end{aligned}$$

If the force is 4000 N, the acceleration is

$$\begin{aligned} a &= \frac{F}{m} = \frac{4000\,\text{N}}{1000\,\text{kg}} \\ &= \frac{4000\,\text{kg}\cdot\text{m/s}^2}{1000\,\text{kg}} = 4\,\text{m/s}^2 \end{aligned}$$

Doubling the force on the same mass simply doubles the acceleration.

Physics problems are often more complicated than these. We don't focus on solving complicated problems in this book; instead, we emphasize equations, such as Newton's second law, in which the relationships among physical quantities are clear. Such equations serve as guides to thinking, rather than recipes for mathematical problem solving. Remember, mastering concepts first makes problem solving more meaningful.

When Acceleration Is g*—Free Fall*

Only a single force acts on something in free fall—the force of gravity.

Although Galileo articulated both the concepts of inertia and acceleration and was the first to measure the acceleration of falling objects, he was unable to explain why objects of various masses fall with equal accelerations. Newton's second law provides the explanation.

We know that a falling object accelerates toward Earth because of the gravitational force of attraction between the object and Earth. As discussed in Chapter 2, when the force of gravity is the only force—that is, when air resistance is negligible—we say that the object is in a state of *free fall.* An object in free fall accelerates toward Earth at 10 m/s^2 (or, more precisely, at 9.8 m/s^2).

The greater the mass of an object, the stronger the gravitational pull between it and the Earth. The double brick in Figure 3.7, for example, has twice the gravitational attraction to Earth as the single brick. Why, then, doesn't the double brick fall twice as fast, as Aristotle supposed it would? The answer is evident in Newton's second law: the acceleration of an object depends not only on the force (weight, in this case) but also on the object's resistance to motion—its inertia (mass). Whereas a force produces an acceleration, inertia is a *resistance* to acceleration. So twice the force exerted on twice the inertia produces the same acceleration as half the force exerted on half the inertia. Both accelerate equally. The acceleration due to gravity is symbolized by *g*. We use the symbol *g*, rather than *a*, to denote that acceleration is due to gravity alone.

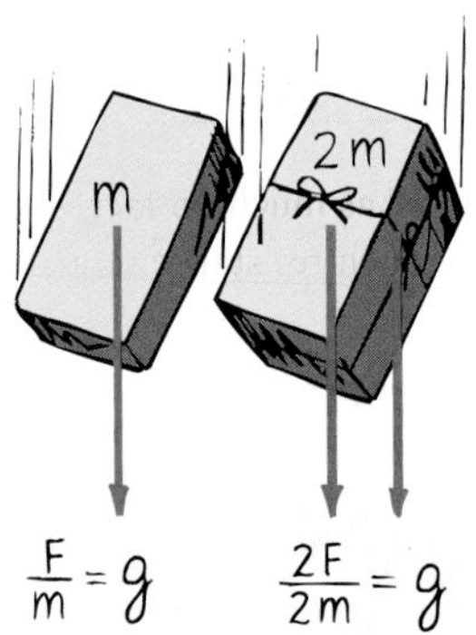

Figure 3.7 Interactive Figure

The ratio of weight (F) to mass (m) is the same for all objects in the same locality; hence, their accelerations are the same in the absence of air resistance.

So the ratio of weight to mass for freely falling objects equals a constant, *g*. This is similar to the constant ratio of circumference to diameter for circles, which equals the constant π. The ratio of weight to mass is identical both for

INSIGHT When Galileo tried to explain why all objects fall with equal accelerations, wouldn't he have loved to know the rule $a = \frac{F}{m}$?

heavy objects and for light objects, just as the ratio of circumference to diameter is the same both for large and small circles (Figure 3.8).

We now understand that the acceleration of free fall is independent of an object's mass. A boulder 100 times more massive than a pebble falls at the same acceleration as the pebble because, although the force on the boulder (its weight) is 100 times that of the pebble, the greater force offsets the equally greater mass.

CHECK YOURSELF

In a vacuum, a coin and a feather fall at an equal rate, side by side. Would it be correct to say that equal forces of gravity act on both the coin and feather in the vacuum?

CHECK YOUR ANSWER

No, no, no—a thousand times no! These objects accelerate equally not because of equal forces of gravity acting on them but because the ratios of their weights to their masses are equal. Although air resistance is not present in a vacuum, gravity is. (You'd know this if you placed your hand in a vacuum chamber and a cement truck rolled over it.) If you answered *yes* to this question, let this be a signal to be more careful when you think about physics.

When Acceleration Is Less Than g*—Non-Free Fall*

Unifying Concept
Friction
Section 2.8

Most often, air resistance is not negligible for falling objects. Then the acceleration of the object's fall is less. Air resistance, which is the force of friction acting between an object and the surrounding air, depends on two things: speed and surface area. When a skydiver steps from a high-flying plane, the air resistance on the skydiver's body builds up as the falling speed increases. The result is reduced acceleration. The acceleration can be reduced further by increasing the surface area. A skydiver does this by orienting his or her body so that more air is encountered by its surface—by spreading out like a flying squirrel. So air resistance depends both on speed and on the surface area encountered by the air.

For free fall, the downward net force is weight—only weight. But, when air is present, the downward net force = weight − air resistance. Can you see that the presence of air resistance reduces net force? And that less force means less acceleration? So, as a skydiver falls faster and faster, the acceleration of fall

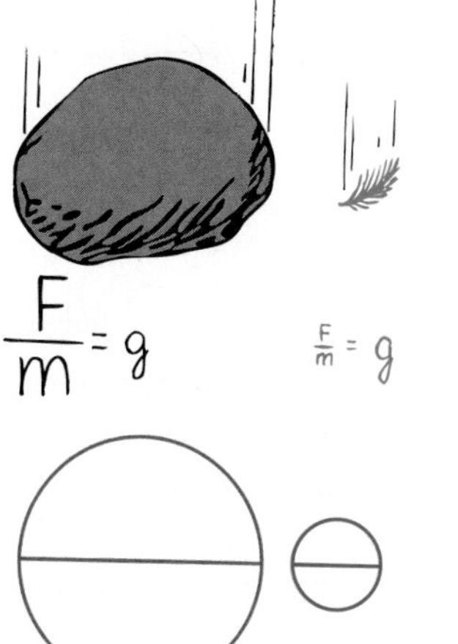

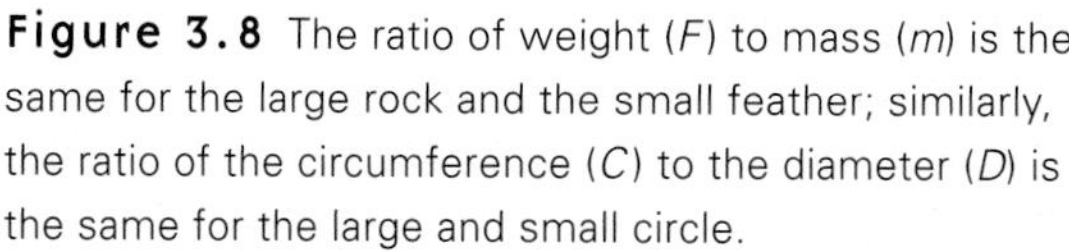

Figure 3.8 The ratio of weight (F) to mass (m) is the same for the large rock and the small feather; similarly, the ratio of the circumference (C) to the diameter (D) is the same for the large and small circle.

Figure 3.9 In a vacuum, a feather and a coin fall with equal accelerations.

Figure 3.10 Interactive Figure
The heavier parachutist must fall faster than the lighter parachutist for air resistance to cancel her greater weight.

Figure 3.11 A flying squirrel increases its frontal area by spreading out. The result is greater air resistance and a slower fall.

When the force of gravity and air resistance act on a falling object, it is not in free fall.

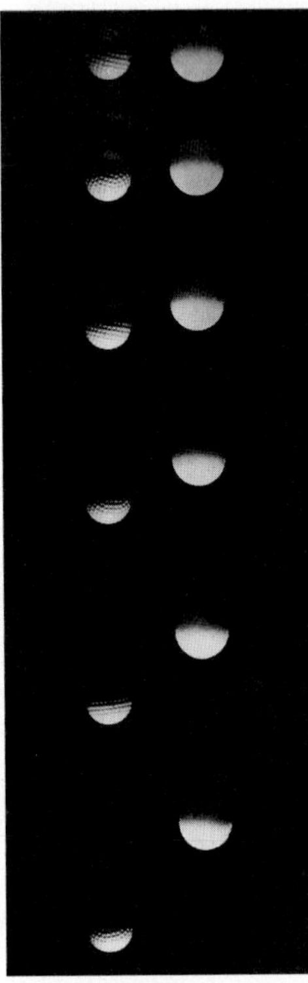

Figure 3.12 Interactive Figure

A stroboscopic study of a golf ball (left) and a Styrofoam ball (right) falling in air. The air resistance is negligible for the heavier golf ball, and its acceleration is nearly equal to *g*. Air resistance is not negligible for the lighter Styrofoam ball, however, and it reaches its terminal velocity sooner.

decreases.* What happens to the net force if air resistance builds up to equal the weight of the skydiver? The answer is, net force becomes zero. Does this mean the skydiver comes to a stop? No! What it means is that the skydiver no longer gains speed. Acceleration terminates—it no longer occurs. We say that the skydiver has reached **terminal speed.** If we are concerned with direction—down, for falling objects—we say that the skydiver has reached *terminal velocity*.

Terminal speed for a human skydiver varies from about 150 to 200 km/h, depending on the weight, size, and orientation of the body. A heavier person has to fall faster for air resistance to balance weight.† The greater weight is more effective in "plowing through" the air, resulting in more terminal speed for a heavier person. Increasing frontal area reduces terminal speed. That's where a parachute is useful. A parachute increases frontal area, which greatly increases air resistance, reducing terminal speed to a safe 15 to 25 km/h.

Consider the interesting demonstration of the falling coin and the feather in the glass tube. When air is inside the tube, we see that the feather falls more slowly because of air resistance. The feather's weight is very small, so it reaches terminal speed very quickly because it doesn't have to fall very far or very fast before air resistance builds up to equal its small weight. The coin, on the other hand, doesn't have a chance to fall fast enough for air resistance to build up to equal its weight. If you were to drop a coin from a very high location, such as from the top of a tall building, its terminal speed would be reached when the speed of the coin would be greater than 100 km/h. This is a much, much higher terminal speed than that of a falling feather.

When Galileo allegedly dropped objects of different weights from the Leaning Tower of Pisa, they didn't actually hit at the same time. They almost did, but, because of air resistance, the heavier one hit slightly before the other. But this contradicted the much longer time difference expected by the followers of Aristotle. The behavior of falling objects was never really understood until Newton announced his second law of motion.

CHECK YOURSELF

Consider two parachutists, a heavy person and a light person, who jump from the same altitude with parachutes of the same size.

1. Which person reaches terminal speed first?
2. Which person has the higher terminal speed?
3. Which person reaches the ground first?
4. If there were no air resistance, like on the Moon, how would your answers to the questions differ?

CHECK YOUR ANSWERS

To answer these questions correctly, think of a coin and a feather falling in air.

1. Just as a feather reaches terminal speed very quickly, the lighter person reaches terminal speed first.
2. Just as a coin falls faster than a feather through air, the heavy person falls faster and reaches a higher terminal speed.
3. Just as in the race between a falling coin and a falling feather, the heavier person falls faster and will reach the ground first.
4. If there were no air resistance, there would be no terminal speed at all. Both would be in free fall, and both would hit the ground at the same time.

*In mathematical notation, $a = \frac{F_{net}}{m} = \frac{mg - R}{m}$, where mg is the weight and R is the air resistance. Note that, when $R = mg$, $a = 0$; then, with no acceleration, the object falls at a constant velocity. With elementary algebra, we can proceed another step and get $a = \frac{net}{m} = \frac{mg - R}{m} = g - R/m$. We see that the acceleration a will always be less than g if air resistance R impedes falling. Only when $R = 0$ does $a = g$.

†A skydiver's air resistance is proportional to speed squared.

MATH CONNECTION

When Air Resistance Slows Acceleration

The effect of air resistance on acceleration can be made clearer with some problem-solving practice. Examine the problems and solutions. You'll have more practice at the end of the chapter.

Problems

1. A skydiver jumps from a high-flying helicopter. As she falls faster and faster through the air, does her acceleration increase, decrease, or remain the same?
2. What will be the acceleration of the skydiver when air resistance builds up to be equal to half her weight?

Solutions

1. Acceleration decreases because the net force on her decreases. Net force is equal to her weight minus her air resistance, and, because air resistance increases with increasing speed, net force and, therefore, acceleration also decrease. According to Newton's second law,

$$a = \frac{F_{net}}{m} = \frac{mg - R}{m}$$

where mg is her weight and R is the air resistance she encounters. As R increases, a decreases. Note that, if she falls fast enough so that $R = mg$, $a = 0$; then, with no acceleration, she falls at terminal speed.

2. We find the acceleration from

$$a = \frac{F_{net}}{m} = \frac{mg - R}{m} = \frac{(mg - mg/2)}{m} = \frac{(mg/2)}{m} = \frac{g}{2}$$

INTEGRATED SCIENCE BIOLOGY

Gliding

INSIGHT Action and reaction are two equal and oppositely directed forces that are coequal parts of a single interaction between two different things.

Only three groups of living organisms—birds, bats, and insects—can truly fly. Gliding, however, has evolved many more times in the biological world. Gliding describes a mode of locomotion in which animals move through the air in a controlled fall. There are gliding squirrels, gliding lizards, gliding snakes, gliding frogs, even gliding ants. Although gliders cannot generate the forward thrust that allows fliers to power through the air, many gliders nevertheless have remarkable control—many, for example, are able to execute sharp turns in midair.

How does gliding work? When an animal jumps out of a tree, it falls toward the ground due to the downward force of gravity. Air resistance slows the animal's fall, just as it slows the motion of any object moving through air. The more air resistance an animal encounters, the slower and more controllable its fall. Since the amount of air resistance a falling object encounters depends on the object's surface area, all gliding animals have evolved special structures that increase their surface area. "Flying" squirrels have large flaps of skin between their front and hind legs. "Flying dragons" (gliding lizards of the genus *Draco*) have long extendable ribs that support large gliding membranes. "Flying" frogs have very long toes with extensive webbing between them. Gliding geckos have skin flaps along their sides and tails in addition to webbed toes. Gliding tree snakes spread out their ribs and suck in their stomachs when they leap off a branch, creating a concave parachute to slow their descent.

Gliding locomotion is particularly common in certain types of habitat. For example, the presence of tall trees without much "clutter" between them has resulted in the evolution of many gliding species in the rainforests of Southeast Asia. Gliding offers a number of advantages. First, it allows rapid, energetically efficient descent, which is useful in many contexts, such as escaping from predators. Second, gliding allows animals to move from one tree to another without descending all the way to the ground and climbing back up. This is, again, energy efficient. And it also allows gliders to avoid potentially dangerous forest understories.

CHECK YOURSELF

Would it be harder or easier for an animal to glide effectively at high altitudes?

CHECK YOUR ANSWER

Because air is thinner at high altitudes, there is less air resistance. So, it would be harder for an animal to glide effectively.

3.3 Forces and Interactions

Conceptual Science Place

Forces and Interactions

So far, we've treated force in its simplest sense—as a push or a pull. In a broader sense, a force is not a thing in itself but an **interaction** between one thing and another. If you push a wall with your fingers, more is happening than you pushing on the wall. You're interacting with the wall, and the wall is simultaneously pushing on you. The fact that your fingers and the wall push on each other is evident in your bent fingers (Figure 3.13). Your push on the wall and the wall's push on you are equal in magnitude (amount) and opposite in direction. The pair of forces constitutes a single interaction. In fact, you can't push on the wall unless the wall pushes back.*

In Figure 3.14, we see a boxer's fist hitting a massive punching bag. The fist hits the bag (and dents it) while the bag hits back on the fist (and stops its motion). This force pair is fairly large. But what if the boxer were hitting a piece of tissue paper? The boxer's fist can exert only as much force on the tissue paper as the tissue paper can exert on the boxer's fist. Furthermore, the fist can't exert any force at all unless what is being hit exerts the same amount of reaction force. An interaction requires a pair of forces acting on two different objects.

When a hammer hits a stake and drives it into the ground, the stake exerts an equal amount of force on the hammer, and that force brings the hammer to an abrupt halt. And when you pull on a cart, the cart pulls back on you, as evidenced, perhaps, by the tightening of the rope wrapped around your hand. One thing interacts with another: the hammer interacts with the stake, and you interact with the cart.

Figure 3.13 When you lean against a wall, you exert a force on the wall. The wall simultaneously exerts an equal and opposite force on you. Hence, you don't topple over.

Figure 3.14 He can hit the massive bag with considerable force. But, with the same punch, he can exert only a tiny force on the tissue paper in midair.

Figure 3.15 In the interaction between the hammer and the stake, each exerts the same amount of force on the other.

*We tend to think only of living things pushing and pulling, but inanimate things can do likewise. So please don't be troubled about the idea of the inanimate wall pushing back at you. It does push back, just as another person pushing back at you would.

3.4 Newton's Third Law of Motion

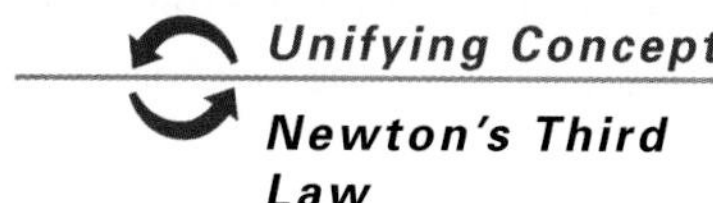

Newton's **third law** states:

> **Whenever one object exerts a force on a second object, the second object exerts an equal and opposite force on the first.**

We can call one force the action force, and we can call the other the reaction force. The important thing is that they are coequal parts of a single interaction and that neither force exists without the other. Action and reaction forces are equal in strength and opposite in direction. They occur in pairs, and they make up a single interaction between two things (Figure 3.16).

When walking, you interact with the floor. Your push against the floor is coupled to the floor's push against you. The pair of forces occurs simultaneously. Likewise, the tires of a car push against the road while the road pushes back on the tires—the tires and the road push against each other. In swimming, you interact with the water that you push backward, while the water pushes you forward—you and the water push against each other. The reaction forces are what account for our motion in these cases. These simultaneous forces depend on friction; a person or a car on ice, for example, may not be able to exert the action force necessary to produce the desired reaction force.

INSIGHT You can't pull on something unless that something simultaneously pulls back on you. That's the law!

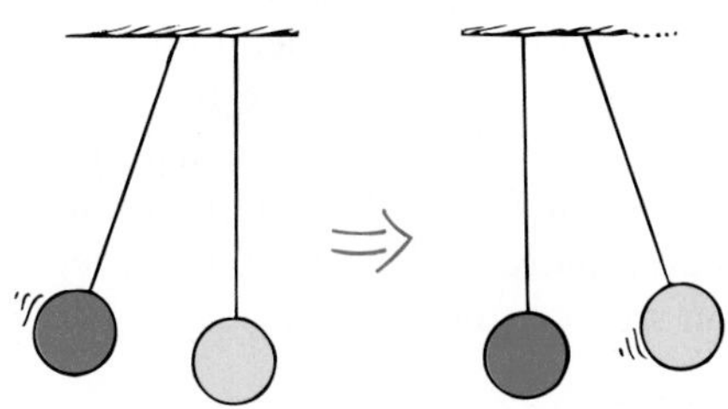

Figure 3.16 The impact forces between the blue and yellow ball move the yellow ball and stop the blue ball.

Simple Rule to Identify Action and Reaction

There is a simple rule for identifying action and reaction forces. First, identify the interaction—one thing (object A) interacts with another (object B). Then, action and reaction forces can be stated in the following form:

> **Action: Object A exerts a force on object B.**
>
> **Reaction: Object B exerts a force on object A.**

The rule is easy to remember. If action is A acting on B, reaction is B acting on A. We see that A and B are simply switched around. Consider the case of your hand pushing on the wall. The interaction is between your hand and the wall. We'll say the action is your hand (object A) exerting a force on the wall (object B). Then the reaction is the wall exerting a force on your hand.

Action: tire pushes on road Reaction: road pushes on tire

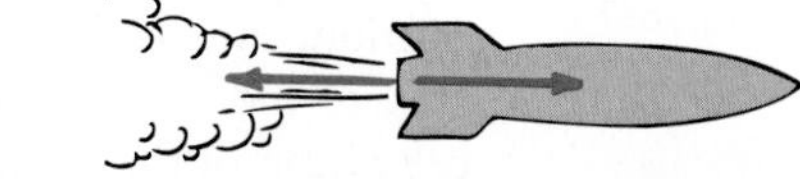

Action: rocket pushes on gas Reaction: gas pushes on rocket

Action: man pulls on spring Reaction: spring pulls on man

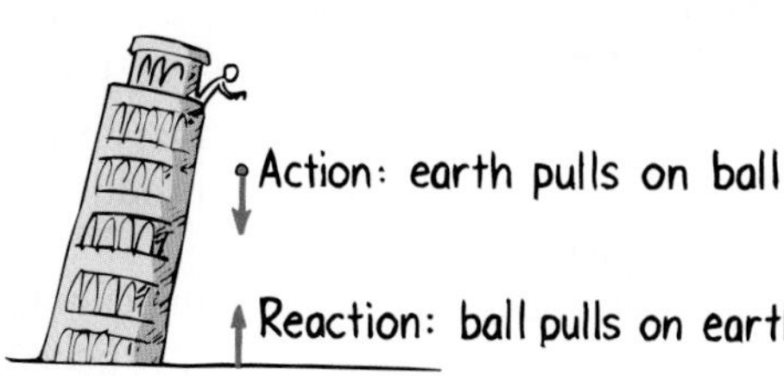

Figure 3.17 Action and reaction forces. Note that, when action is "A exerts force on B," the reaction is then simply "B exerts force on A."

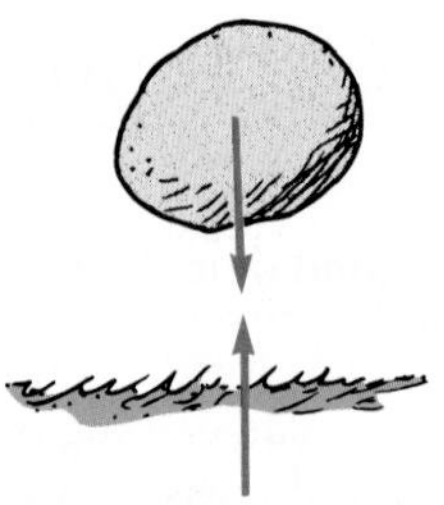

Figure 3.18 Earth is pulled up by the boulder with just as much force as the boulder is pulled down by Earth.

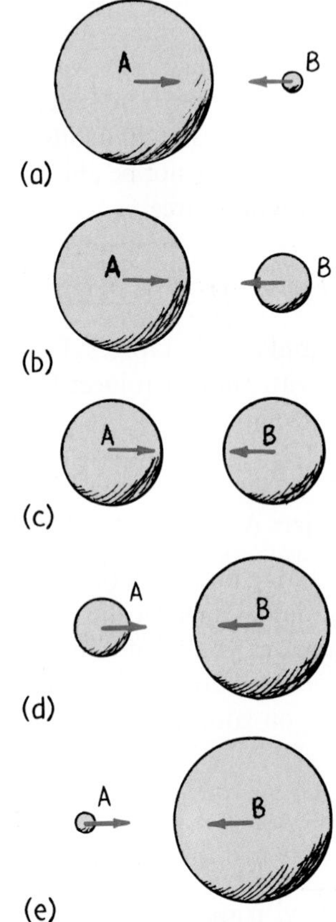

Figure 3.19 Which falls toward the other, A or B? Do the accelerations of each relate to their relative masses?

Figure 3.20 Interactive Figure

The force exerted against the recoiling cannon is just as great as the force that drives the cannonball along the barrel. Why, then, does the cannonball undergo more acceleration than the cannon?

CHECK YOURSELF

1. A car accelerates along a road. Identify the force that moves the car.
2. Identify the action and reaction forces for the case of an object in free fall (with no air resistance).

CHECK YOUR ANSWERS

1. It is the road that pushes the car along. Except for air resistance, only the road provides a horizontal force on the car. How does it do this? The rotating tires of the car push back on the road (action). The road simultaneously pushes forward on the tires (reaction). How about that!
2. To identify a pair of action–reaction forces in any situation, first identify the pair of interacting objects. In this case, Earth interacts with the falling object via the force of gravity. So Earth pulls the falling object downward (call it action). Then reaction is the falling object pulling Earth upward. It is only because of Earth's enormous mass that you don't notice its upward acceleration.

Action and Reaction on Different Masses

Quite interestingly, a falling object pulls upward on Earth with as much force as the Earth pulls downward on it. The resulting acceleration of the falling object is evident, while the upward acceleration of Earth is too small to detect (Figure 3.18).

Consider the exaggerated examples of two planetary bodies, shown in parts (a) through (e) in (Figure 3.19). The forces between bodies A and B are equal in magnitude and oppositely directed in each case. If the acceleration of Planet A is unnoticeable in (a), then it is more noticeable in (b), where the difference between masses is less extreme. In (c), where both bodies have equal mass, the acceleration of Planet A is as evident as it is for Planet B. Continuing, we see that the acceleration of Planet A becomes even more evident in (d) and even more so in (e). So, strictly speaking, when you step off the curb, the street rises ever so slightly to meet you.

When a cannon is fired, there is an interaction between the cannon and the cannonball (Figure 3.20). The sudden force that the cannon exerts on the cannonball is exactly equal and opposite to the force the cannonball exerts on the cannon. This is why the cannon recoils (kicks). But the effects of these equal forces are very different. This is because the forces act on different masses. The different accelerations are evident via Newton's second law,

$$a = \frac{F}{m}$$

Let F represent both the action and reaction forces, m the mass of the cannon, and $\mathcal{m}$ the mass of the cannonball. Different-sized symbols are used here to indicate the relative masses and the resulting accelerations. Then the acceleration of the cannonball and cannon can be represented in the following way.

$$\text{cannonball: } \frac{F}{m} = \mathcal{A}$$

$$\text{cannon: } \frac{F}{\mathcal{M}} = a$$

Thus we see why the change in velocity of the cannonball is so large compared with the change in velocity of the cannon. A given force exerted on a small mass produces a large acceleration, while the same force exerted on a large mass produces a small acceleration.

We can extend the idea of a cannon recoiling from the ball it fires to understanding rocket propulsion. Consider an inflated balloon recoiling when air is expelled. If the air is expelled downward, the balloon accelerates upward. The

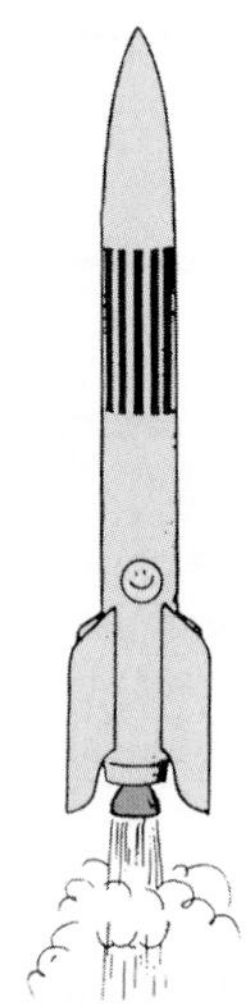

Figure 3.21 Interactive Figure

The rocket recoils from the "molecular cannonballs" it fires, and it rises.

same principle applies to a rocket, which continually "recoils" from the ejected exhaust gas. Each molecule of exhaust gas is like a tiny cannonball shot from the rocket (Figure 3.21).

A common misconception is that a rocket is propelled by the impact of exhaust gases against the atmosphere. In fact, before the advent of rockets, it was commonly thought that sending a rocket to the Moon was impossible. Why? Because there is no air above Earth's atmosphere for the rocket to push against. But this is like saying a cannon wouldn't recoil unless the cannonball had air to push against. Not true! Both the rocket and recoiling cannon accelerate because of the reaction forces by the material they fire, not because of any pushes on the air. In fact, a rocket operates better above the atmosphere, where there is no air resistance.

CHECK YOURSELF

1. We know that Earth pulls on the Moon. Does it follow that the Moon also pulls on Earth?
2. Which pulls harder, Earth on the Moon or the Moon on Earth?
3. A high-speed bus and an unfortunate bug have a head-on collision. The force of the bus on the bug splatters it all over the windshield. Is the corresponding force of the bug on the bus greater than, less than, or the same as the force of the bus on the bug? Is the resulting deceleration of the bus greater than, less than, or the same as that of the bug?

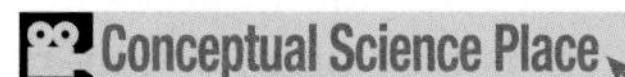

Action and Reaction on Different Mass

CHECK YOUR ANSWERS

1. Yes, both pulls make up an action–reaction pair of forces associated with the gravitational interaction between Earth and the Moon. We can say that (a) Earth pulls on the Moon, and (b) the Moon likewise pulls on Earth; but it is more insightful to think of this as a single interaction—Earth and Moon simultaneously pulling on each other, each with the *same* amount of force.
2. Both pull with the same strengths. This is like asking which distance is greater; from San Francisco to New York, or from New York to San Francisco. Both distances, like both forces in the Moon–Earth pulls, are the same.
3. The magnitudes of the forces are the same, because they constitute an action–reaction force pair that makes up the interaction between the bus and the bug. The accelerations, however, are very different, because the masses are different. The bug undergoes an enormous and lethal deceleration, while the bus undergoes a very tiny deceleration—so tiny that the very slight slowing of the bus is unnoticed by its passengers. But, if the bug were more massive—as massive as another bus, for example—the slowing down would be very apparent.

Figure 3.22 Interactive Figure

A force acts on the orange, and the orange accelerates to the right.

Defining Your System

An interesting question often arises: since action and reaction forces are equal and opposite, why don't they cancel to zero? To answer this question, we must consider the system under consideration. Consider, for example, a system consisting of a single orange, Figure 3.22. The dashed line surrounding the orange encloses and defines the system. The vector that pokes outside the dashed line represents an external force on the system. The system accelerates in accord with Newton's second law. In Figure 3.23, we see that this force is provided by the apple, which doesn't change our analysis. The apple is outside the system. The fact that the orange simultaneously exerts a force on the apple, which is external to the system, may affect the apple (another system), but not the orange. You can't cancel a force on the orange with a force on the apple. So, in this case, the action–reaction forces don't cancel.

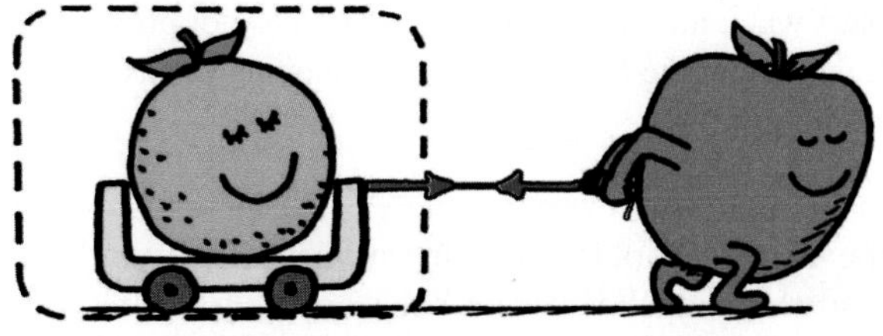

Figure 3.23 Interactive Figure

The force on the orange, provided by the apple, is not cancelled by the reaction force on the apple. The orange still accelerates.

Figure 3.24 Interactive Figure
In the larger system of orange + apple, action and reaction forces are internal and cancel. If these are the only horizontal forces, with no external force, no acceleration of the system occurs.

Figure 3.25 Interactive Figure
An external horizontal force occurs when the floor pushes on the apple (reaction to the apple's push on the floor). The orange–apple system accelerates.

INSIGHT A system may be as tiny as an atom or as large as the universe.

Now let's consider a larger system, enclosing both the orange and the apple. We see the system bounded by the dashed line in Figure 3.24. Notice that the force pair is internal to the orange–apple system. These forces do cancel each other. They play no role in accelerating the system. A force external to the system is needed for acceleration. That's where friction with the floor comes in (Figure 3.25). When the apple pushes against the floor, the floor simultaneously pushes on the apple—an external force on the system. The system accelerates to the right.

Inside a baseball are trillions and trillions of interatomic forces at play. They hold the ball together, but they play no role in accelerating the ball. Although every one of the interatomic forces is part of an action–reaction pair within the ball, they combine to zero, no matter how many of them there are. A force external to the ball, such as a swinging bat provides, is needed to accelerate the ball.

If this is confusing, it may be well to note that Newton had difficulties with the third law himself.

CHECK YOURSELF

1. On a cold, rainy day, your car battery is dead, and you must push the car to move it and get it started. Why can't you move the car by remaining comfortably inside and pushing against the dashboard?
2. Does a fast-moving baseball possess force?

CHECK YOUR ANSWERS

1. In this case, the system to be accelerated is the car. If you remain inside and push on the dashboard, the force pair you produce acts and reacts within the system. These forces cancel out, as far as any motion of the car is concerned. To accelerate the car, there must be an interaction between the car and something external—for example, you on the outside pushing against the road.
2. No, a force is not something an object has, like mass; it is part of an interaction between one object and another. A speeding baseball may possess the capability of exerting a force on another object when interaction occurs, but it does not possess force as a thing in itself. As we will see in the following chapter, moving things possess momentum and kinetic energy.

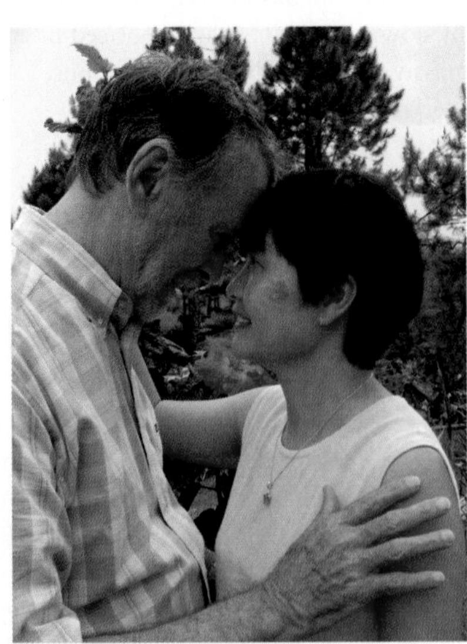

Figure 3.26 You cannot touch without being touched—Newton's third law.

We see Newton's third law in action everywhere. A fish propels water backward with its fins, and the water propels the fish forward. The wind caresses the branches of a tree, and the branches caress back on the wind to produce whistling sounds. Forces are interactions between different things. Every contact requires at least a twoness; there is no way that an object can exert a force on nothing. Forces, whether large shoves or slight nudges, always occur in pairs, each opposite to the other. Thus, we cannot touch without being touched (Figure 3.26).

INTEGRATED SCIENCE BIOLOGY

Animal Locomotion

The study of how animals move, *animal locomotion*, is a branch of *biophysics*. Biophysics draws both on physics and on biology and is one of many crossover science disciplines thriving today.

Much of the study of animal locomotion is based on Newton's third law. To move forward, an animal pushes back on something else; the reaction force pushes the animal forward. For example, birds fly by pushing against air. When they flap their wings, birds push the air downward. The air, in turn, pushes the bird upward. When a bird is soaring, its wing is shaped so that moving air particles are deflected downward; the upward reaction force is lift. A fish swims by pushing against water—the fish propels water backward with its fins, and the water propels the fish forward. Likewise, land animals such as humans push against the ground, and the ground in turn pushes them forward.

Let's consider an example of animal locomotion on land in more detail. Specifically, how does Newton's third law come into play when you walk? First, note that, when you are standing still, you are not accelerating. The forces that act on you, gravity and the normal force, balance as shown in Figure 3.27a. To walk, you must accelerate horizontally—the vertical forces of gravity and the normal force don't help. The forces involved in walking are horizontal *frictional* forces (Figure 3.27b). Because your feet are firmly pressed to the floor, there is friction when you push your foot horizontally against the floor. By Newton's third law, the floor pushes back on you in the opposite direction—forward. So the reaction force that allows you to walk is the friction force that the floor applies to your foot, and thereby to your mass as a whole. (Don't be confused by all the internal forces within your body that are involved in walking, such as the rotation of your bones and stretching of your muscle and tendons. An *external* force must act on your body to accelerate it; friction is that force.)

After friction nudges you forward from a standstill, your step is like a controlled fall. You step forward, and your body drops a short distance until your front foot becomes planted in front of you. Friction, as shown in Figure 3.27c, acts in the opposite direction now as it prevents your front foot from sliding forward.

Locomotion is important for many life functions (eating, finding mates, escaping predators, etc.). Biophysical research in this area, therefore, has beneficial applications for countless animals—human and otherwise—that have impaired locomotion.

CHECK YOURSELF

1. In what way is the study of animal locomotion an integrated science?
2. Why is Newton's third law such a necessary piece of information for understanding animal locomotion?

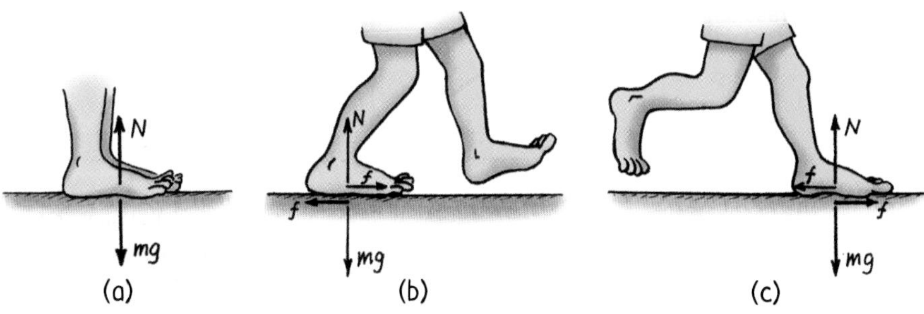

Figure 3.27 (a) Standing still, you push against the floor with a force equal to your weight; the normal force pushes you back equally—action and reaction. (b) Your lifted foot doesn't accelerate you horizontally. It's that frictional force you apply to the floor and its frictional push back on you that moves you forward. (c) When your forward foot lands, friction acts again but in the opposite direction. Friction stops your foot from slipping forward as the rest of your body catches up.

3.2 Newton's Second Law of Motion

3. Cite Newton's second law of motion.
4. (a) Is acceleration *directly* proportional to force, or is it *inversely* proportional to force? Give an example.
(b) Is acceleration *directly* proportional to mass, or is it *inversely* proportional to mass? Give an example.
5. If the mass of a sliding block is tripled at the same time the net force on it is tripled, how does the resulting acceleration compare with the original acceleration?
6. What is the acceleration of a 10-N freely falling object with no air resistance?
7. Why doesn't a heavy object accelerate more than a light object when both are freely falling?
8. What is the acceleration of a falling object that has reached its terminal velocity?
9. What two things does air resistance depend upon?
10. Who falls faster when wearing the same-size parachute—a heavy person, a light person, or both the same?

3.3 Forces and Interactions

11. How many forces are required for a single interaction?
12. When you push against a wall with your fingers, they bend because they experience a force. Identify this force.
13. A boxer can hit a heavy bag with a great force. Why can't he hit a sheet of newspaper in midair with the same amount of force?

3.4 Newton's Third Law of Motion

14. Cite Newton's third law of motion.
15. Consider hitting a baseball with a bat. If we call the force on the bat against the ball the action force, identify the reaction force.
16. Do action and reaction forces act in succession, or simultaneously?
17. If the forces that act on a cannonball and the recoiling cannon from which it is fired are equal in magnitude, why do the cannonball and cannon have very different magnitudes?
18. What is needed to accelerate a system?

3.5 Vectors

19. Cite three examples of a vector quantity. Then cite three example of a scalar quantity.
20. How great is the resultant of two equal-magnitude vectors at right angles to each other?
21. According to the parallelogram rule, what does the diagonal of a constructed parallelogram represent?
22. (a) Can it be said that, when a pair of vectors are at right angles to each other, the resultant is greater than either of the vectors separately? (b) When a vector at an angle is resolved into horizontal and vertical components, can it be said that each component has less magnitude than the original vector?

Multiple Choice Questions

Choose the BEST answer to the following.

1. A heavy rock and a light rock in free fall (zero air resistance) have the same acceleration. The *reason* the heavy rock doesn't have a greater acceleration is that the
(a) force due to gravity is the same on each.
(b) air resistance is always zero in free fall.
(c) inertia of both rocks is the same.
(d) ratio of force to mass is the same.
(e) None of these.
2. A karate chop delivers a force of 3000 N to a board that breaks. The force that the board exerts on the hand during this event is
(a) less than 3000 N.
(b) 3000 N.
(c) greater than 3000 N.
(d) Need more information.
3. Two parachutists, a heavy person and a light person jumping from the same altitude, have the same size parachute. Which reaches the ground first?
(a) The heavy person.
(b) The light person.
(c) They reach the ground at the same time.
(d) Need more information.
4. What kind of path would Earth follow if suddenly its attraction to the Sun no longer existed?
(a) The Earth would continue in its same path, but at a reduced speed.
(b) The Earth would continue traveling in its same path, but at a greater speed.
(c) The Earth would move in a straight line with constant speed.
(d) The Earth would spiral toward the Sun
5. When you push a marble with a 0.5-N force, the marble
(a) accelerates at 10 m/s^2.
(b) resists being pushed with its own 0.5 N.
(c) will likely not move.
(d) pushes on you with a 0.5-N force.

INTEGRATED SCIENCE CONCEPTS

Biology—Gliding

1. What is gliding locomotion?
2. Why is having a large surface area important for effective gliding?
3. Describe some of the structures gliding organisms have evolved to increase their surface area.

Biology—Animal Locomotion

1. Explain how Newton's third law underlies many forms of animal locomotion—from fish, to birds, to humans.
2. A squid propels itself forward by pushing water backward. Why does this occur?
3. When you walk, what is the force that pushes you forward?
4. A duck stuck in an oil spill finds it very difficult to walk. Why?

Active Explorations

1. The net force acting on an object and the resulting acceleration are always in the same direction. You can demonstrate this with a spool. If the spool is pulled horizontally to the right, in which direction will it roll? (Some of your classmates may be surprised.)

2. Hold your hand with the palm down like a flat wing outside the window of a moving automobile. Then slightly tilt the front edge of your hand upward and notice the lifting effect as air bounces from the bottom of your hand. Which of Newton's laws is illustrated here?

Exercises

1. In the orbiting space shuttle, you are handed two identical closed boxes, one filled with sand and the other filled with feathers. How can you tell which is which without opening the boxes?
2. Your empty hand is not hurt when it bangs lightly against a wall. Why is it hurt if it is carrying a heavy load? Which of Newton's laws is most applicable here?
3. Each of the vertebrae forming your spine is separated from its neighbors by disks of elastic tissue. What happens, then, when you jump heavily on your feet from an elevated position? Can you think of a reason why you are a little taller in the morning than you are at the end of the day? (Hint: Think about how Newton's first law of motion applies in this case.)
4. As you stand on a floor, does the floor exert an upward force against your feet? How much force does it exert? Why are you not moved upward by this force?
5. To pull a wagon across a lawn at a constant velocity, you must exert a steady force. Reconcile this fact with Newton's first law, which states that motion with a constant velocity indicates no force.
6. A rocket becomes progressively easier to accelerate as it travels through space. Why is this so? (Hint: About 90% of the mass of a newly launched rocket is fuel.)
7. As you are leaping upward from the ground, how does the force you exert on the ground compare with your weight?
8. A common saying goes, "It's not the fall that hurts you; it's the sudden stop." Translate this into Newton's laws of motion.
9. On which of these hills does the ball roll down with increasing speed and decreasing acceleration along the path? (Use this example if you wish to explain to someone the difference between speed and acceleration.)

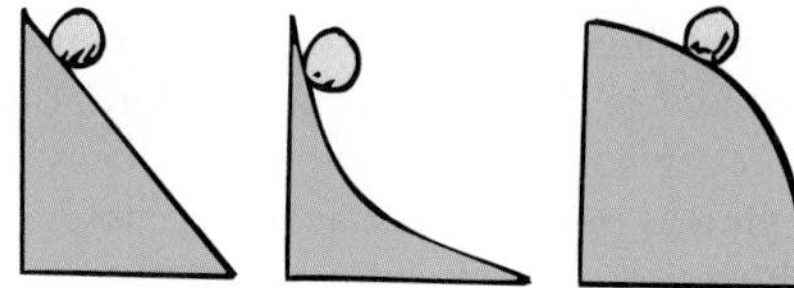

10. Neglecting air resistance, if you drop an object, its acceleration toward the ground is 10 m/s^2. If you throw it down instead, would its acceleration after throwing be greater than 10 m/s^2? Why or why not?
11. Can you think of a reason why the acceleration of the object thrown downward through the air in the preceding exercise would actually be less than 10 m/s^2?
12. You hold an apple over your head. (a) Identify all the forces acting on the apple and their reaction forces. (b) When you drop the apple, identify all the forces acting on it as it falls and the corresponding reaction forces.
13. Does a stick of dynamite contain force? Defend your answer.
14. Can a dog wag its tail without the tail in turn "wagging the dog"? (Consider a dog with a relatively massive tail.)
15. If the Earth exerts a gravitational force of 1000 N on an orbiting communications satellite, how much force does the satellite exert on the Earth?
16. If you exert a horizontal force of 200 N to slide a crate across a factory floor at a constant velocity, how much friction is exerted by the floor on the crate? Is the force of friction equal and oppositely directed to your 200-N push? Does the force of friction make up the reaction force to your push? Why not?
17. If a Mack truck and a motorcycle have a head-on collision, upon which vehicle is the impact force greater? Which vehicle undergoes the greater change in motion? Explain your answers.
18. Two people of equal mass attempt a tug-of-war with a 12-m rope while standing on frictionless ice. When they pull on the rope, each person slides toward the other. How do their accelerations compare, and how far does each person slide before they meet?
19. Suppose that one person in the preceding exercise has twice the mass of the other. How far does each person slide before they meet?
20. Which team wins in a tug-of-war—the team that pulls harder on the rope or the team that pushes harder against the ground? Explain.
21. The photo shows Steve Hewitt and his daughter Gretchen. Is Gretchen touching her dad or is he touching her? Explain.

22. When your hand turns the handle of a faucet, water comes out. Do your push on the handle and the water coming out constitute an action–reaction pair? Defend your answer.
23. Why is it that a cat that falls from the top of a 50-story building will hit the safety net below no faster than if it fell from the twentieth story?

24. Free fall is motion in which gravity is the only force acting. (a) Is a skydiver who has reached terminal speed in free fall? (b) Is a satellite circling the Earth above the atmosphere in free fall?
25. How does the weight of a falling body compare with the air resistance it encounters just before it reaches terminal velocity? Just after?
26. You tell your friend that the acceleration of a skydiver before the chute opening decreases as falling progresses. Your friend then asks if this means the skydiver is slowing down. What is your response?
27. If and when Galileo dropped two balls from the top of the Leaning Tower of Pisa, air resistance was not really negligible. Assuming that both balls were the same size yet one was much heavier than the other, which ball struck the ground first? Why?
28. If you simultaneously drop a pair of tennis balls from the top of a building, they will strike the ground at the same time. If one of the tennis balls is filled with lead pellets, will it fall faster and hit the ground first? Which of the two will encounter more air resistance? Defend your answers.
29. Which is more likely to break, the ropes supporting a hammock stretched tightly between a pair of trees or the ropes supporting a hammock that sags more when you sit on it? Defend your answer.
30. A stone is shown at rest on the ground. (a) The vector shows the weight of the stone. Complete the vector diagram by showing another vector that results in zero net force on the stone. (b) What is the conventional name of the vector you have drawn?

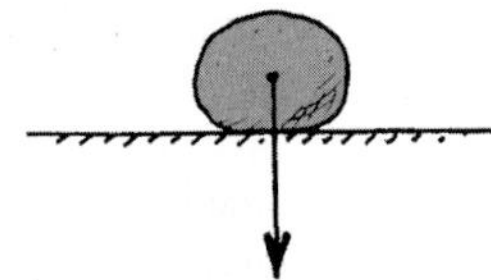

31. Here a stone at rest is suspended by a string. (a) Draw force vectors for all the forces that act on the stone. (b) Should your vectors have a zero resultant? (c) Why, or why not?

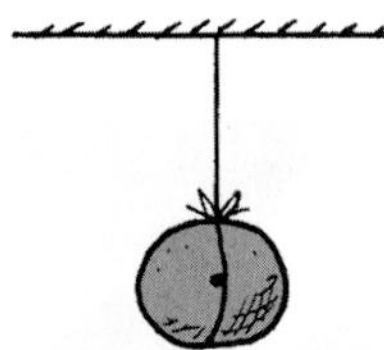

32. Here the same stone is being accelerated vertically upward. (a) Draw force vectors to some suitable scale showing relative forces acting on the stone. (b) Which is the longer vector and why?
33. Suppose that the string in the preceding exercise breaks and that the stone slows in its upward motion. Draw a force-vector diagram of the stone when it reaches the top of the path.

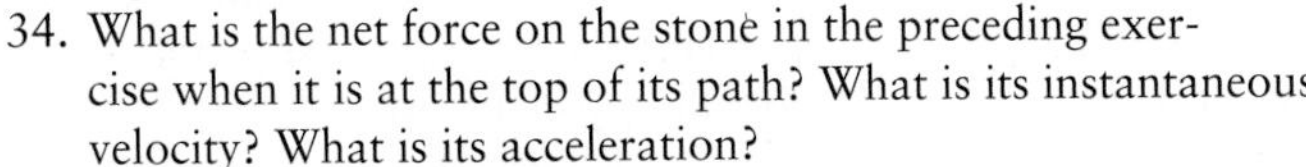

34. What is the net force on the stone in the preceding exercise when it is at the top of its path? What is its instantaneous velocity? What is its acceleration?
35. Here is the stone sliding down a friction-free incline. (a) Identify the forces that act on it, and draw appropriate force vectors. (b) By the parallelogram rule, construct the resultant force on the stone (carefully showing that it has a direction parallel to the incline—the same direction as the stone's acceleration).

36. Here is the stone at rest, interacting with both the surface of the incline and the block. (a) Identify all the forces that act on the stone and draw appropriate force vectors. (b) Show that the net force on the stone is zero. (Hint 1: There are two normal forces *on* the stone. Hint 2: Be sure the vectors you draw are for forces that act on the stone, not for forces that act on the surfaces *by* the stone).

Problems

1. A 400-kg bear grasping a vertical tree slides down at a constant velocity. Show that the friction force that acts on the bear is about 4000 N.
2. When two horizontal forces are exerted on a cart, 600 N forward and 400 N backward, the cart undergoes acceleration. Show that the additional force needed to produce nonaccelerated motion is 200 N.
3. You push with a 20-N horizontal force on a 2-kg mass resting on a horizontal surface. The horizontal friction force is 12 N. Show that the acceleration is 4 m/s^2.
4. You push with a 40-N horizontal force on a 4-kg mass resting on a horizontal surface. The horizontal friction force is 12 N. Show that the acceleration is 7 m/s^2.
5. A cart of mass 1 kg is accelerated 1 m/s^2 by a force of 1 N. Show that a 2-kg cart pushed with a 2-N force, would also accelerate at 1 m/s^2.
6. A rocket of mass 100,000 kg undergoes an acceleration of 2 m/s^2. Show that the force being developed by the rocket engines is 200,000 N.
7. A 747 jumbo jet of mass 30,000 kg experiences a 30,000-N thrust for each of four engines during takeoff. Show that its acceleration is 4 m/s^2.
8. Suppose the jumbo jet in the previous problem flies against an air resistance of 90,000 N while the thrust of all four engines is 100,000 N. Show that its acceleration will be about 0.3 m/s^2. What will the acceleration be when air resistance builds up to 100,000 N?
9. A boxer punches a sheet of paper in midair, bringing it from rest to a speed of 25 m/s in 0.05 second. If the mass of the paper is 0.003 kg, show that the force the boxer exerts on it is only 1.5 N.
10. Suppose that you are standing on a skateboard near a wall and that you push on the wall with a force of 30 N. (a) How hard does the wall push on you? (b) Show that if your mass is 60 kg your acceleration while pushing will be 0.5 m/s^2.
11. If raindrops fall vertically at a speed of 3 m/s and you are running horizontally at 4 m/s, show that the drops will hit your face at a speed of 5 m/s.
12. Horizontal forces of 3 N and 4 N act at right angles on a block of mass 5 kg. Show that the resulting acceleration will be 1 m/s^2.
13. Suzie Skydiver with her parachute has a mass of 50 kg.
    (a) Before opening her chute, show that the force of air resistance she encounters when reaching terminal velocity is about 500 N.
    (b) After her chute is open and she again reaches a smaller terminal velocity, show that the force of air resistance she encounters is also about 500 N.
    (c) Discuss why your answers are the same.
14. An airplane with an air speed of 120 km/h encounters a 90-km/h crosswind. Show that its groundspeed is 150 km/h.

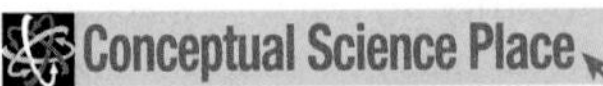

CHAPTER 3 ONLINE RESOURCES

Interactive Figures 3.3, 3.7, 3.12, 3.20, 3.21, 3.22, 3.23, 3.24, 3.25 • ***Tutorials*** Parachutes and Newton's Second Law • ***Videos*** Newton's Law of Inertia, Newton's Second Law, Free Fall: Acceleration Explained, Air Resistance and Falling Objects, Forces and Interactions, Action and Reaction on Different Mass • ***Quiz*** • ***Flashcards*** • ***Links***

Momentum and Energy

The avalanche rips ice, rock, trees, and most anything else in its path off the mountain as it accelerates downhill. Newton's laws predict the overall motion of the avalanche. However, we use derived quantities such as power, work, momentum, and energy, to analyze the complex motion of an avalanche more easily and in better detail.

CHAPTER OUTLINE

Moving objects have a quantity that objects at rest don't have. More than a hundred years ago, this quantity was called *impedo*. A boulder at rest had no impedo, while the same boulder rolling down a steep incline possessed impedo. The faster it moved, the greater the impedo. The change in impedo depended on force and, more importantly, on how long the force acted. Apply a force to a cart and you give it impedo. Apply a long force and you give it more impedo.

But what do we mean by "long?" Does "long" refer to time or distance? When this distinction was made, the term impedo gave way to two more precise ideas—*momentum* and *kinetic energy*. And these two ideas are related to a cluster of other concepts including work, power, efficiency, potential energy, and impulse. What are the distinctions and relationships among these quantities? How can they be used to analyze matter in motion, the workings of machines, and even such complex phenomena as the energy sources that power modern industry and the "machinery" of living organisms?

4.1 Momentum

We all know that a heavy truck is more difficult to stop than a small car moving at the same speed. We state this fact by saying that the truck has more momentum than the car. By **momentum**, we mean inertia in motion—or, more specifically, the product of the mass of an object and its velocity; that is,

$$\textbf{Momentum} = \textbf{mass} \times \textbf{velocity}$$

Or, in shorthand notation,

$$\textbf{Momentum} = mv$$

When direction is not an important factor, we can say momentum = mass × speed, which we still abbreviate *mv*.

We can see from the definition that a moving object can have a large momentum if either its mass or its velocity is large or both its mass and its velocity are large. A truck has more momentum than a car moving at the same velocity because it has a greater mass. We can see that a huge ship moving at a small velocity can have a large momentum, as can a small bullet moving at a high velocity. A massive truck moving down a steep hill with no brakes has a large momentum, whereas the same truck at rest has no momentum at all.

Figure 4.1 The boulder, unfortunately, has more momentum than the runner.

CHECK YOURSELF

When can a 1000-kg car and a 2000-kg truck have the same momentum?

CHECK YOUR ANSWER

They have the same momentum when the car travels twice as fast as the truck. Then (1000 kg × 2v) for the car equals (2000 kg × 2v) for the truck. Or, if they were both at rest, they'd certainly have the same momentum—zero.

4.2 Impulse

Changes in momentum may occur when there is a change in the mass of an object, or a change in its velocity, or both. If momentum changes while the mass remains unchanged, as is most often the case, then the velocity changes. Acceleration occurs. And what produces acceleration? The answer is *force*. The greater the force acting on an object, the greater will be the change in velocity and, hence, the change in momentum.

But something else is important also: time—how long the force acts. Apply a force briefly to a stalled automobile and you produce a small change in its momentum. Apply the same force over an extended period of time, and a greater momentum change results (Figure 4.2). A long sustained force produces more change in momentum than the same force applied briefly. So, for changing an object's momentum, both force and the time interval during which the force acts are important.

The quantity "force × time interval" is called **impulse.**

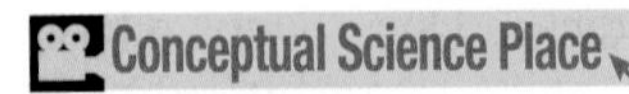

Definition of Momentum; Changing Momentum; Decreasing Momentum Over Short Time

4.3 Impulse–Momentum Relationship

The greater the impulse exerted on something, the greater will be its change in momentum. This is known as the impulse–momentum relationship. Mathematically, the exact relationship is

$$\text{Impulse} = \text{change in momentum}$$

or

$$Ft = \text{change in } mv$$

which reads, "force multiplied by the time-during-which-it-acts equals change in momentum."*

We can express all terms in this relationship in shorthand notation and use the delta symbol Δ (a capital letter in the Greek alphabet signifying "change in" or "difference in"):

$$Ft = \Delta(mv).$$

So, whenever you exert a force on something, you also exert an impulse. Recall that a net force produces acceleration. Now we are also saying that a net force multiplied by the time during which that force acts produces a change in an object's momentum.

Figure 4.2 When you push with twice the same force for twice the time, you impart twice the impulse and produce twice the momentum.

*In Newton's second law ($F/m = a$), we can insert the definition of acceleration (a = change in v/t), and get F/m = (change in v)/t. Then multiplying both sides by mt gives Ft = change in (mv), or, in delta notation, $Ft = \Delta(mv)$.

CHECK YOURSELF

1. Does a moving object possess impulse?
2. Does a moving object possess momentum?

CHECK YOUR ANSWERS

1. No. Recall that an object cannot possess force. Similarly, an object cannot possess impulse. Just as a force is something an object can provide when it changes velocity, an impulse is something an object can provide, or something it can experience, only when it interacts with another object.
2. Yes, a moving object can possess momentum, but, like velocity, only in a relative sense—that is, with respect to a frame of reference, such as the Earth's surface. For example, a fly inside a fast-moving airplane cabin may have a large momentum relative to the Earth below, but it has very little momentum relative to the cabin.

Impulse may be viewed as causing momentum change, or momentum change may be viewed as causing impulse. It doesn't matter which way you think about it. The important thing is that impulse and change of momentum are always linked. Here we will consider some ordinary examples in which impulse is related (1) to increasing momentum and (2) to decreasing momentum.

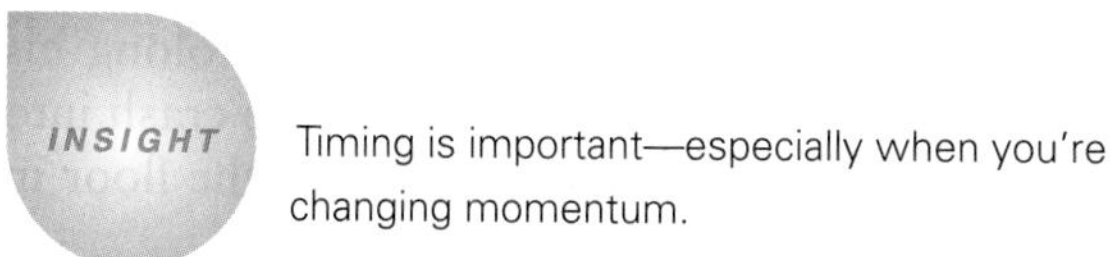

Timing is important—especially when you're changing momentum.

Increasing Momentum

If you wish to produce the maximum increase in the momentum of something, you not only apply the greatest force, you also extend the time of application as much as possible (hence the different results obtained by pushing briefly on a stalled automobile and by giving it a sustained push).

Long-range cannons have long barrels. The longer the barrel, the greater the velocity of the emerging cannonball or shell. Why? The force of exploding gunpowder in a long barrel acts on the cannonball for a longer time, increasing the impulse on it, which increases its momentum. Of course, the force that acts on the cannonball is not steady—it is strong at first and weaker as the gases expand. Most often the forces involved in impulses vary over time. The force that acts on the golf ball in Figure 4.3, for example, increases rapidly as the ball is distorted and then decreases as the ball comes up to speed and returns to its original shape. When we speak of any force that makes up impulse in this chapter, we mean the *average* force.

Figure 4.3 The force of impact on a golf ball varies throughout the duration of the impact.

Decreasing Momentum

Imagine that you are in a car that is out of control, and you're faced with a choice of slamming either into a concrete wall or into a haystack. You don't need much physics knowledge to make the better decision, but knowing some physics aids you in knowing why hitting something soft is entirely different from hitting something hard. Whether you hit the wall or the haystack, your momentum will be decreased by the same amount, and this means that the impulse required to stop you is the same. The same impulse means the same product of force and time, not the same force or the same time. You have a choice. By hitting the haystack instead of the wall, you extend the time of impact—you extend the time during which your momentum is brought to zero. The longer time is compensated for by a lesser force. If you extend the time of impact 100 times, you reduce the force of impact to a hundredth of

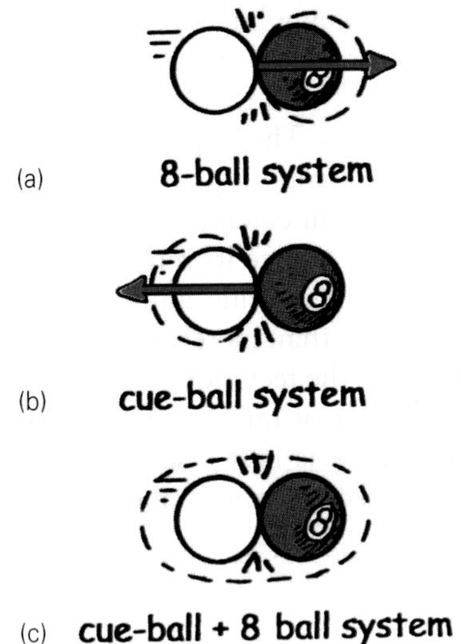

Figure 4.9 A cue ball hits an eight ball head-on. Consider this event in three systems: (a) An external force acts on the eight-ball system, and its momentum increases. (b) An external force acts on the cue-ball system, and its momentum decreases. (c) No external force acts on the cue-ball-plus-eight-ball system, and momentum is conserved (momentum is simply transferred from one part of the system to the other).

Whenever a physical quantity remains unchanged during a process, that quantity is said to be *conserved*. We say momentum is conserved.

The concept that momentum is conserved when no external force acts is so important that it is considered one of the basic laws of mechanics. It is called the **law of conservation of momentum:**

> **In the absence of an external force, the momentum of a system remains unchanged.**

Newton's first law tells us that a body in motion remains in motion when no external forces act. We say the same thing in a different context when we say the momentum of a body doesn't change when no external forces act. Whatever momentum a system may have, in the absence of external force, that momentum remains unchanged (Figure 4.9). For example, the forces involved in an exploding star are internal forces, which means the net momentum of its flying fragments is the same as the star's momentum before the explosion. If the star is initially spinning, the fragments maintain their *rotational equilibrium* (which we'll treat in Appendix B).

CHECK YOURSELF

If you toss a ball horizontally while standing on a skateboard, you'll roll backward with the same amount of momentum that you have given to the ball. Will you roll backward if you go through the motions of tossing the ball but instead hold onto it?

CHECK YOUR ANSWER

No, you'll not roll backward without immediately rolling forward to produce no net rolling. In third-law fashion, if no net force acts on the ball, no net force acts on you. In terms of momentum, if no net momentum is imparted to the ball, no net momentum will be imparted to you. Try it and see.

Collisions

The conservation of momentum is especially useful in collisions, where the forces involved are internal forces. In any collision, we can say that

Net momentum before collision = net momentum after collision.

When a moving billiard ball hits another billiard ball at rest head-on, the first ball comes to rest and the second ball moves with the initial velocity of the first ball. We call this an **elastic collision**; the colliding objects rebound without lasting deformation or the generation of heat. In this collision, momentum is transferred from the first ball to the second (Figure 4.10). Momentum is conserved. Billiard balls approximate perfectly elastic collisions, while collisions between molecules in a gas are perfectly elastic.

Momentum is conserved even when the colliding objects don't rebound. This is an **inelastic collision**, characterized by deformation and/or generation of heat. Sometimes an inelastic collision results in the coupling of colliding objects. Consider, for example, the case of a freight car moving along a track and colliding with another freight car at rest (Figure 4.11). If the freight cars are of equal mass and are coupled by the collision, can we predict the velocity of the coupled cars after impact?

Suppose the moving car has a velocity of 10 meters per second, and we consider the mass of each car to be m. Then, from the conservation of momentum,

$$(\text{net } mv)_{\text{before}} = (\text{net } mv)_{\text{after}}$$

$$(m \times 10)_{\text{before}} = (2m \times v)_{\text{after}}$$

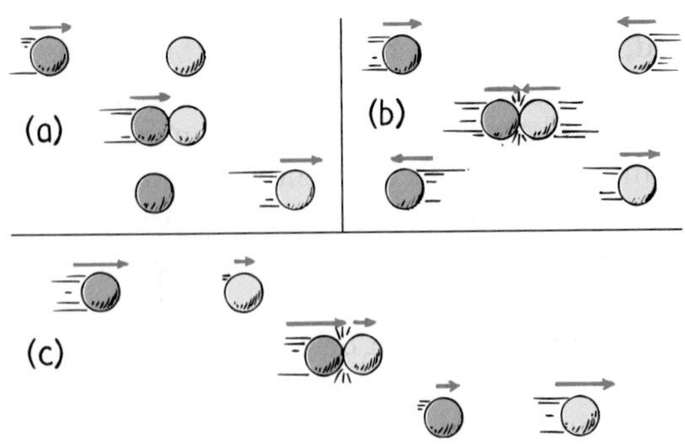

Figure 4.10 Interactive Figure
Elastic collisions of equally massive balls. (a) A green ball strikes a yellow ball at rest. (b) A head-on collision. (c) A collision of balls moving in the same direction. In each case, momentum is transferred from one ball to the other.

MATH CONNECTION

Quantifying Collisions

Billiard balls, cars, molecules, football players—collisions between objects are everywhere. You can use the conservation of momentum to analyze them. For example, consider this practice problem.

Problem Consider the air track in the photo. Suppose that a gliding cart with a mass of 0.5 kg bumps into, and sticks to, a stationary cart that has a mass of 1.5 kg. If the speed of the gliding cart before impact is 4 m/s, how fast will the coupled carts glide after collision?

Solution According to momentum conservation, the momentum of the cart of mass m and velocity v before the collision will equal the momentum of both carts stuck together after the collision.

$$(\text{total } mv)_{\text{before}} = (\text{total } mv)_{\text{after}}$$

$$0.5 \text{ kg } (4 \text{ m/s}) = (0.5 \text{ kg} + 1.5 \text{ kg})\, v$$

$$v = \frac{(0.5 \text{ kg})(4 \text{ m/s})}{(0.5 \text{ kg} + 1.5 \text{ kg})}$$

$$= \frac{(2.0 \text{ kg} \cdot \text{m/s})}{2.0 \text{ kg}} = 1 \text{ m/s}$$

This makes sense, because four times as much mass will be moving after the collision, so the coupled carts will glide more slowly. In keeping the momentum equal, four times the mass glides one-quarter as fast.

INSIGHT Momentum is conserved for all collisions, elastic and inelastic (whenever outside forces don't interfere).

By simple algebra, $v = 5$ m/s. This makes sense because, since twice as much mass is moving after the collision, the velocity must be half as much as the velocity before collision. Both sides of the equation are then equal.

So we see that changes in an object's motion depend both on force and on how long the force acts. When "how long" means time, we refer to the quantity "force × time" as *impulse*. But "how long" can mean distance also. When we consider the quantity "force × *distance*," we are talking about something entirely different—the concept of *energy*.

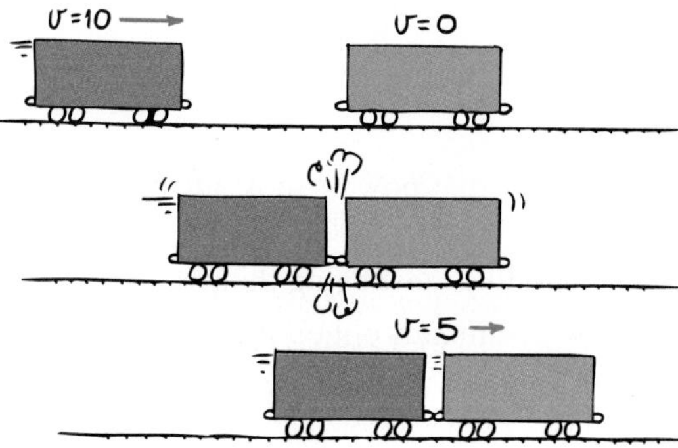

Figure 4.11 Interactive Figure
Inelastic collision. The momentum of the freight car on the left is shared with the freight car on the right after collision.

4.5 Energy

Energy is perhaps the most central concept in science. The combination of energy and matter makes up the universe; matter is substance, and energy is the mover and changer of substance. The idea of matter is easy to grasp—it is stuff that we can see, smell, and feel. It has mass and it occupies space. Energy, on the other hand, is abstract. We cannot see, smell, or feel most forms of energy. It isn't even noticeable unless it is undergoing a change of some kind—being transferred or transformed. Surprisingly, the idea of energy was unknown to Isaac Newton, and its existence was still being debated in the 1850s. Energy comes from the Sun in the form of sunlight, it is in the food you eat, and it sustains all life. It's in heat, sound, electricity, and radiation. Even matter itself is condensed, bottled-up energy, as set forth in Einstein's famous formula $E = mc^2$, which we'll return to in Chapter 10.

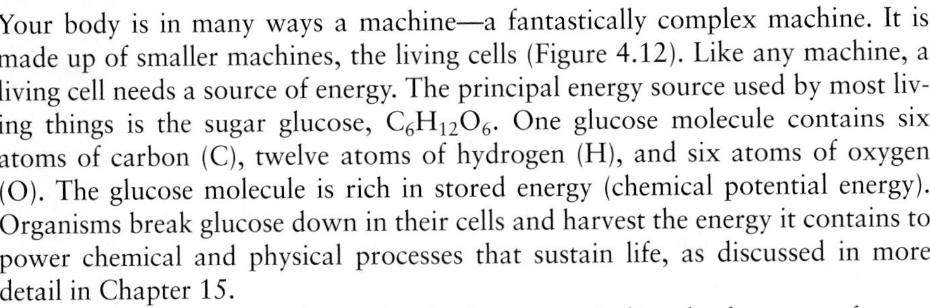

INTEGRATED SCIENCE BIOLOGY AND CHEMISTRY

Glucose: Energy for Life

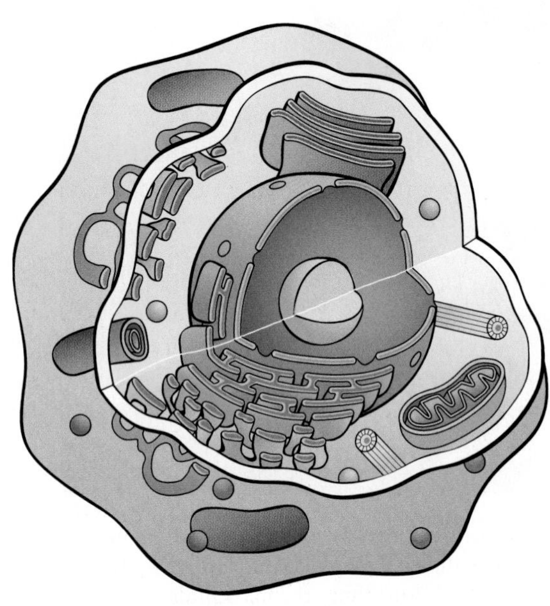

Figure 4.12 This cutaway view of a generalized animal cell shows various specialized structures, including the orange-and-yellow nucleus in the center. The pink structure at the lower right and the others scattered throughout the cell are *mitochondria,* which provide the cell with energy through cellular respiration.

Your body is in many ways a machine—a fantastically complex machine. It is made up of smaller machines, the living cells (Figure 4.12). Like any machine, a living cell needs a source of energy. The principal energy source used by most living things is the sugar glucose, $C_6H_{12}O_6$. One glucose molecule contains six atoms of carbon (C), twelve atoms of hydrogen (H), and six atoms of oxygen (O). The glucose molecule is rich in stored energy (chemical potential energy). Organisms break glucose down in their cells and harvest the energy it contains to power chemical and physical processes that sustain life, as discussed in more detail in Chapter 15.

You obtain glucose from the food you eat indirectly, by way of some rather complex chemical reactions. A few super-sweet foods contain glucose, but most consist of other, more complex carbohydrates, such as starch, or some combination of carbohydrates, fats, and proteins. Your body must break down these nutrients to produce glucose, a raw fuel that is then passed on to your cells. Glucose molecules are taken apart inside your cells, where energy is liberated from them. The actual energy harvesting typically takes place through *cellular respiration,* a process that occurs in specialized structures within the cell—mitochondria—the "power plants" of the cells. Cells use the released energy to do all the tasks they must do to stay alive and to perform their specialized functions.

Green plants, on the other hand, manufacture glucose directly during *photosynthesis.* Photosynthesis is the process by which plants, algae, and certain kinds of bacteria convert light energy from the Sun to chemical energy in sugar molecules. Almost all life on Earth is either directly or indirectly dependent on photosynthesis. The overall chemical reaction for photosynthesis is:

$$6CO_2 + 6H_2O + \text{sunlight} \rightarrow C_6H_{12}O_6 + 6O_2.$$

Carbon dioxide, water, and sunlight go in; glucose and oxygen come out. Glucose is typically converted by plant tissues to complex carbohydrates, which are long molecules built of glucose units. Some plants, of course, don't have the opportunity to consume the glucose they make for themselves, instead donating it to the animals that consume them. A potato, for instance, is crammed with glucose stored in a thicket of starch molecules. The potato's starch molecules are broken down to glucose in your mouth and small intestine, and the glucose is transported to your cells, powering their lives—and yours.

Figure 4.13 Plants capture solar energy and transform it to chemical energy, which is stored in large molecules. When other organisms consume the plants, they obtain the energy they need for life.

CHECK YOURSELF

1. Why do cells need energy?
2. What is the ultimate source of the energy that powers most life on Earth? Explain.
3. What is the principal energy source used by most living things?
4. How do plants obtain glucose? How do animals obtain it?

CHECK YOUR ANSWERS

1. Cells need energy to do all those things that require work. For example, to move, to change their shape, to reproduce, to maintain and repair cellular structures, or to create new cells requires considerable energy.
2. The Sun; it is the source of the energy plants need to perform photosynthesis. Plants form the base of most of food chains, and so the energy they obtain from the Sun is transferred to other organisms that consume them.

INSIGHT First we talked about force × time—impulse. Now we talk about force × distance—work.

3. Glucose.
4. Plants obtain glucose through photosynthesis. Animals obtain it by breaking down nutrients in the foods they eat, such as proteins, carbohydrates, and fats.

Figure 4.14 The man may expend energy when he pushes the wall, but, if it doesn't move, no work is performed on the wall.

Figure 4.15 Work is done in lifting the barbell.

Work

In Sections 4.2 and 4.3, we learned about impulse—force × time. Now we will consider the quantity force × *distance*, an entirely different quantity—**work.**

$$\textbf{Work} = \textbf{force} \times \textbf{distance}$$

In every case in which work is done, two things enter: (1) the exertion of a force and (2) the movement of something by that force.* For example, if we lift two loads one story high, we do twice as much work as we do lifting one load the same distance, because the *force* needed to lift twice the weight is twice as much. Similarly, if we lift a load two stories high instead of one story high, we do twice as much work, because the *distance* is twice as much.

When a weightlifter raises a heavy barbell, he does work on the barbell and gives energy to it. Interestingly, when a weightlifter simply holds a barbell overhead, he does no work on it. Work involves not only force but motion as well. He may get tired holding the barbell still, but, if the barbell is not moved by the force he exerts, he does no work *on the barbell.* Work may be done on his muscles, as they stretch and contract, which is force × distance on a biological scale. But this work is not done *on the barbell. Lifting* the barbell is different from *holding* the barbell.

The unit of work combines the unit of force (N) with the unit of distance (m), the newton-meter (N·m). We call a newton-meter the **joule** (J), which rhymes with cool. One joule of work is done when a force of 1 newton is exerted over a distance of 1 meter in the direction of the force, as in lifting an apple over your head. For larger values, we speak of kilojoules (kJ), which are thousands of joules, or megajoules (MJ), which are millions of joules. The weightlifter in Figure 4.15 does work that can be measured in kilojoules. The work done to vertically hoist a heavily loaded truck can be measured in kilojoules.

4.6 Power

INSIGHT The word *work,* in common usage, means physical or mental exertion. Don't confuse the science definition of work with the everyday notion of work.

The definition of work says nothing about how long it takes to do the work. The same amount of work is done when carrying a load up a flight of stairs, whether we walk up or run up. So why are we more tired after running upstairs in a few seconds than after walking upstairs in a few minutes? To understand this difference, we need to talk about a measure of how fast the work is done—power. **Power** is the rate at which work is done—the amount of work done divided by the amount of time it takes to do it:

$$\textbf{Power} = \frac{\textbf{work done}}{\textbf{time interval}}$$

The work done in climbing stairs requires more power when the worker is running up rapidly than when the worker is climbing up slowly. A high-power automobile engine does work rapidly. An engine that delivers twice the power of another, however, does not necessarily move a car twice as fast or twice as far. Twice the power means that the engine can do twice the work in the same amount of time—or it can do the same amount of work in half the time. A powerful engine can produce greater acceleration.

*Force and distance must be in the same direction. When force is not along the direction of motion, then work equals the component of force in the direction of motion × distance moved.

MATH CONNECTION

Power Practice Problems

Power calculations are straightforward. Try a few to reinforce your understanding of the concept.

Problems

1. How much power is expended when lifting a 1000-N load a vertical distance of 4 m in a time of 2 s?
2. How much power is needed to perform the same job in 1 s?

Solutions

1. $$\text{Power} = \frac{\text{work done}}{\text{time interval}} = \frac{1000\text{ N} \times 4\text{ m}}{2\text{ s}}$$
$$= \frac{4000\text{ J}}{2\text{ s}} = 2000\text{ W (or 2 kW)}.$$
2. Twice the power is needed to do the same job in half the time. That is,
$$\text{Power} = \frac{\text{work done}}{\text{time interval}} = \frac{4000\text{ J}}{1\text{ s}}$$
$$= 4000\text{ W (or 4 kW)}.$$

Power is also the rate at which energy is changed from one form to another. The unit of power is the joule per second, called the watt. This unit was named in honor of James Watt, the eighteenth-century developer of the steam engine. One watt (W) of power is used when one joule of work is done in one second. One kilowatt (kW) equals 1000 watts. One megawatt (MW) equals one million watts.

Figure 4.16 The potential energy of Tenny's drawn bow equals the work (average force × distance) she did in drawing back the arrow into position. When released, most of the potential energy of the drawn bow will become the kinetic energy of the arrow.

4.7 Potential Energy

An object can store energy because of its position, shape, or state. Such stored energy is called potential energy (PE), because, in the stored state, it has the potential to do work. For example, a stretched or compressed spring has the potential for doing work. Or, when an archer draws a bow with an arrow, energy is stored in the fibers of the bent bow. When the bowstring is released, the energy in the bow is transferred to the arrow (Figure 4.16).

The chemical energy in fuels is potential energy, for it is the energy of position from a microscopic point of view. Such energy characterizes fossil fuels, electric batteries, and the food we eat. This energy is available when atoms are rearranged—that is, when a chemical change occurs. Any substance that can do work through chemical action possesses potential energy.

The easiest-to-visualize form of potential energy is when work is done to elevate objects against Earth's gravity. The potential energy due to elevated positions is called gravitational potential energy. Water in an elevated reservoir and the elevated ram of a pile driver have gravitational potential energy. The amount of gravitational potential energy possessed by an elevated object is equal to the work done against gravity in lifting it. The work done equals the force required to move it upward times the vertical distance it has been moved $(W = Fd)$. The

MATH CONNECTION

Calculating Potential Energy

Why are ramps used? Solve these practice problems to appreciate why.

Problems

1. How much work is done in lifting the 100-N block of ice a vertical distance of 2 m, as shown in Figure 4.18?
2. How much work is done in pushing the same block of ice up the 4-m-long ramp? The force needed is only 50 N (which is the reason inclines are used).
3. What is the increase in the block's potential energy in each case?

Solutions

1. $W = F \times d = 100\text{ N} \times 2\text{ m} = 200\text{ J}$.
2. $W = F \times d = 50\text{ N} \times 4\text{ m} = 200\text{ J}$.
3. Either way increases the block's potential energy by 200 J. The ramp simply makes this work easier to perform.

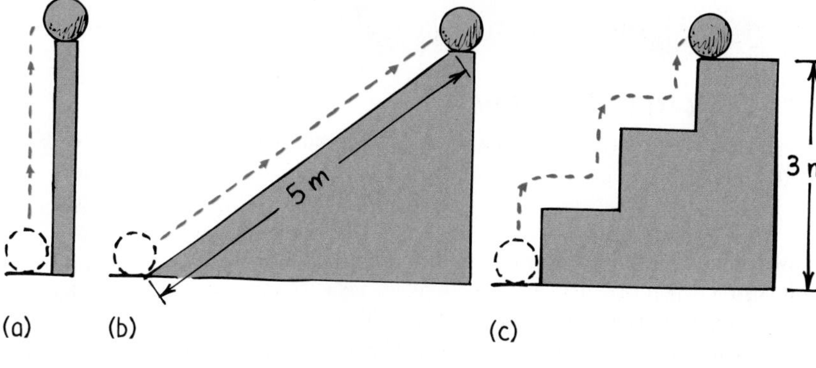

Figure 4.17 The potential energy of the 10-N ball is the same (30 J) in all three cases because the work done in elevating it 3 m is the same whether it is (a) lifted vertically with 10 N of force, (b) pushed with 6 N of force up the 5-m incline, or (c) lifted with 10 N up each 1-m stair. No work is done in moving it horizontally (neglecting friction).

upward force equals the weight mg of the object. So the work done in lifting it to a given height h is given by the product mgh:

$$\text{Gravitational potential energy} = \text{weight} \times \text{height}$$

$$\text{PE} = mgh$$

Figure 4.18 Both do the same work in elevating the same-mass blocks. When both blocks are raised to the same vertical height, both possess the same potential energy.

Note that the height h is the distance above some reference level, such as the ground or the floor of a building. The potential energy mgh is relative to that level, and it depends only on mg and height h. You can see, in Figure 4.17, that the potential energy of the ball at the top of the structure depends on vertical displacement and not on the path taken to get it there.

4.8 Kinetic Energy

When you push on an object, you can make it move. If it moves, the object can apply a force to something else and move it through a distance—the object you pushed can do work. If a moving object has the capacity to do work, it must have energy, but what kind of energy? The energy associated with a moving body, by virtue of its motion alone, is called **kinetic energy (KE)**.

Energy can be *transferred* from one object to another, such as when a rolling bowling ball transfers some of its kinetic energy to the pins and sets them in motion. Energy also *transforms*, or changes form. For example, the gravitational potential energy of a raised ram transforms to kinetic energy when the ram is released from its elevated position, as shown in Figure 4.19. And, when you raise a pendulum bob against the force of gravity, you do work on it. That work is stored as potential energy until you let the pendulum bob go. Its potential energy transforms to kinetic energy as it picks up speed and loses elevation.

Weight = mg, so a 10-kg block of ice weighs 100 N.

The kinetic energy of an object depends on its mass and its speed. It is equal to half the mass multiplied by the square of the speed, multiplied by the constant $\frac{1}{2}$.

$$\text{Kinetic energy} = \frac{1}{2}\,\text{mass} \times \text{speed}^2$$

$$\text{KE} = \frac{1}{2}mv^2$$

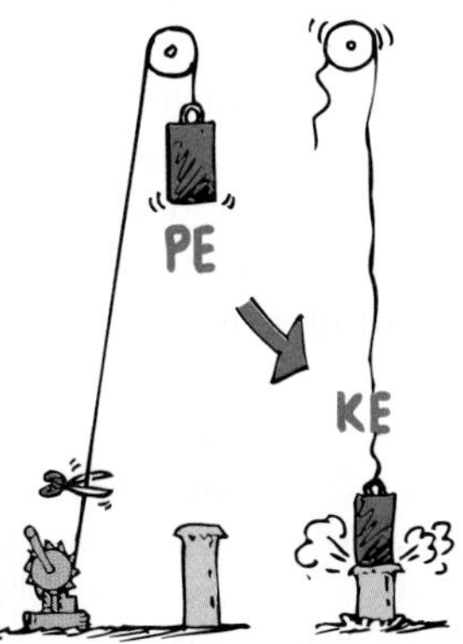

Figure 4.19 The potential energy of the elevated ram is converted to kinetic energy when released.

Potential energy *to* Potential + kinetic *to* Kinetic energy *to* Potential energy
And so on

Figure 4.20 Energy transitions in a pendulum.

MATH CONNECTION

Work Practice Problems

Solve some practice problems relating to work to be sure you can apply this concept.

Problems

1. How much work is needed to lift an object that weighs 500 N to a height of 2 m?
2. How much work is needed to lift it twice as high?
3. How much work is needed to lift a 1000-N object to a height of 4 m?

Solutions

1. $W = F \times d = 500\ \text{N} \times 2\ \text{m} = 1000\ \text{J}$.
2. Twice the height requires twice the work. That is, $W = F \times d = 500\ \text{N} \times 4\ \text{m} = 2000\ \text{J}$.
3. Lifting twice the load twice as high requires four times the work. That is, $F \times d = 1000\ \text{N} \times 4\ \text{m} = 4000\ \text{J}$.

A car moving along a road has kinetic energy. A car that is twice as heavy moving at the same speed has twice the kinetic energy. That's because a car that is twice as heavy has twice the mass. Kinetic energy depends on mass. But note that it also depends on speed—not just plain speed, but speed multiplied by itself—speed squared. If you double the speed of a car, you'll increase its kinetic energy by four ($2^2 = 4$). Or, if you drive three times as fast, you will have nine times the kinetic energy ($3^2 = 9$). The fact that kinetic energy depends on the square of the speed means that small changes in speed can produce large changes in kinetic energy. The squaring of speed means that kinetic energy can only be zero or positive—never negative. Now, let's relate this to work.

Which of these does a speeding baseball not possess? Force, momentum, energy. (Hint: The correct answer begins with an *F*.)

4.9 The Work–Energy Theorem

To increase the kinetic energy of an object, work must be done on it. The change in kinetic energy is equal to the work done. This important relationship is called the work–energy theorem. We abbreviate "change in" with the delta symbol Δ, and say

$$\text{Work} = \Delta KE$$

Work equals change in kinetic energy. The work in this equation is net work—that is, the work based on the net force.

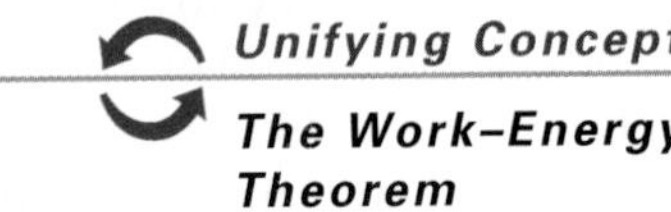

Recall, from Section 4.3, that a cannonball fired from a cannon with a longer barrel has a greater velocity because of the longer time of the impulse. The greater speed is also evident from the work–energy theorem, because of the longer *distance* through which the force acts. The work done on the cannonball is the force exerted on it multiplied by the distance through which the force acts: $Fd = \Delta KE$.

The work–energy theorem emphasizes the role of change. If there were no change in an object's energy, then we know no net work was done on it. This theorem applies to changes in potential energy as well. Recall our previous example of the weightlifter raising the barbell. When work was being done on the barbell, its potential energy was being changed. But, when the barbell was held stationary, no further work was being done on it, as evidenced by no further change in its energy.

Similarly, if you push against a box on a floor and it doesn't slide, then you are not doing work on the box. There is no change in kinetic energy. But, if you push harder and it slides, then you're doing work on it. When the amount of work done to overcome friction is small, the amount of work done on the box is practically matched by its gain in kinetic energy.

Figure 4.21 The work required in raising the roller-coaster car against gravity converts to kinetic energy as it begins to fall.

The work–energy theorem applies to decreasing speed as well. Energy is required to reduce the speed of a moving object and to bring it to a halt. When we apply the brakes to slow a moving car, we do work on it. This work is the friction supplied by the brakes multiplied by the distance over which the friction force acts. The more kinetic energy something has, the more work is required to stop it.

Interestingly, the friction supplied by the brakes is the same whether the car moves slowly or quickly. Friction doesn't depend on speed. The variable that makes

Figure 4.22 Due to the friction, energy is transferred both into the floor and into the tire when the bicycle skids to a stop. An infrared camera reveals (*left*) the heated tire track (*red streak on the floor*) and (*right*) the warmth of the tire. (Courtesy of Michael Vollmer.)

a difference is the braking distance. A car moving at twice the speed of another takes four times ($2^2 = 4$) as much work to stop. Therefore, it takes four times as much distance in which to stop. Accident investigators are well aware that an automobile going 100 kilometers per hour has four times the kinetic energy as it would have if it were going 50 kilometers per hour. So a car going 100 kilometers per hour will skid four times as far when its brakes are applied as it would if it were going 50 kilometers per hour. Kinetic energy depends on speed squared.

CHECK YOURSELF

1. When the brakes of a car are locked, the car skids to a stop. How much farther will the car skid if it's moving three times as fast?
2. Can an object possess energy?
3. Can an object possess work?

CHECK YOUR ANSWERS

1. Nine times farther. The car has nine times as much energy when it travels three times as fast: $\frac{1}{2}m(3v)^2 = \frac{1}{2}m9v^2 = 9(\frac{1}{2}mv^2)$. The friction force will ordinarily be the same in either case. Therefore, to do nine times the work to stop requires nine times as much stopping distance.
2. Yes, but only in a relative sense. For example, an elevated object may possess PE relative to the ground, but none relative to a point at the same elevation. Similarly, the kinetic energy of an object is relative to a frame of reference, usually the Earth's surface.
3. No; unlike energy, work is not something an object has. Work is something an object does to some other object. An object can *do* work only if it has energy. Or, stated another way, an object spends energy when it does work on something else.

Kinetic energy underlies other seemingly different forms of energy, such as heat, sound, and light. Random molecular motion is sensed as heat: when fast-moving molecules bump into the molecules in the surface of your skin, they transfer kinetic energy to your molecules, much as the balls in a game of pool or billiards transfer energy to each other. Sound consists of molecules vibrating in rhythmic patterns: shake a group of molecules in one place and, in cascade fashion, they disturb neighboring molecules that, in turn, disturb others, preserving the rhythm of shaking throughout the medium. Electrons in motion produce electric currents. Even light energy originates from the motion of electrons within atoms. Kinetic energy is far-reaching.

MATH CONNECTION

Applying the Work–Energy Theorem

Determine the work done on an object even though you don't know the forces or distances involved.

Problems

1. Calculate the change in kinetic energy when a 50-kg shopping cart moving at 2 m/s is pushed to a speed of 6 m/s.
2. How much work is required to make this change in kinetic energy?

Solutions

1. $\Delta KE = \frac{1}{2} m (v_f^2 - v_o^2)$

$$= \frac{1}{2} (50 \text{ kg})[(6 \text{ m/s})^2 - (2 \text{ m/s})^2]$$

$$= 800 \text{ J}$$

2. $W = \Delta KE = 800$ J, because the change in KE equals the work done on the shopping cart.

INSIGHT Understanding the distinction between momentum and kinetic energy is high-level physics.

Comparison of Kinetic Energy and Momentum

Momentum and kinetic energy are properties of moving things, but they differ from each other. Like velocity, momentum is a vector quantity and is therefore directional and capable of being cancelled entirely. But kinetic energy is a non-vector (scalar) quantity, like mass, and can never be cancelled. The momenta (plural of momentum) of two firecrackers approaching each other may cancel, but, when they explode, there is no way their energies can cancel. Energies transform to other forms, but momenta do not. Another difference is the velocity dependence of the two. Whereas momentum depends on velocity (mv), kinetic energy depends on the square of velocity ($\frac{1}{2} mv^2$). An object that moves with twice the velocity of another object of the same mass has twice the momentum but four times the kinetic energy. So, when a car traveling twice as fast crashes, it crashes with four times the energy.

If the distinction between momentum and kinetic energy isn't really clear to you, you're in good company. Failure to make this distinction, when impedo was in vogue, resulted in disagreements and arguments between the best British and French physicists for two centuries.

4.10 Conservation of Energy

Whenever energy is transformed or transferred, none is lost and none is gained. In the absence of work input or output, the total energy of a system before some process or event is equal to the total energy after.

Consider the system of a bow, arrow, and target. In the process of drawing the arrow in the bow, we do work in bending the bow, and we give the arrow and the bow potential energy. When the bowstring is released, most of this potential energy is transferred to the arrow as kinetic energy (the rest slightly warms the bow). The arrow, in turn, transfers this energy to its target, perhaps a bale of hay. The distance the arrow penetrates into the hay multiplied by the average force of impact doesn't quite match the kinetic energy of the arrow. The energy score doesn't balance. But, if we investigate further, we discover that both the arrow and the hay are a bit warmer. By how much? By the energy difference. In these transformations of energy, taking the form of thermal energy into account, we find energy transforms without net loss or gain. Quite remarkable!

The study of various forms of energy and their transformations has led to one of the greatest generalizations in physics—the law of **conservation of energy:**

Unifying Concept
Conservation of Energy

In the absence of external work input or output, the energy of a system remains unchanged. Energy cannot be created or destroyed.

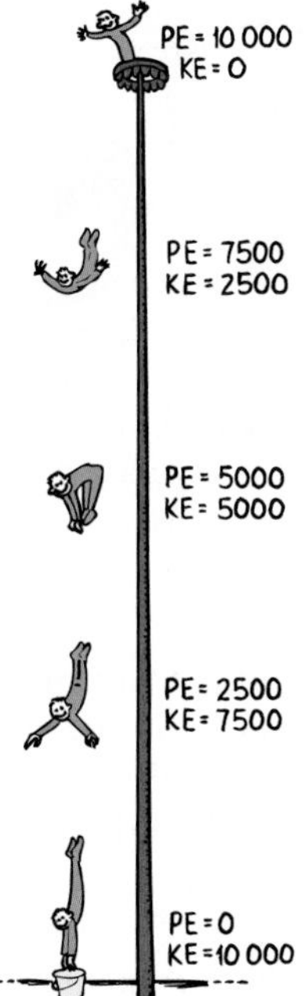

Figure 4.23 Interactive Figure

A circus diver at the top of a pole has a potential energy of 10,000 J. As he dives, his potential energy converts to kinetic energy. Notice that, at successive positions (one-fourth, one-half, three-fourths, and all the way down), the total energy is constant.

When we consider any system in its entirety, whether it be as simple as a swinging pendulum or as complex as an exploding galaxy, there is one quantity that doesn't change: energy. It may change form or it may simply be transferred from one part of the system to another, but, as far as we can tell, the total energy score remains the same. This energy score takes into account the fact that the atoms that make up matter are themselves concentrated bundles of energy. When the nuclei (cores) of atoms rearrange themselves, enormous amounts of energy can be released. The Sun shines because some of this energy is transformed into radiant energy. In nuclear reactors, much of this energy is transformed into heat. Enormous gravitational forces in the deep, hot core of the Sun push hydrogen nuclei together to form helium. This welding together of atomic nuclei is called thermonuclear fusion (Chapter 10). This process produces radiant energy, some of which reaches the Earth. Part of this energy falls on plants, and part of this, in turn, later becomes coal. Another part supports life in the food chain that begins with plants, and part of this energy later becomes petroleum. Part of the energy from the Sun powers the evaporation of water from the ocean, and part of this returns to the Earth as rain that may be trapped behind a dam. By virtue of its position, the water behind the dam has energy that may be used to power a generating plant below, where it will be transformed to electric energy. The energy travels through wires to homes, where it is used for lighting, heating, cooking, and operating electric gadgets. How wonderful that energy is transformed from one form to another!

CHECK YOURSELF

Rows of wind-powered generators are used in various windy locations to generate electric power. Does the power generated affect the speed of the wind? Would locations behind the windmills be windier if they weren't there?

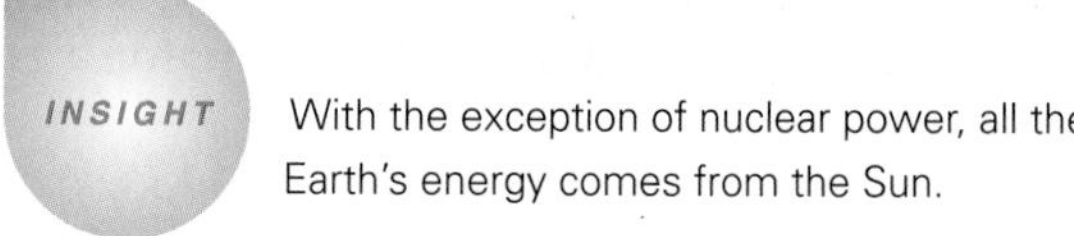

INSIGHT With the exception of nuclear power, all the Earth's energy comes from the Sun.

CHECK YOUR ANSWERS

Windmills generate power by taking kinetic energy from the wind, so the wind is slowed by interaction with the windmill blades. So, yes, it would be windier behind the windmills if they weren't present.

4.11 Machines

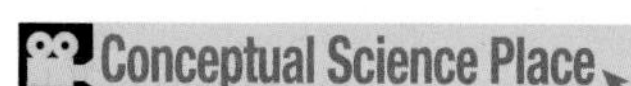

Conservation of Energy; Machines: Pulleys

A machine is a device for multiplying forces or simply changing the direction of forces. Underlying every machine is the conservation of energy. Consider one of the simplest machines, the **lever** (Figure 4.24). At the same time we do work on one end of the lever, the other end does work on the load. We see that the direction of force is changed: if we push down, the load is lifted up. If the heat from friction forces is small enough to neglect, the work input will be equal to the work output.

$$\text{Work input} = \text{work output}$$

Since work equals force times distance,

$$(\text{Force} \times \text{distance})_{\text{input}} = (\text{force} \times \text{distance})_{\text{output}}$$

If the pivot point, or fulcrum, of the lever is relatively close to the load, then a small input force will produce a large output force. This is because the input force is exerted through a large distance and the load is moved over a correspondingly short distance. In this way, a lever can multiply forces. But no machine can multiply work or multiply energy. That's a conservation-of-energy no-no!

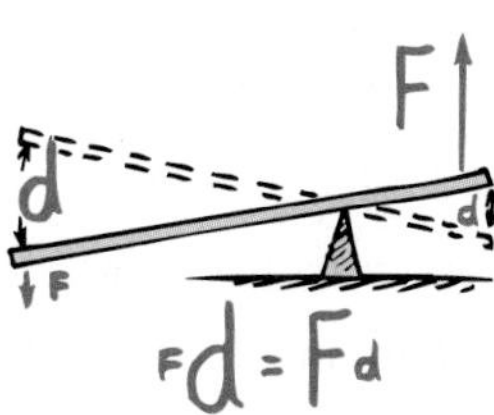

Figure 4.24 The lever.

Figure 4.25 Applied force × applied distance = output force × output distance.

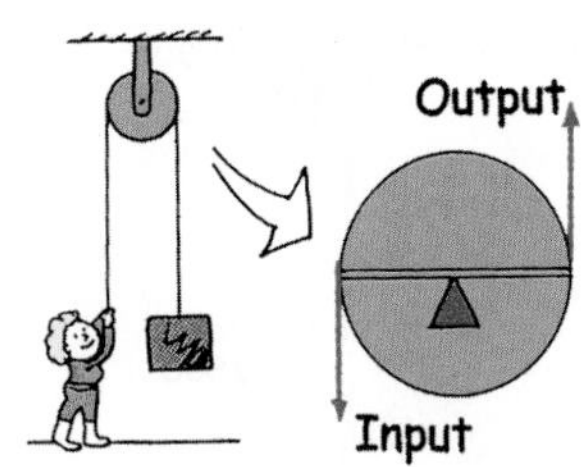

Figure 4.26 The pulley acts like a lever. It changes only the direction of the force.

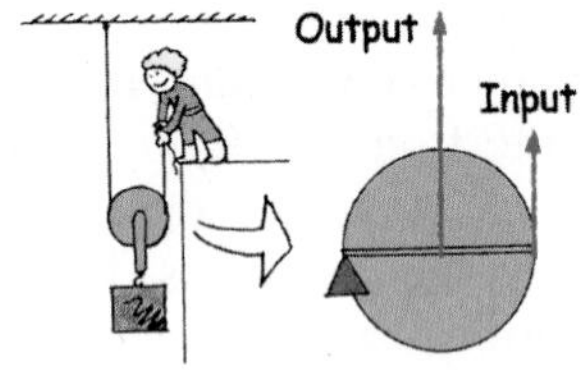

Figure 4.27 In this arrangement, a load can be lifted with half the input force.

You use the principle of the lever in jacking up the front of an automobile. By exerting a small force through a large distance, you are able to provide a large force acting through a small distance. Consider the ideal example illustrated in Figure 4.25. Every time the jack handle is pushed down 25 cm, the car rises 0.25 cm—only a hundredth as far, but with 100 times the force.

A block and tackle, or system of pulleys, is a simple machine that multiplies force at the expense of distance. One can exert a relatively small force through a relatively large distance and lift a heavy load through a relatively short distance. With the ideal pulley shown in Figure 4.28, the man pulls 7 m of rope with a force of 50 N and lifts 350 N through a vertical distance of 1 m. The work done by the man in pulling the rope is numerically equal to the increased potential energy of the 350-N block.

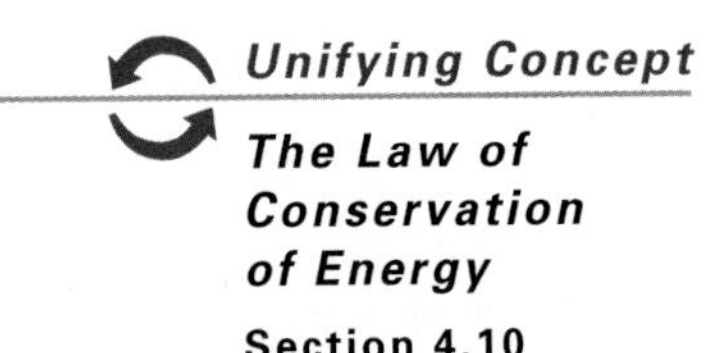

Unifying Concept
The Law of Conservation of Energy

Section 4.10

Any machine that multiplies force does so at the expense of distance. Likewise, any machine that multiplies distance, such as that of your forearm and elbow, does so at the expense of force. No device or machine can put out more energy than is put into it. No machine can create energy; it can only transfer it from one place to another or transform it from one form to another.

Figure 4.28 Applied force × applied distance = output force × output distance.

CHECK YOURSELF

If a lever is arranged so that input distance is twice output distance, can we predict that energy output will be doubled?

CHECK YOUR ANSWER

No, no, a thousand times no! We can predict output force will be doubled, but never output energy. Work and energy remain the same, which means force × distance remains the same. Shorter distance means greater force, and vice versa. Be careful to distinguish between the concepts of *force* and *energy*.

Efficiency

Given the same energy input, some machines can do more work than others. The machines that can perform more work are said to be more efficient. **Efficiency** can be expressed by the ratio,

$$\text{Efficiency} = \frac{\text{work done}}{\text{energy used}}$$

A machine can multiply force, but never energy. No way!

Even a lever converts a small fraction of input energy into heat when it rotates about its fulcrum. We may do 100 J of work but get out only 98 J of

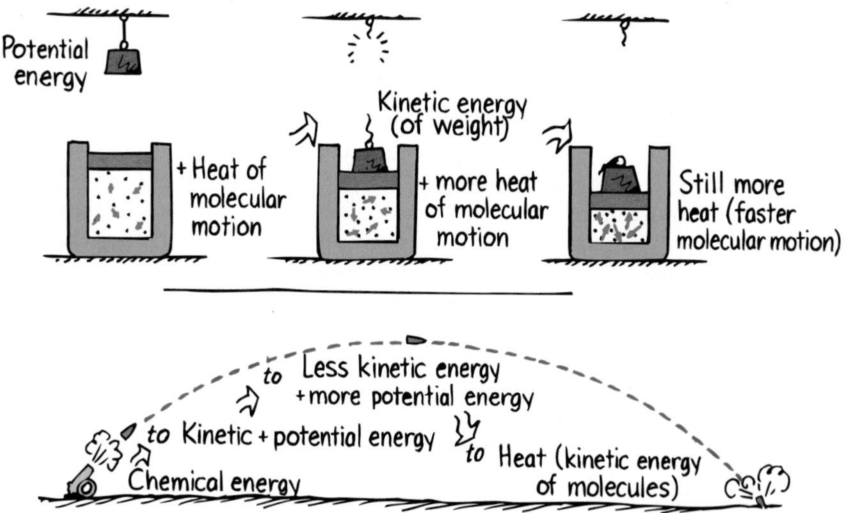

Figure 4.29 Energy transitions. The graveyard of kinetic energy is strewn with heat.

productive work. The lever is then 98% efficient, and we waste 2 J of work input as heat. In a pulley system, a larger fraction of input energy goes into heat. If we do 100 J of work, the forces of friction acting through the distances in which the pulleys turn and rub about their axles may dissipate 60 J of energy as heat. So the work output is only 40 J, and the pulley system has an efficiency of 40%. The lower the efficiency of a machine, the greater the amount of energy wasted as heat.

An automobile engine is a machine that transforms chemical energy stored in gasoline into mechanical energy. But only a fraction of the energy stored in the fuel actually moves the car. Nearly half is wasted in the friction of the moving engine parts. Some goes out in the hot exhaust gases as waste. In addition to these inefficiencies, some of the fuel doesn't even burn completely and goes unused.

MATH CONNECTION

Efficiency Calculations

Energy efficiency is a consideration when choosing appliances, insulating your home, or driving a car, and in countless other practical matters of daily life. Hone your understanding of the concept with a couple practice problems.

Problems

1. Consider an imaginary dream car that has a 100% efficient engine and burns fuel that has an energy content of 40 MJ/L. If the air resistance plus frictional forces on the car traveling highway speed is 500 N, what is the maximum distance the car can go on 1 L of fuel?
2. One can only dream of a car with a 100%-efficient engine. More realistically, a car engine is about 30% efficient. With the same air resistance and same fuel as the car in problem 1, what is the maximum distance per liter for the realistic car?

Solutions

1. From the definition work = force × distance, simple rearrangement gives distance = work/force. If all 40 MJ of energy in 1 L is used to do the work of overcoming air resistance and frictional force, the distance covered is:

$$\text{Distance} = \frac{\text{work}}{\text{force}} = \frac{40{,}000{,}000\ \text{J}}{500\ \text{N}}$$
$$= 80{,}000\ \text{m} = 80\ \text{km}$$

   The important point here is that, even with a perfect engine, there is an upper limit of fuel economy dictated by the conservation of energy.
2. The realistic distance per liter is 30% of 80 km = 0.3 (80 km) = 24 km/L.

TECHNOLOGY

Direct and Indirect Forms of Solar Power

Except for nuclear power, the source of practically all our energy is the Sun. In this sense, all energy technologies, except for nuclear power, are solar. So we make a distinction between direct and indirect solar power. *Direct* solar-power technologies such as photovoltaic cells, involve only one energy transformation from solar radiation to useful energy. Photovoltaic cells convert sunlight to electricity; the simplest of them power watches and calculators. More complex systems of photovoltaic cells bring light and heat to homes and provide power to the electric grid. Solar thermal collectors, another direct solar-power technology, trap solar energy to provide heat. Ingenious solar thermal collectors, from solar ovens to solar panels, provide heat on a relatively small scale.

Indirect solar-power technologies require more than one energy transformation to change sunlight into usable energy. Though they are not usually considered forms of solar power, petroleum, coal, natural gas, and wood are, in a sense, indirect forms of it. That's because these fuels are created by photosynthesis—the process by which plants trap solar energy and store it as plant tissue.

Wind power uses the energy of the wind to turn generator turbines within specially equipped windmills. Since wind is caused by unequal warming of the Earth's surface, wind power is an indirect form of solar power.

Hydroelectric dams generate electricity indirectly as well. Sunlight evaporates water, which later falls as rain; rainwater flows into rivers and into dams, where it is directed to generator turbines. Then it returns to the sea, where the cycle continues.

Ocean thermal energy technology (OTEC) transforms solar radiation to electricity using the ocean's thermal gradient—the variation in temperature with depth. Since the oceans cover 71% of Earth's surface, they are the world's largest solar collector and storage system. If less than one-tenth of 1% of this stored energy could be converted to electric power, it would supply more than 20 times the total amount of electricity consumed in the United States per day. OTEC is one of the promising alternative energy technologies that are ultimately powered by the Sun.

CHAPTER 4 Review

Summary of Terms

Conservation of energy In the absence of external work input or output, the energy of a system remains unchanged. Energy cannot be created or destroyed.

Conservation of energy and machines The work output of any machine cannot exceed the work input. In an ideal machine, where no energy is transformed into heat,

$$\text{work}_{\text{input}} = \text{work}_{\text{output}}$$

and

$$(Fd)_{\text{input}} = (Fd)_{\text{output}}$$

Conservation of momentum In the absence of an external force, the momentum of a system remains unchanged. Hence, the momentum before an event involving only internal forces is equal to the momentum after the event:

$$mv_{(\text{before event})} = mv_{(\text{after event})}$$

Efficiency The percentage of the work put into a machine that is converted into useful work output. (More generally, efficiency is useful energy output divided by total energy input.)

Elastic collision A collision in which colliding objects rebound without lasting deformation or the generation of heat.

Energy The property of a system that enables it to do work.

Impulse The product of the force acting on an object and the time during which it acts.

Inelastic collision A collision in which the colliding objects become distorted, generate heat, and possibly stick together.

Joule The SI unit of energy and work, equivalent to a newton-meter.

Kinetic energy Energy of motion, described by the relationship

$$\text{Kinetic energy} = \frac{1}{2}mv^2$$

Momentum The product of the mass of an object and its velocity.

Potential energy The stored energy that a body possesses because of its position.

Power The time rate of work:

$$\text{Power} = \frac{\text{work}}{\text{time}}$$

(More generally, power is the rate at which energy is expended.)

Relationship of impulse and momentum Impulse is equal to the change in the momentum of the object upon which the impulse acts. In symbolic notation,

$$Ft = \Delta mv$$

Work The product of the force and the distance through which the force moves:

$$W = Fd$$

Work–Energy Theorem The work done on an object equals the change in kinetic energy of the object:

$$\text{Work} = \Delta KE$$

Review Questions

4.1 Momentum

1. Which has a greater momentum, a heavy truck at rest or a moving skateboard?
2. How can a huge ship have an enormous momentum when it moves relatively slowly?

4.2 Impulse

3. How does impulse differ from force?
4. What are the two ways in which the impulse exerted on something can be increased?

4.3 Impulse–Momentum Relationship

5. For the same force, which cannon imparts the greater speed to a cannonball—a long cannon or a short one? Explain.
6. Consider a baseball that is caught and thrown at the same speed. Which case illustrates the greatest change in momentum: The baseball (1) being caught, (2) being thrown, or (3) being caught and then thrown back?
7. In the preceding question, which case requires the greatest impulse?

4.4 Conservation of Momentum

8. Can you produce a net impulse on an automobile by sitting inside and pushing on the dashboard? Defend your answer.
9. What does it mean to say that a quantity is conserved?
10. Distinguish between an elastic collision and an inelastic collision. For which type of collision is momentum conserved?
11. Railroad car A rolls at a certain speed and collides elastically with car B of the same mass. After the collision, car A is at rest. How does the speed of B after the collision compare with the initial speed of A?
12. If the equally massive cars of the previous question stick together after colliding inelastically, how does their speed after the collision compare with the initial speed of car A?

4.5 Energy

13. When is energy most evident?
14. What do we call the quantity force × distance, and what quantity does it change?
15. In what units are work and energy measured?

4.6 Power

16. True or false: One watt is the unit of power equivalent to one joule per second.
17. How many watts of power are expended when a force of 6 N moves a book 2 m in a time interval of 3 s?

4.7 Potential Energy

18. A car is lifted a certain distance in a service station and therefore has potential energy with respect to the floor. If it were lifted twice as high, how much potential energy would it have?
19. Two cars, one twice as heavy as the other, are lifted to the same elevation in a service station. How do their potential energies compare?

4.8 Kinetic Energy

20. When a car travels at 50 km/h, it has kinetic energy. How much more kinetic energy does it have at 100 km/h?

4.9 The Work–Energy Theorem

21. What is the evidence for saying whether or not work is done on an object?
22. The brakes do a certain amount of work to stop a car that is moving at a particular speed. How much work must the brakes do to stop a car that is moving four times as fast?

4.10 Conservation of Energy

23. Cite the law of energy conservation.
24. What is the source of energy that powers a hydroelectric power plant?

4.11 Machines

25. Can a machine multiply input force? Input distance? Input energy? (If your three answers are the same, seek help, for the last question is especially important.)
26. A force of 50 N applied to the end of a lever moves that end a certain distance. If the other end of the lever is moved half as far, how much force does it exert?
27. Is it possible to design a machine that has an efficiency that is greater than 100%? Discuss.

Multiple Choice Questions

Choose the BEST answer to the following

1. When a bullet is fired from a rifle, the force that accelerates the bullet is equal in magnitude to the force that makes the rifle recoil. But, compared with the rifle, the bullet has a greater
   (a) inertia.
   (b) potential energy.
   (c) kinetic energy.
   (d) momentum.
2. When an increase in speed doubles the momentum of a moving body, its kinetic energy
   (a) increases, but less than doubles.
   (b) doubles.
   (c) more than doubles.
   (d) depends on factors not stated.
3. When an increase in speed doubles the kinetic energy of a moving body, its momentum
   (a) increases, but less than doubles.
   (b) doubles.
   (c) more than doubles.
   (d) depends on factors not stated.
4. Car 1 rolls at a certain speed and collides elastically with car 2 at rest of the same mass. The collision brings car 1 to rest. How does the speed of car 2 after the collision compare with the initial speed of car 1?
   (a) Car 2 has twice the initial speed of car 1.
   (b) Car 2 has the same speed as the initial speed of car 1.
   (c) Car 2 has half the speed as car 1.
   (d) Car 2 will also be at rest.
5. You lift a barbell a certain distance from the floor, so the barbell has potential energy relative to the floor. If you lifted the barbell twice as high, how much potential energy would it have?
   (a) twice as much
   (b) half as much
   (c) the same amount with zero potential energy in both cases
   (d) impossible to know from the information provided

INTEGRATED SCIENCE CONCEPTS

Biology—The Impulse–Momentum Relationship in Sports

1. (a) Why is it a good idea to have your hand extended forward when you are getting ready to catch a fast-moving baseball with your bare hand? (b) In boxing, why is it advantageous to roll with the punch? (c) In karate, why is it advantageous to apply a force for a very brief time?
2. Referring to Figure 4.7, how does the force that Cassy exerts on the bricks compare with the force exerted on her hand?
3. How will the impulse differ if her hand bounces back when striking the bricks?

Biology and Chemistry—Glucose: Energy for Life

1. The word "burn" is often used to describe the process of cellular respiration, in which cells release energy from the chemical bonds in food molecules. How is the "burning" that goes on in cells different from literal burning—for example, the burning of a log on a campfire?
2. In what sense are you powered by solar energy?

Active Explorations

1. Pour some dry sand into a tin can with a cover. Compare the temperature of the sand before and after vigorously shaking the can for more than a minute. Explain your observations.
2. Place a small rubber ball on top of a basketball, and then drop them together. How high does the smaller ball bounce? Can you reconcile this with energy conservation? (What if the basketball were not elastic?)

Exercises

1. What is the purpose of a "crumple zone" (which has been manufactured to collapse steadily in a crash) in the front section of an automobile?
2. To bring a supertanker to a stop, its engines are typically cut off about 25 km from port. Why is it so difficult to stop or turn a supertanker?
3. Why might a wine glass survive a fall onto a carpeted floor but not onto a concrete floor?
4. If you throw an egg against a wall, the egg will break. If you throw an egg at the same speed into a sagging sheet, it won't break. Why?
5. Why is a punch more forceful with a bare fist than with a boxing glove?
6. A boxer can punch a heavy bag for more than an hour without tiring but will tire more quickly when boxing with an opponent for a few minutes. Why? (Hint: When the boxer's punches are aimed at the bag, what supplies the impulse to stop them? When aimed at the opponent, what (or who) supplies the impulse to stop the punches that are missed?)
7. Railroad cars are loosely coupled so that there is a noticeable delay from the time the first car is moved and the time the last cars are moved from rest by the locomotive. Discuss the advisability of this loose coupling and slack between cars from an impulse–momentum point of view.

8. A fully dressed person is at rest in the middle of a pond on perfectly frictionless ice and must reach the shore. How can this be accomplished?
9. A high-speed bus and an innocent bug have a head-on collision. The sudden change in momentum of the bus is greater, less, or the same as the change in momentum of the unfortunate bug?
10. Why is it difficult for a firefighter to hold a hose that ejects large quantities of water at high speed?
11. You're on a small raft next to a dock, and you jump from the raft only to fall in the water. What physics principle did you fail to take into account?
12. Your friend says the conservation of momentum is violated when you step off a chair and gain momentum as you fall. What do you say?
13. If a Mack truck and a Honda Civic have a head-on collision, which vehicle will experience the greater force of impact? The greater impulse? The greater change in momentum? The greater acceleration?
14. Would a head-on collision between two cars be more damaging to the occupants if the cars stuck together or if the cars rebounded upon impact?
15. In Chapter 3, rocket propulsion was explained in terms of Newton's third law. That is, the force that propels a rocket is from the exhaust gases pushing against the rocket, the reaction to the force the rocket exerts on the exhaust gases. Explain rocket propulsion in terms of momentum conservation.
16. Suppose there are three astronauts outside a spaceship, and two of them decide to play catch using the third man. All the astronauts weigh the same on Earth and are equally strong. The first astronaut throws the second one toward the third one and the game begins. Describe the motion of the astronauts as the game proceeds. In terms of the number of throws, how long will the game last?

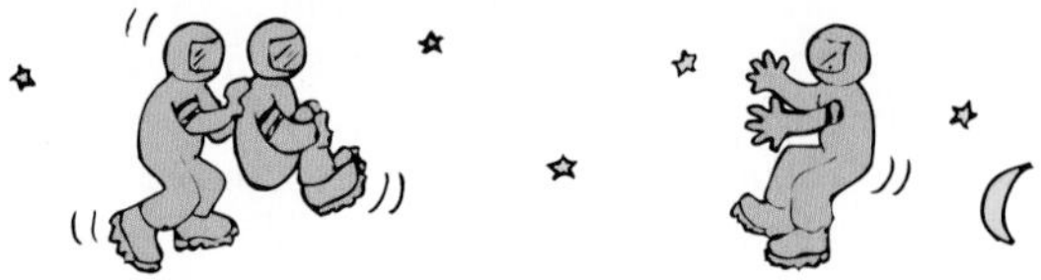

17. How is it possible that a flock of birds in flight can have a momentum of zero but not have zero kinetic energy?
18. When a cannon with a long barrel is fired, the force of expanding gases acts on the cannonball for a longer distance. What effect does this have on the velocity of the emerging cannonball?
19. You and a flight attendant toss a ball back and forth in an airplane in flight. Does the KE of the ball depend on the speed of the airplane? Carefully explain.
20. Can something have energy without having momentum? Explain. Can something have momentum without having energy? Defend your answer.
21. In an effort to combat wasteful habits, we often urge others to "conserve energy" by turning off lights when they are not in use, for example, or by setting thermostats at a moderate level. In this chapter, we also speak of "energy conservation." Distinguish between these two usages.
22. An inefficient machine is said to "waste energy." Does this mean that energy is actually lost? Explain.
23. A child can throw a baseball at 20 mph. Some professional ball players can throw a baseball at 100 mph, which is five times as fast. How much more energy does the pro ball player give to the faster ball?
24. If a golf ball and a Ping-Pong ball both move with the same kinetic energy, can you say which has the greater speed? Explain in terms of KE. Similarly, in a gaseous mixture of massive molecules and light molecules with the same average KE, can you say which have the greater speed?
25. Consider a pendulum swinging to and fro. At what point in its motion is the KE of the pendulum bob at a maximum? At what point is its PE at a maximum? When its KE is half its maximum value, how much PE does it have?
26. A physics instructor demonstrates energy conservation by releasing a heavy pendulum bob, as shown in the sketch, allowing it to swing to and fro. What would happen if, in his exuberance, he gave the bob a slight shove as it left his nose? Why?

27. Discuss the design of the roller coaster shown in the sketch in terms of the conservation of energy.

28. Consider the identical balls released from rest of tracks A and B as shown. When each ball has reached the right end of it track, which will have the greater speed? Why is this question easier to answer than the similar question asked in Exercise 25 back in Chapter 2?

29. Strictly speaking, does a car burn more gasoline when the lights are turned on? Does the overall consumption of gasoline depend on whether or not the engine is running? Defend your answer.
30. If an automobile had an engine that was 100% efficient, would it be warm to your touch? Would its exhaust heat the surrounding air? Would it make any noise? Would it vibrate? Would any of its fuel go unused?

Problems

1. A car with a mass of 1000 kg moves at 20 m/s. Show that the braking force is needed to bring the car to a halt in 10 s is 2000 N.
2. A railroad diesel engine weighs four times as much as a freight car. If the diesel engine coasts at 5 km per hour into a freight car that is initially at rest, show that the two coast at 4 km/h after they couple together?
3. A 5-kg fish swimming at 1 m/s swallows an absent-minded 1-kg fish at rest. (a) Show that the speed of the larger fish after lunch is −5 m/s. What would be its speed if the smaller fish were swimming toward it at 4 m/s?
4. Comic-strip hero Superman meets an asteroid in outer space and hurls it at 800 m/s, as fast as a bullet. The asteroid is a thousand times more massive than Superman. In the strip, Superman is seen at rest after the throw. Taking physics into account, what would be his recoil velocity?
5. Consider the inelastic collision between the two freight cars in Figure 4.11. The momentum before and after the collision is the same. The KE, however, is less after the collision than before the collision. How much less, and what has become of this energy?
6. This question is typical on some driver's license exams: A car moving at 50 km/h skids 15 m with locked brakes. How far will the car skid with locked brakes at 150 km/h?
7. In the hydraulic machine shown, it is observed that, when the small piston is pushed down 10 cm, the large piston is raised 1 cm. If the small piston is pushed down with a force of 100 N, show that the large piston is capable of exerting 1000 N of force.

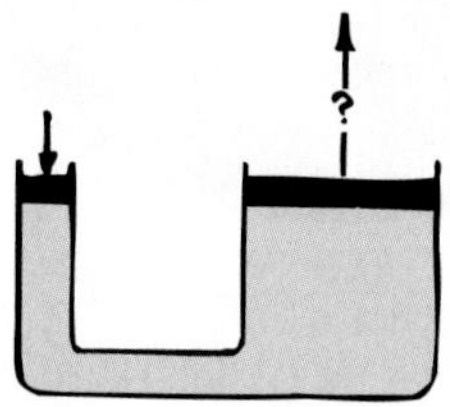

8. Consider a car with a 25% efficient engine that encounters an average retarding force of 1000 N. Assume that the energy content of the gasoline is 40 MJ/L. Show that the car will get 20 km per liter of fuel.
9. When a cyclist expends 1000 W of power to deliver mechanical energy to her bicycle at a rate of 100 W. Show that the efficiency of her body is 10%.
10. The decrease in PE for a freely falling object equals its gain in KE, in accord with the conservation of energy. (a) By simple algebra, find an equation for an object's speed v after falling a vertical distance h. Do this by equating KE to the object's change in PE. (b) Then figure out how much higher a freely falling object must fall to have twice the speed when it hits the ground.

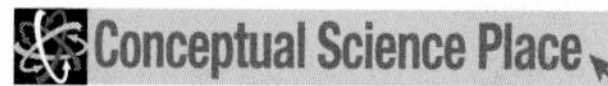

CHAPTER 4 ONLINE RESOURCES

Interactive Figures 4.8, 4.10, 4.11, 4.23 • ***Videos*** Definition of Momentum, Changing Momentum, Decreasing Momentum Over Short Time, Bowling Ball and Conservation of Energy, Conservation of Energy, Machines, Pulleys • ***Quiz*** • ***Flashcards*** • ***Links***

CHAPTER 5

Gravity

The space shuttle is not beyond the reach of Earth's gravity; it stays in orbit because of *Earth's gravity. In fact, because of Earth's gravity, the shuttle doesn't need to burn fuel to stay in orbit, and it can turn off its engines during the 5 to 10 days it spends in space.*

CHAPTER OUTLINE

It would be erroneous to say that Newton discovered gravity. The discovery of gravity goes back much further than Newton's era, to earlier times when Earth dwellers fell from trees and from ledges inside their caves, or when they discovered the consequences of tripping. What Newton discovered was that gravity is universal—that it is not a phenomenon unique to Earth, as his contemporaries had assumed. Further, Newton demonstrated that gravity is universal and can be described by a simple law. What is Newton's universal law of gravitation? How does it unify phenomena as seemingly diverse as falling apples and orbiting planets? What is the relationship between weight and gravity? How is the motion of a thrown basketball essentially similar to that of a satellite? We'll discover answers to these and other gravity-related questions in this chapter.

5.1 The Legend of the Falling Apple

According to popular legend, Newton was sitting under an apple tree when he made a connection that changed the way we see the world. He saw an apple fall. Perhaps he then looked up through the branches toward the origin of the falling apple and noticed the Moon. In any event, Newton had the insight to realize that the force pulling on the apple was the same force that pulls on the Moon. Newton realized that Earth's gravity reaches to the Moon (Figure 5.1).

5.2 The Fact of the Falling Moon

If Earth's gravity is pulling the Moon toward it, why doesn't the Moon fall toward the Earth, like an apple falls from a tree? If an apple or anything else drops from rest, it falls in a vertical straight-line path. To get a better idea of this, consider a tree sitting in the back of a truck (Figure 5.2). If the truck is at rest when the apple falls, we see that the apple's path is vertical. But if the truck is moving when the apple begins to fall, the apple follows a curved path. Can you see that, the faster the truck moves, the wider the curved path of the falling apple? Later in this chapter, we'll see that, if the apple or anything else moves fast enough so that its curved path matches the Earth's curvature, it becomes a satellite.

Figure 5.1 Could the gravitational pull on the apple reach to the Moon?

As the Moon traces out its orbit around the Earth, it maintains a **tangential velocity**—a velocity parallel to the Earth's surface. Newton realized that the Moon's tangential velocity keeps it falling *around* the Earth instead of directly into it. Newton further realized that the Moon's path around the Earth is similar to the paths of the planets around the Sun.

From the time of Aristotle, the circular motions of heavenly bodies were regarded as natural. The ancients believed that stars, planets, and the Moon moved in divine circles, free from the forces that dictate motion here on Earth. They believed there were two sets of laws, one for earthly events and a different set for motions in the heavens. Newton's stroke of intuition—that the force between Earth and apples is the same force that pulls moons, planets and everything else—was a revolutionary break with prevailing notions. Newton synthesized terrestrial and cosmic laws.

5.3 Newton's Law of Universal Gravitation

Figure 5.2 If an apple falls from a tree at rest, it falls straight downward. If it falls from a moving tree, however, it follows a curved path.

Newton further realized that *everything* pulls on *everything* else. He discovered that a **gravitational force** acts on all things in a beautifully simple way—a way that involves only mass and distance. According to Newton, every mass pulls on every other mass with a force that is directly proportional to the product of the two interacting masses. The force is inversely proportional to the square of the distance separating them. This statement is known as the law of universal gravitation.

$$\text{Force} \sim \frac{(\text{mass}_1 \times \text{mass}_2)}{\text{distance}^2}$$

Expressed in symbolic shorthand,

$$F \sim \frac{(m_1 \times m_2)}{d^2}$$

where m_1 and m_2 are the masses, and d is the distance between their centers. Thus, the greater the masses m_1 and m_2, the greater the force of attraction between them. The greater the distance of separation d, the weaker is the force of attraction—weaker as the inverse square of the distance between their centers (Figure 5.4).

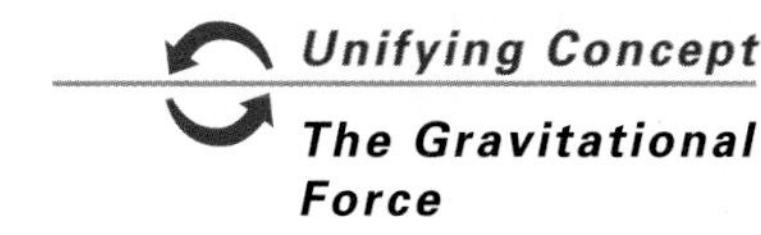

Unifying Concept

The Gravitational Force

CHECK YOURSELF

1. According to the equation for gravity, what happens to the force between two bodies if the mass of only one body is doubled?
2. What happens if the masses of both bodies are doubled?
3. What happens if the mass of one body is doubled and the other is tripled?
4. Gravitational force acts on all bodies in proportion to their masses. Why, then, doesn't a heavy body fall faster than a light body?

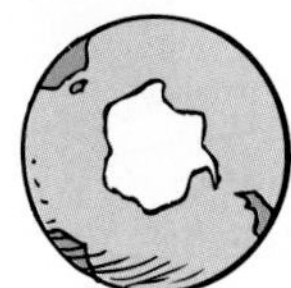

Figure 5.3 The tangential velocity of the Moon about Earth allows it to fall around Earth rather than directly into it. If this tangential velocity were reduced to zero, what would be the fate of the Moon?

CHECK YOUR ANSWERS

1. When one mass is doubled, the force between them doubles.
2. The force is four times as much.
3. Double × triple = six. So the force is six times as much.
4. The answer goes back to Chapter 3. Recall Figure 3.9, in which heavy and light bricks fall with the same acceleration because both have the same ratio of weight to mass. Newton's second law ($a = \frac{F}{m}$) reminds us that greater force acting on greater mass does not result in greater acceleration.

Figure 5.4 As the rocket gets farther from the Earth, the strength of the gravitational force between the rocket and the Earth decreases.

Inverse-Square Law

Unifying Concept

Inverse-Square Law

5.4 Gravity and Distance: The Inverse-Square Law

Gravity gets weaker with distance the same way a light gets dimmer as you move farther from it. Consider the candle flame in Figure 5.5. Light from the flame travels in all directions in straight lines. A patch is shown 1 meter from the flame. Notice that, at a distance of 2 meters away, the light rays that fall on the patch spread to fill a patch twice as tall and twice as wide. In other words, the same light falls on a patch with 4 times the area. If the same light were 3 meters away, it would spread to fill a patch 3 times as tall and 3 times as wide, and it would fill a patch 9 times the area.

As the light spreads out, its brightness decreases. Can you see that when you're twice as far away, it appears $\frac{1}{4}$ as bright? And can you see that when you're 3 times as far away, it appears $\frac{1}{9}$ as bright? There is a rule here: The intensity of the light decreases as the inverse square of the distance. This is the **inverse-square law**.

The inverse-square law also applies to a paint sprayer. Pretend you hold a paint gun at the center of a sphere with a radius of 1 meter (Figure 5.6). Say a burst of paint produces a square patch of paint 1 millimeter thick. How thick would the patch be if the experiment were done in a sphere with twice the radius—that is, with the spray gun twice as far away? The answer is not half as thick, because the paint would spread to a patch twice as tall *and* twice as wide. It would spread over an area *four times* as big, and its thickness would only be $\frac{1}{4}$ millimeter. Can you see that for a sphere of radius 3 meters the thickness of the patch would only be $\frac{1}{9}$ millimeter? Do you see that the thickness of paint decreases as the *square* of the distance? The inverse-square law holds for light, for spray paint, and for gravity. It holds for all phenomena where something from a localized source spreads uniformly throughout the surrounding space—for example, gas molecules from a perfume bottle, radiation from a piece of uranium, and sound from a cricket.

The greater the distance from Earth's center, the less the gravitational force on an object. In Newton's equation for gravity, the distance term d is the distance between the centers of the masses of objects attracted to each other. Note that the girl at the top of the ladder in Figure 5.7 weighs only $\frac{1}{4}$ as much as she weighs at the Earth's surface. That's because she is *twice* the distance from Earth's center. (Do you remember from Chapter 2 that *weight* is the force due to gravity on a body?)

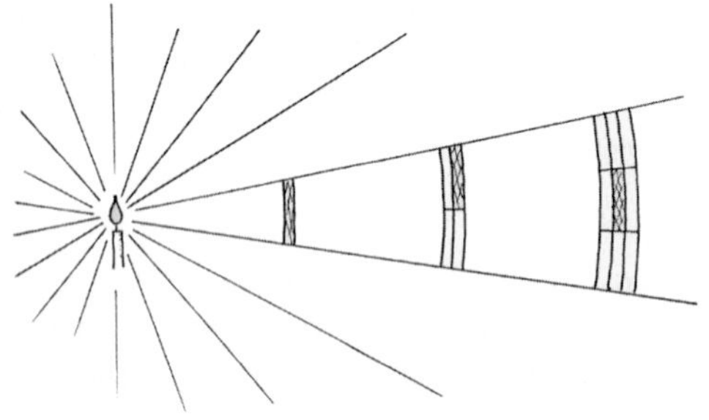

Figure 5.5 Light from the flame spreads in all directions. At twice the distance, the same light is spread over 4 times the area; at 3 times the distance, it is spread over 9 times the area.

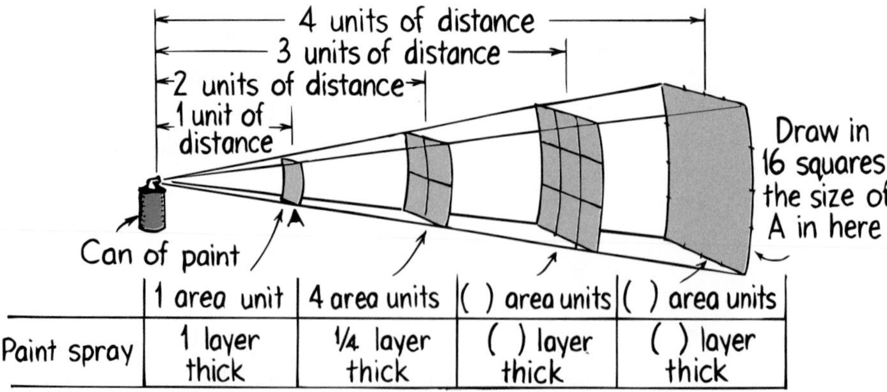

Figure 5.6 The inverse-square law. Paint spray travels radially away from the nozzle of the can in straight lines. Like gravity, the "strength" of the spray obeys the inverse-square law. Check that you understand how this works—fill in the blanks.

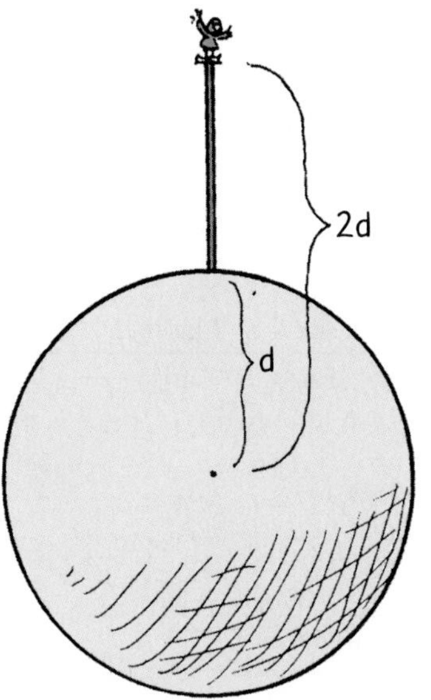

Figure 5.7 At the top of the ladder, the girl is twice as far from Earth's center, and she weighs only $\frac{1}{4}$ as much as she did at the bottom of the ladder.

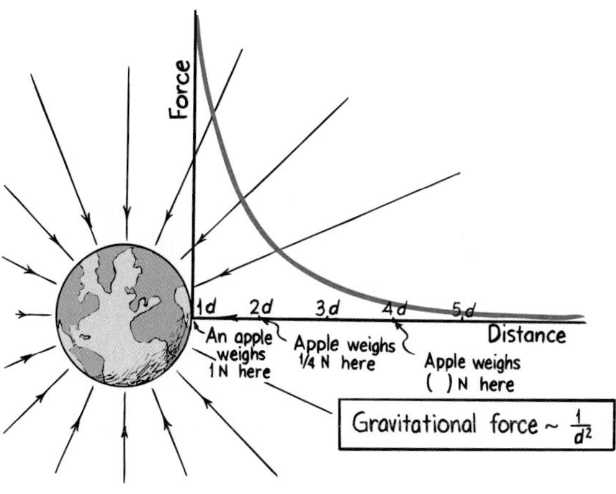

Figure 5.8 Interactive Figure

If an apple weighs 1 N at the Earth's surface, it would weigh only $\frac{1}{4}$ N twice as far from the center of the Earth. At three times the distance, it weighs only $\frac{1}{9}$ N. What would the apple weigh at four times the distance? At five times the distance? Gravitational force versus distance is plotted in color.

INSIGHT Just as π relates the circumference of a circle to its diameter, *G* relates gravitational force with mass and distance. *G*, like π, is a proportionality constant.

CHECK YOURSELF

1. How much does the force of gravity change between the Earth and a receding rocket when the distance between them is doubled? When it is tripled? When it is increased tenfold?
2. Consider an apple at the top of a tree. The apple is pulled by Earth's gravity with a force of 1 N. If the tree were twice as tall, would the force of gravity be only $\frac{1}{4}$ as strong? Defend your answer.

CHECK YOUR ANSWERS

1. When the distance is doubled, the force is only $\frac{1}{4}$ as much. When the distance is tripled, the force is only $\frac{1}{9}$ as much. When the distance is increased tenfold, the force is only $\frac{1}{100}$ as much.
2. No, because the twice-as-tall apple tree is not twice as far from the Earth's center. The taller tree would have to be 6370 km tall (as tall as the Earth's radius) for the apple's weight to reduce to $\frac{1}{4}$ N. For a decrease in weight by 1%, an object must be raised 32 km—nearly four times the height of Mt. Everest. So, as a practical matter, we disregard the effects of everyday changes in elevation for gravity. The apple has practically the same weight at the top of the tree as it has at the bottom.

So the gravitational attraction between two objects gets weaker rather quickly as the objects get farther apart. But no matter how great the distance, gravity approaches, but never quite reaches, zero. There is still a gravitational attraction between any two masses, no matter how far apart they are. Even if you were removed to the far reaches of the universe, the gravitational influence of home would still remain with you. It may be overwhelmed by the influences of nearer and/or more massive bodies, but its presence is still there. The gravitational influence of every material object, however small or however far, is exerted through all of space.

5.5 The Universal Gravitational Constant, G

The universal law of gravitation can be written as an exact equation when the universal constant of gravitation, *G*, is used. Then we have

$$F = \frac{G(m_1 \times m_2)}{d^2}$$

The units of *G* make the force come out in newtons. The magnitude of *G* is the same as the gravitational force between two 1-kilogram masses that are 1 meter apart: 0.0000000000667 newton,

$$G = 6.67 \times 10^{-11} \frac{\text{N} \cdot \text{m}^2}{\text{kg}^2}$$

This is an extremely small number. It shows that gravity is a very weak force compared with electrical forces. The large net gravitational force that we feel as weight occurs is because of the immensity of the number of bits of mass in the planet Earth that are pulling on us.

MATH CONNECTION

Comparing Gravitational Attractions

Every mass gravitationally attracts every other mass. How do different pulls between different pairs of interacting objects compare? Try a few examples.

Problems A 3-kg newborn baby at the Earth's surface is gravitationally attracted to Earth with a force of about 30 N.

a. Calculate the force of gravity with which the baby on Earth is attracted to the planet Mars, when Mars is closest to Earth. (The mass of Mars is 6.4×10^{23} kg and its closest distance is 5.6×10^{10} m.)
b. Calculate the force of gravity between the baby and the physician who delivers her. Assume that the physician has a mass of 100 kg and is 0.5 m from the baby.
c. How do these forces compare?

Solutions

a. Mars: $F = G\dfrac{(m_1 \times m_2)}{d^2}$

$$= \frac{[6.67 \times 10^{-11}\text{N}\cdot\text{m}^2/\text{kg}^2(3\text{ kg})(6.4 \times 10^{23}\text{ kg})]}{(5.6 \times 10^{10}\text{ m})^2}$$

$$= 4.1 \times 10^{-8}\text{ N}$$

b. Physician: $F = G\dfrac{(m_1 \times m_2)}{d^2}$

$$= \frac{[6.67 \times 10^{-11}(3\text{ kg})(10^2\text{ kg})]}{(0.5)^2\text{ m}}$$

$$= 8.0 \times 10^{-8}\text{ N}$$

c. The gravitational force between the baby and the physician is about twice that between the baby and Mars.

INTEGRATED SCIENCE BIOLOGY

Your Biological Gravity Detector

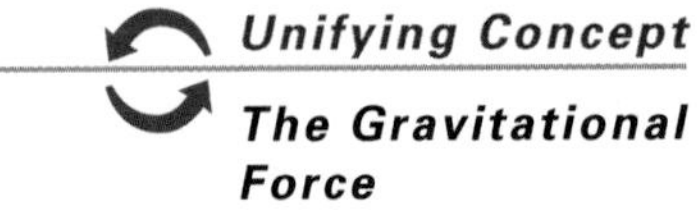

Section 5.3

How many senses do you have? The answer is *not* five. Beyond sight, taste, smell, hearing, and touch, there are other senses—for example, hunger and thirst. You have a sense of how your body is oriented in space, too. This is called your *vestibular sense,* and it depends on your ability to detect both your acceleration and your orientation with respect to Earth's gravitational field. You do this by means of organs in your inner ear called *vestibular organs.*

There are two kinds of vestibular organs, as shown in Figure 5.9—the *semicircular canals* and the *otolith organs.* While it's the job of the semicircular canals to detect rotational motion of the head, the otolith organs detect linear acceleration of the head as well as whether or not the head is tilted with respect to Earth's gravitational field. Otolith organs contain small sensory areas about 2 mm in diameter known as *maculas.* Each macula contains thousands of receptor cells called *hair cells.* The hair cells have stalklike cilia that are embedded in a gelati-

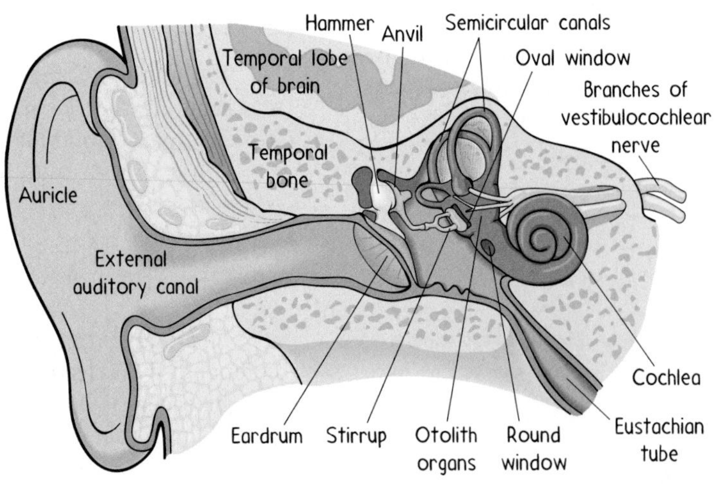

Figure 5.9 The vestibular system consists of the semicircular canals and the otolith organs.

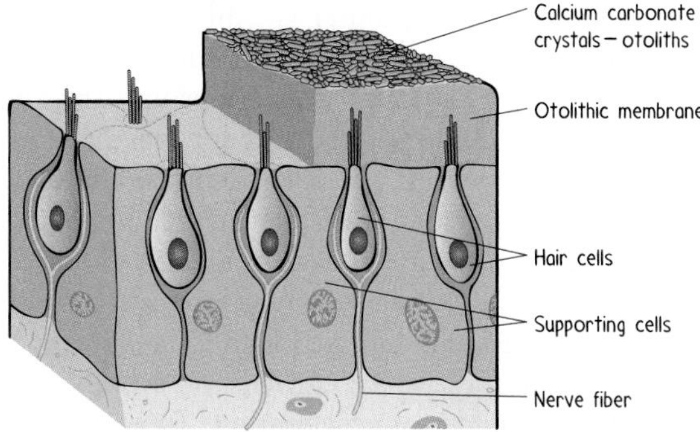

Figure 5.10 Anatomical features of a macula—the structure that contains receptor cells that detect acceleration of the head as well as its orientation with respect to gravity. Each macula contains several thousand hair cells.

nous matrix called the *otolithic membrane*; And, on top of the membrane, there are small piles of calcium carbonate ($CaCO_3$) crystals, which are called *otoliths*. The anatomy of a macula is shown in Figure 5.10.

Because the otoliths are more dense than the surrounding tissue, they have greater inertia. This enables them to indicate linear acceleration: When you accelerate in one direction, the mass of the otolithic membrane and its otoliths causes the hair-cell cilia to bend. This bending of the cilia stimulates the cells to send signals in a particular pattern that your brain interprets as acceleration. When you move with a constant velocity, the otoliths reach equilibrium and you no longer have the perception that you are moving (Figure 5.11).

The otoliths detect head tilt in a way similar to how they sense linear acceleration. When your head tilts, the planes of your maculae change with respect to the direction of gravity's pull, and this bends the cilia of your hair cells. When the cilia bend in one direction, they increase the rate at which they fire nerve signals. When the cilia bend in the opposite direction, they fire fewer nerve signals, as shown in Figure 5.12. The overall pattern of signal firing from tilted macula indicates the direction and degree of head tilt.

By the way, the literal meaning of the word "otoliths" is *ear stones*. So you really do have rocks in your head!

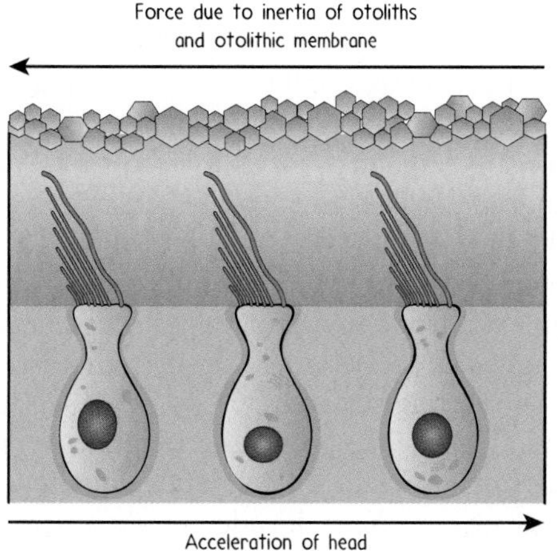

Figure 5.11 The greater density of the otoliths and otolithic memebrane compared with the surrounding tissue pulls and bends the cilia of receptor cells when you accelerate.

CHECK YOURSELF

1. What is the vestibular sense, and what does it have to do with gravity?
2. Microgravity is defined as a state in which gravity is reduced to negligible levels—specifically, below one millionth of Earth's gravity. Would your otolithic organs be able to detect head tilt in microgravity? Why or why not?
3. How do you sense the direction in which your head tilts?

CHECK YOUR ANSWERS

1. It is the sense that indicates body orientation. It depends on knowing how the body is oriented with respect to the downward-directed pull of Earth's gravity.
2. No. The otolith organs could not sense variation with respect to a standard gravitational field direction in microgravity.
3. The rate of firing of nerve signals indicates direction of tilt. When the head tilts in one direction, nerve firing increases. When the head tilts in the opposite direction, nerve firing decreases.

5.6 Weight and Weightlessness

Weight and Weightlessness; Apparent Weightlessness

When you step on a bathroom scale, in effect, you compress a spring inside it that is affixed to a pointer. When the pointer stops moving, the elastic force of the deformed spring balances the gravitational force between you and the

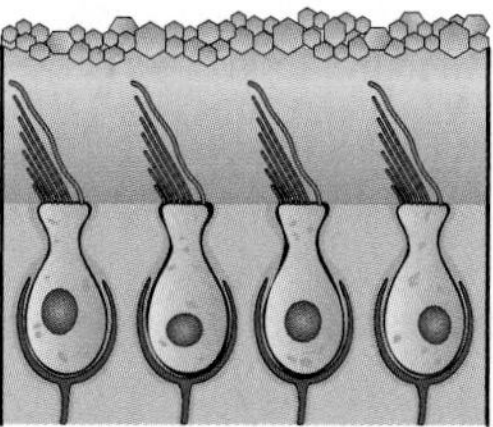

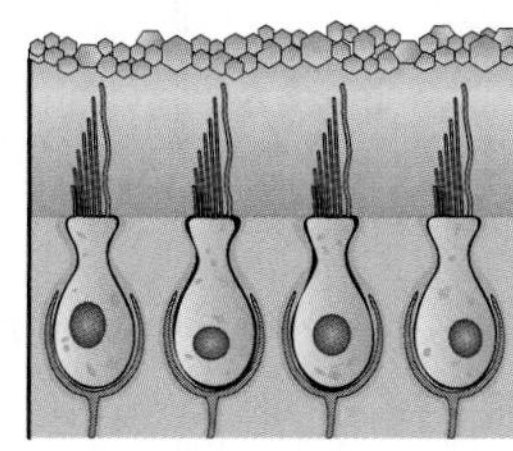

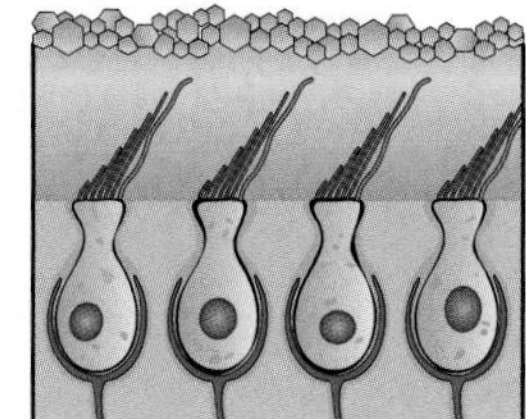

Figure 5.12 Maculae respond to tilting by varying the frequency of neural signals.

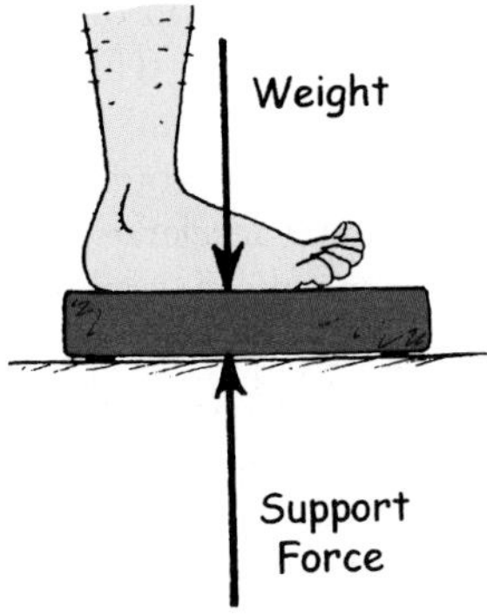

Figure 5.13 When you step on a weighing scale, two forces act on it; a downward force of gravity (your ordinary weight, *mg*, if there is no acceleration) and an upward support force. These equal and opposite forces squeeze a springlike device inside the scale that is calibrated to show your weight.

Earth—you and the scale are in equilibrium. The pointer is calibrated to show your weight (Figure 5.13). If you stand on a bathroom scale in a moving elevator, you'll find variations in your weight. If the elevator accelerates upward, the springs inside the bathroom scale are more compressed and your weight reading is greater. If the elevator accelerates downward, the springs inside the scale are less compressed and your weight reading is less. If the elevator cable breaks and the elevator falls freely, the reading on the scale goes to zero. According to the scale's reading, you would be weightless. Would you really be weightless? We can answer this question only if we agree on what we mean by *weight*.

In Chapter 2, we defined weight as the force due to gravity on a body, *mg*. Your weight does have the value of *mg* if you're not accelerating. To generalize, we now refine this definition by saying that the weight of something is the force it exerts against a supporting floor or weighing scale. According to this definition, you are as heavy as you feel. So, in an elevator that accelerates downward, the supporting force of the floor is less and, therefore, you weigh less. If the elevator is in free fall, your weight is zero (Figure 5.14). Even in this weightless condition, however, there is still a gravitational force acting on you, causing your downward acceleration. But gravity now is not felt as weight because there is no support force.

Consider an astronaut in orbit. The astronaut is weightless because he is not supported by anything (Figure 5.15). There would be no compression in the springs of a bathroom scale placed beneath his feet because the bathroom scale is falling as fast as he is. Any objects that are released fall together with him and remain in his vicinity, unlike what occurs on the ground. All the local effects of gravity are eliminated. The body organs respond as if gravity forces were absent, and this gives the sensation of weightlessness. The astronaut experiences the same sensation in orbit that he would feel falling in an elevator—a state of free fall.

INSIGHT Astronauts inside an orbiting space vehicle have no weight, even though the force of gravity between them and the Earth is only slightly less than it would be if they were at ground level.

On the other hand, if the astronaut were in a spacecraft undergoing acceleration, even in deep space and far removed from any attracting objects, he *would* have weight. Like the girl in the upward accelerating elevator, the astronaut would be pressed against a scale or supporting surface.

The International Space Station in Figure 5.16 provides a weightless environment. The station facility and the astronauts inside all accelerate equally toward Earth, at somewhat less than 1 *g* because of their altitude. This acceleration is not sensed at all. With respect to the station, the astronauts experience zero *g*.

Figure 5.14 Your weight equals the force with which you press against the supporting floor. If the floor accelerates up or down, your weight varies (even though the gravitational force *mg* that acts on you remains the same).

Figure 5.15 Both are weightless.

Figure 5.16 The inhabitants of this laboratory and docking facility continually experience weightlessness. They are in free fall around the Earth. Does a force of gravity act on them?

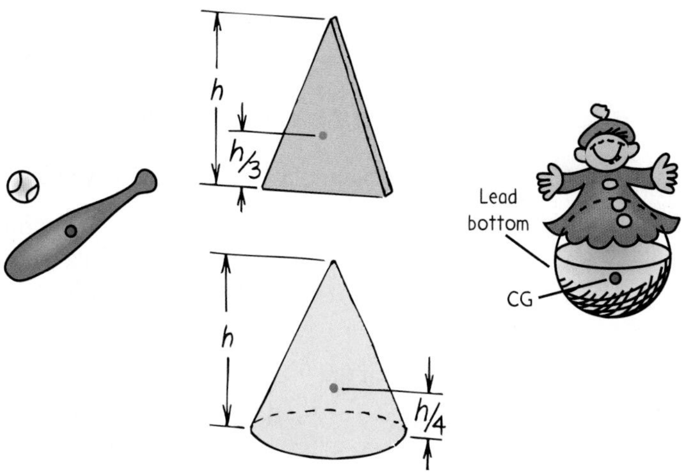

Figure 5.17 The center of gravity for each object is shown by a colored dot.

INTEGRATED SCIENCE BIOLOGY

Center of Gravity of People

Conceptual Science Place

Locating the Center of Gravity; Toppling

The *center of gravity (CG)* of an object is the point located at the object's average position of weight. For a symmetrical object, this point is at the geometric center. But an irregularly shaped object, such as a baseball bat, has more weight at one end, so its CG is toward the heavier end. A piece of tile cut into the shape of a triangle has its CG one third of the way up from its base (Figure 5.17).

The position of an object's CG relative to its base of support determines the object's stability. The rule for stability is this: If the CG of an object is above the area of support, the object will remain upright. If the CG extends outside the area of support, the object will topple.

This is why the Leaning Tower of Pisa doesn't topple. Its CG does not extend beyond its base. If the tower leaned far enough over so that its CG extended beyond its base, it would topple (Figure 5.18).

When you stand erect with your arms hanging at your sides, your CG is typically 2 to 3 cm below your navel and midway between your front and back. The CG is slightly lower in women than it in men because women tend to be proportionally larger in the pelvis and smaller in the shoulders.

When you stand, your CG is somewhere above your support base, the area bounded by your feet (Figure 5.19). In unstable situations, you place your feet farther apart to increase this area. Standing on one foot greatly decreases the area of your support base. A baby must learn to coordinate and position its CG

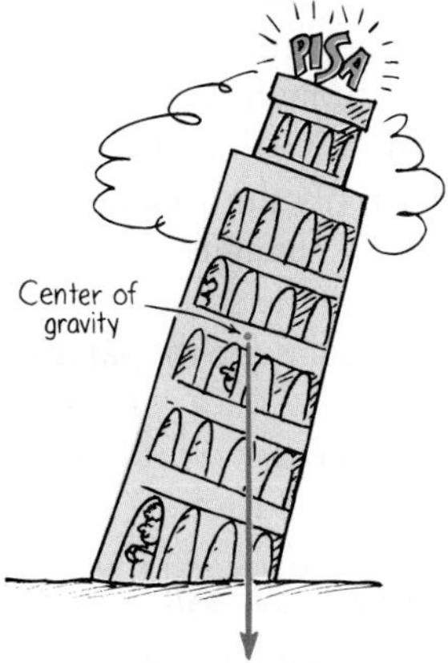

Figure 5.18 The Leaning Tower of Pisa does not topple over because its CG lies above its base.

Figure 5.19 You can lean over and touch our toes only if our CG is above the area bounded by your feet.

above one foot. Many birds—pigeons for example—do this by jerking their heads back and forth with each step.

Conceptual Science Place

Centripetal Force; Simulated Gravity

5.7 Gravity Can Be a Centripetal Force

If you whirl an empty tin can on the end of a string, you find that you must keep pulling on the string (Figure 5.20). You pull inward on the string to keep the can revolving over your head in a circular path. A force of some kind is required for any circular motion, including the nearly circular motions of the planets around the Sun. Any force that causes an object to follow a circular path is called a **centripetal force.** Centripetal means "center-seeking," or "toward the center." Centripetal force is not a new kind of force. It is simply a name given to any force that is directed at right angles to the path of a moving object and that tends to produce circular motion. The gravitational force acting across space is a centripetal force that keeps the Moon in Earth's orbit. Likewise, electrons that revolve about the nucleus of an atom are held by an electrical force that is directed toward the central nucleus.

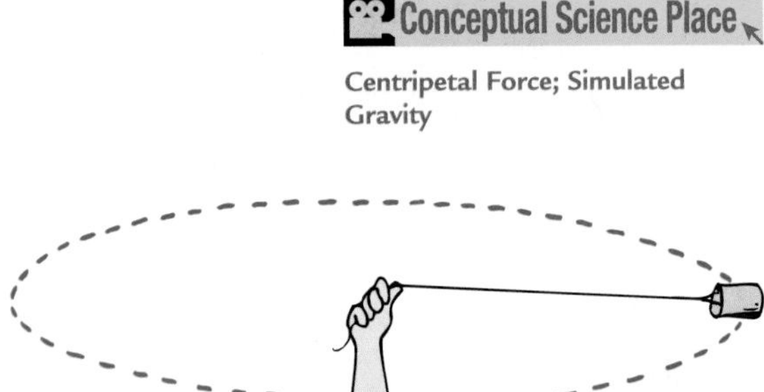

Figure 5.20 The only force that is exerted on the whirling can (neglecting the downward pull of gravity) is directed *toward* the center of circular motion. It is called a centripetal force. No *outward* force is exerted on the can.

Centripetal force could rescue future space travelers from the effects of weightlessness. Today's space habitats may someday be replaced by lazily rotating giant wheels (Figure 5.21). Occupants in the rotating habitat travel in circular paths and feel a centripetal force, which is directed inward toward the axis of rotation. The centripetal force is the support force, and it is sensed as weight. If the space habitat rotates at just the right speed, the support force can simulate normal Earth weight.

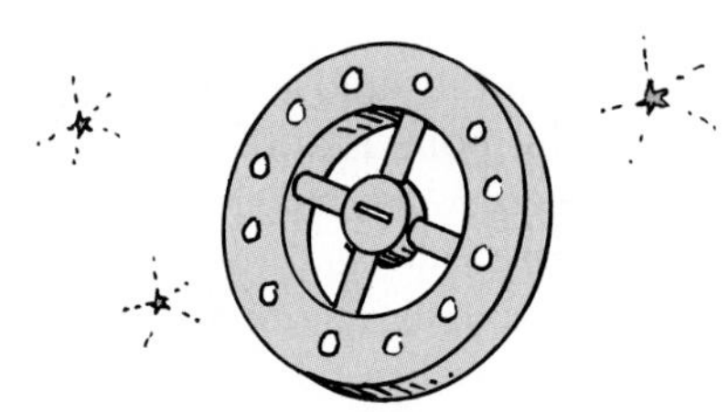

Figure 5.21 Occupants in the rotating space habitat experience simulated weight and can "stand up" inside. In upright positions, their feet press against the outer rim and their heads point toward the center.

5.8 Projectile Motion

Without gravity, you could toss a rock upward at an angle and it would follow a straight-line path. But, due to gravity, the path curves. A tossed rock, a cannonball, or any object that is projected by some means and continues in motion by its own inertia is called a **projectile.** To the cannoneers of earlier centuries, the curved paths of projectiles seemed very complex. Today, we see that these paths are surprisingly simple when the horizontal and vertical components of velocity are considered separately. We'll first consider the vertical part of a projectile's motion, the component that is affected by gravity.

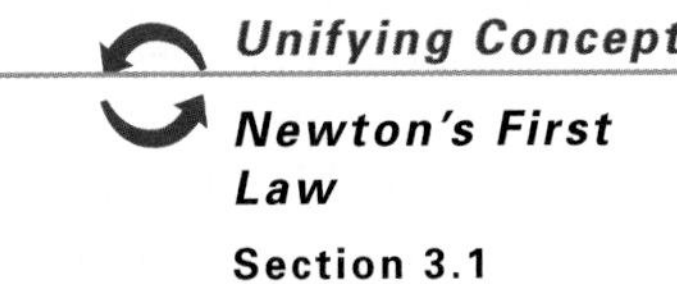

Unifying Concept

Newton's First Law

Section 3.1

A very simple projectile is a falling stone, as shown in Figure 5.22. This is a version of Figure 2.27, which we studied in Chapter 2. The stone gains speed as it falls straight down, as indicated by a speedometer. Remember that a freely falling object gains 10 meters/second during each second of fall. This is the acceleration due to gravity, 10 m/s^2. If it begins its fall from rest, 0 m/s, then at the end of the first second of fall its speed is 10 m/s. At the end of 2 seconds, its speed is 20 m/s—and so on. It keeps gaining 10 m/s each second it falls.

Although the change in speed is the same each second, the distance of fall keeps increasing. That's because the average speed of fall increases each second. Let's apply this to a new situation—throwing the stone horizontally off the cliff.

First, imagine that gravity doesn't act on the stone. In Figure 5.23 we see the positions the stone would have if there were no gravity. Note that the positions each second are the same distance apart. That's because there's no force acting on the stone.

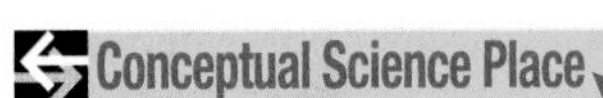

Projectile Motion

In the real world, there is gravity. The thrown stone falls beneath the straight line it would follow with no gravity (Figure 5.24). The stone curves as it falls. Interestingly, this familiar curve is the result of two kinds of motion occurring at the same time. One kind is the straight-down vertical motion of Figure 5.22. The other is the horizontal motion of constant velocity, as imagined in Figure 5.23. Both occur simultaneously. As the stone moves horizontally, it also falls straight downward—beneath the place it would be if there were no gravity. This is indicated in Figure 5.24.

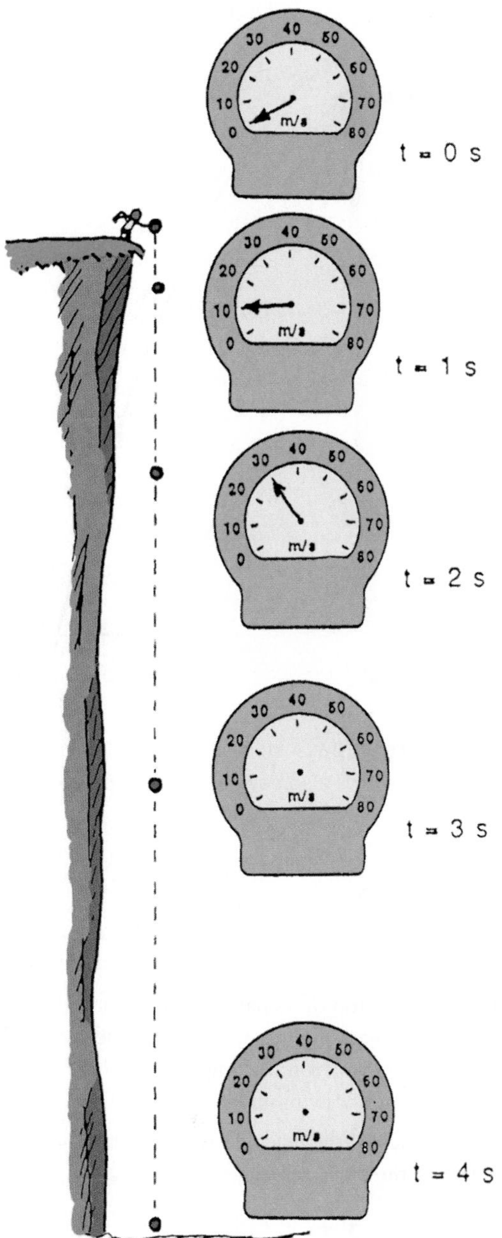

Figure 5.22 The falling stone gains a speed of 10 m/s each second. How would you fill in the speedometer readings for the times 3 seconds and 4 seconds?

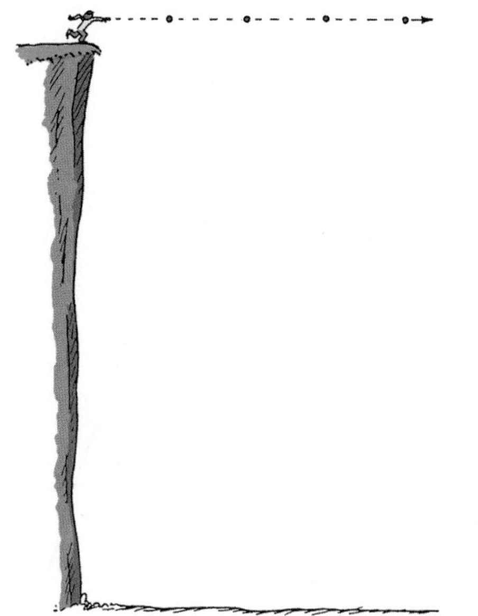

Figure 5.23 If there were no gravity, a stone thrown horizontally would move in a straight-line path and cover equal distances in equal time intervals.

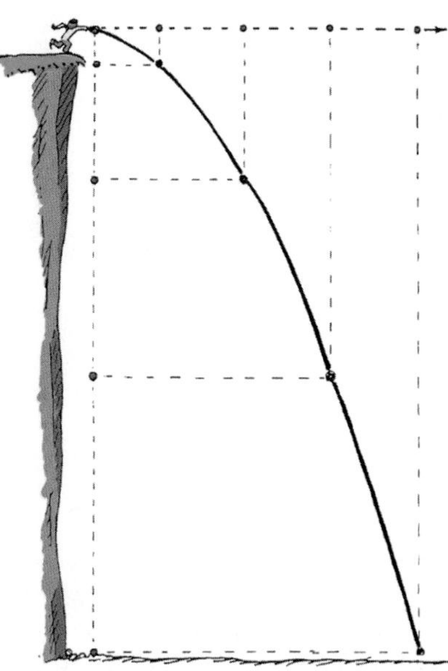

Figure 5.24 The vertical dashed line is the path of a stone dropped from rest. The horizontal dashed line would be its path if there were no gravity. The curved solid line shows the resulting trajectory that combines horizontal and vertical motion.

The curved path of a projectile is the result of constant motion horizontally and accelerated motion vertically under the influence of gravity. This curve is a **parabola**.

In Figure 5.25, we consider a stone thrown upward at an angle. If there were no gravity, the path would be along the dashed line with the arrow. Positions of the stone at 1-second intervals along the line are shown by red dots. Because of gravity, the actual positions (dark dots) are below these points. How far below? The answer is, the same distance an object would fall if it were dropped from the red-dot positions. When we connect the dark dots to plot the path, we get a different parabola.

In Figure 5.26, we consider a stone thrown at a downward angle. The physics is the same. If there were no gravity, it would follow the dashed line with the arrow. Because of gravity, it falls beneath this line, just as in the previous cases. The path is a somewhat different parabola.

The curved path of a projectile is a combination of horizontal and vertical motions. Consider the girl throwing the stone in Figure 5.27. The velocity she gives the stone is shown by the light blue vector. Notice that this vector has horizontal and vertical components. These components, interestingly, are completely independent of each other. Each of them acts as if the other didn't exist. Combined, they produce the curved path.

A typical projectile path would have the velocity vectors and their components as shown in Figure 5.28. Notice that the horizontal component remains the same at all points. That's because no horizontal force exists to change this component of velocity (assuming negligible air drag). The vertical component changes because of the vertical influence of gravity.

CHECK YOURSELF

1. At what part of its trajectory does a projectile have its minimum speed?
2. A tossed ball changes speed along its parabolic path. When the Sun is directly overhead, does the shadow of the ball across the field also change speed?

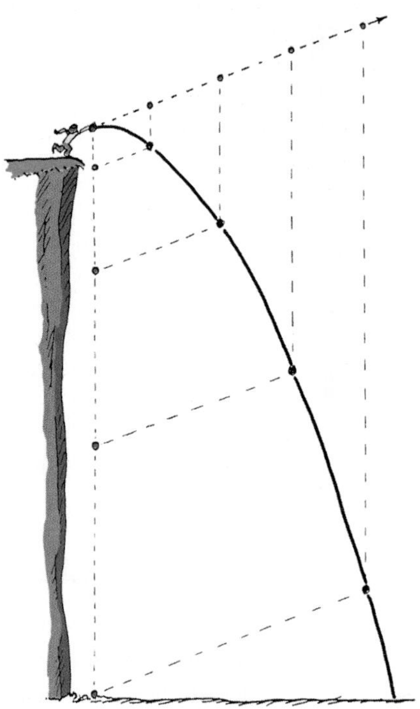

Figure 5.25 A stone thrown at an upward angle would follow the dashed line in the absence of gravity. Because of gravity, however, it falls beneath this line and describes the parabola shown by the solid curve.

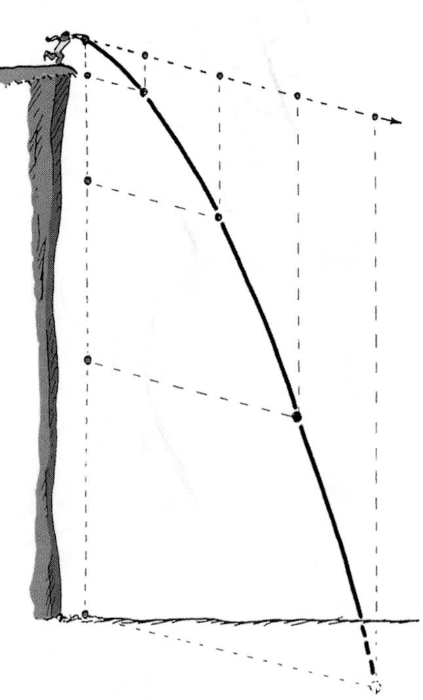

Figure 5.26 A stone thrown at a downward angle follows a somewhat different parabola.

Figure 5.27 The velocity of the ball (light blue vector) has vertical and horizontal components. The vertical component relates to how high the ball will go. The horizontal component relates to the horizontal range of the ball.

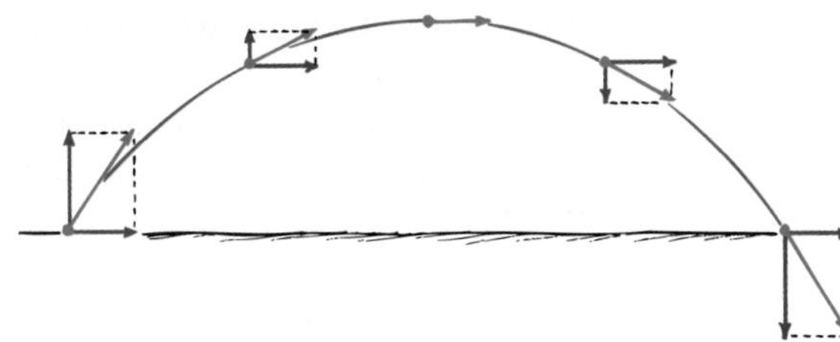

Figure 5.28 Interactive Figure
The velocity of a projectile at various points along its trajectory. Note that the vertical component changes while the horizontal component is the same everywhere.

CHECK YOUR ANSWERS

1. The speed of a projectile is at a minimum at the top of its path. If it is launched vertically, its speed at the top is zero. If it is projected at an angle, the vertical component of speed is zero at the top, leaving only the horizontal component. So the speed at the top is equal to the horizontal component at any point.
2. No, the shadow moves at constant velocity across the field, showing exactly the motion due to the horizontal component of the ball's velocity.

5.9 Projectile Altitude and Range

In Figure 5.29, we see the paths of several projectiles in the absence of air drag. All of them have the same initial speed but different projection angles. Notice that these projectiles reach different *altitudes*, or heights above the ground. They also have different *ranges*, or distances traveled horizontally. The remarkable thing to note is that the same range is obtained from two different projection angles—a pair that add up to 90°! An object thrown into the air at an angle of 60°,

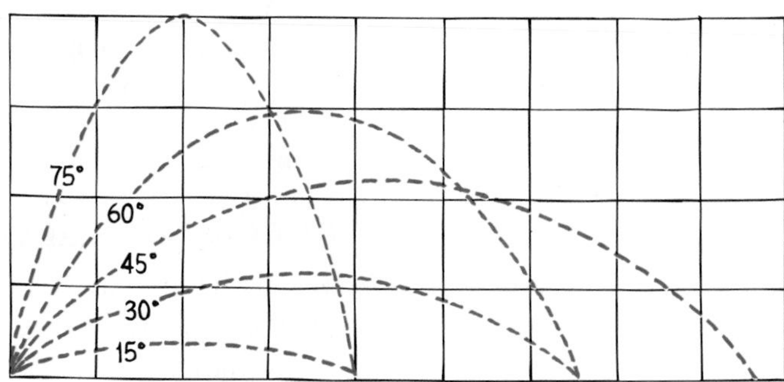

Figure 5.29 The paths of projectiles launched with equal speeds but different projection angles. Note that the same range occurs for pairs of angles that add to 90°.

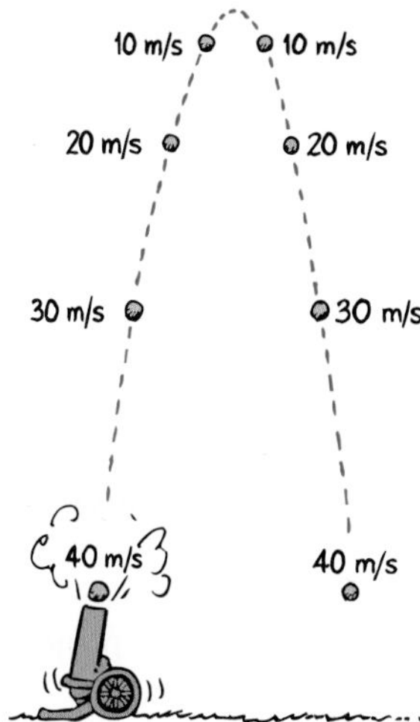

Figure 5.30 Without air drag, speed lost while going up equals speed gained while going down.

for example, will have the same range as if it were thrown at the same speed at an angle of 30°. For the smaller angle, of course, the object remains in the air for a shorter time.

When air drag is low enough to be negligible, a projectile will rise to its maximum height in the same time it takes to fall from that height to the ground. This is because the speed it loses while going up is the same speed it had when it was projected from the ground (Figure 5.30).

5.10 The Effect of Air Drag on Projectiles

We have considered projectile motion without air drag. You can neglect air drag for a ball you toss back and forth with your friends because the speed is low. But at higher speeds, air resistance matters. Air drag is a factor for high-speed projectiles, such as tennis balls, footballs, and basketballs cast airborne in a vigorous game. The result of air drag is that both range and altitude are decreased.

Consider baseball: Air drag greatly affects the range of balls batted and thrown in baseball games. Without air drag, a ball normally batted to the middle of center field would be a home run. If baseball were played on the Moon (not scheduled in the near future!), the range of balls would be considerably farther—about six times the ideal range on Earth. This is because there is no atmosphere on the Moon, so air drag on the Moon is completely absent. In addition, gravity is one-sixth as strong on the Moon, which allows higher and longer paths.

Unifying Concept
Friction
Section 2.8

CHECK YOURSELF

If the boy in Figure 5.31 simply drops a baseball a vertical distance of 5 m, it will hit the ground in 1 s. Suppose instead that he throws the ball horizontally as shown. The ball lands 20 m downrange. What is his pitching speed?

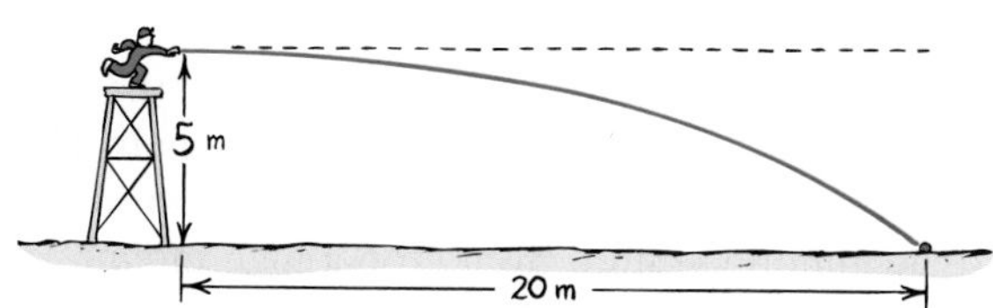

Figure 5.31 How fast is the ball thrown?

CHECK YOUR ANSWER

The ball is thrown horizontally, so the pitching speed is the horizontal distance divided by the time. A horizontal distance of 20 m is given. How about the time? Isn't the time along the parabola the same time it takes to fall vertically 5 m? Isn't this time 1 s? So the pitching speed $v = \frac{d}{t} = \frac{(20\text{ m})}{(1\text{ s})} = 20$ m/s.

Back here on Earth, baseball games normally take place on level ground. Baseballs curve over a flat playing field. The speeds of baseballs are not great enough for the Earth's curvature to affect the ball's path. For very long-range projectiles, however, the curvature of Earth's surface must be taken into account. As we will now see, when an object is projected fast enough, it can fall all the way around the Earth and become a *satellite*.

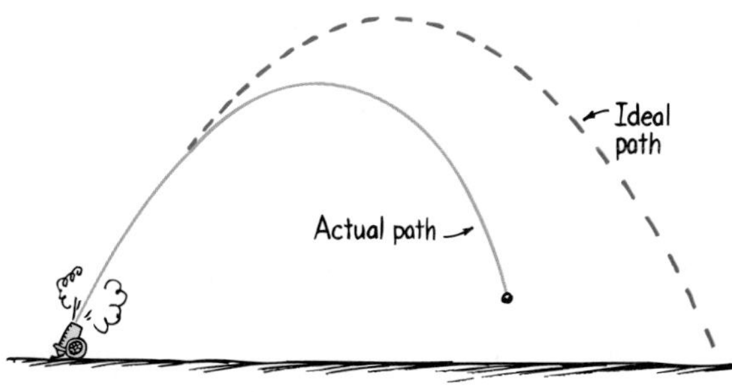

Figure 5.32 Interactive Figure
In the presence of air resistance, a high-speed projectile falls short of a parabolic path. The dashed line shows an ideal path with no air resistance. The solid line indicates an actual path.

5.11 Fast-Moving Projectiles—Satellites

Suppose that a cannon fires a cannonball so fast that its curved path matches the curvature of the Earth. Then, without air drag, it would be an Earth satellite! The same would be true if you could throw a stone fast enough. Any satellite is simply a projectile moving fast enough to fall continually around the Earth.

In Figure 5.33, we see the curved paths of a stone thrown horizontally at different speeds. Whatever the pitching speed, in each case the stone drops the same vertical distance in the same time. For a 1-second drop, that distance is 5 meters. (Perhaps by now you have made use of this fact in lab.) So, if you simply drop a stone from rest, it will fall 5 meters in 1 second of fall. Toss the stone sideways, and in 1 second it will be 5 meters below where it would have been without gravity. To become an Earth satellite, the stone's horizontal velocity must be great enough for its falling distance to match Earth's curvature.

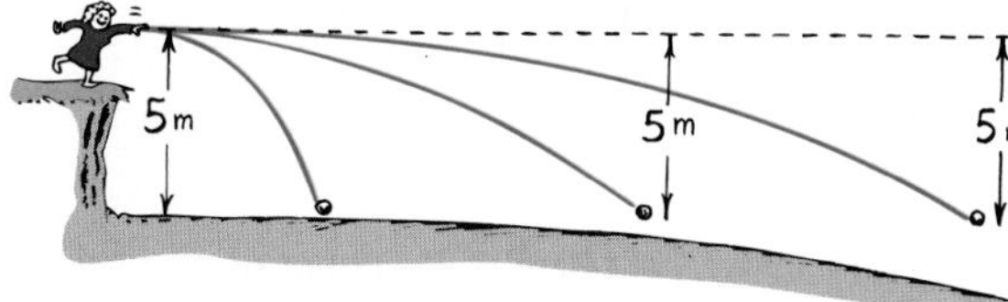

Figure 5.33 If you throw a stone at any speed, one second later it will have fallen 5 m below where it would have been if there were no gravity.

It is a geometrical fact that the surface of the Earth drops a vertical distance of 5 meters for every 8000 meters tangent to the surface (Figure 5.34). A tangent to a circle or to Earth's surface is a straight line that touches the circle or surface at only one place. (So the tangent is parallel to the circle or sphere at any point of contact.) With this amount of Earth's curvature, if you were floating in a calm ocean, you would be able to see only the top of a 5-meter mast on a ship 8000 meters (8 kilometers) away. We live on a round Earth.

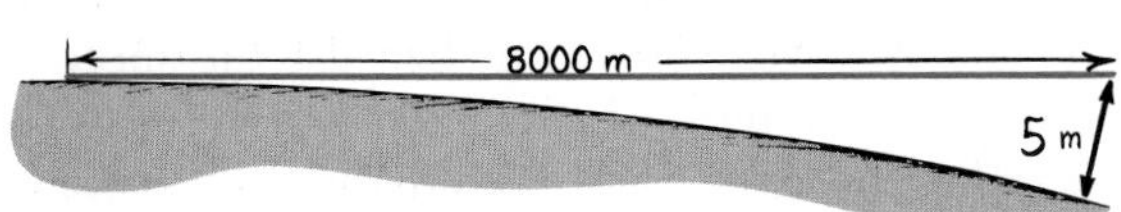

Figure 5.34 The Earth's curvature drops a vertical distance of 5 m for each 8000-m tangent (not to scale).

What do we call a projectile that moves fast enough to travel a horizontal distance of 8 kilometers during 1 second? We call it a satellite. Neglecting air drag, it would follow the curvature of the Earth. A little thought tells you that the minimum required speed is 8 kilometers per second. If this doesn't seem fast, convert it to kilometers per hour, and you will get an impressive 29,000 kilometers per hour (18,000 mi/hr). Fast, indeed!

At this speed, atmospheric friction would incinerate the projectile. This happens to grains of sand and other small meteors that graze the Earth's atmosphere, burn up, and appear as "falling stars." That is why satellites like the space shuttles are launched to altitudes higher than 150 kilometers—to be above the atmosphere.

It is a common misconception that satellites orbiting at high altitudes are free of gravity. Nothing could be farther from the truth. The force of gravity on a satellite 150 kilometers above the Earth's surface is nearly as great as it would be at the surface. If there were no gravity, the motion of the satellite would be along a straight-line path instead of curving around the Earth. High altitude puts the satellite beyond the Earth's *atmosphere*, but not beyond its *gravity*. As you know, Earth's gravity goes on forever, getting weaker and weaker with distance, but never reaching zero.

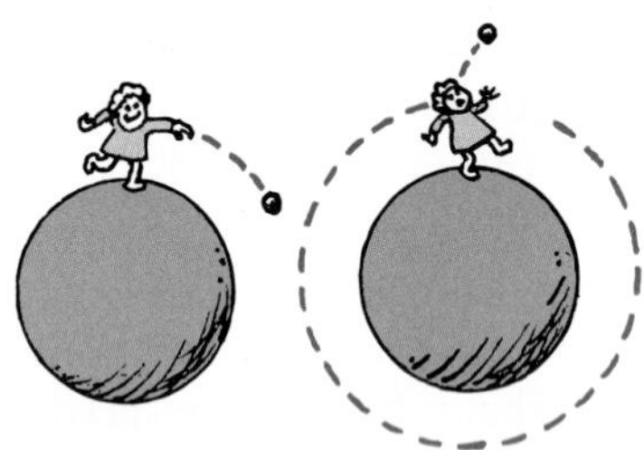

Figure 5.35 If the speed of the object and the curvature of its trajectory are great enough, the stone may become a satellite.

Satellite motion was understood by Isaac Newton. He reasoned that the Moon is simply a projectile circling the Earth under gravitational attraction. This concept is illustrated in Figure 5.36, which is an actual drawing by Newton. He compared the Moon's motion to a cannonball fired from atop a high mountain. He imagined that the mountaintop was above the Earth's atmosphere, so that air drag would not slow the motion of the cannonball. If a cannonball were fired with a low horizontal speed, it would follow a curved path and soon hit the Earth below. If it were fired faster, its path would be wider and it would hit a place on Earth farther away. If the cannonball were fired fast enough, Newton reasoned, the curved path would circle the Earth indefinitely. The cannonball would be in orbit.

Newton calculated the speed for a circular orbit about the Earth. However, since such a cannon-muzzle velocity was clearly impossible, he did not foresee humans launching satellites (and it is quite likely that he didn't foresee multistage rockets).

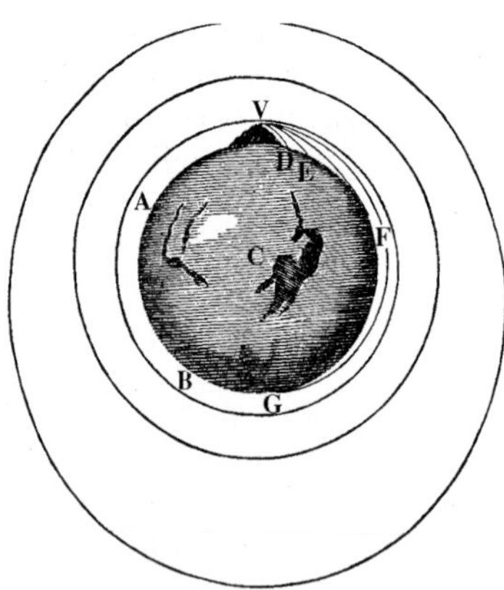

Figure 5.36 "The greater the velocity . . . with which [an object] is projected, the farther it goes before it falls to the Earth. We may therefore suppose the velocity to be so increased that it would describe an arc of 1, 2, 5, 10, 100, 1000 miles before it arrived at the Earth, till at last, exceeding the limits of the Earth, it should pass into space without touching." —Isaac Newton, *System of the World*.

Both the cannonball and the Moon continuously move sideways, with a tangential velocity parallel to Earth's surface. This velocity is enough to ensure motion around the Earth rather than into it. Without air drag to reduce speed, the Moon or any Earth satellite "falls" around and around the Earth indefinitely. Similarly, the planets continually fall around the Sun in closed paths.

CHECK YOURSELF

Can we say that a satellite stays in orbit because it's above the pull of Earth's gravity?

CHECK YOUR ANSWER

No, no, no! No satellite is "above" Earth's gravity. If the satellite were not in the grip of Earth's gravity, it would not orbit but would follow a straight-line path instead.

Why don't the planets crash into the Sun? They don't because of their tangential velocities. What would happen if their tangential velocities were reduced

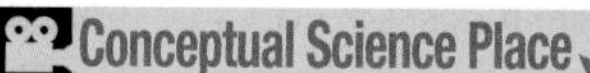

Projectile Motion Demo; More Projectile Motion

to zero? The answer is simple enough: Their motion would be straight toward the Sun, and they would indeed crash into it. Any objects in the solar system without sufficient tangential velocities have long ago crashed into the Sun. What remains is the harmony we observe.

Conceptual Science Place

Orbits and Kepler's Laws

5.12 Elliptical Orbits

If a projectile just above the resistance of the atmosphere is given a horizontal speed somewhat greater than 8 kilometers per second, it will overshoot a circular path and will trace an oval path called an **ellipse**.

An ellipse is a specific curve; the closed path taken by a point that moves in such a way that the sum of its distances from two fixed points (called *foci*, the plural of *focus*) is constant. For a satellite orbiting a planet, one focus is at the center of the planet; the other focus could be inside or outside of the planet. An ellipse can be constructed easily by using a pair of tacks (one at each focus), a loop of string, and a pencil (Figure 5.37). The nearer the foci are to each other, the closer the ellipse is to a circle. When both foci are together, the ellipse *is* a circle. So we see that a circle is a special case of an ellipse.

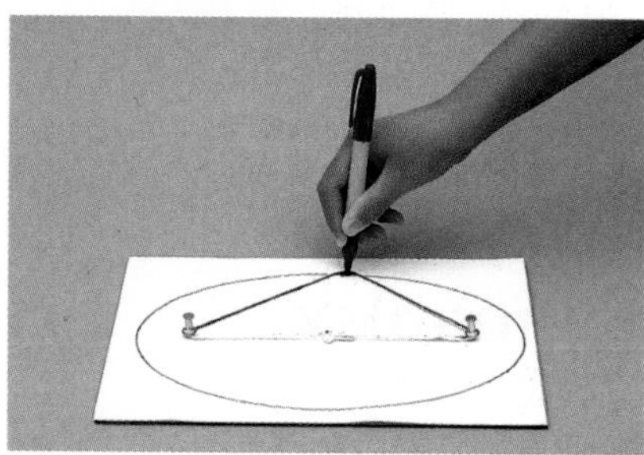

Figure 5.37 A simple method for constructing an ellipse.

Whereas the speed of a satellite is constant in a circular orbit, the speed of a satellite varies in an elliptical orbit. When the initial speed is greater than 8 km/s, the satellite overshoots a circular path and moves away from the Earth, against

TECHNOLOGY

Communications Satellites

Satellites are payloads carried above the atmosphere by rockets. Putting a payload into orbit requires control over the speed and direction of the rocket. A rocket initially fired vertically is intentionally tipped from the vertical course as it rises. Then, once above the drag of the atmosphere, it is aimed horizontally, whereupon the payload is given a final thrust to orbital speed.

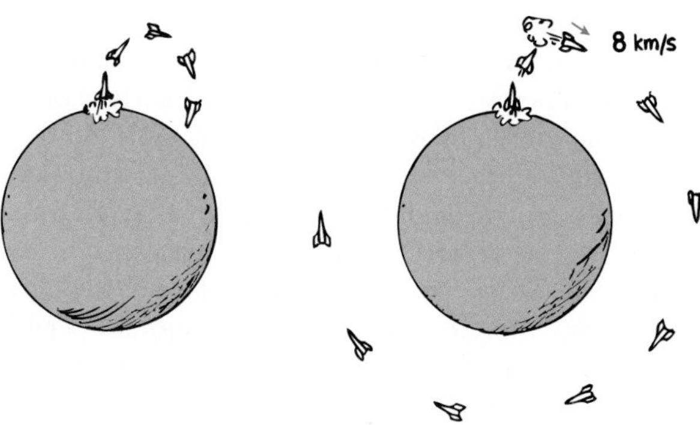

For a satellite close to Earth, the period (the time for a complete orbit about the Earth) is about 90 minutes. For higher altitudes, gravitation is less and the orbital speed is less, so the period is longer. For example, communication satellites located at an altitude of 5.5 Earth radii have a period of 24 hours. This period matches the period of daily Earth rotation. For an orbit around the equator, such a satellite stays above the same point on the ground. That is, it is in *geosynchronous* orbit.

Satellite television employs communications satellites. Satellite TV is much like traditional broadcast television, but it has a larger range. Both systems use electromagnetic signals (radio waves) to send programming to your home. Broadcast stations transmit the waves from powerful land-based antennas, and viewers pick up the signals with smaller antennas.

The problem with this technology is that radio signals travel away from a broadcast antenna in a straight line. To receive the signals, you need to be in the direct "line-of-sight" of the antenna. The Earth's curvature interrupts the line of sight, so the broadcast signals can only be sent over a short distance.

Satellite TV solves the problem by transmitting the signals from satellites in orbit high above Earth. This way, Earth's curvature doesn't interrupt the line of sight. Since the satellite is in geosynchronous orbit, the relative positions of the satellite and receiving dish are fixed. You don't need to readjust your dish—just grab the remote.

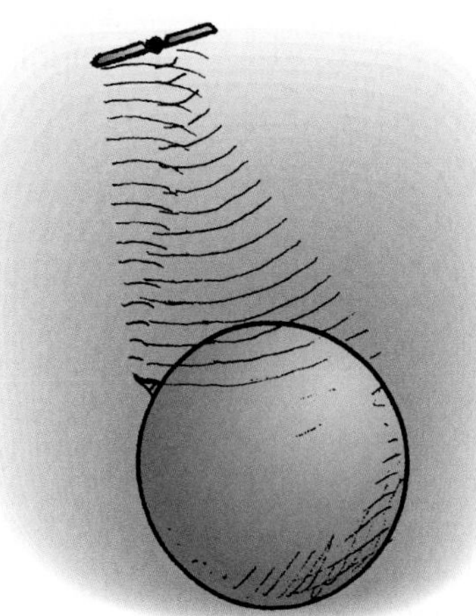

the force of gravity. It therefore loses speed. The speed it loses in receding is gained as it falls back toward the Earth, and it finally rejoins its path with the same speed it had initially (Figure 5.38). The procedure repeats over and over, and the process repeats each cycle.

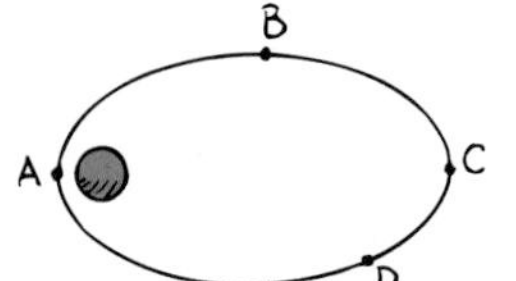

CHECK YOURSELF

The orbital path of a satellite is shown in the sketch. In which of the marked positions A through D does the satellite have the greatest speed? In which position does it have the lowest speed?

CHECK YOUR ANSWER

The satellite has its greatest speed as it whips around A and has its lowest speed at C. After passing C, it gains speed as it falls back to A to repeat the cycle.

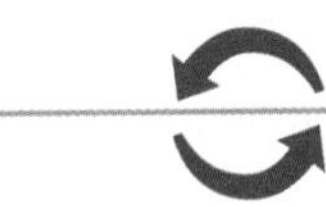

INTEGRATED SCIENCE ASTRONOMY

Escape Speed

We know that a cannonball fired horizontally at 8 kilometers per second from Newton's mountain goes into orbit. But what would happen if the cannonball were instead fired at the same speed vertically? It would rise to some maximum height, reverse direction, then fall back to Earth. Then the old saying "What goes up must come down" would hold true, just as surely as a stone tossed skyward will be returned by gravity (unless, as we shall see, its speed is great enough.)

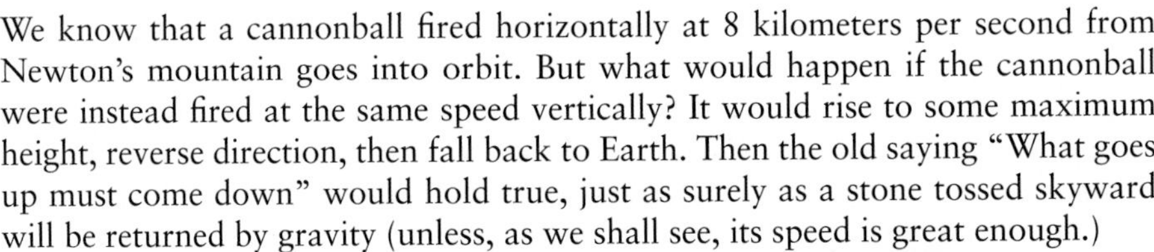

In this age of space travel, it is more accurate to say "What goes up *may* come down," for there is a critical starting speed that allows a projectile to outrun gravity and to escape the Earth. This critical speed is called the **escape speed,** or, if direction is involved, the *escape velocity*. From the surface of the Earth, the escape speed is 11.2 kilometers per second. Launch a projectile at any speed greater than that and it will leave the Earth, traveling slower and slower, but never stopping due to Earth's gravity.* We can understand the magnitude of this speed from an energy point of view.

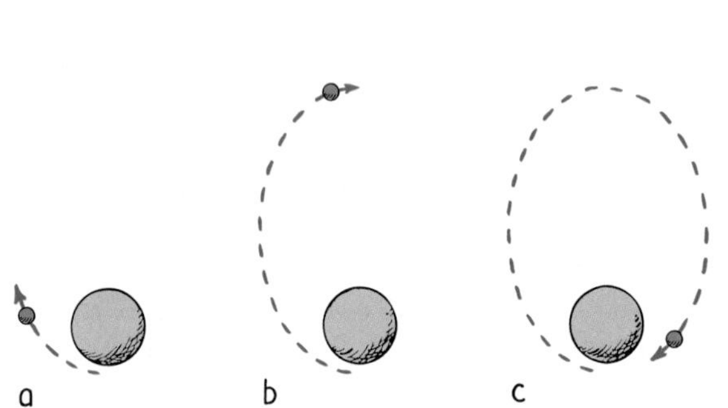

Figure 5.38 Elliptical orbit. An Earth satellite that has a speed somewhat greater than 8 km/s overshoots a circular orbit (a) and travels away from the Earth. Gravitation slows it to a point at which it no longer moves farther away from the Earth (b). It falls toward the Earth, gaining the speed that it lost in receding (c), and it follows the same path as before in a repetitious cycle.

How much work would be required to lift a payload against the force of Earth's gravity to a distance very, very far ("infinitely far") away? We might think that the change of potential energy (PE) would be infinite because the distance is infinite. But gravitation diminishes with distance by the inverse-square law. The force of gravity on the payload would be strong only near the Earth. It turns out that the change of PE of a 1-kilogram body moved from the surface of the Earth to infinite distance is 62 million joules, or 62 megajoules (62 MJ). So to put a payload infinitely far from Earth's surface requires at least 62 megajoules of energy per kilogram of load. We won't go through the calculation here, but 62 megajoules per kilogram corresponds to a speed of 11.2 kilometers per second, whatever the total mass involved. This is the escape speed from the surface of the Earth.†

If we give the payload any more energy than 62 megajoules per kilogram at the surface of the Earth—or, equivalently, any more speed than 11.2 kilometers per second—then, neglecting air resistance, the payload will escape from the Earth, never to return. As it continues outward, its PE increases and its kinetic

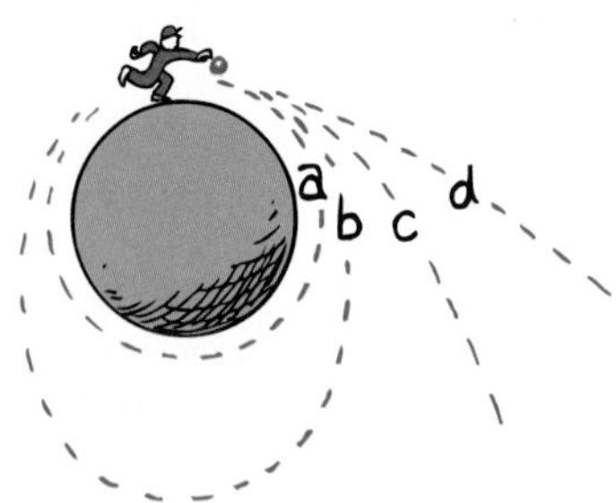

Figure 5.39 Interactive Figure

If Superman were to toss a ball 8 km/s horizontally from the top of a mountain high enough to be just above air resistance (a), then, about 90 minutes later, he could turn around and catch it (neglecting the Earth's rotation). Tossed slightly faster (b), the ball would take an elliptical orbit and return in a slightly longer time. Tossed at more than 11.2 km/s (c), it would escape the Earth's gravitational pull. Tossed at more than 42.5 km/s (d), it would escape from the solar system.

* Escape speed from any planet or any body is given by $v = \sqrt{\frac{2GM}{d}}$, where G is the universal gravitational constant, M is the mass of the attracting body, and d is the distance from its center. (At the surface of the body, d would simply be the radius of the body.) Compare this formula with the one for orbital speed $v = \sqrt{\frac{GM}{d}}$.

† Interestingly enough, this might be called the maximum falling speed. Any object, however far from the Earth, if it is released from rest and allowed to fall to Earth only under the influence of Earth's gravity, would not exceed 11.2 km/s. (With air resistance, it would be somewhat less.)

INSIGHT If you were to drop a candy bar to the Earth from a distance as far away from the Earth as the Earth is from Pluto, its speed of impact would be about 11.2 km/s.

energy (KE) decreases. Its speed becomes less and less, although it is never reduced to zero. The payload outruns the gravity of the Earth. It escapes.

The escape speeds from various bodies in the solar system are shown in Table 5.1. Note that the escape speed from the surface of the Sun is 620 km/s. Even at a 150,000,000-km distance from the Sun (Earth's distance), the escape speed needed to break free of the Sun's influence is 42.5 km/s, which is considerably more than the escape speed of the Earth. An object projected from the Earth at a speed greater than 11.2 km/s but less than 42.5 km/s will escape the Earth, but it will not escape the Sun. Rather than receding forever, it will occupy an orbit around the Sun.

The first probe to escape the solar system, *Pioneer 10*, was launched from Earth in 1972 with a speed of only 15 km/s. The escape was accomplished by directing the probe into the path of oncoming Jupiter. It was whipped about by Jupiter's great gravitational field, gaining speed in the process—similar to the increase in speed of a baseball encountering an oncoming bat. Its speed of departure from Jupiter was increased enough to exceed the escape speed from the Sun at the distance of Jupiter. *Pioneer 10* passed the orbit of Pluto in 1984. Unless it collides with another body, it will wander indefinitely through interstellar space. Like a bottle cast into the sea with a note inside, *Pioneer 10* contains information about Earth that might be of interest to extraterrestrials, information put there in hopes that it will one day wash up and be found on some distant "seashore" (Figure 5.40).

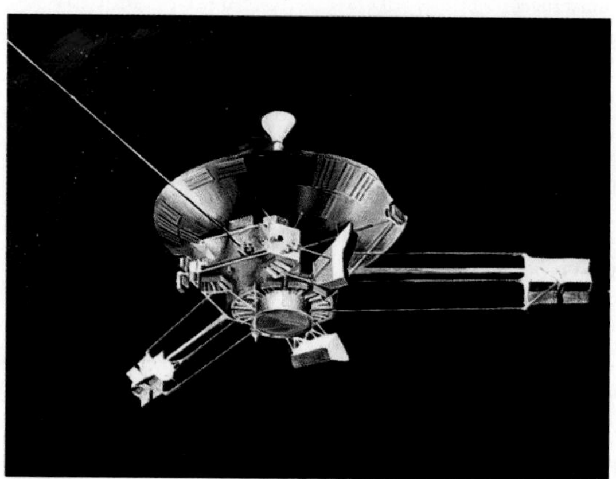

Figure 5.40 *Pioneer 10*, launched from Earth in 1972, passed the outermost planet in 1984 and is now wandering in our galaxy.

It is important to point out that the escape speed of a body is the initial speed given by an initial thrust, after which there is no force to assist motion. One could escape Earth at any sustained speed more than zero, given enough time. For example, suppose a rocket is launched to a destination such as the Moon. If the rocket engines burn out when the rocket is still close to the Earth, the rocket needs a minimum speed of 11.2 kilometers per second. But, if the rocket engines can be sustained for long periods of time, the rocket could go to the Moon without ever attaining 11.2 kilometers per second.

It is interesting to note that the accuracy with which an unmanned rocket reaches its destination is not determined by its remaining on a preplanned path or by its getting back on that path if it strays off course. No attempt is made to return the rocket to its original path. Instead, the control center asks, in effect, "Where is it now and what is its velocity? What is the best way to reach its destination, given its present situation?" With the aid of high-speed computers, the answers to these questions are used in finding a new path. Corrective thrusters

Table 5.1 Escape Speeds from the Surfaces of Bodies in the Solar System

Astronomical Body	Mass (in Earth masses)	Radius (in Earth radii)	Escape Speed (km/s)
Sun	333,000	109	620
Sun (at a distance of Earth's orbit)		23,500	42.2
Jupiter	318.0	11.0	60.2
Saturn	95.2	9.2	36.0
Neptune	17.3	3.47	24.9
Uranus	14.5	3.7	22.3
Earth	1.00	1.00	11.2
Venus	0.82	0.95	10.4
Mars	0.11	0.53	5.0
Mercury	0.055	0.38	4.3
Moon	0.0123	0.27	2.4

INSIGHT Just as planets fall around the Sun, stars fall around the centers of galaxies. Those with insufficient tangential speeds are pulled into, and are gobbled up by, the galactic nucleus, which is usually a black hole.

put the rocket on this new path. This process is repeated over and over again all the way to the goal.*

CHECK YOURSELF

1. What is the minimum speed a molecule in Earth's upper atmosphere must have if it is to escape Earth's gravity and wander into space?
2. Refer to the photograph at the beginning of this chapter, which shows a space shuttle orbiting the Earth. Is the launch speed of the shuttle less than, equal to, or greater than Earth's escape speed? Justify your answer.

CHECK YOUR ANSWERS

1. 11.2 km/s.
2. The launch speed must be less than Earth's escape speed because, when the shuttle's engines are turned off, the shuttle orbits the Earth under the influence of Earth's gravity.

* Is there a personal lesson to be learned here? Suppose you find that you are off course. You may, like the rocket, find it more fruitful to take a course that will lead you to your goal, as best plotted from your present position and circumstances, rather than to try to get back on the course you plotted from a previous position and, perhaps, under different circumstances.

CHAPTER 5 Review

Summary of Terms

Centripetal force Any force that is directed at right angles to the path of a moving object and that tends to produce circular motion.

Ellipse The sum of the distances from any point on the path to two points called foci is a constant; also the oval path followed by a satellite.

Escape speed The speed that a projectile, space probe, or similar object must reach in order to escape the gravitational influence of the Earth or of another celestial body to which it is attracted.

Gravitational force The attractive force between objects due to mass.

Inverse-square law Law relating the intensity of an effect to the inverse square of the distance from the cause: Intensity $\sim \frac{1}{\text{distance}^2}$.

Law of universal gravitation Every body in the universe attracts every other body with a mutually attracting force. For two bodies, this force is directly proportional to the product of their masses and inversely proportional to the square of the distance separating them: $F = \frac{G(m_1 \times m_2)}{d^2}$.

Parabola The curved path followed by a projectile near the Earth under the influence of gravity only.

Projectile Any object that moves through the air or through space under the influence of gravity.

Satellite A projectile or small body that orbits a larger body.

Tangential velocity Velocity that is parallel to (tangent to) a curved path.

Universal constant of gravitation, G The proportionality constant in Newton's law of gravitation.

Weightlessness A condition encountered in free fall wherein a support force is lacking.

Review Questions

5.1 The Legend of the Falling Apple

1. What connection did Newton make between a falling apple and the Moon?

5.2 The Fact of the Falling Moon

2. What does it mean to say that something moving in a curve path has a tangential velocity?
3. In what sense does the Moon "fall"?

5.3 Newton's Law of Universal Gravitation

4. State Newton's law of gravitation in words. Then do the same with one equation.

5.4 Gravity and Distance: The Inverse-Square Law

5. How does the force of gravity between two bodies change when the distance between them is doubled?
6. How does the thickness of paint sprayed on a surface change when the sprayer is held twice as far away?
7. How does the brightness of light on a surface change when a point source of light is brought twice as far away?
8. At what distance from Earth is the gravitational force on an object zero?

5.5 The Universal Gravitational Constant, G

9. What is the magnitude of gravitational force between two 1-kilogram bodies that are 1 meter apart?
10. What is the magnitude of the gravitational force between the Earth and a 1-kilogram body?

5.6 Weight and Weightlessness

11. Would the springs inside a bathroom scale be more compressed or less compressed if you weighed yourself in an elevator that accelerated upward? That accelerated downward?
12. Would the springs inside a bathroom scale be more compressed or less compressed if you weighed yourself in an elevator that moved upward at constant *velocity*? That moved downward at a constant *velocity*?
13. When is your weight equal to mg?
14. Give an example of when your weight is more than mg.
15. Give an example of when your weight is zero.

5.7 Gravity Can Be a Centripetal Force

16. When you whirl a can at the end of a string in a circular path, what is the direction of the force that acts on the can?
17. How can weight be simulated in a space habitat?

5.8 Projectile Motion

18. What exactly is a projectile?
19. How much speed does a freely falling object gain during each second of fall?
20. With no gravity, a horizontally moving projectile follows a straight-line path. With gravity, how far below the straight-line path does it fall compared with the distance of free fall?
21. A ball is batted upward at an angle. What happens to the vertical component of its velocity as it rises?
22. With no air drag, what happens to the horizontal component of velocity for the batted baseball?

5.9 Projectile Altitude and Range

23. A projectile is launched upward at an angle of 75° from the horizontal and strikes the ground a certain distance down range. For what other angle of launch at the same speed would this projectile land just as far away?
24. A projectile is launched vertically at 30 m/s. If air drag can be neglected, at what speed will it return to its initial level?

5.10 The Effect of Air Drag on Projectiles

25. What is the effect of air drag on the height and range of batted baseballs?

5.11 Fast-Moving Projectile—Satellites

26. Why will a projectile that moves horizontally at 8 km/s follow a curve that matches the curvature of the Earth?
27. Why is it important that the projectile in the previous question be above the Earth's atmosphere?
28. Are the planets of the solar system simply projectiles falling around the Sun?

5.12 Elliptical Orbits

29. Why does the force of gravity on a satellite moving in an elliptical orbit vary?
30. Why does the speed of a satellite moving in an elliptical orbit vary?

Multiple Choice Questions

Choose the BEST answer to the following.

1. Imagine you're standing on the surface of a shrinking planet. If it shrinks to one-tenth its original radius with no change in mass, on the shrunken surface you'd weigh
   (a) $\frac{1}{100}$ as much.
   (b) 10 times as much.
   (c) 100 times as much.
   (d) 1000 times as much.
   (e) None of these.
2. A spacecraft on its way from Earth to the Moon is pulled equally by Earth and the Moon when it is
   (a) closer to the Earth's surface.
   (b) closer to the Moon's surface.
   (c) halfway from Earth to the Moon.
   (d) at no point, since Earth always pulls more strongly.
3. Man-made satellites such as the space shuttle orbit at altitudes that are above Earth's
   (a) atmosphere.
   (b) gravitational field.
   (c) Both of these.
4. How does the force of gravity between two bodies change when the distance between them is increased by 4?
   (a) The force of gravity also increases by 4.
   (b) The force of gravity is reduced to $\frac{1}{4}$ its original strength.
   (c) The force of gravity is multiplied by 16.
   (d) None of these.
5. A projectile is launched vertically at 40 m/s. If air drag can be neglected, at what speed will it return to its initial level?
   (a) 0 m/s
   (b) 20 m/s
   (c) 40 m/s
   (d) None of these.

INTEGRATED SCIENCE CONCEPTS

Biology—Your Biological Gravity Detector

1. Do people have sense organs that allow them to sense gravity? If so, where are they located?
2. Speculate on how the vestibular system might be involved in "space sickness"—the feeling of nausea and disorientation suffered by astronuats.

Biology—Center of Gravity of People

1. Why does the Leaning Tower of Pisa not topple?
2. Why does spreading his feet apart help a surfer stay on his board?

Astronomy—Escape Speed

1. What is the minimum speed for orbiting the Earth in a close orbit? What is the maximum speed? What happens above this speed?
2. How was *Pioneer 10* able to escape the solar system with an initial speed less than escape speed?

Active Explorations

1. Hold your hand outstretched with one hand twice as far from your eyes as the other. Make a casual judgment about which hand looks bigger. Most people see them to be about the same size, while many see the nearer hand as slightly bigger. Very few people see the hand as four times as big; but, by the inverse-square law, the nearer hand should appear twice as tall and twice as wide. Twice times twice means four times as big. That's four times as much of your visual field as is occupied by the more distant hand. It is likely that your belief that your hands are the same size is so strong that your brain overrules this information. Try it again, only this time overlap your hands slightly and view them with one eye closed. Aha! Do you now see more clearly that the nearer hand is bigger? This raises an interesting question: What other illusions do you experience that are not so easily checked?

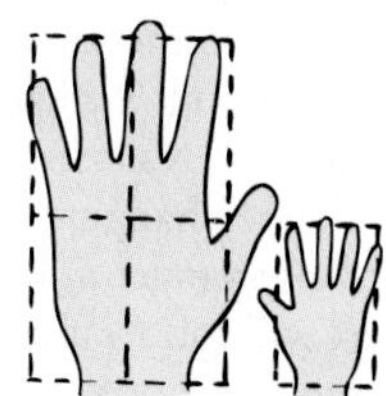

2. Repeat the eye-balling experiment, only this time use two dollar bills—one regular and the other folded in half length-wise, and again width-wise so it has $\frac{1}{4}$ the area. Now hold the two in front of your eyes. Where do you hold the folded bill so that it appears to be of the same size as the unfolded one? Share this experiment with a friend.
3. Ask a friend to stand facing a wall with her toes against the wall. Then ask her to stand on the balls of her feet without toppling backward. Your friend won't be able to do it. Explain to her why it can't be done.

Exercises

1. What would be the path of the Moon if somehow all gravitational force between Earth and the Moon vanished to zero?
2. Is the force of gravity stronger on a piece of iron than it is on a piece of wood if both have the same mass? Defend your answer.
3. Is the force of gravity on a piece of paper stronger when it is crumpled? Defend your answer.
4. What is the magnitude and direction of the gravitational force that acts on a professor who weighs 1000 N at the surface of the Earth?
5. The Earth and the Moon are attracted to each other by gravitational force. Does the more massive Earth attract the less massive Moon with a force that is greater than, smaller than, or the same as the force with which the Moon attracts the Earth?
6. What would you say to a friend who says that, if gravity follows the inverse-square law, the effect of gravity on you when you are on the twentieth floor of a building should be one-fourth as much as it would be if you were on the tenth floor?
7. Why do the passengers of high-altitude jet planes feel the sensation of weight while passengers in an orbiting space vehicle, such as a space shuttle, do not?
8. Is gravitational force acting on a person who falls off a cliff? Is it acting on an astronaut inside an orbiting space shuttle?
9. If you were in a freely falling elevator and you dropped a pencil, it would hover in front of you rather than falling to the floor. Is there a force of gravity that is acting on the pencil? Defend your answer.
10. Are the planets of the solar system simply projectiles falling around the Sun?
11. What path would you follow if you fell off a rotating merry-go-round? What force prevents you from following that path while you're on the merry-go-round?
12. A heavy crate accidentally falls from a high-flying airplane just as it flies over a shiny red sports car parked in a parking lot. Relative to the car, where will the crate crash?
13. How does the vertical component of motion for a ball kicked off a high cliff compare with the motion of vertical free fall?
14. In the absence of air drag, why does the horizontal component of the ball's motion not change, while the vertical component does?
15. At what point in its trajectory does a batted baseball have its minimum speed? If air drag can be neglected, how does this compare with the horizontal component of its velocity at other points?
16. Each of two golfers hits a ball at the same speed, one at 60°, and the other at 30°, above the horizontal. Which ball goes farther? Which hits the ground first? (Ignore air resistance.)
17. A park ranger shoots a monkey hanging from a branch of a tree with a tranquilizing dart. The ranger aims directly at the monkey, not realizing that the dart will follow a parabolic path and, therefore, will fall below the monkey's position. The monkey, however, seeing the dart leave the gun, lets go of the branch to avoid being hit. Will the monkey be hit anyway? Defend your answer.
18. Since the Moon is gravitationally attracted to the Earth, why doesn't it simply crash into the Earth?
19. Does the speed of a falling object depend on its mass? Does the speed of a satellite in orbit depend on its mass? Defend your answers.
20. If you have ever watched the launching of an Earth satellite, you may have noticed that the rocket starts vertically upward, then veers from a vertical course and continues its rise at an angle. Why does it start vertically? Why does it not continue vertically?
21. A satellite can orbit at 5 km above the Moon, but not at 5 km above the Earth. Why?

Problems

1. If you stood atop a ladder that was so tall that you were three times as far from the Earth's center as you would be if you were standing on the Earth's surface, how would your weight compare with it present value?
2. Find the change in the force of gravity between two planets when the masses of both planets are doubled but the distance between them stays the same.
3. Find the change in the force of gravity between two planets when their masses remain the same but the distance between them is increased tenfold.
4. Find the change in the force of gravity between two planets when the distance between them is *decreased* to a tenth of the original distance.
5. Find the change in the force of gravity between two planets when the masses of the planets don't change but the distance between them is decreased to a fifth of the original distance.
6. By what factor would your weight change if the Earth's diameter were doubled and its mass were doubled?
7. Find the change in the force of gravity between two objects when both masses are doubled and the distance between them is also doubled.
8. Consider a bright point light source located 1 m from a square opening with an area of one square meter. Light passing through the opening illuminates an area of 4 m^2 on a wall 2 m from the opening. (a) Find the areas that would be illuminated if the wall were moved to distances of 3 m, 5 m, and 10 m from the opening. (b) How can the same amount of light illuminate more area as the wall is moved farther away?

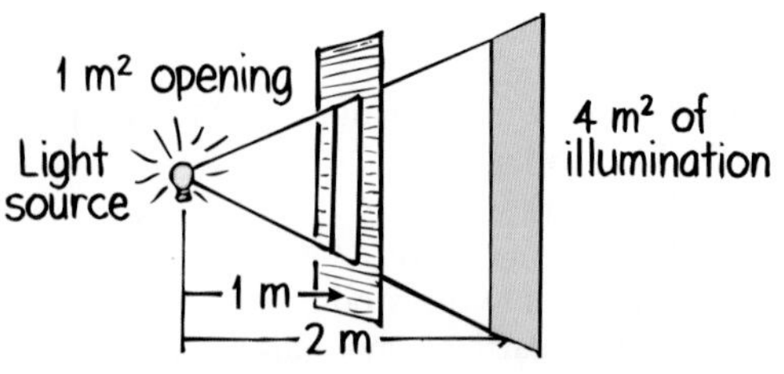

9. Calculate the force of gravity between the Earth (6.4×10^{24} kg) and the Sun (2×10^{30} kg). The average distance between the two is 1.5×10^{11} m.
10. Students in a lab roll a steel ball off the edge of a table. They measure the speed of a horizontally launched ball to be 4.0 m/s. They also know that, if they simply dropped the ball from rest off the edge of the table, it would take 0.5 seconds to hit the floor. Question: How far from the bottom of the table should they place a small piece of paper so that the ball will hit it when it lands?
11. Calculate the speed, in m/s, at which the Earth revolves about the Sun. Assume that the Earth's orbit is nearly circular.

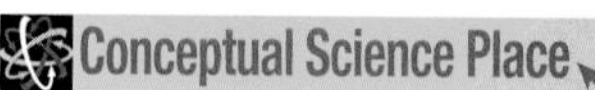

CHAPTER 5 ONLINE RESOURCES

Interactive Figures 5.8, 5.28, 5.32, 5.39 • ***Tutorials*** Projectile Motion, Orbits and Kepler's Laws • ***Videos*** Inverse-Square Law, Weight and Weightlessness, Apparent Weightlessness, Locating the Center of Gravity, Toppling, Centripetal Force, Simulated Gravity, Projectile Motion Demo, More Projectile Motion, Circular Orbits • ***Quiz*** • ***Flashcards*** • ***Links***

state of motion. Specific heat capacity is like thermal inertia since it signifies the resistance of a substance to a change in temperature.

CHECK YOURSELF

Which has a higher specific heat capacity, water or sand? In other words, which takes longer to warm in sunlight (or longer to cool at night)?

CHECK YOUR ANSWER

Water has the higher specific heat capacity. In the same sunlight, the temperature of water increases more slowly than the temperature of sand. And water will cool more slowly at night. The low specific heat capacity of sand and soil, as evidenced by how quickly they warm in the morning Sun and how quickly they cool at night, affects local climates.

Water has a much higher capacity for storing energy than most all other substances. A lot of heat energy is needed to change the temperature of water. This explains why water is very useful in the cooling systems of automobiles and other engines. It absorbs a great quantity of heat for small increases in temperature. Water also takes longer to cool.

INTEGRATED SCIENCE EARTH SCIENCE

Specific Heat Capacity and Earth's Climate

Figure 6.14 Because water has a high specific heat capacity and is transparent, it takes more energy to warm a body of water than to warm the land. Solar energy incident on the land is concentrated at the surface, but solar energy hitting the water extends beneath the surface and is "diluted."

Water's high specific heat capacity changes the world's climate. Look at a globe or a map of the Northern Hemisphere and notice the high latitude of Europe. Water's high specific heat keeps Europe's climate appreciably milder than regions of the same latitude in northeastern regions of Canada. Both Europe and Canada receive about the same amount of sunlight per square kilometer. What happens is that the Atlantic Ocean current known as the Gulf Stream carries warm water northeastward from the Caribbean Sea. It retains much of its thermal energy long enough to reach the North Atlantic Ocean off the coast of Europe. Then it cools, releasing 4.18 joules of energy for each gram of water that cools by 1°C. The released energy is carried by westerly winds over the European continent.

A similar effect occurs in the United States. The winds in North America are mostly westerly. On the West Coast, air moves from the Pacific Ocean to the land. In winter months, the ocean water is warmer than the air. Air blows over the warm water and then moves over the coastal regions. This warms the climate. In summer, the opposite occurs. The water cools the air, and the coastal regions are cooled. The East Coast does not benefit from the moderating effects of water because the direction of the prevailing wind is eastward from the land to the Atlantic Ocean. Land, with a lower specific heat capacity, gets hot in the summer but cools rapidly in the winter.

Islands and peninsulas do not have the extremes of temperatures that are common in the interior regions of a continent. The high summer and low winter temperatures common in the Manitoba and the Dakotas, for example, are largely due to the absence of large bodies of water. Europeans, islanders, and people living near ocean air currents should be glad that water has such a high specific heat capacity. San Franciscans certainly are!

See Chapter 25 for more discussion of the effects of ocean currents on global climate.

CHECK YOURSELF

1. Bermuda is close to North Carolina, but, unlike North Carolina, it has a tropical climate year-round. Why?

9. Calculate the force of gravity between the Earth (6.4×10^{24} kg) and the Sun (2×10^{30} kg). The average distance between the two is 1.5×10^{11} m.
10. Students in a lab roll a steel ball off the edge of a table. They measure the speed of a horizontally launched ball to be 4.0 m/s. They also know that, if they simply dropped the ball from rest off the edge of the table, it would take 0.5 seconds to hit the floor. Question: How far from the bottom of the table should they place a small piece of paper so that the ball will hit it when it lands?
11. Calculate the speed, in m/s, at which the Earth revolves about the Sun. Assume that the Earth's orbit is nearly circular.

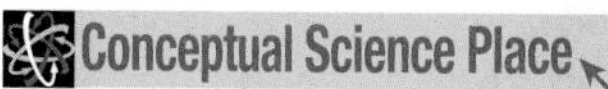

CHAPTER 5 **ONLINE RESOURCES**

Interactive Figures 5.8, 5.28, 5.32, 5.39 • ***Tutorials*** Projectile Motion, Orbits and Kepler's Laws • ***Videos*** Inverse-Square Law, Weight and Weightlessness, Apparent Weightlessness, Locating the Center of Gravity, Toppling, Centripetal Force, Simulated Gravity, Projectile Motion Demo, More Projectile Motion, Circular Orbits • ***Quiz*** • ***Flashcards*** • ***Links***

CHAPTER 6

Heat

Hot materials, such as the lava shown here, consist of fast-moving atoms and molecules.

CHAPTER OUTLINE

Lava has an average temperature of about 1000°C—ten times the boiling point of water. White-hot sparks from a Fourth of July sparkler are about 2000°C, twice as hot as typical lava. Why, then, would lava severely burn your skin on contact, but the sparkler's sparks leave you unhurt? Can an object get colder and colder or hotter and hotter forever—or are there limits to coldness and hotness? Why does a tile floor feel colder than a carpeted floor at the same temperature? Another question . . . why does air in a balloon, the concrete of a sidewalk, and almost everything else expand as it heats up? Why does ice water do the opposite, contracting instead of expanding as its temperature rises? And why does ice form on the top of a pond rather than at the bottom, thus enabling fish to swim comfortably all winter? This chapter will answer these and other questions relating to heat.

6.1 The Kinetic Theory of Matter

According to the *kinetic theory of matter*, matter is made up of tiny particles—atoms or molecules. The particles are always moving, and they move in a number of ways. They rotate, vibrate, and move in straight lines between collisions (Figure 6.1).

The energy that atoms and molecules have relates to their motion and their position. They have translational kinetic energy due to their translational (straight-line) motion as well as rotational and vibrational kinetic energy. Also, the particles have potential energy from the attractions between them or from their mutual repulsion when they are at close range. The *total* of all these forms of energy in a particular substance is its **thermal energy**. (Physicists usually refer to thermal energy as *internal energy*, for it is internal to a substance.)

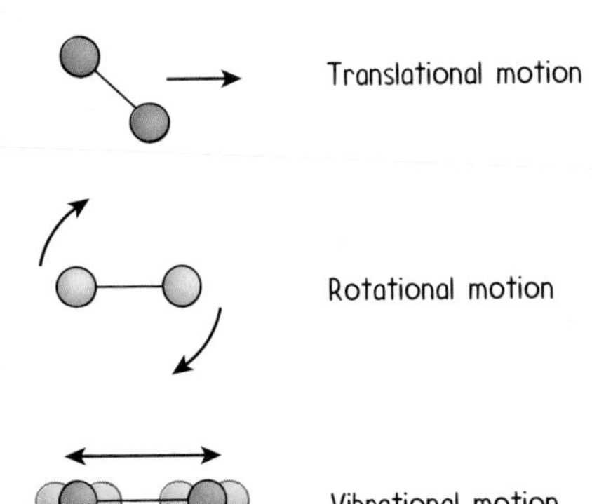

Figure 6.1 Types of motion of particles in matter.

6.2 Temperature

When you strike a nail with a hammer, it becomes warm. Why? Because the hammer's blow makes the nail's atoms move faster. When you put a flame to a liquid, the liquid becomes warmer as its molecules move faster. When you rapidly compress air in a tire pump, the air becomes warmer. In these cases, the molecules are made to race back and forth faster. They gain kinetic energy. In general, the warmer an object, the more kinetic energy its atoms and molecules possess. **Temperature**, the degree of "hotness" or "coldness" of an

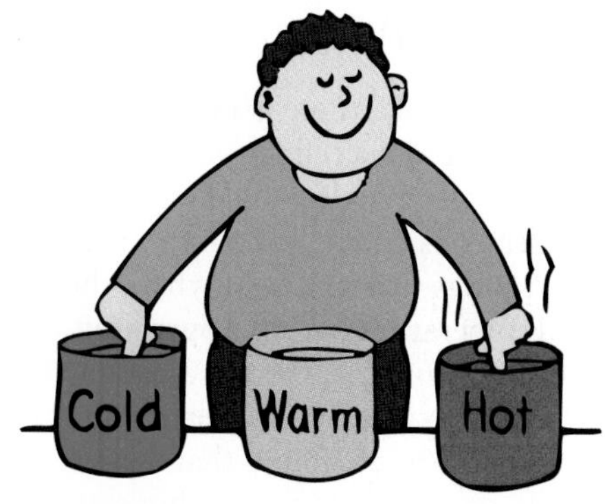

Figure 6.2 Can we trust our sense of hot and cold? Will both fingers feel the same temperature when they are put in the warm water? Try this and see (feel) for yourself.

Figure 6.3 A testament to Fahrenheit outside his home (now in Gdansk, Poland).

Conceptual Science Place

Low Temperatures with Liquid Nitrogen; How a Thermostat Works

INSIGHT Which has a greater temperature, a red-hot tack or a lake? Which has more thermal energy?

object, is proportional to the average translational kinetic energy of the atoms or molecules making it up (Figure 6.2).

It is important to note that temperature is not a measure of total kinetic energy in a substance. For example, there is twice as much molecular kinetic energy in 2 liters of boiling water as in 1 liter of boiling water—but the temperatures of both volumes of water are the same because the *average* kinetic energy per molecule is the same.

We express temperature quantitatively by a number that corresponds to the degree of hotness on some chosen scale. A common thermometer takes advantage of the fact that most substances expand with temperature. A thermometer measures temperature by comparing the expansion and contraction of a liquid (usually mercury or colored alcohol) to increments on a scale.

On a world-wide basis, the thermometer most often used is the Celsius thermometer. This thermometer is named in honor of the Swedish astronomer Anders Celsius (1701–1744), who first suggested the scale of 100 degrees between the freezing point and boiling point of water. A zero (0) is assigned to the temperature at which water freezes, and the number 100 is assigned to the temperature at which water boils (at standard atmospheric pressure). In between freezing and boiling temperatures are 100 equal parts called *degrees*.

In the United States, the number 32 is traditionally assigned to the temperature at which water freezes, and the number 212 is assigned to the temperature at which water boils. Such a scale makes up the Fahrenheit thermometer, named after its inventor, the German physicist G.D. Fahrenheit (1686–1736). Although the Fahrenheit scale is the one most commonly used in the United States, the Celsius scale is standard in scientific applications.

Arithmetic formulas used for converting from one temperature scale to the other are common in classroom exams. Because such arithmetic conversions aren't really physics, we won't be concerned with them here. Besides, the conversion between Celsius and Fahrenheit temperatures is closely approximated in the side-by-side scales of Figure 6.4.*

Interestingly, a thermometer actually registers its own temperature. When a thermometer is in thermal contact with something whose temperature we wish to know, thermal energy flows between the two until their temperatures are equal. At this point, thermal equilibrium is established. So, when we look at the temperature of the thermometer, we learn about the temperature of the substance with which it reached thermal equilibrium.

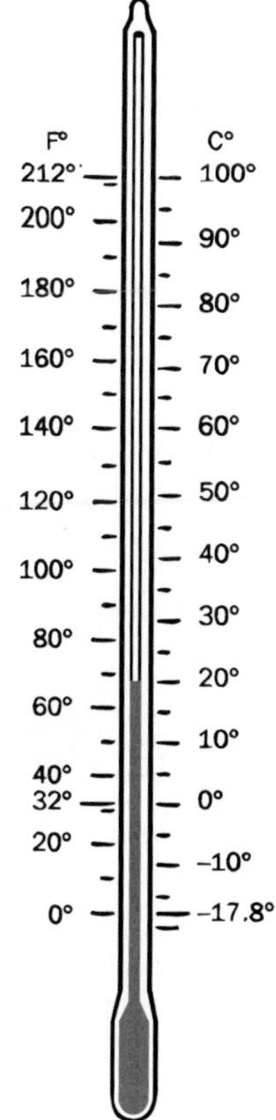

Figure 6.4 Fahrenheit and Celsius scales on a thermometer.

* Okay, if you really want to know, the formulas for temperature conversion are $C = \frac{5}{9}(F - 32)$ and $F = \frac{9}{5}C + 32$, where C is the Celsius temperature and F is the corresponding Fahrenheit temperature.

6.3 Absolute Zero

In principle, there is no upper limit to temperature. As thermal motion increases, a solid object first melts and then vaporizes. As the temperature is further increased, molecules dissociate into atoms, and atoms lose some or all of their electrons, thereby forming a cloud of electrically charged particles—a plasma. Plasmas exist in stars, where the temperature is many millions of degrees Celsius. Temperature has no upper limit.

In contrast, there is a definite limit at the opposite end of the temperature scale. Gases expand when heated, and they contract when cooled. Nineteenth-century experimenters found that all gases, regardless of their initial pressures or volumes, change by $\frac{1}{273}$ of their volume at 0°C for each drop in temperature of 1 degree Celsius, provided that the pressure is held constant. So, if a gas at 0°C were cooled down by 273°C, it would contract $\frac{273}{273}$ volumes and be reduced to zero volume. Clearly, we cannot have a substance with zero volume (Figure 6.5).

The same is true of pressure. The pressure of a gas of fixed volume decreases by $\frac{1}{273}$ for each drop in temperature of 1 degree Celsius. If it were cooled to 273°C below zero, it would have no pressure at all. In practice, every gas converts to a liquid before becoming this cold. Nevertheless, these decreases by $\frac{1}{273}$ increments suggested the idea of lowest temperature: −273°C. That's the lower limit of temperature, **absolute zero**. At this temperature, molecules have lost all available kinetic energy.* No more energy can be removed from a substance at absolute zero. It can't get any colder.

The absolute temperature scale is called the Kelvin scale, named after the famous British mathematician and physicist William Thomson, First Baron Kelvin. Absolute zero is 0 K (short for "zero kelvin"; note that the word "degrees" is not used with Kelvin temperatures). There are no negative numbers on the Kelvin scale. Its temperature divisions are identical to the divisions on the Celsius scale. Thus, the melting point of ice is 273 K, and the boiling point of water is 373 K.

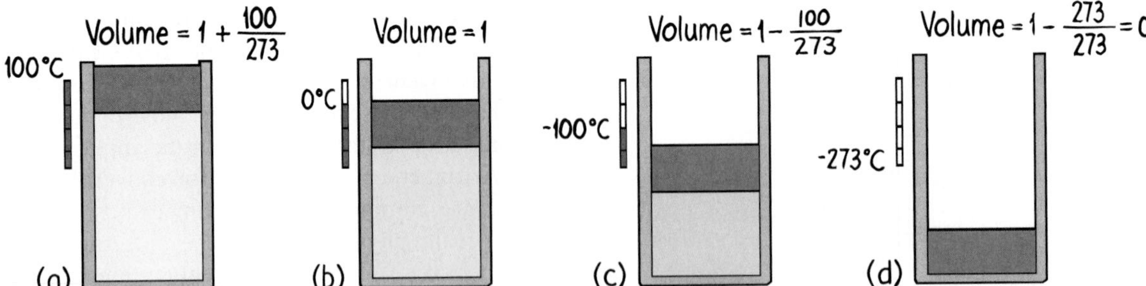

Figure 6.5 The gray piston in the vessel goes down as the volume of gas (blue) shrinks. The volume of gas changes by $\frac{1}{273}$ its volume at 0°C with each 1°C change in temperature when the pressure is held constant. (a) At 100°C, the volume is $\frac{100}{273}$ greater than it is at (b), when its temperature is 0°C. (c) When the temperature is reduced to −100°C, the volume is reduced by $\frac{100}{273}$. (d) At −273°C, the volume of the gas would be reduced by $\frac{273}{273}$ and so would be zero.

* Even at absolute zero, molecules still possess a small amount of kinetic energy, called the zero-point energy. Helium, for example, has enough motion at absolute zero to prevent it from freezing. The explanation for this involves quantum theory.

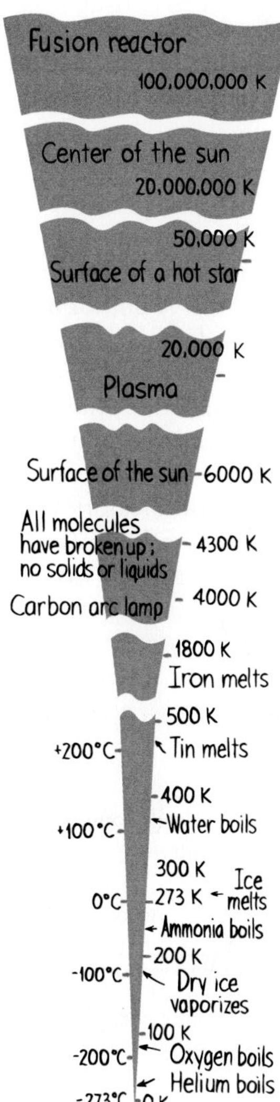

Figure 6.6 Some absolute temperatures.

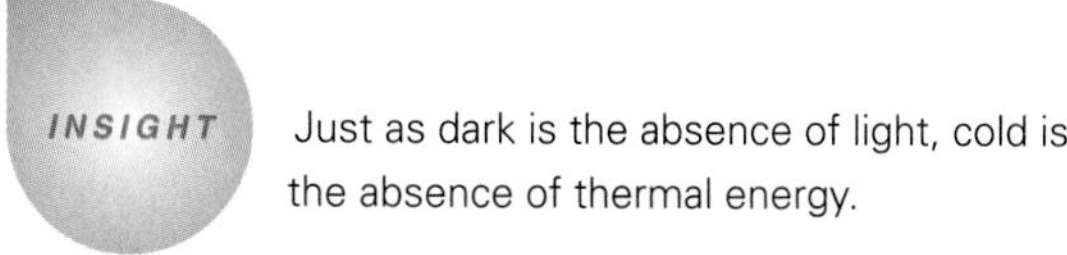

Just as dark is the absence of light, cold is the absence of thermal energy.

Figure 6.7 The temperature of the sparks is very high, about 2000°C. That's a lot of thermal energy per molecule of spark. However, because there are few molecules per spark, internal energy is safely small. Temperature is one thing; transfer of energy is another.

CHECK YOURSELF

1. Which is larger, a Celsius degree or a kelvin?
2. A sample of hydrogen gas has a temperature of 0°C. If the sample were heated until it had twice the thermal energy, what would its temperature be?

CHECK YOUR ANSWERS

1. Neither. They are equal.
2. The 0°C gas has an absolute temperature of 273 K. Twice the thermal energy means that it has twice the absolute temperature, or 2×273 K. This would be 546 K, or 273°C.

6.4 What Is Heat?

If you touch a hot stove, thermal energy enters your hand because the stove is warmer than your hand. When you touch a piece of ice, however, thermal energy passes out of your hand and into the colder ice. The direction of energy flow is always from a warmer thing to a neighboring cooler thing. A scientist defines heat as the thermal energy transferred from one thing to another due to a temperature difference.

According to this definition, matter does not contain heat. Matter contains thermal energy, not heat. Heat is the flow of thermal energy due to a temperature difference. Once thermal energy has been transferred to an object or substance, it ceases to be heat. Heat is thermal energy in transit.

For substances in thermal contact, thermal energy flows from the higher-temperature substance to the lower-temperature one until thermal equilibrium is reached. This does not mean that thermal energy necessarily flows from a substance with more thermal energy into one with less thermal energy. For example, there is more thermal energy in a bowl of warm water than there is in a red-hot thumbtack. If the tack is placed into the water, thermal energy doesn't flow from the warm water to the tack. Instead, it flows from the hot tack to the cooler water. Thermal energy never flows unassisted from a low-temperature substance to a higher-temperature one.

If heat is thermal energy that transfers in a direction from hot to cold, what is cold? Does a cold substance contain something opposite to thermal energy? No. An object is not cold because it contains something, but because it lacks something. It lacks thermal energy. When outdoors on a near-zero winter day, you feel cold not because something called cold gets to you. You feel cold because you lose heat. That's the purpose of your coat—to slow down the heat flow from your body. Cold is not a thing in itself, but the result of reduced thermal energy.

CHECK YOURSELF

1. Suppose you apply a flame to 1 liter of water for a certain time and its temperature rises by 2°C. If you apply the same flame for the same time to 2 liters of water, by how much will its temperature rise?
2. If a fast marble hits a random scatter of slow marbles, does the fast marble usually speed up or slow down? Which lose(s) kinetic energy and which gain(s) kinetic energy, the initially fast-moving marble or the initially slow ones? How do these questions relate to the direction of heat flow?

CHECK YOUR ANSWERS

1. Its temperature will rise by only 1°C, because there are twice as many molecules in 2 liters of water, and each molecule receives only half as much energy on the average. So the average kinetic energy, and thus the temperature, increases by half as much.
2. A fast-moving marble slows when it hits slower-moving marbles. It gives up some of its kinetic energy to the slower ones. Likewise with heat. Molecules with more kinetic

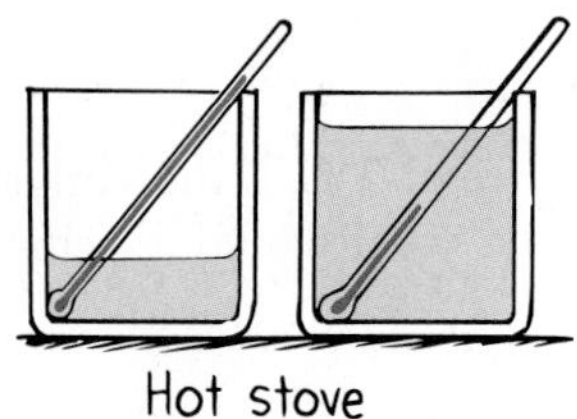

Figure 6.8 Although the same quantity of heat is added to both containers, the temperature increases more in the container with the smaller amount of water.

If you add 1 calorie of heat to 1 gram of water, you'll raise its temperature by 1°C.

Figure 6.9 To the weight watcher, the peanut contains 10 Calories; to the physicist, it releases 10,000 calories (or 41,840 joules) of energy when burned or digested.

energy that make contact with molecules that have lower kinetic energy give up some of their excess kinetic energy to the slower ones. The direction of energy transfer is from hot to cold. For both the marbles and the molecules, however, the total energy before and after contact is the same.

Quantity of Heat

Heat is a form of energy, and it is measured in joules. It takes about 4.2 joules of heat to change 1 gram of water by 1 Celsius degree. A unit of heat still common in the United States is the **calorie.*** A calorie is defined as the amount of heat needed to change the temperature of 1 gram of water by 1 Celsius degree. (The relationship between calories and joules is that 1 calorie = 4.18 joules.)

The energy ratings of foods and fuels are measured by the energy released when they are burned. (Metabolism is really "burning" at a slow rate.) The heat unit for labeling food is the kilocalorie, which is 1000 calories (the heat needed to change the temperature of 1 kilogram of water by 1°C). To differentiate this unit and the smaller calorie, the food unit is usually called a Calorie, with a capital C. So 1 Calorie is really 1000 calories.

CHECK YOURSELF

Which will raise the temperature of water more, adding 4.18 joules or 1 calorie?

CHECK YOUR ANSWER

Both are the same. This is like asking which is longer, a 1-mile long track or a 1.6-kilometer-long track. They're the same in different units.

6.5 The Laws of Thermodynamics

What we've learned thus far about heat and thermal energy is summed up in the laws of thermodynamics. The word *thermodynamics* stems from Greek words meaning "movement of heat."

When thermal energy transfers as heat, it does so without net loss or gain. The energy lost in one place is gained in another. When the conservation of energy (which we discussed back in Chapter 4) is applied to thermal systems, we have the **first law of thermodynamics.**

Whenever heat flows into or out of a system, the gain or loss of thermal energy equals the amount of heat transferred.

Unifying Concept

The Law of Conservation of Energy

Section 4.10

A *system* is any substance, device, or well-defined group of atoms or molecules. The system may be the steam in a steam engine, the entire Earth's atmosphere, or even the body of a living creature. When heat is added to any of these systems, we increase its thermal energy. The added energy enables the system to do work. The first law makes good sense.

The first law is nicely illustrated when you put an airtight can of air on a hot stove and warm it. The energy that is put in increases the thermal energy of the enclosed air, so its temperature rises. If the can is fitted with a movable piston, then the heated air can do *mechanical work* as it expands and pushes the piston outward. This ability to do mechanical work is energy that comes from the energy you put in to begin with. The first law says you don't get energy from nothing.

The **second law of thermodynamics** restates what we've learned about the direction of heat flow.

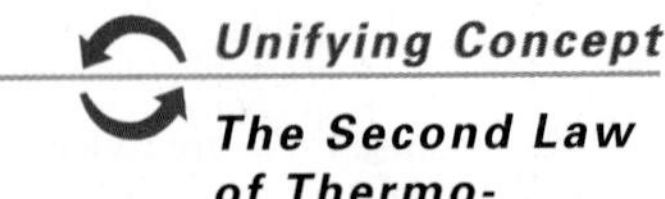

Unifying Concept

The Second Law of Thermodynamics

Heat never spontaneously flows from a lower-temperature substance to a higher-temperature substance.

* Another common unit of heat is the British thermal unit (Btu). The Btu is defined as the amount of heat required to change the temperature of 1 lb of water by 1 degree Fahrenheit.

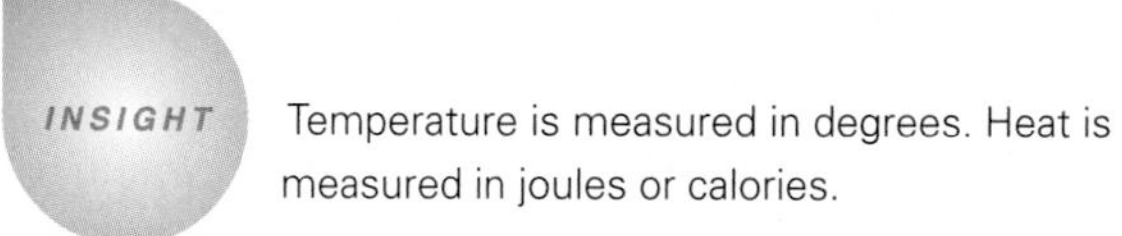

INSIGHT Temperature is measured in degrees. Heat is measured in joules or calories.

When heat flow is spontaneous—that is, without the assistance of external work—the direction of flow is always from hot to cold. In winter, heat flows from inside a warm home to the cold air outside. In summer, heat flows from the hot air outside into the cooler interior. Heat can be made to flow the other way only when work is done on the system or by adding energy from another source. This occurs with heat pumps and air conditioners. In these devices, thermal energy is pumped from a cooler to a warmer region. But without external effort, the direction of heat flow is from hot to cold. The second law, like the first law, makes logical sense.*

The **third law of thermodynamics** restates what we've learned about the lowest limit of temperature:

No system can reach absolute zero.

As investigators attempt to reach this lowest temperature, it becomes more difficult to get closer to it. Physicists have been able to record temperatures that are less than a millionth of 1 kelvin, but never as low as 0 K.

INTEGRATED SCIENCE CHEMISTRY AND BIOLOGY

Entropy

Figure 6.10 When Pete pushes down on the piston, he does work on the air inside. What happens to the air's temperature?

Energy tends to disperse. It flows from where it is localized to where it is spread out. For example, consider a hot pan once you have taken it off the stove. The pan's thermal energy doesn't stay localized in the pan. Instead, it disperses outward, away from the pan into its surroundings. As it's heated, energetic molecules transfer energy to the air by molecular collisions as well as radiation. Consider a second example: The chemical energy in gasoline burns explosively in a car engine when the molecules combust. Some of this energy disperses through the transmission to get the car moving. The rest of the energy disperses as heat into the metal of the engine, into the coolant that flows through the engine and the radiator, and into the gases that flow through the engine and out the exhaust pipe. The energy, once localized in the small volume of the gasoline, is now spread out through a larger volume of space. Or, witness the dispersion of energy when you pick up an object, say a marble for example, and drop it. When you lift the marble, you give it potential energy. Drop it and that potential energy converts to kinetic energy, pushing air aside as it falls (therefore spreading out the marble's kinetic energy a bit), before hitting the ground. When the marble hits, it disperses energy as sound and as heat (when it heats the ground a bit). The potential energy you put into the marble by lifting it, which was once localized in the marble, is now spread out and dispersed in a little air movement plus the heating of the air and ground. The marble bounces before it finally comes to rest; in each bounce, energy spreads out from the marble to its surroundings.

The tendency of energy to spread out is one of the central driving forces of nature. Processes that disperse energy tend to occur spontaneously—they are favored. The opposite hold true as well. Processes that result in the concentration of energy tend not to occur—they are not favored. Heat from the room doesn't spontaneously flow into the frying pan to heat it up. Likewise, the lower-energy molecules of a car's exhaust won't on their own come back together to reform the higher-energy molecules of gasoline. And, needless to say, dropped marbles don't jump back into your hand. The natural flow of energy is always a one-way trip from where it is concentrated to where it is spread out. The second law of thermodynamics states this principle for heat flow. But now we can see that the second law of thermodynamics can be generalized and stated this way:

* The laws of thermodynamics were all the rage back in the 1800s. At that time, horses and buggies were yielding to steam-driven locomotives. There is the story of the engineer who explained the operation of a steam engine to a peasant. The engineer cited in detail the operation of the steam cycle, how expanding steam drives a piston that in turn rotates the wheels. After some thought, the peasant said, "Yes, I understand all that. But where's the horse?" This story illustrates how hard it is to abandon our way of thinking about the world when a newer method comes along to replace established ways. Are we any different today?

Figure 6.11 Entropy.

INSIGHT The laws of thermodynamics can be stated this way: You can't win (because you can't get any more energy out of a system than you put into it), you can't break even (because you can't get as much useful energy out as you put in), and you can't get out of the game (entropy in the universe is always increasing).

Natural systems tend to disperse from concentrated and organized-energy states toward diffuse and disorganized states.

The least concentrated form of energy is thermal energy. So, since organized forms of energy tend to become less organized, they ultimately degrade into the environment as thermal energy. Further, when energy is dispersed, it is less able to do useful work than when it was concentrated—in effect, it becomes diluted. So thermal energy is the graveyard of useful energy.

The measure of energy dispersal is a quantity known as **entropy.*** More entropy means more degradation of energy. Since energy tends to degrade and disperse with time, the total amount of entropy in any system tends to increase with time (Figure 6.11). The same is true for the largest system, the universe. The net entropy in the universe is continually increasing (the universe is continually running "downhill").

We say *net* entropy because there are some regions where energy is actually being organized and concentrated. Work input from outside of an isolated system can decrease entropy in the system, with energy proceeding toward organization and concentration in that system. For example, diffuse thermal energy in the air can be concentrated in a heat pump. And living things seem to defy the second law of thermodynamics with their highly organized and concentrated energy. But, on closer examination, the orderliness we observe among life forms is a result of energy input. Ultimately, the energy that builds and maintains orderly biological systems mostly comes from the Sun, when plants build energy-rich sugar molecules from disorderly gases and liquids during photosynthesis. The *spontaneous* processes that occur within organisms actually do increase entropy—consider for example, the diffusion of nutrients across a cell membrane. Without some outside energy input, processes in which entropy decreases are not observed in nature (Figure 6.12).

Interestingly, the direction of time's passage links to increasing entropy. This includes a leaf falling from a tree, wood burning in a fire, and even the moving hands of a clock. As these occur, energy is dispersed, and we gain the sense that time moves forward. To put it another way, consider the likelihood of a burned log in a fire becoming whole, or a leaf on the ground spontaneously moving upward to join the branch from which it came. These cases involve the opposite of the dispersion of energy, which would be perceived as time moving backward. Hence, entropy is both a gauge for the dispersal of energy and time's arrow.

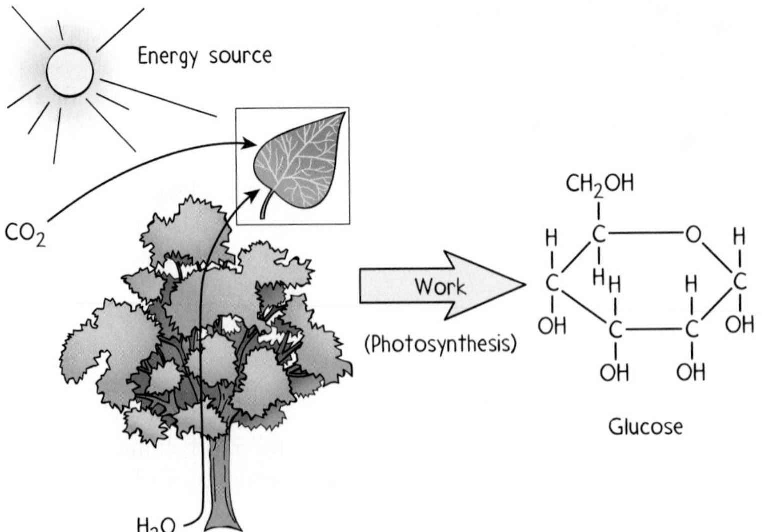

Figure 6.12 Work is needed to transform a state of disorderly, diffuse energy to a state of more concentrated energy. The Sun supplies the energy to do this work when plants transform liquids and vapors into sugar molecules—a plant's storehouse of usable, concentrated energy.

*Entropy can be expressed mathematically. The increase in entropy ΔS of a thermodynamic system is equal to the amount of heat added to the system ΔQ divided by the temperature T at which the heat is added: $\Delta S = \Delta Q/T$.

CHECK YOURSELF

1. As energy disperses, where does it ultimately go?
2. Which has greater entropy: molecules of perfume in a closed perfume bottle, or the molecules of perfume when the perfume bottle is opened? Why?
3. For the formation of molecular hydrogen H_2 from atomic hydrogen H, there is a net increase in entropy. Will this chemical reaction proceed on its own? Justify your answer.
4. A tree takes in carbon dioxide from the air, water from the soil, and a small amount of water vapor from the air. Structures in the tree's leaves, called chloroplasts, convert these disorderly building materials into sugar molecules, which are highly concentrated forms of energy. Does this violate the second law of thermodynamics? Explain your answer.

CHECK YOUR ANSWERS

1. Ultimately, to thermal energy.
2. The molecules diffuse when the bottle is opened, spreading their thermal energy over a larger volume of space. Entropy is increased.
3. Yes; by the second law of thermodynamics, processes that increase entropy are favored.
4. No; the radiant energy of the Sun provides the energy input needed to convert less-concentrated energy to a more concentrated, usable form.

6.6 Specific Heat Capacity

Figure 6.13 The filling of a hot apple pie may be too hot to eat, even though the crust is not.

While eating, you've likely noticed that some foods remain hotter much longer than others. Whereas the filling of hot apple pie can burn your tongue, the crust does not, even when the pie has just been removed from the oven. Or a piece of toast may be comfortably eaten a few seconds after coming from the hot toaster, whereas you must wait several minutes before eating soup that initially had the same temperature (Figure 6.13).

Different substances have different capacities for storing thermal energy. If we heat a pot of water on a stove, we might find that it requires 15 minutes to raise it from room temperature to its boiling temperature. But if we put an equal mass of iron on the same stove, we'd find it would rise through the same temperature range in only about 2 minutes. For silver, the time would be less than a minute. We find that different materials require different quantities heat to raise the temperature of a given mass of the material by a specified number of degrees. This is because different materials absorb energy in different ways. The energy may increase the translational motion of molecules, which raises the temperature; or it may increase the amount of internal vibration or rotation within the molecules and go into potential energy, which does not raise temperature. Generally, there is a combination of both.

A gram of water requires 1 calorie of energy to raise the temperature 1 degree Celsius. It takes only about one-eighth as much energy to raise the temperature of a gram of iron by the same amount. Water absorbs more heat than iron for the same change in temperature. We say water has a higher **specific heat capacity** (sometimes called *specific heat*).

> **The specific heat capacity of any substance is defined as the quantity of heat required to change the temperature of a unit mass of the substance by 1 degree Celsius.**

We can think of specific heat capacity as thermal inertia. Recall that inertia is a term used in mechanics to signify the resistance of an object to a change in its state of motion. Specific heat capacity is like thermal inertia since it signifies the resistance of a substance to a change in temperature.

CHECK YOURSELF

Which has a higher specific heat capacity, water or sand? In other words, which takes longer to warm in sunlight (or longer to cool at night)?

CHECK YOUR ANSWER

Water has the higher specific heat capacity. In the same sunlight, the temperature of water increases more slowly than the temperature of sand. And water will cool more slowly at night. The low specific heat capacity of sand and soil, as evidenced by how quickly they warm in the morning Sun and how quickly they cool at night, affects local climates.

Water has a much higher capacity for storing energy than most all other substances. A lot of heat energy is needed to change the temperature of water. This explains why water is very useful in the cooling systems of automobiles and other engines. It absorbs a great quantity of heat for small increases in temperature. Water also takes longer to cool.

INTEGRATED SCIENCE EARTH SCIENCE

Specific Heat Capacity and Earth's Climate

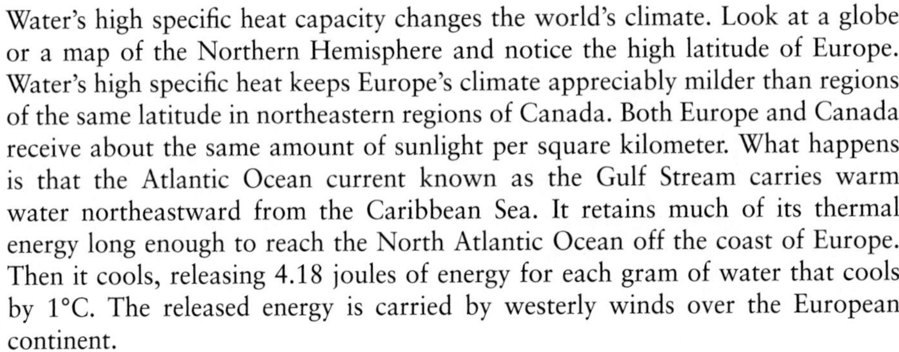

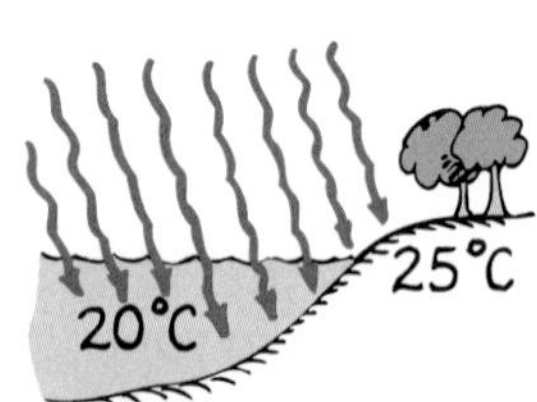

Figure 6.14 Because water has a high specific heat capacity and is transparent, it takes more energy to warm a body of water than to warm the land. Solar energy incident on the land is concentrated at the surface, but solar energy hitting the water extends beneath the surface and is "diluted."

Water's high specific heat capacity changes the world's climate. Look at a globe or a map of the Northern Hemisphere and notice the high latitude of Europe. Water's high specific heat keeps Europe's climate appreciably milder than regions of the same latitude in northeastern regions of Canada. Both Europe and Canada receive about the same amount of sunlight per square kilometer. What happens is that the Atlantic Ocean current known as the Gulf Stream carries warm water northeastward from the Caribbean Sea. It retains much of its thermal energy long enough to reach the North Atlantic Ocean off the coast of Europe. Then it cools, releasing 4.18 joules of energy for each gram of water that cools by 1°C. The released energy is carried by westerly winds over the European continent.

A similar effect occurs in the United States. The winds in North America are mostly westerly. On the West Coast, air moves from the Pacific Ocean to the land. In winter months, the ocean water is warmer than the air. Air blows over the warm water and then moves over the coastal regions. This warms the climate. In summer, the opposite occurs. The water cools the air, and the coastal regions are cooled. The East Coast does not benefit from the moderating effects of water because the direction of the prevailing wind is eastward from the land to the Atlantic Ocean. Land, with a lower specific heat capacity, gets hot in the summer but cools rapidly in the winter.

Islands and peninsulas do not have the extremes of temperatures that are common in the interior regions of a continent. The high summer and low winter temperatures common in the Manitoba and the Dakotas, for example, are largely due to the absence of large bodies of water. Europeans, islanders, and people living near ocean air currents should be glad that water has such a high specific heat capacity. San Franciscans certainly are!

See Chapter 25 for more discussion of the effects of ocean currents on global climate.

CHECK YOURSELF

1. Bermuda is close to North Carolina, but, unlike North Carolina, it has a tropical climate year-round. Why?

2. How is the thermal energy that is lost by the Northern Atlantic Ocean off the coast of Europe carried to the European continent?
3. If the winds at the latitude of San Francisco and Washington, D.C., were from the east rather than from the west, why might San Francisco be able to grow only cherry trees and Washington be able to grow only palm trees?

CHECK YOUR ANSWERS

1. Bermuda is an island. The surrounding water warms it when it might otherwise be too cold, and cools it when it might otherwise be too warm.
2. The heat moves from the Atlantic Ocean into the air and is carried over Europe by westerly winds.
3. As the ocean off the coast of San Francisco cools in the winter, the heat it loses warms the atmosphere it comes in contact with. This warmed air blows over the California coastline to produce a relatively warm climate. If the winds were easterly instead of westerly, the climate of San Francisco would be chilled by winter winds from dry and cold Nevada. The climate would be reversed also in Washington, D.C., because air warmed by the cooling of the Atlantic Ocean would blow over Washington, D.C. and produce a warmer climate there in the winter.

MATH CONNECTION

The Heat-Transfer Equation

We can use specific heat capacity to write a formula for the quantity of heat Q involved when a mass m of a substance undergoes a change in temperature: $Q = cm\Delta T$. In words, heat transferred to or from an object = specific heat capacity of the object × mass of the object × its temperature change. This equation is valid for a substance that gets warmer as well as for one that cools. When a substance is warming up, the heat transferred into it, Q, is positive. When a substance is cooling off, Q has a minus sign.

Let's apply this equation to a few examples.

Problems

1. A 2.0-kg aluminum pan is heated on the stove from 20°C to 110°C. How much heat had to be transferred to the aluminum? The specific heat capacity of aluminum is 900 J/kg·°C.
2. What would be the final temperature of a mixture of 50 g of 20°C water and 50 g of 40°C water?
3. What would be the final temperature when 100 g of 25°C water is mixed with 75 g of 40°C water?
4. Radioactive decay of granite and other rocks in Earth's interior provides enough energy to keep the interior hot, to produce magma, and to provide warmth to natural hot springs. This is due to the average release of about 0.03 J per kilogram each year. How many years are required for a chunk of thermally insulated granite to increase 500°C in temperature, assuming that the specific heat of granite is 800 J/kg·°C?

Solutions

1. $Q = cm\Delta T$ = (900 J/kg·°C)(2.0 kg)(110°C − 20°C) = 1.62×10^5 J. The sign of Q is positive, since the pan is absorbing heat.
2. The heat gained by the cooler water = the heat lost by the warmer water. Since the masses of the water are the same, the final temperature is midway between the two, 30°C. So we'll end up with 100 g of 30°C water.
3. Here we have different masses of water that are mixed together. We equate the heat gained by the cool water to the heat lost by the warm water. We can express this equation formally, and then we can let the expressed terms lead to a solution:

   Heat gained by cool water = heat lost by warm water

   $$cm_1\Delta T_1 = cm_2\Delta T_2$$

   ΔT_1 doesn't equal ΔT_2 as in Problem 2 because of different masses of water. We can see that ΔT_1 will be the final temperature T minus 25°, since T will be greater than 25°; ΔT_2 is 40°C minus T, because T will be less than 40°. Then,

   $$c(100)(T - 25) = c(75)(40 - T)$$
   $$100T - 2500 = 3000 - 75T$$
   $$T = 31.4°\text{C}$$

4. Here, we switch to rock, but the same concept applies. No particular mass is specified, so we'll work with quantity of heat/mass (for our answer should be the same for a small chunk of rock or a huge chunk). From $Q = cm\Delta T$, $\frac{Q}{m} = c\Delta T$ = (800 J/kg·°C) × (500°C) = 400,000 J/kg. The time required is (400,000 J/kg)/(0.03 J/kg·yr) = 13.3 million years. Small wonder it remains hot down there!

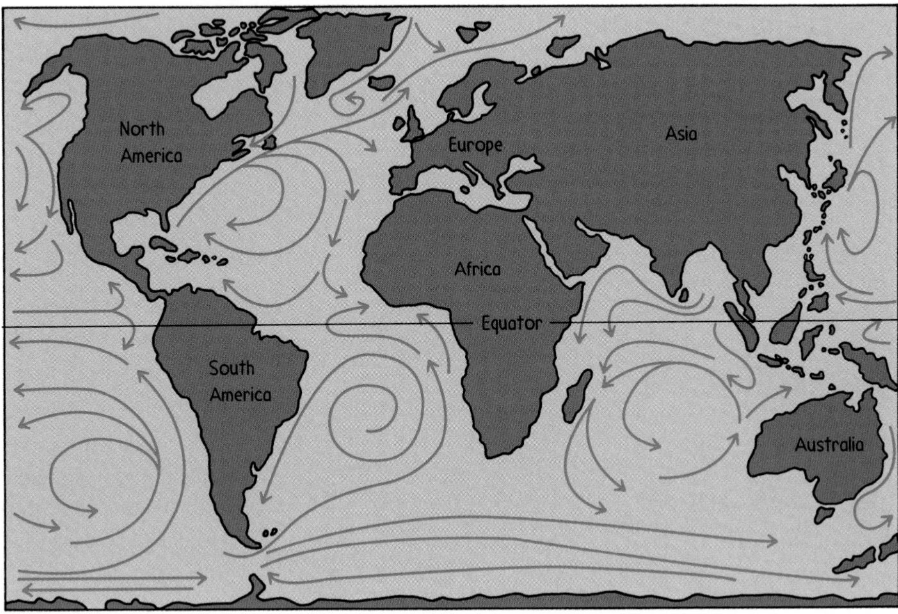

Figure 6.15 Many ocean currents, shown in blue, distribute heat from the warmer equatorial regions to the colder polar regions.

6.7 Thermal Expansion

Molecules in a hot substance jiggle faster and move farther apart. The result is thermal expansion. Most substances expand when heated and contract when cooled. Sometimes the changes aren't noticed and sometimes they are. Telephone wires are longer and sag more on a hot summer day than they do in winter. Railroad tracks laid on cold winter days expand and buckle in the hot summer. Metal lids on glass fruit jars can often be loosened by heating them under hot water. If one part of a piece of glass is heated or cooled more rapidly than adjacent parts, the resulting expansion or contraction may break the glass. This is especially true of thick glass. Pyrex glass is an exception because it is specially formulated to expand very little with increasing temperature.

INSIGHT Thermal expansion accounts for the creaky noises often heard in the attics of old houses on cold nights. The construction materials expand during the day and contract at night, creaking as they grow and shrink.

Thermal expansion must be taken into account in structures and devices of all kinds. A dentist uses filling material that has the same rate of expansion as teeth. A civil engineer uses reinforcing steel with the same expansion rate as concrete. A long steel bridge usually has one end anchored while the other rests on rockers (Figure 6.16). Notice also that many bridges have tongue-and-groove gaps called expansion joints (Figure 6.17). Similarly, concrete roadways and sidewalks are intersected by gaps, which are sometimes filled with tar, so that the concrete can expand freely in summer and contract in winter.

Figure 6.16 One end of the bridge rides on rockers to allow for thermal expansion. The other end (not shown) is anchored.

With increases in temperature, liquids expand more than solids. We notice this when gasoline overflows from a car's tank on a hot day. If the tank and its contents expanded at the same rate, no overflow would occur. This is why a gas tank being filled shouldn't be "topped off," especially on a hot day.

Figure 6.17 The gap in the roadway of a bridge is called an expansion joint. It allows the bridge to expand and contract. (Was this photo taken on a warm day or a cold day?)

Expansion of Water

Water, like most substances, expands when heated, except in the temperature range between 0°C and 4°C. Something fascinating happens in this range. Ice has a crystalline structure, with open-structured crystals. Water molecules in this open structure occupy a greater volume than they do in the liquid phase (Figures 6.18 and 6.19). This means that ice is less dense than water.

TECHNOLOGY

Engineering for Thermal Expansion

An illustration of the fact that different substances expand at different rates can be provided by a bimetallic strip. This device is made of two strips of different metals welded together, one of brass and the other of iron. When heated, the greater expansion of the brass bends the strip. This bending may be used to turn a pointer, to regulate a valve, or to close a switch.

Room temperature

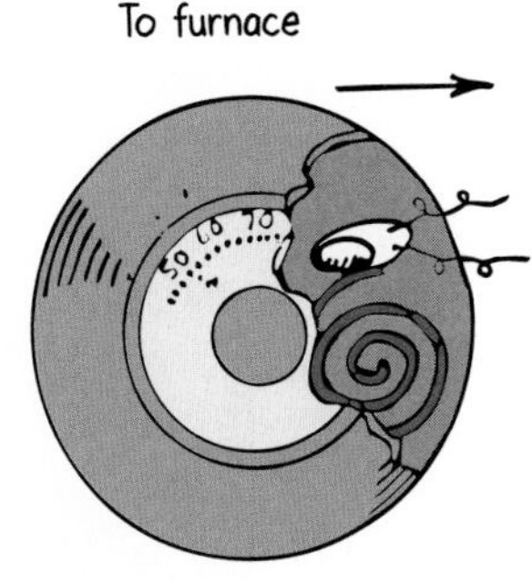

A practical application of a bimetallic strip wrapped into a coil is the thermostat. When a room becomes too warm, the bimetallic coil expands, the drop of liquid mercury rolls away from the electrical contacts and breaks the electrical circuit. When the room is too cool, the coil contracts, the mercury rolls against the contacts and completes the circuit. Bimetallic strips are used in oven thermometers, refrigerators, electric toasters, and various other devices.

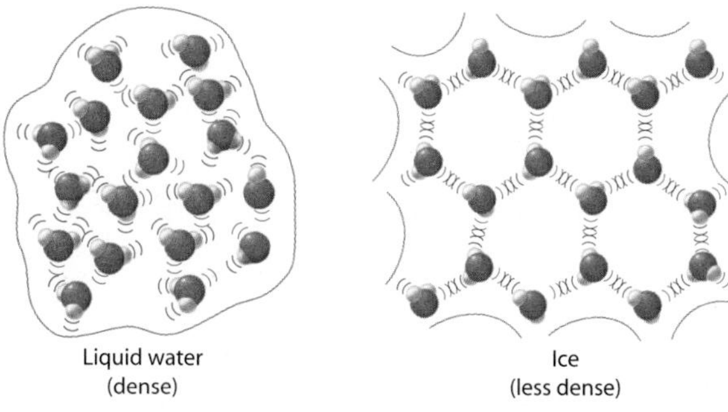

Figure 6.18 Liquid water is more dense than ice because water molecules in a liquid are closer together than water molecules frozen in ice, where they have an open crystalline structure.

When ice melts, not all the six-sided crystals collapse. Some of them remain in the ice-water mixture, making up a microscopic slush that slightly "bloats" the water, increasing its volume slightly (Figure 6.20). This results in ice water being less dense than slightly warmer water. As the temperature of water at 0°C is increased, more of the remaining ice crystals collapse. This further decreases the volume of the water. This contraction occurs only up to 4°C. That's because two things occur at the same time—contraction and expansion. Volume tends to decrease as ice crystals collapse, while volume tends to increase due to greater molecular motion. The collapsing effect dominates until the temperature reaches 4°C. After that, expansion overrides contraction, because most of the ice crystals have melted (Figure 6.21).

When ice water freezes to become solid ice, its volume increases tremendously—and its density is therefore much lower. That's why ice floats on water. Like most other substances, solid ice contracts without further cooling. This behavior of water is very important in nature. If water were most dense at 0°C, it would settle to the bottom of a lake or pond. Because water at 0°C is less dense, it floats at the surface. That's why ice forms at the surface (Figure 6.22).

Figure 6.19 The six-sided structure of a snowflake is a result of the six-sided ice crystals that make it up.

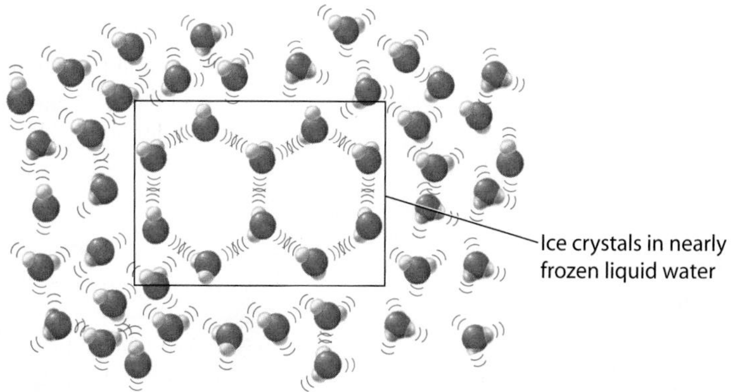

Figure 6.20 Close to 0°C, liquid water contains crystals of ice. The open structure of these crystals increases the volume of the water slightly.

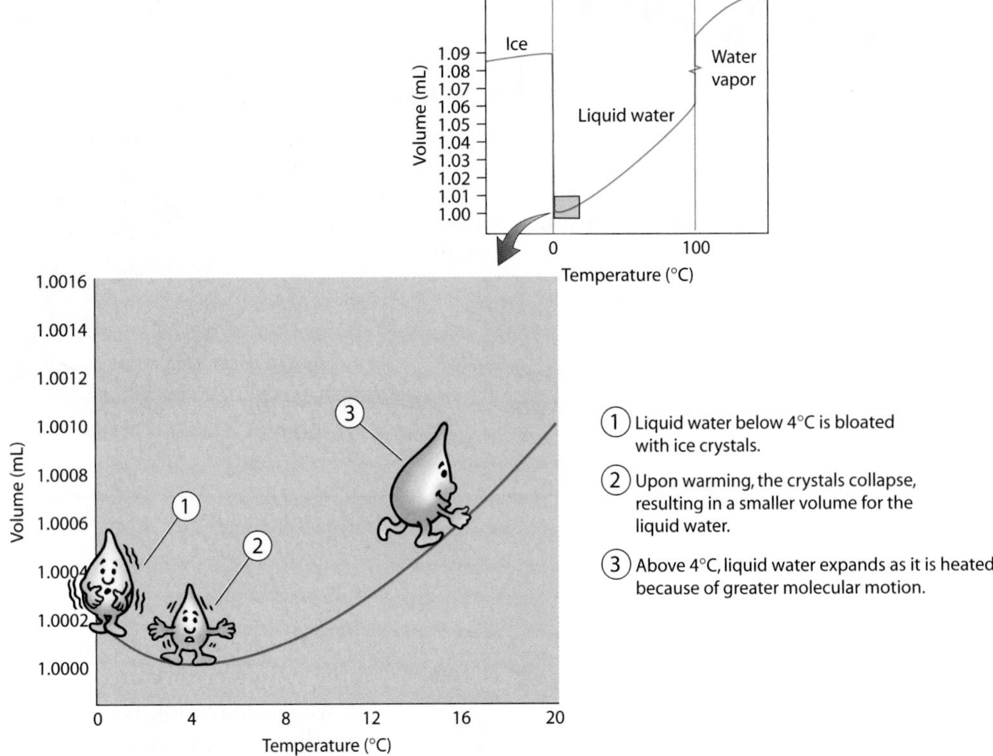

Figure 6.21 Between 0°C and 4°C, the volume of liquid water decreases as temperature increases. Above 4°C, water behaves the way other substances do. Its volume increases as its temperature increases. The volumes shown here are for a 1-gram sample.

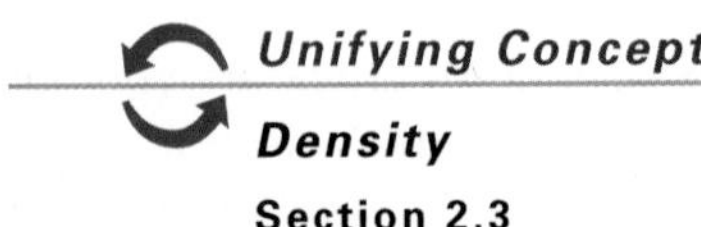

Unifying Concept
Density
Section 2.3

CHECK YOURSELF

1. What was the precise temperature at the bottom of Lake Michigan on New Year's Eve in 1901?
2. What's inside the open spaces of the ice crystals shown in Figure 6.20? Is it air, water vapor, or nothing?

CHECK YOUR ANSWERS

1. The temperature at the bottom of any body of water that has 4°C water in it is 4°C, for the same reason that rocks are at the bottom. Both 4°C water and rocks are denser than water at any temperature. Water is a poor heat conductor, so, if the body of water is deep and in a region of long winters and short summers, the water at the bottom is likely to remain a constant 4°C year round.
2. There's nothing at all in the open spaces. It's empty space—a void. If there were air or vapor in the open spaces, the illustration should show molecules there—oxygen and nitrogen for air and H_2O for water vapor.

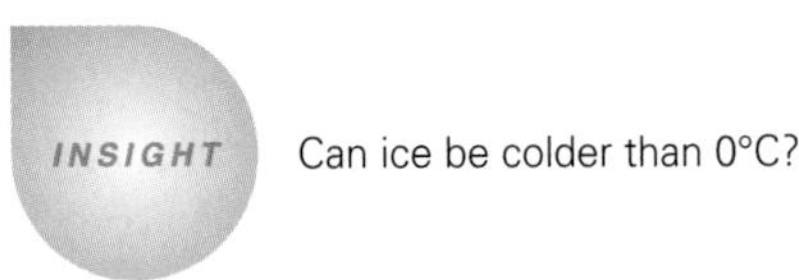

Can ice be colder than 0°C?

Figure 6.22 As water cools, it sinks until the entire lake is 4°C. Then, as water at the surface is cooled further, it floats on top and can freeze. Once ice is formed, temperatures lower than 4°C can extend down into the lake.

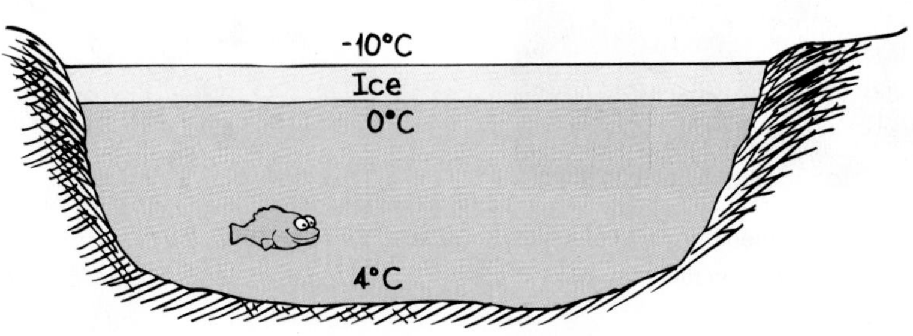

Conceptual Science Place

The Secret to Walking on Hot Coals; Air Is a Poor Conductor

6.8 Heat Transfer: Conduction

Heat transfers from warmer to cooler objects, so that both objects tend to reach a common temperature. This process occurs in three ways: by *conduction*, by *convection*, and by *radiation*.

When you hold one end of an iron nail in a flame, it quickly becomes too hot to hold. Thermal energy at the hot end travels along the nail's entire length. This method of heat transfer is called conduction. Thermal conduction occurs by means of the movement of particles in a material, mainly electrons. Every atom has electrons, and metal atoms have loosely held electrons that are free to migrate in the metal. We shall see, in Chapter 7, that metals are good electrical conductors for the same reason. Thermal conduction occurs by atomic particles colliding inside the heated object.

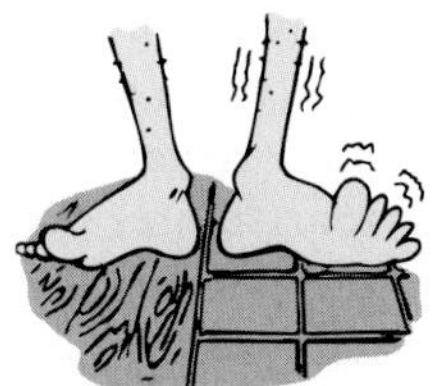

Figure 6.23 The tile floor feels colder than the wooden floor, even though both floor materials are at the same temperature. This is because tile is a better conductor of heat than wood, and so heat is more readily conducted out of the foot touching the tile.

Solids whose atoms or molecules have loosely held electrons are good conductors of heat. Metals have the loosest electrons, and they are excellent conductors of heat. Silver is the best, copper is next, and then, among the common metals, aluminum and iron. Poor conductors include wool, wood, paper, cork, and plastic foam. Molecules in these materials have electrons that are firmly attached to them. Poor conductors are called *insulators*.

Figure 6.24 When you touch a nail stuck in ice, does cold flow from the ice to your hand, or does energy flow from your hand to the ice?

Wood is a great insulator, and it is often used for cookware handles. Even when a pot is hot, you can briefly grasp the wooden handle with your bare hand without harm. An iron handle of the same temperature would surely burn your hand. Wood is a good insulator even when it's red hot. This explains how firewalking coauthor John Suchocki can walk barefoot on red-hot wood coals without burning his feet (Figure 6.25). (CAUTION: Don't try this on your own; even experienced firewalkers sometimes receive bad burns when conditions aren't just right.) The main factor here is the poor conductivity of wood—even red-hot wood. Although its temperature is high, very little energy is conducted to the feet. A firewalker must be careful that no iron nails or other good conductors are among the hot coals. Ouch!

Figure 6.25 Firewalking author John Suchocki isn't burned by the red-hot wooden coals due to the low heat conductivity of wood.

Air is a very poor conductor as well. Hence, you can briefly put your hand in a hot pizza oven without harm. The hot air doesn't conduct thermal energy well. But don't touch the metal in the hot oven. Ouch again! The good insulating properties of such things as wool, fur, and feathers are largely due to the air spaces they contain. Porous substances are also good insulators because of their many small air spaces. Be glad that air is a poor conductor; if it weren't you'd feel quite chilly on a 20°C (68°F) day!

Snow is a poor conductor of thermal energy. Snowflakes are formed of crystals that trap air and provide insulation. That's why a blanket of snow keeps the ground warm in winter. Animals in the forest find shelter from the cold in snow banks and in holes in the snow. The snow doesn't provide them with thermal energy—it simply slows down the loss of body heat generated by the animals. Then there are igloos, Arctic dwellings built from compacted snow to shield those inside from the bitter cold of Arctic winters.

Figure 6.26 Snow patterns on the roof of a house show areas of conduction and insulation. Bare parts show where heat from inside has leaked through the roof and melted the snow.

Convection ovens, now common, are simply ovens with a fan inside. Cooking is speeded up by the circulation of heated air.

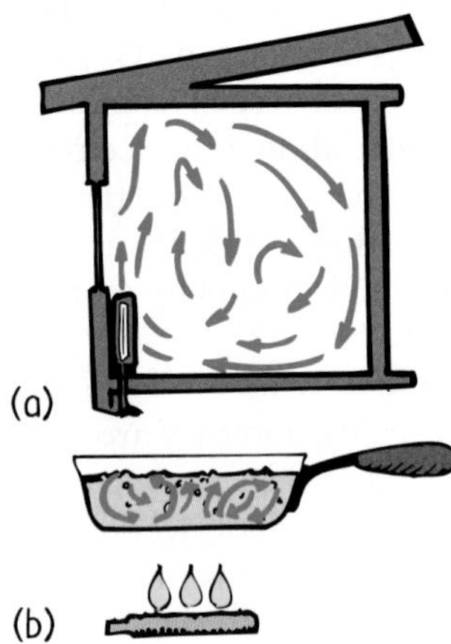

Figure 6.27 (a) Convection currents in air. (b) Convection currents in liquid.

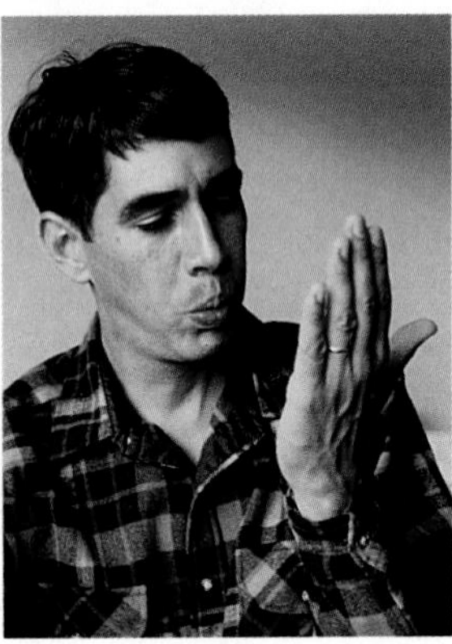

Figure 6.28 Blow warm air onto your hand from your wide-open mouth. Now reduce the opening between your lips so that the air expands as you blow. Try it now. Do you notice a difference in air temperature?

Figure 6.29 The hot steam expands from the pressure cooker and is cool to Millie's touch.

6.9 Heat Transfer: Convection

Liquids and gases transfer thermal energy mainly by convection, which is heat transfer due to the actual motion of the fluid itself. Unlike conduction (in which heat is transferred by successive collisions of electrons and atoms), convection involves motion of a fluid—currents. Convection occurs in all fluids, whether liquids or gases. Whether we heat water in a pan or heat air in a room, the process is the same (Figure 6.27). As the fluid is heated from below, the molecules at the bottom move faster, spread apart more, become less dense, and are buoyed upward. Denser, cooler fluid moves in to take their place. In this way, convection currents keep the fluid stirred up as it heats—warmer fluid moving away from the heat source and cooler fluid moving toward the heat source.

We can see why warm air rises. When warmed, it expands, becomes less dense, and is buoyed upward in the cooler surrounding air like a balloon buoyed upward. When the rising air reaches an altitude at which the air density is the same, it no longer rises. We see this occurring when smoke from a fire rises and then settles off as it cools and its density matches that of the surrounding air. To see for yourself that expanding air cools, right now do the experiment shown in Figure 6.28. Expanding air really does cool.*

A dramatic example of cooling by expansion occurs with steam expanding through the nozzle of a pressure cooker (Figure 6.29).The combined cooling effects of expansion and rapid mixing with cooler air will allow you to hold your hand comfortably in the jet of condensed vapor. (Caution: If you try this, be sure to place your hand high above the nozzle at first and then lower it slowly to a comfortable distance above the nozzle. If you put your hand directly at the nozzle where no steam is visible, watch out! Steam is invisible, and is clear of the nozzle before it expands and cools. The cloud of "steam" you see is actually condensed water vapor, which is much cooler than live steam.)

Cooling by expansion is the opposite of what occurs when air is compressed. If you've ever compressed air with a tire pump, you probably noticed that both air and pump became quite hot. Compression of hot air warms it.

Convection currents stir the atmosphere and produce winds. Some parts of the Earth's surface absorb thermal energy from the Sun more readily than others. This results in uneven heating of the air near the ground. We see this effect at the seashore, as Figure 6.30 shows. In the daytime, the ground warms up more than

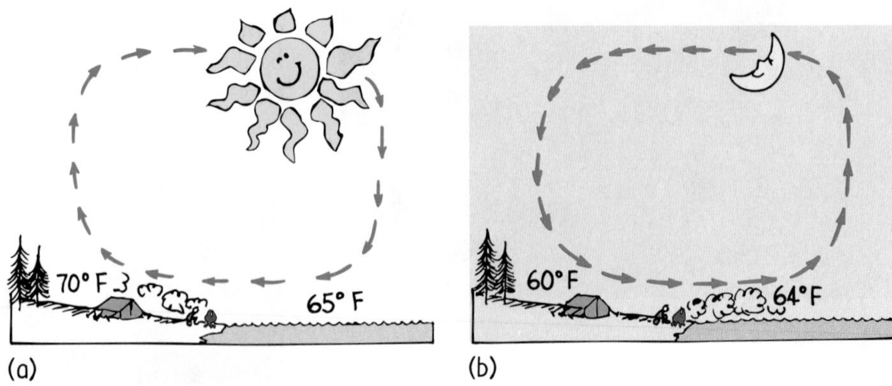

Figure 6.30 Convection currents produced by unequal heating of land and water. (a) During the day, warm air above the land rises, and cooler air over the water moves in to replace it. (b) At night, the direction of the air flow is reversed, because now the water is warmer than the land.

* Where does the energy go in this case? It goes to work done on the surrounding air as the expanding air pushes outward.

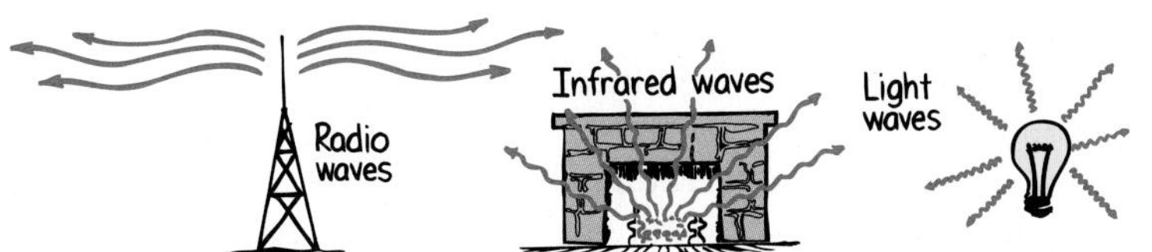

Figure 6.31 Types of radiant energy (electromagnetic waves).

Figure 6.32 A wave of long wavelength is produced when the rope is shaken gently (at a low frequency). When it is shaken more vigorously (at a high frequency), a wave of shorter wavelength is produced.

As something expands, it spreads its energy over a greater area and therefore it cools.

the water. Then warmed air close to the ground rises and is replaced by cooler air that moves in from above the water. The result is a sea breeze. At night, the process reverses because the shore cools off more quickly than the water, and then the warmer air is over the sea. If you build a fire on the beach, you'll see that the smoke sweeps inland during the day and then seaward at night.

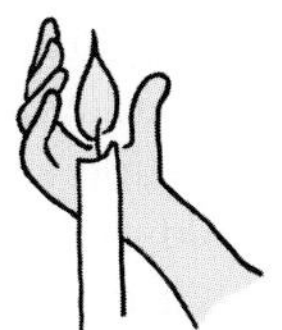

CHECK YOURSELF

Explain why you can hold your fingers beside the candle flame without harm, but not above the flame.

CHECK YOUR ANSWER

Thermal energy travels upward by convection. Since air is a poor conductor, very little energy travels sideways to your fingers.

6.10 Heat Transfer: Radiation

Thermal radiation from the Sun travels through space and then through the Earth's atmosphere and warms the Earth's surface. This energy cannot pass through the empty space between the Sun and Earth by conduction or convection, for there is no medium for doing so. Energy must be transmitted some other way—by radiation.* The energy so radiated is called radiant energy.

Radiant energy exists in the form of electromagnetic waves. It includes a wide span of waves, ranging from longest to shortest: radio waves, microwaves, infrared waves (invisible waves below red in the visible spectrum), visible waves, then to waves that can't be seen by the eye, including ultraviolet waves, X-rays, and gamma rays. We'll treat waves further in Chapter 8, and electromagnetic waves in Chapter 7.

The wavelength of radiation is related to the frequency of vibration. Frequency is the rate of vibration of a wave source. Nellie Newton in Figure 6.32 shakes a rope at a low frequency (left), and a higher frequency (right). Note that shaking at a low frequency produces a long, lazy wave, and the higher-frequency shake produces shorter waves. This is also true with electromagnetic waves. We shall see, in Chapter 8, that vibrating electrons emit electromagnetic waves. Low-frequency vibrations produce long waves, and high-frequency vibrations produce shorter waves (Figure 6.33).

(a) cool
(b) medium
(c) hot

Figure 6.33 (a) A low-temperature (cool) source emits primarily low-frequency, long-wavelength waves. (b) A medium-temperature source emits primarily medium-frequency, medium-wavelength waves. (c) A high-temperature (hot) source emits primarily high-frequency, short-wavelength waves.

Emission of Radiant Energy

All substances at any temperature above absolute zero emit radiant energy. The average frequency f of the radiant energy is directly proportional to the absolute temperature T of the emitter: $f \sim T$.

* The radiation we are talking about here is electromagnetic radiation, including visible light. Don't confuse this with radioactivity, a process of the atomic nucleus that we'll discuss in Chapter 10.

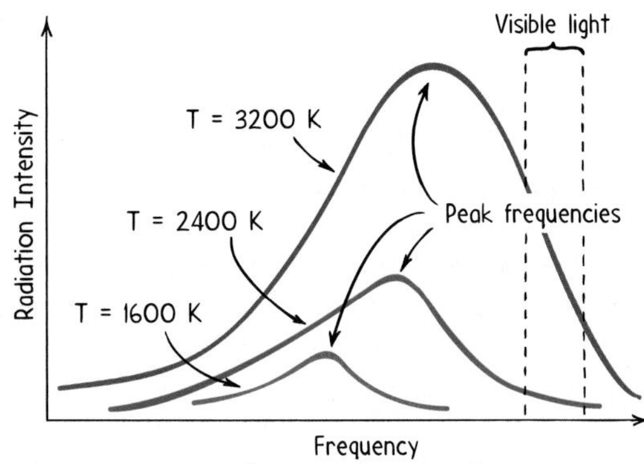

Figure 6.34 Radiation curves for different temperatures. The peak frequency of radiant energy is directly proportional to the absolute temperature of the emitter.

The fact that all objects in our environment continually emit infrared radiation underlies the increasingly common infrared thermometers. Quite remarkably, you simply point the thermometer at something whose temperature you want, press a button, and a digital temperature reading appears. There is no need to touch the thermometer to whatever is being measured. The radiation it emits provides the reading. Typical classroom infrared thermometers operate in the range of about −30°C to 200°C.

If an object is hot enough, some of the radiant energy it emits is in the range of visible light. At a temperature of about 500°C an object begins to emit the longest waves we can see, red light. Higher temperatures produce a yellowish light. At about 1200°C all the different waves to which the eye is sensitive are emitted and we see an object as "white hot." A blue-hot star is hotter than a white-hot star, and a red-hot star is less hot. Since a blue-hot star has twice the light frequency of a red-hot star, it therefore has twice the surface temperature of a red-hot star.

The surface of the Sun has a high temperature (by earthly standards) and therefore emits radiant energy at a high frequency—much of it in the visible portion of the electromagnetic spectrum. The surface of the Earth, by comparison, is relatively cool, and so the radiant energy it emits has a frequency lower than that of visible light. The radiation emitted by the Earth is in the form of infrared waves—below our threshold of sight. As you will learn in Chapter 25, radiant energy emitted by the Earth is called *terrestrial radiation.*

Radiant energy is emitted by both the Sun and the Earth, and it differs only in the range of frequencies and the amount (Figure 6.35). When we study meteorology in Chapter 25, we'll learn how the atmosphere is transparent to the high-frequency solar radiation but opaque to much of the lower-frequency terrestrial radiation. This produces the greenhouse effect, which plays a role in global warming.

Figure 6.35 Both the Sun and the Earth emit the same kind of radiant energy. The Sun's glow consists of longer waves and isn't visible to the eye.

All objects—you, your instructor, and everything in your surroundings—continually emit radiant energy in a mixture of frequencies (because temperature corresponds to a mixture of different molecular kinetic energies). Objects with everyday temperatures mostly emit low-frequency infrared waves. When the higher-frequency infrared waves are absorbed by your skin, you feel the sensation of heat. So it is common to refer to infrared radiation as heat radiation.

Common infrared sources that give the sensation of heat are the burning embers in a fireplace, a lamp filament, and the Sun. All of these emit both infrared radiation and visible light. When this radiant energy falls on other objects, it is partly reflected and partly absorbed. The part that is absorbed increases the thermal energy of the objects on which it falls.

CHECK YOURSELF

Which of these do not emit radiant energy? (a) the Sun; (b) lava from a volcano; (c) red-hot coals; (d) this textbook.

CHECK YOUR ANSWER

All the above emit radiant energy—even your textbook, which, like the other substances listed, has a temperature. According to the rule $f \sim T$, the book therefore emits radiation whose average frequency f is quite low compared with the radiation frequencies emitted by the other substances. Everything with any temperature above absolute zero emits radiant energy. That's right—everything!

Absorption of Radiant Energy

If everything is radiating energy, why doesn't everything finally run out of it? The answer is that everything is also absorbing energy. Good emitters of radiant energy are also good absorbers; poor emitters are poor absorbers. For example, a radio dish antenna constructed to be a good emitter of radio waves is also, by its

Everything around you both radiates and absorbs energy continuously.

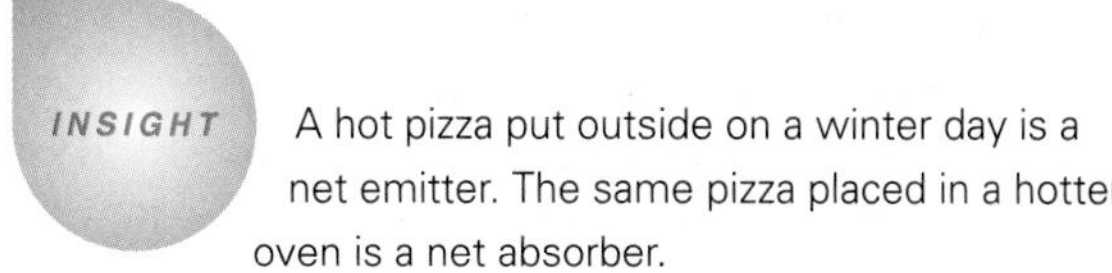

A hot pizza put outside on a winter day is a net emitter. The same pizza placed in a hotter oven is a net absorber.

very design, a good receiver (absorber) of them. A poorly designed transmitting antenna is also a poor receiver.

It's interesting to note that, if a good absorber were not also a good emitter, then black objects would remain warmer than lighter-colored objects, and the two would never reach a common temperature. Objects in thermal contact, given sufficient time, will reach the same temperature. A blacktop pavement may remain hotter than the surroundings on a hot day, but, at nightfall, it cools faster. Sooner or later, all objects come to thermal equilibrium. So a dark object that absorbs a lot of radiant energy must emit a lot as well (Figure 6.36).

Figure 6.36 When the black rough-surfaced container and the shiny polished one are filled with hot (or cold) water, the blackened one cools (or warms) faster.

The surface of any material, hot or cold, both absorbs and emits radiant energy. If the surface absorbs more energy than it emits, it is a net absorber and its temperature rises. If it emits more than it absorbs, it is a net emitter and its temperature drops. Whether a surface plays the role of net absorber or net emitter depends on whether its temperature is above or below that of its surroundings. In short, if it's hotter than its surroundings, the surface will be a net emitter and will cool; if it's colder than its surroundings, it will be a net absorber and will warm.

CHECK YOURSELF

1. If a good absorber of radiant energy were a poor emitter, how would its temperature compare with the temperature of its surroundings?
2. A farmer turns on the propane burner in his barn on a cold morning and heats the air to 20°C (68°F). Why does he still feel cold?

CHECK YOUR ANSWERS

1. If a good radiator were not also a good emitter, there would be a net absorption of radiant energy and the temperature of the absorber would remain higher than the temperature of the surroundings. Things around us approach a common temperature only because good absorbers are, by their very nature, also good emitters.
2. The walls of the barn are still cold. He radiates more energy to the walls than the walls radiate back at him, and he feels chilly. (On a winter day, you are comfortable inside your home or classroom only if the walls are warm—not just the air.)

All matter contains thermal energy. In Chapter 8, you will learn why this means that all objects are emitting radiant energy. The study of heat continues in Chapter 11, where you will learn how heat absorbed by a material can change its *phase* or state—from the solid phase to liquid or the liquid phase to gas. Heat given off by a material can change its phase in the reverse direction—from gas to liquid, or liquid to solid. Since all things contain thermal energy, the study of heat is truly far-reaching.

CHAPTER 6 Review

Summary of Terms

Absolute zero The theoretical temperature at which a substance possesses no thermal energy.

Calorie The amount of heat associated with changing 1 gram of water by 1 degree Celsius.

Conduction The transfer of thermal energy by molecular and electronic collisions within a substance (especially within a solid).

Convection The transfer of thermal energy in a gas or liquid by means of currents in the heated fluid. The fluid flows, carrying energy with it.

Entropy The measure of the energy dispersal of a system. Whenever energy freely transforms from one form to another, the direction of transformation is toward a state of greater disorder and, therefore, toward one of greater entropy.

First law of thermodynamics A restatement of the law of energy conservation, usually as it applies to systems involving changes in temperature: Whenever heat flows into or out of a system, the gain or loss of thermal energy equals the amount of heat transferred.

Heat The thermal energy that flows from a substance of higher temperature to a substance of lower temperature, commonly measured in calories or joules.

Radiation The transfer of energy by means of electromagnetic waves.

Second law of thermodynamics Heat never spontaneously flows from a low-temperature substance to a high-temperature substance. Also, all systems tend to become more and more disordered as time goes by.

Specific heat capacity The quantity of heat per unit mass required to raise the temperature of a substance by 1 degree Celsius.

Temperature A measure of the hotness or coldness of substances, related to the average translational kinetic energy per molecule in a substance; measured in degrees Celsius, or in degrees Fahrenheit, or in kelvins.

Thermal energy (or *internal energy*) The total energy (kinetic plus potential) of the submicroscopic particles that make up a substance.

Thermodynamics The study of heat and its transformation to different forms of energy.

Third law of thermodynamics No system can reach absolute zero.

Review Questions

6.1 The Kinetic Theory of Matter

1. What is thermal energy?
2. What does the kinetic molecular theory of matter state about the composition of matter?
3. Why does a penny become warmer when it is struck by a hammer?

6.2 Temperature

4. What are the temperatures for freezing water on the Celsius and Fahrenheit scales? What are the temperature for boiling water on those scales?
5. Is the temperature of an object a measure of the total translational kinetic energy of molecules in the object or a measure of the average translational kinetic energy per molecule in the object?
6. What is meant by the statement "a thermometer measures its own temperature"?

6.3 Absolute Zero

7. What pressure would you expect in a rigid container of 0°C gas if you cooled it by 273 Celsius degrees?
8. How much energy can be taken from a system at a temperature of 0 K?

6.4 What Is Heat?

9. When you touch a cold surface, does cold travel from the surface to your hand or does energy travel from your hand to the cold surface? Explain.
10. (a) Distinguish between temperature and heat.
    (b) Distinguish between heat and thermal energy.
11. What determines the direction of heat flow?
12. Distinguish between a calorie and a Calorie, and between a calorie and a joule.

6.5 The Laws of Thermodynamics

13. How does the law of conservation of energy relate to the first law of thermodynamics?
14. What happens to the thermal energy of a system when mechanical work is done on the system? What happens to the temperature of the system?
15. How does the second law of thermodynamics relate to the direction of heat flow?

6.6 Specific Heat Capacity

16. Which warms up faster when heat is applied—iron or silver?
17. Does a substance that heats up quickly have a high or low specific heat capacity?
18. How does the specific heat capacity of water compare with specific heat capacities of other common materials?

6.7 Thermal Expansion

19. Which generally expands more for an equal increase in temperature—solids or liquids?
20. What is the reason for ice being less dense than water?
21. Why does ice form at the surface of a pond instead of at the bottom?

6.8 Heat Transfer: Conduction

22. What is the role of "loose" electrons in heat conduction?
23. Distinguish between a heat conductor and a heat insulator.
24. What is the explanation for a barefoot firewalker being able to walk safely on red-hot wooden coals?
25. Why do such materials as wood, fur, and feathers—and even snow—qualify as good insulators?

6.9 Heat Transfer: Convection

26. Describe how convection transfers heat.
27. What happens to the temperature of air when it expands?
28. Why does the direction of coastal winds change from day to night?

6.10 Heat Transfer: Radiation

29. (a) What exactly is radiant energy? (b) What is heat radiation?
30. How does the frequency of radiant energy relate to the absolute temperature of the radiating source?
31. Since all objects continuously radiate energy to their surroundings, why don't the temperatures of all objects continually decrease?

Multiple Choice Questions

Choose the BEST answer to the following.

1. When scientists discuss kinetic energy per molecule, the concept being discussed is
   (a) temperature.
   (b) heat.
   (c) thermal energy.
   (d) entropy.
2. In a mixture of hydrogen gas, oxygen gas, and nitrogen gas, the molecules with the greatest average speed are those of
   (a) hydrogen.
   (b) oxygen.
   (c) nitrogen.
   (d) All have the same speed.
3. Consider a sample of water at 2°C. If the temperature is increased by one degree, the volume of water
   (a) increases.
   (b) decreases.
   (c) remains unchanged.
4. The principal reason one can walk barefoot on red-hot wood coals without burning the feet has to do with
   (a) the low temperature of the coals.
   (b) the low thermal conductivity of the coals.
   (c) mind-over-matter techniques.
5. Which answer best fills in the blanks?
   The ________ law of thermodynamics tells us that without energy input heat ________ flows from a cold substance to a hot substance.
   (a) first; always
   (b) first; never
   (c) second; never
   (d) second; always

INTEGRATED SCIENCE CONCEPTS

Chemistry and Biology—Entropy

1. What does it mean to say that, when energy is transformed, it becomes less useful?
2. What is the physicist's term for the measure of energy dispersal?
3. Consider the decomposition of water (H_2O) to form hydrogen (H_2) and oxygen (O_2). At room temperature, the products of this reaction

have less entropy than the reactants. Is this reaction thermodynamically favored? Justify your answer.

4. A deer is a more concentrated form of energy than the grass it feeds upon. Does this imply that, as the deer converts its food into tissue, it violates the second law of thermodynamics? Explain.

Earth Science—Specific Heat Capacity and Earth's Climate

1. Northeastern Canada and much of Europe receive about the same amount of sunlight per unit area. Why, then, is Europe generally warmer in the winter?
2. Iceland, so named to discourage conquest by expanding empires, is not at all ice-covered like Greenland and parts of Siberia, even though it is nearly on the Arctic Circle. The average winter temperature of Iceland is considerably higher than regions at the same latitude in eastern Greenland and central Siberia. Why is this so?
3. Why does the presence of large bodies of water tend to moderate the climate of nearby land—to make it warmer in cold weather and hotter in warm weather?

Active Explorations

1. Hold the bottom end of a test tube full of cold water in your hand. Heat the top part in a flame until the water boils. The fact that you can still hold the bottom shows that water is a poor conductor of heat. This is even more dramatic when you wedge chunks of ice at the bottom; then the water above can be brought to a boil without melting the ice. Try it and see.

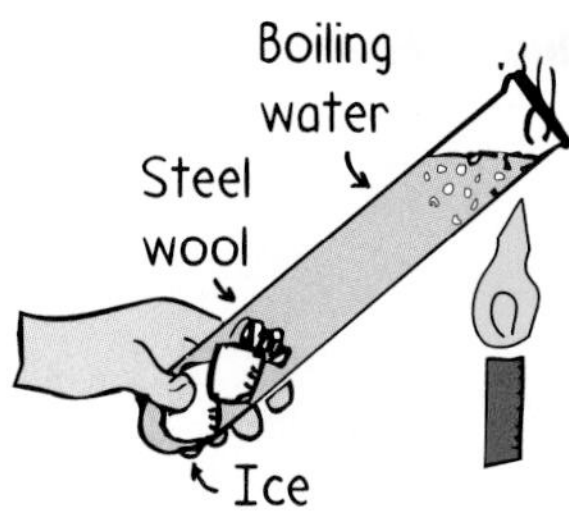

2. Wrap a piece of paper around a thick metal bar and place it in a flame. Note that the paper will not catch fire. Can you figure out why? (Paper will generally not ignite until its temperature reaches 233°C.)
3. Watch the spout of a teakettle of boiling water. Notice that you cannot see the steam that issues from the spout. The cloud that you see farther away from the spout is not steam but condensed water droplets. Now hold the flame of a candle in the cloud of condensed steam. Can you explain your observations?

Exercises

1. In your room, there are such things as tables, chairs, other people, and so forth. Which of these things has a temperature (1) lower than, (2) greater than, (3) equal to the temperature of the air?
2. Why can't you establish whether you are running a high temperature by touching your own forehead?
3. Which is greater, an increase in temperature of 1°C or one of 1°F?
4. Which has the greater amount of internal energy, an iceberg or a hot cup of coffee? Explain.
5. Use the laws of thermodynamics to defend the statement that 100% of the electrical energy that goes into lighting a lamp is converted to thermal energy.
6. When air is rapidly compressed, why does its temperature increase?
7. What happens to the gas pressure within a sealed gallon can when it is heated? What happens to the pressure when the can is cooled? Why?
8. After a car has been driven for some distance, why does the air pressure in the tires increase?
9. If you drop a hot rock into a pail of water, the temperature of the rock and the water will change until both are equal. The rock will cool and the water will warm. Does this hold true if the hot rock is dropped into the Atlantic Ocean? Explain.
10. In the old days, on a cold winter night, it was common to bring a hot object to bed with you. Which would be better to keep you warm through the cold night—a 10-kilogram iron brick or a 10-kilogram jug of hot water at the same temperature? Explain.
11. Why does adding the same amount of heat to two different objects not necessarily produce the same increase in temperature?
12. Why will a watermelon stay cool for a longer time than sandwiches when both are removed from a cooler on a hot day?
13. Cite an exception to the claim that all substances expand when heated.
14. An old method for breaking boulders was to put them in a hot fire, then to douse them with cold water. Why would this fracture the boulders?
15. Would you or the gas company gain by having gas warmed before it passed through your gas meter?
16. A metal ball is just able to pass through a metal ring. When the ball is heated, however, it will not pass through the ring. What would happen if the ring, rather than the ball, were heated? Does the size of the hole increase, stay the same, or decrease?

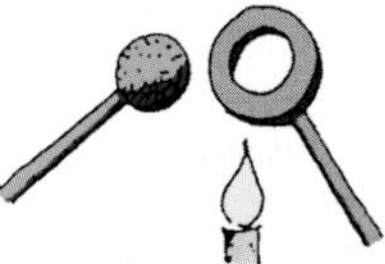

17. After a machinist very quickly slips a hot, snugly fitting iron ring over a very cold glass cylinder, there is no way that the two can be separated intact. Can you explain why this is so?

18. Suppose you cut a small gap in a metal ring. If you heat the ring, will the gap become wider or narrower?

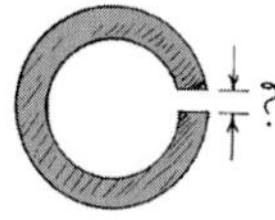

19. Suppose that water is used in a thermometer instead of mercury. If the temperature is at 4°C and then changes, why can't the thermometer indicate whether the temperature is rising or falling?
20. Why is it important to protect water pipes so that they don't freeze?
21. If you wrap a fur coat around a thermometer, will its temperature rise?
22. If you hold one end of a nail against a piece of ice, the end in your hand soon becomes cold. Does cold flow from the ice to your hand? Explain.
23. In terms of physics, why do some restaurants serve baked potatoes wrapped in aluminum foil?

24. Wood is a better insulator than glass, yet fiberglass is commonly used as an insulator in wooden buildings. Explain.
25. Visit a snow-covered cemetery and note that the snow does not slope upward against the gravestones but, instead, forms depressions around them, as shown. Can you think of a reason for this?

26. Why is it that you can safely hold your bare hand in a hot pizza oven for a few seconds, but, if you were to touch the metal inside, you'd burn yourself?
27. In a still room, smoke from a candle will sometimes rise only so far, not reaching the ceiling. Explain why.
28. From the rules that a good absorber of radiation is a good radiator and a good reflector is a poor absorber, state a rule relating the reflecting and radiating properties of a surface.
29. Suppose that, at a restaurant, you are served coffee before you are ready to drink it. In order that it be as warm as possible when you are ready to drink it, would you be wiser to add cream right away or to add it just before you are ready to drink it?

Problems

1. Pounding a nail into wood makes the nail warmer. Consider a 5-gram steel nail 6 cm long and a hammer that exerts an average force of 500 N on the nail when it is being driven into a piece of wood. Show that the increase in the nail's temperature will be 13.3°C. (Assume that the specific heat capacity of steel is 450 J/kg·°C.)
2. If you wish to warm 100 kg of water by 20°C for your bath, show that 8370 kJ of heat is required.
3. The specific heat capacity of copper is 0.092 calories per gram per degree Celsius. Show that the heat required to raise the temperature of a 10-gram piece of copper from 0°C to 100°C is 92 calories.
4. When 100 g of 40°C iron nails are submerged in 100 g of 20°C water, show that the final temperature of the water will be 22.1°C. (The specific heat capacity of iron is 0.12 J/kg·°C. Here, you should equate the heat gained by the water to the heat lost by the nails.)
5. A 10-kg iron ball is dropped onto a pavement from a height of 100 m. If half the heat generated goes into warming the ball, show that the temperature increase of the ball will be 1.1°C. (In SI units, the specific heat capacity of iron is 450 J/kg·°C.) Why is the answer the same for an iron ball of any mass?

Conceptual Science Place

CHAPTER 6 **ONLINE RESOURCES**

Videos Low Temperatures with Liquid Nitrogen, How a Thermostat Works, The Secret to Walking on Hot Coals, Air Is a Poor Conductor • ***Quiz*** • ***Flashcards*** • ***Links***

CHAPTER 7

Electricity and Magnetism

Power lines deliver electric energy to operate the myriad electrical devices found in homes and industry. Lightning is nature's electric-energy delivery system.

CHAPTER OUTLINE

Electricity underlies almost everything around us. It's in the lightning from the sky, it powers devices from computers to flashlights, and it's what holds atoms together to form molecules. But what *is* electricity? Why can you feel a spark when you grab a doorknob after scuffing your feet along the carpet? What's the difference between electric current and voltage—and which of these gives us electric shocks? How do electrical circuits work? Is electricity related to magnetism—if so, how? How do the magnetic strips on credit cards, timing systems for traffic lights, and metal detectors at airports tap both magnetic and electrical forces? We begin our study with a fundamental idea—the concept of electric charge.

7.1 Electric Force and Charge

Suppose that the universe consisted of two kinds of particles—say, positive, and negative. Suppose positives repelled positives but attracted negatives, and that negatives repelled negatives but attracted positives. What would the universe be like? The answer is simple: it would be like the one we are living in. For there *are* such particles and there *is* such a force—the *electrical force*.

The terms *positive* and *negative* refer to electric charge, the fundamental quantity that underlies all electric phenomena. The positively charged particles in ordinary matter are protons, and the negatively charged particles are electrons. Charge is not something added to protons and electrons. It is a basic attribute of them, just as gravitational attractiveness is an attribute of any mass.

The attraction between protons and electrons pulls them together into tiny units—atoms. (Atoms also contain neutral particles called neutrons. More interesting details about atoms are presented in Part 2 of this book). In order to understand the basic principles of electricity, however, we must preview some fundamental facts about atoms:

1. Every atom is composed of a positively charged nucleus surrounded by negatively charged electrons.
2. Each of the electrons in any atom has the same quantity of negative charge and the same mass. Electrons are identical to one another.
3. Protons and neutrons compose the nucleus. (The only exception is the most common form of hydrogen atom, which has no neutrons.) Protons are

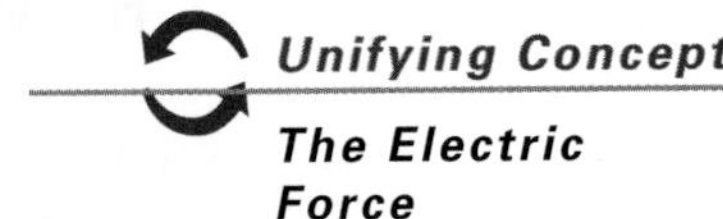

about 1800 times more massive than electrons, but each one carries an amount of positive charge equal to the negative charge of the electrons. Neutrons have slightly more mass than protons and have no net charge.

Normally, an atom has as many electrons as protons. When an atom loses one or more electrons, it has a positive net charge; when it gains one or more electrons, it has a negative net charge. A charged atom is called an *ion*. A *positive ion* has a net positive charge. A *negative ion*, with one or more extra electrons, has a net negative charge.

Material objects are made of atoms, which means they are composed of electrons and protons (and neutrons as well). Although the innermost electrons in an atom are attracted very strongly to the oppositely charged atomic nucleus, the outermost electrons of many atoms are attracted more loosely and can be dislodged. The amount of work required to tear an electron away from an atom varies for different substances. Electrons are held more firmly in rubber or plastic than in your hair, for example. Thus, when you rub a comb against your hair, electrons transfer from the hair to the comb. The comb then has an excess of electrons and is said to be *negatively charged*. Your hair, in turn, has a deficiency of electrons and is said to be *positively charged*.

So protons attract electrons and we have atoms. Electrons repel electrons and we have matter—because atoms don't mesh into one another. This pair of rules is the guts of electricity.

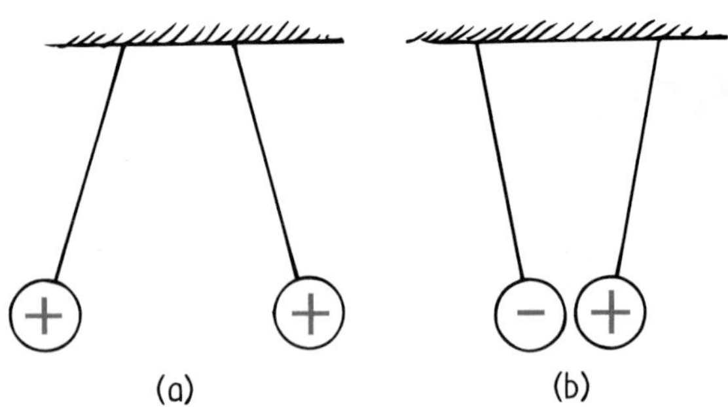

Figure 7.1 Interactive Figure
(a) Like charges repel. (b) Unlike charges attract.

Conservation of Charge

Another basic fact of electricity is that, whenever something is charged, no electrons are created or destroyed. Electrons are simply transferred from one material to another. Charge is *conserved*. In every event, whether large-scale or at the atomic and nuclear level, the principle of **conservation of charge** has always been found to apply. No case of the creation or destruction of charge has ever been found.

CHECK YOURSELF

If you walk across a rug and scuff electrons from your feet, are you negatively or positively charged?

CHECK YOUR ANSWER

You have fewer electrons after you scuff your feet, so you are positively charged (and the rug is negatively charged).

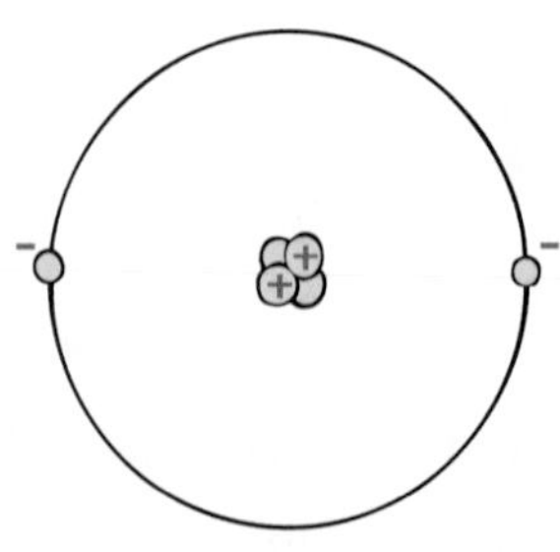

Figure 7.2 Interactive Figure
Model of a helium atom. The atomic nucleus is made up of two protons and two neutrons. The positively charged protons attract two negatively charged electrons. What is the net charge of this atom?

Figure 7.3 Fur has a greater affinity for electrons than plastic. So when a plastic rod is rubbed with fur, electrons are transferred from the fur to the rod. The rod is then negatively charged and the fur is positively charged.

Figure 7.4 Why will you get a slight shock from the doorknob after scuffing across the carpet?

7.2 Coulomb's Law

Electrostatics

The electrical force, like gravitational force, decreases inversely as the square of the distance between the charges. This relationship, which was discovered by Charles Coulomb in the eighteenth century, is called Coulomb's Law. It states that, for two charged objects that are much smaller than the distance between them, the force between them varies directly as the product of their charges and inversely as the square of the separation distance. The force acts along a straight line from one charge to the other. Coulomb's Law can be expressed as

$$F = k\frac{q_1q_2}{d^2}$$

where d is the distance between the charged particles, q_1 represents the charge on one particle, q_2 represents the charge of the second particle, and k is the proportionality constant.

The unit of charge is called the **coulomb**, abbreviated C. It turns out that a charge of 1 C is the charge associated with 6.25 billion billion electrons. This might seem like a great number of electrons, but it only represents the amount of charge that flows through a common 100-watt light bulb in a little more than a second. Dividing the value of 1 C by the number of electrons with this much charge, we have the charge on a single electron: 1.60×10^{-19} C. An electron and proton carry this same magnitude of charge. Because all charged objects carry multiples of this charge, 1.60×10^{-19} C is considered the *fundamental charge.*

INSIGHT Negative and positive are just names given to opposite charges. The names chosen could just as well have been "east and west," or "up and down," or "Mary and Larry." Positive charge is not "better" than "negative" charge—the two kinds of charge are just opposites of each other.

The proportionality constant k in Coulomb's law is similar to G in Newton's law of gravity. Instead of being a very small number, like G, k is a very large number, approximately

$$k = 9{,}000{,}000{,}000 \text{ N}\cdot\text{m}^2/\text{C}^2$$

In scientific notation, $k = 9.0 \times 10^9$ N·m²/C². The unit N·m²/C² is not central to our interest here; it simply converts the right-hand side of the equation to the unit of force, the newton (N). What is important is the large magnitude of k. If, for example, a pair of charges of 1 coulomb each were 1 meter apart, the force of repulsion between the two would be 9 billion N. That would be about ten times the weight of a battleship. Obviously, such quantities of unbalanced charge do not usually exist in our everyday environment.

So Newton's law of gravitation for masses is similar to Coulomb's law for electrically charged bodies. The most important difference between gravitational and electrical forces is that electrical forces may be either attractive or repulsive, whereas gravitational forces are only attractive.

CHECK YOURSELF

1. The proton is the nucleus of the hydrogen atom, and it attracts the electron that orbits it. Relative to this force, does the electron attract the proton with less force, more force, or the same amount of force?
2. If a proton at a particular distance from a charged particle is repelled with a given force, by how much will the force decrease when the proton is three times as distant from the particle? When it is five times as distant?
3. What is the sign of the charge on the particle in this case?

CHECK YOUR ANSWERS

1. The same amount of force, in accord with Newton's third law—basic mechanics! Recall that a force is an interaction between two things—in this case, between the proton and the electron. They pull on each other equally.

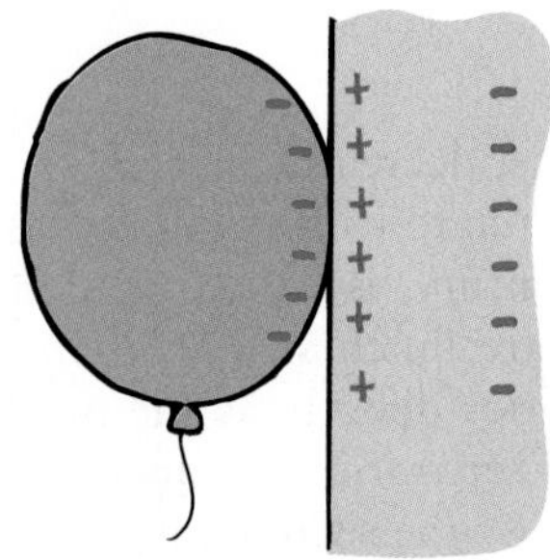

Figure 7.5 The negatively charged balloon polarizes molecules in the wooden wall and creates a positively charged surface, so the balloon sticks to the wall.

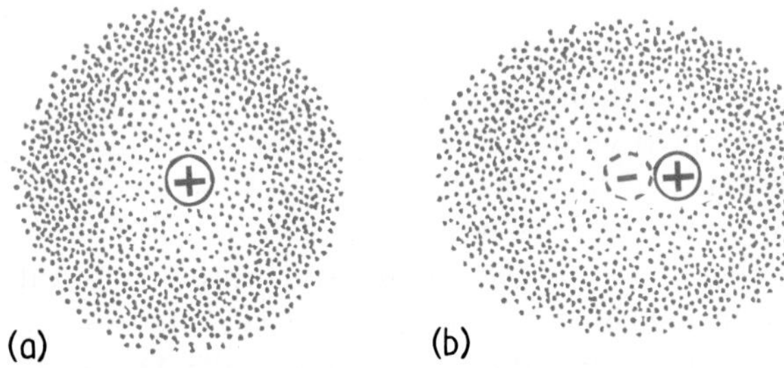

Figure 7.6 An electron buzzing around the atomic nucleus makes up an electron cloud. (a) The center of the negative "cloud" of electrons coincides with the center of the positive nucleus in an atom. (b) When an external negative charge is brought nearby to the right, as on a charged balloon, the electron cloud is distorted so that the centers of negative and positive charge no longer coincide. The atom is electrically polarized.

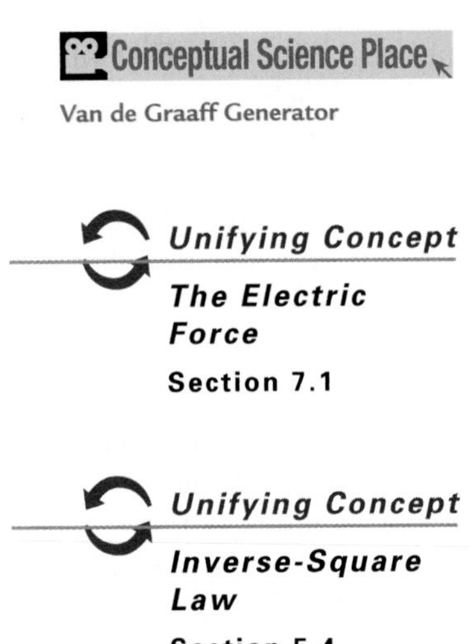

2. In accord with the inverse-square law, it decreases to $\frac{1}{9}$ its original value. To $\frac{1}{25}$ of its original value.
3. Positive.

Charge Polarization

If you charge an inflated balloon by rubbing it on your hair and then place the balloon against a wall, it sticks (Figure 7.5). This is because the charge on the balloon alters the charge distribution in the atoms or molecules in the wall, effectively inducing an opposite charge on the wall. The molecules cannot move from their relatively stationary positions, but their "centers of charge" are moved. The positive part of the atom or molecule is attracted toward the balloon while the negative part is repelled. This has the effect of distorting the atom or molecule (Figure 7.6). The atom or molecule is said to be *electrically polarized.* (We will treat electrical polarization further in Chapter 12.)

7.3 Electric Field

Electrical forces, like gravitational forces, can act between things that are not in contact with each other. How do bodies that are not touching exert forces on one another? To model "action-at-a-distance" forces such as gravity and the electric force, we use the concept of the *force field.* Every mass is surrounded by a gravitational field, while an electric field surrounds any charged object. Since you are more familiar with the gravitational force than the electric force at this point, first consider how the gravitational field works. The space surrounding any mass is altered such that another mass introduced into this region experiences a force. This "alteration in space" is called its *gravitational field.* We can think of any other mass as interacting with the field and not directly with the mass that produces it. For example, when an apple falls from a tree, we say that it is interacting with the mass of the Earth, but we can also think of an apple as responding to the gravitational field of the Earth.

Just as the space around a planet and every other mass is filled with a gravitational field, the space around every electric charge is filled with an **electric field**—an energetic aura that extends through space. An electric field is a vector quantity, having both magnitude and direction. The magnitude of the field at any point is simply the force per unit charge. If a charge q experiences a force F at some point in space, then the electric field at that point is $E = F/q$. The direction of the electric field is away from positive charge and toward negative charge.

If you place a charged particle in an electric field, it will experience a force. The direction of the force on a positive charge is the same direction as the field. The electric field can best be visualized with *field lines.* Field lines show the direction of the electric field—away from positive and toward negative charge. Field lines also show the intensity of the electric field—where the field lines are most tightly bunched together, the field is strongest. The electric field and field lines about an electron point toward the electron (Figure 7.7). The electric field and field lines about a proton point in the opposite direction—radially away from the proton. As with electric force, the electric field about a particle obeys the inverse-square law. Some electric field configurations are shown in Figure 7.8.

7.4 Electric Potential

In our study of energy in Chapter 4, we learned that an object has gravitational potential energy because of its location in a gravitational field. Similarly, a charged object has potential energy by virtue of its location in an electric field. Just as work is required to lift a massive object against the gravitational field of

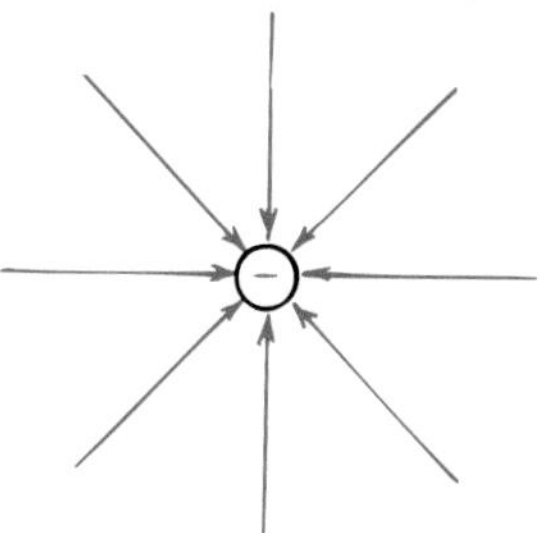

Figure 7.7 Electric-field representations about a negative charge.

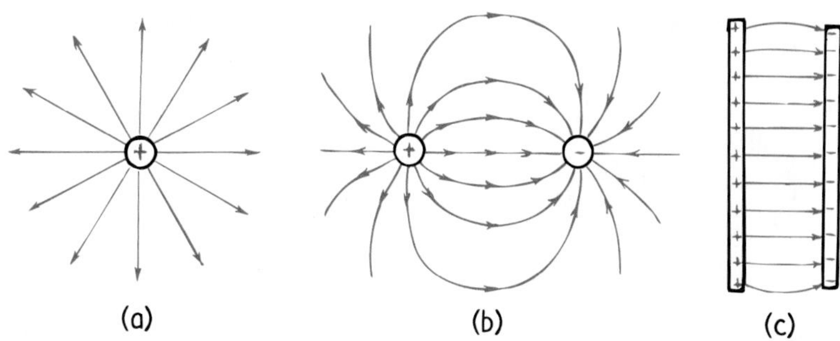

Figure 7.8 Some electric field configurations. (a) Field lines about a single positive charge. (b) Field lines for a pair of equal but opposite charges. (c) Field lines between two oppositely charged parallel plates.

the Earth, work is required to push a charged particle against the electric field of a charged body. This work changes the electric potential energy of the charged particle. Similarly, work done in compressing a spring increases the potential energy of the spring (Figure 7.9a). Likewise, the work done in pushing a positively charged particle closer to the positively charged sphere in Figure 7.9b increases the potential energy of the charged particle. We call this energy possessed by the charged particle that is due to its location **electric potential energy.** If the particle is released, it accelerates in a direction away from the sphere, and its electric potential energy changes to kinetic energy.

If we push a particle with twice the charge, we do twice as much work. Twice the charge in the same location has twice the potential energy; with three times the charge, there is three times as much potential energy; and so on. When working with electricity, rather than dealing with total potential energy of a charged body, it is convenient to consider the electric potential energy per charge. We simply divide the amount of energy by the amount of charge. The concept of electric energy per charge is called **electric potential**; that is,

$$\text{Electric potential} = \frac{\text{electric potential}}{\text{charge}}$$

The unit of measurement for electric potential is the volt, so electric potential is often called *voltage*. A potential of 1 volt (V) equals 1 joule (J) of energy per coulomb (C) of charge.

$$1 \text{ volt} = \frac{1 \text{ joule}}{\text{coulomb}}$$

Thus, a 1.5-volt battery gives 1.5 joules of energy to every 1 coulomb of charge flowing through the battery. Electric potential and voltage are the same thing, and they are commonly used interchangeably.

The significance of electric potential (voltage) is that a definite value for it can be assigned to a location. We can speak about the voltages at different locations in an electric field whether or not charges occupy those locations (Figure 7.10). The same is true of voltages at various locations in an electric circuit. Later in this chapter, we'll see that the location of a positive terminal of a 12-volt battery is maintained at a voltage 12 volts higher than the location of the negative terminal. When a conducting medium connects this voltage difference, any charges in the medium will move between these locations.

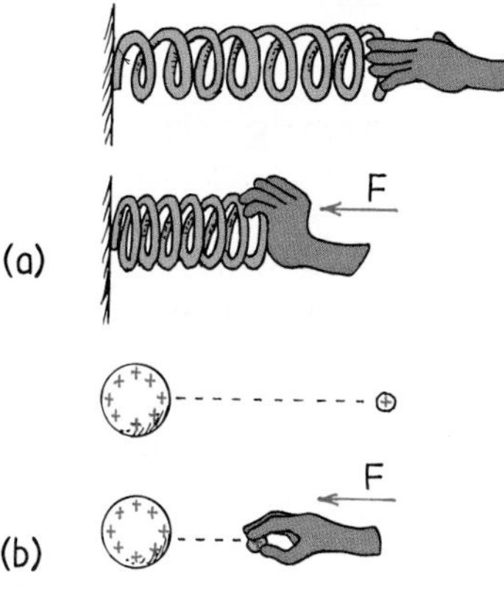

Figure 7.9 (a) The spring has more elastic PE when compressed. (b) The small charge similarly has more PE when pushed closer to the sphere of like charge. In both cases, the increased PE is the result of work input.

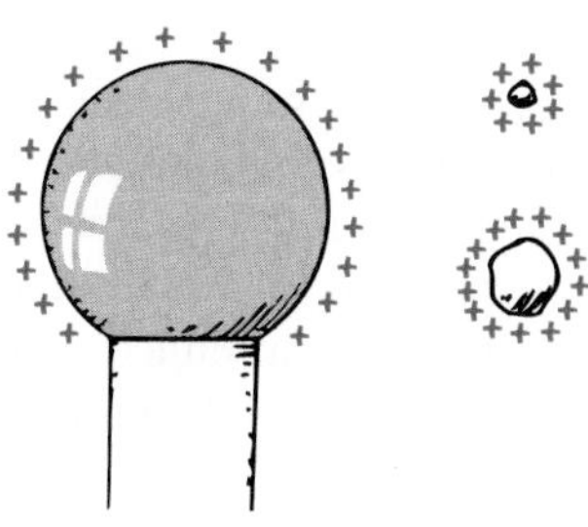

Figure 7.10 The larger test charge has more PE in the field of the charged dome, but the electric potential of any amount of charge at the same location is the same.

CHECK YOURSELF

1. Suppose there were twice as many coulombs in one of the test charges near the charged sphere in Figure 7.10. Would the electric potential energy of this

Figure 7.11 Although the voltage of the charged balloon is high, the electric potential energy is low because of the small amount of charge.

test charge relative to the charged sphere be the same, or would it be twice as great? Would the electric potential of the test charge be the same, or would it be twice as great?

2. What does it mean to say that your car has a 12-volt battery?

CHECK YOUR ANSWERS

1. The result of twice as many coulombs is twice as much electric potential energy because it takes twice as much work to put the charge there. But the electric potential would be the same. Twice the energy divided by twice the charge gives the same potential as one unit of energy divided by one unit of charge. Electric potential is not the same thing as electric potential energy. Be sure you understand this before you study any further.
2. It means that one of the battery terminals is 12 V higher in potential than the other one. Soon we'll see that it also means that, when a circuit is connected between these terminals, each coulomb of charge in the resulting current will be given 12 J of energy as it passes through the battery (and 12 J of energy is "spent" in the circuit).

Rub a balloon on your hair, and the balloon becomes negatively charged—perhaps to several thousand volts. That would be several thousand joules of energy, if the charge were 1 coulomb. However, 1 coulomb is a fairly respectable amount of charge. The charge on a balloon rubbed on hair is typically much less than a millionth of a coulomb. Therefore, the amount of energy associated with the charged balloon is very, very small (Figure 7.11).

INSIGHT High voltage at low energy is very similar to the harmless high-temperature sparks emitted by a Fourth of July sparkler. Temperature, the ratio of energy/molecule, means a lot of energy only if a lot of molecules are involved. High voltage means a lot of energy only if a lot of charge is involved.

7.5 Conductors and Insulators

Electrical **conductors** are materials that allow charged particles (usually electrons) to pass through them easily. Copper, silver, and other metals are good electrical conductors for the same reason they are good heat conductors: atoms of metals have one or more outer electrons that are loosely bound to their nuclei. These are called free electrons. It is these free electrons that conduct through a metallic conductor when an electric force is applied to it, making up a current. (A *current* is a flow of charged particles, usually electrons, as you will learn much more about in Section 7.7).

The electrons in other materials—rubber and glass, for example—are tightly bound and belong to particular atoms. Consequently, it isn't easy to make them flow. These materials are poor electrical conductors for the same reason they are generally poor heat conductors. Such a material is called a good **insulator.**

INSIGHT There are great differences in the electrical conductivity of conductors and insulators. For example, in a common appliance cord, electrons flow through several meters of metal wire rather than through a small fraction of a centimeter of vinyl or rubber insulation.

All substances can be arranged in order of their electrical conductivity. Those at the top of the list are conductors and those at the bottom are insulators. The ends of the list are very far apart. The conductivity of a metal, for example, can be more than a million trillion times greater than the conductivity of an insulator such as glass.

Some materials are neither good conductors nor good insulators: they are **semiconductors.** These materials fall into the midrange of conductivity since they possess few electrons that are free to move. However, the number of free electrons in a semiconductor can be adjusted by introducing small amounts of another element. This process is called *doping*. By doping a semiconductor, a scientist can create a conductor with a specific conductivity. Silicon and germanium are the most common semiconductors. These elements, once they have been doped, serve as the basic material for computer chips and miniaturized electronic components—transistors, for example. Transistors are essential components of computers and other electronic devices. A transistor can act as a conductor or insulator depending on the applied voltage, controlling current flow to different parts of a circuit.

INSIGHT Miniaturized semiconducting electronic components take up less space, and they are faster and require less energy to operate, than old-fashioned circuit components.

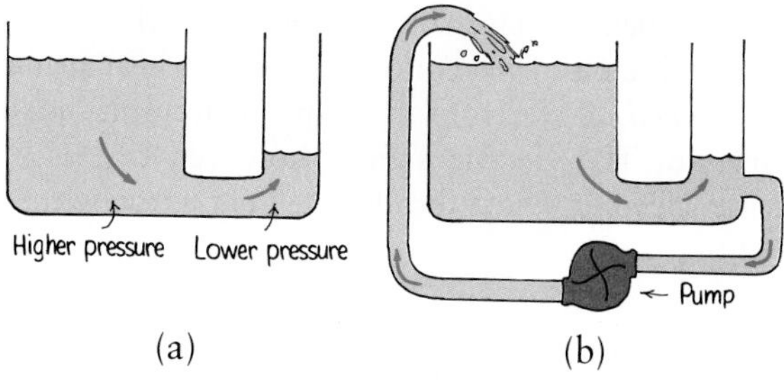

Figure 7.12 (a) Water flows from the reservoir of higher pressure to the reservoir of lower pressure. The flow will cease when the difference in pressure ceases. (b) Water continues to flow because a difference in pressure is maintained with the pump.

Figure 7.13 An unusual source of voltage. The electric potential between the head and tail of the electric eel (*Electrophorus electricus*) can be up to 650 V.

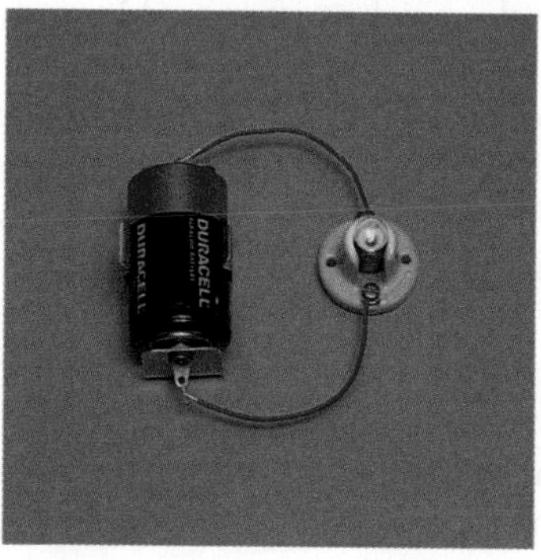

Figure 7.14 Each coulomb of charge that is made to flow in a circuit that connects the ends of this 1.5-V flashlight cell is energized with 1.5 J.

INSIGHT We often think of current flowing through a circuit; but we don't say this around somebody who is picky about grammar, because the expression "current flows" is redundant. More properly, charge flows—which is current.

7.6 Voltage Sources

When the ends of an electrical conductor are at different electric potentials—when there is a **potential difference**—charges in the conductor flow from the higher potential to the lower potential. The flow of charges persists until both ends reach the same potential. Without a potential difference, no flow of charge will occur.

To attain a sustained flow of charge in a conductor, some arrangement must be provided to maintain a difference in potential while charge flows from one end to another. The situation is analogous to the flow of water from a higher reservoir to a lower one (Figure 7.12a). Water will flow in a pipe that connects the reservoirs only as long as a difference in water level exists. The flow of water in the pipe, like the flow of charge in a wire, will cease when the pressures at each end are equal. A continuous flow is possible if the difference in water levels—hence the difference in water pressures—is maintained with the use of a suitable pump (Figure 7.12b).

A sustained electric current requires a suitable pumping device to maintain a difference in electric potential—to maintain a voltage. Chemical batteries or generators are "electrical pumps" that can maintain a steady flow of charge. These devices do work to pull negative charges apart from positive ones. In chemical batteries, this work is done by the chemical disintegration of lead or zinc in acid, and the energy stored in the chemical bonds is converted to electric potential energy. (See Chapter 13 for a discussion of fuel cells, highly efficient devices that convert chemical energy to electricity.) Generators separate charge by electromagnetic induction, a process described later in this chapter.

The work that is done (by whatever means) in separating the opposite charges is available at the terminals of the battery or generator. This energy per charge provides the difference in potential (voltage) that provides the "electrical pressure" to move electrons through a circuit joined to those terminals. A common automobile battery will provide an electrical pressure of 12 volts to a circuit connected across its terminals. Then 12 joules of energy are supplied to each coulomb of charge that is made to flow in the circuit. A simple battery that can be made with a lemon provides about 1 V. (Instructions for making a lemon battery are in "Active Explorations" at the end of the Chapter.)

7.7 Electric Current

Just as water current is a flow of H_2O molecules, electric current is a flow of charged particles. In circuits of metal wires, electrons make up the flow of charge. In metals, one or more electrons from each atom are free to move throughout the atomic lattice. These charge carriers are called *conduction electrons*. Protons, on the other hand, do not move because they are bound within nuclei of atoms that are more or less locked in fixed positions. In fluids, however, positive ions as well as electrons may compose the flow of an electric charge. This occurs inside a common automobile battery.

An important difference between water flow and electron flow has to do with their conductors. If you purchase a water pipe at a hardware store, the clerk doesn't sell you the water to flow through it. You provide that yourself. By contrast, when you buy an "electric pipe" (that is, an electric wire), you also get the electrons. Every bit of matter, wires included, contains enormous numbers of electrons that swarm about in random directions. When a source of voltage sets them moving, we have an electric circuit.

The rate of electrical flow is measured in amperes. An ampere is the rate of flow of 1 coulomb of charge per second. (That's a flow of 6.25 billion billion electrons per second.) In a wire that carries 4 amperes to a car headlight bulb, for example, 4 coulombs of charge flow past any cross section in the wire each second.

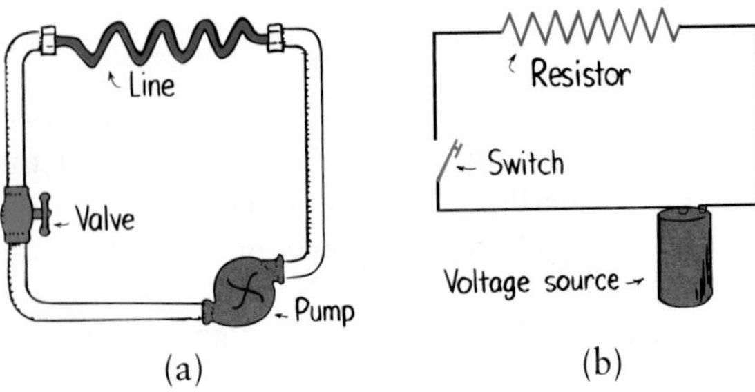

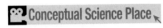

Figure 7.15 Analogy between a simple hydraulic circuit and an electrical circuit. (a) In a hydraulic circuit, a narrow pipe (green) offers resistance to water flow. (b) In an electric circuit, a lamp or other device (shown by the zigzag symbol for resistance) offers resistance to electron flow.

It is interesting to note that the speed of electrons as they drift through a wire is surprisingly slow. This is because electrons continually bump into atoms in the wire. The net speed, or *drift speed*, of electrons in a typical circuit is much less than one centimeter per second. The electric signal, however, travels at nearly the speed of light. That's the speed at which the electric *field* in the wire is established.

Conceptual Science Place

Alternating Current; Caution on Handling Wires

Direct Current and Alternating Current

Electric current may be DC or AC. By DC, we mean **direct current**, which refers to charges flowing in one direction. A battery produces direct current in a circuit because the terminals of the battery always have the same sign. Electrons move from the repelling negative terminal toward the attracting positive terminal, and they always move through the same circuit in the same direction.

Alternating current (**AC**) acts as the name implies. Electrons in the circuit are moved first in one direction and then in the opposite direction, alternating to and fro about relatively fixed positions (Figure 7.16). This is accomplished in a generator or alternator by periodically switching the sign at the terminals. Nearly all commercial AC circuits involve currents that alternate back and forth at a frequency of 60 cycles per second. This is 60-hertz current. (One cycle per second is called a hertz.) In some countries, 25-hertz, 30-hertz, or 50-hertz current is used.

INSIGHT In AC circuits, 120 volts is what is called the "root-mean-square" average of the voltage. The actual voltage in a 120-volt AC circuit varies between +170 volts and –170 volts, delivering the same power to an iron or toaster as a 120-volt DC circuit.

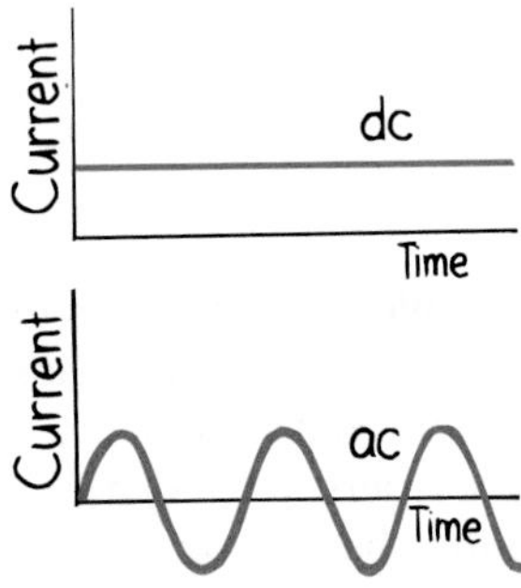

Figure 7.16 Time graphs of DC and AC.

7.8 Electrical Resistance

How much current is in a circuit depends not only on voltage but also on the **electrical resistance** of the circuit. Just as narrow pipes resist water flow more than wide pipes, thin wires resist electrical current more than thicker wires. And length contributes to resistance also. Just as long pipes have more resistance than short ones, long wires offer more electrical resistance. Most important is the material from which the wires were made. Copper has a low electrical resistance, while a strip of rubber has an enormous resistance. Temperature also affects electrical resistance. The greater the jostling of atoms within a conductor (the higher the temperature), the greater its resistance. The resistance of some materials reaches zero at very low temperatures. These materials are referred to as *superconductors*.

Electrical resistance is measured in units called *ohms*. The Greek letter *omega*, Ω, is commonly used as a symbol for the ohm. The unit was named after Georg Simon Ohm, a German physicist who, in 1826, discovered a simple and very important relationship among voltage, current, and resistance.

HISTORY OF SCIENCE

History of 110 Volts

In the early days of electrical lighting, high voltages burned out electric filaments, so low voltages were more practical. The hundreds of power plants built in the United States prior to 1990 adopted 110 volts (or 115 volts, or 120 volts) as their standard. The tradition of 110 volts was decided upon because it made bulbs of the day glow as brightly as a gas lamp. By the time electrical lighting became popular in Europe, engineers had figured out how to make light bulbs that would not burn out so fast at higher voltages. Power transmission is more efficient at higher voltages, so Europe adopted 220 volts as their standard. The U.S. remained with 110 volts (today, it is officially 120 volts) because of the initial huge expense in the installation of 110-volt equipment.

MATH CONNECTION

Ohm's Law

Rearranging Ohm's law shows that $I = \frac{V}{R}$, can be re-expressed $R = \frac{V}{I}$. If any two variables are known, the third can be found.

Problems

1. How much current flows through a lamp with a resistance of 60 Ω when the voltage across the lamp is 12 V?
2. What is the resistance of a toaster that draws a current of 12 A when connected to a 120-V circuit?

Solutions

1. From Ohm's law: $\text{Current} = \frac{\text{voltage}}{\text{resistance}} = \frac{12\text{ V}}{12\ \Omega} = 0.2\text{ A}$.
2. Rearranging Ohm's law:

$$\text{Resistance} = \frac{\text{voltage}}{\text{current}} = \frac{120\text{ V}}{12\text{ A}} = 10\Omega.$$

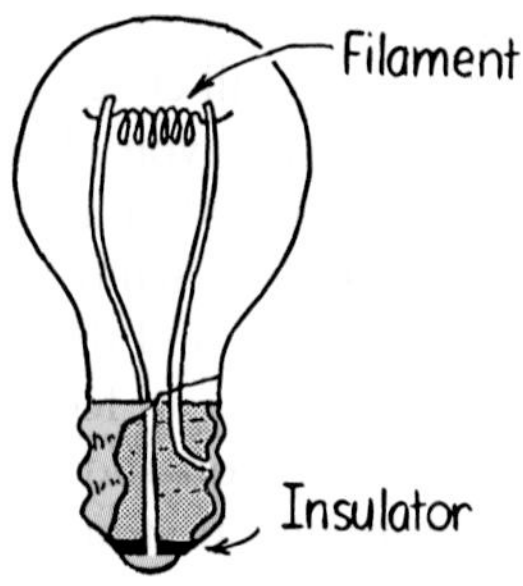

Figure 7.17 The conduction electrons that surge to and fro in the filament of the lamp do not come from the voltage source. They are in the filament to begin with. The voltage source simply provides them with surges of energy. When switched on, the resistance of the very thin tungsten filament heats up to 3000°C and roughly doubles its resistance.

7.9 Ohm's Law

The relationship between voltage, current, and resistance is summarized by a statement known as **Ohm's Law.** Ohm discovered that the amount of current in a circuit is directly proportional to the voltage established across the circuit and is inversely proportional to the resistance of the circuit:

$$\text{Current} = \frac{\text{voltage}}{\text{resistance}}$$

Or, in the form of units,

$$\text{Amperes} = \frac{\text{volts}}{\text{ohms}}$$

And, in symbolic form [since V stands for voltage (in volts), I for current (in amperes), and R for resistance (in ohms)], we express Ohm's law as

$$I = \frac{V}{R}.$$

Ohm's Law

So, for a given circuit of constant electrical resistance, current and voltage are proportional to each other. This means that we'll get twice the current for twice the voltage. The greater the voltage, the greater the current. But, if the resistance is doubled for a circuit, the current will be half what it would have been otherwise. The greater the resistance, the smaller the current. Ohm's law makes good sense.

The resistance of a typical lamp cord is much less than 1 ohm, and a typical light bulb has a resistance of more than 100 ohms. An iron or an electric toaster has a resistance of 15 to 20 ohms. The current inside these and all other electrical devices is regulated by circuit elements called *resistors* (Figure 7.18), whose resistance may be a few ohms or millions of ohms.

Current is a flow of charge, pressured into motion by voltage and hampered by resistance.

INTEGRATED SCIENCE BIOLOGY

Electric Shock

The damaging effects of electric shock are the result of current passing through the human body. But what causes electric shock in the body—current or voltage? From Ohm's law, we can see that this current depends on the voltage that is

CHECK YOURSELF

1. An electric razor completes 60 cycles every second. (a) What is its frequency? (b) What is its period?
2. Gusts of wind cause the Sears Building in Chicago to sway back and forth, completing a cycle every ten seconds. (a) What is its frequency? (b) What is its period?

CHECK YOUR ANSWERS

1. (a) 60 cycles per second, or 60 Hz. (b) $\frac{1}{60}$ second.
2. (a) $\frac{1}{10}$ Hz. (b) 10 seconds.

8.2 Wave Motion

INSIGHT A wave transfers energy without transferring matter. If matter were to move along with the energy in a wave, oceans would be emptied as ocean waves travel to the shore.

If you drop a stone into a calm pond, waves will travel outward in expanding circles. Energy is carried by the wave, traveling from one place to another. The water itself goes nowhere. This can be seen by waves encountering a floating leaf. The leaf bobs up and down, but it doesn't travel with the waves. The waves move along, not the water. The same is true of waves of wind over a field of tall grass on a gusty day. Waves travel across the grass, while the individual grass plants remain in place; instead, they swing to and fro between definite limits, but they go nowhere. When you speak, wave motion through the air travels across the room at about 340 meters per second. The air itself doesn't travel across the room at this speed. In these examples, when the wave motion ceases, the water, the grass, and the air return to their initial positions. It is characteristic of wave motion that the medium transporting the wave returns to its initial condition after the disturbance has passed. Putting all the information about waves together, we can now specifically define what a wave is: A **wave** is a disturbance that travels from one place to another transporting energy, but not necessarily matter, along with it.

Figure 8.2 Water waves.

Wave Speed

The speed of periodic wave motion is related to the frequency and wavelength of the waves. Consider the simple case of water waves (Figures 8.2 and 8.3). Imagine that we fix our eyes on a stationary point on the water's surface and observe the waves passing by that point. We can measure how much time passes between the arrival of one crest and the arrival of the next one (the period), and we can also observe the distance between crests (the wavelength). We know that speed is defined as distance divided by time. In this case, the distance is one wavelength and the time is one period, so wave speed = wavelength/period.

For example, if the wavelength is 10 meters and the time between crests at a point on the surface is 0.5 second, the wave is traveling 10 meters in 0.5 seconds and its speed is 10 meters divided by 0.5 seconds, or 20 meters per second. Since period is equal to the inverse of frequency, the formula wave speed = wavelength/period can also be written

$$\text{wave speed} = \text{wavelength} \times \text{frequency}$$

This relationship applies to all kinds of waves, whether they are water waves, sound waves, or light waves.

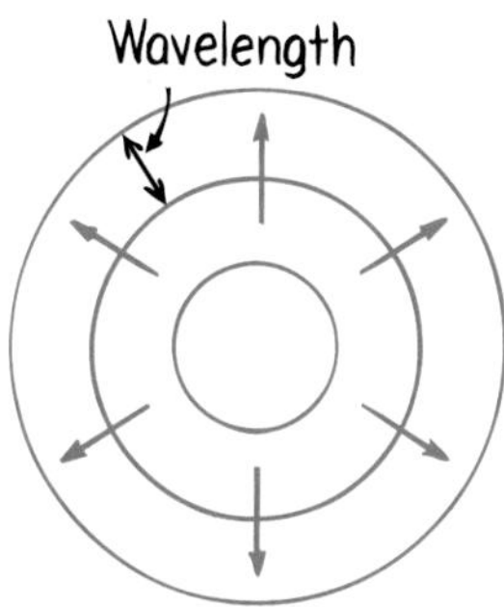

Figure 8.3 A top view of water waves.

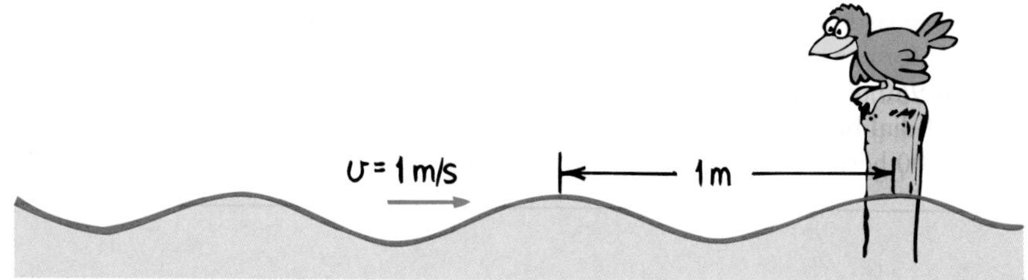

Figure 8.4 Interactive Figure
If the wavelength is 1 m, and one wavelength per second passes the pole, then the speed of the wave is 1 m/s.

MATH CONNECTION

Ohm's Law

Rearranging Ohm's law shows that $I = \frac{V}{R}$, can be re-expressed $R = \frac{V}{I}$.

If any two variables are known, the third can be found.

Problems

1. How much current flows through a lamp with a resistance of 60 Ω when the voltage across the lamp is 12 V?
2. What is the resistance of a toaster that draws a current of 12 A when connected to a 120-V circuit?

Solutions

1. From Ohm's law: $\text{Current} = \frac{\text{voltage}}{\text{resistance}} = \frac{12\text{ V}}{12\ \Omega} = 0.2\text{ A}$.
2. Rearranging Ohm's law:

$$\text{Resistance} = \frac{\text{voltage}}{\text{current}} = \frac{120\text{ V}}{12\text{ A}} = 10\Omega.$$

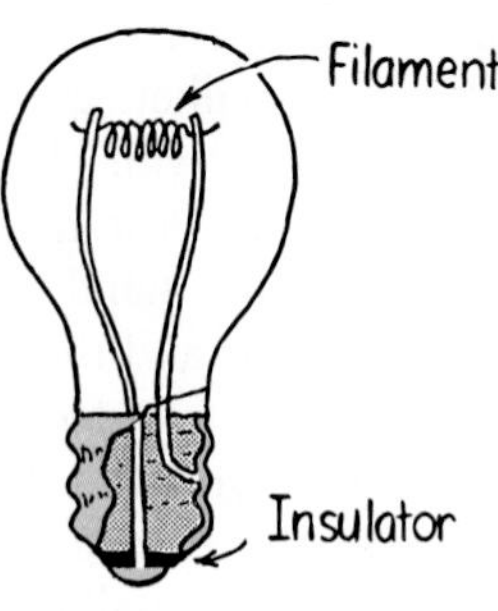

Figure 7.17 The conduction electrons that surge to and fro in the filament of the lamp do not come from the voltage source. They are in the filament to begin with. The voltage source simply provides them with surges of energy. When switched on, the resistance of the very thin tungsten filament heats up to 3000°C and roughly doubles its resistance.

Ohm's Law

Current is a flow of charge, pressured into motion by voltage and hampered by resistance.

7.9 Ohm's Law

The relationship between voltage, current, and resistance is summarized by a statement known as **Ohm's Law.** Ohm discovered that the amount of current in a circuit is directly proportional to the voltage established across the circuit and is inversely proportional to the resistance of the circuit:

$$\text{Current} = \frac{\text{voltage}}{\text{resistance}}$$

Or, in the form of units,

$$\text{Amperes} = \frac{\text{volts}}{\text{ohms}}$$

And, in symbolic form [since V stands for voltage (in volts), I for current (in amperes), and R for resistance (in ohms)], we express Ohm's law as

$$I = \frac{V}{R}.$$

So, for a given circuit of constant electrical resistance, current and voltage are proportional to each other. This means that we'll get twice the current for twice the voltage. The greater the voltage, the greater the current. But, if the resistance is doubled for a circuit, the current will be half what it would have been otherwise. The greater the resistance, the smaller the current. Ohm's law makes good sense.

The resistance of a typical lamp cord is much less than 1 ohm, and a typical light bulb has a resistance of more than 100 ohms. An iron or an electric toaster has a resistance of 15 to 20 ohms. The current inside these and all other electrical devices is regulated by circuit elements called *resistors* (Figure 7.18), whose resistance may be a few ohms or millions of ohms.

INTEGRATED SCIENCE BIOLOGY

Electric Shock

The damaging effects of electric shock are the result of current passing through the human body. But what causes electric shock in the body—current or voltage? From Ohm's law, we can see that this current depends on the voltage that is

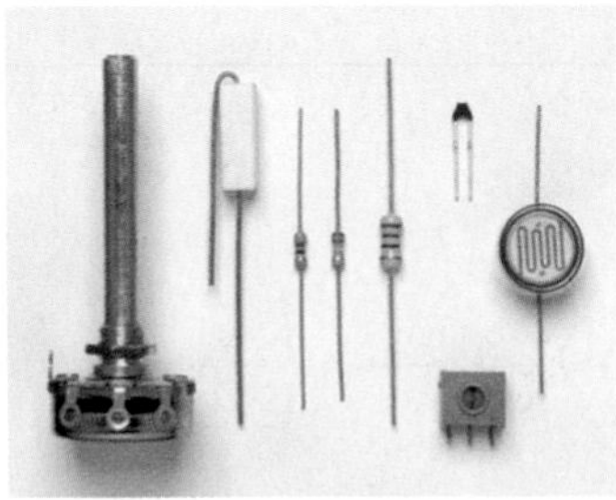

Figure 7.18 Resistors. The graphic symbol for a resistor in an electric circuit is -\/\/\/-.

applied and also on the electrical resistance of the human body. The resistance of one's body depends on its condition, and it ranges from about 100 ohms, if it is soaked with salt water, to about 500,000 ohms if the skin is very dry. If we touch the two electrodes of a battery with dry fingers, completing the circuit from one hand to the other, we offer a resistance of about 100,000 ohms. We usually cannot feel 12 volts, and 24 volts just barely tingles. If our skin is moist, 24 volts can be quite uncomfortable. Table 7.1 describes the effects that different amounts of current have on the human body.

To receive a shock, there must be a difference in electric potential between one part of your body and another part. Most of the current will pass along the path of least electrical resistance connecting these two points. Suppose you fell from a bridge and managed to grab onto a high-voltage power line, halting your fall. So long as you touch nothing of different potential, you will receive no shock at all. Even if the wire is a few thousand volts above ground potential and you hang by it with two hands, no appreciable charge will flow from one hand to the other. This is because there is no appreciable difference in potential between your two hands. If, however, you reach over with one hand and grab onto a wire of different potential—zap! We have all seen birds perched on high-voltage wires. Every part of their bodies is at the same potential as the wire, so they feel no ill effects (Figure 7.19).

Figure 7.19 The bird can stand harmlessly on one wire of high potential, but it had better not reach over and touch a neighboring wire. Why not?

Many people are killed each year from common 120-volt electric circuits. If your hand touches a faulty 120-volt light fixture while your feet are on the ground, there's likely a 120-volt "electrical pressure" between your hand and the ground. Resistance to current is usually greatest between your feet and the ground, and so the current is usually not enough to do serious harm. But if your feet and the ground are wet, there is a low-resistance electrical path between you and the ground. Pure water is not a good conductor. But the ions that normally are found in water make it a fair conductor. Dissolved materials in water, especially small quantities of salt, lower the resistance even more. There is usually a layer of salt remaining on your skin from perspiration, which, when wet, lowers your skin resistance to a few hundred ohms or less. Handling electrical devices when wet is a definite no-no.

Injury by electric shock occurs in three forms: (1) burning of tissues by heating, (2) contraction of muscles, and (3) disruption of cardiac rhythm. These conditions are caused by the delivery of excessive power for too long a time in critical regions of the body.

Electric shock can upset the nerve center that controls breathing. In rescuing shock victims, the first thing to do is remove them from the source of the electricity. Use a dry wooden stick or some other nonconductor so that you don't get electrocuted yourself. Then apply artificial respiration. It is important to continue artificial respiration. There have been cases of victims of lightning who did not breathe without assistance for several hours, but who were eventually revived and who completely regained good health.

CHECK YOURSELF

1. What causes electric shock—current or voltage?
2. At 100,000 Ω, how much current will flow through your body if you touch the terminals of a 12-V battery?

Table 7.1 Effect of Electric Currents on the Body

Current (A)	Effect
0.001	Can be felt
0.005	Is painful
0.010	Causes involuntary muscle contractions (spasms)
0.15	Causes loss of muscle control
0.70	Goes through the heart; causes serious disruption; probably fatal if current lasts for more than 1 s

3. If your skin is very moist, so that your resistance is only 1000 Ω, and you touch the terminals of a 12-V battery, how much current do you receive?

CHECK YOUR ANSWERS

1. Electric current occurs when current is produced in the body, but the current is *caused* by impressed voltage.
2. $\text{Current} = \dfrac{\text{Voltage}}{\text{resistance}} = \dfrac{12\ \text{V}}{100{,}000\ \Omega} = 0.00012\ \text{A}$
3. $\text{Current} = \dfrac{\text{Voltage}}{\text{resistance}} = \dfrac{12\ \text{V}}{1000\ \Omega} = 0.012\ \text{A}$. Ouch!

Most electric plugs and sockets today are wired with three, instead of two, connections. The principal two flat prongs on an electrical plug are for the current-carrying double wire, one part "live" and the other neutral, while the third round prong is grounded—connected directly to the Earth (Figure 7.20). Appliances such as stoves, washing machines, and dryers are connected with these three wires. If the live wire accidentally comes into contact with the metal surface of the appliance and you touch the surface of the appliance, you could receive a dangerous shock. This won't occur when the appliance casing is grounded via the ground wire, which assures that the appliance casing is at zero ground potential. A lamp has an insulating body and doesn't need the third (ground) wire.

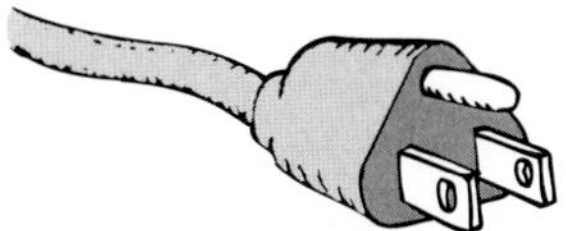

Figure 7.20 The third prong connects the body of the appliance directly to ground. Any charge that builds up on an appliance is therefore conducted to the ground.

7.10 Electric Circuits

Any path along which electrons can flow is a circuit. For a continuous flow of electrons, there must be a complete circuit with no gaps. A gap is usually provided by an electric switch that can be opened or closed either to cut off energy or to allow energy to flow. Most circuits have more than one device that receives electric energy. These devices are commonly connected to a circuit in one of two ways, in *series* or in *parallel.*

A simple series circuit is shown in Figure 7.21. Three lamps are connected in series with a battery. The same current exists almost immediately in all three lamps when the switch is closed. The current does not "pile up" or accumulate in any lamp but flows through each lamp. Electrons that make up this current leave the negative terminal of the battery, pass through each of the resistive filaments in the lamps in turn, and then return to the positive terminal of the battery. (The same amount of current passes through the battery.) This is the only path of the electrons through the circuit. A break anywhere in the path results in an open circuit, and the flow of electrons ceases. Such a break occurs when the switch is opened, when the wire is accidentally cut, or when one of the lamp filaments burns out.

It is easy to see the main disadvantage of the series circuit; if one device fails, current in the entire circuit ceases. Some cheap Christmas-tree lights are connected in series. When one bulb burns out, it's fun and games (or frustration) trying to locate which one to replace.

Most circuits are wired so that it is possible to operate several electrical devices, each independently of the other. In your home, for example, a lamp can be turned on or off without affecting the operation of other lamps or electrical devices. This is because these devices are connected not in series but in parallel with one another. A simple parallel circuit is shown in Figure 7.22. Three lamps are connected to the same two points, A and B. Electrical devices connected to the same two points of an electrical circuit are said to be *connected in parallel.*

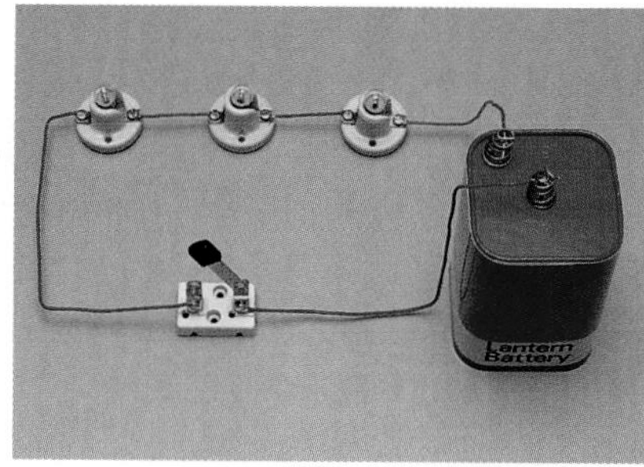

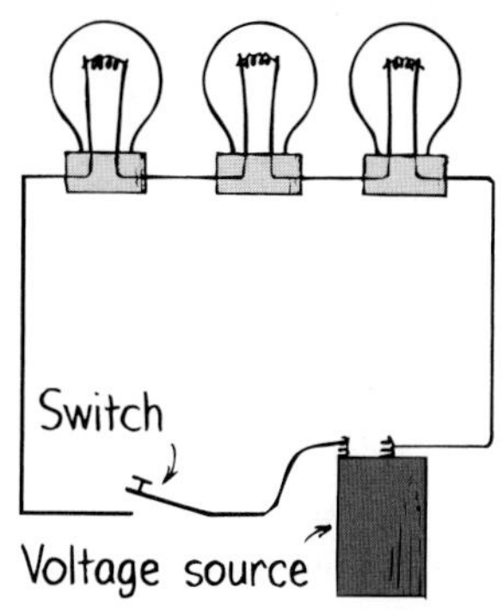

Figure 7.21 Interactive Figure

A simple series circuit. The 6-V battery provides 2 V across each lamp.

Electric Circuits

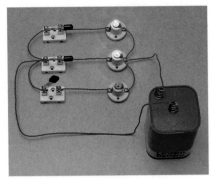

Electrons leaving the negative terminal of the battery need travel only through one lamp filament before returning to the positive terminal of the battery. In this case, current branches into three separate pathways from A to B. A break in any one path does not interrupt the flow of charge in the other paths. Each device operates independently of the other devices.

CHECK YOURSELF

1. In a circuit consisting of two lamps connected in series, if the current in one lamp is 1 A, what is the current in the other lamp?
2. In a circuit consisting of two lamps connected in parallel, if there is 6 V across one lamp, what is the voltage across the other lamp?

CHECK YOUR ANSWERS

1. 1 A; current does not "pile up" anywhere in the circuit.
2. 6 V; the voltage in all branches of a parallel circuit are the same.

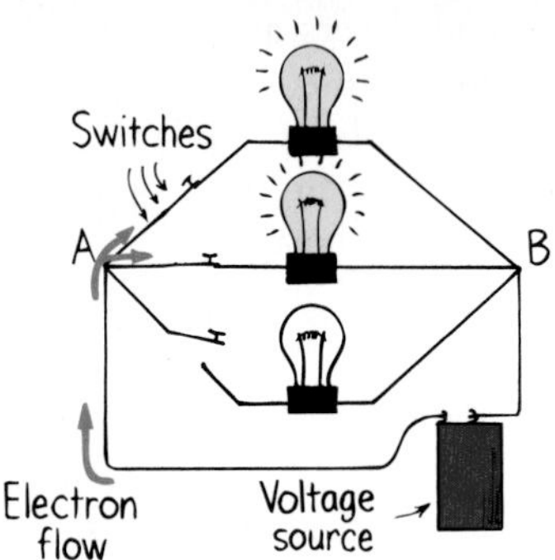

Figure 7.22 Interactive Figure
A simple parallel circuit. A 6-V battery provides 6 V across each lamp.

7.11 Electric Power

The moving charges in an electric current do work. This work, for example, can heat a circuit or turn a motor. The rate at which this work is done—that is, the rate at which electric energy is converted to another form, such as mechanical energy, heat, or light—is called **electric power**. Electric power is equal to the product of current and voltage.*

$$\text{Power} = \text{current} \times \text{voltage} = IV$$

If a lamp rated at 120 volts operates on a 120-volt line, you can figure that it will draw a current of 1 ampere (120 watts = 1 ampere × 120 volts). A 60-watt lamp draws $^1/_2$ ampere on a 120-volt line. This relationship becomes a practical matter when you wish to know the cost of electrical energy, which is usually a small fraction of a dollar per kilowatt-hour, depending on the locality.

MATH CONNECTION

Solving Power Problems

You will find that there are many practical, everyday questions related to electricity that you can answer if you know the relationship between power, voltage, and current. Here are a couple examples.

Problems

1. If a 120-V line to a socket is limited to 15 A by a safety fuse, will it operate a 1200-W hair dryer?
2. At 30¢/kWh, show that it costs 36¢ to operate the 1200-W hair dryer for 1 h.

Solutions

1. Yes. From the expression watts = amperes × volts, we can see that current $= \frac{1200 \text{ W}}{120 \text{ V}} = 10$ A, so the hair dryer will operate when connected to the circuit. But two hair dryers on the same circuit will blow the fuse.
2. 1200 W = 1.2 kW; 1.2 kW × 1 h × 30¢/kWh = 36¢.

*Recall from Chapter 4 that Power $= \frac{\text{work}}{\text{time}}$; 1 Watt = 1 J/s. Note that the units for mechanical power and electrical power agree (work and energy are both measured in joules):

$$\text{Power} = \frac{\text{charge}}{\text{time}} \times \frac{\text{energy}}{\text{charge}} = \frac{\text{energy}}{\text{time}}$$

If the voltage is expressed in volts and the current is expressed in amperes, then the power is expressed in watts. So, in units form,

$$\text{Watts} = \text{amperes} \times \text{volts}$$

Using Ohm's Law to substitute IR for V, we have an alternate statement for power: Power = I^2R.

Figure 7.23 The power-and-voltage designation on the light bulb reads "100 W 120 V." How many amperes will flow through the bulb?

A kilowatt is 1,000 watts, and a kilowatt-hour represents the amount of energy consumed in 1 hour at the rate of 1 kilowatt.* Therefore, in a locality where electric energy costs 25 cents per kilowatt-hour, a 100-watt electric light bulb can operate for 10 hours at a cost of 25 cents, or a half nickel for each hour (Figure 7.23).

7.12 Magnetic Force

Anyone who has played around with magnets knows that magnets exert forces on one another. Magnetic forces are similar to electrical forces, in that magnets can both attract and repel without touching (depending on which ends of the magnets are held near one another), and the strength of their interaction depends on the distance between them. Whereas electric charges produce electrical forces, regions called *magnetic poles* give rise to magnetic forces.

If you suspend a bar magnet at its center by a piece of string, you've got a compass. One end, called the north-seeking pole, points northward. The opposite end, called the south-seeking pole, points southward. More simply, these are called the *north* and *south* poles. All magnets have both a north and a south pole (some have more than one of each). Refrigerator magnets have narrow strips of alternating north and south poles. These magnets are strong enough to hold sheets of paper against a refrigerator door, but they have a very short range because the north and south poles cancel a short distance from the magnet. In a simple bar magnet, the magnetic poles are located at the two ends. A common horseshoe magnet is a bar magnet bend into a U shape. Its poles are also located at its two ends (Figure 7.24).

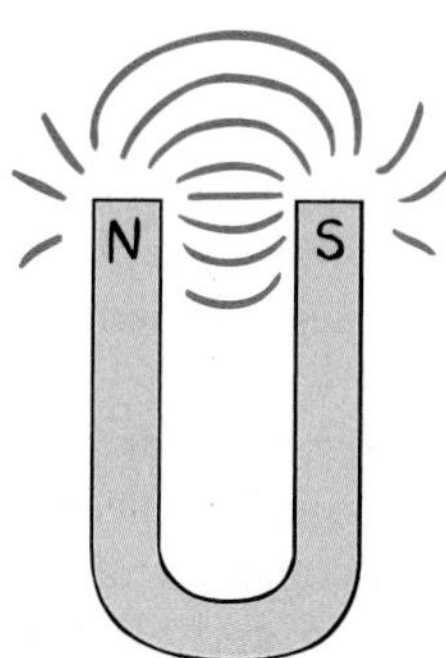

Figure 7.24 A horseshoe magnet.

When the north pole of one magnet is brought near the north pole of another magnet, they repel each other. The same is true of a south pole near a south pole. If opposite poles are brought together, however, attraction occurs. We find the following rule:

Like poles repel; opposite poles attract.

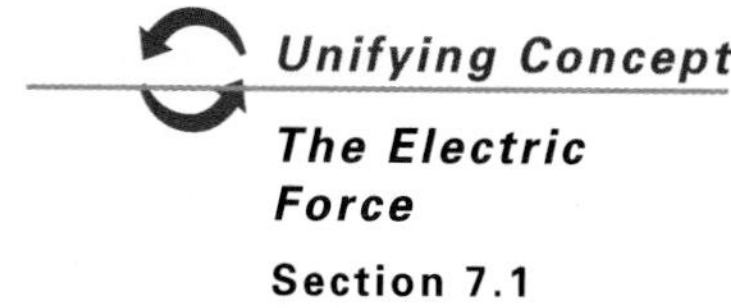

This rule is similar to the rule for the forces between electric charges, where like charges repel one another and unlike charges attract. But there is a very important difference between magnetic poles and electric charges. Whereas electric charges can be isolated, magnetic poles cannot. Electrons and protons are entities by themselves. But the north and south poles of a magnet are like the head and tail of the same coin.

If you break a bar magnet in half, each half still behaves as a complete magnet. Break the pieces in half again, and you have four complete magnets. You can continue breaking the pieces in half and never isolate a single pole. Even if your pieces were one atom thick, there would still be two poles on each piece, which suggests that the atoms themselves are magnets (Figure 7.25).

Figure 7.25 If you break a magnet in half, you will have two magnets. If you break these two magnets in half, you will have four magnets, each with a north and a south pole. If you continue breaking the pieces further and further, you will find that you always get the same results. Magnetic poles exist in pairs.

7.13 Magnetic Fields

You have learned that the space around a planet and every other mass is filled with a gravitational field, and that the space around every electric charge is filled with an electric field. Similarly, the space around a magnet contains a **magnetic field**—an energetic aura that extends through space. If you sprinkle some iron filings on a sheet of paper placed on a magnet, you'll see that the filings move in response to the magnetic field and trace out an orderly pattern of lines that surround the magnet. The shape of the field is revealed by magnetic field lines that spread out from one pole and return to the other pole. It is interesting to

*Since power = energy/time, simple rearrangement gives energy = power × time; thus energy can be expressed in the unit *kilowatt-hours* (kWh).

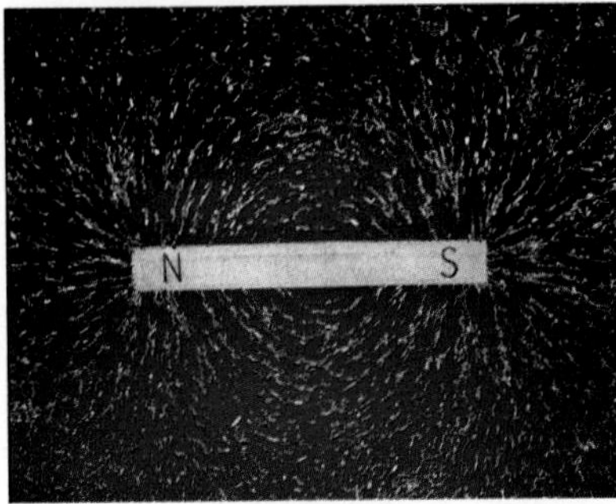

Figure 7.26 Interactive Figure
Top view of iron filings sprinkled on a sheet of paper on top of a magnet. The filings trace out a pattern of magnetic field lines in the surrounding space. Interestingly enough, the magnetic field lines continue inside the magnet (not revealed by the filings) and form closed loops.

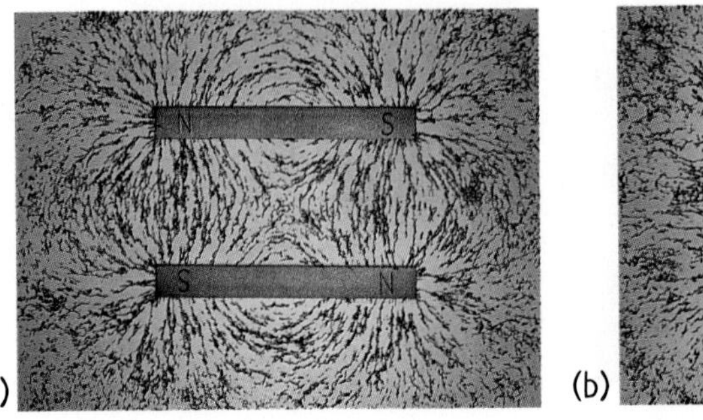

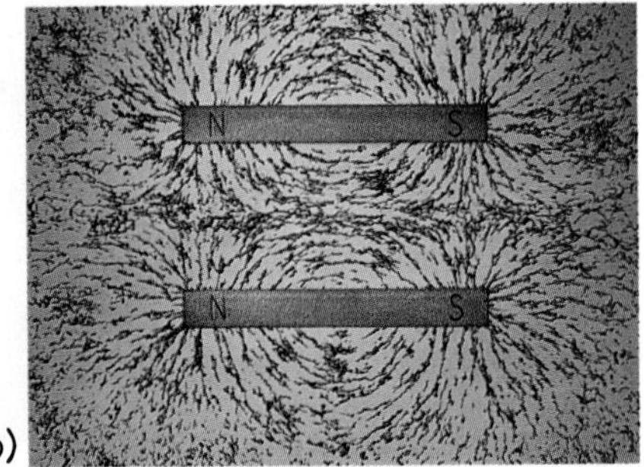

Figure 7.27 The magnetic field patterns for a pair of magnets. (a) Opposite poles are nearest each other, and (b) like poles are nearest each other.

compare the field patterns in Figures 7.26 and 7.27 with the electric field patterns in 7.8b.

The direction of the field outside the magnet is from the north pole to the south pole. Where the lines are closer together, the field is stronger. We can see that the magnetic field strength is greater at the poles. If we place another magnet or a small compass anywhere in the field, its poles will tend to align with the magnetic field (Figure 7.28).

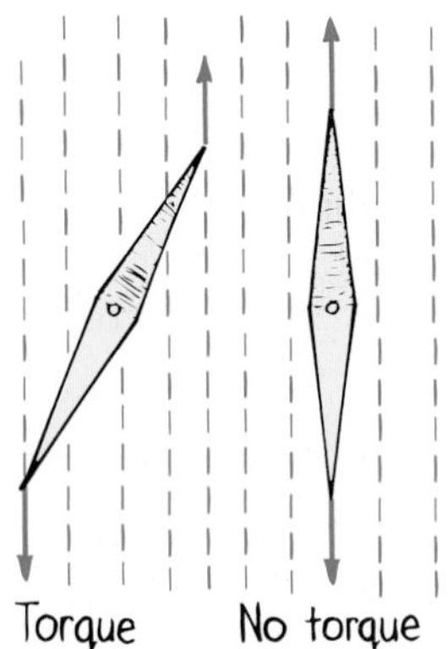

Figure 7.28 When the compass needle is not aligned with the magnetic field, the oppositely directed forces produce a pair of *torques* (a torque is a turning force) that twist the needle into alignment.

A magnetic field is produced by moving electric charges. Where, then, is this motion in a common bar magnet? The answer is, in the electrons of the atoms that make up the magnet. These electrons are in constant motion. Two kinds of electron motion produce magnetism: electron spin and electron revolution. In most common magnets, electron spin is the main contributor to magnetism.

Every spinning electron is a tiny magnet. A pair of electrons spinning in the same direction creates a stronger magnet. A pair of electrons spinning in opposite directions, however, work against each other, and the magnetic fields cancel. This is why most substances are not magnets. In most atoms, the various fields cancel one another because the electrons spin in opposite directions. In such materials as iron, nickel, and cobalt, however, the fields do not cancel entirely. Each iron atom has four electrons whose spin magnetism is uncanceled. Each iron atom, then, is a tiny magnet. The same is true, to a lesser extent, of nickel and cobalt atoms. Most common magnets are therefore made from alloys containing iron, nickel, cobalt, and aluminum in various proportions.

Magnetic Domains

The magnetic field of an individual iron atom is so strong that interactions among adjacent atoms cause large clusters of them to line up with one another. These clusters of aligned atoms are called **magnetic domains**. Each domain is perfectly magnetized and is made up of billions of aligned atoms. The domains are microscopic (Figure 7.29), and there are many of them in a crystal of iron.

Not every piece of iron is a magnet, because the domains in ordinary iron are not aligned. In a common nail, for example, the domains are randomly oriented. But, when you bring a magnet nearby, they can be induced into alignment. When you remove the nail from the magnet, ordinary thermal motion causes most or all of the domains in the nail to return to a random arrangement.

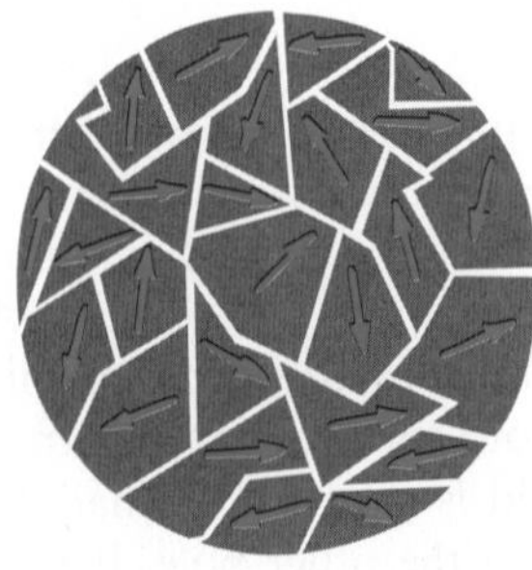

Figure 7.29 A microscopic view of magnetic domains in a crystal of iron. Each domain consists of billions of aligned atoms.

Electric Currents and Magnetic Fields

A moving electric charge produces a magnetic field. A current of charges, then, also produces a magnetic field. The magnetic field that surrounds a current-carrying wire can be demonstrated by arranging an assortment of compasses around the

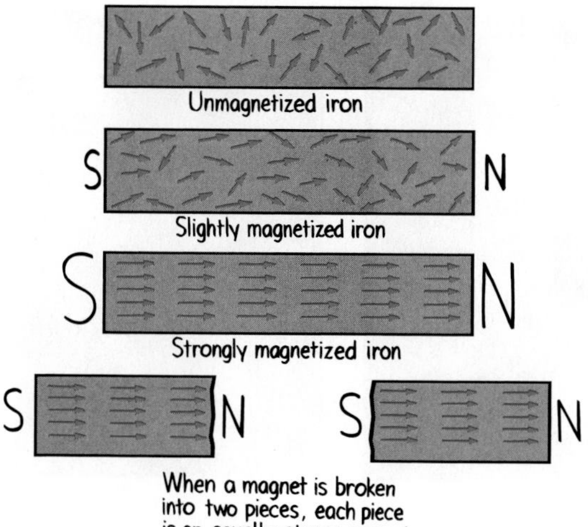

Figure 7.30 Pieces of iron in successive stages of magnetization. The arrows represent domains; the head is a north pole and the tail is a south pole. Poles of neighboring domains neutralize each other's effects, except at the ends.

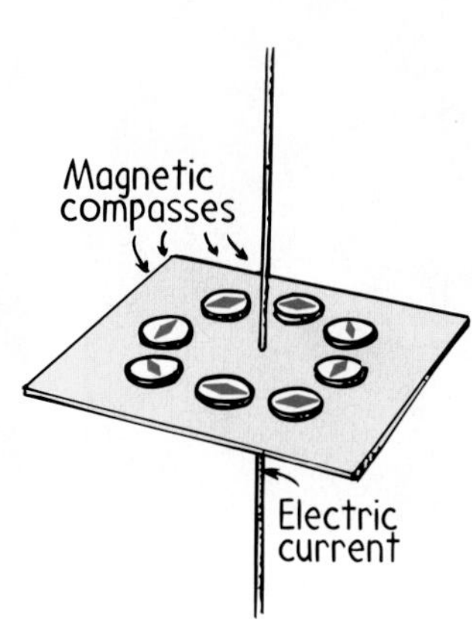

Figure 7.31 The compasses show the circular shape of the magnetic field surrounding the current-carrying wire.

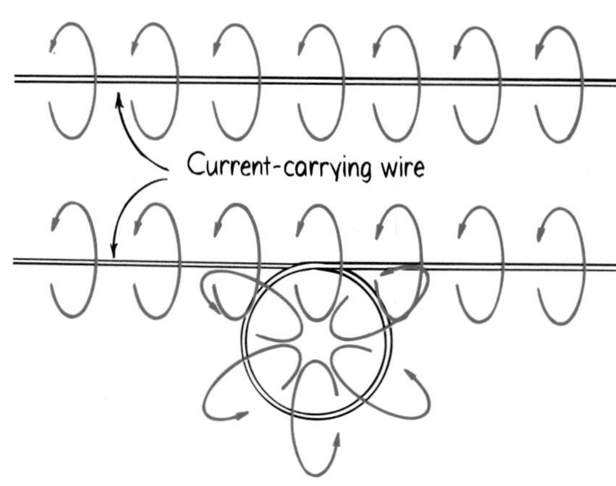

Figure 7.32 Magnetic field lines about a current-carrying wire become bunched up when the wire is bent into a loop.

INSIGHT A magstripe on a credit card contains millions of tiny magnetic domains held together by a resin binder. Data is encoded in binary code, with zeroes and ones distinguished by the frequency of domain reversals. It is quite amazing how quickly your name pops up when an airline reservationist swipes your card.

wire (Figure 7.31). The magnetic field about the current-carrying wire makes up a pattern of concentric circles. When the current reverses direction, the compass needle turns around, showing that the direction of the magnetic field changes also.

If the wire is bent into a loop, the magnetic field lines become bunched up inside the loop (Figure 7.32). If the wire is bent into another loop that overlaps the first, the concentration of magnetic field lines inside the loops is doubled. It follows that the magnetic field intensity in this region is increased as the number of loops is increased. The magnetic field intensity is appreciable for a current-carrying coil that has many loops.

If a piece of iron is placed in a current-carrying coil of wire, the alignment of magnetic domains in the iron produces a particularly strong magnet known as an **electromagnet.** The strength of an electromagnet can be increased simply by increasing the current through the coil. Electromagnets powerful enough to lift automobiles are a common sight in junkyards.

INTEGRATED SCIENCE BIOLOGY AND EARTH SCIENCE

Earth's Magnetic Field and the Ability of Organisms to Sense It

A suspended magnet or compass points northward because Earth itself is a huge magnet and the compass aligns with the magnetic field of the Earth (Figure 7.34).

The configuration of Earth's magnetic field is similar to that of a strong bar magnet placed near the center of the Earth. But Earth is not a magnetized chunk of iron like a bar magnet. Temperatures rise quickly underground, and they are too high for individual atoms to stabilize in magnetic domains with the proper orientation. Random thermal motion would destroy such an organized alignment of atoms. Rather, the explanation of Earth's magnetic field has to do with the convection cells that occur in Earth's interior, especially in the liquid outer core. Accelerating electric charges produce magnetic fields, as we know. As the electrically charged, iron-rich material of the hot outer core circulates in convection patterns, it creates a magnetic field. (More on this in Chapter 22).

Many organisms are able to sense Earth's magnetic field. They use it, rather than the maps and street signs that humans employ, to figure out how to get where they need to go—and they often use their magnetic sense with remarkable accuracy.

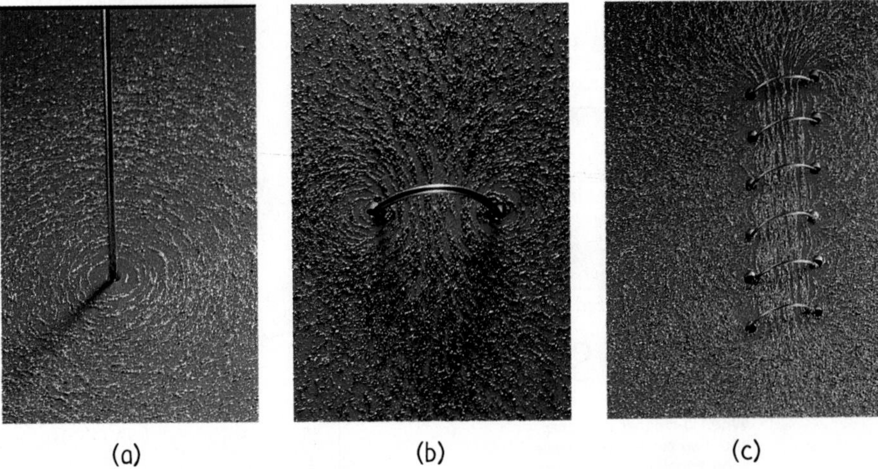

Figure 7.33 Iron fillings sprinkled on paper reveal the magnetic-field configurations about (a) a current-carrying wire, (b) a current-carrying loop, and (c) a current-carrying coil of loops.

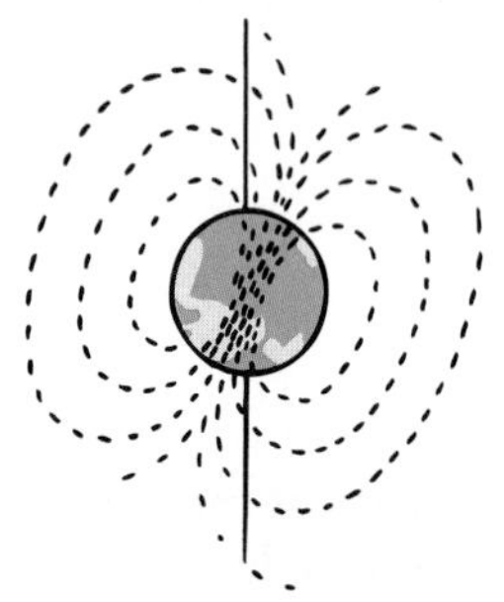

Figure 7.34 The Earth is a magnet.

Figure 7.35 The pigeon may well be able to sense direction because of a built-in magnetic "compass" in its skull.

For example, certain bacteria biologically produce single-domain-sized grains of magnetite (an iron-containing magnetic material) that they string together to form magnetic compasses. They use these compasses to detect the dip in the Earth's magnetic field. Equipped with a sense of direction, the organisms are able to locate food supplies. Pigeons have been found to have multiple-domain magnetite magnets within their skulls that are connected by a large number of nerves to their brains. Magnetic material has also been found in the abdomens of bees, whose behavior is affected by small magnetic fields and fluctuations in Earth's magnetic field. Monarch butterflies and sea turtles are known to employ their magnetic sense to migrate vast distances. Organisms as diverse as hamsters, salamanders, sparrows, rainbow trout, spiny lobsters, and many bacteria have demonstrated a magnetic sense in laboratory experiments. Magnetite crystals resembling those found in bacteria have been found in human brains. No one knows how these may be useful, but it seems likely that humans, too, have a magnetic sense.

CHECK YOURSELF

1. Is Earth a magnet? Justify your answer.
2. How does the presence of magnetite in its body help an organism navigate?

CHECK YOUR ANSWERS

1. Yes. The Earth is a magnet: it is an object that produces a net magnetic field outside itself.
2. Magnetite is a magnetic material; it is composed of domains that line up in an external magnetic field much like a compass. Thus organisms that contain magnetite contain, in effect, an internal compass that indicates the organism's orientation with respect to Earth's magnetic field.

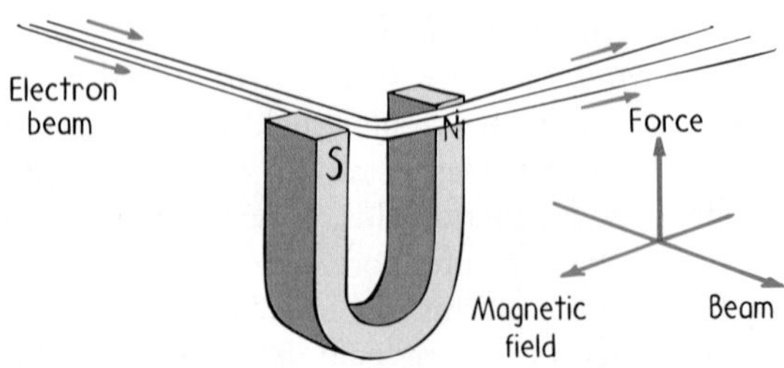

Figure 7.36 A beam of electrons is deflected by a magnetic field.

7.14 Magnetic Forces on Moving Charges

A charged particle has to be moving to interact with a magnetic field. Charges at rest don't respond to magnets. But, when they are moving, charged particles experience a deflecting force.* The force is greatest when the particles move at right

*When particles of electric charge q and velocity v move perpendicularly into a magnetic field of strength B, the force F on each particle is simply the product of the three variables: $F = qvB$. For non-perpendicular angles, v in this relationship must be the component of velocity perpendicular to B.

angles to the magnetic field lines. At other angles, the force is less, and it becomes zero when the particles move parallel to the field lines. The force is always perpendicular to the magnetic field lines and perpendicular to the velocity of the charged particle (Figure 7.36). So a moving charge is deflected when it crosses through a magnetic field, but, when it travels parallel to the field, no deflection occurs.

This deflecting force is very different from the forces that occur in other interactions. Gravitation acts in a direction parallel to the line between masses, and electrical forces act in a direction parallel to the line between charges. But magnetic force acts at right angles to the magnetic field and the velocity of the charged particle.

We are fortunate that charged particles are deflected by magnetic fields. This fact is employed in guiding electrons onto the inner surface of a television picture tube to produce a picture. Also, charged particles from outer space are deflected by the Earth's magnetic field. Otherwise the harmful cosmic rays bombarding the Earth's surface would be much more intense (Figure 7.37).

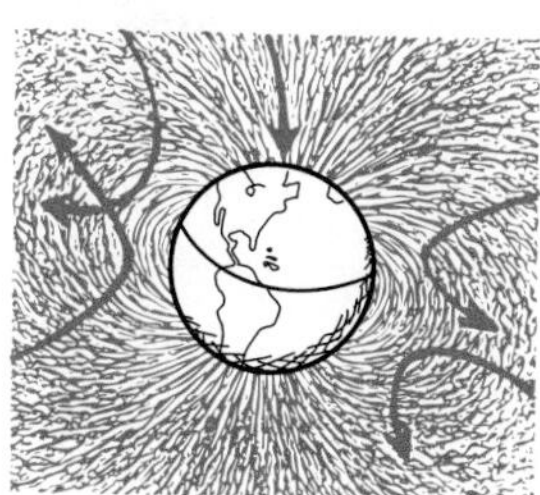

Figure 7.37 The magnetic field of the Earth picks up many charged particles that make up cosmic radiation.

Magnetic Force on Current-Carrying Wires

Magnetic Forces on a Current-Carrying Wire

Simple logic tells you that, if a charged particle moving through a magnetic field experiences a deflecting force, then a current of charged particles moving through a magnetic field also experiences a deflecting force. If the particles are deflected while moving inside a wire, the wire is also deflected (Figure 7.38).

If we reverse the direction of current, the deflecting force acts in the opposite direction. The force is strongest when the current is perpendicular to the magnetic field lines. The direction of force is neither along the magnetic field lines nor along the direction of current. The force is perpendicular both to the field lines and to the current. It is a sideways force.

We see that, just as a current-carrying wire will deflect a magnet, such as a compass needle, a magnet will deflect a current-carrying wire. When discovered, these complementary links between electricity and magnetism created much excitement. Almost immediately, people began harnessing the electromagnetic force for useful purposes—with great sensitivity in electric meters and with great force in electric motors.

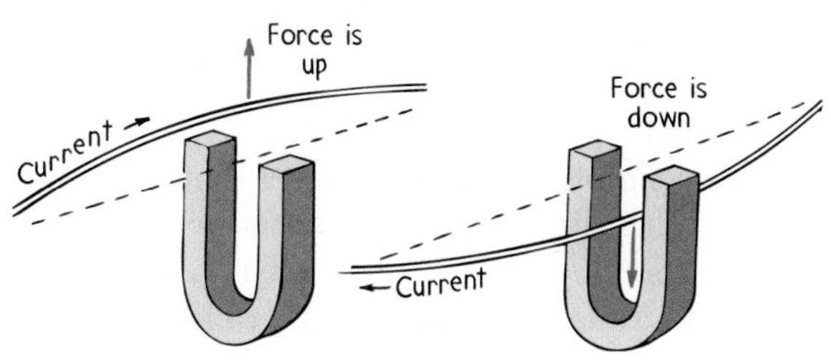

Figure 7.38 A current-carrying wire experiences a force in a magnetic field.

Electric Meters

The simplest meter to detect electric current is a magnetic compass. The next simplest meter is a compass in a coil of wires (Figure 7.39). When an electric current passes through the coil, each loop produces its own effect on the needle, so even a very small current can be detected. Such a current-indicating instrument is called a galvanometer.

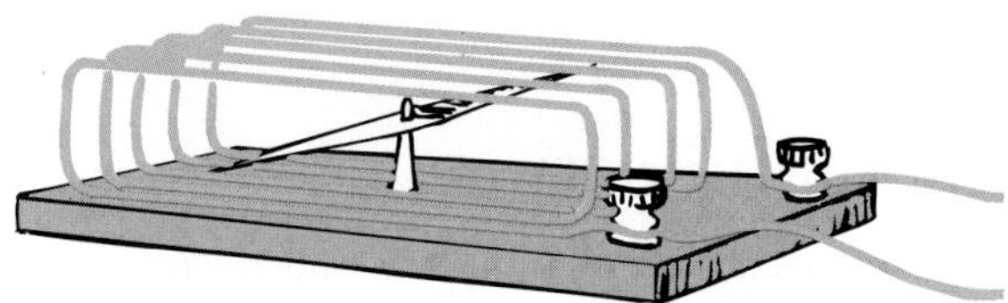

Figure 7.39 A very simple galvanometer.

A more common design is shown in Figure 7.40. It employs more loops of wire and is therefore more sensitive. The coil is mounted for movement, and the magnet is held stationary. The coil turns against a spring, so the greater the current in its windings, the greater its deflection. A galvanometer may be calibrated to measure current (amperes), in which case it is called an *ammeter*, or it may be calibrated to measure electric potential (volts), in which case it is called a *voltmeter* (Figure 7.41).

Electric Motors

If we change the design of the galvanometer slightly, so that deflection makes a complete turn rather than a partial rotation, we have an electric motor. The principal difference is that the current in a motor is made to change direction each time the coil makes a half rotation. This happens in a cyclic fashion to produce continuous rotation, which has been used to run clocks, operate gadgets, and lift heavy loads.

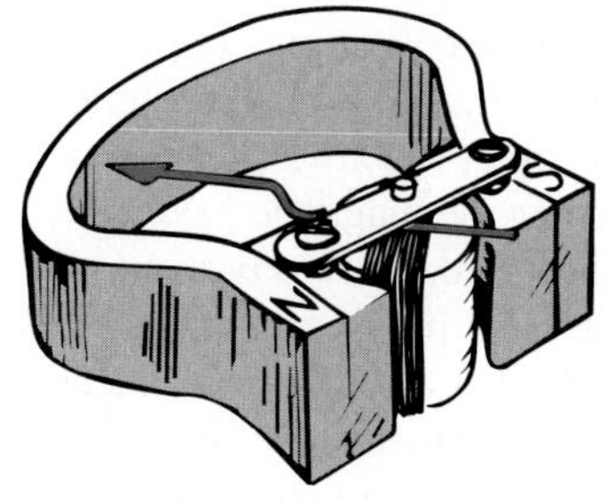

Figure 7.40 A common galvanometer design.

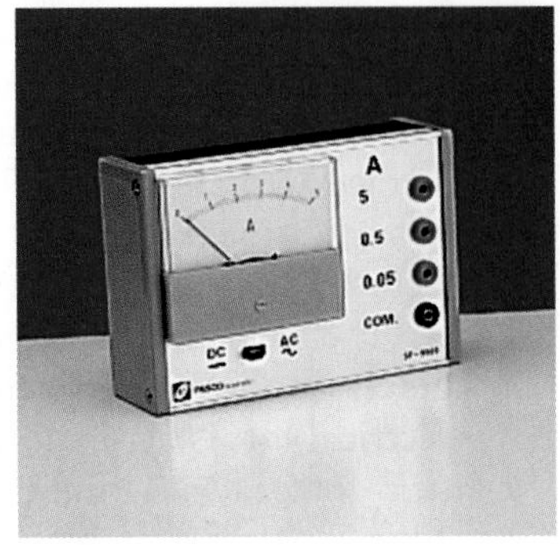

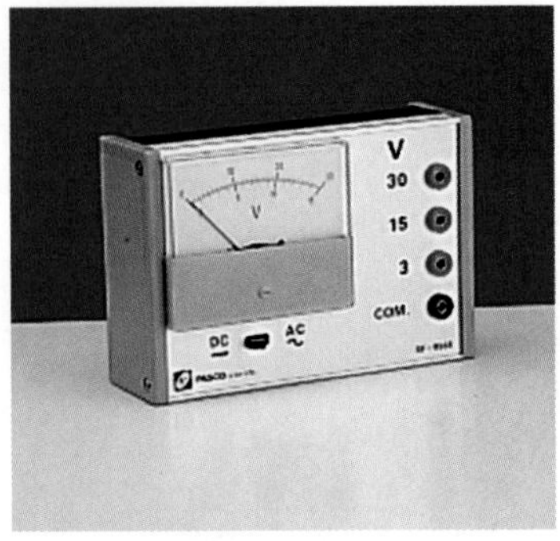

Figure 7.41 Both the ammeter and the voltmeter are basically galvanometers. (The electrical resistance of the instrument is designed to be very low for the ammeter and very high for the galvanometer.)

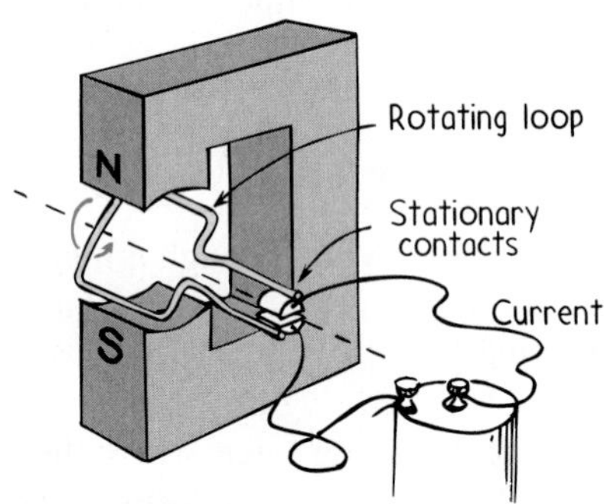

Figure 7.42 Interactive Figure
A simple motor.

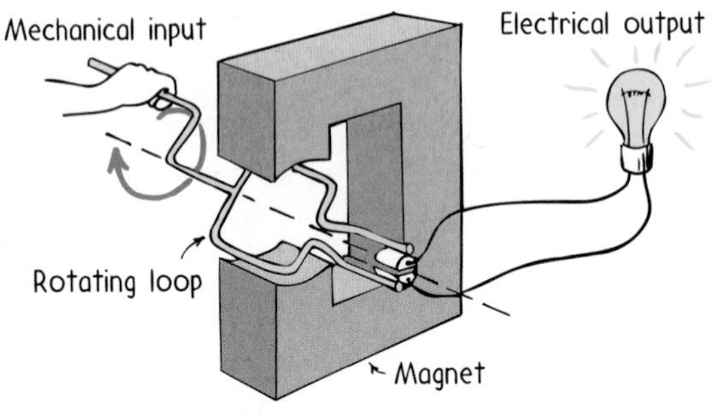

Figure 7.43 Interactive Figure
A simple generator. Voltage is induced in the loop when it is rotated in the magnetic field.

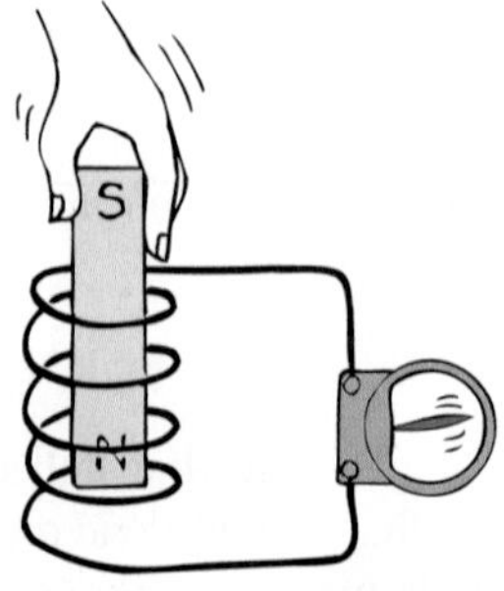

Figure 7.44 When the magnet is plunged into the coil, charges in the coil are set in motion, and voltage is induced in the coil.

In Figure 7.42, we see the principle of the electric motor in bare outline. A permanent magnet produces a magnetic field in a region where a rectangular loop of wire is mounted to turn about the axis shown by the dashed line. When a current passes through the loop, it flows in opposite directions in the upper and lower sides of the loop. (It must do this because, if charge flows into one end of the loop, it must flow out the other end.) If the upper portion of the loop is forced to the left, then the lower portion is forced to the right, as if it were a galvanometer. But, unlike a galvanometer, the current is reversed during each half revolution by means of stationary contacts on the shaft. The parts of the wire that brush against these contacts are called brushes. In this way, the current in the loop alternates so that the forces in the upper and lower regions do not change directions as the loop rotates. The rotation is continuous as long as current is supplied.

We have described here only a very simple DC motor. Larger motors, DC or AC, are usually manufactured by replacing the permanent magnet by an electromagnet that is energized by the power source. Of course, more than a single loop is used. Many loops of wire are wound about an iron cylinder, called an armature, which then rotates when the wire carries current.

A generator is a motor in reverse (Figure 7.43). In a motor, electrical energy is the input and mechanical energy is the output. In a generator, mechanical energy is the input and electric energy is the output. Both devices transform energy from one form to another. To understand how a generator works, you need to understand electromagnetic induction, the subject of the next section.

7.15 Electromagnetic Induction

In the early 1800s, when electricity and magnetism were topics of much scientific research, the question arose as to whether electricity could be produced from magnetism. The answer was provided in 1831 by two physicists, Michael Faraday in England and Joseph Henry in the United States—each working without knowledge of the other. Their discovery changed the world by making electricity commonplace.

Faraday and Henry both discovered that electric current could be produced in a wire simply by moving a magnet into or out of a coil of wire (Figure 7.44). No battery or other voltage source was needed—only the motion of a magnet in a coil or a wire loop. They discovered that voltage is caused, or *induced*, by the relative motion between a wire and a magnetic field. Whether the magnetic field moves near a stationary conductor or vice versa, voltage is induced either way (Figure 7.45).

The greater the number of loops of wire moving in a magnetic field, the greater the induced voltage (Figure 7.46). Pushing a magnet into a coil with twice as many loops induces twice as much voltage; pushing into a coil with ten times as many loops induces ten times as much voltage; and so on. Electromagnetic induction can be summarized by Faraday's Law:

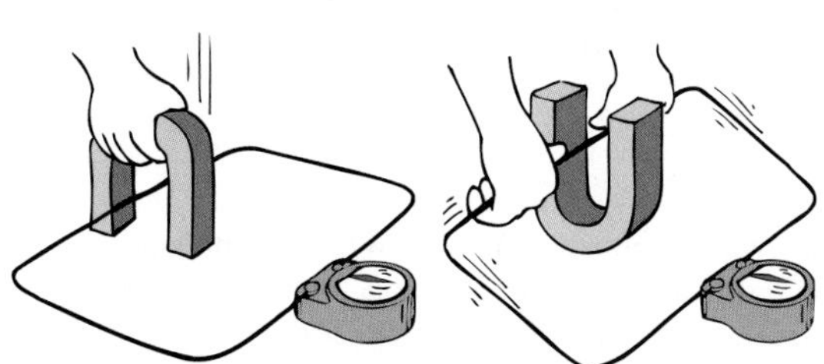

Figure 7.45 Voltage is induced in the wire loop whether the magnetic field moves past the wire or the wire moves through the magnetic field.

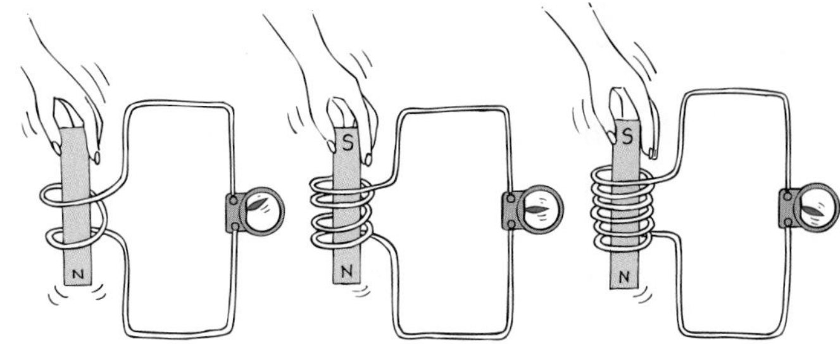

Figure 7.46 When a magnet is plunged into a coil with twice as many loops as another, twice as much voltage is induced. If the magnet is plunged into a coil with three times as many loops, three times as much voltage is induced.

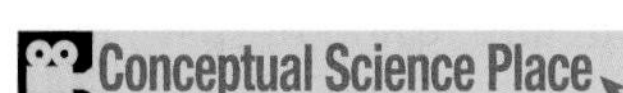

Faraday's Law; Applications of Electromagnetic Induction

The induced voltage in a coil is proportional to the number of loops multiplied by the rate at which the magnetic field changes within those loops.

Note that the amount of current produced by electromagnetic induction depends on the resistance of the coil and the circuit it connects, as well as the induced voltage.

We have mentioned two ways in which voltage can be induced in a loop of wire: by moving the loop near a magnet, or by moving a magnet near the loop. There is a third way: by changing the current in a nearby loop. All three of these cases possess the same essential ingredient—a changing magnetic field in the loop.

Electromagnetic induction explains the induction of voltage in a wire, as we have seen. However, the more basic concept of fields is at the root of the induced voltages and currents we observe in coils. The modern view of electromagnetic induction states that electric and magnetic *fields* are induced—and these in turn produce the voltages we have considered. So induction occurs whether or not a conducting wire or any material medium is present. In this more general sense, Faraday's law states:

An electric field is induced in any region of space in which a magnetic field is changing with time.

There is a second effect, an extension of Faraday's law. It is the same except that the roles of electric and magnetic fields are interchanged. It is one of nature's many symmetries. This effect, which was introduced by British physicist James Clerk Maxwell in about 1860 is known as **Maxwell's counterpart to Faraday's Law:**

HISTORY OF SCIENCE

Generators and Power Production

Fifty years after Faraday and Henry discovered electromagnetic induction, Nikola Tesla and George Westinghouse put those findings to practical use. They showed the world that electricity could be generated reliably and in sufficient quantities to light entire cities.

Tesla built generators much like those still in use. Tesla's generators were more complicated than the simple model we have discussed. His generators had armatures made up of bundles of copper wires. The armatures were forced to spin within strong magnetic fields by a turbine, which, in turn, was spun by the energy of either steam or falling water. The rotating loops of wire in the armature cut through the magnetic field of the surrounding electromagnets. In this way, they induced alternating voltage and currents.

Today, the energy to turn the turbine is still often delivered by steam. In traditional power plants, the water is heated to steam by burning fossil fuels. In nuclear power plants, the energy to heat water to make steam is supplied by nuclear fission reactions.

It's important to know that generators don't create energy—they simply convert energy from some other form into electric energy. Energy from a source—whether fossil or nuclear fuel, or wind, or water—is converted to mechanical energy to drive the turbine, and then converted to electricity, which then carries the energy to where it can be used.

TECHNOLOGY

Electromagnetic Induction in Everyday Technologies

We see electromagnetic induction all around us. On the road, we see it operate when we drive over buried coils of wire to activate a nearby traffic light. When iron parts of a car move over the buried coils, the effect of Earth's magnetic field is changed, inducing a voltage to trigger the changing of the traffic lights. Similarly, when you walk through the upright coils in the security system at an airport, any metal you carry slightly alters the magnetic field in the coils. This change induces voltage, which sounds an alarm. When the magnetic strip on the back of a credit card is scanned, induced voltage pulses identify the card. Something similar occurs in the recording head of a tape recorder: magnetic domains in the tape are sensed as the tape moves past a current-carrying coil. Electromagnetic induction is everywhere. As we shall see in the next chapter, it underlies the electromagnetic waves we call light.

A magnetic field is induced in any region of space in which an electric field is changing with time.

In each case, the strength of the induced field is proportional to the rates of change of the inducing field. The induced electric and magnetic fields are at right angles to each other.

Maxwell saw the link between electromagnetic waves and light. If electric charges are set into vibration in the range of frequencies that match those of light, waves are produced that *are* light! Maxwell discovered that light is simply electromagnetic waves in the range of frequencies to which the eye is sensitive. You will learn much more about electromagnetic waves in the following chapter.

CHAPTER 7 Review

Summary of Terms

Alternating current (AC) Electric current that repeatedly reverses its direction; the electric charges vibrate about relatively fixed positions. In the United States, the vibrational rate is 60 Hz.

Conductor Any material having free charged particles that easily flow through it when an electric force acts on them.

Coulomb The SI unit of electrical charge. One coulomb (symbol C) is equal in magnitude to the total charge of 6.25×10^{18} electrons.

Coulomb's Law The relationship among force, charge, and distance:

$$F = k\left(\frac{q_1 q_2}{d^2}\right)$$

If the charges are alike in sign, the force is repelling; if the charges are unlike, the force is attractive.

Direct current (DC) An electric current flowing in one direction only.

Electric current The flow of electric charge that transports energy from one place to another. It is measured in amperes, where 1 A is the flow of 6.25×10^{18} electrons per second, or 1 coulomb per second.

Electric field Defined as force per unit charge, it can be considered to be an energetic "aura" surrounding charged objects. About a charged point, the field decreases with distance according to the inverse-square law, like a gravitational field. Between oppositely charged parallel plates, the electric field is uniform.

Electric potential The electric potential energy per amount of charge, measured in volts, and often called *voltage*:

$$\text{Voltage} = \frac{\text{electric energy}}{\text{amount of charge}}$$

Electric potential energy The energy a charge possesses by virtue of its location in an electric field.

Electric power The rate of energy transfer, or rate of doing work; the amount of energy per unit time, which can be measured by the product of current and voltage:

$$\text{Power} = \text{current} \times \text{voltage}$$

It is measured in watts (or in kilowatts).

Electric resistance The property of a material that resists the flow of electric charge through it. It is measured in ohms (Ω).

Electrically polarized Term applied to an atom or molecule in which the charges are aligned so that one side has a slight excess of positive charge and the other side a slight excess of negative charge.

Electromagnet A magnet whose field is produced by an electric current. It is usually in the form of a wire coil with a piece of iron inside the coil.

Electromagnetic induction The induction of voltage when a magnetic field changes with time.

Electrostatics The study of electric charge at rest (not *in motion*, as in electric currents).

Faraday's law An electric field is induced in any region of space in which a magnetic field is changing with time. The magnitude of the induced electric field is proportional to the rate at which the magnetic field changes. The direction of the induced field is at right angles to the changing magnetic field.

Insulator Any material without free charged particles and through which current does not easily flow.

Magnetic domains Clustered regions of aligned magnetic atoms. When these regions themselves are aligned with one another, the substance containing them is a magnet.

Magnetic field The region of magnetic influence around a magnetic pole or around a moving charged particle.

Magnetic force (1) Between magnets, it is the attraction of unlike magnetic poles for each other and the repulsion between like magnetic poles. (2) Between a magnetic field and a moving charge, it is a deflecting force due to the motion of the charge; the deflecting force is perpendicular to the velocity of the charge and perpendicular to the magnetic field lines.

Maxwell's counterpart to Faraday's Law A magnetic field is induced in any region of space in which an electric field is changing with time. The magnitude of the induced magnetic field is proportional to the rate at which the electric field changes. The direction of the induced magnetic field is at right angles to the changing electric field.

Ohm's Law The statement that the current in a circuit varies in direct proportion to the potential difference or voltage and inversely with the resistance:

$$\text{Current} = \frac{\text{voltage}}{\text{resistance}}$$

Parallel circuit An electric circuit with two or more devices connected in such a way that the same voltage acts across each one, and any single one completes the circuit independently of all the others.

Potential difference The difference in potential between two points, measured in volts, and often called *voltage difference*.

Semiconductor A material that can be made to behave sometimes as an insulator and sometimes as a conductor.

Series circuit An electric circuit with devices connected in such a way that the same electric current exists in all of them.

Review Questions

7.1 Electrical Force and Charge

1. Which part of an atom is positively charged, and which part is negatively charged?
2. What is meant by saying that charge is conserved?

7.2 Coulomb's Law

3. How is Coulomb's law similar to Newton's law of gravitational force? How is it different?
4. How does a coulomb of charge compare with the charge of a single electron?

7.3 Electric Field

5. Give two examples of common force fields.
6. How is the direction of an electric field defined?

7.4 Electric Potential

7. In terms of the units that measure them, distinguish between *electric potential energy* and *electric potential.*
8. A balloon may easily be charged to several thousand volts. Does that mean it has several thousand joules of energy? Explain.

7.5 Conductors and Insulators

9. What is the difference between a conductor and an insulator? Between a conductor and a semiconductor?
10. What kind of materials are the best conductors? Why are they so good at conducting electricity?

7.6 Voltage Sources

11. What condition is necessary for heat energy to flow from one end of a metal bar to another? For electric charge to flow?
12. What condition is necessary for a sustained flow of electric charge through a conducting medium?

7.7 Electric Current

13. Why do electrons, rather than protons, make up the flow of charge in a metal wire?
14. Distinguish between DC and AC.

7.8 Electrical Resistance

15. Which has the greater resistance, a thick wire or a thin wire of the same length?
16. What is the unit of electrical resistance?

7.9 Ohm's Law

17. What is the effect on current through a circuit of steady resistance when the voltage is doubled? What is the effect on current if both voltage and resistance are doubled?
18. How much current flows through a radio speaker that has a resistance of 8 Ω when 12 V is impressed across the speaker?

7.10 Electric Circuits

19. Which type of circuit is favored for operating several electrical devices, each independently of the other; series, or parallel? Defend your answer.
20. How does the sum of the currents though the branches of a simple parallel circuit compare with the current that flows through the voltage source?

7.11 Electric Power

21. What is the relationship among electric power, current, and voltage?
22. Between a kilowatt and a kilowatt-hour, which is a unit of energy and which is a unit of power?

7.12 Magnetic Force

23. In what way is the rule for the interaction between magnetic poles similar to the rule for the interaction between electric charges?
24. In what way are magnetic poles very different from electric charges?

7.13 Magnetic Fields

25. What produces a magnetic field?
26. Why is iron magnetic and wood not?

7.14 Magnetic Forces on Moving Charges

27. What is the direction of the magnetic force on a moving charged particle?
28. What is a galvanometer? What is a galvanometer called when it is calibrated to read current? When it is calibrated to read voltage?

7.15 Electromagnetic Induction

29. What are the three ways in which voltage can be induced in a wire?
30. (a) What is induced by the rapid alternation of a magnetic field?
    (b) What is induced by the rapid alternation of an electric field?

Multiple Choice Questions

Choose the BEST answer to the following.

1. The electrical force of attraction between an electron and a proton is greater on the
   (a) proton.
   (b) electron.
   (c) neither; both are the same.
2. Immediately after two separated charged particles are released from rest, both increase in speed. The particles therefore have
   (a) the same sign of charge.
   (b) opposite signs of charge.
   (c) either the same or opposite signs of charge.
   (d) Need more information.
3. The current through a 12-ohm hair dryer connected to a 120-volt power source is
   (a) 1 A.
   (b) 10 A.
   (c) 12 A.
   (d) 120 A.
   (e) None of these.

4. Compared with the amount of current in the filament of a lamp, the amount of current in the connecting wire is
   (a) definitely less.
   (b) often less.
   (c) actually more.
   (d) the same.
   (e) incredibly, all of these.
5. If you change the magnetic field in a closed loop of wire, you induce a
   (a) current.
   (b) voltage.
   (c) electric field.
   (d) All of these.
   (e) None of these.
6. The mutual induction of electric and magnetic fields can produce
   (a) light.
   (b) sound.
   (c) Both of these.
   (d) Neither of these.

INTEGRATED SCIENCE CONCEPTS

Biology—Electric Shock

1. High voltage by itself does not produce electric shock. What does?
2. What is the source of electrons that shock you when you touch a charged conductor?
3. If a current of one- or two-tenths of an ampere were to flow into one of your hands and out the other, you would probably be electrocuted. But if the same current were to flow into your hand and out at the elbow above the same hand, you could survive, even though the current might be large enough to burn your flesh. Explain.

Biology and Earth Science—Earth's Magnetic Field and the Ability of Organisms to Sense It

1. People have wondered about the "mystery" of animal migration for countless generations. Give one possible, very general explanation as to how animals find their way during migration.
2. Does Earth's magnetic field arise because Earth's interior consists of aligned magnetic domains? If not, what is the likely cause of Earth's magnetic field?

Active Explorations

1. An electric cell is made by placing two plates of different materials that have different affinities for electrons in a conducting solution. You can make a simple 1.5-V cell by placing a strip of copper and a strip of zinc in a tumbler of salt water. The voltage of a cell depends on the materials and the solution they are placed in, not the size of the plates. A battery is actually a series of cells.

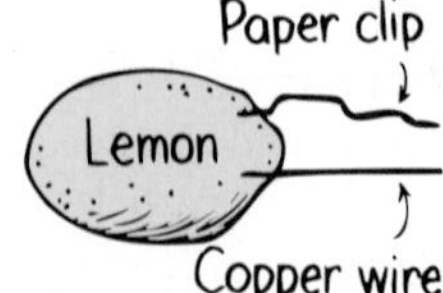

An easy cell to construct is the lemon cell. Stick a straightened paper clip and a piece of copper wire into a lemon. Hold the ends of the paper clip and the wire close together, but not touching, and place the ends on your tongue. The slight tingle you feel and the metallic taste you experience result from a slight current of electricity pushed by the citrus cell through the paper clip and the wire when your moist tongue closes the circuit.

2. An iron bar can be magnetized easily by aligning it with the magnetic field lines of Earth and striking it lightly a few times with a hammer. This works best if the bar is tilted down to match the dip of Earth's magnetic field. The hammering jostles the magnetic domains in the bar so that they can fall into a better alignment with Earth's magnetic field. The bar can be demagnetized by striking it when it is oriented in an east–west direction.

Exercises

1. With respect to forces, how are electric charge and mass alike? How are they different?
2. When combing your hair, you scuff electrons from your hair onto the comb. Is your hair then positively or negatively charged? How about the comb?
3. An electroscope is a simple device consisting of a metal ball that is attached by a conductor to two thin leaves of metal foil protected from air disturbances in a jar, as shown. When the ball is touched by a charged body, the leaves, which normally hang straight down, spread apart. Why?

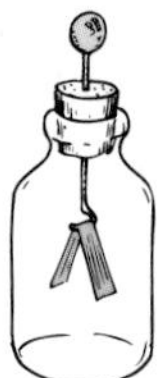

4. The five thousand billion billion freely moving electrons in a penny repel one another. Why don't they fly out of the penny?
5. Two equal charges exert equal forces on each other. What if one charge has twice the magnitude of another? How do the forces they exert on each other compare?
6. Suppose that the strength of the electric field about an isolated point charge has a certain value at a distance of 1 m. How will the electric field strength compare at a distance of 2 m from the point charge? What law guides your answer?
7. Why is a good conductor of electricity also a good conductor of heat?
8. When a car is moved into a painting chamber, a mist of paint is sprayed around it. When the body of the car is given a sudden electric charge and the mist of paint is attracted to it, presto—the car is quickly and uniformly painted. What does the phenomenon of polarization have to do with this?
9. If you place a free electron and a free proton in the same electric field, how will the forces acting on them compare? How will their accelerations compare? How will their directions of travel compare?
10. You are not harmed by contact with a charged metal ball, even though its voltage may be very high. Is the reason for this similar to the reason why you are not harmed by the sparks from a Fourth-of-July sparkler, even though the temperature of each of those sparks is greater than 1000°C? Defend your answer in terms of the energies that are involved.
11. In which of these circuits does a current exist to light the bulb?

12. Sometimes you hear someone say that a particular appliance "uses up" electricity. What is it that the appliance actually uses up, and what becomes of it?

13. Will the current in a light bulb connected to a 220-V source be greater or less than the current in the same bulb when it is connected to a 110-V source?
14. Are automobile headlights wired in parallel or in series? What is your evidence?
15. A car's headlights dissipate 40 W on low beam and 50 W on high beam. Is there more resistance or less resistance in the high-beam filament?
16. To connect a pair of resistors so that their equivalent resistance will be greater than the resistance of either one, should you connect them in series or in parallel?
17. Why might the wingspans of birds be a consideration in determining the spacing between parallel wires on power poles?
18. In the circuit shown, how do the brightnesses of the individual bulbs compare? Which light bulb draws the most current? What will happen if bulb A is unscrewed? If bulb C is unscrewed?

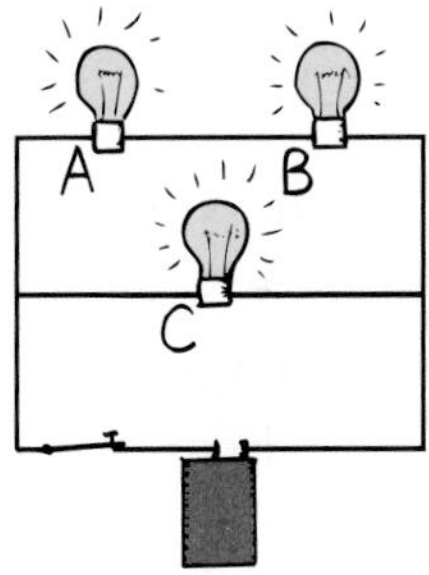

19. As more and more bulbs are connected in series to a flashlight battery, what happens to the brightness of each bulb? Assuming that heating inside the battery is negligible, what happens to the brightness of each bulb when more and more bulbs are connected in parallel?
20. Since every iron atom is a tiny magnet, why aren't all things made of iron also magnets?
21. What surrounds a stationary electric charge? A moving electric charge?
22. A strong magnet attracts a paper clip to itself with a certain force. Does the paper clip exert a force on the strong magnet? If not, why not? If so, does it exert as much force on the magnet as the magnet exerts on it? Defend your answers.
23. Wai Tsan Lee shows iron nails that have become induced magnets. Is there similar physics here with the sticking balloon of Figure 7.5? Defend your answer.

24. Can an electron at rest in a magnetic field be set into motion by the magnetic field? What if it were at rest in an electric field?
25. Residents of northern Canada are bombarded by more intense cosmic radiation than are residents of Mexico. Why is this so?
26. A magician places an aluminum ring on a table, underneath which is hidden an electromagnet. When the magician says "abracadabra" (and pushes a switch that starts current flowing through the coil under the table), the ring jumps into the air. Explain his "trick."
27. A friend says that changing electric and magnetic fields generate one another, and this gives rise to visible light when the frequency of change matches the frequency of light. Do you agree? Explain.
28. Write a letter to Grandma and convince her that whatever electric shocks she may have received over the years have been due to the movement of electrons already in her body—not electrons coming from somewhere else.

Problems

1. Two point charges are separated by 6 cm. The attractive force between them is 20 N. Show that the force between them when they are separated by 12 cm is 5 N.
2. A droplet of ink in an industrial ink-jet printer carries a charge of 1.6×10^{-10} C and is deflected onto paper by a force of 3.2×10^{-4} N. Show that the strength of the electric field that is required to produce this force is 2×10^6 N/C.
3. Find the voltage change when (a) an electric field does 12 J of work on a charge of 0.0001 C, and when (b) the same electric field does 24 J of work on a charge of 0.0002 C.
4. Rearrange this equation

$$\text{Current} = \frac{\text{voltage}}{\text{resistance}}$$

to express resistance in terms of current and voltage. Then consider the following: A certain device in a 120-V circuit has a current rating of 20 A. Show that the resistance of the device is 6 Ω.

5. Using the formula

$$\text{Power} = \text{current} \times \text{voltage}$$

Find the current drawn by a 1200-W hair dryer connected to 120 V is 10 A. Then, using the method you used in the previous problem, show that the resistance of the hair dryer is 12 Ω.

6. Show that it costs $3.36 to operate a 100-W lamp continuously for a week if the power utility rate is 20¢/kWh.

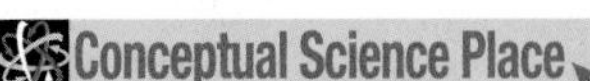

CHAPTER 7 **ONLINE RESOURCES**

Interactive Figures 7.1, 7.8, 7.21, 7.22, 7.26, 7.42, 7.43
Tutorials Electrostatics • ***Videos*** Van de Graaff Generator, Alternating Current, Caution on Handling Wires, Ohm's Law, Electric Circuits, Magnetic Forces on a Current-Carrying Wire, Faraday's Law, Application of Electromagnetic Induction • ***Quiz*** • ***Flashcards*** • ***Links***

CHAPTER 8

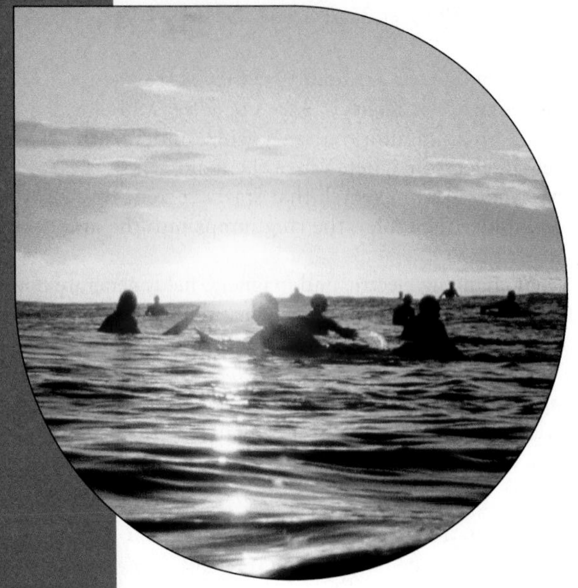

Waves—Sound and Light

These surfers ride waves of water and are immersed in an undulating "sea" of electromagnetic waves as well.

CHAPTER OUTLINE

Light is the only thing we can really see. But what *is* light? You may know that light is an electromagnetic wave and that it is part of the electromagnetic spectrum—a continuum of waves including X-rays, radio waves, microwaves and others. Where do electromagnetic waves come from? What are their properties? Do electromagnetic waves of various sizes really permeate our environment like vibrating strands of invisible spaghetti? Sound, like light, is a wave phenomenon. How do sound waves differ from light waves? Does the speed of sound differ in various materials? Can it travel in a vacuum? Can one sound wave cancel another, so that two loud noises combine to make zero noise? Are there technological applications for this idea? Many things—from musical instruments to atoms to vocal chords—vibrate, and when they do, they produce waves. Waves, the subject of this chapter, are everywhere!

8.1 Vibrations and Waves

Anything—from your vocal chords to a pendulum—that moves back and forth, to and fro, in and out, or up and down is vibrating. A vibration is a wiggle. A wiggle that travels is a **wave**. A wave extends from one location to another. Light and sound are both vibrations that propagate throughout space as waves, but as waves of two very different kinds. Sound is the propagation of vibrations through a material medium—a solid, a liquid, or a gas. If there is no medium to vibrate, then no sound is possible. Sound cannot travel in a vacuum. But light can, because light is a vibration of nonmaterial electric and magnetic fields—a vibration of pure energy. Although light can pass through many materials, it needs none. This is evident when it propagates through the near vacuum between the Sun and the Earth.

The relationship between a vibration and a wave is shown in Figure 8.1. A marking pen on a bob attached to a vertical spring vibrates up and down and traces a wave form on a sheet of paper that is moved horizontally at constant speed. The wave form is actually a *sine curve*, a graphical representation of a wave. Like a water wave, the high points of a sine wave are called *crests* and the low points are the *troughs*. The straight dashed line represents the "home" position, or midpoint, of the vibration. The term **amplitude** refers to the distance from the midpoint to the crest (or to the trough) of the wave. So the amplitude

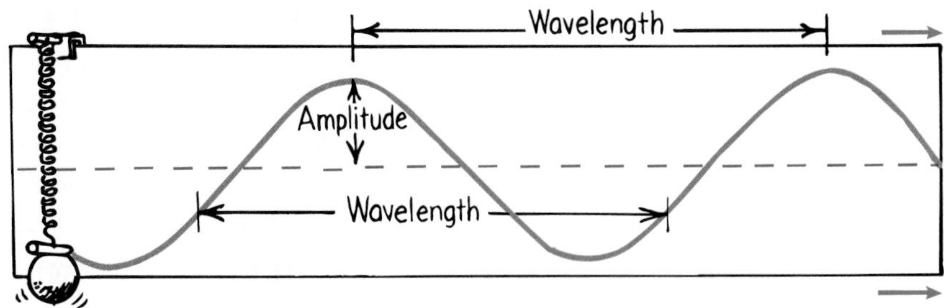

Figure 8.1 Interactive Figure
When the weight vibrates up and down, a marking pen traces out a sine curve on paper that is moved horizontally at constant speed.

equals the maximum displacement from equilibrium. Waves carry energy from one place to another. The amount of energy a wave carries depends on its amplitude. The larger the amplitude of a wave, the more energy it has.

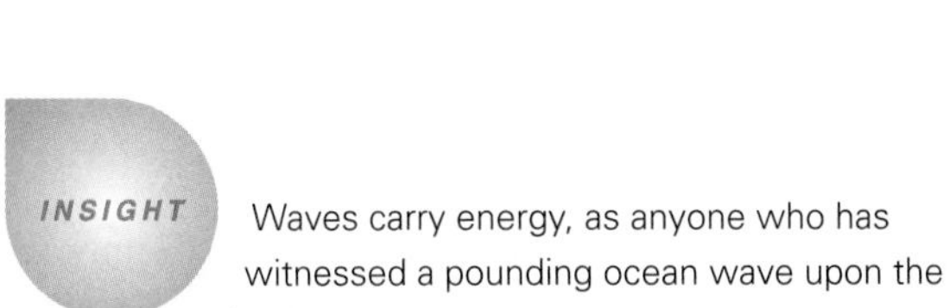

INSIGHT Waves carry energy, as anyone who has witnessed a pounding ocean wave upon the shore can see.

The **wavelength** of a wave is the distance from the top of one crest to the top of the next one, or, equivalently, the distance between successive identical parts of the wave. The wavelengths of waves at the beach are measured in meters, the wavelengths of ripples in a pond in centimeters, and the wavelengths of light in billionths of a meter (nanometers). All waves have a vibrating source.

How frequently a vibration occurs is described by its frequency. The **frequency** of a vibrating pendulum, or of an object on a spring, specifies the number of to-and-fro vibrations it makes in a given time (usually in one second). A complete to-and-fro oscillation is one vibration. If it occurs in one second, the frequency is one vibration per second. If two vibrations occur in one second, the frequency is two vibrations per second.

The unit of frequency is called the **hertz** (Hz), after Heinrich Hertz, who demonstrated the existence of radio waves in 1886. One vibration per second is 1 hertz; two vibrations per second is 2 hertz; and so on. Higher frequencies are measured in kilohertz (kHz), and still higher frequencies in megahertz (MHz).

INSIGHT The frequency of a "classical wave"—such as a sound wave, water wave, or radio wave—matches the frequency of the vibrating source. (In the quantum world of atoms and photons, the rules are different.)

The **period** of a wave or vibration is the time it takes for a complete vibration—for a complete cycle. Period can be calculated from frequency, and vice versa. Suppose, for example, that a pendulum makes two vibrations in one second. Its frequency is 2 Hz. The time needed to complete one vibration—that is, the period of vibration—is $\frac{1}{2}$ second. Or, if the vibration frequency is 3 Hz, then the period is $\frac{1}{3}$ second. The frequency and period are the inverse of each other:

$$\text{Frequency} = \frac{1}{\text{period}}$$

or, vice versa,

$$\text{Period} = \frac{1}{\text{frequency}}$$

TECHNOLOGY

Broadcasting Radio Waves

AM radio waves are usually measured in kilohertz, while FM radio waves are measured in megahertz. A station at 960 kHz on the AM radio dial, for example, broadcasts radio waves that have a frequency of 960,000 vibrations per second. A station at 101.7 MHz on the FM dial broadcasts radio waves with a frequency of 101,700,000 hertz. These radio-wave frequencies are the frequencies

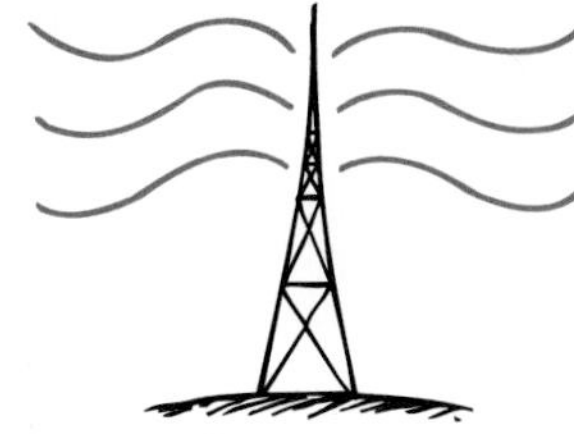

at which electrons are forced to vibrate in the antenna of a radio station's transmitting tower. The frequency of the vibrating electrons and the frequency of the wave produced are the same.

CHECK YOURSELF

1. An electric razor completes 60 cycles every second. (a) What is its frequency? (b) What is its period?
2. Gusts of wind cause the Sears Building in Chicago to sway back and forth, completing a cycle every ten seconds. (a) What is its frequency? (b) What is its period?

CHECK YOUR ANSWERS

1. (a) 60 cycles per second, or 60 Hz. (b) $\frac{1}{60}$ second.
2. (a) $\frac{1}{10}$ Hz. (b) 10 seconds.

8.2 Wave Motion

INSIGHT A wave transfers energy without transferring matter. If matter were to move along with the energy in a wave, oceans would be emptied as ocean waves travel to the shore.

If you drop a stone into a calm pond, waves will travel outward in expanding circles. Energy is carried by the wave, traveling from one place to another. The water itself goes nowhere. This can be seen by waves encountering a floating leaf. The leaf bobs up and down, but it doesn't travel with the waves. The waves move along, not the water. The same is true of waves of wind over a field of tall grass on a gusty day. Waves travel across the grass, while the individual grass plants remain in place; instead, they swing to and fro between definite limits, but they go nowhere. When you speak, wave motion through the air travels across the room at about 340 meters per second. The air itself doesn't travel across the room at this speed. In these examples, when the wave motion ceases, the water, the grass, and the air return to their initial positions. It is characteristic of wave motion that the medium transporting the wave returns to its initial condition after the disturbance has passed. Putting all the information about waves together, we can now specifically define what a wave is: A **wave** is a disturbance that travels from one place to another transporting energy, but not necessarily matter, along with it.

Figure 8.2 Water waves.

Wave Speed

The speed of periodic wave motion is related to the frequency and wavelength of the waves. Consider the simple case of water waves (Figures 8.2 and 8.3). Imagine that we fix our eyes on a stationary point on the water's surface and observe the waves passing by that point. We can measure how much time passes between the arrival of one crest and the arrival of the next one (the period), and we can also observe the distance between crests (the wavelength). We know that speed is defined as distance divided by time. In this case, the distance is one wavelength and the time is one period, so wave speed = wavelength/period.

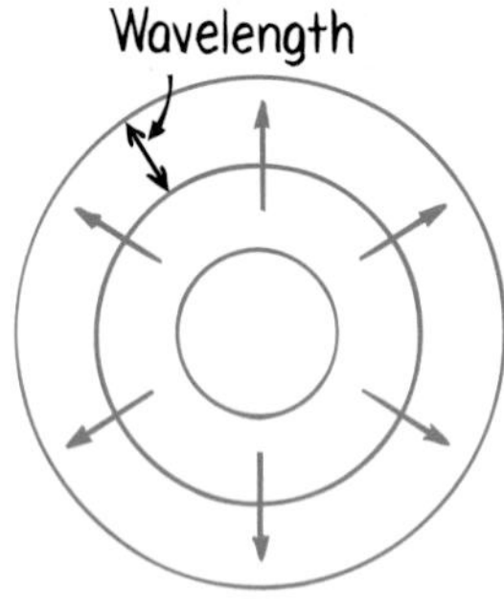

Figure 8.3 A top view of water waves.

For example, if the wavelength is 10 meters and the time between crests at a point on the surface is 0.5 second, the wave is traveling 10 meters in 0.5 seconds and its speed is 10 meters divided by 0.5 seconds, or 20 meters per second. Since period is equal to the inverse of frequency, the formula wave speed = wavelength/period can also be written

$$\text{wave speed} = \text{wavelength} \times \text{frequency}$$

This relationship applies to all kinds of waves, whether they are water waves, sound waves, or light waves.

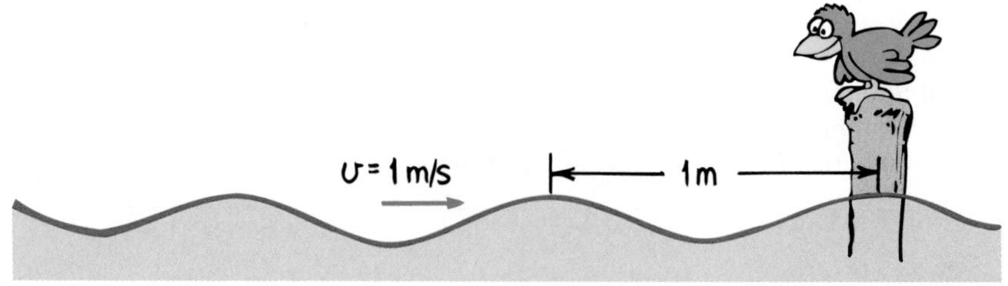

Figure 8.4 Interactive Figure

If the wavelength is 1 m, and one wavelength per second passes the pole, then the speed of the wave is 1 m/s.

INSIGHT The speed of light waves in a vacuum, approximately 3.0×10^8 m/s and denoted as *c*, is nature's speed limit. No material objects in the universe travel faster than this.

INSIGHT It is customary to express the speed of a wave by the equation $v = f\lambda$, where v is wave speed, f is wave frequency, and λ (the Greek letter lambda) is wavelength.

CHECK YOURSELF

1. If a train of freight cars, each 10 m long, rolls by you at the rate of three cars each second, what is the speed of the train?
2. If a water wave oscillates up and down three times each second and the distance between wave crests is 2 m, (a) what is its frequency? (b) What is its wavelength? (c) What is its wave speed?
3. The sound from a 60-Hz electric razor spreads out at 340 meters per second. (a) What is the frequency of the sound waves? (b) What is their period? (c) What is their speed? (d) What is their wavelength?

CHECK YOUR ANSWERS

1. (a) 30 m/s. We can see this in two ways. According to the definition of speed in Chapter 2, $v = \frac{d}{t} = \frac{3 \times 10\text{ m}}{1\text{ s}} = 30$ m/s, since 30 m of train passes you in 1 s.
   (b) If we compare our train to wave motion, where wavelength corresponds to 10 m and frequency is 3 Hz, then

$$\text{Speed} = \text{frequency} \times \text{wavelength} = 3\text{ Hz} \times 10\text{ m} = 30\text{ m/s}.$$

2. (a) 3 Hz. (b) 2 m. (c) Wave speed = frequency × wavelength = $\frac{3}{\text{s}} \times 2$ m = 6 m/s.
3. (a) 60 Hz. (b) $\frac{1}{60}$ s. (c) 340 m/s. (d) 5.7 m.

8.3 Transverse and Longitudinal Waves

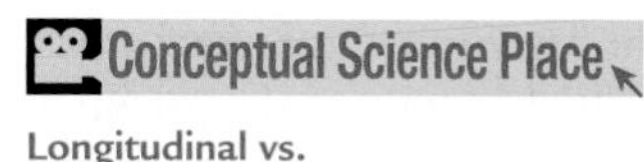

Longitudinal vs. Transverse Waves

Fasten one end of a Slinky to a wall and hold the free end in your hand. If you shake the free end up and down, you will produce vibrations that are at right angles to the direction of wave travel. The right-angled, or sideways, motion is called *transverse motion*. This type of wave is called a **transverse wave**. Waves in the stretched strings of musical instruments are transverse waves. We will see later that electromagnetic waves, such as radio waves and light waves, are also transverse waves.

A **longitudinal wave** is one in which the direction of wave travel is along the direction in which the source vibrates. You produce a longitudinal wave with your Slinky when you shake it back and forth along the Slinky's axis (Figure 8.5). The vibrations are then parallel to the direction of energy transfer. Part of the Slinky is compressed, and a wave of compression travels along it. In between successive compressions is a stretched region, called a rarefaction. Both compressions and rarefactions travel in the same direction along the Slinky. Together they make up the longitudinal wave.

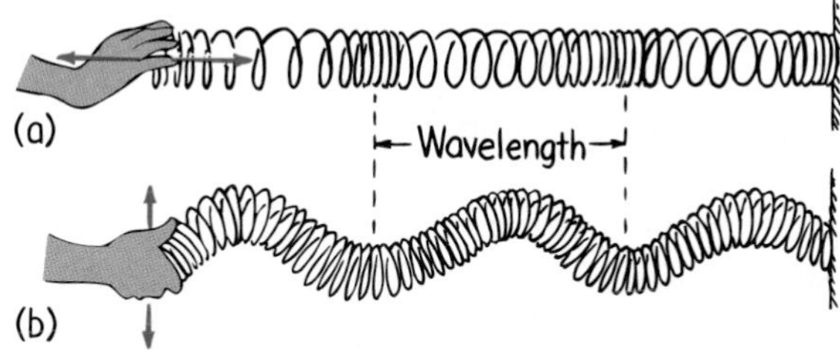

Figure 8.5 Interactive Figure
Both waves transfer energy from left to right. (a) When the end of the Slinky is pushed and pulled rapidly along its length, a longitudinal wave is produced. (b) When its end is shaken up and down (or from side to side), a transverse wave is produced.

8.4 The Nature of Sound

Think of the air molecules in a room as tiny randomly moving Ping-Pong balls (Figure 8.6). If you vibrate a Ping-Pong paddle in the midst of the balls, you'll set them vibrating to and fro. The balls will vibrate in rhythm with your vibrating paddle. In some regions, they will be momentarily bunched up (compressions), and in other regions in between, they will be momentarily spread out (rarefactions). The vibrating prongs of a tuning fork do the same to air molecules. Vibrations made up of compressions and rarefactions spread from the tuning fork throughout the air, and a sound wave is produced.

The wavelength of a sound wave is the distance between successive compressions or, equivalently, the distance between successive rarefactions. Each molecule in the air vibrates to and fro about some equilibrium position as the waves move by.

Figure 8.6 If you vibrate a Ping-Pong paddle in the midst of a lot of Ping-Pong balls, the balls will vibrate also.

INTEGRATED SCIENCE BIOLOGY

Sensing Pitch

Our subjective impression about the frequency of sound is described as *pitch*. A high-pitched sound, like that from a tiny bell, has a high vibration frequency. Sound from a large bell has a low pitch because its vibrations are of a low frequency.

The human ear can normally hear pitches from sound ranging from about 20 hertz to about 20,000 hertz. As we age, this range shrinks. Sound waves of frequencies below 20 hertz are called *infrasonic* waves, and those of frequencies above 20,000 hertz are called *ultrasonic* waves. We cannot hear infrasonic or ultrasonic sound waves. But dogs and some other animals can.

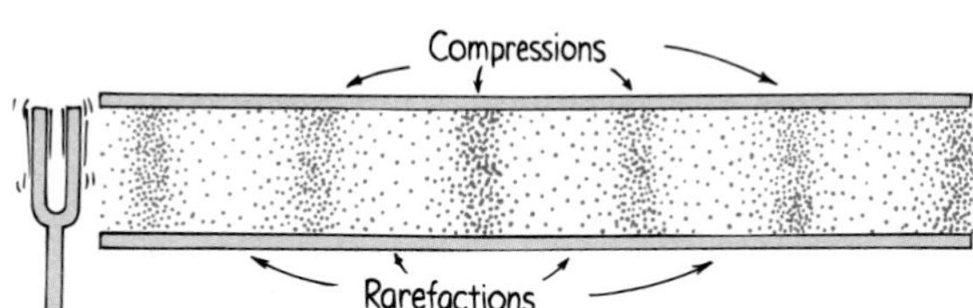

Figure 8.7 Compressions and rarefactions travel (both at the same speed and in the same direction) from the tuning fork through the air in the tube.

Hearing any sound occurs because air molecules next to a vibrating object are themselves set into vibration. These, in turn, vibrate against neighboring molecules, which, in turn, do the same, and so on. As a result, rhythmic patterns of compressed and rarefied air emanate from the sound source. The resulting vibrating air sets your eardrum into vibration, which, in turn, sends cascades of rhythmic electrical impulses along nerves in the cochlea of your inner ear and into the brain. Thus, when you hear a high-pitched sound, a high-frequency wave from a quickly vibrating source sets your eardrum into fast vibration. Bass guitars, foghorns, and deep-throated bullfrogs vibrate slowly, making low-pitched waves that set your eardrums into slow vibration.

CHECK YOURSELF

1. A singer sings a high-pitched note and then a low-pitched note. For which note are her vocal chords vibrating faster? For which note is your eardrum vibrating faster? Which note sets the air into higher-frequency vibrations?
2. Would you consider infrasonic waves to be sound waves? Are ultrasonic waves sound waves?

CHECK YOUR ANSWERS

1. A high-pitched note is a high-frequency sound wave arising from a rapidly vibrating source. It would set the air into oscillations at a higher frequency than a low-pitched sound, and it would likewise vibrate your eardrum at a higher frequency.
2. "Sound" requires perception. Since infrasonic and ultrasound waves are outside the range of human hearing, they would not be considered to be sound waves.

INSIGHT In hospitals, concentrated beams of ultrasound are used to break up kidney stones and gallstones, eliminating the need for surgery.

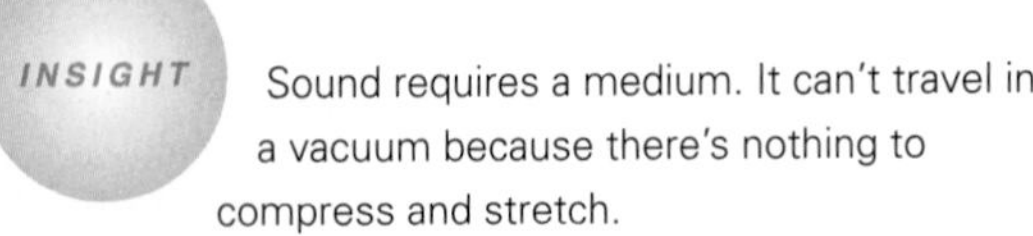

INSIGHT Sound requires a medium. It can't travel in a vacuum because there's nothing to compress and stretch.

Most sound is transmitted through air, but any elastic substance—solid, liquid, or gas—can transmit sound.* Air is a poor conductor of sound compared with solids and liquids. You can hear the sound of a distant train clearly by placing your ear against the rail. When swimming, have a friend at a distance click two rocks together beneath the surface of water while you are submerged. You will observe how well water conducts the sound.

Speed of Sound

If, from a distance, we watch a person chopping wood or hammering, we can easily see that the blow occurs a noticeable time before its sound reaches our ears. Thunder is often heard seconds after a flash of lightning is seen. These common experiences show that sound requires time to travel from one place to

* An elastic substance is "springy," has resilience, and can transmit energy with little loss. Steel, for example, is elastic, while lead and putty are not.

another. The speed of sound depends on wind conditions, temperature, and humidity. It does not depend on the loudness or the frequency of the sound; all sounds travel at the same speed in a given medium. The speed of sound in dry air at 0°C is about 330 meters per second, which is nearly 1200 kilometers per hour. Water vapor in the air increases this speed slightly. Sound travels faster through warm air than cold air. This is to be expected, because the faster-moving molecules in warm air bump into each other more frequently and, therefore, can transmit a pulse in less time.* For each degree rise in temperature above 0°C, the speed of sound in air increases by 0.6 meter per second. Thus, in air at a normal room temperature of about 20°C, sound travels at about 340 meters per second. In water, sound speed is about four times its speed in air; in steel, it's about fifteen times its speed in air.

CHECK YOURSELF

1. Do compressions and rarefactions in a sound wave travel in the same direction or in opposite directions from one another?
2. What is the approximate distance of a thunderstorm when you note a 3-s delay between the flash of lightning and the sound of thunder?

CHECK YOUR ANSWERS

1. They travel in the same direction.
2. Assuming that the speed of sound in air is about 340 m/s, in 3 s it will travel 340 m/s × 3 s = 1020 m. There is no appreciable time delay for the flash of light, so the storm is slightly more than 1 km away.

8.5 Resonance

If you strike an unmounted tuning fork, its sound is rather faint. Repeat with the handle of the tuning fork held against a table after striking it, and the sound is louder. This is because the table is forced to vibrate, and its larger surface sets more air in motion. The table is forced into vibration by a fork of any frequency. This is an example of **forced vibration.** The vibration of a factory floor caused by the running of heavy machinery is another example of forced vibration. A more pleasing example is given by the sounding boards of stringed instruments.

If you drop a wrench and a baseball bat on a concrete floor, you will easily notice the difference in their sounds. This is because each vibrates differently when striking the floor. They are not forced to vibrate at a particular frequency, but, instead, each vibrates at its own characteristic frequency. Any object composed of an elastic material will, when disturbed, vibrate at its own special set of frequencies, which together form its special sound. We speak of an object's **natural frequency**, which depends on such factors as the elasticity and the shape of the object. Bells and tuning forks, of course, vibrate at their own characteristic frequencies. Interestingly, most things, from atoms to planets and almost everything else in between, have a springiness to them, and they vibrate at one or more natural frequencies.

When the frequency of forced vibrations on an object matches the object's natural frequency, a dramatic increase in amplitude occurs. This phenomenon is called **resonance.** Putty doesn't resonate, because it isn't elastic, and a dropped handkerchief is too limp to resonate. In order for something to resonate, it needs both a force to pull it back to its starting position and enough energy to maintain its vibration.

INSIGHT Resonance occurs on a swing. When pumping, you pump in rhythm with the natural frequency of the swing. More important than the force with which you pump is the timing. Even small pumps, or small pushes from someone else, if delivered in rhythm with the frequency of the swinging motion, produce large amplitudes.

Resonance is not restricted to wave motion. It occurs whenever successive impulses are applied to a vibrating object in rhythm with its natural frequency. Cavalry troops marching across a footbridge near Manchester, England, in 1831 inadvertently caused the bridge to collapse when they marched in rhythm with

* The speed of sound in a gas is about $\frac{3}{4}$ the average speed of its molecules.

Figure 8.8 In 1940, four months after being completed, the Tacoma Narrows Bridge in the state of Washington was destroyed by wind-generated resonance. A mild gale produced a fluctuating force in resonance with the natural frequency of the bridge, steadily increasing the amplitude until the bridge collapsed.

the bridge's natural frequency. Since then, it has been customary to order troops to "break step" when crossing bridges. A more recent bridge disaster was caused by wind-generated resonance, as shown in Figure 8.8.

Figure 8.9 If you shake an electrically charged object to and fro, you will produce an electromagnetic wave.

8.6 The Nature of Light

During the day, the primary source of light is the Sun. Other common sources are flames, white-hot filaments in light bulbs, and glowing gases in glass tubes. What these sources emit, and what we perceive as light, are electromagnetic waves with frequencies that fall within a certain range. Recall, from Chapter 7, that an **electromagnetic wave** is a wave of energy produced when an electric charge accelerates (Figure 8.9). Light is only a tiny part of a larger whole—a wide range of electromagnetic waves called the **electromagnetic spectrum** (Figure 8.10).

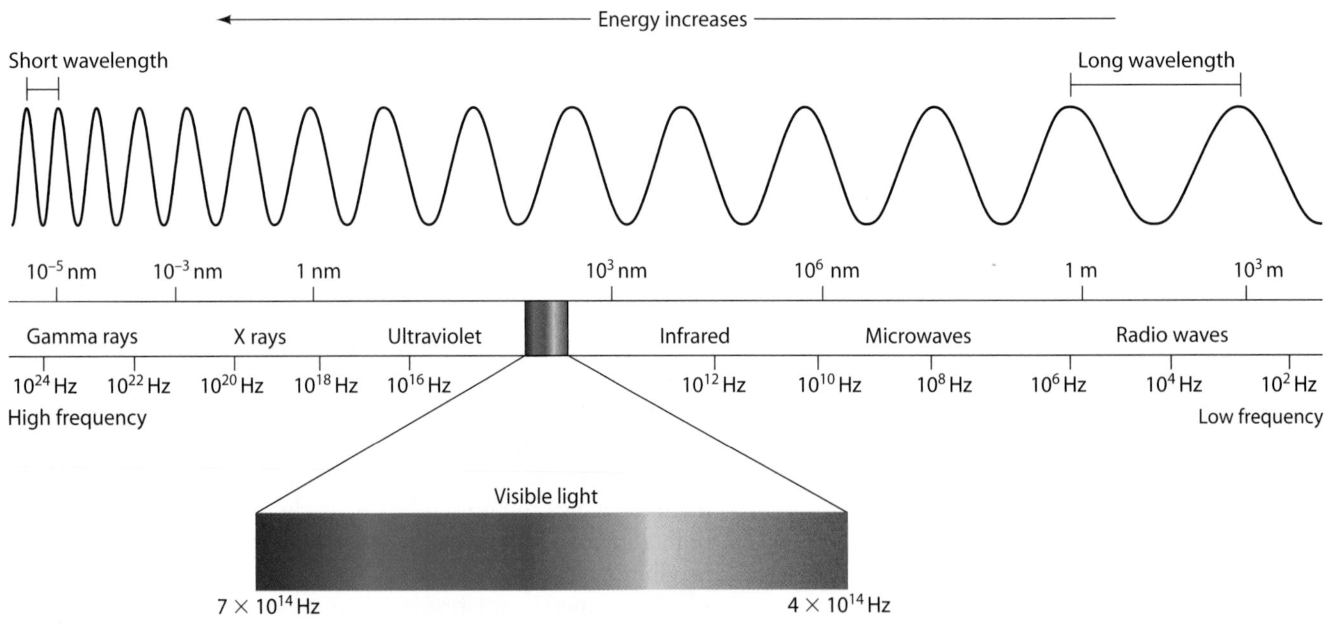

Figure 8.10 Interactive Figure
The electromagnetic spectrum is a continuous range of electromagnetic waves extending from radio waves to gamma rays. The descriptive names of the sections are merely a historical classification, for all the waves are the same in nature, differing principally in frequency and wavelength; all travel at the same speed.

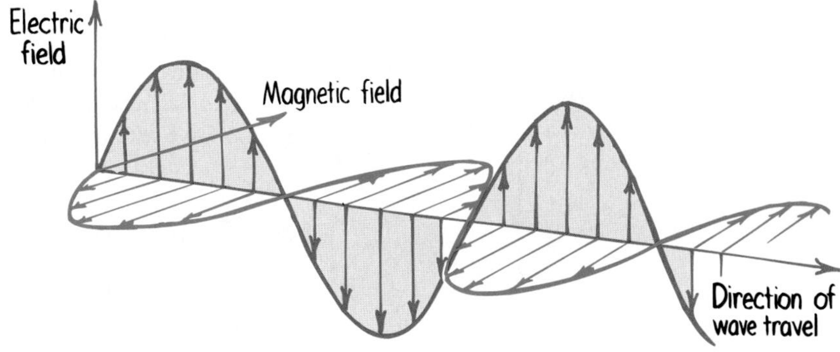

Figure 8.11 Interactive Figure
The electric and magnetic fields of an electromagnetic wave are perpendicular to each other and to the direction to the wave.

Light originates from the accelerated motion of electrons. If you shake the end of a stick back and forth in still water, you'll create waves on the water's surface. Similarly, if you shake an electrically charged rod to and fro in empty space, you'll create electromagnetic waves in space. This is because the moving charge on the rod is an electric current. Recall, from Chapter 7, that a magnetic field surrounds an electric current and that the field changes as the current changes. Recall also that a changing magnetic field induces an electric field—electromagnetic induction. And what does the changing electric field do? It induces a changing magnetic field. The vibrating electric and magnetic fields regenerate each other to make up an electromagnetic wave (Figure 8.11).

INSIGHT Light is the only thing we see. Sound is the only thing we hear.

In a vacuum, all electromagnetic waves move at the same speed—the speed of light, *c*. How they differ from one another is in terms of their frequency. Electromagnetic waves have been detected with a frequency as low as 0.01 hertz (Hz). Others, with frequencies of several thousand hertz (kHz), are classified as low-frequency radio waves. One million hertz (1 MHz) lies in the middle of the AM radio band. The very high frequency (VHF) television band of waves begins at about 50 million hertz (MHz), and FM radio frequencies are from 88 to 108 MHz. Then come ultrahigh frequencies (UHF), followed by microwaves, beyond which are infrared waves. Further still is visible light, which makes up less than a millionth of 1% of the electromagnetic spectrum.

The lowest frequency of light we can see with our eyes appears red. The highest visible frequencies, which are nearly twice the frequency of red light, appear violet. Still higher frequencies are ultraviolet. These higher-frequency waves are more energetic and can cause sunburns. Beyond ultraviolet light are the X-ray and gamma-ray regions. There is no sharp boundary between regions of the spectrum, for they actually grade continuously into one another. The spectrum is divided into these arbitrary regions for the sake of classification.

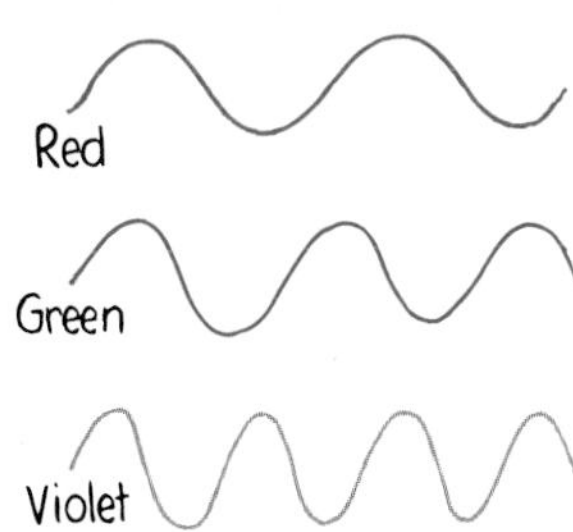

Figure 8.12 Interactive Figure
Relative wavelengths of red, green, and violet light. Violet light has nearly twice the frequency of red light and half the wavelength.

The frequency of the electromagnetic wave as it vibrates through space is identical with the frequency of the oscillating electric charge that generates it. Different frequencies result in different wavelengths—low frequencies produce long wavelengths and high frequencies produce short wavelengths. The higher the frequency of the vibrating charge, the shorter the wavelength of the radiation.*

CHECK YOURSELF

Is it correct to say that a radio wave is a low-frequency light wave? Is a radio wave also a sound wave?

CHECK YOUR ANSWER

Both radio waves and light waves are electromagnetic waves, which originate in the vibrations of electrons. Radio waves have lower frequencies than light waves, so a radio

* The relationship is $c = f\lambda$, where c is the speed of light (constant), f is the frequency, and λ is the wavelength. It is common to describe sound and radio by frequency and light by wavelength. In this book, however, we'll favor the single concept of frequency in describing light.

INSIGHT Light is energy carried in an electromagnetic wave emitted by vibrating electrons in atoms. In air, light travels a million times faster than sound.

wave might be considered to be a low-frequency light wave (and a light wave might be considered to be a high-frequency radio wave). But a sound wave is a mechanical vibration of matter and is not electromagnetic. A sound wave is fundamentally different from an electromagnetic wave. So a radio wave is definitely not a sound wave. (Don't confuse a radio wave with the sound that a loudspeaker emits.)

8.7 Reflection

If you look carefully at the waves next time you are in a swimming pool, you will see that they bounce back when they hit the side of the pool. This is an example of **reflection**—the returning of a wave to the medium from which it came when it hits a barrier.

Water waves reflect off a surface in much the same way that a ball bounces back when it strikes a surface. When the ball hits a surface and bounces back, the angle of incidence (the angle at which the ball strikes the surface) equals the angle of rebound. Waves behave the same way. This is the **law of reflection**, and it holds for all angles:

The angle of reflection equals the angle of incidence.

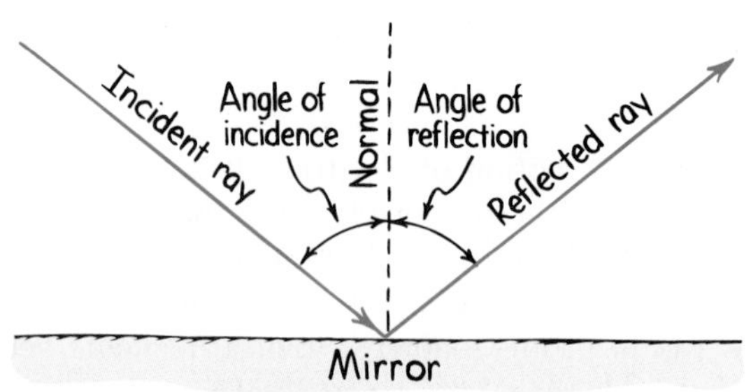

Figure 8.13 Interactive Figure
The law of reflection.

The law of reflection is illustrated with light rays (with arrows representing the direction of light wave travel) in Figure 8.13. Instead of measuring the angles of incident and reflected rays from the reflecting surface, it is customary to measure them from a line perpendicular to the plane of the reflecting surface. This imaginary line is called the *normal*. The incident ray, the normal, and the reflected ray all lie in the same plane.

If you place a candle in front of a mirror, rays of light radiate from the flame in all directions. Figure 8.14 shows only four of the infinite number of rays leaving one of the infinite number of points on the candle. When these rays meet the mirror, they reflect at angles equal to their angles of incidence. The rays diverge from the flame. Note that they also diverge when reflecting from the mirror. These divergent rays appear to emanate from behind the mirror (dashed lines). You see an image of the candle at this point. The light rays do not actually come from this point, so the image is called a *virtual image*. The image is as far behind the mirror as the object is in front of the mirror, and image and object have the same size. When you view yourself in a mirror, for example, the size of your image is the same as the size your twin would appear to be, if located as far behind the mirror as you are in front, as long as the mirror is flat.

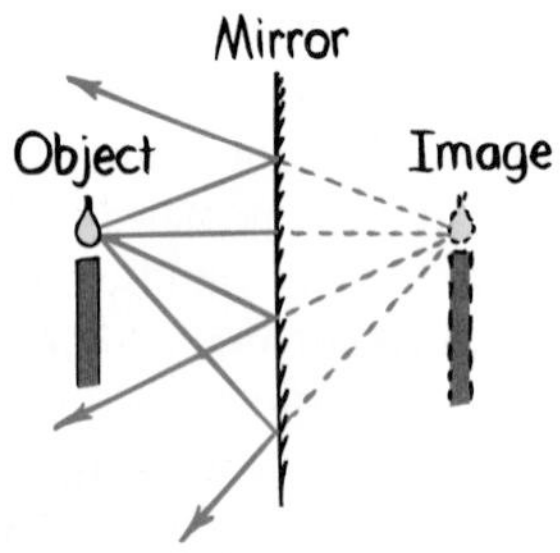

Figure 8.14 A virtual image is located behind the mirror and is located at the position where the extended reflected rays (dashed lines) converge.

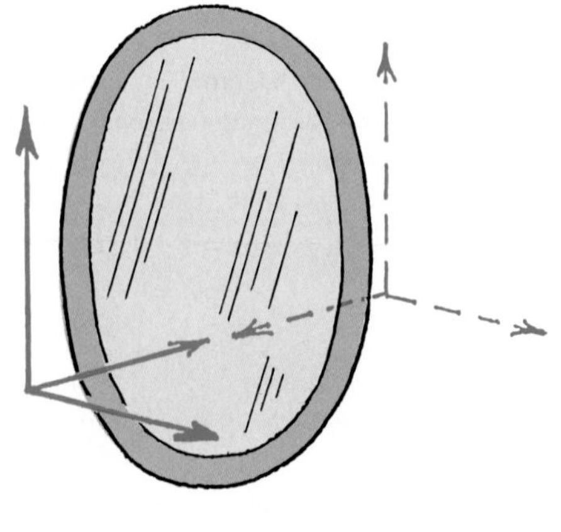

Figure 8.15 Marjorie is as far behind the mirror as she is in front of it. Note that she and her image have the same color of clothing—evidence that light doesn't change frequency upon reflection. Interestingly, her left and right axis is no more reversed than her up and down axis. The axis that is reversed, as shown to the right, is her front and back axis. That's why it appears that her left hand faces the right hand of her image.

Only part of the light that strikes a surface is reflected. For example, on a surface of clear glass and for normal incidence (light perpendicular to the surface), only about 4% is reflected from each surface. On a clean and polished aluminum or silver surface, however, about 90% of the incident light is reflected. The light that is not reflected by a surface penetrates the surface and is either absorbed there (if the material is opaque) or *transmitted*—absorbed and reemitted (if the material is transparent).

Diffuse Reflection

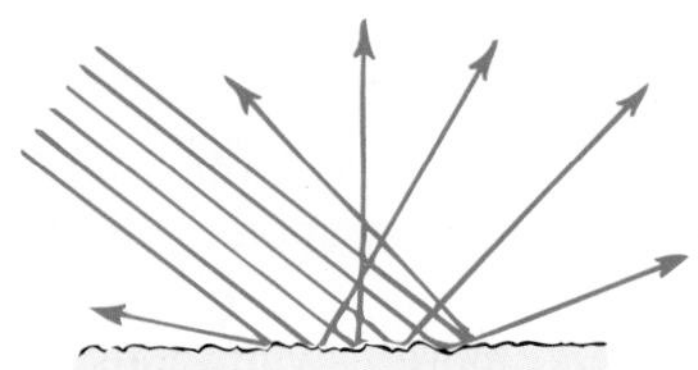

Figure 8.16 Diffuse reflection. Although reflection of each single ray obeys the law of reflection, the many different surface angles that light rays encounter in striking a rough surface produce reflection in many directions.

When light is incident on a rough surface, it is reflected in many directions. This is called *diffuse reflection* (Figure 8.16). If the surface is so smooth that the distances between successive elevations on the surface are less than about one-eighth the wavelength of the light, there is very little diffuse reflection, and the surface is said to be polished. A surface therefore may be polished for radiation of long wavelengths but rough for light of short wavelengths. The wire-mesh "dish" shown in Figure 8.17 is very rough for light waves and is hardly mirrorlike. But, for long-wavelength radio waves, it is "polished" and is an excellent reflector.

This page is smooth to a radio wave, but, to a light wave, it is rough so light reflecting from this page is diffuse. Rays of light striking this page encounter millions of tiny flat surfaces facing in all directions. The incident light, therefore, is reflected in all directions. This is a desirable circumstance. It enables us to see objects from any direction or position. You can see the road ahead of your car at night, for instance, because of diffuse reflection by the rough road surface. When the road is wet, however, it is smoother with less diffuse reflection and therefore more difficult to see. Most of our environment is seen by diffuse reflection.

Figure 8.17 The open-mesh parabolic dish is a diffuse reflector for short wavelength light but a polished reflector for long-wavelength radio waves.

Reflection of Sound

We call the reflection of sound an *echo*. The fraction of sound energy reflected from a surface is large if the surface is rigid and smooth, but it is less if the surface is soft and irregular. The sound energy that is not reflected is transmitted or absorbed.

Sound reflects from a smooth surface in the same way that light does—the angle of incidence is equal to the angle of reflection. Sometimes, when sound reflects from the walls, ceiling, and floor of a room, the surfaces are too reflective and the sound becomes garbled. This is due to multiple reflections called reverberations. On the other hand, if the reflective surfaces are too absorbent, the sound level is low and the room may sound dull and lifeless. Reflected sound in a room makes it sound lively and full, as you have probably experienced while singing in the shower. In the design of an auditorium or concert hall, a balance must be found between reverberation and absorption. The study of sound properties is called *acoustics*.

8.8 Transparent and Opaque Materials

Light is energy carried in an electromagnetic wave emitted by vibrating electrons in atoms. When light is incident upon matter, some of the electrons in the matter are forced into vibration. In this way, vibrations in the emitter are transformed to vibrations in the receiver. This is similar to the way in which sound is transmitted (Figure 8.20).

Thus the way a receiving material responds when light is incident upon it depends on the frequency of the light and on the natural frequency of the electrons in the material. Visible light vibrates at a very high rate, some 100 trillion times per second (10^{14} hertz). If a charged object is to respond to these ultrafast vibrations, it must have very, very little inertia. Electrons are light enough to vibrate at this rate.

Figure 8.18 A magnified view of the surface of ordinary paper.

Such materials as glass and water allow light to pass through in straight lines. We say they are **transparent** to light. To understand how light penetrates a

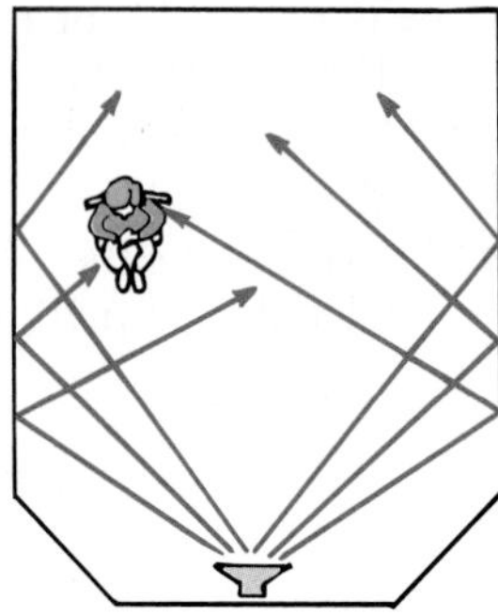

Figure 8.19 The angle of incident sound is equal to the angle of reflected sound.

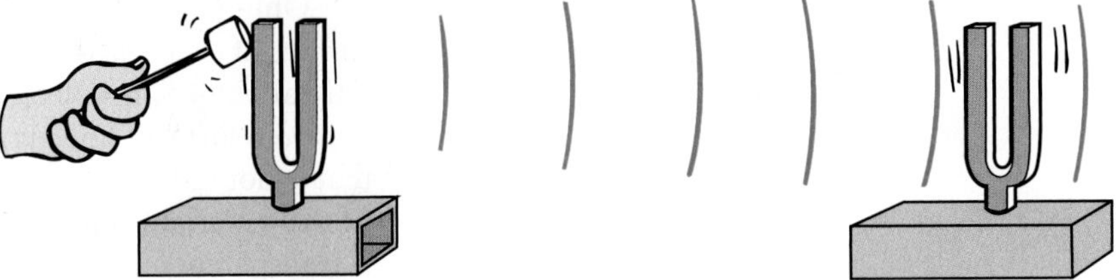

Figure 8.20 Just as a sound wave can force a sound receiver into vibration, a light wave can force the electrons in materials into vibration.

transparent material, visualize the electrons in an atom as if they were connected by springs (Figure 8.21).* When a light wave is incident upon them, the electrons are set into vibration.

Materials that are springy (elastic) respond more to vibrations at some frequencies than they do to vibrations at other frequencies. Bells ring at a particular frequency, tuning forks vibrate at a particular frequency, and so do the electrons of atoms and molecules. The natural vibration frequencies of an electron depend on how strongly it is attached to its atom or molecule. Different atoms and molecules have different "spring strengths." Electrons in glass have a natural vibration frequency in the ultraviolet range. When ultraviolet rays shine on glass, resonance occurs as the wave builds and maintains a large amplitude of vibration of the electron, just as pushing someone at the resonant frequency on a swing builds a large amplitude. The energy that atoms in the glass receive may be passed on to neighboring atoms by collisions, or the energy may be reemitted. Resonating atoms in the glass can hold onto the energy of the ultraviolet light for quite a long time (about 100 millionths of a second). During this time, the atom makes about 1 million vibrations and collides with neighboring atoms and transfers absorbed energy as heat. Thus, glass is not transparent to ultraviolet. Glass absorbs ultraviolet.

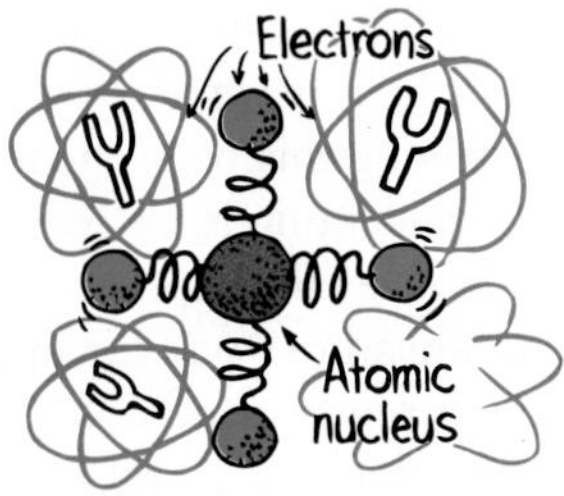

Figure 8.21 The electrons of atoms have certain natural frequencies of vibration, and these can be modeled as particles connected to the atomic nucleus by springs. As a result, atoms and molecules behave somewhat like optical tuning forks.

At lower wave frequencies, such as those of visible light, electrons in the glass are forced into vibration at a lower amplitude. The atoms or molecules in the glass hold the energy for less time, with less chance of collision with neighboring atoms and molecules, and with less of the energy being transformed to heat. The energy of vibrating electrons is reemitted as light. Glass is transparent to all the frequencies of visible light. The frequency of the reemitted light that is passed from molecule to molecule is identical to the frequency of the light that produced the vibration originally. However, there is a slight time delay between absorption and reemission.

INSIGHT When light falls on the surface of a material, it is usually either reemitted without change in frequency or is absorbed into the material and turned into heat. Usually, both of these processes occur in varying degrees.

It is this time delay that results in a lower average speed of light through a transparent material (Figure 8.22). Light travels at different average speeds through different materials. We say average speeds, for the speed of light in a vacuum, whether in interstellar space or in the space between molecules in a

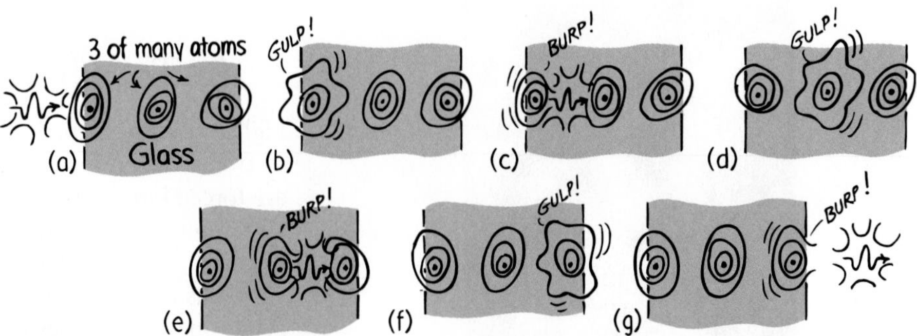

Figure 8.22 A light wave incident upon a pane of glass sets up vibrations in the molecules that produce a chain of absorptions and reemissions, which pass the light energy through the material and out the other side. Because of the time delay between absorptions and reemissions, the light travels through the glass more slowly than through empty space.

* Electrons, of course, are not really connected by springs. We are simply presenting a "spring model" of the atom to help us understand the interaction of light and matter. Scientists devise such conceptual models to understand nature, particularly at the submicroscopic level. The worth of a model lies not in whether it is "true" but in whether it is useful—in explaining observations and in predicting new ones. If predictions are contrary to new observations, the model is usually either refined or abandoned.

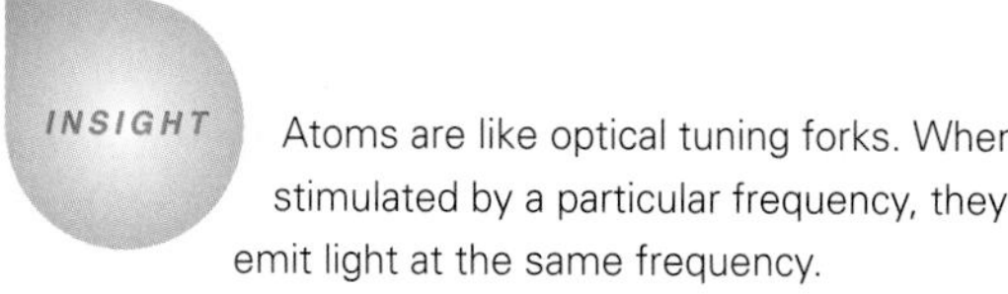

INSIGHT Atoms are like optical tuning forks. When stimulated by a particular frequency, they emit light at the same frequency.

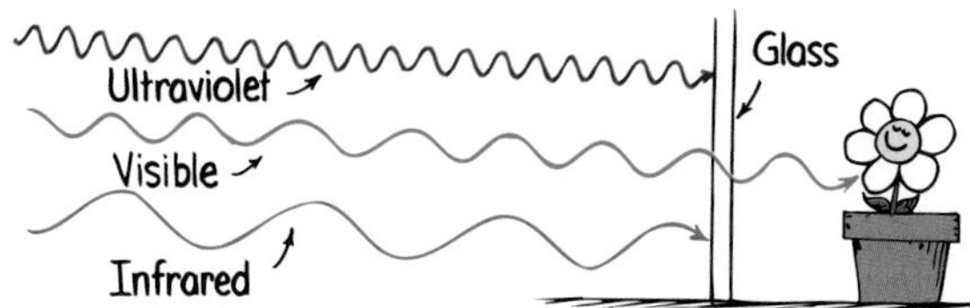

Figure 8.23 Clear glass blocks both infrared and ultraviolet, but it is transparent to all frequencies of visible light.

piece of glass, is a constant 300,000 kilometers per second. We call this speed of light *c*. (This corresponds to 186,000 m/s.) The speed of light in the atmosphere is slightly less than it is in a vacuum, but is usually rounded off as *c*. In water, light travels at 75% of its speed in a vacuum, or 0.75 *c*. In glass, light travels about 0.67 *c*, depending on the type of glass. In a diamond, light travels at less than half its speed in a vacuum, only 0.41 *c*. When light emerges from these materials into the air, it travels at its original speed.

Infrared waves, which have frequencies lower than those of visible light, vibrate not only the electrons but the entire molecules in the structure of the glass and in many other materials. This molecular vibration increases the thermal energy and temperature of the material, which is why infrared waves produce temperature increases in these materials. Glass is transparent to visible light, but not to ultraviolet and infrared light (Figure 8.23).

CHECK YOURSELF

1. Pretend that, while you are at a social gathering, you make several momentary stops across the room to greet people who are "on your wavelength." How is this analogous to light traveling through glass?
2. In what way is it not analogous?

CHECK YOUR ANSWERS

1. Your average speed across the room would be less because of the time delays associated with your momentary stops. Likewise, the speed of light in glass is less because of the time delays in interactions with atoms along its path.
2. In the case of walking across the room, it is you who begin the walk and you who complete the walk. This is not analogous to the similar case of light because (according to our model for light passing through a transparent material) the light that is absorbed by an electron that has been made to vibrate is not the same light that is reemitted—even though the two, like identical twins, are indistinguishable.

Most things around us are **opaque**—they absorb light without reemission. Books, desks, chairs, and people are opaque. Vibrations given by light to their atoms and molecules are turned into random kinetic energy—into thermal energy. They become slightly warmer.

Figure 8.24 The rose appears red because it reflects light in this frequency range.

8.9 Color

Roses are red and violets are blue; colors intrigue artists and scientists too. The colors we see depend on the frequency of the light we see. Different frequencies of light are perceived as different colors; the lowest frequency we detect appears, to most people, as the color red, and the highest appears as violet. Between them range the infinite number of hues that make up the color spectrum of the rainbow. By convention, these hues are grouped into seven colors: red, orange, yellow, green, blue, indigo, and violet. These colors together appear white. The white light from the Sun is a composite of all the visible frequencies.

Selective Reflection

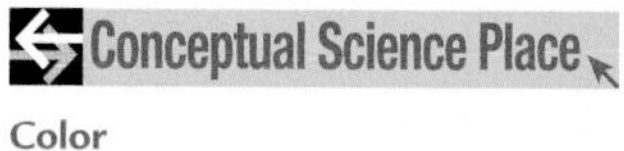

Color

A rose, for example, doesn't emit light; it reflects light. If we pass sunlight through a prism and then place a deep-red rose in various parts of the spectrum, the rose will appear brown or black in all regions of the spectrum except in the red region. In the red part of the spectrum, the petals also will appear red, but the green stem and leaves will appear black. This shows that the red rose has the ability to reflect red light, but it cannot reflect other colors; the green leaves have the ability to reflect green light and, likewise, cannot reflect other colors. When the rose is held in white light, the petals appear red and the leaves appear green,

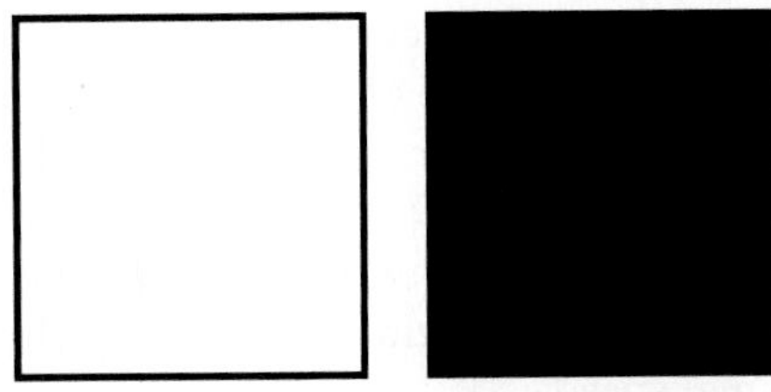

Figure 8.25 The square on the left reflects all the colors illuminating it. In sunlight, it is white. When illuminated with blue light, it is blue. The square on the right absorbs all the colors illuminating it. In sunlight, it is warmer than the white square.

because the petals reflect the red part of the white light and the leaves reflect the green part of the white light.

Usually, a material will absorb light of some frequencies and reflect the rest. If a material absorbs most of the light and reflects red, for example, the material appears red. If it reflects light of all the visible frequencies, like the white part of this page, it will be the same color as the light that shines on it. If a material absorbs light and reflects none, then it is black (Figures 8.25 and 8.26).

Interestingly, the petals of most yellow flowers, like daffodils, reflect red and green as well as yellow. Yellow daffodils reflect a broad band of frequencies. The reflected colors of most objects are not pure single-frequency colors but are a mixture of frequencies.

An object can reflect only those frequencies present in the illuminating light. The appearance of a colored object, therefore, depends on the kind of light that illuminates it. An incandescent lamp, for instance, emits light of lower average frequencies than sunlight, enhancing any reds viewed in this light. In a fabric having only a little bit of red in it, the red is more apparent under an incandescent lamp than it is under a fluorescent lamp. Fluorescent lamps are richer in the higher frequencies, and so blues are enhanced in their light. For this reason, it is difficult to tell the true color of objects viewed in artificial light (Figure 8.27).

Figure 8.26 Rosie's dark fur absorbs all of the radiant energy in incident sunlight and therefore appears black. Fur on her chest reflects all frequencies and therefore appears white.

Selective Transmission

The color of a transparent object depends on the color of the light it transmits. A red piece of glass appears red because it absorbs all the colors of white light except red, so red light is transmitted. Similarly, a blue piece of glass appears blue because it transmits primarily blue and absorbs the other colors that illuminate it. These pieces of glass contain dyes or pigments—fine particles that selectively absorb light of particular frequencies and selectively transmit others. Ordinary window glass doesn't have a color because it transmits light of all visible frequencies equally well.

CHECK YOURSELF

1. When illuminated with green light, why do the petals of a red rose appear black?
2. If you hold a match, a candle flame, or any small source of white light between you and a piece of red glass, you'll see two reflections from the glass: one from the front surface of the glass and one from the back surface. What color reflections will you see?

Figure 8.27 Color depends on the light source.

CHECK YOUR ANSWERS

1. The petals absorb rather than reflect the green light. Since green is the only color illuminating the rose, and green contains no red to be reflected, the rose reflects no color at all and appears black.
2. You see white reflected from the top surface. You see red reflected from the back surface because only red reaches the back surface and reflects from there.

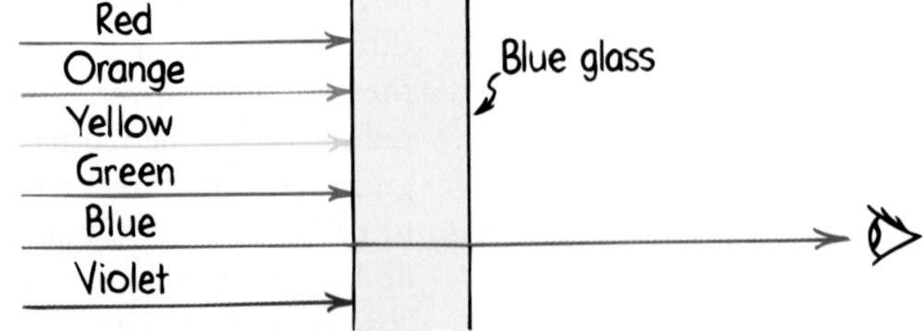

Figure 8.28 Only energy having the frequency of blue light is transmitted; energy of the other frequencies is absorbed and warms the glass.

INTEGRATED SCIENCE BIOLOGY

Mixing Colored Lights

Conceptual Science Place

Colored Shadows; Yellow-Green Peak of Sunlight

You can see that white light from the Sun is composed of all the visible frequencies when you pass sunlight through a prism. The white light is dispersed into a rainbow-colored spectrum. The distribution of solar frequencies (Figure 8.29) is uneven, and the light is most intense in the yellow-green part of the spectrum. How fascinating it is that our eyes have evolved to have maximum sensitivity in this range. That's why many fire engines are painted yellow-green, particularly at airports, where visibility is vital. Our sensitivity to yellow-green light is also why we see better under the illumination of yellow sodium-vapor lamps at night than we do under incandescent lamps of the same brightness.

All the colors combined produce white. Interestingly, we also see white from the combination of only red, green, and blue light. We can understand this by dividing the solar radiation curve into three regions, as in Figure 8.30. Three types of cone-shaped receptor cells in our eyes ("cones") perceive color. (For more on this, see Chapter 19.) Each cone cell is stimulated only by certain frequencies of light. Light of lower visible frequencies stimulates the cones sensitive to low frequencies and appears red. Light of middle frequencies stimulates the midfrequency-sensitive cones and appears green. Light of higher frequencies stimulates the higher-frequency-sensitive cones and appears blue. When all three types of cones are stimulated equally, we see white.

Project red, green, and blue lights on a screen and, where they all overlap, white is produced. If two of the three colors overlap, or are added, then another color sensation will be produced (Figure 8.31). By adding various amounts of red, green, and blue, the colors to which each of our three types of cones are sensitive, we can produce any color in the spectrum. For this reason, red, green, and blue are called the *additive primary colors*. Here's what happens when any two of the primary colors are combined:

INSIGHT Every artist knows that if you mix red, green, and blue paint, the result will not be white but a muddy dark brown. The mixing of pigments in paints and dyes is a process called subtractive color mixing and is entirely different from mixing light of different colors.

Red + Blue = Magenta

Red + Green = Yellow

Blue + Green = Cyan

We say that magenta is the opposite of green; cyan is the opposite of red; and yellow is the opposite of blue. The addition of any color to its opposite color results

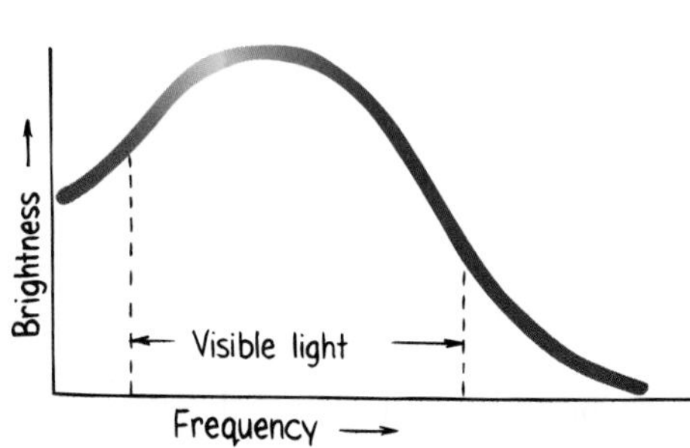

Figure 8.29 The radiation curve of sunlight is a graph of brightness against frequency. Sunlight is brightest in the yellow-green region, which is in the middle of the visible range.

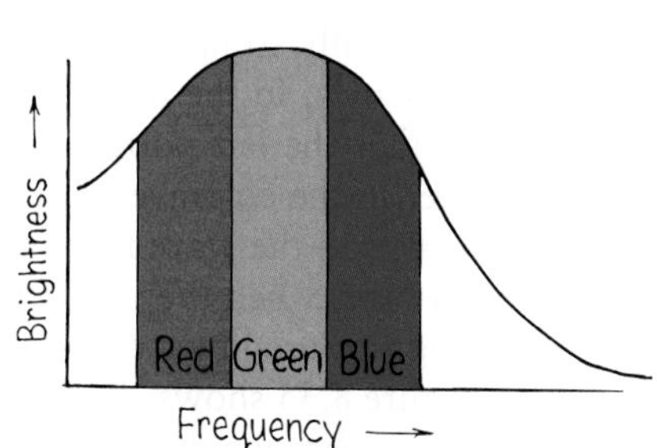

Figure 8.30 The radiation curve of sunlight divided into three regions—red, green, and blue. Red, green, and blue are the additive primary colors, so-called because every color of light can be produced by adding them together in different proportions.

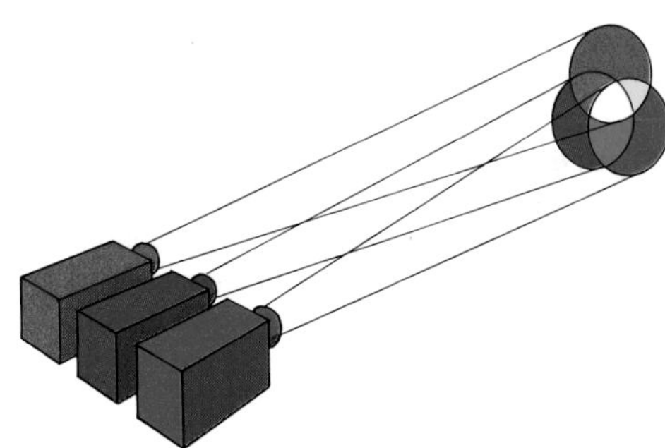

Figure 8.31 Interactive Figure

Color addition. When three projectors shine red, green, and blue light on a white screen, the overlapping parts produce different colors. Red and blue light combine to make magenta; red and green light combine to make yellow; and blue and green light combine to make cyan. White is produced where all three overlap.

farther a galaxy is from Earth, the faster it is moving away. This is the basis of our current belief that the universe is ever-expanding. You will learn more about Hubble and his fundamentally important contributions to astronomy in Chapter 28.

CHECK YOURSELF

1. If Hubble had observed that the light from distant galaxies was blue-shifted, would this be evidence for an expanding or a shrinking universe? Explain.
2. Explain why, in terms of the bunching together of waves, why light from a receding source is red-shifted.

CHECK YOUR ANSWERS

1. Blue-shifted atomic spectra would indicate that the source was approaching the observer; if all surrounding galaxies showed blue-shifted atomic spectra, this would be evidence for a shrinking universe.
2. Light from a receding source should be red-shifted because, just like the bug on the pond, wave crests grow successively farther apart and thus have a lower frequency.

8.14 The Wave–Particle Duality

We have described light as a wave. The earliest ideas about the nature of light, however, were that light was composed of tiny particles. In ancient times, Plato and other Greek philosophers held to a particle view of light—as, in the early 1700s, did Isaac Newton, who first became famous for his experiments with light. A hundred years later, the wave nature of light was demonstrated by Thomas Young in his interference experiments. The wave view was reinforced in 1862 by Maxwell's finding that light is energy carried in oscillating electric and magnetic fields of electromagnetic waves. The wave view of light was confirmed experimentally by Heinrich Hertz 25 years later.

Then, in 1905, Albert Einstein published a Nobel Prize–winning paper that challenged the wave theory of light. Einstein stated that light in its interactions with matter was confined not in continuous waves, as Maxwell and others had envisioned, but in tiny particles of energy called *photons*. Einstein's particle model of light explained a perplexing phenomenon of that time—*the photoelectric effect*.

The Photoelectric Effect

When light shines on certain metal surfaces, electrons are ejected from those surfaces. This is the photoelectric effect, which is put to use in electric eyes, in light meters, and in motion-picture sound tracks. What perplexed investigators at the turn of the twentieth century was that ultraviolet and violet light imparted sufficient energy to knock electrons from those metal surfaces, while lower-frequency light did not—even when the lower-frequency light was very bright. Ejection of electrons depended only on the frequency of light, and the higher the frequency of the light, the greater the kinetic energy of the ejected electrons. Very dim high-frequency light ejected fewer electrons, but it ejected each of them with the same kinetic energy of electrons ejected in brighter light of the same frequency (Figure 8.51).

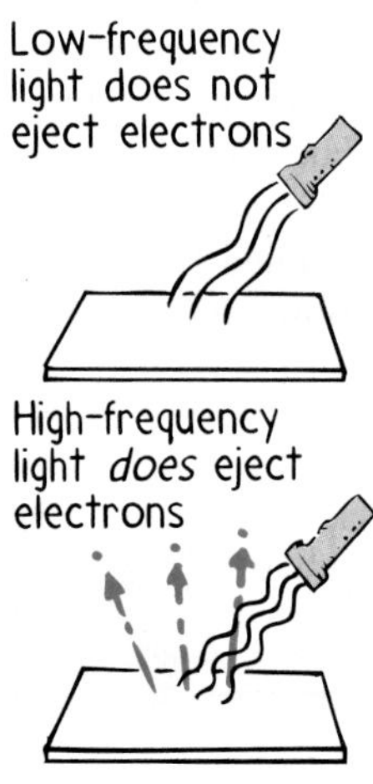

Low-frequency light does not eject electrons
High-frequency light *does* eject electrons

Figure 8.51 The photoelectric effect depends on frequency.

Einstein's explanation was that the electrons in the metal were being bombarded by "particles of light"—by photons. Einstein stated that the energy of each photon was proportional to its frequency: That is,

$$E \sim f$$

So Einstein viewed light as a hail of photons, each carrying energy proportional to its frequency. One photon is completely absorbed by each electron ejected from the metal.

All attempts to explain the photoelectric effect by waves failed. A light wave has a broad front, and its energy is spread out along this front. For the light wave to eject a single electron from a metal surface, all its energy would somehow have to be concentrated on that one electron. But this is as improbable as an ocean wave hitting a beach and knocking only one single seashell far inland with an energy equal to the energy of the whole wave. Therefore, the photoelectric effect suggests that, instead of thinking of light encountering a surface as a continuous train of waves, we should conceive of light encountering a surface, or any detector, as a succession of particle-like photons. The energy of each photon is proportional to the frequency of light, and that energy is given completely to a single electron in the metal's surface. The number of ejected electrons has to do with the number of photons—the brightness of the light.

Experimental verification of Einstein's explanation of the photoelectric effect was made 11 years later by the American physicist Robert Millikan. Every aspect of Einstein's interpretation was confirmed. The photoelectric effect proves conclusively that light has particle properties. A wave model of light is inconsistent with the photoelectric effect. On the other hand, interference demonstrates convincingly that light has wave properties, and a particle model of light is inconsistent with interference.

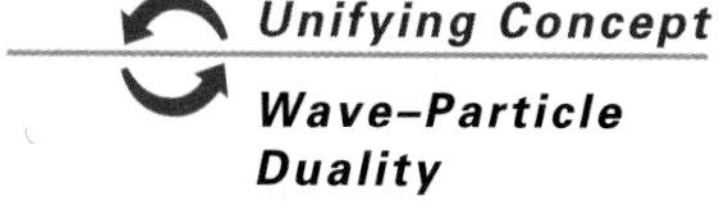

Evidently, light has both a wave nature and a particle nature—a **wave–particle duality.** It reveals itself as a wave or particle depending on how it is being observed. Simply stated, light behaves as a stream of photons when it interacts with a sheet of metal or other detector and it behaves as a wave in traveling from a source to the place where it is detected. Light travels as a wave and hits as a stream of photons. The fact that light exhibits both wave and particle behavior is one of the most interesting surprises that physicists discovered in the twentieth century.

CHAPTER 8 Review

Summary of Terms

Amplitude The distance from the midpoint to the crest (or to the trough) of a wave.

Diffraction Any bending of light by means other than reflection and refraction.

Dispersion The separation of light into colors arranged by frequency.

Doppler effect The change in frequency of a wave due to the motion of the source (or due to the motion of the receiver).

Electromagnetic spectrum The continuous range of electromagnetic waves that extends in frequency from radio waves to gamma rays.

Electromagnetic wave An energy-carrying wave produced when an electric charge accelerates.

Forced vibration The setting up of vibrations in an object by a vibrating source.

Frequency For a vibrating body, the number of vibrations per unit time. For a wave, the number of crests that pass a particular point per unit time.

Interference The combined effect of two or more waves overlapping.

Law of reflection The angle of reflection equals the angle of incidence.

Longitudinal wave A wave in which the medium vibrates in a direction parallel (longitudinal) with the direction in which the wave travels.

Natural frequency A frequency at which an elastic object naturally tends to vibrate.

Opaque The term applied to materials that absorb light without reemission.

Period The time required for a vibration or a wave to make a complete cycle.

Reflection The returning of a wave to the medium from which it came when it hits a barrier.

Refraction The bending of waves due to a change in the medium.

Resonance A dramatic increase in the amplitude of a wave that results when the frequency of forced vibrations matches an object's natural frequency.

Transparent The term applied to materials through which light can pass in straight lines.

Transverse wave A wave in which the medium vibrates in a direction perpendicular (transverse) to the direction in which the wave travels.

Wave A disturbance that travels from one place to another transporting energy, but not necessarily matter, along with it.

Wavelength The distance from the top of one crest to the top of the next one, or, equivalently, the distance between successive identical parts of the wave.

Review Questions

8.1 Vibrations and Waves

1. Distinguish between these different parts of a wave: amplitude, wavelength, frequency, and period.
2. What is the source of all waves?

8.2 Wave Motion

3. In one word, what is it that moves from source to receiver in wave motion?
4. Does the medium in which a wave travels move with the wave?
5. What is the relationship among frequency, wavelength, and wave speed?

8.3 Transverse and Longitudinal Waves

6. (a) In what direction are the vibrations in a transverse wave, relative to the direction of wave travel?
   (b) In what direction are the vibrations in a longitudinal wave, relative to the direction of wave travel?
7. Distinguish between a compression and a rarefaction.

8.4 The Nature of Sound

8. Define the wavelength of sound in terms of successive compressions of air.
9. Can sound travel through a vacuum? Why or why not?

8.5 Resonance

10. Why does a struck tuning fork sound louder when its handle is held against a table?
11. Distinguish between forced vibrations and resonance.

8.6 The Nature of Light

12. What is the principal difference between a radio wave and light? Between light and X-rays?
13. How does the frequency of an electromagnetic wave compare with the frequency of the vibrating electrons that produces it?

8.7 Reflection

14. What is the law of reflection?
15. Does the law of reflection hold for diffuse reflection? Explain.

8.8 Transparent and Opaque Materials

16. The sound coming from one tuning fork can force another to vibrate. What is the analogous effect for light?
17. (a) What is the fate of the energy in ultraviolet light incident on glass?
    (b) What is the fate of the energy in visible light incident on glass?
18. How does the average speed of light in glass compare with its speed in a vacuum?

8.9 Color

19. What is the relationship between the frequency of light and its color?
20. Distinguish between the white of this page and the black of this ink, in terms of what happens to the white light that falls on both.
21. Why does common window glass have no color?

8.10 Refraction

22. What causes the bending of light in refraction?
23. Is refraction a property of sound, light, or both?
24. Distinguish between a converging lens and a diverging lens.

8.11 Diffraction

25. For an opening of a given size, is diffraction more pronounced for a longer wavelength or a shorter wavelength?
26. What are some of the ways in which diffraction can be useful or troublesome?

8.12 Interference

27. What kinds of waves exhibit interference?
28. Distinguish between constructive interference and destructive interference.

8.13 The Doppler Effect

29. Why does an observer measure waves from an approaching source as having a higher frequency than if the source were stationary?
30. Give an example of how you might observe the Doppler Effect in your everyday life.

8.14 The Wave–Particle Duality

31. What evidence can you cite for the particle nature of light?
32. Which are more successful in dislodging electrons from a metal surface, photons of violet light or photons of red light? Why?
33. When does light behave as a particle? When does it behave as a wave?

Multiple Choice Questions

Choose the BEST answer to the following.

1. As a water wave passes by, a portion of water vibrates up and down two complete cycles in 1 second. What is the wave's speed for a wavelength of 5 meters?
   (a) 2 m/s
   (b) 5 m/s
   (c) 10 m/s
   (d) 15 m/s
   (e) none of these
2. When a source of sound approaches you, you detect an increase in its
   (a) speed.
   (b) wavelength.
   (c) frequency.
   (d) All of these.
3. A singer holds a high note and shatters a distant crystal glass. This phenomenon best demonstrates
   (a) forced vibrations.
   (b) the Doppler effect.
   (c) interference.
   (d) resonance.
4. Which of the following does not belong to the same family?
   (a) light wave
   (b) radio wave
   (c) sound wave
   (d) microwave
   (e) X-ray
5. If water naturally absorbed blue and violet light rather than infrared, water would appear
   (a) greenish blue, as it presently appears.
   (b) a more intense greenish blue.
   (c) orange-ish yellow.
   (d) to have no color at all.
6. Light has both a wave nature and a particle nature. Light behaves primarily as a particle when it
   (a) travels from one place to another.
   (b) interacts with matter.
   (c) does neither.

INTEGRATED SCIENCE CONCEPTS

Biology—Sensing Pitch

1. How does the pitch of sound relate to its frequency?
2. A cat can hear sound frequencies up to 70,000 Hz. Bats send and receive ultrahigh-frequency squeaks up to 120,000 Hz. Which hears sound of shorter wavelengths, cats or bats?

Biology—Mixing Colored Lights

1. Explain how you are able to see a wide range of colors although there are only three kinds of light-sensitive cells ("cones") in your eyes?
2. Stare at a piece of colored paper for 45 seconds or so. Then look at a plain white surface. What do you see now? Can you explain what

you see? (Hint: To figure this out, you need to know that the cones in your retina that are receptive to the color of the paper have become fatigued.)

Astronomy—The Doppler Effect and the Expanding Universe

1. Swing a buzzer of any kind over your head in a circle. Ask some friends to stand off the side, listen to the buzzer, and report their observations. Then switch places so you can hear the buzzer as it moves too. How does the sound of the buzzer change? Why?
2. How does the Doppler effect provide evidence that we live in an expanding universe?

Active Explorations

1. Suspend a wire grille from a refrigerator or an oven from a string, holding the ends of the string to your ears. Let a friend gently stroke the grille with pieces of broom straw and with other objects. The effect is best appreciated when you are in a relaxed condition with your eyes closed. Be sure to try this!

2. Set up two pocket mirrors at right angles, and place a coin between them. You'll see four coins. Change the angle of the mirrors, and see how many images of the coin you can see. With the mirrors at right angles, look at your face. Then wink. What do you see? You now see yourself as others see you. Hold a printed page up to the double mirrors and compare its appearance with the reflection of a single mirror.

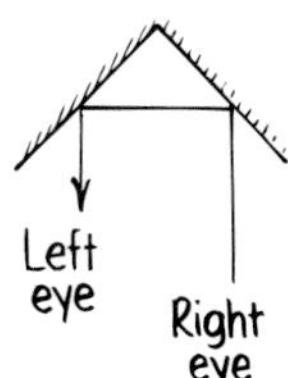

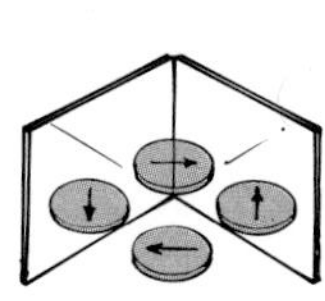

Exercises

1. What kind of motion should you impart to a stretched coiled spring (or to a Slinky) to produce a transverse wave? To produce a longitudinal wave?
2. What does it mean to say that a radio station is "at 101.1 on your FM dial?"
3. At the stands of a race track, you notice smoke from the starter's gun before you hear it fire. Explain.
4. What is the danger posed by people in the balcony of an auditorium stamping their feet in a steady rhythm?
5. The sitar, an Indian musical instrument, has a set of strings that vibrate and produce music, even though they are never plucked by the player. These "sympathetic strings" are identical to the plucked strings and are mounted below them. What is your explanation?
6. A railroad locomotive is at rest with its whistle shrieking, and then it starts moving toward you. (a) Does the frequency that you hear increase, decrease, or stay the same? (b) How about the wavelength reaching your ear? (c) How about the speed of sound in the air between you and the locomotive?
7. What is the fundamental source of electromagnetic radiation?
8. Which has the shorter wavelengths, ultraviolet or infrared? Which has the higher frequencies?
9. Do radio waves travel at the speed of sound, at the speed of light, or at some speed in between?
10. What determines whether a material is transparent or opaque?
11. You can get a sunburn on a cloudy day, but you can't get a sunburn even on a sunny day if you are behind glass. Explain.
12. Suppose that sunlight falls on both a pair of reading glasses and a pair of dark sunglasses. Which pair of glasses would you expect to become warmer? Defend your answer.
13. Fire engines used to be red. Now many of them are yellow-green. Why the change of color?
14. The radiation curve of the Sun (Figure 8.29) shows that the brightest light from the Sun is yellow-green. Why then do we see the Sun as whitish instead of yellow-green?
15. Her eye at point P looks into the mirror. Which of the numbered cards can she see reflected in the mirror?

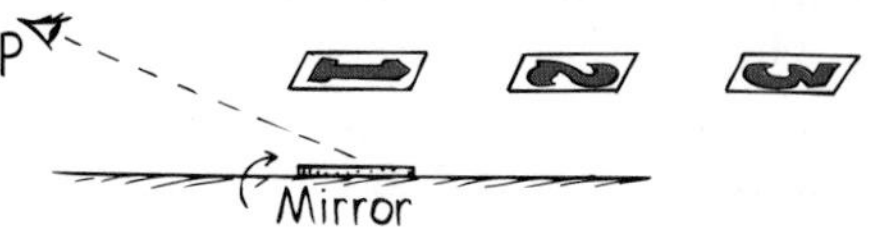

16. Cowboy Joe wishes to shoot his assailant by ricocheting a bullet off a mirrored metal plate. To do so, should he simply aim at the mirrored image of his assailant? Explain.
17. If, while standing on the bank of a stream, you wished to spear a fish swimming in the water out in front of you, would you aim above, below, or directly at the observed fish to make a direct hit? If you decided instead to zap the fish with a laser, would you aim above, below, or directly at the observed fish? Defend your answers.
18. What happens to light of a certain frequency when it is incident on a material whose natural frequency is the same as the frequency of the light?
19. The ocean wave is cyan. What color(s) of light does it absorb? What colors does it reflect?

20. A rule of thumb for estimating the distance in kilometers between an observer and a lightning strike is to divide the number of seconds in the interval between the flash and the sound by 3. Is this rule correct?
21. If a single disturbance some unknown distance away sends out both transverse and longitudinal waves that travel at distinctly different speeds in the medium, such as the ground during earthquakes, how could the origin of the disturbance be located?
22. A bat flying in a cave emits a sound and receives its echo 0.1 s later. How far away is the cave wall?
23. Why do radio waves diffract around buildings whereas light waves do not?
24. Suntanning produces cellular damage in the skin. Why is ultraviolet radiation capable of producing this damage whereas visible radiation is not?
25. Explain briefly how the photoelectric effect is used in the operation of at least two of the following: an electric eye, a photographer's light meter, and the sound track of a motion picture.
26. Does the photoelectric effect prove that light is made of particles? Do interference experiments prove that light is composed of waves? (Is there a distinction between what something is and how it behaves?)
27. Write a letter to Grandpa explaining why we now say that light is not just a particle, and is not just a wave, but in fact is both—a "wavicle"!

Problems

1. What is the frequency, in hertz, that corresponds to each of the following periods: (a) 0.10 s, (b) 5 s, (c) $\frac{1}{60}$ s?
2. The nearest star beyond the Sun is Alpha Centauri, which is 4.2×10^{16} meters away. If we were to receive a radio message from this star today, show that it would have been sent 1.4×10^{8} seconds ago (4.4 years ago).
3. Blue-green light has a frequency of about 6×10^{14} Hz. Use the relationship $c = f\lambda$ to show that the wavelength of this light in air is 5×10^{-7} meters. How does this wavelength compare with the size of an atom, which is about 10^{-10} m?
4. The wavelength of light changes as light goes from one medium to another, while the frequency remains the same. Is the wavelength longer or shorter in water than in air? Explain in terms of the equation: Speed = frequency × wavelength.
5. A certain blue-green light has a wavelength of 600 nm in air. Show that its wavelength in water, where light travels at 75% of its speed in air is 450 nm. Show that its speed in Plexiglas, where light travels at 67% of its speed in air, is 400 nm. (1 nm = 10^{-9}m).
6. A certain radar installation that is used to track airplanes transmits electromagnetic radiation with a wavelength of 3 cm. (a) Show that the frequency of this radiation, measured in billions of hertz, is 10 GHz. (b) Show that the time required for a pulse of radar waves to reach an airplane 5 km away and return is 3.3×10^{-5} seconds.
7. Suppose that you walk toward a mirror at 2 m/s. Show that you and your image approach each other at a speed of 4 m/s (and not 2 m/s.)

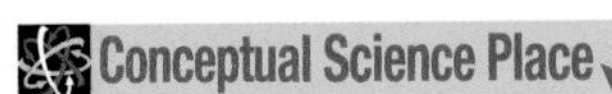

CHAPTER 8 ONLINE RESOURCES

Tutorials Color, The Doppler Effect • ***Videos*** Longitudinal vs. Transverse Waves, Colored Shadows, Yellow-Green Peak of Sunlight, Refraction of Sound, The Doppler Effect • ***Quiz*** • ***Flashcards*** • ***Links***

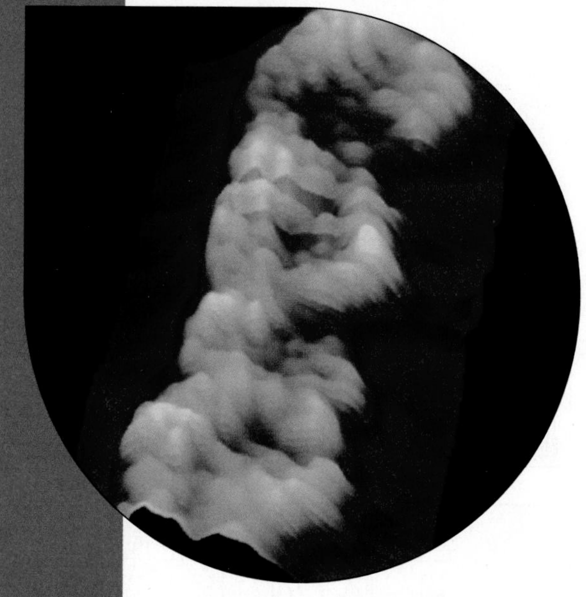

CHAPTER 9

The Atom

This false-color image of a DNA double-helix was obtained with a scanning tunneling microscope (STM). An STM image is made by dragging an instrument with an ultra-fine point over the contours of a specimen and recording the height of the point as it moves. This image of DNA is magnified 2,000,000×.

CHAPTER OUTLINE

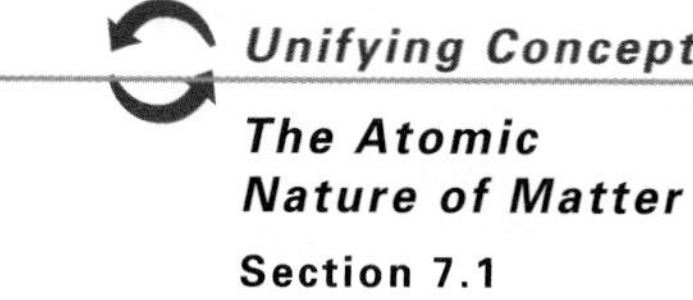

Unifying Concept
The Atomic Nature of Matter
Section 7.1

Imagine you are falling off a chair in slow motion. And while you're falling, you are also shrinking in size. By the time you hit the floor, you are the size of an atom, just one-billionth of a meter across. What does the submicroscopic world that you now inhabit look like? Are the atoms that make up the floor solid structures or are they mostly empty space? What particles make up the atom's core, the nucleus? And the tiny, electrically charged particles whizzing around the nucleus—these can only be found in certain regions of space. Why? Do electrons have wavelike properties, just as light has a wave–particle dual nature? How old are these undulating blobs, atoms, that surround you in your submicroscopic environment? Where do they come from? How do atoms differ from one another? How much do we know about what atoms really look like? These are just a few of the questions we will explore as we investigate the world of the atom.

9.1 The Elements

You know that atoms make up the matter around you, from stars to steel to chocolate ice cream. Given all these various materials, you might think that there must be many different kinds of atoms. But the number of different kinds of atoms is surprisingly small. The great variety of substances results from the many ways in which a few kinds of atoms can be combined. Just as the three colors red, green, and blue can be combined to form any color on a television screen, or just as the 26 letters of the alphabet make up all the words in a dictionary, only a few kinds of atoms combine in different ways to produce all of the countless substances in the universe. To date, we know of 115 distinct kinds of atoms. Of these, about 90 are found in nature. The remaining kinds of atoms have been created in the laboratory.

Any material that is made up of only one type of atom is classified as an element. A few examples are shown in Figure 9.1. Pure gold, for example, is an element—it contains only gold atoms. Nitrogen gas is an element because it contains only nitrogen atoms. Likewise, the graphite in your pencil is an element—carbon. Graphite is made up solely of carbon atoms. All of the elements are listed in a chart called the **periodic table**, which is shown in Figure 9.2.

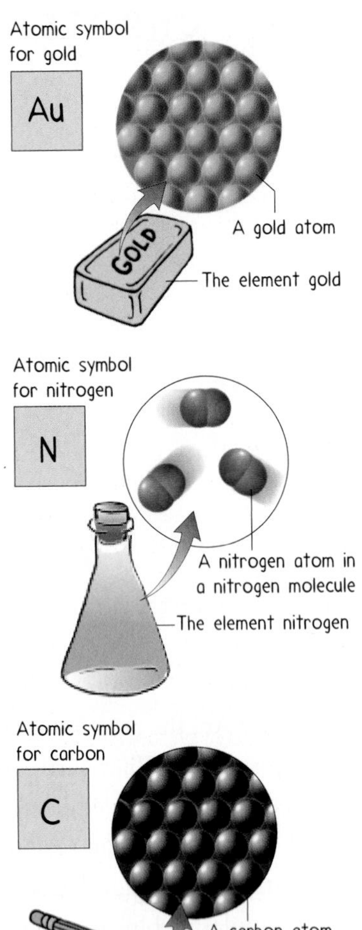

Figure 9.1 Any element consists only of one kind of atom.

1 H																	2 He
3 Li	4 Be											5 B	6 C	7 N	8 O	9 F	10 Ne
11 Na	12 Mg											13 Al	14 Si	15 P	16 S	17 Cl	18 Ar
19 K	20 Ca	21 Sc	22 Ti	23 V	24 Cr	25 Mn	26 Fe	27 Co	28 Ni	29 Cu	30 Zn	31 Ga	32 Ge	33 As	34 Se	35 Br	36 Kr
37 Rb	38 Sr	39 Y	40 Zr	41 Nb	42 Mo	43 Tc	44 Ru	45 Rh	46 Pd	47 Ag	48 Cd	49 In	50 Sn	51 Sb	52 Te	53 I	54 Xe
55 Cs	56 Ba	57 La	72 Hf	73 Ta	74 W	75 Re	76 Os	77 Ir	78 Pt	79 Au	80 Hg	81 Tl	82 Pb	83 Bi	84 Po	85 At	86 Rn
87 Fr	88 Ra	89 Ac	104 Rf	105 Db	106 Sg	107 Bh	108 Hs	109 Mt	110 Ds	111 Uuu	112 Uub	113 Uut	114 Uuq	115 Uup	116 Uuh		

58 Ce	59 Pr	60 Nd	61 Pm	62 Sm	63 Eu	64 Gd	65 Tb	66 Dy	67 Ho	68 Er	69 Tm	70 Yb	71 Lu
90 Th	91 Pa	92 U	93 Np	94 Pu	95 Am	96 Cm	97 Bk	98 Cf	99 Es	100 Fm	101 Md	102 No	103 Lr

Figure 9.2 The periodic table of the elements.

In the periodic table, each element is designated by its **atomic symbol,** which comes from the letters of the element's name. For example, the atomic symbol for carbon is C and that for chlorine is Cl. In many cases, the atomic symbol is derived from the element's Latin name. Gold has the atomic symbol Au, which comes from its Latin name, *aurum.* Lead has the atomic symbol Pb, which derives from its Latin name, *plumbum.* Hence, "plumbers" are so named because they once worked with lead pipes. Some elements are named after people. Curium is named after Pierre and Marie Curie, discoverers of radioactivity. Still other elements are named after particular places—californium (Cf), francium (Fr), germanium (Ge), and berkelium (Bk) are tributes to the locations where they were discovered (Figure 9.3).

Note that only the first letter of an atomic symbol is capitalized. The symbol for the element cobalt, for instance, is Co; CO denotes a combination of two elements: carbon, C, and oxygen, O.

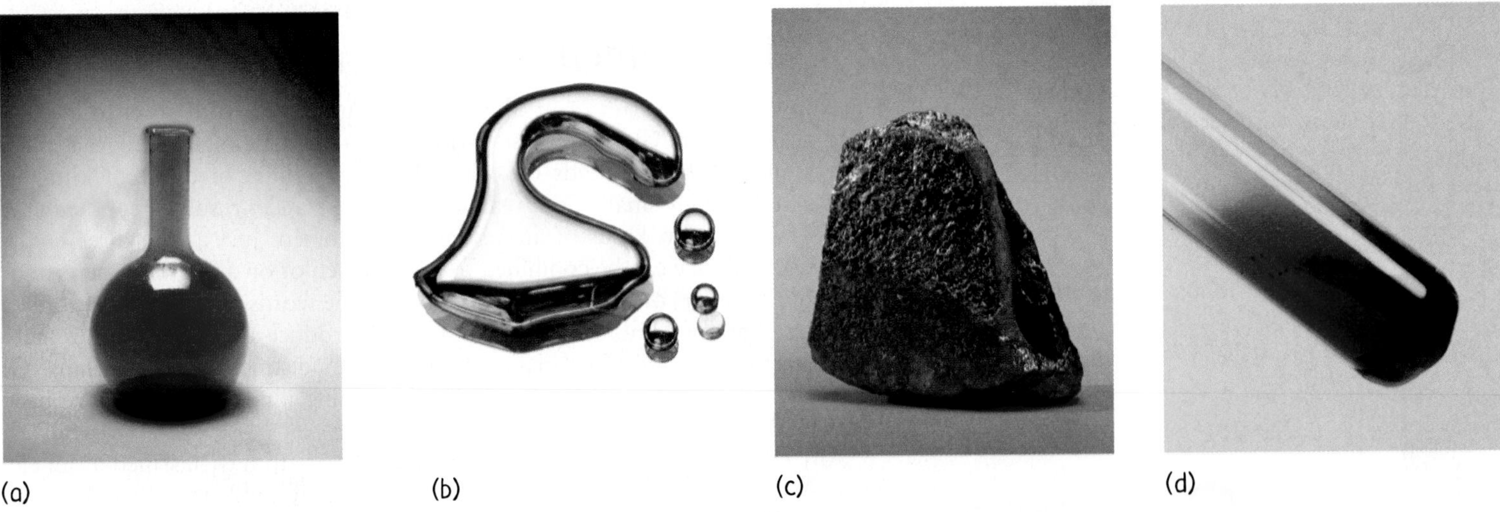

Figure 9.3 Some chemical elements. (a) Bromine (Br) is a reddish-brown liquid at room temperature that emits a red vapor. (b) Mercury (Hg), sometimes called "quicksilver" is a liquid metal. (c) Silicon (Si) is the second most abundant element in the Earth's crust (after oxygen). (d) Iodine (I) is a black solid at room temperature but with gentle heating it sublimes—transitions directly—to a purple gas.

INSIGHT All things are made of atoms—little particles that move around in perpetual motion, attracting each other when they are a little distance apart, but repelling upon being squeezed into one another.—*Richard P. Feynman*

9.2 Atoms Are Ancient, Tiny, and Empty

The origin of most atoms goes back to the birth of the universe. Hydrogen, H, the lightest atom, was also the original atom, and it still makes up more than 90% of the atoms in the known universe. The next most abundant element is helium, He. Heavier atoms were produced in stars, which are massive collections of hydrogen and helium atoms pulled together by gravitational forces. Great pressures deep in a star's interior cause light atoms to fuse and become heavier elements. With the exception of hydrogen, all the atoms that occur naturally on Earth—including those in your body—are the products of stars. A tiny fraction of these atoms came from our own star, the Sun, but most are from stars that perished long before our solar system came into existence. You are made of stardust, as is all matter that surrounds you.

Conceptual Science Place

Evidence for Atoms; Atoms Are Recyclable

So most atoms are ancient. They have existed through imponderable ages, recycling through the universe in innumerable forms, both nonliving and living. In this sense, you don't "own" the atoms that make up your body—you are simply their present caretaker. There will be many caretakers to follow.

INSIGHT A nice way to state the smallness of atoms: The diameter of an atom is to the diameter of an apple as the diameter of an apple is to the diameter of Earth. So, to imagine the number of atoms in an apple, think of Earth solid-packed with apples. Both have about the same number.

Atoms are in a state of perpetual motion. You can see evidence of this motion when you place a drop of ink into a glass of water. It soon spreads to color the entire glassful. Likewise, if a cupful of the atoms making up the pesticide DDT is thrown into an ocean, those atoms disperse and are later found in every part of the world's oceans. The same is true of materials released into the atmosphere.

Atoms are so small that there are more than 10 billion trillion of them in each breath you exhale. This is more than the number of breaths in Earth's atmosphere. Within a few years, the atoms of your breath are uniformly mixed throughout the atmosphere. What this means is that anyone anywhere on Earth inhaling a breath of air takes in numerous atoms that were once part of you. And, of course, the reverse is true: you inhale atoms that were once part of everyone who has ever lived. We are literally breathing one another.

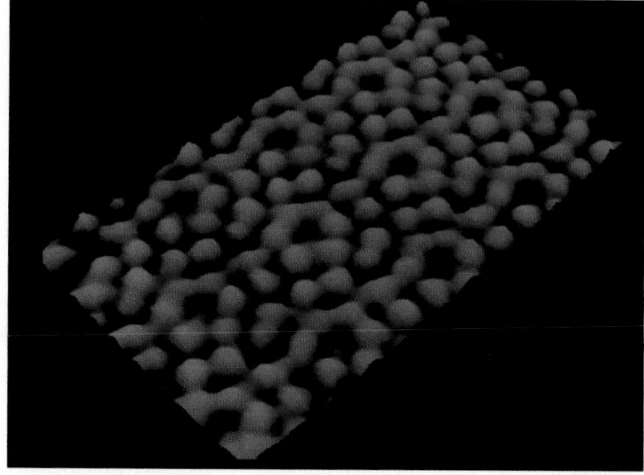

Figure 9.4 An image of carbon atoms obtained with a scanning tunneling microscope.

Atoms are so small that they can't be seen with visible light. That's because they are smaller than the wavelengths of visible light. We could stack microscope on top of microscope and never "see" an atom. Photographs of atoms, such as in Figure 9.4, are obtained with a scanning tunneling microscope (STM). Discussed further in Chapter 10, this is an imaging device that bypasses the use of light and optics altogether.

Today we know the atom is made up of subatomic particles—electrons, protons, and neutrons. We also know that atoms differ from one another only in the number of these subatomic particles. **Protons** and **neutrons** are bound together at the atom's center to form the atomic nucleus. The nucleus is much smaller than the atom but it contains most of an atom's mass. Surrounding the nucleus are the tiny **electrons**, as shown in Figure 9.5. When you turn on a light switch, you make electrons flow. As we learned in Chapter 9, these make up the current flowing through a light bulb. The electrons that are so important in electricity are the same electrons that dictate chemical reactions and flow as nerve impulses in animals. You will learn more about this in following chapters.

INSIGHT If a typical atom were expanded to a diameter of 3 km, about as big as a medium-sized airport, the nucleus would be about the size of a basketball. Atoms are mostly empty space.

We and all materials around us are mostly empty space. How can this be? Electrons move about the nucleus in an atom, defining the volume of space that the atom occupies. But since electrons are very small, and because they are widely separated from each other and from the nucleus, atoms are indeed mostly empty space.

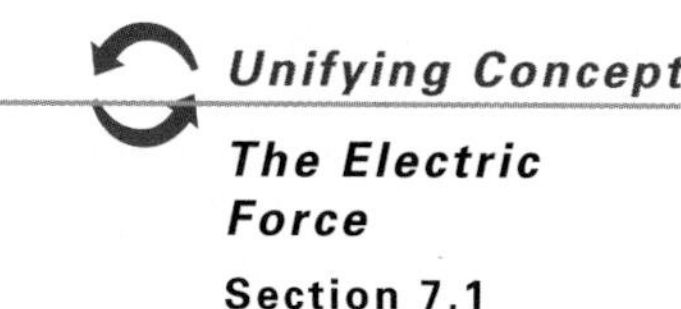

Unifying Concept

The Electric Force

Section 7.1

So what keeps atoms from oozing through one another? How are we supported by a floor despite the empty nature of its atoms? The answer is that although the space within the atom is mostly devoid of matter, it is filled with an electric field. The range of this field is several times larger than the atomic

TECHNOLOGY

The Scanning Tunneling Microscope (STM)

Although we cannot see atoms directly, we can generate images of them indirectly. In the mid-1980s, researchers developed the scanning tunneling microscope (STM), not to be confused with the scanning electron microscope (SEM) profiled in Chapter 15. The STM (a) is based on quantum physics. It produces images by dragging an ultra-thin needle back and forth over the surface of a sample. Bumps the size of atoms on the surface cause the needle to move up and down. This vertical motion is detected and translated by a computer into a topographical image that corresponds to the positions of atoms on the surface. An STM can also be used to push individual atoms into desired positions. This ability opened up the field of *nanotechnology*, in which incredibly small electronic circuits and motors are built atom by atom.

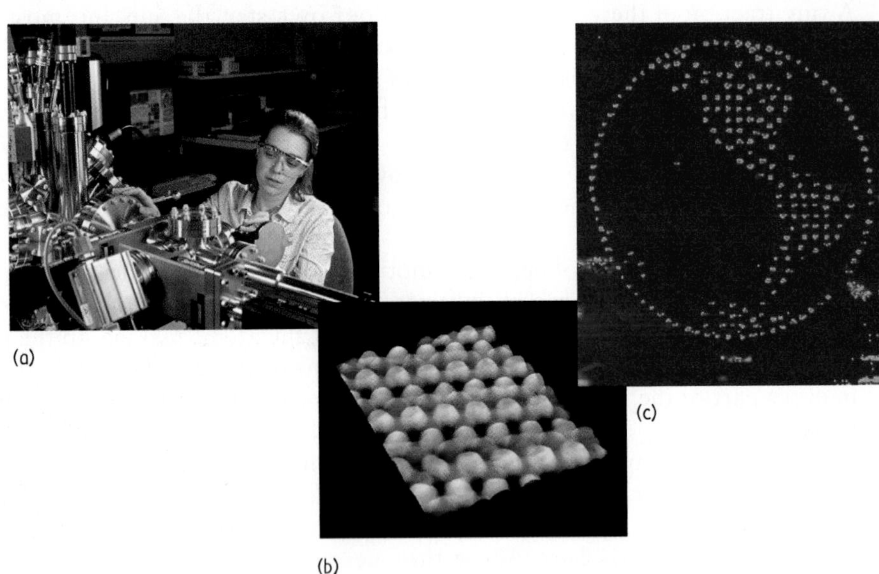

(a) (b) (c)

(b) An image of gallium and arsenic atoms obtained with an STM. (c) Each dot in the world's tiniest map consists of a few thousand gold atoms, each atom moved into its proper location by an STM.

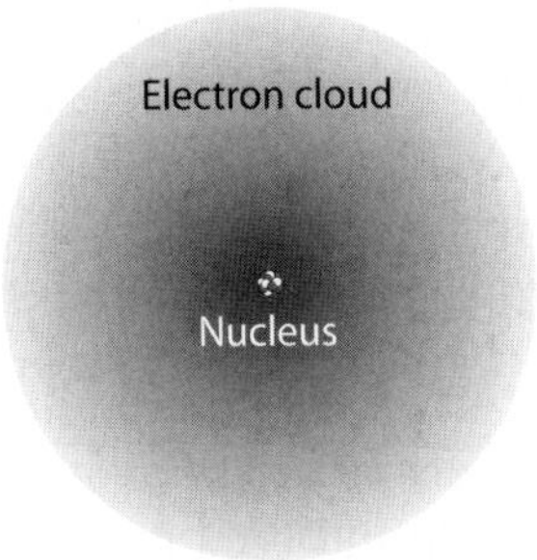

Figure 9.5 Electrons whiz around the atomic nucleus, forming what can be best described as a cloud that is more dense where the electrons tend to spend most of their time. Electrons, however, are invisible to us. Hence, such a cloud can only be imagined. Furthermore, if this illustration were drawn to scale, the atomic nucleus would be too small to be seen. In short, atoms are not well suited to visual depictions.

volume. Electrons in the outer regions of any atom repel the electrons of neighboring atoms. Any two atoms, therefore, can get only so close to each other before repulsion dominates (provided they don't join in a chemical bond, as discussed in Chapter 12).

When your hand pushes against a wall, you make contact with the wall—at the macroscopic level. At the atomic level, electrical repulsion between electrons in your hand and electrons in the wall prevent contact in the conventional sense—and, furthermore, prevent your hand from passing through the wall. This same electrical repulsion prevents us from falling through a solid floor. So, quite interestingly, when you touch someone, your atoms and those of the other person do not meet. Instead, atoms from the two of you get close enough so that you sense an electrical repulsion. There is still a tiny, though imperceptible, gap between the two of you.

CHECK YOURSELF

What kind of force prevents atoms from meshing into one another?

CHECK YOUR ANSWER

The kind of force that prevents atoms from meshing with one another is electrical.

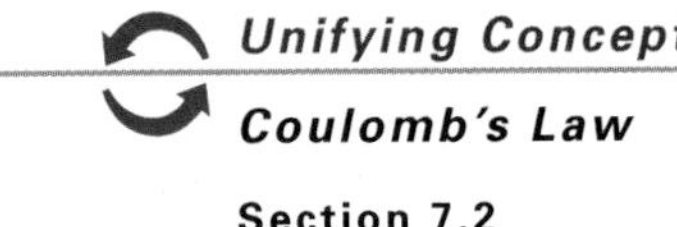

Unifying Concept

Coulomb's Law

Section 7.2

9.3 Protons and Neutrons

Let's take a closer look at the atom and investigate the particles found in the atomic nucleus. First, consider protons. A proton carries a positive charge and is relatively heavy—nearly 2000 times as massive as the electron. The proton and electron have the same quantity of charge, but the opposite sign. The number of

HISTORY OF SCIENCE

The Atomic Hypothesis

The idea that matter is composed of atoms goes back to the Greeks in the fifth century BC. Investigators of nature back then wondered whether matter was continuous or not. We can break a rock into pebbles, and the pebbles into fine gravel. The gravel can be broken into fine sand, which then can be pulverized into powder. Perhaps it seemed to the fifth-century Greeks that there was a smallest bit of rock, an "atom," that could not be divided any further.

Aristotle, the most famous of the early Greek philosophers, didn't agree with the idea of atoms. In the fourth century BC, he taught that all matter is composed of various combinations of four elements—earth, air, fire, and water. This view seemed reasonable because, in the world around us, matter is seen in only four forms: solids (earth), gases (air), liquids (water), and in the state of flames (fire). The Greeks viewed fire as the element of change, since fire was observed to produce changes on substances that burned. Aristotle's ideas about the nature of matter held sway for more than 2000 years.

The atomic idea was revived in the early 1800s by an English meteorologist and schoolteacher, John Dalton. He successfully explained the natures of chemical reactions by proposing that matter is made of atoms. He and others of the time, however, had no direct evidence of their existence. Then, in 1827, a Scottish botanist named Robert Brown noticed something very unusual in his microscope. He was studying grains of pollen suspended in water, and he saw that the grains were perpetually moving and jumping about. At first he thought that the grains were some kind of moving life forms, but later he found that dust particles and grains of soot suspended in water moved the same way. The perpetual jiggling of particles—now called Brownian motion—results from random collisions between visible particles and invisible atoms in air or water. The atoms are invisible because they're so small. Although he couldn't see the atoms, Brown could see the effect they had on the visible specks. The pollen grains moved because they were constantly being jostled by the atoms (actually, by the atomic combinations known as molecules) that made up the water surrounding them. The drawing shows the random path taken by a dust speck experiencing Brownian motion in air.

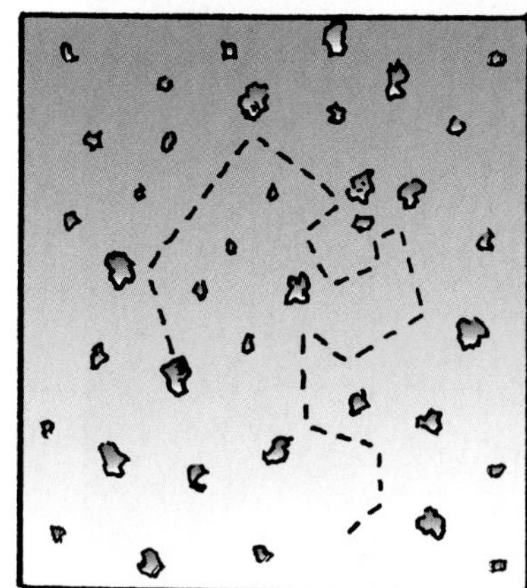

All this was explained in 1905 by Albert Einstein. Einstein observed that, if the kinetic theory of matter was right (Chapter 6), then molecules of water would move randomly so a small speck would receive random impacts and move randomly as Brown described. Until Einstein's explanation—which made it possible to find the masses of atoms—many prominent physicists remained skeptical about the existence of atoms. So we see that the reality of the atoms was not firmly established until the twentieth century.

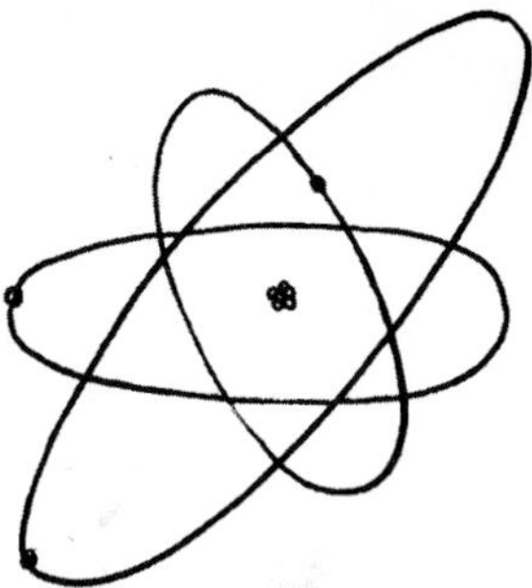

An early model of the atom consists of a central nucleus and orbiting electrons, much like a solar system with orbiting planets.

protons in the nucleus of any atom is equal to the number of electrons whirling about the nucleus. So the opposite charges of protons and electrons balance each other, producing a zero net charge. The atom is electrically neutral. For example, an electrically balanced oxygen atom has a total of eight electrons and eight protons.

Scientists have agreed to identify elements by **atomic number**, which is the number of protons each atom of a given element contains. The periodic table lists the elements in order of increasing atomic number. Hydrogen, with one proton per atom, has atomic number 1; helium, with two protons per atom, has atomic number 2; and so on.

CHECK YOURSELF

How many protons are there in an iron atom, Fe (atomic number 26)?

CHECK YOUR ANSWER

The atomic number of an atom and its number of protons are the same. Thus, there are 26 protons in an iron atom. Another way to put this is that all atoms that contain 26 protons are, by definition, iron atoms.

Table 9.1 Subatomic Particles

	Mass Compared	Particle	Charge to Electron	Actual Mass* (kg)
	Electron	−1	1	9.11×10^{-31}†
Nucleons	Proton	+1	1836	1.673×10^{-21}
Nucleons	Neutron	0	1841	1.675×10^{-27}

*Not measured directly but calculated from experimental data.
†9.11×10^{-31} kg = 0.000000000000000000000000000000911 kg.

Figure 9.6 As close as they are in this photograph, none of their atoms meet.

If we compare the electric charges and masses of different atoms, we see that the atomic nucleus must be made up of more than just protons. Helium, for example, has twice the electric charge of hydrogen but four times the mass. The additional mass is due to another subatomic particle found in the nucleus, the neutron. The neutron has about the same mass as the proton, but it has no electric charge. Any object that has no net electric charge is said to be electrically neutral, which accounts for the neutron's name. In Chapter 10, we will discuss the important role that neutrons play in holding the nucleus of each atom together.

Both protons and neutrons are called **nucleons**, a term that denotes their location in the atomic nucleus. Table 9.1 summarizes the basic facts about our three subatomic particles.

9.4 Isotopes and Atomic Mass

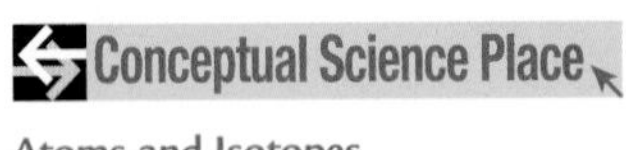

Atoms and Isotopes

For any element, there is no set number of neutrons in the nucleus. For example, most hydrogen atoms (atomic number 1) have no neutrons. A small percentage, however, have one neutron, and a smaller percentage have two neutrons. Similarly, most iron atoms (atomic number 26) have 30 neutrons, but a small percentage of them have 29 neutrons. Atoms of the same element that contain different numbers of neutrons are referred to as **isotopes** of that element.

We identify an isotope by its **mass number**, which is the total number of protons and neutrons (nucleons) in the nucleus. As Figure 9.7 shows, a hydrogen isotope with only one proton is called hydrogen-1, where 1 is the mass number. A hydrogen isotope with one proton and one neutron is therefore hydrogen-2, and a hydrogen isotope with one proton and two neutrons is hydrogen-3. Similarly, an iron isotope with 26 protons and 30 neutrons is called iron-56, and one with only 29 neutrons is iron-55.

INSIGHT Protons and neutrons are made of still smaller particles, the fundamental particles called quarks. Some recent theories suggest that quarks may be made of infinitely thin string loops.

An alternative method of indicating isotopes is to write the mass number as a superscript and the atomic number as a subscript to the left of the atomic symbol. For example, an iron isotope with a mass number of 56 and atomic number of 26 is written:

Mass number—56
Fe—Atomic symbol
Atomic number—26

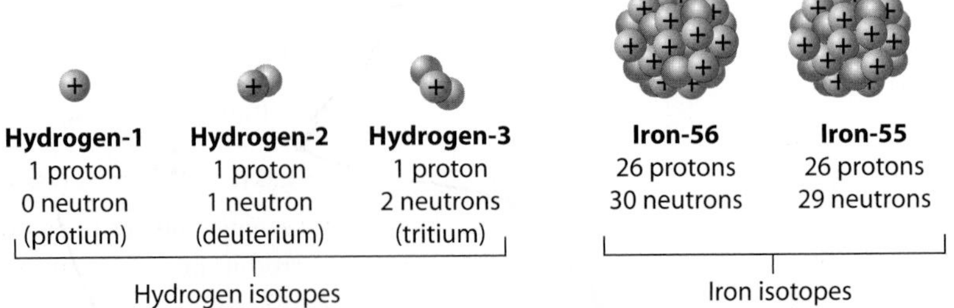

Figure 9.7 Isotopes of an element have the same number of protons but different numbers of neutrons and, therefore, different mass numbers. The three hydrogen isotopes have special names: protium for hydrogen-1 (the most common isotope), deuterium for hydrogen-2, and tritium for hydrogen-3. The isotopes of most elements have no special names and are indicated merely by mass number.

The total number of neutrons in an isotope can be calculated by subtracting its atomic number from its mass number:

$$\begin{array}{r} \text{Mass number} \\ -\ \text{Atomic number} \\ \hline \text{Number of neutrons} \end{array}$$

For example, uranium-238 has 238 nucleons. The atomic number of uranium is 92, which tells us that 92 of these 238 nucleons are protons. The remaining 146 nucleons must be neutrons:

$$\begin{array}{r} \text{238 protons and neutrons} \\ -\ \text{92 protons} \\ \hline \text{146 neutrons} \end{array}$$

Atoms interact with one another electrically. Therefore, the way any atom behaves in the presence of other atoms is determined largely by the charged particles it contains, especially its electrons. Isotopes of an element differ only in mass, not in electric charge. For this reason, isotopes of an element share many characteristics—in fact, as chemicals they cannot usually be distinguished from one another. For example, a sugar molecule containing seven neutrons per carbon nucleus is digested no differently from a sugar molecule containing six neutrons per carbon nucleus. Interestingly, about 1 percent of the carbon we eat is the carbon-13 isotope, which contains seven neutrons per nucleus. An even smaller amount is carbon-14 (which we will discuss further in Chapter 10). The remaining nearly 99 percent of the carbon in our diet is the more common carbon-12 isotope, which contains six neutrons per nucleus.

The total mass of an atom is called its **atomic mass**. This is the sum of the masses of all the atom's components (electrons, protons, and neutrons). Because electrons are so much less massive than protons and neutrons, their contribution to atomic mass is negligible. A special unit has been developed for atomic masses. This is the **atomic mass unit**, amu, where 1 atomic mass unit is equal to 1.661×10^{-24}g, which is slightly less than the mass of a single proton. As shown in Figure 9.8, the atomic masses listed in the periodic table are in atomic mass units. As explained in the Math Connection, the atomic mass of an element as presented in the periodic table is actually the average atomic mass of its various isotopes.

INSIGHT Most water molecules, H_2O, consist of hydrogen atoms with no neutrons. The few that do, however, are heavier and because of this difference they can be isolated. Such water is appropriately called "heavy water."

MATH CONNECTION

Calculating Atomic Mass

Most elements have a variety of isotopes, each with its own atomic mass. For this reason, the atomic mass listed in the periodic table for any given element is the average of the masses of all the element's isotopes based on their relative abundance. For example, about 99% of all carbon atoms are atoms of the isotope carbon-12, and most of the remaining 1% are atoms of the heavier isotope carbon-13. This small amount of carbon-13 raises the average mass of carbon from 12.000 atomic mass units to the slightly greater value of 12.011 atomic mass units.

To arrive at the atomic mass presented in the periodic table, you must first multiply the mass of each naturally occurring isotope of an element by the fraction of its abundance, and then add up all the fractions.

Problem Carbon-12 has a mass of 12.0000 atomic mass units, and it makes up 98.89% of naturally occurring carbon. Carbon-13 has a mass of 13.0034 atomic mass units, and it makes up 1.11% of naturally occurring carbon. Use this information to show that the atomic mass of carbon shown in the periodic table, 12.011 atomic mass units, is correct.

Solution Recognize that 98.89% and 1.11% expressed as fractions are 0.9889 and 0.0111, respectively.

	Contributing Mass of C-12	Contributing Mass of C-13
Fraction of Abundance	0.9889	0.0111
Step 1		
Mass (amu)	× 12.0000	× 13.0034
Step 2		
Atomic mass =	11.867 + 0.144 = 12.011	

You will find similar problems to try in the Problems section at the end of the chapter.

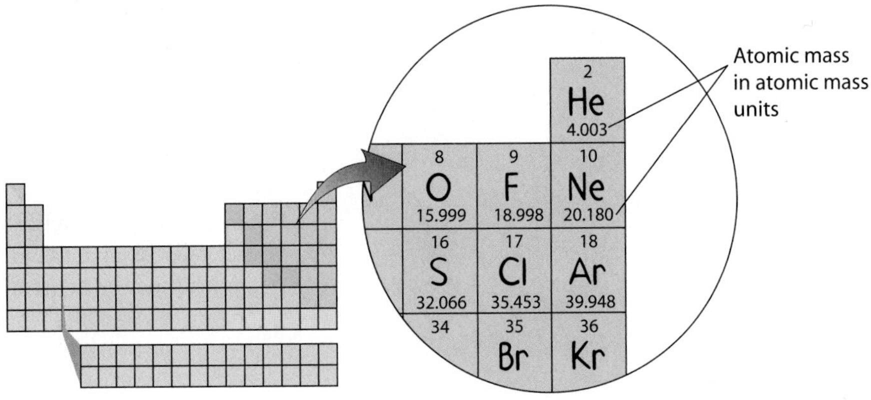

Figure 9.8 Helium, He, has an atomic mass of 4.003 atomic mass units, and neon, Ne, has an atomic mass of 20.180 atomic mass units.

CHECK YOURSELF

1. Distinguish between mass number and atomic mass.
2. If two atoms are isotopes of the same element, do they have the same atomic number? The same atomic mass?

CHECK YOUR ANSWERS

1. Both terms include the word *mass* and so are easily confused. Focus your attention on the second word of each term, however, and you'll be correct every time. Mass number is a count of the number of nucleons in an isotope and requires no units because it is simply a count. Atomic mass, however, is a measure of the total mass of an atom, which is given in atomic mass units. For example, the mass number of helium-4 is simply 4, but its atomic mass is 4.003 amu.
2. Isotopes of the same element have the same atomic number, but they have different atomic masses (because they have the same number of protons in the nucleus, but different numbers of neutrons).

INTEGRATED SCIENCE CHEMISTRY, BIOLOGY, EARTH SCIENCE

Physical and Conceptual Models

Visible objects, whether very small or very large, can be represented with a physical model, which is a model that replicates the object at a more convenient scale. Figure 9.9a, for instance, shows a large-scale **physical model** of a cell that a biology student uses to study the cell's internal structure. Applying a physical model to invisible atoms, however, is ineffective. We cannot simply scale up the atom to a larger size, as we might with a microorganism. (An STM merely shows the positions of atoms and not the actual images of atoms, which do not have the solid surfaces implied in STM images.) So, rather than describing the atom with a physical model, scientists use what is known as a **conceptual model** to describe it. The more accurate a conceptual model is, the more accurately it predicts the behavior of the system it is meant to describe. Climate is best described using a conceptual model like the one shown in Figure 9.9b. Such a model shows how the various components of the system—concentrations of gases, atmospheric pressure, temperature, sunlight, the motion of large masses of air—interact with one another. Other systems that can be described effectively by conceptual models include the economy, population growth, the spread of diseases, and team sports.

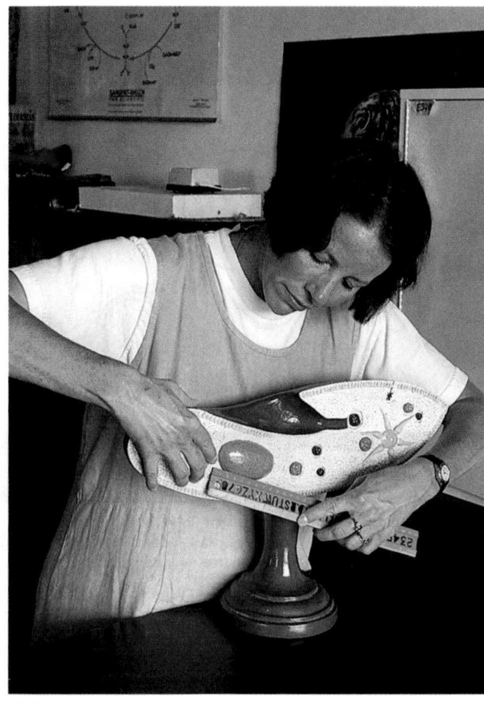

(a)

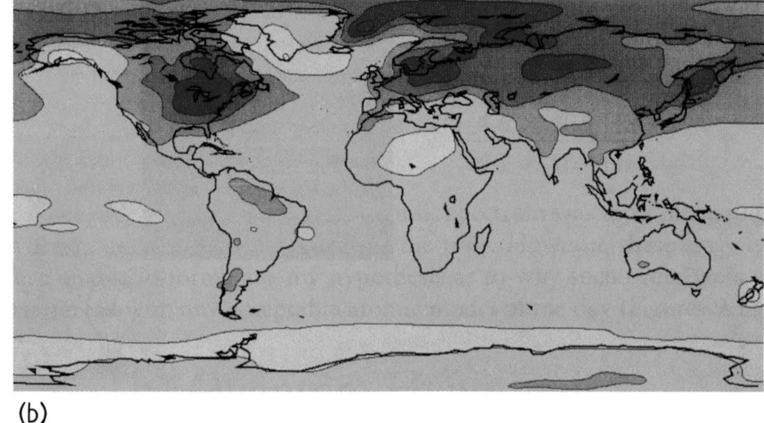

(b)

Figure 9.9 (a) This large-scale model of a cell is a physical model. (b) If human carbon dioxide emissions were to double, this climate model predicts areas in pale yellow would warm by 0–2 °C and red areas would warm by 8–12 °C during the winter months.

INSIGHT We can't "see" an atom because it's too small. We can't see the farthest star either. There's much that we can't see. But that doesn't prevent us from thinking about such things or even finding indirect evidence.

Like the climate, the atom is a complex system of interacting components, and it is best described with a conceptual model. You should therefore be careful not to interpret any visual representation of an atomic conceptual model as a recreation of an actual atom. In the following sections of this chapter, for example, you will be introduced to the planetary model of the atom, wherein electrons are shown orbiting the atomic nucleus much as planets orbit the sun. This planetary model is limited, however, in that it fails to explain many properties of atoms. Hence, newer and more accurate (and more complicated) conceptual models of the atom have since been introduced. In these models, electrons appear as a cloud hovering around the atomic nucleus, but even these models have their limitations. Ultimately, the best models of the atom are ones that are purely mathematical.

9.5 Atomic Spectra

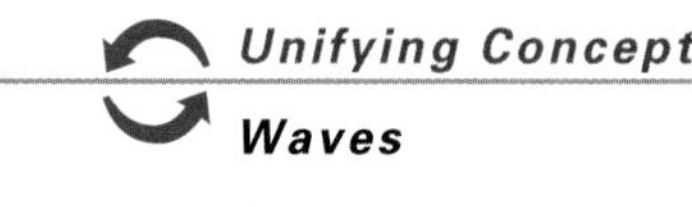

Unifying Concept
Waves
Section 8.1

Recall (from Chapter 8) that when all frequencies of visible light reach our eye at the same time, we see white light. By passing white light through a prism or through a diffraction grating, the color components of the light can be separated,

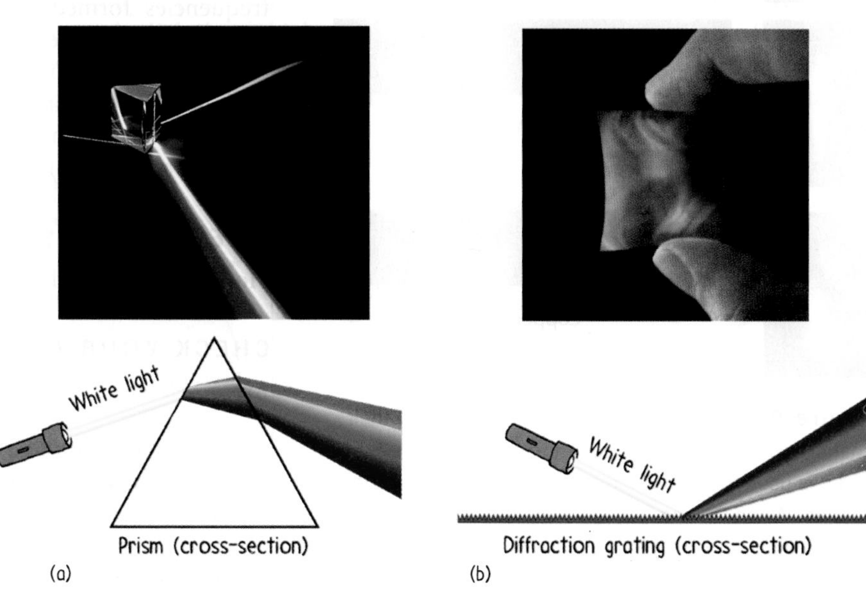

Figure 9.10 White light is separated into its color components by (a) a prism and (b) a diffraction grating.

Table 9.2 The Four Major Types of Orbitals: *s, p, d, f*

Orbital Type	Spatial Orientations
The **s orbital** has only one *s* shape which is spherical.	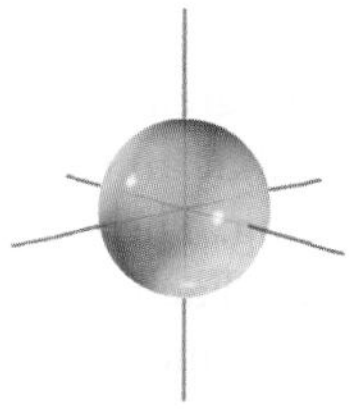
There are three ***p* orbitals.** They differ by orientation.	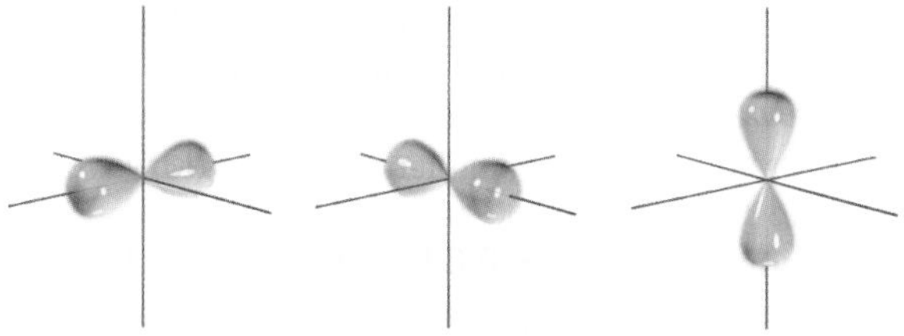
There are five ***d* orbitals**	
There are seven ***f* orbitals**	

another only by their orientation in three-dimensional space. The more complex *d* orbitals have five possible shapes, and the *f* orbitals have seven. Please do not feel compelled to memorize all the orbital shapes, especially the *d* and *f* ones. However, you should understand that each orbital represents a different region in which an electron of a given energy is most likely to be found.

CHECK YOURSELF

What is the relationship between an electron wave and an atomic orbital?

CHECK YOUR ANSWER

The atomic orbital is an approximation of the shape of the standing electron wave surrounding the atomic nucleus.

In addition to a variety of shapes, atomic orbitals also come in a variety of sizes that correspond to different energy levels. In general, highly energized electrons are able to extend themselves farther away from the attracting nucleus,

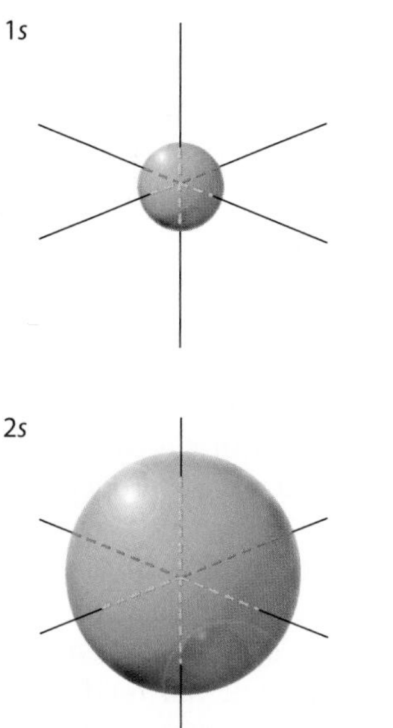

Figure 9.23 The 2*s* orbital is larger than the 1*s* orbital because the 2*s* accommodates electrons of greater energy.

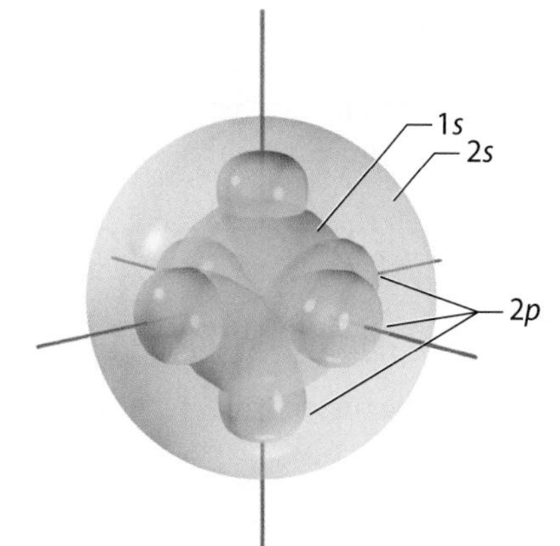

Figure 9.24 The fluorine atom has five overlapping atomic orbitals that contain its nine electrons, which are not shown.

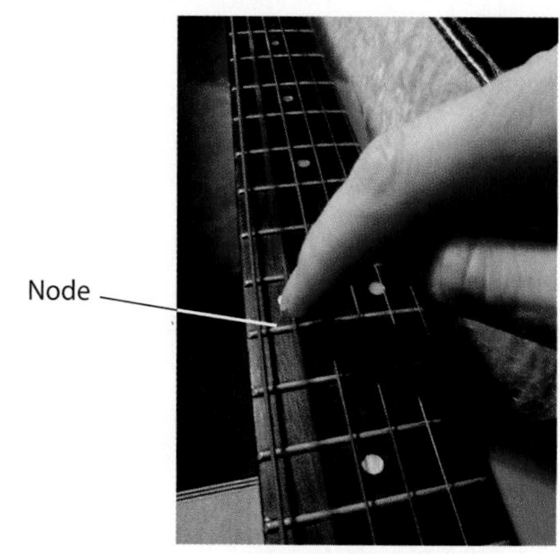

Figure 9.25 The guitar string can oscillate on both sides of the 12th-fret node even when the string is plucked on only one side of the node.

which means they are distributed over a greater volume of space. The higher the energy of the electron therefore, the larger its atomic orbital. Because electron energies are quantized, however, the possible sizes of the atomic orbitals are quantized. The size of an orbital is thus indicated by Bohr's principal quantum number $n = 1, 2, 3, 4, 5, 6, 7$, or greater. The first two *s* orbitals are shown in Figure 9.23. The smallest *s* orbital is 1*s*, where 1 is the principal quantum number. The next largest *s* orbital is 2*s*, and so forth.

So we see that an atomic orbital is simply a volume of space within which an electron may reside. Orbitals may overlap one another in an atom. As shown in Figure 9.24, the electrons of a fluorine atom are distributed among its 1*s*, 2*s*, and three 2*p* orbitals.*

The hourglass-shaped *p* orbital illustrates the significance of the wave nature of the electron. Unlike the case with a real hourglass, the two lobes of this orbital are not open to each other, and yet an electron freely moves from one lobe to the other. To understand how this can happen, consider an analogy from the macroscopic world. A guitar player can gently tap a guitar string at its midpoint (the 12th fret) and pluck it elsewhere at the same time to produce a high-pitched tone called a harmonic. Close inspection of this string, shown in Figure 9.25, reveals that it oscillates everywhere along the string except at the point directly above the 12th fret. This point of zero oscillation is called a node. Although there is no motion at the node, waves nonetheless travel through it. Thus, the guitar string oscillates on both sides of the node when only one side is plucked. Similarly, the point between the two lobes of a *p* orbital is a node through which the electron may pass—but only by virtue of its ability to take on the form of a wave.

CHECK YOURSELF

Distinguish between an orbital and one of Bohr's orbits.

CHECK YOUR ANSWER

An orbit is a distinct path followed by an object in its revolution around another object. In Bohr's planetary model of the atom, he proposed an analogy between electrons orbiting the atomic nucleus and planets orbiting the Sun. What orbits and orbitals have in common is that they both use Bohr's principal quantum number to indicate energy levels in an atom.

Bohr's planetary atomic model postulated discrete energy values for electrons in order to account for spectral data. The electron-wave model goes further and shows that discrete electron energy values are a natural consequence of the electron's confinement to the atom. While Bohr's planetary model accounts for the generation of

* For reasons beyond the scope of this text, the 1*p* orbital does not exist. The smallest *p* orbital is therefore the 2*p*.

light quanta, the wave model takes things a step further by treating light and matter in the same way—both behaving sometimes like a wave and sometimes like a particle. As abstract as the wave model may be, these successes indicate that it presents a more fundamental description of the atom than does Bohr's shell model.

INTEGRATED SCIENCE CHEMISTRY

The Shell Model

To begin to understand how atoms interact with one another in chemical processes, we need not take into account the shapes of electron orbitals. For a good basic understanding of chemical reactions and atomic properties, we utilize the shell model of the atom.

According to the shell model, electrons behave as if they are arranged in a series of concentric shells. A **shell** is defined as a region of space about the atomic nucleus within which electrons may reside. An important aspect of this model is that there are at least seven shells and that each shell can hold only a limited number of electrons. As shown in Figure 9.26, the innermost shell can hold two; the second and third shells, eight electrons each; the fourth and fifth shells, 18 each; and the sixth and seventh shells, 32 each.

Recall from Section 9.1 that the chemical elements are organized in a chart called the *periodic table*. The periodic table contains a wealth of information about the elements; learning to read the table and extract this information is a topic of the Chemistry portion of this book. For now, we will simply note that elements in the periodic table are arranged in order of increasing atomic number. Elements in the same column or *group* of the periodic table generally share similar chemical behavior. In the same row or *period* of the table, the properties of elements tend to vary gradually from one element to the next, moving from left to right across the table.

The series of seven such concentric shells described by the shell model accounts for the seven periods of the periodic table, with each period adding a shell. Furthermore, the number of elements in each period is equal to the shell's capacity for electrons. The first shell, for example, has a capacity for only two electrons. That's why we find only two elements, hydrogen and helium, in the first period (Figure 9.27). The second and third shells each have a capacity for eight electrons, and so eight elements are found in both the second and third periods.

How does this shell model relate to atomic orbitals? Briefly, each shell represents a grouping of orbitals of similar energy levels. The first shell represents only the 1*s* orbital. The second shell represents the 2*s* and three 2*p* orbitals. The third

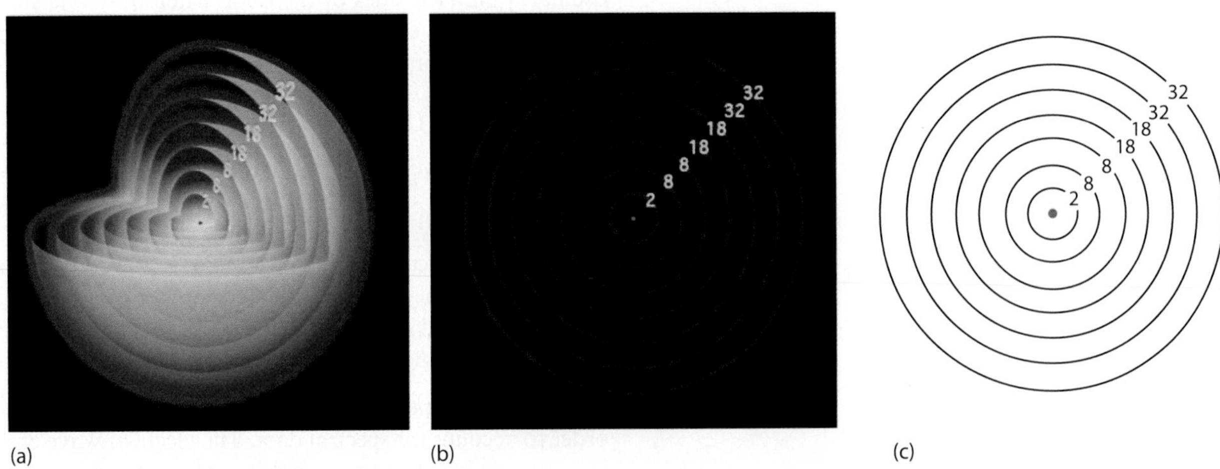

Figure 9.26 (a) A cutaway view of the seven shells, indicating the number of electrons each shell can hold. (b) A two-dimensional, cross-sectional view of the shells. (c) An easy-to-draw cross-sectional view that resembles Bohr's planetary model.

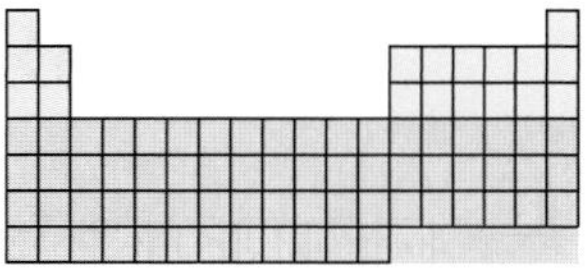

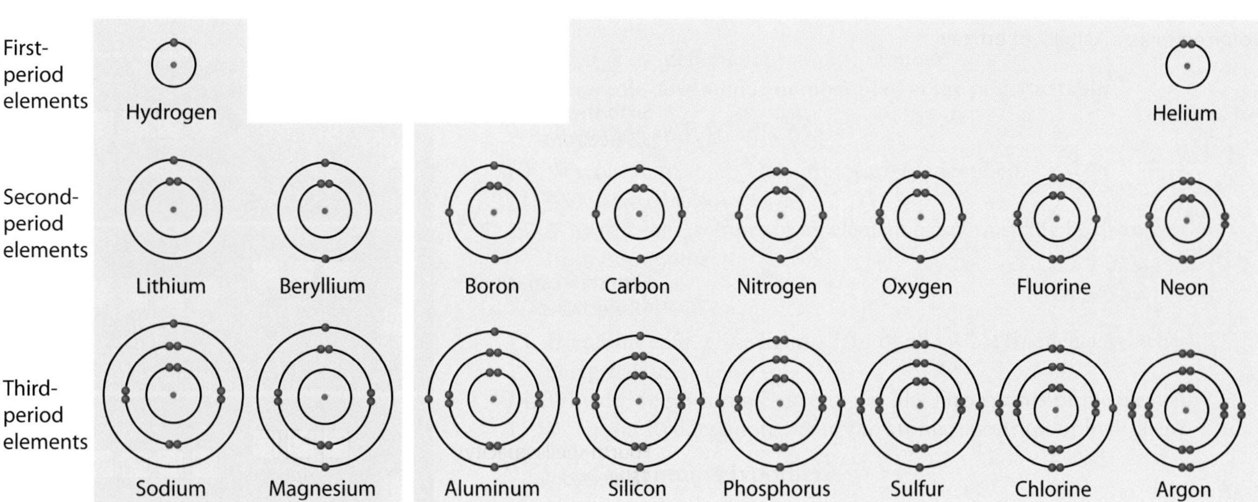

Figure 9.27 The first three periods of the periodic table, according to the shell model. Elements in the same period have electrons in the same shells. Elements in the same period differ from one another by the number of electrons in the outermost shell. Elements in the same group of the periodic table have the same number of valence electrons, in general. This is why elements of the same group tend to have very similar chemical properties.

shell represents the 3*s* and three 3*p* orbitals, and so on, as described in Figure 9.28. The limited capacity of each shell arises from the fact that each orbital can hold no more than two electrons. The first shell, representing only one orbital, therefore has a capacity for two electrons. The second and third shells each represent four orbitals, and so each has a capacity to hold eight electrons.

The electrons of the outermost occupied shell in any atom are directly exposed to the external environment and are the first to interact with other atoms. Most notably, they are the ones that participate in chemical bonding, as we shall be discussing in Chapter 12. The electrons in the outermost shell, therefore, are quite important. They are called *valence electrons*. The term *valence* is derived from the Latin *valentia*, "strength," and it refers to the "combining power" of an atom.

Look carefully again at Figure 9.28. Can you see that the elements in the same period have electrons in the same shells and that atomic number increases moving from left to right across a period? This is the reason that the properties of elements change gradually moving from left to right across the periodic table. Can you also see that valence electrons of atoms above and below one another within the same group are similarly organized? For example, atoms of the first group, which includes hydrogen, lithium, and sodium, each have a single valence electron. The atoms of the second group, including beryllium and magnesium, each have two valence electrons. Similarly, atoms of the last group, including helium, neon, and argon, each have their outermost shells filled to capacity with valence electrons—two for helium, and eight each for neon and argon. In general, the valence electrons of atoms in the same group of the periodic table are similarly organized. This explains why elements of the same group have similar properties—another concept we will explore in the Chemistry part of this book.

CHECK YOURSELF

1. The fourth shell represents the 4*s* orbital, along with five 3*d* orbitals and three 4*p* orbitals. How many orbitals is this altogether? What, then, is the

3. What is the difference between a physical model and a conceptual model?
4. What is the function of an atomic model?

Chemistry—The Shell Model

1. Which electrons are most responsible for the properties of an atom?
2. The shell model presented in this book is not very accurate. Why, then, is it presented here?
3. How many orbitals are present in the third shell?

Active Explorations

Spectral Patterns

Purchase some "rainbow" glasses from a nature, toy, or hobby store. The lenses of these glasses are diffraction gratings. Looking through them, you will see light separated into its color components. Certain light sources, such as the Moon or a car's headlights, are separated into a continuous spectrum—in other words, all the colors of the rainbow appear in a continuous sequence from red to violet.

Other light sources, however, emit a distinct number of discontinuous colors. Examples include streetlights, neon signs, sparklers, and fireworks. The spectral patterns you see from any of these light sources are the atomic spectra of elements that are heated in the light source. You'll be able to see the patterns best when you are at least 50 meters away from the light source. This distance makes the spectrum appear as a series of dots similar to the series of lines shown in Figure 9.13.

To the naked eye, a glowing element appears as only a single color. However, this color is an average of the many different visible frequencies the element is emitting. Only with a device such as a spectroscope are you able to discern the different frequencies. So when you look at an atomic spectrum, don't get confused and think that each frequency of light (color) corresponds to a different element. Instead, remember that what you are looking at is all the frequencies of light emitted by a single element as its electrons make transitions back and forth between energy levels.

Not all elements produce discrete line patterns in the visible spectrum. Tungsten, for example, produces the full spectrum of colors (white light). This property makes it useful as the glowing component of a car's headlights, as shown in the photograph. Also, the sunlight reflecting off the Moon is so bright and contains the glow of so many different elements that it, too, appears as a broad spectrum.

Exercises

1. A cat strolls across your backyard. An hour later, a dog with its nose to the ground follows the trail of the cat. Explain what is going on in terms of atoms.
2. Which are older, the hydrogen atoms in a young star or those in an old star?
3. In what sense can you truthfully say that you are a part of every person around you?
4. Where were the atoms that make up a newborn baby manufactured?
5. Considering how small atoms are, what are the chances that at least one of the atoms exhaled in your first breath will be in your last breath?
6. Name ten elements you have access to in macroscopic quantities as a consumer here on Earth.
7. Which of the following diagrams best represent the size of the atomic nucleus relative to the size of the atom?

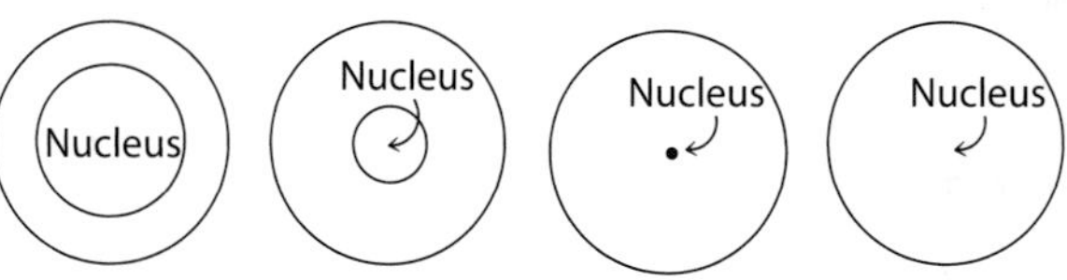

8. Which contributes more to an atom's mass: electrons or protons? Which contributes more to an atom's size?
9. If two protons and two neutrons are removed from the nucleus of an oxygen atom, a nucleus of which element remains?
10. What element results if one of the neutrons in a nitrogen nucleus is converted by radioactive decay into a proton?
11. The atoms that constitute your body are mostly empty space, and structures such as the chair you're sitting on are composed of atoms that are also mostly empty space. So why don't you fall through the chair?
12. If an atom has 43 electrons, 56 neutrons, and 43 protons, what is its approximate atomic mass? What is the name of this element?
13. The nucleus of an electrically neutral iron atom contains 26 protons. How many electrons does this iron atom have?
14. Why are the atomic masses listed in the periodic table not whole numbers?
15. Where did the carbon atoms in Leslie's hair originate?

16. Would you use a physical model or a conceptual model to describe the following: brain, mind, solar system, birth of universe, stranger, best friend, gold coin, dollar bill, car engine, virus, spread of a cold virus?
17. How might you distinguish a sodium-vapor lamp from a mercury-vapor lamp?
18. How can a hydrogen atom, which has only one electron, create so many spectral lines?
19. Which color of light comes from a greater energy transition, red or blue? Explain.
20. Which has the greatest energy, a photon of infrared light, a photon of visible light, or a photon of ultraviolet light?
21. If we take a piece of metal at room temperature and begin to heat it continuously in a dark room, it will soon begin to glow visibly. What will be its first visible color, and why?
22. Figure 9.20 shows three energy-level transitions that produce three spectral lines in a spectroscope. Note that the distance between the

$n = 1$ and $n = 2$ levels is slightly greater than the distance between the $n = 2$ and $n = 3$ levels. Would the number of spectral lines produced change if the distance between the $n = 1$ and $n = 2$ levels were exactly the same as the distance between the $n = 2$ and $n = 3$ levels?

23. What is the evidence for the claim that iron exists in the relatively cool outer layer of the Sun?
24. What does it mean to say that something is *quantized*?
25. The frequency of violet light is about twice that of red light. Compare the energy of a violet photon with the energy of a red photon.
26. If a beam of red light and a beam of violet light have equal energies, which beam has the greater number of photons?
27. How does the wave model of electrons orbiting the nucleus account for the fact that the electrons can have only discrete energy values?
28. How might the spectrum of an atom appear if the atom's electrons were not restricted to particular energy levels?
29. How does an electron move from one lobe of a p orbital to the other?
30. Light is emitted as an electron transition from a higher-energy orbital to a lower-energy orbital. How long does it take for the transition to take place? At what point in time is the electron found between the two orbitals?
31. Why is there only one spatial orientation for the s orbital?
32. Place the proper number of electrons in each shell:

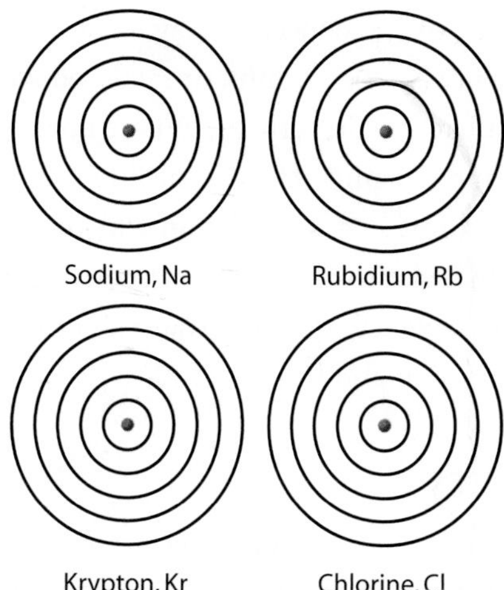

33. Which element is represented in Figure 9.28 if all seven shells are filled to capacity?
34. Does an orbital or shell have to contain electrons in order to exist?
35. Use the shell model to explain why a potassium atom, K, is larger than a sodium atom, Na.
36. Light has been described as being a wave, and then as being a particle, and then again as a wave. Does this indicate that light's true nature probably lies somewhere in between these models?
37. Write a letter to Grandma telling her to what extent we can now "see" atoms.

Problems

1. Chlorine (atomic number 17) is composed of two principal isotopes, chlorine-35, which has a mass of 34.9689 atomic mass units, and chlorine-37, which has a mass of 36.9659 atomic mass units. Assume that 75.77 percent of all chlorine atoms are the chlorine-35 isotope and 24.23 percent are the chlorine-37 isotope. Show that the atomic mass of natural chlorine is 35.45.
2. Lithium (atomic number 3) is composed of two principle isotopes. The isotope lithium-7 has a mass of 7.0160 atomic mass units, and the isotope lithium-6 has a mass of 6.0151 atomic mass units. Assume that 92.58 percent of all lithium atoms found in nature are lithium-7 and 7.42 percent are lithium-6. Show that the atomic mass of lithium is 6.94.
3. The element bromine, Br (atomic number 35), has two major isotopes of similar abundance, both approximately 50 percent. The atomic mass of bromine is reported in the periodic table as 79.904 atomic mass units. Choose the most likely set of mass numbers for these two bromine isotopes: (a) ^{80}Br, ^{81}Br; (b) ^{79}Br, ^{80}Br; (c) ^{79}Br, ^{81}Br.
4. Gas A is composed of diatomic molecules (two atoms to a molecule) of a pure element. Gas B is composed of monatomic molecules (one atom to a "molecule") of another pure element. Gas A has three times the mass of an equal volume of gas B at the same temperature and pressure. Show that the atomic mass of Element A is $\frac{3}{2}$ the mass of Element B.

The following problems require some knowledge of exponents. See Appendix D for help on this if necessary.

5. The diameter of an atom is about 10^{-10} m. (a) Show that 10^4 atoms make a line a millionth of a meter (10^{-6} m) long? (b) Show that 10^8 atoms cover a square a millionth of a meter on a side. (c) Show that 10^{12} atoms fill a cube a millionth of a meter on a side. (d) If a dollar were attached to each atom, what could you buy with your line of atoms? With your square of atoms? With your cube of atoms?
6. Assume that the present world population of about 6×10^9 people is about $\frac{1}{20}$ of the people who ever lived on Earth. Show that the number of people who ever lived is incredibly small compared with the number of air molecules in a single breath.

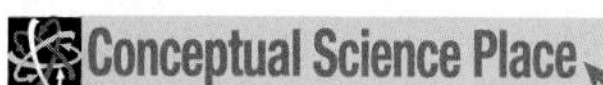

CHAPTER 9 ONLINE RESOURCES

Tutorials Bohr's Shell Model • ***Videos*** Evidence for Atoms, Atoms Are Recyclable • ***Quiz*** • ***Flashcards*** • ***Links***

CHAPTER 10

Nuclear Physics

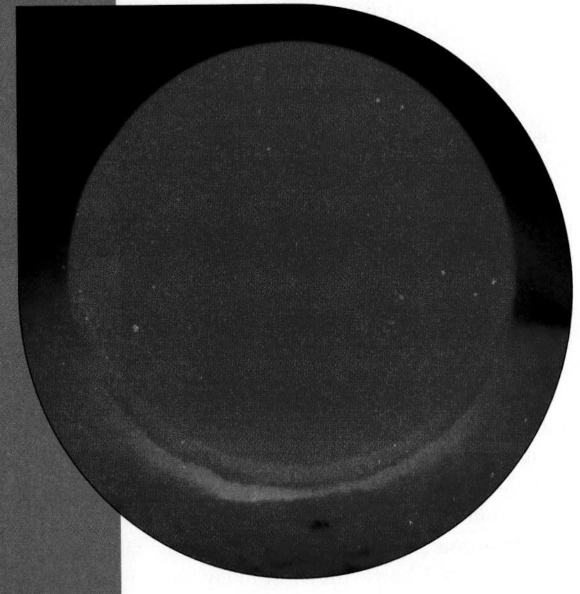

A pellet of pure radioactive plutonium-238. All isotopes of plutonium are radioactive because their nuclei are unstable. Plutonium-238 is used as a power source in space applications. For example, the interplanetary space probe Pioneer-10 *is powered by plutonium-238.*

CHAPTER OUTLINE

Radioactivity exists naturally in our environment; in fact, the human body itself is radioactive. We are exposed to additional radioactivity when we fly in an airplane, watch TV, or burn coal for fuel. You may know that radioactivity is an important part of some medical diagnostics and therapy. At the same time, excess radioactivity can be harmful or fatal and we are warned not to expose ourselves to it. What is radioactivity? Where does it come from? How can it be both beneficial and harmful? How is the Sun like a nuclear reactor? Does Einstein's famous equation $E = mc^2$ explain why nuclear reactions can produce so much energy? In order to make the best possible decisions about nuclear technologies, we should have a good understanding of nuclear energy and how it arises from nuclear fission and fusion.

10.1 Radioactivity

As we learned in Chapter 9, atoms are made up of electrons, neutrons, and protons. The neutrons and protons lie at the heart of the atom—in the nucleus. Some atoms have stable nuclei. These nuclei have the right balance of neutrons to protons and the right amount of energy to remain unchanged for a long time. Other atoms don't have the right mix of protons and neutrons or have the wrong amount of energy. The nuclei of these atoms are unstable. Atoms with unstable nuclei are said to be radioactive. Sooner or later, they break down and eject energetic particles and emit electromagnetic radiation. This process is **radioactivity**, which, because it involves the decay of the atomic nucleus, is often called *radioactive decay*.

A common misconception is that radioactivity is something new in the environment. Actually, it has been around far longer than the human race. It has always been in the soil we walk on and in the air that we breathe, and it warms Earth's interior. In fact, radioactive decay in Earth's interior heats the water that spurts from a geyser or wells up from a natural hot spring. The helium in a child's balloon is nothing more than the products of radioactive decay.

As Figure 10.1 shows, most of the radiation we encounter is natural background radiation that originates in Earth and in space and was present long before we humans arrived. Even the cleanest air we breathe is radioactive as a result of bombardment by cosmic rays. These rays originate in the Sun and other stars and make up the background radiation in space. Cosmic rays are of two types: high-energy particles or high-frequency electromagnetic radiation

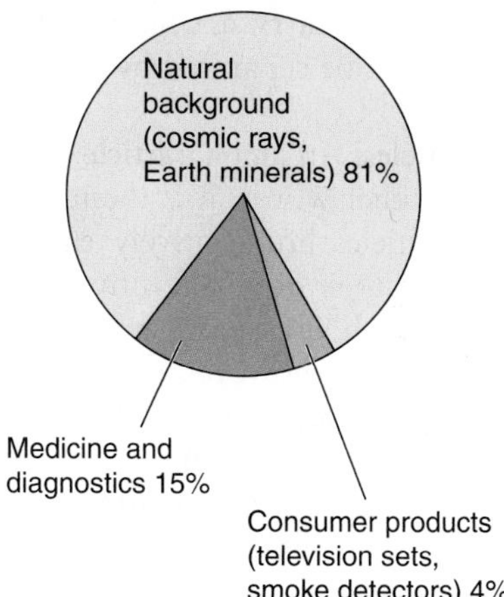

Figure 10.1 Origins of radiation exposure for an average individual in the United States.

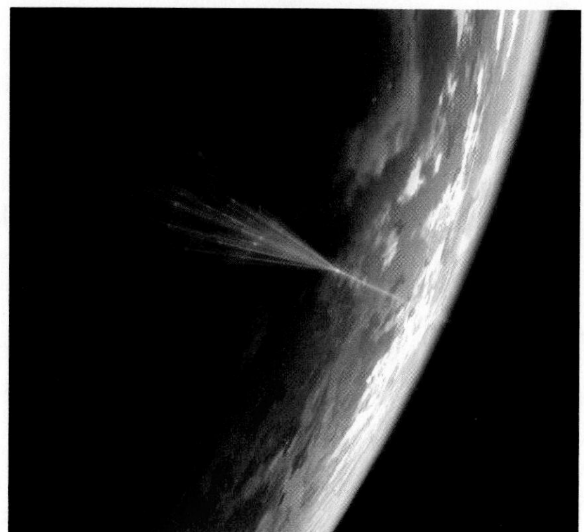

Figure 10.2 This illustration depicts cosmic rays (*lower right*) from an energetic astronomical event (such as a quasar, supernova, or gamma-ray burst) hitting Earth's atmosphere and triggering an "air shower" of subatomic particles (left and center). The shower occurs as the cosmic ray collides with a particle high in Earth's atmosphere, creating a stream of new particles, some of which in turn collide with other atmospheric particles.

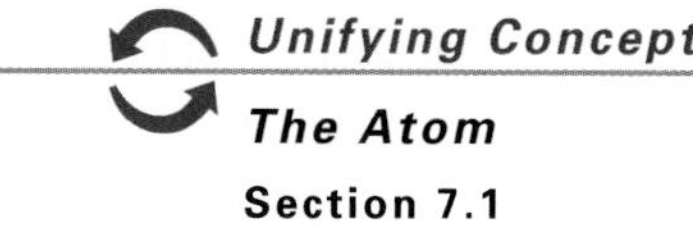

Radioactivity has been around since Earth's beginning.

(gamma rays). At sea level, the protective blanket of the atmosphere reduces this background radiation, but radiation is more intense at higher altitudes. In Denver, the "mile-high city," a person receives more than twice as much radiation from cosmic rays as someone at sea level. A couple of round-trip flights between New York and San Francisco exposes us to as much radiation as we receive in a chest X-ray at the physician's office. Because extended exposure to this level of radiation is dangerous, the flight time of airline personnel is limited.

Cosmic radiation also affects us indirectly, by transforming the nitrogen atoms in the air to radioactive carbon-14, which is incorporated through photosynthesis into the plants we consume. Radioactive potassium from the earth is in our bones. So even our bodies are radioactive.

10.2 Alpha, Beta, and Gamma Rays

Radiation and Biological Effects

All elements having an atomic number greater than 82 (lead) are radioactive. These elements, and others, emit three distinct types of radiation, named by the first three letters of the Greek alphabet, α, β, γ—alpha, beta, and gamma. Alpha

SCIENCE AND SOCIETY

Nuclear Technology

When household electricity made its way across the country more than a century ago, it represented a new technology with great potential. But it was not without its hazards. While electric grids could provide cities with a quantity and quality of energy previously unheard of, they could also kill people in ways previously unheard of. Many people opposed the adoption of household electricity because of the inherent dangers. But safeguards were engineered, and society determined that the benefits of electricity outweighed the risks. Similar debates continue today; not over electricity, but over nuclear energy and radioactivity.

Without nuclear technology, we would not have X-rays, radiation treatments for fighting cancer, smoke detectors, nuclear power as a source of electricity, nor hundreds of other useful applications. Then again, we would not have nuclear bombs. Along with any technology comes responsibility, in this case for safeguarding nuclear material and disposing of it in such a way as to avoid endangering future generations. As a member of society, you will have to make important decisions in these matters. You should do so with an adequate understanding of the atomic nucleus and its inner processes.

rays carry a positive electrical charge, beta rays carry a negative charge, and gamma rays carry no charge. The three rays can be separated by placing a magnetic field across their paths (Figure 10.3).

INSIGHT Once an alpha particle slows by collisions, it becomes a harmless helium atom.

Alpha radiation is a stream of alpha particles. An **alpha particle** is the combination of two protons and two neutrons (in other words, it is the nucleus of the helium atom, atomic number 2). Alpha particles are relatively easy to shield against because of their relatively large size and their double positive charge (+2). For example, they do not normally penetrate such lightweight materials as paper or clothing. Because of their great kinetic energies, however, alpha particles can cause significant damage to the surface of a material, especially living tissue. When traveling through only a few centimeters of air, alpha particles pick up electrons and become nothing more than harmless helium. As a matter of fact, that's where the helium in a child's balloon comes from—practically all Earth's helium atoms were at one time energetic alpha particles.

INSIGHT Once a beta particle slows by collisions, it becomes just another electron in the environment.

Beta radiation is a stream of beta particles. A **beta particle** is an electron ejected from a nucleus. Once ejected, it is indistinguishable from an electron in a cathode ray or in an electrical circuit, or from an electron orbiting the atomic nucleus. The difference is that a beta particle originates inside the nucleus—from a neutron. As we shall soon see, the neutron becomes a proton once it loses the electron that has become a beta particle. A beta particle is normally faster than an alpha particle, and it carries only a single negative charge (−1). Beta particles are not as easy to stop as alpha particles are, and they are able to penetrate such low-mass materials as paper or clothing. They can penetrate fairly deeply into skin, where they have the potential for harming or killing living cells. But they are unable to penetrate even thin sheets of denser materials, such as aluminum. Beta particles, once stopped, simply become a part of the material they are in, like any other electron.

Gamma rays are the high-frequency electromagnetic radiation emitted by radioactive elements. Like photons of visible light, a gamma ray is pure energy. The amount of energy in a gamma ray, however, is much greater per photon than in visible light, ultraviolet light, or even X-rays. Because they have no mass or electric charge, and because of their high energies, gamma rays are able to penetrate through most materials. However, they cannot easily penetrate very dense materials, such as lead. Lead is commonly used as a shielding material in laboratories or hospitals where there can be much gamma radiation. Delicate molecules inside cells throughout our bodies that are zapped by gamma rays suffer structural damage. Hence, gamma rays are generally more harmful to us than are

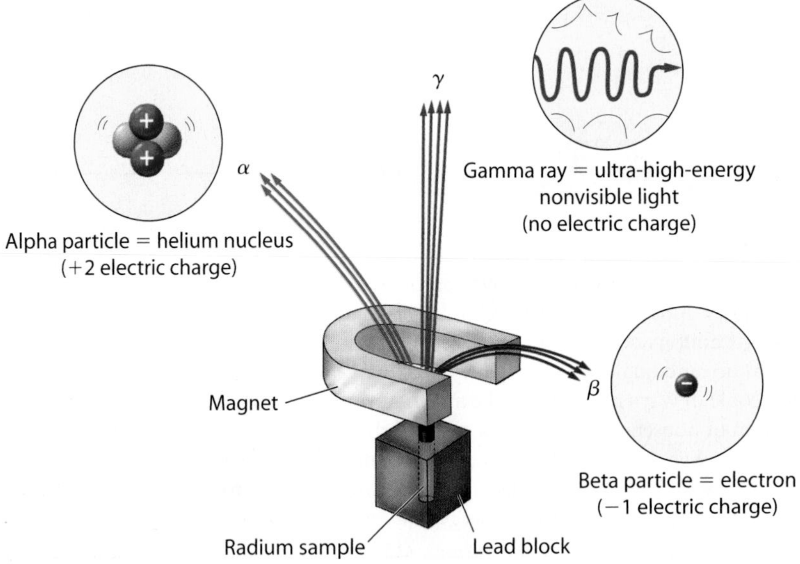

Figure 10.3 Interactive Figure
In a magnetic field, alpha rays bend one way, beta rays bend the other way, and gamma rays don't bend at all. Note that the alpha rays bend less than beta rays. This occurs because alpha particles have more inertia (mass) than beta particles. The combined beam comes from a radioactive material placed at the bottom of a hole drilled in a block of lead.

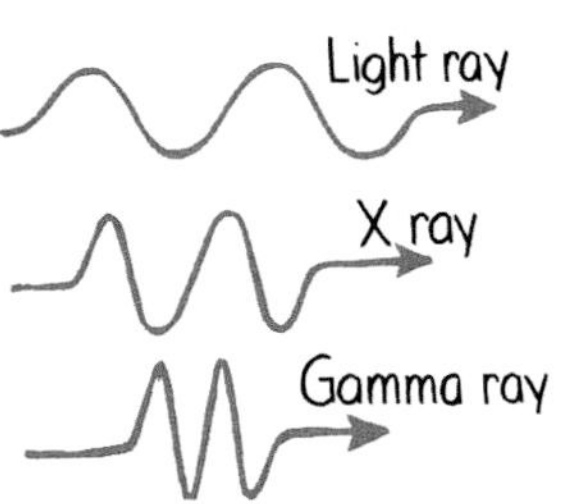

Figure 10.4 A gamma ray is simply electromagnetic radiation, much higher in frequency and energy than light and X-rays.

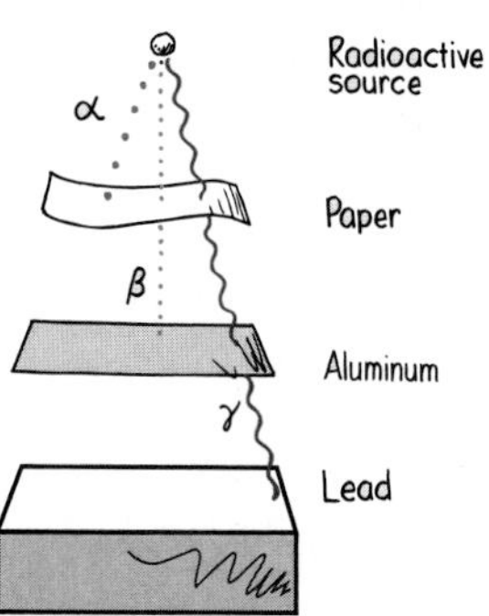

Figure 10.5 Interactive Figure
Alpha particles are the least penetrating and can be stopped by a few sheets of paper. Beta particles will readily pass through paper, but not through a sheet of aluminum. Gamma rays penetrate several centimeters into solid lead.

Figure 10.6 The shelf-life of fresh strawberries and other perishables is markedly increased when the food is subjected to gamma rays from a radioactive source. The strawberries on the right were treated with gamma radiation, which kills the microorganisms that normally lead to spoilage. The food is only a receiver of radiation and is in no way transformed into an emitter of radiation, as can be confirmed with a radiation detector.

alpha or beta particles. However, radioactive materials that emit alpha particles and beta particles do not pose a significant health risk if they are ingested.

CHECK YOURSELF

Pretend that you are given three radioactive rocks. One is an alpha emitter, one is a beta emitter, and one is a gamma emitter, and you know which is which. You can throw away one, but, of the remaining two, you must hold one in your hand and place one in your pocket. What can you do to minimize your exposure to radiation?

CHECK YOUR ANSWER

Hold the alpha emitter in your hand, because the skin on your hand will shield you. Put the beta emitter in your pocket, because beta particles will likely be stopped by the combined thickness of your clothing and skin. Throw away the gamma emitter, because it would penetrate your body from any of these locations. (Ideally, of course, you should distance yourself as much as possible from all of the rocks.)

10.3 Environmental Radiation

Common rocks and minerals in our environment contain significant quantities of radioactive isotopes, because most of them contain trace amounts of uranium. As a matter of fact, people who live in brick, concrete, or stone buildings are exposed to greater amounts of radiation than people who live in wooden buildings.

The leading source of naturally occurring radiation is radon-222, an inert gas arising from uranium deposits. Radon is a heavy gas that tends to accumulate in basements after it seeps up through cracks in the floor. Levels of radon vary from region to region, depending upon local geology. You can check the radon level in your home with a radon detector kit (Figure 10.7). If levels are high, corrective measures, such as sealing the basement floor and walls and maintaining adequate ventilation, should be taken. Radon gas poses a serious health risk.

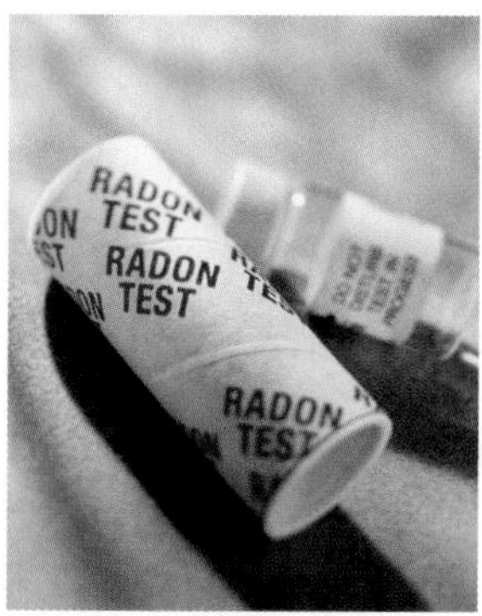

Figure 10.7 A commercially available radon test kit.

About 20 percent of our annual exposure to radiation comes from sources outside of nature, primarily medical procedures. Televisions, computer monitors, cell phones, fallout from nuclear testing, and the coal and nuclear power industries

INSIGHT The average coal-burning power plant is a far greater source of airborne radioactive material than is a nuclear power plant.

are also contributors. The coal industry far outranks the nuclear power industry as a source of radiation. The global combustion of coal annually releases about 13,000 tons of radioactive thorium and uranium into the atmosphere (in addition to other environmentally damaging molecules). Both of these elements are found naturally in coal deposits, so their release is a natural consequence of burning coal. Worldwide, the nuclear power industries generate about 10,000 tons of radioactive waste each year. Most of this waste, however, is contained and *not released into the environment.*

Units of Radiation

Radiation dosage is commonly measured in *rads* (*r*adiation *a*bsorbed *d*ose), a unit of absorbed energy. One **rad** is equal to 0.01 joule of radiant energy absorbed per kilogram of tissue.

The capacity for nuclear radiation to cause damage is not just a function of its level of energy, however. Some forms of radiation are more harmful than others. For example, suppose you have two arrows, one with a pointed tip and one with a suction cup at its tip. If you shoot each of them at an apple at the same speed, both will have the same kinetic energy. The arrow with the pointed tip, however, will invariably do more damage to the apple than the one with the suction cup. Similarly, some forms of radiation cause greater harm than other forms, even when we receive the same number of rads from both forms.

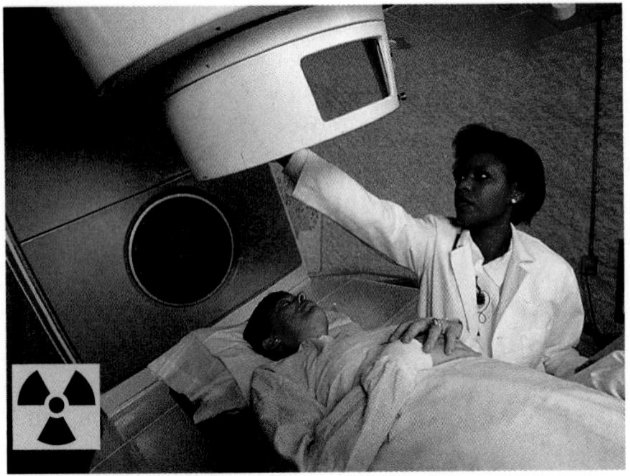

Figure 10.8 Nuclear radiation is focused on harmful tissue, such as a cancerous tumor, to kill cells selectively or to shrink the tissue in a technique known as radiation therapy. This application of nuclear radiation has saved millions of lives—a clear-cut example of the benefits of nuclear technology. The inset shows the internationally used symbol indicating an area where radioactive material is being handled or produced.

The unit of measure for radiation dosage based on potential damage is the **rem** (roentgen equivalent man).* In calculating the dosage in rems, we multiply the number of rads by a factor that corresponds to different health effects of different types of radiation as determined by clinical studies. For example, 1 rad of alpha particles has the same biological effect as 10 rads of beta particles.† We call both of these dosages 10 rems:

Particle	Radiation Dosage		Factor		Health Effect
Alpha	1 rad	×	10	=	10 rems
Beta	10 rad	×	1	=	10 rems

CHECK YOURSELF

If you have to choose between the two, would you prefer exposure to 1 rad of alpha particles or 1 rad of beta particles?

CHECK YOUR ANSWER

Multiply these quantities of radiation by the appropriate factor to get the dosages in rems. Alpha: 1 rad × 10 = 10 rems. Beta: 1 rad × 1 = 1 rem. The factors show us that, physiologically speaking, alpha particles are 10 times more damaging than beta particles.

INTEGRATED SCIENCE BIOLOGY

Doses of Radiation

Lethal doses of radiation begin at 500 rems. A person has about a 50 percent chance of surviving a dose of this magnitude received over a short period of time. During radiation therapy, a patient may receive localized doses in excess of 200 rems daily for a period of weeks (Figure 10.8).

* This unit is named for the discoverer of X-rays, Wilhelm Roentgen.

† This is true even though beta particles have more penetrating power, as discussed in Section 16.2.

Table 10.1 Annual Radiation Exposure

Typical Annual Dose Source	(mrem)
Natural Origin	
Cosmic radiation	30
Ground	30
Air (radon-222)	200
Human tissues (K-40; Ra-226)	40
Human Origin	
Smoking	280
Medical procedures	
Diagnostic X rays	40
Nuclear medicine	15
TV tubes, other consumer products	11
Weapons-test fallout	1
Coal-burning power plants	<1
Commercial nuclear power plants	$\ll 1$

All the radiation we receive from natural sources and from medical procedures is only a fraction of 1 rem. For convenience, the smaller unit millirem is used, where 1 millirem (mrem) is $\frac{1}{1000}$ of a rem. The average person in the United States is exposed to about 360 mrem a year, as Table 10.1 indicates. About 80 percent of this radiation comes from natural sources, such as cosmic rays and Earth itself. A typical chest X-ray exposes a person to 5 to 30 mrem (0.005 to 0.030 rem), less than one ten-thousandth of the lethal dose. As mentioned earlier, the human body is a significant source of natural radiation, primarily from the potassium we ingest. Our bodies contain about 200 grams of potassium. Of this quantity, about 20 milligrams is the radioactive isotope potassium-40, which is a gamma-ray emitter. Between every heartbeat, about 50,000 potassium-40 isotopes in the average human body undergo spontaneous radioactive decay. Radiation is indeed everywhere.

When radiation encounters the intricately structured molecules in our cells, it can create chaos on the atomic scale. Some molecules are broken, and this change alters other molecules, which can be harmful to our life processes. Cells are able to repair most kinds of molecular damage caused by radiation, if the radiation is not too severe. A cell can survive an otherwise lethal dose of radiation if the dose is spread over a long period of time to allow intervals for healing. When radiation is sufficient to kill cells, the dead cells can be replaced by new ones (except for most nerve cells, which are irreplaceable). Sometimes a radiated cell will survive with a damaged DNA molecule. DNA damage can alter the genetic information contained in a cell, producing one or more mutations (Chapter 16). Although the effects of many mutations are insignificant, others affect the functioning of cells. Genetic mutations are the cause of most cancers, for example. In addition, if mutations occur in an individual's reproductive cells, they can be passed to the individual's offspring. In this case, the mutation will be present in every cell in the offspring organism's body—and may well have an effect on the functioning of the organism.

TECHNOLOGY

Radioactive Tracers

In scientific laboratories, all the elements have been made radioactive by bombardment with neutrons and other particles. Radioactive materials are extremely useful in scientific research and industry. In order to check the action of a fertilizer, for example, researchers combine a small amount of radioactive material with the fertilizer and then apply the combination to a few plants. The amount of radioactive fertilizer absorbed by the plants can be easily measured with radioactive detectors. From such measurements, scientists can inform farmers about proper usages of fertilizer. When used in this way, radioactive isotopes are called *tracers*.

Perhaps the simplest tracer is pouring radioactive isotopes into a faulty pipe system to locate leaks. Tracers are widely used in medicine to diagnose disease. Small quantities of particular radioactive isotopes, after being injected into the bloodstream, concentrate at trouble spots, such as bone fractures and tumors. Using radiation detectors, medical staff can locate isotope concentrations.

Engineers can study how the parts of an automobile engine wear away during use by making the cylinder walls radioactive. While the engine is running, the piston rings rub against the cylinder walls. The tiny particles of metal that are worn away fall into the lubricating oil, where they can be measured with a radiation detector. By repeating this test with different oils, the engineer can determine which oil provides the least wear and longest life to the engine.

There are hundreds more examples of the use of trace amounts of radioactive isotopes. The important thing is that this technique provides a way to detect and count atoms in samples of materials too small to be seen with a microscope.

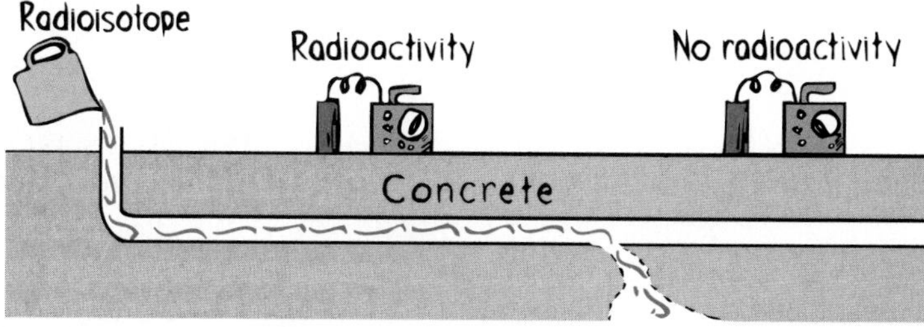

CHECK YOURSELF

1. Is the human body radioactive?
2. How can radioactivity produce a genetic mutation?
3. How can radioactivity produce birth defects?

CHECK YOUR ANSWERS

1. Yes. The human body is radioactive because of all the radioactive isotopes it contains.
2. Radioactivity produces genetic mutations by damaging a cell's DNA—its genetic blueprint—without killing the cell.
3. Radioactivity can cause birth defects by producing genetic mutations in sex cells (sperm or eggs) that later develop into a new organism, or by producing mutations in a developing fetus. A developing fetus is particularly vulnerable to DNA damage, which is why pregnant women should be careful to avoid exposure to radiation.

10.4 The Atomic Nucleus and the Strong Nuclear Force

As described in Chapter 9, the atomic nucleus occupies only a few quadrillionths of the volume of the atom, leaving most of the atom as empty space. The nucleus is composed of **nucleons**, which is the collective name for protons and neutrons.

Just as there are energy levels for the orbital electrons of an atom, there are energy levels within the nucleus. Whereas orbiting electrons emit photons when making transitions to lower energy levels, similar energy changes occur in radioactive nuclei that result in the emission of gamma-ray photons. This is gamma radiation.

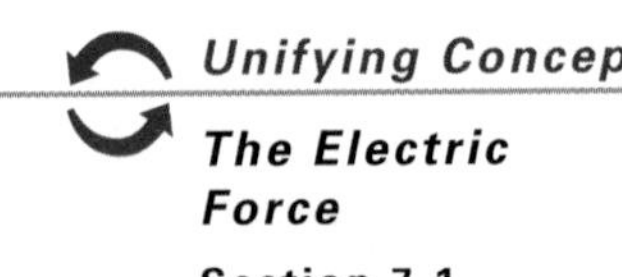
Unifying Concept
The Electric Force
Section 7.1

We know that electrical charges of like signs repel one another. So how is it possible that positively charged protons in the nucleus remain clumped together? This question led to the discovery of an attraction called the **strong nuclear force**, which acts between all nucleons. This force is very strong, but only over extremely short distances (about 10–15 meters, the diameter of a typical atomic nucleus). Repulsive electrical interactions, on the other hand, have a relatively long range. Figure 10.9 compares the strength of these two forces over distance. For protons that are close together, as in small nuclei, the attractive strong nuclear force easily overcomes the repulsive electrical force. But for protons that are far apart, such as those on opposite edges of a large nucleus, the attractive strong nuclear force may be weaker than the repulsive electrical force.

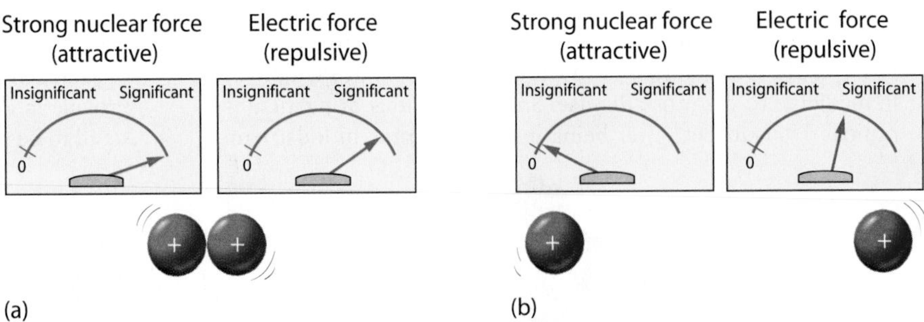

Figure 10.9 (a) Two protons near each other experience both an attractive strong nuclear force and a repulsive electric force. At this tiny separation distance, the strong nuclear force overcomes the electric force, resulting in their remaining together. (b) When the two protons are relatively far from each other, the electric force is more significant. The protons repel each other. This proton–proton repulsion in large atomic nuclei reduces nuclear stability.

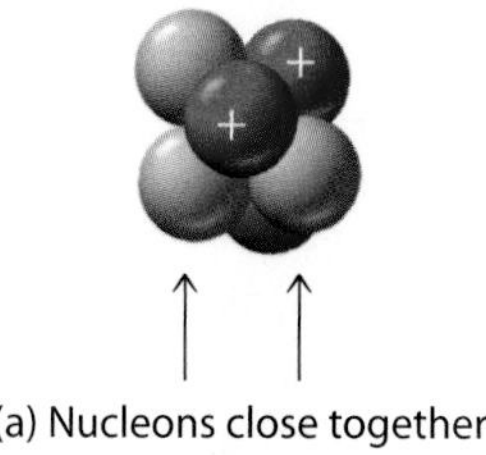

(a) Nucleons close together

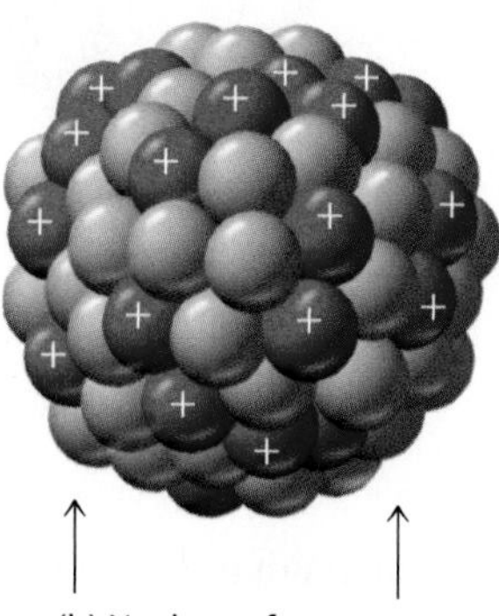

(b) Nucleons far apart

Figure 10.10 (a) All nucleons in a small atomic nucleus are close to one another; hence, they experience an attractive strong nuclear force. (b) Nucleons on opposite sides of a larger nucleus are not as close to one another, and so the attractive strong nuclear forces holding them together are much weaker. The result is that the large nucleus is less stable.

A large nucleus is not as stable as a small one. In a helium nucleus, for example, each of the two protons feels the repulsive effect of the other. In a uranium nucleus, each proton feels the repulsive effects of the other 91 protons! The nucleus is unstable. We see that there is a limit to the size of the atomic nucleus. It is for this reason that all nuclei having more than 82 protons are radioactive (Figure 10.10).

CHECK YOURSELF

Two protons in the atomic nucleus repel each other, but they are also attracted to each other. Why?

CHECK YOUR ANSWER

While two protons repel each other by the electric force, they also attract each other by the strong nuclear force. Both of these forces act simultaneously. So long as the attractive strong nuclear force is stronger than the repulsive electric force, the protons will remain together. Under conditions in which the electric force overcomes the strong nuclear force, however, the protons fly apart from each other.

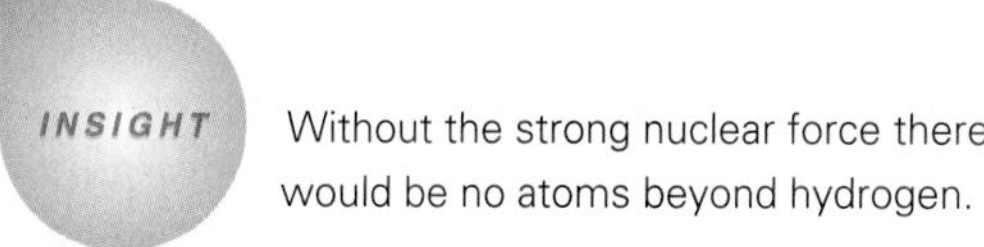

INSIGHT Without the strong nuclear force there would be no atoms beyond hydrogen.

Neutrons serve as a "nuclear cement" holding the atomic nucleus together. Protons attract both protons and neutrons by the strong nuclear force. Protons also repel other protons by the electric force. Neutrons, on the other hand, have no electric charge and so only attract other protons and neutrons by the strong nuclear force. The presence of neutrons therefore adds to the attraction among nucleons and helps to hold the nucleus together (Figure 10.11).

The more protons there are in a nucleus, the more neutrons are needed to help balance the repulsive electric forces. For light elements, it is sufficient to have about as many neutrons as protons. For example, the most common isotope of carbon, C-12, has equal numbers of each—six protons and six neutrons. For large nuclei, more neutrons than protons are needed. Because the strong nuclear force diminishes rapidly over distance, nucleons must be practically touching in order for the strong nuclear force to be effective. Nucleons on opposite sides of a large atomic nucleus are not attracted to one another. The electric force, however, diminishes very little across the diameter of a large nucleus, and so it wins out over the strong nuclear force. To compensate for the near absence of the strong nuclear force across the diameter of the nucleus, large nuclei have more neutrons than protons. Lead, for example, has about one and one-half times as many neutrons as protons.

So we see that neutrons have a stabilizing effect, and large nuclei require an abundance of them. But neutrons are not always successful in keeping a nucleus intact. Interestingly, neutrons are unstable when they are by themselves. A lone neutron is radioactive, and it spontaneously transforms to a proton and an electron (Figure 10.12a). A neutron seems to need protons around to keep this from occurring. After the size of a nucleus reaches a certain point, the neutrons so outnumber the protons that there are not sufficient protons in the mix to prevent the neutrons from turning into protons. As neutrons in a nucleus change into protons, the stability of the nucleus decreases because the repulsive electric force

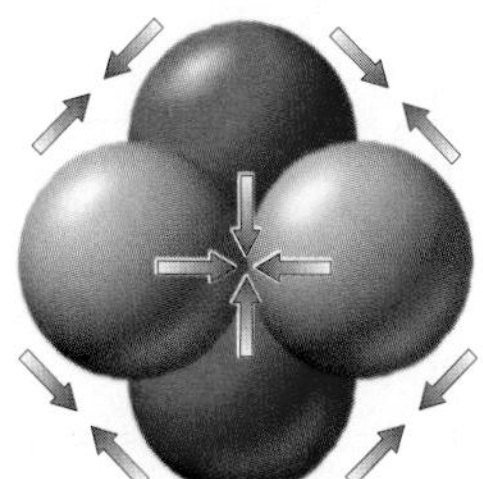

All nucleons, both protons and neutrons, attract one another by the strong nuclear force.

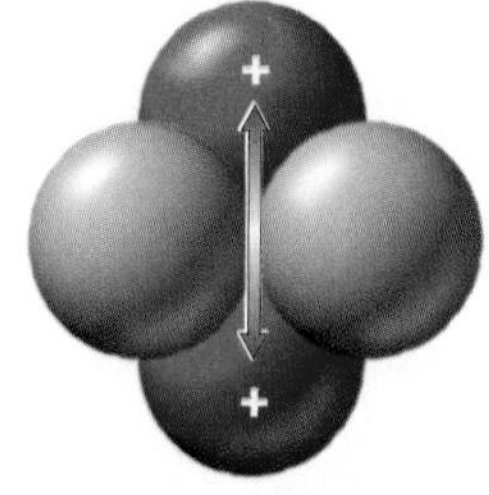

Only protons repel one another by the electric force.

Figure 10.11 The presence of neutrons helps hold the nucleus together by increasing the effect of the strong nuclear force, represented by the single-headed arrows.

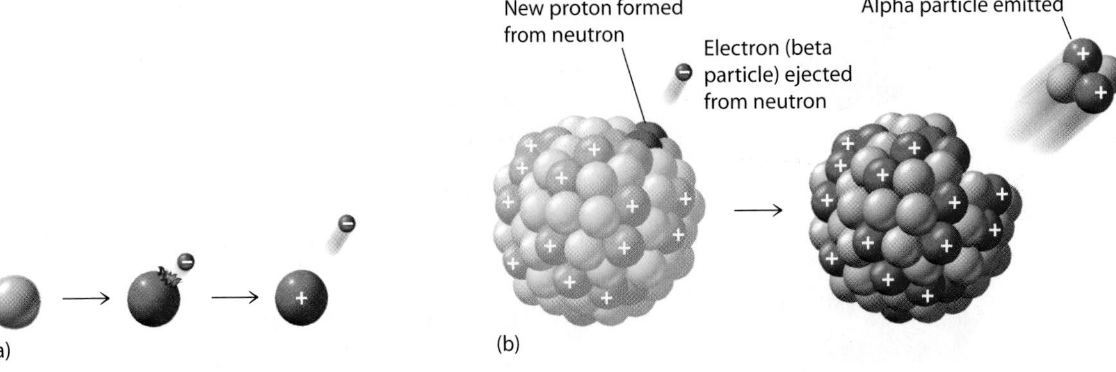

Figure 10.12 (a) A neutron near a proton is stable, but a neutron by itself is unstable and decays to a proton by emitting an electron. (b) Destabilized by an increase in the number of protons, the nucleus begins to shed fragments, such as alpha particles.

becomes increasingly significant. The result is that pieces of the nucleus fragment away in the form of radiation, as indicated in Figure 10.12b.

CHECK YOURSELF

What role do neutrons serve in the atomic nucleus? What is the fate of a neutron when alone or distant from one or more protons?

CHECK YOUR ANSWER

Neutrons serve as a nuclear cement in nuclei, and they add to nuclear stability. But when alone or away from protons, a neutron becomes radioactive, and it spontaneously transforms to a proton and an electron.

10.5 Half-Life

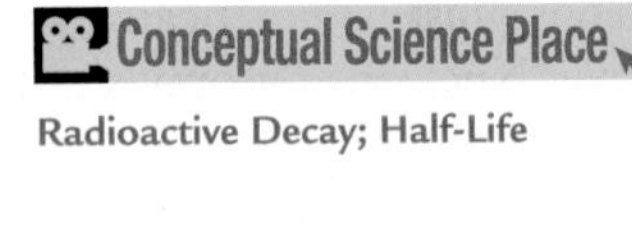

Radioactive Decay; Half-Life

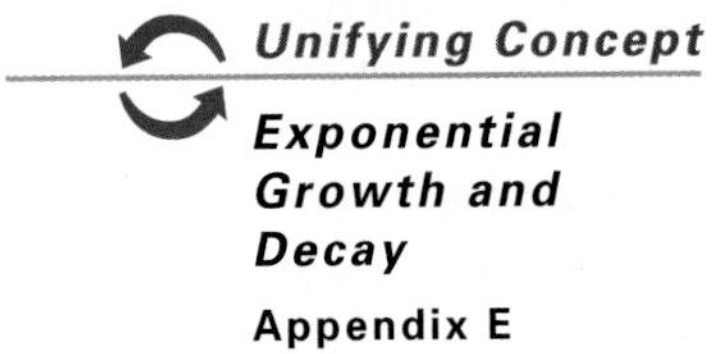

The rate of decay for a radioactive isotope is measured in terms of a characteristic time, the **half-life.** This is the time it takes for half of an original quantity of an element to decay. For example, radium-226 has a half-life of 1620 years, which means that half of a pure radium-226 sample will be converted to other elements by the end of 1620 years. In the next 1620 years, half of the remaining radium will decay, leaving only one-fourth the original amount of radium. (After 20 half-lives, the initial quantity of radium-226 will be diminished by a factor of about 1 million; Figure 10.13.)

The half-life of an element is remarkably constant and not affected by external conditions. Some radioactive isotopes have half-lives that are less than a millionth of a second, while others have half-lives of more than a billion years.

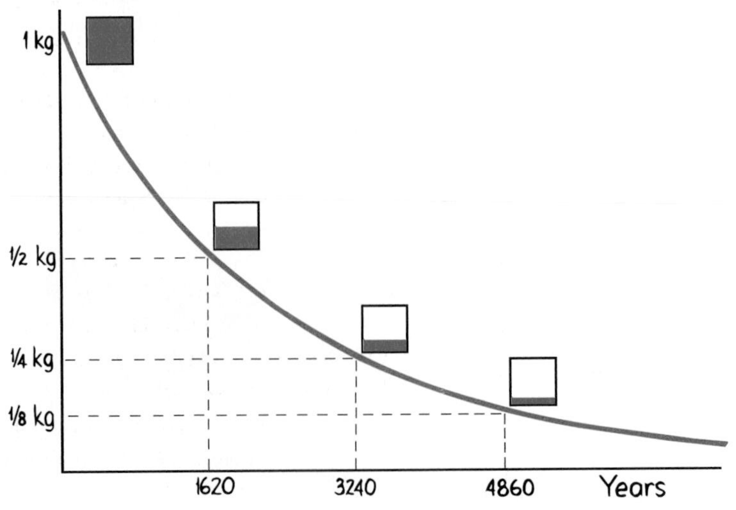

Figure 10.13 Interactive Figure
Every 1620 years, the amount of radium in a given sample decreases by half.

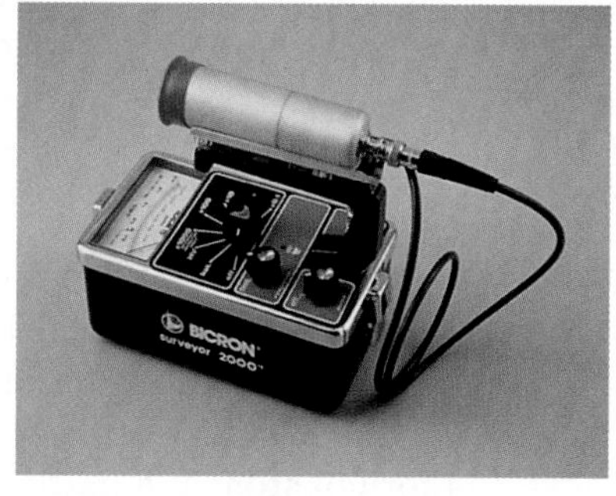

(a)

(b)

Figure 10.14 Radiation detectors. (a) A Geiger counter detects incoming radiation by its ionizing effect on enclosed gas in the tube. (b) A scintillation counter detects incoming radiation by flashes of light that are produced when charged particles or gamma rays pass through it.

Uranium-238 has a half-life of 4.5 billion years. All uranium isotopes eventually decay in a series of steps to become lead.

It is not necessary to wait through the duration of a half-life in order to measure it. The half-life of an element can be calculated at any given moment by measuring the rate of decay of a known quantity. This is easily done using a radiation detector (Figure 10.14). In general, the shorter the half-life of a substance, the faster it disintegrates, and the more radioactivity per amount is detected.

The radioactive half-life of a material is also the time for its decay rate to reduce to half.

CHECK YOURSELF

1. If a sample of radioactive isotopes has a half-life of 1 day, how much of the original sample will remain at the end of the second day? The third day?
2. Which will give a higher counting rate on a radiation detector—radioactive material that has a short half-life or a long half-life?

CHECK YOUR ANSWERS

1. One-fourth of the original sample will be left—the three-fourths that underwent decay is then a different element altogether. At the end of 3 days, one-eighth of the original sample will remain.
2. The material with the shorter half-life is more active and will show a higher counting rate on a radiation detector.

10.6 Transmutation of Elements

Conceptual Science Place
Nuclear Chemistry

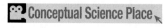

Conceptual Science Place
Plutonium

When a radioactive nucleus emits an alpha or a beta particle, there is a change in atomic number—a different element is formed. The changing of one chemical element to another is called **transmutation**. Transmutation occurs in natural events and is also initiated artificially in the laboratory.

Natural Transmutation

Consider uranium-238, the nucleus of which contains 92 protons and 146 neutrons. When an alpha particle is ejected, the nucleus loses two protons and two neutrons. Because an element is defined by the number of protons in its nucleus, the 90 protons and 144 neutrons left behind are no longer identified as being uranium. What we have is the nucleus of a different element—thorium. This transmutation can be written as a nuclear equation:

92 146 → 90 144 +

$$^{238}_{92}\text{U} \rightarrow {}^{234}_{90}\text{Th} + {}^{4}_{2}\text{He}$$

We see that $^{238}_{92}\text{U}$ transmutes to the two elements written to the right of the arrow. When this transmutation occurs, energy is released, partly in the form of kinetic energy of the alpha particle ($^{4}_{2}\text{He}$), partly in the kinetic energy of the thorium atom and partly in the form of gamma radiation. In this and all such equations, the mass numbers at the top balance ($238 = 234 + 4$) and the atomic numbers at the bottom also balance ($92 = 90 + 2$). Thorium-234, the product of this reaction, is also radioactive. When it decays, it emits a beta particle.* Since a beta particle is an electron, the atomic number of the resulting nucleus is *increased* by 1. So, after beta emission by thorium with 90 protons, the resulting element has 91 protons (a neutron ejects the beta and becomes a proton). It is no longer thorium, but has become the element protactinium. Although the atomic number has increased by 1 in this process, the mass number (protons + neutrons) remains the same. The nuclear equation is:

90 144 → 91 143 +

$$^{234}_{90}\text{Th} \rightarrow {}^{234}_{91}\text{Pa} + {}^{0}_{-1}\text{e}$$

We write an electron as $^{0}_{-1}\text{e}$. The superscript 0 indicates that the electron's mass is insignificant relative to that of protons and neutrons. The subscript -1 is the electric charge of the electron.

So we see that, when an element ejects an alpha particle from its nucleus, the mass number of the resulting atom is decreased by 4, and its atomic number is decreased by 2. The resulting atom is an element two spaces back in the periodic table of the elements. When an element ejects a beta particle from its nucleus, the mass of the atom is practically unaffected, meaning there is no change in its mass number, but its atomic number increases by 1. The resulting atom belongs to an element one place forward in the periodic table. Gamma emission results in no change in either the mass number or the atomic number. So we see that radioactive elements can decay backward or forward in the periodic table.†

The successions of radioactive decays of $^{238}_{92}\text{U}$ to $^{206}_{82}\text{Pb}$, an isotope of lead, is shown in Figure 10.15. Each green arrow shows an alpha decay, and each red arrow shows a beta decay. Notice that some of the nuclei in the series can decay in both ways. This is one of several similar radioactive series that occur in nature.

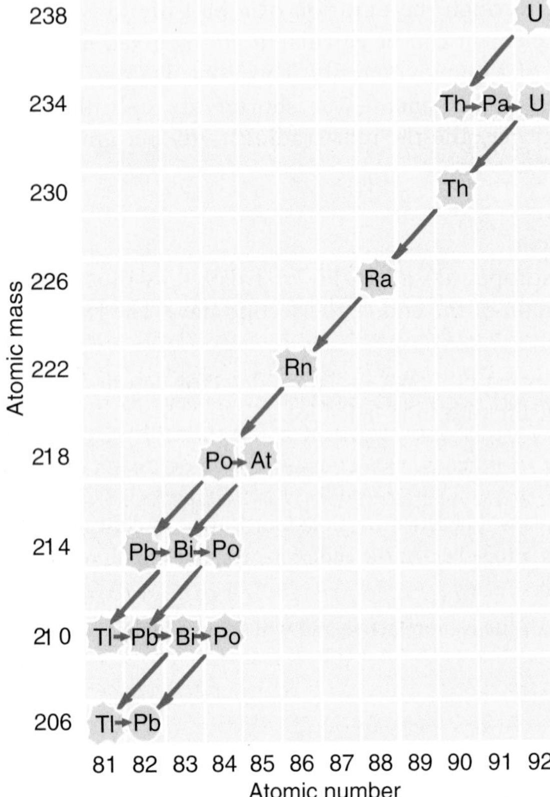

Figure 10.15 U-238 decays to Pb-206 through a series of alpha and beta decays.

* Beta emission is always accompanied by the emission of a neutrino (actually an antineutrino), a neutral particle with nearly zero mass that travels at about the speed of light. The neutrino ("little neutral one") was postulated by Wolfgang Pauli in 1930 and detected in 1956. Neutrinos are difficult to detect because they interact very weakly with matter. A piece of solid lead a few centimeters thick will stop most gamma rays from a radium source, whereas a piece of lead about 8 light-years thick would be needed to stop half the neutrinos produced in typical nuclear decays. Thousands of neutrinos are flying through you every second of every day, because the universe is filled with them. Only one or two times a year does a neutrino or two interact with the matter of your body.

At this writing, the mass of the neutrino is unknown. Neutrinos are so numerous in the universe that, if they have even the tiniest mass, they might constitute most of the mass of the universe. Neutrinos may be the "glue" that holds the universe together.

† Sometimes a nucleus emits a positron, which is the "antiparticle" of an electron. In this case, a proton becomes a neutron, and the atomic number is decreased.

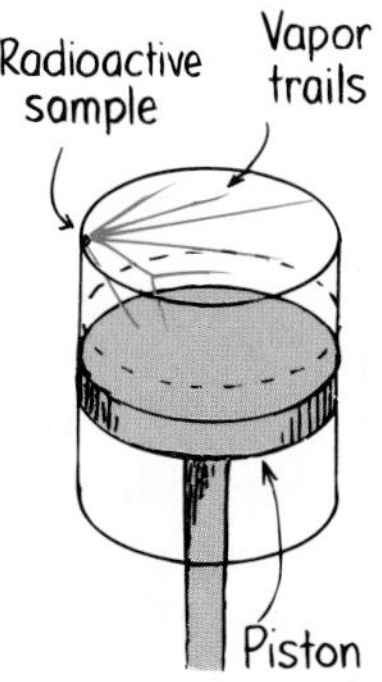

Figure 10.16 A simplified cloud chamber. Charged particles moving through supersaturated vapor leave trails. When the chamber is in a strong electric or magnetic field, bending of the tracks provides information about the charge, mass, and momentum of the particles.

CHECK YOURSELF

1. Complete the following nuclear reactions.
   a. $^{226}_{88}\text{Ra} \rightarrow ??? + ^{0}_{-1}\text{e}$
   b. $^{209}_{84}\text{Po} \rightarrow ^{205}_{82}\text{Pb} + ???$
2. What finally becomes of all the uranium that undergoes radioactive decay?

CHECK YOUR ANSWERS

1. a. $^{226}_{88}\text{Ra} \rightarrow ^{226}_{89}\text{Ac} + ^{0}_{-1}\text{e}$
   b. $^{209}_{84}\text{Po} \rightarrow ^{205}_{82}\text{Pb} + ^{4}_{2}\text{He}$
2. All uranium will ultimately become lead. On the way to becoming lead, it will exist as a series of elements, as shown in Figure 10.15.

Artificial Transmutation

Ernest Rutherford, in 1919, was the first of many investigators to succeed in transmuting a chemical element. He bombarded nitrogen gas with alpha particles from a piece of radioactive ore. The impact of an alpha particle on a nitrogen nucleus transmutes nitrogen into oxygen:

$$^{14}_{7}\text{N} + ^{4}_{2}\text{He} \rightarrow ^{17}_{8}\text{O} + ^{1}_{1}\text{H}$$

INSIGHT The alchemists of old tried vainly for more than 2000 years to cause the transmutation of one element to another. Despite their fervent efforts and rituals, they never came close to succeeding. Ironically, nuclear transmutations were going on all around them, which they never detected. And if they had detected them, they would not have recognized the transmutation process, because they lacked an adequate theory to guide them.

Rutherford used a device called a *cloud chamber* to record this event (Figure 10.16). In a cloud chamber, moving charged particles leave a visible trail of ions along their path, much as a jet plane high in the sky leaves a contrail of water vapor and ice crystals in its wake. From a quarter-million cloud-chamber tracks photographed on movie film, Rutherford showed seven examples of atomic transmutation. Analysis of tracks bent by a strong external magnetic field showed that when an alpha particle collided with a nitrogen atom, a proton bounced out and the heavy atom recoiled a short distance. The alpha particle disappeared. The alpha particle was absorbed in the process, transforming nitrogen to oxygen.

Since Rutherford's announcement in 1919, experimenters have carried out many other nuclear reactions, first with natural bombarding projectiles from radioactive ores and then with still more energetic projectiles—protons and electrons hurled by huge particle accelerators. Artificial transmutation is what produces the previously unknown synthetic elements from atomic numbers 93 to 115. All these artificially made elements have short half-lives. If they ever existed naturally when Earth was formed, they have long since decayed.

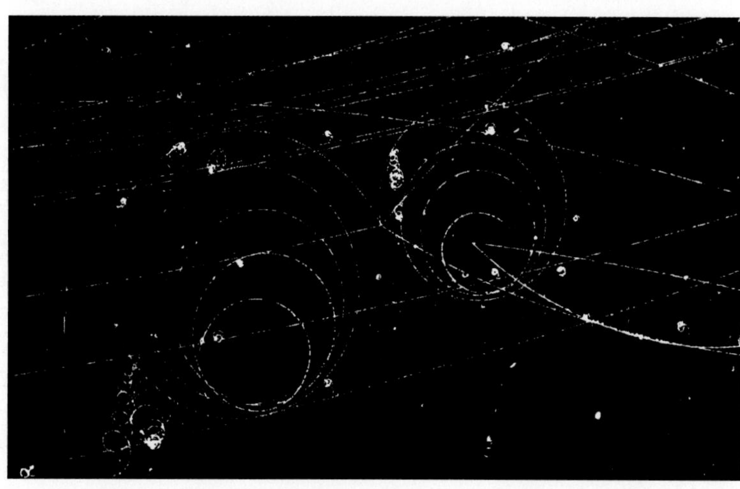

Figure 10.17 Tracks of elementary particles in a bubble chamber, a similar yet more complicated device than a cloud chamber. Two particles have been destroyed at the points where the spirals emanate, and four others have been created in the collision.

INTEGRATED SCIENCE BIOLOGY, EARTH SCIENCE

Isotopic Dating

Earth's atmosphere is continuously bombarded by cosmic rays, and this bombardment causes many atoms in the upper atmosphere to transmute. These transmutations result in many protons and neutrons being "sprayed out" into the environment. Most of the protons are stopped as they collide with the atoms of the upper atmosphere, stripping electrons from these atoms to become hydrogen atoms. The neutrons, however, keep going for longer distances because they have no electrical charge and therefore do not interact electrically with matter. Eventually, many of them collide with the nuclei in the denser lower atmosphere. A nitrogen atom that captures a neutron, for example, becomes an isotope of carbon by emitting a proton:

$$^{14}_{7}\text{N} + ^{1}_{0}n \rightarrow ^{14}_{6}\text{C} + ^{1}_{1}\text{H}$$

INSIGHT The terms *isotopic dating* and *radiometric dating* are interchangeable. Both terms refer to the nuclear decay of naturally occurring radioactive isotopes. Although the terms are interchangeable, for geologic age determination we generally use the term *radiometric dating*.

This carbon-14 isotope, which makes up less than one-millionth of 1 percent of the carbon in the atmosphere, is radioactive and has eight neutrons. (The most common isotope, carbon-12, has six neutrons and is not radioactive.) Because both carbon-12 and carbon-14 are forms of carbon, they have the same chemical properties. Both these isotopes can react chemically with oxygen to form carbon dioxide, which is taken in by plants. This means that all plants contain a tiny bit of radioactive carbon-14. All animals eat plants (or at least plant-eating animals) and therefore have a little carbon-14 in them. In short, all living things on Earth contain some carbon-14.

Carbon-14 is a beta emitter, and it decays back to nitrogen by the following reaction:

$$^{14}_{6}\text{C} \rightarrow ^{14}_{7}\text{N} + ^{0}_{-1}e$$

Because plants continue to absorb carbon dioxide as long as they live, any carbon-14 lost by decay is immediately replenished with fresh carbon-14 from the atmosphere. In this way, a radioactive equilibrium is reached where there is a constant ratio of about 1 carbon-14 atom to every 100 billion carbon-12 atoms. When a plant dies, replenishment of carbon-14 stops. Then the percentage of carbon-14 decreases at a constant rate given by its half-life.* The longer a plant or other organism is dead, therefore, the less carbon-14 it contains relative to the constant amount of carbon-12.

The half-life of carbon-14 is about 5730 years. This means that half of the carbon-14 atoms that are now present in a plant or animal that dies today will decay in the next 5730 years. Half of the remaining carbon-14 atoms will then decay in the following 5730 years, and so forth (Figure 10.18).

With this knowledge, scientists are able to calculate the age of carbon-containing artifacts or remains, such as wooden tools or skeletons, by measuring their current level of radioactivity. This process, known as carbon-14 dating, enables us to probe as much as 50,000–60,000 years into the past. Beyond this time span, there is too little carbon-14 remaining to permit accurate analysis.

* A 1-g sample of contemporary carbon contains about 5×10^{22} atoms, 6.5×10^{10} of which are C-14 atoms. Carbon-14 has a beta disintegration rate of about 13.5 decays per minute.

22,920 years ago

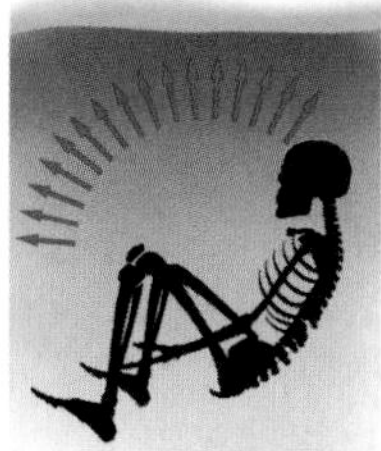

17,190 years ago

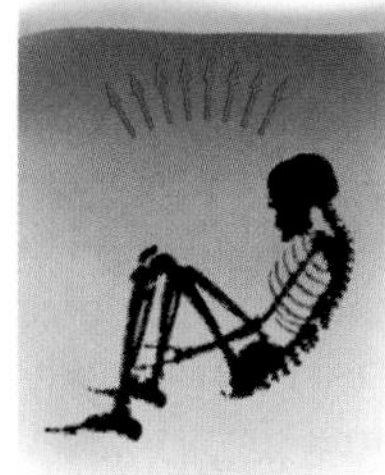

11,460 years ago

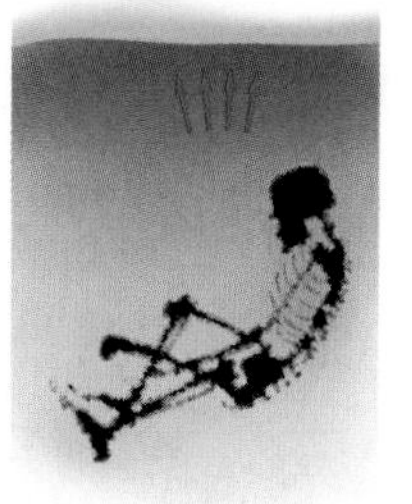

5730 years ago

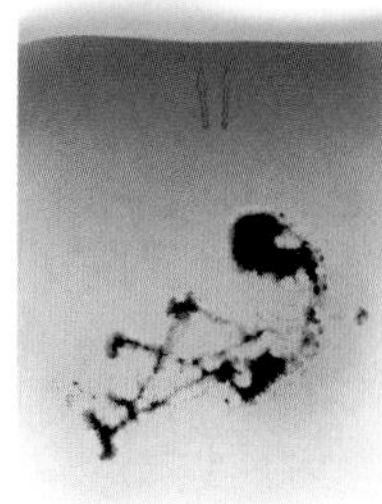

Present

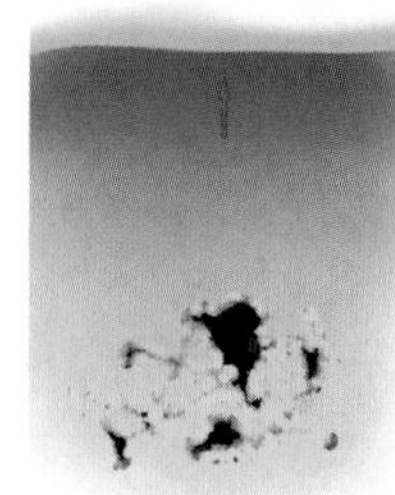

Figure 10.18 The amount of radioactive carbon-14 in the skeleton diminishes by one-half every 5730 years, with the result that the skeleton today contains only a fraction of the carbon-14 it originally had. The red arrows symbolize relative amounts of carbon-14.

Carbon-14 dating would be an extremely simple and accurate dating method if the amount of radioactive carbon in the atmosphere had been constant over the ages. But it hasn't been. Fluctuations in the Sun's magnetic field, as well as changes in the strength of Earth's magnetic field, affect cosmic-ray intensities in Earth's atmosphere, which, in turn, produce fluctuations in the production of C-14. In addition, changes in Earth's climate affect the amount of carbon dioxide in the atmosphere. The oceans are great reservoirs of carbon dioxide. When the oceans are cold, they release less carbon dioxide into the atmosphere than when they are warm.

The dating of older, but inorganic, things is accomplished with radioactive minerals, such as uranium. The naturally occurring isotopes U-238 and U-235 decay very slowly and ultimately become isotopes of lead—but not the common lead isotope Pb-208. For example, U-238 decays through several stages to finally become Pb-206, whereas U-235 finally becomes the isotope Pb-207. Lead isotopes 206 and 207 that now exist were at one time uranium. The older the uranium-bearing rock, the higher the percentage of these remnant isotopes.

From the half-lives of uranium isotopes, and the percentage of lead isotopes in uranium-bearing rock, it is possible to calculate the date at which the rock was formed.

CHECK YOURSELF

1. Suppose that an archaeologist extracts a gram of carbon from an ancient ax handle and finds that it is one-fourth as radioactive as a gram of carbon extracted from a freshly cut tree branch. About how old is the ax handle?
2. If it were known that cosmic ray intensity was much greater thousands of years ago, how would this affect the ages assigned to ancient samples of once-living matter?
3. The age of the Dead Sea Scrolls was found by carbon dating. Could this technique have worked if they were carved in stone tablets? Explain.

CHECK YOUR ANSWERS

1. Assuming the ratio of C-14/C-12 was the same when the ax was made, the ax handle is as old as two half-lives of C-14, or about 11,460 years old.
2. If the cosmic ray intensity were much greater thousands of years ago, then the proportion of carbon-14 in tissues living at that time would be greater. If such an increased intensity in cosmic rays were not taken into account, then the once-living matter would appear to be younger than it really is.
3. Stone tablets cannot be dated by the carbon-dating technique. Nonliving stone does not ingest carbon and transform that carbon by radioactive material. Carbon dating pertains to organic material.

10.7 Nuclear Fission

Conceptual Science Place
Nuclear Fission

In 1938, two German scientists, Otto Hahn and Fritz Strassmann, made an accidental discovery that was to change the world. While bombarding a sample of uranium with neutrons in the hope of creating new, heavier elements, they were astonished to discover chemical evidence for the production of barium, an element having about half the mass of uranium. Hahn wrote of this news to his former colleague Lise Meitner, who had fled from Nazi Germany to Sweden because of her Jewish ancestry. From the evidence given to her by Hahn, Meitner concluded that the uranium nucleus, activated by neutron bombardment, had split in half. Soon thereafter, Meitner, working with her nephew, Otto Frisch, also a physicist, published a paper in which the term *nuclear fission* was first coined.

In the nucleus of every atom, there exists a delicate balance between attractive nuclear forces and repulsive electric forces between protons. In all known nuclei, the nuclear forces dominate. In uranium, however, this domination is tenuous. If a uranium nucleus stretches into an elongated shape (Figure 10.19), the electrical forces may push it into an even more elongated shape. If the elongation passes a certain point, electrical forces overwhelm strong nuclear forces and the nucleus splits. This is **nuclear fission**.

The energy released by the fission of one U-235 nucleus is relatively enormous—about seven million times the energy released by the combustion of one TNT molecule. This energy is mainly in the form of the kinetic energy of the fission fragments that fly apart from one another, with some energy given to ejected neutrons and the remainder to gamma radiation.

A typical uranium fission reaction is:

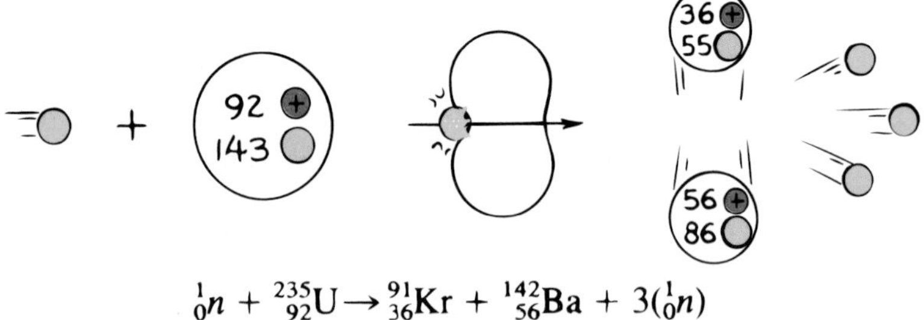

92
143
36
55
56
86

$$ {}^{1}_{0}n + {}^{235}_{92}\mathrm{U} \rightarrow {}^{91}_{36}\mathrm{Kr} + {}^{142}_{56}\mathrm{Ba} + 3({}^{1}_{0}n) $$

Note that, in this reaction, 1 neutron starts the fission of a uranium nucleus, and the fission produces 3 neutrons. (A fission reaction may produce fewer or more than 3 neutrons.) These product neutrons can cause the fissioning of three other uranium atoms, releasing 9 more neutrons. If each of these 9 neutrons succeeds in splitting a uranium atom, the next step in the reaction produces 27 neutrons, and so on. Such a sequence, illustrated in Figure 10.20 is called a **chain reaction**—a self-sustaining reaction in which the products of one reaction event stimulate further reaction events.

If a chain reaction were to occur in a baseball-sized chunk of pure U-235, an enormous explosion would result. If the chain reaction were started in a smaller chunk of pure U-235, however, no explosion would occur. This is because of geometry: The ratio of surface area to mass is larger in a small piece than in a

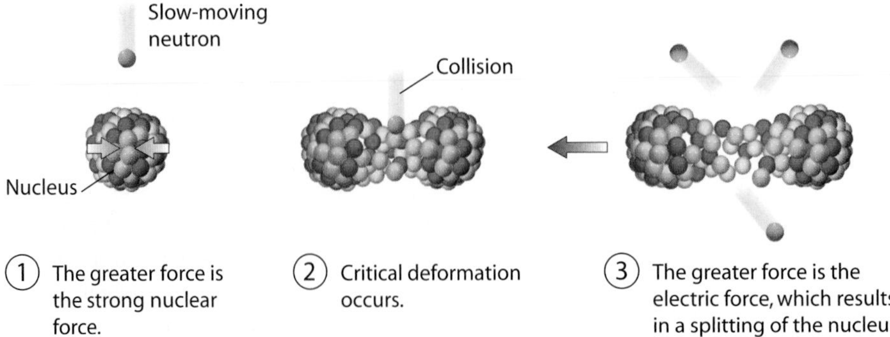

Slow-moving neutron
Collision
Nucleus
1 The greater force is the strong nuclear force.
2 Critical deformation occurs.
3 The greater force is the electric force, which results in a splitting of the nucleus.

Figure 10.19 Nuclear deformation may result in repulsive electrical forces overcoming attractive nuclear forces, in which case fission takes place.

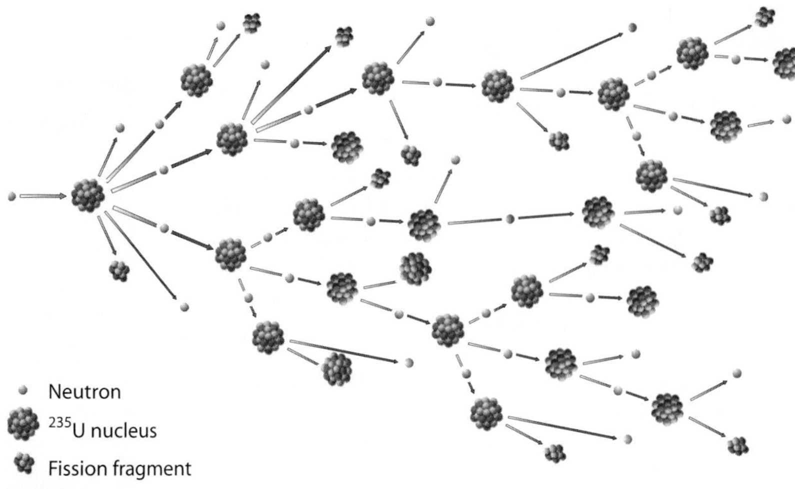

Figure 10.20 A chain reaction.

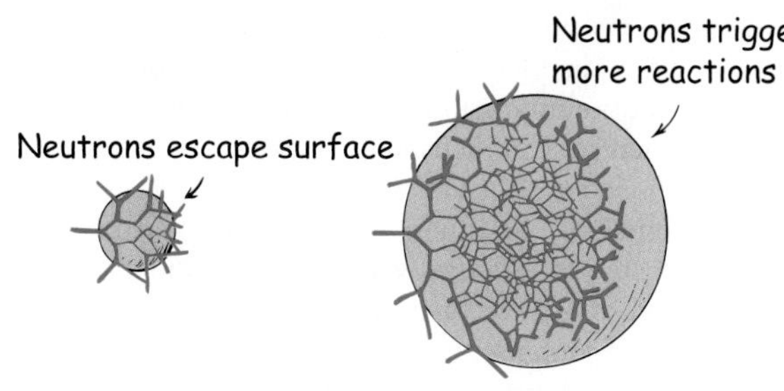

Figure 10.21 The exaggerated view shows that a chain reaction in a small piece of pure U-235 runs its course before it can cause a large explosion because neutrons leak from the surface too soon. The surface area of the small piece is large relative to the mass. In a larger piece, more uranium and less surface is presented to the neutrons.

large one (just as there is more skin on six small potatoes having a combined mass of 1 kilogram than there is on a single 1-kilogram potato). So there is more surface area on a bunch of small pieces of uranium than on a large piece. In a small piece of U-235, neutrons leak through the surface before an explosion can occur. In a bigger piece, the chain reaction builds up to enormous energies before the neutrons reach the surface and escape (Figure 10.21). For masses greater than a certain amount, called the **critical mass**, an explosion of enormous magnitude may occur.

Consider a large quantity of U-235 divided into two pieces, each having a mass less than critical. The pieces are *subcritical*. Neutrons in either piece readily reach a surface and escape before a sizable chain reaction builds up. But if the pieces are suddenly driven together, the total surface area decreases. If the timing is right, and if the combined mass is greater than critical, a violent explosion occurs. This is what happens in a nuclear fission bomb (Figure 10.22).

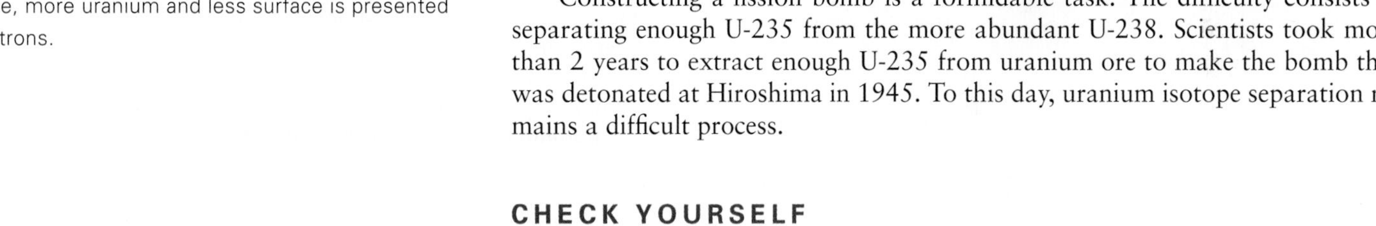

Constructing a fission bomb is a formidable task. The difficulty consists in separating enough U-235 from the more abundant U-238. Scientists took more than 2 years to extract enough U-235 from uranium ore to make the bomb that was detonated at Hiroshima in 1945. To this day, uranium isotope separation remains a difficult process.

CHECK YOURSELF

A 1-kilogram ball of U-235 is critical, but the same ball broken up into small chunks is not. Explain.

CHECK YOUR ANSWER

The small chunks have more combined surface area than the ball from which they came. Neutrons escape via the surface before a sustained chain reaction can build up.

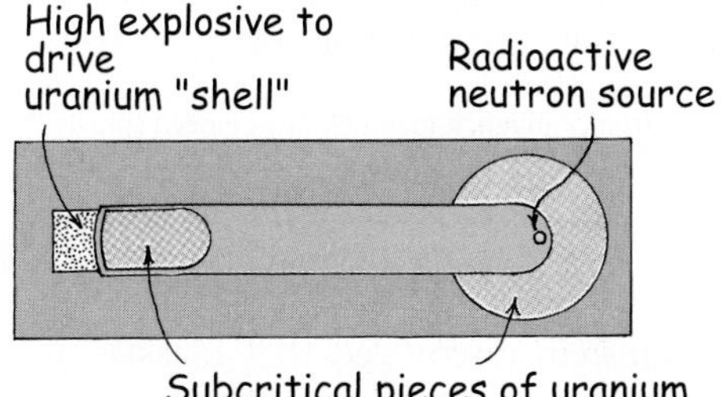

Figure 10.22 Simplified diagram of a uranium-fission bomb.

10.8 The Mass–Energy Relationship—$E = mc^2$

Clearly, a lot of energy comes from every gram of nuclear fuel that is fissioned. What is the source of this energy? As we will see, it comes from nucleons losing mass as they undergo nuclear reactions.

TECHNOLOGY

Nuclear Fission Reactors

The awesome energy of nuclear fission was introduced to the world in the form of nuclear bombs, and this violent image still colors our thinking about nuclear power, making it difficult for many people to recognize its potential usefulness. Currently, about 20 percent of electric energy in the United States is generated by nuclear fission reactors (whereas most electric power is nuclear in some other countries—more than 70 percent in France). These reactors are simply nuclear furnaces. They, like fossil-fuel furnaces, do nothing more elegant than boil water to produce steam for a turbine. The greatest practical difference is in the amount of fuel required: A mere 1 kilogram of uranium fuel, less than the size of a baseball, yields more energy than 30 freight car loads of coal.

The most common type of fission reactor contains four components: nuclear fuel, control rods, a moderator (to slow the neutrons, which are required for fission),* and a liquid (usually water) to transfer heat from the reactor to the turbine and generator. The nuclear fuel is primarily U-238, plus about 3 percent U-235. Because the U-235 isotopes are so highly diluted with U-238, an explosion like that of a nuclear bomb is not possible.† The reaction rate, which depends on the number of neutrons that initiate the fission of other U-235 nuclei, is controlled by rods inserted into the reactor. The control rods are made of a neutron-absorbing material, usually the metal cadmium or boron.

Heated water around the nuclear fuel is kept under high pressure to keep it at a high temperature without boiling. It transfers heat

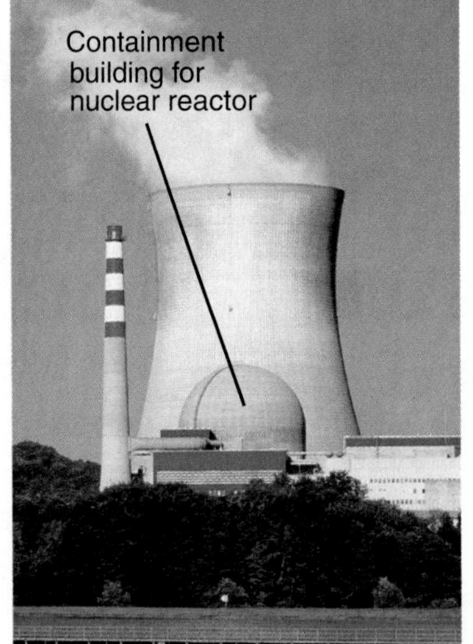

to a second water system with lower pressure, which operates the turbine and electric generator in a conventional fashion. In this design, two separate water systems are used so that no radioactivity reaches the turbine or the outside environment.

One disadvantage of fission power is the generation of radioactive waste products. Light atomic nuclei are most stable when composed of equal numbers of protons and neutrons, and mainly heavy nuclei need more neutrons than protons for stability. For example, there are 143 neutrons but only 92 protons in U-235. When uranium fissions into two medium-weight elements, the extra neutrons in their nuclei make them unstable. These elements are radioactive. Most of these elements have very short half-lives, but some have half-lives of thousands of years. Safely disposing of these waste products (as well as materials made radioactive in the production of nuclear fuels) requires special storage casks and procedures. Although fission has been successfully producing electricity for a half century, the search for satisfactory ways of disposing of radioactive wastes in the United States has remained unsuccessful.

Risks attend nuclear power, but so do benefits. The potential benefits of fission power are: plentiful electricity; conservation of the many billions of tons of fossil fuels that every year are literally turned to heat and smoke (which fuels, in the long run, may be far more precious as sources of organic molecules than as sources of heat); reduction of the carbon dioxide emissions that occur with the combustion of fossil fuels and are linked to global warming; and the elimination of the megatons of sulfur oxides and other poisons that are put into the air each year by the burning of fossil fuels. Yet, these benefits do not come risk-free.

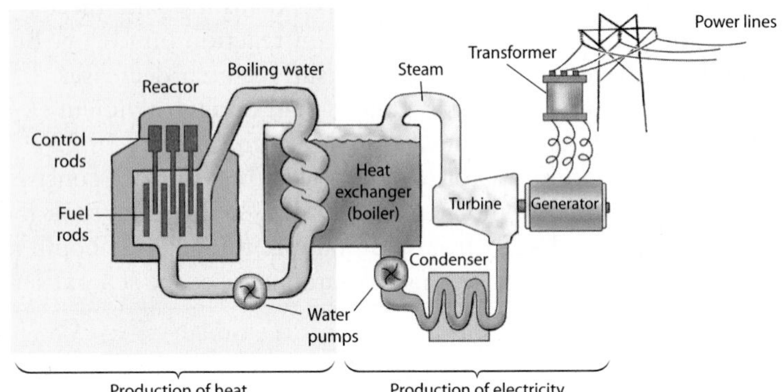

* Moderators are graphite, heavy water, or other substances that decrease the speeds of neutrons so they can be captured by the fissionable isotope. Interestingly, although slow neutrons in a reactor maintain the fission process, in a detonated nuclear bomb, slow neutrons can't keep up with the explosion and it would fizzle out. So one of the safeguards of commercial reactors is that slow neutrons can't sustain a substantial explosion. Even the 1986 Chernobyl accident was an incomplete explosion in a primitive reactor of a type no longer being manufactured.

† In a worst-case accident, however, heat sufficient to melt the reactor core is possible—and, if the reactor building is not strong enough, a meltdown can indeed spread radioactivity into the environment. Such an accident occurred at the Chernobyl reactor.

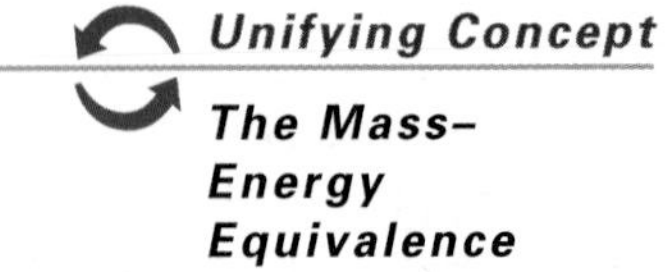

Early in the early 1900s, Albert Einstein discovered that mass is actually "congealed" energy. Mass and energy are two sides of the same coin, as stated in his celebrated equation $E = mc^2$. In this equation, E stands for the energy that any mass has at rest, m stands for mass, and c is the speed of light. The quantity c^2 is the proportionality constant of energy and mass. This relationship between energy and mass is the key to understanding why and how energy is released in nuclear reactions.

The more energy associated with a particle, the greater the mass of the particle. Is the mass of a nucleon inside a nucleus the same as that of the same nucleon

Figure 10.23 Albert Einstein (1879–1955)—German-American physicist Albert Einstein is revered as the scientist who contributed most to the 20th century understanding of physical reality.

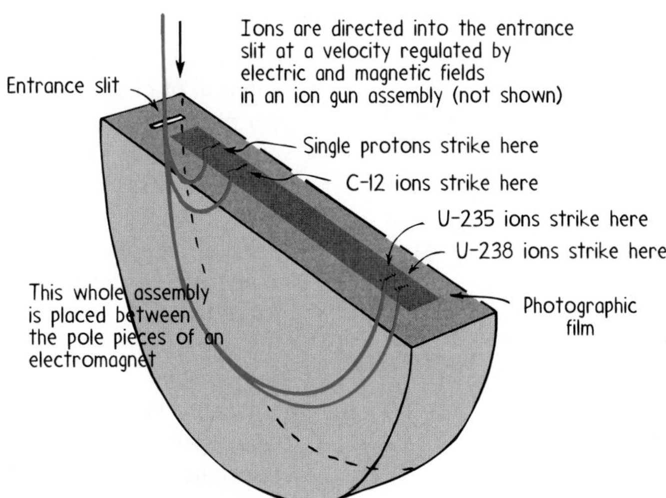

Figure 10.24 The mass spectrometer. Electrically charged isotopes are directed into the semicircular "drum," where they are forced into semicircular paths by a strong magnetic field. Lighter isotopes have less inertia (less mass), change direction more easily, and are pulled into curves of smaller radii. Heavier isotopes have greater inertia (more mass) and are pulled into curves of larger radii. The mass of an isotope, therefore, is directly proportional to its striking distance from the slit.

outside a nucleus? This question can be answered by considering the work that would be required to separate nucleons from a nucleus. From physics, we know that work, which is expended energy, is equal to force × distance. Think of the amount of force required to pull a nucleon out of the nucleus through a sufficient distance to overcome the attractive strong nuclear force. Enormous work would be required. This work is energy that is added to the nucleon that is pulled out.

According to Einstein's equation, this newly acquired energy reveals itself as an increase in the nucleon's mass. The mass of a nucleon outside a nucleus is greater than the mass of the same nucleon locked inside a nucleus. As discussed in Chapter 9, the nucleus of a carbon-12 atom, composed of six protons and six neutrons, has a mass of exactly 12.00000 atomic mass units (amu). Therefore, on average, each nucleon contributes a mass of 1 amu. However, outside the nucleus, a proton has a mass of 1.00728 amu and a neutron has a mass of 1.00867 amu. Thus, we see that the combined mass of six free protons and six free neutrons—$(6 \times 1.00728) + (6 \times 1.00867) = 12.09570$—is greater than the mass of one carbon-12 nucleus. The greater mass reflects the energy required to pull the nucleons apart from one another. Thus, what mass a nucleon has depends on where the nucleon is.

INSIGHT Miniature mass spectrometers are used for detecting molecules associated with explosives at airport security stations. The security agent swabs luggage with a soft cloth that is placed in the device. Molecules on the swab are ionized and scrutinized.

The masses of the isotopes of various elements can be very accurately measured with a mass spectrometer (Figure 10.24). This important device uses a magnetic field to deflect ions of these isotopes into circular arcs. The greater the inertia (mass) of the ion, the harder it is to deflect, and the greater the radius of its curved path. The magnetic force sweeps lighter ions into more sharply curved arcs and heavier ions into larger arcs.

A graph of the nuclear masses of the elements ranging from hydrogen through uranium is shown in Figure 10.25. The graph slopes upward with increasing atomic number, as expected—elements are more massive as atomic number increases. The slope curves because there are proportionally more neutrons in the more massive atoms.

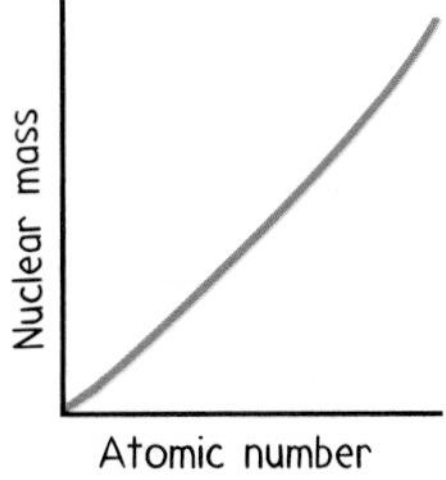

Figure 10.25 The plot shows how nuclear mass increases with increasing atomic number.

A more important graph results from the plot of nuclear mass per nucleon inside each nucleus ranging from hydrogen through uranium (Figure 10.26). This graph is the key to understanding the energy associated with nuclear processes.

INSIGHT The graphs shown below reveal the energy of the atomic nucleus, the primary source of energy in the universe—which is why they can be considered the most important graphs in this book.

To obtain the average mass per nucleon, you divide the total mass of a nucleus by the number of nucleons in the nucleus. (Similarly, if you divide the total mass of a roomful of people by the number of people in the room, you get the average mass per person.)

Note that the masses of the nucleons are different when they are combined in different nuclei. The greatest mass per nucleon occurs for the proton alone, hydrogen, because it has no binding energy to pull its mass down. Progressing beyond hydrogen, the mass per nucleon is smaller, and it is least for a nucleon in the nucleus of the iron atom. Beyond iron, the process reverses itself, as nucleons have progressively more and more mass in atoms of increasing atomic number. This continues all the way to uranium and the transuranic elements.

From Figure 10.26, we can see how energy is released when a uranium nucleus splits into two nuclei of lower atomic number. Uranium, being towards the right-hand side of the graph, is shown to have a relatively large amount of mass per nucleon. When the uranium nucleus splits in half, however, smaller nuclei of lower atomic numbers are formed. As shown in Figure 10.27, these nuclei are lower on the graph than uranium, which means that they have a smaller amount of mass per nucleon. Thus, nucleons lose mass in their transition from being in a uranium nucleus to being in one of its fragments. When this decrease in mass is multiplied by the speed of light squared (c^2 in Einstein's equation), the product is equal to the energy yielded by each uranium nucleus as it undergoes fission.

CHECK YOURSELF

Correct the following incorrect statement: When a heavy element such as uranium undergoes fission, there are fewer nucleons after the reaction than before.

CHECK YOUR ANSWER

When a heavy element such as uranium undergoes fission, there aren't fewer nucleons after the reaction. Instead, there's less mass in the same number of nucleons.

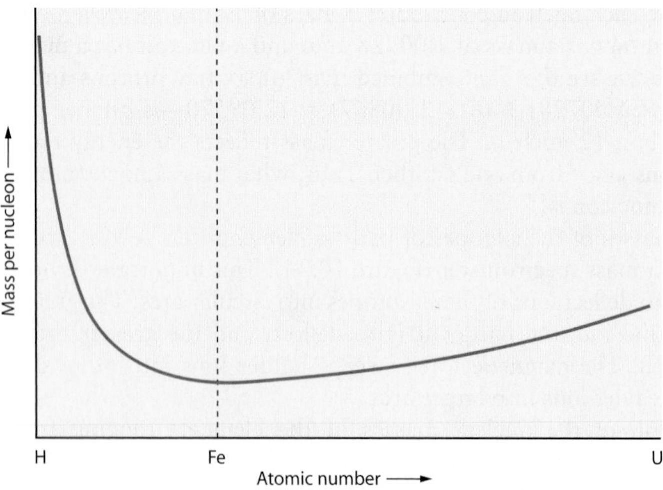

Figure 10.26 This graph shows that the average mass of a nucleon depends on which nucleus it is in. Individual nucleons have the most mass in the lightest (hydrogen) nuclei, the least mass in iron, and intermediate mass in the heaviest (uranium) nuclei.

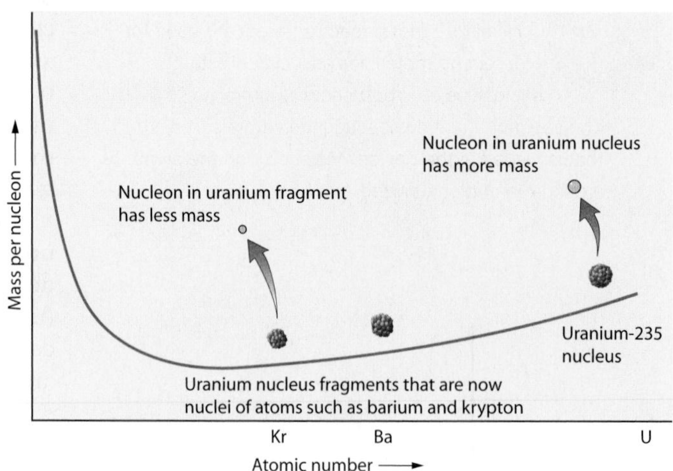

Figure 10.27 The mass of each nucleon in a uranium nucleus is greater than the mass of each nucleon in any one of its nuclear fission fragments. This lost mass is mass that has been transformed into energy, which is why nuclear fission is an energy-releasing process.

We can think of the mass-per-nucleon graph as an energy valley that starts at hydrogen (the highest point) and slopes steeply to iron (the lowest point), then slopes gradually up to uranium. Iron is at the bottom of the energy valley and is the most stable nucleus. It is also the most tightly bound nucleus; more energy per nucleon is required to separate nucleons from its nucleus than from any other nucleus.

All nuclear power today is produced by nuclear fission. A more promising long-range source of energy is to be found on the left side of the energy valley in a process known as nuclear fusion.

MATH CONNECTION

Mass–Energy Conversion

A proton bombards a lithium nucleus and produces two helium nuclei in the following nuclear reaction:

$$P + {}^{7}_{3}\text{Li} \rightarrow {}^{4}_{2}\text{He} + {}^{4}_{2}\text{He}$$

Problem Calculate the amount of energy released, given the following masses:

mass of proton = 1.0073 amu
mass of ${}^{4}_{2}$He = 4.0016 amu
mass of ${}^{7}_{3}$Li = 7.016 amu

where 1 amu = 1.66×10^{-27} kg.

Solution The mass of the proton and lithium nucleus combined equals 1.0073 + 7.016 amu = 8.023 amu. The mass of the two helium nuclei is 8.003 amu. The mass lost in the reaction is 8.023 amu − 8.003 amu = 0.020 amu.

Convert the lost mass to energy using the mass–energy equivalence:

$$E = mc^2$$
$$= 0.020 \text{ amu} \times (1.66 \times 10^{-27} \text{ kg/amu}) \times (3 \times 10^8 \text{ m/s})^2$$
$$= 2.99 \times 10^{-12} \text{ J}$$

There is a similar problem for you to try in the Problem section at the end of this chapter.

10.9 Nuclear Fusion

Conceptual Science Place
Controlling Nuclear Fusion

Notice, in the graphs of Figures 10.26 and 10.27, that the steepest part of the energy valley goes from hydrogen to iron. Energy is gained as light nuclei combine. This combining of nuclei is **nuclear fusion**—the opposite of nuclear fission. We can see from Figure 10.28 that as we move along the list of elements from hydrogen to iron, the average mass per nucleon decreases. Thus, when two small nuclei fuse—say, a pair of hydrogen isotopes—the mass of the resulting helium nucleus is less than the mass of the two small nuclei before fusion. Energy is released as smaller nuclei fuse.

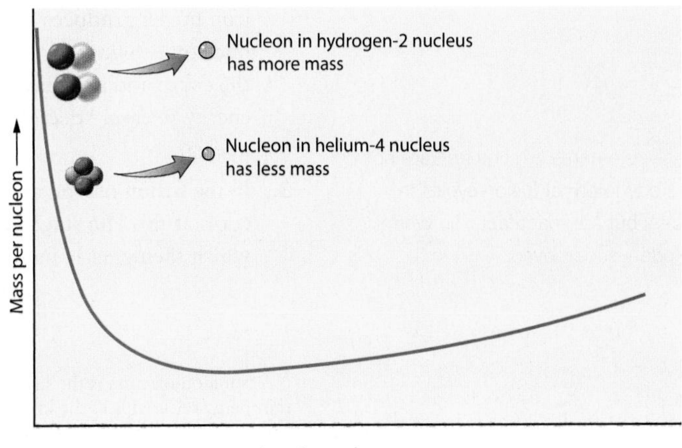

Figure 10.28 The mass of each nucleon in a hydrogen-2 nucleus is greater than the mass of each nucleon in a helium-4 nucleus, which results from the fusion of two hydrogen-2 nuclei. This lost mass is mass that has been converted to energy, which is why nuclear fusion is a process that releases energy.

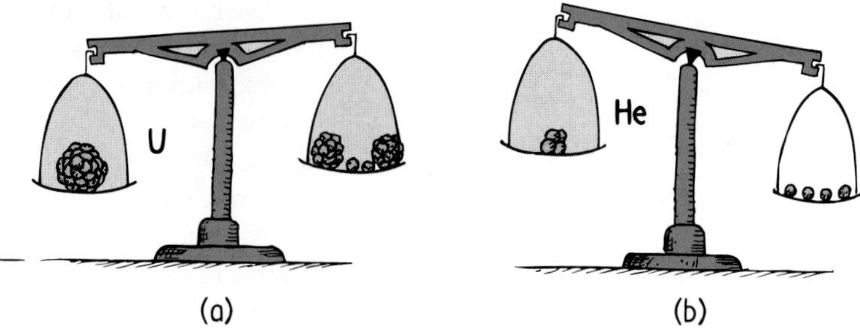

Figure 10.29 The mass of a nucleus is not equal to the sum of the mass of its parts. (a) The fission fragments of a uranium nucleus are less massive than the uranium nucleus. (b) Two protons and two neutrons are more massive in their free states than when they are combined to form a helium nucleus.

For a fusion reaction to occur, the nuclei must collide at a very high speed in order to overcome their mutual electric repulsion.* The required speeds correspond to the extremely high temperatures found in the Sun and other stars. Fusion brought about by high temperatures is called **thermonuclear fusion.** In the high temperatures of the Sun approximately 657 million tons of hydrogen is converted into 653 million tons of helium each second. The missing 4 million tons of mass is converted to energy—a tiny bit of which reaches Planet Earth as sunshine.

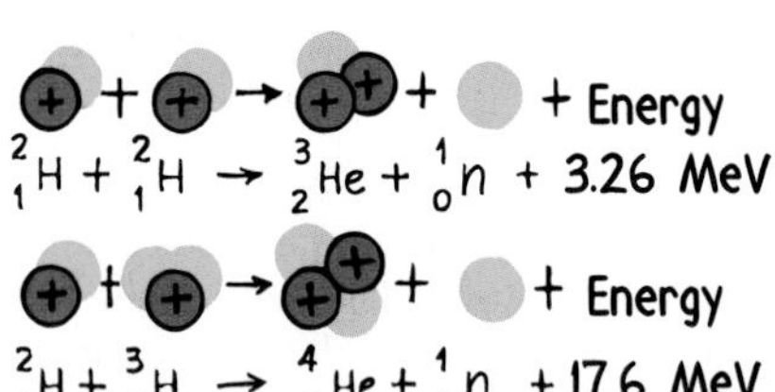

Figure 10.30 Fusion reactions of hydrogen isotopes. Most of the energy released is carried by the neutrons ejected at high speeds.

Such reactions are, quite literally, nuclear burning. Thermonuclear fusion is analogous to ordinary chemical combustion. In both chemical and nuclear burning, a high temperature starts the reaction; the release of energy by the reaction maintains a high enough temperature to spread the fire. The net result of the chemical reaction is a combination of atoms into more tightly bound molecules. In nuclear fusion reactions, the net result is more tightly bound nuclei. In both cases, mass decreases as energy is released.

INSIGHT With each passing day, the Sun is less massive, and we benefit by receiving its radiant warmth and welcome light.

CHECK YOURSELF

1. Fission and fusion are opposite processes, yet each releases energy. Isn't this contradictory?
2. To get a release of nuclear energy from the element iron, should iron undergo fission or fusion?
3. Predict whether the temperature of the core of a star increases or decreases when iron and elements of higher atomic number than iron in the core are fused.

CHECK YOUR ANSWERS

1. No, no, no! This is contradictory only if the same element is said to release energy by both fission and fusion. Only the fusion of light elements and the fission of heavy elements result in a decrease in nucleon mass and a release of energy.
2. Neither, because iron is at the very bottom of the "energy valley." Fusing a pair of iron nuclei produces an element to the right of iron on the curve, where mass per nucleon is higher. If you split an iron nucleus, the products lie to the left of iron on the curve and also have a higher mass per nucleon. So no energy is released. For energy release, "decrease mass" is the name of the game—any game, chemical or nuclear.
3. In the fusion of iron and any nuclei beyond, energy is absorbed and the star core cools at this late stage of its evolution. This, however, leads to the star's collapse, which then greatly increases its temperature.

INSIGHT Interestingly, elements beyond iron are not manufactured in normal fusion cycles in stellar sources, but are manufactured when stars violently explode—supernovae.

* A common reaction is the fusion of a pair of H-2 and H-3 nuclei to become He-4 plus a neutron. Most of the energy released is in the kinetic energy of the ejected neutron. The rest of the energy is in the kinetic energy of the recoiling He-4 nucleus. Interestingly, without the neutron energy carrier, a fusion reaction won't occur. The intensity of fusion reactions is measured by the accompanying neutron flux.

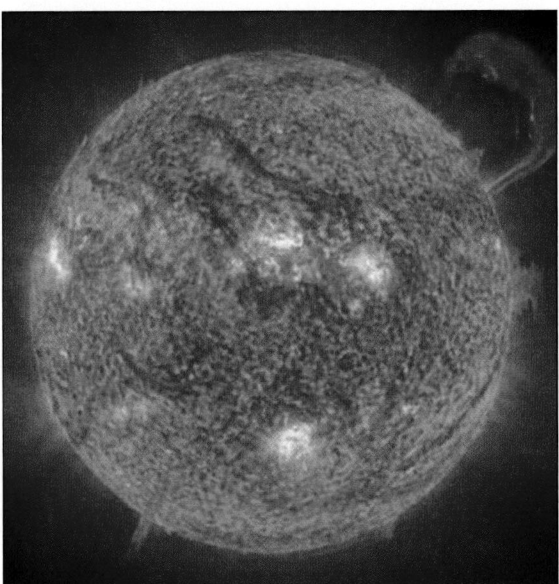

Figure 10.31 Thermonuclear fusion takes place in stars, such as the Sun. You will learn much more about this in Chapters 27 and 28. Some day, humans may produce vast quantities of energy through thermonuclear fusion as the stars have always done.

Carrying out thermonuclear fusion reactions under controlled conditions requires temperatures of millions of degrees. There are a variety of techniques for attaining high temperatures. But no matter how the temperature is produced, a problem is that all materials melt and vaporize at the temperatures required for fusion. One solution to this problem is to confine the reaction in a nonmaterial container, such as a magnetic field.

At this writing, fusion by magnetic confinement has only been partially successful. A sustained and controlled reaction has so far been out of reach. We await "break-even day," when one of the techniques for controlled nuclear fusion produces sustained energy. If and when this occurs, a vast energy supply will become available. The fuel for fusion—hydrogen—is found in every part of the universe, not only in the stars but also in the space between them. About 91 percent of the atoms in the universe are estimated to be hydrogen. If people are one day to dart about the universe in the same way we jet about Earth today, their supply of fuel is assured!

CHAPTER 10 Review

Summary of Terms

Alpha particle The nucleus of a helium atom, which consists of two neutrons and two protons, ejected by certain radioactive elements.

Beta particle An electron (or positron) emitted during the radioactive decay of certain nuclei.

Chain reaction A self-sustaining reaction in which the products of one reaction event stimulate further reaction events.

Critical mass The minimum mass of fissionable material in a reactor or nuclear bomb that will sustain a chain reaction.

Gamma ray High-frequency electromagnetic radiation emitted by the nuclei of radioactive atoms.

Half-life The time required for half the atoms in a sample of a radioactive isotope to decay.

Nuclear fission The splitting of the nucleus of a heavy atom, such as uranium-235, into two main parts, accompanied by the release of much energy.

Nuclear fusion The combining of nuclei of light atoms to form heavier nuclei, with the release of much energy.

Nucleon A nuclear proton or neutron.

Rad Equal to 0.01 joule of radiant energy absorbed per kilogram of tissue.

Radioactivity The process whereby unstable atomic nuclei break down and emit radiation.

Rem Unit of measure for radiation dosage (roentgen equivalent man).

Strong force The powerful force that attracts nucleons to one another over a short distance.

Thermonuclear fusion Nuclear fusion produced by high temperature.

Transmutation The conversion of an atomic nucleus of one element into an atomic nucleus of another element through a loss or gain in the number of protons.

Review Questions

10.1 Radioactivity

1. Where does most of the radiation you encounter originate?
2. What sorts of atoms are radioactive?

10.2 Alpha, Beta, and Gamma Rays

3. How do the electric charges of alpha, beta, and gamma rays differ?
4. What is the difference between a beta particle and an ordinary electron?

10.3 Environmental Radiation

5. Approximately what proportion of the radioactivity we absorb annually comes from sources outside nature?
6. What is the unit of radiation dosage based on potential biological damage?

10.4 The Atomic Nucleus and the Strong Nuclear Force

7. Which acts over a longer distance: the electrical force or the strong nuclear force?
8. Why is a larger nucleus generally less stable than a smaller nucleus?
9. What is the role of neutrons in the atomic nucleus?
10. Why does the transformation of neutrons to protons in a nucleus reduce the stability of the nucleus?

10.5 Half-Life

11. What is meant by radioactive half-life?
12. What is the half-life of Ra-226? Of uranium-238?
13. True or false: in general, the shorter the half-life of a radioactive element, the more radioactivity per amount can be detected.

10.6 Transmutation of Elements

14. What is the result of a radioactive element emitting an alpha or beta particle?
15. When an atom decays by emitting a beta particle, does its mass change significantly?
16. By what process are the synthetic elements with atomic numbers from 93 to 115 produced?
17. How is the mass number of an element affected when it decays by ejecting an alpha particle? How is atomic number affected?
18. Is it possible for a nitrogen atom to change into an oxygen atom? If so, how?

10.7 Nuclear Fission

19. How much energy is released in the fissioning of one Uranium-235 nucleus?
20. If two subcritical pieces of Uranium-235 are combined rapidly and their combined mass is greater than critical, what results?
21. Why can a reactor not explode like a fission bomb?

10.8 The Mass–Energy Relationship—$E = mc^2$

22. What is the source of the energy obtained when nuclear fuel is fissioned?
23. A uranium nucleus splits in half. Do the products have more mass per nucleon or less mass per nucleon than the original uranium nucleus?
24. What is the most stable nucleus? Why is it so stable?

10.9 Nuclear Fusion

25. If the graph in Figure 10.26 is seen as an energy valley, what can be said of nuclear transformations that progress toward iron?
26. In what ways is thermonuclear fusion analogous to chemical combustion?

Multiple Choice Questions

Choose the BEST answer to the following.

1. The discovery of radioactivity was a boost to earth scientists, who were then able to know more about
   (a) the ages of various rocks.
   (b) why Earth's interior is hot.
   (c) Both.
   (d) Neither.
2. There is a greater ratio of C-14 to C-12 in
   (a) old bones.
   (b) new bones.
   (c) It depends on the organism.
   (d) Not enough information to say.
3. Most of the radioactivity we personally encounter comes from
   (a) fallout from past and present testing of nuclear weapons.
   (b) nuclear power.
   (c) medical X-ray examinations.
   (d) the natural environment.
4. Fusing a pair of iron nuclei yields a net
   (a) absorption of energy.
   (b) release of energy.
   (c) Neither a nor b.
   (d) Both a and b.
5. When uranium or another nuclear fuel is fissioned to liberate energy, the source of the energy obtained is
   (a) electrons being boosted to a higher quantum level.
   (b) atoms traveling at the speed of light in accordance with $E = mc^2$.
   (c) nucleons losing mass as they undergo nuclear reactions.
   (d) iron nuclei forming.

INTEGRATED SCIENCE CONCEPTS

Biology—Doses of Radiation

1. Humans receive more radiation from natural sources than artificial sources. Does this imply that the amount of radiation received from artificial sources is not important, in terms of one's health?
2. Name one application of radioactivity that saves lives. Name another that kills.
3. What is the most significant artificial source of radioactivity? What is the best way to remove this source? Is it common knowledge that this source provides high doses of radioactivity? Why not?
4. Describe the mechanism by which radioactivity destroys cells.
5. Why is radioactivity more dangerous when delivered in a single dose rather than gradually over time?

Biology, Earth Science—Isotopic Dating

1. What does the proportion of lead and uranium tell us about the age of a rock?
2. What occurs when a nitrogen nucleus captures an extra neutron?
3. How are radioactive isotopes produced?
4. Does the amount of carbon-14 in a fossil tell the amount of time that has elapsed since the organism was born or since it died? Explain.

Active Explorations

Some watches and clocks have luminous hands that glow continuously. Some of these have traces of radium bromide mixed with zinc sulfide. (Safer clock faces use light rather than radioactive disintegration as a means of excitation, and they become progressively dimmer in the dark.) If you have a luminous watch or clock available, take it into a completely dark room and, after your eyes have become adjusted to the dark, examine the luminous hands with a very strong magnifying glass or the eyepiece of a microscope or telescope. You should be able to see individual tiny flashes, which together seem to be a steady source of light to the unaided eye. Each flash occurs when an alpha particle ejected by a radium nucleus strikes a molecule of zinc sulfide.

Exercises

1. Is radioactivity in the world something relatively new? Defend your answer.
2. Why is a sample of radium always a little warmer than its surroundings?
3. Some people say that all things are possible. Is it at all possible for a hydrogen nucleus to emit an alpha particle? Defend your answer.
4. Why are alpha and beta rays deflected in opposite directions in a magnetic field? Why are gamma rays undeflected?
5. In bombarding atomic nuclei with proton "bullets," why must the protons be accelerated to high energies to make contact with the target nuclei?
6. Just after an alpha particle leaves the nucleus, would you expect it to speed up? Defend your answer.
7. Within the atomic nucleus, which interaction tends to hold it together and which interaction tends to push it apart?
8. What evidence supports the contention that the strong nuclear force is stronger than the electrical force at short internuclear distances?

9. A friend asks if a radioactive substance with a half-life of 1 day will be entirely gone at the end of 2 days. What is your answer?
10. Elements with atomic numbers greater than that of uranium do not exist in any appreciable amounts in nature because they have short half-lives. Yet there are several elements with atomic numbers smaller than that of uranium that have equally short half-lives and that do exist in appreciable amounts in nature. How can you account for this?
11. You and your friend journey to the mountain foothills to get closer to nature and to escape such things as radioactivity. While bathing in the warmth of a natural hot spring, she wonders aloud how the spring gets its heat. What do you tell her?
12. Coal contains minute quantities of radioactive materials, yet there is more environmental radiation surrounding a coal-fired power plant than a fission power plant. What does this indicate about the shielding that typically surrounds these power plants?
13. When we speak of dangerous radiation exposure, are we customarily speaking of alpha radiation, beta radiation, or gamma radiation? Defend your answer.
14. People who work around radioactivity wear film badges to monitor the amount of radiation that reaches their bodies. These badges consist of small pieces of photographic film enclosed in a light-proof wrapper. What kind of radiation do these devices monitor?
15. A friend produces a Geiger counter to check the local background radiation. It ticks. Another friend, who normally fears most that which is understood least, makes an effort to keep away from the region of the Geiger counter and looks to you for advice. What do you say?
16. When food is irradiated with gamma rays from a cobalt-60 source, does the food become radioactive? Defend your answer.
17. Why will nuclear fission probably not be used directly for powering automobiles? How could it be used indirectly?
18. Why does a neutron make a better nuclear bullet than a proton or an electron?
19. U-235 releases an average of 2.5 neutrons per fission, while Pu-239 releases an average of 2.7 neutrons per fission. Which of these elements might you therefore expect to have the smaller critical mass?
20. Why is lead found in all deposits of uranium ores?
21. Why does plutonium not occur in appreciable amounts in natural ore deposits?
22. Why does a chain reaction not occur in uranium mines?
23. A friend makes the claim that the explosive power of a nuclear bomb is due to static electricity. Do you agree or disagree? Defend your answer.
24. If a nucleus of ${}^{232}_{90}\text{Th}$ absorbs a neutron and the resulting nucleus undergoes two successive beta decays (emitting electrons), what nucleus results?
25. How does the mass per nucleon in uranium compare with the mass per nucleon in the fission fragments of uranium?
26. How is chemical burning similar to nuclear fusion?
27. To predict the approximate energy release of either a fission or a fusion reaction, explain how a physicist makes use of the curve of Figure 10.26 or a table of nuclear masses and the equation $E = mc^2$.
28. Which process would release energy from gold, fission or fusion? From carbon? From iron?
29. If uranium were to split into three segments of equal size instead of two, would more energy or less energy be released? Defend your answer in terms of Figure 10.26.
30. Explain how radioactive decay has always warmed Earth from the inside and nuclear fusion has always warmed Earth from the outside.
31. Write a letter to Grandma to dispel any notion she might have about radioactivity being something new in the world. Tie this to the idea people sometimes have the strongest views about that which they know the least.

Problems

1. Radiation from a point source obeys the inverse-square law. If a Geiger counter 1 meter from a small sample reads 360 counts per minute, what will be its counting rate at 2 meters from the source? At 3 meters from the source?
2. If a sample of a radioactive isotope has a half-life of 1 year, how much of the original sample will be left at the end of the second year? At the end of the third year? At the end of the fourth year?
3. A certain radioactive substance has a half-life of 1 hour. If you start with 1 gram of the material at noon, how much will be left at 3:00 P.M.? At 6:00 P.M.? At 10:00 P.M.?
4. A sample of a particular radioisotope is placed near a Geiger counter, which is observed to register 160 counts per minute. Eight hours later, the detector counts at a rate of 10 counts per minute. Show that the half-life of the material is 2 hours.
5. The isotope cesium-137, which has a half-life of 30 years, is a product of nuclear power plants. Show that it will take 120 years for this isotope to decay to about one-sixteenth its original amount.
6. Suppose that you measure the intensity of radiation from carbon-14 in an ancient piece of wood to be 6 percent of what it would be in a freshly cut piece of wood. Show that this artifact is 22,920 years old.
7. Suppose that you want to find out how much gasoline is in an underground storage tank. You pour in 1 gallon of gasoline that contains some radioactive material with a long half-life that gives off 5000 counts per minute. The next day, you remove a gallon from the underground tank and measure its radioactivity to be 10 counts per minute. How much gasoline is in the tank?
8. The kiloton, which is used to measure the energy released in an atomic explosion, is equal to 4.2×10^{12} J (approximately the energy released in the explosion of 1000 tons of TNT). Recall that 1 kilocalorie of energy raises the temperature of 1 kilogram of water by 1°C and that 4184 joules is equal to 1 kilocalorie, show that 4.0×10^8 kilograms of water can be heated 50°C by a 20-kiloton atomic bomb.
9. An atom of uranium (m = 232.03174 amu) radioactively decays into an atom of thorium (m = 228.02873 amu) plus an atom of helium (m = 4.00260 amu). Show that about 6.1×10^{-14} J of energy is released in this decay.

Conceptual Science Place

CHAPTER 10 ONLINE RESOURCES

Interactive Figures 10.3, 10.5, 10.13 • ***Tutorials*** Radiation and Biological Effects, Nuclear Chemistry • ***Videos*** Radioactive Decay, Half-Life, Plutonium, Nuclear Fission, Controlling Nuclear Fusion • ***Quiz*** • ***Flashcards*** • ***Links***

PART TWO

CHEMISTRY

Like everyone and everything, I'm made of atoms. My atoms are energized by the food I eat, and they are constantly moving, coming together in clusters called *molecules,* then breaking apart again. This energetic dance of my atoms underlies much about me, including how I grow. It's amazing that chemistry can reveal so much about the complicated and invisible dance of atoms, giving us the power to understand, care for, and improve the material world!

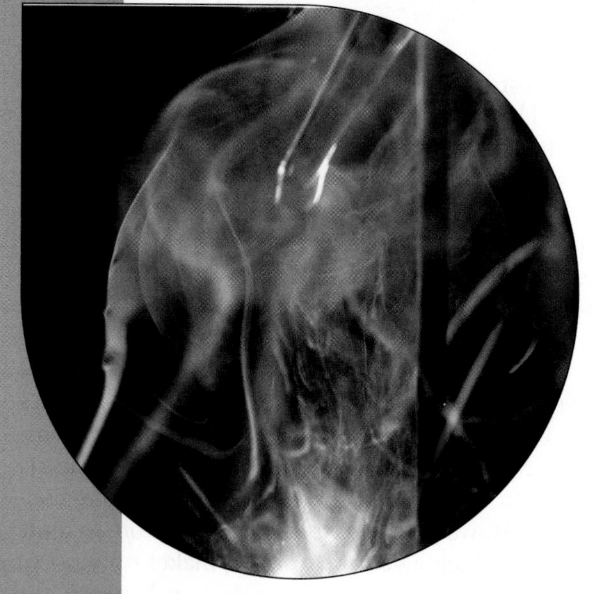

Investigating Matter

Water is dripped into a dish containing the element potassium. Potassium, a silver-white-metal, undergoes a violent chemical reaction with water in which the compound potassium hydroxide and the element hydrogen gas are produced, along with heat.

CHAPTER OUTLINE

We know that two oxygen atoms joined together make an oxygen molecule—which is good for life. So why does adding another oxygen atom to the pair make a poison? Similarly, we know that common table salt is also good for life. How is it, then, that the two elements composing salt are poisonous by themselves? Why does a strong iron bar eventually become a crumbling pile of rust when left out in the rain? Is rusting the same kind of change that occurs when water freezes or glass breaks? The answers to these questions make up the science of matter: chemistry!

11.1 Chemistry: Integral to Our Lives

When you wonder what the Earth, sky, or ocean is made of, you are thinking about chemistry. When you wonder how a rain puddle dries up, how a car gets energy from gasoline, or how your body gets energy from the food you eat, you are again thinking about chemistry. By definition, **chemistry** is the study of matter and the transformations it can undergo. **Matter** is anything that occupies space. It is the stuff that makes up all material things—anything you can touch, taste, smell, see, or hear is matter. Chemistry, therefore, is very broad in scope.

Chemistry is often described as a central science because it touches all the other sciences. It builds up from physics and serves as the foundation for the most complex science of all—biology. Chemistry, along with physics, forms the foundation for the Earth sciences, including geology, oceanography, and meteorology. It is also an important component of space science, as described in Figure 11.1. Just as we learned about the origin of the Moon from the chemical analysis of Moon rocks in the early 1970s, we are now learning about the history of Mars and other planets from the chemical information gathered by space probes.

INTEGRATED SCIENCE ASTRONOMY

Origin of the Moon

Conceptual Science Place
What Is Chemistry?

Chemistry plays an integral part in solving mysteries both large and small and in verifying or disproving theories. For instance, while humans have wondered for centuries about the origin of the Moon, only recently did we gather chemical evidence to support a widely accepted theory called the Giant Impact Theory. This

Figure 11.1 This illustration shows a Mars-sized object colliding with Earth approximately 4.5 billion years ago, when the Earth was newly formed. According to the Giant Impact Theory, such a collision generated the debris that later aggregated as Earth's Moon. Chemical analysis of rocks obtained from the Moon provides crucial evidence for this theory.

theory holds that, early in Earth's history, our planet was bombarded with a continuous rain of debris from space. The rocky bodies hitting Earth generally ranged from sand-grain to boulder size, but some were larger, tens of kilometers or more in diameter. One particular collision was much more catastrophic than the others. According to the Giant Impact Theory, a huge rocky mass the size of the planet Mars collided with Earth about 20 million years after Earth had formed. Material blasted out from the Earth and fragments from the impactor itself formed a transient ring-shaped cloud around Earth. The particles in the cloud later aggregated through mutual gravitational attraction to form a spherical body, which, captured by Earth's gravity, continues to orbit our planet today—the Moon.

Some of the most convincing evidence for the Giant Impact Theory comes from chemical analysis of lunar rocks that were brought to Earth by Apollo astronauts. The rocks show that the Moon's chemical composition is unlike the average chemical composition of Earth as a whole (the Moon, for example, contains much less iron than Earth). However, the Moon's composition is quite similar to that of Earth's mantle—the thick layer of earth that makes up most of the outer portion of the planet (see Chapter 22). This matchup between the Moon and Earth's mantle supports the idea that the Moon formed from a collision between Earth and an impactor. Earth's iron is concentrated in its core. The iron therefore would be shielded during the blast and be retained by Earth, while the mantle material would be blasted into space and later aggregate as the Moon. Further, lunar rock chemistry also indicates that the lunar surface was once tremendously heated. For example, the Moon lacks water and other volatile materials. This is what we would expect from the tremendous heating due to friction that would have occurred in a giant impact.

CHECK YOURSELF

1. Besides Earth, where did material that comprises the Moon come from?
2. What is one explanation for the Moon's lack of water?

CHECK YOUR ANSWERS

1. The Mars-sized body that impacted Earth was fragmented in the collision and provided much of the material that would become the Moon.
2. One explanation is that any volatile materials, such as water, would have evaporated from the heat generated by a giant impact.

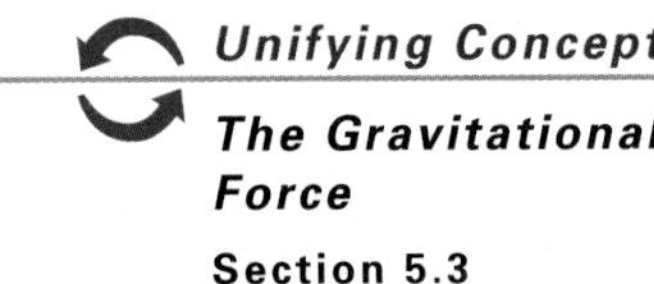

Scientific research is activity aimed at discovering and interpreting new knowledge. **Basic research** leads to greater understanding of how the natural world operates. Many scientists focus on basic research. The foundation of knowledge laid down by basic research often leads to useful applications. **Applied research** focuses on developing these applications. While physicists tend to focus on basic research, most chemists focus on applied research. Applied research in chemistry has provided us with medicine, food, water, shelter, and so many of the material goods that characterize modern life. Just a few of a myriad of examples are shown in Figure 11.2.

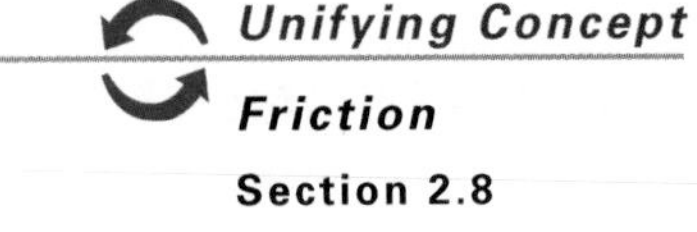

In the past century, we excelled at creating materials to suit our needs. However, we did not always excel in caring for the environment. Waste products were dumped into rivers, buried in the ground, or vented into the air without regard for possible long-term consequences. Many people believed that Earth was so large that its resources were virtually unlimited and that it could absorb wastes without being significantly harmed. People minimized the seriousness of the effects of these wastes on human health, as well.

Most nations now recognize this as a dangerous attitude. As a result, government agencies, industries, and concerned citizens are involved in extensive efforts to clean up toxic-waste sites. Such regulations as the international ban on ozone-destroying chlorofluorocarbons (CFCs) have been enacted to protect the

Figure 11.2 Most of the material items in any modern house are shaped by some human-devised chemical process.

environment. Members of the American Chemistry Council, who as a group produce 90 percent of the chemicals manufactured in the United States, have adopted a program called *Responsible Care*, in which they have pledged to manufacture without causing environmental damage. The Responsible Care program—its emblem shown in Figure 11.3—is based on the understanding that modern technology can be used to both harm and protect the environment. By using chemistry wisely, most waste products can be minimized, recycled, or engineered into sellable ones that are environmentally benign.

Figure 11.3 The Responsible Care symbol of the American Chemistry Council.

CHECK YOURSELF

Chemists have learned how to produce aspirin using petroleum as a starting material. Is this an example of basic or applied research?

CHECK YOUR ANSWER

This is an example of applied research because the primary goal was to develop a useful commodity. However, the ability to produce aspirin from petroleum depended on an understanding of atoms and molecules, an understanding that came from many years of basic research.

SCIENCE AND SOCIETY

Chemistry and Public Policy

Chemistry has influenced our lives in profound ways and will continue to do so in the future. In fact, more than 70 percent of all legislation placed before the Congress of the United States addresses science-related questions and issues, and many of these pertain to chemistry. Understanding of the science behind the issues is therefore important for everyone in a democracy—especially political leaders and policy makers.

Consider, for example, just a small sample of the chemistry-related questions and issues we face as a society. Is it safe to radiate food to retard spoilage? (More on this in Chapter 10.) Should the gasoline additive MTBE be prohibited even though it improves a car's performance (Chapter 24)? What is happening to stratospheric ozone, and how does this problem differ from global warming (Chapter 25)? Are disease-causing bacteria becoming immune to antibiotics—if so, what can be done about it (Chapter 16)? At some point either we or the people we elect will be considering questions such as these, as the scene in the figure illustrates. The more informed we are, the greater the likelihood the decisions we make will be good ones.

In a democracy, it's important that everyone understands the scientific facts pertinent to any political debate. How else can rational choices be made?

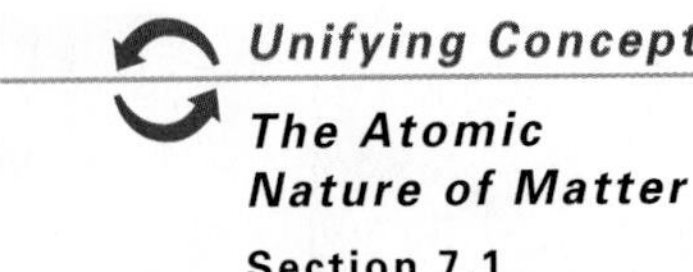

11.2 The Submicroscopic World

From afar, a sand dune looks like it's made of a smooth, continuous material. But up close, you can tell the dune is granular—it's made of tiny particles of sand. Similarly, all matter is granular rather than continuous. Everything around us, no matter how smooth it may appear, is made of basic units called *atoms*, as you learned in Chapter 9. Atoms are so small that a single grain of sand contains about 125 million trillion of them. There are roughly a quarter million times more atoms in a single grain of sand than there are grains of sand in the dunes shown in Figure 11.4.

Some atoms link together to form larger but still incredibly small basic units of matter. These are molecules. As shown in Figure 11.4, two hydrogen atoms and one oxygen atom link to form a single molecule of water, which we know as H_2O. Water molecules are so small that an 8-oz glass of water contains about a trillion trillion of them. Molecules are made of atoms.

The world can be studied at different levels of magnification. At the *macroscopic* level, matter is large enough to be seen, measured, and handled. A handful of sand and a glass of water are macroscopic samples of matter. At the *microscopic* level, physical structure is so fine that it can be seen only with a microscope. A biological cell is microscopic, as is the detail on a dragonfly's wing. Beyond the microscopic level is the **submicroscopic**—the realm of atoms and molecules. Only recently has the STM (scanning tunneling microscope) rendered the submicroscopic world visible. (Look back at Figure 9.11 to see an actual STM image). This realm—the realm of the atom—is the major focus of chemistry.

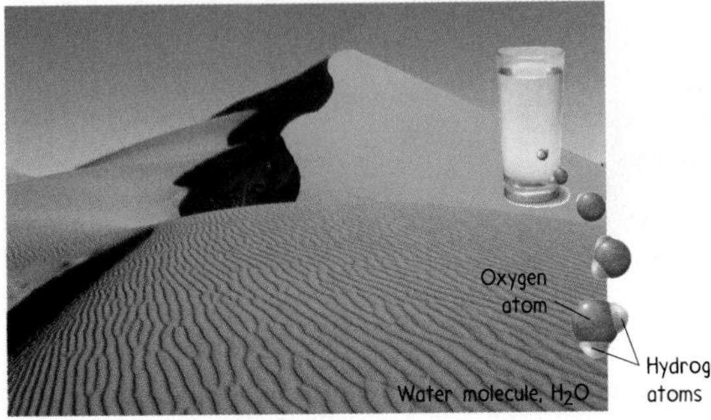

Figure 11.4 From a distant perspective, you can't tell that a sand dune is composed of individual particles of sand. It looks continuous. Likewise, a glass of water seems to be continuous on the macroscopic level. But if you could see it on the submicroscopic level, you would realize that water is made up of discrete particles.

11.3 Phases of Matter

One of the most obvious ways that matter varies is in its **phase** (also called its physical *state*). The three common phases of matter are: *solid*, *liquid*, and *gas*. A **solid** material, such as a rock, occupies a constant amount of space and does not readily deform when pressure is applied. In other words, a solid has both definite volume and definite shape. A **liquid** also occupies a constant amount of space (it has a definite volume), but its form changes readily (it has an indefinite shape). A liter of milk, for example, may take the shape of its carton or the shape of a puddle, but its volume is the same in both cases. A **gas** has neither definite volume nor definite shape. Any sample of gas assumes both the shape and the volume of the container it occupies. A given amount of air, for example, may assume the volume and shape of a toy balloon or the volume and shape of a bicycle tire. Released from its container, a gas spreads out into the atmosphere, which is a collection of various gases held to our planet only by the force of gravity.

On the submicroscopic level, the solid, liquid, and gaseous phases are distinguished by how well the particles—atoms or molecules—hold together. This is illustrated in Figure 11.5. You learned in Chapter 6 that, according to the kinetic theory of matter, the particles within any sample of matter are always moving. In solid matter, the attractions between particles are strong enough to hold them in some fixed three-dimensional arrangement despite their random, thermal motion. The particles are able to vibrate about fixed positions, but they cannot move past one another. Adding heat causes these vibrations to increase until, at a certain temperature, the vibrations are strong enough to disrupt the fixed arrangement. The particles can then slip past one another and tumble around, much like a bunch of marbles in a bag. This is the liquid phase of matter, and it is the mobility of atoms or molecules in this state that gives rise to the liquid's fluid character—its ability to flow and take on the shape of its container.

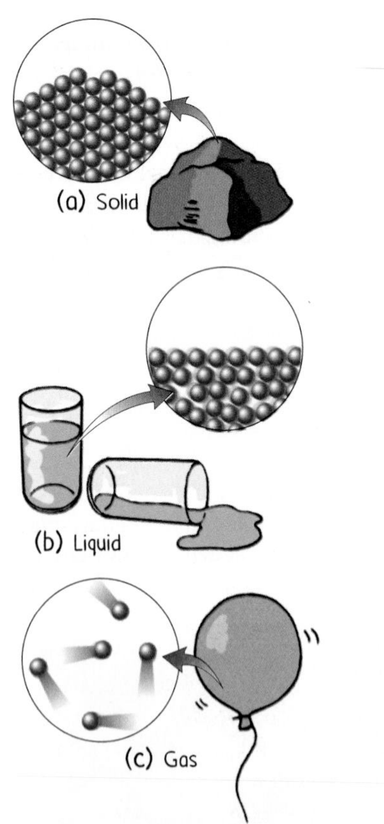

Figure 11.5 The familiar bulk properties of a solid, liquid, and gas. (a) The submicroscopic particles of the solid phase vibrate about fixed positions. (b) The submicroscopic particles of the liquid phase slip past one another. (c) The fast-moving submicroscopic particles of the gaseous phase are separated by large average distances.

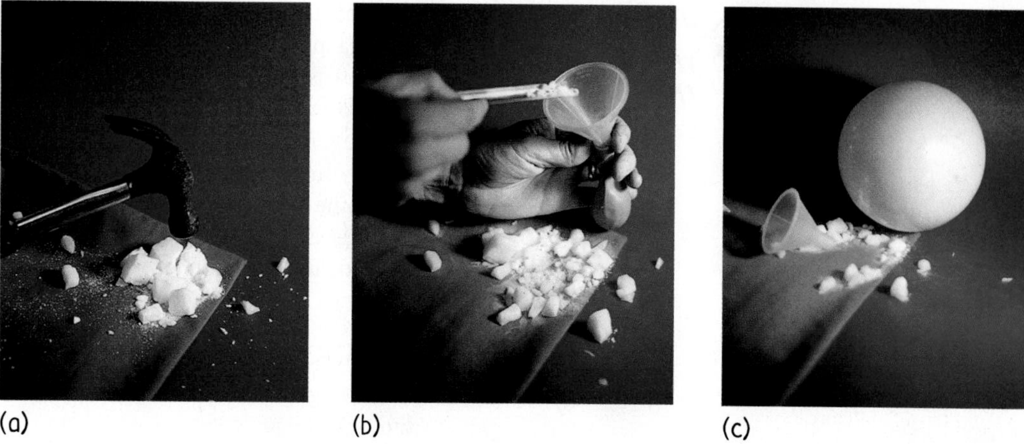

Figure 11.6 The gaseous phase of any material occupies significantly more volume than either its solid or liquid phase. (a) Solid carbon dioxide (dry ice) is broken up into powder form. (b) The powder is funneled into a balloon. (c) The balloon expands as the contained carbon dioxide becomes a gas as the powder warms up.

Further heating causes the particles in the liquid to move with such high-amplitude vibrations that the attractions they have for one another can't hold them together. They then separate from one another, forming a gas. The speed of the atoms or molecules in a gas depends on the temperature of the gas. At room temperature (25°C), the atoms or molecules move at an average speed of some 500 meters per second (1100 miles per hour). The particles move in straight lines until they collide with one another or with the walls of their container, then they bounce away from one another like billiard balls in a game of pool. Because they bounce back after colliding, the particles of a gas become widely separated. Matter in the gaseous phase therefore occupies much more volume than it does in the solid or liquid phase, as Figure 11.6 shows. Applying pressure to a gas squeezes the gas particles closer together, which makes for a smaller volume. Enough air for an underwater diver to breathe for many minutes, for example, can be squeezed (compressed) into a tank small enough to be carried on the diver's back.

Although gas particles move at high speeds, the speed at which they can travel from one side of a room to the other is relatively slow. This is because the gas particles are continually hitting one another, and so they move in an erratic, zigzag pattern as Figure 11.7 shows. At home, you get a sense of how long it takes for gas particles to migrate each time someone opens the oven door after baking. A shot of aromatic gas particles escapes the oven, but there is a notable delay before the aroma reaches the nose of someone sitting in the next room.

Figure 11.7 In traveling from point A to point B, the typical gas particle travels a circuitous path because of numerous collisions with other gas particles—about 8 billion collisions every second! The changes in direction shown here represent only a few of these collisions. Although the particle travels at very high speeds, it takes a relatively long time to cross between two distant points because of these numerous collisions. This net movement of gas molecules from an area of high concentration to low concentration is called *diffusion*.

The net migration of aromatic gas molecules away from a baking pie to a person in an adjacent room is an example of **diffusion**. Diffusion is the tendency of molecules to move from an area of high concentration to one of low concentration. Over time, a gas will diffuse to completely fill its container. Because all gases do this, a mixture of gases eventually will become thoroughly, evenly mixed.* This is not surprising when you think about it. The gases making up the air in your classroom, for example, have an even composition throughout the classroom. You wouldn't expect the oxygen molecules to

* All gases do not diffuse at the same rate, however. At a constant temperature, molecules with a small mass diffuse more rapidly than molecules of large mass. Graham's Law states gives the quantitative relationship: The rate of diffusion of a gas is inversely proportional to the square root of its formula mass.

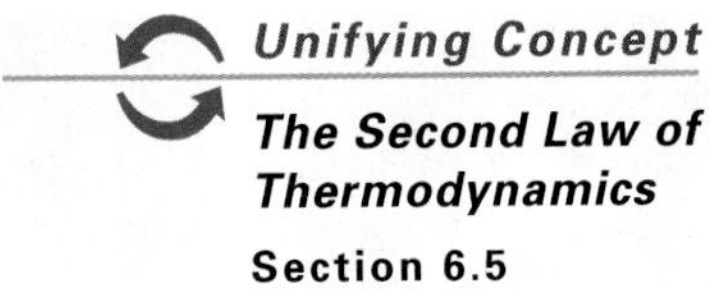

concentrate on one side of the classroom while leaving the other half of the class gasping from breath!

Diffusion is not limited to gases. Liquids diffuse readily as well and even solids diffuse gradually. If you put a bar of gold and a bar of silver side-by-side and leave them alone for several months, then chemically analyze their compositions, you will find that some of the gold particles have diffused into the silver and some of the silver has diffused into the gold. In Chapter 15, you will learn how the diffusion of materials across cellular membranes is a principal mechanism for moving nutrients in and wastes out of a cell.

CHECK YOURSELF

Why are gases so much easier to compress into smaller volumes than are solids and liquids?

CHECK YOUR ANSWER

Because there is a lot of space between gas particles. The particles of a solid or liquid, on the other hand, are already close to one another, meaning there is little room left for a further decrease in volume.

INSIGHT Diffusion occurs as a result of the random motion of molecules, which in turn is a direct result of the dispersion of energy as spelled out by the Second Law of Thermodynamics.

11.4 Change of Phase

Figure 11.8 illustrates that you must either add heat to a substance or remove heat from it if you want to change its phase. **Melting** occurs when a substance changes from a solid to a liquid. To visualize what happens during melting, imagine you are holding hands with a group of people and each of you start jumping around randomly. The more violently you jump, the more difficult it is to hold onto one another. If everyone jumps violently enough, keeping hold is impossible. Something like this happens to the submicroscopic particles of a solid when it is heated. As heat is added to the solid, the particles vibrate more and more violently. If enough heat is added, the attractive forces between the particles are no longer able to hold them together. The solid melts.

A liquid changes to a solid when enough heat is removed. This process is called **freezing**, and it is the reverse of melting. As heat is withdrawn from the liquid, particle motion diminishes until the particles, on average, are moving slowly enough for attractive forces between them to take permanent hold. The only motion the particles are capable of then is vibration about fixed positions, which means the liquid has solidified, or frozen.

INSIGHT Because a gas results from evaporation, this phase is also sometimes referred to as *vapor*. Water in the gaseous phase, for example, may be referred to as water vapor.

A liquid can be heated so that it becomes a gas—**evaporation**. As heat is added, the particles of the liquid acquire more kinetic energy and move faster. Particles at the liquid surface eventually gain enough energy to jump out of the

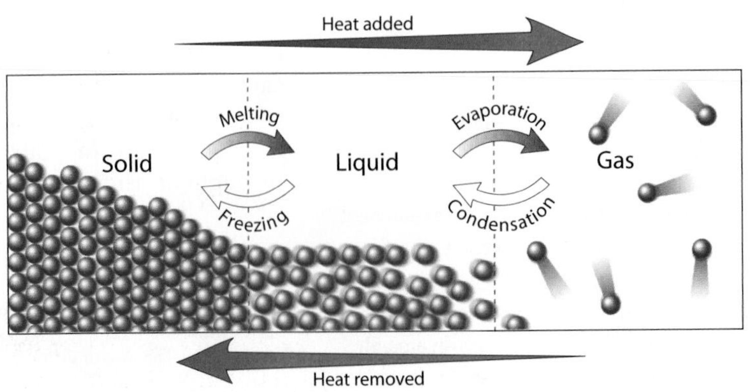

Figure 11.8 Melting and evaporation involve the addition of heat; condensation and freezing involve the removal of heat.

liquid and enter the air. In other words, they enter the gaseous phase. Even frozen water "evaporates." In this form of evaporation, called **sublimation**, molecules jump directly from the solid to the gaseous phase. Mothballs are well known for their sublimation. Ice also sublimes. Because water molecules are so tightly held in a solid, frozen water sublimes much more slowly than liquid water evaporates. Sublimation accounts for the loss of much snow and ice, especially on high, sunny mountain tops. Sublimation also explains why ice cubes left in the freezer for a long time get smaller, and how dry ice goes from a solid directly to a gas.

The rate at which a liquid evaporates increases with temperature. A puddle of water, for example, evaporates from a hot pavement more quickly than it does from your cool kitchen floor. When the temperature is hot enough, evaporation occurs beneath the surface of the liquid. As a result, bubbles form and are buoyed up to the surface. When these bubbles form, we say that the liquid is **boiling**. A substance is often characterized by its *boiling point*, which is the temperature at which it boils. At sea level, the boiling point of fresh water is 100°C.

The transformation from gas to liquid—the reverse of evaporation—is **condensation**. This process can occur when the temperature of a gas decreases. The water vapor held in the warm daylight air, for example, may condense to form a wet dew in the cool of the night.

Note that the underlying cause of phase changes is always the transfer of energy. Energy must be added to melt ice into water or vaporize water into steam. Energy must be removed to condense steam back into water or freeze water into ice. We call the energy that is released or absorbed in a change of phase **latent heat**. The word *latent* means "hidden." The heat energy involved in changing the phase of a material is *hidden* in the sense that the material does not change its temperature as it absorbs or releases this heat. For example, water boiling in a pot on the stove will remain at 100°C during its phase change to water vapor. The heat pumped into the pot of water goes into disrupting the attractive forces between molecules in the liquid state, rather than changing the water's temperature. A thermometer would register 100°C until all the water has evaporated.

More specifically, the amount of energy needed to change any substance from solid to liquid (and vice versa) is called the **heat of fusion** for the substance. For water, this is 334 joules per gram. And the amount of energy required to change any substance from liquid to gas (and vice versa) is called the **heat of vaporization** for the substance. For water, this is a whopping 2256 joules per gram.

Water's high heat of vaporization allows you to briefly touch your wetted finger to a hot skillet on a hot stove without harm. You can even touch it a few times in succession as long as your finger remains wet. Energy that ordinarily would burn your finger goes instead into changing the phase of the moisture on your finger. Similarly, you are able to judge the hotness of a clothes iron with a wetted finger. Better work fast though—when the water on your finger vaporizes, the iron's hot, highly conductive metal surface will transfer plenty of thermal energy to your finger, quickly, in its attempt to reach thermal equilibrium with your skin—ouch!

INTEGRATED SCIENCE PHYSICS AND BIOLOGY

Evaporation Cools You Off and Condensation Warms You Up

Evaporation is a cooling process. That is to say, whenever a liquid evaporates from the surface of a material, that material is cooled. Our sweat glands take advantage of this by producing perspiration to cool us down when we begin to overheat. As the water molecules in perspiration evaporate from the surface of our skin, they carry away unwanted energy and thereby cool us down.

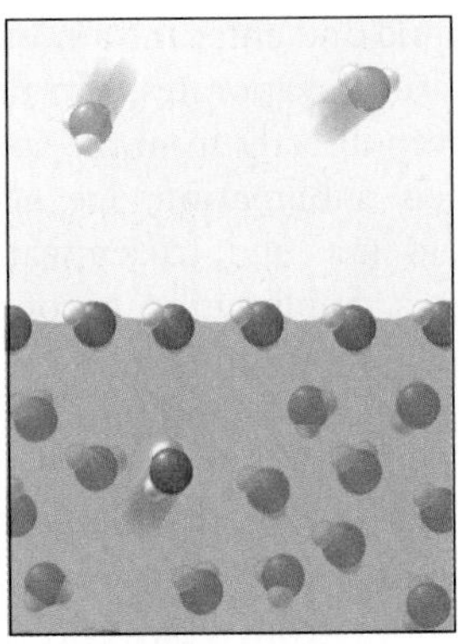
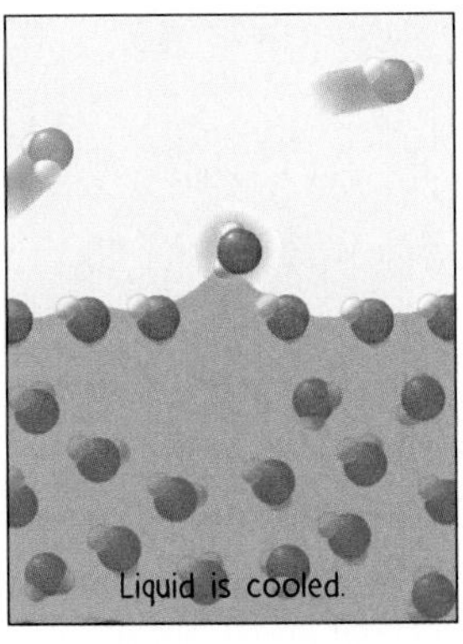

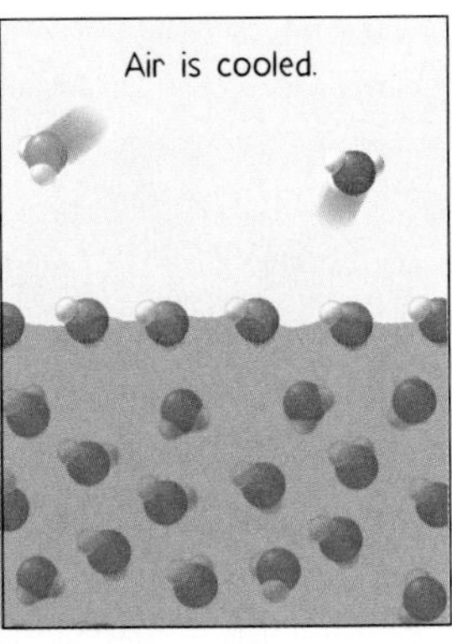

(1) Liquid water molecule having sufficient kinetic energy to overcome attractions to surrounding molecules approaches liquid surface.

(2) Liquid water is cooled as it loses this high-speed water molecule.

(3) Air is cooled as it collects slowly moving gaseous particles.

Figure 11.9 Evaporation is a cooling process.

Figure 11.10 Sam has no sweat glands (except between his toes). He cools himself by panting. In this way, evaporation occurs in the mouth and within the bronchial tract.

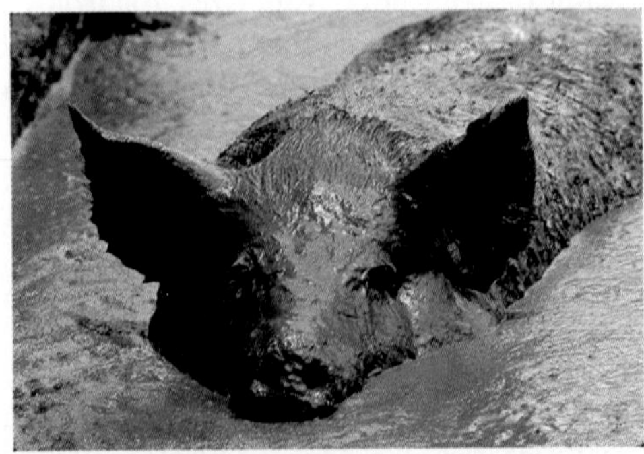

Figure 11.11 Pigs have no sweat glands and therefore cannot cool by the evaporation of perspiration. Instead, they wallow in the mud to cool themselves.

Think of how this works on the submicroscopic level. Liquid molecules are like tiny billiard balls, moving helter-skelter, continually bumping into one another. During their bumping, some gain kinetic energy while some lose kinetic energy. Molecules at the surface of the liquid gain kinetic energy by being bumped from below. Given sufficient energy to overcome attractions to their molecular neighbors, they can break free from the liquid. They leave the surface and escape into the space above the liquid to become a gas (Figure 11.9). When fast-moving molecules leave the liquid, the molecules left behind are the slow-moving ones. In this way, a layer of perspiration on your skin continually moves the high-temperature molecules away from your body. Many animals, however, do not have sweat glands and must cool themselves by other means (Figures 11.10 and 11.11).

Condensation works in just the opposite way. When gas molecules near the surface of a liquid are attracted to the liquid, they strike the surface with increased kinetic energy and become part of the liquid. Their kinetic energy is absorbed by the liquid. The result is increased temperature. Condensation is a warming process (Figure 11.12). A dramatic example of warming by condensation is the energy released by steam when it condenses on your skin. The steam gives up a lot of energy when it condenses to a liquid and moistens the skin. That's why a burn from 100°C steam is much more damaging than 100°C boiling water.

When taking a shower, you may have noticed that you feel warmer in the moist shower region than outside the shower (Figure 11.13). This difference is quickly sensed when you step outside. Away from the moisture, the rate of evaporation is much higher than the rate of condensation, and you feel chilly. When you remain in the shower stall where the humidity is higher, the rate of condensation is increased, so that you feel warmer. So now you know why you can dry yourself with a towel more comfortably if you remain in the shower stall. If you're in a hurry and don't mind the chill, dry yourself off in the hallway.

CHECK YOURSELF

1. Would evaporation be a cooling process if there were no transfer of molecular kinetic energy from water to the air above?
2. On a July afternoon in dry Phoenix or Santa Fe, you'll feel a lot cooler than in New York City or New Orleans, even when the temperatures are the same. Can you explain why?

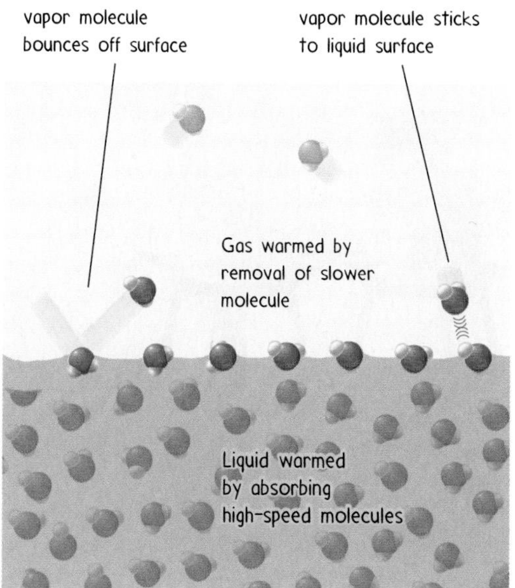

Figure 11.12 Condensation is a warming process.

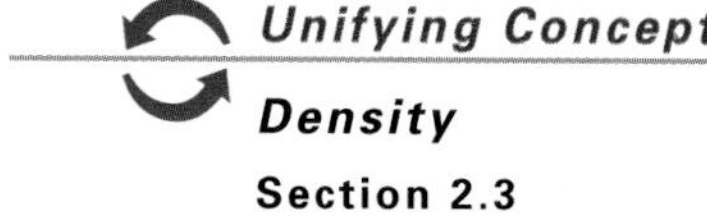

Figure 11.13 If you're chilly outside the shower stall, step back inside and be warmed by the condensation of the excess water vapor there.

Figure 11.14 Gold, diamond, and water are distinguishable on the basis of physical properties such as color, phase, and density.

CHECK YOUR ANSWERS

1. No, a liquid cools only when kinetic energy is carried away by evaporating molecules. This is similar to billiard balls that gain speed at the expense of others that lose speed. Those that leave (evaporate) are gainers, while losers remain behind and lower the temperature of the water.
2. In the drier cities, the rate of evaporation is much greater than the rate of condensation. In humid locations, the rate of condensation is greater than the rate of evaporation. You feel the warming effect as vapor in the air condenses on your skin. You are literally being bombarded by the impact of H_2O molecules in the air slamming into you.

11.5 Physical and Chemical Properties

Properties that describe the look or feel of a substance, such as color, hardness, density, texture, and phase, are called **physical properties.** Every substance has its own set of characteristic physical properties that we can use to identify that substance. For example, as Figure 11.14 shows, gold is an opaque, yellowish substance that is a solid at room temperature and has a density of 19.3 grams per milliliter. Diamond is a transparent substance that is a solid at room temperature and has a density of 3.5 grams per milliliter. Water is a transparent substance that is a liquid at room temperature and has a density of 1.0 gram per milliliter.

The physical properties of a substance can change when conditions change, but this does not mean a different substance is created. Cooling liquid water to below 0°C causes the water to transform to solid ice, but the substance is still water. H_2O is H_2O no matter which phase it is in. The only difference when water changes phase is how the H_2O molecules are oriented relative to one another. In the liquid, the water molecules tumble around one another, whereas in the ice, they vibrate about fixed positions. The freezing of water is an example of what chemists call a **physical change.** During a physical change, a substance changes its phase or some other physical property but *not* its chemical composition, as Figure 11.15 shows.

CHECK YOURSELF

The melting of gold is a physical change. Why?

CHECK YOUR ANSWER

During a physical change, a substance changes only one or more of its physical properties; its identity does not change. Because melted gold is still gold, but in a different form, this change is a physical change.

Gold
Opacity: opaque
Color: yellowish
Phase at 25°C: solid
Density: 19.3 g/mL

Diamond
Opacity: transparent
Color: colorless
Phase at 25°C: solid
Density: 3.5 g/mL

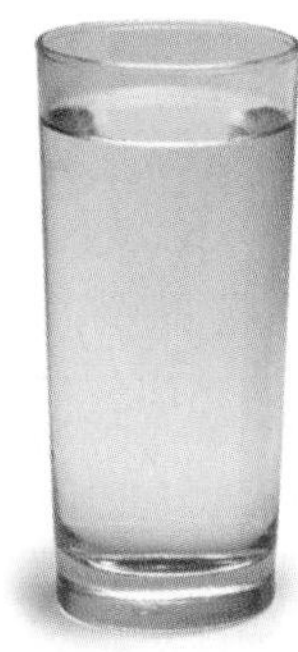

Water
Opacity: transparent
Color: colorless
Phase at 25°C: liquid
Density: 1.0 g/mL

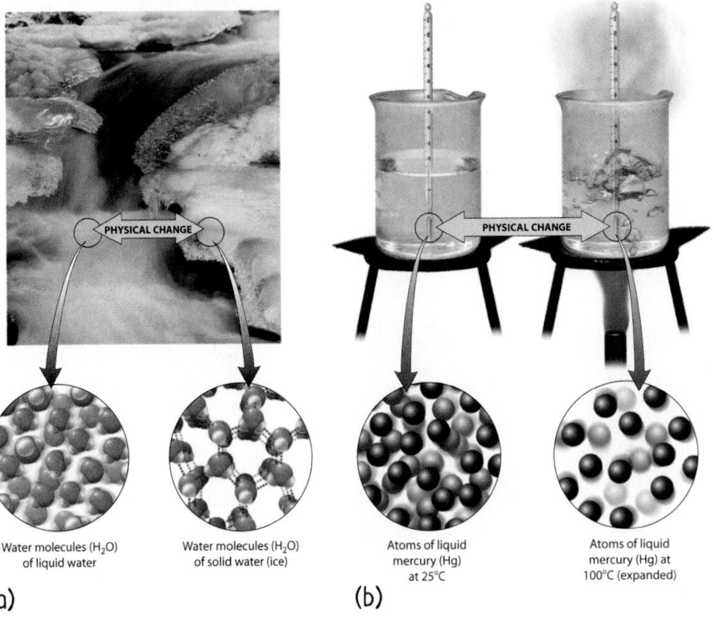

Figure 11.15 Two physical changes. (a) Liquid water and ice might look like different substances, but at the submicroscopic level, it is evident that both consist of water molecules. (b) At 25°C, the atoms in a sample of mercury are a certain distance apart, yielding a density of 13.53 grams per milliliter. At 100°C, the atoms are farther apart, meaning that each milliliter now contains fewer atoms than at 25°C, and the density is now 13.35 grams per milliliter. The physical property we call density has changed with the temperature, but the identity of the substance remains unchanged.

Chemical properties relate to how one substance reacts with others or how a substance transforms. Figure 11.16 shows three examples. The methane of natural gas has the chemical property of reacting with oxygen to produce carbon dioxide and water, along with lots of heat energy. Similarly, it is a chemical property of baking soda to react with vinegar to produce carbon dioxide and water while absorbing a small amount of heat energy. Copper has the chemical property of reacting with carbon dioxide and water to form a greenish-blue solid known as *patina*. Copper statues exposed to the carbon dioxide and water in the air become coated with patina. The patina is not copper, it is not carbon dioxide, and it is not water. It is a new substance formed by the reaction of these chemicals with one another.

All three of these transformations involve a change in the way the atoms in the molecules are *chemically bonded* to one another. A **chemical bond** is the attraction between two atoms that holds them together in a compound. A methane molecule, for example, is made of a single carbon atom bonded to four hydrogen atoms. An oxygen molecule is made of two oxygen atoms bonded to each other. Figure 11.17 shows the chemical change in which the atoms in a methane molecule and those in two oxygen molecules first pull apart and then form new bonds with different partners, resulting in the formation of molecules of carbon dioxide and water.

Methane
Reacts with oxygen to form carbon dioxide and water, giving off lots of heat during the reaction.

Baking soda
Reacts with vinegar to form carbon dioxide and water, absorbing heat during the reaction.

Copper
Reacts with carbon dioxide and water to form the greenish-blue substance called patina.

Figure 11.16 The chemical properties of substances allow them to transform to new substances. Natural gas and baking soda transform to carbon dioxide, water, and heat. Copper transforms to patina.

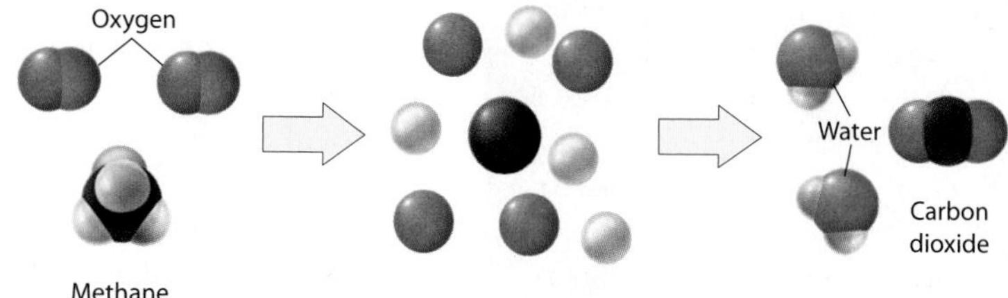

Figure 11.17 The chemical change in which molecules of methane and oxygen transform to molecules of carbon dioxide and water as atoms break old bonds and form new ones. The actual mechanism of this transformation is more complicated than depicted here. However, the idea that new materials are formed by the rearrangement of atoms is accurate.

Any change in a substance that involves a rearrangement of the way its atoms are bonded is called a **chemical change.** Thus the transformation of methane to carbon dioxide and water is a chemical change, as are the other two transformations shown in Figure 11.16.

The chemical change shown in Figure 11.18 occurs when an electric current is passed through water. The energy of the current causes the water molecules to split into atoms that then form new chemical bonds. Thus, water molecules are changed into molecules of hydrogen and oxygen, two substances that are very different from water. The hydrogen and oxygen are both gases at room temperature; they can be seen as bubbles rising to the surface.

In the language of chemistry, materials undergoing a chemical change are said to be *reacting*. Methane *reacts* with oxygen to form carbon dioxide and water. Water *reacts* when it's exposed to electricity to form hydrogen gas and oxygen gas. Thus the term *chemical change* means the same thing as *chemical reaction*. During a **chemical reaction**, new materials are formed by a change in the way atoms are bonded together. We shall explore chemical bonds and reactions in which they are formed and broken in later chapters.

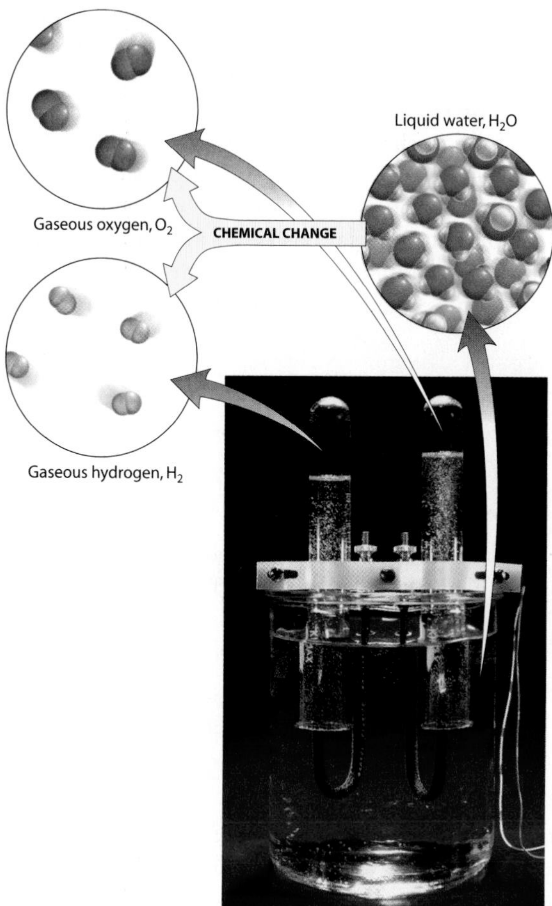

Figure 11.18 Water can be transformed to hydrogen gas and oxygen gas by the energy of an electric current. This is a chemical change because new materials (the two gases) are formed as the atoms originally in the water molecules are rearranged.

CHECK YOURSELF

Each sphere in the following diagrams represents an atom. Joined spheres represent molecules. One set of diagrams shows a physical change, and the other shows a chemical change. Which is which?

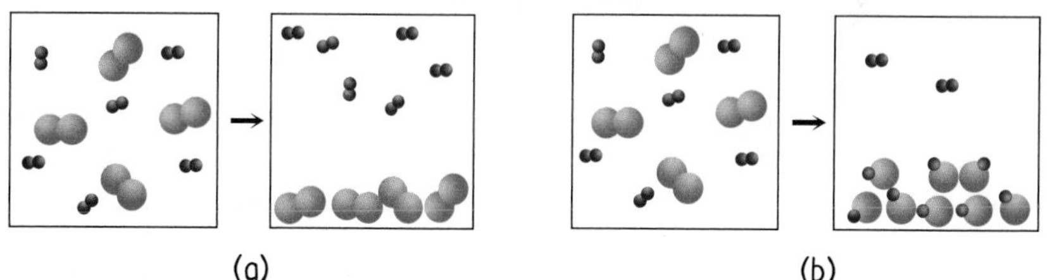

CHECK YOUR ANSWER

Remember that a chemical change (also known as a chemical reaction) involves molecules breaking apart so that the atoms are free to form new bonds with new partners. You must be careful to distinguish this breaking apart from a mere change in the relative positions of a group of molecules. In set A, the molecules before and after the change are the same. They differ only in their positions relative to one another. Set A therefore represents a physical change. In set B, new molecules consisting of bonded red and blue spheres appear after the change. These molecules represent a new material, and so B is a chemical change.

How can you tell whether a change you observe is physical or chemical? It can be tricky because in both cases there are changes in physical appearance. Water, for example, looks quite different after it freezes, just as a car looks quite different after it rusts (Figure 11.19). The freezing of water results from a change in how water molecules are oriented relative to one another. This is a physical change because liquid water and frozen water are both forms of water. The rusting of a car, by contrast, is the result of the transformation of iron to rust. This is a chemical change because iron and rust are two different materials, each consisting

Figure 11.19 The transformation of water to ice and the transformation of iron to rust both involve a change in physical appearance. The formation of ice is a physical change, and the formation of rust is a chemical change.

of a different arrangement of atoms. As we shall see in following sections, iron is an *element* and rust is a *compound* consisting of iron and oxygen atoms.

By studying this chapter, you can expect to learn the difference between a physical change and a chemical change. However, you cannot expect to have a firm handle on how to categorize an observed change as physical or chemical. Doing so requires a knowledge of the chemical identity of the materials involved as well as an understanding of how their atoms and molecules behave. This sort of insight builds over many years of study and laboratory experience.

There are, however, two powerful guidelines that can assist you in assessing physical and chemical changes. First, in a physical change, a change in appearance is the result of a new set of conditions imposed on the *same* material. Restoring the original conditions restores the original appearance; we say a physical change is easy to reverse. The freezing and melting of water is a good example. Second, in a chemical change, a change in appearance is the result of the formation of a *new* material that has its own unique set of physical properties. Most chemical changes, therefore, are not so easy to reverse. The more evidence you have suggesting that a different material has been formed, the greater the likelihood that the change is a chemical change. Iron is a material that can be used to build cars. Rust is not. This suggests that the rusting of iron is a chemical change.

CHECK YOURSELF

Evan has grown an inch in height over the past year. Is this best described as a physical or a chemical change?

CHECK YOUR ANSWER

Are new materials being formed as Evan grows? Absolutely—created out of the food he eats. His body is very different from, say, the peanut butter sandwich he ate yesterday. Yet through some very advanced chemistry, his body is able to take the atoms of that peanut butter sandwich and rearrange them into new materials. Biological growth, therefore, is best described as a chemical change.

11.6 Elements and the Periodic Table

Conceptual Science Place

Atoms and Isotopes

The terms *element* and *atom* are often used in a similar context. You might hear, for example, that gold is an element made of gold atoms. Generally, *element* is used in reference to an entire macroscopic or microscopic sample, and *atom* is used when speaking of the submicroscopic particles in the sample. The important distinction is that elements are made of atoms and not the other way around.

How many atoms are bound together in an element is shown by an **elemental formula**. For elements in which the basic units are individual atoms, the elemental formula is simply the chemical symbol: Au is the elemental formula for gold, and Li is the elemental formula for lithium, to name just two examples. For elements in which the basic units are two or more atoms bonded into molecules, the elemental formula is the chemical symbol followed by a subscript indicating the number of atoms in each molecule. For example, elemental nitrogen, as was shown in Figure 9.1, commonly consists of molecules containing two nitrogen atoms per molecule. Thus N_2 is the usual elemental formula given for nitrogen. Similarly, O_2 is the elemental formula for oxygen (two oxygen atoms per molecule), and S_8 is the elemental formula for sulfur (eight atoms per molecule).

The periodic table is a listing of all the known elements with their atomic masses, atomic numbers, and chemical symbols. Recall from our introduction to the atom in Chapter 9 that the total mass of an atom is called its *atomic mass*. This is the sum of the masses of all the atom's components (electrons, protons, and neutrons). A special unit is used for atomic masses. This is the *atomic mass*

Please put to rest any fear you may have about needing to memorize the periodic table, or even parts of it—better to focus on the many great concepts behind its organization.

Figure 11.20 The periodic table provides a variety of information about the elements.

unit, amu, where 1 atomic mass unit is equal to 1.661×10^{-24} g, which is slightly less than the mass of a single proton. Also recall from Chapter 9 that the atomic number of an element is the number of protons in an atom of that element. It is also equal to the number of electrons in the neutral atom.

Besides atomic numbers, atomic masses, and chemical symbols, there is much more information about the elements in the periodic table (Figure 11.20). The way the table is organized in groups tells you a lot about the elements' structures and how they behave.

Look carefully at Figure 11.21. It shows that metals make up most elements. **Metals** are defined as those elements that are shiny, opaque, and good conductors of electricity and heat. Metals are *malleable*, which means they can be hammered into different shapes or bent without breaking. They are also *ductile*, which means they can be drawn into wires. All but a few metals are solid at room temperature. The exceptions include mercury, Hg; gallium, Ga; cesium, Cs; and francium, Fr. These metals are all liquids at a warm room temperature of 30°C (86°F). Another interesting exception is hydrogen, H. Hydrogen acquires the properties of a liquid metal only at very high pressures (Figure 11.22). Under normal conditions, hydrogen behaves as a nonmetallic gas.

The nonmetallic elements, with the exception of hydrogen, are on the right of the periodic table. **Nonmetals** are very poor conductors of electricity and heat, and they may also be transparent. Solid nonmetals are neither malleable nor ductile. Rather, they are brittle and shatter when hammered. At 30°C (86°F), some

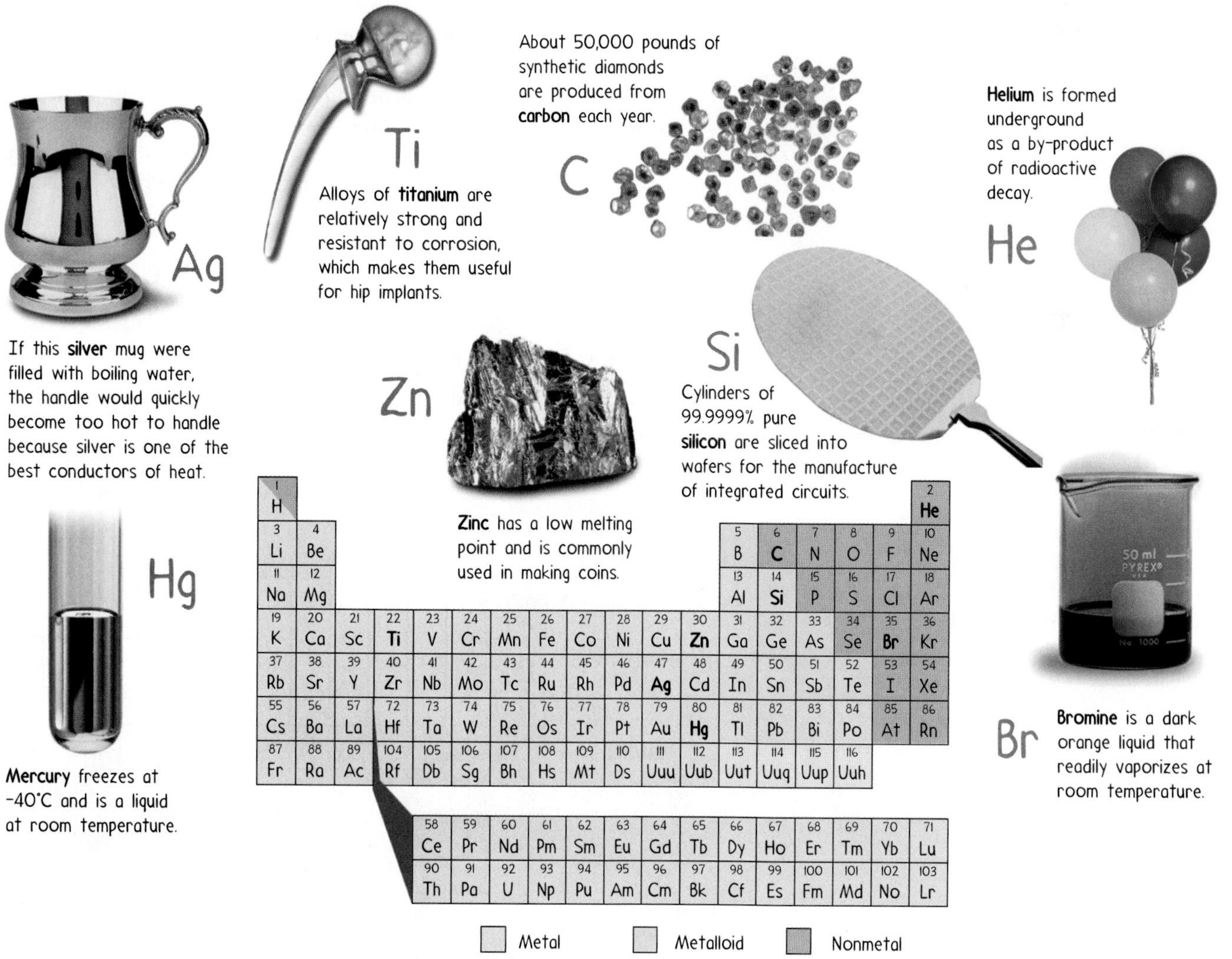

Ag

If this **silver** mug were filled with boiling water, the handle would quickly become too hot to handle because silver is one of the best conductors of heat.

Ti

Alloys of **titanium** are relatively strong and resistant to corrosion, which makes them useful for hip implants.

C

About 50,000 pounds of synthetic diamonds are produced from **carbon** each year.

He

Helium is formed underground as a by-product of radioactive decay.

Si

Cylinders of 99.9999% pure **silicon** are sliced into wafers for the manufacture of integrated circuits.

Zn

Zinc has a low melting point and is commonly used in making coins.

Hg

Mercury freezes at -40°C and is a liquid at room temperature.

Br

Bromine is a dark orange liquid that readily vaporizes at room temperature.

1 H																	2 **He**
3 Li	4 Be											5 B	6 **C**	7 N	8 O	9 F	10 Ne
11 Na	12 Mg											13 Al	14 **Si**	15 P	16 S	17 Cl	18 Ar
19 K	20 Ca	21 Sc	22 **Ti**	23 V	24 Cr	25 Mn	26 Fe	27 Co	28 Ni	29 Cu	30 **Zn**	31 Ga	32 Ge	33 As	34 Se	35 **Br**	36 Kr
37 Rb	38 Sr	39 Y	40 Zr	41 Nb	42 Mo	43 Tc	44 Ru	45 Rh	46 Pd	47 **Ag**	48 Cd	49 In	50 Sn	51 Sb	52 Te	53 I	54 Xe
55 Cs	56 Ba	57 La	72 Hf	73 Ta	74 W	75 Re	76 Os	77 Ir	78 Pt	79 Au	80 **Hg**	81 Tl	82 Pb	83 Bi	84 Po	85 At	86 Rn
87 Fr	88 Ra	89 Ac	104 Rf	105 Db	106 Sg	107 Bh	108 Hs	109 Mt	110 Ds	111 Uuu	112 Uub	113 Uut	114 Uuq	115 Uup	116 Uuh		

58 Ce	59 Pr	60 Nd	61 Pm	62 Sm	63 Eu	64 Gd	65 Tb	66 Dy	67 Ho	68 Er	69 Tm	70 Yb	71 Lu
90 Th	91 Pa	92 U	93 Np	94 Pu	95 Am	96 Cm	97 Bk	98 Cf	99 Es	100 Fm	101 Md	102 No	103 Lr

Metal   Metalloid   Nonmetal

Figure 11.21 The periodic table, color-coded to show metals, nonmetals, and metalloids.

Figure 11.22 Geoplanetary models suggest that hydrogen exists as a liquid metal deep beneath the surfaces of Jupiter (shown here) and Saturn. These planets are composed mostly of hydrogen. Interior pressures exceed 3 million times the Earth's atmospheric pressure. At this tremendously high pressure, hydrogen is pressed to a liquid-metal phase. Back here on Earth, at our relatively low atmospheric pressure, hydrogen exists as a nonmetallic gas.

nonmetals are solid (such as carbon, C). Other nonmetallic elements are liquid (such as bromine, Br). Still other nonmetals are gaseous (like helium, He).

Six elements are classified as **metalloids**: boron, B; silicon, Si; germanium, Ge; arsenic, As; tin, Sn; and antimony, Sb. You'll see them between the metals and the nonmetals in the periodic table. The metalloids have both metallic and nonmetallic characteristics. For example, they are weak conductors of electricity. This makes them useful as semiconductors in the integrated circuits of computers. Note in the periodic table how germanium, Ge (number 32), is closer to the metals than to the nonmetals. Because of this positioning, we can tell that germanium has more metallic properties than silicon, Si (number 14), and is a slightly better conductor of electricity. So we find that integrated circuits fabricated with germanium operate faster than those fabricated with silicon. Because silicon is much more abundant and less expensive to obtain, however, silicon computer chips remain the industry standard.

11.7 Organization of the Periodic Table

Two other important ways in which the elements are organized in the periodic table are by horizontal rows and vertical columns. Each horizontal row is called a **period**, and each vertical column is called a **group** (or sometimes a *family*). As shown in Figure 11.23, there are 7 periods and 18 groups.

Across any period, the properties of elements gradually change. This gradual change is called a **periodic trend.** As Figure 11.24 shows, one periodic trend is that atomic size becomes smaller as you move from left to right across any period. Note that the trend repeats from one horizontal row to the next. This repeating of trends is called *periodicity*, a term used to indicate that the trends recur in cycles. Each horizontal row is called a *period* because it corresponds to one full cycle of a trend. There are many other properties of elements that change gradually in moving from left to right across the periodic table.

CHECK YOURSELF

Which are larger: atoms of cesium, Cs (number 55), or atoms of radon, Rn (number 86)?

CHECK YOUR ANSWER

Perhaps you tried looking to Figure 11.24 to answer this question and quickly became frustrated because the sixth-period elements are not shown. Well, relax. Look at the trends and you'll see that all atoms to the left are larger than those to the right.

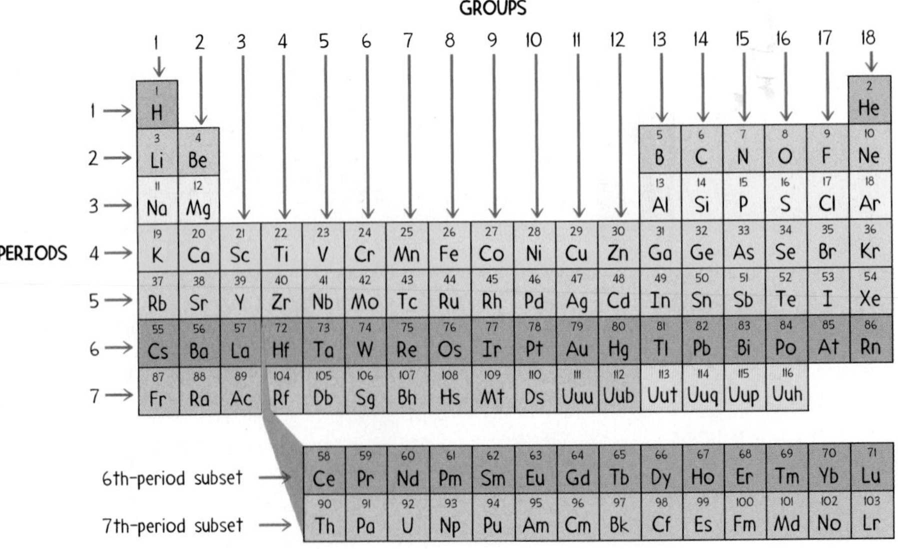

Figure 11.23 The 7 periods (horizontal rows) and 18 groups (vertical columns) of the periodic table. Note that not all periods contain the same number of elements. Also note that, for reasons explained later, the sixth and seventh periods each include a subset of the elements, which are listed apart from the main body.

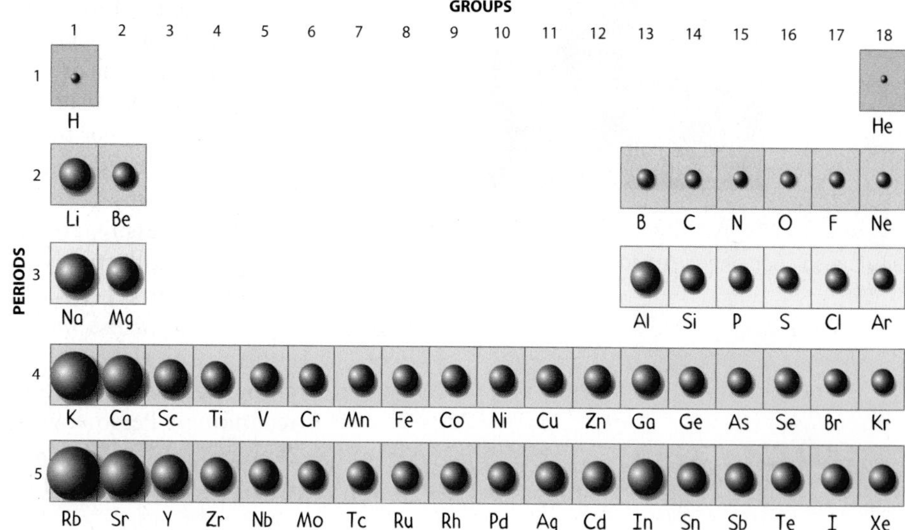

Figure 11.24 The size of atoms gradually decreases in moving left to right across any period. Atomic size is a periodic (repeating) property.

Accordingly, cesium is positioned at the far left of period 6, and so you can reasonably predict that its atoms are larger than those of radon, which is positioned at the far right of period 6. The periodic table is a road map to understanding the elements.

Moving down any group (vertical column), the properties of elements tend to be remarkably similar. This is why these elements are said to be "grouped" or arranged "in a family." As Figure 11.25 shows, several groups have traditional names that describe the properties of their elements. Early in human history, people discovered that ashes mixed with water produce a slippery solution useful for removing grease. By the Middle Ages, such mixtures were described as being *alkaline*, a term derived from the Arabic word for ashes, *al-qali*. Alkaline mixtures found many uses, particularly in the preparation of soaps. We now know that alkaline ashes contain compounds of group 1 elements, most notably potassium carbonate, also known as *potash*. Because of this history, group 1 elements, which are metals, are called the *alkali metals*.

Elements of group 2 also form alkaline solutions when mixed with water. Furthermore, medieval alchemists noted that certain minerals (which we now know are made up of group 2 elements) do not melt or change when placed in fire. These fire-resistant substances were known to the alchemists as "earth." As a holdover from these ancient times, group 2 elements are known as the *alkaline-earth metals*.

Toward the right side of the periodic table elements of group 16 are known as the *chalcogens* ("ore-forming" in Greek) because the top two elements of this group, oxygen and sulfur, are so commonly found in ores. Elements of group 17 are known as the *halogens* ("salt-forming" in Greek) because of their tendency to form various salts. Interestingly, a small amount of the halogens iodine or bromine

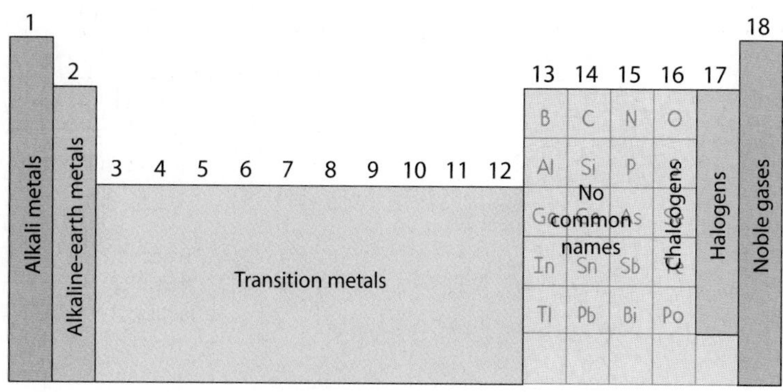

Figure 11.25 The common names for various groups of elements.

Figure 11.26 Ashes and water make a slippery alkaline solution once used to clean hands.

inside a lamp allows the tungsten filament of the lamp to glow brighter without burning out so quickly. Such lamps are commonly referred to as halogen lamps. Group 18 elements are all unreactive gases that tend not to combine with other elements. For this reason, they are called the *noble gases* (presumably because people of nobility of earlier times were above interacting with "common folk"!).

The elements of groups 3 through 12 are all metals that do not form alkaline solutions with water. These metals tend to be harder than the alkali metals and less reactive with water. Hence these metals are used for structural purposes. Collectively they are known as the *transition metals*, a name that denotes their central position in the periodic table. The transition metals include some of the most familiar and important elements. They are iron, Fe; copper, Cu; nickel, Ni; chromium, Cr; silver, Ag; and gold, Au. They also include many lesser-known elements that are nonetheless important in modern technology. Persons with hip replacements appreciate the transition metals titanium (Ti), molybdenum (Mo), and manganese (Mn) because these noncorrosive metals are used in implant devices.

CHECK YOURSELF

The elements copper (Cu), silver (Ag), and gold (Au) are three of the few metals that can be found naturally in their elemental state. These three metals have found great use as currency and jewelry for a number of reasons, including their resistance to corrosion and their remarkable colors. How is the fact that these metals have similar properties indicated in the periodic table?

CHECK YOUR ANSWER

Copper (number 29), silver (number 47), and gold (number 79) are all in the same group in the periodic table (group 11). This suggests they should have similar—though not identical—physical and chemical properties.

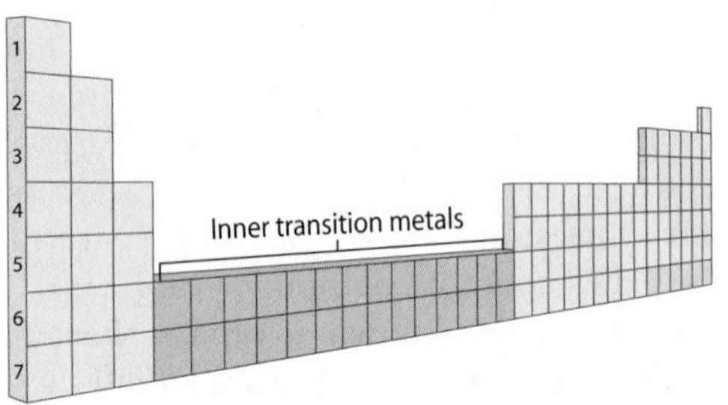

Figure 11.27 Inserting the inner transition metals between atomic groups 3 and 4 results in a periodic table that is not easy to fit on a standard sheet of paper.

In the sixth period is a subset of 14 metallic elements (numbers 58 to 71) that are quite unlike any of the other transition metals. A similar subset (numbers 90 to 103) is found in the seventh period. These two subsets are the *inner transition metals*. Inserting the inner transition metals into the main body of the periodic table, as in Figure 11.27, results in a long and cumbersome table. So that the table can fit nicely on a standard paper size, these elements are commonly placed below the main body of the table, as shown in Figure 11.28.

The sixth-period inner transition metals are called the *lanthanides* because they follow lanthanum, La. Because of their similar physical and chemical properties, they tend to occur mixed together in the same locations in the earth. Also

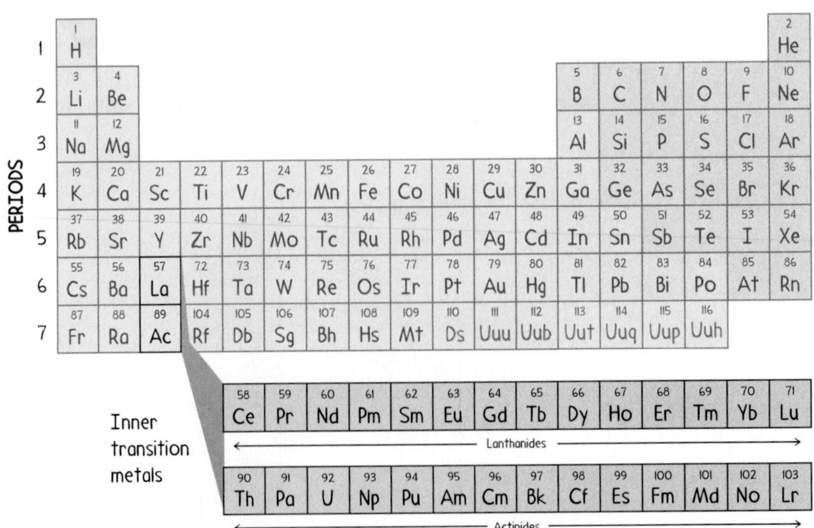

Figure 11.28 The typical display of the inner transition metals. The count of elements in the sixth period goes from lanthanum (La, 57) to cerium (Ce, 58) on through to lutetium (Lu, 71) and then back to hafnium (Hf, 72). A similar jump is made in the seventh period.

because of their similarities, lanthanides are unusually difficult to purify. Recently, the commercial use of lanthanides has increased. Several lanthanide elements, for example, are used in the fabrication of the light-emitting diodes (LEDs) of laptop computer monitors.

The seventh-period inner transition metals are called the *actinides* because they follow actinium, Ac. They, too, all have similar properties and hence are not easily purified. The nuclear power industry faces this obstacle because it requires purified samples of two of the most publicized actinides: uranium, U, and plutonium, Pu. Actinides heavier than uranium are not found in nature but are synthesized in the laboratory.

HISTORY OF SCIENCE

Dimitri Mendeleev

The mid-nineteenth century was an exciting time in the history of chemistry. New elements were being discovered almost every year and innovative laboratory apparatus and techniques enabled chemists to look at matter in more detail than ever before. The most decisive moment in the history of chemistry occurred during this period of intellectual ferment. The discovery of unexpected patterns in nature—the realization that order exists where none was previously seen—is pivotal in the progress of science. Thus, when Dimitri Mendeleev developed the first periodic table and revealed the underlying order of the elements, he brought the science of chemistry into a new era of understanding.

— 70 —

но въ ней, мнѣ кажется, уже ясно выражается примѣнимость выставляемаго мною начала ко всей совокупности элементовъ, пай которыхъ извѣстенъ съ достовѣрностію. На этотъ разъ я и желалъ преимущественно найдти общую систему элементовъ. Вотъ этотъ опытъ:

			Ti=50	Zr=90	?=180.
			V=51	Nb=94	Ta=182.
			Cr=52	Mo=96	W=186.
			Mn=55	Rh=104,4	Pt=197,4
			Fe=56	Ru=104,4	Ir=198.
			Ni=Co=59	Pl=106,6,	Os=199.
H=1			Cu=63,4	Ag=108	Hg=200.
	Be=9,4	Mg=24	Zn=65,2	Cd=112	
	B=11	Al=27,4	?=68	Ur=116	Au=197?
	C=12	Si=28	?=70	Sn=118	
	N=14	P=31	As=75	Sb=122	Bi=210
	O=16	S=32	Se=79,4	Te=128?	
	F=19	Cl=35,5	Br=80	I=127	
Li=7	Na=23	K=39	Rb=85,4	Cs=133	Tl=204
		Ca=40	Sr=87,6	Ba=137	Pb=207.
		?=45	Ce=92		
		?Er=56	La=94		
		?Yt=60	Di=95		
		?In=75,6	Th=118?		

а потому приходится въ разныхъ рядахъ имѣть различное измѣненіе разностей, чего нѣтъ въ главныхъ числахъ предлагаемой таблицы. Или же придется предполагать при составленіи системы очень много недостающихъ членовъ. То и другое мало выгодно. Мнѣ кажется притомъ, наиболѣе естественнымъ составить кубическую систему (предлагаемая есть плоскостная), но и попытки для ея образованія не повели къ надлежащимъ результатамъ. Слѣдующія двѣ попытки могутъ показать то разнообразіе сопоставленій, какое возможно при допущеніи основнаго начала, высказаннаго въ этой статьѣ.

Li	Na	K	Cu	Rb	Ag	Cs	—	Tl
7	23	39	63,4	85,4	108	133		204
Be	Mg	Ca	Zn	Sr	Cd	Ba	—	Pb
B	Al	—	—	—	Ur	—	—	Bi?
C	Si	Ti	—	Zr	Sn	—	—	—
N	P	V	As	Nb	Sb	—	Ta	—
O	S	—	Se	—	Te	—	W	—
F	Cl	—	Br	—	J	—	—	—
19	35,5	58	80	190	127	160	190	220.

An early draft of Mendeleev's periodic table.

Dimitri Mendeleev (1834–1907) was a popular chemistry professor at the Technological Institute of St. Petersburg in Russia. In 1869, he wrote the properties of the 63 known elements along with their atomic weights on small paper cards. He arranged the cards in various ways to see if he could find order among them; some say Mendeleev was inspired by the card game known as solitaire. By arranging the cards in order of increasing atomic mass as well as by their properties, he found a way to reveal one set of relationships when the cards were read up and down and another set of relationships when the cards were read side-to-side. Elements in the same column had similar properties—for example copper, gold, and silver are all metals while helium, argon, and neon are all nonreactive gases. Across each horizontal row, were elements of repeating properties. Medeleev found however, that in order to align elements properly in a column, he had to shift elements right or left occasionally. This left gaps—blank spaces that could not be filled in by any element. Instead of looking at the gaps as defects, Mendeleev predicted the existence of elements that had not yet been discovered.

In the years after the periodic table was published, more elements were discovered and gaps in Mendeleev's table were filled in according to his predictions. The newly discovered elements had just the masses and properties anticipated. Of course, as more has been learned about atomic structure, Mendeleev's original table has been modified and improved. But because it laid the groundwork for our understanding of atomic behavior in such a fundamental way, the periodic table is recognized as one of the most important achievements of modern science.

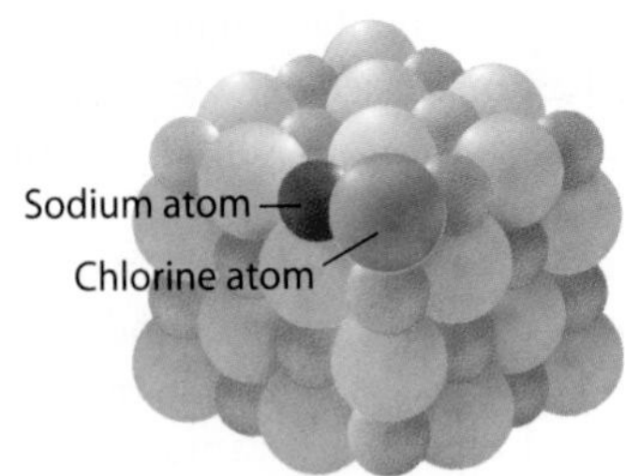

Sodium chloride, NaCl

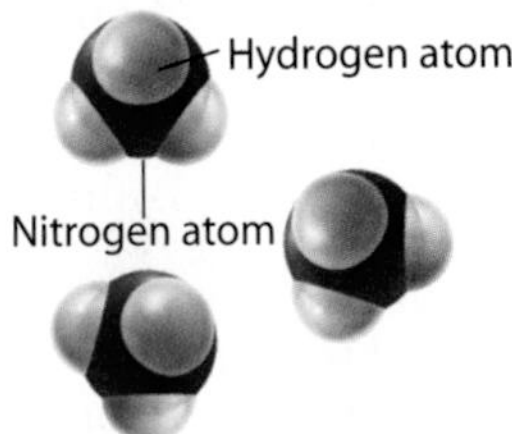

Ammonia, NH_3

Figure 11.29 The compounds sodium chloride and ammonia are represented by their chemical formulas, NaCl and NH_3. A chemical formula shows the ratio of atoms used to make the compound.

11.8 Elements to Compounds

When atoms of *different* elements bond to one another, they make a **compound.** Sodium atoms and chlorine atoms, for example, bond to make the compound sodium chloride, commonly known as table salt. Nitrogen atoms and hydrogen atoms join to make the compound ammonia, a common household cleaner.

A compound is represented by its **chemical formula,** in which the symbols for the elements are written together. The chemical formula for sodium chloride is NaCl, and that for ammonia is NH_3. Numerical subscripts indicate the ratio in which the atoms combine. By convention, the subscript 1 is understood and omitted. So the chemical formula NaCl tells us that in the compound sodium chloride there is one sodium for every one chlorine. The chemical formula NH_3 tells us that in the compound ammonia there is one nitrogen atom for every three hydrogen atoms, as Figure 11.29 shows.

Compounds have physical and chemical properties that are different from the properties of their elemental components. The sodium chloride, NaCl, shown in Figure 11.30 is very different from the elemental sodium and elemental chlorine used to form it. Elemental sodium, Na, consists of nothing but sodium atoms, which form a soft, silvery metal that can be cut easily with a knife. Its melting point is 97.5°C, and it reacts violently with water. Elemental chlorine, Cl_2, consists of chlorine molecules. This material, a yellow-green gas at room temperature, is very toxic and was used as a chemical warfare agent during World War I. Its boiling point is 234°C. The compound sodium chloride, NaCl, is a translucent, brittle, colorless crystal having a melting point of 800°C. Sodium chloride does not chemically react with water the way sodium does, and not only is it not toxic to humans the way chlorine is, but the very opposite is true: It is an essential component of all living organisms. Sodium chloride is not sodium, nor is it chlorine; it is uniquely sodium chloride, a tasty chemical when sprinkled lightly over popcorn.

CHECK YOURSELF

Hydrogen sulfide, H_2S, is one of the smelliest compounds. Rotten eggs get their characteristic bad smell from the hydrogen sulfide they release. Can you infer from this information that elemental sulfur, S_8, is just as smelly?

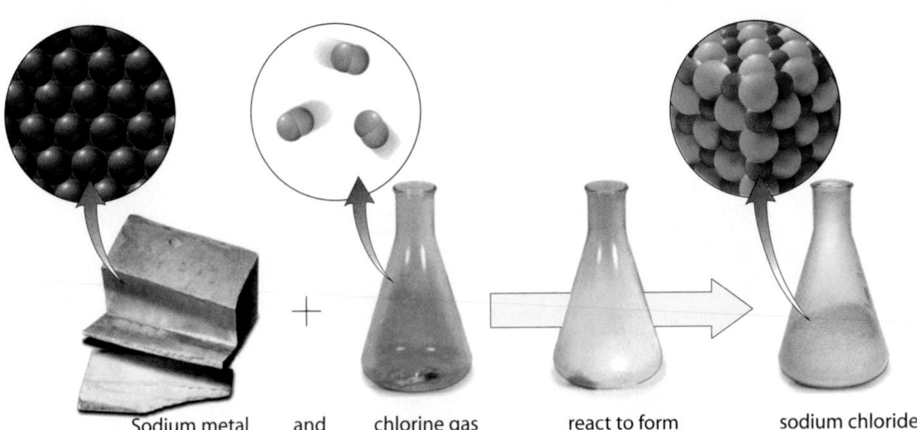

Figure 11.30 The chemical properties of sodium metal and chlorine gas react together to form sodium chloride. Although the compound sodium chloride is composed of sodium and chlorine, the physical and chemical properties of sodium chloride are very different from the physical and chemical properties of either sodium metal or chlorine gas.

CHECK YOUR ANSWER

No, you cannot. In fact, the odor of elemental sulfur is negligible compared with that of hydrogen sulfide. Compounds are truly different from the elements from which they are formed. Hydrogen sulfide, H_2S, is as different from elemental sulfur, S_8, as water, H_2O, is from elemental oxygen, O_2.

11.9 Naming Compounds

A system for naming the countless number of possible compounds has been developed by the International Union for Pure and Applied Chemistry (IUPAC). This system is designed so that a compound's name reflects the elements it contains and how those elements are put together. Anyone familiar with the system, therefore, can deduce the chemical identity of a compound from its *systematic name*.

As you might imagine, this system is very intricate. However, you need not learn all its rules. At this point, learning some guidelines will prove most helpful. These guidelines alone will not enable you to name every compound; however, they will acquaint you with how the system works for many simple compounds consisting of only two elements.

Guideline 1: The name of the element farther to the left in the periodic table is followed by the name of the element farther to the right, with the suffix *-ide* added to the name of the latter:

NaCl	Sodium chloride	HCl	Hydrogen chloride
Li_2O	Lithium oxide	MgO	Magnesium oxide
CaF_2	Calcium fluoride	Sr_3P_2	Strontium phosphide

Guideline 2: When two or more compounds have different numbers of the same elements, prefixes are added to remove the ambiguity. The first four prefixes are *mono-* ("one"), *di-* ("two"), *tri-* ("three"), and *tetra-* ("four"). The prefix *mono-*, however, is commonly omitted from the beginning of the first word of the name:

Carbon and Oxygen

CO	Carbon monoxide
CO_2	Carbon dioxide

Nitrogen and Oxygen

NO_2	Nitrogen dioxide
N_2O_4	Dinitrogen tetroxide

Sulfur and Oxygen

SO_2	Sulfur dioxide
SO_3	Sulfur trioxide

Guideline 3: Many compounds are not usually referred to by their systematic names. Instead, they are assigned *common names* that are more convenient or have been used traditionally for many years. Some common names we use in *Conceptual Science* are *water* for H_2O, *ammonia* for NH_3, and *methane* for CH_4.

CHECK YOURSELF

What is the systematic name for NaF?

CHECK YOUR ANSWER

This compound is a cavity-fighting substance added to some toothpastes—sodium fluoride.

HISTORY OF SCIENCE

Chemical Discovery

The history of chemical discovery is full of interesting stories. Here are just a few examples. The first is a tale of serendipity; the next an account of deliberate, long-term research and development. The third story is an example of one chemical discovery leading researchers in a new direction, which can often produce a treasure of other chemical finds.

Coal tar dyes For over five thousand years, people have been dying fabric with natural dyes from plants, animals, and mineral sources. In 1856, a teenage chemistry student named William Henry Perkin was doing a little experimenting in his home chemistry lab during spring vacation. Perkin's teacher had suggested that the young lad try to make synthetic quinine, a medicine, with coal tar. Perkin didn't make the quinine—instead his chemical combining created a gooey, black substance that stuck to his reaction flask. When Perkin tried to rinse the mess out of the flask, he saw that it produced a beautiful purple color when mixed with alcohol or water. Dazzled by the color, Perkin experimented with the purple liquid and found it to be an excellent fabric dye. He set about finding a way to extract the dye from the black solid, and soon thereafter he patented his discovery—the first coal tar dye. Within months of his initial discovery, Perkin had built a factory and was producing coal tar dyes on a large scale. Today, coal tar dyes are used in products ranging from food coloring to hair dyes.

Aspirin One of the medicines used by the ancient Greek physician Hippocrates to help heal headaches, pains, and fevers was a powder made from the bark and leaves of the willow tree. By 1829, scientists knew that the compound salicin was the ingredient in willow plants that provided the relief. Soon thereafter, salicylic acid, derived from salicin, was produced and prescribed on a large scale despite the fact that it was irritating to patients' stomachs. In 1899, a German chemist named Felix Hoffman sought to find a way to modify or "buffer" salicylic acid to reduce stomach irritation—Hoffman's father was taking salicylic acid for his arthritis and suffering severe side effects. Hoffman searched the literature for information about compounds that were similar to salicylic acid and tested some on his father. He found that the compound acetylsalicylic acid worked best. Hoffman shared his findings with the president of the company where he worked, Bayer Company. The Bayer Company recognized the value of Hoffman's work and marketed the new drug as aspirin; the "a" in the name comes from acetyl; and "spir" refers to the *spiraea* plant salicylic acid is derived from. The first commercial aspirin was marketed by Bayer Company in 1899.

Buckyballs Also known as molecules of buckminsterfullerene or C_{60}, are hollow spheres built of 60 linked carbon atoms. Buckyballs are so named because they resemble the geodesic dome, designed by architect R. Buckminster Fuller.

Buckyballs, which do literally bounce when hurled against a hard substance such as steel, were discovered in 1985. British chemist Harry Kroto was investigating strange chains of carbon atoms that were detected many light-years from Earth by radio telescopes. Kroto thought that the carbon chains might form under the physical conditions that exist near red-giant stars (Chapter 25). Kroto, working with colleagues Richard Smalley and Robert Curl in their U.S. laboratory, attempted to create the exotic carbon chains. They vaporized graphite (which consists of sheets of carbon atoms) with a powerful laser in a helium gas atmosphere. When Kroto, Smalley, and Curl analyzed the reaction products, they found a spherical structure of interlocking hexagons and pentagons with carbon atoms at each vertex—a Buckyball.

The discovery caused a great stir when it was announced in the prestigious scientific journal *Nature*. The hollow, spherical structure of buckyballs was unlike compounds previously known. Chemists were intrigued by the stability of buckyballs at high temperatures and pressures as well as their ability to react with other molecules, thus producing a wide range of carbon-based spheres, tubes, and ellipsoid molecules known collectively as fullerenes.

Medical researchers believe that fullerenes may someday be used as tiny chemical sponges to absorb dangerous chemicals created in injured brain tissue. Buckyballs may also be put to use as "molecular ball bearings"—lubricants allowing surfaces to glide over one another. Also, fullerenes with added potassium act as superconductors. Research into fullerenes is ongoing as potential applications continue to be discovered. In 1996, Curl, Kroto, and Smalley were awarded the Nobel Prize for their discovery.

(a)

(b)

(a) This computer graphic of buckminsterfullerene (C_{60}) shows the soccer ball-like structure of this molecule. The carbon (C) atoms (pink spheres) are bonded together as pentagons or hexagons, which are in turn bonded together to form the hollow sphere. (b) Geodesic dome designed by R. Buckminster Fuller, which uses interlocking triangular pieces and supports to spread the load evenly across the entire dome.

CHAPTER 11 Review

Summary of Terms

Applied research Research dedicated to the development of useful products and processes.

Basic research Research dedicated to the discovery of the fundamental workings of nature.

Boiling Evaporation in which bubbles form beneath the liquid surface.

Chemical bond The attraction between two atoms that holds them together in a compound.

Chemical change During this kind of change, atoms in a substance are rearranged to give a new substance having a new chemical identity.

Chemical formula A notation used to indicate the composition of a compound, consisting of the atomic symbols for the different elements of the compound and numerical subscripts indicating the ratio in which the atoms combine.

Chemical properties A property that relates to how a substance changes its chemical identity.

Chemical reaction A chemical change.

Chemistry The study of matter and the transformations it can undergo.

Compound A material in which atoms of different elements are bonded to one another.

Condensation A transformation from a gas to a liquid.

Diffusion The tendency of molecules to move from an area of high concentration to one of low concentration.

Elemental formula A notation that uses the atomic symbol and (sometimes) a numerical subscript to denote how atoms are bonded in an element.

Evaporation A transformation from a liquid to a gas.

Freezing A transformation from a liquid to a solid.

Gas Matter that has neither a definite volume nor a definite shape, always filling any space available to it.

Group A vertical column in the periodic table, also known as a family of elements.

Heat of fusion The heat energy released by a substance as it transforms from liquid to solid. This term is also used to indicate the heat energy absorbed as the substance transforms from a solid to liquid.

Heat of vaporization The heat absorbed by a substance as it vaporizes from liquid to gas. This term is also used to indicate the heat energy released as the gas condenses back into the liquid.

Latent heat Energy that is released or absorbed in a change of phase.

Liquid Matter that has a definite volume but no definite shape, assuming the shape of its container.

Matter Anything that occupies space.

Melting A transformation from a solid to a liquid.

Metal An element that is shiny, opaque, and able to conduct electricity and heat.

Metalloid An element that exhibits some properties of metals and some properties of nonmetals.

Nonmetal An element located toward the upper right of the periodic table and that is neither a metal nor a metalloid.

Period A horizontal row in the periodic table.

Periodic trend The gradual change of any property in the elements across a period.

Physical change A change in which a substance alters its physical properties without changing its chemical identity.

Physical property Any physical attribute of a substance, such as color, density, or hardness.

Solid Matter that has a definite volume and a definite shape.

Sublimation The change of phase of a solid directly to a gas.

Submicroscopic The realm of atoms and molecules, where objects are smaller than can be detected by optical microscopes.

Review Questions

11.1 Chemistry: Integral to Our Lives

1. Why is chemistry often called the central science?
2. What do members of the Chemical Manufacturers Association pledge in the Responsible Care program?

11.2 The Submicroscopic World

3. Are atoms made of molecules, or are molecules made of atoms?
4. At what three levels of organization do we view matter?

11.3 Phases of Matter

5. What determines the shape of a gas? What determines its volume?
6. On a submicroscopic level, what determines whether a sample of matter is in the solid, liquid, or gaseous phase?

11.4 Change of Phase

7. Explain how, in molecular terms, adding enough heat to a solid makes the solid melt.
8. What is it called when evaporation takes place beneath the surface of a liquid?
9. Distinguish between heat of fusion and heat of vaporization.

11.5 Physical and Chemical Properties

10. What doesn't change during a physical change?
11. What changes during a chemical reaction?
12. What does it mean to say that a physical change is reversible?

11.6 Elements and the Periodic Table

13. Are atoms made of elements or are elements made of atoms?
14. How many atoms are in a sulfur molecule that has the elemental formula S_8?
15. How do the physical properties of nonmetals differ from the physical properties of metals?

11.7 Organization of the Periodic Table

16. How many periods are there in the periodic table? How many groups?
17. (a) Why are group 2 elements called alkaline-earth metals? (b) Why are group 17 elements called halogens?
18. Why are the inner transition metals not listed in the main body of the periodic table?

11.8 Elements to Compounds

19. What is the difference between an element and a compound?
20. Are the physical and chemical properties of a compound necessarily similar to those of the elements from which it is composed?

11.9 Naming Compounds

21. What does its systematic (IUPAC) name tell you about a compound?
22. What is the chemical formula for the compound magnesium oxide?
23. Why are common names often used for chemical compounds instead of systematic names?

Multiple Choice Questions

Choose the BEST answer to the following.

1. Is chemistry the study of the submicroscopic, the microscopic, the macroscopic, or all three?
   (a) submicroscopic, because it deals with atoms and molecules, which can't be seen with a microscope.
   (b) microscopic, because it pertains to the formation of crystals.
   (c) macroscopic, because it deals with powders, liquids, and gases that fill beakers and flasks.
   (d) all of the above, because most everything is made of atoms and molecules.
2. Of the following three sciences—physics, chemistry, and biology—the most complex is
   (a) physics, because it involves many mathematical equations.
   (b) chemistry, because there are so many possible combinations of atoms.
   (c) biology, because it is based upon both the laws of chemistry and physics.
   (d) all three, because these sciences are equally complex.
3. What type of phase change does the following figure best describe?

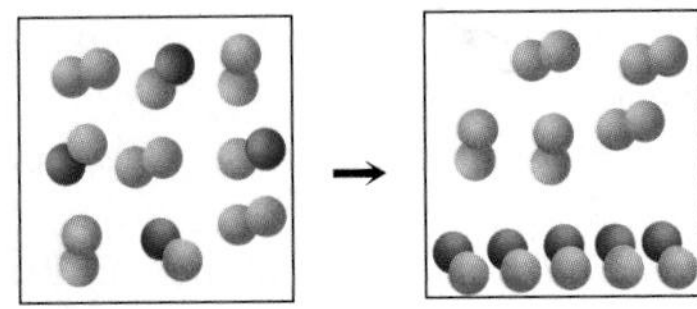

   (a) melting
   (b) condensation
   (c) evaporation
   (d) freezing
4. The phase in which atoms and molecules no longer move is the
   (a) solid phase.
   (b) liquid phase.
   (c) gas phase.
   (d) None of the above.
5. Oxygen, O_2, has a boiling point of 90 K (−183°C), and nitrogen, N_2, has a boiling point of 77 K (−196°C). Which is a liquid and which is a gas at 80 K (−193°C)?
   (a) Oxygen is a liquid and nitrogen is a gas.
   (b) Nitrogen is a liquid and oxygen is a gas.
   (c) They are both liquids at 80 K (−193°C).
   (d) They are both gases at 80 K (−193°C).
6. Is the following transformation representative of a physical change or a chemical change?

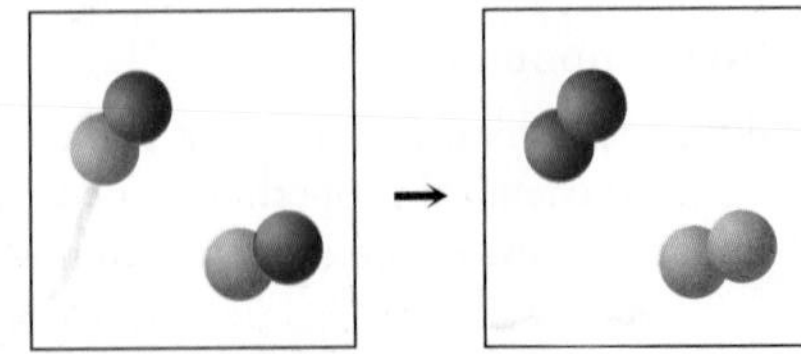

   (a) chemical, because of the formation of elements
   (b) physical, because a new material has been formed
   (c) chemical, because the atoms are connected differently
   (d) physical, because of a change in phase
7. Helium (He) is a nonmetallic gas and the second element in the periodic table. Rather than being placed adjacent to hydrogen (H), however, helium is placed on the far right of the table. Why?
   (a) because hydrogen and helium repel one another
   (b) because the sizes of their atoms are vastly different
   (c) because they come from different sources
   (d) because helium is most similar to other group 18 elements
8. Strontium, Sr (number 38), is especially dangerous to humans because it tends to accumulate in calcium-dependent bone marrow tissues [calcium (Ca), number 20]. This fact relates to the organization of the periodic table in that strontium and calcium are both
   (a) metals.
   (b) in group 2 of the periodic table.
   (c) made of relatively large atoms.
   (d) soluble in water.
9. The systematic names for water, ammonia, and methane are dihydrogen monoxide (H_2O); trihydrogen nitride (NH_3); and tetrahydrogen carbide (CH_4). Why do most people, including chemists, prefer to use the common names for these compounds?
   (a) The common names are shorter and easier to pronounce.
   (b) These compounds are encountered frequently.
   (c) The common names are more widely known.
   (d) All of the above.
10. The systematic name for the compound N_2O_4 is
   (a) nitrogen oxide.
   (b) dintrogen oxide.
   (c) nitrogen tetroxide.
   (d) dinitrogen tetroxide.

INTEGRATED SCIENCE CONCEPTS

Astronomy—Origin of the Moon

1. Summarize the Giant Impact Theory. What sort of chemical evidence supports it?
2. Has chemical analysis of lunar rocks shown that chemical elements not listed in the periodic table exist on the Moon?

Physics, Biology—Evaporation Cools You Off and Condensation Warms You Up

1. What is evaporation, and why is it a cooling process? Exactly what is it that cools?
2. What is condensation, and why is it a warming process? Exactly what is it that warms?
3. Why is a steam burn more damaging than a burn from boiling water of the same temperature?
4. Why do you feel uncomfortably warm on a hot and humid day?

Active Explorations

1. **Fire Water** This activity is for those of you with access to a gas stove.

Safety Note

Tie long hair back and roll up long, loose sleeves. Of course, watch the pot carefully while it is on the burner. Protect your hands with oven mitts if you pick up the pot.

Now, place a large pot of cool water on top of the stove, and set the burner on high. What product from the combustion of the natural gas do you see condensing on the outside of the pot? Where did it come from? Would more or less of this product form if the pot contained ice water? Where does this product go as the pot gets warmer? What physical and chemical changes can you identify?

2. **Oxygen Bubble Bursts** Compounds can be broken down to their component elements. For example, when you pour a solution of the compound hydrogen peroxide, H_2O_2, over a cut, an enzyme in your blood decomposes it to produce oxygen gas, O_2. You see the oxygen gas as the bubbling that takes place. This oxygen at high concentrations at the site of injury kills off microorganisms and prevents your cut from getting infected. A similar enzyme is found in baker's yeast.

What You Need

Packet of baker's yeast; 3 percent hydrogen peroxide solution; short, wide drinking glass; tweezers; matches

Safety Note

Wear safety glasses, and remove all materials that could burn (such as paper towels, etc.) from the area where you are working. Keep your fingers well away from the flame because it will glow brighter as it is exposed to the oxygen.

Procedure

1. Pour the yeast into the glass. Add a couple of capfuls of the hydrogen peroxide and watch the oxygen bubbles form.
2. Test for the presence of oxygen. Hold a lighted match with the tweezers and put the flame near the bubbles. Look for the flame to glow brighter as the escaping oxygen passes over it.

Describe oxygen's physical and chemical properties.

Exercises

1. In what sense is a color computer monitor or television screen similar to our view of matter? Place a drop (and only a drop) of water on your computer monitor or television screen for a closer look.
2. Is chemistry the study of the submicroscopic, the microscopic, the macroscopic, or all three? Defend your answer.
3. Which has stronger attractions among its submicroscopic particles: a solid at 25°C or a gas at 25°C? Explain.
4. Gas particles travel at speeds of up to 500 meters per second. Why, then, does it take so long for gas molecules to travel the length of a room?
5. Humidity is a measure of the amount of water vapor in the atmosphere. Why is humidity always very low inside your kitchen freezer?
6. A cotton ball is dipped in alcohol and wiped across a tabletop. Explain what happens to the alcohol molecules deposited on the tabletop. Is this a physical or chemical change? Would the resulting smell of the alcohol be more or less noticeable if the tabletop were much warmer? Explain.
7. Alcohol wiped across a tabletop rapidly disappears. What happens to the temperature of the tabletop? Why?
8. Try to explain how alcohol evaporates from the surface of a tabletop assuming that matter is continuous and NOT made of tiny atoms and molecules.
9. A skillet is lined with a thin layer of cooking oil followed by a layer of unpopped popcorn kernels. Upon heating, the kernels all pop, thereby escaping the skillet. Identify any physical or chemical changes.
10. Red-colored Kool-Aid crystals are added to a still glass of water. The crystals sink to the bottom. Twenty-four hours later, the entire solution is red even though no one stirred the water. Explain.
11. Gas molecules move 500 meters per second at room temperature; yet there is a noticeable delay in the time it takes for you to smell perfume when someone walks into the room. Explain.
12. Oxygen, O_2, has a boiling point of 90 K (−183°C), and nitrogen, N_2, has a boiling point of 77 K (−196°C). Which is a liquid and which is a gas at 80 K (−193°C)?
13. What happens to the properties of elements across any period of the periodic table?
14. Each sphere in the following diagrams represents an atom. Joined spheres represent molecules. Which box contains a liquid phase? Why can you not assume that box B represents a lower temperature?

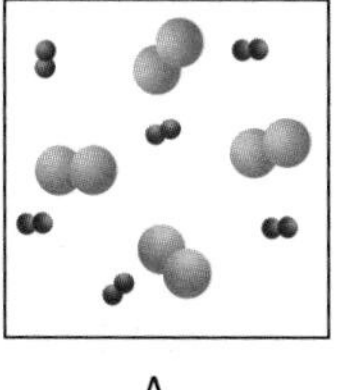

A

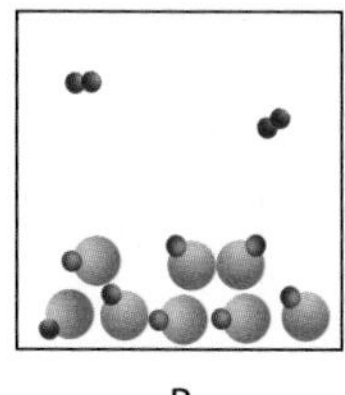

B

15. Based on the information given in the following diagrams, which substance has the lower boiling point?

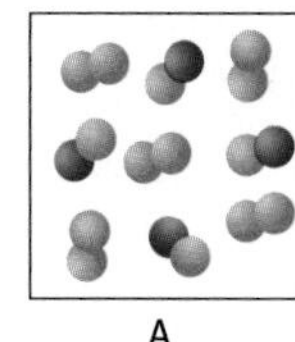

A

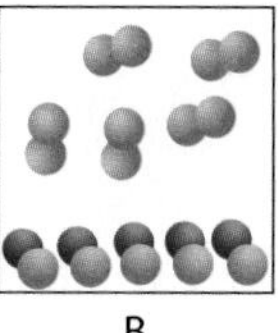

B

16. What physical and chemical changes occur when a wax candle burns?
17. Germanium, Ge (number 32), computer chips operate faster than silicon, Si (number 14), computer chips. So how might a gallium, Ga (number 31), chip compare with a germanium chip?
18. Helium, He, is a nonmetallic gas and the second element in the periodic table. Rather than being placed adjacent to hydrogen, H, however, helium is placed on the far right of the table. Why?
19. Strontium, Sr (number 38), is especially dangerous to humans because it tends to accumulate in calcium-dependent bone marrow tissues (calcium, Ca, number 20). How does this fact relate to what you know about the organization of the periodic table?
20. Do all the molecules in a liquid have about the same speed, or do they have a wide variety of speeds? Likewise, do all the molecules in a gas have the same speeds?
21. Why does increasing the temperature of a solid make it melt?
22. (a) How many atoms are there in one molecule of H_3PO_4? (b) How many atoms of each element are there in one molecule of H_3PO_4?
23. Why does decreasing temperature of a liquid make it freeze?
24. Write a letter to Grandpa to explain to him, in molecular terms, why he will stay warmer if he pats himself dry in the stall after a shower.

Problem

Calculate the height from which a block of ice at 0°C must be dropped for it to completely melt upon impact. Assume that there is no air resistance and that all the energy goes into melting the ice. [*Hint*: Equate the joules of gravitational potential energy to the product of the mass of ice and its heat of fusion (in SI units, 335,000 J/kg). Do you see why the answer doesn't depend on mass?]

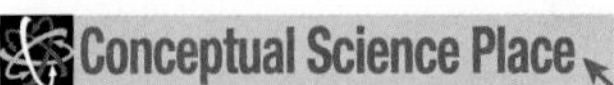

CHAPTER 11 ONLINE RESOURCES

Tutorials • What Is Chemistry?, Atoms and Isotopes • ***Quiz*** • ***Flashcards*** • ***Links***

The Nature of Chemical Bonds

Calcium fluoride is made up of calcium and fluoride ions. These ions are held together by strong chemical bonds, so calcium fluoride has a high melting point. Calcium fluoride occurs naturally as the mineral fluorite in the drinking water of some communities, where it is a good source of the teeth-strengthening fluoride ion, F^-.

CHAPTER OUTLINE

In previous chapters, you learned about the structure of the atom and the organization of the periodic table. This knowledge of atomic structure and periodicity will help you understand the forces of attraction that hold atoms together—chemical bonds. For instance, what is it about the electron structure of calcium and fluoride atoms that makes them bond so strongly, as shown in the photograph? On the other hand, why are the noble gases such as neon and helium almost never combined with other elements? Why is it that metallic elements lose electrons while nonmetallic atoms readily absorb them? What sorts of forces, operating between molecules, account for the clinginess of plastic wrap, the spherical shape of water drops, and the ability of geckos to walk on walls? Questions such as these and more will be answered in this chapter.

12.1 Electron Shells and Chemical Bonding

We know from our previous study of atoms that an atom consists of a positively charged nucleus surrounded by moving, negatively charged electrons. According to the *shell model* of atom, introduced in Chapter 9, electrons behave as if they were arranged in concentric shells around the nucleus. As shown in Figure 12.1, there are at least seven shells available to the electrons in any atom. Each of these shells has a limited capacity for the number of electrons it can hold. The first shell can hold up to 2 electrons, while the second and third shells can each hold up to 8 electrons. The fourth and fifth shells can each hold 18 electrons, and the

Figure 12.1 The shell model of the atom. The numbers indicate the maximum number of electrons in each shell.

sixth and seventh shells can each hold 32 electrons. Figure 12.2 shows how this model applies to the first four elements of group 18, the noble gases.

Electrons, being negatively charged, are attracted to the positively charged nucleus. They occupy the innermost shells first, where they are closest to the nucleus and possess minimum potential energy. Outer shells only get filled once the inner shells have reached their capacity for electrons.

It is the exposed electrons in the outermost occupied shell that are the ones most responsible for an atom's chemical properties, including its ability to form bonds with other atoms. As you may recall from Chapter 7, a **chemical bond** is an electrostatic force of attraction between atoms that holds them together. To indicate their importance, an atom's outer-shell electrons are called **valence electrons** and they are said to occupy the atom's **valence shell.**

If we are going to keep track of the bonding behavior of an atom, we need to keep track of its valence electrons. We do this by depicting valence electrons as a series of dots surrounding an atomic symbol. The atomic symbol represents the nucleus and the atom's inner-shell electrons. This notation is called the **electron-dot structure** or, sometimes, a *Lewis dot symbol* in honor of the American chemist G. N. Lewis who first proposed the concept of shells. Figure 12.3 shows the electron-dot structures for the representative elements of the periodic table.

When you look at the electron-dot structure of an atom, you immediately know two important things about that element that relate to its bonding behavior. You know how many valence electrons it has and how many of these are paired. Chlorine, for example, has three sets of paired electrons and one unpaired electron, and carbon has four unpaired electrons:

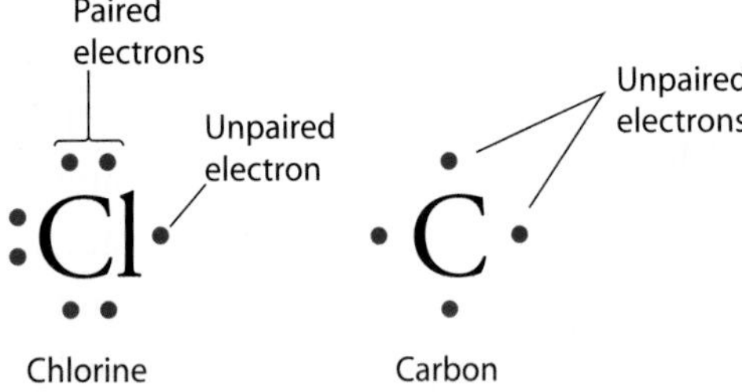

Paired valence electrons are relatively *stable*, or resistant to change. They usually do not form chemical bonds with other atoms. For this reason, electron

Figure 12.2 Occupied shells in the group 18 elements helium through krypton. Each of these elements has a filled outermost occupied shell.

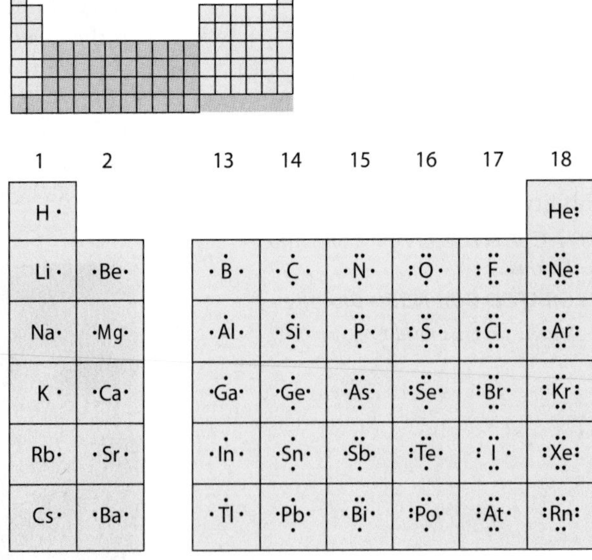

Figure 12.3 The valence electrons of an atom are shown in its electron-dot structure.

INSIGHT What takes more work: lifting a rock a centimeter off the ground or lifting it a meter off the ground? Similarly, what takes more work: pulling an electron a short distance away from the attracting nucleus or pulling it far away from the nucleus? Potential energy can be thought of as "stored work"—does this help you see why an electron closer to the nucleus has lower potential energy than one farther away?

pairs in an electron-dot structure are called *nonbonding pairs*. Valence electrons that are unpaired, by contrast, have a strong tendency to participate in chemical bonding. By doing so, they become paired with an electron from another atom. The most stable electron arrangement for an atom is reached when all its valence electrons are paired so that its outermost occupied shell is filled to capacity.

How can an atom with an unfilled valence shell attain a completely filled valence shell? It can *share* electrons with another atom or *transfer* electrons to another atom through bonding. The three types of chemical bonds discussed in this chapter—ionic bonds, covalent bonds, and metallic bonds—all result from either a transfer or a sharing of unpaired valence electrons.

You were introduced to the group 18 elements, the noble gases, in Chapters 9 and 11. The noble gases are referred to as the *inert gases* because they are chemically nonreactive—they almost never bond with other atoms. What is the reason for this? The valence shells of all noble gases are already filled to capacity, so they do not increase their stability by bonding. Atoms tend to bond with one another so that they achieve the same type of electron configuration as the noble gases. This tendency is called the **octet rule**:

> **Atoms tend to form chemical bonds so that they each have eight electrons in their valence shells, similar to the electron configuration of a noble gas.**

Other arrangements of electrons can also increase an atom's stability, especially for the transition metals. Nevertheless, the octet rule is the best indicator of stability for the main-block elements of the periodic table.

Unifying Concept
The Atomic Nature of Matter
Section 7.1

Consider the bonding behavior of sodium, Na, as an example of the octet rule. We can see from Figure 12.3 that sodium, being a group 1 element, has one valence electron. If an atom has only one or only a few electrons in its valence shell, it will tend to lose its outer-shell electrons so that the next shell inward, which is filled, becomes the outermost occupied shell. Then, the atom will have a filled valence shell. Sodium readily gives up the single electron in its third shell. This makes the second shell, which is already filled to capacity, the outermost occupied shell.

CHECK YOURSELF

1. Where are valence electrons located, and why are they important?
2. Why are noble gas elements chemically inert?

CHECK YOUR ANSWERS

1. Valence electrons are located in the outermost occupied shell of an atom. They are important because they play a leading role in determining the chemical properties of the atom.
2. The valence shells of noble gas elements are all filled to capacity.

12.2 The Ionic Bond

Recall from Chapter 7 that atoms are electrically neutral because they have equal numbers of protons and electrons. However, to gain stability, an atom may give up electrons from its outermost energy level. Other atoms accept additional electrons to fill their valence shells. If an atom loses or accepts electrons, it will have unbalanced electric charges and will no longer be electrically neutral. Any atom having a net electric charge is referred to as an **ion**. If electrons are lost, protons outnumber electrons and the ion's net charge is positive. If electrons are gained, electrons outnumber protons and the ion's net charge is negative.

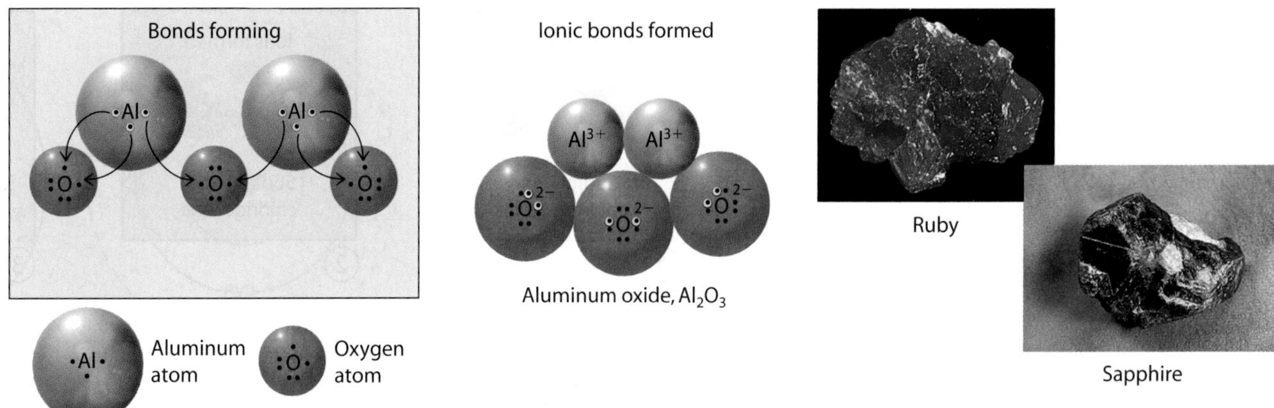

Figure 12.6 Two aluminum atoms lose a total of six electrons to form two aluminum ions, Al^{3+}. These six electrons may be picked up by three oxygen atoms, transforming the atoms to three oxide ions, O^{2-}. The aluminum and oxide ions then join to form the ionic compound aluminum oxide, Al_2O_3.

ride ion carries a charge of only 1^-. Because two fluoride ions are needed to balance each calcium ion, the formula for calcium fluoride is CaF_2. Also, an aluminum ion carries a 3^+ charge, and an oxide ion carries a 2^- charge. Together, these ions make the ionic compound aluminum oxide, Al_2O_3, the mineral that composes rubies and sapphires. The three oxide ions in Al_2O_3 carry a total charge of 6^-, which balances the total 6^+ charge of the two aluminum ions (Figure 12.6).

An ionic compound typically contains a multitude of ions grouped together in a highly ordered, three-dimensional array. In sodium chloride, for example, each sodium ion is surrounded by six chloride ions and each chloride ion is surrounded by six sodium ions (Figure 12.7). Overall, there is one sodium ion for each chloride ion, but there are no identifiable sodium–chloride pairs. Such an orderly array of ions is known as an *ionic crystal.* On the atomic level, the crystalline structure of sodium chloride is cubic, which is why macroscopic crystals of table salt are also cubic. Smash a large cubic sodium chloride crystal with a hammer, and what do you get? Smaller cubic sodium chloride crystals! (You will find out much more about crystals when you study minerals in Chapter 23.)

Sodium ion, Na^+
Chloride ion, Cl^-
(a) (b)

Figure 12.7 (a) Sodium chloride, as well as other ionic compounds, forms ionic crystals in which every internal ion is surrounded by ions of the opposite charge. (For simplicity, only a small portion of the ion array is shown here. A typical NaCl crystal involves millions and millions of ions.) (b) A view of crystals of table salt through a microscope shows their cubic structure. The cubic shape is a consequence of the cubic arrangement of sodium and chloride ions.

12.3 The Metallic Bond

Metals conduct electricity and heat, are opaque to light, and are ductile (easily drawn into a wire or hammered thin). Because of these properties, metals have found numerous applications. We use them to build homes, appliances, cars, bridges, airplanes, and skyscrapers. We stretch metal wire across poles to transmit communication signals and electricity. We wear metal jewelry, exchange metal currency, and drink from metal cans. Yet, what is it that gives a metal its metallic properties? We can answer this question by looking at the behavior of the atoms of the metallic elements.

The outer electrons of most metal atoms tend to be weakly held to the atomic nucleus. Consequently, these outer electrons are easily dislodged, leaving behind positively charged metal ions. The many electrons easily dislodged from a large group of metal atoms flow freely through the resulting metal ions, as is depicted in Figure 12.8. This "fluid" of electrons holds the positively charged metal ions together in the type of chemical bond known as a **metallic bond.**

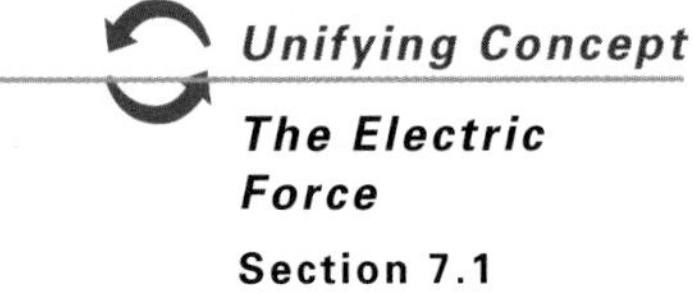

Unifying Concept

The Electric Force

Section 7.1

The mobility of electrons in a metal accounts for the metal's ability to conduct electricity and heat. Also, metals are opaque and shiny because the free

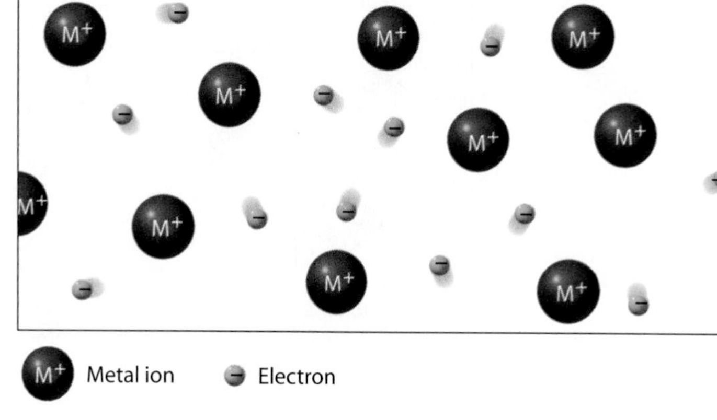

Figure 12.8 Metal ions are held together by freely flowing electrons. These loose electrons form a kind of "electronic fluid" that flows through the lattice of positively charged ions.

electrons easily vibrate to the oscillations of any light falling on them, reflecting most of it. Furthermore, the metal ions are not rigidly held to fixed positions, as ions are in an ionic crystal. Rather, because the metal ions are held together by a "fluid" of electrons, these ions can move into various orientations relative to one another, which is what happens when a metal is pounded, pulled, or molded into a different shape.

Two or more metals can be bonded to each other by metallic bonds. This occurs, for example, when molten gold and molten palladium are blended to form the homogeneous solution known as white gold. The quality of the white gold can be modified simply by changing the proportions of gold and palladium. White gold is an example of an *alloy*, which is any mixture composed of two or more metallic elements. By playing around with proportions, metal workers can readily modify the properties of an alloy. For example, in designing the Sacagawea dollar coin, shown in Figure 12.9, the U.S. Mint needed a metal that had a gold color—so that it would be popular—and also had the same electrical characteristics as the Susan B. Anthony dollar coin—so that the new coin could substitute for the Anthony coin in vending machines.

Figure 12.9 The gold color of the Sacagawea U.S. dollar coin is achieved by an outer surface made of an alloy of 77 percent copper, 12 percent zinc, 7 percent manganese, and 4 percent nickel. The interior of the coin is pure copper.

CHECK YOURSELF

In which material do the sodium nuclei have access to 11 electrons: sodium chloride, NaCl, or metallic sodium, Na?

CHECK YOUR ANSWER

In sodium chloride each sodium nucleus has access to only 10 electrons, which is why this sodium is called an ion. Within metallic sodium each sodium nucleus has access to 11 electrons. Due to the nature of the metallic bond, however, the outermost 11th electron tends to migrate from one sodium to the next.

12.4 The Covalent Bond

Imagine two children playing together and sharing their toys. A force that keeps the children together is their mutual attraction to the toys they share. In a similar fashion, two atoms can be held together by their mutual attraction for electrons they share. A fluorine atom, for example, has a strong attraction for one additional electron to fill its outermost occupied shell. As shown in Figure 12.10, a fluorine atom can obtain an additional electron by holding on to the unpaired valence electron of another fluorine atom. This results in a situation in which the two fluorine atoms are mutually attracted to the same two electrons. This type of electrical attraction in which atoms are held together by their mutual attraction for shared electrons is called a **covalent bond**, where *co-* signifies sharing and *-valent* refers to the fact that valence electrons are being shared.

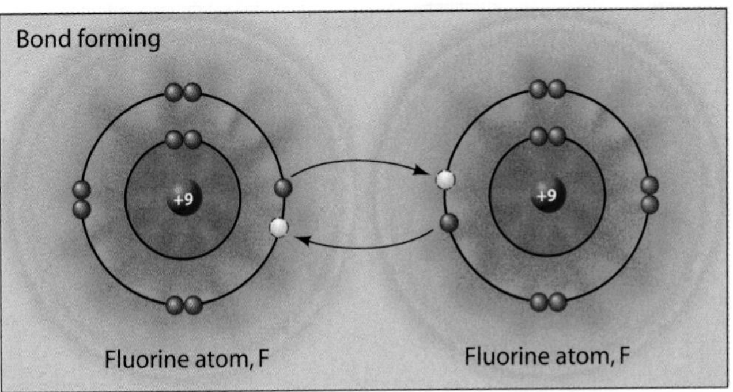

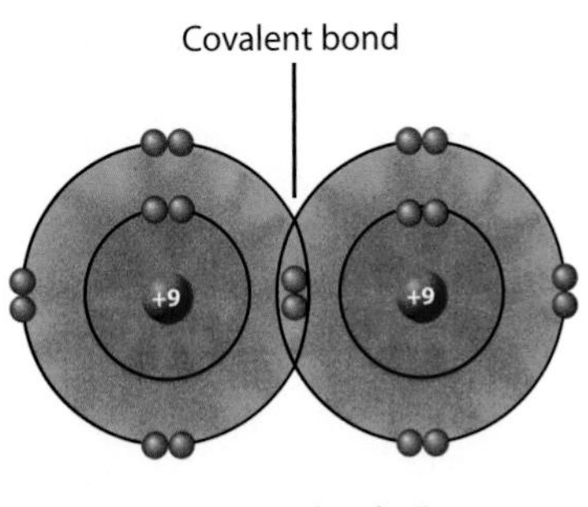

Figure 12.10 The effect of the positive nuclear charge (represented by red shading) of a fluorine atom extends beyond the atom's outermost occupied shell. This positive charge can cause the fluorine atom to become attracted to the unpaired valence electron of a neighboring fluorine atom. Then the two atoms are held together in a fluorine molecule by the attraction they both have for the two shared electrons. Each fluorine atom achieves a filled valence shell.

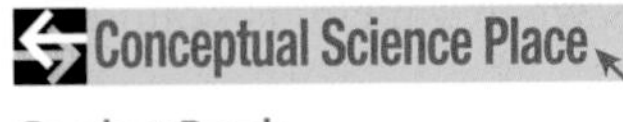

Covalent Bonds

A substance composed of atoms held together by covalent bonds is a **covalent compound.** The fundamental unit of most covalent compounds is a **molecule,** which we can now formally define as any group of atoms held together by covalent bonds. Figure 12.11 uses the element fluorine to illustrate this principle.

When writing electron-dot structures for covalent compounds, chemists often use a straight line to represent the two electrons involved in a covalent bond. In some representations, the nonbonding electron pairs are left out. This is done in instances where these electrons play no significant role in the process being illustrated. Following are two frequently used ways of showing the electron-dot structure for a fluorine molecule without using spheres to represent the atoms. Remember: the straight line in both versions represents two electrons, *one from each atom.*

$$:\ddot{\underset{\cdot\cdot}{F}}-\ddot{\underset{\cdot\cdot}{F}}: \qquad F-F$$

You've learned that an ionic bond is formed when an atom that tends to lose electrons is placed in contact with an atom that tends to gain them. A covalent bond, by contrast, is formed when two atoms that tend to gain electrons are brought into contact with each other. Atoms that tend to form covalent bonds are therefore primarily atoms of the nonmetallic elements in the upper right corner of the periodic table (with the exception of the noble gas elements, which are very stable and tend not to form bonds at all). Hydrogen tends to form covalent bonds because, unlike the other group 1 elements, it has a fairly strong attraction for an additional electron. Two hydrogen atoms, for example, covalently bond to form a hydrogen molecule, H_2, as shown in Figure 12.12.

The number of covalent bonds an atom can form is equal to the number of additional electrons it can attract, which is the number it needs to fill its valence shell. Hydrogen attracts only one additional electron, and so it forms only one

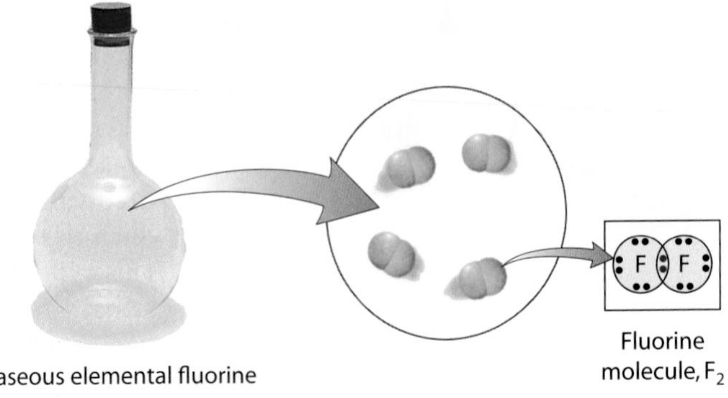

Figure 12.11 Molecules are the fundamental units of the gaseous covalent compound fluorine, F_2. Notice that in this model of a fluorine molecule, the spheres overlap, whereas the spheres shown earlier for ionic compounds do not. Now you know that this difference in representation is because of the difference in bond types.

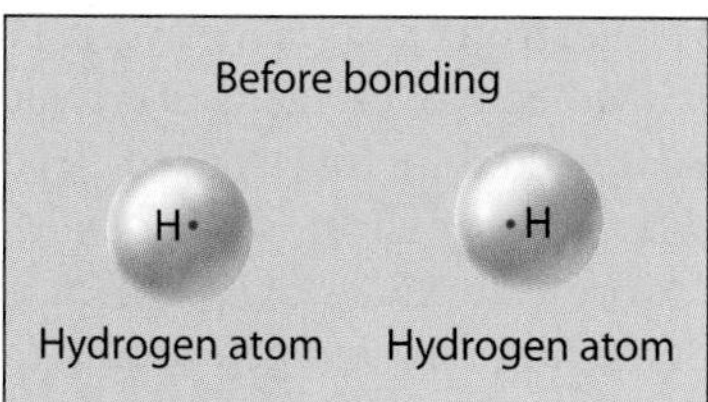

Covalent bond formed

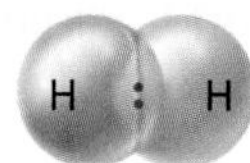

Hydrogen molecule, H_2

Figure 12.12 Two hydrogen atoms form a covalent bond as they share their unpaired electrons.

Before bonding

Oxygen atom

Hydrogen atom

Hydrogen atom

Covalent bonds formed

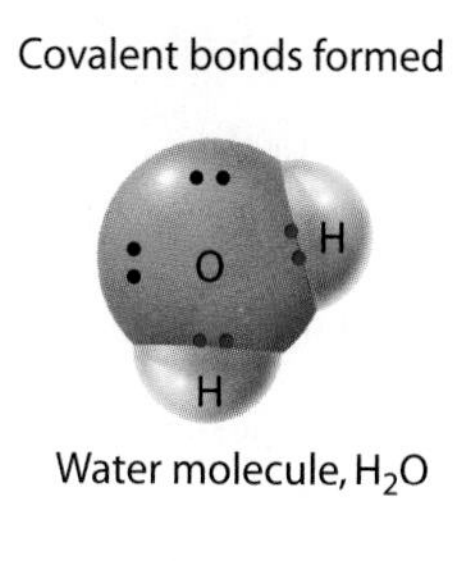

Water molecule, H_2O

Figure 12.13 The two unpaired valence electrons of oxygen pair with the unpaired valence electrons of two hydrogen atoms to form the covalent compound water.

covalent bond. Oxygen, which attracts two additional electrons, finds them when it encounters two hydrogen atoms and reacts with them to form water, H_2O, as Figure 12.13 shows. In water, not only does the oxygen atom have access to two additional electrons by covalently bonding to two hydrogen atoms but each hydrogen atom has access to an additional electron by bonding to the oxygen atom. Each atom thus achieves a filled valence shell.

Nitrogen attracts three additional electrons and is thus able to form three covalent bonds, as occurs in ammonia, NH_3, shown in Figure 12.14. Likewise, a carbon atom can attract four additional electrons and is thus able to form four covalent bonds, as occurs in methane, CH_4.*

CHECK YOURSELF

How many electrons make up a covalent bond?

CHECK YOUR ANSWER

Two—one from each participating atom.

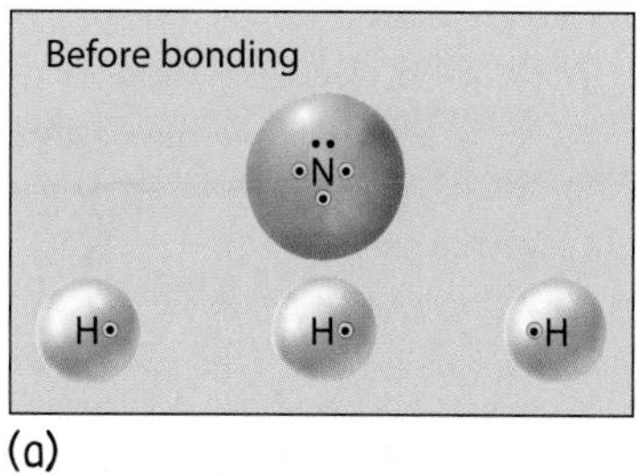

(a)

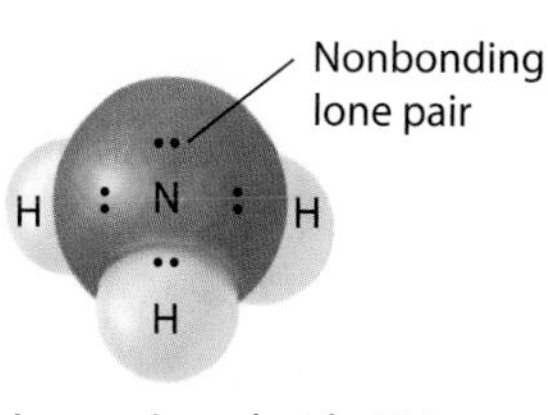

Ammonia molecule, NH_3

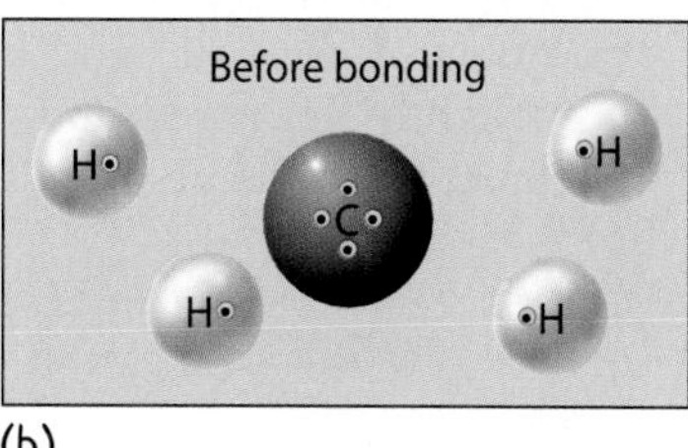

(b)

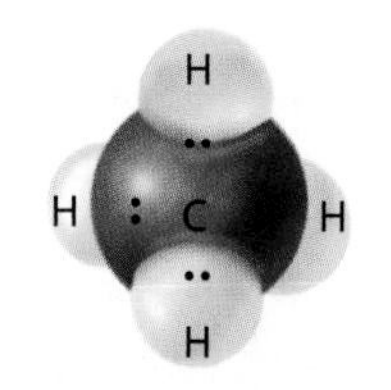

Methane molecule, CH_4

Figure 12.14 (a) A nitrogen atom attracts the three electrons in three hydrogen atoms to form ammonia, NH_3, a gas that can dissolve in water to make an effective cleanser. (b) A carbon atom attracts the four electrons in four hydrogen atoms to form methane, CH_4, the primary component of natural gas. In these and most other cases of covalent bond formation, the result is a filled valence shell for all the atoms involved.

* Electron-dot structures, such as those for methane and ammonia shown in Figure 12.1, can be used to determine the three-dimensional shape of a molecule. To do this, we use the valence shell electron-pair repulsion (VSEPR) model. According to this model, electron pairs strive to get as far away as possible from all other electron pairs in the shell. The VSEPR model shows that methane is a tetrahedral molecule while ammonia is pyramidal. In the laboratory part of this course, you may have the opportunity to build three-dimensional molecular models from electron-dot structures using the VSEPR model.

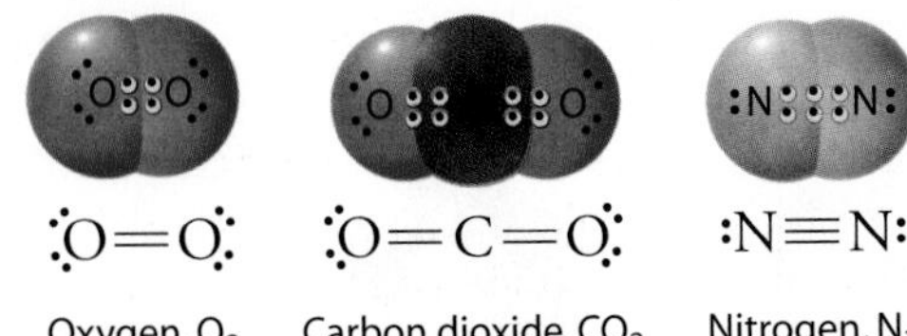

Figure 12.15 Double covalent bonds in molecules of oxygen, O_2, and carbon dioxide, CO_2, and a triple covalent bond in a molecule of nitrogen, N_2.

It is possible to have more than two electrons shared between two atoms, and Figure 12.15 shows a few examples. Molecular oxygen, O_2, which is what we inhale, consists of two oxygen atoms connected by four shared electrons. This arrangement is called a *double covalent bond* or, for short, a *double bond*. As another example, the covalent compound carbon dioxide, CO_2, which is what we exhale, consists of two double bonds connecting two oxygen atoms to a central carbon atom.

Some atoms can form *triple covalent bonds*, in which six electrons—three from each atom—are shared. One example is molecular nitrogen, N_2. Most of the air surrounding you right now (about 78 percent) is gaseous molecular nitrogen, N_2. Multiple bonds higher than triple bonds, such as the quadruple covalent bond, are not commonly observed.

12.5 Polar Bonds and Polar Molecules

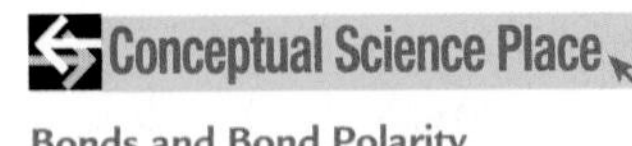

Bonds and Bond Polarity

If the two atoms in a covalent bond are identical, their nuclei have the same positive charge, and therefore the electrons are shared *evenly*. We can represent these electrons as being centrally located by using an electron-dot structure in which the electrons are situated exactly halfway between the two atomic symbols. Alternatively, we can draw a probability cloud (Chapter 9) in which the positions of the two bonding electrons over time are shown as a series of dots. Where the dots are most concentrated is where the electrons have the greatest probability of being located:

H : H   H H

In a covalent bond between nonidentical atoms, the nuclear charges are different, and consequently the bonding electrons may be shared *unevenly*. This occurs in a hydrogen–fluorine bond, in which electrons are more attracted to fluorine's greater nuclear charge:

H :F   H F

The bonding electrons spend more time around the fluorine atom. For this reason, the fluorine side of the bond is slightly negative and, because the bonding electrons have been drawn away from the hydrogen atom, the hydrogen side of the bond is slightly positive. This separation of charge is called a **dipole** (pronounced die-pole) and is represented either by the characters δ^- and δ^+, read "slightly negative" and "slightly positive," respectively, or by a crossed arrow pointing to the negative side of the bond.

$$\overset{\delta+}{\text{H}}-\overset{\delta-}{\text{F}} \qquad \overset{+\!\!\longrightarrow}{\text{H}-\text{F}}$$

So, atoms forming a chemical bond engage in a tug-of-war for electrons. How strongly an atom is able to tug on bonding electrons has been measured experimentally and quantified as the atom's **electronegativity**. The greater an atom's electronegativity, the greater its ability to pull electrons toward itself when bonded. Thus, in hydrogen fluoride, fluorine has a greater electronegativity, or pulling power, than does hydrogen.

INSIGHT The term *electronegativity* refers to the *negative* charge that an atom in a molecule picks up from the *electrons* it may attract.

Electronegativity is greatest for elements (such as fluorine and iodine) at the upper right of the periodic table and lowest for elements at the lower left (such as barium and cesium). Noble gases are not considered in electronegativity discussions because, with only a few exceptions, they do not participate in chemical bonding because their outermost shells are already filled and because electrons in those shells are tightly held.

No dipole forms when the two atoms in a covalent bond have the same electronegativity (as is the case with H_2) and the bond is classified as a **nonpolar bond.** When the electronegativities of the atoms differ, a dipole may form (as with HF) and the bond is classified as a **polar bond.** Just how polar a bond is depends on the difference between the electronegativity values of the two atoms—the greater the difference, the more polar the bond.

CHECK YOURSELF

How might a chemist predict which bonds are more polar than others?

CHECK YOUR ANSWER

The chemist should compare their electronegativities. In general, all the chemist need do is look at the relative positions of the atoms in the periodic table.

What is important to understand here is that there is no black-and-white distinction between ionic and covalent bonds. Rather, there is a gradual change from one to the other as the atoms that bond are located farther and farther apart in the periodic table. This continuum is illustrated in Figure 12.16, where the bonds are shown in order of decreasing polarity from left to right with the size of the crossed arrow representing the degree of polarity. Atoms on opposite sides of the periodic table have great differences in electronegativity, and hence the bonds between them are highly polar—in other words, ionic. Nonmetallic atoms of the same type have the same electronegativities, and so their bonds are nonpolar covalent. The polar covalent bond with its uneven *sharing* of electrons and slightly *charged* atoms is between these two extremes.

If all the bonds in a molecule are nonpolar, the molecule as a whole is also nonpolar—as is the case with H_2, O_2, and N_2. If a molecule consists of only two atoms and the bond between them is polar, the polarity of the molecule is the same as the polarity of the bond—as with HF, HCl, and ClF.

INSIGHT A polar molecule always contains polar bonds, but some molecules containing polar bonds—for example, carbon dioxide—are nonpolar.

Complexities arise when assessing the polarity of a molecule containing more than two atoms. Consider carbon dioxide, CO_2, shown in Figure 12.17. The cause of the dipole in either one of the carbon–oxygen bonds is oxygen's

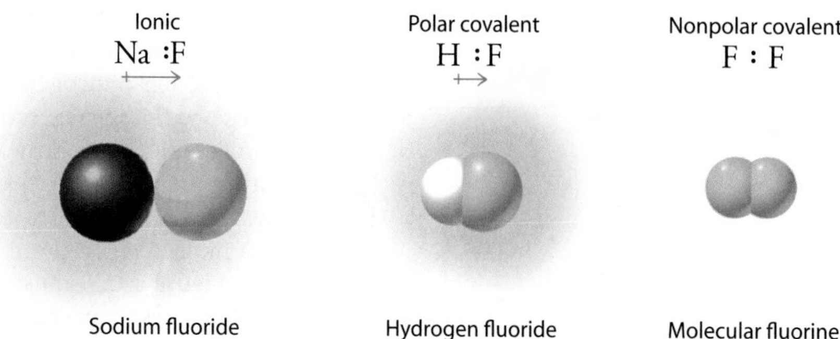

Figure 12.16 The ionic bond and the nonpolar covalent bond represent the two extremes of chemical bonding. The ionic bond involves a transfer of one or more electrons, and the nonpolar covalent bond involves the equitable sharing of electrons. The character of a polar covalent bond falls between these two extremes.

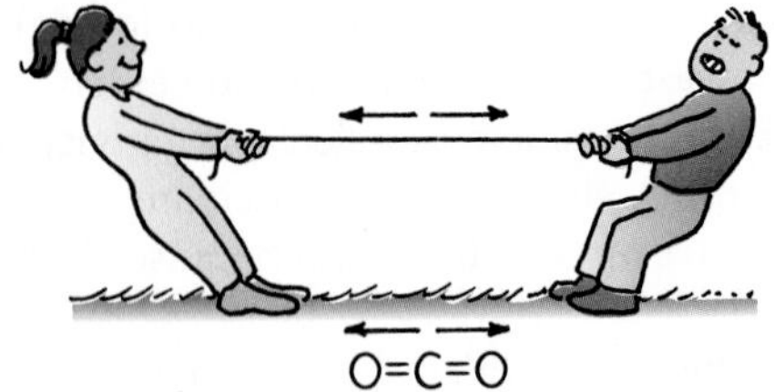

Figure 12.17 There is no net dipole in a carbon dioxide molecule, and so the molecule is nonpolar. This is analogous to two people in a tug-of-war. When they pull with equal forces but in opposite directions, the rope remains stationary.

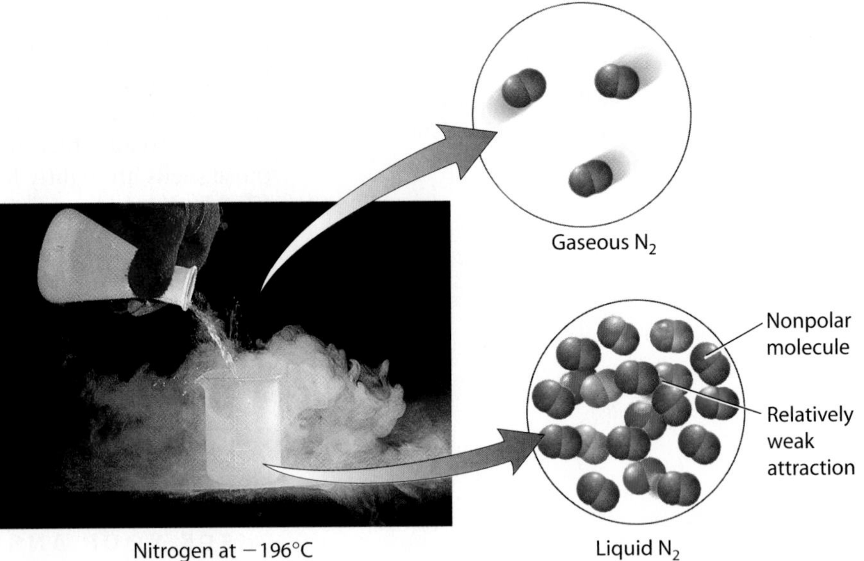

Figure 12.18 Nitrogen is a liquid at temperatures below its chilly boiling point of −196°C. Nitrogen molecules are not very attracted to one another because they are nonpolar. As a result, the small amount of heat energy available at −196°C is enough to separate them and allow them to enter the gaseous phase.

greater pull (because oxygen is more electronegative than carbon) on the bonding electrons. At the same time, however, the oxygen atom on the opposite side of the carbon pulls those electrons back to the carbon. The net result is an even distribution of bonding electrons around the whole molecule. So, dipoles that are of equal strength but pull in opposite directions in a molecule effectively cancel each other, with the result that the molecule as a whole is nonpolar.

Unifying Concept

The Electric Force

Section 7.1

Nonpolar molecules have only relatively weak attractions to other nonpolar molecules. The covalent bonds in a carbon dioxide molecule, for example, are many times stronger than any forces of attraction that might occur between two adjacent carbon dioxide molecules. This lack of attraction between nonpolar molecules explains the low boiling points of many nonpolar substances. Recall from Chapter 11 that boiling is a process wherein the molecules of a liquid separate from one another as they go into the gaseous phase. When there are only weak attractions between the molecules of a liquid, less heat energy is required to liberate the molecules from one another and allow them to enter the gaseous phase. This translates into a relatively low boiling point for the liquid, as, for instance, in the nitrogen, N_2, shown in Figure 12.18. The boiling points of hydrogen, H_2; oxygen, O_2; and carbon dioxide, CO_2; are also quite low for the same reason.

There are many instances in which the dipoles of different bonds in a molecule do not cancel each other. Perhaps the most relevant example is water, H_2O. Each hydrogen–oxygen covalent bond has a relatively large dipole because of the great electronegativity difference. Because of the bent shape of the molecule, however, the two dipoles, shown in Figure 12.19, do not cancel each other the way the C=O dipoles in Figure 12.19 do. Instead, the dipoles in the water molecule work together to give an overall dipole, shown in purple, for the molecule.

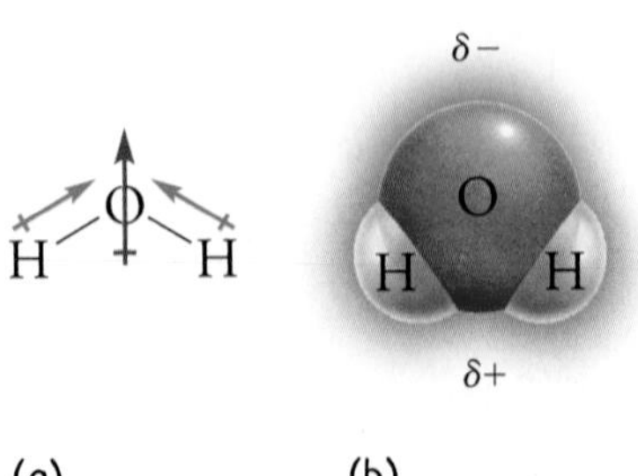

Figure 12.19 The individual dipoles in a water molecule add together to give a large overall dipole for the whole molecule. The region around the oxygen atom is therefore slightly negative, and the region around the two hydrogens is slightly positive.

Figure 12.20 illustrates how polar molecules electrically attract one another and as a result are relatively difficult to separate. In other words, polar molecules can be thought of as being "sticky," which is why it takes more energy to separate them and let them enter the gaseous phase. For this reason, substances composed of polar molecules typically have higher boiling points than substances composed of nonpolar molecules. Water, for example, boils at 100°C, whereas carbon dioxide boils at −79°C.

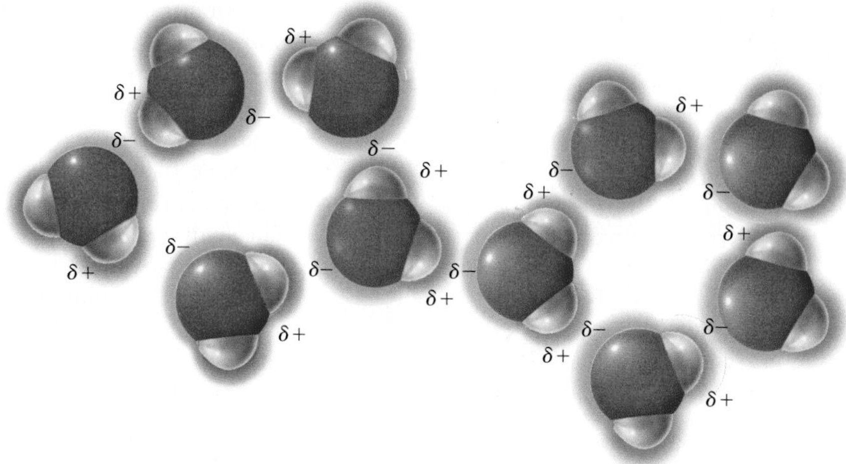

Figure 12.20 Water molecules attract one another because each contains a slightly positive (red) side and a slightly negative (blue) side. The molecules position themselves such that the positive side of one faces the negative side of a neighbor.

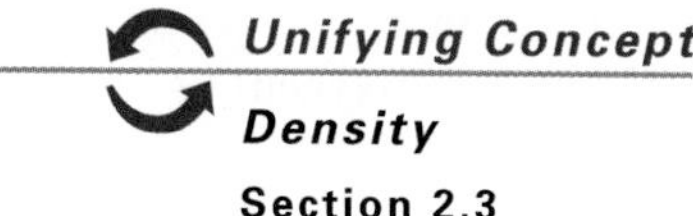

Unifying Concept

Density

Section 2.3

Because molecular "stickiness" can play a lead role in determining a substance's macroscopic properties, molecular polarity is a central concept of chemistry. Figure 12.21 shows an interesting example.

12.6 Interparticle Attractions

Polar Attraction; Intermolecular Forces

We can think of any pure substance as being made up of one type of submicroscopic particle. For an ionic compound, that particle is an ion; for a covalent compound, it is a molecule; and for an element, it is an atom. Table 12.1 lists three types of electrical attractions that can occur between these particles. The strength of even the strongest of these attractions is many times weaker than any chemical bond, however. Although particle-to-particle attractions are relatively weak compared to ionic, covalent, or metallic bonds, you can see their profound effect on the substances around you. Intermolecular forces, particularly the weakest ones not involving ions, are known as *Van der Waals forces*.

Unifying Concept

The Electric Force

Section 7.1

An attraction between an ion and the dipole of a polar molecule is called an *ion–dipole attraction*. Ion–dipole attractions between polar H_2O molecules and the ionic compound NaCl are responsible for the solubility of NaCl in water, as you will learn more about in Section 12.7.

Figure 12.21 Oil and water are difficult to mix, as is evident from this 1989 oil spill of the *Exxon Valdez* oil tanker in Alaska's Prince William Sound. It's not, however, that oil and water repel each other. Rather, water molecules are so attracted to themselves because of their polarity that they pull themselves together. The nonpolar oil molecules are thus excluded and left to themselves. Being less dense than water, oil floats on the surface, where it poses great danger to wildlife.

Table 12.1 Electrical Attractions Between Submicroscopic Particles

Attraction	Relative Strength
Ion–dipole	Strongest
Dipole–dipole	
Dipole–induced dipole	Weakest

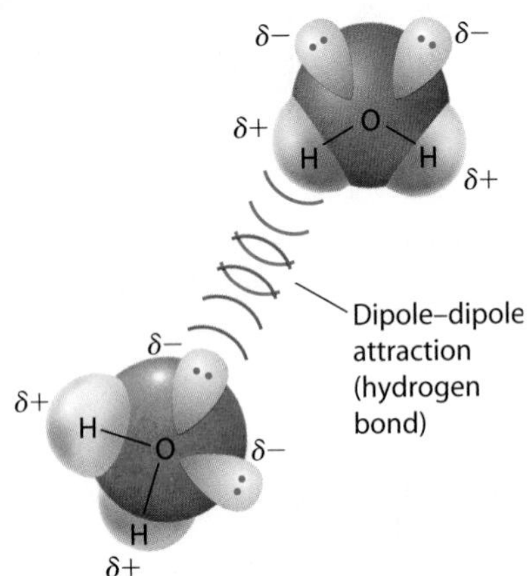

Figure 12.22 The dipole–dipole attraction between two water molecules is a hydrogen bond because it involves hydrogen atoms bonded to highly electronegative oxygen atoms.

An attraction between two polar molecules is called a *dipole–dipole attraction*. An unusually strong dipole–dipole attraction is the **hydrogen bond.** This attraction occurs between molecules that have a hydrogen atom covalently bonded to a highly electronegative atom, usually nitrogen, oxygen, or fluorine. Look at Figure 12.22 to see how hydrogen bonding works. The hydrogen side of a polar molecule (water in this example) has a positive charge because the more electronegative oxygen the hydrogen is bonded to tugs on the hydrogen's electron. This hydrogen is therefore electrically attracted to a pair of nonbonding electrons on the negatively charged atom of another molecule (in this case, another water molecule). This mutual attraction between hydrogen and the negatively charged atom of another molecule is a hydrogen bond. Many, if not most, of the molecules discussed in this textbook owe their "stickiness" to the hydrogen bonds they are able to form.

In many molecules, the electrons are distributed evenly, and so there is no dipole. The oxygen molecule, O_2, is an example. Such a nonpolar molecule can be induced to become a temporary dipole, however, when it is brought close to a water molecule or any other polar molecule, as Figure 12.23 illustrates. The slightly negative side of the water molecule pushes the electrons in the oxygen molecule away. Thus oxygen's electrons are pushed to the side that is farthest from the water molecule. The result is a temporary uneven distribution of electrons called an **induced dipole.** The resulting attraction between the permanent dipole (water) and the induced dipole (oxygen) is a *dipole–induced dipole attraction.*

CHECK YOURSELF

How does the electron distribution in an oxygen molecule change when the hydrogen side of a water molecule is nearby?

CHECK YOUR ANSWER

Because the hydrogen side of the water molecule is slightly positive, the electrons in the oxygen molecule are pulled *toward* the water molecule, inducing in the oxygen molecule a temporary dipole in which the larger side is nearest the water molecule (rather than as far away as possible as it was in Figure 12.23).

Remember, induced dipoles are only temporary. If the water molecule in Figure 12.23b were removed, the oxygen molecule would return to its normal, nonpolar state. As a consequence, dipole–induced dipole attractions are weaker than dipole–dipole attractions. Nevertheless, these productions produce many macroscopic effects. For example, dipole–induced dipole attractions occur between molecules of carbon dioxide, which are nonpolar, and water. It is these attractions that help keep carbonated beverages (which are mixtures of carbon dioxide in water) from losing their fizz too quickly after they've been opened. Dipole–induced dipole attractions are also responsible for holding plastic wrap to glass, as shown in Figure 12.24. These wraps are made of very long nonpolar

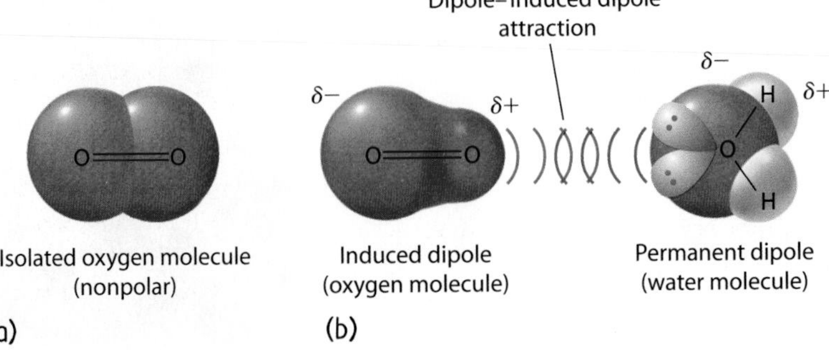

Figure 12.23 (a) An isolated oxygen molecule has no dipole; its electrons are distributed evenly. (b) An adjacent water molecule induces a redistribution of electrons in the oxygen molecule. (The slightly negative side of the oxygen molecule is shown larger than the slightly positive side because the slightly negative side contains more electrons.)

Figure 12.24 Temporary dipoles induced in the normally nonpolar molecules in plastic wrap makes it stick to glass.

molecules that are induced to have dipoles when placed in contact with glass, which is highly polar. The molecules of a nonpolar material, such as plastic wrap, can also induce dipoles among themselves. This explains how plastic wrap sticks not only to polar materials such as glass but also to itself.

CHECK YOURSELF

Distinguish between a dipole–dipole attraction and a dipole–induced dipole attraction.

CHECK YOUR ANSWER

The dipole–dipole attraction is stronger and involves two permanent dipoles. The dipole–induced dipole attraction is weaker and involves a permanent dipole and a temporary one.

INTEGRATED SCIENCE PHYSICS AND BIOLOGY

How Geckos Walk on Walls—The Adhesive Force

The forces of attraction between molecules of a single substance are called *cohesive forces*. For water, the cohesive forces are hydrogen bonds. Cohesive forces are what pull water molecules into spherical drops.

Van der Waals forces of attraction between molecules of two different substances are called *adhesive forces*. Adhesive forces are the "glue" that stick one material to a different material. For example, if you have ever looked at water in a test tube or other slender glass cylinder, you may have noticed that the water is curved upward so that it looks like it is climbing up the sides of the glass. This is an example of adhesion.

Recently, scientists have discovered another interesting case of adhesion. Geckos are probably nature's most spectacular climbers. A gecko can race up a glass wall and support its entire body weight with only a single toe! Gecko climbing has puzzled observers since Aristotle first took note of it in the fourth century B.C.

How do geckos do it? A gecko's feet are covered with billions of microscopic hairs called *spatulae*. Each spatulae is only 0.3-millionth of a meter wide; about 300 times thinner than a human hair. Adhesive forces between the spatulae and a climbing surface keep the gecko "glued" to the surface since individual molecules on the spatulae and surface attract one another. Because there is so much surface area on all the tiny spatulae, the total adhesive force is enough to keep the gecko clinging to walls and ceilings, seemingly defying gravity (Figure 12.25).

The sort of adhesive forces between the gecko's feet and a wall include dipole-induced dipole attractions. Also involved are even weaker *van der Waal* dispersion forces in which the random motion of electrons allows for brief moments of polarity within a normally nonpolar surface.

Research is currently underway to develop a synthetic, dry glue based on gecko adhesion. Velcro, watch out!

Figure 12.25 The gecko's feet are covered by billions of microscopic hairs called spatulae. Adhesive forces between the molecules in the spatulae and the surface the gecko is climbing on enable the gecko to "stick."

CHECK YOURSELF

How are the size and number of the spatulae an important part of the gecko's strategy?

CHECK YOUR ANSWER

Adhesive forces are weak. To support a gecko, there must be a significant amount of surface area contact between the surface and the gecko's feet. A large number of tiny hairs increases the surface area of contact between gecko and climbing surface.

INTEGRATED SCIENCE BIOLOGY AND EARTH SCIENCE

Mixtures

A *pure substance* consists of only one type of atom, molecule, or ion. Methane, sodium chloride, and hydrogen are examples of pure substances. A **mixture**, on the other hand, is a collection of two or more pure substances that are physically mixed and in which each of the pure substances retains its properties. A mixture cannot be represented by a chemical formula because the proportions of the constituent substances in a mixture can vary.

Stainless steel, for example, is a mixture of the elements iron, chromium, nickel, and carbon. Seltzer water is a mixture of the liquid compound water and the gaseous compound carbon dioxide. Our atmosphere, as Figure 12.26 illustrates, is a mixture of the elements nitrogen, oxygen, and argon plus small amounts of such compounds as carbon dioxide and water vapor.

Tap water is a mixture containing mostly water but also many other compounds. Depending on your location, your water may contain compounds of calcium, magnesium, fluorine, iron, and potassium; chlorine disinfectants; trace amounts of compounds of lead, mercury, and cadmium; organic compounds; and dissolved oxygen, nitrogen, and carbon dioxide. Although it is surely important to minimize any toxic components in your drinking water, it is unnecessary, undesirable, and impossible to remove all other substances from it. Some of the dissolved solids and gases give water its characteristic taste, and many of them promote human health. For instance, chlorine destroys bacteria, and as much as 10 percent of our daily requirement for iron, potassium, calcium, and magnesium is obtained from drinking water.

Mixtures can be classified as heterogeneous or homogeneous. A *homogeneous mixture* is one in which the substances are evenly distributed. Tap water is a homogeneous mixture for example; a sip from the bottom of a glass or water tastes the same as your first sip. A heterogeneous mixture contains substances that are not evenly distributed, so different regions of the mixture have different properties. A bowl of cereal is an obvious heterogeneous mixture, but some substances that appear uniform to the naked eye are in fact heterogeneous mixtures (Figure 12.27).

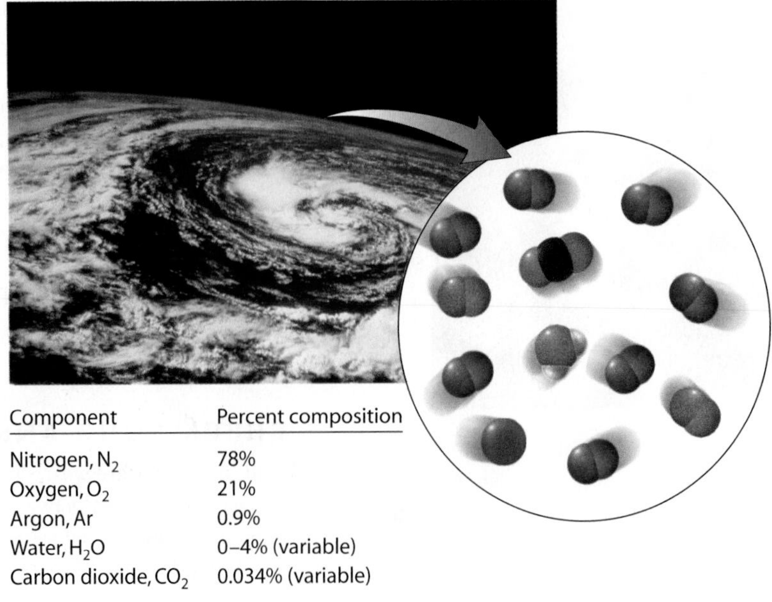

Component	Percent composition
Nitrogen, N_2	78%
Oxygen, O_2	21%
Argon, Ar	0.9%
Water, H_2O	0–4% (variable)
Carbon dioxide, CO_2	0.034% (variable)

Figure 12.26 Earth's atmosphere is a mixture of gaseous elements and compounds. Some of them are shown here. You will find out much more about the composition of the atmosphere and its effect on the weather in Chapter 25.

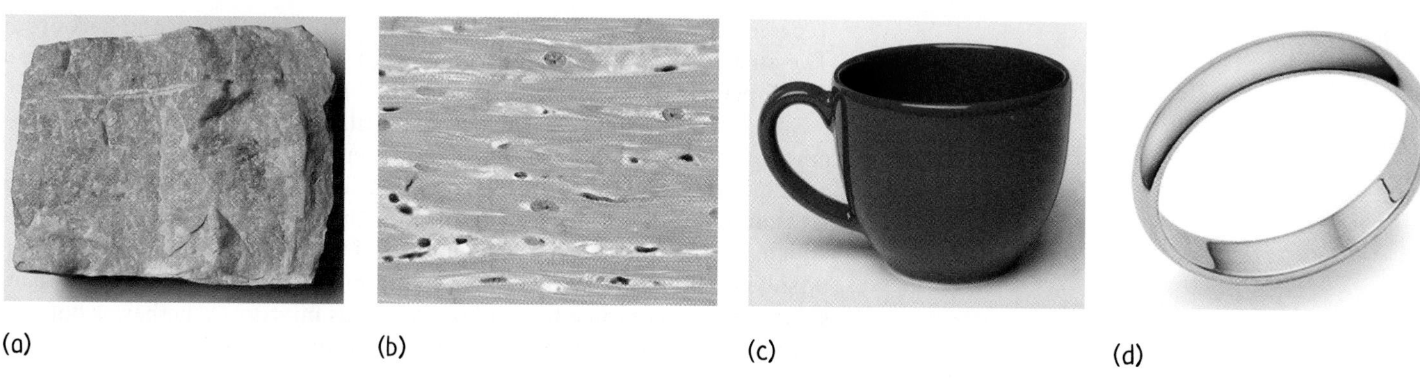

Figure 12.27 (a) Marble, (b) a muscle cell, and (c) most forms of plastic are heterogeneous mixtures. (d) Rose gold is a homogeneous mixture. It is not always possible to spot a heterogeneous mixture at the macroscopic level.

CHECK YOURSELF

Identify each of the following as a heterogeneous mixture; a homogeneous mixture; a pure substance:

(a) a green salad **(b)** chlorine gas **(c)** a cookie **(d)** sugar-water **(e)** nerve cell

CHECK YOUR ANSWERS

Heterogeneous mixture: (a) green salad, (c) cookie, (e) nerve cell
Homogeneous mixture: (d) sugar-water
Pure substance: (b) chlorine gas

12.7 Solutions

A homogeneous mixture consisting of ions or molecules is called a **solution**. Because a solution is a mixture at the submicroscopic level of atoms and molecules, a solution the most finely mixed form of homogeneous mixture. Sugar in water is a solution in the liquid phase (Figure 12.28). But solutions are not always liquids. They can also be solid or gaseous. Metal alloys are solid solutions, mixtures of different metallic elements. The alloy brass is a solid solution of copper and zinc, for instance, and the alloy stainless steel is a solid solution of iron, chromium, nickel, and carbon. Rose gold and white gold are other examples of solutions in the solid state. An example of a gaseous solution is the air we inhale. By volume, this solution is 78 percent nitrogen gas, 21 percent oxygen gas, and

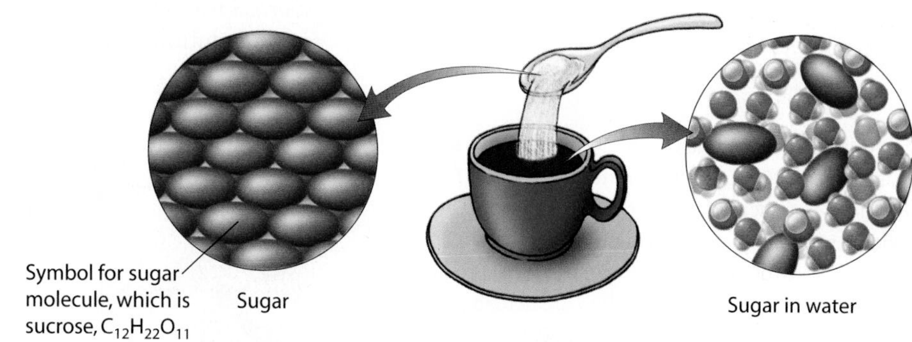

Figure 12.28 Table sugar is a compound consisting of only sucrose molecules. Once these molecules are mixed into hot tea, they become interspersed among the water and tea molecules and form a sugar-tea-water mixture. No new compounds are formed, and so this is an example of a physical change.

INSIGHT All gases are generally infinitely soluble in other gases because they can mix together in just about any proportion.

1 percent other gaseous materials, including water vapor and carbon dioxide. The air we *exhale* is a gaseous solution of 75 percent nitrogen, 14 percent oxygen, 5 percent carbon dioxide, and around 6 percent water vapor.

In describing solutions, it is usual to call the component present in the largest amount the **solvent** and the other component(s) the **solute(s)**. For example, when a teaspoon of table sugar is mixed with 1 liter of water, we identify the sugar as the solute and the water as the solvent. An *aqueous* solution is any solution in which water is the solvent. The process of a solute mixing in a solvent is called **dissolving.** To make a solution, a solute must *dissolve* in a solvent; that is, the solute and solvent must form a homogeneous mixture. Whether or not one material dissolves in another is a function of electrical attractions.

CHECK YOURSELF

What is the solvent in the gaseous solution we call air?

CHECK YOUR ANSWER

Nitrogen is the solvent because it is the component present in the greatest quantity.

There is typically a limit to how much of a given solute can dissolve in a given solvent, as Figure 12.29 illustrates. When you add table sugar to a glass of water, for example, the sugar rapidly dissolves. As you continue to add sugar, however, there comes a point when it no longer dissolves. Instead, it collects at the bottom of the glass, even after stirring. At this point, the water is *saturated* with sugar, meaning the water cannot accept any more sugar. When this happens, we have what is called a **saturated** solution, defined as one in which no more solute can dissolve. A solution that has not reached the limit of solute that will dissolve is called an **unsaturated** solution.

However, when the molecule-to-molecule attractions among solute molecules are comparable to the molecule-to-molecule attractions among solvent molecules, the result can be that there is no practical point of saturation. For example, the hydrogen bonds among water molecules are about as strong as those between molecules. These two liquids can therefore mix together quite well in just about any proportion. A solute that has no practical point of saturation in a given solvent is said to be infinitely soluble in that solvent. Ethanol, for example, is infinitely soluble in water.

The quantity of solute dissolved in a solution is described in mathematical terms by the solution's **concentration**, which is the amount of solute dissolved per amount of solution:

$$\text{concentration of solution} = \frac{\text{amount of solute}}{\text{amount of solution}}$$

For example, a sucrose–water solution may have a concentration of 1 gram of sucrose for every liter of solution. This can be compared with concentrations of other solutions. A sucrose–water solution containing 2 grams of sucrose per

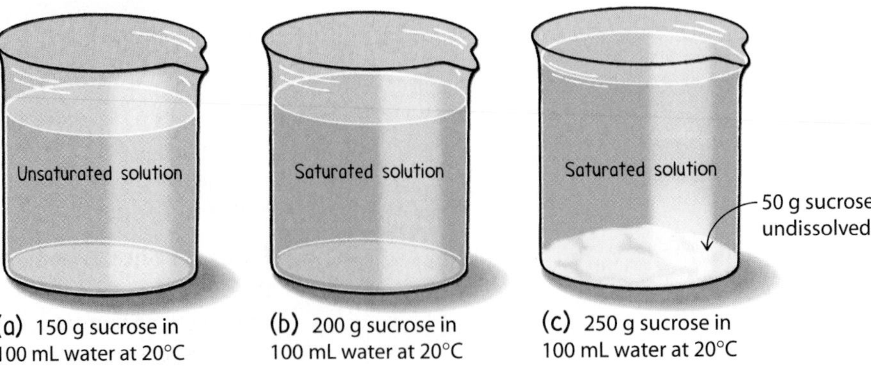

Figure 12.29 A maximum of 200 grams of sucrose dissolves in 100 milliliters of water at 20°C. (a) Mixing 150 grams of sucrose in 100 milliliters of water at 20°C produces an unsaturated solution. (b) Mixing 200 grams of sucrose in 100 milliliters of water at 20°C produces a saturated solution. (c) If 250 grams of sucrose is mixed with 100 milliliters of water at 20°C, 50 grams of sucrose remain undissolved. (As we will discuss later, the concentration of a saturated solution varies with temperature.)

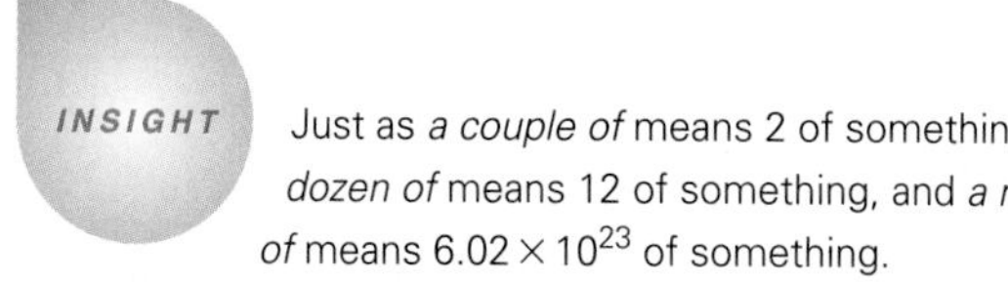

Just as *a couple of* means 2 of something, *a dozen of* means 12 of something, and *a mole of* means 6.02×10^{23} of something.

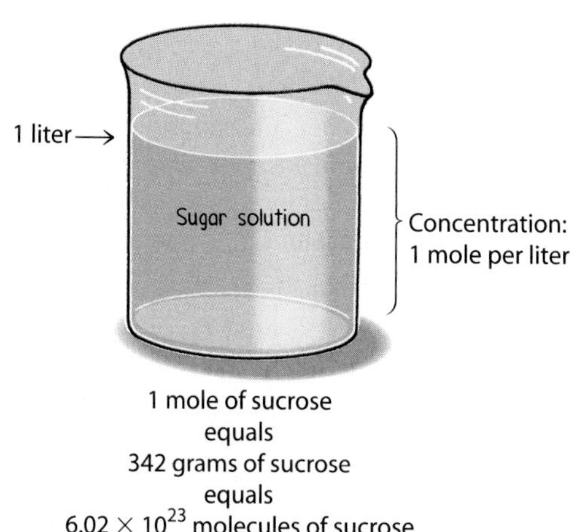

Figure 12.30 An aqueous solution of sucrose that has a concentration of 1 mole of sucrose per liter of solution contains 6.02×10^{23} sucrose molecules (342 grams) in every liter of solution.

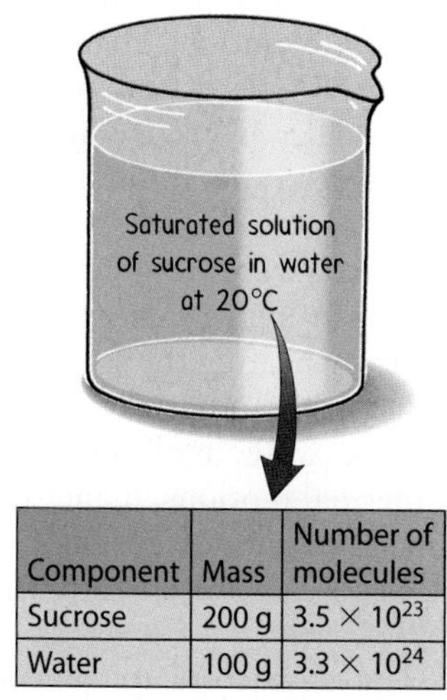

Component	Mass	Number of molecules
Sucrose	200 g	3.5×10^{23}
Water	100 g	3.3×10^{24}

Figure 12.31 Although 200 grams of sucrose is twice as massive as 100 grams of water, there are about 10 times as many water molecules in 100 grams of water as there are sucrose molecules in 200 grams of sucrose. How can this be? Each water molecule is about 20 times less massive (and smaller) than each sucrose molecule, which means that about 10 times as many water molecules can fit within half the mass.

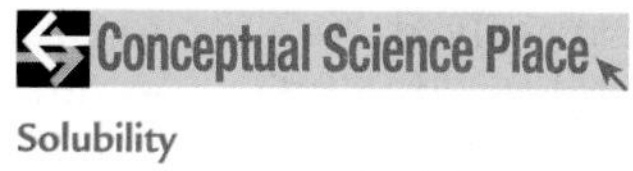

liter of solution, for example, is more *concentrated*, and one containing only 0.5 gram of sucrose per liter of solution is less concentrated, or more *dilute*.

Chemists are often more interested in the number of solute particles in a solution rather than the number of grams of solute. Submicroscopic particles, however, are so very small that the number of them in any observable sample is incredibly large. To get around having to use awkwardly large numbers, scientists use a unit called the mole. One **mole** of any type of particle is, by definition, 6.02×10^{23} particles (this superlarge number is about 602 billion trillion):

$$1 \text{ mole} = 6.02 \times 10^{23} \text{ particles}$$
$$= 602{,}000{,}000{,}000{,}000{,}000{,}000{,}000 \text{ particles}$$

One mole of pennies, for example, is 6.02×10^{23} pennies, 1 mole of marbles is 6.02×10^{23} marbles, and 1 mole of sucrose molecules is 6.02×10^{23} sucrose molecules. The number 6.02×10^{23} is known as **Avogadro's number**, after the nineteenth-century Italian scientist and lawyer who first proposed this unit.

A stack containing 1 mole of pennies would reach a height of 860 quadrillion kilometers, which is roughly equal to the diameter of our Milky Way galaxy. Sucrose molecules are so small, however, that there are 6.02×10^{23} of them in only 342 grams of sucrose, which is about a cupful. Thus because 342 grams of sucrose contains 6.02×10^{23} molecules of sucrose, we say that 342 grams of sucrose contains 1 mole of sucrose. As Figure 12.30 shows, therefore, an aqueous solution that has a concentration of 342 grams of sucrose per liter of solution also has a concentration of 6.02×10^{23} sucrose molecules per liter of solution or, by definition, a concentration of 1 mole of sucrose per liter of solution.

A common unit of concentration used by chemists is **molarity**, which is the solution's concentration expressed in moles of solute per liter of solution:

$$\text{molarity} = \frac{\text{number of moles of solute}}{\text{liters of solution}}$$

A solution that contains 1 mole of solute per liter of solution has a concentration of 1 *molar*, which is often abbreviated 1 *M*. A more concentrated, 2-molar (2 *M*) solution contains 2 moles of solute per liter of solution.

The difference between referring to the number of molecules of solute and referring to the number of grams of solute can be illustrated by the following question. A saturated aqueous solution of sucrose contains 200 grams of sucrose and 100 grams of water. Which is the solvent: sucrose or water?

As shown in Figure 12.31, there are 3.5×10^{23} molecules of sucrose in 200 grams of sucrose but almost 10 times as many molecules of water in 100 grams of water—3.3×10^{24} molecules. As defined earlier, the solvent is the component present in the largest amount, but what do we mean by *amount*? If amount means number of molecules, then water is the solvent. If amount means mass, then sucrose is the solvent. So, the answer depends on how you look at it. From a chemist's point of view, *amount* typically means the number of molecules, and so water is the solvent in this case.

Another unit of concentration is **parts per million (ppm)**. A concentration of 1 ppm means that there is 1 particle of substance for every 999,999 other particles. A drop of orange juice in a 40-gallon drum of water would have a concentration of 1 ppm, for example. This unit is used where low concentrations are significant, for example, when quantifying trace elements in the Earth's crust or pollutants in the environment.

12.8 Solubility

The **solubility** of a solute is its ability to dissolve in a solvent. As you might expect, this ability depends in great part on the submicroscopic attractions between solute particles and solvent particles. If a solute has any appreciable solubility in a solvent, then that solute is said to be **soluble** in that solvent.

MATH CONNECTION

Concentration Calculations

Use what you have learned about concentration to solve the following problems.

Problems

1. How much sucrose, in moles, is there in 0.5 liter of a 2-molar solution? How many molecules of sucrose is this?
2. How much sucrose, in grams, is there in 3 liters of an aqueous solution that has a concentration of 2 grams of sucrose per liter of solution?

Solutions

1. First you need to understand that 2-molar means 2 moles of sucrose per liter of solution. Then you should multiply solution concentration by amount of solution to obtain amount of solute:

$$\left(\frac{2\text{ moles}}{\text{L}}\right)(0.5\text{ L}) = 1\text{ mole}$$

which is the same as 6.02×10^{23} molecules.

2. Start with the defintion of concentration: concentration of solution = amount of solute/amount of solution. Since the question asks for amount of solute, rearrange the equation as: amount of solute = concentration of solution × amount of solution:

$$\frac{\text{amount of solute}}{1\text{ L}} = 2\text{ g} \times (3\text{ L}) = 6\text{ g}$$

Solubility depends on attractions of solute particles for one another and attractions of solvent particles for one another. For example, there are many polar hydrogen–oxygen bonds in a sucrose molecule. Sucrose molecules, therefore, can form multiple hydrogen bonds with one another. These hydrogen bonds are strong enough to make sucrose a solid at room temperature and give it a relatively high melting point of 185°C. In order for sucrose to dissolve in water, the water molecules must first pull sucrose molecules away from one another. This puts a limit on the amount of sucrose that can dissolve in water—eventually a point is reached where there are not enough water molecules to separate the sucrose molecules from one another. This is the point of saturation, and any additional sucrose added to the solution does not dissolve.

You probably know from experience that water-soluble solids usually dissolve better in hot water than in cold water. A highly concentrated solution of sucrose in water, for example, can be made by heating the solution almost to the boiling point. This is how syrups and hard candy are made. Solubility increases with increasing temperature because water molecules have greater kinetic energy. Therefore, in their random thermal motions, higher-temperature molecules are able to collide with the solid solute more vigorously. The vigorous collisions facilitate the disruption of electrical particle-to-particle attractions in the solid.

Although the solubilities of some solid solutes—sucrose, to name just one example—are greatly affected by temperature changes, the solubilities of other solid solutes, such as sodium chloride, are only mildly affected, as Figure 12.32 shows. This difference has to do with a number of factors, including the strength of the chemical bonds in the solute molecules and the way those molecules are packed together.

When a solution saturated at a high temperature is allowed to cool, some of the solute usually comes out of solution and forms what is called a **precipitate**. When this happens, the solute is said to have *precipitated* from the solution. For example, at 100°C, the solubility of sodium nitrate, $NaNO_3$, in water is 165 grams per 100 milliliters of water. As we cool this solution, the solubility of $NaNO_3$ decreases as shown in Figure 12.33, and this change in solubility causes some of the dissolved $NaNO_3$ to precipitate (come out of solution). At 20°C, the solubility of $NaNO_3$ is only 87 grams per 100 milliliters of water. So if we cool the 100°C solution to 20°C, 78 grams (165 grams − 87 grams) precipitates.

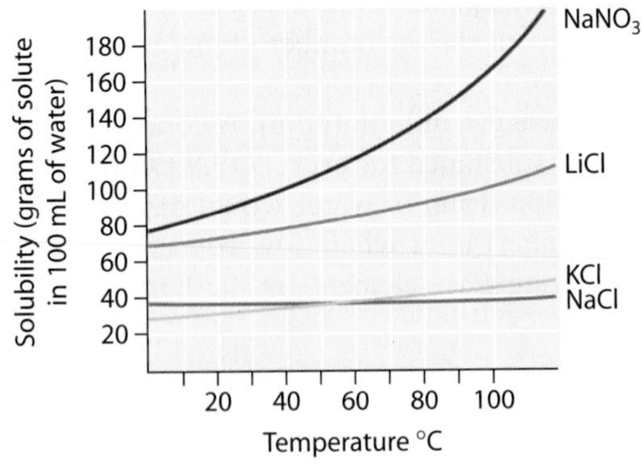

Figure 12.32 The solubility of many water-soluble solids increases with temperature, whereas the solubility of others is only very slightly affected by temperature.

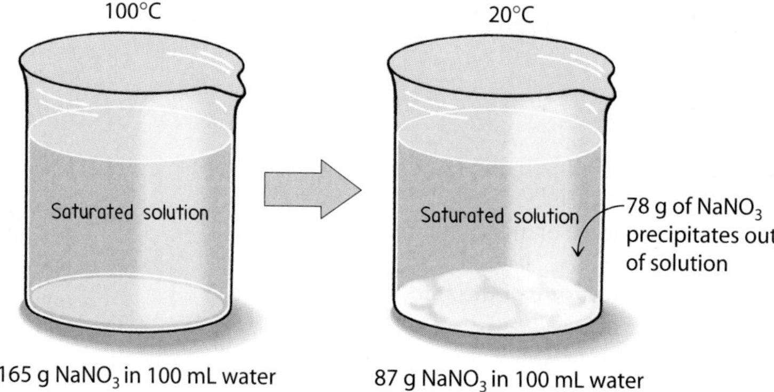

Figure 12.33 The solubility of sodium nitrate is 165 grams per 100 milliliters of water at 100°C but only 87 grams per 100 milliliters at 20°C. Cooling a 100°C saturated solution of $NaNO_3$ to 20°C causes 78 grams of the solute to precipitate.

INTEGRATED SCIENCE BIOLOGY

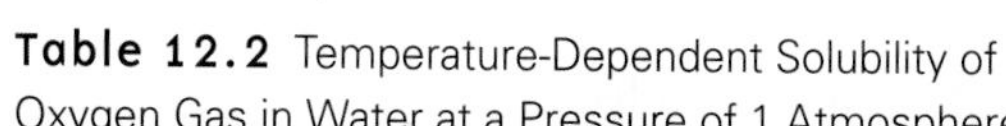

Fish Depend on Dissolved Oxygen

In contrast to the solubilities of most solids, the solubilities of gases in liquids *decrease* with increasing temperature, as Table 12.2 shows. Note, for example, that oxygen gas is almost twice as soluble near 0°C as it is around 30°C. This is true because, with an increase in temperature, the solvent molecules have more kinetic energy. This makes it more difficult for a gaseous solute to stay in solution because the solute molecules are literally being kicked out by the high-energy solvent molecules.

Table 12.2 Temperature-Dependent Solubility of Oxygen Gas in Water at a Pressure of 1 Atmosphere

Temperature (°C)	O_2 Solubility (g O_2/L H_2O)
0	0.0141
10	0.0109
20	0.0092
25	0.0083
30	0.0077
35	0.0070
40	0.0065

All animals require oxygen to live. Aquatic animals cannot obtain the oxygen they need from the water molecules they are surrounded by because the oxygen atom in H_2O is strongly bonded to the hydrogen atoms. So fish and other aquatic organisms must have an adequate supply of oxygen gas dissolved in the water they live in to survive. Some of this oxygen comes from the air above the water surface. Oxygen gas is also supplied to aquatic environments by photosynthesizing aquatic plants.

The minimum concentration of dissolved oxygen needed to support aquatic life varies from organism to organism. Fish, for example, require at least 0.004 g of dissolved oxygen per 1000 g of solution. This is equivalent to a concentration of 4 ppm. However, the internal body temperature of most fish rises and falls with the temperature of their surroundings. In warm water, the body temperature and metabolic activity of fish is higher than it is in cold water. As a result, in warm water they require more oxygen. In warm water, the solublity of oxygen decreases—yet this is precisely when fish need oxygen most. For this reason, some streams experience large fish kills during long stretches of hot summer weather.

Figure 12.34 Fish depend on dissolved oxygen to breathe.

CHECK YOURSELF

Thermal pollution is the discharge of excess heat into rivers, streams, and lakes by industry. Do you think thermal pollution can lead to fish kills? If so, how?

CHECK YOUR ANSWER

Since the solubility of gases, including oxygen, O_2, generally decreases with increasing temperature, overly warm water is threatening to fish regardless of the source of excess heat. So yes, thermal pollution can lead to fish kills.

CHAPTER 12 Review

Summary of Terms

Avogadro's number The number of particles (atoms, ions, or molecules) in one mole; 6.02×10^{23} particles.

Chemical bond The electrical force of attraction holding two atoms together.

Concentration A quantitative measure of the amount of solute in a solution.

Covalent bond A chemical bond in which atoms are held together by their mutual attraction for two electrons they share.

Dipole A separation of charge that occurs in a chemical bond because of differences in the electronegativities of the bonded atoms.

Dissolving The process of mixing a solute in a solvent.

Electron-dot structure A shorthand notation of the shell model of the atom in which valence electrons are shown around an atomic symbol.

Electronegativity The ability of an atom to attract a bonding pair of electrons to itself when bonded to another atom.

Hydrogen bond A strong dipole–dipole attraction between a slightly positive hydrogen atom on one molecule and a pair of nonbonding electrons on another molecule.

Induced dipole A dipole temporarily created in an otherwise nonpolar molecule, induced by a neighboring charge.

Ion An electrically charged particle created when an atom either loses or gains one or more electrons.

Ionic bond A chemical bond in which an attractive electric force holds ions of opposite charge together.

Ionic compound Any chemical compound containing ions.

Metallic bond A chemical bond in which the metal ions in a piece of solid metal are held together by their attraction to a "fluid" of electrons in the metal.

Mixture A combination of two or more substances in which each substance retains its properties.

Molarity A unit of concentration equal to the number of moles of a solute per liter of solution.

Mole 6.02×10^{23} of anything.

Molecule A group of atoms held tightly together by covalent bonds.

Nonpolar bond A chemical bond that has no dipole.

Octet rule A rule stating that atoms gain or lose electrons to acquire the outer shell electron configuration of a noble gas, usually neon or argon, which each have eight electrons in their outermost shell.

Polar bond A chemical bond that has a dipole.

Polyatomic ion Molecules that carry a net electric charge.

Precipitate A solute that has come out of solution.

Saturated A solution containing the maximum amount of solute that will dissolve.

Solubility The ability of a solute to dissolve in a given solvent.

Soluble Capable of dissolving to an appreciable extent in a given solvent.

Solute Any component in a solution that is not the solvent.

Solution A homogeneous mixture in which all components are dissolved in the same phase.

Solvent The component in a solution present in the largest amount.

Unsaturated Said of a solution in which more solute can dissolve.

Valence electron An electron that is located in the outermost occupied shell in an atom and can participate in chemical bonding.

Valence shell The outermost occupied shell of an atom.

Review Questions

12.1 Electron Shells and Chemical Bonding

1. How many shells are needed to account for the seven periods of the periodic table?
2. How many shells are completely filled in an argon atom, Ar (atomic number 18)?
3. Which electrons are represented by an electron-dot structure?

12.2 The Ionic Bond

4. How does an ion differ from an atom?
5. To become a negative ion, does an atom lose or gain electrons?
6. Which elements tend to form ionic bonds?
7. What is the electric charge on the calcium ion in the compound calcium chloride, $CaCl_2$?

12.3 The Metallic Bond

8. What type of bond holds sodium atoms together?
9. Why do metals conduct electricity so well?

12.4 The Covalent Bond

10. Which elements tend to form covalent bonds?
11. What force holds two atoms together in a covalent bond?
12. How many electrons are shared in a double covalent bond?

12.5 Polar Bonds and Polar Molecules

13. What is a dipole?
14. Which element of the periodic table has the greatest electronegativity? Which has the smallest?
15. How is a polar covalent bond similar to an ionic bond?
16. How can a molecule be nonpolar when it consists of atoms that have different electronegativities?

12.6 Interparticle Attractions

17. What is the primary difference between a chemical bond and an attraction between two molecules?
18. Why are water molecules attracted to sodium chloride?
19. What is a hydrogen bond?

12.7 Solutions

20. Distinguish between a solute and a solvent.
21. What does it mean to say a solution is concentrated?
22. Distinguish between a saturated solution and an unsaturated solution.

12.8 Solubility

23. What kind of electrical attraction is responsible for oxygen's ability to dissolve in water?
24. What is the relationship between a precipitate and a solute?

Multiple Choice Questions

Choose the BEST answer to the following.

1. Magnesium ions carry a 2^+ charge, and chloride ions carry a 1^- charge. What is the chemical formula for the ionic compound magnesium chloride?
   (a) $MgCl$
   (b) Mg_2Cl
   (c) $MgCl_2$
   (d) Mg_2Cl_2

2. The image shown represents which kind of matter?

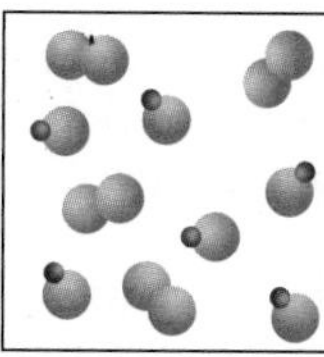

   (a) an element
   (b) a mixture
   (c) a compound
   (d) none of the above
   (e) all of the above
3. Which would you expect to have a higher melting point: sodium chloride (NaCl), or aluminum oxide (Al_2O_3)?
   (a) The aluminum oxide has a higher melting point because it is a larger molecule and has a greater number of molecular interactions.
   (b) The sodium chloride has a higher melting point because it is a solid at room temperature.
   (c) The aluminum oxide has a higher melting point because of the greater charges of the ions, and hence the greater force of attractions between them.
   (d) The aluminum oxide has a higher melting point because of the covalent bonds within the molecule.
4. Distinguish between a metal and a metal-containing compound.
   (a) There is no difference between the two.
   (b) Only one of these contains ionic bonds.
   (c) Only one of these contains covalent bonds.
   (d) Only one of these occurs naturally.
5. A hydrogen atom does not form more than one covalent bond because it
   (a) has only one shell of electrons.
   (b) has only one electron to share.
   (c) loses its valence electron so readily.
   (d) has such a strong electronegativity.
6. What drives an atom to form a covalent bond: its nuclear charge or the need to have a filled outer shell?
   (a) its nuclear charge
   (b) its need to have a filled outer shell
   (c) both
   (d) neither
7. Classify the following bonds as ionic, covalent, or neither (O, atomic number 8; F, atomic number 9; Na, atomic number 11; Cl, atomic number 17; U, atomic number 92).
   O with F   Ca with Cl   Na with Na
   (a) covalent, ionic, covalent
   (b) ionic, covalent, neither
   (c) neither, ionic, covalent
   (d) covalent, ionic, neither
8. Which of the following molecules is the most polar?
   (a) HCl
   (b) BrF
   (c) CO
   (d) Br_2
9. How can you tell whether a sugar solution is saturated or not?
   (a) Add more sugar; if it dissolves, it is saturated.
   (b) There will be a precipitate if the water is heated.
   (c) Cool the solution to see if there is a precipitate.
   (d) As long as there are more water molecules than sugar molecules, there is a saturated solution.
10. Hydrogen chloride (HCl) is a gas at room temperature. Would you expect this material to be very soluble or not very soluble in water?
   (a) HCl is very soluble in water by virtue of the dipole–dipole attractions occurring between the HCl and H_2O molecules.
   (b) It is not very soluble because it is a gas, and all gases have very low solubility in water at room temperature.
   (c) HCl is very soluble in water because it is such a small molecule, there is little electrical attraction to other HCl molecules.
   (d) It is not very soluble because as a gas with low density, it floats to the surface of the water and then into the surrounding atmosphere.

INTEGRATED SCIENCE CONCEPTS

Physics and Biology—How Geckos Walk on Walls—The Adhesive Force

1. What is the difference between adhesive and cohesive forces? Give an example of each.
2. Describe what is happening at the molecular level when a gecko clings to a wall.

Biology and Earth Science—Mixtures

1. Would you expect a skin cell to be a heterogeneous or homogeneous mixture? Would you expect Earth's atmosphere to be a heterogeneous or homogeneous mixture?
2. Are the particles making up a mixture held together by chemical bonds? By intermolecular forces?

Biology—Fish Depend on Dissolved Oxygen

1. What effect does temperature have on the solubility of a gas solute in a liquid solvent?
2. How do fish obtain the oxygen they need to breathe?
3. Why do fish sometimes suffocate when the water they are living in warms up in hot summer months?

Active Explorations

1. View crystals of table salt with a magnifying glass or, better yet, a microscope if one is available. If you do have a microscope, crush the crystals with a spoon and examine the resulting powder. Purchase some sodium-free salt, which is potassium chloride, KCl, and examine these ionic crystals, both intact and crushed. Sodium chloride and potassium chloride both form cubic crystals, but there are significant differences. What are they?
2. Black ink contains pigments of many different colors. Acting together, these pigments absorb all the frequencies of visible light. Because no light is reflected, the ink appears black. We can use electrical attractions to separate the components of black ink with a technique called *paper chromatography*.

   Place a concentrated dot of water-soluble ink (black felt-tip pens work well) at the center of a piece of porous paper, such as a napkin. Carefully place one drop of water on top of the dot, and watch the ink spread radially with the solvent. Because the different components of the ink have different affinities for the solvent, they travel with the solvent at different rates. Just after the drop of water is completely absorbed, add a second drop at the same place you put the first one, then a third, and so on until the ink components have separated to your satisfaction.

Exercises

1. What happens when hydrogen's electron gets close to the valence shell of a fluorine atom?
2. An atom loses an electron to another atom. Is this an example of a physical or a chemical change?
3. Why doesn't the neon atom tend to gain any electrons? Why doesn't it tend to lose any electrons?
4. Which should be larger, the potassium atom, K, or the potassium ion, K^+?
5. Which should have a higher melting point, sodium chloride, NaCl, or aluminum oxide, Al_2O_3?
6. Two fluorine atoms join together to form a covalent bond. Why don't two potassium atoms do the same thing?
7. Why doesn't a hydrogen atom form more than one covalent bond?
8. Is there an abrupt or gradual change between ionic and covalent bonds? Explain.
9. Atoms of metallic elements can form ionic bonds, but they are not very good at forming covalent bonds. Why?
10. What is the source of an atom's electronegativity?
11. Which molecule is most polar:
    (a) S=C=S (b) O=C=O (c) O=C=S
12. Which is more polar, a sulfur–bromine (S–Br) bond or a selenium–chlorine (Se—Cl) bond?
13. Water, H_2O, and methane, CH_4, have about the same mass and differ by only one type of atom. Why is the boiling point of water so much higher than that of methane?
14. Three kids sitting equally apart around a table are sharing jelly beans. One of the kids, however, tends only to take jelly beans and only rarely gives one away. If each jelly bean represents an electron, who ends up being slightly negative? Who ends up being slightly positive? Is the negative kid just as negative as one of the positive kids is positive? Would you describe this as a polar or nonpolar situation? How about if all three kids were equally greedy?
15. Which is stronger: the covalent bond that holds atoms together within a molecule or the dipole–dipole attraction between two neighboring molecules?
16. Why is a water molecule more attracted to a calcium ion than a sodium ion?
17. The charges within sodium chloride are all balanced—for every positive sodium ion, there is a corresponding negative chloride ion. Since its charges are balanced, how can sodium chloride be attracted to water, and vice versa?
18. The volume of many liquid solvents expands with increasing temperature. What happens to the concentration of a solution made with such a solvent as the temperature of the solution is increased?
19. Suggest why sodium chloride, NaCl, is insoluble in gasoline. Consider the electrical attractions.
20. Would you expect to find more dissolved oxygen in ocean water around Alaska or in ocean water close to the equator? Why?
21. Why are the melting points of most ionic compounds far higher than the melting points of most covalent compounds?

Problems

1. Show that there are 2.5 grams of sucrose in 5 liters of an aqueous solution of sucrose having a concentration of 0.5 gram of sucrose per liter of solution.
2. Show that 45 grams of sodium chloride is needed to make 15 L of a solution having a concentration of 3.0 grams of sodium chloride per liter of solution.
3. If water is added to 1 mole of sodium chloride in a flask until the volume of the solution is 1 liter, show that the molarity of this solution is 1 M. Show that a 4-M solution results when water is added to 2 moles of sodium chloride to make 0.5 liter of solution.
4. Show that one mole of sugar equals 342 grams.

Conceptual Science Place

CHAPTER 12 ONLINE RESOURCES

Tutorials • Covalent Bonds, Bonds and Bond Polarity, Polar Attraction, Intermolecular Forces, Solubility • ***Quiz*** • ***Flashcards*** • ***Links***

CHAPTER 13

Chemical Reactions

The heat of a lightning bolt causes many chemical reactions in the atmosphere, including one in which nitrogen and oxygen react to form nitrogen monoxide, NO. The NO then reacts with atmospheric oxygen and water vapor to form nitric acid, HNO_3, and nitrous acid, HNO_2. These acids are carried by rain into the ground, where they form ions, which plants need for growth. Chemical reactions such as these make up the nitrogen cycle, one of Earth's essential systems.

CHAPTER OUTLINE

Countless chemical reactions take place continually in the outside environment, such as the reactions of the nitrogen cycle. Millions more occur within our bodies, such as the chemical reactions that help us digest our food. What are the most common types of reactions? How do you write and "read" a chemical equation? What is an acid–base reaction? What is pH and how do we measure it? How are oxidation–reduction reactions different from acid–base reactions, and how are they similar? What do our bodies have in common with the burning of a campfire or the rusting of old farm equipment? How do chemicals store and release energy? We will address such questions in this chapter as we investigate the dynamic submicroscopic world of reacting chemicals.

13.1 Chemical Reactions and Chemical Equations

During a **chemical reaction**, one or more new compounds are formed as a result of the rearrangement of atoms. To represent a chemical reaction, we can write a **chemical equation**, which shows the substances about to react, called **reactants**, to the left of an arrow that points to the newly formed substances, called **products**:

$$\text{reactants} \rightarrow \text{products}$$

Typically, reactants and products are represented by their atomic or molecular formulas, but molecular structures or simple names may be used instead. Phases are also often shown: (s) for solid, (l) for liquid, and (g) for gas. Compounds dissolved in water are designated (aq) for aqueous. Lastly, numbers are placed in front of the reactants or products to show the ratio in which they either combine or form. These numbers are called **coefficients**, and they represent numbers of individual atoms and molecules. For instance, two hydrogen gas molecules, H_2, react with one oxygen gas molecule, O_2, to produce two molecules of water, H_2O, in the gaseous phase:

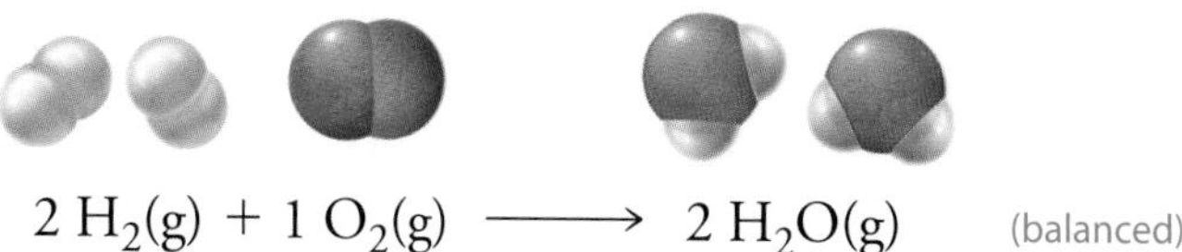

$$2\,H_2(g) + 1\,O_2(g) \longrightarrow 2\,H_2O(g) \quad \text{(balanced)}$$

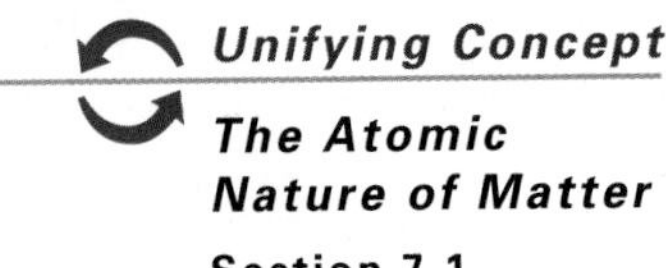

One of the great discoveries of chemistry is that matter is neither created nor destroyed during a chemical reaction. The atoms present at the beginning of a reaction merely rearrange to form new molecules. This means that no atoms are lost nor gained during any reaction. This idea is known as the **Law of Mass Conservation**. According to this law, the chemical equation must be *balanced*, which means each atom shown in the equation must appear on both sides of the arrow the same number of times. The equation for the formation of water is balanced—there are four hydrogen and two oxygen atoms before and after the arrow. You can count the number of atoms in the space-filling models to see this for yourself.

A coefficient in front of a chemical formula tells us the number of times that element or compound must be counted. For example, 2 H_2O indicates two water molecules, which contain a total of four hydrogen atoms and two oxygen atoms.

By convention, the coefficient 1 is omitted so that the chemical equation above is typically written

$$2\ H_2(g) + O_2(g) \rightarrow 2\ H_2O(g)\ \text{(balanced)}$$

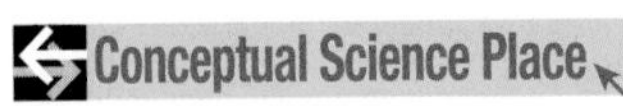

Chemical Reactions and Equations

CHECK YOURSELF

How many oxygen atoms are indicated by the balanced equation

$$3\ O_2(g) \rightarrow 2\ O_3(g)$$

CHECK YOUR ANSWER

Six. Before the reaction these six oxygen atoms are found in three O_2 molecules. After the reaction these same six atoms are found in two O_3 molecules.

An unbalanced chemical equation shows the reactants and products without the correct coefficients. For example, the equation

$$NO(g) \rightarrow N_2O(g) + NO_2(g)\ \text{(not balanced)}$$

is not balanced because there is one nitrogen atom and one oxygen atom before the arrow but three nitrogen atoms and three oxygen atoms after the arrow.

You can balance unbalanced equations by adding or changing coefficients to produce correct ratios. (It's important not to change subscripts, however, because to do so changes the compound's identity—H_2O is water, but H_2O_2 is hydrogen peroxide!) For example, to balance the preceding equation, add a 3 before the NO:

$$3\ NO(g) \rightarrow N_2O(g) + NO_2(g)\ \text{(balanced)}$$

Now there are three nitrogen atoms and three oxygen atoms on each side of the arrow, and the Law of Mass Conservation is not violated.

CHECK YOURSELF

Write a balanced equation for the reaction showing hydrogen gas, H_2, and nitrogen gas, N_2, forming ammonia gas, NH_3:

CHECK YOUR ANSWER

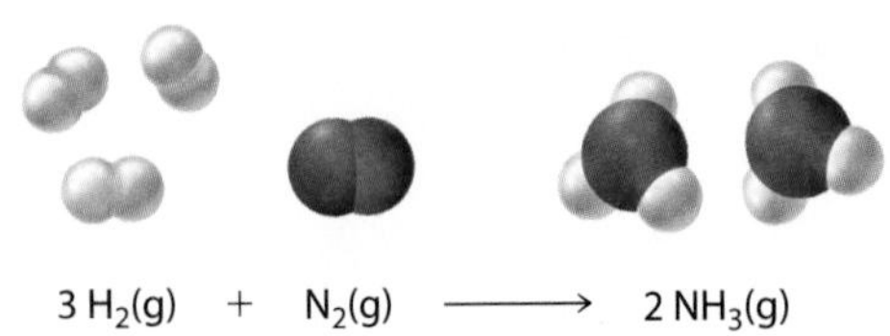

$$3\ H_2(g) + N_2(g) \longrightarrow 2\ NH_3(g)$$

INSIGHT Chemical explosions involve the transformation of solid or liquid chemicals, which occupy relatively little volume, into gases which occupy much greater volumes. For example, upon detonation, one mole of nitroglycerin, $C_3H_5N_3O_9$, produces 7.25 moles of gases, including carbon dioxide, CO_2; nitrogen, N_2; oxygen, O_2; and water vapor, H_2O. The volume change is dramatic—from less than 0.3 liter to about 170 liters, which is an increase of about 600 percent.

You can see that there are equal numbers of each kind of atom before and after the arrow. For more practice balancing equations, see the Exercises at the end of this chapter.

Chemical equations must be balanced or they do not reflect the reality of the chemistry occurring in the reaction. The Law of Mass Conservation tells us that atoms are neither created nor destroyed in a chemical reaction—they are simply rearranged. So every atom present before the reaction must be present after the reaction, even though the groupings of atoms are different.

Two common types of chemical reactions occur over and over again in your everyday world: acid–base reactions and oxidation–reduction reactions. *Acid–base reactions* involve the transfer of protons from one reactant to another. Acid–base reactions are responsible for the sharp taste of a cola drink and the digestion of food in your stomach, as well as the formation and removal of atmospheric carbon dioxide. *Oxidation–reduction reactions* involve the transfer of one or more electrons from one reactant to another. Oxidation–reduction reactions underlie the workings of batteries, fuel cells, and the creation of energy from food in animals and are also responsible for the corrosion of metals and combustion of nonmetallic materials such as wood.

We will begin our examination of chemical reactions by first looking at acids and bases.

13.2 Acid–Base Reactions

Conceptual Science Place
The Nature of Acids and Bases

The term *acid* comes from the Latin *acidus*, which means "sour." The sour taste of vinegar and citrus fruits is due to the presence of acids. Food is digested in the stomach with the help of strong acids, and acids are also essential in the chemical industry. Today, for instance, more than 85 billion pounds of sulfuric acid is produced annually in the United States. Sulfuric acid is used to produce fertilizers, detergents, paint dyes, plastics, pharmaceuticals, storage batteries, iron, and steel. Figure 13.1 shows only a very few of the acids we commonly encounter.

Bases are characterized by their bitter taste. Many pharmaceuticals, for example, are bitter because of their basic nature, which is why they are swallowed whole, rather than chewed. A solution of a base also tends to be slippery. Interestingly, bases themselves are not slippery. Rather, they cause skin oils to transform

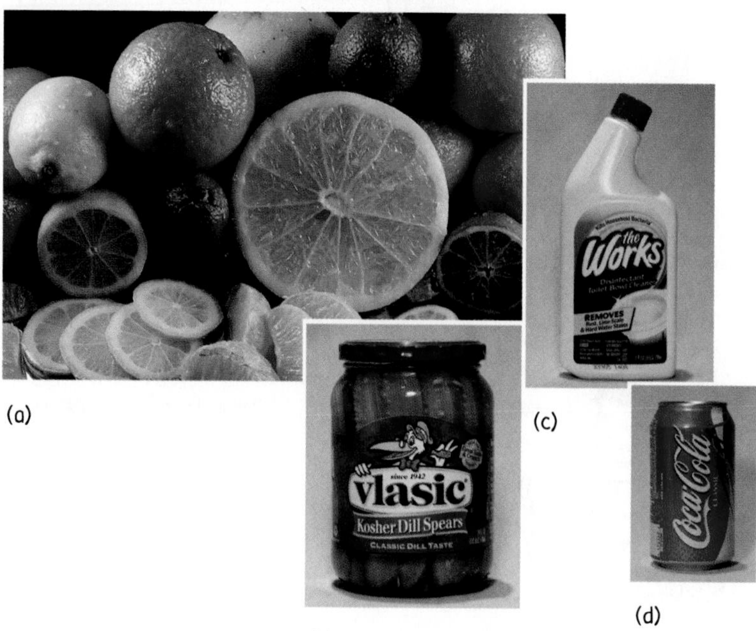

Figure 13.1 Examples of acids. (a) Citrus fruits contain many types of acids, including ascorbic acid, $C_6H_8O_8$, which is vitamin C. (b) Vinegar contains acetic acid, $C_2H_4O_2$, and can be used to preserve foods. (c) Many toilet-bowl cleaners are formulated with hydrochloric acid, HCl. (d) All carbonated beverages contain carbonic acid, H_2CO_3, while many also contain phosphoric acid, H_3PO_4.

Figure 13.2 Examples of bases. (a) Reactions involving sodium bicarbonate, $NaHCO_3$, make baked goods rise. (b) Ashes contain potassium carbonate, K_2CO_3. (c) Soap is made by reacting bases with animal or vegetable oils. The soap itself, then, is slightly alkaline. (d) Powerful bases such as sodium hydroxide, NaOH, are used in drain cleaners. Warning: Don't touch concentrated acid or base solutions, as they will corrode your skin.

INSIGHT Common definitions of acids and bases involve how they taste and feel. But don't taste and feel chemicals to determine their acidity or basicity! Many acids and bases are corrosive and dangerous. Battery acid and lye are common examples.

into slippery solutions of soap. Common household baking soda is a mild base. Most commercial preparations for unclogging drains are composed of sodium hydroxide, NaOH (also known as lye), which is extremely basic and hazardous when concentrated. Bases such as sodium hydroxide are also heavily used in industry. Solutions containing bases are often called *alkaline*, a term derived from the Arabic word for ashes (*al-qali*). Ashes are slippery when wet because of the presence of the base potassium carbonate, K_2CO_3. Figure 13.2 shows some common bases with which you are probably familiar.

We can define an **acid** as any chemical that donates a hydrogen ion, H^+, and a **base** as any chemical that accepts a hydrogen ion.* Recall from Chapter 12 that, because a hydrogen atom consists of one electron surrounding a one-proton nucleus, a hydrogen ion formed from the loss of an electron is nothing more than a lone proton. Thus, it is also sometimes said that an acid is a chemical that donates a proton and a base is a chemical that accepts a proton.

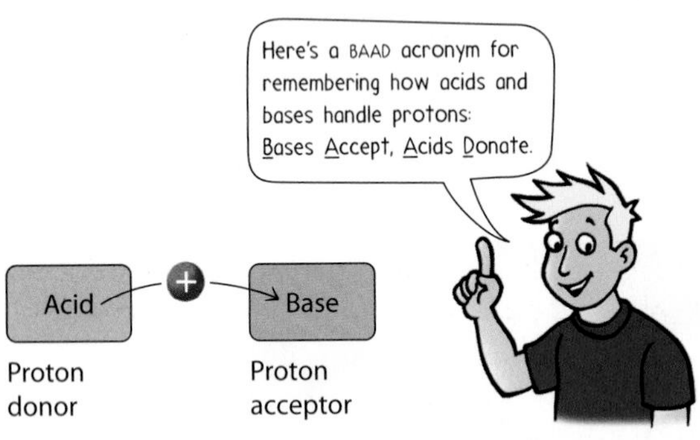

Here's a BAAD acronym for remembering how acids and bases handle protons: Bases Accept, Acids Donate.
Acid
Base
Proton donor
Proton acceptor

Consider what happens when hydrogen chloride is mixed into water:

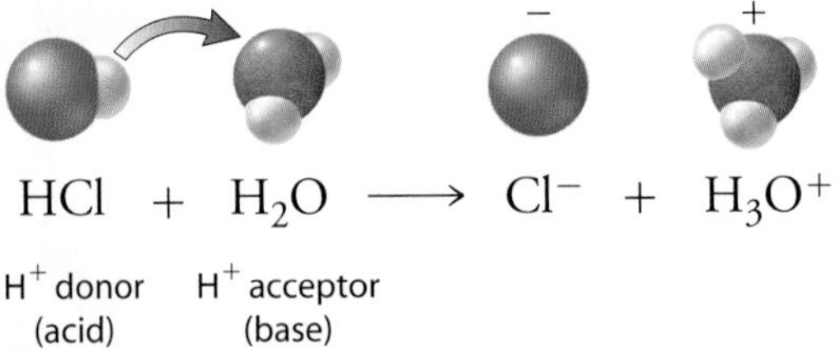

$$HCl + H_2O \longrightarrow Cl^- + H_3O^+$$

H^+ donor (acid) H^+ acceptor (base)

Hydrogen chloride donates a hydrogen ion to one of the nonbonding electron pairs on a water molecule, resulting in a third hydrogen bonded to the oxygen. In this case, hydrogen chloride behaves as an acid (proton donor) and water behaves as a base (proton acceptor). The products of this reaction are a chloride

Recall that a hydrogen ion with a positive charge is simply a lone proton.
Hydrogen atom
Positive hydrogen ion (lone proton)

* There are several definitions of acids and bases. The one we use in this book is known as the Brønsted–Lowry definition. The earliest, simplest, but narrowest definition was offered in 1887, by the Swedish Chemist Arrhenius. The Arrhenius definition states that acids are compounds containing hydrogen ions that ionize to yield hydrogen ions (H^+) in aqueous solutions while bases are compounds that ionize to yield hydroxide ions (OH^-). On the other hand, a very broad and more recent definition is the Lewis definition, which you may encounter in a follow-up chemistry course. A Lewis acid is a substance that can accept a pair of electrons to form a covalent bond. A Lewis base is a substance that can donate a pair of electrons to form a covalent bond.

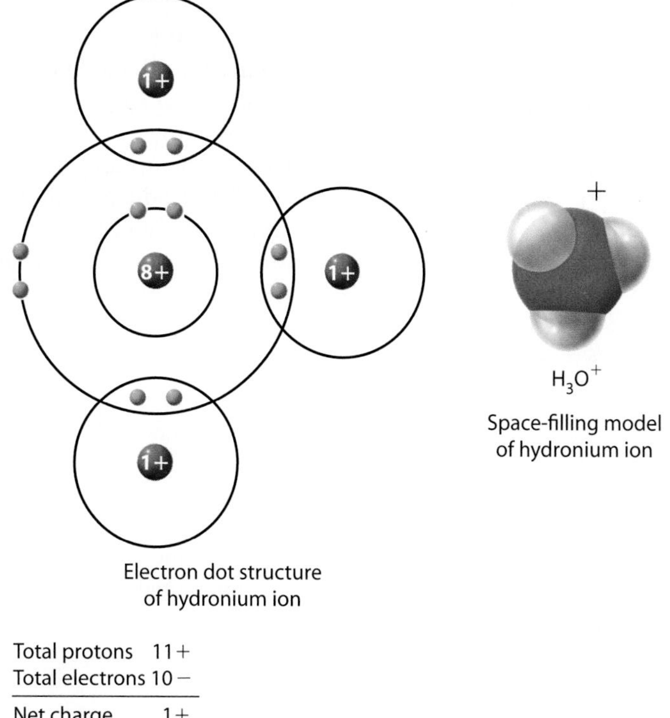

Figure 13.3 The hydronium ion's positive charge is a consequence of the extra proton this molecule has acquired. Hydronium ions, which play a part in many acid–base reactions, are polyatomic ions, which, as mentioned in Chapter 12, are molecules that carry a net electric charge.

ion, Cl^-, and a **hydronium ion**, H_3O^+, which, as Figure 13.3 shows, is a water molecule with an extra proton.

When added to water, ammonia (NH_3) behaves as a base as its nonbonding electrons accept a hydrogen ion from water, which, in this case, behaves as an acid:

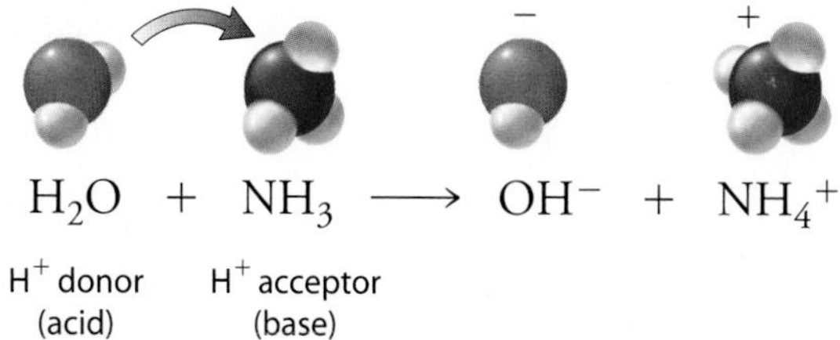

This reaction results in the formation of an ammonium ion and a **hydroxide ion**, which, as shown in Figure 13.4, is a water molecule without the nucleus of one of the hydrogen atoms.

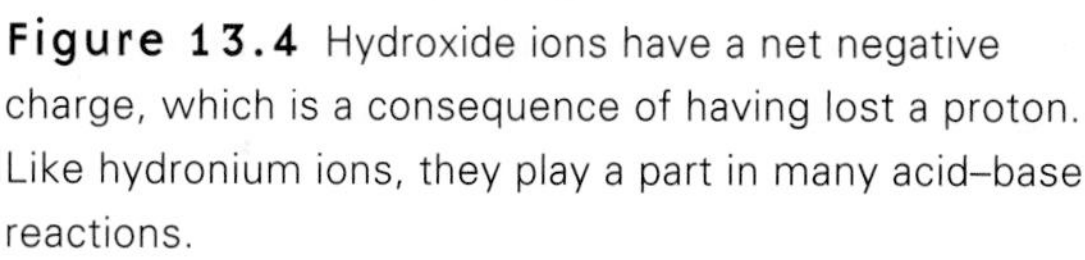

Figure 13.4 Hydroxide ions have a net negative charge, which is a consequence of having lost a proton. Like hydronium ions, they play a part in many acid–base reactions.

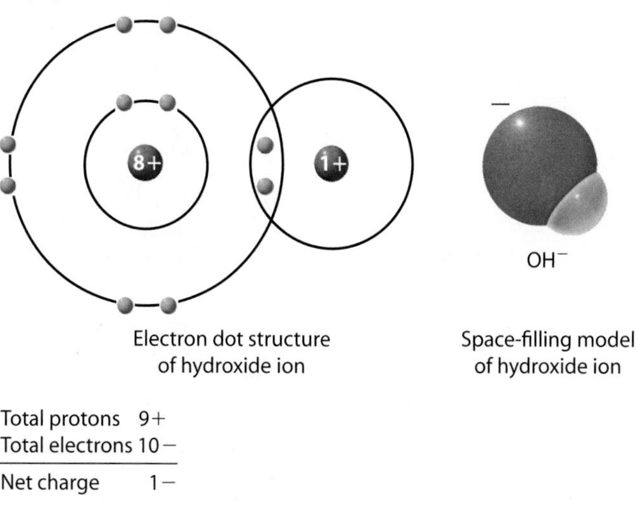

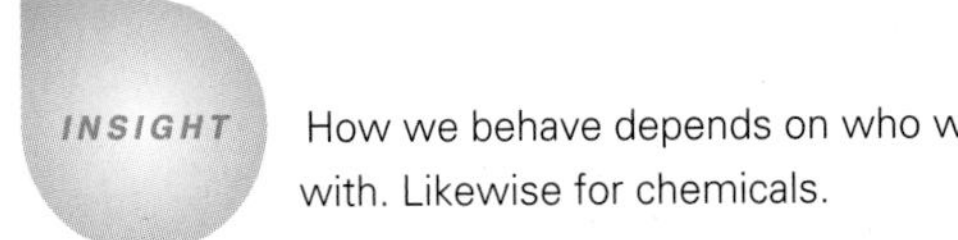

How we behave depends on who we're with. Likewise for chemicals.

It is important to recognize acid–base reactions as a *behavior*. We say, for example, that hydrogen chloride *behaves* as an acid when mixed with water, which *behaves* as a base. Similarly, ammonia *behaves* as a base when mixed with water, which under this circumstance *behaves* as an acid. Because acid–base is seen as a behavior, there is really no contradiction when a chemical like water behaves as a base in one instance but as an acid in another instance.

The products of an acid–base reaction can also behave as acids or bases. An ammonium ion, for example, may donate a hydrogen ion back to a hydroxide ion to re-form ammonia and water:

$$H_2O + NH_3 \longleftarrow OH^- + NH_4^+$$

H^+ acceptor (base) — H^+ donor (acid)

Forward and reverse acid–base reactions proceed simultaneously and can therefore be represented as occurring at the same time by using two oppositely facing arrows:

$$H_2O + NH_3 \rightleftharpoons OH^- + NH_4^+$$

H^+ donor (acid) — H^+ acceptor (base) — H^+ acceptor (base) — H^+ donor (acid)

When the equation is viewed from left to right, the ammonia behaves as a base because it accepts a hydrogen ion from the water, which therefore acts as an acid. Viewed in the reverse direction, the equation shows that the ammonium ion behaves as an acid because it donates a hydrogen ion to the hydroxide ion, which therefore behaves as a base.

CHECK YOURSELF

Identify the acid or base behavior of each participant in the reaction

$$H_2PO_4^- + H_3O^+ \rightarrow H_3PO_4 + H_2O$$

CHECK YOUR ANSWER

In the forward reaction (left to right), $H_2PO_4^-$ gains a hydrogen ion to become H_3PO_4. In accepting the hydrogen ion, $H_2PO_4^-$ is behaving as a base. It gets the hydrogen ion from the H_3O^+, which is behaving as an acid. In the reverse direction, H_3PO_4 loses a hydrogen ion to become $H_2PO_4^-$ and is thus behaving as an acid. The recipient of the hydrogen ion is the H_2O, which is behaving as a base as it transforms to H_3O^+.

13.3 Salts

In everyday language, the word *salt* implies sodium chloride, NaCl, table salt. In the language of chemistry, however, **salt** is a general term meaning any ionic compound formed from the reaction between an acid and a base. Hydrogen chloride and sodium hydroxide, for example, react to produce the salt sodium chloride and water:

$$HCl + NaOH \rightarrow NaCl + H_2O$$

Similarly, the reaction between hydrogen chloride and potassium hydroxide yields the salt potassium chloride and water:

$$HCl + KOH \rightarrow KCl + H_2O$$

Figure 13.5 "Salt-free" table salt substitutes contain potassium chloride in place of sodium chloride. Some people with high blood pressure are advised to substitute "salt-free" salt for NaCl. However, before you switch to KCl, you should check with your doctor. Excessive quantities of potassium salts can lead to serious disorders of the nervous system. Furthermore, sodium ions are a vital component of our diet and should never be totally excluded. For a good balance of these two important ions, you might inquire about commercially available half-and-half mixtures of sodium chloride and potassium chloride, such as the one shown here.

Table 13.1 Acid–Base Reactions and Salts Formed

Acid		Base		Salt		Water
HCN	+	NaOH	→	NaCN	+	H_2O
HNO_3	+	KOH	→	KNO_3	+	H_2O
2 HCl	+	$Ca(OH)_2$	→	$CaCl_2$	+	2 H_2O
HF	+	NaOH	→	NaF	+	H_2O

Potassium chloride is the main ingredient in "salt-free" table salt, as noted in Figure 13.5.

Salts are generally far less corrosive than the acids and bases from which they are formed. A *corrosive* chemical has the power to disintegrate a material or wear away its surface. Hydrogen chloride is a remarkably corrosive acid, which makes it useful for cleaning toilet bowls and etching metal surfaces. Sodium hydroxide is a very corrosive base used for unclogging drains. Mixing hydrogen chloride and sodium hydroxide together in equal portions, however, produces an aqueous solution of sodium chloride—saltwater, which is nowhere near as destructive as either starting material.

There are as many salts as there are acids and bases. Sodium cyanide, NaCN, is a deadly poison. "Salt peter," which is potassium nitrate, KNO_3, is useful as a fertilizer and in the formulation of gun powder. Calcium chloride, $CaCl_2$, is commonly used to deice roads, and sodium fluoride, NaF, is believed to prevent tooth decay. The acid–base reactions forming these salts are shown in Table 13.1.

The reaction between an acid and a base is called a **neutralization** reaction. As can be seen in the neutralization reactions in Table 13.1, the positive ion of a salt comes from the base and the negative ion comes from the acid. The remaining hydrogen and hydroxide ions join to form water.

Not all neutralization reactions result in the formation of water. In the presence of hydrogen chloride, for example, the drug pseudoephedrine behaves as a base by accepting a hydrogen ion, H^+. The negative Cl^- then joins the pseudoephedrine H^+ ion to form the salt pseudoephrine hydrochloride, which is a common nasal decongestant, shown in Figure 13.6. This salt is soluble in water and can be absorbed through the digestive system.

Hydrogen chloride (salt)

Pseudoephedrine (base)

Pseudoephedrine · HCl (salt)

Figure 13.6 Hydrogen chloride and pseudoephedrine react to form the salt pseudoephedrine HCl, which, because of its solubility in water, is readily absorbed into the body.

CHECK YOURSELF

Is a neutralization reaction best described as a physical change or a chemical change?

CHECK YOUR ANSWER

New chemicals are formed during a neutralization reaction, meaning the reaction is a chemical change.

13.4 Solutions: Acidic, Basic, or Neutral

A substance whose ability to behave as an acid is about the same as its ability to behave as a base is said to be **amphoteric**. Water is a good example. Because it is amphoteric, water has the ability to react with itself. In behaving as an acid, a water molecule donates a hydrogen ion to a neighboring water molecule, which in accepting the hydrogen ion is behaving as a base. This reaction produces a hydroxide ion and a hydronium ion, which react together to re-form the water:

INSIGHT As an analogy, consider a large room packed full of people, each wearing one hat. At the sound of a bell, five of these people take off their hats and give them to someone nearby who is still wearing one hat. There are now five people with no hats and five people with two hats. No matter how many people gave their hats away, the number of people with no hats will always be the same as the number of people with two hats—no matter how many times the bell rings. Likewise, with pure water, the concentration of hydronium and hydroxide ions will always be the same.

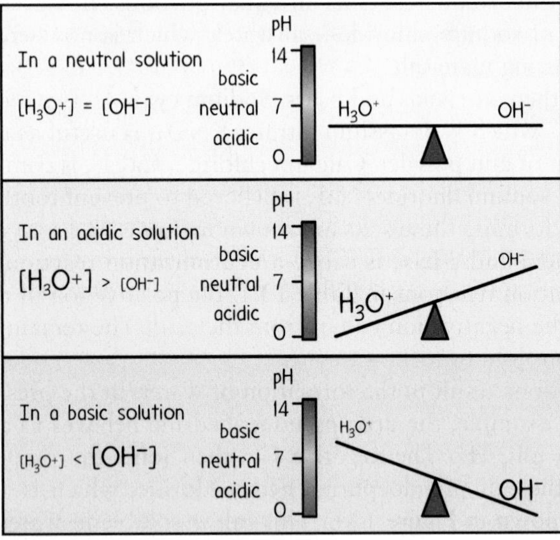

Figure 13.7 The relative concentrations of hydronium and hydroxide ions determines whether a solution is neutral, acidic, or basic.

From this reaction we can see that, in order for a water molecule to gain a hydrogen ion, a second water molecule must lose a hydrogen ion. This means that for every one hydronium ion formed, there is also one hydroxide ion formed. In pure water, therefore, the total number of hydronium ions must be the same as the total number of hydroxide ions. Experiments reveal that the concentration of hydronium and hydroxide ions in pure water is extremely low—about 0.0000001 *M* for each, where *M* stands for molarity or moles per liter (Chapter 9). Water by itself, therefore, is a very weak acid as well as a very weak base.

CHECK YOURSELF

Do water molecules react with one another?

CHECK YOUR ANSWER

Yes, but not to any large extent. When they do react, they form hydronium and hydroxide ions. (*Note*: Make sure you understand this point because it serves as a basis for understanding the pH scale, which we will cover in the next section.)

An aqueous solution can be described as neutral, acidic, or basic. A **neutral solution** is one in which the hydronium ion concentration equals the hydroxide ion concentration. Pure water is an example of a neutral solution—not because it contains so few hydronium and hydroxide ions, but because it contains equal numbers of them. A neutral solution is also obtained when equal quantities of acid and base are combined, which is why acids and bases are said to *neutralize* each other.

An **acidic solution** is one in which the hydronium ion concentration is higher than the hydroxide ion concentration. An acidic solution is made by adding an acid to water. The effect of this addition is to increase the concentration of hydronium ions. Interestingly, the excess amounts of hydronium ions have the effect of neutralizing the hydroxide ions. Thus, as the hydronium ion concentration increases, the hydroxide ion concentration necessarily decreases, as is depicted in Figure 13.7.

A **basic solution** is one in which the hydroxide ion concentration is higher than the hydronium ion concentration. A basic solution is made by adding a base to water. This addition increases the concentration of hydroxide ions. Notably, the excess amounts of hydroxide ions have the effect of neutralizing the hydronium ions. Thus, as the hydroxide ion concentration increases, the hydronium ion concentration necessarily decreases, as is depicted in Figure 13.7.

CHECK YOURSELF

How does adding ammonia, NH_3, to water make a basic solution when there are no hydroxide ions in the formula for ammonia?

CHECK YOUR ANSWER

Ammonia indirectly increases the hydroxide ion concentration by reacting with water:

$$NH_3 + H_2O \rightarrow NH_4^+ + OH^-$$

This reaction raises the hydroxide ion concentration. With the hydroxide ion concentration now higher than the hydronium ion concentration, the solution is basic.

13.5 The pH Scale

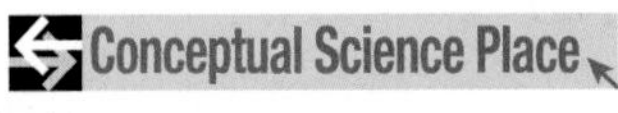

The pH Scale

The **pH scale** is a numeric scale used to express the acidity of a solution. The scale is a logarithmic value based on the concentration of hydronium ions in the solution.

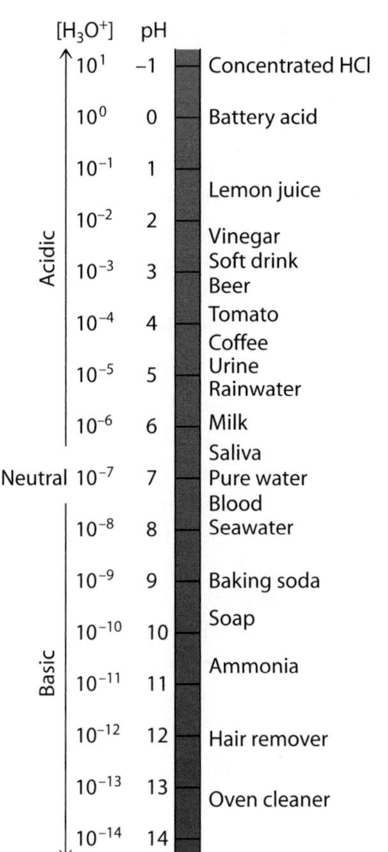

Figure 13.8 The pH values of some common solutions.

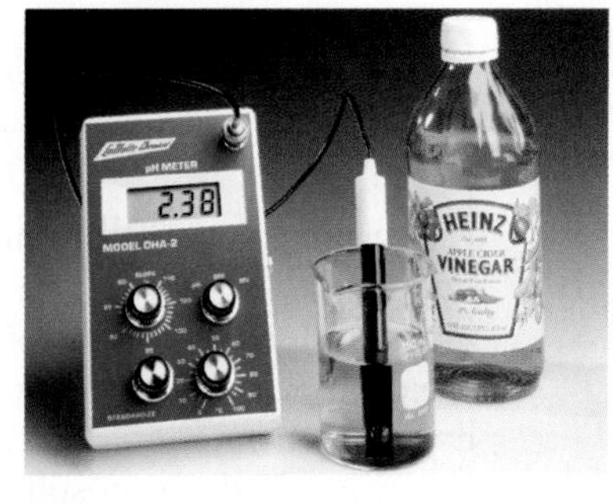

(a)

(b)

Figure 13.9 (a) The pH of a solution can be measured electronically using a pH meter. (b) A rough estimate of the pH of a solution can be obtained with litmus paper, which is coated with a dye that changes color with pH.

As shown in Figure 13.7, acidic solutions have pH values less than 7. An acidic solution in which the hydronium ion concentration is 1.0×10^{-4} *M*, for example, the pH $= -\log(1.0 \times 10^{-4}) = 4$. The more acidic a solution is, the greater its hydronium ion concentration and the lower its pH. *Strong acids have low pH.*

Basic solutions have pH values greater than 7. For a basic solution in which the hydronium ion concentration is 1.0×10^{-8} *M*, for example, pH $= -\log(1.0 \times 10^{-8}) = 8$. The more basic a solution is, the smaller its hydronium ion concentration and the higher its pH. *Strong bases have high pH.*

Figure 13.8 shows typical pH values of some familiar solutions, and Figure 13.9 shows two common ways of determining pH values.

MATH CONNECTION

Logarithms and pH

The pH of a solution is calculated from the concentration of hydronium ions, H_3O^+. Mathematically, **pH** is equal to the negative of the base-10 logarithm of the hydronium ion concentration:

$$\text{pH} = -\log[H_3O^+]$$

Note that brackets are used to represent molar concentrations, meaning $[H_3O^+]$ is read "the molar concentration of hydronium ions."

So, what the heck is a "base-10 logarithm"? This logarithm is simply a fancy way of asking "To what power is 10 raised?" The logarithm of 10^2, for example, is 2 because that is the power to which 10 is raised. If you know that 10^2 is equal to 100, then you'll understand that the logarithm of 100 also is 2, which is the power to which 10 is raised to give you 100. Similarly, the log of 10^3 equals 3, because 3 is the power to which 10 is raised. Do you recognize that 10^3 equals $10 \times 10 \times 10$, which equals 1000? If so, then you'll recognize that the log of 1000 also equals 3. The logarithm of 1000 is 3 because 10 raised to the third power, 10^3, equals 1000.

Any positive number, including a very small one, has a logarithm. The logarithm of $0.0001 = 10^{-4}$, for example, is -4 (the power to which 10 is raised to equal this number).

Problem What is the logarithm of 0.01?

Solution The number 0.01 is 10^{-2}, the logarithm of which is -2 (the power to which 10 is raised).

The concentration of hydronium ions in most solutions is typically much less than 1 *M*. Recall, for example, that in neutral water the hydronium ion concentration is 0.0000001 *M* (10^{-7} *M*). The logarithm of any number smaller than 1 (but greater than zero) is a negative number. The definition of pH includes the minus sign so as to transform the logarithm of the hydronium ion concentration to a positive number.

When a solution has a hydronium ion concentration of 1 *M*, the pH is 0 because 1 *M* = 10^0 *M*. A 10 *M* solution has a pH of -1 because 10 *M* = 10^1 *M*.

Problem What is the pH of a solution that has a hydronium ion concentration of 0.001 *M*?

Solution The number 0.001 is 10^{-3}, and so

$$\begin{aligned} \text{pH} &= -\log[H_3O^+] \\ &= -\log 10^{-3} \\ &= -(-3) = 3 \end{aligned}$$

INTEGRATED SCIENCE EARTH SCIENCE

Acid Rain

Rainwater is naturally acidic. One source of this acidity is carbon dioxide, the same gas that gives fizz to soda drinks. There are about 810 billion tons of CO_2 in the atmosphere, most of it (about 675 billion tons) from natural sources such as volcanoes and decaying organic matter but a growing amount (about 135 billion tons) from human activities such as the burning of fossil fuels.

Water in the atmosphere reacts with carbon dioxide to form *carbonic acid*:

$$CO_2(g) + H_2O(\ell) \rightarrow H_2CO_3(aq)$$

Carbonic acid, as its name implies, behaves as an acid and lowers the pH of water. The CO_2 in the atmosphere brings the pH of rainwater to about 5.6—noticeably below the neutral pH value of 7. Because of local fluctuations, the normal pH of rainwater varies between 5 and 7. This natural acidity of rainwater may accelerate the erosion of land and, under the right circumstances, can lead to the formation of underground caves, such as those found at Carlsbad Caverns, New Mexico, or Mammoth Cave, Kentucky.

By convention, *acid rain* is a term used for rain having a pH lower than 5. Acid rain is created when airborne pollutants such as sulfur dioxide are absorbed by atmospheric moisture. Sulfur dioxide is readily converted to sulfur trioxide, which reacts with water to form *sulfuric acid*:

$$2\,SO_2(g) + O_2(g) \rightarrow 2\,SO_3(g)$$

$$SO_3(g) + H_2O(\ell) \rightarrow H_2SO_4(aq)$$

When we burn fossil fuels, the reactants that produce sulfuric acid are emitted into the atmosphere. Each year, about 20 million tons of SO_2 are released into the atmosphere by the combustion of sulfur-containing coal and oil. Sulfuric acid is much stronger than carbonic acid, and as a result rain laced with sulfuric acid eventually corrodes metal, paint, and other exposed substances. The damage costs billions of dollars. The cost to the environment is also high (Figure 13.10).

(a)

(b)

Figure 13.10 The two photographs in (a) show the same obelisk before and after the effects of acid rain. (b) Many forests downwind from heavily industrialized areas, such as in the northeastern United States and in Europe, have been noticeably hard hit by acid rain.

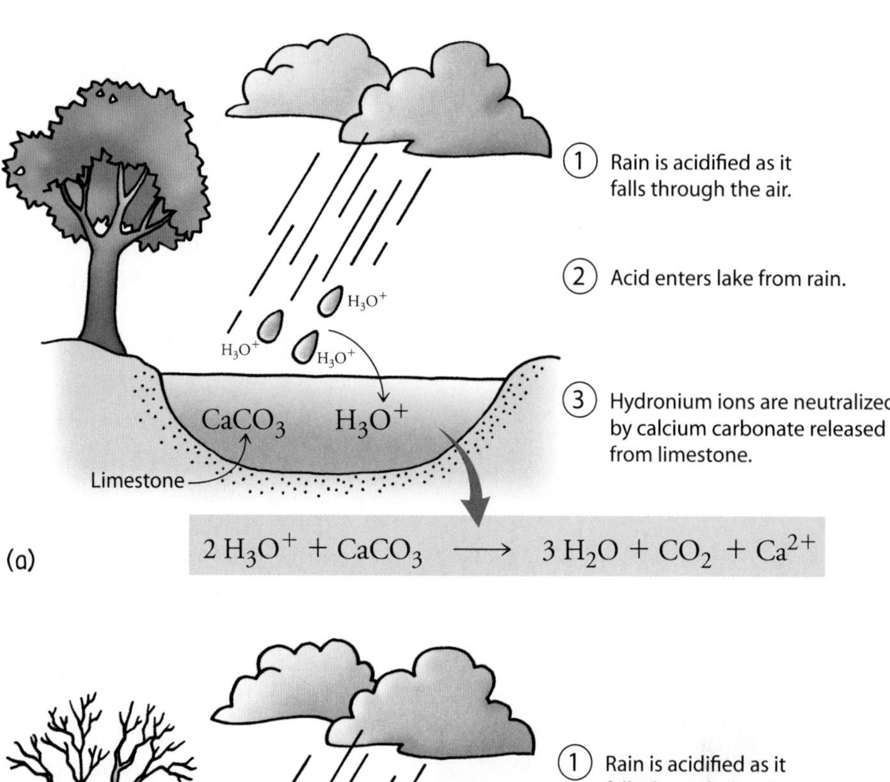

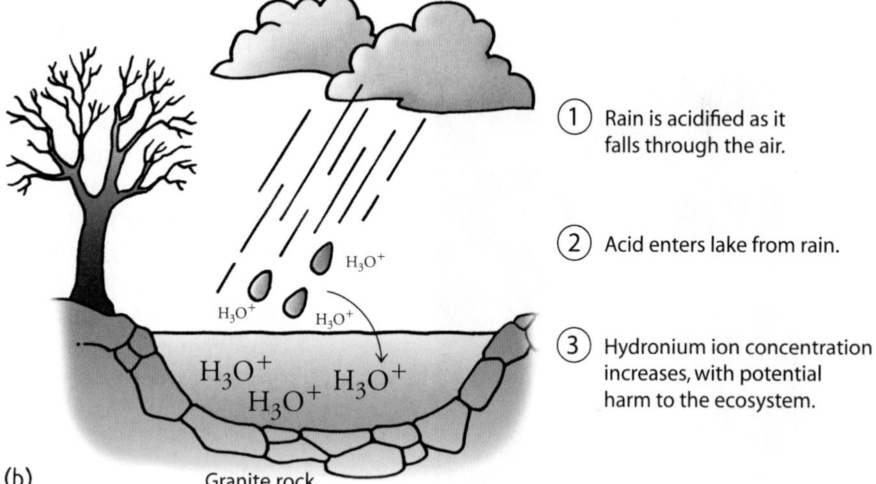

Figure 13.11 (a) The damaging effects of acid rain do not appear in bodies of fresh water lined with calcium carbonate, which neutralizes any acidity. (b) Lakes and rivers lined with inert materials are not protected.

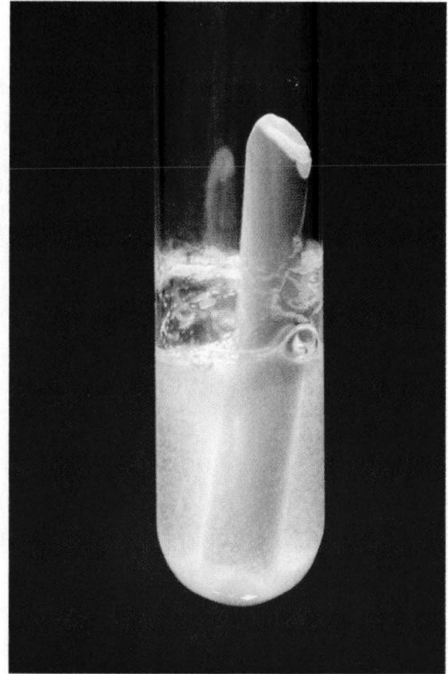

Figure 13.12 Most chalks are made from calcium carbonate, which is the same chemical found in limestone. The addition of even a weak acid, such as the acetic acid of vinegar, produces hydronium ions that react with the calcium carbonate to form several products, the most notable being carbon dioxide, which rapidly bubbles out of solution.

Many rivers and lakes receiving acid rain become less capable of sustaining life because the greater acidity is harmful to living tissues. Much vegetation that receives acid rain doesn't survive for the same reason. This is particularly evident in heavily industrialized regions and in areas downwind of these areas, where acid rain travels.

The environmental impact of acid rain depends on local geology, as Figure 13.11 illustrates. In certain regions, such as the midwestern United States, the ground contains significant quantities of the alkaline compound calcium carbonate (limestone), deposited when these lands were submerged under oceans 200 million years ago. Acid rain pouring into these regions is often neutralized by the calcium carbonate before any damage is done. (Figure 13.12 shows calcium carbonate neutralizing an acid.) In the northeastern United States and many other regions, however, the ground contains very little calcium carbonate and is composed primarily of chemically less reactive materials, such as granite. In these regions, the effect of acid rain on lakes and rivers accumulates.

A long-term solution to acid rain is to prevent most of the generated sulfur dioxide and other pollutants from entering the atmosphere in the first place. Toward this end, smokestacks have been designed or retrofitted to minimize the quantities of pollutants released. Though costly, the positive effects of these adjustments have been realized in many regions, such as the northeastern United

States. An ultimate long-term solution to acid rain, however, would be a shift from fossil fuels to energy sources that do not emit sulfur, such as nuclear and solar energy.

CHECK YOURSELF

1. When sulfuric acid, H_2SO_4, is added to water, what makes the resulting aqueous solution corrosive?
2. What kind of lakes are protected against the negative effects of acid rain?

CHECK YOUR ANSWERS

1. Because H_2SO_4 is a strong acid, it readily forms hydronium ions when dissolved in water. Hydronium ions are responsible for the corrosive action.
2. Lakes that have a floor consisting of basic minerals, such as limestone, are more resistant to acid rain because the chemicals of the limestone (mostly calcium carbonate, $CaCO_3$) neutralize any incoming acid.

13.6 Oxidation–Reduction Reactions

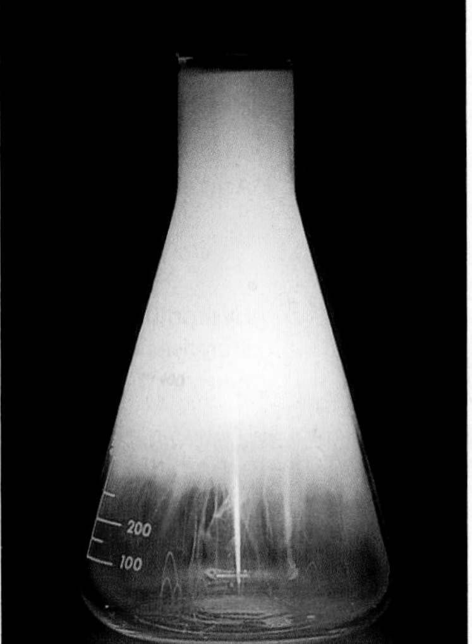

Figure 13.13 In the formation of sodium chloride, sodium metal is oxidized by chlorine gas and chlorine gas is reduced by sodium metal.

What do campfires, the rusting of iron, the tarnishing of silver, the creation of metals from metal ores, batteries, fuel cells, and your car's engine have in common? They all involve oxidation–reduction reactions. **Oxidation** is the process whereby a reactant loses one or more electrons. **Reduction** is the opposite process whereby a reactant gains one or more electrons. Oxidation and reduction are complementary processes that occur at the same time. They always occur together; you cannot have one without the other. The electrons lost by one chemical in an oxidation reaction don't simply disappear; they are gained by another chemical in a reduction reaction.

An oxidation–reduction reaction occurs when sodium and chlorine react to form sodium chloride, as shown in Figure 13.13. The equation for this reaction is

$$2\ Na + Cl_2 \rightarrow 2\ NaCl$$

To see how electrons are transferred in this reaction, we can look at each reactant individually. Each electrically neutral sodium atom changes to a positively charged ion. We can also say that each atom loses an electron and is therefore oxidized:

$$2\ Na \rightarrow 2\ Na^+ + 2\ e^-\ \text{(oxidation)}$$

Each electrically neutral chlorine molecule changes to two negatively charged ions. Each of these atoms gains an electron and is therefore reduced:

$$Cl_2 + 2\ e^- \rightarrow 2\ Cl^-\ \text{(reduction)}$$

The net result is that the two electrons lost by the sodium atoms are transferred to the chlorine atoms. Therefore, each of the two equations shown above actually represents one-half of an entire process, which is why they are each called a **half-reaction.** In other words, an electron won't be lost from a sodium atom without there being a chlorine atom available to pick up that electron. Both half-reactions are required to represent the *whole* oxidation–reduction process. Half-reactions are useful for showing which reactant loses electrons and which reactant gains them, which is why half-reactions are used throughout this chapter.

Because the sodium causes reduction of the chlorine, the sodium is acting as a *reducing agent.* A reducing agent is any reactant that causes another reactant to be reduced. Note that sodium is oxidized when it behaves as a reducing agent—it loses electrons. Conversely, the chlorine causes oxidation of the sodium and so is

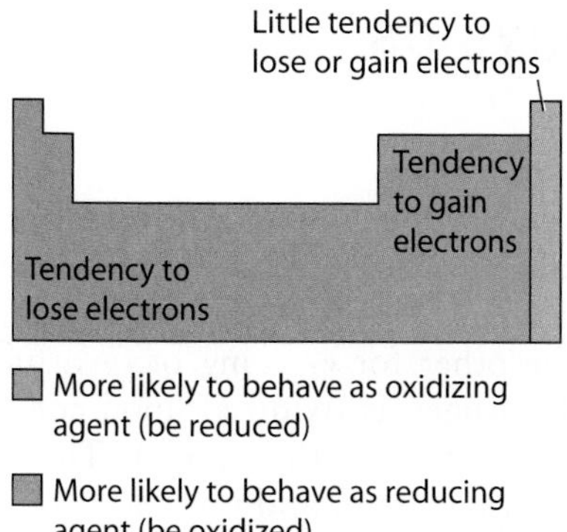

Figure 13.14 The ability of an atom to gain or lose electrons is a function of its position in the periodic table. Those at the upper right tend to gain electrons, and those at the lower left tend to lose them.

INSIGHT When we say a substance was oxidized, we're saying that it lost electrons. When we say a substance was reduced, we're saying that it gained electrons.

acting as an *oxidizing agent*. Because it gains electrons in the process, an oxidizing agent is reduced. Just remember that *loss of* electrons is *oxidation*, and *gain of* electrons is *reduction*. Here is a helpful mnemonic adapted from a once-popular children's story: *Leo the lion went "ger."*

Different elements have different oxidation and reduction tendencies—some lose electrons more readily, while others gain electrons more readily, as Figure 13.14 illustrates. The tendency to do one or the other is a function of how strongly the atom's nucleus holds electrons. The greater the electronegativity (Section 12.6), the greater the tendency of the atom to *gain* electrons. Because the atoms of elements at the upper right of the periodic table have the strongest electronegativities (with the noble gases excluded), these atoms have the greatest tendency to gain electrons and hence behave as oxidizing agents. The atoms of elements at the lower left of the periodic table have the weakest electronegativities and therefore the greatest tendency to *lose* electrons and behave as reducing agents.

CHECK YOURSELF

True or false:

1. Reducing agents are oxidized in oxidation–reduction reactions.
2. Oxidizing agents are reduced in oxidation–reduction reactions.

CHECK YOUR ANSWERS

Both statements are true.

Whether a reaction classifies as an oxidation–reduction reaction is not always immediately apparent. The chemical equation, however, can provide some important clues. First, look for changes in the ionic states of elements. Sodium metal, for example, consists of neutral sodium atoms. In the formation of sodium chloride, these atoms transform into positively charged sodium ions, which occurs as sodium atoms lose electrons (oxidation). A second way to identify a reaction as an oxidation–reduction reaction is to look to see whether an element is gaining or losing oxygen atoms. As the element gains the oxygen it is losing electrons to that oxygen because of the oxygen's high electronegativity. The gain of oxygen, therefore, is oxidation (loss of electrons), while the loss of oxygen is reduction (gain of electrons). For example, hydrogen, H_2, reacts with oxygen, O_2, to form water, H_2O, as follows:

$$\mathrm{H{-}H + H{-}H + O{=}O \rightarrow H{-}O{-}H + H{-}O{-}H}$$

Note that the element hydrogen becomes attached to an oxygen atom through this reaction. The hydrogen, therefore, is oxidized.

A third way to identify a reaction as an oxidation–reduction reaction is to see whether an element is gaining or losing hydrogen atoms. The gain of hydrogen is reduction, while the loss of hydrogen is oxidation. For the formation of water shown above, we see that the element oxygen is gaining hydrogen atoms, which means that the oxygen is being reduced—that is, the oxygen is gaining electrons from the hydrogen, which is why the oxygen atom within water is slightly negative, as discussed in Chapter 12. The three ways of identifying a reaction as an oxidation–reduction type of reaction are summarized in Figure 13.15.

Oxidation (Ionic state becomes more positive)	Reduction (Ionic state becomes more negative)
Loses electrons	Gains electrons
Gains oxygen	Loses oxygen
Loses hydrogen	Gains hydrogen

Figure 13.15 Oxidation results in a greater positive charge, which can be achieved by losing electrons, gaining oxygen atoms, or losing hydrogen atoms. Reduction results in a greater negative charge, which can be achieved by gaining electrons, losing oxygen atoms, or gaining hydrogen atoms.

CHECK YOURSELF

In the following equation, is carbon oxidized or reduced?

$$CH_4 + 2\ O_2 \rightarrow CO_2 + 2\ H_2O$$

CHECK YOUR ANSWER

As the carbon of methane, CH_4, forms carbon dioxide, CO_2, it is losing hydrogen and gaining oxygen, which tells us that the carbon is being oxidized.

INTEGRATED SCIENCE PHYSICS

Fuel Cells

A *fuel cell* is a device that changes the chemical energy of a fuel to electrical energy. Fuel cells are by far the most efficient means of generating electricity. A hydrogen–oxygen fuel cell is shown in Figure 13.16. It has two compartments, one for entering hydrogen fuel and the other for entering oxygen fuel, separated by a set of porous electrodes. Hydrogen is oxidized upon contact with hydroxide ions at the hydrogen-facing electrode (the anode). The electrons from this oxidation flow through an external circuit and provide electric power before meeting up with oxygen at the oxygen-facing electrode (the cathode). The oxygen readily picks up the electrons (in other words, the oxygen is reduced) and reacts with water to form hydroxide ions. To complete the circuit, these hydroxide ions migrate across the porous electrodes and through an ionic paste of potassium hydroxide, KOH, to meet up with hydrogen at the hydrogen-facing electrode.

As the oxidation equation shown at the top of Figure 13.16 demonstrates, the hydrogen and hydroxide ions react to produce energetic water molecules that arise in the form of steam. This steam may be used for heating or to generate electricity in a steam turbine. Furthermore, the water that condenses from the steam is pure water, suitable for drinking!

Fuel cells are similar to dry-cell batteries, but fuel cells don't run down as long as fuel is supplied. The space shuttle uses hydrogen–oxygen fuel cells to meet its electrical needs. The cells also produce more than 100 gallons of drinking water for the astronauts during a typical week-long mission. Back on Earth, researchers are developing fuel cells for buses and automobiles. As shown in Figure 13.17, experimental fuel-cell buses are already operating in several cities, such as Vancouver, British Columbia, and Chicago, Illinois. These vehicles produce very few pollutants and can run much more efficiently than vehicles that run on fossil fuels.

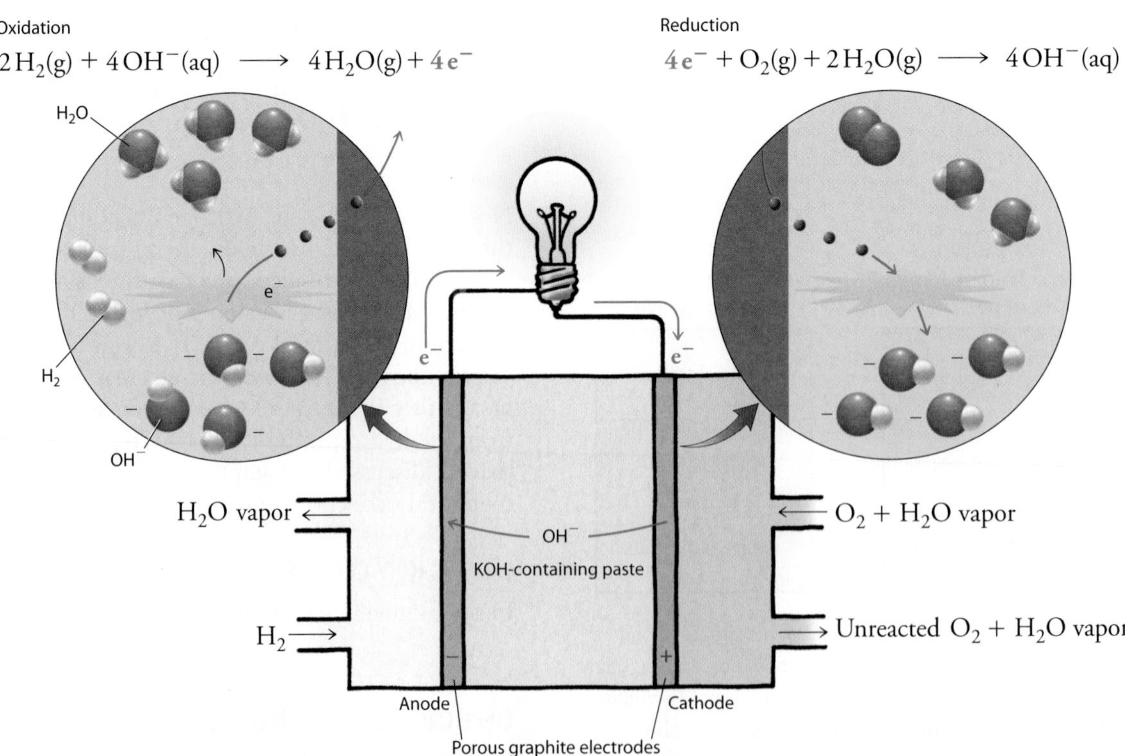

Figure 13.16 The hydrogen–oxygen fuel cell.

Figure 13.17 Because this bus is powered by a fuel cell, its tailpipe emits mostly water vapor.

In the future, commercial buildings as well as individual homes may be outfitted with fuel cells as an alternative to receiving electricity (and heat) from regional power stations. Researchers are also working on miniature fuel cells that could replace the batteries used for portable electronic devices, such as cellular phones and laptop computers. Such devices could operate for extended periods of time on a single "ampule" of fuel available at your local supermarket.

Amazingly, a car powered by a hydrogen–oxygen fuel cell requires only about 3 kilograms of hydrogen to travel 500 kilometers. However, this much hydrogen gas at room temperature and atmospheric pressure would occupy a volume of about 36,000 liters, the volume of about four midsize cars! Thus the major hurdle to the development of fuel-cell technology lies not with the cell, but with the fuel. This volume of gas could be compressed to a much smaller volume, as it is on the experimental buses in Vancouver.

Compressing a gas takes energy, however, and as a consequence the inherent efficiency of the fuel cell is lost. Chilling hydrogen to its liquid phase, which occupies much less volume, poses similar problems. Instead, researchers are looking for novel ways of providing fuel cells with hydrogen. In one design, hydrogen is generated within the fuel cell from chemical reactions involving liquid hydrocarbons, such as methanol, CH_3OH. Alternatively, certain porous materials, including the recently developed carbon nanofibers, shown in Figure 13.18, can hold large volumes of hydrogen on their surfaces, behaving in effect like hydrogen "sponges." The hydrogen is "squeezed" out of these materials on demand by controlling the temperature—the warmer the material, the more hydrogen released.

CHECK YOURSELF

As long as fuel is available to it, a given fuel cell can supply electrical energy indefinitely. Why can't batteries do the same?

CHECK YOUR ANSWER

Batteries generate electricity as the chemical reactants they contain are reduced and oxidized. Once these reactants are consumed, the battery can no longer generate electricity. A rechargeable battery can be made to operate again, but only after the energy flow is interrupted so that the reactants can be replenished.

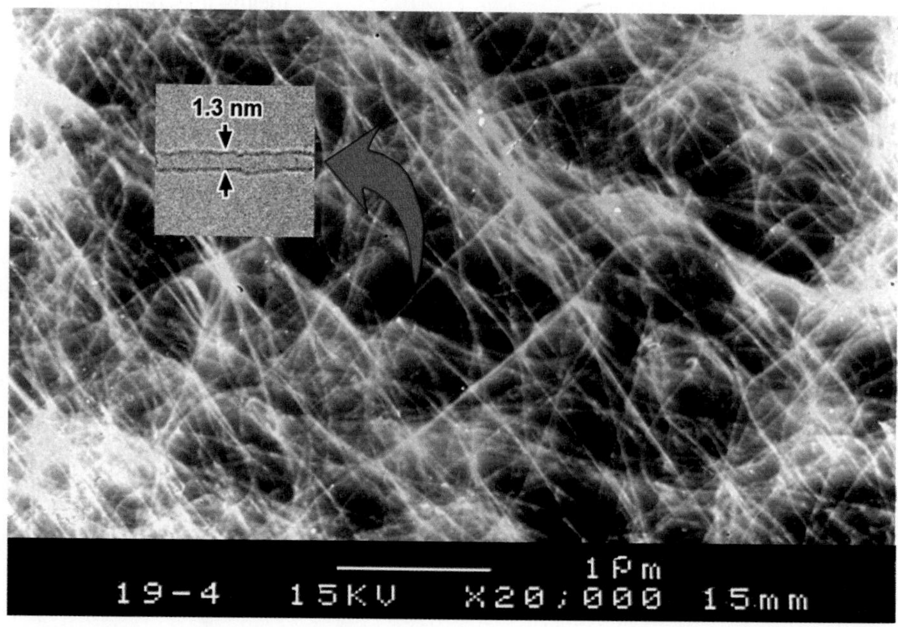

Figure 13.18 Carbon nanofibers consist of near-submicroscopic tubes of carbon atoms. They outclass almost all other known materials in their ability to absorb hydrogen molecules. With carbon nanofibers, for example, a volume of 36,000 liters of hydrogen can be reduced to a mere 35 liters. Carbon nanofibers are a recent discovery, however, and much research is still required to confirm their applicability to hydrogen storage and to develop the technology.

Figure 13.19 Rust itself is not harmful to the iron structures on which it forms. It is the loss of metallic iron that ruins the structural integrity.

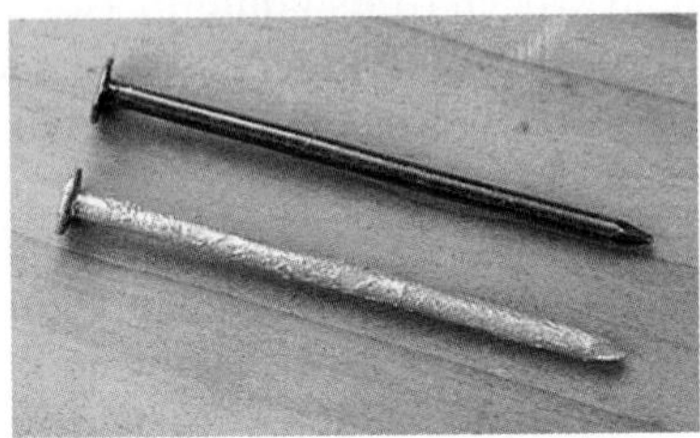

Figure 13.20 The galvanized nail (bottom) is protected from rusting by the sacrificial oxidation of zinc.

Figure 13.21 Zinc strips help protect the iron hull of an oil tanker from oxidizing. The zinc strip shown here is attached to the hull's interior surface.

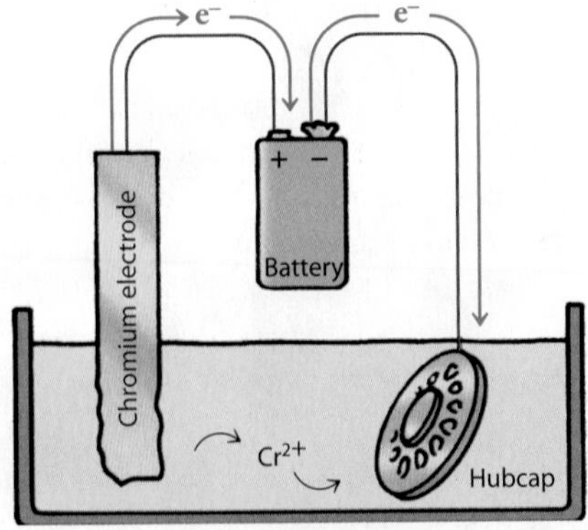

Figure 13.22 As electrons flow into the hubcap and give it a negative charge, positively charged chromium ions move from the solution to the hubcap and are reduced to chromium metal, which deposits as a coating on the hubcap. The solution is kept supplied with ions as chromium atoms in the cathode are oxidized to Cr^{2+} ions.

13.7 Corrosion and Combustion

Look to the upper right of the periodic table, and you will find one of the most common oxidizing agents—oxygen. In fact, if you haven't guessed already, the term *oxidation* is derived from this element. Oxygen is able to pluck electrons from many other elements, especially those that lie at the lower left of the periodic table. Two common oxidation–reduction reactions involving oxygen as the oxidizing agent are *corrosion* and *combustion.*

CHECK YOURSELF

Oxygen is a good oxidizing agent, but so is chlorine. What does this tell you about their relative positions in the periodic table?

CHECK YOUR ANSWER

Chlorine and oxygen must lie in the same area of the periodic table. Both have strong effective nuclear charges, and both are strong oxidizing agents.

Corrosion is the process whereby a metal deteriorates. Corrosion caused by atmospheric oxygen is a widespread and costly problem. About one-quarter of the steel produced in the United States, for example, goes into replacing corroded iron at a cost of billions of dollars annually. Iron corrodes when it reacts with atmospheric oxygen and water to form iron oxide trihydrate, which is the naturally occurring reddish-brown substance you know as *rust*, shown in Figure 13.19.

Another common metal oxidized by oxygen is aluminum. The product of aluminum oxidation is aluminum oxide, Al_2O_3, which is water-insoluble. Because of its water insolubility, aluminum oxide forms a protective coat that shields aluminum from further oxidation. This coat is so thin that it's transparent, which is why aluminum maintains its metallic shine.

A protective water-insoluble oxidized coat is the principle underlying a process called *galvanization.* Zinc has a slightly greater tendency to oxidize than does iron. For this reason, many iron articles, such as the nails pictured in Figure 13.20 are *galvanized* by coating them with a thin layer of zinc. The zinc oxidizes to zinc oxide, an inert, insoluble substance that protects the inner iron from rusting.

In a technique called *cathodic* protection, iron structures can be protected from oxidation by placing them in contact with metals, such as zinc or magnesium, that have a greater tendency to oxidize. This forces the iron to accept electrons, which means it is behaving as a cathode—recall that rusting occurs only where iron behaves as an anode. Ocean tankers, for example, are protected from corrosion by strips of zinc affixed to their hulls, as shown in Figure 13.21.

Yet another way to protect iron and other metals from oxidation is to coat them with a corrosion-resistant metal, such as chromium, platinum, or gold. *Electroplating* is the operation of coating one metal with another by electrolysis, and it is illustrated in Figure 13.22. The object to be electroplated is connected to a negative battery terminal and then submerged in a solution containing ions of the metal to be used as the coating. The positive terminal of the battery is connected to an electrode made of the coating metal. The circuit is completed when this electrode is submerged in the solution. Dissolved metal ions are attracted to the negatively charged object, where they pick up electrons and are deposited as metal atoms. The ions in solution are replenished by the forced oxidation of the coating metal at the positive electrode.

Combustion is an oxidation–reduction reaction between a nonmetallic material and molecular oxygen. Combustion reactions are characteristically exothermic (energy-releasing). A violent combustion reaction is the formation of water from hydrogen and oxygen. The energy from this reaction is used to lift the space shuttle into outer space. More common examples of combustion include the burning of wood and fossil fuels. The combustion of these and other carbon-

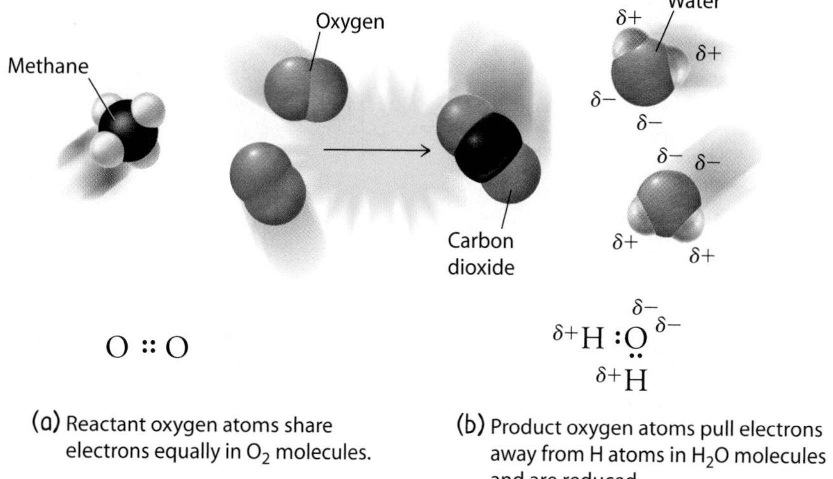

(a) Reactant oxygen atoms share electrons equally in O_2 molecules.

(b) Product oxygen atoms pull electrons away from H atoms in H_2O molecules and are reduced.

Figure 13.23 (a) Neither atom in an oxygen molecule is able to preferentially attract the bonding electrons. (b) The oxygen atom of a water molecule pulls the bonding electrons away from the hydrogen atoms on the water molecule, making the oxygen slightly negative and the two hydrogens slightly positive.

based chemicals forms carbon dioxide and water. Consider, for example, the combustion of methane, the major component of natural gas:

$$CH_4 + 2\,O_2 \rightarrow CO_2 + 2\,H_2O + \text{energy}$$

This gain of electrons by oxygen and loss of electrons by hydrogen is an energy-releasing process. Typically, the energy is released either as molecular kinetic energy (heat) or as light (the flame).

Combustion oxidation–reduction reactions occur throughout your body. You can visualize a simplified model of your metabolism by reviewing Figure 13.23 and substituting a food molecule for the methane. Food molecules relinquish their electrons to the oxygen molecules you inhale. The products are carbon dioxide, water vapor, and energy. You exhale the carbon dioxide and water vapor, but much of the energy from the reaction is used to keep your body warm and to drive the many other biochemical reactions necessary for living.

13.8 The Rate of Chemical Reactions

Some chemical reactions, such as the rusting of iron, are slow, whereas others, such as the burning of gasoline, are fast. The speed of any reaction is determined by its **reaction rate**, the rate at which the concentration of products increases (or equivalently, the rate at which reactants decrease).

What determines the rate of a reaction? The answer is complex, but one important factor is that reactant molecules must physically come together, or collide. We can illustrate the relationship between molecular collisions and reaction rate by considering Figure 13.24. Because reactant molecules must collide in order for a reaction to occur, the rate of a reaction can be increased by increasing the rate of collisions. One effective way to increase the rate of collisions is to increase the concentration of reactants. The reason is simply that, with higher concentrations, there are more molecules in a given volume, which makes collisions more probable (Figure 13.25).

However, not all collisions lead to product formation. In order for reactant molecules to convert to product molecules, reactant molecules must collide at the right orientation to one another ("head-on" collisions are better at

INSIGHT The life sciences involve fantastic applications of the physical sciences, the chemistry of cellular respiration being just one example. Others include nitrogen fixation, photosynthesis, and molecular genetics. So there are distinct advantages to learning about the physical sciences *before* advancing to the life sciences.

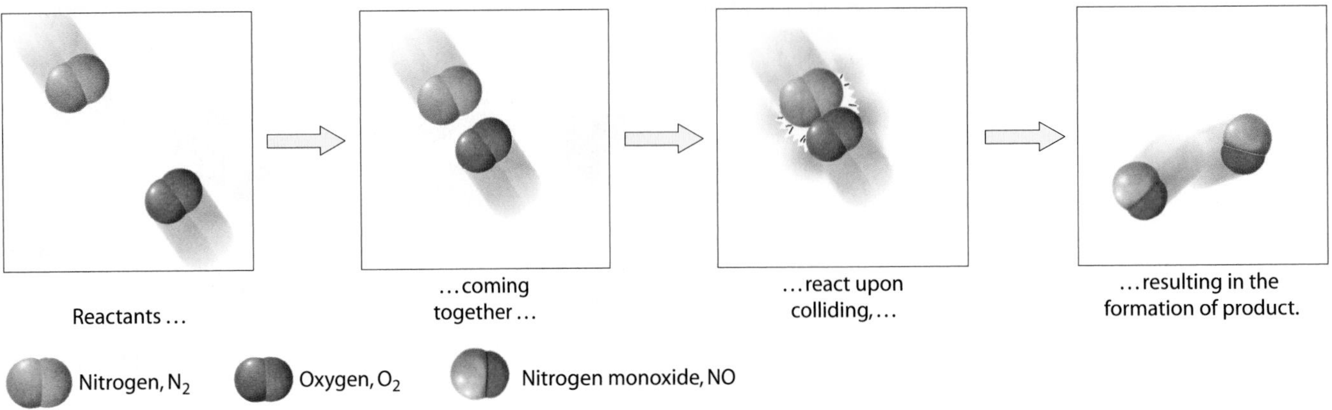

Figure 13.24 During a reaction, reactant molecules collide with one another.

breaking bonds than glancing blows). Also, for products to form, colliding molecules must possess enough kinetic energy to break their bonds. Only then is it possible for the atoms in the reactant molecules to change bonding partners and form product molecules. The bonds in N_2 and O_2 molecules, for example, are quite strong. In order for these bonds to be broken, collisions between the molecules must contain enough energy to break the bonds. As a result, collisions between slow-moving N_2 and O_2 molecules may not form NO, as is shown in Figure 13.26.

The energy required to break bonds sometimes comes from the absorption of electromagnetic radiation. As the radiation is absorbed by reactant molecules, atoms in the molecules may start to vibrate with so much energy that the bonds between them are easily broken. For example, the common atmospheric pollutant nitrogen dioxide, NO_2, may transform to nitrogen monoxide and atomic oxygen merely upon exposure to sunlight:

$$NO_2 + \text{sunlight} \rightarrow NO + O$$

The energy required to break bonds is also often supplied by heat, as explored in the following *Integrated Science* section.

INTEGRATED SCIENCE PHYSICS, BIOLOGY, AND EARTH SCIENCE

The Effect of Temperature on Reaction Rate

Higher temperatures tend to increase reaction rates. The reason is that the higher the temperature of a material, the faster its molecules are moving, and the more forceful the collisions between them so the more likely that these collisions will break bonds within reactant molecules.

Whether the result of collisions, absorption of electromagnetic energy, or both, broken bonds are a necessary first step in most chemical reactions. The

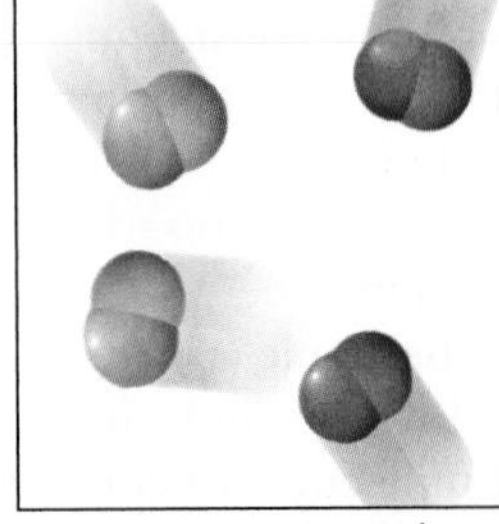

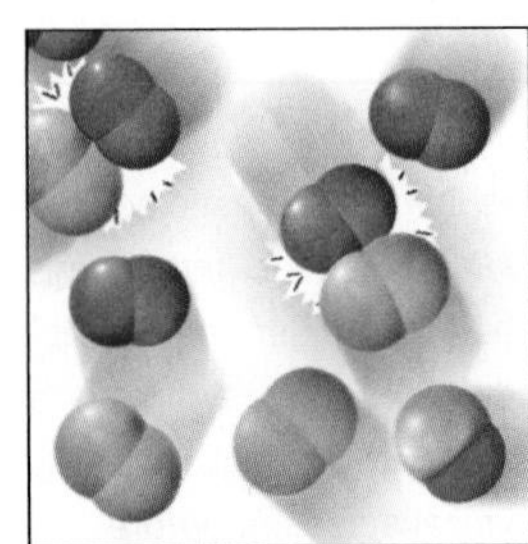

Figure 13.25 The more concentrated a sample of nitrogen and oxygen, the greater the likelihood that N_2 and O_2 molecules will collide and form nitrogen monoxide.

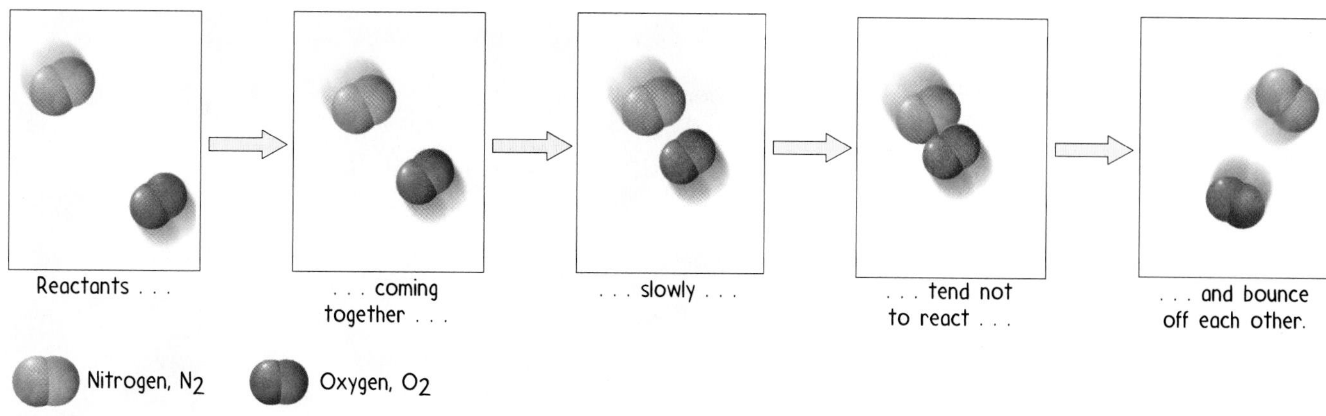

Figure 13.26 Slow-moving molecules may collide without enough force to break the bonds. In this case, they cannot react to form product molecules.

energy required to get this bond-breaking process moving can be viewed as an energy barrier. The minimum energy required to overcome the energy barrier is known as the *activation energy*, E_a. In the reaction between nitrogen and oxygen to form nitrogen monoxide, the energy barrier is so high (because the bonds in N_2 and O_2 are so strong) that only the fastest nitrogen and oxygen molecules possess enough energy to react. Figure 13.27 shows the energy barrier in this reaction as a vertical hump.

At any given temperature, there is a wide distribution of kinetic energies among reactant molecules. Some are moving slowly and others quickly. As discussed in Chapter 6, the temperature of a material is simply the *average* of all these kinetic energies. The few fast-moving reactant molecules in Figure 13.27 are the first to transform to product molecules because they are the molecules that have enough energy to pass over the energy barrier. When the temperature of reactants is increased, the number of reactant molecules having sufficient energy to pass over the energy barrier also increases, which is why reactions are generally faster at higher temperatures.

Most chemical reactions are influenced by temperature in this manner. For example, the nitrogen and oxygen molecules that make up our atmosphere are always colliding with one another. But at the ambient temperatures of our atmosphere, these molecules generally do not have sufficient kinetic energy to allow for the formation of the molecule nitrogen monoxide. The heat of a lightning bolt, however, dramatically increases the kinetic energy of these molecules, to the point that a large portion of the collisions in the vicinity of the bolt result in the formation of nitrogen monoxide, as shown at the beginning of this chapter. The

Figure 13.27 Reactant molecules must gain a minimum amount of energy, called the activation energy E_a, in order to transform into product molecules.

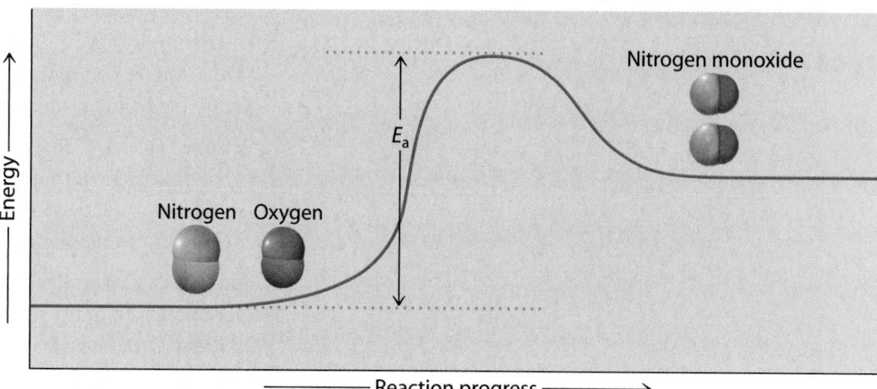

Figure 13.28 This alligator became immobilized on the pavement after being caught in the cold night air. By midmorning, shown here, the temperature had warmed sufficiently to allow the alligator to get up and walk away.

nitrogen monoxide formed in this manner undergoes further atmospheric reactions to form chemicals known as *nitrates* that plants depend on to survive.

The chemical reactions occurring in living bodies are no exception to the rule; these also tend to proceed faster at higher temperatures. The body temperature of animals that regulate their own temperature, (called *endotherms*, as you will learn more about in Chapter 18) is fairly constant. However, the body temperature of some animals (*ectotherms*), such as the alligator shown in Figure 13.28, rises and falls with the temperature of the environment. On a warm day, the chemical reactions occurring in the alligator are "up to speed" and the animal can be most active. On a chilly day, however, the chemical reactions proceed at a lower rate, and as a consequence, the alligator's movements are unavoidably sluggish.

CHECK YOURSELF

1. What can you deduce about the activation energy of a reaction that takes billions of years to go to completion? How about a reaction that takes only a fraction of a second?
2. What kitchen device is used to lower the rate at which microorganisms grow on food?
3. What effect would increased pressure tend to have on the rate of a chemical reaction? Why do you think so?

CHECK YOUR ANSWERS

1. If reactants and products are in physical contact yet the reaction takes billions of years to proceed, you can conclude that rarely do reactant molecules have enough energy to get over the energy barrier. Either high activation energy is required for the reaction, or there is little energy available to the molecules from ambient heat or radiation. A fast reaction has low activation energy in comparison with energy available to the molecules.
2. The refrigerator! Microorganisms, such as bread mold, are everywhere and difficult to avoid. By lowering the temperature of microorganism-contaminated food, the refrigerator decreases the rate of the chemical reactions that these microorganisms depend on for growth, thereby increasing the food's shelf life.
3. Increased pressure generally corresponds to increased reaction rate for the same reason increased temperature does—molecules strike one another with greater force and therefore chemical bonds are more likely to break.

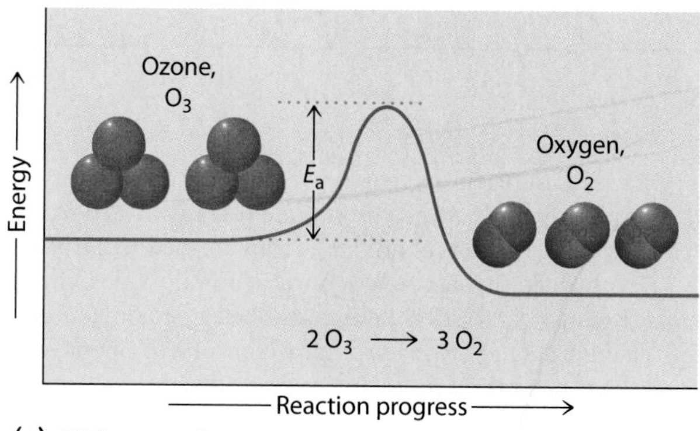

(a) Without catalyst

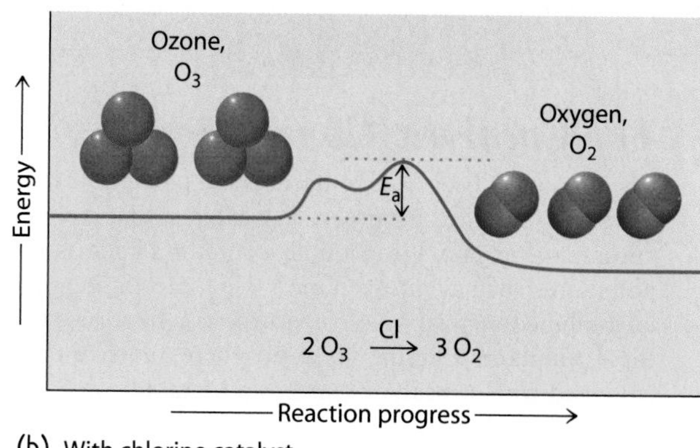

(b) With chlorine catalyst

Figure 13.29 (a) The relatively high energy barrier indicates that only the most energetic ozone molecules can react to form oxygen molecules. (b) Chlorine atoms lower the energy barrier, which means more reactant molecules have sufficient energy to form product. The chlorine allows the reaction to proceed in two steps, and the two smaller energy barriers correspond to these steps.

A chemical reaction can be made to go faster if the reactants absorb heat or electromagnetic radiation, as you have seen. There is a third way to increase the rate of a reaction: adding a catalyst. A **catalyst** is a substance that increases the rate of a chemical reaction by lowering its activation energy. The catalyst may participate as a reactant but it is then regenerated as a product and is thus available to catalyze subsequent reactions.

The conversion of ozone, O_3, to oxygen O_2, is normally sluggish because the reaction has a relatively high energy barrier, as shown in Figure 13.29a. However, when chlorine atoms act as a catalyst, the energy barrier is lowered, as shown in Figure 13.29b, and the reaction is able to proceed faster. Atomic chlorine lowers the energy barrier of this reaction by providing an alternative pathway involving intermediate reactions, each having a lower activation energy than the uncatalyzed reaction. Chlorine atoms in the atmosphere catalyze the destruction of Earth's protective stratospheric ozone shield layer, an issue you will investigate further in Chapter 25.

Figure 13.30 For the chemical reaction taking place in burning wood, there is a net release of energy. For those taking place in a photosynthetic plant, there is a net absorption of energy.

13.9 Endothermic Versus Exothermic Reactions

All chemical reactions involve transfer of energy. For some reactions there is a *production* of energy. This energy can be considered as a reaction product and it is sometimes written along with the actual chemical products as follows:

$$\text{reactants} \rightarrow \text{products} + \text{energy}$$

In such a reaction, the energy produced is released to the environment and the reaction is called **exothermic**. Rocket ships lift off into space and campfires glow red hot as a result of exothermic reactions.

If a chemical reaction doesn't produce energy, then it *consumes* energy. In such a case, the energy is a necessary ingredient for the reactants to transform into products. The energy itself, therefore, can be depicted as a reactant:

$$\text{energy} + \text{reactants} \rightarrow \text{products}$$

In such a reaction, the consumed energy is absorbed from the environment and the reaction is called **endothermic**. Photosynthesis, for example, involves a series of endothermic reactions that are driven by the energy of sunlight. Both exothermic and endothermic reactions, illustrated in Figure 13.30, can be understood through the concept of bond energy.

TECHNOLOGY

The Catalytic Converter

Chemists have been able to harness the power of catalysts for numerous practical purposes. The exhaust that comes from an automobile engine, for example, contains a wide assortment of pollutants, such as nitrogen monoxide, carbon monoxide, and uncombined fuel vapors (hydrocarbons). To reduce the amount of these pollutants entering the atmosphere, most automobiles are equipped with a catalytic converter. Metal catalysts in a converter speed up reactions that convert exhaust pollutants to less toxic substances. Nitrogen monoxide is transformed to nitrogen and oxygen, carbon monoxide is transformed to carbon monoxide, and unburned fuel is converted to carbon dioxide and water vapor. Because catalysts are not consumed by the reactions they facilitate, a single catalytic converter may continue to operate effectively for the lifetime of the car.

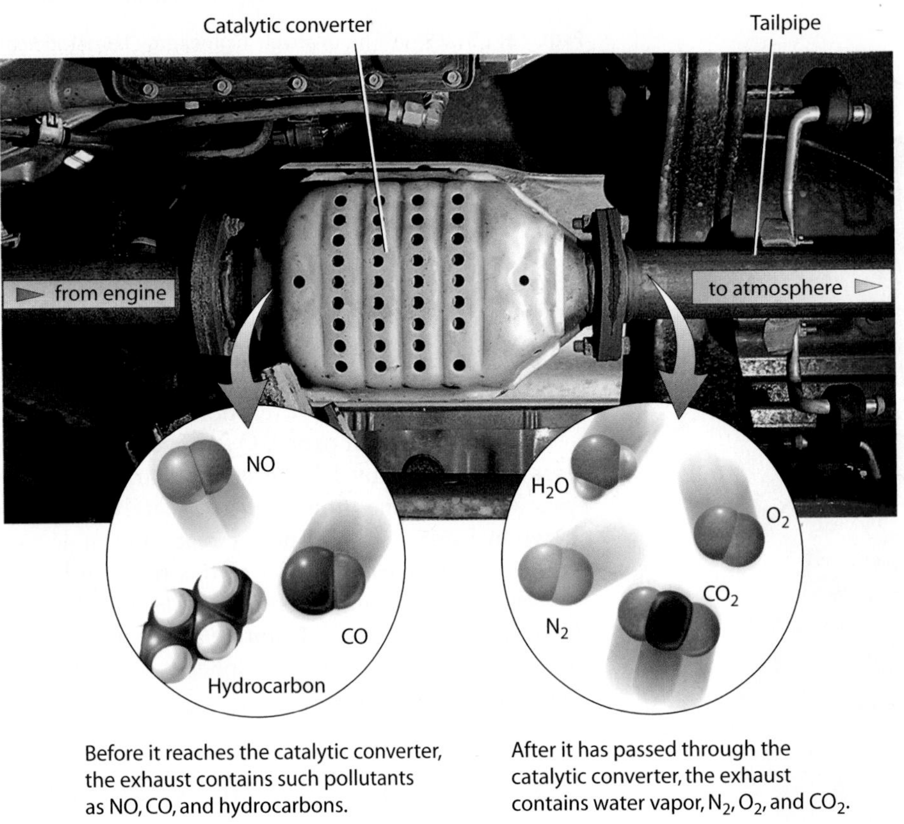

Before it reaches the catalytic converter, the exhaust contains such pollutants as NO, CO, and hydrocarbons.

After it has passed through the catalytic converter, the exhaust contains water vapor, N_2, O_2, and CO_2.

INSIGHT Most exothermic reactions, once started, proceed on their own, ceasing only when there are no more reactants to react. A campfire, for example, will continue to burn only so long as there is a supply of wood and atmospheric oxygen. Most endothermic reactions, by contrast, require the continual input of energy in order to proceed. Plants, for example, cannot photosynthesize without sunlight.

During a chemical reaction, as you learned in the previous section, chemical bonds are broken and atoms rearrange to form new chemical bonds. Such breaking and forming of chemical bonds involves changes in energy. As an analogy, consider a pair of magnets. To separate them requires an input of "muscle energy." Conversely, when the two separated magnets collide, they become slightly warmer than they were, and this warmth is evidence of energy released. Energy must be absorbed by the magnets if they are to break apart, and energy is released as they come together. The same principle applies to atoms. To pull bonded atoms apart requires an energy input. When atoms combine, there is an energy output, usually in the form of faster-moving atoms and molecules, electromagnetic radiation, or both.

The amount of energy required to pull two bonded atoms apart is the same as the amount released when they are brought together. This energy absorbed as a bond breaks or released as one forms is called bond energy. Each chemical bond has its own characteristic bond energy. The hydrogen–hydrogen bond energy, for example, is 436 kilojoules per mole. This means that 436 kilojoules of energy are absorbed as 1 mole of hydrogen–hydrogen bonds break apart, and

Table 13.2 Selected Bond Energies

Bond	Bond Energy (kJ/mole)	Bond	Bond Energy (kJ/mole)
H—H	436	O—O	138
H—C	414	Cl—Cl	243
H—N	389	N—N	159
H—O	464	N=O	631
H—F	569	O=O	498
H—Cl	431	O=C	803
H—S	339	N≡N	946
C—C	347	C≡C	837

436 kilojoules of energy are released upon the formation of 1 mole of hydrogen–hydrogen bonds. Different bonds involving different elements have different bond energies, as Table 13.2 shows. You can refer to the table as you study this section, but please do not memorize these bond energies. Instead, focus on understanding what they mean.

By convention, a positive bond energy represents the amount of energy absorbed as a bond breaks and a negative bond energy represents the amount of energy released as a bond forms. Thus when you are calculating the net energy released or absorbed during a reaction, you'll need to be careful about plus and minus signs. It is standard practice when doing such calculations to assign a plus sign to energy absorbed and a minus sign to energy released. For instance, when dealing with a reaction in which 1 mole of H—H bonds are broken, you'll write +436 kilojoules to indicate energy absorbed, and when dealing with the formation of 1 mole of H—H bonds, you'll write −436 kilojoules to indicate energy released. We'll do some sample calculations in a moment.

CHECK YOURSELF

Do all covalent single bonds have the same bond energy?

CHECK YOUR ANSWER

Bond energy depends on the types of atoms bonding. The H—H single bond, for example, has a bond energy of 436 kilojoules per mole, but the H—O single bond has a bond energy of 464 kilojoules per mole. All covalent single bonds do not have the same bond energy.

Exothermic Reactions

For any chemical reaction, the total amount of energy absorbed in breaking bonds in reactants is always different from the total amount of the energy released as bonds form in the products. Consider the reaction in which hydrogen and oxygen react to form water:

$$\text{H—H} + \text{H—H} + \text{O=O} \longrightarrow \text{H—O—H} + \text{H—O—H}$$

In the reactants, hydrogen atoms are bonded to hydrogen atoms and oxygen atoms are double-bonded to oxygen atoms. The total amount of energy absorbed as these bonds break is as follows:

Type of Bond	Number of Moles	Bond Energy	Total Energy
H—H	2	+436 kJ/mole	+872 kJ
O=O	1	+498 kJ/mole	+498 kJ
Total energy absorbed +1370 kJ			

Figure 13.31 A space shuttle uses exothermic chemical reactions to lift off from the Earth's surface.

In the products there are four hydrogen–oxygen bonds. The total amount of energy released as these bonds form is:

Type of Bond	Number of Moles	Bond Energy	Total Energy
H—O	4	−464 kJ/mole	−1856 kJ
Total energy released −1856 kJ			

For this reaction the amount of energy released exceeds the amount of energy absorbed. The net energy of the reaction is found by adding the two quantities:

$$\begin{aligned}\text{net energy of reaction} &= \text{energy absorbed} + \text{energy released}\\ &= +1370\text{ kJ} + (-1856\text{ kJ})\\ &= -486\text{ kJ}\end{aligned}$$

The negative sign on the net energy indicates that there is a net *release* of energy, and so the reaction is exothermic:

$$2\,H_2 + O_2 \rightarrow 2\,H_2O + \text{energy}$$

The amount of energy released in an exothermic reaction depends on the amount of reactants. The reaction of large amounts of hydrogen and oxygen, for example, provides the energy to lift the space shuttle shown in Figure 13.31 into orbit. There are two compartments in the large central tank to which the orbiter is attached—one filled with liquid hydrogen and the other with liquid oxygen. Upon ignition, these two liquids mix and react chemically to form water vapor, which produces the needed thrust as it is expelled out the rocket cones.

It is important to realize that the energy released by an exothermic reaction is not created by the reaction. This is in accord with the Law of Conservation of Energy, which states that energy is neither created nor destroyed in a chemical reaction. Instead, it is merely converted from one form to another. During an exothermic reaction, energy that was once in the form of the potential energy of the chemical bonds is released as the kinetic energy of fast-moving molecules and/or electromagnetic radiation.

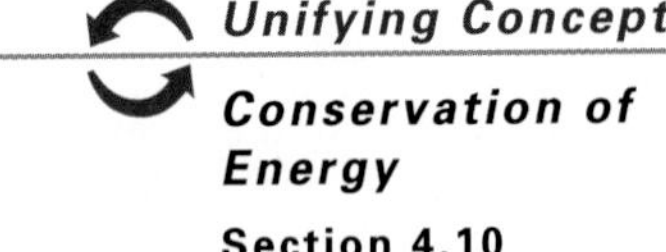

Unifying Concept

Conservation of Energy

Section 4.10

CHECK YOURSELF

Where does the net energy released in an exothermic reaction go?

CHECK YOUR ANSWER

This energy goes into making atoms and molecules move faster and/or into the formation of electromagnetic radiation.

Endothermic Reactions

Many chemical reactions are endothermic, such that the amount of energy released as products form is *less* than the amount of energy absorbed in the breaking of bonds in the reactants. An example is the reaction of atmospheric nitrogen and oxygen to form nitrogen monoxide, which is a reaction that occurs wherever air is heated to very high temperatures, such as within your car's engine:

$$N{=}N + O{=}O \rightarrow N{=}O + N{=}O$$

The amount of energy absorbed as the chemical bonds in the reactants break is

Type of bond	Number of moles	Bond Energy	Total Energy
N═N	1	+946 kJ/mole	+946 kJ
O═O	1	+498 kJ/mole	+498 kJ
Total energy absorbed +1444 kJ			

The amount of energy released upon the formation of bonds in the products is

Type of Bond	Number of Moles	Bond Energy	Total Energy
N═O	2	−631 kJ/mole	−1262 kJ
Total energy released −1262 kJ			

As before, the net energy of the reaction is found by adding the two quantities:

$$\begin{aligned}\text{net energy of reaction} &= \text{energy absorbed} + \text{energy released}\\ &= +1444\text{ kJ} + (-1262\text{ kJ})\\ &= +182\text{ kJ}\end{aligned}$$

The positive sign indicates that there is a net *absorption* of energy, meaning the reaction is endothermic:

$$\text{energy} + N_2 + O_2 \rightarrow 2\text{ NO}$$

13.10 Entropy and Chemical Reactions

The energy in a hot cup of coffee doesn't stay in the cup—it spreads out into the cooler surroundings. Similarly, the concentrated chemical energy found in gasoline disperses as heat of many smaller, lower-energy molecules upon combustion. Some of this heat is converted to mechanical energy as the car moves. The rest spreads into the engine block, radiator fluid, or out the exhaust pipe. As these and countless everyday examples show, heat tends to move away from where it is concentrated.

Unifying Concept

The Second Law of Thermodynamics

Section 6.5

As you recall from Chapter 6, the tendency of heat and other forms of energy to become less localized is known as the *second law of thermodynamics*. The term **entropy** describes the degree to which energy has become dispersed. For example, an ice cube gains entropy as it melts and a puddle gains entropy as it evaporates. So the second law of thermodynamics can be stated in terms of entropy rather than heat: *the entropy of an isolated system tends to increase with time*.

Applied to chemistry, the concept of entropy helps us to answer a most fundamental question: If you take two materials and put them together, will they react to form new materials? If the reaction results in an overall increase in entropy, then the answer is yes. Conversely, if the reaction results in an overall decrease in entropy, then the reaction will *not* occur by itself, without a net input of energy.

A quick way to determine whether or not a reaction might be favorable is to assess whether the reaction leads to an overall increase or decrease in entropy.
Because it is natural for energy to disperse, a reaction that leads to an increase in entropy will likely occur, while a reaction that leads to a decrease in entropy will not likely occur.

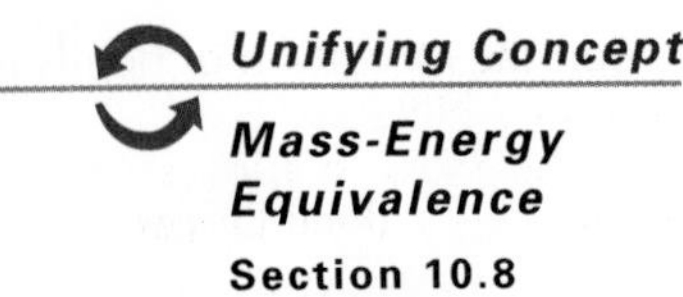

Can you now see why exothermic reactions are self-sustaining, while most endothermic reactions need a continual prodding? Exothermic reactions spread energy out to the surroundings, much like a cooling cup of coffee. This is an increase in entropy; hence, exothermic reactions are favored to occur. An endothermic reaction, by contrast, requires that energy from the surroundings be absorbed by the reactants. This is a concentration of energy, which is counter to energy's natural tendency to disperse. Endothermic reactions, therefore, can only be sustained with the continual input of some external source of energy.*

Photosynthesis is an endothermic reaction of fundamental importance to life. The entropy of the products (complex sugars) is less than that of the reactants (carbon dioxide and water). So in order to occur, photosynthesis requires the input of energy. What is the source of this energy? The Sun—it provides energy for life-sustaining endothermic reactions on Earth by undergoing exothermic nuclear reactions itself (Figure 13.32). (See Chapters 10 and 27 for more on how the Sun produces usable energy by converting its own mass to radiant energy in accord with $E = mc^2$.)

Figure 13.32 Some of the Sun's dispersed energy is used to drive endothermic reactions that allow for the functioning of living organisms.

* There are examples of endothermic reactions that proceed spontaneously, absorbing heat from the environment and increasing in entropy. A classic example is the mixing of a salt in water. In such cases, entropy increases not by a release of energy, but by the dispersion of energy-containing atoms and molecules into solution.

CHAPTER 13 Review

Summary of Terms

Acid A substance that donates hydrogen ions.
Acidic solution A solution in which the hydronium ion concentration is higher than the hydroxide ion concentration.
Amphoteric A description of a substance that can behave as either an acid or a base.
Base A substance that accepts hydrogen ions.
Basic solution A solution in which the hydroxide ion concentration is higher than the hydronium ion concentration.
Bond energy The amount of energy either absorbed as a chemical bond breaks or released as a bond forms.
Catalyst A substance that increases the rate of a chemical reaction by lowering its activation energy.
Chemical equation A representation of a chemical reaction.
Chemical reaction A rearrangement of atoms so that one or more new compounds are formed from preexisting compounds or elements.
Coefficient A number used in a chemical equation to indicate either the number of atoms/molecules or the number of moles of a reactant or product.
Combustion An exothermic oxidation–reduction reaction between a nonmetallic material and molecular oxygen.
Corrosion The deterioration of a metal, typically caused by atmospheric oxygen.
Endothermic A chemical reaction in which heat energy is absorbed from the environment.
Exothermic A chemical reaction in which heat energy is released to the environment.
Half-reaction One portion of an oxidation–reduction reaction, represented by an equation showing electrons as either reactants or products.
Hydronium ion A water molecule after accepting a hydrogen ion.
Hydroxide ion A water molecule after losing a hydrogen ion.
Law of Mass Conservation Matter is neither created nor destroyed during a chemical reaction.
Neutral solution A solution in which the hydronium ion concentration is equal to the hydroxide ion concentration.
Neutralization A reaction in which an acid and a base combine to form a salt.
Oxidation The process whereby a reactant loses one or more electrons.
pH A measure of the acidity of a solution, equal to the negative of the base-10 logarithm of the hydronium ion concentration.
Product A new material formed in a chemical reaction, appearing after the arrow in a chemical equation.
Reactant A starting material in a chemical reaction, appearing before the arrow in a chemical equation.
Reaction rate A measure of how quickly the concentration of products in a chemical reaction increases or the concentration of reactants decreases.
Reduction The process whereby a reactant gains one or more electrons.
Salt An ionic compound formed from the reaction between an acid and a base.

Review Questions

13.1 Chemical Reactions and Chemical Equations

1. What do the letters (s), (l), (g), and (aq) stand for in a chemical equation?
2. Why is it important that a chemical equation be balanced?
3. Why is it important never to change a subscript in a chemical formula when balancing a chemical equation?

13.2 Acid–Base Reactions

4. What is an acid? What is a base?
5. What is a hydronium ion? What is a hydroxide atom?
6. When ammonia is added to water, does ammonia behave as an acid or a base? Does water behave as an acid or a base?

13.3 Salts

7. How is the term *salt* as used in chemistry differ from how it is used in everyday life?
8. Which is more corrosive—a salt or the acid and base from which it formed?

13.4 Solutions: Acidic, Basic, or Neutral

9. How does water behave as an acid and a base at the same time?
10. What is true about the relative concentrations of hydronium and hydroxide ions in an acidic solution? How about a neutral solution? A basic solution?

13.5 The pH Scale

11. What pH values indicate that a solution is basic? That it is acidic?
12. As the hydronium ion concentration of a solution increases, does the pH of the solution increase or decrease?

13.6 Oxidation–Reduction Reactions

13. What happens to the electrons lost by one chemical in an oxidation reaction?
14. Does the gain of hydrogen atoms indicate an oxidation or a reduction reaction? Does the gain of oxygen atoms indicate an oxidation or a reduction reaction?

13.7 Corrosion and Combustion

15. What metal coats a galvanized nail?
16. What are some differences between corrosion and combustion?

13.8 The Rate of Chemical Reactions

17. Why does increasing the concentration of reactants increase the rate of a chemical reaction?
18. What are three ways to increase the rate at which a chemical reaction goes to completion?

13.9 Endothermic Versus Exothermic Reactions

19. If it takes 436 kilojoules to break a bond, how many kilojoules are released when the same bond is formed?
20. What is released by an exothermic reaction?
21. What is absorbed by an endothermic reaction?

13.10 Entropy and Chemical Reactions

22. As energy disperses, where does it go?
23. If you take two materials and put them together, will they react to form new materials?

Multiple Choice Questions

Choose the BEST answer to the following.

1. What coefficients balance the following equation?

$$\_\_\_\_ P_4(s) + \_\_\_\_ H_2(g) \_\_\_\_ PH_3(g)$$

2. Sodium hydroxide (NaOH) is a strong base, which means that it readily accepts hydrogen ions. What products are formed when sodium hydroxide accepts a hydrogen ion from a water molecule?
   (a) water and sodium hydroxide
   (b) sodium hydroxide and hydronium ions
   (c) sodium ions and hydronium ions
   (d) sodium ions and water
3. What happens to the pH of an acidic solution as water is added?
   (a) The pH is not influenced by the addition of water.
   (b) The pH will decrease as the solution becomes more dilute.
   (c) The pH will increase as the solution becomes more dilute.
   (d) The pH will decrease since more hydronium ions are produced from the water.
4. Lakes lying in granite basins, such as those in the northeastern United States, tend to become acidified by acid rain more readily than lakes lying in limestone basins, such as those found in the midwestern United States. Why is this so?
   (a) The granite contains acidic minerals that dissolve in the rainwater and are added to the lakes.
   (b) Since the acidified water cannot percolate through the granite, the acid in the lakes becomes more concentrated.
   (c) The limestone, which is calcium carbonate, serves to neutralize the acid in the rain.
   (d) There is more acid rain in the northeast than in the midwestern regions of the United States.
5. What element is oxidized in the following equation and what element is reduced?

$$I_2 + 2\ Br^- \rightarrow 2\ I^- + Br_2$$

   (a) Iodine (I) is oxidized, while the bromine ion (Br^-) is reduced.
   (b) Iodine (I) is reduced, while the bromine ion (Br^-) is oxidized.
   (c) Both the iodine (I) and the bromine ion (Br^-) are reduced.
   (d) Both the iodine (I) and the bromine ion (Br^-) are oxidized.
6. The general chemical equation for photosynthesis is shown here. Through this reaction, are the oxygens of the water molecules (H_2O) oxidized or reduced?

$$6\ CO_2 + 6\ H_2O \rightarrow C_6H_{12}O_6 + 6\ O_2$$

   (a) The oxygens of the water molecules are oxidized.
   (b) The oxygens of the water molecules are reduced.
   (c) The oxygens of some of these water molecules are oxidized while others are reduced.
   (d) The oxygens of the water molecules are neither oxidized nor reduced.
7. One of the products of combustion is water. Why doesn't this water extinguish the combustion?
   (a) While areas of combustion are being extinguished, new areas are combusting.
   (b) Combustion only produces micro amounts of water.
   (c) The chemical combustion reaction is happening too fast for the water to have an effect on the fire.
   (d) This water is in the gaseous phase and merely floats away from the fire.
8. The yeast in bread dough feeds on sugar to produce carbon dioxide. Why does the dough rise faster in a warmer area?
   (a) There are a greater number of effective collisions among reacting molecules.
   (b) Atmospheric pressure decreases with increasing temperature.
   (c) The yeast tends to "wake up" in warmer temperatures, which is why baker's yeast is best stored in the refrigerator.
   (d) The rate of evaporation increases with increasing temperature.
9. Is the synthesis of ozone (O_3) from oxygen (O_2) an example of an exothermic or endothermic reaction?
   (a) Exothermic, because ultraviolet light is emitted during its formation.
   (b) Endothermic, because ultraviolet light is emitted during its formation.

(c) Exothermic, because ultraviolet light is absorbed during its formation.
(d) Endothermic, because ultraviolet light is absorbed during its formation.

10. How is it possible to cause an endothermic reaction to proceed when the reaction causes energy to become less dispersed?
(a) The reaction should be placed in a vacuum.
(b) The reaction should be cooled down.
(c) The concentration of the reactants should be increased.
(d) The reaction should be heated.

INTEGRATED SCIENCE CONCEPTS

Earth Science—Acid Rain

1. Why do lakes lying in granite basins tend to become acidified by acid rain more readily than lakes lying in limestone basins?
2. What is the definition of acid rain?
3. How does the burning of coal and oil contribute to acid rain?

Physics—Fuel Cells

1. Why don't the electrodes of a fuel cell deteriorate the way the electrodes of a battery do?
2. In one type of fuel cell the following oxidation–reduction reaction is taking place: $2\ H_2 + O_2 \rightarrow 2\ H_2O$. What is the fuel?
(a) H_2
(b) O_2
(c) H_2O
(d) all of the above
(e) none of the above

Physics, Biology, and Earth Science—The Effect of Temperature on Reaction Rate

1. In terms of reaction rates, why is lightning good for plants?
2. Why do ectotherms move sluggishly in cold weather?

Active Exploration

Rainbow Cabbage

The pH of a solution can be approximated with a *pH indicator*, which is any chemical whose color changes with pH. Many pH indicators are found in plants; the pigment of red cabbage is a good example. This pigment is red at low pH values (1 to 5), light purple around neutral pH values (6 to 7), light green at moderately alkaline pH values (8 to 11), and dark green at very alkaline pH values (12 to 14).

Safety Note

Wear safety glasses. Do not use bleach products because they will oxidize the pigment, rendering it insensitive to any changes in pH. You also do not want to run the risk of accidentally mixing a bleach solution with the toilet-bowl cleaner because the resulting solution would generate harmful chlorine gas.

What You Need

Head of red cabbage, small pot, water, four colorless plastic cups or drinking glasses, toilet-bowl cleaner, vinegar, baking soda, ammonia cleanser

Procedure

1. Shred about a quarter of the head of red cabbage and boil the shredded cabbage in 2 cups of water for about 5 minutes. Strain and collect the broth, which contains the pH-indicating pigment.
2. Pour one-fourth of the broth into each cup. (If the cups are plastic, either allow the broth to cool before pouring or dilute with cold water.)
3. Add a small amount of toilet-bowl cleaner to the first cup, a small amount of vinegar to the second cup, baking soda to the third, and ammonia solution to the fourth.
4. Use the different colors to estimate the pH of each solution.
5. Mix some of the acidic and basic solutions together and note the rapid change in pH (indicated by the change in color).

Exercises

1. Balance these equations:
a. ___ $Fe(s)$ + ___ $O_2(g)$ → ___ $Fe_2O_3(s)$
b. ___ $H_2(g)$ + ___ $N_2(g)$ → ___ $NH_3(g)$
c. ___ $Cl_2(g)$ + ___ $KBr(aq)$ → ___ $Br_2(l)$ + ___ $KCl(aq)$
d. ___ $CH_4(g)$ + ___ $O_2(g)$ → ___ $CO_2(g)$ + ___ $H_2O(l)$
2. Is the following chemical equation balanced?

$$4\ C_6H_7N_5O_{16}(s) + 19\ O_2(g) \rightarrow 24\ CO_2(g) + 20\ NO_2(g) + 14H_2O(g)$$

Use the following illustration to answer Exercises 3–5.

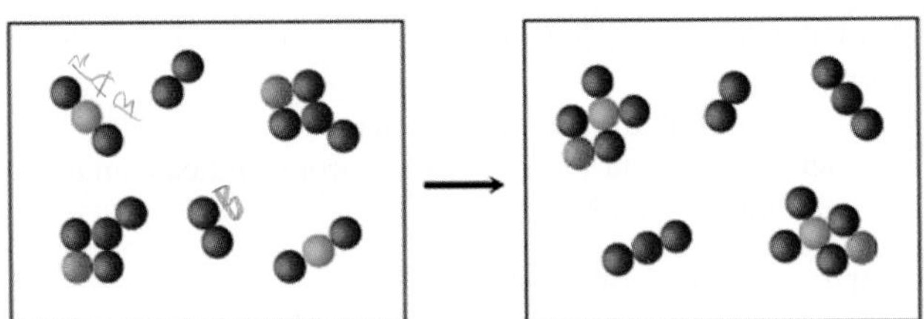

3. Is this reaction balanced?
4. How many diatomic molecules are represented?
5. Which equation best describes this reaction?
a. $2\ AB_2 + 2\ DCB_3 + B_2 \rightarrow 2\ DBA_4 + 2\ CA_2$
b. $2\ AB_2 + 2\ CDA_3 + B_2 \rightarrow 2\ C_2A_4 + 2\ DBA$
c. $2\ AB_2 + 2\ CDA_3 + A_2 \rightarrow 2\ DBA_4 + 2\ CA_2$
d. $2\ BA_2 + 2\ DCA_3 + A_2 \rightarrow 2\ DBA_4 + 2\ CA_2$
6. What is the relationship between a hydroxide ion and a water molecule?
7. What atom in the hydronium ion, H_3O^+, bears the positive charge?
8. Identify the acid or base behavior of each substance in these reactions:
a. $H_3O^+ + Cl^- \rightleftarrows H_2O + HCl$

____ ____ ____ ____

b. $H_2PO_4 + H_2O \rightleftarrows H_3O^+ + HPO_4^-$

____ ____ ____ ____

9. What happens to the corrosive properties of an acid and a base after they neutralize each other? Why?
10. Why do we use the pH scale to indicate the acidity of a solution rather than simply stating the concentration of hydronium ions?
11. What is the concentration of hydronium ions in a solution that has a pH of −3? Why is such a solution impossible to prepare?
12. What happens to the pH of an acidic solution as pure water is added?
13. A weak acid is added to a concentrated solution of hydrochloric acid. Does the solution become more or less acidic?
14. Many of the smelly molecules of cooked fish are alkaline compounds. How might these smelly molecules be conveniently transformed into less smelly salts just prior to eating the fish?

15. What elements are oxidized in the following equations and what elements are reduced?
    a. $Sn_2^+ + 2\ Ag \rightarrow Sn + 2\ Ag^+$
    b. $I_2 + 2\ Br^- \rightarrow 2\ I^- + Br_2$
16. What element behaves as the oxidizing agent in each of the following equations and what element behaves as the reducing agent?
    a. $Sn_2^+ + 2\ Ag \rightarrow Sn + 2\ Ag^+$
    b. $I_2 + 2\ Br^- \rightarrow 2\ I^- + Br_2$
17. The general chemical equation for photosynthesis is shown below. Through this reaction is the carbon oxidized or reduced? Are the oxygen atoms of the water molecules being oxidized or reduced?

$$6\ CO_2 + 6\ H_2O \rightarrow C_6H_{12}O_6 + 6\ O_2$$

18. During strenuous exercise there is little oxygen, O_2, available for muscle cells. Under these conditions, the muscle cells derive most of their energy from the anaerobic conversion of pyruvic acid, $C_3H_4O_3$, into lactic acid, $C_3H_6O_3$. The buildup of lactic acid makes the muscles ache and fatigue quickly. Is the pyruvic acid oxidized or reduced as it transforms into lactic acid?
19. As we digest and subsequently metabolize food, is the food gradually oxidized or reduced? What evidence do you have?
20. Are the chemical reactions that take place in a disposable battery exothermic or endothermic? What evidence supports your answer? Is the reaction going on in a rechargeable battery while it is recharging exothermic or endothermic?
21. Why do exothermic reactions typically favor the formation of products?
22. As the Sun shines on a snow-capped mountain, much of the snow sublimes instead of melts. How is this favored by entropy?

Problems

1. When the hydronium ion concentration of a solution equals 1 mole per liter, what is the pH of the solution? Is the solution acidic or basic?
2. When the hydronium ion concentration of a solution equals 10 moles per liter, what is the pH of the solution? Is the solution acidic or basic?
3. Use the bond energies in Table 13.2 and the accounting format shown in Section 13.9 to determine whether these reactions are exothermic or endothermic:

$$H_2 + Cl_2 \rightarrow 2\ HCl$$
$$2\ HC{\equiv}CH + 5\ O_2 \rightarrow 4\ CO_2 + 2\ H_2O$$

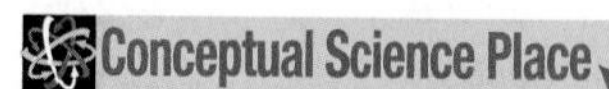

CHAPTER 13 ONLINE RESOURCES

Tutorials Chemical Reactions and Equations, The Nature of Acids and Bases, The pH Scale • ***Quiz*** • ***Exercises*** • ***Flashcards*** • ***Links***

Organic Chemistry

You perceive the flavor of vanilla when the sensory organs of your mouth and nose detect the organic compound vanillin. The flavor of chocolate, on the other hand, derives from a wide assortment of carbon-based molecules. One of the major flavor ingredients in chocolate is the organic compound tetramethylpyrazine, a ringlike structure built of nitrogen, carbon, and hydrogen atoms.

CHAPTER OUTLINE

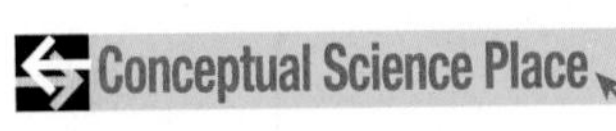

Organic Molecules and Isomers

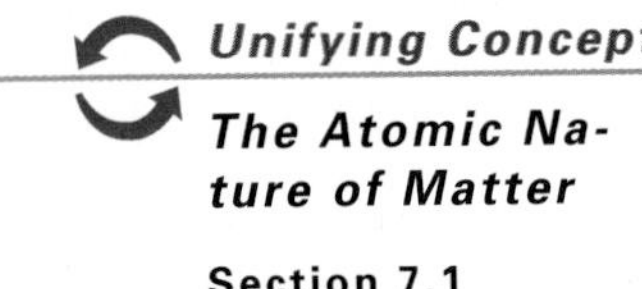

Like tiny toy Lego pieces, carbon atoms can link together in countless different orientations to form an endless diversity of molecules. Interestingly, carbon is the only element of the periodic table to have this property. Life itself is based upon this unique ability of carbon. Reflecting this fact, the branch of chemistry that is the study of carbon-containing compounds has come to be known as **organic chemistry**. (The term *organic* is derived from *organism* and is not necessarily related to the environment-friendly form of farming.) Organic chemicals can be isolated from nature, but they are also readily produced in the laboratory. Today, more than 13 million organic compounds are known, and about 100,000 new ones are added to the list each year. By contrast, there are only 200,000 to 300,000 known *inorganic* compounds, those based on elements other than carbon.

In this chapter we will explore some of the many forms and applications of organic molecules. Along the way, we will discover the answers to some important questions: What are the simplest examples of organic compounds? Is it possible for two different compounds to have the same chemical formula? How is gasoline made? Where does petroleum come from? How do drugs work? What is a polymer and how have polymers affected society?

14.1 Hydrocarbons

Organic compounds that contain only carbon and hydrogen are **hydrocarbons** (Figure 14.1). These differ from one another by the number of carbon and hydrogen atoms they contain. The simplest hydrocarbon is methane, CH_4, with only one carbon per molecule. Methane is the main component of natural gas. The hydrocarbon octane, C_8H_{18}, has eight carbons per molecule and is a component of gasoline. The hydrocarbon polyethylene contains hundreds of carbon and hydrogen atoms per molecule. Polyethylene is a plastic used to make many items, including milk containers and plastic bags.

Hydrocarbons also differ from one another in the way the carbon atoms connect to each other. Figure 14.2 shows the three hydrocarbons *n*-pentane, *iso*-pentane, and *neo*-pentane. These hydrocarbons all have the same molecular formula, C_5H_{12}, but are structurally different from one another. The carbon

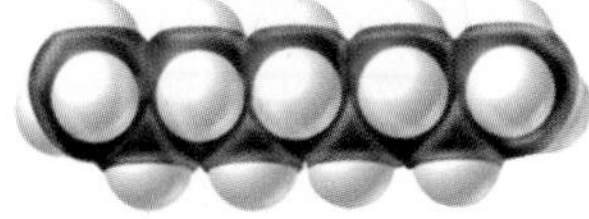

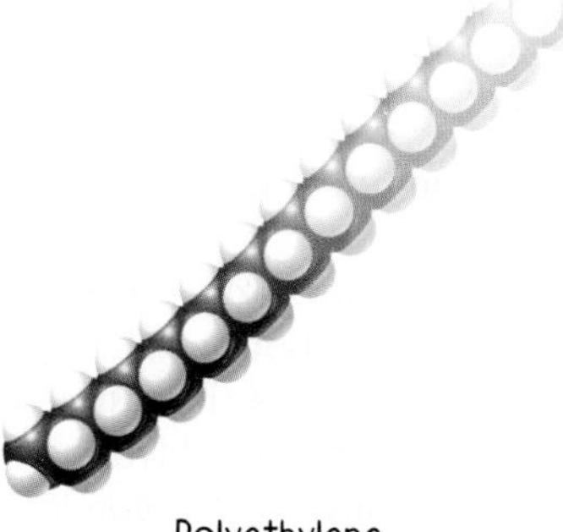

Figure 14.1 Three common hydrocarbons: methane, octane, and polyethylene.

framework of *n*-pentane is a chain of five carbon atoms. In *iso*-pentane, the carbon chain branches, so that the framework is a *four*-carbon chain branched at the second carbon. In *neo*-pentane, a central carbon atom is bonded to four surrounding carbon atoms.

We can see the different structural features of *n*-pentane, *iso*-pentane, and *neo*-pentane more clearly by drawing the molecules in two dimensions, as shown in the middle row of Figure 14.2. Alternatively, we can represent them by the *stick structures* shown in the bottom row. A stick structure is a commonly used, shorthand notation for representing an organic molecule. Each line (stick) represents a covalent bond, and carbon atoms are understood to be wherever two or more straight lines meet and at the end of any line (unless another type of atom is drawn at the end of the line). Any hydrogen atoms bonded to the carbons are also typically not shown. Instead, their presence is implied so that the focus can remain on the skeletal structure formed by the carbon atoms.

Molecules such as *n*-pentane, *iso*-pentane, and *neo*-pentane, which have the same molecular formula but different structures, are known as **structural isomers**. Structural isomers have different physical and chemical properties. For example, *n*-pentane has a boiling point of 36°C, *iso*-pentane's boiling point is 30°C, and *neo*-pentane's is 10°C. The number of possible structural isomers for a chemical formula increases rapidly as the number of carbon atoms increases. There are three structural isomers for compounds having the formula C_5H_{12}, 18 for C_8H_{18}, 75 for $C_{10}H_{22}$, and a whopping 366,319 for $C_{20}H_{42}$!

The hydrocarbons we use are obtained primarily from coal and petroleum. These fossil fuels are both formed from the remains of organisms that decayed under water in the absence of oxygen millions of years ago. (The formation of coal is discussed in more depth in Chapter 23.) Coal is a solid material containing many large, complex hydrocarbon molecules. Most of the coal mined today is used for the production of steel and for generating electricity at coal-burning power plants. Petroleum, or "crude oil," is a liquid readily separated into its hydrocarbon components through a process known as *fractional distillation*, shown in Figure 14.3. Distillation is a method of separating mixtures often used in chemistry. During distillation, a liquid is boiled to produce a vapor that is then

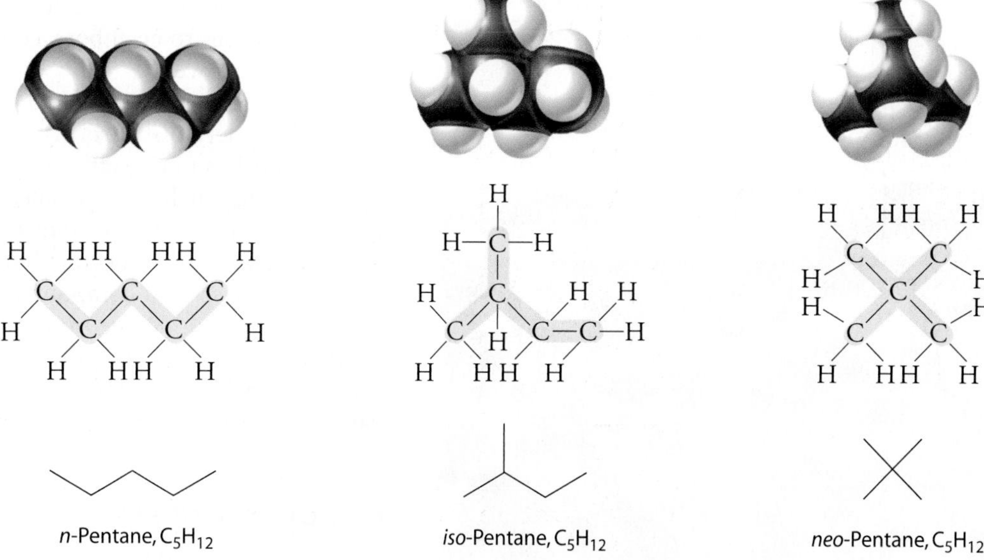

Figure 14.2 These three hydrocarbons all have the same molecular formula. We can see their different structural features by highlighting the carbon framework in two dimensions. Easy-to-draw stick structures that use lines for all carbon–carbon covalent bonds can also be used.

Figure 14.3 A schematic for the fractional distillation of petroleum into its useful hydrocarbon components.

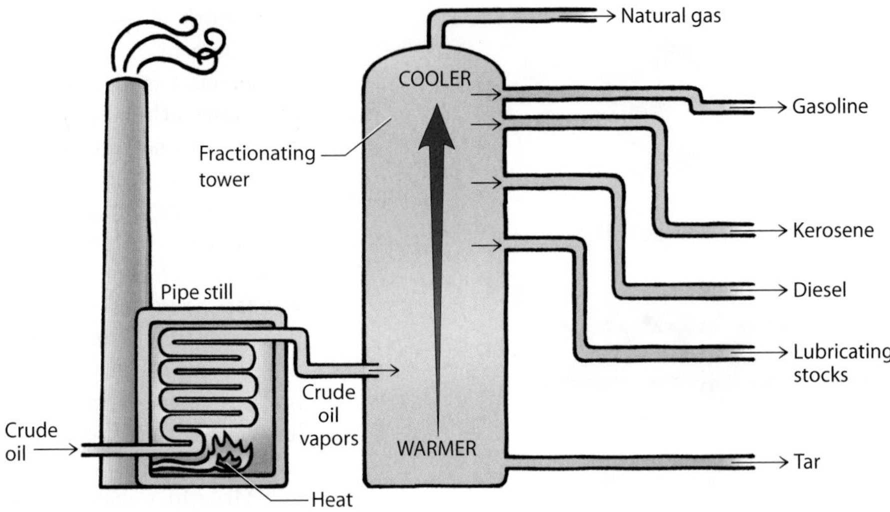

condensed again to a liquid. Fractional distillation refers specifically to the distillation of petroleum.

Petroleum is heated in a pipe still to a temperature high enough to vaporize most of the components. The hot vapor flows into the bottom of a fractionating tower, which is warmer at the bottom than at the top. As the vapor rises in the tower and cools, the various components begin to condense. Hydrocarbons that have high boiling points, such as tar and lubricating stocks, condense first at warmer temperatures. Hydrocarbons that have low boiling points, such as gasoline, travel to the cooler regions at the top of the tower before condensing. Pipes drain the various liquid hydrocarbon fractions from the tower. Natural gas, which is primarily methane, does not condense. It remains a gas and is collected at the top of the tower.

Section 7.1

Differences in the strength of molecular attractions explain why different hydrocarbons condense at different temperatures. Larger hydrocarbons experience many more attractions among themselves than smaller hydrocarbons do. For this reason, the larger hydrocarbons condense readily at high temperatures and so are found at the bottom of the tower. Smaller molecules, because they experience fewer attractions to neighbors, condense only at the cooler temperatures found at the top of the tower.

The gasoline obtained from the fractional distillation of petroleum consists of a wide variety of hydrocarbons having similar boiling points. Some of these components burn more efficiently than others in a car engine. The straight-chain hydrocarbons, such as *n*-hexane, tend to burn quickly, upsetting the timing of the engine and causing *engine knock*. Gasoline hydrocarbons that have more branching, such as *iso*-octane, burn slowly, and as a result the engine runs more smoothly. These two compounds, *n*-hexane and *iso*-octane, are used as standards in assigning *octane ratings* to gasoline. An octane number of 100 is arbitrarily assigned to *iso*-octane, and *n*-hexane is assigned an octane number of 0. The antiknock performance of a particular gasoline is compared with that of various mixtures of *iso*-octane and *n*-hexane, and an octane number is assigned. Figure 14.4 shows octane information on a typical gasoline pump.

Figure 14.4 Octane ratings are posted on gasoline pumps.

CHECK YOURSELF

Which structural isomer in Figure 14.1 should have the highest octane rating?

CHECK YOUR ANSWER

The structural isomer with the greatest amount of branching in the carbon framework will likely have the highest octane rating, making *neo*-pentane the clear winner.

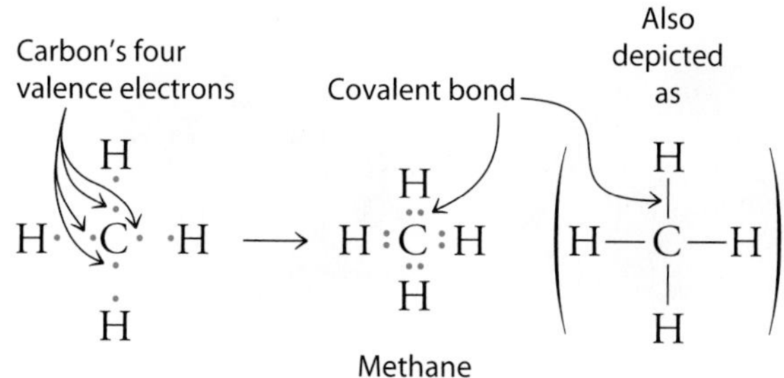

Figure 14.5 Carbon has four valence electrons. Each electron pairs with an electron from a hydrogen atom in the four covalent bonds of methane.

14.2 Unsaturated Hydrocarbons

Recall from Section 12.1 that carbon, as a group 4 element of the periodic table, has four unpaired valence electrons. As shown in Figure 14.5, each of these electrons is available for pairing with an electron from another atom, such as hydrogen, to form a covalent bond.

In all the hydrocarbons discussed so far, including the methane shown in Figure 14.4, each carbon atom is bonded to four neighboring atoms by four single covalent bonds. Such hydrocarbons are known as **saturated hydrocarbons**. The term *saturated* means that each carbon has as many atoms bonded to it as possible—four. We now explore cases where one or more carbon atoms in a hydrocarbon are bonded to fewer than four neighboring atoms. This occurs when at least one of the bonds between a carbon and a neighboring atom is a multiple bond. (See page 250 of Chapter 12 for a review of multiple bonds.)

A hydrocarbon containing a multiple bond—either double or triple—is known as an **unsaturated hydrocarbon**. Because of the multiple bond, two of the carbons are bonded to fewer than four other atoms. These carbons are thus said to be *unsaturated*.

Figure 14.6 compares the saturated hydrocarbon *n*-butane with the unsaturated hydrocarbon 2-butene. The number of atoms bonded to each of the two middle carbons of *n*-butane is four, whereas each of the two middle carbons of 2-butene is bonded to only three other atoms—a hydrogen and two carbons.

An important unsaturated hydrocarbon known as benzene, C_6H_6, once used as a solvent, may be drawn as three double bonds contained within a flat hexagonal ring, as is shown in Figure 14.7a. Unlike the double-bond electrons in most other unsaturated hydrocarbons, however, the electrons of the double bonds in benzene are not fixed between any two carbon atoms. Instead, these electrons are able to move freely around the ring. This is commonly represented by drawing a circle within the ring, as shown in Figure 14.7b, rather than the individual double bonds.

Many organic compounds contain one or more benzene rings in their structure. Because many of these compounds are fragrant, any organic molecule containing a benzene ring is classified as an **aromatic compound** (even if it is not particularly fragrant). Figure 14.8 shows a few examples. Toluene, a common solvent used as paint thinner, is toxic and gives model airplane glue its distinctive odor. Some aromatic compounds, such as naphthalene, contain two or more benzene rings fused together. At one time, mothballs were made of naphthalene. Most mothballs sold today, however, are made of the less toxic 1,4-dichlorobenzene.

CHECK YOURSELF

Prolonged exposure to benzene has been found to increase the risk of developing certain cancers. The structure of aspirin contains a benzene ring. Does this necessarily mean that prolonged exposure to aspirin will increase a person's risk of developing cancer?

Saturated hydrocarbon

n-Butane, C_4H_{10}

Unsaturated hydrocarbon

2-Butene, C_4H_8

Figure 14.6 The carbons of the hydrocarbon *n*-butane are saturated, each being bonded to four other atoms. Because of the double bond, two of the carbons of the unsaturated hydrocarbon 2-butene are bonded to only three other atoms, which makes the molecule an unsaturated hydrocarbon.

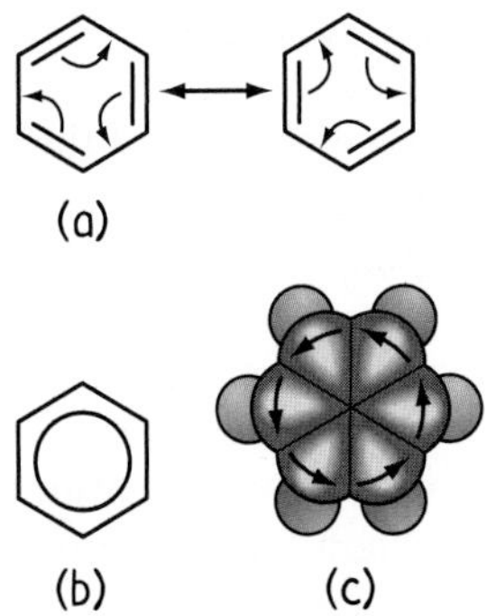

Figure 14.7 (a) The double-bond electrons of benzene, C_6H_6, are able to migrate around the ring. (b) For this reason, they are often represented by a circle within the ring. (c) Three-dimensional model of benzene.

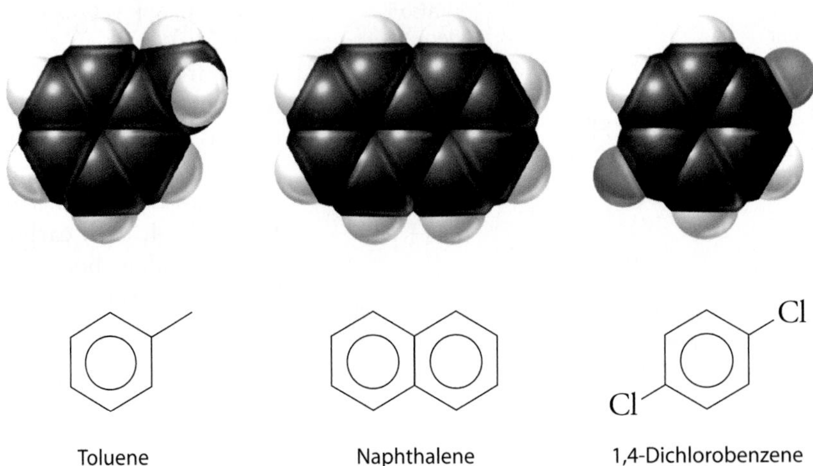

Figure 14.8 The structures for three odoriferous organic compounds containing one or more benzene rings: toluene, naphthalene, and 1,4-dichlorobenzene.

Introduction to Organic Molecules

CHECK YOUR ANSWER

No. Although benzene and aspirin both contain a benzene ring, these two molecules have different overall structures, which means the properties of one are quite different from the properties of the other. Each organic compound has its own set of unique physical, chemical, and biological properties. Although benzene may cause cancer, aspirin is a safe remedy for headaches.

14.3 Functional Groups

Carbon atoms can bond to one another and to hydrogen atoms in many ways, which results in an incredibly large number of hydrocarbons. But carbon atoms can bond to atoms of other elements as well, further increasing the number of possible organic molecules. In organic chemistry, any atom other than carbon or hydrogen in an organic molecule is called a **heteroatom**, where *hetero-* signifies "different from either carbon or hydrogen."

A hydrocarbon structure can serve as a framework to which various heteroatoms can be attached. This is analogous to a Christmas tree serving as the scaffolding on which ornaments are hung. Just as the ornaments give character to the tree, so do heteroatoms give character to an organic molecule. In other words, heteroatoms can have profound effects on the properties of an organic molecule.

Consider ethane, C_2H_6, and ethanol, C_2H_6O, which differ from each other by only a single oxygen atom (Figure 14.9). Ethane has a boiling point of −88°C, making it a gas at room temperature, and it does not dissolve in water very well. Ethanol, by contrast, has a boiling point of +78°C, making it a liquid at room temperature. It is infinitely soluble in water and is the active ingredient of alcoholic beverages. Consider further ethylamine, C_2H_7N, which has a nitrogen atom on the same basic two-carbon framework. This compound is a corrosive, pungent, highly toxic gas—most unlike either ethane or ethanol.

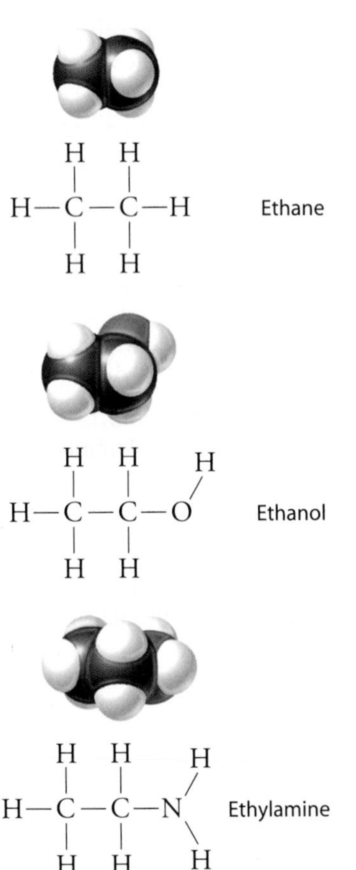

Figure 14.9 Ethane, ethanol, and ethylamine share the same basic two-carbon framework. However, you can see that ethanol has an alcohol functional group, and ethylamine has an amide functional group in place of one of the hydrogen atoms in ethane.

Organic molecules are classified according to the functional groups they contain; a **functional group** is defined as a combination of atoms that behave as a unit. Most functional groups are distinguished by the heteroatoms they contain; some common groups are listed in Table 14.1.

The following two sections, 14.4 and 14.5, introduce the classes of organic molecules shown in Table 14.1. The role heteroatoms play in determining the properties of each class is the underlying theme. As you study this material, focus on understanding the chemical and physical properties of the various classes of

Table 14.1 Functional Groups in Organic Molecules

General Structure	Name	Class
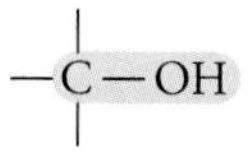 —C—OH	Hydroxyl group	Alcohols
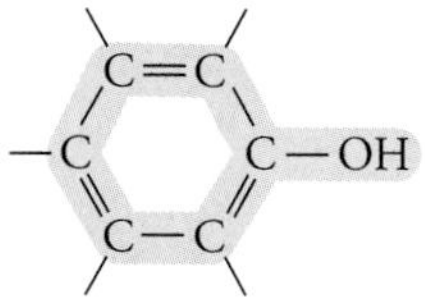 —C₆ ring—OH (C=C, C—OH, C—C)	Phenolic group	Phenols
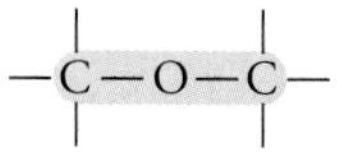 —C—O—C—	Ether group	Ethers
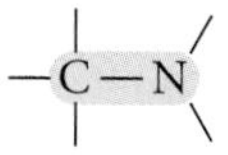 —C—N	Amine group	Amines
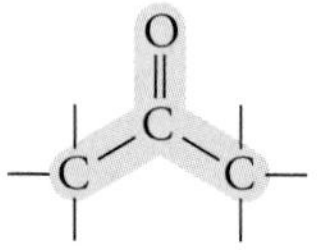 —C—C(=O)—C—	Ketone group	Ketones
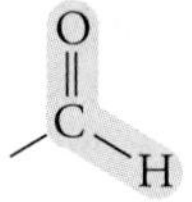 —C(=O)—H	Aldehyde group	Aldehydes
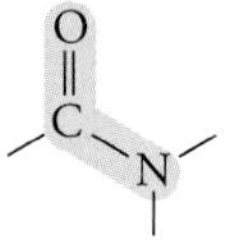 —C(=O)—N	Amide group	Amides
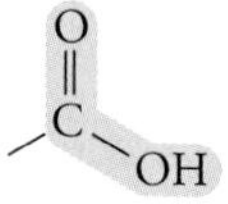 —C(=O)—OH	Carboxyl group	Carboxylic acids
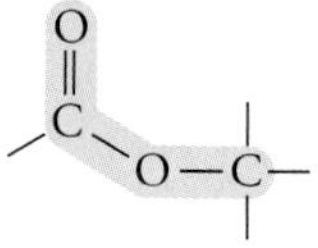 —C(=O)—O—C—	Ester group	Esters

compounds, for doing so will give you a greater appreciation of the remarkable diversity of organic molecules and their many applications.

CHECK YOURSELF

What is the significance of heteroatoms in an organic molecule?

CHECK YOUR ANSWER

Heteroatoms largely determine an organic molecule's "personality."

14.4 Alcohols, Phenols, Ethers, and Amines

Alcohols are organic molecules in which a *hydroxyl group* is bonded to a saturated carbon. The hydroxyl group consists of an oxygen bonded to a hydrogen. Because of the polarity of the oxygen–hydrogen bond, alcohols are often soluble in water, which is itself very polar. Some common alcohols are listed in Table 14.2.

More than 11 billion pounds of methanol, CH_3OH, are produced annually in the United States. Most of it is used for making formaldehyde and acetic acid, important starting materials in the production of plastics. In addition, methanol is used as a solvent, an octane booster, and an anti-icing agent in gasoline. Sometimes called wood alcohol because it can be obtained from wood, methanol should never be ingested because in the body it is metabolized to formaldehyde and formic acid. Formaldehyde is harmful to the eyes, can lead to blindness, and was once used to preserve dead biological specimens. Formic acid, the active ingredient in an ant bite, can lower the pH of the blood to dangerous levels. Ingesting only about 15 milliliters (about 3 tablespoons) of methanol may lead to blindness, and about 30 milliliters can cause death.

Ethanol, C_2H_5OH, is one of the oldest chemicals manufactured by humans. The "alcohol" of alcoholic beverages, ethanol is prepared by feeding the sugars of various plants to certain yeasts, which produce ethanol through a biological process known as *fermentation*. Ethanol is widely used as an industrial solvent. For many years, ethanol intended for this purpose was made by fermentation, but today industrial-grade ethanol is more cheaply manufactured from petroleum by-products.

A third well-known alcohol is 2-propanol, commonly known as isopropyl alcohol. This is the rubbing alcohol you buy at the drugstore. Although 2-propanol has a relatively high boiling point, it readily evaporates, leading to a pronounced cooling effect when applied to skin—an effect once used to reduce fevers. (Isopropyl alcohol is very toxic if ingested. Washcloths wetted with cold water are nearly as effective in reducing fever and far safer.) You are probably most familiar with the use of isopropyl alcohol as a topical disinfectant.

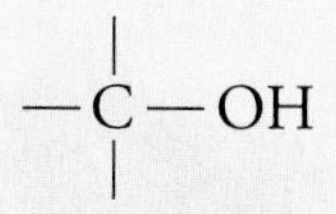

—C—OH

Hydroxyl group

While alcohols are compounds containing an alcohol group—a hydroxyl group attached to a saturated carbon atom—**phenols** are compounds containing a phenolic group, a hydroxyl group attached to a benzene ring. Because of the presence of the benzene ring, the hydrogen of the hydroxyl group is readily lost in an acid–base reaction, which makes the phenolic group mildly acidic. The reason for this acidity is illustrated in Figure 14.10.

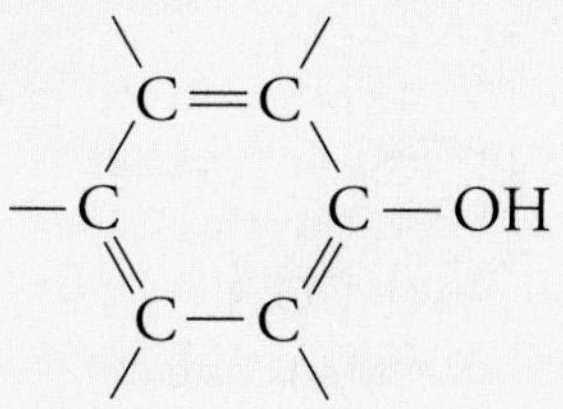

The simplest phenol, shown in Figure 14.11, is called phenol. In 1867, Joseph Lister (1827–1912) discovered the antiseptic value of phenol, which, when applied to surgical instruments and incisions, greatly increased surgery survival rates. Phenol was the first purposefully used antibacterial solution, or *antiseptic*. Phenol damages healthy tissue, however, and so a number of milder phenols have since been introduced. The phenol 4-*n*-hexylresorcinol, for example, is commonly used in throat lozenges and mouthwashes. This compound has even greater antiseptic properties than phenol, and yet it does not damage tissue. Listerine brand mouthwash (named after Joseph Lister) contains the antiseptic phenols thymol and methyl salicylate.

C=C / —C C—OH / C—C

Phenolic group

Table 14.2 Some Simple Alcohols

Structure	Scientific Name
H–C(H)(H)–OH	Methanol
H–C(H)(H)–C(H)(H)–OH	Ethanol
H–C(H)(H)–C(H)(OH)–C(H)(H)–H	2-Propanol

Phenol (acidic) → Phenoxide ion + Hydrogen ion (H^+)

Figure 14.10 The negative charge of the phenoxide ion is able to migrate to select positions on the benzene ring. This mobility helps to accommodate the negative charge, which is why the phenolic group readily donates a hydrogen ion.

CHECK YOURSELF

Why are alcohols less acidic than phenols?

CHECK YOUR ANSWER

An alcohol does not contain a benzene ring adjacent to the hydroxyl group. If the alcohol were to donate the hydroxyl hydrogen, the result would be a negative charge on the oxygen. Without an adjacent benzene ring, this negative charge has nowhere to go. As a result, an alcohol behaves only as a very weak acid, much the way water does.

–C–O–C–

Ether group

Ethers are organic compounds structurally related to alcohols. The oxygen atom in an ether group, however, is bonded not to a carbon and a hydrogen, but rather to two carbons. As we see in Figure 14.12, ethanol and dimethyl ether have the same chemical formula, C_2H_6O, but their physical properties are vastly different. Whereas ethanol is a liquid at room temperature (boiling point 78°C)

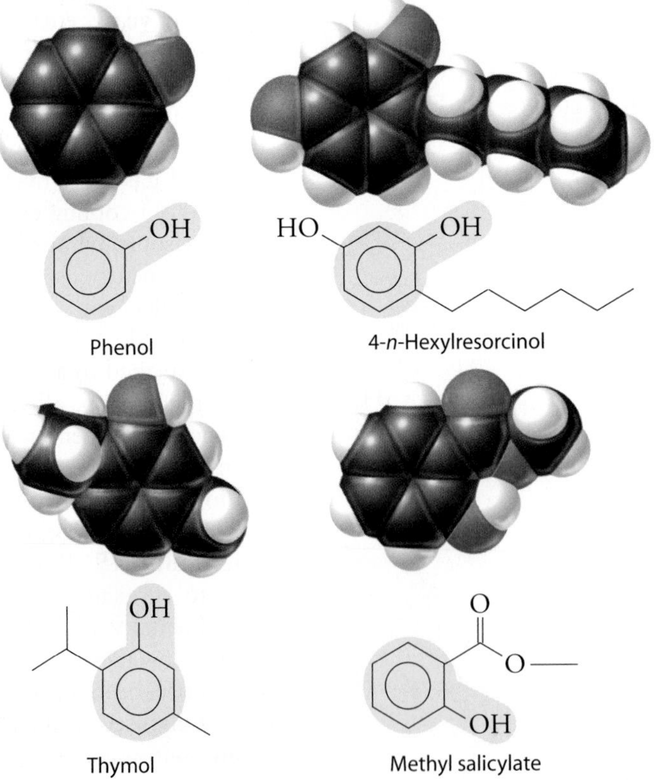

Figure 14.11 Every phenol contains a phenolic group (highlighted in blue).

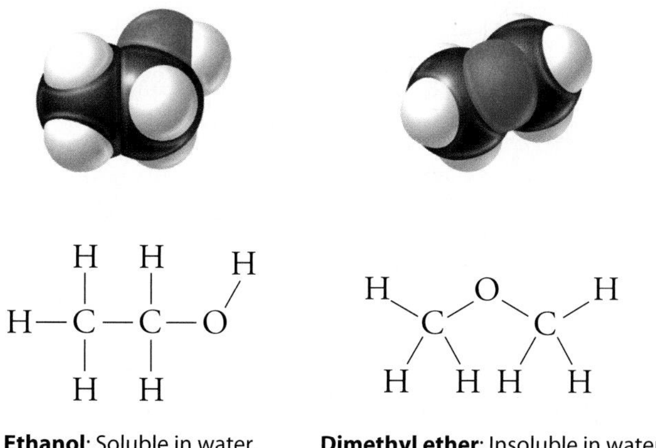

Ethanol: Soluble in water, boiling point 78°C

Dimethyl ether: Insoluble in water, boiling point −25°C

Figure 14.12 The oxygen in an alcohol, such as ethanol, is bonded to one carbon atom and one hydrogen atom. The oxygen in an ether, such as dimethyl ether, is bonded to two carbon atoms. Because of this difference, alcohols and ethers of similar molecular mass have vastly different physical properties.

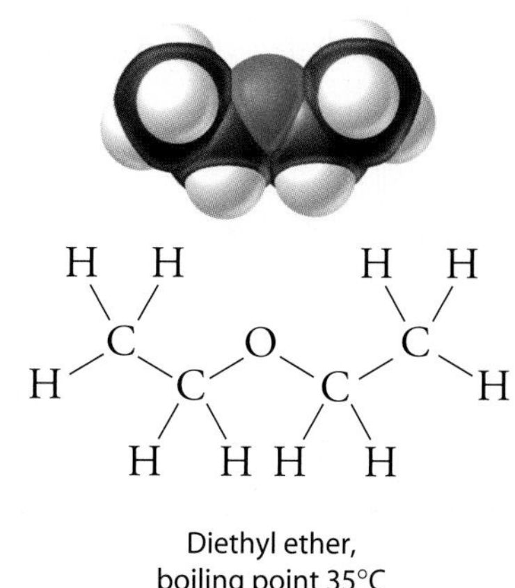

Diethyl ether, boiling point 35°C

Figure 14.13 Diethyl ether is the systematic name for the "ether" historically used as an anesthetic.

and mixes quite well with water, dimethyl ether is a gas at room temperature (boiling point −25°C) and is much less soluble in water.

Ethers are not very soluble in water because, without the hydroxyl group, they are unable to form strong hydrogen bonds with water. Furthermore, without the polar hydroxyl group, the molecular attractions among ether molecules are relatively weak. As a result, it does not take much energy to separate ether molecules from one another. This is why ethers have relatively low boiling points and evaporate so readily.

Diethyl ether, shown in Figure 14.13, was one of the first anesthetics. The anesthetic properties of this compound, discovered in the early 1800s, revolutionized the practice of surgery. Because of its high volatility at room temperature, inhaled diethyl ether rapidly enters the bloodstream. Because this ether has low solubility in water and high volatility, it quickly leaves the bloodstream once introduced. Because of these physical properties, a surgical patient can be brought in and out of anesthesia (a state of unconsciousness) simply by regulating the gases breathed. Modern-day gaseous anesthetics have fewer side effects than diethyl ether but work on the same principle.

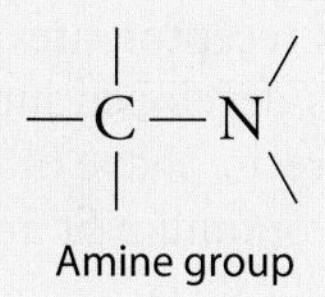

Amine group

Amines are organic compounds that contain the amine group—a nitrogen atom bonded to one, two, or three saturated carbons. Amines are typically less soluble in water than are alcohols because the nitrogen–hydrogen bond is not quite as polar as the oxygen–hydrogen bond. The lower polarity of amines also means their boiling points are typically somewhat lower than those of alcohols of similar formula mass. One of the most notable physical properties of many low-formula-mass amines is their offensive odor. For example, two appropriately named amines, putrescine and cadaverine, are responsible for the odor of decaying flesh. Table 14.3 lists three simple amines.

INTEGRATED SCIENCE BIOLOGY

Drug Action and Discovery

To find new and more effective medicines, chemists use various models that describe how drugs work. By far, one of the most useful models of drug action is the **lock-and-key model**. The basis of this model is that there is a connection between a drug's chemical structure and its biological effect. For example, related pain-relieving opioids, such as codeine and morphine, have the T-shaped structure shown in Figure 14.14.

T-shaped three-dimensional structure found in all opioids

Morphine

Codeine

Oxycodone

Figure 14.14 All drugs that act like morphine have the same basic three-dimensional shape as morphine.

According to the lock-and-key model, illustrated in Figure 14.15, biologically active molecules function by fitting into *receptor sites* on proteins in the body, where they are held by molecular attractions, such as hydrogen bonding (Chapter 12). When a drug molecule fits into a receptor site the way a key fits into a lock, a particular biological event is triggered, such as a nerve impulse, a change in the shape of a protein, or even a chemical reaction. In order to fit into a particular receptor site, however, a molecule must have the proper shape, just as a key must have properly shaped notches in order to fit a lock.

Another facet of this model is that the molecular attractions holding a drug to a receptor site are easily broken. (Recall from Chapter 12 that most molecular attractions are many times weaker than chemical bonds.) A drug is therefore held to a receptor site only temporarily. Once the drug is removed from the receptor site, body metabolism destroys the drug's chemical structure and the effects of the drug are said to have worn off. Using this model, we can understand why some drugs are more potent than others. Oxycodone, for example, is a more potent painkiller than is morphine because its chemical structure allows for tighter and longer binding to its receptor sites.

The lock-and-key model has developed into one of the central tenets of pharmaceutical study. Knowing the precise shape of a target receptor site allows chemists to design molecules that have an optimum fit and a specific biological effect.

Table 14.3 Three Simple Amines

Structure	Scientific Name
	Ethylamine
	Diethylamine
	Triethylamine

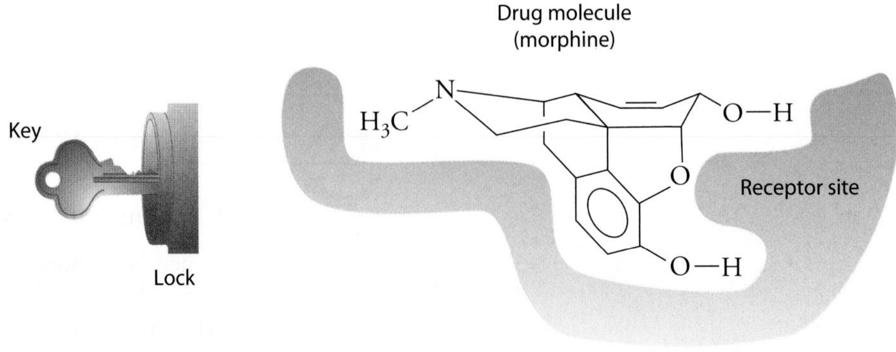

Figure 14.15 Many drugs act by fitting into receptor sites on molecules found in the body, much as a key fits in a lock.

Biochemical systems are so complex, however, that our knowledge is still limited, as is our capacity to design effective medicinal drugs. For this reason, most new medicinal drugs are still discovered instead of designed. One important avenue for drug discovery is ethnobotany. An *ethnobotanist* is a researcher who learns about the medicinal plants used in indigenous cultures, as discussed in greater detail in Chapter 18.

CHECK YOURSELF

Why are organic chemicals so suitable for making drugs?

CHECK YOUR ANSWER

Because their vast diversity permits the manufacture of the many different types of medicines needed to combat the many different types of illnesses humans are subject to.

14.5 Ketones, Aldehydes, Amides, Carboxylic Acids, and Esters

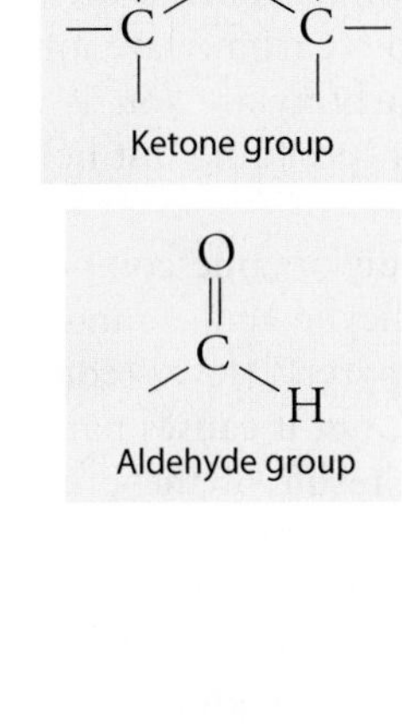

The organic compounds known as the ketones, aldehydes, amides, carboxylic acids, and esters are all similar in that they contain the *carbonyl group*. The carbonyl group consists of a carbon atom double-bonded to an oxygen atom.

A **ketone** is a carbonyl-containing organic molecule in which the carbonyl carbon is bonded to two carbon atoms. A familiar example of a ketone is *acetone*, shown in Figure 14.16a, which is often used in fingernail polish remover. Like ketones, **aldehydes** contain a carbonyl group, but in an aldehyde, the carbonyl carbon is bonded either to one carbon atom and one hydrogen atom, as in Figure 14.16b, or, in the special case of formaldehyde, to two hydrogen atoms.

Many aldehydes are particularly fragrant. A number of flowers, for example, owe their pleasant odor to the presence of simple aldehydes. The smells of lemons, cinnamon, and almonds are due to the aldehydes citral, cinnamaldehyde, and benzaldehyde, respectively. The structures of these aldehydes are shown in Figure 14.17. The aldehyde vanillin, introduced at the beginning of this chapter, is the key flavoring molecule derived from the vanilla orchid. You

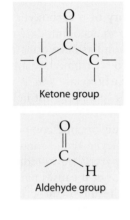

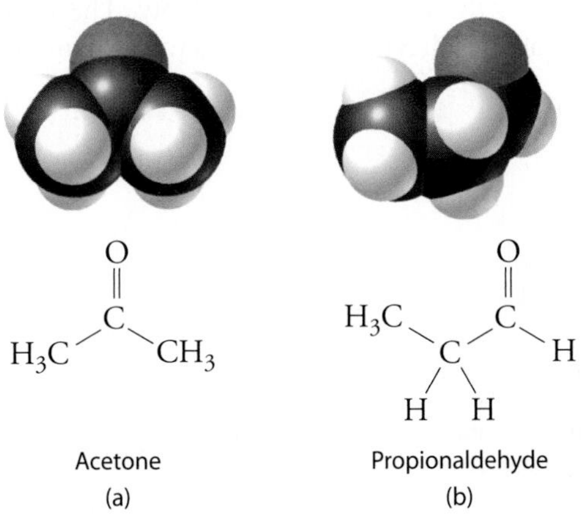

Figure 14.16 (a) When the carbon of a carbonyl group is bonded to two carbon atoms, the result is a ketone. An example is acetone. (b) When the carbon of a carbonyl group is bonded to at least one hydrogen atom, the result is an aldehyde. An example is propionaldehyde.

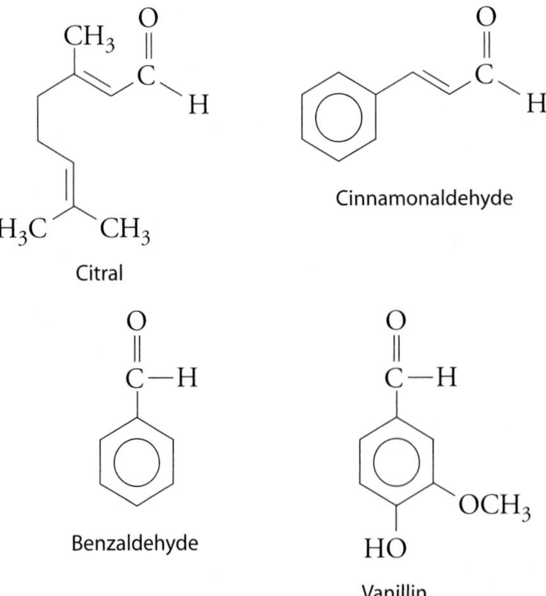

Figure 14.17 Aldehydes are responsible for many familiar fragrances.

may have noticed that vanilla seed pods and vanilla extract are fairly expensive. Imitation vanilla flavoring is less expensive because it is merely a solution of the compound vanillin, which is economically synthesized from the waste chemicals of the wood pulp industry. Although imitation vanilla tastes much like natural vanilla extract, it is not exact. This is because, in addition to vanillin, many other flavorful molecules contribute to the complex taste of natural vanilla. Many books made in the days before "acid-free" paper smell of vanilla because of the vanillin formed and released as the paper ages, a process that is accelerated by the acids the paper contains.

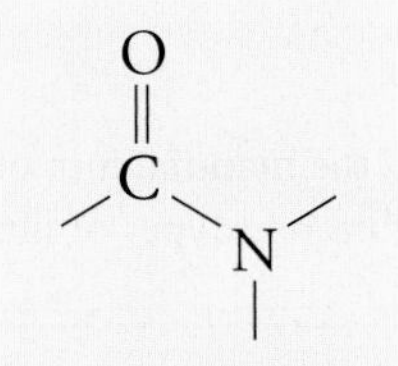

Amide

An **amide** is a carbonyl-containing organic molecule in which the carbonyl carbon is bonded to a nitrogen atom. The active ingredient of most mosquito repellents is an amide whose chemical name is *N*,*N*-diethyl-*m*-toluamide but is commercially known as DEET, shown in Figure 14.18. This compound is actually not an insecticide. Rather, it causes certain insects, especially mosquitoes, to lose their sense of direction, which effectively protects DEET wearers from being bitten.

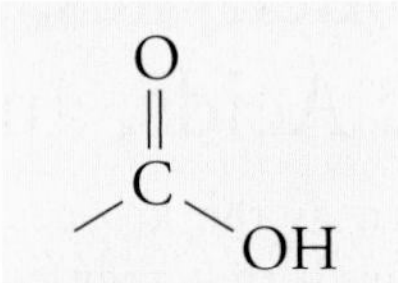

Carboxyl group

A **carboxylic acid** is a carbonyl-containing organic molecule in which the carbonyl carbon is bonded to a hydroxyl group. As its name implies, this functional group is able to donate hydrogen ions, and as a result organic molecules containing it are acidic. An example is acetic acid, $C_2H_4O_2$, the main ingredient of vinegar.

As with phenols, the acidity of a carboxylic acid results in part from the ability of the functional group to accommodate the negative charge of the ion that forms after the hydrogen ion has been donated. As shown in Figure 14.19, a carboxylic acid transforms to a carboxylate ion as it loses the hydrogen ion. The negative charge of the carboxylate ion is able to pass back and forth between the two oxygens. This spreading out helps to accommodate the negative charge.

An interesting example of an organic compound that contains both a carboxylic acid and a phenol is salicylic acid, found in the bark of the willow tree. At one time brewed for its antipyretic (fever-reducing) effect, salicylic acid is an important analgesic (painkiller), but it causes nausea and stomach upset because of its relatively high acidity, a result of the presence of two acidic functional groups. In 1899, Friederich Bayer and Company, in Germany, introduced a chemically modified version of salicylic acid in which the phenolic group was transformed to an ester functional group. Because both the carboxyl group and the phenolic group contribute to the high acidity of salicylic acid, getting rid of the phenolic group reduced the acidity of the molecule considerably. The result was the less acidic and more tolerable acetylsalicylic acid, the chemical name for aspirin, shown in Figure 14.20b.

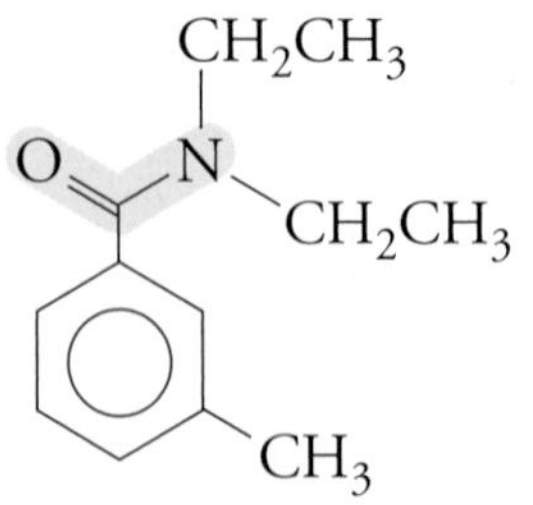

N,*N*-Diethyl-*m*-toluamide
(DEET)

Figure 14.18 *N*,*N*-diethyl-*m*-toluamide is an example of an amide. Amides contain the amide group, shown highlighted in blue.

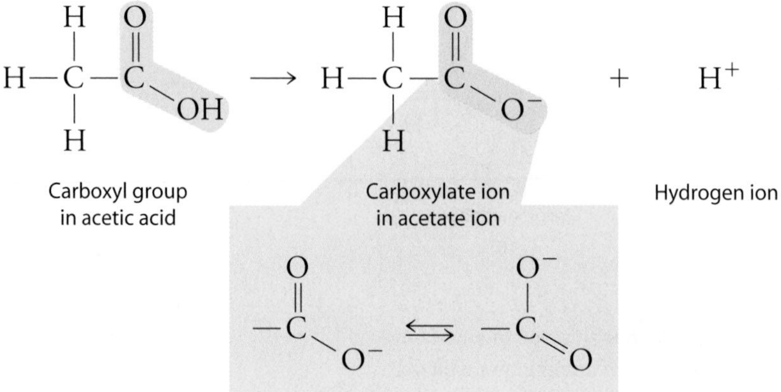

Carboxyl group in acetic acid → Carboxylate ion in acetate ion + Hydrogen ion (H^+)

Figure 14.19 The negative charge of the carboxylate ion is able to pass back and forth between the two oxygen atoms of the carboxyl group.

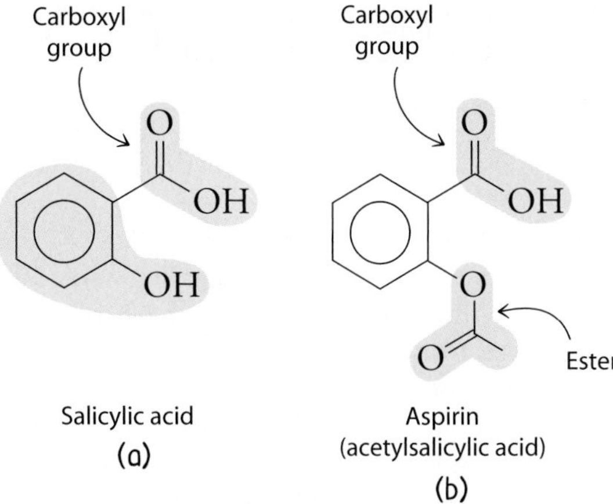

Figure 14.20 (a) Salicylic acid is an example of a molecule containing both a carboxyl group and a phenolic group. (b) Aspirin, acetylsalicylic acid, is less acidic than salicylic acid because it no longer contains the acidic phenolic group, which has been converted to an ester.

Table 14.4 Some Familiar Esters

Structure	Scientific Name	Flavor/odor
$H-C(=O)-O-CH_2CH_3$	Ethyl formate	Rum
$H_3C-C(=O)-O-CH_2CH_2-C(CH_3)_2-H$	Isopentyl acetate	Banana
$H_3C-C(=O)-O-CH_2(CH_2)_6CH_3$	Octyl acetate	Orange
$CH_3CH_2CH_2-C(=O)-O-CH_3$	Methyl butyrate	Apple
$C_6H_4(OH)-C(=O)-O-CH_3$	Methyl salicylate	Wintergreen

Ester group

An **ester** is an organic molecule similar to a carboxylic acid except that in the ester the hydroxyl hydrogen is replaced by a carbon. Unlike carboxylic acids, esters are not acidic because they lack the hydrogen of the hydroxyl group. Like aldehydes, many simple esters have notable fragrances and are used as flavorings. Some familiar ones are listed in Table 14.4.

CHECK YOURSELF

Identify all the functional groups in these four molecules (ignore the sulfur group in penicillin G):

Acetaldehyde   Penicillin G   Testosterone   Morphine

CHECK YOUR ANSWERS

Acetaldehyde: aldehyde; penicillin G: amide (two amide groups), carboxylic acid; testosterone: alcohol and ketone; morphine: alcohol, phenol, ether, and amine.

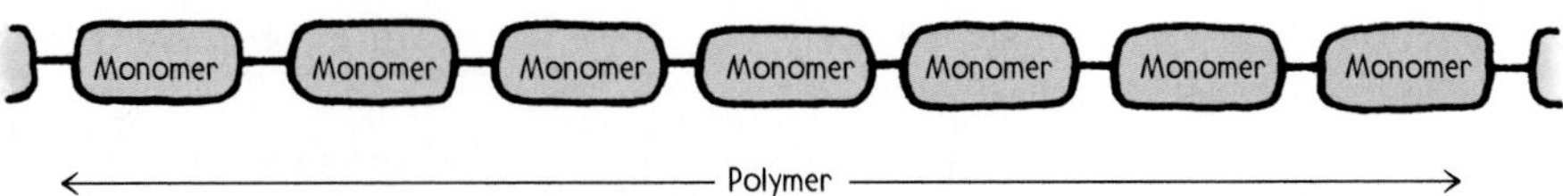

Figure 14.21 A polymer is a long molecule consisting of many smaller monomer molecules linked together.

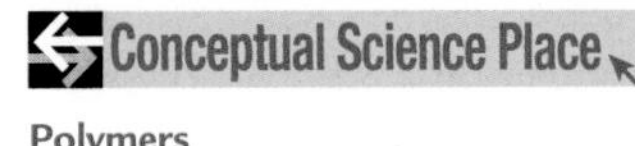

Polymers

14.6 Polymers

Polymers are exceedingly long molecules that consist of repeating molecular units called **monomers**, as Figure 14.21 illustrates. Monomers have relatively simple structures consisting of anywhere from 4 to 100 atoms per molecule. When chained together, they can form polymers consisting of hundreds of thousands of atoms per molecule. These large molecules are still too small to be seen with the unaided eye. They are, however, giants in the world of the submicroscopic—if a typical polymer molecule were as thick as a kite string, it would be 1 kilometer long.

Many of the molecules that make up living organisms are polymers, including DNA, proteins, the cellulose of plants, and the complex carbohydrates of starchy foods. We leave a discussion of these important biological molecules to Chapter 15. For now, we focus on the human-made polymers, also known as synthetic polymers, that make up the class of materials commonly known as plastics.

We begin by exploring the two major types of synthetic polymers used today—*addition polymers* and *condensation polymers*. As shown in Table 14.5,

Table 14.5 Addition and Condensation Polymers

Addition polymers	Repeating unit	Common uses	Recycling codes
Polyethylene (PE)	$\cdots CH_2-CH_2 \cdots$	Plastic bags, bottles	2, 4 HDPE
Polypropylene (PP)	$\cdots CH_2-CH(CH_3) \cdots$	Indoor-outdoor carpets	5 PP
Polystyrene (PS)	$\cdots CH_2-CH(C_6H_5) \cdots$	Plastic utensils, insulation	6 PS
Polyvinyl chloride (PVC)	$\cdots CH_2-CHCl \cdots$	Shower curtains, tubing	3 V
Polyvinylidene chloride (Saran)	$\cdots CH_2-CCl_2 \cdots$	Plastic wrap	
Condensation polymers			
Polyethylene terephthalate (Dacron, Mylar)	$\cdots C(=O)-C_6H_4-C(=O)-O-CH_2CH_2-O \cdots$	Clothing, plastic bottles	1 PET

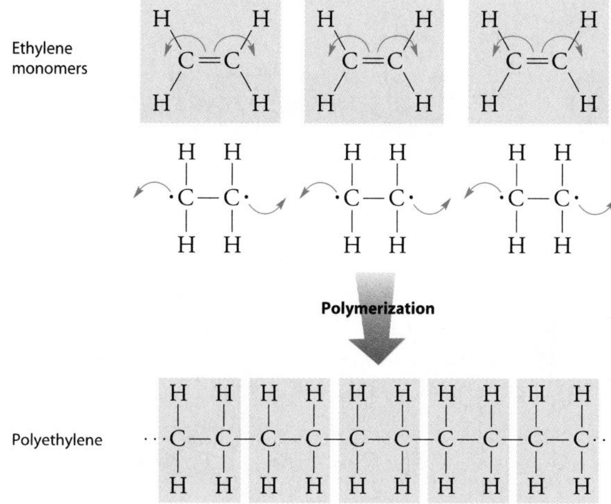

Figure 14.22 The addition polymer polyethylene is formed as electrons from the double bonds of ethylene monomer molecules split away and become unpaired valence electrons. Each unpaired electron then joins with an unpaired electron of a neighboring carbon atom to form a new covalent bond that links two monomer units together.

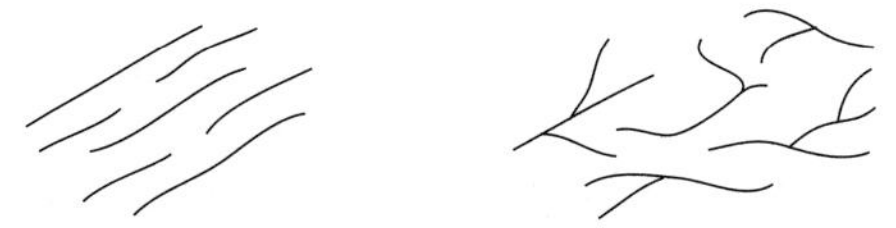

(a) Molecular strands of HDPE (b) Molecular strands of LDPE

Figure 14.23 (a) The polyethylene strands of HDPE are able to pack closely together, much like strands of uncooked spaghetti. (b) The polyethylene strands of LDPE are branched, which prevents the strands from packing well.

addition and condensation polymers have a wide variety of uses. Solely the product of human design, these polymers pervade modern living. In the United States, for example, synthetic polymers have surpassed steel as the most widely used material.

Addition Polymers

Addition polymers form simply by the joining together of monomer units. For this to happen, each monomer must contain at least one double bond. As shown in Figure 14.22, polymerization occurs when two of the electrons from each double bond split away from each other to form new covalent bonds with neighboring monomer molecules. During this process, no atoms are lost, meaning that the total mass of the polymer is equal to the sum of the masses of all the monomers.

Nearly 12 million tons of polyethylene are produced annually in the United States; that's about 90 pounds per U.S. citizen. The monomer from which it is synthesized, ethylene, is an unsaturated hydrocarbon produced in large quantities from petroleum.

Two principal forms of polyethylene are produced by using different catalysts and reaction conditions. High-density polyethylene (HDPE), shown schematically in Figure 14.23a, consists of long strands of straight-chain molecules packed closely together. The tight alignment of neighboring strands makes HDPE a relatively rigid, tough plastic useful for such things as bottles and milk jugs. Low-density polyethylene (LDPE), shown in Figure 14.23b, is made of strands of highly branched chains, an architecture that prevents the strands from packing closely together. This makes LDPE more bendable than HDPE and gives it a lower melting point. HDPE holds its shape in boiling water, LDPE deforms. It is most useful for such items as plastic bags, photographic film, and electrical-wire insulation.

Other addition polymers are created by using different monomers. The only requirement is that the monomer must contain a double bond. The monomer propylene, for example, yields polypropylene, a tough plastic material useful for pipes, hard-shell suitcases, and appliance parts. Fibers of polypropylene are used for upholstery, indoor–outdoor carpets, and even thermal underwear. Using styrene as the monomer yields polystyrene. Transparent plastic cups are made of polystyrene, as are thousands of other household items. Blowing gas into liquid polystyrene generates Styrofoam, widely used for coffee cups, packing material, and insulation.

The addition polymer polyvinylidene chloride (trade name Saran), shown in Figure 14.24, is used as plastic wrap for food. Interestingly, induced dipoles are

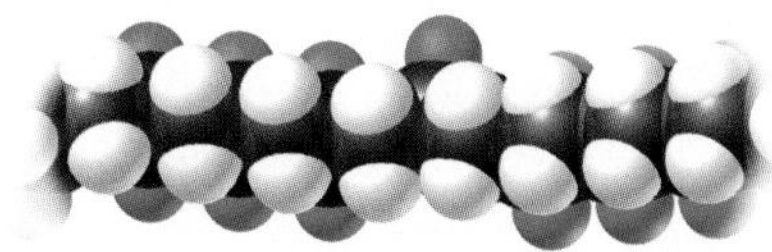

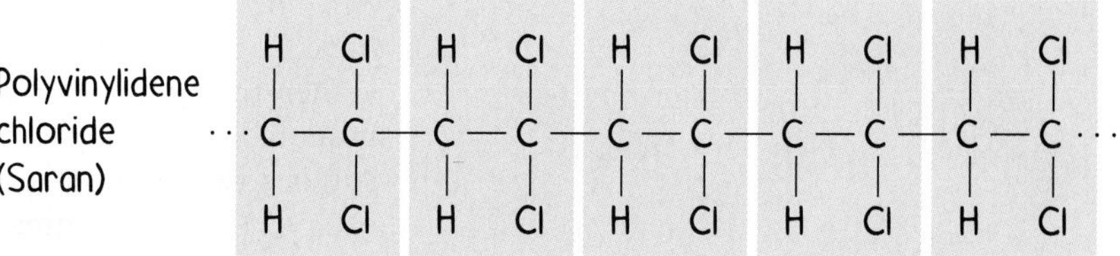

Figure 14.24 The large chlorine atoms (shown in green) in polyvinylidene chloride make this addition polymer sticky.

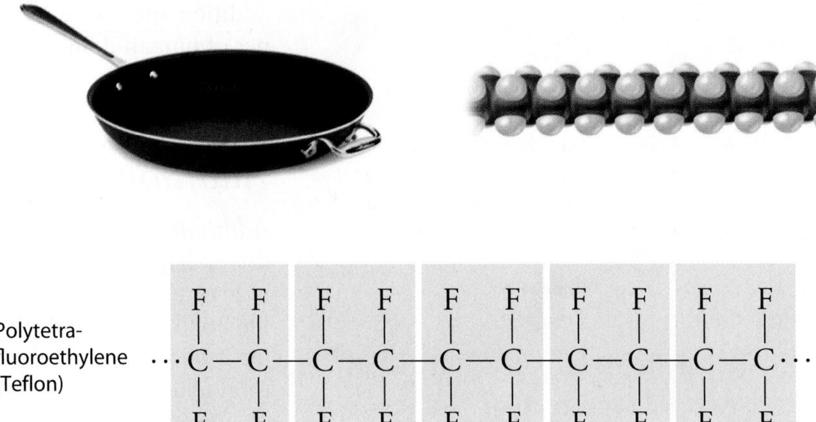

Figure 14.25 The fluorine atoms in polytetrafluoroethylene tend not to experience molecular attractions, which is why this addition polymer is used as a nonstick coating and lubricant.

INSIGHT Pure polyvinylchloride, PVC, is a tough material great for making pipes. Mixed with small molecules called *plasticizers*, however, the PVC becomes soft and flexible and thus useful for making shower curtains, toys, and many other products now found in most households. One of the more commonly used plasticizers are the phthalates, some of which have been shown to disrupt the development of reproductive organs, especially in the fetus and in growing children. Governments and manufacturers are now working to phase out these plasticizers.

more easily formed in larger atoms because the electrons can more easily congregate to one side (more space is available to them). The large chlorine atoms in this polymer help it stick to surfaces such as glass by dipole–induced dipole attractions, as we saw in Section 12.7.

The addition polymer polytetrafluoroethylene, shown in Figure 14.25, is what you know as Teflon. In contrast to the chlorine-containing Saran, fluorine-containing Teflon has a nonstick surface because the fluorine atoms tend not to experience any molecular attractions. (Fluorine atoms are relatively small and so they don't readily form induced dipoles.) In addition, because carbon–fluorine bonds are unusually strong, Teflon can be heated to high temperatures before decomposing. These properties make Teflon an ideal coating for cooking surfaces. It is also relatively inert, which is why many corrosive chemicals are shipped or stored in Teflon containers.

CHECK YOURSELF

What do all monomers of addition polymers have in common?

CHECK YOUR ANSWER

A double covalent bond between two carbon atoms.

Condensation Polymers

A *condensation polymer* forms when the joining of monomer units is accompanied by the loss of a small molecule, such as water or hydrochloric acid. Any monomer capable of becoming part of a condensation polymer must have a functional group on each end. When two such monomers come together to form a condensation polymer, one functional group of the first monomer links up with one functional group of the other monomer. The result is a two-monomer unit that has two terminal functional groups, one from each of the two original monomers. Each of these terminal functional groups in the two-monomer unit is now free to link up with one of the functional groups of a third monomer, and then a fourth, and so on. In this way a polymer chain is built.

SCIENCE AND SOCIETY

Polymers Win World War II

The search for a lightweight, nonbreakable, moldable material began with the invention of vulcanized rubber. This material is derived from natural rubber, which is a semisolid, elastic, natural polymer. In the 1700s, natural rubber was noted for its ability to rub off pencil marks, which is the origin of the term *rubber*. Natural rubber has few other uses, however, because it turns gooey at warm temperatures and brittle at cold temperatures. Then, in 1839, an American inventor, Charles Goodyear, discovered *rubber vulcanization*, a process in which natural rubber and sulfur are heated together. The product, vulcanized rubber, is harder than natural rubber and retains its elastic properties over a wide range of temperatures. With the invention of vulcanized rubber came modern products such as waterproof boots, rain jackets, and, most importantly, the rubber tire.

In the 1930s, more than 90 percent of the natural rubber used for manufacturing in the United States came from Malaysia. In the days after Pearl Harbor was attacked in December 1941 and the United States entered World War II, however, Japan captured Malaysia. As a result, the United States—the land with plenty of everything, except rubber—faced its first natural resource crisis. The military implications were devastating because without rubber tires, military airplanes and jeeps were useless. Petroleum-based synthetic rubber had been developed in 1930 by DuPont chemist Wallace Carothers but was not widely used because it was much more expensive than natural rubber. With Malaysian rubber impossible to get and a war on, however, cost was no longer an issue. Synthetic rubber factories were constructed across the nation, and within a few years, the annual production of synthetic rubber rose from 2000 tons to about 800,000 tons.

Also in the 1930s, British scientists developed radar as a way to track thunderstorms. With war approaching, these scientists turned their attention to the idea that radar could be used to detect enemy aircraft. Their radar equipment was massive, however. A series of ground-based radar stations could be built, but placing massive radar equipment on aircraft was not feasible. The great mass of the equipment was due to the large coils of wire needed to generate the intense radio waves. The scientists knew that if they could coat the wires with a thin, flexible electrical insulator, they would be able to design a radar device that was much less massive. Fortunately, the recently developed polymer polyethylene turned out to be an ideal electrical insulator. This permitted British radar scientists to construct equipment light enough to be carried by airplanes. These planes were slow, but flying at night or in poor weather, they could detect, intercept, and destroy enemy aircraft. Midway through the war, the Germans developed radar themselves, but without polyethylene, their radar equipment was inferior, and the tactical advantage stayed with the Allied forces.

Nylon was invented in 1937, which was just prior to World War II. Aside from its use in hosiery, nylon also found great use in the manufacture of parachutes, important to the U.S. military. Up to that time, parachutes were made mainly of silk. The world's foremost supplier of silk, however, was Japan. By the time World War II began, Japan had stopped exporting silk to the United States. The United States, however, now had nylon, which in many regards was better than silk. Over the course of the war practically all the nylon that DuPont could produce went to the military for the manufacture of a wide variety of nylon-based commodities suited for military purposes, such as parachutes, ropes, and clothing.

Four other polymers that had a significant impact on the outcome of World War II were Plexiglas, polyvinyl chloride, Saran, and Teflon. Plexiglas, or poly(methyl methacrylate), is a glasslike but moldable and lightweight material that made excellent domes for the gunner's nests on fighter planes and bombers. Both Allied and German chemists used poly(methyl methacrylate), but only the Allied chemists learned how small amounts of this polymer in solution could prevent oil or hydraulic fluid from becoming too thick at low temperatures. Equipped with only a few gallons of a poly(methyl methacrylate) solution, Soviet forces were able to keep their tanks operational in the Battle of Stalingrad during the winter of 1943. While Nazi equipment halted in the bitter cold, Soviet tanks and artillery functioned perfectly, resulting in victory and an important turning point in the war.

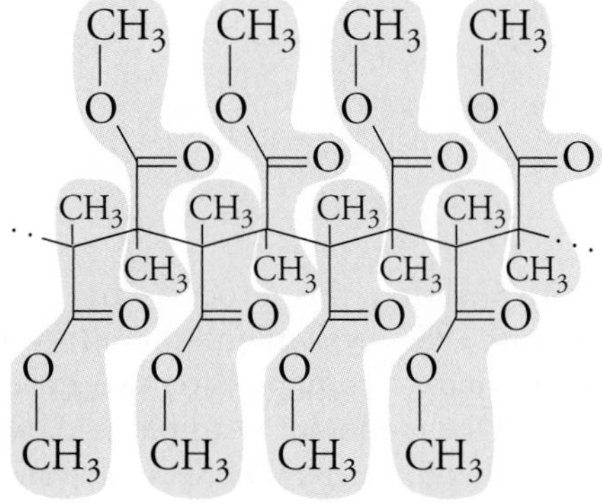

Poly(methyl methacrylate)

The bulky side groups in poly(methyl methacrylate) prevent the polymer chains from aligning with one another. This makes it easy for light to pass through the material, which is tough, transparent, lightweight, and moldable. (Plexiglas® is a registered trademark belonging to ATOFINA.)

continued

Polyvinyl chloride (PVC) had been developed by a number of chemical companies in the 1920s. The problem with this material, however, was that it lost resiliency when heated. In 1929, Waldo Semon, a chemist at BFGoodrich, found that PVC could be made into a workable material by the addition of a plasticizer. In World War II, this material became recognized as an ideal waterproof material for tents and rain gear.

The now-familiar plastic food wrap carton with a cutting edge was introduced in 1953 by Dow Chemical for its brand of Saran wrap.

Originally designed as a covering to protect theater seats from chewing gum, Saran found great use in World War II as a protective wrapping for artillery equipment during sea voyages. (Before Saran, the standard operating procedure had been to disassemble and grease the artillery to avoid corrosion.) After the war, the polymer was reformulated to eliminate the original formula's unpleasant odor and soon pushed cellophane aside to become the most popular food wrap of all time.

In the 1930s, Teflon was discovered serendipitously by chemists working at DuPont. Initially, the discoverers of Teflon were impressed by the long list of things this new material would *not* do. It would not burn, and it would not completely melt. Instead, at 620°F it congealed into a gel that could be conveniently molded. It would not conduct electricity, and it was impervious to attack by mold or fungus. No solvent, acid, or base could dissolve or corrode it. And most remarkably, nothing would stick to it, not even chewing gum.

Because of all the things Teflon would not do, DuPont was not quite sure what to do with it. Then, in 1944 the company was approached by governmental researchers in desperate need of a highly inert material to line the valves and ducts of an apparatus being built to isolate uranium-235 in the manufacture of the first nuclear bomb. Thus Teflon found its first application, and 1 year later, World War II came to a close with the nuclear bombing of Japan.

With a record of wartime successes, plastics were readily embraced in the postwar years. In the 1950s, Dacron polyester was introduced as a substitute for wool. Also, the 1950s were the decade during which the entrepreneur Earl Tupper created a line of polyethylene food containers known as Tupperware.

By the 1960s, a decade of environmental awakening, many people began to recognize the negative attributes of plastics. Being cheap, disposable, and nonbiodegradable, plastic readily accumulated as litter and as landfill. With petroleum so readily available and inexpensive, however, and with a growing population of plastic-dependent baby boomers, little stood in the way of an ever-expanding array of plastic consumer products. By 1977, plastics surpassed steel as the number-one material produced in the United States. Environmental concerns also continued to grow, and in the 1980s plastics-recycling programs began to appear. Although the efficiency of plastics recycling still holds room for improvement, we now live in a time when sports jackets made of recycled plastic bottles are a valued commodity.

Figure 14.26 shows this process for the condensation polymer called nylon, created in 1937 by DuPont chemist Wallace Carothers (1896–1937). Because this polymer is composed of two different monomers, it is classified as a *copolymer*. One monomer is adipic acid, which contains two reactive end groups, both carboxyl groups. The second monomer is hexamethylenediamine, in which two amine groups are the reactive end groups. One end of an adipic acid molecule and one end of a hexamethylamine molecule can be made to react with each other, splitting off a water molecule in the process. After two monomers have joined, reactive ends still remain for further reactions, which leads to a growing polymer chain. Aside from its use in hosiery, nylon also finds great use in the manufacture of ropes, parachutes, clothing, and carpets.

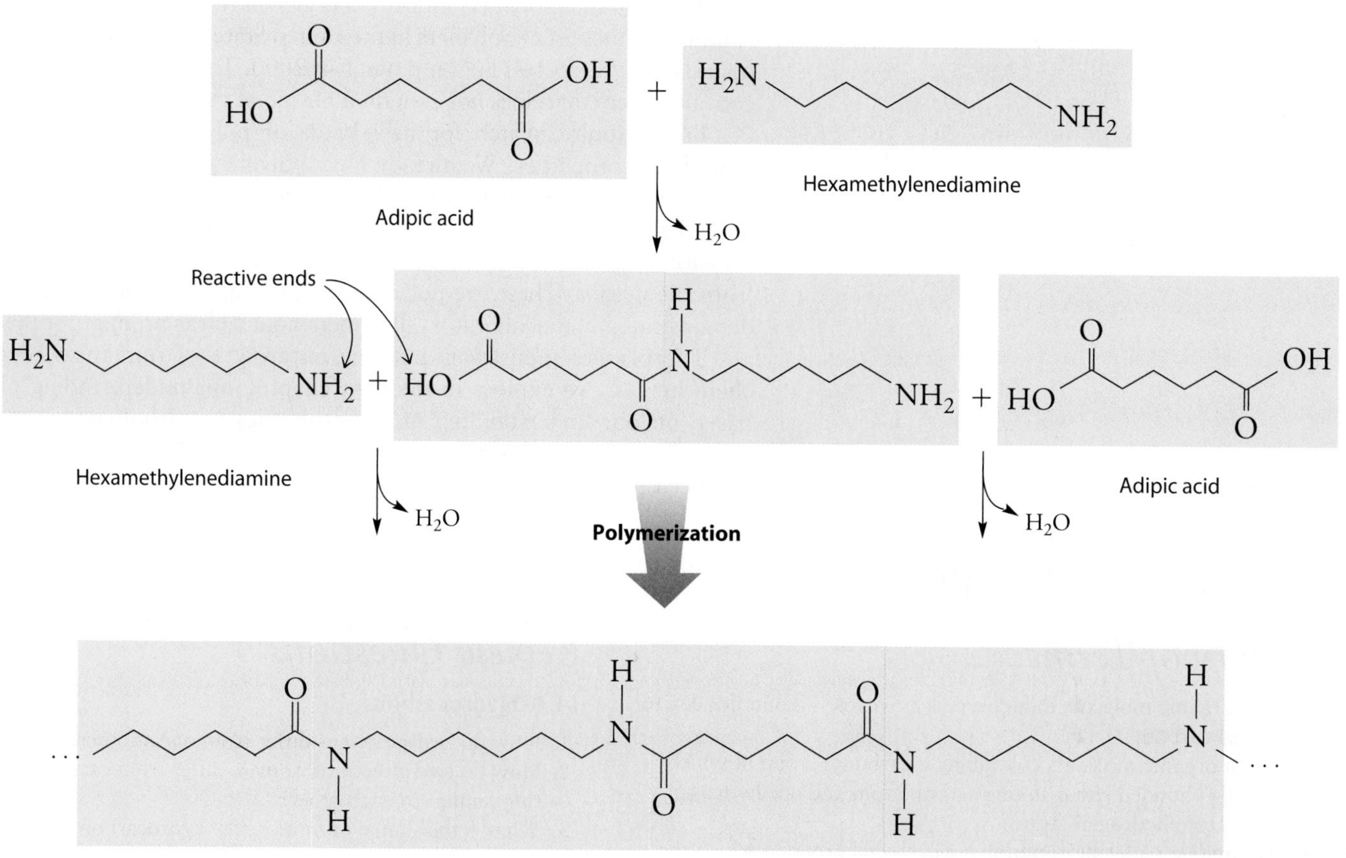

Figure 14.26 Adipic acid and hexamethylenediamine polymerize to form the condensation copolymer nylon.

CHECK YOURSELF

The structure of 6-aminohexanoic acid is:

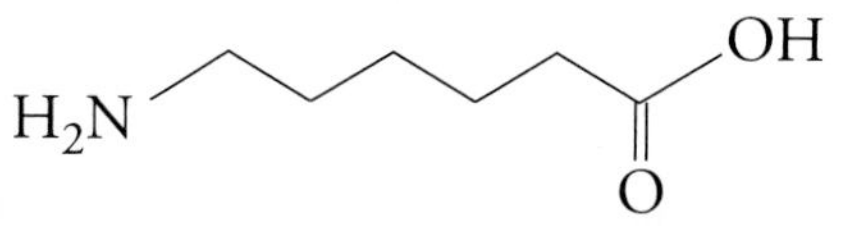

Is this compound a suitable monomer for forming a condensation polymer? If so, what is the structure of the polymer formed, and what small molecule is split off during the condensation?

CHECK YOUR ANSWER

Yes, because the molecule has two reactive ends. You know both ends are reactive because they are the ends shown in Figure 14.26. The only difference here is that both types of reactive ends are on the same molecule. Monomers of 6-aminohexanoic acid combine by splitting off water molecules to form the polymer known as nylon-6:

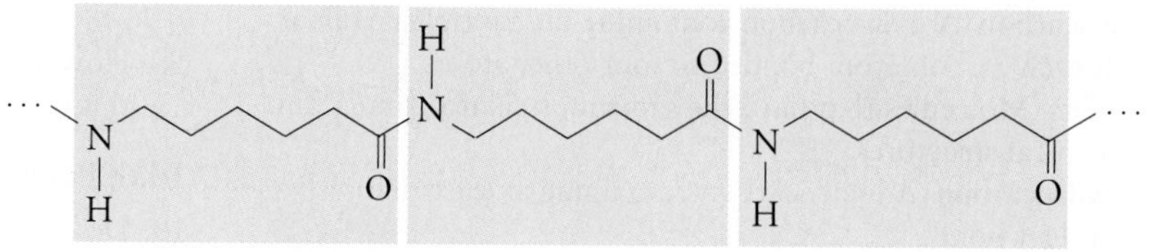

The synthetic-polymers industry has grown remarkably over the past 50 years. Annual production of polymers in the United States alone has grown from 3 billion pounds in 1950 to 100 billion pounds in 2000. Today, it is a challenge to find any consumer item that does *not* contain a plastic of one sort or another.

In the future, watch for new kinds of polymers having a wide range of remarkable properties. We already have polymers that conduct electricity, others that emit light, others that replace body parts, and still others that are stronger but much lighter than steel. Imagine synthetic polymers that mimic photosynthesis by transforming solar energy to chemical energy or efficiently separate fresh water from the oceans. These are not dreams. They are realities chemists have already demonstrated in the laboratory. Polymers hold a clear promise for the future.

The plastics industry is but one outgrowth of our knowledge of organic chemistry. As we explore in the next chapter, our understanding of life itself is based on our understanding of the properties of carbohydrates, fats, proteins, and nucleic acids, all of which are polymers containing the functional groups introduced in this chapter.

CHAPTER 14 Chapter Review

Summary of Terms

Alcohol An organic molecule that contains a hydroxyl group bonded to a saturated carbon.

Aldehyde An organic molecule containing a carbonyl group in which the carbon is bonded either to one carbon atom and one hydrogen atom or to two hydrogen atoms.

Amide An organic molecule containing a carbonyl group in which the carbon is bonded to a nitrogen atom.

Amine An organic molecule containing a nitrogen atom bonded to one or more saturated carbon atoms.

Aromatic compound Any organic molecule containing a benzene ring.

Carboxylic acid An organic molecule containing a carbonyl group in which the carbon is bonded to a hydroxyl group.

Ester An organic molecule containing a carbonyl group in which the carbon is bonded to one carbon atom and one oxygen atom is bonded to another carbon atom.

Ether An organic molecule containing an oxygen atom bonded to two carbon atoms.

Functional group A specific combination of atoms that behave as a unit in an organic molecule.

Heteroatom Any atom other than carbon or hydrogen in an organic molecule.

Hydrocarbon A chemical compound containing only carbon and hydrogen atoms.

Ketone An organic molecule containing a carbonyl group in which the carbon is bonded to two carbon atoms.

Lock-and-key model A conceptual model that explains how drugs interact with receptor sites.

Monomer The small molecular unit from which a polymer is formed.

Organic chemistry The study of carbon-containing compounds.

Phenol An organic molecule in which a hydroxyl group is bonded to a benzene ring.

Polymer A long organic molecule made of many repeating units.

Saturated hydrocarbon A hydrocarbon containing no multiple covalent bonds, with each carbon atom bonded to four other atoms.

Structural isomers Molecules that have the same molecular formula but different chemical structures.

Unsaturated hydrocarbon A hydrocarbon containing at least one multiple covalent bond.

Review Questions

14.1 Hydrocarbons

1. How do hydrocarbons differ from one another?
2. How do two structural isomers differ from each other? How are they similar to each other?
3. What is the source of most of the hydrocarbons people use?

14.2 Unsaturated Hydrocarbons

4. In a saturated hydrocarbon, how many atoms are bonded to each carbon atom?
5. What type of bond must a hydrocarbon have in order to be classified as unsaturated?
6. Aromatic compounds contain what kind of ring?

14.3 Functional Groups

7. What is a heteroatom?
8. How are organic molecules classified?

14.4 Alcohols, Phenols, Ethers, and Amines

9. Why are alcohols often soluble in water?
10. What distinguishes an alcohol from a phenol?
11. What distinguishes an alcohol from an ether?
12. What functional group do amines contain? What atoms make up this functional group?
13. What are two appropriately named amines and why are these names appropriate?

14.5 Ketones, Aldehydes, Amides, Carboxylic Acids, and Esters

14. Which elements make up the carbonyl group?
15. How are ketones and aldehydes related to each other? How are they different?
16. To what class of organic molecules does vanillin, the substance mainly responsible for vanilla flavor, belong to?
17. How are amides and carboxylic acids related to each other? How are they different from each other?
18. How are salicylic acid molecules modified to make aspirin? Why is this modification needed?

14.6 Polymers

19. Describe the general structure of a polymer.

20. What happens to the double bond of a monomer participating in the formation of an addition polymer?
21. What is released in the formation of a condensation polymer?

Multiple Choice Questions

Choose the BEST answer to the following.

1. How many structural isomers are there for hydrocarbons having the molecular formula C_4H_{10}.
   (a) none (b) one (c) two (d) three
2. Which of the following two stick structures are structural isomers?

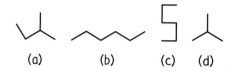

   (a) (b) (c) (d)

   (a) a and c (b) b and d (c) a and b
   (d) b and c (e) none of the above
3. Which of the following is a saturated molecule?

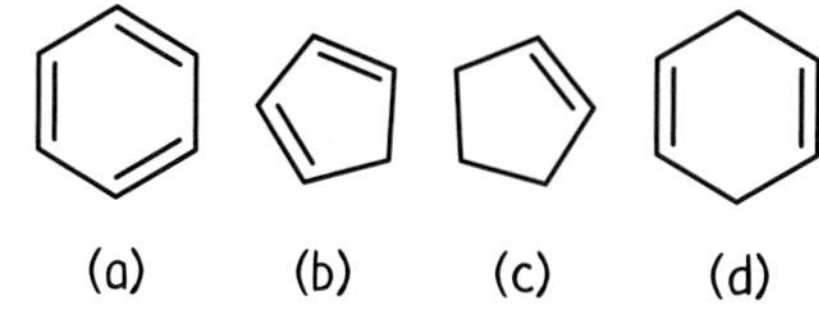

   (a) a (b) c (c) d (d) none of the above
4. How are monomers and polymers related?
   (a) A monomer is one strand of a polymer.
   (b) A monomer is made when you condense a polymer through polymerization.
   (c) A polymer is chemically identical to a monomer, only larger.
   (d) A monomer is a small, repeating unit that makes up a polymer.
   (e) None of the above.
5. Heteroatoms make a difference in the physical and chemical properties of an organic molecule because
   (a) they add extra mass to the hydrocarbon structure.
   (b) each heteroatom has its own characteristic chemistry.
   (c) they can enhance the polarity of the organic molecule.
   (d) all of the above.
6. Suggest an explanation for why aspirin has a sour taste.
   (a) It is the acidic nature of aspirin that gives rise to its sour taste.
   (b) The sour flavor is added to help prevent overdosing.
   (c) Aspirin is made sour as a mandated child safety feature.
   (d) It is the basic nature of aspirin that gives rise to its sour taste.
7. Alkaloid salts are not very soluble in the organic solvent diethyl ether. What might happen to the free-base form of caffeine dissolved in diethyl ether if gaseous hydrogen chloride (HCl) were bubbled into the solution?
   (a) A second layer of water would form.
   (b) Nothing would happen; the HCl gas would merely bubble out of solution.
   (c) The diethyl ether insoluble caffeine salt would form as a white precipitate.
   (d) The acid–base reaction would release heat, which would cause the diethyl ether to start evaporating.
8. Organic chemicals are suitable for making drugs because
   (a) our bodies are also made of organic chemicals.
   (b) they can be produced fairly inexpensively from petroleum.
   (c) they can form a vast diversity of structures.
   (d) they tend to dissolve well within the bloodstream.
9. Aspirin can cure a headache, but when you pop an aspirin tablet, how does the aspirin know to go to your head rather than your big toe?
   (a) The body is able to "pool" medicines around areas of increased neuronal activity.
   (b) The aspirin doesn't "know" to go to your head rather than your big toe. Rather, it gets distributed throughout your body.
   (c) The blood vessels within your head are generally more accessible than those within your big toe.
   (d) Most commercially available aspirins are compounded with buffer agents that direct the aspirin to regions of lower pH, which includes the cerebral hemispheres.
10. Correctly identify the following functional groups in this organic molecule-amide, ester, ketone, ether, alcohol, aldehyde, amine:

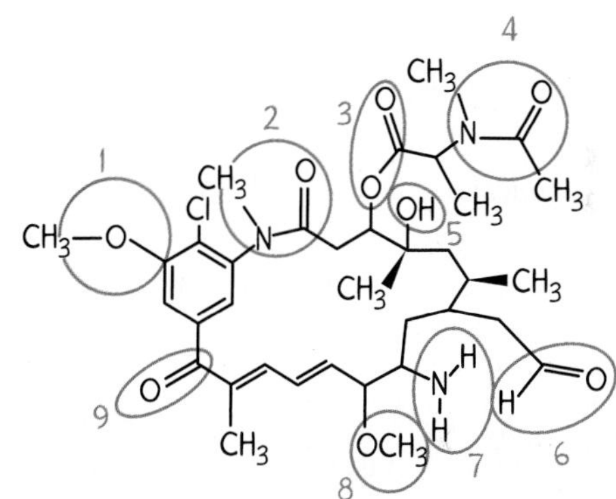

   (a) 1 = ether, 3 = ester, 6 = aldehyde, 9 = alcohol
   (b) 2 = amide, 4 = ester, 7 = amine, 8 = ether
   (c) 1 = ester, 5 = alcohol, 8 = ether, 9 = ketone
   (d) 2 = amide, 6 = aldehyde, 7 = amine, 8 = ether

INTEGRATED SCIENCE CONCEPTS

Biology—Drug Action and Discovery

1. In the lock-and-key model, is a drug viewed as the lock or the key?
2. What holds a drug to its receptor site?
3. Which is better for you: a drug that is a natural product or one that is synthetic? Why do you think so?

Active Exploration

Racing Water Drops

The chemical composition of a polymer has a significant effect on its macroscopic properties. To see this for yourself, place a drop of water on a new plastic sandwich bag, and then tilt the bag vertically so that the drop races off. Observe the behavior of the water carefully. Now race a drop of water off a freshly pulled strip of plastic food wrap. How does the behavior of the drop on the wrap compare with the behavior of the drop on the sandwich bag?

You may need to play around with the drops for a while in order to see the differing affinities that the bag and wrap have for water. One way to do this is to tape the polymers side-by-side stretched out on a sturdy piece of cardboard. Tilt the cardboard to various angles, testing for the speed with which water drops roll down the incline on the two surfaces.

Most brands of sandwich bags are made of polyethylene terephthalate, and most brands of food wrap are made of polyvinylidene chloride. Which shows evidence of stronger dipole–induced dipole interactions with water? Can you explain your results based upon the chemical structures shown in Table 14.5?

Exercises

1. According to Figure 14.3, which has a higher boiling point: gasoline or kerosene?
2. According to Figure 14.3, which consists of smaller molecules: kerosene or diesel?
3. There are five atoms in the methane molecule, CH_4. One out of these five is a carbon atom, which is $\frac{1}{5} \times 100 = 20$ percent carbon. What is the percent carbon in ethane, C_2H_6? Propane, C_3H_8? Butane, C_4H_{10}?
4. What is the chemical formula of the following structure?
5. What is the chemical formula of the following structure?
6. Of the structures shown in Exercises 4 and 5, which is more oxidized?
7. List the following compounds in order of least oxidized to most oxidized:

(a) (b) (c) (d)

8. Circle the longest chain of carbon atoms in the following structure. How many carbon atoms are in this chain?
9. Carbon–carbon single bonds can rotate while carbon–carbon double bonds cannot rotate. How many different structures are shown below?
10. Heteroatoms make a difference in the physical and chemical properties of an organic molecule because
    (a) they add extra mass to the hydrocarbon structure.
    (b) each heteroatom has its own characteristic chemistry.
    (c) they can enhance the polarity of the organic molecule.
    (d) all of the above.
11. One of the skin-irritating components of poison oak is tetrahydrourushiol:

The long, nonpolar hydrocarbon tail embeds itself in a person's oily skin, where the molecule initiates an allergic response. Scratching the itch spreads tetrahydrourushiol molecules over a greater surface area, causing the zone of irritation to grow. Is this compound an alcohol or a phenol? Defend your answer.

12. Explain why caprylic acid, $CH_3(CH_2)_6COOH$, dissolves in a 5 percent aqueous solution of sodium hydroxide but caprylaldehyde, $CH_3(CH_2)_6CHO$, does not.
13. Suggest an explanation for why aspirin has a sour taste.
14. Identify the following functional groups in this organic molecule—amide, ester, ketone, ether, alcohol, aldehyde, amine:

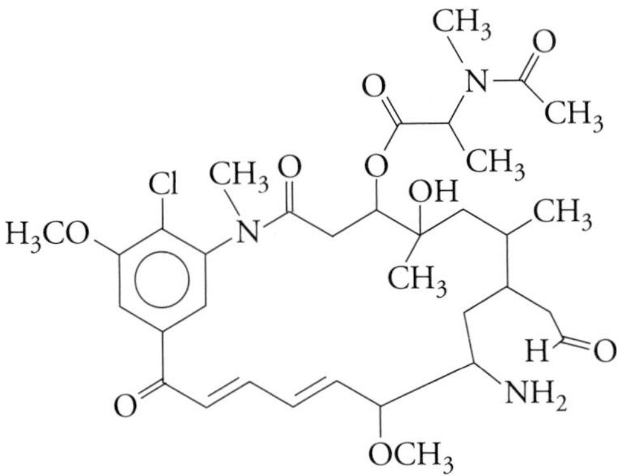

15. Benzaldehyde is a fragrant oil. If stored in an uncapped bottle, this compound will slowly tranform into benzoic acid along the surface. Is this an oxidation or a reduction?

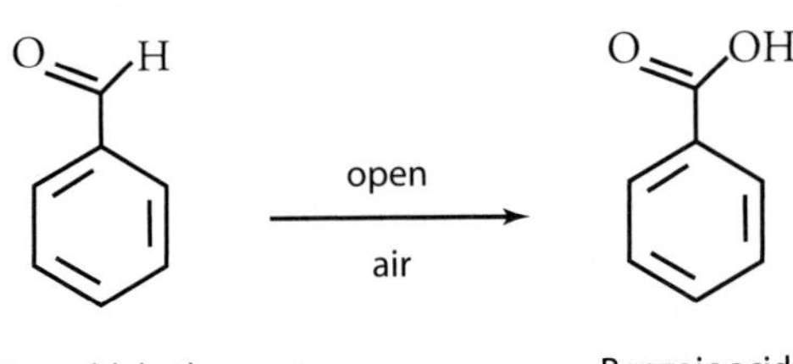

16. Which would you expect to be more viscous, a polymer made of long molecular strands or one made of short molecular strands? Why?
17. Hydrocarbons release a lot of energy when ignited. Where does this energy come from?
18. What type of polymer would be best to use in the manufacture of stain-resistant carpets?
19. The copolymer styrene-butadiene rubber (SBR), shown here, is used for making tires as well as bubble gum. Is it an addition polymer or a condensation polymer?

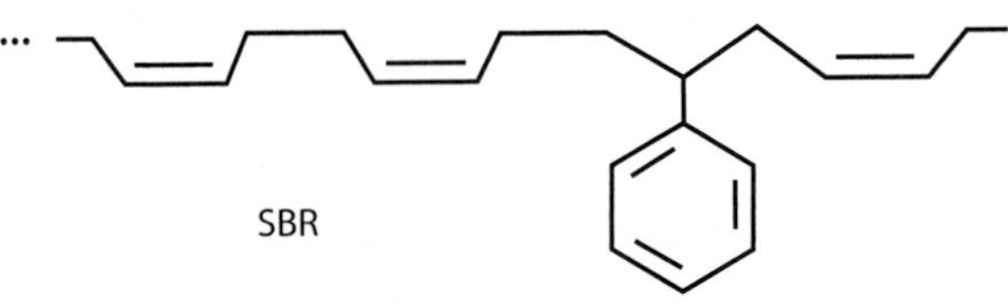

20. Many of the natural product molecules synthesized by plants are formed by the joining together of isoprene monomers via an addition polymerization. A good example is the nutrient beta-carotene, which consists of eight isoprene units. Find and circle these units within the structure shown here.

Isoprene Beta-carotene

21. Write a letter to Grandma summarizing how polymers were important in winning World War II. If Grandma lived through World War II, ask her if she was aware of the role polymers played!

Problems

1. Draw all the structural isomers for hydrocarbons having the molecular formula C_4H_{10}.
2. Draw all the structural isomers for hydrocarbons having the molecular formula C_6H_{14}.
3. Draw all the structural isomers for amines having the molecular formula C_3H_9N.
4. Cetyl alcohol, $C_{16}H_{33}OH$, is a common ingredient of soaps and shampoos. It was once commonly obtained from whale oil, which is where it gets its name (*cetyl* is derived from *cetacean*). Draw the chemical structure for this compound.

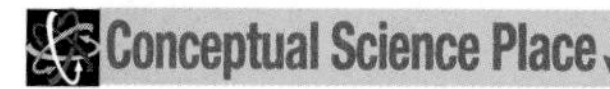

CHAPTER 14 ONLINE RESOURCES

Tutorials Organic Molecules and Isomers, Introduction to Organic Molecules, Polymers • ***Quiz*** • ***Flashcards*** • ***Links***

PART THREE

BIOLOGY

[Io:] "Pico, did you know that every human being is made up of more than 10 trillion little units called cells? Cells allow us to do everything we do—jump, yell, see things... even thinking happens because of little electric signals moving through cells in our brains."

[Pico:] "Wow, Io, that is pretty amazing. But do you know what else is cool? We are what we are because of little strands in our cells called DNA. That's where our genes are, and there's 7 feet of DNA in every cell in the body. What's even more amazing is that more than 99.9% of our DNA is the same as that of every other person on Earth!"

CHAPTER 15

The Basic Unit of Life—The Cell

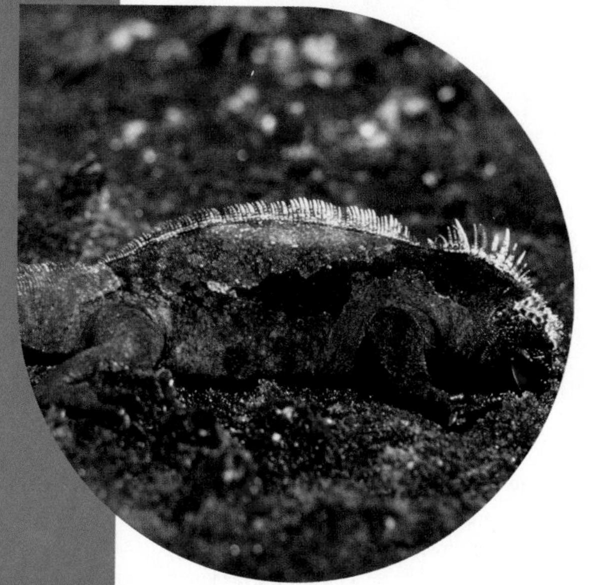

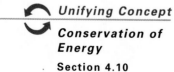

A marine iguana eats algae on Fernandina Island in the Galápagos.

CHAPTER OUTLINE

Unifying Concept
Conservation of Energy
Section 4.10

Instinctively, we know that a marine iguana and the algae it eats are living things. At the same time, it's obvious that the rock beneath them is not living. What are living things? Do living things all share characteristics that differentiate them from nonliving things? Do they all reproduce? Use energy? Evolve? Are all living things made of one or more cells? What are cells? How do cells "talk" with each other? How do cells make new cells? How do cells obtain energy, and how do we take advantage of this process when we bake bread and brew beer? Can a cell live forever? In this chapter, we will explore the nature of life and the world of the cell.

15.1 Characteristics of Life

Biology is the study of life and living organisms. But what is a living thing? What distinguishes living things from nonliving things? We tend to know living things when we see them, although sometimes our eyes can fool us (Figure 15.1). What is it about living things that makes them living?

Living things all share a set of characteristics. For one thing, they all use energy. Living things, such as the sunflowers and lions in Figure 15.2, take energy from the environment and convert it to other forms of energy for their own use. For example, plants take electromagnetic energy from sunlight and use it to build stems and leaves. Animals eat, converting the energy they get from food to chemical energy, which they store in their bodies. This chemical energy is eventually converted again to kinetic and potential energy and heat as they crawl, or fly, or grow. Of course, all the ways in which living things convert energy are consistent with the laws of physics. This means, first, that energy is always conserved and, second, that in any energy conversion, some energy is lost to the environment as heat.

Another characteristic of living things is that they develop and grow. A living organism grows and changes over time, as shown in Figure 15.3. When chicks hatch, they are small and covered with downy yellow feathers. Over time, they grow bigger, and their downy feathers are replaced by stiff adult feathers.

Living things maintain themselves. They generate structures, such as stems and leaves or skin and bones, and they repair damage done to those structures. When you scrape your knee, your blood clots to stop the bleeding, and the wounded skin scabs over and heals. Living things also maintain their internal environment, keeping it stable in the face of changing external conditions. Whether it is freezing cold or blisteringly hot, your body temperature hovers right around 98.6°F (37°C). And

Figure 15.1 Is this a rock or is it a Red Sea scorpion stonefish waiting for prey to pass?

(a)

(b)

Figure 15.2 Living things take energy from the environment and convert it to other forms of energy. (a) Plants such as these sunflowers convert energy from sunlight to plant tissue. (b) Animals such as these lions convert the chemical energy stored in food to motion or other activity and to growth and reproduction.

under a tremendous variety of conditions, your heart continues to pump, your blood carries oxygen and nutrients to your tissues, and your kidneys filter wastes.

Living things have the capacity to reproduce. They make offspring that are exact or inexact copies of themselves. Figure 15.4 shows the two primary ways living things reproduce, asexually and sexually. *Asexual reproduction* occurs when a living organism reproduces all by itself, such as by dividing into two. Bacteria and sea anemones are examples of organisms that are able to reproduce asexually. *Sexual reproduction* occurs when organisms form sperm and eggs that join to develop into new individuals. Humans, penguins, bumblebees, and pine trees reproduce sexually.

Finally, living things are parts of populations that evolve. Populations do not remain constant from one generation to the next but change over time, across generations. During the industrial revolution, when habitats became polluted and blackened with soot, populations of peppered moths evolved so that better-camouflaged dark-colored moths became more common than light-colored moths. Once antipollution laws were passed and habitats were cleaned up, light-colored moths became more prevalent again.

Figure 15.3 Living things grow and develop over time.

CHECK YOURSELF

1. We know that cars are nonliving things. Check this against the list of attributes of living organisms. Which attributes do cars exhibit and which don't they exhibit?
2. Do all living things have all the characteristics of life listed above?

(a)

(b)

Figure 15.4 Living things have the capacity to reproduce. (a) A sea anemone reproduces asexually by dividing. (b) Penguins reproduce sexually. A sperm and egg join and develop into a new individual. These are two emperor penguins with their chick in Antarctica.

CHECK YOUR ANSWERS

1. Cars use energy, converting the energy in gasoline to motion. We might be able to argue that cars "develop" over time, acquiring nicks and dents and wearing down the treads on their tires. However, cars definitely do not maintain themselves, do not reproduce, and do not evolve.
2. Not necessarily. For example, mules are sterile and unable to reproduce, but they are certainly alive.

INTEGRATED SCIENCE CHEMISTRY

Macromolecules Needed for Life

On a typical day of the grueling bicycle race known as the Tour de France, champion Lance Armstrong bikes two hundred kilometers and burns some 7,000 Calories. (The Calorie, you will recall, is a unit of energy equivalent to 4,180 joules.) To get the Calories he needs, Armstrong eats lots of rice, pasta, and baked potatoes—foods with plenty of carbohydrates, or "carbs," as they're known in the world of sport and nutrition. Other diets—high-protein, low-carb, low-fat, and so on—are used for other purposes. But what are these substances, exactly? What is a carbohydrate? What is a protein? Carbohydrates and proteins are examples of *macromolecules*, large molecules made up of smaller molecules linked together. There are four major types of macromolecules in living organisms—proteins, carbohydrates, lipids, and nucleic acids.

Proteins perform a wide range of functions in living organisms. The protein keratin provides structure in the form of skin, hair, and feathers. Insulin is a protein that acts as a hormone, allowing one type of cell to communicate with other types. Actin and myosin are proteins that allow muscles to contract. Hemoglobin, a protein found in red blood cells, transports oxygen to body tissues. Antibodies are proteins that protect the body from disease. And proteins known as digestive enzymes break down food during digestion.

What are proteins? Proteins are folded strings of organic molecules called *amino acids*. All amino acids include a central carbon (C) atom bonded to an amino group (NH_2), a carboxyl group (COOH), a hydrogen atom (H), and a side chain (R) that varies from one amino acid to another. The amino acid leucine is shown in Figure 15.5a. Although only 20 different amino acids are found in living organisms, these can be combined and folded in practically countless ways to create proteins with unique three-dimensional structures. The unique structures of proteins allow them to perform very specific functions in living organisms, as we will see later in this chapter.

Carbohydrates store energy in living organisms. Simple sugars, such as glucose (Figure 15.5b) and fructose, are carbohydrates. More complex carbohydrates are made up of chains of simple sugars. Starch and glycogen—the primary energy-storage substances in plants and animals, respectively—consist of linked glucose molecules. Carbohydrates can also have structural functions. Cellulose, found in plant cell walls, is a structural carbohydrate built from glucose subunits—and it is the most abundant organic compound in the world. Carbohydrates are made up of carbon, hydrogen, and oxygen atoms, generally in the form of $(CH_2O)_n$.

Lipids serve diverse functions in living organisms. As fats or oils, lipids are used by many living organisms to store energy. Lipids store energy much more efficiently than carbohydrates—that is, one gram of fat or oil contains a lot more energy than one gram of carbohydrate. For this reason, lipids are used for long-term energy storage by many organisms, including humans. Lipids can serve structural purposes as well. For example, phospholipids are an essential component of cell membranes. One of the most familiar lipids is cholesterol, which the body uses to make such hormones as estrogen and testosterone. Lipids have a variety of chemical structures, but many include fatty acids—strings of hydrocarbons (carbon and

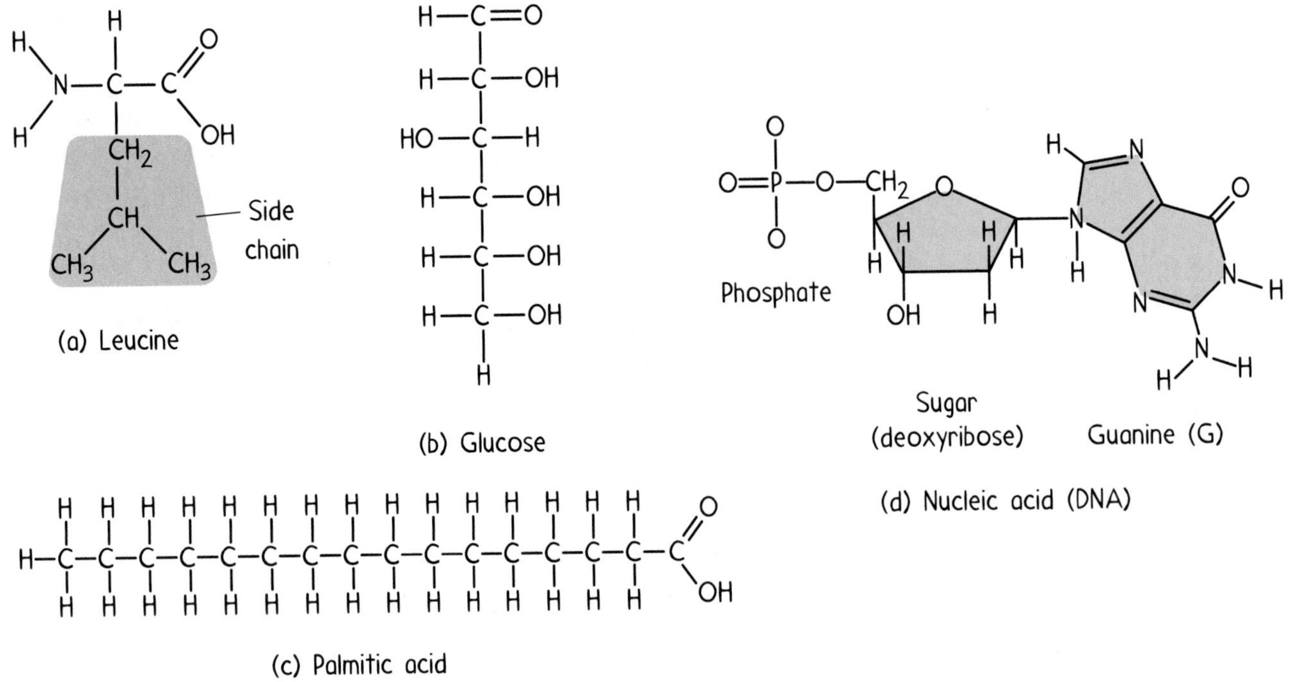

Side chain

(a) Leucine

(b) Glucose

Phosphate

Sugar (deoxyribose)

Guanine (G)

(d) Nucleic acid (DNA)

(c) Palmitic acid

Figure 15.5 (a) The amino acid leucine is one of the twenty amino acids found in living organisms. (b) The carbohydrate glucose is the primary energy source for vertebrates. (c) Fatty acids are an important component of many lipids. This is palmitic acid, a fatty acid found in lard and butter. (d) Nucleic acids are made up of a sugar-and-phosphate backbone attached to a series of nitrogenous bases. Guanine is one of four nitrogenous bases found in DNA.

hydrogen atoms) with a carboxyl (COOH) group at one end—as a major component (Figure 15.5c). Lipids are hydrophobic; that is, they are not soluble in water.

The role of the fourth type of macromolecule, **nucleic acids**, is to store genetic information in living organisms. There are two kinds of nucleic acids, *deoxyribonucleic acid* (DNA) and *ribonucleic acid* (RNA). A nucleic acid strand consists of a sugar-and-phosphate backbone attached to a series of nitrogenous bases. DNA consists of two nucleic acid strands twisted into a spiral, which is why it is sometimes called a double helix. Four nitrogenous bases are used in DNA—adenine, cytosine, guanine, and thymine, or A, C, G, T for short. All the genetic information in living organisms is expressed using this four-letter alphabet. The chemical structure of guanine is shown in Figure 15.5(d). RNA resembles DNA except that it is single-stranded rather than double-stranded, uses a different sugar molecule, and uses the nitrogenous base uracil (U) in place of thymine (T). DNA and RNA will be described in greater detail in Chapter 16.

Figure 15.6 Geckos store fat in their tails. This energy supply helps them get through lean periods.

CHECK YOURSELF

1. Many species of gecko lizards store fat in their tails (see Figure 15.6), using this energy supply to get through lean periods. What is the advantage of storing energy as fat rather than as carbohydrate?
2. Living organisms contain thousands of different kinds of proteins. How is it possible to make so many different kinds of proteins from only 20 different amino acids?

CHECK YOUR ANSWERS

1. Because a gram of fat contains more energy than a gram of carbohydrate, storing the energy as fat allows geckos to store the energy they need without weighing themselves down as much as if they used carbohydrates.

2. The 20 amino acids that make up proteins can be combined and then folded in a huge number of different ways, creating many different proteins.

15.2 Cell Types: Prokaryotic and Eukaryotic

Conceptual Science Place
Comparing Prokaryotic and Eukaryotic Cells

Cells are the basic units of life, the same way that atoms are the basic units of matter. Cells have all the characteristics of life, and all living organisms are made of one or more cells.*

You yourself have more than 10 trillion (10^{13}) cells in your body, and the diversity of things they do is amazing. Right now, muscle cells are moving your eyeballs as you follow the text on this page, sensory cells in your eyes are taking in the shapes of the letters, cells in your ears are absorbing nearby sounds, red blood cells are carrying oxygen to all the other cells in your body, and gut cells are making digestive juices to process your morning oatmeal. And some very impressive cells—the neurons in your brain—are producing your thoughts about how amazing cells are.

Two distinct types of cells are found in different living organisms today—prokaryotic cells and eukaryotic cells. These are distinguished primarily by the presence or absence of a **nucleus**, a distinct structure within the cell that contains the cell's DNA. *Prokaryotic cells* lack a nucleus (*pro* means "before" and *karyote* refers to "nut" or "nucleus"). *Eukaryotic cells* ("true nucleus") have a nucleus as well as other structures not present in prokaryotic cells. Organisms with prokaryotic cells are called **prokaryotes**, and organisms with eukaryotic cells are called **eukaryotes**. Figure 15.7 compares typical prokaryotic and eukaryotic cells.

Prokaryotes have existed on Earth far longer than eukaryotes. In fact, prokaryotes first evolved 3.5 to 4 billion years ago, and they were the only living things on Earth until about two billion years ago, when the first eukaryotes

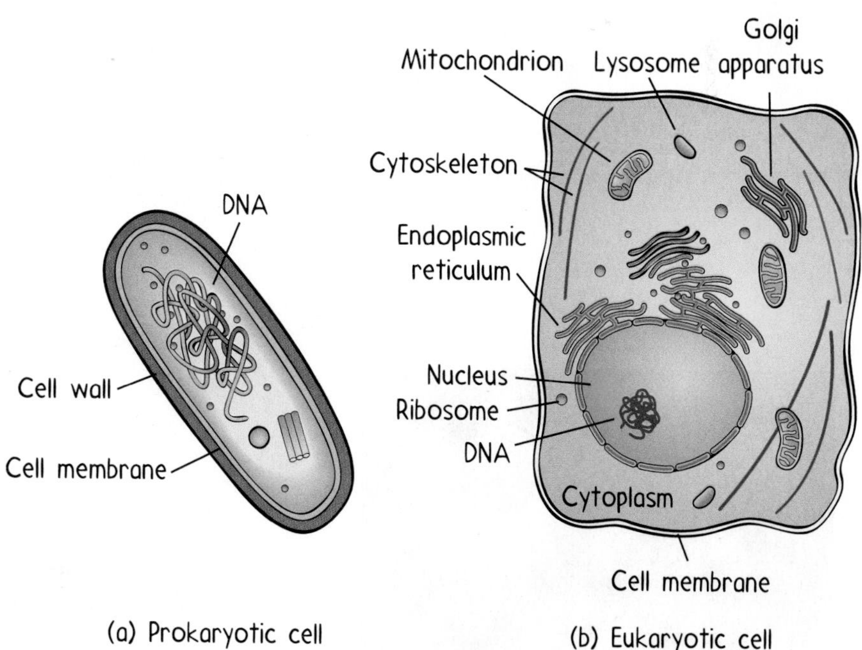

Figure 15.7 (a) Prokaryotic cells have no nucleus. (b) Eukaryotic cells have a nucleus as well as structures called organelles not found in prokaryotes. The organelles shown here will be discussed later in the chapter.

*Viruses, which possess some of the characteristics of life and straddle the line between living and nonliving, are not composed of cells. We'll discuss viruses further in Chapter 18.

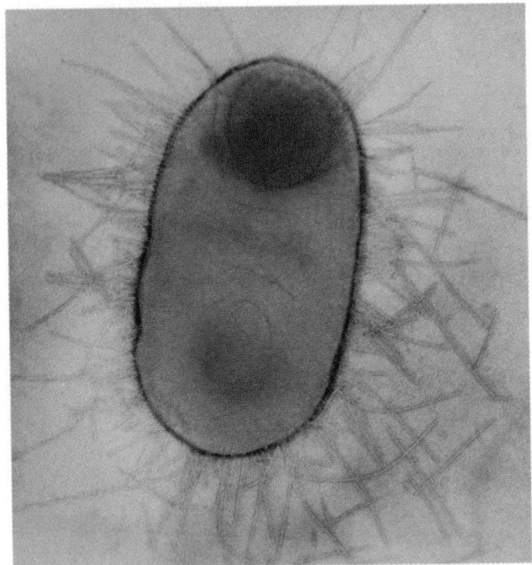

Figure 15.8 *Escherichia coli* (commonly referred to as *E. coli*) is a prokaryote that lives in the human digestive tract.

appeared. Prokaryotes now include two major lineages, the bacteria and the archaea (Chapter 18). Prokaryotes are single-celled organisms and are very small, ranging from about 0.1 to 10 micrometers (10^{-6} meter) in diameter. Their structure is considerably simpler than that of eukaryotes. The DNA of prokaryotes is found in a single circular structure called a **chromosome** and is not contained within a nucleus. Most prokaryotes have an outer *cell wall* that helps protect the cell. The prokaryote *Escherichia coli*, an occupant of the human digestive tract and one of the best-studied organisms in the world, is shown in Figure 15.8.

Eukaryotes can be single-celled, like prokaryotes, or they can be composed of many cells. The fungus known as baker's yeast, commonly used in baking and brewing, is a single-celled eukaryote (Figure 15.9). Plants, animals, and most other fungi are multicellular eukaryotes. Eukaryotic cells have their DNA in a distinct nucleus, a feature that distinguishes them from prokaryotes. In addition, the DNA of eukaryotic cells is found in linear, rather than circular, chromosomes. Eukaryotic cells also have numerous **organelles**, structures that perform specific functions for the cell. Finally, eukaryotic cells are larger than prokaryotic cells—where prokaryotic cells measure 0.1 to 10 micrometers, eukaryotic cells generally measure 10 to 100 micrometers. Some eukaryotic cells are even larger than that, however.

Some of the organelles in eukaryotic cells look suspiciously like whole prokaryotes. **Mitochondria** (Figure 15.10), organelles that obtain energy for the cell's use, are contained within their own membrane and have their own DNA, just like prokaryotes. Furthermore, mitochondrial DNA, like that of prokaryotes, exists in the form of a single circular chromosome. This has led to the hypothesis that certain prokaryotes started to live within early eukaryotes and eventually evolved into organelles, a theory we will discuss further in Chapter 17.

INSIGHT An egg yolk is a single cell. That makes the ostrich egg yolk, which measures 10 to 15 centimeters in diameter, the largest cell in the world.

INSIGHT Ostrich eggs may be the largest cells in the world, but nerve cells are the longest. The longest cell in the human body is a neuron that runs all the way from the spinal cord to the toes. You can try this cell out right now by wiggling your toes.

CHECK YOURSELF

Which of the following organisms are prokaryotes and which are eukaryotes? a. The bacterium that causes tuberculosis. b. A humpback whale. c. A honey mushroom.

CHECK YOUR ANSWERS

a. The tuberculosis bacterium, like all bacteria, is a prokaryote.

b. A humpback whale, like all animals, is a eukaryote.

c. A honey mushroom, like all fungi, is a eukaryote.

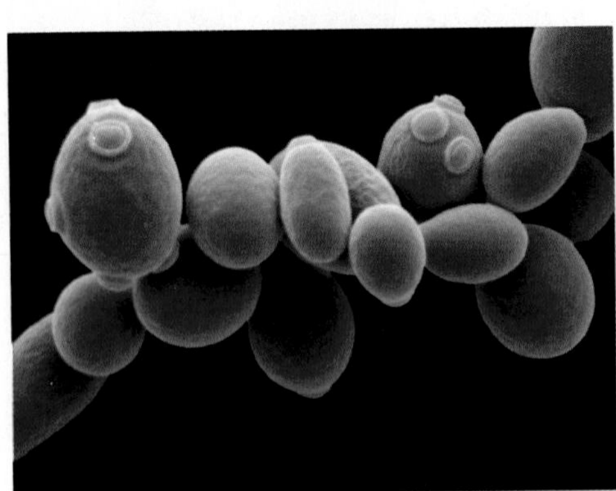

Figure 15.9 *Saccharomyces cerevisiae*, commonly known as baker's yeast or brewer's yeast, is a single-celled eukaryote.

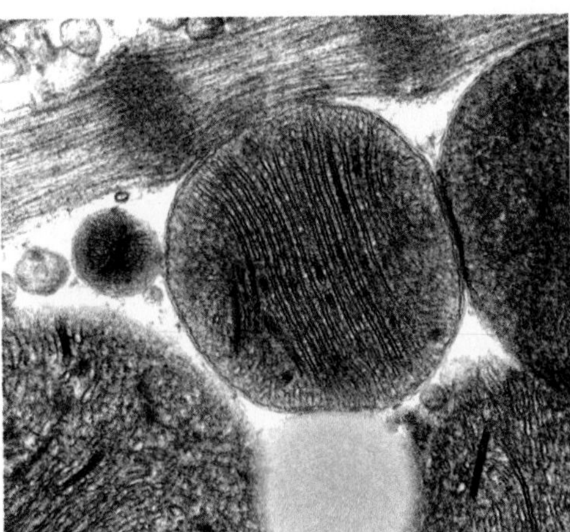

Figure 15.10 The mitochondria in eukaryotic cells have a great deal in common with prokaryotes and probably evolved from them.

HISTORY OF SCIENCE

Cell Theory

Cell theory, the idea that the cell is the basic unit of life, was several centuries in the making. In 1665, Robert Hooke, an English scientist, coined the term *cell* and published the first description of cells in his book *Micrographia.* Hooke examined a piece of cork under a microscope and saw a series of small boxlike chambers. He called these chambers "cells" because they reminded him of monks' cells. We now know that what Hooke saw were not actually living cells but the cell walls that remain in dried plant tissue.

In the late 1600s, the Dutch naturalist Anton van Leeuwenhoek became the first person to describe many types of living cells, including bacteria, sperm cells, and blood cells. After examining plaque scraped from his teeth under a microscope, van Leeuwenhoek wrote, in a 1683 letter to the Royal Society of London, about the amazing "animalcules" he saw: "I then most always saw, with great wonder, that in the said matter there were many very little living animalcules, very prettily a-moving. The biggest sort . . . had a very strong and swift motion, and shot through the water (or spittle) like a pike does through the water. The second sort . . . oft-times spun round like a top."* We now know that these animalcules were actually bacteria.

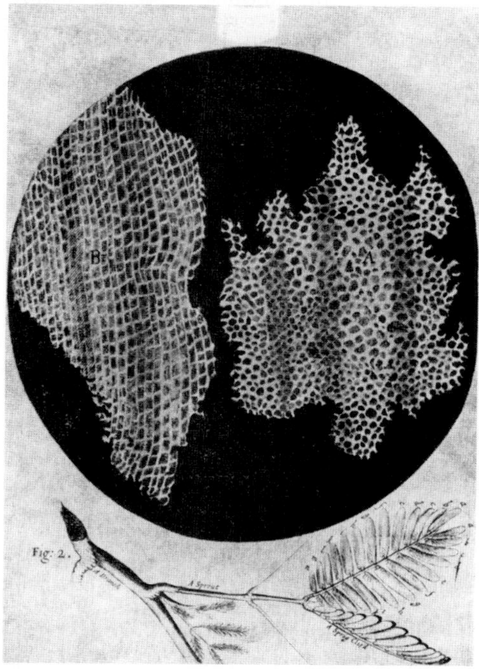

Robert Hooke examined cork tissue under a microscope and called the small chambers he saw "cells." This is Hooke's original drawing of what he saw.

It was not until the 1800s that the central importance of the cell was established. In 1838, careful studies of plants led German scientist Matthias Schleiden to conclude that all plants were made of cells. The following year, another German scientist, Theodor Schwann, came to the same conclusion about animals. The cell theory was finally completed in 1855 with German scientist Rudolph Virchow's observation that all living cells come from other living cells.

In summary, the cell theory states:

1. All living things are made up of one or more cells.
2. All cells come from other cells.

* Dobell, Clifford, *Antony van Leeuwenhoek and His "Little Animals."* Dover Publications, New York, 1960.

INTEGRATED SCIENCE PHYSICS

The Microscope

Microscopes are high-tech magnifying glasses that allow us to see very small objects with a fine level of detail. One type of microscope, the light microscope, has been around for centuries—in fact, Robert Hooke discovered the existence of cells while using one. Light microscopes work by passing visible light through a specimen and then through a series of lenses. The lenses refract, or bend, the light in order to produce a magnified image of the specimen (Figure 15.11a). Often, specimens are treated with stains that make their features easier to see.

Light microscopes are able to resolve objects on the order of a micrometer (10^{-6} meters) in size. (For comparison, the resolving power of the human eye is about 1/10 millimeter, or 10^{-4} meters—two lines closer together than that appear to us as a single line.) With a resolution of 10^{-6} meters, light microscopes allow us to view cells and to make out the larger features within them, such as the nucleus and mitochondria. However, they do not really allow us to see organelles and other cellular structures in detail. This is because diffraction blurs the image if the size of an object is about the same as the wavelength of light (Chapter 8). Further, if an object is smaller than the wavelength of light, no structure can be seen at all. The entire image is lost, due to diffraction. No amount of magnification or perfection of microscope design can defeat this fundamental diffraction limit.

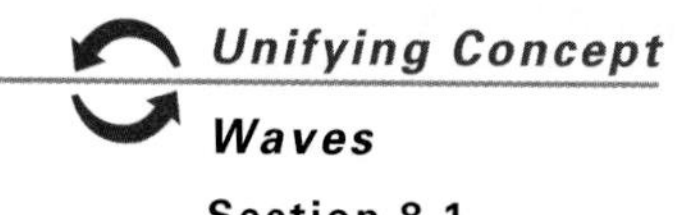

Unifying Concept
Waves
Section 8.1

To get around this problem, scientists illuminate very tiny objects with electron beams rather than with light. All matter has wave properties, and, compared with light waves, electron beams have extremely short wavelengths. In an electron microscope, electric and magnetic fields, rather than optical lenses, are used to focus electron beams (Figure 15.11b). In practice, electron microscopes are able to resolve objects about a nanometer (10^{-9} meter) in size, which covers just about everything of biological interest (Figure 15.12).

Eye
Ocular lens
Objective lens
Specimen
Condenser lens
Light source

(a)

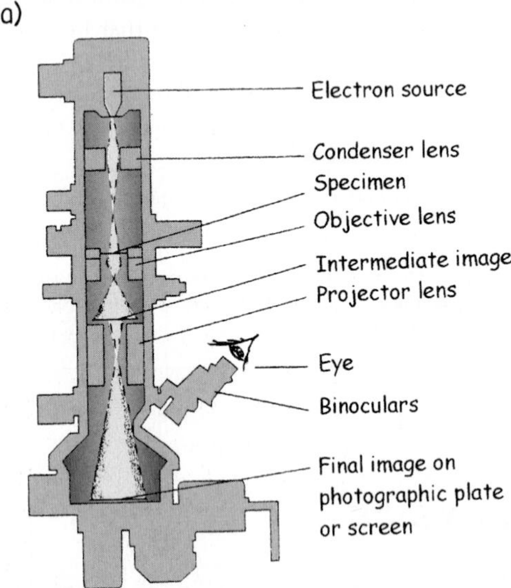

(b)

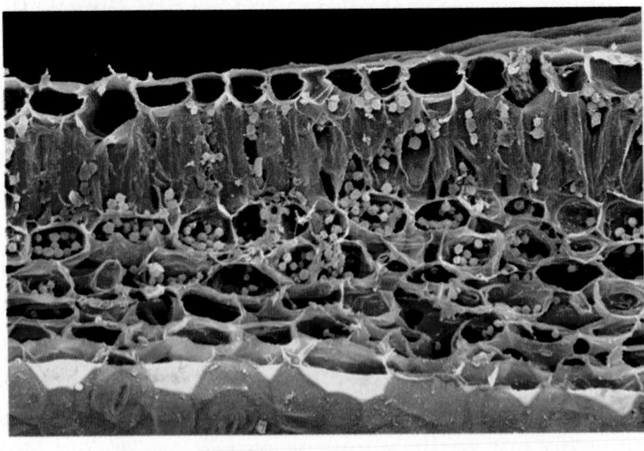

(c)

Figure 15.11 Microscopes allow us to examine objects that are too small for the human eye to see. (a) Light microscopes use lenses to refract light and magnify specimens. (b) Electron microscopes use electric and magnetic fields to focus beams of electrons on specimens. (c) This is a scanning electron microscope image of a leaf in cross section. Chloroplasts and other organelles can be seen inside the cells.

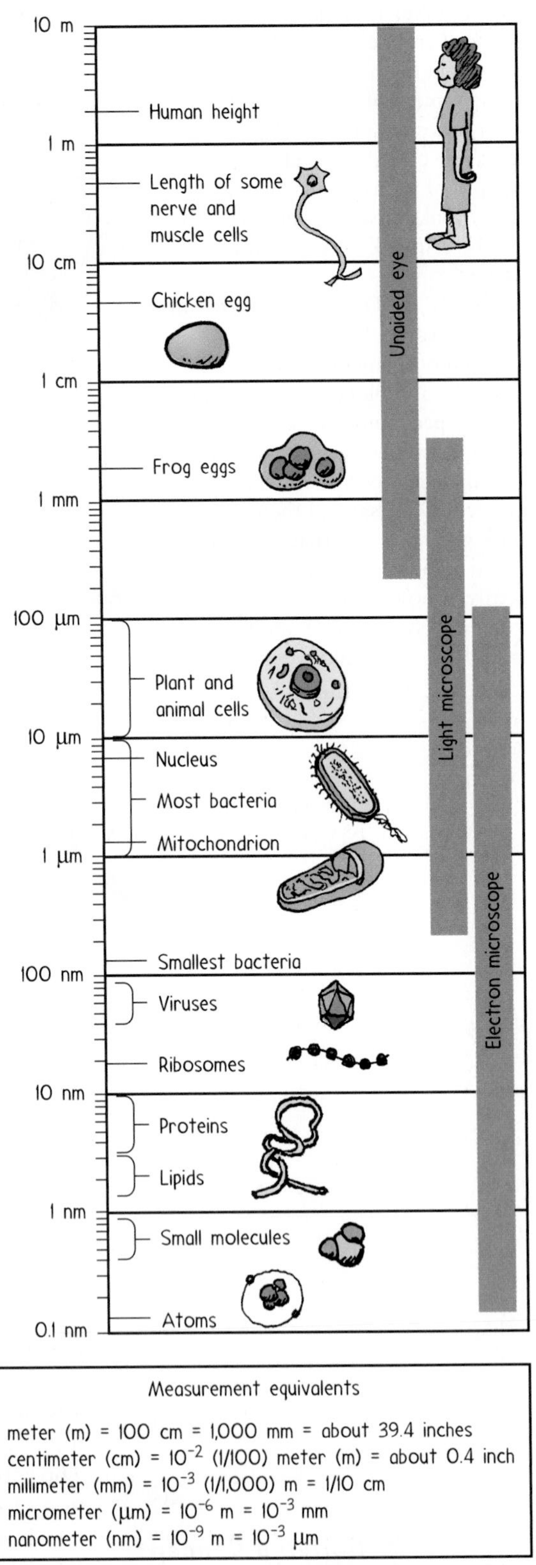

Measurement equivalents

1 meter (m) = 100 cm = 1,000 mm = about 39.4 inches
1 centimeter (cm) = 10^{-2} (1/100) meter (m) = about 0.4 inch
1 millimeter (mm) = 10^{-3} (1/1,000) m = 1/10 cm
1 micrometer (μm) = 10^{-6} m = 10^{-3} mm
1 nanometer (nm) = 10^{-9} m = 10^{-3} μm

Figure 15.12 Microscopes allow us to look into the world of the cell. Depending on the size of a feature, the naked eye, a light microscope, or an electron microscope may be used. (Adapted from Campbell, Reece, Simon, *Essential Biology with Physiology,* © 2004.)

There are two types of electron microscopes. Scanning electron microscopes create a three-dimensional image of the surface of a specimen (Figure 15.11c). Transmission electron microscopes image thin sections. Specimens used in electron microscopy have to be prepared carefully. Typically, a scanning electron microscope specimen is dried and coated with a thin layer of gold. It is then placed in a vacuum column, and the electron beam is moved across it row by row. As the beam hits the specimen, electrons are knocked from the gold layer, collected, and counted in order to put together the image. A specimen prepared for transmission electron microscopy is stained with heavy-metal atoms. The electron beam is projected on the specimen, and the pattern of deflection and transmission of electrons is recorded in order to create the image.

CHECK YOURSELF

Tiny bacteria called mycoplasmas have diameters as small as 0.1 micrometers. Can they be studied with light microscopes? Can they be studied with electron microscopes?

CHECK YOUR ANSWER

Mycoplasmas are too small to be examined with light microscopes, but they can be studied with electron microscopes.

15.3 Tour of a Eukaryotic Cell

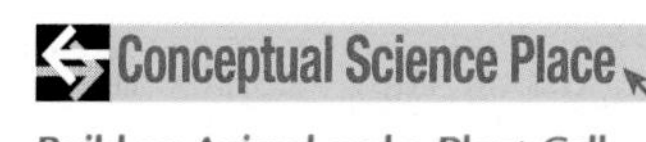

Build an Animal and a Plant Cell

You, your potted geranium, and your Labrador retriever are all eukaryotes, organisms composed of eukaryotic cells. In fact, most of the living things we encounter on a daily basis—all the plants, animals, and fungi—are eukaryotes.

All eukaryotic cells are surrounded by a cell membrane. The **cell membrane** separates the inside of the cell from the outside and is responsible for controlling what goes in and out of the cell. Plant cells also have a rigid cell wall outside the cell membrane made of cellulose and other materials. Typical plant and animal cells are shown in Figure 15.13.

Eukaryotic cells have a nucleus, a distinct structure within the cell surrounded by a double membrane. The nucleus contains the cell's DNA, or genetic material, in the form of linear chromosomes. The portion of the cell that is outside the nucleus is called the **cytoplasm**. The cytoplasm is crisscrossed by fibers of the cytoskeleton, which helps the cell hold its shape and also plays a role in cell movement.

The cytoplasm of eukaryotic cells contains many organelles attached to and suspended from the cytoskeleton. These are called organelles because, like the organs of the body, each plays a specific role in the functioning of the cell. Most organelles—all but the ribosomes—are surrounded by a membrane. *Ribosomes* are organelles that assemble proteins. Some ribosomes are suspended in the fluid of the cytoplasm. These make proteins that will remain inside the cell. Other ribosomes are attached to an organelle called the *rough endoplasmic reticulum*, which assembles proteins either destined to go to the cell membrane or to be exported from the cell. The rough endoplasmic reticulum appears "rough" because of the ribosomes embedded within it. The *smooth endoplasmic reticulum* assembles membranes and, depending on the cell, may have additional functions. For example, the smooth endoplasmic reticulum of liver cells detoxifies drugs and other poisons. In cells that secrete such steroid hormones as estrogen or testosterone, it is the smooth endoplasmic reticulum that synthesizes the hormones. The *Golgi apparatus* is sometimes described as the "post office" of a cell. It receives products from the endoplasmic reticulum, modifies them, and packages them in membrane-bound vesicles for transport within or out of the cell. *Lysosomes* are the garbage disposals of a cell. These organelles break down

INSIGHT Why does our tolerance for certain drugs and medications increase over time, so that higher doses are required to obtain the same effect? Regular exposure to a drug causes the amount of smooth endoplasmic reticulum in liver cells to increase. This allows the liver to detoxify, or break down, the drug molecules more efficiently. As a result, a higher dose is necessary to keep the same amount of medication available to the body.

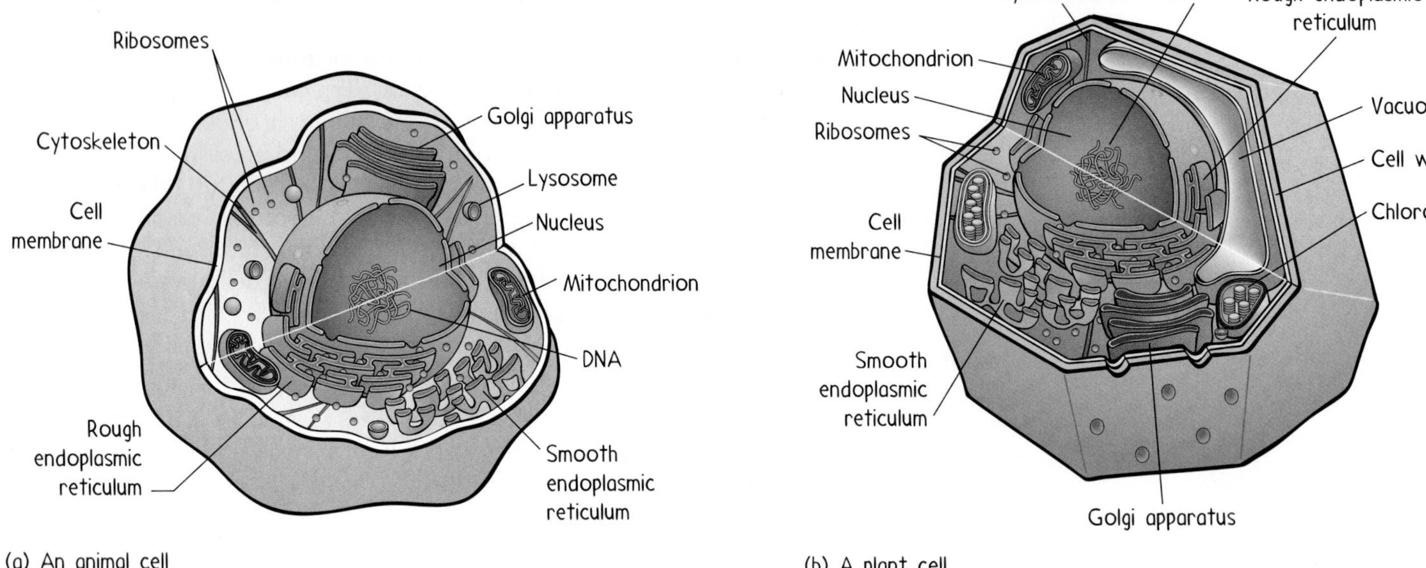

(a) An animal cell

(b) A plant cell

Figure 15.13 Typical plant and animal cells have cell membranes, nuclei, and several different types of organelles. (a) This is a typical animal cell. The lysosomes in animal cells are not found in plant cells. (b) This is a typical plant cell. Plants cells have a cell wall outside the cell membrane, a large central vacuole, and chloroplasts, features that are absent from animal cells.

organic materials. Among other tasks, they destroy damaged organelles. Lysosomes are also used for other purposes. Certain white blood cells of the immune system use lysosomes to destroy the bacteria they have engulfed. *Vacuoles* are sacs surrounded by membrane. Plant cells usually have a single large vacuole that can be used to store nutrients or other materials. In flowers, vacuoles are used to store the pigments that provide color. Animal cells typically have smaller vacuoles, sometimes called *vesicles*, that are used to hold or transport a wide array of products. For example, nerve cells have many vesicles that contain the chemicals they use to signal other nerve cells. Mitochondria, as described above, are organelles that break down organic molecules to obtain energy for cells. Finally, in plants, organelles called **chloroplasts** capture energy from sunlight in order to build organic molecules. Table 15.1 summarizes the major organelles and features of eukaryotic cells.

Table 15.1 Major Features of Eukaryotic Cells

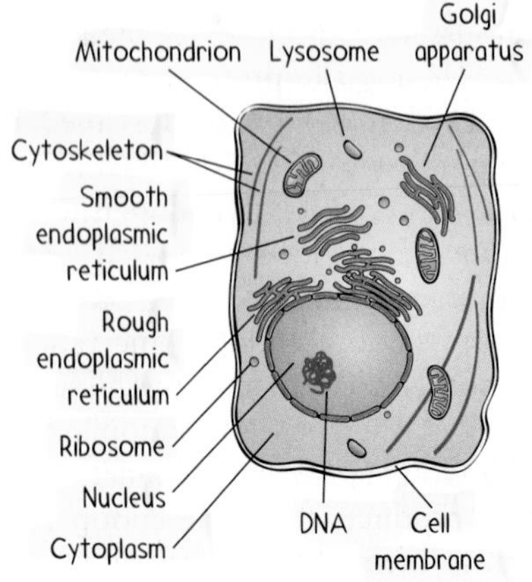

Feature	Function
Nucleus	Contains the cell's DNA
Ribosome	Assembles proteins for the cell
Rough endoplasmic reticulum	Assembles proteins destined either to go to the cell membrane or to leave the cell
Smooth endoplasmic reticulum	Assembles membranes and performs other specialized functions in specific cells
Golgi apparatus	Receives products from the endoplasmic reticulum and packages them for transport
Lysosome	Breaks down organic material
Mitochondrion	Obtains energy for the cell to use
Chloroplast	In plant cells, captures energy from sunlight to build organic molecules
Cytoskeleton	Helps cell hold its shape

15.4 Cell Membrane: Structure and Function

The cell membrane defines a cell's boundary, separating the inside of the cell from the outside. One of its main functions is to serve as a gatekeeper, controlling what goes into the cell and what comes out of it. The cell membrane is also responsible for communicating with other cells. The key to the cell membrane's function is its structure, which is illustrated in Figure 15.14. The three primary components of the cell membrane are phospholipids, proteins, and short carbohydrates.

Phospholipids are part hydrophilic and part hydrophobic. Have you ever noticed the way oil and water separate after they have been combined? The oil floats on top of the water in a distinct layer, rather than mixing with it. This is because oil is *hydrophobic*, or insoluble in water. (Hydrophobic literally means "afraid of water.") The opposite of hydrophobic is *hydrophilic*, or soluble in water. (Hydrophilic literally means "water-loving.") Phospholipids have hydrophilic "heads" and hydrophobic "tails." The hydrophilic heads are naturally drawn to the watery environment inside and outside the cell, whereas the hydrophobic tails naturally try to avoid it. The result is that the phospholipids form a double layer, or bilayer, with the hydrophobic tails pointing in and the hydrophilic heads pointing out.

Conceptual Science Place

Membrane Structure

The cell membrane also includes a large number of *membrane proteins*. These are embedded somewhat like toothpicks in the phospholipid sandwich. Membrane proteins serve a variety of functions—they help cells communicate with other cells, control transport into and out of cells, control the chemical reactions that occur in cells, and join cells to one another. Because different cells have different requirements for various functions, the proteins found in the cell membrane differ from one cell type to another.

Finally, short carbohydrates are attached to the membrane proteins and phospholipids on the outside surface of the cell. These carbohydrates play an important role in cell recognition, the ability to distinguish one type of cell from another. For example, certain immune-system cells use these short carbohydrates to identify foreign cells, such as disease-causing bacteria (Chapter 20).

Because the cell membrane includes a mosaic of phospholipids and proteins, and because the phospholipids and many membrane proteins slide freely around the cell surface, the cell membrane is often described as *fluid mosaic*. The fluidity

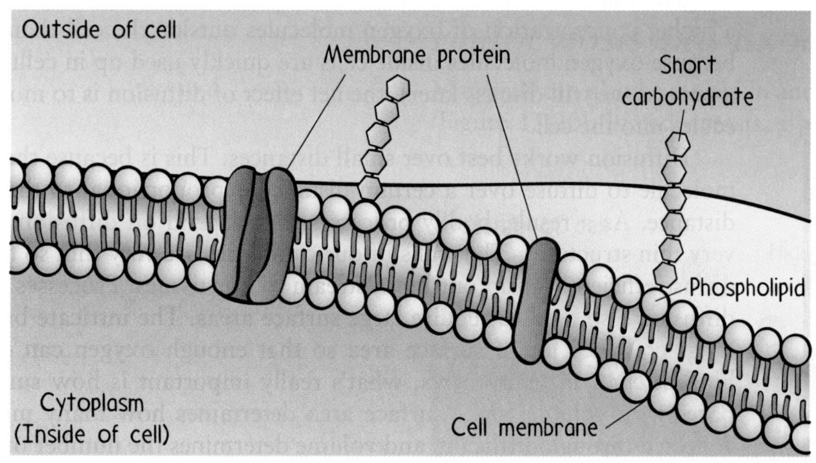

(a)

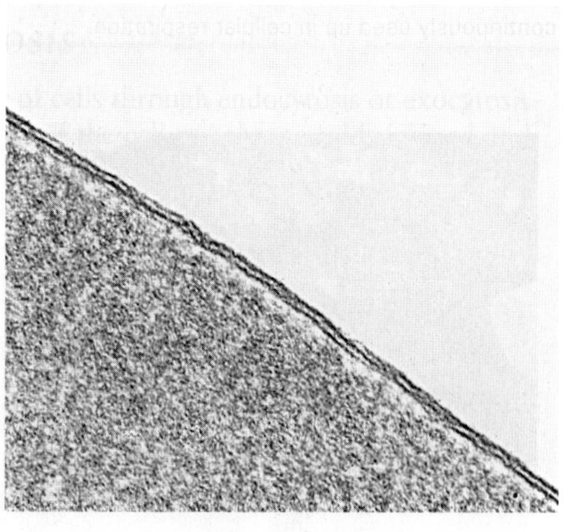

(b)

Figure 15.14 (a) The cell membrane includes a phospholipid bilayer, with hydrophobic tails pointed inward and hydrophilic heads pointing towards the inside and outside of the cell. Proteins are embedded in the membrane, and short carbohydrates are attached to the outside of the membrane. (b) This photograph shows the cell membrane of a red blood cell. Note the double layer of phospholipids. (Adapted from Campbell, Reece, Simon, *Essential Biology with Physiology*, © 2004.)

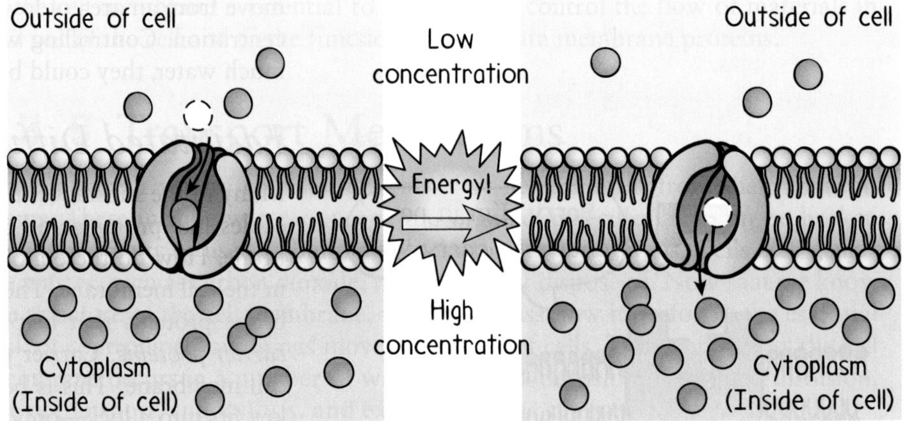

Figure 15.19 In active transport, molecules move from an area of low concentration to an area of high concentration—that is, against a concentration gradient. This requires both a carrier protein and energy.

pinches off, enclosing material within a vesicle inside the cell. Endocytosis is used by some white blood cells of the human immune system to engulf invading bacteria. In exocytosis, the opposite process occurs—a vesicle fuses its membrane with the cell membrane and dumps its contents outside the cell. Exocytosis in used by certain endocrine cells to release hormones into the bloodstream. Neurotransmitters—the chemicals that neurons use to signal one another—are also released through exocytosis (Chapter 19).

CHECK YOURSELF

1. Insects don't have lungs. Instead, they get oxygen from a series of small branching tubules in their bodies that are connected to the outside air. Oxygen diffuses through the tubules to reach their tissues. How might this relate to the fact that there are no 12-foot mosquitoes?
2. While observing the behavior of certain molecules, you notice that they always move from an area of higher concentration to an area of lower concentration when they cross cell membranes. You conclude that they cross the cell membrane either by diffusion or by facilitated diffusion. How can you distinguish between these possibilities? (*Hint:* There are only a limited number of carrier proteins available to move any given type of molecule. How could this affect the speed at which transport takes place?)

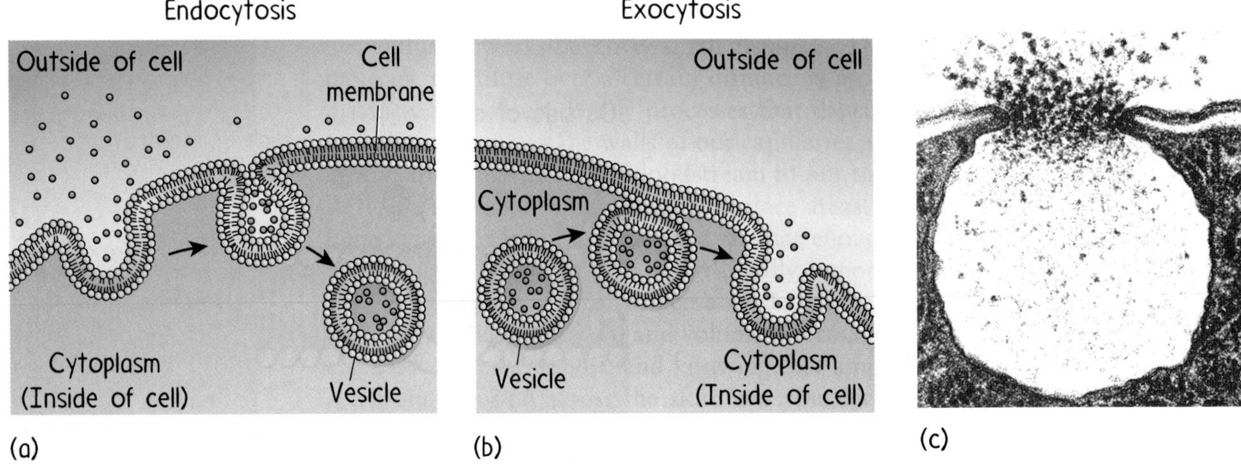

Figure 15.20 (a) In endocytosis, a portion of the cell membrane pinches off to form a vesicle that brings materials into the cell. (b) In exocytosis, a vesicle inside the cell fuses with the cell membrane, dumping its contents outside the cell. (c) This photograph shows a vesicle dumping its contents through exocytosis.

MATH CONNECTION

Why Does Diffusion Limit the Size of Cells?

Cells rely on diffusion to obtain a number of crucial resources. The amount of any given resource that a cell requires generally depends on the cell's volume. The larger the volume, the more resources the cell requires. But, the rate at which molecules diffuse into the cell depends on the cell's surface area—the total area of the cell membrane. This means that for diffusion to work well, a cell needs to have a lot of surface area relative to its volume. What implication does this have for the size of cells?

Let's look at how well diffusion works in cells of different sizes. We'll assume that cells are spherical in shape. The surface area of a sphere is $4\pi r^2$, where π is the constant equal to approximately 3.14 and r is the radius of the sphere. The volume of a sphere is $\frac{4}{3}\pi r^3$. Let's look at how surface area compares to volume in three cells with radii 1, 2, and 3 micrometers respectively. For a cell with a radius of 1 micrometer,

$$\frac{\text{Surface area}}{\text{Volume}} = \frac{4\pi r^2}{4/3\pi r^3} = \frac{4\pi(1)^2}{4/3\pi(1)^3} = \frac{4\pi}{4/3\pi} = 3$$

That is, the amount of surface area the cell has is three times greater than the volume of the cell.

For a cell with a radius of 2 micrometers,

$$\frac{\text{Surface area}}{\text{Volume}} = \frac{4\pi r^2}{4/3\pi r^3} = \frac{4\pi(2)^2}{4/3\pi(2)^3} = \frac{4\pi(4)}{4/3\pi(8)} = 1.5$$

For a cell with a radius of 3 micrometers,

$$\frac{\text{Surface area}}{\text{Volume}} = \frac{4\pi r^2}{4/3\pi r^3} = \frac{4\pi(3)^2}{4/3\pi(3)^3} = \frac{4\pi(9)}{4/3\pi(27)} = 1$$

What's happening here? Even though a cell's surface area increases as it gets bigger, its volume increases also—and more quickly. As a result, bigger cells have a *smaller* surface-area-to-volume ratio, making it harder for them to meet their needs through diffusion.

For this reason, it was long thought that bacteria, which rely on diffusion to obtain nutrients, could not grow very large. The discovery of not just one but two species of bacteria large enough to be visible to the naked eye was startling. A close examination of these organisms reveals how they do it. One giant, *Thiomargarita namibiensis*, positions all its cytoplasm in a thin layer just under its cell membrane. Most of the interior of the cell is occupied by a giant vacuole. The other giant, *Epulopiscium fishelsoni*, has an extremely wrinkled and convoluted cell membrane, increasing the surface area available for diffusion. In order to use diffusion effectively for transport, these species have evolved ways of compensating for their large size.

Problem The smallest bacteria, called mycoplasmas, have a radius of about 0.1 micrometers, and the largest bacteria (the giant *Thiomargarita namibiensis* mentioned above) have a radius close to 500 micrometers. What is the surface-area-to-volume ratio for each of these?

Solution For the mycoplasma:

$$\frac{\text{Surface area}}{\text{Volume}} = \frac{4\pi r^2}{4/3\pi r^3} = \frac{4\pi(0.1)^2}{4/3\pi(0.1)^3} = \frac{4\pi(0.01)}{4/3\pi(0.001)} = 30$$

For the giant bacteria:

$$\frac{\text{Surface area}}{\text{Volume}} = \frac{4\pi r^2}{4/3\pi r^3} = \frac{4\pi(500)^2}{4/3\pi(500)^3} = \frac{4\pi\ (250{,}000)}{4/3\pi\ (125{,}000{,}000)} = 0.006$$

CHECK YOUR ANSWERS

1. Because diffusion works well only at small distances, the insect respiratory system constrains insects to relatively small body sizes. (See Chapter 18, however, for a discussion of the giant insects that lived during the Carboniferous Period of Earth's history.)
2. Because there is only a limited number of carrier proteins available, facilitated diffusion hits a maximal rate of transport when all available carrier proteins are being used. Diffusion is not limited in this way. So, you could flood the cell with molecules to see if the transport rate appears to hit a ceiling.

15.6 Cellular Communication

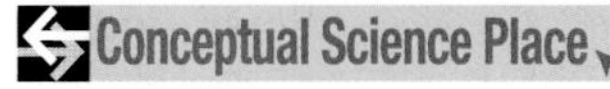

Signal Transduction

The cells of multicellular organisms communicate with one another in order to coordinate their activities. The "messages" they send each other take the form of molecules. For example, nerve cells send molecules to muscle cells that tell them to contract, and the pituitary gland sends molecules to a wide array of cells in the body telling them to grow.

In animals and plants, special structures allow very local messages to pass directly from one cell to an adjacent cell. This mode of communication is like

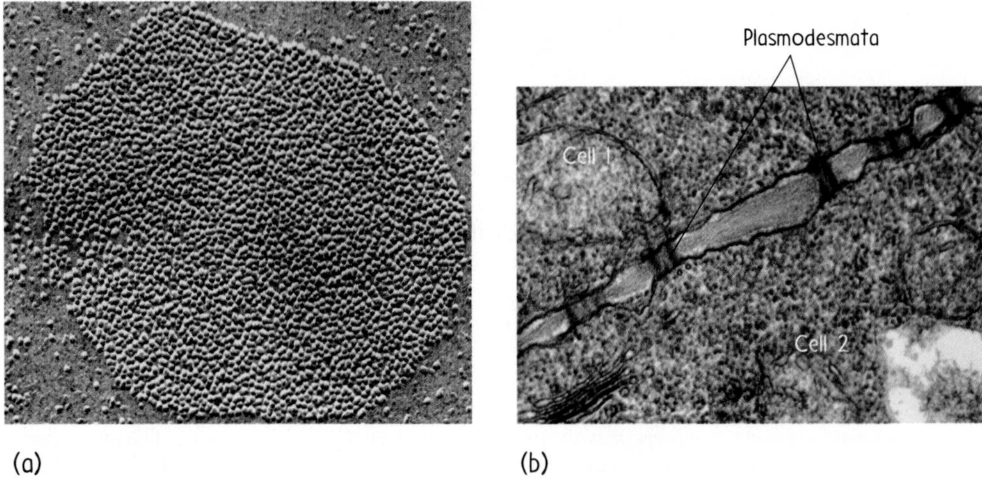

Figure 15.21 (a) Gap junctions are tiny channels between adjacent animal cells that allow small molecules to pass. This is a cross-sectional view of the channels between adjacent liver cells. (b) Plasmodesmata are slender threads of cytoplasm linking adjacent plant cells.

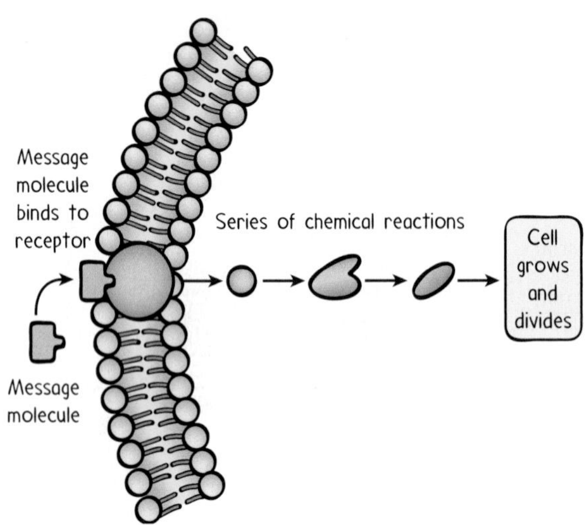

Figure 15.22 Here, a cell receives a message when a molecule binds to a receptor in its cell membrane. This begins a chain reaction that ends with the cell's response to the message.

tossing a note from one room of your house to another. Special "doorways" allow the message molecules to move from one cell to the next. In animal cells, the doorways are gap junctions, tiny channels surrounded by specialized proteins. Gap junctions are found in almost every cell of the body. In the heart, for example, communication via gap junctions allows muscle cells to contract simultaneously to produce the heartbeat. Plasmodesmata in plant cells serve a function similar to gap junctions in animal cells. Plasmodesmata are slender threads of cytoplasm that link adjacent plant cells (Figure 15.21).

For communication over longer distances, message molecules travel to target cells through the bloodstream or some other medium. This mode of communication is like sending a letter through the mail, except that, instead of having mail carriers with organized routes, the letter floats around until, eventually, it finds its way to the right mailbox. When a message molecule reaches a target cell, it binds to a protein called a receptor. Some receptors are membrane proteins, whereas others are found inside the cell. Either way, receptors are extremely specific about the molecules they bind to. This is because a message molecule and its receptor fit together like a key in a lock—only the right combination will work. The binding of a message molecule to its receptor sets off a series of chemical reactions that results ultimately in the target cell's response to the message (Figure 15.22). As just one example, a cell may receive a message that tells it to grow and divide. This normally controlled process is disrupted in many cancer cells, which, because of a problem in the communication process, receive the *grow and divide* message continuously. As a result, they divide out of control. We will see several more examples of message molecules and receptors when we discuss the nervous, sensory, and endocrine systems in Chapter 19.

15.7 How Cells Reproduce

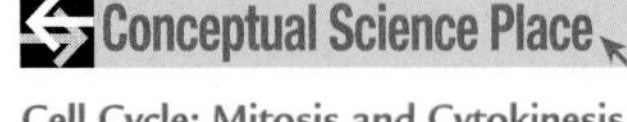

Cell Cycle; Mitosis and Cytokinesis

Cells reproduce by dividing. Cell division allows single-celled organisms to reproduce themselves and multicellular organisms to develop, grow, and maintain their tissues (Figure 15.23).

Mitosis is a form of cell division in which one parent cell divides into two daughter cells, each of which contains the same genetic information as the parent cell. Cells that are preparing to divide enter the *cell cycle*, shown in Figure 15.24. The cell cycle is divided into four stages—gap 1, synthesis, gap 2, and mitosis. Gap 1, synthesis, and gap 2 are collectively known as interphase. During interphase, the cell makes the necessary preparations for division. During mitosis, the cell divides.

During gap 1 (G_1), a cell prepares to divide by growing to approximately double its original size. All the important components of the cytoplasm, including the mitochondria and other organelles, also double in number. (Incidentally, calling this stage a "gap" is a little misleading in that it suggests that nothing is going on. In fact, important events occur during both "gap" stages. They are gaps only from the point of view of someone focused exclusively on whether the cell's DNA is doing anything interesting.)

During synthesis (S), the cell creates an exact copy of its genetic material—its DNA. The process of duplicating DNA will be described in Chapter 16.

During gap 2 (G_2), the cell builds the machinery necessary for division. This includes the structures that will separate the two copies of the genetic material and divide the cell into two daughter cells.

During mitosis (M), the cell divides. First, the nucleus divides in four phases (Figure 15.25). During *prophase*, the normally loosely packed chromosomes condense and the membranes surrounding the nucleus break down. When the chromosomes condense, it becomes clear that each consists of two identical sister chromatids attached at a point called the centromere. The mitotic spindle also forms during prophase. The mitotic spindle, which consists of a series of fibers that attach to the duplicated chromosomes, is responsible for splitting the genetic material between the two daughter cells. During *metaphase*, the chromosomes line up at the equatorial plane, the plane that passes through the imaginary "equator" of the cell. During *anaphase*, the two sister chromatids are pulled

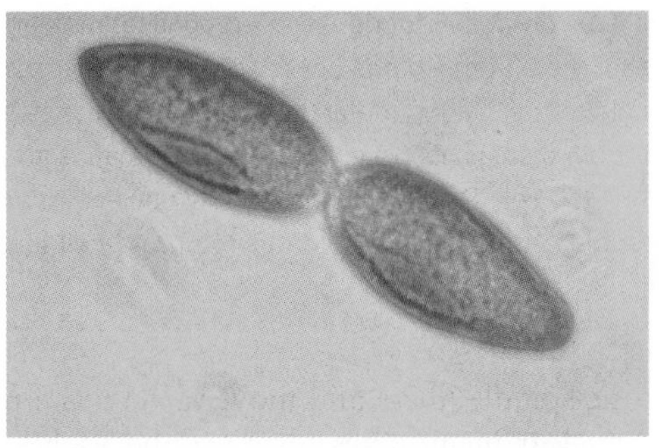

(a)

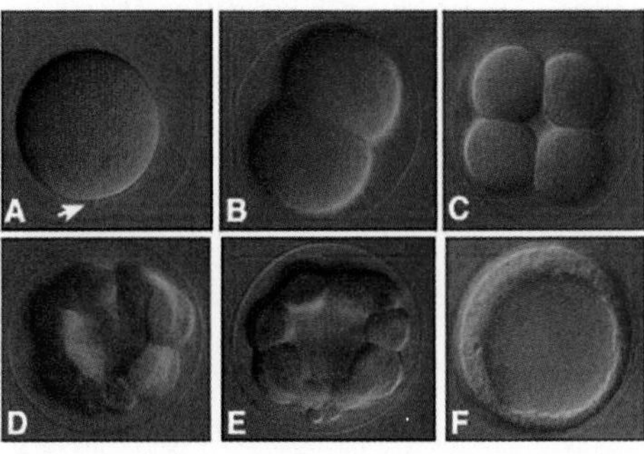

(b)

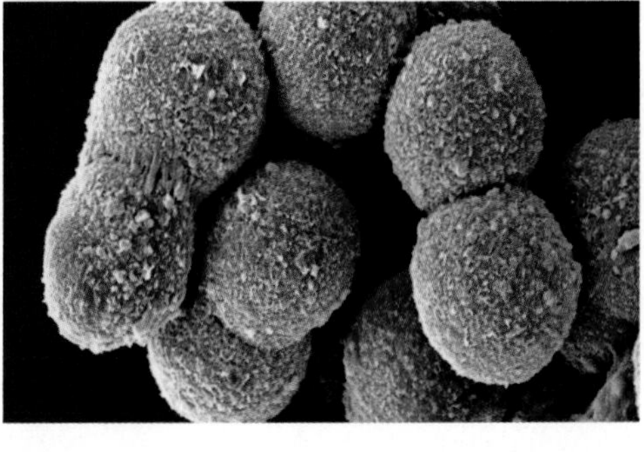

(c)

Figure 15.23 Cell division is essential for reproduction, growth and development, and maintenance. (a) A paramecium, a single-celled organism, reproduces itself by dividing into two. (b) The early development of a sea-urchin embryo involves multiple divisions of the fertilized egg. (c) Cell division in the liver produces new cells to replace old, worn-out cells.

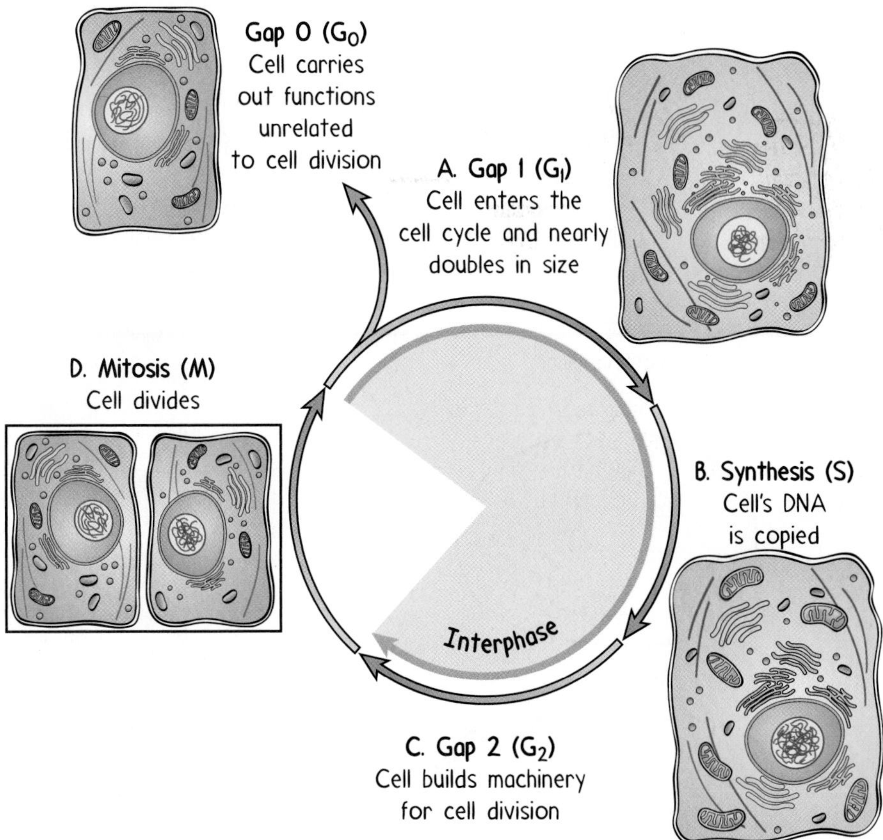

Figure 15.24 The cell cycle has four stages. (a) During G_1, the cell grows to about double its original size. (b) During S, an exact copy of the cell's DNA is made. (c) During G_2, the cell builds the machinery required for mitosis. (d) During mitosis, the cell divides.

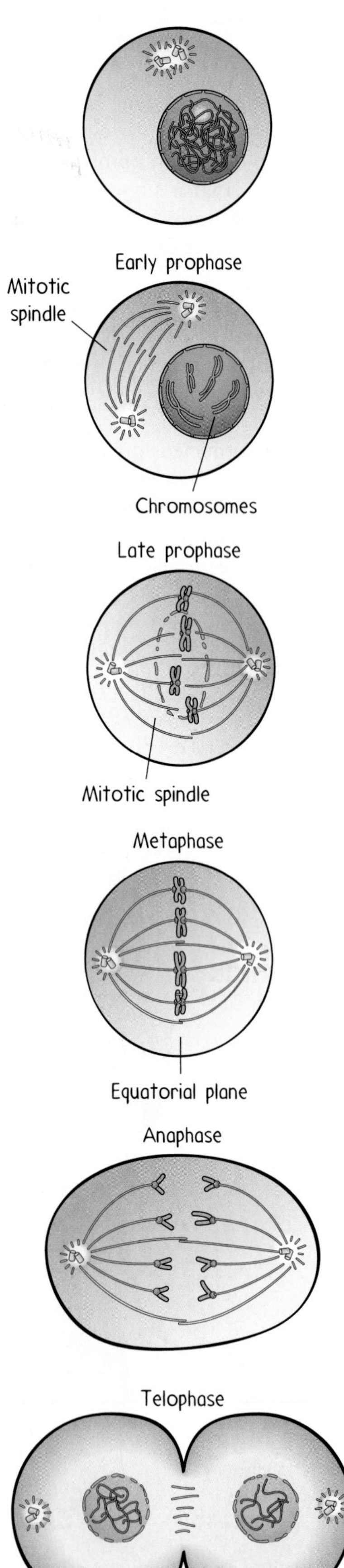

Figure 15.25 During mitosis, the nucleus divides in four phases—prophase, metaphase, anaphase, and telophase. During prophase, the chromosomes condense, the nuclear membranes break down, and the mitotic spindle forms. During metaphase, the chromosomes line up along the equatorial plane of the cell. During anaphase, the two sister chromatids are pulled apart and move to opposite poles of the cell. During telophase, new nuclear membranes form around each set of chromosomes, and the chromosomes return to their loosely packed state.

apart by the shortening of the mitotic-spindle fibers and move to opposite poles of the cell. During *telophase*, new nuclear membranes form around each set of chromosomes, and the chromosomes return to their loosely packed state. The division of the nucleus is followed by *cytokinesis*, the division of the cytoplasm to yield two separate daughter cells. Mitosis in an onion root cell is shown in Figure 15.26.

Cells are not always in the cell cycle. Many cells are neither dividing nor preparing to divide, but simply carrying out their regular functions. These cells are said to be in gap 0 (G_0). Some cells are in G_0 temporarily and will eventually reenter the cell cycle. Other cells, such as many neurons, are in permanent G_0 and will never divide again.

CHECK YOURSELF

During which stages of the cell cycle does the cell have twice the genetic material it normally has?

CHECK YOUR ANSWER

The cell has twice the normal amount of genetic material during G_2 and M stages—that is, after S stage and continuing up until cytokinesis is complete.

15.8 How Cells Use Energy

At any moment, countless chemical reactions are occurring in cells. These reactions sustain life by allowing cells to carry out such essential functions as building macromolecules, transporting ions across membranes, and dividing.

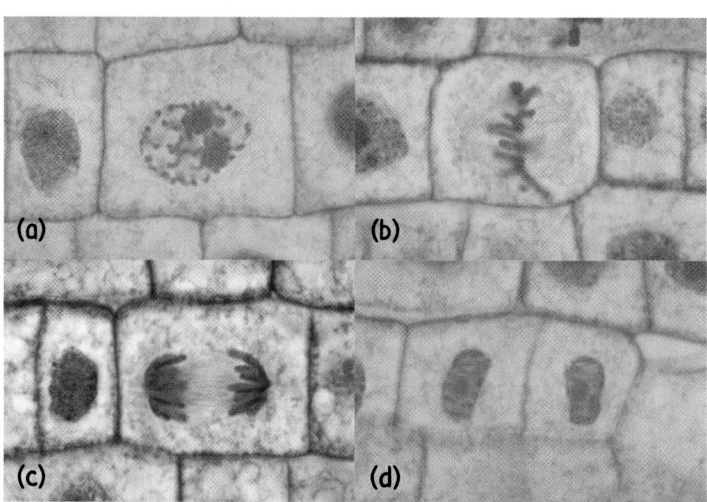

Figure 15.26 These photographs show mitosis in an onion root. DNA is stained red. (a) During prophase, the chromosomes are condensed. Note the diffuse appearance of the chromosomes in the surrounding, non-dividing cells. (b) At metaphase, the chromosomes are lined up along the equatorial plane of the cell. (c) During anaphase, sister chromatids separate. (d) Telophase and cytokinesis are complete, and two daughter cells have been formed.

SCIENCE AND SOCIETY

Stem Cells

What are stem cells? Why do some people consider stem-cell research the most promising avenue of medical research today? Why do other people oppose it adamantly?

Humans are made up of a huge number of different kinds of cells—skin cells, muscle cells, liver cells, brain cells, and so on. Embryonic stem cells come from human embryos that have yet to differentiate into distinct types of cells. As a result, embryonic stem cells have the capacity to develop into all the different kinds of cells in the body. The hope of stem-cell research is to develop techniques for growing healthy cells from embryonic stem cells that can then be used to replace defective or diseased cells in the body. Because embryonic stem cells are easily grown in the lab, this source of healthy cells is potentially quite large. Embryonic stem cells provide great promise for treating a variety of conditions, including diabetes, Parkinson's disease, heart disease, Alzheimer's disease, arthritis, strokes, spinal-cord injuries, and burns. Although stem cells also exist in adults, adult stem cells can only give rise to a limited range of cell types. They are also much harder to grow in the lab. Despite these drawbacks, certain therapies make use of adult stem cells, including the bone-marrow transplants we will discuss in Chapter 20.

Stem-cell research has only just begun. Scientists are still in the process of figuring out how to induce stem cells to develop into specific types of body cells. They have already made some progress—for example, scientists recently figured out how to make embryonic stem cells develop into insulin-producing cells. These could one day be used to treat Type 1 diabetes, a disease in which the insulin-producing cells of the pancreas are destroyed by the body's immune system. Beyond the challenge of directing stem-cell differentiation, scientists will also have to figure out how to prevent a patient's immune system from recognizing transplanted cells as foreign and attacking them. Finally, the great capacity for division that characterizes stem cells has to be carefully controlled in order to be sure they do not form tumors once they have been transplanted.

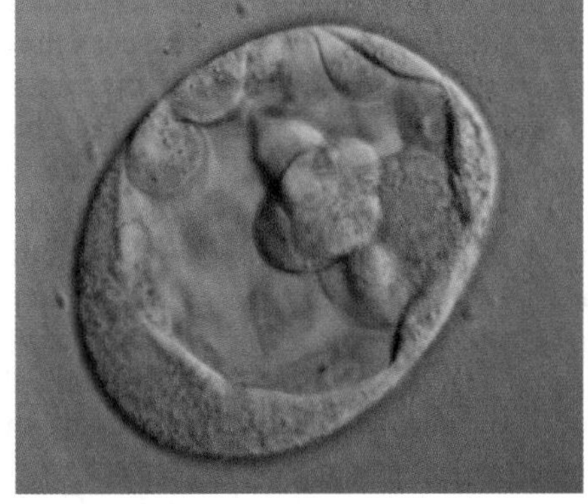

Embryonic stem cells are isolated from 3- to 5-day-old human embryos, such as the one shown here.

Why is stem-cell research controversial? Embryonic stem cells are grown from cells that are removed from three- to five-day-old human embryos. These embryos were created at fertility clinics to aid infertile couples and were then donated to stem-cell research, with donor consent, when they were no longer needed. Because removing the stem cells kills the embryo, however, some people are opposed to stem-cell research. (The extra embryos generated at fertility clinics are discarded if they are not donated to another couple or used for research.) In 2001, President George W. Bush banned federal funding for stem-cell research unless the cells had already been established in laboratories prior to that time. This effectively restricted research to a small number of stem-cell lines. To circumvent these limits, California voters passed a landmark 3-billion-dollar ballot initiative in 2004 to support embryonic-stem-cell research. The author of the initiative was a real-estate developer whose son suffers from Type 1 diabetes and whose mother has Alzheimer's disease. The passage of the bill marked the first time voters approved the use of tax dollars to support a specific type of scientific research.

Because of the law of conservation of energy, however, the only chemical reactions that can occur spontaneously are ones in which energy is released. For all other reactions, energy must be provided (Chapter 13). In cells, usable energy is found in molecules of adenosine triphosphate, or **ATP.**

Unifying Concept
Conservation of Energy
Section 4.10

Even with energy-releasing reactions, however, the reacting molecules need to have a certain activation energy in order for the reaction actually to happen (Chapter 13). Unfortunately for living organisms, the activation energy for many essential chemical reactions is very high. Because of this, cells rely on catalysts to lower the activation energy of reactions and allow them to happen more quickly. The catalysts in cells are large, complex proteins called **enzymes**. An enzyme binds the reactants of a reaction—the enzyme's substrate—at its active site and then releases the products of the reaction. An example of how an enzyme might work is shown in Figure 15.27. Note that enzymes, like other catalysts, are not altered or destroyed in the reactions they catalyze. They are therefore available to catalyze the same reaction over and over again. Enzymes are involved in most of the chemical reactions in cells—in fact, several thousand unique enzymes have been identified in living organisms. Each of these enzymes is highly specific for a certain reaction. That is, each enzyme is very picky about the substrate to which it binds and therefore about the reaction it catalyzes.

As you might imagine, cells regulate enzymes very carefully in order to control the chemical reactions that occur. Regulation takes place in a number of

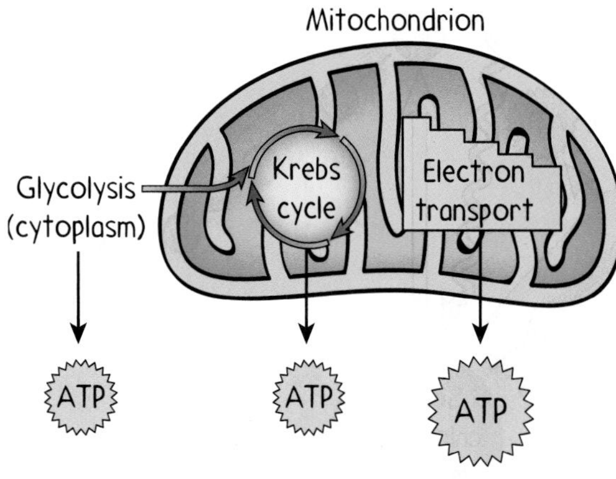

Figure 15.35 Cellular respiration takes place in three stages—glycolysis, the Krebs cycle, and electron transport. Glycolysis occurs in the cytoplasm. The Krebs cycle and electron transport occur in the mitochondria. (Courtesy of Campbell, Reece, Simon, *Essential Biology with Physiology*, © 2004.)

respiration takes place in three stages—glycolysis, the Krebs cycle, and electron transport (Figure 15.35). The equation for cellular respiration is:

$$C_6H_{12}O_6 + 6\,O_2 + \text{about 38 molecules of ADP} \rightarrow 6\,CO_2 + 6\,H_2O + \text{about 38 molecules of ATP}$$

Glucose, oxygen, and ADP go in, and carbon dioxide, water, and ATP are released. About 38 molecules of ATP are produced from a single molecule of glucose.

Glycolysis

The first step in breaking down glucose is **glycolysis** (literally, "sugar splitting"). Glycolysis takes place in the cytoplasm of cells. During glycolysis, the six-carbon glucose molecule is split into two molecules of pyruvic acid, each of which contains three carbon atoms. Two molecules of ATP are produced in the process.

The Krebs Cycle and Electron Transport

The Krebs cycle and electron transport occur in the mitochondria. Before entering the Krebs cycle, the pyruvic acid that comes out of glycolysis is converted to acetic acid and bound to a molecule of coenzyme A. This entire complex is called acetyl-CoA. During the Krebs cycle, acetyl-CoA is broken down to carbon dioxide. Two molecules of ATP are harvested, and additional energy is stored in two other molecules, NADH and $FADH_2$.

During electron transport, electrons carried by NADH and $FADH_2$ are sent down electron transport chains. As electrons are passed from one carrier in the transport chain to the next, they lose energy. The energy released is used to pump hydrogen ions (H^+) across a membrane inside the mitochondria. At the end of the electron transport chain, the electrons combine with an oxygen molecule to generate water. (This is what happens to all the oxygen we breathe—and it is also why humans and so many other organisms need oxygen to survive.) The concentration gradient of hydrogen ions across the inner mitochondrial membrane is then used to make ATP. As hydrogen ions move back across the inner mitochondrial membrane (down the concentration gradient), they pass through a protein complex called ATP synthase, turning ADP into ATP (Figure 15.36). This process generates the bulk of the ATP harvested during cellular respiration.

Fermentation

In certain cells, such as baker's yeast and animal muscle cells, glycolysis is sometimes followed not by the Krebs cycle but by an **anaerobic** (non-oxygen-using) process known as **fermentation.** Fermentation yields no ATP, but it does regenerate the molecules necessary to keep glycolysis going, allowing cells to continue to obtain ATP through glycolysis.

Alcoholic fermentation is an anaerobic process used by yeast. It takes place in the cytoplasm. In alcoholic fermentation, the pyruvic acid from glycolysis is broken down into ethanol and carbon dioxide. Alcoholic fermentation is essential both to winemaking and to breadmaking. Yeast cells ferment the sugar in grape juice to turn it into wine. The same process makes bread rise—yeast cells ferment the sugars in bread dough, releasing tiny bubbles of carbon dioxide gas. (And, yes, fermentation also produces ethanol in bread dough, but all of it evaporates during baking. This is why you don't have to be 21 to purchase dinner rolls.)

Lactic acid fermentation is an anaerobic process that occurs in some animal cells as well as in certain species of bacteria and fungi. It takes place in the cytoplasm, and it breaks pyruvic acid down to lactic acid. In animal muscle cells,

INSIGHT Why do we continue to breathe hard after stopping strenuous activity? The liver, which receives all the lactic acid that has been built up in the muscles, requires oxygen to convert lactic acid back to pyruvic acid. Pyruvic acid can then be converted back to glucose.

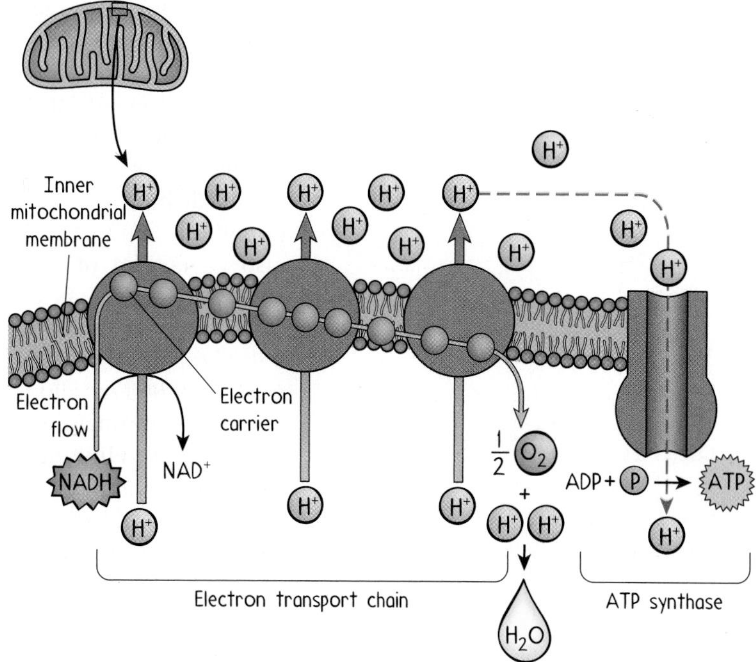

Figure 15.36 As electrons move down electron transport chains, hydrogen ions are pumped across the membrane inside mitochondria. When the hydrogen ions move back across the membrane, they generate ATP. (Courtesy of Campbell, Reece, Simon, *Essential Biology with Physiology*, © 2004.)

lactic acid fermentation occurs during strenuous exercise, when the oxygen supply—despite hard breathing—can't quite meet the demand. By regenerating the molecules required for glycolysis, lactic acid fermentation allows muscle cells to continue to make ATP without oxygen. The lactic acid produced during strenuous exercise causes a burning sensation in the muscles. Red blood cells, which lack mitochondria, also rely on lactic acid fermentation to obtain ATP. Finally, lactic acid fermentation by certain species of bacteria and fungi is used to make cheese and yogurt.

CHECK YOURSELF

How are the chemical reactions of photosynthesis and cellular respiration similar? How are they different?

CHECK YOUR ANSWER

They are similar in that the products of one are the reactants of the other, and vice versa. They are different in that photosynthesis takes energy and small molecules to build a larger molecule (glucose), whereas cellular respiration takes a large molecule and breaks it down into smaller molecules, releasing energy in the process.

15.11 The Life Spans of Cells

Most of the cells in multicellular eukaryotes are not immortal. Instead, they reach a point when they can no longer divide and cell structure and function begin to deteriorate. Interestingly, the life span of a cell appears to be described not by a fixed length of time but by a fixed number of cell divisions. In other words, most cells have the capacity to divide a certain number of times and no more. In the case of most human cells, this number appears to be between 40 and 60 divisions.

The finite life span of cells is explained by a part of the chromosome called the telomere. Telomeres are lengths of DNA at the ends of chromosomes that protect them from damage (Figure 15.37). Every time DNA is copied for cell division, the telomeres get a little shorter. This is because the machinery for copying DNA is unable to copy the linear chromosomes of eukaryotes all the way to their ends. It is sort of like peeling a carrot—because you have to hold the carrot at one end, that end doesn't get peeled all the way. When telomere shortening

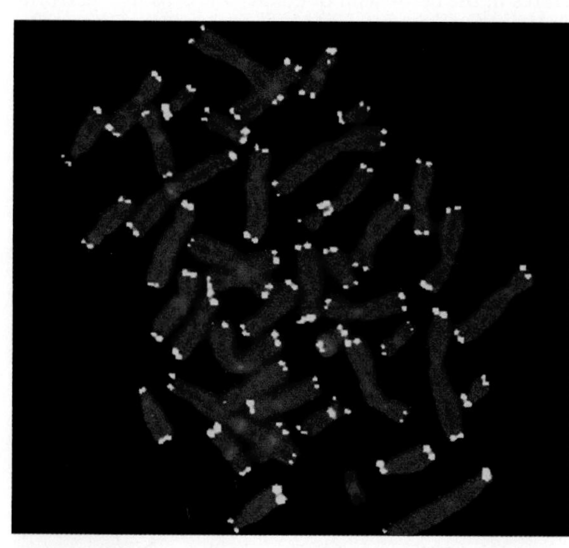

Figure 15.37 Telomeres protect the ends of chromosomes. Because they get shorter with every cell division, they place a natural limit on the life spans of cells. In the photo above, the chromosomes are in blue, the telomeres in yellow.

reaches a critical stage, the cell can no longer divide without losing critical genetic information from the ends of its chromosomes. (Note that prokaryotes do not have this problem because their chromosomes are circular.)

Not all eukaryotic cells are vulnerable to telomere shortening. Some cells have a special enzyme called telomerase that lengthens telomeres. Cells that produce a lot of telomerase are able to divide indefinitely. For example, the germ-cell lines that produce eggs and sperm have a lot of telomerase and are potentially immortal. Some abnormal cells, such as the tumor cells in cancers, also have a lot of telomerase, enabling them to divide indefinitely. This becomes dangerous as masses of rapidly dividing tumor cells prevent the body from functioning normally.

The role of telomeres in setting cellular life spans has made some people wonder whether telomerase could be used to extend human life spans. In laboratory settings, telomerase does allow human cells to divide beyond their normal limit. However, the implication of this for the aging process is still unclear. For example, although some studies have linked shorter telomeres with shorter lives, short telomeres may merely be a sign of aging, not a cause of it. And the fact that mice have short life spans and very long telomeres (much longer than those of humans) suggests that telomerase is unlikely—by itself, anyway—to represent the ever-elusive fountain of youth.

CHAPTER 15 Review

Summary of Terms

Active transport Energy-requiring movement of molecules across the cell membrane.

Aerobic Refers to a cellular process that uses oxygen.

Anaerobic Refers to a cellular process that does not use oxygen.

ATP Adenosine triphosphate, the basic energy molecule used in cellular processes.

Biology The study of life and living organisms.

Carbohydrates Sugars, starches, cellulose, and other organic molecules consisting of carbon, hydrogen, and oxygen atoms, usually in a 1:2:1 ratio.

Cell membrane The membrane that separates the inside of the cell from the outside.

Cells The basic units of life that make up all living organisms.

Cellular respiration The oxygen-using process of breaking down organic molecules (such as glucose) to obtain ATP.

Chlorophyll The green pigment in plant chloroplasts that absorbs light energy during the light-dependent reactions of photosynthesis.

Chloroplasts The organelles in plant cells in which photosynthesis occurs.

Chromosomes The DNA-containing structures found in eukaryotic and prokaryotic cells.

Cytoplasm The contents of a cell, excluding the nucleus.

Diffusion The tendency for molecules to move from an area of high concentration to an area of low concentration.

Enzyme A protein that catalyzes a chemical reaction in a living organism.

Eukaryotes Organisms whose cells have a true nucleus, including protists, animals, plants, and fungi.

Fermentation The anaerobic breakdown of glucose that results in the production of ethanol and carbon dioxide gas (alcoholic fermentation) or lactic acid (lactic acid fermentation).

Glycolysis The first step in the breakdown of glucose, during which glucose is split into two molecules of pyruvic acid.

Lipids Hydrophobic organic compounds, many of which include fatty acids as a primary component.

Mitochondria Eukaryotic organelles that break down organic molecules to obtain ATP.

Mitosis A form of cell division in which one cell divides into two daughter cells, each of which has the same genetic content as the original cell.

Nucleic acid An organic molecule composed of a series of nitrogenous bases attached to a sugar-and-phosphate backbone.

Nucleus A structure in eukaryotic cells that is surrounded by a double membrane and that contains the cell's genetic material.

Organelles Structures in the cytoplasm of eukaryotic cells that perform specific functions for the cell.

Passive transport The movement of molecules across the cell membrane without the use of energy.

Photosynthesis The process in plants and some other organisms in which light energy from the Sun is converted to energy that is stored in organic molecules.

Prokaryotes Single-celled organisms, including bacteria and archaea, whose cells lack a nucleus.

Proteins Organic molecules composed of folded chains of amino acids.

Review Questions

15.1 Characteristics of Life

1. What are some of the characteristics that living organisms share?
2. What does it mean that living things are parts of populations that evolve?

15.2 Cell Types: Prokaryotic and Eukaryotic

3. Describe three or more differences between prokaryotic cells and eukaryotic cells.
4. How is the DNA of prokaryotes packaged differently from the DNA of eukaryotes?
5. What evidence is there for the idea that some eukaryotic organelles evolved from prokaryotes living inside eukaryotic cells?

15.3 Tour of a Eukaryotic Cell

6. What is the nucleus of a cell? What does the nucleus contain?
7. Describe the functions of the following organelles: mitochondria, ribosomes, lysosomes, chloroplasts.

15.4 Cell Membrane: Structure and Function

8. What are three components of the cell membrane?
9. How are phospholipids arranged in the cell membrane?
10. What are some of the functions of membrane proteins?

15.5 Transport Mechanisms

11. Give three examples of molecules that are able to move directly across the cell membrane.
12. What is diffusion?
13. How do carrier proteins allow only very specific molecules to cross the cell membrane?
14. How do endocytosis and exocytosis move materials into and out of cells?

15.6 Cellular Communication

15. What are plasmodesmata? What function do they serve?
16. Describe what happens when a message molecule binds to a receptor on the cell membrane.
17. Do all message molecules use the same receptor?

15.7 How Cells Reproduce

18. What are the stages of the cell cycle? What happens during synthesis (S)?
19. What are the phases of mitosis? What happens during each phase?
20. What are the end products of mitosis?

15.8 How Cells Use Energy

21. Why do cells need catalysts? What are the catalysts in cells?
22. What is the difference between competitive and noncompetitive inhibition of an enzyme?
23. How does penicillin kill bacteria?

15.9 Photosynthesis

24. Why is almost all life on Earth dependent either directly or indirectly on photosynthesis?
25. What happens during the light-dependent reactions of photosynthesis? What happens during the light-independent reactions?
26. What are the end products of photosynthesis?

15.10 Cellular Respiration

27. What happens during glycolysis? Is ATP produced during glycolysis?
28. About how many ATP molecules does a cell obtain from one molecule of glucose in cellular respiration? What other products are released as a result of cellular respiration?
29. What are the products of alcoholic fermentation?
30. Give two examples of cells in the human body that use lactic acid fermentation. Why do each of these use lactic acid fermentation?

15.11 The Life Spans of Cells

31. What is a telomere? Explain how telomere shortening determines the life span of a cell.
32. Do all cells have a finite life span? What are some exceptions?

Multiple Choice Questions

Choose the BEST answer to the following.

1. Which of the following is not a characteristic of living things?
   (a) Living things use energy.
   (b) Living things maintain themselves.
   (c) Living things have the capacity to reproduce.
   (d) Living things are part of populations that remain constant from one generation to the next.
2. The macromolecules made from carefully folded strings of amino acids are
   (a) proteins.
   (b) carbohydrates.
   (c) lipids.
   (d) nucleic acids.
3. One difference between prokaryotic and eukaryotic cells is
   (a) prokaryotic cells have a nucleus, whereas eukaryotic cells do not.
   (b) eukaryotic cells have existed on Earth far longer than prokaryotic cells.
   (c) the DNA of eukaryotic cells is found in linear chromosomes, whereas the DNA of prokaryotic cells is found in a single circular chromosome.
   (d) eukaryotic cells are generally smaller than prokaryotic cells.
4. In plant cells, the organelles that break down organic molecules to obtain energy for the cell's use are
   (a) mitochondria.
   (b) ribosomes.
   (c) chloroplasts.
   (d) lysosomes.
5. Which of the following is not a component of the cell membrane?
   (a) phospholipids
   (b) proteins
   (c) nucleic acids
   (d) carbohydrates
6. Transport across the cell membrane that requires energy is known as
   (a) diffusion.
   (b) facilitated diffusion.
   (c) active transport.
   (d) endocytosis.
7. During which stage of the cell cycle does a cell duplicate its genetic material?
   (a) gap 1
   (b) gap 2
   (c) synthesis
   (d) mitosis
8. Which of the following statements about enzymes is true?
   (a) Enzymes provide energy for specific chemical reactions in cells.
   (b) Enzymes are catalysts that allow specific chemical reactions in cells to happen faster than they would otherwise.
   (c) Both of the above are true.
   (d) Neither of the above is true.
9. The products of photosynthesis are
   (a) carbon dioxide and water.
   (b) carbon dioxide, water, and sunlight.
   (c) glucose.
   (d) glucose and oxygen.
10. Which of the following processes requires oxygen?
    (a) glycolysis
    (b) Krebs cycle and electron transport
    (c) fermentation
    (d) none of the above

INTEGRATED SCIENCE CONCEPTS

Chemistry—Macromolecules Needed for Life

1. What are some of the different functions of proteins?

2. Give an example of (a) a carbohydrate that functions in energy storage and (b) a carbohydrate that has a structural function.
3. How many different nitrogenous bases are used in DNA? What are they?

Physics—The Microscope

4. Why are light microscopes of limited use to cell biologists?
5. Why are electron microscopes particularly useful to cell biologists?

Chemistry—Chemical Reactions in Cells

6. How does ATP provide energy for cells?
7. Is the amount of energy it takes to make ATP the same as the amount of energy that cells eventually get out of ATP?
8. What step in the active transport of sodium and potassium ions by the sodium–potassium pump requires ATP?

Active Explorations

1. Does a pound of fat or oil really store more energy than a pound of carbohydrates? Examine the nutrition labels on a bag of sugar and on a pound of butter or a bottle of cooking oil. How many calories are there in a pound of sugar *versus* a pound of butter?
2. Carefully place a teabag in a mug of very still hot water. Why is the water darker closer to the teabag than farther away from it? If you carefully remove the teabag from the mug and wait, will your tea eventually be uniformly colored? If so, explain why. Also explain how this experiment relates to cellular transport mechanisms.

Exercises

1. What are the characteristics of life? Discuss how these are evident in human beings.
2. During their annual migrations, many birds fly hundreds or even thousands of miles over a relatively short period of time. Why do birds put on a layer of fat before their annual migration? Why don't they store this energy in the form of carbohydrates?
3. Is it true that *all* the DNA contained in a eukaryotic cell is in the nucleus? If not, why not? How does this support the argument that there were once prokaryotes living inside eukaryotic cells?
4. What organelle is found only in plants? What does it do? Does this explain why animals have to eat but plants don't?
5. Certain cells in the body, including nerve cells, muscle cells, and liver cells, have large numbers of mitochondria. Bone cells and fat cells generally have few mitochondria. What can you tell about a cell from the number of mitochondria it contains?
6. In this chapter, we had three examples of molecules fitting together like a "lock and key." What were these? Why do you think it is important, in each of these contexts, to have such a specific fit?
7. In all the instances of molecules fitting together like "lock and key," you'll notice that the molecules involved are proteins. How are proteins able to achieve the specificity required?
8. We mentioned that controlling the flow of water into and out of cells is a problem that all cells face. We also know that water is able to cross the cell membrane directly. Are organisms that occupy freshwater habitats likely to have the problem of too little water entering their cells or too much water entering their cells? Why?
9. Glucose gets into certain cells through facilitated diffusion. Why isn't active transport of glucose necessary? That is, why is there usually a higher concentration of glucose molecules outside the cell than inside the cell?
10. Why does oxygen diffuse into cells rather out of them? Why does carbon dioxide diffuse out of cells rather than into them?
11. In plants, roots absorb water (among other functions). Why are the roots of many plants highly branched?
12. What is the difference between using endocytosis and using a carrier protein to cross the cell membrane? How does each process work?
13. How are gap junctions and plasmodesmata similar? How do they differ?
14. If a cell goes through all the stages of the cell cycle but for some reason fails to undergo cytokinesis, how will that cell differ from normal cells?
15. The figure below shows a cell in the process of cell division. In which stage of the cell cycle is it?

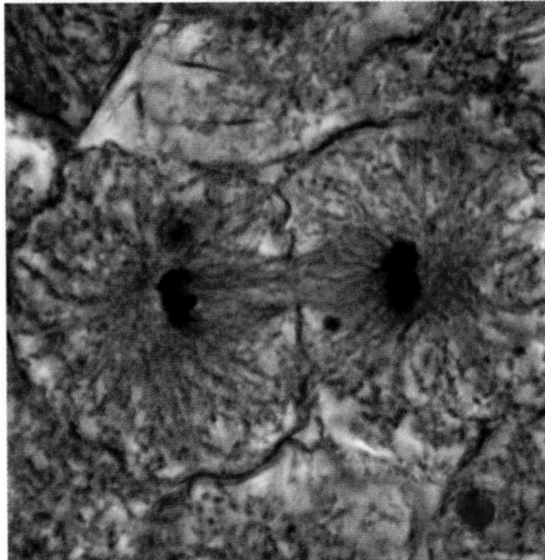

16. The lethal nerve gas sarin binds to an enzyme called acetylcholinesterase, which breaks down acetylcholine in the body. If acetylcholine is not broken down, muscles are unable to relax after contracting. Without prompt treatment, sarin exposure leads to respiratory collapse and death. Sarin works by binding to acetylcholinesterase at the site where acetylcholine normally binds. What form of enzyme regulation does this represent?
17. Global warming has occurred because of the large amounts of carbon dioxide released by burning fossil fuels. Carbon dioxide traps heat. Why might the process of deforestation (the large-scale destruction of forests) also contribute to global warming?
18. What are some differences between fermentation and cellular respiration? Which process produces more ATP? Why do some cells in the human body use fermentation?
19. Where do the bubbles in champagne come from? *Hint:* Unlike still wines, champagne goes through an extra round of fermentation in the bottle, during which the bottles are capped tight.
20. Some animals that live in desert environments, like the kangaroo rat shown in the figure below, go their entire lives without drinking a drop of water. Kangaroo rats subsist entirely on the starches and fats found in the dry seeds they eat. Yet we know that all living organisms need water, and, in fact, the bodies of kangaroo rats have about the same water content as those of other animals. How do kangaroo rats get their water?

21. Do prokaryotes have telomeres? Why don't they need them?
22. Write a letter to Grandma telling her about stem cells. Be sure to tell her how stem cells are different from all the other cells in our bodies and why scientists believe they may be helpful for treating many diseases.

Problems

1. As energy-storage substances, carbohydrates produce about 4 kilocalories of energy per gram, whereas fats produce about 9 kilocalories of energy per gram. The American black bear may hibernate for as long as seven months in the winter, during which it does not eat. Before hibernating, black bears put on a lot of weight, often spending 20 hours a day eating and storing as much as 50 kilograms of fat. Show that the bear would have to gain 112.5 kilograms if it stored energy as carbohydrate instead of as fat.
2. A typical cell in the body makes about 10 million molecules of ATP per second. Show that the cell breaks down about 263,158 molecules of glucose per second.
3. Two different bacteria have radii of 1 micrometer and 5 micrometers, respectively. What is the surface area of each cell? How does surface area compare with volume for each cell—that is, what is the surface-area-to-volume ratio? Why is the larger cell able to obtain more molecules through diffusion? Why is it nonetheless more challenging for the larger cell to meet its needs through diffusion? Recall that the surface area of a sphere is $4\pi r^2$ and that the volume of a sphere is $\frac{4}{3}\pi r^3$.
4. We mentioned that diffusion works best over small distances. This is because the average time it takes a molecule to diffuse a certain distance is proportional to the square of that distance. If two cells have diameters of 1 micrometer and 5 micrometers, respectively, show that it takes a molecules 25 times longer to diffuse across the larger cell than the smaller cell.
5. Proteins are folded chains of amino acids. All the proteins in living organisms are made up of only twenty different amino acids. Show that there are 400 different ways to make a string of two amino acids, and 8000 different ways to make a string of three amino acids. How many different ways are there to make a string of ten amino acids? Do you see why the number of proteins living organisms can make is practically countless?

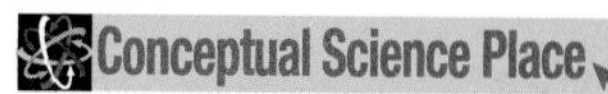

CHAPTER 15 ONLINE RESOURCES

Tutorials Comparing Prokaryotic and Eukaryotic Cells, Build an Animal and a Plant Cell, Membrane Structure, Diffusion, Facilitated Diffusion, Osmosis and Water Balance in Cells, Active Transport, Signal Transduction, Cell Cycle, Mitosis and Cytokinesis, Overview of Photosynthesis, Overview of Cellular Respiration
• ***Quiz*** • ***Flashcards*** • ***Links***

CHAPTER 16

Genetics

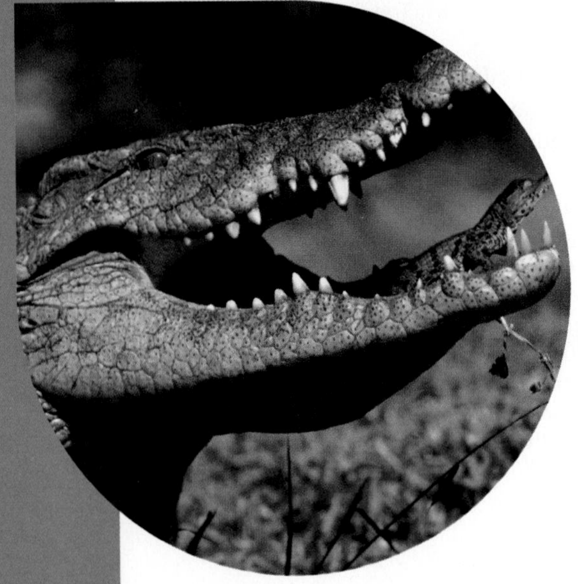

A Nile crocodile carries her young in her mouth in the province of KwaZula Natal, South Africa.

CHAPTER OUTLINE

DNA and RNA Structure; DNA Double Helix

Pine seeds grow into pine trees, and Nile crocodiles raise baby Nile crocodiles—in all living organisms, offspring resemble their parents because they inherit their parents' genes. But what are genes, and how do genes determine the traits of organisms? How can the traits of living things as different as mushrooms and free-tailed bats be coded using the same genetic material? Why do traits sometimes "skip" generations? How are mutations in genes responsible both for the remarkable diversity of life on Earth and for such devastating diseases as cancer? How can genetics research revolutionize everything from what we eat to how long we live? We'll learn the answers to these questions in this chapter.

16.1 What Is a Gene?

Genes determine all sorts of traits in organisms—the colors of an orchid's flowers, the length of a cat's tail, the substances that make up a crab's carapace or a bacterium's cell wall. In humans, genes affect what color eyes we have, whether we are tall or short, and whether our hair is straight or curly. Genes are even believed to influence human personality traits—at least to some degree. But what is a gene, and how does it determine a trait? It may surprise you to learn that a **gene** is simply a section of DNA that contains the instructions for making a protein. The genetic makeup of an organism, contained in its DNA, is known as the organism's **genotype.** The observable physical and biochemical characteristics, or traits, of an organism are known as the organism's **phenotype.** How genotype becomes phenotype—how genes become traits—is one of the subjects of this chapter.

But why do so many of an organism's traits depend on genes and therefore, presumably, on proteins? Because, as we learned in the previous chapter, proteins play a huge variety of roles in living organisms—they provide structure, they act as hormones, they transport molecules, they function in cell signaling, and they protect organisms from disease. In addition, the all-important enzymes, required for practically every chemical reaction that occurs in cells, are proteins as well.

16.2 Chromosomes: Packages of Genetic Information

Let's begin by asking where the genes are. In eukaryotic organisms, DNA is found in the cell nucleus, where it is packaged in linear structures known as *chromosomes* (Figure 16.1). Each chromosome consists of a single long piece of

DNA as well as small proteins called histones. DNA is wrapped around histone "spools" like thread. Because the DNA in each chromosome is so long—a single human cell contains 7 feet of DNA—the histones help to keep it from becoming hopelessly tangled.

Most cells have two of each kind of chromosome, like a pair of matched shoes. These cells are referred to as **diploid**, and their matched chromosomes are known as *homologous chromosomes*. Some cells—such as sperm and egg cells—have only one of each kind of chromosome—these cells are referred to as **haploid**. Different organisms have different numbers of chromosomes. Chickens have 78 (39 pairs), mosquitoes have 6 (3 pairs), lettuces have 18 (9 pairs), and yeast have 32 (16 pairs). In humans, there are 46 chromosomes, or 23 pairs, as shown in Figure 16.2. One pair of *sex chromosomes* determines sex. Females have two X sex chromosomes, whereas males have one X and one Y sex chromosome. The other chromosomes are known as *autosomal chromosomes*.

INTEGRATED SCIENCE CHEMISTRY

The Structure of DNA

Now that we know where the genes are, let's look more closely at what genes are made of, at DNA itself. What does DNA look like? A molecule of **deoxyribonucleic acid**, or **DNA,** consists of two strands that, when put together, resemble a spiraling ladder with two "sides" and a series of regularly spaced "rungs" (Figure 16.3). Because DNA consists of two strands twisted into a spiral or helix, it is often described as a *double helix*. Let's examine a single strand of DNA, and then consider how the two strands fit together.

Each DNA strand has a backbone (or a "side" of the ladder) that is made up of alternating molecules of deoxyribose sugar and phosphate (Chapter 15). Attached to this backbone is a series of *nitrogenous bases* (each represents one half of a "rung"). Only four different nitrogenous bases are used in DNA—adenine (A), guanine (G), cytosine (C), and thymine (T). A single nitrogenous base bound to a single molecule of sugar and a single phosphate is called a *nucleotide*. So, a strand of DNA can also be described as a string of nucleotides.

Now let's put the two strands of DNA together. Each base binds with a base on the opposite strand using hydrogen bonds (Chapter 12). The binding occurs in a very specific way—adenine always pairs with thymine (A–T), and guanine always pairs with cytosine (G–C).

DNA double helix

Histones

Loosely packed DNA, in cells not preparing to divide

Tight helical fiber

Condensed chromosome in dividing cells

Sister chromatids

Centromere

(a)

(b)

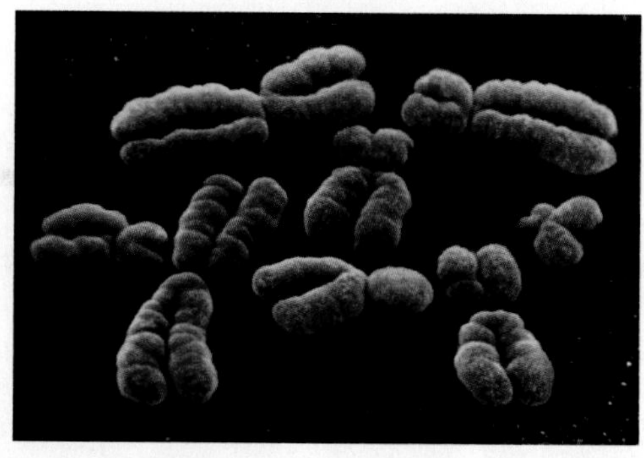

Figure 16.1 (a) Chromosomes are linear structures containing DNA as well as proteins called histones. Histones keep DNA packed in an orderly way. Chromosomes are loosely packed most of the time but become condensed during cell division. (b) These chromosomes are condensed in preparation for cell division. Recall that each consists of two identical sister chromatids attached at the centromere. (Adapted from Campbell, Reece, Simon, *Essential Biology with Physiology*, © 2004.)

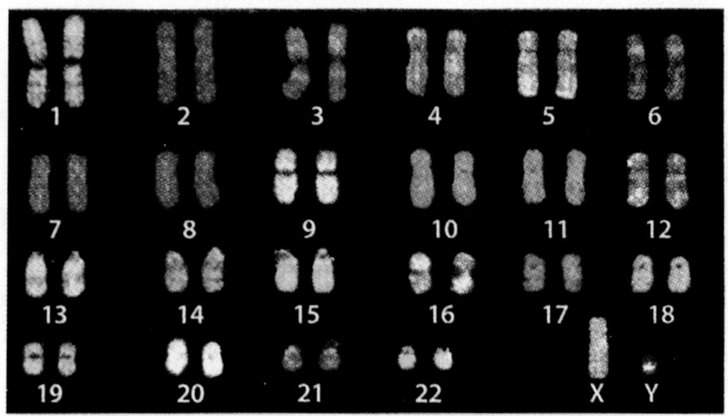

Figure 16.2 Humans have 23 pairs of chromosomes. These are the chromosomes of a human male.

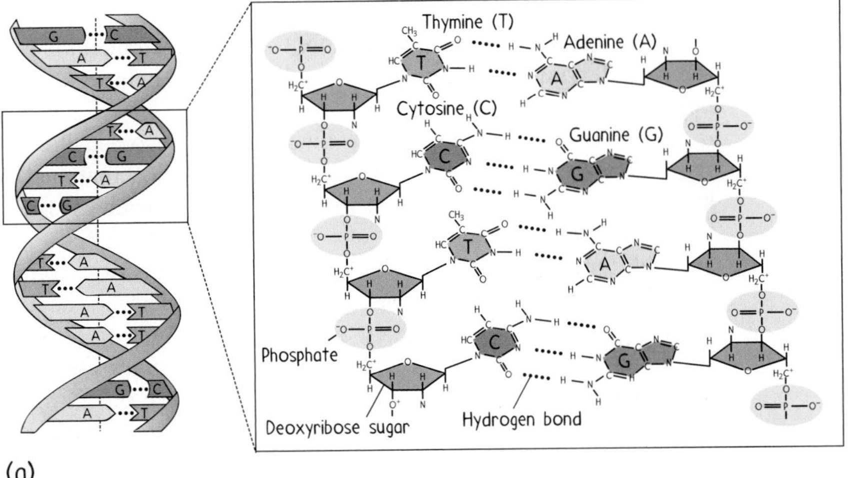

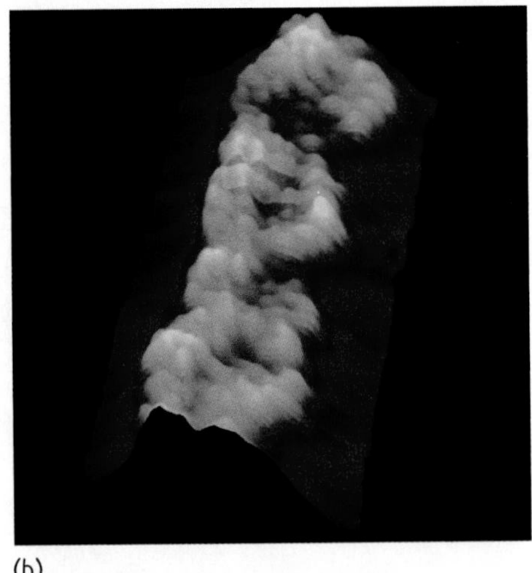

Figure 16.3 (a) DNA is shaped like a spiraling ladder with two sugar-phosphate strands as the "sides" of the ladder and paired bases as the "rungs." (b) This photograph of DNA shows that it is a double helix. (Adapted from Campbell, Reece, Simon, *Essential Biology with Physiology*, © 2004.)

HISTORY OF SCIENCE

Discovery of the Double Helix

By the early 1950s, scientists knew that the genetic material of eukaryotic organisms was contained in their chromosomes. Scientists also knew that, of the two types of molecules found in chromosomes, proteins and DNA, the genetic material was DNA. The structure of DNA was still a mystery, however, and it represented the biggest unsolved problem in biology. In 1951, Francis Crick and James D. Watson began tackling the problem at Cambridge University in England. Their strategy was to build a model of DNA that would be consistent with available experimental evidence. Meanwhile, both Rosalind Franklin and Maurice Wilkins of King's College in London were taking X-ray photos of DNA.

By early 1951, Franklin had gathered data on two forms of DNA. These differed in their water content—there was a dry form, A, and a wet form, B. Franklin photographed the B form, and the photograph very clearly revealed a helix. However, she did not publish this finding and chose instead to concentrate on the A form. Later in 1951, Watson attended a seminar in which Franklin spoke about her findings. He and Crick used what they learned there to develop their first model of DNA, a triple helix that turned out to be incorrect. The real breakthrough, which came in January 1953, was sparked by two key events—Wilkins shared Franklin's photograph of the B form of DNA with Watson, and a copy of a report on Franklin's experimental findings on DNA made its way to Watson and Crick.

Watson and Crick began to build DNA models and to test them against Franklin's photographs and data. Within a few weeks, they had figured it out—DNA was a double helix. Furthermore, adenine and thymine, and guanine and cytosine, paired up between the strands. The Watson–Crick model of DNA was published in April 1953, along with two papers offering supporting evidence, one by Wilkins and his collaborators and one by Franklin and her assistant. Acceptance of the Watson–Crick model was immediate and widespread.

For their discovery of the double helix, Watson, Crick, and Wilkins were awarded the 1962 Nobel Prize in Physiology or Medicine. Franklin had died by then, of ovarian cancer that probably resulted from radiation exposure related to her work. She never knew how important her results had been in allowing Watson and Crick to develop their model. In fact, Franklin's importance to the discovery of DNA's structure was unknown until Watson told the story in his memoir *The Double Helix*.

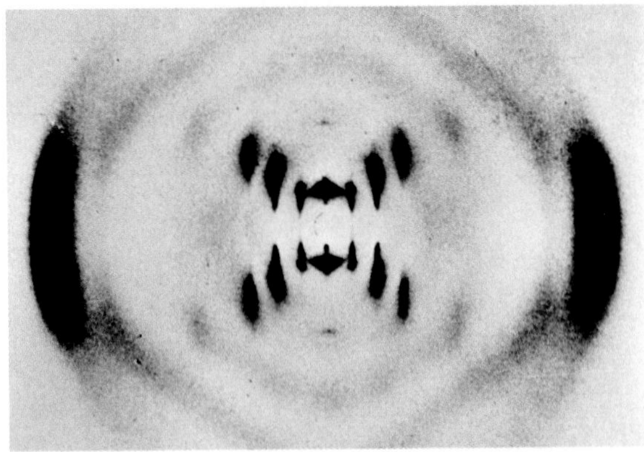

Rosalind Franklin's X-ray photo may look like a blurry "X" to you, but to Watson and Crick it showed that DNA is a double helix.

(a) (b)

(a) James D. Watson and Francis Crick figured out the structure of DNA in 1953. (b) Rosalind Franklin took the famous X-ray photo that led Watson and Crick to the structure of DNA.

INSIGHT Most human cells contain the usual 23 pairs of chromosomes, but red blood cells lack both nuclei and DNA. This allows them to be smaller in size and move more easily through blood vessels as they carry oxygen to tissues. Because red blood cells lack genetic material, however, they are unable to repair and maintain themselves and have a short life span.

Conceptual Science Place

DNA Replication

INSIGHT In humans and other eukaryotes, about 50 nucleotides are added every second during DNA replication. This sounds fast—and it is—but we have about 3 billion base pairs in each cell. DNA replication would take far too long if it didn't occur simultaneously at many different points in each chromosome. As it is, all the DNA in a human cell can be copied in a few hours. Prokaryotes, incidentally, are much faster than eukaryotes—they can add about 500 nucleotides per second.

Nucleic Acid Structure; Transcription; Translation

CHECK YOURSELF

If one strand of DNA contains the base sequence ACCTGA, what is the base sequence on the opposite strand?

CHECK YOUR ANSWER

TGGACT.

16.3 DNA Replication

DNA has to be copied in order for cells to divide and reproduce. When James D. Watson and Francis Crick published their model of the double helix, they made the comment, now famous, "It has not escaped our notice that the specific pairing we have postulated immediately suggests a possible copying mechanism for the genetic material." What did they mean? How is DNA copied?

DNA is copied in a process called **replication.** During replication, DNA's two strands are separated as if the spiral ladder were unzipped down the middle. Because of the specific way the bases pair with other bases—because A always goes with T and G always goes with C—each strand can serve as a template for building a new partner. This is shown in Figure 16.4. Free nucleotides bind to their complementary nucleotides on the template DNA strand and are then attached to the new, growing DNA strand by enzymes called *DNA polymerases.* Because of the way replication occurs, every new DNA molecule includes one old strand and one new strand. The new DNA molecules are identical to the original.

DNA replication always begins at fixed spots within chromosomes. Bacteria, which have only one small chromosome, can get away with starting replication at a single point and copying all around the chromosome. In eukaryotes, however, there is so much DNA that replication begins simultaneously at many different points in a chromosome, allowing the job to be completed efficiently (Figure 16.5).

16.4 Transcription and Translation

Now that we know how DNA is structured and copied, we can turn to the question of how it performs its primary function—how does DNA provide instructions for cells to build proteins? It does this through the processes of transcription and translation.

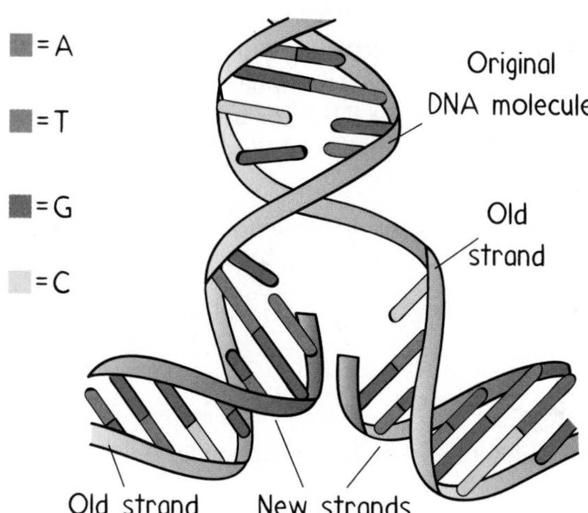

Figure 16.4 When DNA is copied, the two strands are separated and each serves as a template for putting together a new strand. The new DNA molecules are identical to the original.

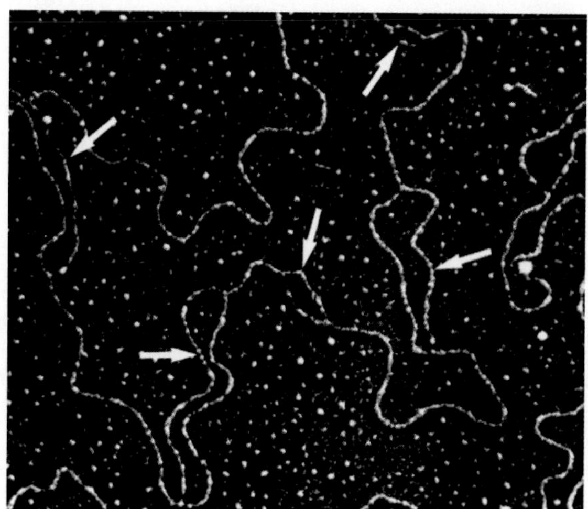

Figure 16.5 This photo shows DNA replication in the cell of a fruitfly (*Drosophila melanogaster*) embryo. Arrows mark where the strands of DNA have been separated and replicated.

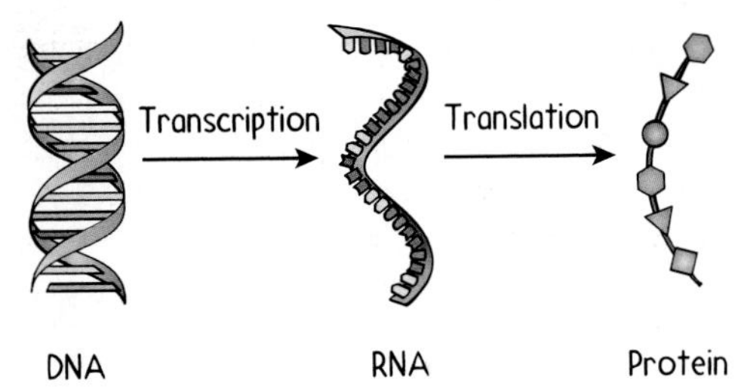

Figure 16.6 Proteins are made from instructions contained in DNA through the processes of transcription and translation.

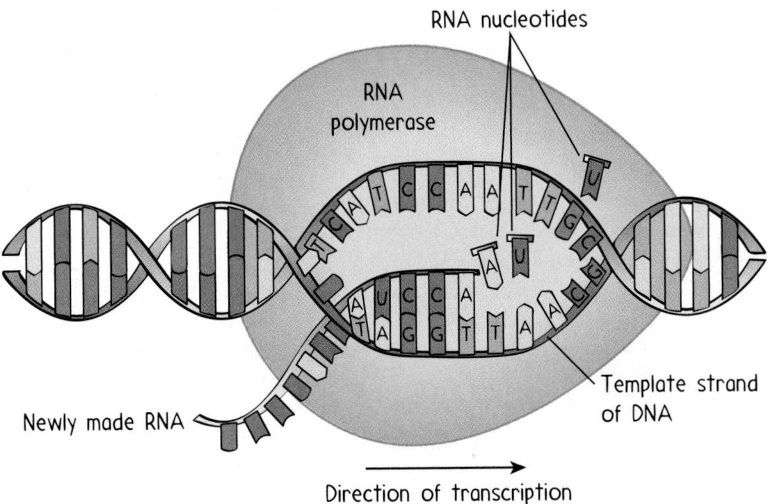

Figure 16.7 During transcription, DNA is used as a template for making mRNA. (Courtesy of Campbell, Reece, Simon, *Essential Biology with Physiology,* © 2004.)

A key player in transcription and translation is *ribonucleic acid*, or RNA. Like DNA, **RNA** is a molecule consisting of a sugar-phosphate backbone attached to a series of nitrogenous bases. However, RNA differs from DNA in three ways—RNA is single-stranded rather than double-stranded, it uses the sugar ribose rather than deoxyribose, and it uses the base uracil (U) instead of thymine (T). During **transcription,** DNA is used as a template for building a molecule of RNA. During **translation,** this RNA molecule is used to assemble a protein (Figure 16.6).

Transcription

In eukaryotes, transcription occurs in the cell nucleus. The two strands of DNA are separated, and one strand serves as a template for constructing the RNA transcript, as shown in Figure 16.7. Special sequences in the DNA indicate where transcription should begin and end, marking the beginning and end of a single gene. The construction of the RNA transcript makes use of the same base-pairing rules we saw before, except that RNA uses uracil instead of thymine. So, where DNA has bases A, C, G, and T, the RNA transcript has U, G, C, and A respectively.

Transcription requires the enzyme *RNA polymerase.* As free nucleotides bind to their complementary nucleotides on the DNA strand, RNA polymerase adds them to the growing RNA molecule. Once the RNA molecule is completed, the DNA zips back up and the RNA begins a processing phase. The RNA molecule made during transcription is called **messenger RNA,** or **mRNA** for short.

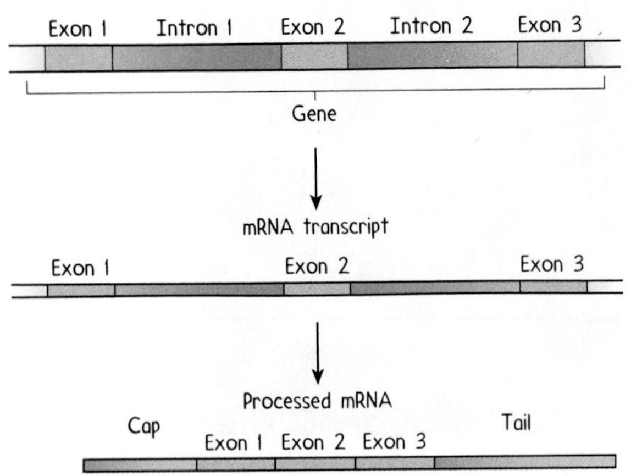

Figure 16.8 After the mRNA molecule is transcribed, it goes through a processing phase—a cap and tail are added, and introns are removed.

CHECK YOURSELF

1. If DNA being transcribed contains the nucleotide sequence ACCTGAT, what sequence will the mRNA contain?
2. What is the role of RNA in transcription?

CHECK YOUR ANSWERS

1. UGGACUA
2. RNA is used to capture information about the sequence of nucleotides from the DNA molecule.

Once the mRNA molecule has been transcribed, it begins a processing phase in which two things happen (Figure 16.8). First, a cap is added to the beginning of the molecule and a tail is added to the end—these allow the cell to recognize the mRNA molecule as mRNA. Second, stretches of nucleotides not relevant to building protein are removed from the mRNA molecule—these stretches are called **introns.** The nucleotides that remain, which do contribute to building protein, are called **exons.** It is as if a preliminary transcript of Shakespeare's *Hamlet* contained the line "*aggfr uidosa to be dfjklsdf or rewerwe not to be*," and the irrelevant parts had to be removed in order to leave the coherent "*to be or not to be.*" No one knows for sure why introns exist. Once processing is complete, the mRNA molecule moves from the nucleus of the cell to the cytoplasm, where it will be translated to assemble a protein.

CHECK YOURSELF

Suppose you have the following mRNA transcript. Exons and introns are labeled. What will the mRNA molecule look like after processing?

UAGCCUGUAUGGACUUGUCAACGGGUCAUACCGAUUCGAUCAGAAUUCG

intron	exon	intron	exon	intron
UAGCCUG	AUGGACUUG	UCAACGGGUCAUACCGA	UUCGAUCA	GAAUUCG

CHECK YOUR ANSWER

The introns will be removed, and a cap and tail will be added, resulting in:
cap-AUGGACUUGUUCGAUCA-tail

Translation

> **INSIGHT** With a very few exceptions, the genetic code is universal—that is, it is the same for all living organisms, including animals, fungi, plants, protists, bacteria, and archaea. Even viruses use the same genetic code. The universality of the genetic code suggests that it arose very early in the evolution of life and was then passed on to all living species.

Translation takes place at ribosomes (Chapter 15) within the cytoplasm. There, mRNA is translated into protein through a simple, elegant mechanism: a triplet of nucleotides along the mRNA strand forms a **codon** that stands for one of the twenty amino acids that make up proteins. During translation, codons are "read" from the mRNA molecule like words and the corresponding amino acids are strung together to build the protein. The correspondence between codons and amino acids, which is called the *genetic code*, is shown in Table 16.1. For example, the codon AGU translates to the amino acid serine, and the codon GUG translates to the amino acid valine. Because there are more triplet codons than there are amino acids, multiple codons translate to the same amino acid. For example, CGU, CGG, and CGC all translate to arginine. Certain codons, such as UAA, are called *stop codons* because they tell the ribosome that there will be no more amino acids in the protein. In

Table 16.1 Genetic Code

First base	Second base: U	Second base: C	Second base: A	Second base: G	Third base
U	UUU Phenylalanine	UCU	UAU Tyrosine	UGU Cysteine	U
	UUC (Phe)	UCC Serine	UAC (Tyr)	UGC (Cys)	C
	UUA Leucine	UCA (Ser)	UAA Stop	UGA Stop	A
	UUG (Leu)	UCG	UAG Stop	UGG Tryptophan (Trp)	G
C	CUU	CCU	CAU Histidine	CGU	U
	CUC Leucine	CCC Proline	CAC (His)	CGC Arginine	C
	CUA (Leu)	CCA (Pro)	CAA Glutamine	CGA (Arg)	A
	CUG	CCG	CAG (Gln)	CGG	G
A	AUU Isoleucine	ACU	AAU Asparagine	AGU Serine	U
	AUC (Ile)	ACC Threonine	AAC (Asn)	AGC (Ser)	C
	AUA	ACA (Thr)	AAA Lysine	AGA Arginine	A
	AUG Met or start	ACG	AAG (Lys)	AGG (Arg)	G
G	GUU	GCU	GAU Aspartic	GGU	U
	GUC Valine	GCC Alanine	GAC acid(Asp)	GGC Glycine	C
	GUA (Val)	GCA (Ala)	GAA Glutamic	GGA (Gly)	A
	GUG	GCG	GAG acid (Glu)	GGG	G

eukaryotes, the first codon to be translated from a mRNA molecule is usually AUG—methionine. So, you read the codons in a molecule of mRNA, string together amino acids in the right order, and (presto!) you have a protein.

Translation requires two other forms of RNA—**transfer RNA** or **tRNA** and **ribosomal RNA** or **rRNA**. tRNA molecules transfer amino acids to the protein being assembled. Each tRNA molecule carries a single, specific amino acid and includes a sequence of three nucleotides called an **anticodon**. rRNA is a type of RNA that, along with proteins, make up ribosomes.

During translation, the mRNA molecule is bound to a ribosome. The codon being translated is positioned at the ribosome's A site (Figure 16.9a). A tRNA

INSIGHT Why a triplet genetic code? Two nucleotides wouldn't be enough—with two nucleotides, there are only 16 possible codons (4 possible codons in the first position × 4 possible codons in the second position), and there are 20 different amino acids found in living organisms. For three nucleotides, there are 4 × 4 × 4 = 64 possible codons. A triplet codon is the shortest that can code for all the amino acids.

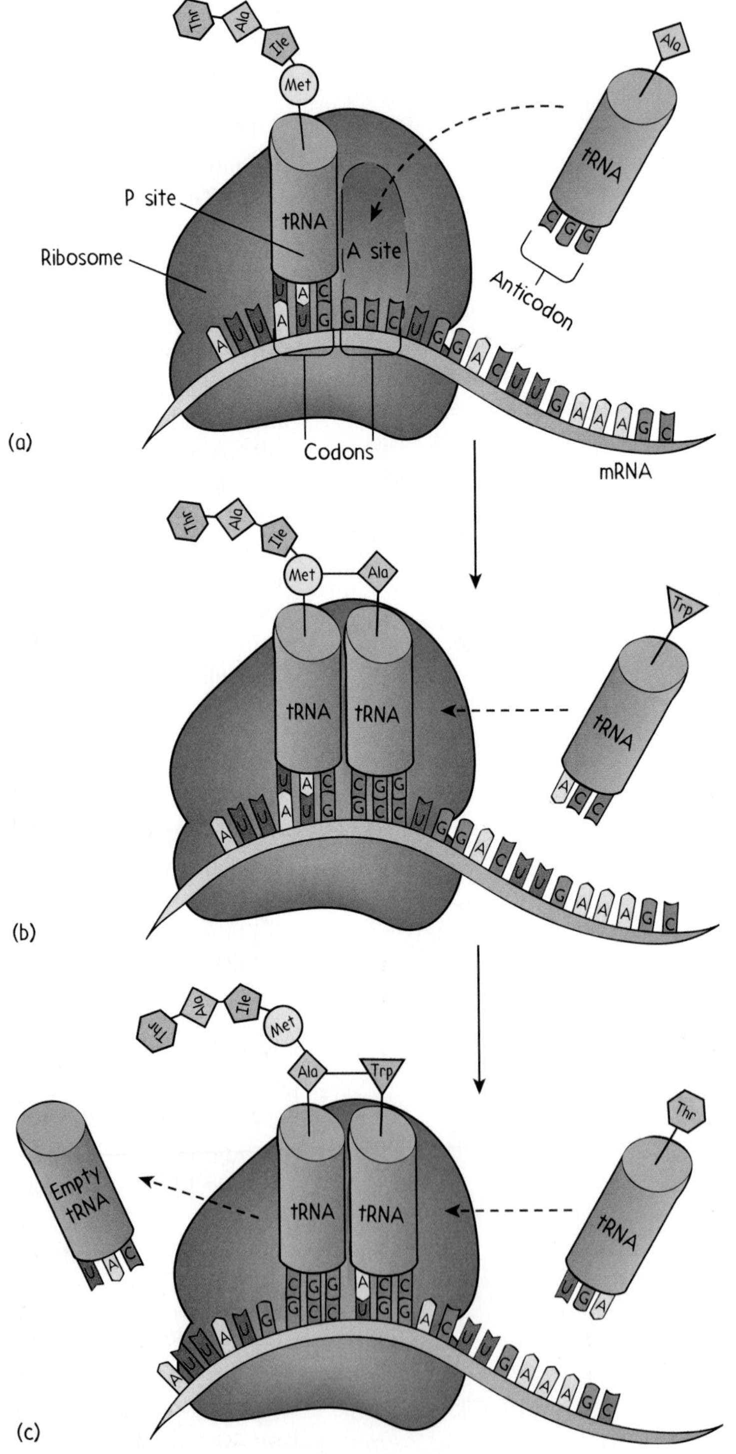

Figure 16.9 During translation, the mRNA molecule is translated to a string of amino acids. (a) A tRNA molecule carrying an amino acid binds to the codon at the ribosome's A site. (b) The new amino acid is added to the growing protein. (c) The mRNA molecule is shifted so that the codon that has just been translated moves to the P site. A new codon occupies the A site and is translated.

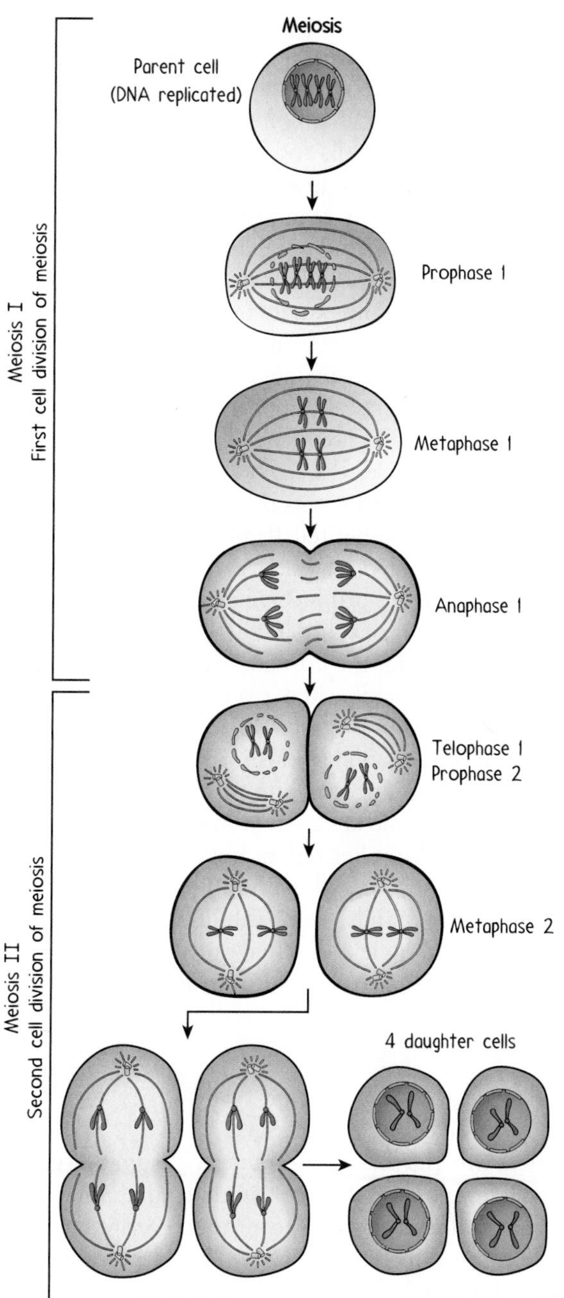

Figure 16.10 Meiosis produces four daughter cells, each of which has one copy of each kind of chromosome.

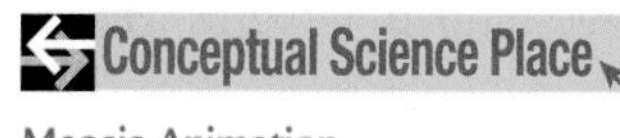

Meosis Animation

INSIGHT Meiosis in human egg cells is sometimes described as "unequal meiosis" because one of the four daughter cells—the future egg—receives almost all of the cytoplasm. The other three daughter cells receive almost no cytoplasm and quickly degenerate. This unequal division assures that the egg will have all the resources it needs for development.

molecule bearing the complementary anticodon and carrying its amino acid binds to the codon. The binding of codon and anticodon follows the usual base-pairing rules—that is, A binds with U, and G binds with C. The new amino acid, carried by the tRNA, is then added to the growing protein (Figure 16.9b). After this, the ribosome shifts the mRNA molecule so that the codon that has just been translated moves to the P site (Figure 16.9c). A new codon occupies the A site, a new tRNA molecule binds to the codon, and the process is repeated until a stop codon is reached.

CHECK YOURSELF

1. Suppose a molecule of mRNA has the codon GUC. What will the anticodon sequence be on the tRNA that binds to it? What amino acid will the tRNA be carrying? (Remember it's *codons* that appear in the genetic code table!)
2. Consider the mRNA sequence: AUGAGCCUGUAC. What string of amino acids does this sequence code for?

CHECK YOUR ANSWERS

1. The tRNA will have the anticodon sequence CAG, and it will be carrying the amino acid valine.
2. Dividing the sequence into codons and using the genetic code table, we have: AUG = methionine, AGC = serine, CUG = leucine, UAC = tyrosine. So, the amino acid string is methionine–serine–leucine–tyrosine.

16.5 Meiosis: Genetic Diversity

In Chapter 15, we described how cells reproduce through the process of mitosis. **Meiosis** is a special form of cell division used to make haploid cells, such as the egg cells and sperm cells of animals. In meiosis, one diploid parent cell, with two of each kind of chromosome, divides into four haploid daughter cells, each with only one of each kind of chromosome. The usual diploid chromosome number is restored during sexual reproduction when sperm and egg fuse at fertilization.

At the start of meiosis, the cell has already replicated its DNA, and so it has twice the normal amount of genetic material (that is, four, rather than two, of each kind of chromosome). Meiosis takes place in two stages, *meiosis I* and *meiosis II*. Each stage includes several phases. During meiosis I, the original cell divides into two cells, each of which has two of each kind of chromosome. During meiosis II, the two cells produced during meiosis I divide again, producing four cells that have one of each kind of chromosome. Meiosis is illustrated in Figure 16.10.

Meiosis I begins with *prophase I*, during which the chromosomes condense, the membrane of the nucleus breaks down, and the spindle apparatus appears. As with mitosis, each chromosome consists of two sister chromatids. During *metaphase I*, homologous chromosomes line up at the equatorial plane. (Note the difference between meiotic metaphase I and mitotic metaphase. In mitosis, the chromosomes line up individually at the equatorial plane; in meiosis, homologous pairs of chromosomes line up opposite each other.) While the homologous chromosomes are lined up, *crossing over* occurs—one chromosome exchanges corresponding parts with its homologue. As a result, the chromosomes of the dividing cell are no longer identical to the ones found in the original parent cell. Instead, many of the chromosomes are now composites of the two homologous parental chromosomes (Figure 16.11). On a genetic level, crossing over results in **recombination**, the production of new combinations of genes different from those found in the parental chromosomes. Because crossing over occurs at different points along the chromosomes in every meiotic division, it produces tremendous genetic variation among the sex cells of a single individual.

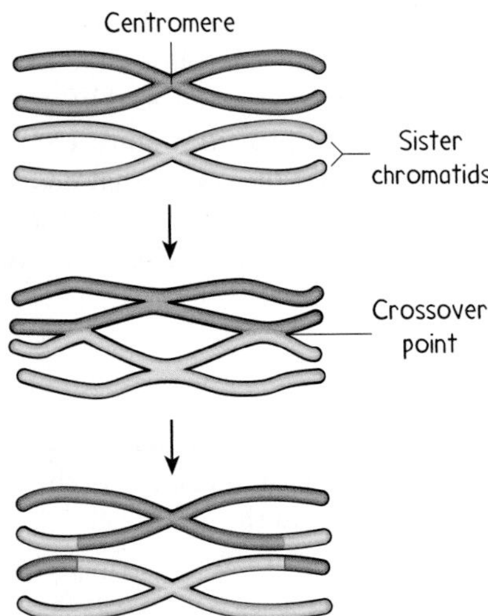

Figure 16.11 During meiosis, homologous chromosomes exchange genetic material in a process called crossing over.

INSIGHT Every once in a while, a mistake occurs during meiosis so that a sperm or egg ends up with two of some chromosome. In humans, embryos produced by these sex cells usually do not survive, but there are exceptions. One of the most common chromosomal abnormalities is having three copies of chromosome 21—trisomy 21. Trisomy 21 causes Down syndrome, a condition characterized by mental retardation and defects of the heart and respiratory system.

During *anaphase I*, the chromosome pairs are separated by the spindle apparatus. During *telophase I*, the chromosomes move to the poles of the cell. Cytokinesis occurs, producing two daughter cells, each with two sets of chromosomes.

Meiosis II begins immediately after meiosis I is complete. During *prophase II*, the chromosomes move to the center of the cell, and a new spindle apparatus forms. During *metaphase II*, the chromosomes line up along the equatorial plane. During *anaphase II*, the sister chromatids separate. During *telophase II*, the sister chromatids move to opposite poles of the cell. The nuclear membranes reform, and the chromosomes revert to their loosely packed, noncondensed form. Cytokinesis occurs, producing four haploid daughter cells.

Meiosis produces great genetic diversity among an organism's sex cells for two reasons. To see how, let's consider the eggs produced by a human female. Like all humans, this woman has two of each kind of chromosome, one that she inherited from her mother and one that she inherited from her father. First, as we have seen, crossing over during meiosis I causes the woman's homologous chromosomes to exchange parts, so that most of the chromosomes in her eggs are unique composites of the chromosome she inherited from her mother and the chromosome she inherited from her father. Even if there were no such thing as crossing over, however, the woman would still produce a huge number of genetically different eggs. This is because each pair of homologous chromosomes separates independently during meiosis I. If we ignore crossing over, each egg would receive either the chromosome the woman inherited from her mother or the chromosome she inherited from her father. One egg could receive chromosomes 1, 3, 4, 5, 7, 10, 13, etc., from her mother and chromosomes 2, 6, 8, 9, 11, 12, etc., from her father. A second egg she produces is almost certain to receive a different set of chromosomes, perhaps chromosomes 2, 3, 5, 6, 7, etc., from her mother and chromosomes 1, 4, 8, 9, etc., from her father. It is clear that the independent separation of homologous chromosomes alone produces a huge number of possible egg cells. Crossing over only expands the possibilities. It's no wonder that no two eggs or sperm produced by a single individual are alike. This genetic diversity produced during meiosis is crucial to evolution, as we will see in Chapter 17.

16.6 Mendelian Genetics

Until 1900, the dominant theory of *inheritance*—how traits pass from one generation to the next—was blending inheritance. Under blending inheritance, the mixing of parental hereditary material was thought to produce offspring intermediate between the parents. Thus, breeding a white cat with a black cat should result in a litter of gray kittens. But does it?

In fact, biologists were hard-pressed to explain a number of hereditary patterns that seemed to contradict blending inheritance: Why do brown-eyed parents sometimes have blue-eyed children? Why do traits so often skip generations? Then, in 1900—three and a half decades after its original publication—the work of an Austrian monk named Gregor Mendel (Figure 16.12) was rediscovered.

Figure 16.12 Gregor Mendel was the founder of modern genetics. Here he is shown examining a plant.

Mendel had completed a series of breeding experiments using pea plants, which vary in a number of traits—they have round seeds or wrinkled seeds, yellow seeds or green seeds, purple flowers or white flowers, and so on. Mendel performed the simple experiment of breeding two plants that differed in a single trait. Did he see blending inheritance? No. In fact, all the offspring resembled *one* of the two parents. For example, when a pea plant with round seeds was bred with a pea plant with wrinkled seeds, all the offspring had round seeds. When a purple-flowered plant was bred with a white-flowered plant, all the offspring had purple flowers. Traits that were expressed in the offspring—such as round seeds or purple flowers—Mendel called *dominant*. Traits not expressed in the offspring—such as wrinkled seeds or white flowers—Mendel called *recessive*. For

	Seed shape	Seed color	Flower color
One form of trait (dominant)	Round	Yellow	Purple
A second form of trait (recessive)	Wrinkled	Green	White

Figure 16.13 Mendel bred pea plants that varied in a number of traits. He found that, of a pair of traits, one was dominant and the other was recessive.

Figure 16.14 Dimples are a dominant human trait. So are widow's peak hairlines.

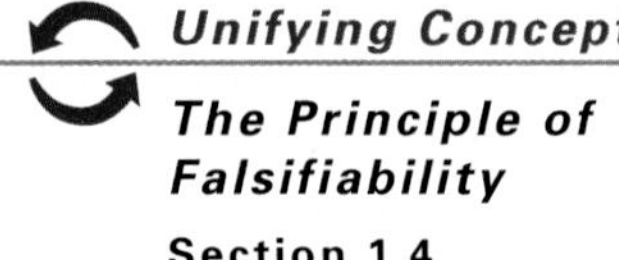

Unifying Concept
The Principle of Falsifiability
Section 1.4

every pair of traits Mendel examined, one trait was dominant and the other was recessive (Figure 16.13). Humans have dominant and recessive traits as well—a few of these are shown in Figure 16.14.

CHECK YOURSELF

1. If blending inheritance were correct, what would Mendel have gotten when he bred pea plants with green seeds and pea plants with yellow seeds?
2. In fact, Mendel found that all the offspring had yellow seeds. Which seed-color trait is dominant? Which seed-color trait is recessive?

CHECK YOUR ANSWERS

1. Plants with yellow-green seeds.
2. Yellow seeds are dominant. Green seeds are recessive.

Mendel then allowed offspring plants from the first generation to breed with themselves. (Pea plants, like many other plants and some animals, can self-fertilize.) What did he find? Mendel found that the recessive trait, which had disappeared in the first generation, reappeared in the second generation. Moreover, in the second generation, the ratio of plants expressing the dominant trait to plants expressing the recessive trait was 3:1. That is, there were three times as many plants expressing the dominant trait as there were plants expressing the recessive trait.

How did Mendel explain his results? Mendel postulated that the heritable factors (which we now call genes) that determine traits consist of two separate **alleles**, or versions, one inherited from each parent. When an individual produces sex cells (sperm cells or egg cells), half the sex cells carry one allele and the other half carry the other allele. This is Mendel's first law, his *principle of segregation.*

Let's see how the principle of segregation accounts for Mendel's breeding results. Mendel looked at plants that differ in a single trait—say, round seeds *versus* wrinkled seeds. The plant with round seeds has two alleles for roundness, *R* and *R*, or genotype *RR* for short. The plant with wrinkled seeds has two

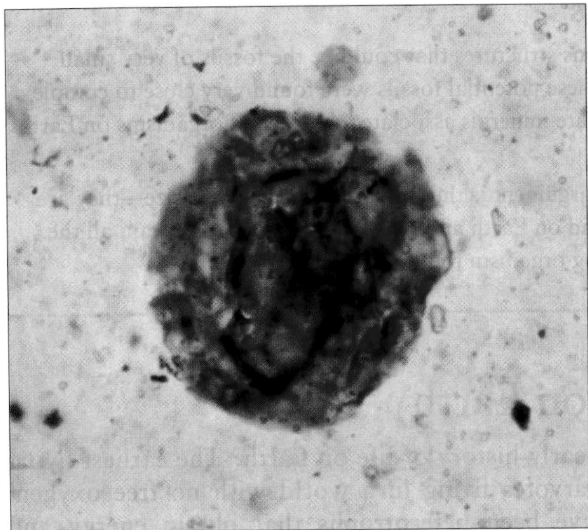

Figure 17.7 The earliest known eukaryote fossils are of acritarchs, photosynthesizing protists. The fossil above is 1.5 billion years old and was collected in northern Australia.

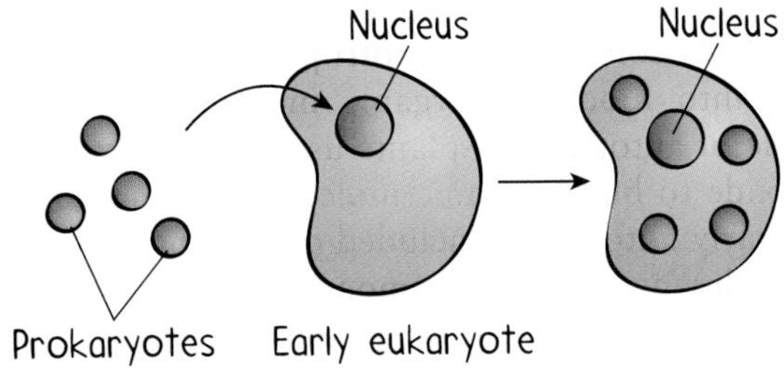

Figure 17.8 The mitochondria and chloroplasts in eukaryotic cells likely evolved from prokaryotes that started living inside early eukaryotes.

however, appear to have a different, and quite fascinating, origin. Mitochondria, organelles that obtain energy for cells, exist in most eukaryotic cells. Chloroplasts are found in plant cells and are responsible for conducting photosynthesis. Scientists believe mitochondria and chloroplasts evolved from prokaryotes living inside the earliest eukaryotic cells (Figure 17.8). This **endosymbiotic theory** (*endo* means "in" and *symbiotic* means "to live with") is supported by several observations. First, mitochondria and chloroplasts have their own membranes and their own DNA. Further, this DNA is in the form of a circular chromosome, just like prokaryotic DNA. Both mitochondria and chloroplasts also make their own proteins, using ribosomes that resemble those of prokaryotes. So, which prokaryotes did mitochondria and chloroplasts evolve from? By studying their structures, scientists have concluded that mitochondria are most likely descended from a group of oxygen-breathing bacteria and that chloroplasts most likely originated from our old friends, the photosynthesizing cyanobacteria. Given that mitochondria use oxygen to obtain energy for eukaryotic cells, and that chloroplasts perform photosynthesis in plant cells, these origins make good sense.

17.4 Charles Darwin and *The Origin of Species*

For thousands of years, ever since the time of Aristotle, it was believed that species were *static*—they had always been the way they were and would continue to be that way until the end of time. A horse was a horse was a horse. Then fossil organisms were discovered, and people began to wonder. Here were ancient creatures clearly different from any species that exist today, yet they also showed an unmistakable resemblance to present-day species. Could these fossils, such as the one shown in Figure 17.9, be the ancestors of modern species?

French naturalist Jean Baptiste Lamarck (1744–1829) was one of the first to argue that this was the case. Lamarck believed that modern species were descended from ancestors that had evolved—changed over time—to become better adapted to the environments they lived in. How did change occur? According to Lamarck's theory, organisms acquired new characteristics over a lifetime of activity and then passed these characteristics onto their offspring. So, modern giraffes have long necks because their ancestors stretched their necks to reach for the high leaves on a tree. These ancestors then passed longer necks onto their offspring, which reached for even higher leaves, stretching their necks even further, and so on (Figure 17.10a). Lamarck's theory for how change occurs, called the *inheritance of acquired characteristics*, proved to be incorrect—organisms cannot pass characteristics acquired during their lifetimes to their offspring because these acquired characteristics are not genetic. However, Lamarck's fierce support for the idea that organisms evolve set the stage for the work of Charles Darwin.

English naturalist Charles Darwin (1809–1882), shown in Figure 17.11, set forth the theory of evolution in his book *The Origin of Species by Means of Natural Selection*, published in 1859. Darwin proposed that **evolution**—heritable changes in organisms over time, or "descent with modification"—had produced all the living forms on Earth. Much of Darwin's theory grew out of the observations he made as the official naturalist aboard the H.M.S. *Beagle*, which voyaged around South America and mapped its coastline between 1831 and 1836. For 5 years, Darwin studied South American species, collecting large numbers of plants, animals, and fossils. He became increasingly intrigued by the question of how species got to be the way they were. Darwin was particularly struck by the life-forms he encountered on the Galápagos Islands, 950 kilometers from the South American continent. Among the organisms Darwin took particular note of were the 13 species of Galápagos finches now known as Darwin's finches. Some

Darwin and the Galápagos Islands

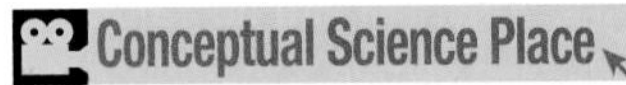

Galápagos Islands Overview;
Galápagos Marine Iguana

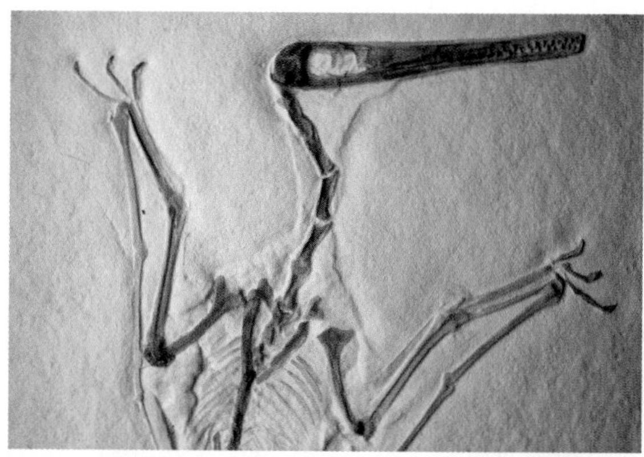

Figure 17.9 Could fossils be the ancestors of modern species?

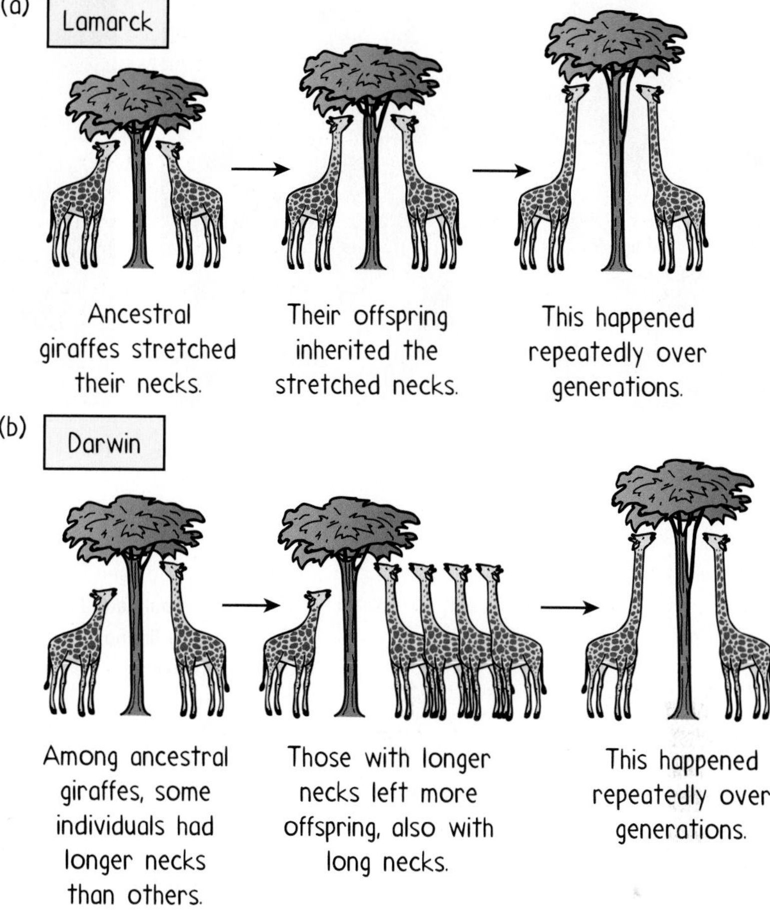
(a)
Lamarck
Ancestral giraffes stretched their necks.
Their offspring inherited the stretched necks.
This happened repeatedly over generations.
(b)
Darwin
Among ancestral giraffes, some individuals had longer necks than others.
Those with longer necks left more offspring, also with long necks.
This happened repeatedly over generations.

Figure 17.10 (a) Under Lamarck's theory of evolution, offspring inherit the characteristics that their parents acquire over a lifetime of activity. So, Lamarck argued that ancestral giraffes stretched their necks after ever-higher leaves on trees and passed these longer necks to their offspring. (b) Under Darwin's theory of evolution by natural selection, organisms with advantageous traits leave more offspring than organisms with other traits, causing advantageous traits to become more common in a population. Darwin argued that modern giraffes have long necks because ancestral giraffes with long necks left more offspring than ancestral giraffes with short necks.

Figure 17.11 Charles Darwin set forth the theory of evolution by natural selection.

of these are shown in Figure 17.12. Darwin's finches showed remarkable variation in the size and shape of their beaks, with each beak being suited to, and used for, a different diet. How had the beaks of these finches come to differ in this way? Darwin later wrote, "Seeing this gradation and diversity of structure in one small, intimately related group of birds, one might really fancy that from an original paucity of birds in this archipelago, one species had been taken and modified for different ends."*

In addition to his *Beagle* observations, Darwin was inspired by the work of two of his contemporaries, Charles Lyell and Thomas Malthus. Lyell, a geologist, argued that the geological features of the Earth were created not by major catastrophic events—the favored theory of the time—but by gradual processes that produced their effects over long time periods. A deep canyon, for example, did not require a cataclysmic flood but could be the result of a river's slow erosion of rock over millennia. Darwin realized this could be true for biological

* Charles Darwin, *The Voyage of the Beagle*, 1909.

Figure 17.12 The finches Darwin saw on the Galápagos Islands—now called Darwin's finches—show remarkable variation in the size and shape of their beaks. Each beak type is suited to, and used for, a different diet. (a) The cactus finches have pointy beaks used to eat cactus pulp and flowers. (b) The ground finches have blunt beaks used to crack seeds. (c) The tree finches have slender beaks used to catch insects. (d) The woodpecker finch has a woodpecker-like beak which it uses to drill holes in wood. It then uses a cactus spine to pry out insects.

organisms as well—the accumulation of gradual changes over long periods could produce all the diversity of living organisms, as well as all their remarkable features.

The economist Thomas Malthus was a second important influence for Darwin, and the one who led Darwin to his great idea on the cause of evolutionary change. Malthus observed that human populations grow much faster than available food supplies and concluded, with despair, that famine was an inevitable feature of human existence. Darwin applied Malthus's idea to the natural world and argued that, because there are not enough resources for all organisms to survive and to reproduce as much as they can, living organisms are involved in an intense "struggle for existence." As a result, organisms with advantageous traits leave more offspring than organisms with other traits, causing populations to change over time (Figure 17.10b). This process, which Darwin called **natural selection,** is the major driving force behind evolution.

INSIGHT If genius is that well-known combination of inspiration and perspiration, Darwin got his share of the latter—during his voyage on the *Beagle*, he collected 1529 alcohol-preserved specimens and 3907 skins, bones, and other dried specimens. He also took close to 2000 pages of notes on the flora, fauna, and geology of the places he saw. It's no wonder that when he finally set forth his theory, he was able to support it with a myriad of well-considered examples.

CHECK YOURSELF

1. When people live at high altitude, where oxygen is scarce, their red blood cell count increases. Is this an example of evolution?
2. If Lamarck had been correct and evolutionary change occurred through the inheritance of acquired characteristics, what trait might a bodybuilder pass to his offspring?

CHECK YOUR ANSWERS

1. No—evolution describes heritable changes in organisms over time—that is, genetic changes that can be passed onto offspring. The adjustments the body makes to high altitude do not affect the genes that are passed on to a person's offspring.
2. If Lamarck were correct, the bodybuilder's children would inherit the increased muscle mass that the bodybuilder had acquired over a lifetime of weightlifting. However, because Lamarck's theory turned out to be incorrect, the children will have to do their own bodybuilding.

17.5 How Natural Selection Works

Conceptual Science Place
Causes of Microevolution

Rabbits were introduced into Australia in 1859, when Thomas Austin released 24 individuals onto his property in southern Victoria, in the southeastern part of the continent. They quickly became pests, devastating farmlands and natural habitats, as shown in Figure 17.13. Breeding "like rabbits," they spread across the continent in such large numbers that they were described as a "gray blanket." Many attempts were made to control the rabbit population, including the laborious construction of a 1822-kilometer-long "rabbitproof" fence across the continent—still the longest fence in the world. Unfortunately, by the time the fence was completed in 1907, the rabbits had already moved through. (It probably wouldn't have worked anyway—even after the fence was in place, rabbits piled up so thickly behind it that some were eventually able to walk right over their companions' backs to the other side!)

In the early 1950s, the government pinned its hopes on a biological approach to population control—introducing the deadly myxoma virus into the rabbit population. Initially, the virus was a wonder, killing over 99.9 percent of infected rabbits. Within a few years, however, it was obvious that the virus was not working as well as it once had. Many fewer rabbits were dying. What had happened? The rabbit population had evolved disease resistance through the process of natural selection. Within the original rabbit population, a small number of individuals just happened to be resistant to myxoma virus. These resistant individuals survived the epidemics and reproduced, producing yet more disease-resistant offspring, as illustrated in Figure 17.14. Over time, the number of disease-resistant rabbits increased, and the virus became less and less effective. Today, fresh outbreaks of myxoma virus kill only about half the infected rabbits. Disease resistance is just one example of natural selection at work.

Natural selection occurs when organisms with certain advantageous traits leave more offspring than organisms with other traits, causing populations to

(a)

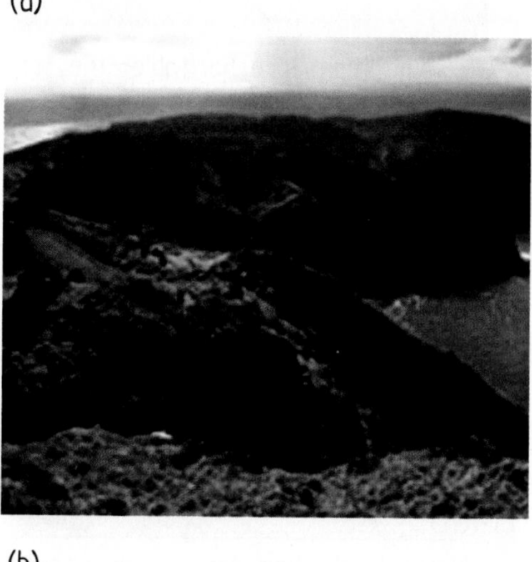

(b)

Figure 17.13 (a) Rabbits introduced into Australia caused widespread devastation, including here on Phillip Island. (b) This photo shows the same area after rabbits were eradicated. Note that vegetation has grown back.

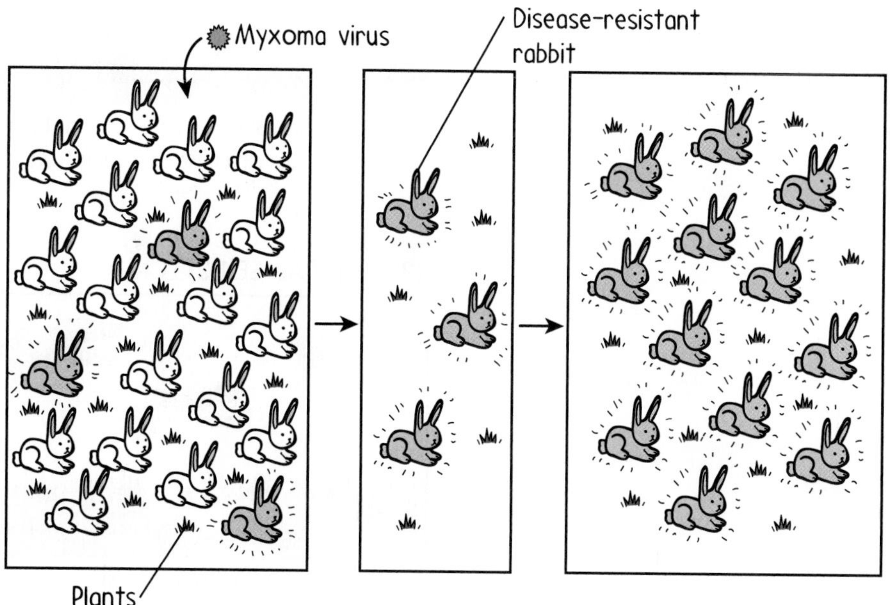

Figure 17.14 The initial outbreak of myxoma virus killed 99.9 percent of infected rabbits, but a small number of naturally disease-resistant rabbits survived. These reproduced. The next outbreak of disease found a more resistant population.

Figure 17.15 Natural selection can only act on traits where there is variation. These beetle specimens in the National Biodiversity Institute, Costa Rica may look nearly identical at first glance, but actually show many subtle differences on closer inspection.

change over time. Let's look at how natural selection works. First, in any population of organisms, individuals have traits, many of which show **variation**—that is, they vary from individual to individual (Figure 17.15). In humans, for example, individuals vary in height, hair color, hairstyle, nose shape, foot size, blood type, and so on. Second, many of these variable traits are determined at least partly by genes and therefore will be **heritable**—passed from parents to offspring. All of the human traits just listed are heritable except hairstyle, which is not heritable because it is not genetically determined. (Note that all heritable variation has its ultimate origin in genetic mutations (Chapter 16). In addition, sexual reproduction plays an important role in generating heritable variation by causing alleles to be brought together in varying combinations with each generation of organisms.) Third, some of these variable, heritable traits will be advantageous and allow the organisms possessing them to leave more offspring than other organisms. The **fitness** of an organism describes the number of offspring it leaves over its lifetime compared to other individuals in the population. An organism that leaves more offspring than other individuals in the population is said to have greater fitness. Finally, because organisms with advantageous traits have greater fitness—and so reproduce more—advantageous traits are "selected for" and become more common in a population. What is the result? The population evolves to become better adapted to its environment. Figure 17.16 shows a summary of the process of natural selection. Note that, although natural selection acts on individuals within a population, allowing some individuals to leave more offspring than others, it is the population that evolves.

(1) VARIATION
Organisms have lots of traits, many of which show variation.

(2) HERITABILITY
Some traits are heritable—they are determined by genes and so are passed from parents to offspring.

(3) NATURAL SELECTION
Variation in heritable traits sometimes results in some organisms leaving more offspring than others—that is, in natural selection.

(4) ADAPTATION
Natural selection causes advantageous traits to become more common in a population, producing the adaptation of organisms to their environments.

Figure 17.16 How natural selection works.

CHECK YOURSELF

1. Which of the following traits are variable in cats? Fur color. Tail length. Number of eyes.
2. Are these cat traits heritable?
3. In elephant seals, males fight to control and mate with large "harems" of female seals. The outcome of such a fight usually depends on factors such as size and strength. Could this lead to natural selection among elephant seals? If so, what evolutionary change might be seen in the elephant seal population?

CHECK YOUR ANSWERS

1. There is lots of variation in fur color among cats—there are tabby cats, black cats, white cats, gray cats, and so on. There is variation in tail length among cats—not all cat tails are exactly the same length. There is no variation in the number of eyes among cats—all cats have two eyes.
2. Yes, all three traits are heritable because they are all determined genetically. (Note that not all heritable traits are necessarily variable—here, having two eyes is a heritable trait but not a variable one.)
3. Yes, this could lead to natural selection. There is variation in the fighting abilities of males, and this variation is due to heritable traits such as size and strength. And, because the winners of fights have more mates, winning males are likely to leave more offspring than losing males. The effect of natural selection would be to cause elephant seals to become stronger, like their fathers.

Adaptation

Natural selection acts as the driving force behind evolution because it leads to the evolution of **adaptations**—traits that make organisms well suited to living and reproducing in their environments. Adaptations can relate to many different aspects of an organism's life. Many of the adaptations organisms evolve help them survive. Survival is, after all, usually a requirement for leaving offspring. Figure 17.17 shows three different survival adaptations that have evolved in butterflies. All are the result of natural selection favoring organisms that were better able to avoid

(a)

(b)

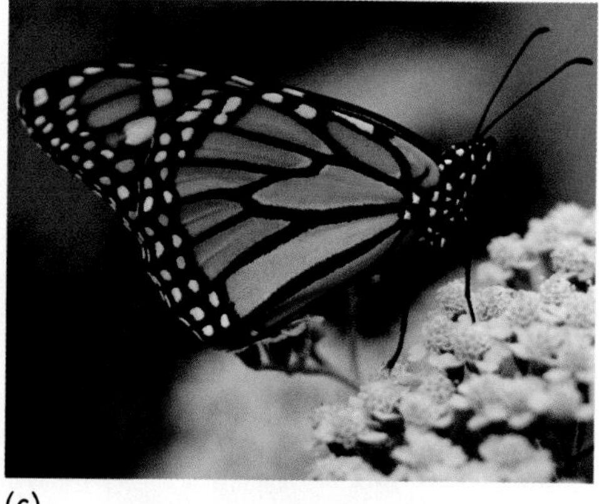

(c)

Figure 17.17 Butterflies have evolved a variety of ways to avoid predators. (a) The Painted Lady flies in an erratic, unpredictable manner, making it hard to catch. (b) The Monarch eats plants that are toxic to other animals so that its tissues become toxic. Birds that try to eat Monarchs vomit, and remember to avoid the striking black and orange pattern in the future. (c) The Viceroy is not toxic but is a mimic of the toxic Monarch, resembling it closely in appearance. As a result, the Viceroy is also avoided by birds.

Figure 17.18 Parental care occurs in many species and aids in the survival of offspring. In poison dart frogs, adults care for their young through the first dangerous phase of life. Here, the male parent is shown carrying two tadpoles on his back.

being eaten by birds. Other adaptations in organisms have evolved to help them acquire mates. These include the beautiful feathers of male peacocks and birds of paradise, the sexy "rib-bits" of males in many species of frogs, and the enchanting songs of many male birds, all of which females of the species find attractive. On the other hand, fighting structures, such as antlers in deer, are adaptations for winning mates in species where males compete with other males for mates. Natural selection that is related to acquiring mates is also called *sexual selection.* Finally, many adaptations relate to bearing and raising young. Figure 17.18 shows a behavioral adaptation related to raising young in a poison dart frog—parental care. Parental care evolved in this species as a result of natural selection favoring organisms that were better able to protect their young from predators.

Natural selection has produced remarkable adaptations over time. Nature does not plan ahead—it does not plan to make a falcon or a polar bear. Instead, adaptations are built step-by-step, through the never-ending selection of the most successful forms that arise from chance mutations.

INSIGHT How did natural selection produce something as spectacular-looking as a peacock? The peacock's magnificent plumage allows it to attract mates—that's right, peahens! In fact, you may have noticed that males are more brightly colored than females in many species of birds. Bright males are believed to result from a female preference for colorful mates. Females, not subject to these pressures, have been selected to be less bright because that makes them less conspicuous to predators.

INSIGHT And speaking of bright colors—the bold colors of organisms such as coral snakes, wasps, and poison dart frogs are believed to have evolved to warn potential predators that they are dangerous.

INTEGRATED SCIENCE PHYSICS

Animal Adaptations to Heat and Cold

Animals that live in extremely hot or extremely cold habitats need to be able to maintain appropriate body temperatures in those environments—to *thermoregulate.* In deserts, animals have to be able to dissipate heat to avoid overheating. In cold habitats, animals have to be able to retain heat. In both types of environments, animals have evolved behavioral, physiological, and anatomical adaptations relating to heat balance. We'll focus on anatomical adaptations.

A key factor in heat balance is an animal's surface area-to-volume ratio. (This ratio, reviewed in Figure 17.19, comes up repeatedly in different biological contexts—its importance in cells was discussed in Chapter 15.) The heat an animal generates is proportional to its volume. The heat an animal dissipates is proportional to its surface area, since heat is lost to the environment through its body surface. Consequently, animals are better able to lose heat if they have a high surface area-to-volume ratio and better able to retain heat if they are have a low surface area-to-volume ratio. This influences both the size and shape of animals that occupy extreme habitats.

Larger organisms, whether they are cells or animals, tend to have *smaller* surface area-to-volume ratios. This is because volume increases more quickly than surface area as organisms get bigger (Chapter 15). For this reason, animals found in cold habitats are often larger than related forms in warm habitats. This pattern is so common that it is called Bergmann's Rule. An example of Bergmann's Rule is seen in bears. The smallest bear in the world is the sun bear, found in the tropical forests of Southeast Asia (Figure 17.20). Adult sun bears weigh between 27 and 65 kilograms (60 to 140 pounds). The largest bear in the world is the polar bear, which ranges throughout the Arctic. Adult polar bears weigh between 200 and 800 kilograms (440 to 1760 pounds).

Figure 17.19 The amount of caramel in this caramel-covered apple is determined by its surface area. The amount of apple in this caramel-covered apple is determined by its volume. If you were the sort of person who wanted to eat a lot more caramel than apple, you would want to choose one with a *high* surface area-to-volume ratio.

Animals adapted to hot versus cold climates also vary in shape. Desert species typically have long legs and large ears that increase the surface area available for heat dissipation. These parts of the body are also covered with extensive blood vessels that carry heat from the core of the body to the skin, where convection, the transfer of heat by moving air, cools the animal. Arctic species typically have short appendages and small ears that help conserve heat. The difference in appendage size between hot- and cold-climate species is again so common that it is known as Allen's Rule. An example of Allen's Rule in rabbits is shown in Figure 17.21.

Coloration also helps animals adapt to extreme environments. Many desert species are paler than their relatives in other habitats. Their coloration

(a)

(b)

Figure 17.20 (a) The small sun bear is found in tropical forests in Southeast Asia. Small animals are better able to dissipate heat because of their large surface area-to-volume ratios. (b) The polar bear, the largest terrestrial carnivore in the world, is found throughout the Arctic. Large animals are better able to retain heat because of their small surface area-to-volume ratios.

simultaneously allows for less heat absorption and provides camouflage in the bright environment. At the other extreme of temperature, polar bear fur appears white, but is actually fairly transparent, effectively transmitting sunlight down to the animal's mottled black, heat-absorbing skin below.

Finally, cold-climate animals have insulating layers that conduct heat slowly and help retard heat loss. Both fur and feathers trap air, a very effective insulator. Many animals also have blubber, a layer of fat under the skin. In polar bears, the blubber layer can be more than 4 inches thick! Blubber is also used by marine mammals such as whales and dolphins.

(a)

(b)

Figure 17.21 (a) This is a black-tailed jackrabbit in Anza-Borrego Desert State Park, California. Extensive blood vessels in the ears help the animal dissipate heat. (b) This is an arctic hare, a relative of the black-tailed jackrabbit, on Ellesmere Island in the Canadian Arctic. Its ears are much smaller.

CHECK YOURSELF

1. If natural selection led to the evolution of light colors in many desert species, why doesn't it appear to have selected for dark colors in many Arctic species? Why are polar bears, Arctic hares, and Arctic foxes all white during the winter?
2. On cold days, people often bundle up babies and small children carefully. Are babies more likely to need the extra bundling than adults? Why or why not?

CHECK YOUR ANSWERS

1. Because camouflage against the snow has also been selected for—camouflage allows prey to avoid being seen by predators, and predators to hide from the prey they're stalking. A trait's overall contribution to an animal's survival and reproduction is what is relevant for natural selection. After all, if thermoregulation was all that mattered, polar bears could reduce their surface area-to-volume ratio further by having no legs at all, but that wouldn't be good for survival and reproduction.
2. Babies have higher surface area-to-volume ratios than adult humans because they are smaller in size. So yes, they are likely to appreciate the extra bundling.

Modes of Selection

Natural selection is sometimes classified into three different modes depending on how it affects populations. These are illustrated in Figure 17.22. In *directional selection*, natural selection favors organisms with a trait that is different from the population average. Over time, directional selection causes the population average to shift towards the favored trait. For example, a population of beetles may have an average length of 1 centimeter, and natural selection may favor organisms that are longer than that. Over time, this directional selection would cause the average length in the beetle population to increase. Directional selection

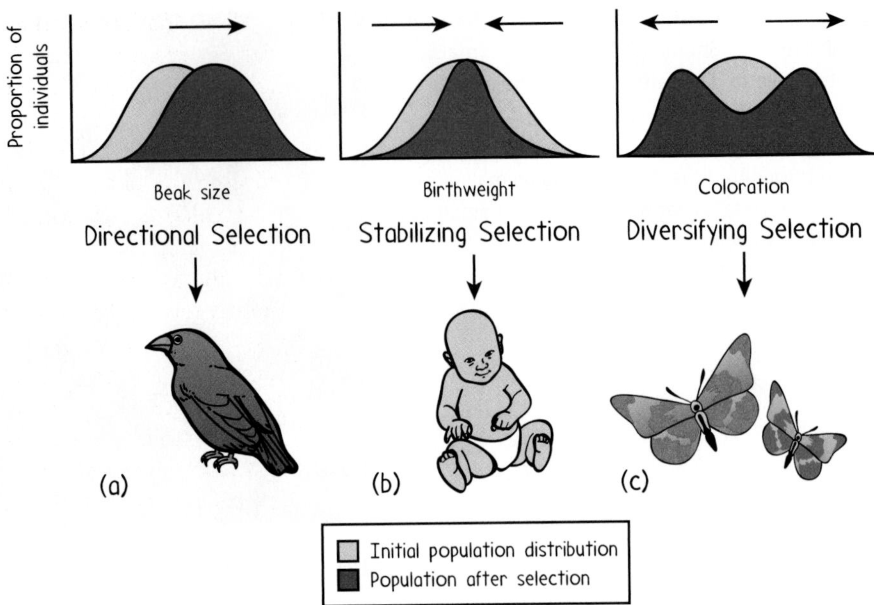

Figure 17.22 Natural selection is sometimes classified into different modes depending on the effect it has on populations. (a) Directional selection selects for organisms that differ from the population average. It causes the average trait in a population to shift. An example of directional selection is the increase in beak size in a finch population on the Galápagos Islands following a drought. (b) Stabilizing selection selects for organisms that have a trait at the population average. It causes the distribution of traits in the population to become narrower. An example of stabilizing selection is birthweight in humans. (c) Diversifying selection selects for traits at two extremes within a population. It causes the population to diverge in the trait. Diversifying selection occurs in the coloration of butterflies that mimic the appearance of two different toxic species.

occurred in a finch population on the Galápagos Islands during a severe drought in 1977. The drought made seeds scarce, with the result that small seeds were quickly eaten up, leaving only larger, tougher seeds. Birds with larger, stronger beaks were better at cracking these larger seeds and so were more likely to survive. As a result, the finch population shifted towards larger beak size.

In *stabilizing selection*, natural selection favors organisms with the average trait in the population over organisms with traits that differ from it. Over time, stabilizing selection causes the distribution of traits in the population to become narrower. Stabilizing selection in our beetle example would favor beetles with a length of 1 centimeter (the population average), causing beetle length in the population to become more tightly clustered around this favored value. An example of stabilizing selection is birthweight in human babies—babies born at average weight survive better than either heavier or lighter babies. Over time, stabilizing selection causes human babies to be born within a narrower weight range.

In *diversifying selection*, natural selection favors traits at two extremes on either side of the population average. Over time, diversifying selection causes the population to diverge into two groups, clustered around the two optimum traits. Diversifying selection in our beetle example would simultaneously favor individuals that are smaller than 1 centimeter and individuals that are larger than 1 centimeter. Over time, diversifying selection would produce a population of beetles consisting of small individuals and large individuals. There would be few medium-sized individuals. An example of diversifying selection occurs in the coloration of butterflies that mimic the appearance of two different toxic species. In these butterfly populations, natural selection favors individuals that closely resemble either of the two toxic species, but not individuals with an intermediate appearance—intermediate butterflies don't look toxic and so are eaten. Diversifying selection produces a butterfly population with many individuals that resemble one toxic species and many individuals that resemble the other toxic species, but few butterflies with an intermediate appearance.

The Modern Synthesis: A Genetic View of Natural Selection and Evolution

When Darwin published his theory of evolution by natural selection, scientists did not yet know exactly how traits were inherited. The incorporation of modern genetics (Chapter 16) into Darwin's theory of evolution took place in the middle of the twentieth century and is known as the *Modern Synthesis*.

The Modern Synthesis focuses on evolution as *changes in the allele frequencies of genes over time*. We can recast our description of natural selection in this light as well: First, there is variation in a gene when alternate alleles (Chapter 16) exist within a population. Second, a specific allele may give an organism some advantage that allows it to reproduce more than other organisms in the population. Third, as a result, more copies of the advantageous allele are passed to the next generation, causing the frequency of the allele to increase in the population.

During the Modern Synthesis, it was also recognized that, although natural selection is the driving force behind evolution and the mechanism that causes populations to become adapted to their environments, it is not the only mechanism that causes populations to evolve. Populations also change over time because of mutation pressure, genetic drift, and migration.

In order to understand these processes, let's imagine a population of lizards in which coloration is variable and heritable and determined by whether individuals carry green or brown alleles. Natural selection occurs if one allele confers greater fitness than the other. For example, if brown lizards have greater fitness than green lizards, brown lizards leave more offspring (also carrying brown alleles), and the population shifts towards a greater frequency of brown alleles.

Mutation pressure exists if the alleles responsible for color are more likely to mutate in one direction than the other. For example, a brown allele may be more likely to mutate into a green allele than vice versa. This would cause the lizard population to evolve to a higher frequency of green alleles.

Genetic drift occurs when, by chance rather than because it confers greater fitness, more alleles of one color are transmitted to the next generation than alleles of the other color. For example, even if brown and green lizards have equal fitness, green lizards might just happen to leave more offspring (and therefore more green alleles) than brown lizards one year, causing the population to evolve to a greater frequency of green alleles. Genetic drift operates in a way that is similar to a coin flip—brown and green lizards have equal fitness the way you are equally likely to get heads or tails when you flip a coin. However, if you flip a coin 100 times you won't necessarily get exactly 50 heads and exactly 50 tails. Similarly, of 100 baby lizards that hatch in the next generation, there won't necessarily be exactly 50 brown ones and 50 green ones. Genetic drift is a particularly important mechanism of evolution in small populations. This is because chance is more likely to change allele frequencies significantly in small populations. To see why this is the case, consider flipping a coin 10 times (akin to a small population) versus 1000 times (akin to a large population). With 10 flips, it is not at all unlikely that you'll get heads 60 percent of the time—that is, 6 heads and 4 tails. On the other hand, a similar result with 1000 flips—600 heads and 400 tails—would be unusual indeed.

Migration occurs if there is a net movement of brown or green alleles either into or out of the population. For example, there could be another population of lizards nearby that consists entirely of brown individuals, and migrants from that population could move to our brown-and-green population. This would cause our population to evolve towards a greater frequency of brown alleles.

CHECK YOURSELF

1. If genetic drift caused our lizard population to evolve towards a greater frequency of brown alleles one year, would it have the same effect the following year?
2. If 20 brown lizards migrated into our lizard population, and 30 green lizards migrated out, what would be the net effect of migration on allele frequencies?

CHECK YOUR ANSWERS

1. Not necessarily. Because genetic drift depends on some individuals leaving more offspring than other individuals by chance, there's no guarantee that chance would

Figure 17.23 Can you find the moths? Light peppered moths are well camouflaged on lichen-covered trees.

produce the same result the next year. It's the same as flipping coins—if you flip a coin 100 times and then repeat the process, you may get more heads the first time and more tails the second time.

2. Migration would cause the population to evolve towards a greater frequency of brown alleles.

A Case Study in Natural Selection: The Peppered Moth

During the Industrial Revolution, coal was the primary fuel in England. Burning coal slathered dark soot on trees, rocks, and ground. And then a startling thing happened to the moths. Peppered moths in England had always been light in color, with the scattering of dark peppery flecks that gave them their name. Their coloration made them hard to see against a habitat of lichen-covered trees and rocks (Figure 17.23). (Lichens are fungi that grow in close association with photosynthetic algae or bacteria—they form crustlike growths on rocks, trees, and other surfaces.) It was believed that this camouflage protected the moths from birds, their primary predators. As the Industrial Revolution progressed, pollution killed the lichens, leaving the trees first bare and then soot-darkened. In 1848, the first dark peppered moth was found in the industrial center of Manchester, England. Over the next decades, as more coal burned and the environment became increasingly sooty, more and more dark moths were seen. By 1895, 98 percent of peppered moths in industrialized areas were dark. Dark moths remained the prevalent form until the second half of the twentieth century, when antipollution laws took effect and soot disappeared. Soon thereafter, light peppered moths increased in number, and today, the dark moths have all but disappeared.

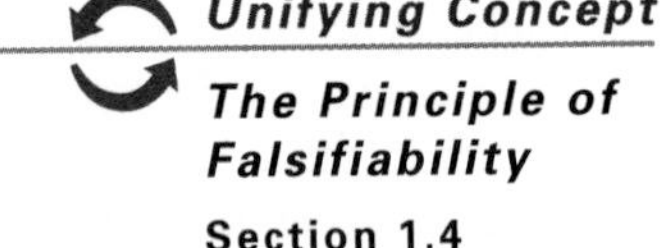

Unifying Concept
The Principle of Falsifiability
Section 1.4

What happened in the peppered moth population? Biologists hypothesized that natural selection caused the shifts in the coloration of the peppered moth. In lichen-covered habitats, light moths were better camouflaged and so were selected for. In sooty habitats, dark moths were selected for. To test the hypothesis that the shift in coloration was caused by natural selection, biologist Bernard Kettlewell performed a series of experiments showing that camouflage and bird predation were in fact crucial pieces to the peppered moth puzzle.

Kettlewell started by releasing equal numbers of marked dark and light moths in polluted and unpolluted areas. After waiting for a set time, he started recapturing marked moths. In polluted habitats, he recaptured many more dark moths than light moths, which suggested that dark moths survived better. The opposite was true in unpolluted habitats, where he recaptured more light moths. Kettlewell also tested the bird predation hypothesis directly by placing differently colored moths on tree trunks and using movie cameras to film birds eating the moths. Kettlewell found that birds ate what they could see, eating more light moths in polluted habitats and more dark moths in unpolluted habitats.

In recent years, certain aspects of Kettlewell's work have come under question. For example, moth experts point out that peppered moths don't usually sit on tree trunks, where Kettlewell placed them, instead preferring the more hidden undersides of branches. Kettlewell's mark-and-recapture experiments have also been criticized because the normally nocturnal moths were released during the day, which may have affected their ability to find resting spots. In addition, Kettlewell used a mix of lab-raised and wild-caught moths, which may differ in their behavior. Despite these flaws, most biologists believe that Kettlewell's conclusions are unlikely to be overturned. For one thing, a shift from light to dark forms in polluted areas (and back again, as pollution is cleaned up) has been reported in over 70 other moth species in England and the United States alone. Kettlewell's experiments are now being repeated, without the flaws, by other scientists. Stay tuned for the results.

SCIENCE AND SOCIETY

Antibiotic-Resistant Bacteria

A patient is ill with pneumonia and gets a prescription for penicillin. After 3 days, he feels much better and stops taking his pills. A few days later, his symptoms return. He quickly finds his pills and starts taking them again, but this time they have no effect. What happened? This frightening phenomenon is called *antibiotic resistance*. Antibiotic resistance is caused by natural selection. Penicillin kills most of the pneumonia bacteria, but a few penicillin-resistant bacteria survive. These bacteria then multiply, and eventually the patient's infection comes back—only this time, the bacteria are resistant to penicillin.

When penicillin, the first antibiotic, became widely available, it was rightfully lauded as a wonder drug and dramatically cut the number of illnesses and deaths due to bacterial infections. After only a decade of use, however, the first penicillin-resistant bacterial strains appeared. Antibiotic resistance has only increased since then, with more and more bacterial populations becoming resistant to more and more different antibiotics. Diseases once easy to treat—tuberculosis, pneumonia, even common childhood ailments such as ear infections—are now often resistant to multiple antibiotics. And in some hospitals, there are infectious bacteria that are resistant to every antibiotic on the market.

Because it results from natural selection, antibiotic resistance is inevitable—*all* antibiotic use contributes to resistance. However, resistance has been greatly accelerated by the overuse of antibiotics. Under pressure from patients, physicians often prescribe antibiotics for illnesses that are not caused by bacteria—many common illnesses, such as colds, flus, and most sore throats, for example, are caused by viruses. These antibiotics select for resistance in the normal (non-disease-causing) bacterial populations in our bodies, making it possible for resistance genes to be transferred to disease-causing bacteria that later invade the body. The fact that patients sometimes stop taking their medications too soon only exacerbates the problem by selecting for antibiotic-resistant strains without providing the sustained dose that would actually kill all the bacteria. Antibiotics also see heavy usage in the livestock industry, where animals are sometimes given antibiotics regularly, even when they are healthy, in an attempt to prevent illness. Unfortunately, this practice only increases antibiotic resistance, a fact highlighted in 1983 by the development of antibiotic-resistant *Salmonella* poisoning in 18 people who ate meat from cows raised on antibiotics.

So, what can be done about antibiotic resistance? First, humans must learn to use antibiotics wisely, taking them only when they are needed—that is, for bacterial infections—and then taking the entire course of treatment. Physicians and veterinarians can also promote a socially responsible approach to antibiotics by educating patients and agriculturalists on the proper application of these drugs. Finally, since many antibiotics are less effective now because of resistance, scientists must continue to search for new antibiotics that will take the place of those that no longer do the job.

CHECK YOURSELF

1. Although dark moths were prevalent in polluted industrial centers, the countryside has always been inhabited primarily by light peppered moths. Is this surprising?
2. Is the evolution of peppered moth coloration an example of directional, stabilizing, or diversifying selection?
3. You release 100 marked light moths and 100 marked dark moths in a polluted industrial center. Do you expect to recapture more light or dark moths?

CHECK YOUR ANSWERS

1. This is what we expect—light moths are selected for in environments that are not polluted.
2. Directional selection—moths shifted towards dark coloration when habitats were polluted and then shifted towards light coloration when habitats were cleaned up following the passage of pollution-control laws.
3. You expect the dark moths to survive better in a polluted environment. So, you expect to recapture more dark moths.

Conceptual Science Place

Polyploid Plants

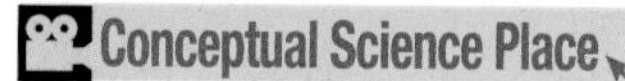

Albatross Courtship;
Booby Courtship

17.6 How Species Form

We have seen how evolution through natural selection allows populations to become adapted to their environments. But how does evolution produce biological diversity—that is, how has it produced all the different species that exist in the world? Where do new species come from?

Figure 17.24 A prezygotic reproductive barrier: Differing courtship rituals prevent individuals of different species from mating. In these red-crowned cranes, courtship involves an elaborate display—birds dance around each other, bob their heads, stretch their necks, extend their wings, and leap straight into the air, singing in unison. Unless you can do all that just right, you have little hope of convincing a red-crowned crane to mate with you.

We can begin by considering what species are. A **species** is a group of organisms whose members can interbreed among themselves but not with members of other species.* This means that the key to **speciation**—the formation of new species—is the evolution of *reproductive barriers* that prevent two groups of organisms from interbreeding.

There are two kinds of reproductive barriers, prezygotic reproductive barriers and postzygotic reproductive barriers. **Prezygotic reproductive barriers** prevent members of different species from mating in the first place or prevent fertilization from occurring if they do mate. (A *zygote* is a fertilized egg, so *prezygotic* means "before fertilization.") There are many types of prezygotic barriers—organisms may differ in when they breed, where they breed, or in the details of their courtship rituals. Their sex organs may not fit together properly, preventing successful sperm transfer, or other factors may prevent fertilization even if sperm is transferred. Figures 17.24 and 17.25 show examples of prezygotic reproductive barriers. **Postzygotic reproductive barriers** act after fertilization has taken place. Postzygotic barriers occur when mating produces hybrids that either don't survive or are sterile—unable to breed themselves. The mule, the offspring of a horse and a donkey, is sterile and cannot reproduce. Likewise, a liger (Figure 17.26), the product of the mating of a lion and a tiger, is sterile.

Now let's consider how reproductive barriers—and therefore new species—evolve. In **allopatric speciation**, shown in Figure 17.27, new species are formed after a geographic barrier divides a single population into two. The two populations,

Figure 17.25 A prezygotic reproductive barrier: The flowers of this and a related primrose species open at different times of day. In this species, flowers open in the late afternoon. In the other species, flowers open in the morning. So, even though the two species occupy the same western North American deserts and both are pollinated by bees, hybrids are rarely produced.

Figure 17.26 A postzygotic reproductive barrier: The liger, a lion-tiger hybrid, is sterile.

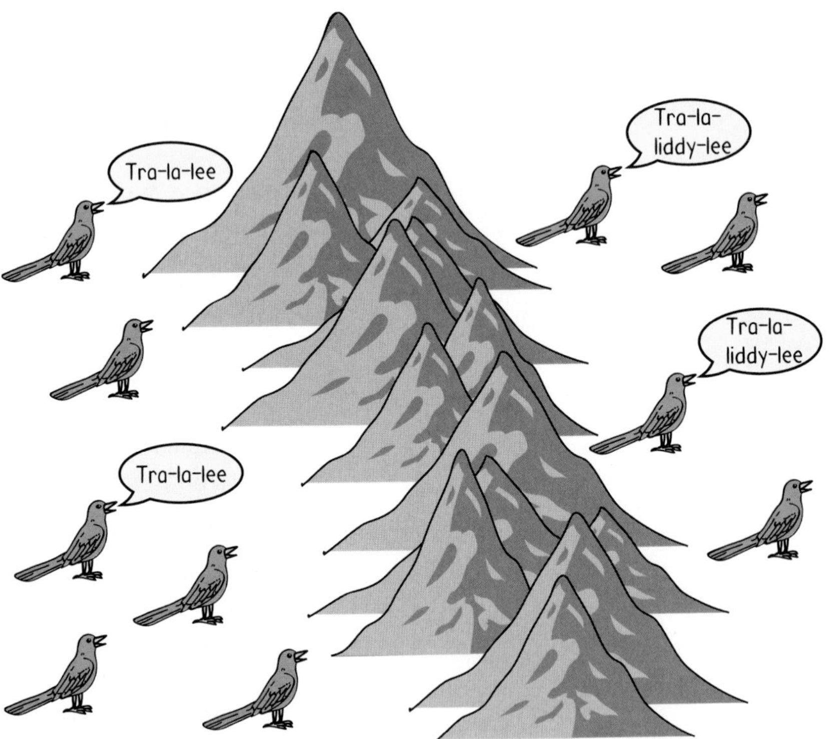

Figure 17.27 Geographic barriers isolate populations and allow them to evolve independently. Sometimes, a reproductive barrier will evolve, producing allopatric speciation. In this example, the courtship song of birds divided by a mountain range diverges, resulting in a prezygotic reproductive barrier.

* This definition of species applies to organisms that reproduce sexually. It doesn't work very well for bacteria and other asexually reproducing organisms. For these groups, species are generally recognized by their similar characteristics and ways of life.

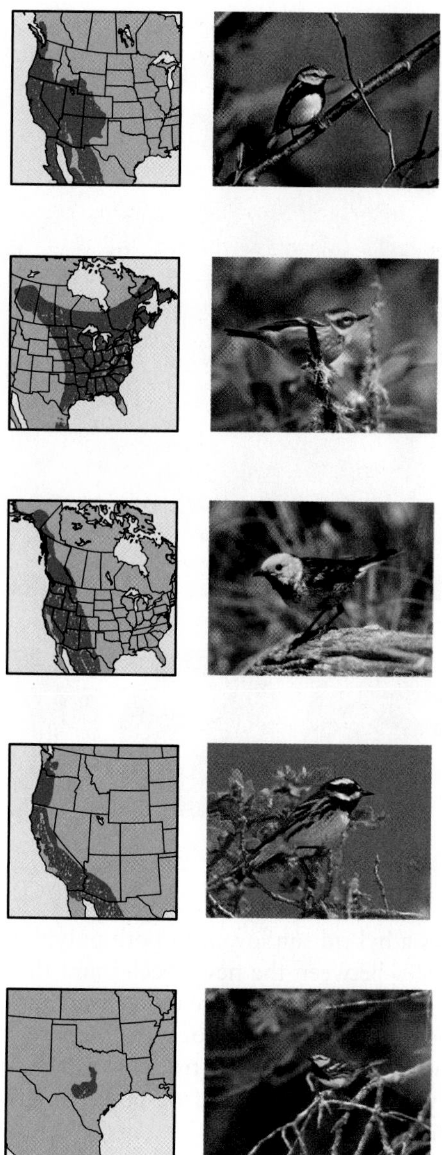

Figure 17.28 Allopatric speciation: Advancing glaciers isolated warbler populations during the Pleistocene Ice Age, causing the evolution of five unique warbler species. The maps show the current breeding ranges of the species.

now that they are isolated from one another, begin to evolve independently, accumulating differences due to natural selection (which is likely to differ in the two environments) as well as genetic drift. Over time, the populations will have evolved key differences that prevent organisms from interbreeding. What type of geographic barrier is sufficient for allopatric speciation to occur? That depends on the organism—a mountain range imposes a great barrier for a tortoise, but not necessarily for an eagle. Geographic barriers may include phenomena such as mountain ranges, rivers, glaciers, oceans, or, in the case of aquatic organisms, land.

Numerous instances of allopatric speciation have been recorded. A series of glacial advances during the Pleistocene Ice Age, 700,000 to 1 million years ago, isolated multiple populations of North American warblers, ultimately producing five distinct species (Figure 17.28). Similarly, the rise of the Isthmus of Panama, 3 million years ago, divided the Caribbean Sea from the Pacific Ocean, splitting hundreds of types of marine organisms into separate Caribbean and Pacific populations (Figure 17.29). Most of these subsequently speciated by evolving reproductive barriers. For example, similar-looking snapping shrimps from the Caribbean and Pacific will not mate when put together. Instead, they snap aggressively at each other.

Island archipelagos, with abundant opportunities for geographic isolation, are the sites of some of the most spectacular examples of speciation, called adaptive radiations. An **adaptive radiation** is the evolution of a large number of new species, each adapted to a distinct way of life, from a single ancestor. Adaptive radiations are most often seen after a few members of a species colonize a new habitat. One example is the 13 species of Darwin's finches on the Galápagos Islands, all descended from a single South American species. Several impressive adaptive radiations also characterize the Hawaiian Archipelago, including the Hawaiian honeycreepers and Hawaiian fruitflies (Figure 17.30).

In **sympatric speciation**, speciation occurs without the introduction of a geographic barrier. Sympatric speciation is much less common than allopatric speciation. In plants, sympatric speciation is often the result of sudden chromosomal changes. One such chromosomal change is *polyploidy*, which occurs when organisms inherit more than the usual two sets of chromosomes, usually as a result of improper meiosis (Chapter 16). In Figure 17.31a, we see that polyploidy has produced a new species of anemone with four, rather than two, copies of each chromosome. Another instance of sympatric speciation through chromosomal change is *hybridization*, which occurs when two species interbreed and produce

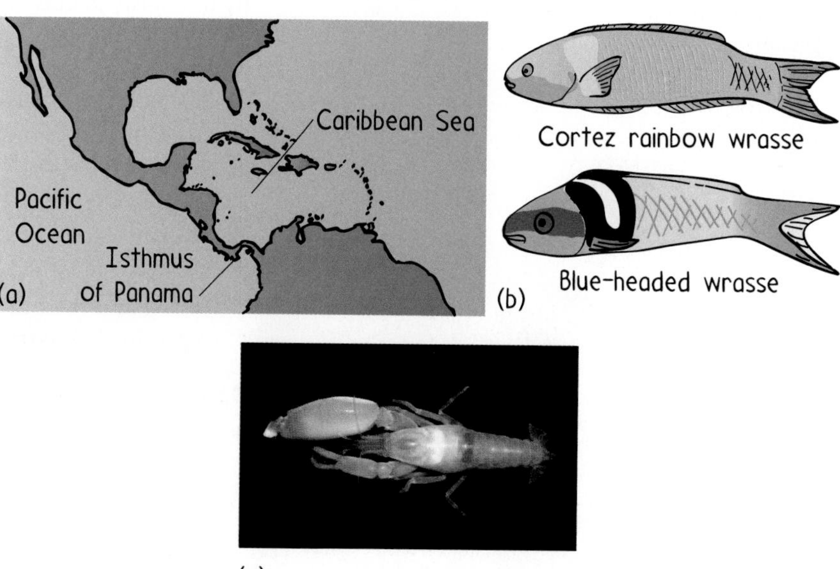

Figure 17.29 (a) The formation of the Isthmus of Panama 3 million years ago isolated Pacific and Caribbean marine populations, producing numerous instances of allopatric speciation. (b) The blue-headed wrasse (Caribbean) and Cortez rainbow wrasse (Pacific) are descended from a single ancestral species that formerly spanned Pacific and Caribbean waters. (c) Caribbean and Pacific snapping shrimps are now distinct species that do not interbreed. This specimen is from the Pacific.

(a)

(b)

Figure 17.30 The Hawaiian Archipelago is characterized by multiple adaptive radiations. (a) The Hawaiian honeycreepers represent an adaptive radiation consisting of more than 30 species that differ in plumage, beak shape and size, and diet. Many honeycreepers are already extinct or are currently endangered because of habitat destruction and the introduction of nonnative species such as rats, pigs, mongooses, and cats. Nonnative mosquitoes are particularly damaging to honeycreeper populations because they carry avian malaria.(b) Hawaiian fruitflies have undergone an impressive radiation—these are only a few of the more than 500 species! Lava flows on the archipelago repeatedly created geographically isolated vegetation "islands"that contributed to speciation in the group.

fertile offspring. Figure 17.31b shows a hybrid sunflower. In both polyploidy and hybridization, chromosomal differences between the new species and the parent species prevent interbreeding. These types of speciation are much more rare in animals than in plants but are not unheard of. An example of polyploidy in animals can be seen in the common gray treefrog and Cope's gray treefrog. These amphibians differ in chromosome number, with Cope's gray treefrog possessing two sets of chromosomes and the common gray treefrog four sets. The two species cannot interbreed.

In Chapter 18, we will examine the diversity of life produced by speciation and evolution, and we will also see how the history of speciation events is used by biologists to classify the world's living organisms.

INSIGHT Humans may vary in significant ways from one part of the world to another, but we all belong to the same species—all humans are able to interbreed!

CHECK YOURSELF

1. A small river forms, dividing a group of moles into two isolated populations. After many years, a biologist puts moles from opposite sides of the river together and finds that they will not mate. Has speciation occurred? If so, what type of speciation is it?
2. Do you think the same river would cause birds to speciate?
3. Two species of frogs do not interbreed because one species breeds in the spring and the other breeds in the fall. Is this a prezygotic or postzygotic reproductive barrier?

CHECK YOUR ANSWERS

1. Yes, the moles on the two sides of the river now represent two different species because they don't interbreed. This is allopatric speciation, because it occurred after a geographic barrier (the river) separated the populations.
2. Probably not, since a river is not much of a geographic barrier for flying organisms.
3. Prezygotic—it prevents mating.

Figure 17.31 Sympatric speciation in plants frequently occurs through abrupt chromosomal changes. (a) This photo shows a new species of anemone produced through polyploidy. Note the doubling of chromosomes. (b) The sunflower species *Helianthus anomalus* originated through the hybridization of two other sunflower species.

17.7 Evidence of Evolution

Reconstructing Forelimbs

The description of evolution as a *theory* may lead some people to believe that evolution is controversial within the scientific community. It is not. All scientific theories—sets of ideas framed to explain a group of natural phenomena—win general acceptance when their predictions are borne out by what is observed in nature (Chapter 1). The theory of evolution is a theory in the way that Sir Isaac Newton's law of universal gravitation is a theory and that Albert Einstein's theory of relativity is a theory. That is, the theory of evolution has been tested repeatedly against observations of the natural world, and the evidence for evolution has met the most rigorous standards of the scientific community.

Many lines of evidence support the argument that Earth's diversity of life is the result of evolution. Populations evolve on human time scales when we see natural selection in action or when organisms change as a result of human-imposed artificial selection. Evolution is also evident in living creatures themselves, in the form of shared anatomy, shared patterns of development, and shared DNA sequences. Finally, the evidence in the rocks—the evidence of fossils and biogeography—argues for the evolution of organisms.

There is plentiful evidence of natural selection in action. This includes some of the examples we have discussed—coloration evolution in peppered moths, resistance to myxoma virus in Australian rabbits, beak size increase in a Galápagos

Figure 17.32 Artificial selection has produced great diversity in dogs.

finch, and antibiotic resistance in bacteria. All these populations underwent distinct, measurable evolutionary changes in response to natural selection.

The evolution of organisms is also visible during human-imposed artificial selection. **Artificial selection** is the selective breeding of organisms with desirable traits in order to produce offspring with the same traits. Humans artificially select for desirable traits in domesticated animals and crops all the time—we breed fast racehorses to try to get faster racehorses; different types of dogs to produce superior hunters, herders, or sled-pullers (Figure 17.32); and varieties of strawberries to grow the largest and sweetest fruit. Through artificial selection, humans have brought about dramatic evolutionary changes in these organisms. Think how much a Chihuahua differs from the animal it is descended from, the wolf. Or look at Figure 17.33 to see the difference between the corn we eat today and teosinte, the plant from which corn was bred. In fact, artificial selection has produced countless forms of domestic animals and crops, all with traits valued by humans.

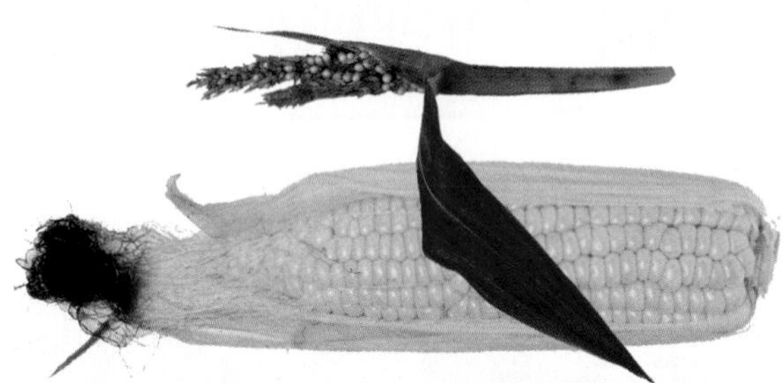

Figure 17.33 Corn, now one of the most important agricultural crops in the world, was laboriously bred through artificial selection from teosinte, which has tiny cobs, only a few rows of kernels, and inedible hard coverings on its seeds. Average cob size increased only slowly over time, from 2 cm in 5000 B.C. to 4.3 cm in 3000 B.C. to 13 cm by 1500 A.D.

In many cases, the evolutionary histories of species are preserved in the structures of their bodies. For example, some snakes actually retain tiny, incomplete hind legs, evidence of their evolution from legged vertebrates. We can also see evidence of shared ancestry whenever we compare the bodies of related species. Consider, for example, the limbs of mammals, some of which are shown in Figure 17.34. Limbs are used by different mammals for different purposes—humans walk, bats fly, whales swim, and moles dig, just to name a few. If each of these mammals had originated independently, we would expect their limbs to look completely different. Yet, in fact, all mammalian limbs resemble each other and are made up of the same set of bones. This suggests that mammals inherited their limbs from a common ancestor and then modified them for different purposes.

As with body structures, the macromolecules of organisms retain evidence of their shared evolutionary history. For example, the DNA of related species have similar ACGT sequences. This is true not only for sequences in the DNA molecule that tell cells how to build proteins but even for sequences that have no obvious function (Chapter 16). This sequence similarity, and the fact that DNA sequences tend to be more similar in more closely related species, is logically explained by shared evolutionary history. Similarly, almost all organisms on Earth share the same genetic code, the rules for translating codons into amino acids during the construction of proteins (Chapter 16). The universality of the genetic code suggests that it arose early in the evolution of life and was then passed on to all living species.

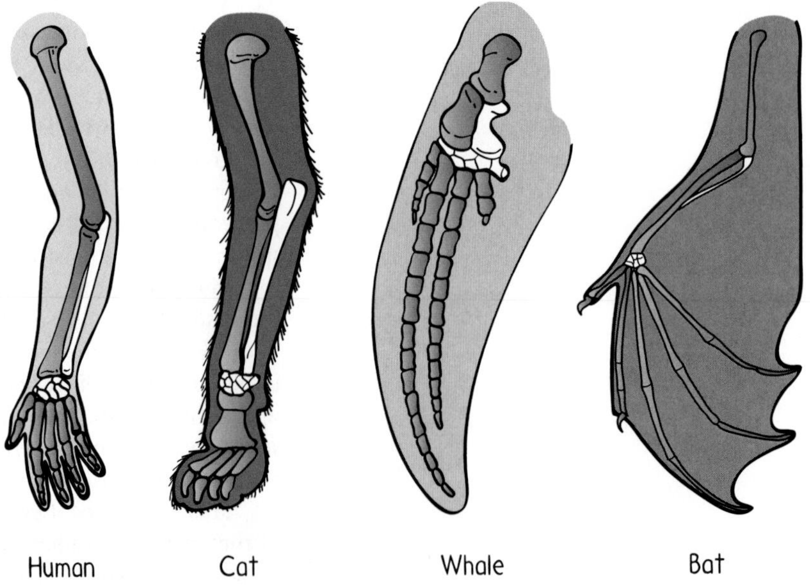

Figure 17.34 Despite the fact that these vertebrate limbs are used for different activities, they are composed of the same set of bones, evidence that they were inherited from a common ancestor. (Courtesy of Campbell, Reece, Simon, *Essential Biology with Physiology*, © 2004.)

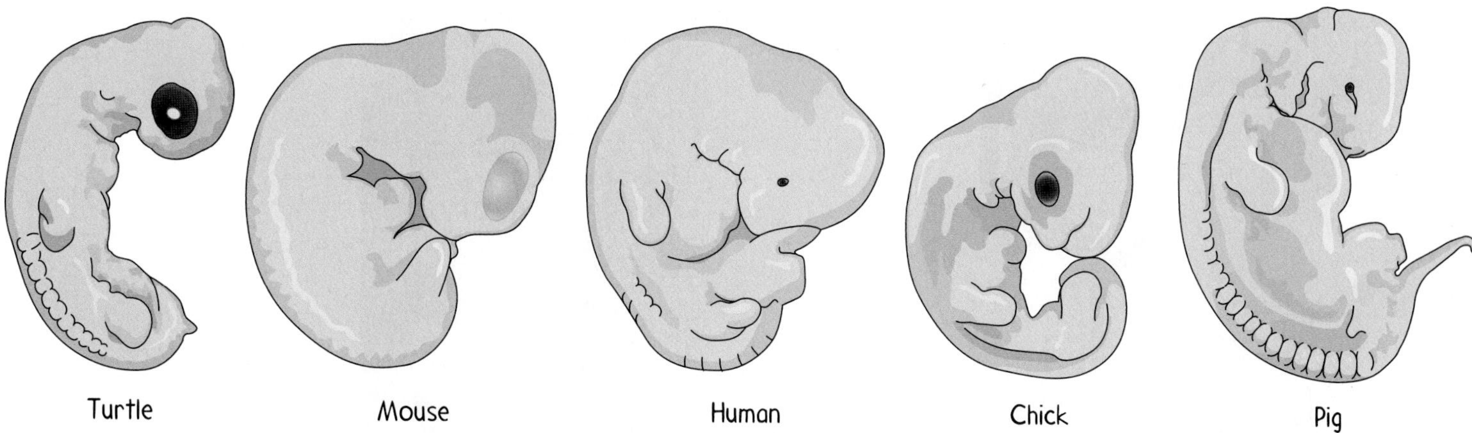

Figure 17.35 Related species go through similar stages in their development. The embryos above are of different vertebrates: a turtle, a mouse, a human, a chick, and a pig. The human embryo goes through a tailed stage just like the other vertebrates, even though humans don't have tails.

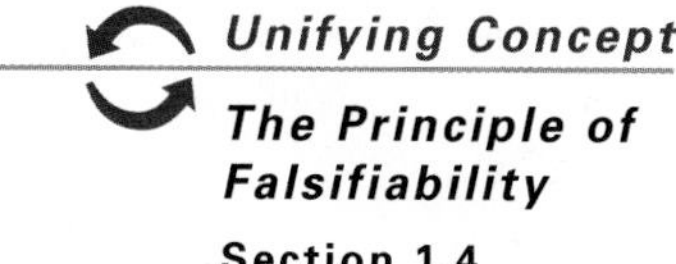

Unifying Concept
The Principle of Falsifiability
Section 1.4

Related species also develop in similar ways. For example, human embryological development resembles that of other vertebrates in that we go through a stage where we have gill slits and a tail (see Figure 17.35), even though these structures aren't actually present in humans. If each species on Earth had originated independently, we wouldn't expect these similarities in development.

INTEGRATED SCIENCE EARTH SCIENCE

Earth's Tangible Evidence of Evolution

Evolution has also left a record in Earth's rocks in the form of fossils. Because we can date fossils from the age of the rock formations they belong to, we can follow the evolution of organisms over time. For example, fossils of now-extinct relatives of the horse (Figure 17.36) show that species grew larger in size over time as well as more specialized for eating grass and running. Some fossil whales exhibit some of the characteristics of the hoofed animals they evolved from—hind limbs, nostrils on their noses rather than blowholes, and different types of teeth (incisors, molars, etc.) rather than rows of uniform teeth. *Archaeopteryx*, the famous 150-million-year-old fossil bird shown in Figure 17.37, has many birdlike features—feathers, wings, a wishbone—but also has dinosaur-like features absent in modern birds, including claws on its wings, bones in its tail, and teeth.

Biogeography, the study of how species are distributed on Earth, is consistent with evolution rather than with the idea that organisms were purposefully distributed around the planet. For example, the argument that each organism was specially designed to fit into its habitat is undermined by the observation that similar habitats are often home to completely different species. New World tropical forests and Old World tropical forests are occupied by entirely different life-forms, as are the similar environments of the Arctic and Antarctic (Figure 17.38). In addition, closely related species tend to be found close together. All of Darwin's finches are found in or near the Galápagos, and all the honeycreepers in Hawaii. Most of the world's marsupials (pouched mammals, such as koalas and kangaroos) are found in Australia. Island species are most closely related to species found on the closest mainland. This pattern holds for fossils as well—fossil armadillos are found only in the New World, where modern armadillos also occur, and fossil apes are found only in Asia and Africa, where modern apes reside. Islands tend to be occupied by many flying animals,

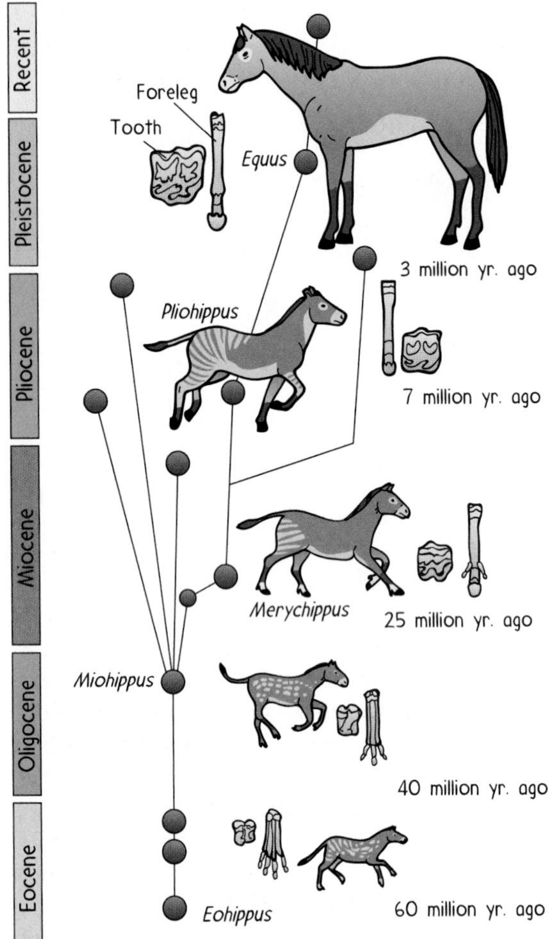

Figure 17.36 Many early relatives of horses have been preserved as fossils. Over time, species grew larger in size, their legs became specialized for running (the lower leg bones fused, and the number of toes was reduced), and their teeth became specialized for grinding tough grasses rather than browsing on soft leaves.

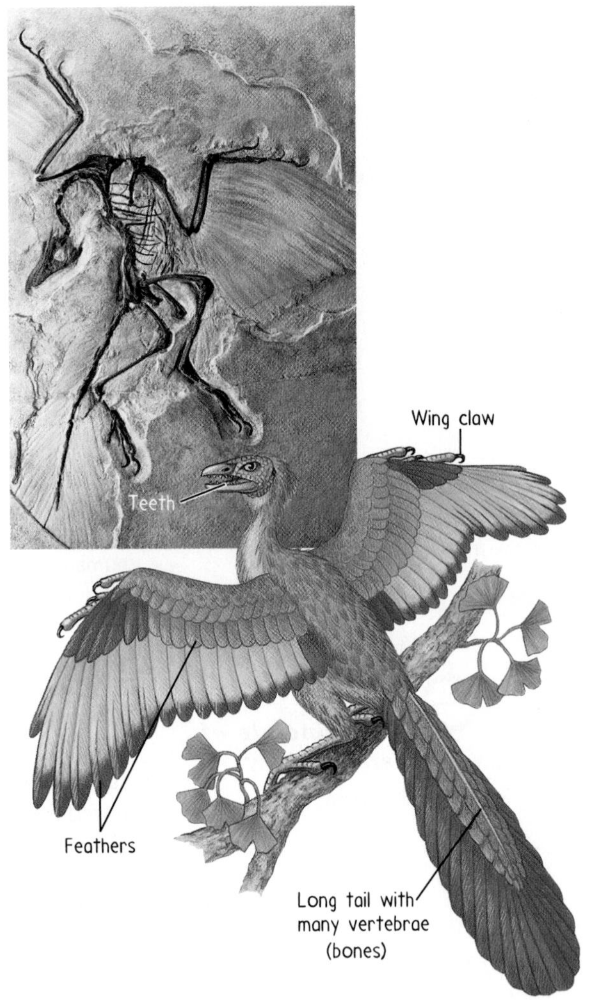

Figure 17.37 *Archaeopteryx*, an early bird, has features of both the dinosaurs it evolved from and modern birds. (Courtesy of Campbell, Reece, Simon, *Essential Biology with Physiology*, © 2004.)

Figure 17.38 The Arctic and Antarctic, which have similar habitats, are occupied by very different species. Polar bears are found in the Arctic but not the Antarctic, penguins in the Antarctic but not the Arctic.

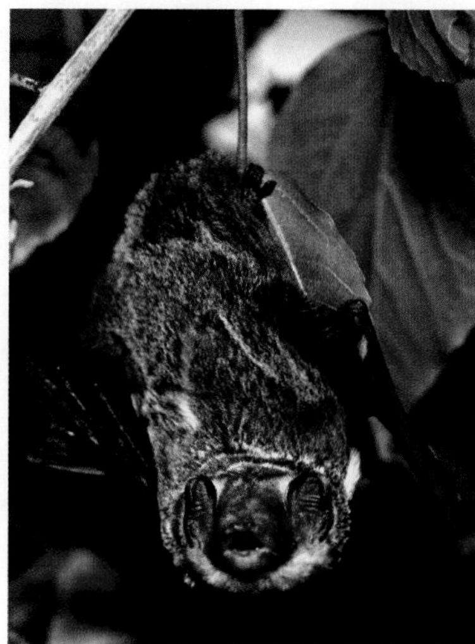

Figure 17.39 Why are terrestrial vertebrates rare or absent from isolated islands, whereas flying species are common? This is the Hawaiian hoary bat, the only mammal found on Hawaii prior to human colonization of the islands.

but few or no terrestrial ones (Figure 17.39). All these biogeographical patterns suggest that organisms were not dispersed purposefully, but instead evolved in a certain place and then spread and left descendants where they could.

CHECK YOURSELF

Why is the fact that many species found on islands resemble species found on nearby mainlands evidence for evolution?

CHECK YOUR ANSWER

It suggests that island species evolved from mainland species, rather than that species were distributed purposefully around the Earth.

17.8 Does Evolution Occur Gradually or in Spurts?

In 1977, biologists Stephen Jay Gould and Niles Eldredge proposed an audacious idea—evolution does not occur gradually, but in spurts. They cited the fossil record, which frequently shows species appearing suddenly, not changing much for a long time, and then disappearing. Scientists had always attributed this observation to the "incompleteness" of the fossil record—that is, because only a tiny fraction of living organisms are fossilized, the probability of seeing a series of gradually changing intermediate forms is slim.

Gould and Eldredge argued that the fossil record should be accepted somewhere closer to face value. Their **punctuated equilibrium** theory hypothesizes that species maintain *equilibrium*, or stability, for long periods of time. These periods of equilibrium, in which species do not change very much, are *punctuated* by bouts of rapid change. Moreover, Gould and Eldredge asserted, the rapid change occurs during speciation. The difference between a punctuated equilibrium model of evolution and a more gradual model is illustrated in Figure 17.40.

Figure 17.40 This diagram illustrates alternate ideas on the pace of evolutionary change. (a) Under a model of gradual evolution, coloration in these butterflies changes slowly over time. (b) Under a model of punctuated equilibrium, coloration changes quickly, in a spurt during speciation. (Courtesy of Campbell, Reece, Simon, *Essential Biology with Physiology*, © 2004.)

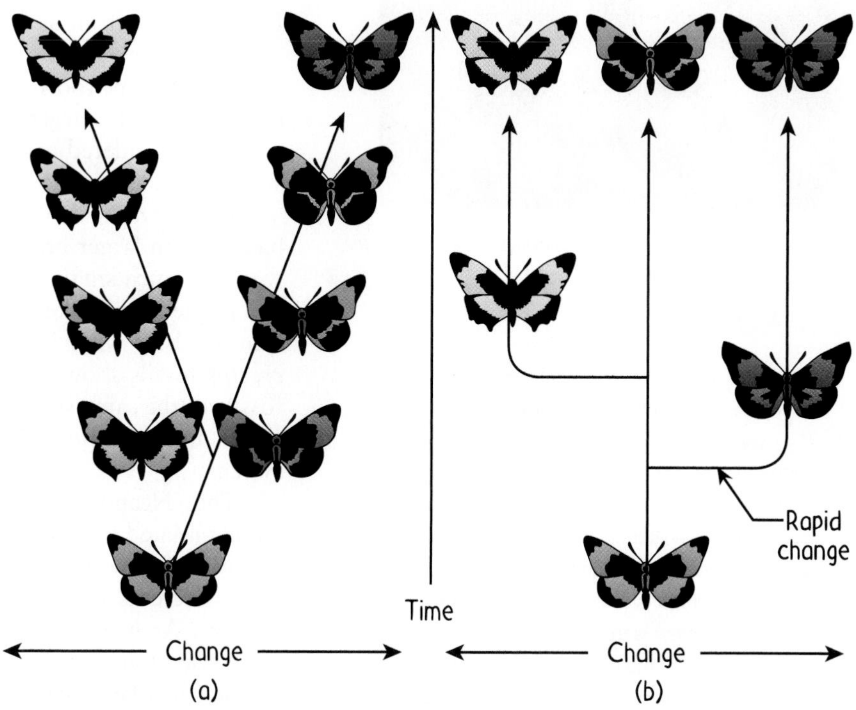

Figure 17.41 This timeline shows when certain hominid species existed on Earth. The skulls are all drawn to the same scale to show relative brain sizes. (Courtesy of Campbell, Reece, Simon, *Essential Biology with Physiology*, © 2004.)

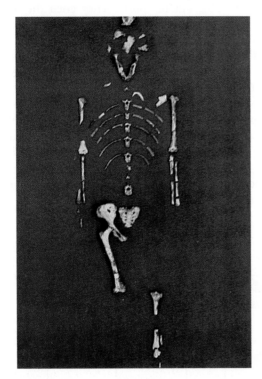

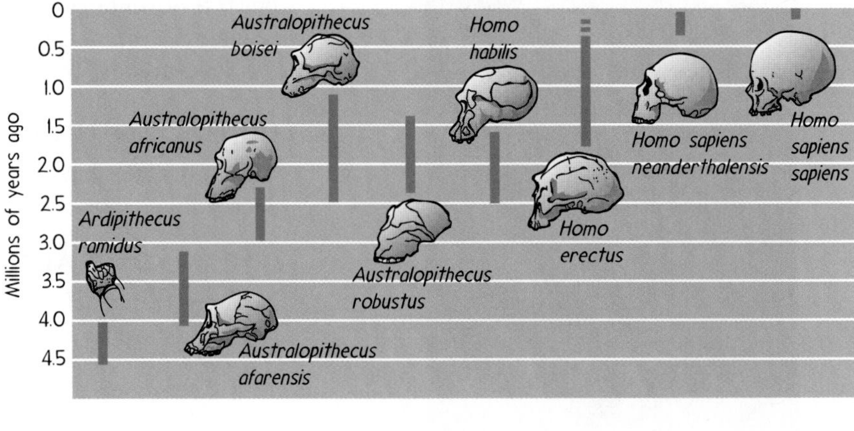

17.9 The Evolution of Humans

Humans are *primates*, a group of mammals that also includes the monkeys and apes. This does not mean we are descended from any modern species of monkey or ape, just that we share a common more recent ancestor with these species than we do with a dog, or frog, or plant. Humans are also *hominids*, the group within the primates that includes modern *Homo sapiens* as well as some of our extinct relatives. Although humans are the only hominids in existence today, fossil hominids provide clues as to how humans evolved. A timeline of human evolution is shown in Figure 17.41.

Some of the earliest hominids known belong to the genus *Australopithecus*. "Lucy," the famous *Australopithecus afarensis* fossil shown in Figure 17.42, dates from 3.2 million years ago. When she was alive, Lucy stood 3 feet 8 inches tall and had a brain about the size of a chimpanzee's. However, the bones of Lucy's pelvis make it clear that she was bipedal, walking upright on two legs. Older *Australopithecus* fossils show that an upright posture dates to at least 4 million years ago and therefore evolved well before increased brain size and intelligence.

Homo habilis is the earliest known species that belongs to the group *Homo*, which includes the species most closely related to modern humans. Fossils date *Homo habilis* from as far back as 2.4 million years. Brain size was significantly larger in *Homo habilis* than in *Australopithecus*. In addition, *Homo habilis* was a maker of stone tools—its scientific name means "handy man." Male *Homo habilis* were much larger than females. In other mammals, such as the elephant seals we considered earlier, a large size difference between the sexes is usually a sign that males battle each other to obtain large numbers of mates.

Homo erectus lived from about 2 million years ago to 400,000 years ago and had an even larger brain than *Homo habilis*. In fact, the brain of *Homo erectus* was not much smaller than that of modern humans. *Homo erectus* was an accomplished toolmaker as well as the first species to migrate out of Africa and into much of what is now Europe and Asia. Like *Homo habilis*, older *Homo erectus* fossils show that males were much larger than females. However, later fossils of the same species show a male-female size difference closer to that present in modern humans, suggesting the development of a more humanlike social system.

The Neanderthals—*Homo sapiens neanderthalensis*—are a subspecies closely related to modern humans. They lived from about 200,000 years ago to about 30,000 years ago. Neanderthals had very thick arms and legs, and their brains were as large as those of modern humans. Neanderthals were effective hunters. Archaeological finds show that they had complex burial rituals and made use of medicinal plants. However, we still do not know whether the Neanderthals had language.

Figure 17.42 "Lucy," a fossil *Australopithecus afarensis*, stood upright and walked on two feet.

INSIGHT In one of the most spectacular archaeological finds in centuries, skeletons of a tiny hominid were discovered on a remote Indonesian island in 2004. Nicknamed "hobbits," *Homo floresiensis* adults had skulls the size of grapefruits and were no bigger than three-year old modern children. *Homo floresiensis* lived alongside pygmy elephants, giant rodents, and Komodo dragons. Not all scientists agree that *Homo florensiensis* is truly a species distinct from humans, however. Some argue that the fossils represent only an anatomically unusual human population.

The earliest fossils of modern humans, *Homo sapiens sapiens*, were found in Ethiopia and are 195,000 years old. Although anatomically modern humans are therefore quite old, the cultural traits we associate with humans—things like art, music, and religion—are a more recent phenomenon, appearing only about 50,000 years ago. The reason for this gap between modern anatomy and modern behavior is the subject of continued debate.

CHECK YOURSELF

1. Have multiple species of hominids ever coexisted on Earth? Do any hominids other than humans survive to this day?
2. What is the significance of the transition from a large male-female size difference in early *Homo erectus* fossils to a size difference closer to that of modern humans in later fossils of the same species?

CHECK YOUR ANSWERS

1. Yes, the timeline of hominid evolution shows that multiple species of hominids coexisted during much of hominid history. Today, however, humans are the only species of hominids in existence.
2. A large size difference between the males and females of a species is often a sign that males battle each other to control large numbers of female mates. This may have been the mating pattern in early *Homo erectus*. The transition to more equivalent body sizes may be evidence that *Homo erectus* had moved to a mating pattern involving longer-term bonds between males and females, perhaps as they raised offspring together.

CHAPTER 17 Review

Summary of Terms

Adaptation An evolved trait that makes organisms more suited to living and reproducing in their environments.

Adaptive radiation The evolution of multiple species, each adapted to a distinct way of life, from a single ancestor.

Allopatric speciation Speciation that occurs after a geographic barrier divides a group of organisms into two isolated populations.

Artificial selection The selective breeding of organisms with desirable traits in order to produce offspring with the same traits.

Autotrophs Living organisms that make their own food and organic materials.

Chemoautotrophs Autotrophs that use energy from inorganic chemicals to make their food and organic materials.

Endosymbiotic theory The theory that the mitochondria and chloroplasts of eukaryotic cells evolved from prokaryotes living inside early eukaryotic cells.

Evolution Heritable changes in living organisms over time—or, as Darwin put it, "descent with modification."

Fitness The number of offspring an organism produces in its lifetime compared to other organisms in the population.

Heritable Traits that are passed from parents to offspring because they are at least partially determined by genes.

Heterotrophs Living organisms that obtain their energy and organic materials from other living organisms or other outside sources.

Natural selection Organisms with heritable, advantageous traits leave more offspring than organisms with other traits, causing advantageous traits to become more common in populations over time.

Postzygotic reproductive barrier A barrier that prevents members of different species from successfully reproducing because the hybrids produced are either unable to survive or sterile.

Prezygotic reproductive barrier A barrier that prevents members of different species from mating in the first place or that keeps fertilization from occurring if they do mate.

Punctuated equilibrium The theory that species do not change very much over long periods of time and then change a lot suddenly, during speciation.

Speciation The formation of new species.

Species A group of organisms whose members can interbreed among themselves, but not with members of other species.

Sympatric speciation Speciation that occurs without the introduction of a geographic barrier.

Variation Differences in a trait from one individual to another.

Review Questions

17.1 The Origin of Life

1. How did Pasteur disprove the idea of spontaneous generation?
2. What experiment did Miller and Urey perform? What were their results?
3. How do liposomes resemble real cells?
4. Why is RNA, rather than DNA, believed to be the first genetic material?

17.2 Early Life on Earth

5. How did the earliest autotrophs make their own food and organic materials?
6. What fundamental change in Earth's environment is attributed to the cyanobacteria?

17.3 The First Eukaryotic Cells

7. How did the mitochondria and chloroplasts originate?
8. Did the nucleus and other organelles come from endosymbiosis as well?

17.4 Charles Darwin and *The Origin of Species*

9. What was Lamarck's theory on how evolutionary change occurred?
10. How did Charles Lyell's work influence Darwin?
11. How did the work of Thomas Malthus influence Darwin?
12. What impressed Darwin about the finches on the Galápagos Islands?

17.5 How Natural Selection Works

13. What is variation?
14. What are heritable traits?
15. Describe how natural selection occurs.
16. What types of traits might be "selected for" in a population—that is, might allow some organisms to leave more offspring than others?

17.6 How Species Form

17. What is a species?
18. What's the difference between a prezygotic reproductive barrier and a postzygotic reproductive barrier? Give an example of each.
19. What are two ways sympatric speciation can occur?
20. Why did the formation of the Isthmus of Panama cause many marine organisms to speciate?
21. What is an adaptive radiation and when is it most commonly seen?

17.7 Evidence of Evolution

22. What is artificial selection? Why does artificial selection provide evidence for evolution?
23. Why does the similarity of the mammalian limb in all different species of mammals provide evidence for evolution?

17.8 Does Evolution Occur Gradually or in Spurts?

24. What feature of the fossil record led some biologists to argue that evolution occurs in spurts?
25. When, according to punctuated equilibrium, do spurts of evolutionary change occur?

17.9 The Evolution of Humans

26. What important features of modern humans can already be seen in 4-million-year-old *Australopithecus* fossils?
27. What was the first species of hominid to leave Africa and spread across Europe and Asia?

Multiple Choice Questions

Choose the BEST answer to the following.

1. Which of the following statements regarding the origin of life is false?
   (a) Life originated on an Earth whose atmosphere contained high levels of oxygen.
   (b) Miller and Urey generated amino acids and many other organic molecules by shooting electric sparks through a model of Earth's early atmosphere.
   (c) The first genes were probably made of RNA, not DNA.
   (d) When certain lipids are added to water, they spontaneously form structures that resemble cell membranes.
2. The primary problem with the hypothesis that life on Earth originated on Mars is
   (a) Mars has never had water.
   (b) the proposed Martian fossils are much smaller than the tiniest bacteria on Earth.
   (c) life on Mars would have had no way to get to Earth.
   (d) life has never been found on Mars.
3. Photosynthesizing plants are
   (a) heterotrophs.
   (b) autotrophs.
   (c) chemoautotrophs.
   (d) archaeans.
4. Which of the following scientists did not play a role in the development of Darwin's theory of evolution by natural selection?
   (a) Lamarck
   (b) Lyell
   (c) Malthus
   (d) Gould
5. Which of the following is an example of a sexually selected trait?
   (a) the erratic flight of the Painted Lady butterfly
   (b) parental care by male poison dart frogs
   (c) large size and strength in male elephant seals
   (d) antibiotic resistance in bacteria
6. If we compare related bear species in tropical and arctic environments, we would expect
   (a) the tropical species to have shorter legs.
   (b) the arctic species to be smaller in size.
   (c) the tropical species to have larger ears.
   (d) the tropical species to have a thicker layer of fat.
7. If natural selection favors a body size smaller than the population average within a group of squirrels, you expect to see
   (a) directional selection.
   (b) stabilizing selection.
   (c) diversifying selection.
   (d) either a or b.
8. Which of the following mechanisms of evolution most effectively causes populations to become more adapted to their environments?
   (a) natural selection
   (b) mutation pressure
   (c) genetic drift
   (d) migration
9. When a lion mates with a tiger, the offspring are sterile. This is an example of
   (a) allopatric speciation.
   (b) sympatric speciation.
   (c) speciation by hybridization.
   (d) a postzygotic reproductive barrier.

10. Which of the following provides evidence for evolution?
    (a) changes in the coloration of peppered moth populations over time
    (b) the presence of tiny incomplete hind legs on snakes such as boa constrictors
    (c) the fact that island species tend to most closely resemble species found on the nearest mainland
    (d) all of the above

INTEGRATED SCIENCE CONCEPTS

Astronomy—Did Life on Earth Originate on Mars?

1. Why do some NASA scientists think life on Earth could originally have come from Mars?
2. How do scientists know the meteorite containing the potential fossils is from Mars?
3. If life on Earth did originate on Mars, how did it get here?
4. Why are some people skeptical that the supposed fossils are of bacteria?

Physics—Animal Adaptations to Heat and Cold

5. Why is surface area-to-volume ratio important in thermoregulation?
6. What is Bergmann's Rule? How does Bergmann's Rule relate to surface area-to-volume ratio? Give an example of Bergmann's Rule in animals.
7. What is Allen's Rule? Provide an example of Allen's Rule in animals.
8. What are some examples of insulators in animals? How does an insulator work?

Earth Science—Earth's Tangible Evidence of Evolution

9. What are some fossils that provide insight into how organisms have evolved over time?
10. How does biogeography provide evidence for evolution?

Active Explorations

1. Look at the following photos, taken from Kettlewell's original publication on peppered moths. If, rather than birds, *you* were the primary predator of peppered moths, would there be natural selection for color? Why or why not? (How many moths do you see? Which one did you see first?)

2. Take a hike, or a walk in your neighborhood, and examine some of the plants, insects, birds, and other organisms that you come across. For each organism, note one or two traits that make it adapted to its environment. Did you notice any adaptations that keep organisms from being eaten by potential predators? What types of adaptations were these—ones that allowed them to flee? Or ones that allow them to remain camouflaged? Did you notice any adaptations related to finding mates? What about adaptations related to raising offspring?
3. Here's one more thing to think about during your walk. The species you see all share the ability to coexist with humans in a human-created environment. What adaptations allow these species (whether they are pigeons, squirrels, mice, robins, sow bugs, or other species) to thrive in human communities?
4. How effective is blubber—a layer of fat—in insulating polar bears, marine mammals, and penguins from their cold environments? Try some yourself by making a "blubber glove." Fill a plastic bag with shortening. Place your hand into a second bag and then put it in the bag filled with shortening. Now fill a bucket or large bowl with ice water, and place your blubber-gloved hand in the water. When does it begin to feel cold? As a control, place your other hand in two plastic bags, this time without the shortening. How quickly does that hand feel cold?

Exercises

1. What types of experiments were necessary to show that living organisms were not spontaneously generated in nonliving matter? Why do you think the idea of spontaneous generation survived so long—why was it so difficult to disprove?
2. Why didn't Miller and Urey include oxygen in their model of the young Earth?
3. How are liposomes similar to cells? How are they different from real cells?
4. How do scientists know that the first living organisms used anaerobic processes to obtain energy? Do any aerobic organisms predate cyanobacteria? Why or why not?
5. How might Lamarck have explained the streamlined shape of fish, which makes them effective swimmers? How would Darwin's theory of evolution by natural selection explain the same phenomenon?
6. How is the story of myxoma virus and Australian rabbits similar to the story of antibiotic resistance in bacteria?
7. What are some human traits that do not show variation? What are some that do show variation?
8. What are some heritable human traits? Some nonheritable human traits?
9. Nancy Burley of the University of California, Irvine, ran the following experiment—she placed red color bands on the feet of some male birds and green color bands on the feet of other male birds. Females preferred to mate with males that had red color bands. Could this lead to natural selection? Why or why not?
10. How would you determine whether a trait you were interested in studying is heritable?
11. You are studying a population of beetles that include some red individuals and some yellow individuals. You know that color is a heritable trait in the population. By counting the number of red and yellow beetles over a period of 5 years, you notice that the population is evolving towards more red individuals. How could you determine whether this is a result of natural selection? Are there other potential explanations?
12. On islands, many large animals—such as elephants—evolve to become miniaturized in size. On the other hand, many small animals—including some rodents—evolve to be exceptionally large in size. Why might natural selection produce these results? Do you think this phenomenon sheds light on *Homo floresiensis*, the miniature human relative?
13. In recent decades, average human height has increased in many parts of the world. Do you think this is an example of evolution?
14. In a population of mice that you are studying, tail length appears to be increasing over time. However, you find no evidence that natural selection is acting on tail length. What are two alternate explanations for your observation?

15. Two species of foxes are shown below. One is a kit fox in the Mojave Desert, California. The other is an arctic fox in Manitoba, Canada. Which is which? How can you tell? Describe at least two traits that make each animal well-adapted to thermoregulating in its habitat.

16. Individuals of two different fish species sometimes mate, but the offspring die soon after hatching. Is this a prezygotic or postzygotic reproductive barrier?
17. Finches on two closely situated islands look noticeably different from one another—individuals on one island have brown tail feathers, whereas individuals on the other island have black tail feathers. Can you conclude that they are two different species? What would you do to determine if they are in fact distinct species?
18. At your field site, there are butterflies with yellow wings and butterflies with orange wings. After observing them carefully, you conclude that the yellow butterflies always mate in shady areas under trees, whereas the orange butterflies always mate in sunny meadows. Can you conclude that they are different species?
19. Many of the living organisms on Hawaii are found nowhere else on Earth. This includes species of plants, birds, insects, mammals, mushrooms, and so forth. Why do you think this is?
20. What are some examples of artificial selection? How are artificial selection and natural selection similar? How are they different?
21. Islands tend to have fewer species than the mainlands they resemble. Furthermore, island species often include many flying organisms and few terrestrial ones. Do these biogeographical patterns support evolution or the purposeful distribution of organisms? Why?
22. Write a letter to Grandma telling her about drug resistance in living organisms. Explain to her why drug resistance is such a common phenomenon—including why insects become more resistant to pesticides over time, and why diseases such as tuberculosis and malaria have become harder to treat in recent years.

Problems

1. Let's look at how natural selection causes advantageous traits to become more common in populations. Suppose there is a population of bugs in which some individuals are green and some individuals are brown. Suppose that, because brown bugs are better camouflaged against predators, each brown bug leaves two brown offspring per generation and each green bug leaves one green offspring per generation. (Is this natural selection? Why?) You start with two brown bugs and two green bugs in Generation 1. How many brown and green bugs are there in Generation 2? Calculate the number of brown and green bugs there are in Generations 1 to 10. Show that 50 percent of the bugs in Generation 1 are brown, 94 percent in Generation 5 are brown, and over 99 percent in Generation 10 are brown. What is happening here?
2. Suppose that, in the population above, migration is operating in addition to natural selection. Suppose that three green bugs migrate into the population each generation. Again, calculate the number of brown and green bugs in Generations 1 to 10 and the fraction of brown bugs in each generation. Show that, with migration, 50 percent of the bugs in Generation 1 are brown, 70 percent in Generation 5 are brown, and 97 percent in Generation 10 are brown. What effect does migration have on how quickly this population is adapting to its environment?
3. Let's consider a very small population of snapdragons, one with only two individuals. One snapdragon has two red alleles for flower color—that is, it is RR. The other snapdragon has a red allele and a white allele for flower color—it is RW. (You may wish to read about the inheritance of flower color in snapdragons in Chapter 16.) Show that the frequency of the red allele R in the population is 0.75 and that the frequency of the white allele W is 0.25.
4. Now let's assume the two snapdragons in our tiny population mate and produce a single offspring. We now have a snapdragon population with only one individual. What are the two possibilities for the genotype of this individual? For each of the two possibilities, calculate the allele frequencies of the red and white alleles in the population. Show that, in either case, the allele frequencies are different from those found in the parental population (calculated in Problem 3). Is this an example of genetic drift?

Conceptual Science Place

CHAPTER 17 ONLINE RESOURCES

Tutorials Darwin and the Galápagos Islands, Causes of Microevolution, Polyploid Plants, Reconstructing Forelimbs • ***Videos*** Galápagos Islands Overview, Galápagos Marine Iguana, Albatross Courtship, Booby Courtship • ***Quiz*** • ***Exercises*** • ***Flashcards*** • ***Links***

Biological Diversity

Life on Earth has evolved into countless forms. Clockwise from the upper right, these are seeds from tropical legume plants, euphorbia flowers, jewel beetles, cockle shells, foliose lichen, a fossilized fish, tropical butterflies, sprats, and pheasant feathers.

CHAPTER OUTLINE

Over 1.5 million species have been described and named on Earth, and scientists believe that many more—probably somewhere between 10 and 100 million—have yet to be found and named. Earth's living species show remarkable diversity in their adaptations and ways of life. What are the major groups of living things? How can we classify them? What do we know about their evolutionary histories? Is it true that we humans are more closely related to bread mold than to daisies? How is it adaptive for certain flowers to mimic the smell of a dead horse? And for certain salamanders to have no lungs? Should we worry that evolution may one day produce a 10-foot mosquito? And are there really thousands of species of dinosaurs flying around on Earth today?

18.1 Classifying Living Things

The drive to classify living things is an ancient one. Thousands of years ago, Aristotle arranged a "Chain of Being" that proceeded from minerals to plants, animals, man, and God. Other thinkers built on Aristotle's idea, arranging organisms in a hierarchy from "simple" things likes plants and worms to the most "complex"—you guessed it!—humans at the top (Figure 18.1).

Linnaean Classification

In the eighteenth century, Swedish naturalist Carolus Linnaeus developed a new system of classification that emphasized the shared similarities of organisms. The Linnaean system of classification makes use of multiple levels, which from largest to smallest are the domain, kingdom, phylum, class, order, family, genus, and species. Each domain contains one or more kingdoms, each kingdom contains one or more phyla, each phylum contains one or more classes, and so on, until you get to each genus containing one or more species. Every living species belongs to one domain, one kingdom, one phylum, one class, one order, one family, one genus, and one species. You can think of each successive Linnaean level as allowing you to "home in" on a particular species, much the way successive geographical levels allow you to home in on a particular house and human resident, as shown in Figure 18.2. At every level in the Linnaean system, species are grouped together based on shared similarities. For example, species in the class Mammalia (all mammals) are grouped together based on shared features such as their possession of hair and their production of milk. Furthermore, under the

Figure 18.1 Early classifications of living organisms placed them in a hierarchy from simple to complex.

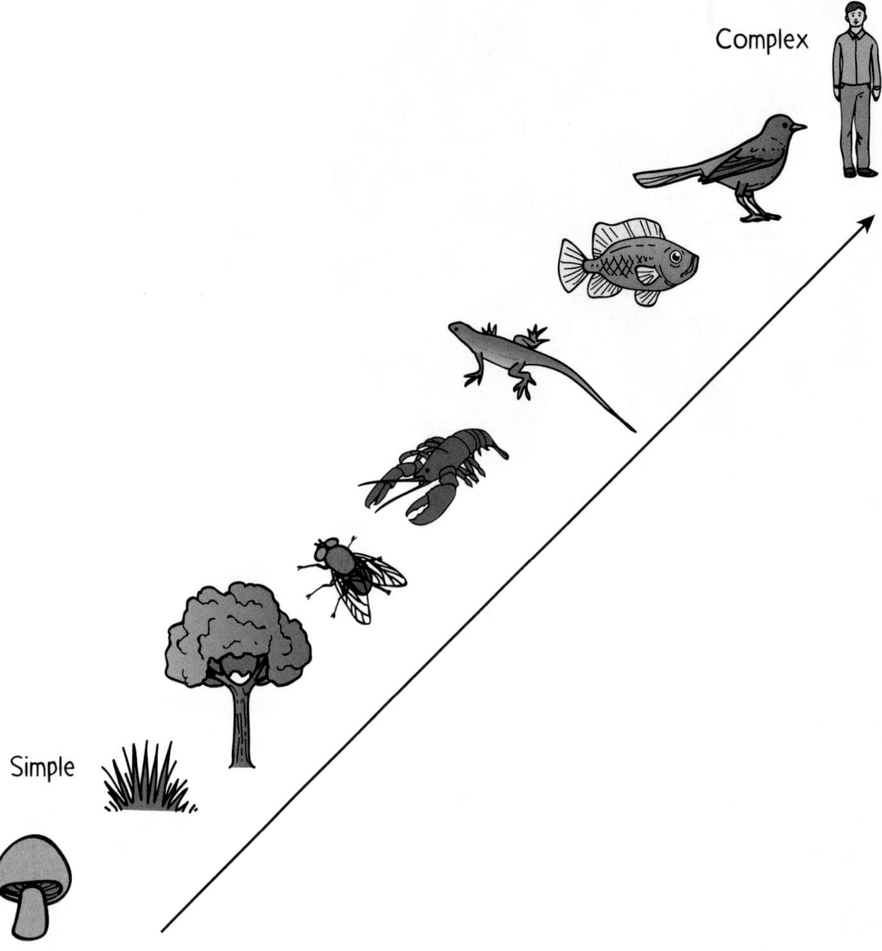

INSIGHT A variety of mnemonics have been invented to help people remember the Linnaean levels of classification. This one has the advantage that each word sounds similar to the corresponding Linnaean level: "*King Phillip* called the *class* to *order*—the *family genius* will now *speak.*"

Linnaean system, all species have a two-part scientific name consisting of the genus name and species name. Some examples of scientific names include *Homo sapiens* ("wise man") for humans and *Canis familiaris* ("intimate dog") for humankind's best friend. Genus and species names are always Latinized and italicized, with the genus capitalized. Sometimes the genus name is abbreviated as a single letter, as in *E. coli* for the human gut bacterium *Escherichia coli.*

Cladistic Classification

Since Linnaeus's time, science has expanded our understanding of the history of life on Earth. In particular, Darwin's theory of evolution showed that the wealth of species on Earth is the result of numerous instances of speciation followed by the independent evolution of new lineages (Chapter 17). Biologists now aim to classify living organisms based on this evolutionary history, a strategy that has several advantages over Linnaean classification. First, while making judgments about the "shared similarities" of different organisms is inherently subjective, a classification based on evolutionary history is not at all arbitrary. Second, classifying organisms based on their evolutionary history allows biologists to make predictions about the characteristics of organisms that are unstudied or poorly known. Finally, a classification system based on evolutionary history allows biologists to study more effectively how specific features of organisms evolved. For example, knowing that birds and bats are distinct lineages with separate evolutionary histories allows us to infer that flight evolved independently in each group. Someone interested in studying the evolution of flight would proceed to ask questions very different from the ones they might formulate if, say, birds evolved from bats or vice versa.

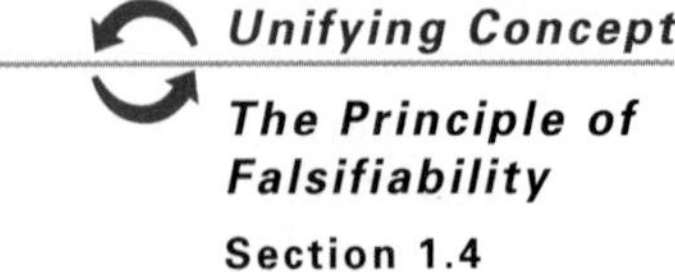

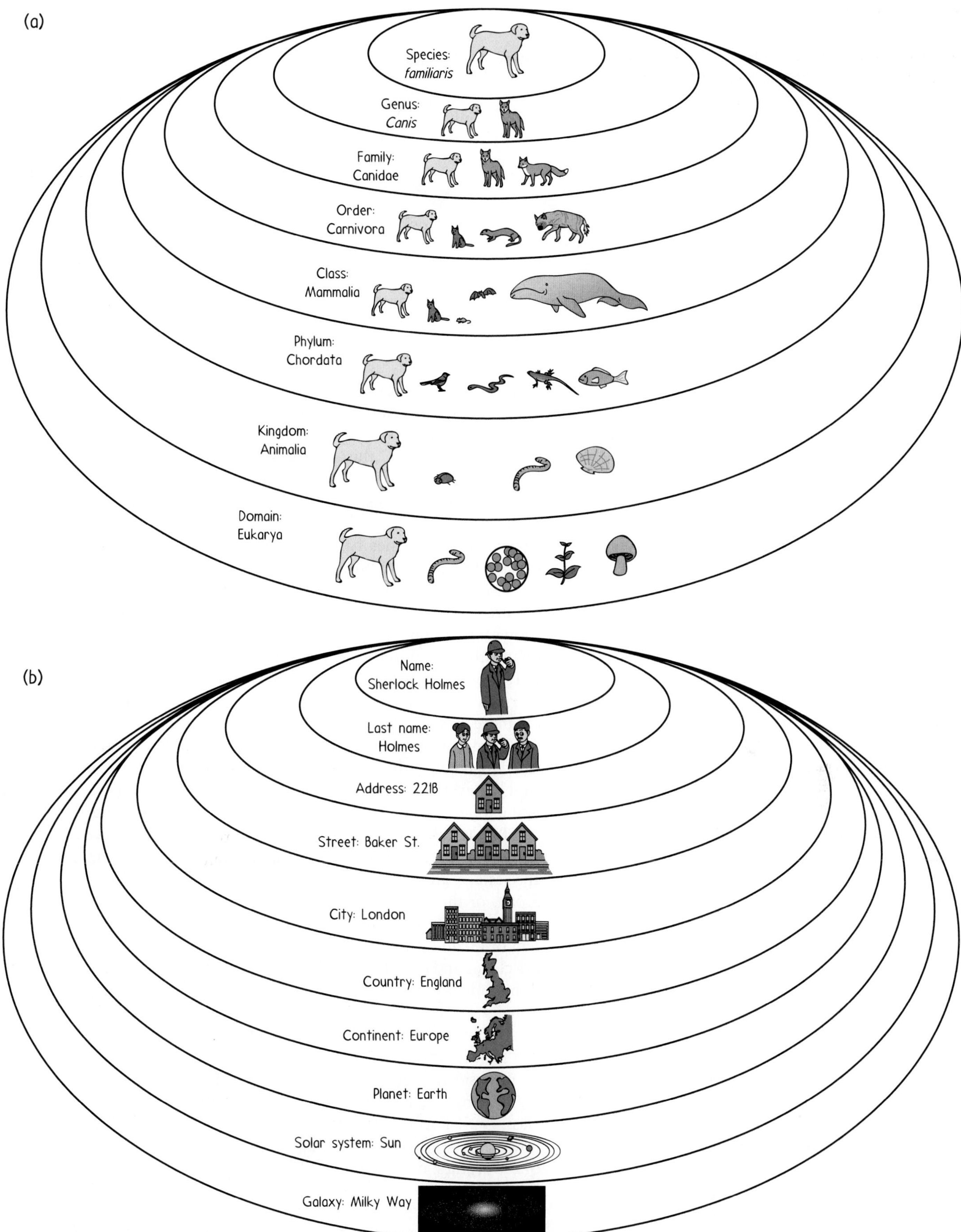

Figure 18.2 (a) Under the Linnaean system, the dog belongs to the domain Eukarya (eukaryotes), kingdom Animalia (animals), phylum Chordata (chordates), class Mammalia (mammals), order Carnivora (carnivores), family Canidae (dogs, wolves, foxes, etc.), genus *Canis*, and species *familiaris*. (b) Sherlock Holmes belongs to the galaxy Milky Way, solar system Sun, planet Earth, continent Europe, country England, city London, street Baker Street, address 221B, last name Holmes, name Sherlock Holmes.

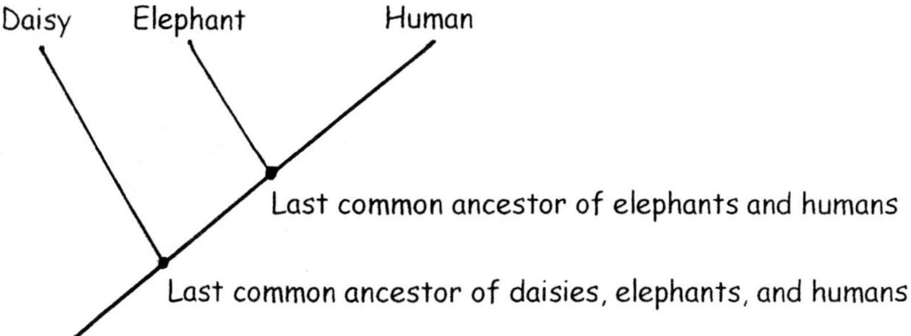

Figure 18.3 Cladograms are used to diagram the evolutionary relationships among species or other biological groups. The cladogram for humans, elephants, and daisies shows that humans and elephants are more closely related to each other than either is to daisies.

In order to actually accomplish the goal of classifying organisms based on evolutionary history, biologists have to reconstruct the history of speciation events among a group of organisms. This can be challenging. Typically, biologists make use of fossils as well as information on the anatomical, behavioral, and molecular traits of existing organisms to try to reconstruct this history. DNA sequences have proved to be a particularly valuable source of information.*

Once biologists have a hypothesis for how speciation occurred within a group of organisms, this can be diagrammed in a **cladogram**, or *evolutionary tree*. A simple cladogram is shown in Figure 18.3. This cladogram includes only three species—humans, elephants, and daisies. The last common ancestor of the three species is indicated with a dot. Up to this last common ancestor, humans, elephants, and daisies had a shared history. Then a speciation event occurred, causing two separate lineages to emerge. One of the lineages eventually gave rise to daisies (and many other organisms, which are not shown), and the other lineage eventually gave rise to humans and elephants (as well as other organisms not shown). Later on, another speciation event caused the lineage that produced elephants to split from the lineage that produced humans. From this cladogram, we infer that humans and elephants have a longer period of shared history than either has with daisies. Another way of saying this is that humans and elephants are more closely related to each other than to daisies. In order for biological classification to reflect this evolutionary history, humans and elephants must be classified together to the exclusion of daisies. On the other hand, humans and daisies should never be classified together to the exclusion of elephants, and elephants and daisies should never be classified together to the exclusion of humans.

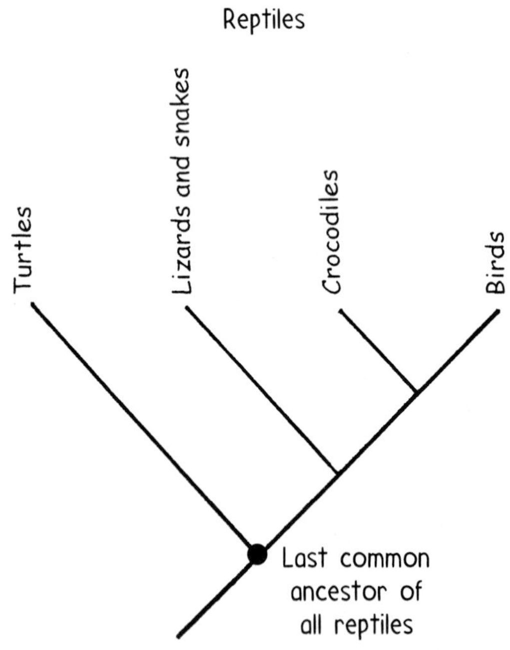

Figure 18.4 This cladogram shows that birds are descended from the last common ancestor of reptiles, and so are crocodiles.

Biological groups that are constructed based on evolutionary history are called clades. A **clade** (rhymes with "made") is a group that includes an ancestral species and all of its descendants. Clades can be small groups, such as the genus *Homo* or the species *Homo sapiens*, or broad groups, such as mammals or animals or eukaryotes. How does a cladistic classification relate to the Linnaean system of kingdoms, classes, orders, and families? In many cases, Linnaean groups *are* clades and continue to be used by biologists. For example, the Linnaean kingdom Animalia—all animals—is a clade. The Linnaean kingdoms Plantae and Fungi are also clades. Mammals, amphibians, primates, rodents, birds, frogs—these familiar Linnaean groups are all clades.

As it turns out, however, some Linnaean groups are not clades and have had to be reconsidered. For example, the Linnaean class Reptilia—the reptiles—grouped together turtles, lizards, snakes, and crocodiles. Birds were placed in a separate class, Aves. Reconstructing the evolutionary relationships among these organisms, however, made it apparent that birds are descended from the last common ancestor of the reptiles, and so are reptiles too! As shown in Figure 18.4, birds are closely

* Interestingly, DNA can also provide evidence for how long it has been since particular speciation events occurred. This gives scientists an estimate for when particular groups of organisms diverged evolutionarily from one another. How does this work? Certain genes appear to function as *molecular clocks*, evolving (accumulating changes in their DNA sequence) at a fairly constant rate over time. By looking at how much the DNA sequence of a gene differs between two organisms, scientists can estimate how long ago the organisms diverged. Multiple genes, as well as information from the fossil record, are usually used to estimate divergence times.

related to crocodiles. Classifying turtles, lizards, snakes, and crocodiles together to the exclusion of birds is just like classifying elephants and daisies together to the exclusion of humans. Why this discrepancy between Linnaean and cladistic classification? Recall that, under the Linnaean system, species are grouped together based on shared similarities. Turtles, lizards, snakes, and crocodiles had been grouped together based on shared features such as "cold-bloodedness" and the possession of scales. Birds, because they are "warm-blooded" and have feathers, were placed in a different group. A cladistic classification groups species together based on their evolutionary relationships and places birds squarely among the reptiles.

INSIGHT Birds aren't only reptiles—they're dinosaurs! This is because birds are descended from the last common ancestor of all the dinosaurs, too. So dinosaurs didn't *all* go extinct—birds survived, and they are certainly alive and well today, with nearly 10,000 described species.

Biological classification is always a work in progress. As biologists learn more about the evolutionary history of species, whether this occurs through the discovery of new fossils or the reexamination of existing species, they sometimes need to redraw cladograms. Classification then changes to reflect this new understanding of species relationships.

CHECK YOURSELF

1. The scientific name of the endangered orangutan of Sumatra is *Pongo abelii*. What is its genus name? Its species name?
2. Why is a cladistic classification, the classification of life by evolutionary history, more useful for biologists studying the evolution of traits than the Linnaean classification system?

CHECK YOUR ANSWERS

1. The genus name is *Pongo*. The species name is *abelii*.
2. A cladistic classification reveals, rather than obscures, the evolutionary history of species. Knowing that birds are reptiles, for example, allows biologists to ask appropriate questions about the evolution of their traits. For example, birds are "warm-blooded" rather than "cold-blooded" like the other reptiles—how did they evolve warm-bloodedness? Birds have feathers, rather than the scales seen in other reptiles—could scales have been modified to form feathers? Under Linnaean classification, where reptiles and birds are separate classes, there is no clear relationship between the two groups.

18.2 The Three Domains of Life

Life is classified into three domains—Bacteria, Archaea, and Eukarya. Early in the history of life—probably 2.5 to 3.5 billion years ago—living organisms split into two separate lineages, one that produced the Bacteria and one that produced the Archaea and the Eukarya (Figure 18.5). Of the three domains, Bacteria and Archaea consist of prokaryotic organisms, organisms whose cells do not have a nucleus (Chapter 15). Eukarya includes all eukaryotic organisms, organisms with nucleated cells.

The domain Eukarya is further divided into four kingdoms—Protists, Plants, Fungi, and Animals. The Protist kingdom is problematic because it includes all eukaryotes that aren't plants, animals, or fungi—in other words, it's a hodgepodge of species that don't represent a clade. Amoebas, kelp, and diatoms are all protists, although they have little in common other than the fact that they are eukaryotes. Until a more accurate classification emerges, however, we are stuck with the term Protists.

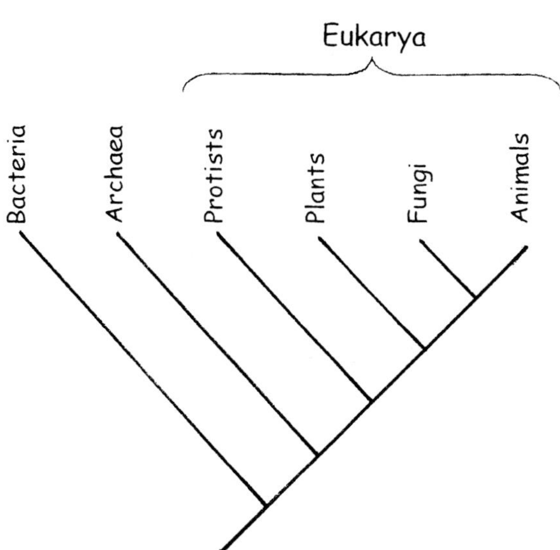

Figure 18.5 The three domains of life are Bacteria, Archaea, and Eukarya.

CHECK YOURSELF

Are humans more closely related to button mushrooms or to celery?

CHECK YOUR ANSWER

Humans are animals, button mushrooms are fungi, and celery is a plant. In the cladogram shown in Figure 18.5, we see that animals and fungi are more closely related to each other than either is to plants. So, we are more closely related to button mushrooms.

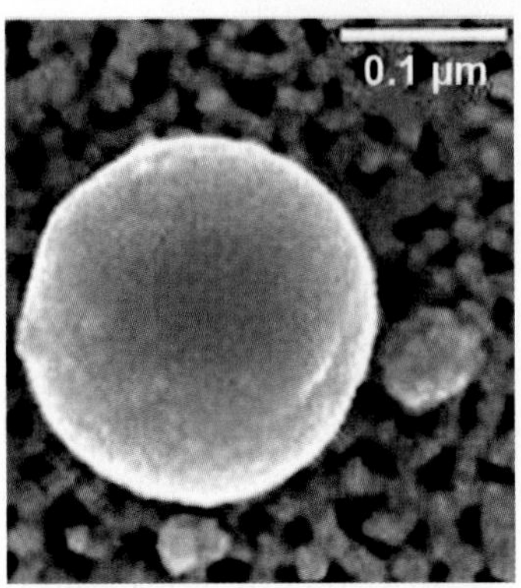

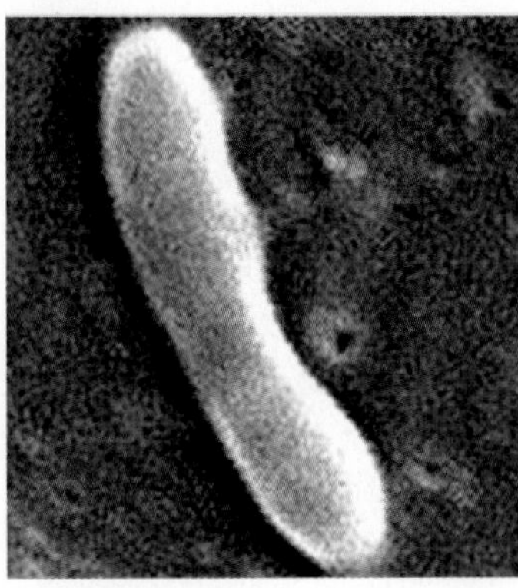

Figure 18.6 Bacteria can live in habitats where no other organisms can survive. These bacteria were found thriving in Antarctica's Lake Vostok, which is thousands of meters below the ice sheet.

18.3 Bacteria

They live on your body by the hundreds of trillions, occupy habitats where no other organisms can survive (Figure 18.6), and devastate human populations with diseases such as plague and tuberculosis—yet life on Earth would quickly end without them. *They* are bacteria, one of the three domains of life and one of the most ancient lineages on Earth. Earth's oldest fossils, 3.5 billion years old, are of bacteria (Chapter 17).

Bacteria are prokaryotes so diverse that it is hard to make generalizations about them. Some are autotrophs that, like plants, make their own food through photosynthesis. Others are heterotrophs that obtain food from other organisms. Most bacteria are single-celled, but others gather in multicellular clusters. Bacteria come in varied shapes, including spheres, rods, and spirals. Many are mobile, propelling themselves with whiplike structures called *flagella*. Bacteria typically reproduce asexually by dividing. However, most species exchange genetic material at least occasionally—when they take up small pieces of naked DNA from the environment, when bacterial viruses inadvertently transfer DNA between organisms, or when two bacteria join together and one passes DNA to the other. Under favorable conditions, bacteria can divide very quickly, as often as every 20 minutes. This allows bacterial populations to grow rapidly when resources are plentiful (Figure 18.7). In poor conditions, many bacteria form hardy spores that remain dormant until conditions improve.

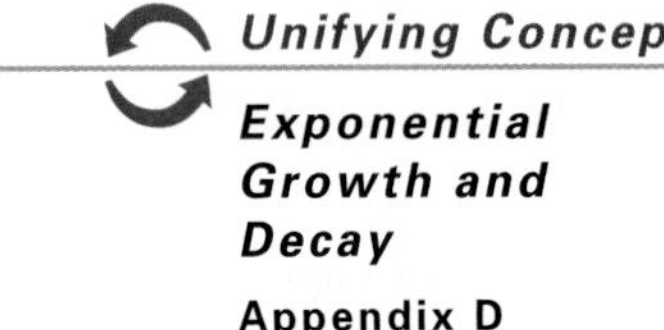

Unifying Concept

Exponential Growth and Decay

Appendix D

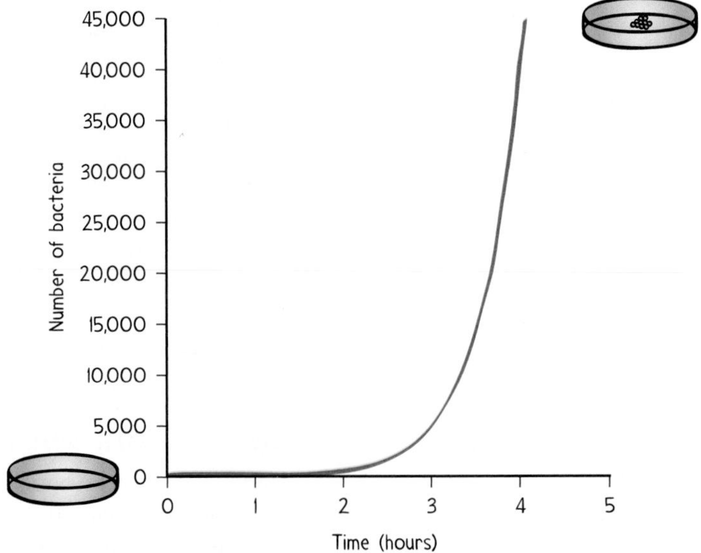

Figure 18.7 When resources are plentiful, bacterial populations grow exponentially. (This is true for populations of other organisms as well, but bacteria more commonly encounter such favorable conditions.) The population shown here begins with 10 bacteria at time 0 and doubles every 20 minutes.

Life would be impossible without bacteria, which are essential in decomposition, the breaking down of organic matter. Without bacterial decomposition, carbon (C) would remain trapped in dead organic matter and there eventually would be none available for photosynthesis. Bacteria also help cycle other nutrients; some bacteria fix nitrogen (N), transforming it from its inorganic atmospheric form to organic varieties that can be used by living organisms (Chapter 21).

Countless bacteria live in and on our bodies, particularly on the skin and in the mouth, respiratory tract, intestine, and urethra. A few of these are potentially harmful, but others benefit us by producing vitamins and by keeping more dangerous bacteria from invading our bodies. Bacteria are important for making foods such as cheese and yogurt, and genetically engineered strains have been used to produce human insulin and other medically important molecules. Of course, other bacteria are familiar to us because they cause diseases such as tuberculosis, syphilis, and Lyme disease. The development of *antibiotics*, substances that kill bacteria, was a significant step forward in medicine.

INSIGHT Most bacteria are very small because they rely on diffusion to obtain nutrients (Chapter 15). However, a few are big enough to be seen with the naked eye! The largest bacteria known belong to the species *Thiomargarita namibiensis*, which means "Sulfur Pearls of Namibia." They grow in long strands like strings of pearls and were discovered in sediment collected off the coast of Namibia. These giant bacteria measure $\frac{3}{4}$ of a millimeter across, or about the size of the period at the end of this sentence. If a typical bacterium were the size of a mouse, scientists say, these bacteria would be as big as blue whales!

CHECK YOURSELF

Some people contract yeast infections after taking a course of antibiotics. Why?

CHECK YOUR ANSWER

Antibiotics kill both the targeted "bad" bacteria as well as normal "good" bacteria in our bodies. The normal bacteria help keep yeast in check. Without them, yeast proliferates and causes an infection.

18.4 Archaea

Once considered odd-looking bacteria, **archaea** ("OUR-kee-uh") are now recognized as a distinct domain of prokaryotic organisms more closely related to eukaryotes than to bacteria. Many features of archaean genetics in particular link

TECHNOLOGY

Bacteria Power

Will our cars run on bacteria power one day? Mud-dwelling bacteria of the genus *Geobacter* release electrons as they consume organic pollutants and decaying plant and animal matter. (Humans and other organisms also produce electrons during cellular respiration, but these combine with oxygen and hydrogen atoms to form water (Chapter 15). *Geobacter*, on the other hand, releases its electrons directly.) By designing a simple system for catching these electrons, biologist Derek Lovley and his colleagues at the University of Massachusetts, Amherst, were able to create a battery that runs on pure bacteria power. Lovley collected polluted mud and seawater from Boston Harbor and placed them in a fish tank. He then stuck an electrode in the mud and connected it with copper wire to a second electrode in the seawater. Over time, *Geobacter* bacteria gathered on the mud-embedded electrode and passed electrons out their cell membranes as they ate. These electrons moved up the copper wire, producing an electric current.

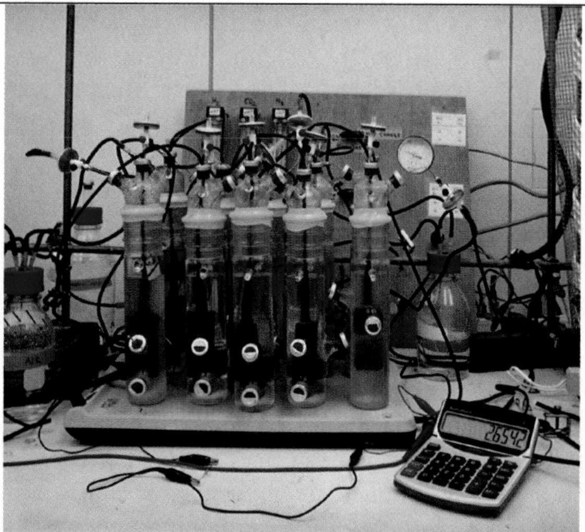

A series of *Geobacter* "batteries" (the tubes in the background) are powering this calculator.

The next challenge will be to make *Geobacter* energy generation more efficient. At present, it would take many square miles of muddy seafloor to generate 60 watts, the amount of power required to run a household light-bulb—an unrealistic proposition. Scientists are tackling this problem now and have taken the first step of sequencing the *Geobacter* genome. Studying *Geobacter*'s genes may just help them figure out how to get the bacteria to eat faster. In fact, it appears that *Geobacter* has more than 100 genes that are related to its ability to transfer electrons. If scientists are successful at manipulating these, we may soon have microbial fuel cells that run on garbage and other organic wastes. In the meantime, *Geobacter* is already making itself useful running weather sensors and deep-sea mapping instruments, devices that require only 1 watt of power.

Figure 18.8 The orange and yellow encrustations are colonies of extremophile archaea that occupy the scalding waters of this Nevada geyser.

archaea to eukaryotes—their ribosomes and tRNA resemble those of eukaryotes, their genes contain introns like those of eukaryotes, and their DNA is associated with histone proteins, like that of eukaryotes.

Many archaea are adapted to extreme environments, such as very salty ponds or the scalding waters of hot springs and hydrothermal vents (Figure 18.8). These archaea are called "extremophiles"—lovers of the extreme—and are of particular interest to biologists because they live in conditions thought to be similar to those found on the early Earth. However, not all archaea are extremophiles. Many occur in more familiar locales such as the open ocean or the digestive tracts of termites and cows.

Some archaea are *chemoautotrophs*. They make food using chemical energy rather than energy from sunlight. Archaea in hydrothermal vent habitats (Chapter 22), for example, obtain energy from a chemical abundant there, hydrogen sulfide. These archaea form the basis of remarkable biological vent communities that are entirely independent of sunlight.

18.5 Protists

Conceptual Science Place

Paramecium **Cilia;** ***Paramecium*** **Vacuole; Diatoms Moving; Dinoflagellates;** ***Amoeba; Amoeba*** **Pseudopodia**

Eukaryotes that are not plants, animals, or fungi are **protists**. This somewhat miscellaneous category includes autotrophs, heterotrophs, and even species that use both strategies to obtain nutrition. Protists may be single-celled or multicellular. Certain protists, the slime molds, actually hover somewhere between single-celled and multicellular—they go from one condition to the other during the course of their lives. By studying this process, scientists have obtained clues about how multicellularity might have evolved in other biological groups. Although many protists reproduce asexually, others undergo sexual reproduction.

Many protists are autotrophs that get their food from photosynthesis. Single-celled *diatoms* float in the open ocean and perform the bulk of oceanic photosynthesis. Diatoms have elaborate, often very beautiful shells made of silica (Figure 18.9). These shells are sometimes used in manmade products—for example, they provide the gritty texture of toothpaste. The single-celled *dinoflagellates* are also important oceanic photosynthesizers—although there are also many species of heterotrophic dinoflagellates. Dinoflagellate population explosions are responsible for the "red tides" that occur when sunlight and nutrients are plentiful. The

Figure 18.9 This microscopic view of a diatom shows its elaborate silica shell.

INSIGHT Red tides aren't always red—the color of the water may also be pink, purple, green, orange, brown, or blue, depending on the dinoflagellate responsible and which pigment it uses for photosynthesis.

discoloration that gives red tides their name is caused, believe it or not, by the sheer number of dinoflagellates in the water! Some red tides are toxic—shellfish that eat the dinoflagellates become contaminated and poisonous to humans.

Other photosynthetic protists are multicellular and can grow quite large. Many familiar seaweeds are protists. Kelp forms huge oceanic "forests." Red algae is the source of Japanese nori. Photosynthetic green algae likely gave rise to terrestrial plants.

Most heterotrophic protists are mobile, active hunters, and all have special cell vacuoles for digesting prey. *Amoebas* move by extending part of their body forward as a pseudopodium, a temporary protrusion of the cell, and then pulling the rest of the body behind. Amoebas surround and engulf prey (Figure 18.10). *Ciliates* move by beating numerous hairlike projections called cilia. *Flagellates* move by whipping long flagella. Both ciliates and flagellates have openings that function as "mouths." One group of flagellates, the *choanoflagellates*, are thought to include the ancestors of animals.

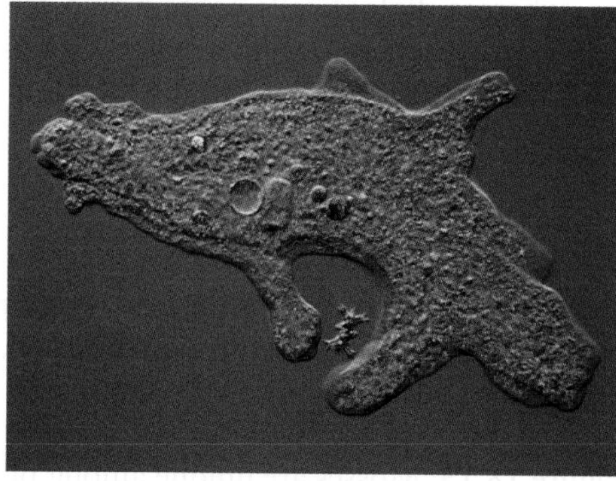

Figure 18.10 An amoeba captures another protist (a ciliate) with pseudopodia.

Protists are responsible for a number of serious human diseases, including malaria, African sleeping sickness, and amoebic dysentery. Malaria is caused by *Plasmodium* protists that divide their life cycle between mosquitoes and humans. Humans contract the disease when they are bitten by infected mosquitoes. Plasmodium protists infect human red blood cells (Figure 18.11), where they reproduce in huge numbers. The synchronized emergence of protists from host red blood cells causes periodic bouts of chills, fever, and vomiting.

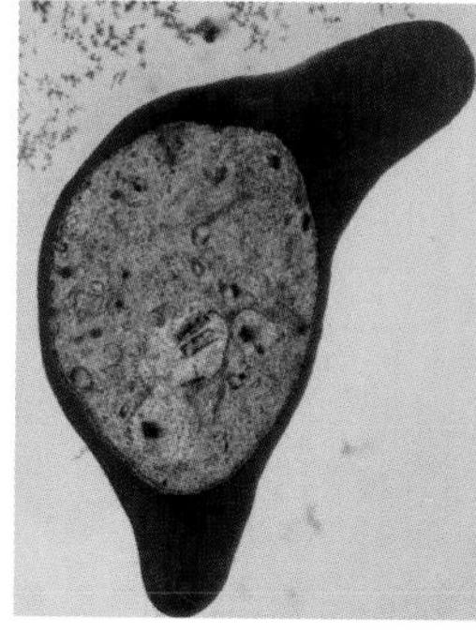

Figure 18.11 This misshapen red blood cell has been infected by the malaria-causing protist *Plasmodium*.

18.6 Plants

Pictures of Earth from space show large green patches stretching across wide areas of the continents (see photo on page 505). The reason much of Earth's land surface is green is because vast areas are covered with plants. In fact, in almost any natural terrestrial environment you might explore, the majority of life forms you will come across are plants. **Plants** are autotrophic, multicellular, terrestrial eukaryotes that obtain energy through photosynthesis. The green in plants comes from chlorophyll, the pigment they use in photosynthesis (Chapter 15).

Plants have a variety of adaptations that make them well suited to their terrestrial environments. Roots anchor them to the ground and absorb water and nutrients from the soil. Shoots, which include stems and leaves, are a plant's above-ground structures and conduct photosynthesis. Leaves are the site of the bulk of photosynthesis. Sunlight is caught on their surfaces, and carbon dioxide is taken in through small pores known as *stomata* (Figure 18.12). Most plants also have a vascular system, a sort of plant "circulatory system" that distributes water and

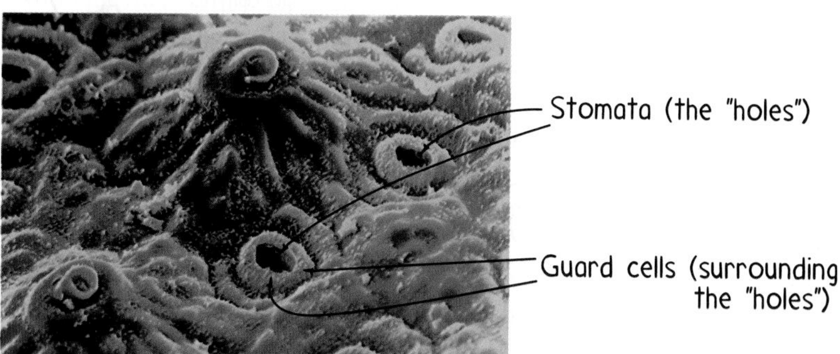

Figure 18.12 Stomata are tiny pores in plant leaves used to take in carbon dioxide. To prevent too much water loss, the size of stomata is controlled by flanking "guard" cells. Most stomata are found on the shaded undersides of leaves, again to help reduce water loss.

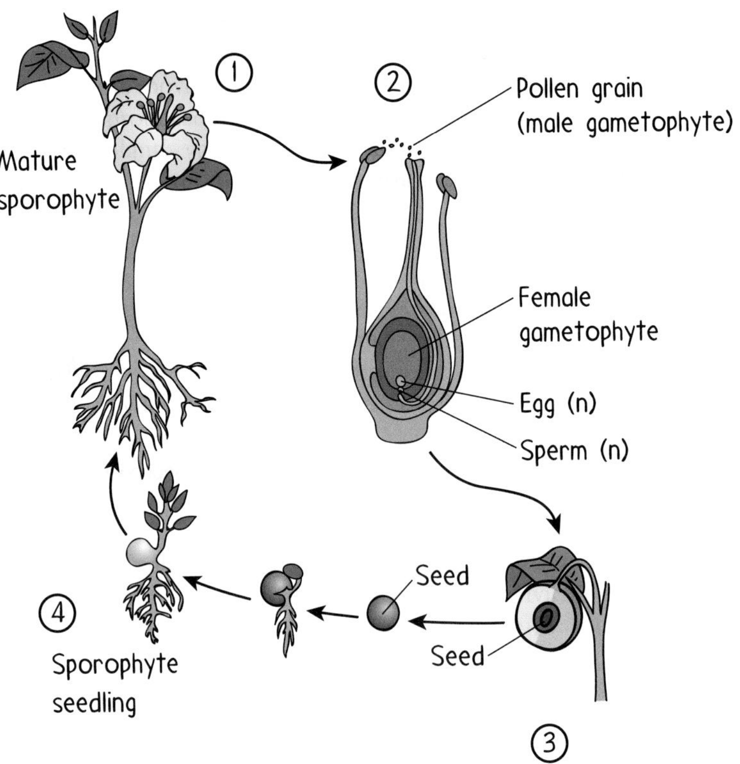

Figure 18.15 In the alternation of generations in seed plants, the gametophyte is small and dependent on the sporophyte for protection and survival. (1) A mature sporophyte produces pollen, which contains male gametophytes. (2) A sperm from a pollen grain fertilizes an egg found in the female gametophyte. (3) After fertilization, the diploid sporophyte embryo is encased in a tough outer coating with a food supply—this is the seed. (4) The seed grows into a new sporophyte.

and grow into tiny, but independent, haploid gametophytes. The gametophytes then form eggs and sperm, which fuse and grow into new sporophytes.

Seed Plants

Seed plants are the largest group of plants. Seed plants have two key features that have made them successful in a wide variety of land habitats—pollen and seeds. **Pollen** consists of numerous tiny grains, each of which is an immature male gametophyte wrapped in a protective coating. Pollen can be transported to female gametophytes by wind or (as we will see) by animals. As a result, seed plants do not require the swimming sperm that limit mosses and ferns to moist habitats. The fertilized eggs of seed plants grow into small embryonic sporophytes that are then encased in a tough outer coating along with a food supply—this entire structure is called a **seed.** Seeds are able to survive in a dormant state, during which growth and development are suspended until conditions are favorable for growth. In seed plants, the diploid sporophyte is larger than the haploid gametophyte, which is small and completely dependent on the sporophyte for protection and survival—all the seed plants you see are diploid sporophytes. The seed plant life cycle is shown in Figure 18.15.

The two major groups of seed plants are conifers and flowering plants. *Conifers* include well-known species such as redwoods, pines, cedars, spruces, and firs. Conifers have waxy, needlelike leaves and reproductive structures called *cones* (Figure 18.16). Pollen released by male cones is carried by wind to female cones. Because of the haphazard nature of wind pollination, conifers typically produce huge amounts of pollen. Fertilization takes place in the female cones and is followed by seed development. Seeds are eventually shed from female cones.

Figure 18.16 Conifers are seed plants with reproductive structures called cones. These are cones from a bristlecone pine.

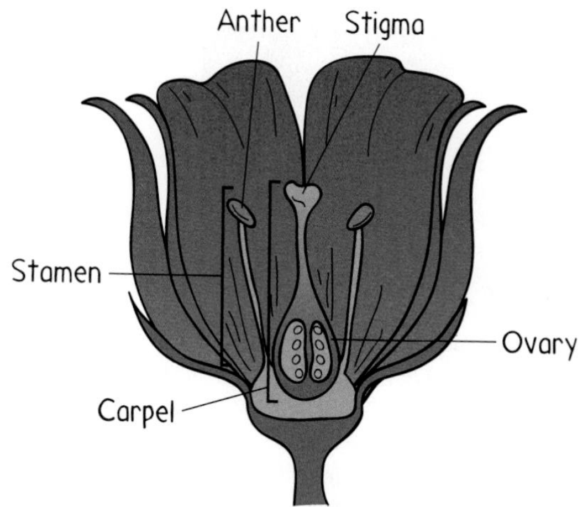

Figure 18.17 Flowers contain the reproductive structures of flowering plants.

INSIGHT The tallest tree in the world is a conifer, a coast redwood named the "Stratosphere Giant." It stands 112.7 meters tall. According to one team of researchers, no tree can grow past a theoretical limit of 130 meters, the height of a 35-story skyscraper. Because of gravity and friction between water and the vessels of a tree's vascular system, it's just too hard to effectively transport water any higher. As it is, the highest leaves in the tallest coast redwoods exist in a state of perpetual drought, despite plentiful moisture at ground level.

Flowering plants are the largest and most successful group of seed plants. Flowering plants have two important features absent in other seed plants—flowers and fruit. **Flowers** function in reproduction. The *stamen* of a flower is the male reproductive structure. It consists of a stalk capped with an *anther* where pollen develops. The *carpel* is the female reproductive structure. It includes an *ovary* where eggs develop and a stalk capped by the *stigma*, a sticky structure that traps pollen. Figure 18.17 shows these structures in a typical flower. In flowering plants, pollen is often transported from one flower to another by insects or other animals. The features of these flowers, including their petals, scent, and nectar, have often evolved to attract specific animal pollinators. In fact, flowers and their pollinators can be so well suited to each other that the features of one can be predicted by examining the other. After studying a species of night-blooming orchid in Madagascar, for example, Darwin predicted the existence of a nocturnal moth with a tongue 30 centimeters long. Forty years later, the moth was discovered! Figure 18.18 provides additional examples of flowers and their pollinators.

Flowering plants surround their seeds with a structure called a **fruit**. Fruits help spread seeds around—when fruits are eaten by birds or mammals, for example, the seeds pass unharmed through the digestive tract and eventually emerge far from the parent plants. Tasty fruits evolved in certain plants because they were more likely to be eaten by animals and so were more likely to be dispersed. Seeds themselves, on the other hand, often taste bad, discouraging animals from eating them—something you know if you've ever bitten into one. Some fruits help plants spread their seeds using other strategies—the burrs that catch on your socks during a hike are also fruits. These hitch a ride until you pull them off and drop them on the ground, where the seeds inside may just grow and take root.

Seed plants are important to human societies in many ways. Aside from their essential role in photosynthesis, many trees provide lumber, and a huge part of the human diet comes from the shoots, leaves, and fruits of flowering plants.

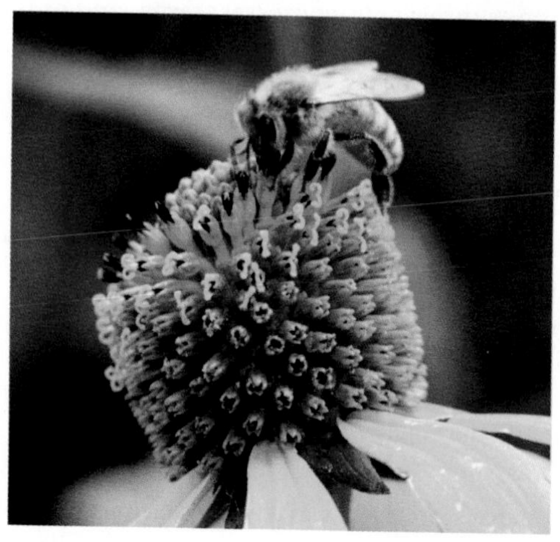
(a)

(b)

(c)

Figure 18.18 Flowering plants make use of a wide variety of pollinators, primarily, but certainly not exclusively, insects. (a) Flowers pollinated by bees are frequently blue or yellow, the colors bees see best. Bee-pollinated flowers may also give off a pleasant scent and provide suitable bee landing spots. (b) Hummingbirds are attracted to red flowers. Birds have a poor sense of smell, and hummingbird-pollinated flowers are usually odorless. The trumpetlike shape of many hummingbird-pollinated flowers is well suited to the hovering flight hummingbirds use while feeding. (c) Flowers pollinated by bats—which are active at night—tend to be white, so more easily visible in darkness, and pungent. They also often grow at the tops of plants for easy access.

SCIENCE AND SOCIETY

Ethnobotany

Human societies use plants in many different ways—for food, medicine, shelter, clothing, tools, ceremonial functions, and other purposes. Ethnobotany is the study of how people use plants, and ethnobotanists generally conduct their work by interviewing local peoples and studying their traditions and habits. Paleoethnobotanists focus on plant use in prehistoric times by examining seeds, pollen, wood, and other plant remains found at archaeological sites.

Although ethnobotanists are interested in all types of plant use, the study of medicinal plants, with its potential for social and economic benefit, has always been of particular interest. In fact, ethnobotanical investigations have led to the development of numerous important medicines. Quinine, the first antimalarial drug, comes from the bark of cinchona trees, long used in native Peruvian medicine to treat fever, digestive ailments, and malaria. Aspirin originally came from willow bark, which has been used for thousands of years to relieve pain (Chapter 11). More recently, Madagascar periwinkle, a plant used by several native peoples for diabetes and other conditions, provided two new cancer drugs.

Scientists pursuing drug development have traditionally relied on the knowledge of local healers in their search for promising plants. However, the development of modern drugs from medicinal plant compounds leads to a difficult ethical issue. What are the rights of indigenous peoples, and how can these be protected? Critics are quick to point to the fact that cancer drugs developed from Madagascar periwinkle produced over a billion dollars in profit for the pharmaceutical giant Eli Lilly, but nothing for traditional societies in Madagascar. Some drug developers now attempt to ensure that local peoples also benefit. For example, when the AIDS Research Alliance found an anti-HIV compound in a Samoan medicinal tree, the group made direct contributions to the village where the tree was known and also promised 20 percent of the profits from drugs that are eventually developed. Unfortunately, because of the ongoing loss of native cultures and the extinction of plant species through deforestation and habitat destruction, it is all too likely that many medically useful species will never be known.

Ethnobotanists are interested in how different peoples use plants. Here, a Matses Indian shaman points out a medicinal plant in the rainforest surrounding his village in the Galvez River area, Amazon Basin, Peru.

INSIGHT Not all flowers smell sweet! Flowers of the dead horse arum lily mimic the smell of rotting flesh to attract certain flies that normally lay their eggs in carrion.

INSIGHT Apple seeds contain cyanide, as do those of cherries, apricots, and peaches. In small quantities, these seeds are unlikely to harm you. But don't get in the habit of eating them.

CHECK YOURSELF

1. How does the alternation of generations differ among mosses, ferns, and seed plants?
2. Why don't mosses and ferns need pollinators?
3. Fruits evolved to help flowering plants disperse their seeds. How is dispersal adaptive?

CHECK YOUR ANSWERS

1. In mosses, the gametophyte is much larger than the sporophyte, and the sporophyte is completely dependent on the gametophyte for water and nutrients. In ferns, the sporophyte is larger than the gametophyte. The gametophyte is very small but independent. In seed plants, the sporophyte is larger than the gametophyte, and the gametophyte is completely dependent on the sporophyte—in other words, seed plants are sort of the "opposites" of mosses.
2. In both groups, gametophytes release sperm that swim directly to the eggs to fertilize them.
3. Dispersal allows plants to spread their seeds to a variety of environments, some of which may be well suited to their survival and reproduction.

Conceptual Science Place

Phlyctochytrium Zoospore Release

18.7 Fungi

When you keep a loaf of bread too long, or leave a lemon in the fruit bowl for weeks, you often end up with some fuzzy stuff called mold. Mold is a fungus, a living organism belonging to the group Fungi, and it is doing what you once meant to do—it is eating your food!

Figure 18.19 Many fungi are decomposers and obtain their nutrients from dead organic material. This fungus is growing on a dead grasshopper. The mushrooms are spore-producing structures.

Fungi were once grouped with plants because of their stationary way of life but are actually more closely related to animals. Like animals, fungi are heterotrophs that obtain food from other organisms. Some fungi, like yeast, are single-celled, but most species are multicellular. Multicellular fungi are composed of masses of small threadlike filaments. They obtain food by secreting digestive enzymes over organic matter and then absorbing the nutrients. Many fungi are decomposers that obtain the bulk of their nutrients from dead organic material, as the one in Figure 18.19 is doing. Fungi, along with bacteria, are the most important decomposers in terrestrial ecosystems.

Fungi reproduce either sexually or asexually. Most species use both strategies at some point in their life cycles. Reproduction in fungi occurs through the formation of spores, tiny reproductive bodies that can exist in a dormant state until conditions become favorable for growth. The spore-producing structures of some fungi are familiar to us as mushrooms. Fungal spores spread far and wide by floating through air or water—this is why mold finds your leftovers no matter where you hide them.

INSIGHT The largest living organism in the world is a fungus—an underground honey mushroom in Oregon's Blue Mountains that measures 5.6 kilometers (3.5 miles) across! Scientists have collected DNA from different parts of this organism to confirm that it is all one individual.

Fungi are essential to the survival and growth of many, perhaps most, plants. This is because in most plant species, roots form close associations with fungi known as **mycorrhizae** ("my-kuh-RYE-zuh"). The fungus receives nutrients from the plant while helping roots absorb water and minerals from soil.

Fungi important to humans include yeast, which is used in baking and brewing, and edible mushrooms and truffles. Fungi are also used to make blue cheeses such as Roquefort and gorgonzola (the blue stuff is actually a colossal number of minuscule fungal spores—enjoy!). Penicillin, the first antibiotic, was originally isolated from a fungus in 1928. Human fungal diseases include yeast infections, ringworm, and athlete's foot. Dutch elm disease, which devastated elm trees across the United States, is also caused by a fungus.

Animals

Sponges
Cnidarians
Flatworms
Roundworms
Arthropods
Mollusks
Annelids
Echinoderms
Chordates

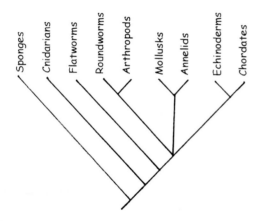

Figure 18.20 This cladogram shows relationships among the major groups of animals.

18.8 Animals

Animals include everything from sea stars and beetles to coral and antelope. **Animals** are multicellular, heterotrophic eukaryotes that obtain nutrients by eating other organisms. Animals *ingest* food, taking it into their bodies for digestion (unlike fungi, those other multicellular heterotrophs, which secrete digestive enzymes *out over* their food). Most animals reproduce sexually and are diploid during most of their life cycle, with the gametes—sperm and eggs—being the only haploid stage. Many animals go through a juvenile period as a **larva** that is markedly different from adults in form and ecology. Think, for example, of the caterpillars of butterflies or the tadpoles of frogs. Most animals also have muscles for moving, sense organs for taking in information from their environments, and nervous systems for controlling their actions. The cladogram in Figure 18.20 shows evolutionary relationships among major animal groups.

Sponges

Sponges (Figure 18.21) are sedentary marine animals. Most possess a tubelike structure with a large central cavity. Special cells in a sponge beat their flagella to produce a constant water current through the animal. Water enters through numerous pores, flows into the sponge's central cavity, and goes out the top. The purpose of this current is to catch food—sponge cells trap bacteria from the moving water, digest them, and then distribute the nutrients to other cells. Sponges are the only animals that lack tissues, groups of similar cells that perform a certain function. This allows sponges to do unusual things—for example, if you separate a sponge's cells by passing it through a sieve, the cells will reassemble on the other side, forming a new sponge. No other animal can do that. The sponge's ability to reassemble appears to rely on a protein similar to human fibronectin, which is found on cell surfaces and helps to promote cellular adhesion.

Figure 18.21 This is a purple tube sponge.

Figure 18.22 This purple-banded jellyfish is a medusa-stage cnidarian.

Cnidarians

Cnidarians ("nye-DARE-ee-uhns") include animals such as jellyfish, sea anemones, corals, and hydras. Unlike sponges, cnidarians have two distinct tissue layers—an outer layer that protects the body and an inner layer that digests food. These are separated by a jellylike middle layer. Cnidarians catch prey using tentacles armed with barbed stinging cells. In many species, the stinging cells release potent toxins. (This is why jellyfish can be a danger to ocean swimmers.) Prey are digested in a *gastrovascular cavity* that has a single opening that serves as both mouth and anus. Many cnidarians alternate between a sedentary polyp stage and a mobile, bell-shaped, medusa stage. In cnidarians such as sea anemones and corals, the polyp stage dominates the life cycle. In cnidarians such as jellyfish (Figure 18.22), the medusa form dominates.

INTEGRATED SCIENCE EARTH SCIENCE

Coral Bleaching

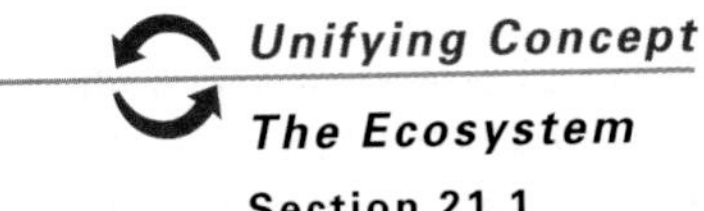

Unifying Concept

The Ecosystem

Section 21.1

Coral reefs occur in tropical oceans, where they are found in clear, shallow waters with temperatures between 20 and 28 °C (68 and 82°F) (Figure 18.23). They are among the most diverse, and most important, ecosystems in the world. Numerous marine species, including commercially important fish, spend all or part of their life cycles in coral reefs. Reefs also help protect shorelines from the effects of ocean waves. Unfortunately, many scientists fear that we may lose all our coral reefs by the end of the century due to a phenomenon called coral bleaching.

Corals are cnidarians that form colonies of tiny polyps encased in calcium carbonate skeletons. Unlike other cnidarians, corals obtain the bulk of their nutrients from photosynthesizing dinoflagellates that live within their cells. Coral bleaching occurs when corals evict their dinoflagellates. The corals quite literally turn white (Figure 18.24), since it is the dinoflagellates that give them their colors.

Coral bleaching is most often triggered by increases in seawater temperature. A rise of as little as a few degrees Celsius can start bleaching. However, only extended periods of warming—not rapid temperature fluctuations—produce bleaching. Scientists have shown that elevated temperatures interfere with dinoflagellate photosynthesis, damaging certain proteins that help convert carbon dioxide to glucose.

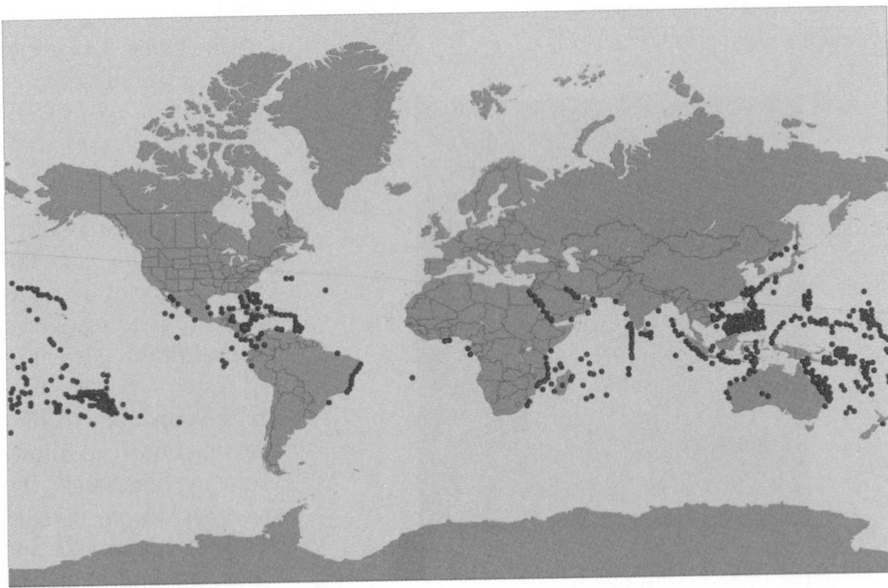

Figure 18.23 Coral reefs (red dots) are found in shallow, warm, tropical waters. Most occur between 30 degrees north latitude and 30 degrees south latitude.

Figure 18.24 This bleaching brain coral is evicting its dinoflagellates.

The inability to complete photosynthesis causes toxins to build up in the dinoflagellates' bodies, which causes the corals to eject them. In the laboratory, scientists have seen polyps expel dinoflagellates into their gastrovascular cavities and then "spit" them out. In addition to elevated water temperatures, coral bleaching has been associated with UV radiation and pollution. Corals can survive for a short time without their dinoflagellates. If water temperatures decrease again, dinoflagellates recolonize bleached corals, and the corals survive. However, if warm temperatures or other stressful conditions continue for too long, the corals starve to death.

Mass coral bleaching events have become common in recent years, almost certainly due to the higher water temperatures that have resulted from global warming (Chapter 25). In 2002, high ocean temperatures affected coral reefs worldwide, killing more than 90 percent of corals in some places. With global temperatures expected to continue their rise, many scientists fear that bleaching will only become more widespread and more devastating. Will coral reefs disappear? That is still uncertain. Scientists do note that corals vary in their preferred water temperatures and their susceptibility to bleaching. For example, certain corals have special fluorescent pigments that they use, like sunscreen, to shield their dinoflagellates. These fluorescent corals have survived mass bleaching episodes better than nonfluorescent corals.

CHECK YOURSELF

1. The temperature increases associated with global warming devastate coral reefs through bleaching. A second effect of global warming is rising sea levels (due primarily to the expansion of seawater as its temperature increases). How would rising sea levels affect corals?
2. If global warming continues, how might the species composition of coral reefs change over time?

CHECK YOUR ANSWERS

1. Corals are found only in shallow waters, where there is enough sunlight for their dinoflagellates to photosynthesize. Rising sea levels will require them to either shift their ranges to shallower ground, or to grow upward quickly. Whether they are able to find appropriate habitat and keep pace with sea-level changes remains to be seen.
2. Coral species that are currently adapted to warmer water temperatures may spread as global warming continues. In addition, coral reefs are likely to include more species that are better able to resist bleaching. For example, many reefs may come to be dominated by species that have fluorescent pigments, as species that lack these sunscreens die out.

Flatworms

Unlike sponges and cnidarians, *flatworms* have distinct "head" and "tail" ends as well as "back" and "belly" sides. A single body opening serves as both mouth and anus, and an elaborately branched digestive tract transports nutrients to the entire body. The flat shape of flatworms allows oxygen to be absorbed efficiently across the skin via diffusion.

Flatworms include free-living (nonparasitic) species as well as parasites such as tapeworms and blood flukes. Tapeworms are long, ribbonlike flatworms that live as intestinal parasites, absorbing food from the digestive tracts of their hosts. Blood flukes live in the blood vessels of humans and other species. In humans, they cause the disease schistosomiasis, which is characterized by diarrhea and intense abdominal pain.

Roundworms

C. elegans Crawling

Roundworms (not to be confused with the more familiar earthworms) occupy almost every conceivable habitat, and often in staggering numbers—thousands of individuals can sometimes be found in a single handful of soil. Roundworms have small, slender bodies with tapered ends and a round cross section. They have both a mouth and an anus. Like their relatives the arthropods, roundworms have a tough outer cuticle that is shed periodically during growth. Roundworms may eat a variety of things—bacteria, plants, fungi, other animals—but many specialize on decaying organic material. In many habitats, they are important decomposers. The muscles of roundworms all run longitudinally (from head to tail) down the body. As a result, roundworms move like flailing whips as muscles on alternate sides of the body contract. Roundworms are responsible for several human diseases, including hookworm, pinworm, elephantitis, and trichinosis. One roundworm of particular significance to biologists is *Caenorhabditis elegans*, which, for historical reasons, is used in many studies of genetics and development. *C. elegans* was in fact the first species to have its entire genome sequenced.

Arthropods

Conceptual Science Place

Lobster Mouthparts

Arthropods include species as diverse as lobsters, barnacles, spiders, scorpions, ticks, centipedes, and insects. Arthropods are found in just about every known habitat on Earth. All arthropods have an external **exoskeleton** made of chitin that protects and supports the body. The exoskeleton is incapable of growth and must be shed periodically as animals grow. Arthropods also have segmented bodies and jointed legs. However, in most species, some segments are fused and some legs have been modified during the course of evolution to form mouthparts, antennae, or reproductive structures. Arthropods have a brain and a number of highly developed sense organs. Many arthropods pass through a distinct larval stage during their growth and development.

The major groups of arthropods are the crustaceans, the chelicerates, and the uniramians (Figure 18.25). The *crustaceans* include lobsters, crabs, shrimp, krill, and barnacles. The *chelicerates* include horseshoe crabs, spiders, scorpions, ticks, and mites. Chelicerates have four pairs of legs and one pair of mouthparts. The

(a)

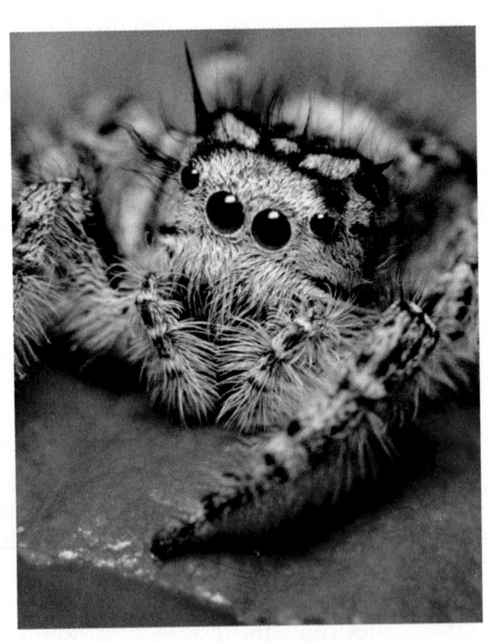
(b)

(c)

Figure 18.25 Arthropods have segmented bodies and jointed legs. (a) Barnacles belong to the group of arthropods known as crustaceans. Although the shells of barnacles may lead some people to confuse them with mollusks, their jointed legs are an arthropod feature. (b) Spiders belong to the group of arthropods known as chelicerates. This is a metaphid jumping spider. (c) Insects belong to the uniramians. This damselfly is shedding its exoskeleton.

INSIGHT The earliest flying insects weren't able to fold their wings back on their bodies. That evolutionary innovation came later in insect history. Dragonflies are an example of a group that still retains the old trait—have you noticed how their wings always stick straight out?

uniramians include centipedes, millipedes, and insects. Insects are noteworthy as the most diverse group of living organisms on Earth—there are over a million known species and perhaps as many as ten times that number waiting to be discovered. All insects have three body parts—a head, thorax, and abdomen—and three pairs of legs. Most also have two pairs of wings. Many insects are important to humans as plant pollinators. Others impact humans as disease carriers (mosquitoes carry malaria and West Nile virus) or as agricultural pests.

For all their success, the one thing insects don't seem to have is size—why are there no giant insects? Much of the answer has to do with the way insects breathe. Insects rely on *trachea*, series of branched tubules connected to the outside air, for oxygen. Oxygen must diffuse through the trachea to reach the tissues, a strategy that works well only in extremely small bodies. (This is because the time it takes a molecule to diffuse a certain distance is proportional to the square of that distance, as we discussed in Chapter 15.) Interestingly, there were much bigger insects during the Carboniferous Period, some 300 million years ago, when atmospheric oxygen levels were much higher than they are today. Those insects included a 5-foot millipede and a dragonfly with a $2\frac{1}{2}$-foot wingspan!

Unifying Concept
The Second Law of Thermodynamics
Section 6.5

Mollusks

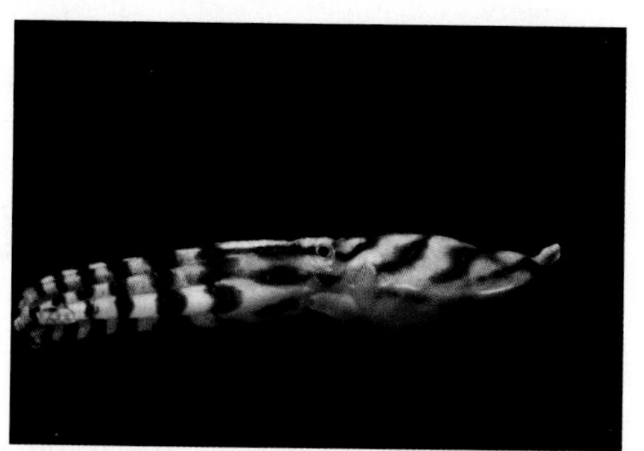

Figure 18.26 The blue-ringed octopus is a cephalopod, a type of mollusk.

Mollusks are soft-bodied animals such as clams, oysters, squids, octopuses, snails, and slugs. Most mollusks have a protective shell, although the shell is tiny in some species (squids) and entirely absent in others (octopuses and slugs). All mollusks have a muscular "foot" responsible for locomotion, a visceral mass that holds the digestive and reproductive organs, and a mantle that secretes the shell. There are three main groups of mollusks. *Bivalves* have two hinged shells and include species such as clams, oysters, mussels, and scallops. Most bivalves are sedentary and feed by filtering small particles from the water. *Cephalopods* such as squids and octopuses (Figure 18.26) are active predators that use arms (eight in octopuses and ten in squids) to capture prey. Cephalopods also have well-developed brains and eyes. *Gastropods* have a single, spiral shell and include species such as snails, abalone, and limpets. Most gastropods are herbivores.

Annelids

Earthworm Locomotion

Annelids are segmented worms. The group includes earthworms, leeches, and the less familiar marine bristleworms. The muscles of annelids are arranged in both circular (around the body) and longitudinal (head-to-tail) orientations, allowing for great flexibility of motion. Unlike roundworms, for example, annelids are able to contract one part of the body while keeping the rest of the body still. The familiar earthworms feed by passing large amounts of soil through their digestive tracts and absorbing the available nutrients. Earthworms play an essential role in decomposing organic materials, and their burrowing activity helps to aerate soil, supplying it with oxygen. Leeches are external parasites that feed off the blood of their hosts. Leeches cut through skin using their bladelike teeth and secrete anticoagulants to keep blood from clotting while they feed.

Echinoderms

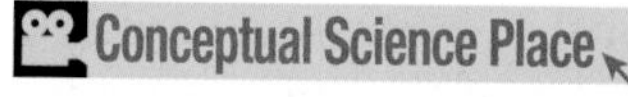

Echinoderm Tube Feet

Echinoderms are a group of marine animals with spiny surfaces. This group includes the sea stars, sea urchins, and sea cucumbers (Figure 18.27). All echinoderms have internal skeletons, or **endoskeletons**, made of small interlocking plates. They use small, suckerlike appendages called tube feet to move, stick to rocks and other surfaces, and, in sea stars, to pry open the shells of bivalves. Most echinoderms are bottom-dwellers and move very slowly.

Figure 18.27 A close encounter between two echinoderms: A sunflower star attempts to prey on a sea cucumber, which manages to wriggle out of its captor's arms.

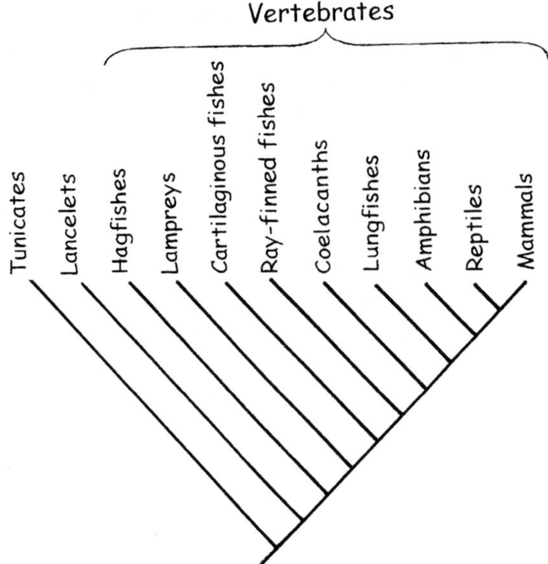

Figure 18.28 The chordates consist of tunicates, lancelets, and vertebrates.

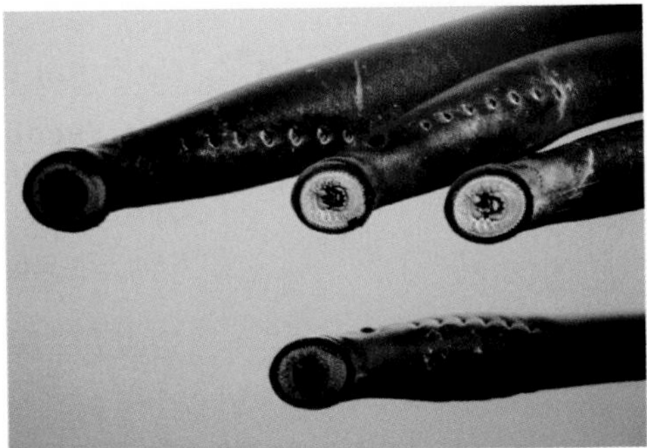

Figure 18.29 The earliest vertebrates did not have hinged jaws. Modern lampreys retain this condition, as is apparent from this photograph of several individuals clinging to glass.

INSIGHT At one time, "bleeding" a patient with leeches was standard medical treatment for a wide variety of ailments. Then this practice was viewed as medical quackery. Now, the leech is back! Leeches release powerful anticoagulants—substances that keep blood from clotting—when they feast on their hosts. And leech anticoagulants, it turns out, are still better than anything in the arsenal of modern medicine.

Chordates

The *chordates* include tunicates, lancelets, and vertebrates, the group that includes humans. Chordates share four key features: a brain and a spinal cord that runs along the back of the body; a *notochord*—a stiff but bendable rod that supports the back; gill slits; and a tail that extends beyond the anus. These features are not apparent in all adult chordates (for example, humans don't have tails), but are generally present at some stage of development (human embryos do go through a tailed stage). The evolutionary relationships among chordates are shown in Figure 18.28.

Tunicates are sedentary marine species also known as sea squirts. They feed by filtering small particles from the water. *Lancelets* are small, swimming, blade-shaped marine species that live buried in sand. Like tunicates, lancelets filter food from the water.

Vertebrates are animals with backbones. The vertebrates include several groups of fishes as well as amphibians, reptiles, and mammals. The earliest vertebrates had mouths lacking hinged jaws. The only jawless vertebrates still in existence are hagfishes, slimy marine species that eat marine worms and other animals, and lampreys, parasites that suck blood from ray-finned fishes (Figure 18.29).

Cartilaginous fishes have skeletons made of cartilage rather than bone. Cartilaginous fishes include sharks, skates, and rays. Sharks are known for being frighteningly effective predators. To aid them in hunting, sharks have *electroreceptive organs* that are able to detect the electric currents given off by the muscles and nerves of nearby prey. (We'll see why nerves and muscles give off electric currents in Chapter 19.)

Ray-finned fishes include most of the animals we think of as "fish"—tuna, salmon, perch, bass, and so forth. Ray-finned fishes have a gas-containing sac called a *swim bladder* that they use to adjust their buoyancy. The swim bladder is filled with just enough gas that the overall density (mass/volume) of the fish is the same as the density of water—because of this, a ray-finned fish neither sinks nor

Figure 18.30 Lungfishes may look like other ray-finned fishes, but they are more closely related to terrestrial vertebrates. Note the unusual shape of the fins.

INSIGHT Hundreds of species of salamanders, found in the family Plethodontidae, have neither lungs nor gills. Instead, they get all their oxygen through their skins. Scientists hypothesize that lunglessness originally evolved in fast-flowing mountain streams where lungs—more or less big bags of air—made animals too buoyant to swim effectively.

floats, allowing it great flexibility of motion in the water. Cartilaginous fishes, which do not have swim bladders, sink if they stop swimming.

Lungfishes (Figure 18.30) and *coelacanths* resemble ray-finned fishes superficially, but they are actually more closely related to terrestrial vertebrates such as amphibians, reptiles, and mammals. The bones in their fins are arranged serially, like those in the limbs of terrestrial vertebrates, rather than as rays emanating from the base of the fin, as in ray-finned fishes. Although there are only a few species, lungfishes and coelacanths are of particular interest to evolutionary biologists because they provide clues about how the limbs of terrestrial vertebrates evolved.

Amphibians include salamanders, frogs, and caecilians. Salamanders have elongated bodies and tails, frogs have compact bodies suited to jumping, and caecilians, the least familiar group, are limbless amphibians that live primarily as burrowers. The name "amphibian" refers to the fact that many species make use of both aquatic and terrestrial habitats. In many amphibians, larvae are aquatic and eventually metamorphose into terrestrial adults. However, some amphibians are entirely aquatic and many are entirely terrestrial. All amphibians are restricted to moist environments because their skins are composed of living cells that are vulnerable to drying out. Amphibian eggs, which have no shells, also require moisture (Figure 18.31).

In recent decades, dozens of amphibian species have suddenly gone extinct, and hundreds have suffered drastic, unexplained population declines. Particularly disturbing is the fact that many of these, including the golden toads shown in Figure 18.32, have disappeared from protected, supposedly pristine habitats. Amphibians are believed to be particularly vulnerable to pollution and environmental contamination because water and other substances are easily absorbed through their skins.

Reptiles (including the feathered ones—birds) and mammals are *amniotes*. Amniotes have skin that is made of dead cells, helping to prevent water loss. Amniotes also have shelled eggs that are not vulnerable to drying out. Both these traits make them better adapted than amphibians to diverse terrestrial habitats. *Reptiles* include turtles, lizards, crocodilians, and birds. Turtles have protective shells that are actually, believe it or not, modified ribs. (Imagine squeezing your entire body inside your ribcage!) Lizards are a diverse group found in a wide array of terrestrial environments. Snakes are a group of lizards that have lost their legs and evolved adaptations for subduing large prey and swallowing them whole (Figure 18.33a). Birds are reptiles (a group of dinosaurs, in fact) that have evolved the ability to fly (Figure 18.33b). Flight in birds is associated with a suite of adaptations including wings, feathers, hollow bones, air sacs in the body, and a four-chambered heart.

Figure 18.31 Amphibian eggs have no shells and so require moisture to keep from drying out. Because the eggs are transparent, amphibians have long been favorite organisms for studies of animal development.

Figure 18.32 The striking golden toad is only one of dozens of amphibian species to disappear in the last few decades. Particularly disturbing is the fact that it had inhabited a protected and supposedly pristine reserve in the Monteverde Cloud Forest, Costa Rica.

(a)

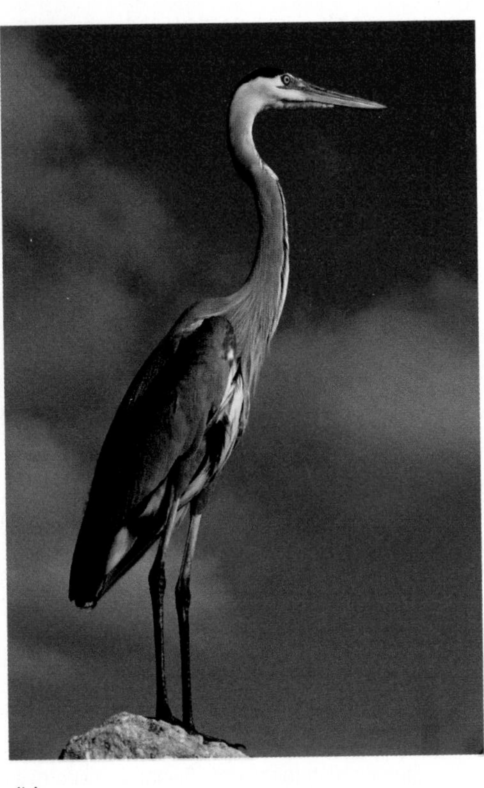

(b)

Figure 18.33 Reptiles include turtles, lizards, crocodilians, and birds. (a) Many snakes have adaptations for swallowing large prey whole. This rat snake is eating a bird. (b) Birds are reptiles that have evolved the ability to fly. This is a great blue heron.

Birds and crocodiles are each other's closest living relatives. Both show parental care of young (the photo on page 348 shows a female crocodile taking care of her offspring) in addition to sharing skeletal features.

INSIGHT Flight has evolved three separate times in vertebrates—in birds, bats, and the extinct pterodactyls.

With the exception of birds, reptiles are **ectotherms**, animals that regulate their body temperature behaviorally by seeking sunlight when they need to warm up and shade when they need to cool down. The body temperature of ectotherms tends to vary, to some degree, depending on environmental conditions. Birds (as well as mammals) are **endotherms**, animals that maintain a relatively constant and relatively high body temperature by metabolizing large amounts of food. During metabolism, when organic molecules are broken down to make ATP, energy is lost to the environment as heat—this energy helps warm the body. (In Chapter 21, we will calculate exactly how much energy is lost as heat when glucose is broken down to make ATP.) Because of this, endotherms have to eat much more food than equivalently sized ectotherms.

Unifying Concept
The Second Law of Thermodynamics
Section 6.5

Mammals are amniotes that have hair and feed their young milk. All mammals are endotherms. The majority of mammals live on the ground, but bats fly and two groups, seals and whales, are partly or fully aquatic. There are three major groups of mammals. Monotremes such as the platypus and spiny echidna differ from other mammals in that they lay eggs. Monotremes feed their young milk but do not have nipples. Instead, milk is secreted over the skin and lapped up by the young. Marsupials such as possums, koalas, and kangaroos give birth to live young at a very early developmental stage. Upon birth, the young crawl up to the mother's pouch and attach to a nipple where they feed and continue development. Placentals, which include the vast majority of living mammal species, also give birth to live young, but these are born at a more advanced stage of development than in marsupials.

INSIGHT Ectotherms and endotherms were once called "cold-blooded" and "warm-blooded" respectively, but these terms just aren't accurate. For example, some desert lizards have body temperatures well over 100 degrees Fahrenheit, hotter than the body temperatures of most birds and mammals.

HISTORY OF SCIENCE

Classifying the Platypus

The first dried platypus specimen reached England from Australia in 1799 and created an immediate sensation. This animal, native to eastern Australia and Tasmania, has a ducklike bill, webbed front legs, clawed hind legs, and a covering of thick fur. It also possesses a cloaca—a single opening for the reproductive, excretory, and digestive systems—and lacks nipples, two attributes that characterize birds but not mammals. To all appearances, the platypus was an impossible cross between a bird and a mammal. How to classify it? Everyone had a different answer. Very quickly, however, attention fastened on one key question—how did this odd animal reproduce? Did it lay eggs? Did it give birth to live young?

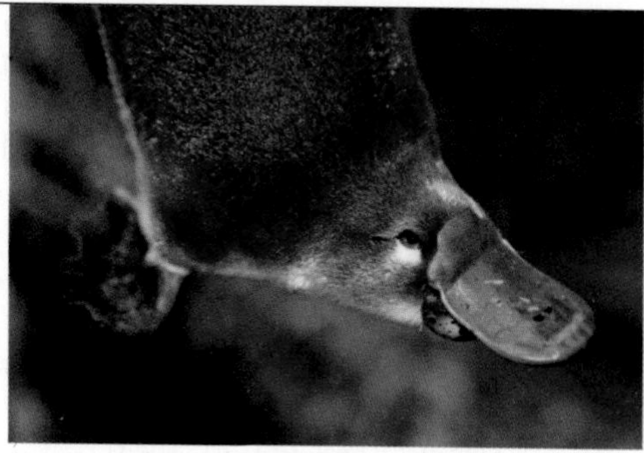
The platypus is an egg-laying mammal.

The first breakthrough came in 1821, when naval surgeon Patrick Hill reported that certain Australian Aborigines were familiar with the platypus and knew that it laid eggs in nests on the water surface. Then, in 1824, the anatomist Johann Meckel found mammary glands on a specimen—these opened directly to the skin without nipples. But this led to another conundrum—how could a platypus with a ducklike bill suckle? In 1831, platypus nests containing broken eggshells were found, and milk was seen to ooze from the skin of a female platypus's abdomen. In 1834, the famous anatomist Richard Owen examined the mouth of a baby platypus and concluded it could suckle in the usual fashion. Plus, he found milk in its stomach. Finally, in 1884, a female platypus was shot while in the process of laying eggs. To those who had yet to be convinced, this was decisive. The platypus was indeed a mammal and it did indeed lay eggs.

Today, the platypus is classified in the mammalian clade Monotremata, which means "one hole" and refers to the animals' single exit for reproductive, excretory, and digestive systems. Only one other creature—the spiny echidna, the platypus's closest relative left on Earth—belongs to the group.

CHECK YOURSELF

Dispersal is important for most living organisms. How do animals disperse? How do animal dispersal strategies compare with those of plants?

CHECK YOUR ANSWER

Most animals are able to move. The ones that are sedentary (sponges, some bivalves, barnacles, and so forth) typically release their eggs into the water, where water currents disperse them. Some sedentary animals also have a mobile larval stage. Plants, which are sedentary, are able to disperse by forming spores or seeds that are carried by wind, water, or mobile animals. Flowering plants form special structures called fruit that help them disperse their seeds.

INTEGRATED SCIENCE PHYSICS

How Birds Fly

Humans have always been interested in how birds fly. Without birds as a model, would we ever have dreamed up something as crazy as an airplane? How can something heavier than air stay aloft?

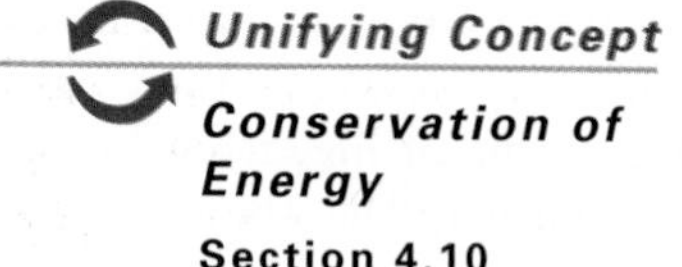
Unifying Concept
Conservation of Energy
Section 4.10

It's all about shape. As shown in Figure 18.34, the wings of birds are *airfoils*. The curved shape of the wing causes air to flow faster over the top of the wing than under the wing. This is because, as a bird cuts through the air, air molecules have a greater distance to travel over the wing and so must move faster. Bernoulli's Principle (Appendix E) shows that when the speed of air over a surface increases, its pressure decreases. The reason for this is that, when air molecules accelerate horizontally, energy conservation means there is less energy available for them to move vertically. And it is the vertical movement of air against the wing that determines air pressure on the wing. The result of greater air speed over the wing than under it is that air pressure above the wing is lower than air pressure below the wing. This produces *lift*, an upward force that counters gravity and keeps birds aloft.

Unifying Concept
Newton's Third Law
Section 3.4

Birds move forward through the air by flapping their wings. During the downstroke, the wings push against the air and the air pushes back, creating *thrust*, a force that propels the animal forward. As birds pull their wings back up, however, they produce *drag*, a force that drives them backwards. Birds twist their wings during the upstroke to reduce drag. (You do something similar when you swim the breaststroke—on the downstroke, you push your fully extended arms back, and your body moves forward. On the upstroke, however, you pull your arms in to reduce drag.)

Some birds manage to fly for long periods without flapping their wings—this is called soaring. You often see eagles and vultures soar. Soaring is possible because the birds have located a thermal, a pocket of rising hot air, and are floating on it. It's sort of like sitting on top of a geyser but (luckily for the birds) considerably more controlled.

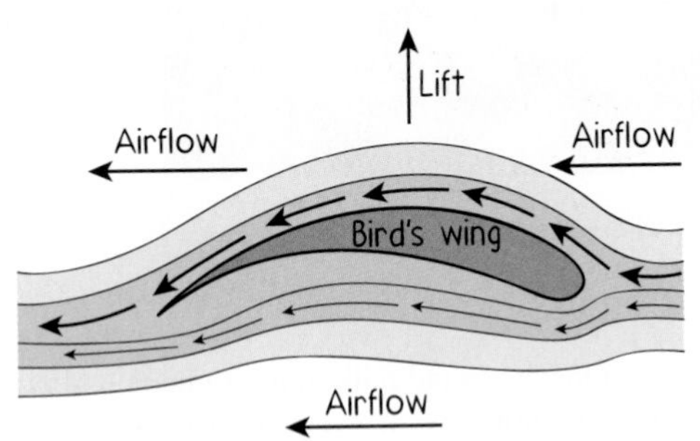

Figure 18.34 Bird wings function as airfoils. Lift is produced by the difference in air pressure above and below the wing.

CHECK YOURSELF

1. Does an eagle or a sparrow produce more lift as it flies?
2. Given the way lift is produced, do you think it is a greater challenge for large birds or small birds to keep themselves aloft?

CHECK YOUR ANSWERS

1. An eagle produces more lift. Birds must produce enough lift to counter gravity in order to stay in the air. An eagle weighs more than a sparrow, and so produces more lift.
2. Large birds have a more difficult time keeping themselves aloft. This is because lift is generated by air pressure under the wings, and so is proportional to the surface area of the wings. Lift must counter gravity, which is proportional to the volume of the bird. Since the surface area-to-volume ratio is smaller in larger animals (Chapters 15 and 17), larger birds have more difficulty generating the necessary lift. This explains why the wings of large birds are relatively larger than the wings of small birds. It also explains why small birds are more adept fliers than larger birds. For example, tiny hummingbirds have tremendous maneuverability in the air and can even hover, an ability that is rare among birds and critical to the way they feed. On the other hand, many large birds have trouble getting off the ground and then actually spend much of their time soaring on thermals.

18.9 Viruses and Infectious Molecules

Viruses straddle the line between living and nonliving. They are not made of cells and they can reproduce only within a host cell, but they have genes and they evolve. **Viruses** are small pieces of genetic material wrapped in a protein coat. They range from about 20 to 400 nanometers in size. Many viruses have normal, double-stranded DNA genomes, but others use single-stranded DNA, single-stranded RNA, or double-stranded RNA. Viruses reproduce by infecting a host cell and then using the cell's enzymes and ribosomes to copy their genetic material and build viral proteins. These are then assembled to form new viruses. Viruses infect all other forms of life, from bacteria to plants and animals. Human immunodeficiency virus (HIV), the virus that causes acquired immunodeficiency syndrome (AIDS) in humans, is shown infecting an immune cell in Figure 18.35.

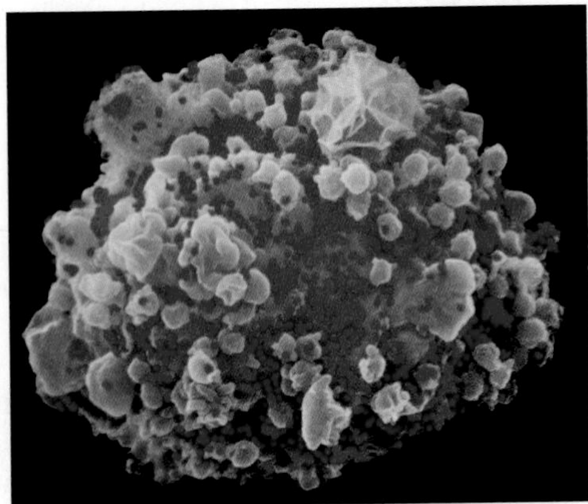

Figure 18.35 This photo shows a human immune cell infected with HIV, the virus that causes AIDS. The cell is covered with viral particles that will soon spread and infect other cells.

Where did viruses come from? Most scientists believe viruses originate when little pieces of host DNA or RNA evolve the ability to move from one cell to another. This would explain why viral genomes often resemble the DNA of their hosts. This also means that viruses have almost certainly originated numerous times.

INSIGHT Are viruses ever beneficial to their hosts? Not generally. However, in recent years, scientists have manipulated the genomes of certain human cold viruses to make them preferentially attack cancer cells in the body. These viruses are known as oncolytic viruses (*onco* means "cancer" and *lytic* means "killing"). Where radiation and chemotherapy generally kill many normal cells in addition to tumor cells, oncolytic viruses have been designed to target cancer cells specifically. As a result, treatment with oncolytic viruses has fewer negative side effects than traditional therapies. A number of oncolytic viruses are now in clinical trials.

Viruses are responsible for many human diseases, including the common cold, flu, AIDS, herpes, and smallpox. One feature of viruses that makes them hard to deal with from the point of view of disease control is that they mutate very quickly. This is particularly true of viruses with RNA genomes, since there is no error-checking and repair system for copying RNA, as there is with DNA. The rapid mutation of viral genomes explains why there's never a shortage of colds to catch and why the flu comes back, in a different form, year after year.

Viroids are circular molecules of RNA that infect plants. They are smaller even than viruses, and lack a protein coat. Like viruses, however, they infect cells and then use their host's cellular machinery to make copies of themselves. Viroids typically spread through seeds or pollen.

Prions are unique among infectious agents in that they lack genetic material altogether. Prions are misfolded proteins believed to cause mad cow disease and the related Creutzfeldt-Jakob disease in humans. Both of these conditions are characterized by fatal brain degeneration. Prions infect cells, where they "reproduce" by converting normal proteins to the misfolded, prion variety. Disease-causing prions are most likely transmitted through the consumption of infected animal products. Cooking, which destroys the nucleic acids found in all other types of infectious agents, has no effect on prions.

SCIENCE AND SOCIETY

Bird Flu

One of the great fears of scientists at the World Health Organization and elsewhere is bird flu. Bird flu, which has devastated populations of domesticated birds, is caused by a virus that occurs naturally among wild birds. In 1997, the first case of a human infected by bird flu was reported in Hong Kong, and dozens of additional cases have been seen since then. So far, however, the virus cannot be transmitted easily from person to person. The evolution of this capability is the event scientists await with trepidation.

This fear turns out to be more than justified. Scientists recently discovered that the infamous "Spanish flu" epidemic of 1918—which killed more people than any other disease over a similar length of time—was a bird flu that became easily transmissible among humans. Using virus fragments removed from three individuals who succumbed during the 1918 epidemic (two soldiers whose tissues had been saved by the U.S. Army and an Alaskan woman buried in permanently frozen ground), scientists painstakingly reconstructed the genetic sequence of the virus. They were startled to discover that this was no human flu virus—it was the bird flu virus, changed in critical ways that allowed it to spread easily among humans. (Interestingly, the viruses responsible for the two other devastating flu epidemics of the century, in 1957 and 1968, appear from their genetic sequences to be hybrid viruses—human flu viruses that picked up certain genes from the bird flu virus.)

Many people had opposed research into the 1918 virus, fearing the accidental release of the virus or its eventual use as a biological weapon. Most scientists, however, felt that the benefits outweighed the dangers. Knowing the genetic characteristics of the 1918 virus allows scientists to track the evolution of current bird flu viruses and to determine whether they are acquiring the genetic changes that made the 1918 virus so deadly. It appears that current bird flu strains have some, but not all, the crucial genetic characteristics of the 1918 virus. Reconstructing the 1918 virus may also help with the development of vaccines that will be effective against any eventual outbreak.

CHAPTER 18 Review

Summary of Terms

Alternation of generations The plant life cycle, which alternates between a haploid gametophyte stage and a diploid sporophyte stage.

Animals A clade of multicellular heterotrophic eukaryotes that take food into their bodies for digestion.

Archaea One of the three domains of life, consisting of prokaryotic organisms, many of which are adapted to extreme environments.

Bacteria One of the three domains of life, consisting of a wide range of generally single-celled prokaryotic organisms.

Clade A group of species that includes an ancestor and all its descendants.

Cladogram A diagram that shows the history of speciation events among a group of organisms.

Ectotherms Organisms that regulate their body temperature behaviorally, by seeking either warm or cool areas—their body temperature tends to fluctuate depending on environmental conditions.

Endoskeleton An internal skeleton, such as that found in echinoderms and chordates.

Endotherms Organisms that rely on food metabolism to maintain a relatively constant and relatively high body temperature.

Exoskeleton An external skeleton, such as that found in arthropods.

Flower The reproductive structure of flowering plants, which may include stamens (male reproductive structures) and/or carpels (female reproductive structures), as well as petals.

Fruit In flowering plants, a structure surrounding the seeds that typically contributes to seed dispersal.

Fungi A clade of heterotrophic eukaryotes that obtain food by secreting digestive enzymes over organic matter and then absorbing the nutrients.
Larva A stage in the growth and development of animals that is distinct from the adult in form and ecology.
Mycorrhizae Close associations between fungi and the roots of plants in which the fungi obtain nutrients from the plant while helping the roots absorb water and minerals from the soil.
Plants A clade of autotrophic, multicellular, terrestrial eukaryotes that obtain energy through photosynthesis.
Pollen In seed plants, immature male gametophytes wrapped in protective coatings.
Protists Eukaryotic organisms that are not plants, animals, or fungi.
Seed In seed plants, a structure consisting of a sporophyte plant embryo, a food supply, and a tough outer coating.
Viruses Small pieces of genetic material wrapped in protein coats that infect and reproduce within host cells.

Review Questions

18.1 Classifying Living Things

1. How are species grouped together under the Linnaean system?
2. What is a clade?
3. Why are birds reptiles?

18.2 The Three Domains of Life

4. What are the three domains of life? Which domain do eukaryotes belong to?

18.3 Bacteria

5. How do bacteria reproduce? Do they ever exchange genetic material?
6. How do some bacteria respond to poor conditions?
7. Why is bacterial decomposition important?

18.4 Archaea

8. Why are some archaea referred to as "extremophiles"?
9. Some archaea are chemoautotrophs. Explain what this means.

18.5 Protists

10. What are protists?
11. Heterotrophic protists have to be able to move to capture prey. Describe some of the different ways they do this.

18.6 Plants

12. What are the two components of the plant vascular system? What is the function of each?
13. How does the "alternation of generations" differ between mosses and all other plants?
14. How is pollen transferred from one plant to another in conifers? In flowering plants?
15. How does fruit contribute to the dispersal of plants?

18.7 Fungi

16. How do fungi obtain food?
17. What are fungal spores?
18. Why are fungi essential to the growth and survival of so many species of plants?

18.8 Animals

19. Animals and fungi are both heterotrophs—what is different about the way they obtain nutrients?
20. How many body openings do cnidarians have? What about flatworms?
21. What are some features shared by arthropods?
22. The mouth parts, antennae, and reproductive structures of arthropods are modified _______.
23. Describe the three major groups of mollusks.
24. What do all annelids have in common?
25. Why are amphibians generally restricted to moist environments?
26. What is the difference between an ectotherm and an endotherm? Which vertebrates are endotherms?
27. How do monotremes differ from other mammals?

18.9 Viruses and Infectious Molecules

28. What is the structure of a virus?
29. How do viruses reproduce?
30. How did viruses most likely originate?
31. What are prions? How do they "reproduce"?

Multiple Choice Questions

Choose the BEST answer to the following.

1. Under the Linnaean classification system, organisms are grouped together based on
   (a) their shared similarities.
   (b) their evolutionary history.
   (c) their position in a hierarchy from "simple" to "complex."
   (d) none of the above.
2. The scientific name of a species
   (a) consists of two parts, the family name and the species name.
   (b) is always underlined.
   (c) is always Latinized.
   (d) none of the above.
3. Why are birds considered reptiles under a cladistic classification system?
   (a) because feathers evolved from scales
   (b) because birds are more similar to reptiles than was previously realized
   (c) because birds are descended from the last common ancestor of all reptiles
   (d) because they share certain similarities with crocodiles
4. Life would be impossible without bacteria because
   (a) they photosynthesize.
   (b) they function in decomposition.
   (c) they reproduce quickly.
   (d) they occupy habitats where no other organisms can survive.
5. All protists are
   (a) eukaryotes.
   (b) autotrophs.
   (c) heterotrophs.
   (d) single-celled.
6. Which of the following is a characteristic of all plants?
   (a) seeds
   (b) pollen
   (c) swimming sperm
   (d) alternation of generations
7. Fruits help plants
   (a) attract animal pollinators.
   (b) obtain nutrients.
   (c) disperse seeds.
   (d) both a and b.
8. All fungi
   (a) are heterotrophs.
   (b) are multicellular.
   (c) reproduce asexually.
   (d) none of the above.

9. Animals
   (a) are multicellular heterotrophs.
   (b) are more closely related to plants than fungi.
   (c) often go through a haploid larval stage during development.
   (d) include some stationary, nonmobile organisms such as sponges, corals, and certain mushrooms.
10. Which of the following is true?
    (a) All viruses have DNA.
    (b) A few viruses have evolved the ability to reproduce outside a host cell.
    (c) Viruses infect all other forms of life, from bacteria to plants and animals.
    (d) Viroids are circular proteins that infect plants.

INTEGRATED SCIENCE CONCEPTS

Earth Science—Coral Bleaching

1. What single factor is most frequently responsible for coral bleaching events?
2. Why do corals turn white during bleaching episodes?
3. Why is coral bleaching expected to become more widespread and more devastating?

Physics—How Birds Fly

4. How are birds able to stay in the air?
5. How do birds propel themselves forward during flight?
6. How do birds soar?

Active Explorations

1. Let's grow mold! Obtain some bread slices and plastic sandwich bags. Keep the bread in the bags so that you won't breathe in too many spores during the course of the experiment. Now try growing mold under different conditions. How does mold grow best? Does it grow better at room temperature or in the refrigerator? Does it grow better dry, or if the bread has been moistened with water? Does it grow better in sunlight or darkness? Come up with hypotheses to explain your results.
2. Go for a walk and examine some of the flowers you see. Can you guess at the pollinators used by some of these flowers? Look for yellow and blue flowers that may be pollinated by bees and red trumpet-shaped flowers that are likely to be pollinated by hummingbirds. Do you see any white or pale flowers that are closed during the day? These may be pollinated by nocturnal species such as moths or bats. If you wait long enough, you may see pollinators visit a flower! Is it the same species each time?
3. We saw that birds stay in the air due to lift, an upward force produced by the difference in air pressure above and below the wings. This air pressure difference is, in turn, the result of air flowing more quickly over the top of the wing than below the wing. Let's experiment with air flow and air pressure using a tissue. Hold the tissue vertically and blow on one side of it—the side you blow on is analogous to the top of the wing, where air flows faster. What happens to the tissue? Does it do what you expect? Why or why not? Can you produce enough lift to keep the tissue horizontal?
4. Imagine that you are being interviewed by an ethnobotanist about how you use plants in your everyday life. What would you tell her? Remember that ethnobotanists are interested in everything related to plant use, so don't stop at the plants you eat. What are other ways you use plants and plant products? Clothing? Tools? Medicine? Shelter? Decoration?

Exercises

1. Of the three domains of life, Bacteria and Archaea both consist of prokaryotes, whereas Eukarya consists of eukaryotes. Why can't we lump Bacteria and Archaea together and call them all Bacteria?
2. What is the difference between a heterotroph and an autotroph? Name two groups of living organisms that include both heterotrophs and autotrophs.
3. What is the advantage of being able to produce spores, as many bacteria and fungi do?
4. Why is decomposition important? What are some important groups of decomposers?
5. We saw that life on Earth would be impossible without bacteria. Would life on Earth be impossible without eukaryotes?
6. Do all autotrophs use photosynthesis to make food? If not, what do they do?
7. Of the three major plant groups we discussed, which is most dependent on living in a moist habitat? Why? Which is least dependent?
8. You may have heard that moss most often grows on the north sides of trees—a potentially useful fact if you are ever lost in a forest! Why do mosses do best on the north sides?
9. Which plants produce pollen? What are some strategies plants use for pollination? What strategy do most flowering plants employ?
10. Some plants have green flowers. An artichoke, for example, is a green flower. How do you think green flowers are pollinated?
11. Some people are allergic to pollen. Do you think bee-pollinated plants or wind-pollinated plants are more likely to cause allergies? Why?
12. What do fungi and animals have in common? How do they differ?
13. What are two different strategies used by cnidarians to obtain food?
14. Although roundworms and arthropods are different in many ways, they are related groups. Name an important feature they have in common.
15. How do the muscles of roundworms and earthworms differ? What does this mean about the way each type of worm moves?
16. What features do insects and crustaceans share? How are they different?
17. Why are amphibians more dependent on living in a moist habitat than amniotes?
18. Many snakes can survive eating just once every few weeks. Why can't birds do this?
19. Birds and mammals are both endotherms, and they both have four-chambered hearts. Why are birds classified as reptiles rather than as mammals?
20. What are some organisms that reproduce asexually? What are some advantages of asexual reproduction?
21. Most living organisms reproduce sexually sometimes, or have some other mechanism for exchanging genetic material. What is the advantage of sexual reproduction or genetic exchange?
22. Scientists have never been sure whether to classify viruses as living things or nonliving things. In what ways do viruses resemble living things? In what ways do they resemble nonliving things?
23. Write a letter to Grandpa telling him about the current scare over bird flu. Be sure to explain what viruses are and why they sometimes mutate so quickly.

Problems

1. Suppose a species of bacteria divides once every 20 minutes. You start with a single bacterium on your unrefrigerated egg-and-baloney sandwich at 8:00 A.M. Show that when you sit down to lunch at noon, there will be 4096 bacteria on your sandwich.
2. The lightest and heaviest flying birds are the bee hummingbird of Cuba, which weighs about 1.6 grams, and the great bustard of Europe and Asia, which can weigh as much as 21 kilograms.

Show that the bee hummingbird produces about 0.016 newtons of lift when it flies, whereas the great bustard produces about 205.8 newtons of lift. Which species would you expect to have proportionally larger wings? Why?

The bee hummingbird and the great bustard are the lightest and heaviest flying birds respectively. The bee hummingbird weighs less than a penny!

3. Draw a diagram showing how the following organisms are related: broad beech fern, cherry tree, shitake mushroom, hammerhead shark, red-eyed treefrog, sidewinder, albatross, duck-billed platypus, Albert Einstein.

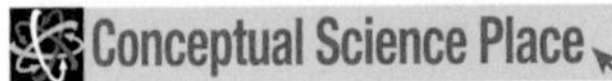

CHAPTER 18 ONLINE RESOURCES

Videos *Paramecium* Cilia, *Paramecium* Vacuole, Diatoms Moving, Dinoflagellates, *Amoeba, Amoeba* Pseudopodia, *Phlyctochytrium* Zoospore Release, *C. elegans* Crawling, Lobster Mouthparts, Earthworm Locomotion, Echinoderm Tube Feet • ***Quiz*** • ***Flashcards*** • ***Links***

A performer juggles axes while walking a tightrope at an Atlanta, Georgia fair.

CHAPTER 19

Human Biology I—Control and Development

CHAPTER OUTLINE

Every feat of the human body requires careful coordination. This is obvious during the performance of a world-class acrobat, but it is no less true of the smallest things we do, such as read a word or take a breath or a step. In the next two chapters, we'll see how the body maintains exquisite control over its systems and allows us to function in a complex world. Why is it that we can ride a bike, juggle, or fold laundry without thinking about what we're doing? What part of the brain is at work when we contemplate Plato or do a math problem? How do our eyes tell our brains what we're seeing? What is "taste" and "smell"? How do our bones and muscles interact with each other and with our other systems?

19.1 Organization of the Human Body

At any moment, a lot of things are going on in our bodies. Some of it we are aware of—we look at a traffic light, step off the curb, think about whether we want noodles or a sandwich for lunch. Other things happen without our knowledge but are no less essential to life—our lungs inhale and exhale, our blood transports oxygen all around the body, our intestines digest food and drink, our kidneys filter wastes, our immune cells dispatch bacteria at the site of a paper cut, we make sperm or release an egg in preparation for reproduction. All these processes require great coordination among participating cells. Cells achieve this coordination by working together in structures with varying levels of complexity—tissues, organs, and organ systems (Figure 19.1).

Multiple cells make up a **tissue**, a group of similar cells that performs a certain function. Skin and muscle are examples of tissues. Multiple tissues combine to make an **organ**, a structure in the body that has a certain function. The heart, stomach, ovaries, and brain are all organs. Finally, multiple organs make up an **organ system** that is responsible for performing particular bodily functions. The digestive system, consisting of the mouth, esophagus, stomach, small and large intestines, as well as the pancreas and liver, is an organ system that is responsible for processing the food we eat. In fact, the human body functions through ten major organ systems—nervous, endocrine, reproductive, sensory, muscular and skeletal, circulatory, respiratory, digestive, excretory, and immune. How these organ systems work is the subject of the next two chapters.

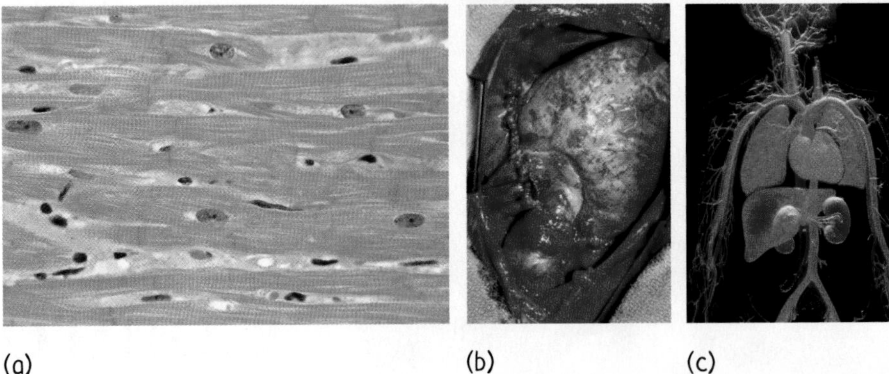

Figure 19.1 Tissues, organs, and organ systems are three levels of organization in the human body. (a) Muscle tissue is composed of multiple muscle cells. This is cardiac muscle from a human heart. (b) The heart is an organ. This heart has just been transplanted and is now beating in the body of the recipient. (c) The circulatory system is made up of the heart, blood, and blood vessels. It is responsible for transporting nutrients, gases, and wastes to different parts of the body.

19.2 Homeostasis

Whether we are swimming in icy waters or hiking through scorching heat, our body temperature stays pretty much the same, at 37° Celsius (98.6° Fahrenheit). This consistency in body temperature is an example of **homeostasis**, the maintenance of a stable internal environment. Homeostasis is a characteristic of all living organisms (Chapter 15), and a huge amount of the body's activity contributes to maintaining it.

As just one example, our cells need a certain amount of oxygen to function, and our lungs and heart maintain a normal level of activity that supplies that oxygen. If our activity level increases, say because we run for the bus or sprint quickly up a long flight of stairs, our cells use up more oxygen. What happens? The body responds by breathing harder to take in more oxygen and by increasing heart rate to get that oxygen to our cells. Once our activity level returns to normal and oxygen use decreases, our breathing and heart rate slow again.

Figure 19.2 Our bodies respond to changing conditions in ways that help maintain homeostasis. For example, when we're cold, shivering warms us up.

To go back to body temperature, when it's cold outside, we *feel* cold and pile on more clothes, or wrap our arms around our bodies to reduce heat loss. In addition, less blood is sent to our limbs and extremities, which lose heat faster than the core of the body. (This explains why our fingers and toes often feel most cold when we're cold.) We may also shiver to generate heat (Figure 19.2). On the other hand, when it's hot outside, we take off our clothes, look for shade, and sweat to cool off. Also, more blood goes to the extremities and to the face, which are good at shedding heat. (This explains why our faces get red when we're hot.)

INSIGHT How does increased activity trigger an increase in oxygen intake? Interestingly, the body doesn't respond to low oxygen levels directly. Instead, increased oxygen use in cellular respiration produces an increase in blood carbon dioxide levels. A higher concentration of carbon dioxide causes blood to become more acidic. And this acidity is detected by sensors, causing the body to increase respiration and heart rate.

Oxygen supply and body temperature are only two of the many variables the body carefully maintains. The amount of water in the body, the concentration of nutrients such as glucose and of waste products in the blood, the concentrations of important ions inside and outside cells, and blood pH—all these variables are carefully controlled as part of maintaining homeostasis.

19.3 The Brain

The brain is the origin of our thoughts, feelings, and desires, and the keeper of our memories. The brain also controls all our activities, from conscious, high-level endeavors such as writing poetry to basic functions essential for life such as breathing, blood circulation, and digestion. Although many functions of the brain are still not completely understood, we do know that certain activities are associated with specific parts of the brain. The parts of the brain are shown in Figure 19.3.

The *brainstem*, located at the base of the brain, connects the spinal cord to the thalamus and cerebrum. The brainstem controls many of the body's involuntary activities, including heartbeat, respiration, and digestion. It is also the brainstem that wakes you up every morning, bringing your body from sleep to a state of conscious wakefulness (Figure 19.4).

The *cerebellum*, located in the posterior portion of the head, controls balance, posture, coordination, and fine motor movements. Any time you decide to move a

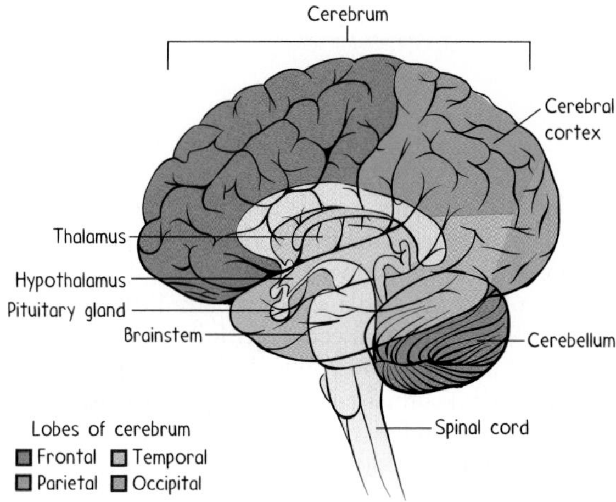

Figure 19.3 The major parts of the human brain include the cerebrum, thalamus, hypothalamus, cerebellum, and brainstem.

Figure 19.4 The brainstem controls the transition between sleep and wakefulness.

part of your body, whether it be to walk down the street or take a sip of milk, the cerebellum checks what you intended to do against how your body is actually responding and corrects any problems. The cerebellum is also involved in performing automated movements such as riding a bicycle or hitting a baseball. When we first learn to pump bicycle pedals or swing a baseball bat, we employ awkward, voluntary movements, but with practice, the cerebellum takes over and we do these things easily and automatically—"without thinking."

The *cerebrum* is found in the top and front part of the head and is the largest part of the brain. The cerebrum takes in information from the senses and controls all the conscious, voluntary activities of the body. The right hemisphere (right side) of the cerebrum receives information from and controls the left side of the body and vice versa, which is why damage to one side of the brain (such as from a tumor or stroke) affects functioning on the opposite side of the body.

Most of the information processing that occurs in the cerebrum takes place in the cerebral cortex, the thin layer that covers the surface of the cerebrum. "Wrinkles" in the cerebral cortex give the brain its familiar convoluted appearance and increase the area available for information processing. Each cerebral hemisphere consists of four lobes that are responsible for different activities. The *frontal lobes* deal with reasoning, control of voluntary movements, and speech. The *parietal lobes* hold sensory areas for temperature, touch, taste, and pain. The *occipital lobes* process visual information. The *temporal lobes* interpret sound and play an essential role in comprehending language.

The control of certain cognitive functions is dominated by either the right or left cerebral hemisphere. The left hemisphere is more adept with math, reasoning, language, and detail-oriented activities. The right hemisphere is more adept with spatial relations, emotional processing, and music. This distinction has led to the popular conception of "left-brained" people, often described as organized, analytical, and attentive to detail, and "right-brained" people, described as intuitive, flexible, and creative.

The *thalamus*, located centrally in the brain, receives information from many different parts of the brain. It sorts and filters these data and then passes them on to the cerebral cortex.

The *hypothalamus* is found below the thalamus. The hypothalamus is responsible for emotions such as pleasure and rage, controls bodily drives such as hunger, thirst, and sex drive, and regulates body temperature and blood pressure. Another function of the hypothalamus is to control the body's internal clock, which tells us when it is day (and therefore time to be awake and alert) and when it is night (and therefore time to sleep). Light cues that the hypothalamus receives from the eyes help set the clock. This is why when flying across time zones, we get over jet lag more quickly if we can spend time in the sun at our destination. The hypothalamus performs many of its activities by controlling the release of chemical molecules called hormones, which we will discuss later in this chapter.

CHECK YOURSELF

1. What are some parts of the brain that you use when you read a book?
2. In sports, it's called *choking*. It's a tense, decisive moment in a game, and the pressure's on. The ball comes towards you—it's an easy play, something you've done a million times before. But suddenly you become superaware of all your movements, and then it happens—the ball rolls right between your legs. What part of the brain normally controls practiced, automated movements? What part of your brain supersedes that part when you choke?

CHECK YOUR ANSWERS

1. You use the occipital lobes to process visual information (the words in the book) and the temporal lobes to comprehend language.
2. Automated movements are normally controlled by the cerebellum. When you consciously control your movements, you are using the frontal lobes of the cerebrum.

TECHNOLOGY

Mapping the Brain in Action: Functional MRIs

Traditional lie-detector tests are not always dependable. Relying as they do on pulse, blood pressure, breathing rate, and skin conductance (sweating), they incriminate the nervously innocent while exonerating liars who keep their cool. But tellers of tall tales may find a new technique harder to fool. *Functional magnetic resonance imaging* (fMRI) reveals which parts of the brain are activated during different types of activity—and it shows that the brains of liars are doing different things from those of truth-tellers.

How does fMRI work? fMRI builds on the earlier technology of magnetic resonance imaging (MRI). MRI makes use of the fact that every hydrogen atom in the body—and there are two in every single water molecule inside us—is a tiny magnet. (This is because, as we learned in Chapter 7, every accelerating charged particle, including the spinning proton in the nucleus of a hydrogen atom, produces a magnetic field.) Because of this, when living tissue is placed in the field of a very strong magnet—typically one that is over 10,000 times more powerful than Earth's magnetic field—all the hydrogen atoms line up in the field much the way a compass lines up with Earth's magnetic field. A radio wave is then used to perturb the atoms, knocking them slightly out of line. As the atoms bounce back to their natural alignment within the field, they release a small amount of energy that can be detected and recorded. Because body tissues vary in water concentration, different tissues release different amounts of energy, allowing a very detailed image to be constructed. Like MRI, fMRI constructs images based on different concentrations of water molecules in different parts the body. With fMRI, however, the focus is on blood oxygen levels. Like all cells, neurons in the brain require more energy when they're active, and so require more oxygen for cellular respiration. In order to accommodate this need, blood flow to active areas of the brain is increased. This increased flow is always in excess of what the active tissue requires, resulting in high local blood oxygen levels that can be detected and converted to an image of active brain areas.

So what happens in the brain when people lie? In one study, volunteers were given a playing card—the five of clubs—and a $20 bill. They were told they could keep the money if they managed to fool the computer into thinking they had a different card. fMRI maps were made of the brain while volunteers deceitfully denied having the five of clubs and while they truthfully denied having other cards. The maps were then compared. Lying caused increased activity in several areas of the brain (see figure), including those responsible for attention, inhibiting actions, and monitoring errors. This suggests that lying requires the inhibition of a natural tendency toward truth-telling as well as increased effort and attention. Interestingly, truth-telling did not increase activity in any part of the brain. Lying appears, overall, to be much harder work than telling the truth.

fMRI is much too costly to be used regularly for lie detection. However, it has proven invaluable in studies that look at the areas of the brain responsible for different sensations, emotions, and activities. fMRI has already contributed to our understanding of how we remember information, feel love, gamble, recognize faces, and respond to pain.

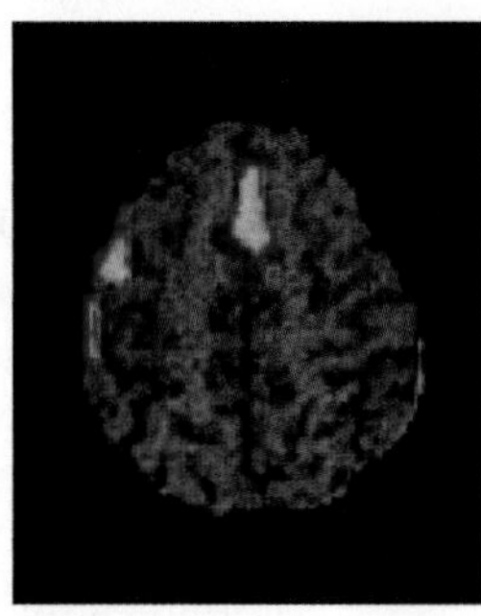

Functional magnetic resonance imaging (fMRI) allows scientists to compare activity levels in different parts of the brain during different activities. This image shows the areas of the brain that are activated during lying.

19.4 The Nervous System

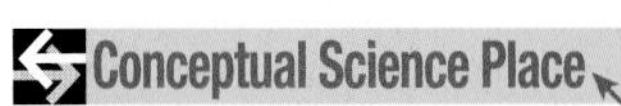

Neuron Structure

Pam looks up from eating cereal and reading the sports page and sees on the clock that she's 5 minutes late for class. Immediately, her body goes into hyperdrive. Her heart starts pounding and she begins to sweat. She dashes out the door, down the steps of the Student Union—and trips. Pain shoots through her knee, but she pulls herself up and continues her race to class. Once there, she drops into a seat and feels her heart rate gradually slow and her body relax.

The *nervous system* is responsible for collecting information about the body's internal and external environments and for controlling the body's activities. It includes the brain, the spinal cord, and all the other nerves in the body. The nervous system has two parts—the **central nervous system**, which includes the brain and spinal cord, and the **peripheral nervous system**, which includes all the other nerves in the body. The central nervous system is like a general who receives information from the field, mulls it over, and shouts out orders. The peripheral nervous system functions as an array of messengers, carrying information from the sense organs to the general, and carrying the general's orders to the foot soldiers—muscles and other organs that respond.

The nervous system is made up of two cell types—**neurons**, specialized cells that receive and transmit electrical impulses from one part of the body to another,

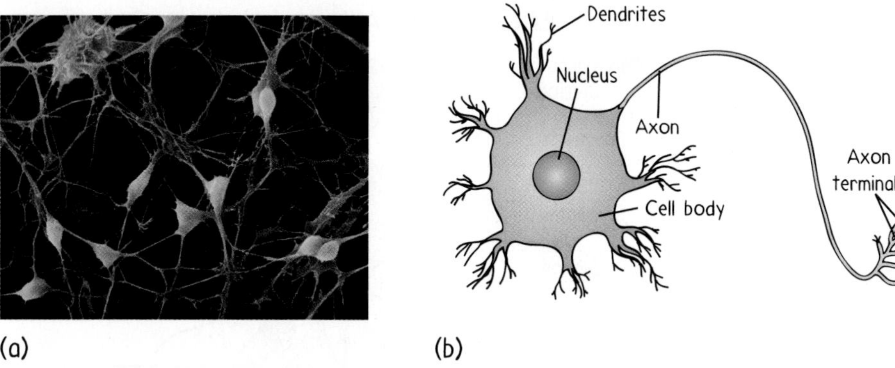

Figure 19.5 (a) Neurons transmit electrical impulses from one part of the body to another. The yellow cells shown here are neurons. The green cell is a glial, or support, cell. (b) Neurons consist of a cell body, dendrites that receive information from other cells, and an axon that transmits information to other neurons or cells.

and *glial cells*, cells with no electrical impulses that serve to support, insulate, and protect neurons (Figure 19.5a). Interestingly, there are actually 10 to 50 times more glial cells than neurons in the human brain.

The structure of a typical neuron is shown in Figure 19.5b. **Dendrites** receive information from other neurons or cells. The *cell body* contains the nucleus and organelles. A cell extension called the **axon** transmits information to other neurons or cells.

The neurons of the nervous system are divided into three categories depending on the origin and destination of their messages—these are sensory neurons, interneurons, and motor neurons. **Sensory neurons** carry messages from the senses to the central nervous system. The neurons that transmit visual information from our eyes to our brain are sensory neurons. So are the neurons that go from our fingertips to our central nervous system, telling us whether we are touching something cool or something warm. Sensory neurons inform Pam's brain about the time on the clock and the pain of hitting her knee on hard concrete. **Interneurons** connect neurons to other neurons. They are found exclusively within the central nervous system. **Motor neurons** carry messages from the central nervous system to muscles or other organs. The neurons that travel from the central nervous system to Pam's legs, making her run, are motor neurons, as is the neuron to Pam's heart that accelerates her heartbeat.

Motor neurons are further divided into two groups, the *somatic nervous system*, which controls voluntary actions by stimulating our voluntary muscles, and the *autonomic nervous system*, which controls involuntary actions by stimulating our involuntary muscles and other internal organs. Of the motor neurons activated during Pam's morning race, the neurons to her legs are part of her somatic nervous system, and the neuron to her heart is part of her autonomic nervous system.

The autonomic nervous system, which controls involuntary activities, includes a *sympathetic* and a *parasympathetic* division, shown in Figure 19.6. The sympathetic nervous system promotes a "fight or flight" response, preparing our bodies to respond to danger—or to a perceived emergency, such as being late to class. For example, the sympathetic nervous system speeds Pam's heartbeat, so that her muscles will get the oxygen they need to run. The sympathetic nervous system also slows digestion and other processes that are of secondary importance during emergencies. The parasympathetic nervous system operates in times of relaxation. Its effects are the direct opposite of those of the sympathetic nervous system. Pam's parasympathetic nervous system takes over after she finds a seat in class.

CHECK YOURSELF

1. You stand at the edge of a pool, wondering whether the water is too cold for a swim. What kinds of neurons will it take to find out?
2. What type of motor neuron is required to find out if the water is swimmable? Somatic or autonomic?

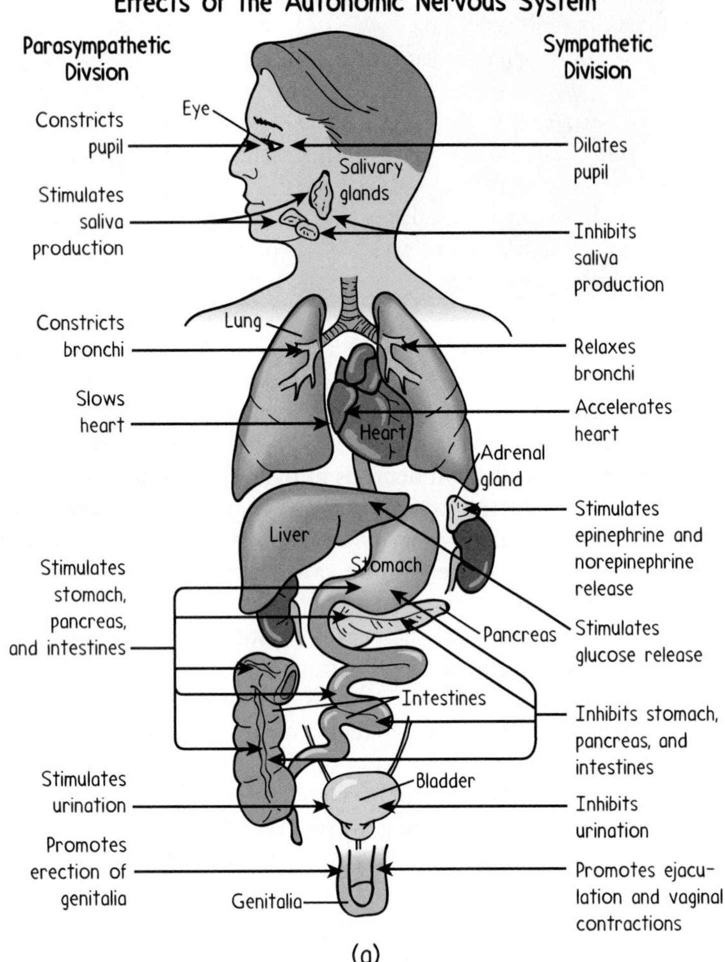

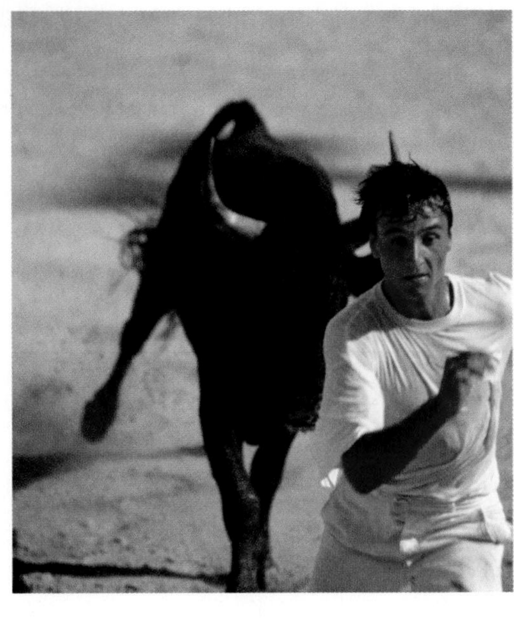

Figure 19.6 (a) The autonomic nervous system controls involuntary activities of various internal organs. It has a sympathetic division and a parasympathetic division. The sympathetic division prepares the body for danger—the "fight or flight" response. The parasympathetic division acts in a directly opposed way, and operates in times of calm. (b) The sympathetic nervous system is responsible for the "fight or flight" response that is activated when we are in danger. (Courtesy of Campbell, Reece, Simon, *Essential Biology with Physiology*, © 2004.)

CHECK YOUR ANSWERS

1. It takes a motor neuron to dip your toe into the pool. It takes a sensory neuron to tell you whether the water is cold.
2. Somatic, since dipping your toe in the pool is a voluntary action.

19.5 How Neurons Work

Nerve Signals; Neuron Communication

Our bodies are essentially electric appliances. Like computers, digital clocks, or telephones, neurons rely on electrical signals do their work. In neurons, the electrical signals are changes in the voltage, or electric potential (Chapter 7), across the cell membrane.

How a Neuron Fires

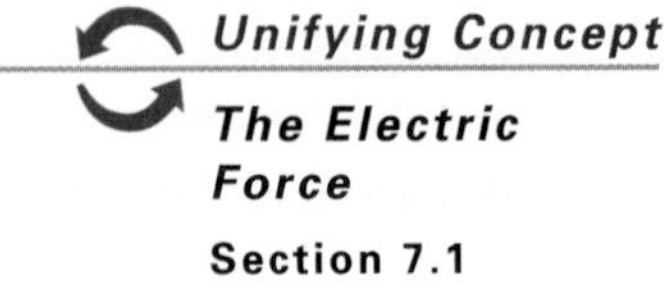

Neurons maintain an electric potential across their cell membranes by keeping differently charged particles on opposite sides. This electric potential is also known as a *membrane potential* because it is the cell membrane that keeps the charged particles apart. The normal membrane potential of a neuron, the neuron's *resting potential*, is created by two phenomena. First, like other cells, neurons have more potassium ions inside the cell than outside and more sodium ions outside the cell than inside. You may recall from Chapter 15 that maintaining this state of affairs requires constant work by the sodium-potassium pump as well as energy in the form of ATP. Second, neurons contain many other negatively charged ions, including proteins and other organic molecules. As a result of these two factors, the inside of a neuron is normally negatively charged and the outside of a neuron is normally positively charged, creating a resting potential of about −70 millivolts (mV) across the cell membrane.

In order to send information, a neuron must signal or "fire." In order to fire, it must first be stimulated. A neuron is stimulated when its membrane potential is increased, typically by allowing positively charged ions to enter the neuron. If a neuron's membrane potential reaches a certain *threshold* value—typically around −55 mV—sodium channels in the neuron's cell membrane suddenly open, allowing positively charged sodium ions to flow into the neuron. (Sodium ions flow into the neuron because there are more sodium ions outside the cell than inside). This influx of positively charged ions causes the membrane potential to spike and become positive. This spike is called an **action potential.** The action potential is the neuron's way of signaling, or "firing." Once the spike occurs, the sodium channels quickly close and potassium channels open. Positively charged potassium ions flow out of the neuron (they flow out because there are more potassium ions inside the cell than outside), causing the membrane potential to return to its resting value. The events that occur during an action potential are shown in Figure 19.7.

Unifying Concept
The Second Law of Thermodynamics
Section 6.5

Action potentials are all-or-nothing events—a neuron either fires or it doesn't. So, a neuron can't fire "harder" when a stimulus is more intense. But, what it can do is fire more frequently. For example, touch receptors in our skin fire slowly when they feel a little pressure (tick… tick… tick…) and more often when they feel lots of pressure (tick-tick-tick-tick-tick).

How an Action Potential Is Propagated

An action potential doesn't spike everywhere along a neuron's cell membrane at once. Instead, it begins near the neuron's cell body and travels neatly down the axon. How does this happen? When an action potential begins at the cell body end of the axon, sodium ions enter the axon there. These ions then diffuse into adjacent areas where, because of their positive charge, they cause the local membrane potential to increase. When this local membrane potential hits threshold, a new action potential, further along the axon, begins. In this way, the action potential travels down the entire axon. This process is shown in Figure 19.8.

Because sodium ions diffuse both forward and backward along the axon, a special mechanism is necessary to keep the action potential from also being propagated backwards. Specifically, once an action potential has occurred in one part of the axon membrane, there's a brief time period when that area can't be stimulated again. This guarantees that the signal travels in only one direction—down the axon. You might compare this to the functioning of an exhausted fire brigade, where the brigade is like an axon and the bucket of water is like an action potential—after a person passes the bucket, there's a brief delay before she can pass another one (see Figure 19.9).

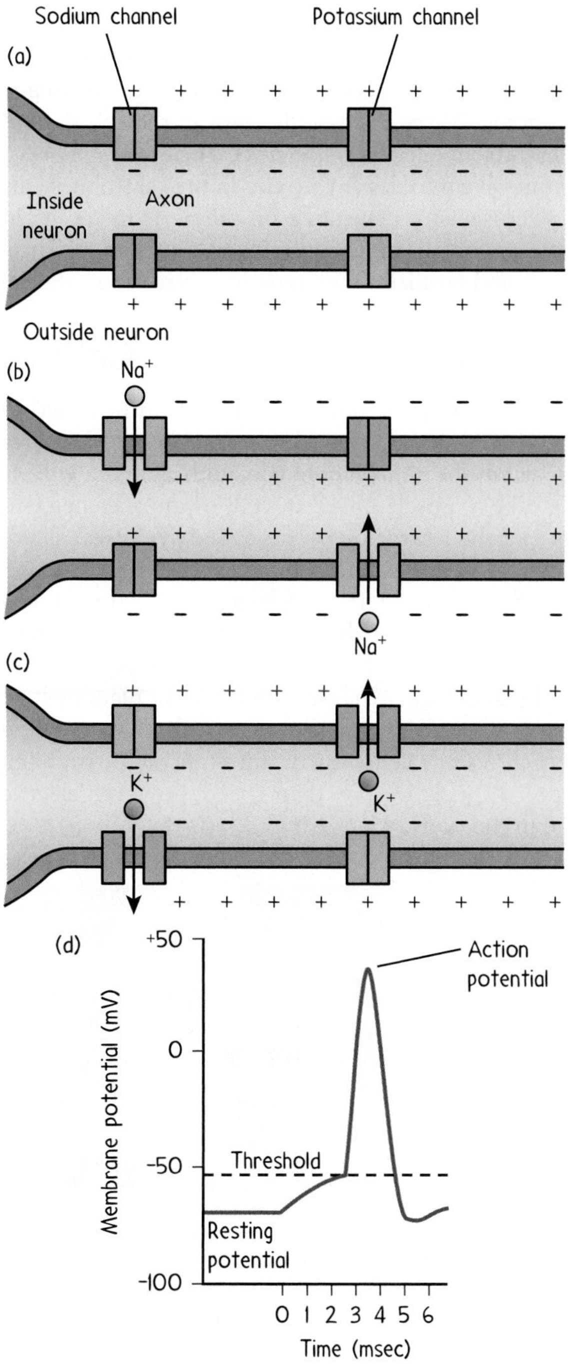

Figure 19.7 The action potential is the neuron's way of signaling. (a) A neuron at rest has a negative membrane potential, called its resting potential. (b) During an action potential, sodium channels in the neuron's cell membrane open, allowing sodium ions to flow into the neuron. This causes the membrane potential to spike and become positive. (c) Soon after the sodium channels open, they close. Potassium channels open, allowing potassium ions to flow out of the neuron. This causes the membrane potential to decrease and return to its resting potential. (d) This graph shows how the membrane potential changes during an action potential. As the neuron is stimulated, the membrane potential increases gradually until it reaches threshold. At this point, the action potential occurs—sodium channels open, causing the membrane potential to spike. The membrane potential decreases again when potassium channels open and returns to resting potential. (Courtesy of Campbell, Reece, Simon, *Essential Biology with Physiology*, © 2004.)

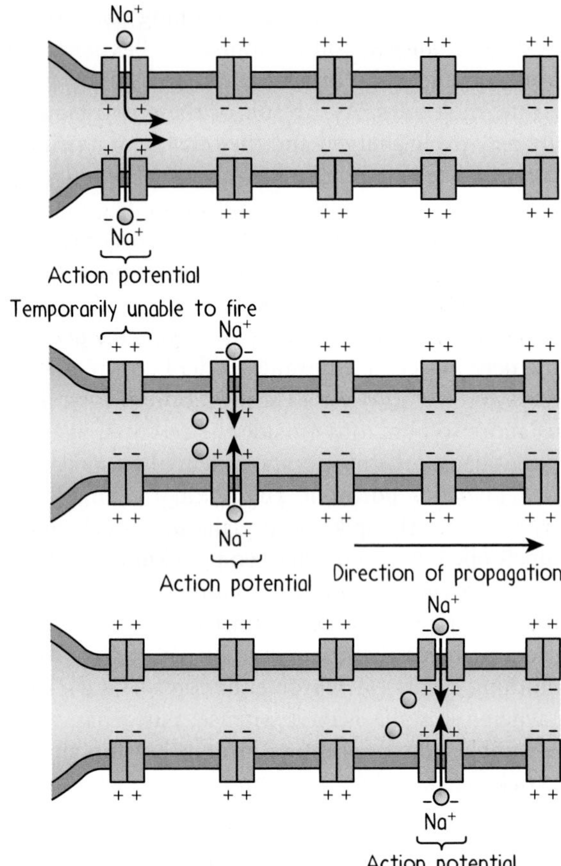

Figure 19.8 Action potentials are propagated down a neuron's axon. The action potential does not move backward along the axon because once an action potential has passed through one part of the axon, there's a brief period when that area cannot be stimulated again.

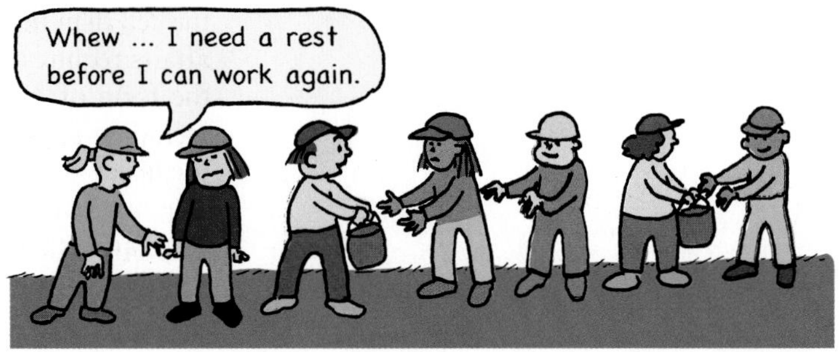

Figure 19.9 Once an action potential has moved through one part of an axon, there's a brief time period when that area can't be stimulated again. This is similar to the way the workers on an exhausted fire brigade might need a break before passing another bucket of water.

Some axons, such as the one shown in Figure 19.10, are surrounded by a *myelin sheath* that allows the neuron's signal to be transmitted more quickly. The sheath consists of glial cells wrapped around and around the axon, insulating it with multiple layers of cell membrane. There are periodic gaps in the sheath, and the action potential essentially leaps from one gap to the next. The disease *multiple sclerosis* causes the body's immune system to destroy myelin in the central nervous system. Symptoms of multiple sclerosis vary depending on the parts of the brain and spinal cord affected, but they can include fatigue, dizziness, bladder and bowel control problems, vision problems, and problems with muscle control and balance.

How a Neuron Communicates with Other Cells

So far, we have seen what happens during an action potential and how action potentials travel down axons. What happens when the action potential arrives at the end of the axon? This is where the neuron connects with a target cell, either a cell that does something, like a muscle or organ cell, or another neuron. Connections between neurons and their target cells are called **synapses.** There are two types of synapses, electrical synapses and chemical synapses.

In *electrical synapses*, ions flow directly from a neuron to a target cell through gap junctions (Chapter 15), and immediately initiate an action potential. The advantage of an electrical synapse is that the transmission of an action potential occurs extremely rapidly. Electrical synapses are found in places where this rapidity is essential. For example, in heart muscle, electrical synapses allow the contraction command to reach muscle cells nearly simultaneously, enabling the heart chambers to contract all at once. Despite their speed, however, electric synapses are actually fairly rare in the body.

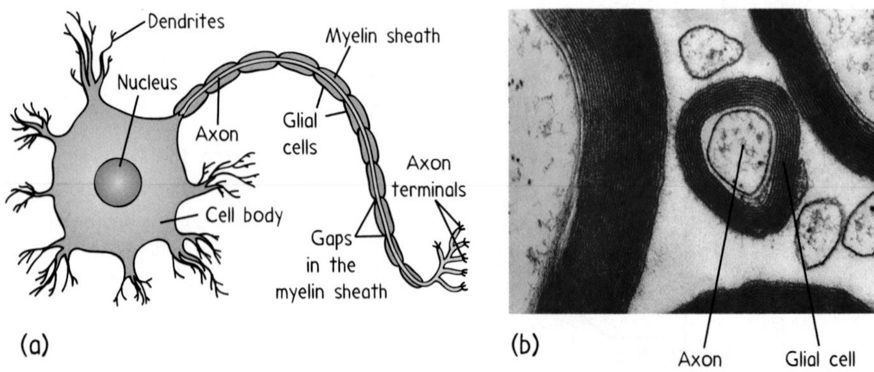

Figure 19.10 (a) Myelin sheaths surround some axons, allowing the signals they transmit to go faster. (b) The myelin sheath consists of glial cells wrapped around and around an axon, as shown in this cross-sectional view.

Most synapses are *chemical synapses*. Chemical synapses aren't as fast as electrical synapses, but they allow for a finer degree of control that is essential for most neural activity. In a chemical synapse, there is a narrow space between the signaling neuron and its target cell (Figure 19.11). When the action potential arrives at the end of the axon, the neuron releases chemical messengers called *neurotransmitters* into this space. The neurotransmitters are released through exocytosis (Chapter 15); that is, small vesicles containing the neurotransmitters fuse to the neuron's cell membrane and release their contents outside the cell. The neurotransmitters then diffuse across the space between the neuron and its target cell and bind to receptor proteins on the cell membrane of the target cell. The binding of neurotransmitters causes the receptor proteins to change, opening ion channels and allowing ions to flow into the target cell. Depending on how many ions flow into the target cell and what kind of ions they are, the target cell's membrane potential either increases or decreases. An increase in membrane potential makes the target cell more likely to fire, whereas a decrease in membrane potential makes it less likely to fire. A target cell may receive signals from many other neurons (see Figure 19.12), each of which either increases or decreases its membrane potential. Depending on the sum total of these effects, the neuron will either reach its threshold and fire, or not. The effect of a neurotransmitter on a target cell ends when the neurotransmitter is removed—the molecules can be degraded by enzymes, taken up by the signaling neuron and repackaged into vesicles, or collected and broken down by glial cells.

Unifying Concept
The Second Law of Thermodynamics
Section 6.5

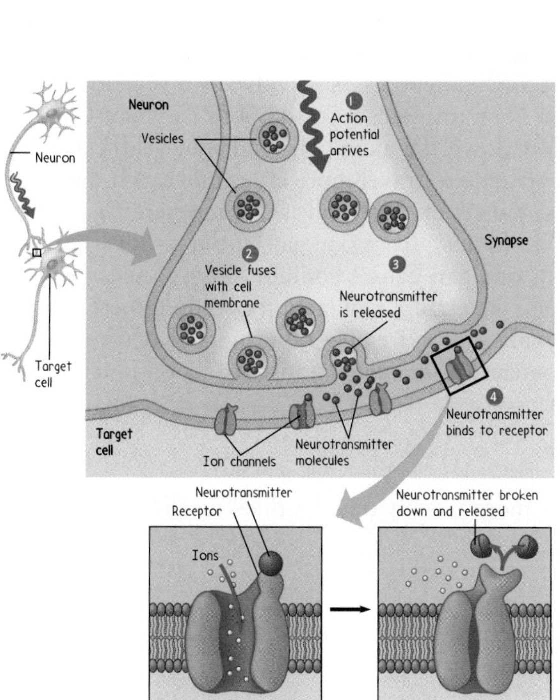

Figure 19.11 In a chemical synapse, a neuron uses chemicals called neurotransmitters to transmit information to another cell. When an action potential arrives at the end of the axon, vesicles release neurotransmitters into the space between the neuron and the target cell. The neurotransmitters diffuse to the target cell and bind to receptors on the target cell's membrane, causing ion channels to open and making the neuron either more or less likely to fire. (Courtesy of Campbell, Reece, Taylor, Simon, *Biology: Concepts and Connections, 5e*, © 2006.)

CHECK YOURSELF

1. What causes sodium channels to open, initiating an action potential?
2. What causes the action potential to end?
3. What is the advantage of an electrical synapse over a chemical synapse? Why don't all parts of the body use electrical synapses?

CHECK YOUR ANSWERS

1. The membrane potential reaching its threshold value.
2. The opening of potassium channels, which returns the membrane potential to its resting potential.
3. Electrical synapses are faster than chemical synapses. However, because they allow ions to flow freely from one cell to the next, they don't permit the fine degree of control that chemical synapses permit. With a chemical synapse, the binding of neurotransmitters to receptors on the target cell can have a gradation of effects on the target cell's membrane potential. For example, it can cause the target cell to immediately fire, make it impossible for the target cell to fire, or anywhere in between.

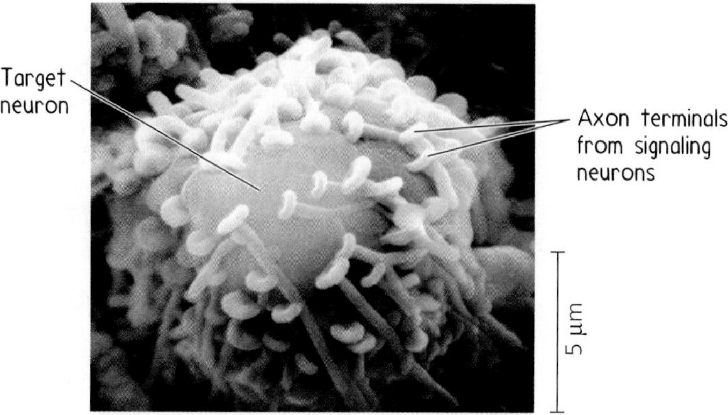

Figure 19.12 Some neurons receive signals from many other neurons. Axons from a large number of neurons have synapses with this target cell.

INTEGRATED SCIENCE PHYSICS

How Fast Can Action Potentials Travel?

INSIGHT Some cases of depression are believed to be caused by the inability of signaling neurons to stimulate target neurons. Prozac, Paxil, and Zoloft are among a group of drugs known as selective serotonin reuptake inhibitors (SSRIs). SSRIs work by preventing signaling neurons from taking up the neurotransmitter serotonin once it has been released at synapses. That way, more serotonin remains to bind to and stimulate target neurons.

For a squid escaping from a hungry shark, speed is of the essence. How quickly the signal to move travels from the brain to the muscles could be the difference between life and death. An action potential's speed depends in part on how quickly successive parts of the axon's cell membrane (that is, parts further down the axon) can be induced to increase to threshold. How quickly the membrane reaches threshold is in turn dependent on how fast sodium ions flow downstream to increase the membrane potential there. And how fast sodium ions flow down an axon depends on Ohm's Law (Chapter 7). Ohm's Law tells us that current = voltage/resistance, so the lower the resistance, the more current (that is, ions) flows and the faster the action potential travels. Like any other material, an axon has lower resistance if it is thicker around—a thick axon resists current less than a thin axon the same way a wide pipe resists water flow less than a thin pipe. Consequently, one way to get an action potential to travel quickly is to build a really thick axon. Thick axons have in fact evolved numerous times in different animals, including cockroaches, crayfish, earthworms, and squid. The squid giant axon, which can measure nearly a millimeter in diameter, conducts action potentials extremely quickly—at a speed close to 100 meters/second. Compare this to a more typical axon that is only 1 micrometer in diameter—it might conduct action potentials at a speed of 2 meters/second.

Unfortunately, it isn't practical to have large numbers of giant axons—they simply take up too much space. In vertebrates and certain other animals, an alternative way to achieve fast action potentials has evolved—myelination. The myelin sheath that surrounds an axon insulates it so that ions cannot escape out the cell membrane, but must flow down the axon. The end result is the same as for a giant axon—sodium ions are able to travel more efficiently down the axon. Moreover, in myelinated axons, the action potential is not regenerated at every point along the axon; instead, it "jumps" from one gap in the myelin sheath to the next. An action potential at one gap causes sodium ions to move into the axon, flow down to the next gap, generate a new action potential there, and so on. This jumping propagation makes for extremely rapid signal transmission—a myelinated axon only 20 micrometers in diameter transmits action potentials as quickly as the squid giant axon.

CHECK YOURSELF

Would an action potential move more quickly down a thick myelinated axon or a thin myelinated axon?

CHECK YOUR ANSWER

A thick myelinated axon. A thick myelinated axon has the best of both worlds—decreased resistance and insulation.

INTEGRATED SCIENCE CHEMISTRY

Endorphins

What do chocoholics and long-distance runners have in common? Just maybe, an addiction to endorphins. *Endorphins* are proteins made by the brain in times of stress or pain. Endorphins are neurotransmitters that bind to opiate receptors on neurons—the same receptors targeted by drugs such as morphine, codeine, opium, and heroin (Figure 19.13). In fact, endorphins were discovered after scientists identified the morphine receptor and hypothesized that the body must make

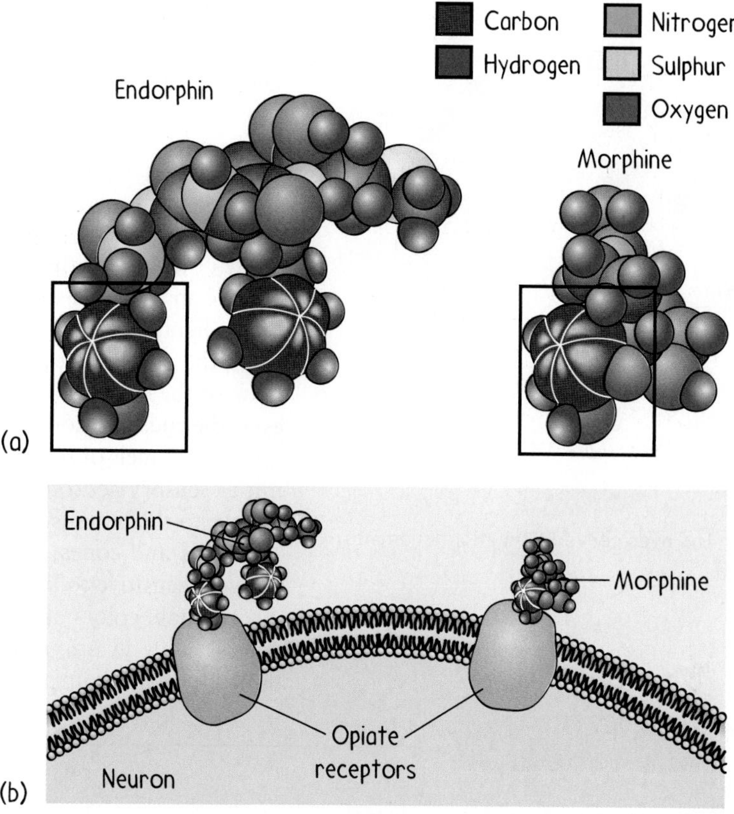

Figure 19.13 (a) Structures of endorphin and morphine. The boxed portion of the endorphin molecule (left) binds to receptors on neurons in the brain. The boxed portion of the morphine molecule (right) is a close match. Morphine has a structure similar enough to allow it to bind to the same receptors. (b) Endorphin receptors on the surface of a neuron can bind to both endorphin and morphine. (Courtesy of Campbell, Reece, *Biology* 7e, © 2005.)

a substance of its own that bound to the same receptor. When this molecule was found, it was called an endorphin after "endogenous morphine."

Like opiates, endorphins decrease sensitivity to pain and bring on a feeling of euphoria. It is this euphoria that runners call "runner's high." Besides sports, endorphin release has also been associated with activities such as laughter, orgasm, acupuncture, massage, and deep meditation. Finally, certain foods—notably chocolate and chili peppers—increase the release of endorphins. Comfort food indeed.

CHECK YOURSELF

Do you expect other opiates, such as codeine and heroin, to resemble endorphins in their chemical structure? What do you expect them to have in common with endorphins?

CHECK YOUR ANSWER

Codeine and heroin would not be expected to resemble endorphins in their overall chemical structure, just as morphine does not. However, you would expect each to have a region that closely resembles the portion of the endorphin molecule that binds to the opiate receptor, since they bind to the same receptor.

19.6 The Senses

Our senses are our connection to the world. Where would we be if we couldn't see, hear, smell, touch, or taste? If we couldn't tell, without looking down at them, where our hands were? If pain didn't warn us of danger or injury? These are just some of the critical functions our senses perform. But how do the senses work?

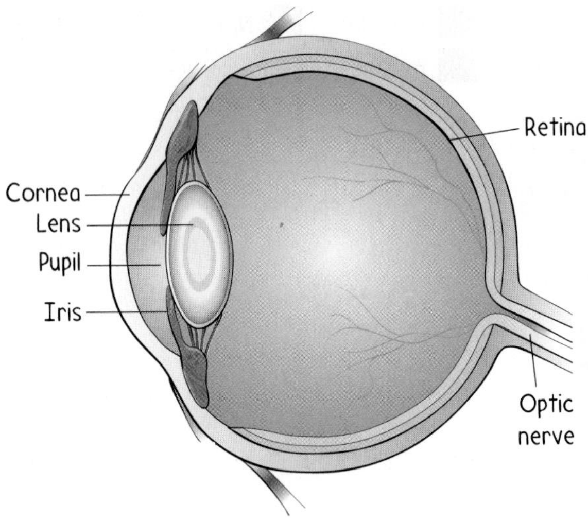

Figure 19.14 The eyes convert light information into action potentials.

Each of our senses takes information from the environment—light, sound, touch, or molecules in the air—and translates it into action potentials that the brain can understand. Each sense accomplishes this in its own way.

Vision

Our eyes are responsible for our sense of sight. Light enters each eye (Figure 19.14) through a tough, transparent layer called the *cornea*, which is continuous with the white of the eye. It then passes through a small hole, the *pupil*. The *iris*, the part of the eye that gives us our eye color, surrounds the pupil and controls its size. In bright light, the pupil is small. In dim light, the pupil expands to let in more light. From the pupil, light passes through the *lens*, which focuses it on the **retina** at the back of the eyeball. It is the retina that holds the actual light-sensitive cells of the eyes, the rods and cones. When light hits the rods and cones, it produces changes in the ion channels of these cells and, consequently, in the action potentials they transmit to sensory neurons. The bundle of neurons that takes visual information from the eye to the brain is called the *optic nerve*.

Rods and cones, shown in Figure 19.15, form a mosaic in the retina. *Rods* are very sensitive to light and are responsible for vision in dim light. Rods cannot discriminate colors and so allow us to see only black, white, and shades of gray. This is why, in our unlighted bedrooms at night, we often can't distinguish a navy-blue shirt from a maroon shirt. (Interestingly, however, we are so accustomed

TECHNOLOGY

Visual Prostheses for the Blind

Will blindness one day be as easy to treat as a toothache? Scientists are beginning to develop visual prostheses that bring vision to the blind. The most promising strategy at the moment involves implanting an array of *electrodes*—devices that conduct electric currents—in front of the retina. The electrodes receive signals from a chip connected to a camera and then stimulate intact neurons at the back of the retina. Retinal prostheses currently being tested use 16 electrodes. With the implants, volunteers were able to detect light, locate and count objects, and distinguish between simple objects. The next generation of retinal prostheses will have 64 electrodes—probably enough, judging by preliminary tests, to offer patients significant mobility. Although this technology is far from reproducing what the human eye can do, it does promise to bring activities such as navigating unfamiliar locations, reading text, and recognizing faces within reach.

For patients without intact retinal neurons, researchers are pursuing other strategies for visual prostheses. Specifically, some are exploring the possibility of directly stimulating the optic nerve, or even the parts of the brain that deal with vision. Stay tuned.

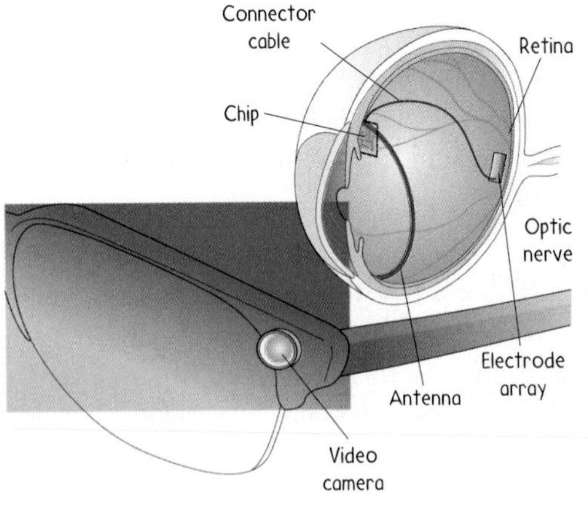

(a)

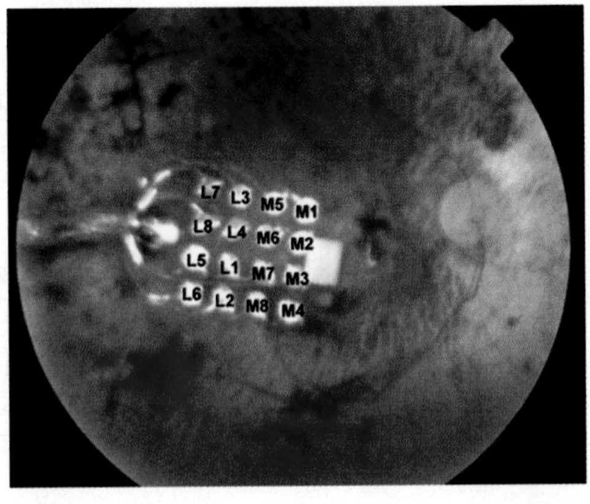

(b)

(a) Retinal prostheses work by artificially stimulating retinal neurons. A camera located in a pair of glasses sends visual data to a chip implanted in the eye. The chip processes the data and sends information to an electrode array in front of the retina, which then stimulates retinal neurons. This technology, although still being tested, may be commercially available in a few years. (b) This photo shows an electrode array in the retina of a patient.

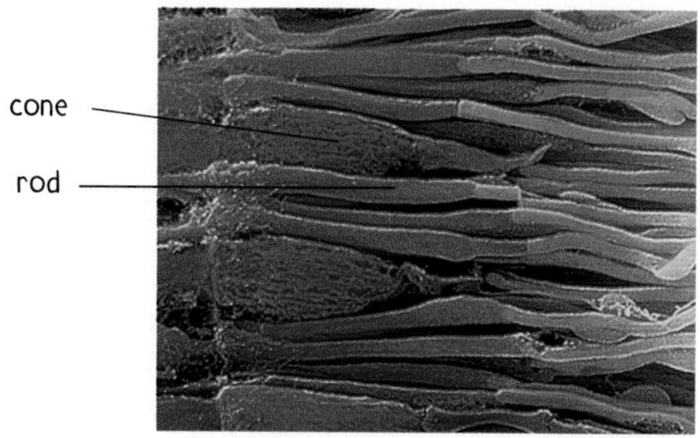

Figure 19.15 The light-sensitive cells in the eyes are the rods and cones. As you can see, both rods and cones are named for their shape.

to this that many of us don't even realize that we lose our color vision in dim light!) Rods are so light-sensitive they respond to the tiniest amount of light possible—a single photon, or light particle. However, rods aren't very good at making out fine details. This is why, in addition to lacking color, our night vision isn't very sharp. *Cones* detect color. Our eyes have three types of cones that respond most strongly to red, green, and blue light, respectively. All the shades we see are made up of different combinations of these three colors (see the discussion of the electromagnetic spectrum in Chapter 8). Color blindness is the result of having a nonfunctioning version of one or more cone type.

Hearing

The ears consist of three parts: the outer, middle, and inner ear (Figure 19.16). Sound waves move through the air to the outer ear, or *pinna*, which funnels them in. Inside the ear, the waves hit a thin membrane of skin—the *eardrum*—and

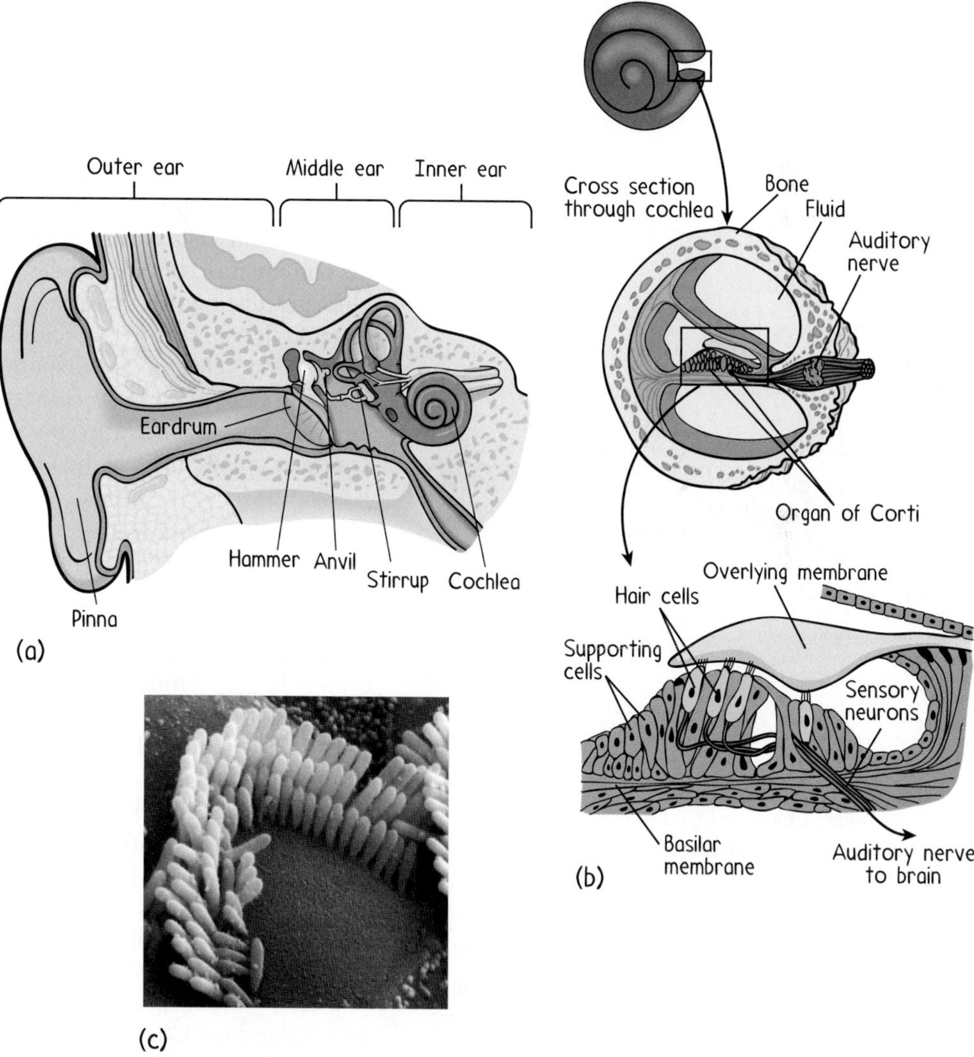

Figure 19.16 (a) The structures of the ear convert sound waves to action potentials. (b) Sound-sensitive cells are contained in the organ of Corti. Fluid vibrations in the inner ear vibrate the basilar membrane of the organ of Corti, causing sensory "hairs" embedded in the organ to bend against the overlying membrane. The bending opens ion channels, starting action potentials that are transmitted to the brain. (Courtesy of Campbell, Reece, Simon, *Essential Biology with Physiology*, © 2004.) (c) These are sensory "hairs" from a cell in the organ of Corti. Note that some of the "hairs" on the left side of the photo are in slight disarray, indicating that the cell has suffered mild damage.

INSIGHT If you ever accidentally hit your eye, you may "see" a flash of light—this is because the impact affects ion channels in the rods and cones of the retina. You "see" light because your brain interprets any signal it receives from the rods and cones as light.

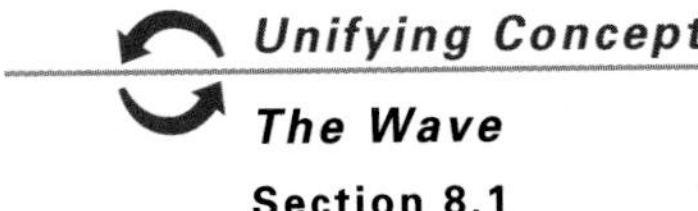

Section 8.1

INSIGHT Vision is arguably our best-developed sense—the rods and cones account for 70 percent of all the sensory receptors in the body.

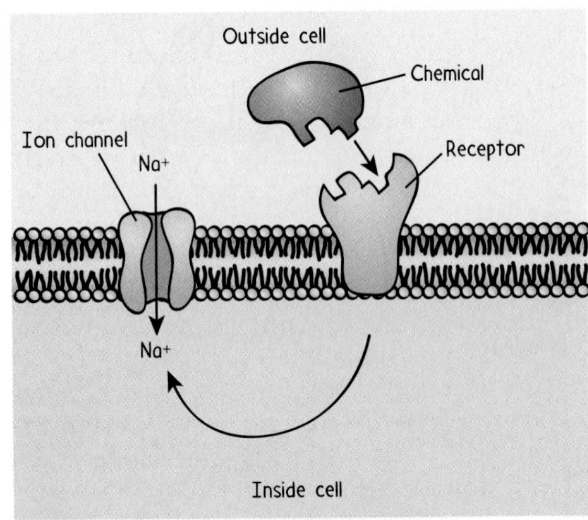

Figure 19.17 Smell and taste are examples of chemoreception. Chemicals bind to specific receptors on the cell membrane, causing ion channels to open.

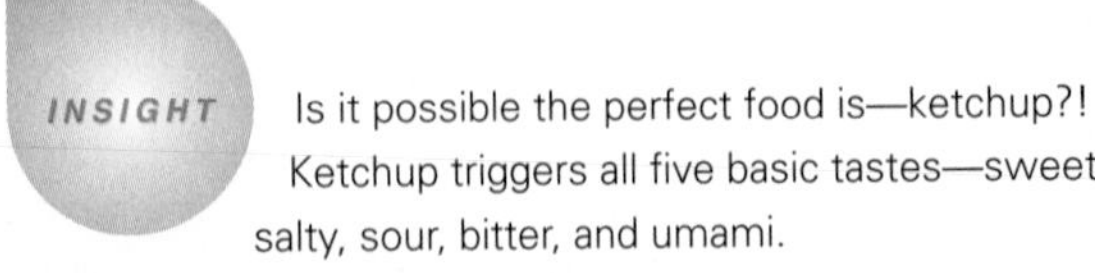
INSIGHT Is it possible the perfect food is—ketchup?! Ketchup triggers all five basic tastes—sweet, salty, sour, bitter, and umami.

INSIGHT When we're hungry, our taste buds become more sensitive to salt and sugar, making food taste better. The ability to taste bitterness, however, which plays the important role of helping us identify toxic foods, is unaffected by hunger.

cause it to vibrate, just the way blowing on a piece of paper causes it to shake. The eardrum's vibrations move three middle ear bones—the hammer, the anvil, and the stirrup—in sequence. These bones amplify the vibrations, making them more pronounced. The final bone, the stirrup, then transfers the vibrations on to the fluid-filled inner ear. In the inner ear, sound vibrations enter the cochlea, a coiled tube holding the *organ of Corti*, which contains the sensory cells responsible for hearing. Fluid vibrations in the inner ear move the organ of Corti's basilar membrane, causing sensory "hairs" (actually extensions of sensory cells) embedded in it to bend against an overlying membrane. This bending opens ion channels, starting action potentials that are transmitted to the brain. We can distinguish different noises—the high pitch of a siren versus the low pitch of a street drill—because different parts of the organ of Corti respond to different pitches.

Smell and Taste

Smell and taste rely on **chemoreception**, a process in which chemicals bind to receptors on the surface of special chemosensory cells. The binding causes ion channels to open and action potentials to happen (Figure 19.17).

The sensory cells for smell lie in two patches, each about the size of a dime, at the top of our nasal passages. We have over 1000 different kinds of chemosensory cells for smell, each containing only one or a few different kinds of receptors. Because different chemicals trigger different combinations of sensory cells, however, we can distinguish well over 10,000 distinct odors.

Our taste cells cluster in small bumps called taste buds. Taste buds are found on our tongues, on the insides of our cheeks, and on the roofs of our mouths. Humans distinguish five basic tastes—sweet, salty, sour, bitter, and umami. Umami, Japanese for "delicious," is the flavor found in foods such as meat, mushrooms, cheese, and asparagus. Monosodium glutamate, or MSG, has a strong umami taste. Interestingly, our experience of food—what we think of as "taste"—comes largely from our sense of smell. This is why food doesn't have nearly as much "taste" when we have stuffy noses.

Touch

The sense of touch is actually several different senses, telling us about stimuli as diverse as pressure, temperature, and pain. Sensory receptors in the skin are illustrated in Figure 19.18. Pressure, like hearing, causes the "hairs" on sensory cells to bend, opening ion channels and starting action potentials. We have separate sensory cells for detecting light touch and heavy pressure. Temperature sensing relies on cells with ion channels that are directly affected by temperature. Some temperature-sensing cells respond to cold, others to warmth. Interestingly, the chemical menthol (found in peppermint) also stimulates cold receptors—it is this coincidence, not an actual cold temperature, which brings on the cool feeling you get from eating a mint. Pain receptors respond to stimuli that cause damage to the body. These sensory cells generally require strong stimulation before they will respond. However, damaged tissues release chemicals called *prostaglandins* that increase the sensitivity of pain receptors. (Aspirin, you may recall, provides pain relief by interfering with the production of prostaglandins (Chapter 15)). Also, whereas most sensory cells become less sensitive with repeated stimulation—this is why you stop noticing the funny smells in your house, or feeling the weight of your backpack—pain receptors actually become more sensitive with continued stimulation. Some types of chronic pain may in fact result from pain receptors that have become abnormally sensitive.

Other Senses

We have other senses too, in addition to the big five. *Proprioceptors* in our muscles, tendons, and joints tell us where different parts of our body are. These allow

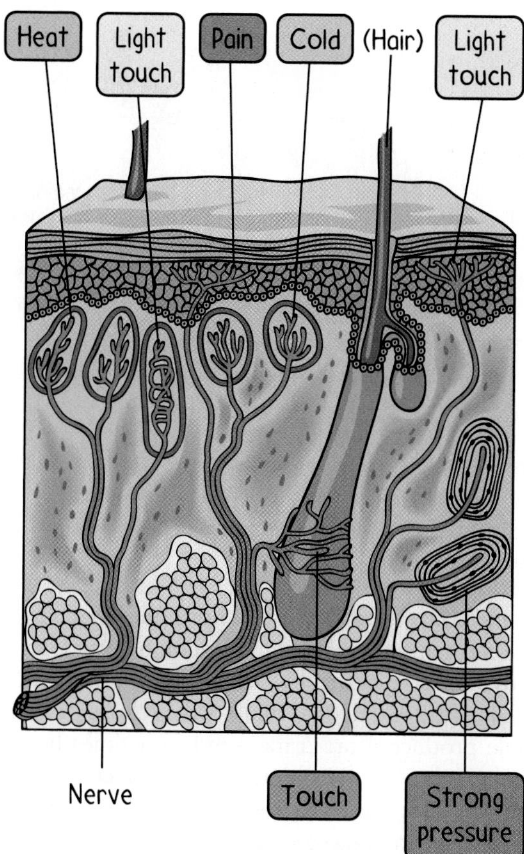

Figure 19.18 Sensory receptors in the skin are responsible for our various senses of touch. (Courtesy of Campbell, Reece, Simon, *Essential Biology with Physiology*, © 2004.)

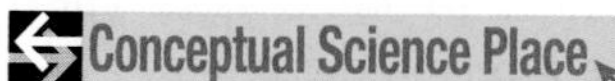

Overview of Cell Signaling; Nonsteroid Hormone Action; Steroid Hormone Action

us to do things like touch our noses with our eyes shut. (You may be too used to this ability to be much impressed, but consider the fact that you can't easily touch another person's nose with your eyes shut!) And the *vestibular senses*, dependent on the vestibular organs in the inner ear, provide information on body rotation and movement and also tell us which way is up (Chapter 5).

CHECK YOURSELF

1. When you are outside at night, looking at a starry sky, which light receptors are you using?
2. Deafness can result from a number of problems, including a ruptured eardrum or damaged sensory cells in the organ of Corti. Why would each of these problems make you unable to hear?

CHECK YOUR ANSWERS

1. Rods.
2. If the eardrum is ruptured, sound waves cannot be conducted to the middle ear and no vibrations make it to the organ of Corti. If sensory cells in the organ of Corti are damaged, they are unable to send signals to the brain.

19.7 Hormones

Where the nervous system handles rapid actions, the hormones of the endocrine system control activities that take place over longer periods. Among other activities, hormones control our growth and development, prepare us for reproduction, determine how quickly we metabolize food, and tell us whether it is night or day. A number of hormones come in pairs with opposing effects and play an important role in maintaining homeostasis in the body. So, what are hormones and how do they work?

Hormones are chemical messengers that give instructions to the body. They are produced in one place in the body, released into the bloodstream, and then received by target cells elsewhere in the body. Hormones come in two types. Most are proteins or modified amino acids. Others—the *steroid hormones*—are derived from cholesterol. Protein hormones bind to receptors on the cell membrane of their target cells. This binding initiates a series of chemical reactions that end with the cell's response to the hormone. Steroid hormones cross the cell membrane and bind to receptors in either the cytoplasm or nucleus of target cells. The hormone-receptor complex then binds to DNA in the nucleus and directly impacts gene transcription. The functioning of protein and steroid hormones is illustrated in Figure 19.19.

Our endocrine organs include the hypothalamus, anterior and posterior pituitary glands, thyroid gland, parathyroid glands, adrenal glands, pancreas, ovaries and testes, and pineal gland (Figure 19.20). The hypothalamus (see also Figure 19.3) acts as a control center for the endocrine system. A number of the hypothalamus's hormones are released directly to the anterior pituitary, located beneath it. These hormones control the release of anterior pituitary hormones. The hypothalamus also makes several hormones that are stored and released by the posterior pituitary.

The anterior pituitary is sometimes called the "master gland" because many of its hormones control the function of other endocrine organs. For example, anterior pituitary hormones stimulate the thyroid, sex organs, and adrenal glands to release their hormones. The anterior pituitary also makes growth hormone and prolactin. *Growth hormone* does just that—promote growth. Too little growth hormone results in dwarfism, and too much of it results in gigantism. *Prolactin* stimulates milk production in nursing mothers.

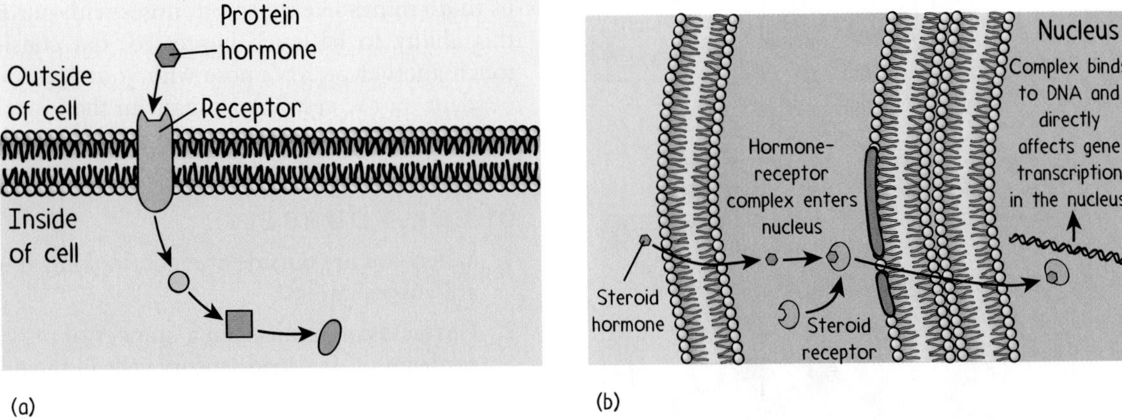

Figure 19.19 (a) Protein hormones bind to membrane receptors on target cells, beginning a sequence of events that culminates in the cell's reaction to the hormone. (b) Steroid hormones pass directly into cells and bind to receptors in the cytoplasm or nucleus. They directly affect gene transcription in the nucleus.

The posterior pituitary stores and controls the release of hormones made by the hypothalamus. *Antidiuretic hormone* helps regulate the amount of water in the body. Specifically, it helps the body conserve water by producing a more concentrated urine. (We will see how in Chapter 20.) Alcohol inhibits the release of antidiuretic hormone. This is why people produce more urine—and sometimes become dehydrated—when they consume alcohol. *Oxytocin* stimulates uterine contractions during childbirth. Women whose labor does not progress, or who need labor induced, may be given pitocin, a synthetic form of oxytocin, to stimulate contractions.

The thyroid gland produces *thyroid hormones* that are involved in metabolism, growth, and development. The thyroid hormones are particularly important for proper brain and nervous system development during childhood. Thyroid hormones contain iodine, and a deficiency in dietary iodine can result in problems associated with insufficient amounts of thyroid hormones. It is for this reason that iodine is often added to table salt. The thyroid also produces *calcitonin*, which lowers calcium levels in the blood.

The parathyroid glands lie next to the thyroid. They produce *parathyroid hormone*, which raises calcium levels in the blood. Parathyroid hormone does this in three ways—it causes calcium to be released from bones, increases calcium absorption in the intestine, and decreases calcium excretion in the urine.

The adrenal glands, located above the kidneys, secrete *epinephrine* (adrenaline) and *norepinephrine*. These hormones are involved in the "fight or flight" response. Signals from the sympathetic nervous system trigger their release. The adrenal glands also produce *glucocorticoids*, which raise glucose levels in the blood, and *mineralocorticoids*, which help control water and salt balance in the body.

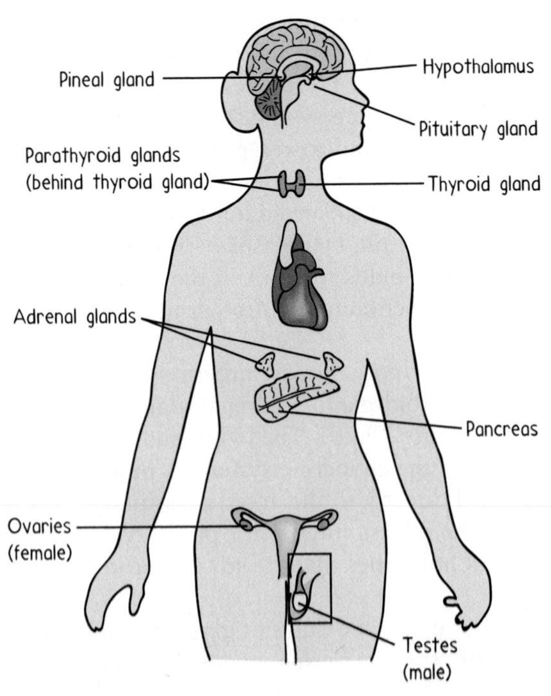

Figure 19.20 Endocrine organs include the hypothalamus, pituitary gland, thyroid gland, parathyroid glands, adrenal glands, pancreas, ovaries and testes, and pineal gland.

The pancreas produces *insulin* and *glucagon*. These hormones regulate the amount of glucose in the blood. Insulin lowers blood glucose levels by directing muscle and other cells to take in glucose and by stimulating the liver to convert glucose to the storage substance glycogen. Glucagon increases blood glucose levels by causing the liver to break down glycogen and release glucose. Diabetes results when the body doesn't make enough insulin or when cells do not respond to insulin. In either case, blood glucose levels become abnormally high. People with diabetes must control their diets and monitor their blood glucose levels. Many require regular injections of insulin.

The sex organs, ovaries in women and testes in men, produce sex hormones. The three types of sex hormones—estrogens, progestins, and androgens—are produced by both women and men, though in different quantities. Women make more *estrogens* and *progestins*. Estrogen is involved in the development of female secondary sexual characteristics. It promotes breast development and fat storage in the hips and thighs. Estrogen and progesterone control ovulation and the menstrual

INSIGHT A special group of light-sensitive cells in the retina tells the pineal gland whether it is day or night, allowing the gland to set the body's internal clock. These cells are intact in some blind people but not in others. In people who are totally blind, the internal clock may shift out of sync with day and night, so that they are sleepy during the day and wide awake at night.

cycle and are also involved in pregnancy. Men produce more *androgens*. Androgens such as testosterone are required for sperm production and for the development of male secondary sexual characteristics such as facial hair and increased muscle mass. It is this last effect of androgens that tempts some athletes to use anabolic steroids—synthetic versions of testosterone—to improve performance.

The pineal gland produces the hormone *melatonin*. Melatonin regulates the body's internal clock, telling us when it is day and when it is night. Using light cues from the eyes, the pineal gland releases melatonin during the night hours. This is why some people use it as a sleeping pill and why overseas travelers take it to help them adjust to a new time zone.

The major endocrine organs and the hormones they produce are summarized in Table 19.1.

CHECK YOURSELF

Some of the hormones that contribute to maintaining homeostasis in the body come in pairs with opposing effects. What are two examples?

Table 19.1 Major Endocrine Organs and the Hormones They Produce

Endocrine Organ	Hormone	Hormone Type	Effect
Hypothalamus	Hormones released by the posterior pituitary and hormones that regulate the anterior pituitary (see below)		
Pituitary gland			
Anterior pituitary	Growth hormone	Protein	Stimulates growth
	Prolactin	Protein	Stimulates milk production
	Follicle-stimulating hormone	Protein	Stimulates production of eggs and sperm
	Luteinizing hormone	Protein	Stimulates ovaries and testes
	Thyroid-stimulating hormone	Protein	Stimulates thyroid gland
	Adrenocorticotropic hormone	Protein	Stimulates adrenal glands to secrete glucocorticoids
Posterior pituitary (releases hormones made by hypothalamus)	Antidiuretic hormone	Protein	Promotes retention of water by kidneys
	Oxytocin	Protein	Stimulates contraction of uterus
Thyroid gland	Thyroid hormones	Protein	Stimulate and maintain metabolic processes, growth, and development
	Calcitonin	Protein	Lowers blood calcium level
Parathyroid glands	Parathyroid hormone	Protein	Raises blood calcium level
Adrenal glands	Epinephrine (adrenaline) and norepinephrine	Protein	Promote "fight or flight" response
	Glucocorticoids	Steroid	Raise blood glucose level
	Mineralocorticoids	Steroid	Control water and salt balance
Pancreas	Insulin	Protein	Lowers blood glucose level
	Glucagon	Protein	Raises blood glucose level
Sex organs			
Testes	Androgens (testosterone)	Steroid	Support sperm formation; promote development and maintenance of male secondary sex characteristics
Ovaries	Estrogens	Steroid	Stimulate uterine lining growth; promote development and maintenance of female secondary sex characteristics
	Progesterone	Steroid	Promotes uterine lining growth
Pineal gland	Melatonin	Protein	Regulates internal clock

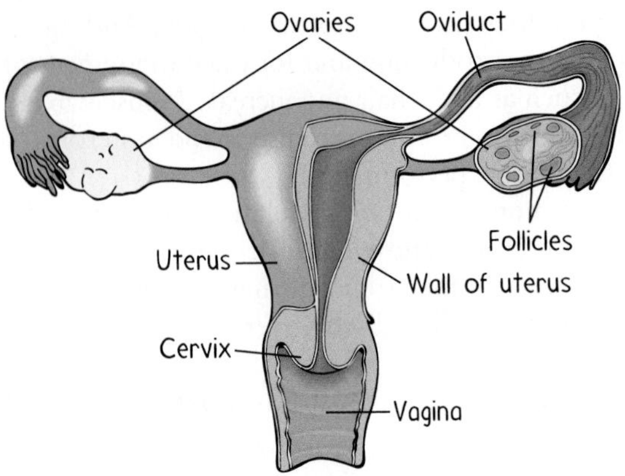

Figure 19.21 Female reproductive anatomy.

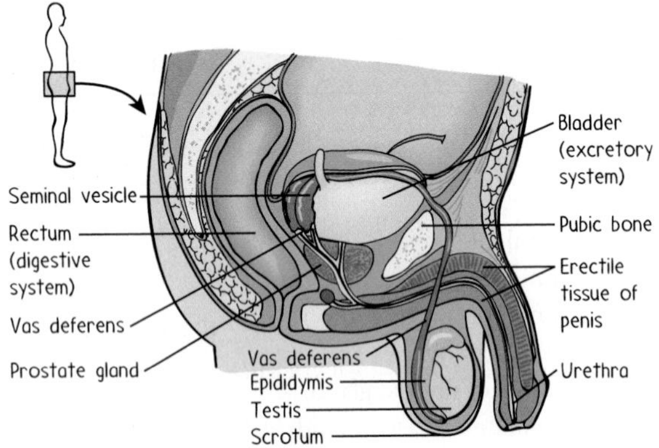

Figure 19.22 Male reproductive anatomy. (Courtesy of Campbell, Reece, Simon, *Essential Biology with Physiology*, © 2004.)

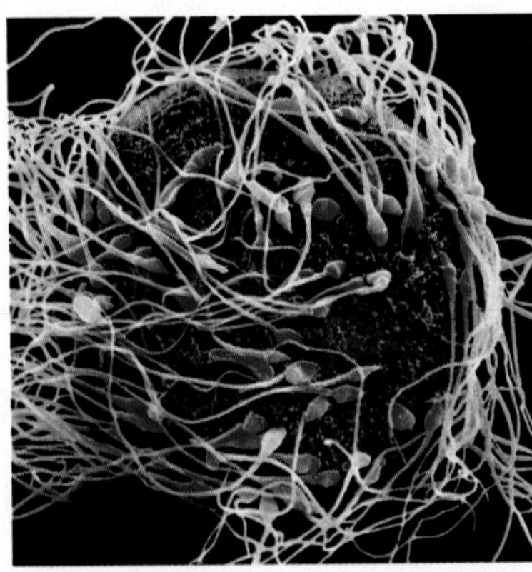

Figure 19.23 Sperm swim on the surface of a human egg.

CHECK YOUR ANSWER

Calcitonin and parathyroid hormone together regulate blood calcium levels—calcitonin lowers blood calcium levels, and parathyroid hormone raises blood calcium levels. Insulin and glucagon together regulate blood glucose levels—insulin reduces glucose levels in the blood and glucagon increases glucose levels.

19.8 Reproduction and Development

Where do babies come from? Without reproduction, humans—or any other species—would quickly die out. Sexual reproduction begins with the production of gametes, eggs in women and sperm in men. Eggs and sperm are haploid cells produced through meiosis (Chapter 16). At fertilization, egg and sperm join to form a diploid cell that develops ultimately into a new human being.

The *ovaries* (Figure 19.21) are made up of follicles, developing eggs surrounded by support cells. During each menstrual cycle, a single follicle matures and releases an egg in a process called **ovulation**. The egg is a large cell, with lots of stored nutrients and resources in its cytoplasm. Eggs end up big because they are the result of unequal meiosis—during cell division, the future egg gets almost all the cytoplasm, while the other cells (which quickly degenerate) receive almost nothing (Chapter 16). Following ovulation, the egg makes its way down the *oviduct*. It is in the oviduct that fertilization, the joining of an egg and a sperm, takes place. The egg is actually in the middle of meiosis II at this point—it only completes meiosis if fertilized. The fertilized egg then continues to the *uterus*, where it implants and completes development.

Sperm are made in the *testes*, which are located in the scrotum (Figure 19.22). The scrotum hangs from the body, keeping the testes at a temperature lower than body temperature. This is essential for sperm production. From the testes, sperm move to the *epididymis*, where they complete development and become mobile. Each mature sperm cell has a head that contains DNA, mitochondria, and enzymes for penetrating the egg, as well as a tail for swimming. During sexual intercourse, sperm travel along the *vas deferens* and are ejaculated from the urethra in *semen*. In addition to sperm, semen contains fluids from the seminal vesicles and prostate gland that nourish sperm and protect them from the acidic environment of the vagina. There are about half a billion sperm in each ejaculate.

Development begins with the fertilization of the egg by a single sperm (Figure 19.23). Sperm swim up the oviduct to the egg and encounter a jellylike layer surrounding the egg called the *zona pellucida*. Enzymes released from the head of the sperm eat away at this layer. A single sperm finally reaches the egg's cell membrane, and the membranes of the egg and sperm fuse. Once this occurs, the zona pellucida undergoes changes that make it impenetrable to additional sperm, assuring that the fertilized egg doesn't end up with too many chromosomes. This doesn't always happen quickly enough, however—fertilization by multiple sperm apparently occurs regularly and represents one of the reasons why couples often take several months to conceive successfully.

Following fertilization, the egg begins to divide. By the time implantation occurs in the uterus, about 6 days after fertilization, the developing egg has become a ball of cells with an inner cavity—the *blastocyst*. Part of the blastocyst forms the embryo. The rest forms structures that protect and nourish the embryo. The fluid-filled **amnion** is the membrane that immediately surrounds the embryo, cushioning and protecting it. The amnion is what ruptures when a pregnant woman's "water breaks." The **placenta** provides oxygen and nutrients to the developing embryo and carries away wastes. It consists of both embryonic and maternal tissues. Maternal and embryonic blood are never in direct contact in the placenta but come close enough to allow for the exchange of nutrients and wastes. The placenta also produces the sex hormones estrogen and progesterone. These hormones prevent further ovulation and maintain the uterus in its nurturing condition throughout pregnancy.

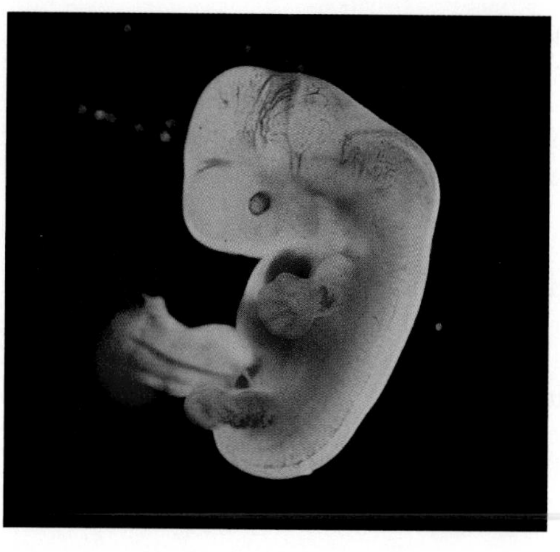
(a)

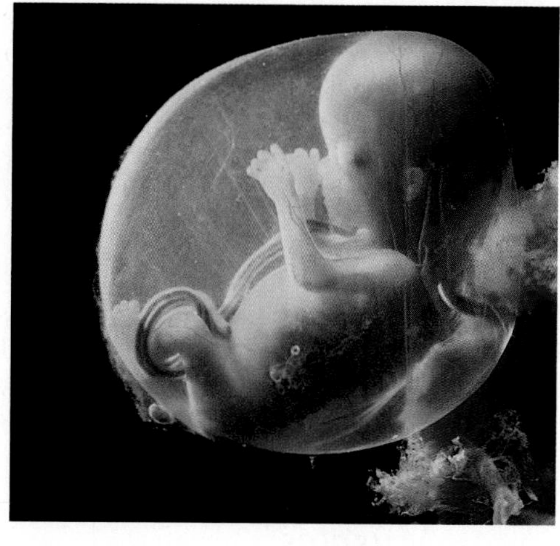
(b)

(c)

Figure 19.24 (a) This is a human embryo at 5 weeks. (b) This is a human fetus at 14 weeks. By the end of the first trimester, all the major organs and body parts have developed. (c) This is a human fetus at 20 weeks.

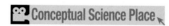

Ultrasound of Human Fetus 1; Ultrasound of Human Fetus 2

The 9 months of pregnancy are divided into three 3-month trimesters. By the end of the first trimester, all the fetus's major organs and body parts have developed. Further development, as well as most of the fetus's growth, occurs in the second and third trimesters. Some stages in human development are shown in Figure 19.24.

INSIGHT You may have heard that for couples trying to conceive, men should wear boxers rather than briefs. It's not just superstition—briefs sometimes hold the scrotum too close to the body and keep the testes at a temperature higher than what is ideal for sperm production.

CHECK YOURSELF

If the zona pellucida were removed from an unfertilized egg and the egg were placed in a petri dish with lots of sperm, what would be the likely result?

CHECK YOUR ANSWER

More than one sperm would fertilize the egg. This is because, when fertilization occurs, the zona pellucida undergoes changes making it impenetrable to additional sperm. Embryos that have more than two sets of chromosomes are unable to survive.

19.9 The Skeleton and Muscles

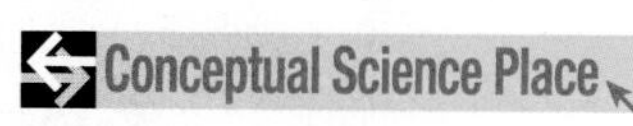

Muscle Contraction

Can you imagine what we'd be like without our skeleton and muscles? We'd be no more than shapeless blobs of tissue, unable to move or even stand. And there wouldn't be a whole lot of us, period! Muscles make up about half our body weight, bones another 14 percent.

The Skeleton

The skeleton, shown in Figure 19.25, consists of the bones and cartilages that protect and support the body. Human adults have 206 bones in all. Babies are born with more, but many of these fuse during growth. Bones vary in size. The largest bone in the body is the femur, or thigh bone. The smallest is the stirrup, a middle ear bone about a quarter of a centimeter long. Besides bones, the skeleton includes several cartilages, including our external ears (pinnas) and the tips of our noses.

One function of the skeleton is to protect the body. The skull surrounds and protects the brain, the vertebrae protect the spinal cord, and the ribs protect the heart and lungs. The skeleton also supports the body and, in cooperation with the muscles, moves it. *Joints* allow movable connections between bones. Some joints, like the elbow and knee, act like hinges, only bending in one direction.

Others, like the articulation of the hip and thigh, resemble a ball and socket and allow for a greater range of motion. At a joint, the ends of connecting bones are covered with smooth cartilage and enclosed in a fluid-filled capsule. The fluid lubricates the joint so that the bones can move smoothly, without rubbing against each other. *Arthritis* is a condition where the tissues of the joint become inflamed and produce too much fluid. Both the inflammation and excess fluid cause painful bone damage.

What would we see if we looked inside a bone? Bones have a strong, hard outer layer of *compact bone* that surrounds a lighter inner layer of *spongy bone*. Inside the spongy bone is the jellylike material bone marrow (Figure 19.26). *Red bone marrow* produces red and white blood cells. *Yellow bone marrow* stores fat. Like other parts of the body, bones are made of living cells. Bone cells secrete the hard calcium-containing matrix that gives bones both strength and flexibility.

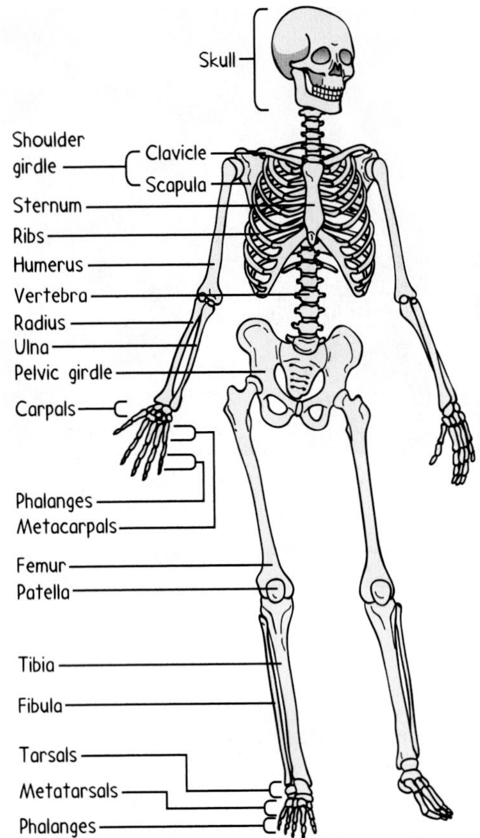

Figure 19.25 The human skeleton includes 206 bones (in adults) as well as a number of cartilages.

Muscles

Muscles allow us to move. We have over 600 muscles in our bodies, 30 in our faces alone. These muscles allow us to make the remarkably fine movements required when we do things like perform brain surgery or build a tower of cards. Our muscles also allow us to smile at our friends, make a peanut butter sandwich, or run down the street. In fact, just about every time we say we're "doing" something, we're doing it with muscles.

Muscles accomplish everything they do by contracting, that is, by shortening. Many of our muscles are connected to bones via *tendons*. When these muscles contract, they pull at our bones, moving us. Because muscles can only pull, not push, we often have pairs of muscles that have opposite effects. For example, the biceps muscle pulls on the inner part of the forearm, causing the forearm to bend up towards the shoulder (Figure 19.27). The triceps muscle pulls on the back end of the forearm (it attaches to the "funny bone" at the end of the elbow) and causes the arm to straighten.

How do muscles contract? First, a muscle has to receive a signal from a motor neuron telling it to contract (Figure 19.28). Motor neurons connect to muscles through a chemical synapse that uses the neurotransmitter *acetylcholine*. When acetylcholine binds to receptors on muscle cells, it starts an action potential in the cells, which respond by contracting. A number of well-known toxins work by interfering with the neuron-to-muscle connection. Curare, an arrow

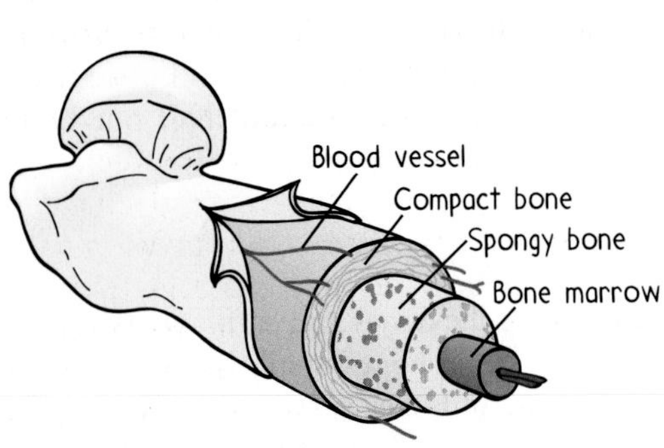

Figure 19.26 In bones, a layer of compact bone surrounds a layer of spongier bone, which surrounds the bone marrow.

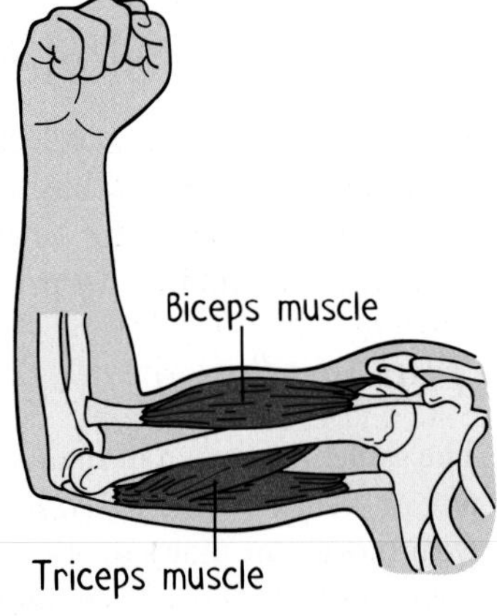

Figure 19.27 The biceps and triceps move the forearm in opposite directions. The biceps bends the forearm, and the triceps straightens it.

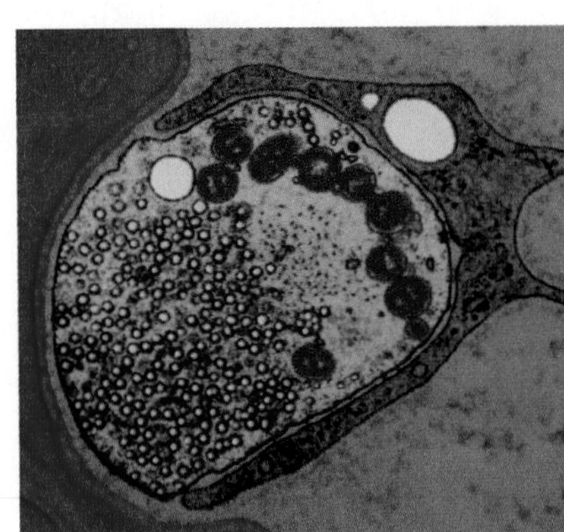

Figure 19.28 This photo shows the junction between a motor neuron (in blue) and a muscle cell (in red). Numerous small vesicles containing the neurotransmitter acetylcholine can be seen clustered near the muscle cell. The larger round structures in the neuron are mitochondria. A glial cell (shown in green) surrounds the neuron.

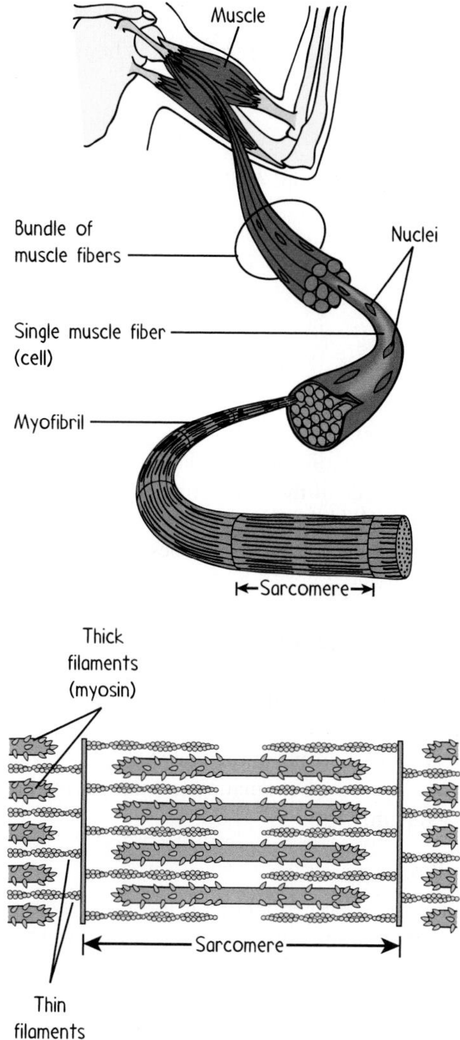

Figure 19.29 Muscles consist of contractile units called sarcomeres. Two proteins found in sarcomeres, actin and myosin, are responsible for muscle contraction. (Courtesy of Campbell, Reece, Simon, *Essential Biology with Physiology*, © 2004.)

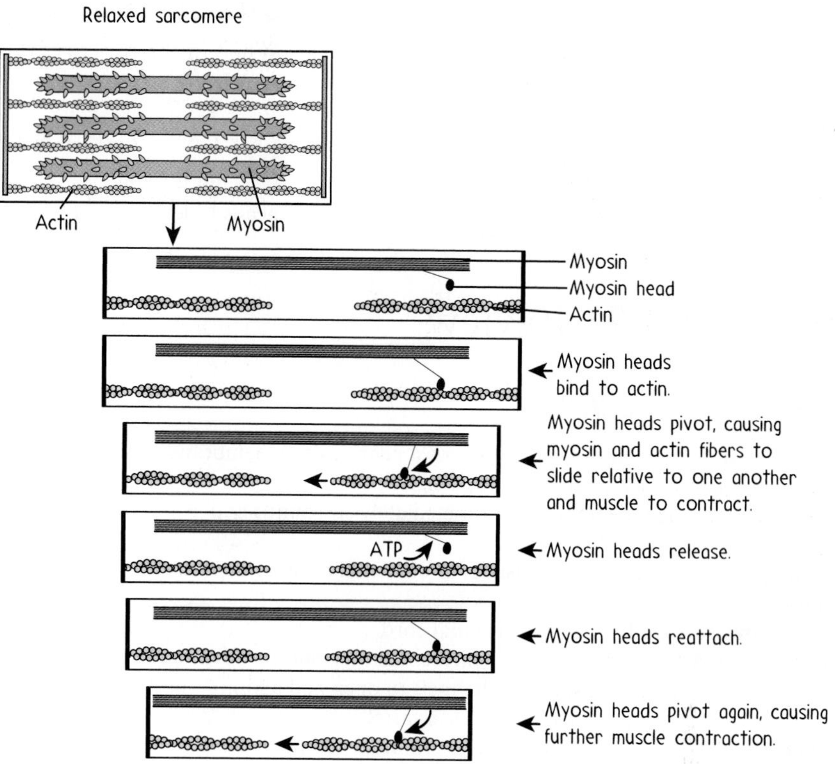

Figure 19.30 During contraction, the myosin heads bind to actin, pivot (causing the actin to slide), release, reattach, and pull again.

poison used in the South American tropics for hunting, binds to acetylcholine receptors on muscle cells, preventing acetylcholine itself from binding. Curare causes paralysis and then death as the respiratory muscles become paralyzed. The powerful nerve gas sarin prevents acetylcholine from being broken down after muscles contract. Muscles are stimulated continuously and soon become exhausted. Again, death occurs through asphyxiation as the respiratory muscles stop working. Sarin is perhaps best known as the agent used in a terrorist attack on the Tokyo subway in 1995.

How does an action potential in a muscle cell lead to muscle contraction? Let's start by looking at how muscles are organized (Figure 19.29). A muscle consists of a bundle of elongated muscle fibers. Each muscle fiber is actually a single cell with multiple nuclei and can be several centimeters long. Each muscle fiber contains bundles of smaller elements called *myofibrils*, which in turn are made up of a series of contractile units called **sarcomeres.** Sarcomeres are made up of carefully arranged fibers of two proteins, thin filaments called *actin* and thick filaments called *myosin*. When an action potential occurs in a muscle cell, calcium ions are released from the cell's endoplasmic reticulum. Calcium ions allow a series of pivoting heads on the myosin fibers to attach to actin. The myosin heads attach and pivot, pulling on the actin as shown in Figure 19.30. Each pull shortens the length of the sarcomere a tiny bit—about 10 nanometers, to be exact—and, consequently, the length of the muscle as a whole. After pulling, the myosin heads release, recock, reattach, and pull again. This cycle repeats until the signal to contract ends or until the muscle has fully contracted. Muscle contractions require energy, of course. ATP is required for the myosin heads to release actin, an essential step in the contraction cycle. This accounts for the phenomenon of rigor mortis, the stiffness of the body that sets in after death—calcium ions leak from the endoplasmic reticulum of muscle cells, causing myosin to bind to actin and muscles to contract. When the available ATP is depleted, the myosin heads are unable to disengage, and the muscles remain contracted.

INSIGHT The number of muscle fibers we have in a given muscle is fixed. The increased muscle mass that develops through activities such as weightlifting comes from a thickening of existing fibers.

CHECK YOURSELF

When in the process of muscle contraction is calcium required? When is ATP required?

CHECK YOUR ANSWER

Calcium is required for the myosin heads to bind to actin. ATP is required for the myosin heads to let go of actin.

CHAPTER 19 Review

Summary of Terms

Action potential A spike in the voltage across a neuron's cell membrane that represents the way a neuron sends a signal.
Amnion The fluid-filled membrane that immediately surrounds and protects a developing embryo.
Axon The part of a neuron that transmits information to other neurons or target cells.
Central nervous system The brain and spinal cord.
Chemoreception A form of sensing in which chemicals bind to receptors on chemosensory cells, causing ion channels to open and action potentials to happen; pertains to smell and taste.
Dendrites The parts of a neuron that receive information from other cells, either sensory cells or other neurons.
Homeostasis The maintenance of a stable internal environment.
Hormones Chemical messengers produced in one place in the body, released into the bloodstream, and received by target cells elsewhere in the body.
Interneuron A neuron that connects one neuron to another neuron.
Motor neuron A neuron that carries messages from the central nervous system to muscle cells, organs, or other responsive cells.
Neuron A cell that receives and transmits information from one part of the body to another.
Organ A structure in the body that has a certain function.
Organ system A set of organs that work together to perform certain bodily functions.
Ovulation The release of a mature egg cell that occurs once during each menstrual cycle.
Peripheral nervous system All the nerves in the body that are not part of the central nervous system.
Placenta The organ that provides for nutrient and waste exchange between the mother and a developing embryo.
Retina The structure of the eye that contains the light-sensitive cells for sight, rods and cones.
Sarcomere The contractile unit of muscle cells.
Sensory neuron A neuron that carries messages from sense receptors to the central nervous system.
Synapse A connection between a neuron and its target cell.
Tissue A group of similar cells that performs a certain function.

Review Questions

19.1 Organization of the Human Body

1. A group of similar cells that performs a certain function is a ___.
2. Multiple tissues combine to make ___, which are organized into an ____ that is responsible for performing particular bodily functions.

19.2 Homeostasis

3. What is homeostasis?
4. What are some bodily characteristics the body keeps stable?

19.3 The Brain

5. What does the brainstem do?
6. Which part of the brain is responsible for balance and posture?
7. Why does damage to one side of the brain often affect functioning on the opposite side of the body?
8. What are the functions of each of the four lobes of the cerebrum?

19.4 The Nervous System

9. What are the two parts of the nervous system, and what structures does each include?
10. A typical neuron has dendrites, a cell body, and an axon. What is the function of each part?
11. What are the three types of neurons in the body?
12. What is the difference between the somatic nervous system and the autonomic nervous system?

19.5 How Neurons Work

13. Why is the resting membrane potential of a neuron negative?
14. What happens to the membrane potential of a neuron during an action potential?
15. What is the significance of a neuron's threshold for firing?
16. Why is an action potential described as an "all-or-nothing" event?
17. How does an action potential move down an axon?
18. How do electrical synapses work? Chemical synapses?

19.6 The Senses

19. Describe the path that light takes to reach the retina.
20. What are the two types of light-sensitive cells in the eyes? How do they differ from one another?
21. Describe how sound waves enter the ear and ultimately cause you to hear.
22. How are action potentials generated in chemosensory cells?
23. How do pain receptors work?

19.7 Hormones

24. What are the two types of hormones, and how do they differ in the way they work?
25. Why is the anterior pituitary sometimes called the "master gland"?
26. What hormones are associated with the "fight or flight" response and where are they made?
27. What hormone made by the hypothalamus helps control the amount of water in the body? What is its effect?

19.8 Reproduction and Development

28. What is unusual about the meiosis that produces the egg? What does this unusual form of meiosis accomplish?
29. How do sperm get through the zona pellucida that surrounds the egg?
30. What is the function of the placenta?
31. When in pregnancy do the major organs of the body develop?

19.9 The Skeleton and Muscles

32. What are three functions of the skeleton?
33. How do motor neurons tell muscles to contract?
34. How do the proteins actin and myosin bring about muscle contraction?
35. At what point in the process of muscle contraction is ATP required?

Multiple Choice Questions

Choose the BEST answer to the following.

1. Which part of the brain controls posture, balance, and fine motor movements?
   (a) brainstem
   (b) cerebellum
   (c) cerebrum
   (d) thalamus
2. Which part of the brain is responsible for reasoning, controlling voluntary movement, and comprehending language?
   (a) brainstem
   (b) cerebellum
   (c) cerebrum
   (d) thalamus
3. This part of a neuron receives information from another cell or neuron:
   (a) dendrite.
   (b) cell body.
   (c) axon.
   (d) myelin sheath.
4. During an action potential
   (a) the membrane potential becomes more negative until it hits threshold.
   (b) the membrane potential increases as potassium ions flow out.
   (c) the membrane potential increases as sodium ions flow in, then decreases as potassium ions flow out.
   (d) the membrane potential may or may not reach threshold.
5. Chemoreception characterizes
   (a) vision.
   (b) hearing.
   (c) touch.
   (d) taste.
6. The hormone insulin
   (a) helps regulate blood glucose level.
   (b) is involved in the "fight-or-flight" response.
   (c) helps regulate the amount of water in the body.
   (d) acts in conjunction with glucagon to regulate calcium levels in the blood.
7. Which of the following statements is false?
   (a) Fertilization takes place in the oviduct.
   (b) The human egg only completes meiosis once it is fertilized.
   (c) Semen contains fluids that help nourish sperm.
   (d) Multiple sperm never fertilize a single egg.
8. The structure that provides oxygen and nutrients to the developing embryo is the
   (a) zona pellucida.
   (b) amnion.
   (c) blastocyst.
   (d) placenta.
9. The connection between motor neurons and muscles
   (a) occurs through an electrical synapse.
   (b) determines whether muscles contract or extend.
   (c) is the target of toxins such as curare and sarin.
   (d) uses an endorphin neurotransmitter.
10. During muscle contraction
    (a) sodium ions are released by the endoplasmic reticulum.
    (b) calcium allows the myosin heads to bind to actin.
    (c) ATP causes the myosin heads to pivot, pulling on actin.
    (d) actin heads repeatedly bind to myosin, pivot, release, recock, reattach, and pivot again.

INTEGRATED SCIENCE CONCEPTS

Physics—How Fast Can Action Potentials Travel?

1. Why do action potentials travel more quickly down thicker axons than thinner ones?
2. What is the problem with achieving rapidly traveling action potentials through large numbers of giant axons?
3. How does myelination speed the propagation of action potentials?

Chemistry—Endorphins

4. What are endorphins?
5. Why do endorphins have effects similar to those of drugs like morphine and heroin?
6. What are some of the activities associated with endorphin release by the brain?

Active Explorations

1. We all have a *blind spot* in our vision where the optic nerve exits the retina because in that spot there are no rods or cones. However, we don't notice the blind spot because the brain cleverly fills it in using visual information from surrounding areas. Explore your blind spot. Draw a small dot on the left side of a piece of paper and a small x on the right side. The x should be about 6 inches from the dot. Now hold the paper in front of you at arm's length. Close your right eye and look at the x with your left eye. Slowly bring the sheet of paper closer to your face. At some point, the dot will disappear. What is happening in the brain that prevents you from seeing the dot?
2. Try this: Place your index finger in front of you and look at it. Now move your head rapidly back and forth, as if you were shaking your head no, keeping your eyes on your finger. What happens? Now hold your head still, and shake your finger quickly back and forth. What happens? In which case is your finger in focus and in which case is it a blur? The vestibular organs in your inner ear, which are responsible for sensing motion, are taking in the movements of your head and reporting them to the brain. The brain then adjusts your eyeballs, moving them as necessary to correct for the motion of your head. Explain why this works when your eyes are moving but not when the object you are viewing is moving.
3. Is what we think of as "taste" largely smell? Cut up an apple and a pear, hold your nose, and chew. Can you distinguish one from the other? Now release your nose and allow yourself to smell, too, as you eat. Now can you tell the difference? This exploration can also be performed with different flavors of jarred baby food (these work well because they all have the same texture).
4. This is a pretty neat trick: Stand in a doorway with your arms at your sides. Now move your arms out so that the backs of your hands are pushing against the sides of the doorframe. Push out as hard as you can for about 40 seconds. Now step away from doorway and relax. Your arms should float upwards, even though you aren't pushing anymore! As you push, signals from the motor neurons to your arm muscles cause calcium ions to be released from the endoplasmic reticulum in your muscle cells. Normally, as soon as your neurons stop sending the push signal, the calcium ions are cleared away, back into the endoplasmic reticulum. But because you push so hard for so long, lots of calcium gets released—too much to clear away quickly.

Exercises

1. You have a conversation with a friend on the telephone. What parts of your brain are you using?
2. The following figure shows a map of the primary motor control area of the brain, found in the frontal lobes of the cerebrum. Why do you think controlling the actions of body parts such as the hands and lips requires such large portions of the brain? Why does controlling the back (trunk) require only a small portion?

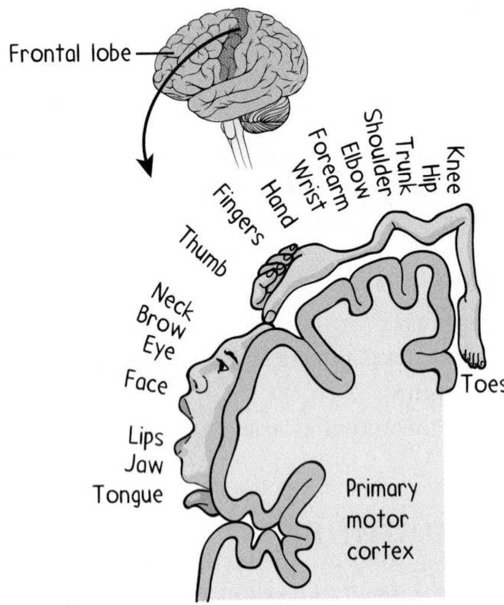

(Courtesy of Campbell, Reece, *Biology* 7e, © 2005.)

3. Of the three types of neurons—sensory neurons, motor neurons, and interneurons—which type goes to your biceps muscle and tells you to bend your elbow? Which type transmits information from your feet as to whether they feel cold?
4. Is a neuron that slows your heartbeat part of the somatic or autonomic nervous system? Is this neuron part of the sympathetic or parasympathetic division?
5. What would happen if you artificially excited a neuron, initiating an action potential in the middle of the axon? How and why is this different from how action potentials actually move along axons?
6. If the myelin sheath were removed from the axon of a neuron, what effect would that have on the neuron's action potential?
7. Do neurotransmitters enter the target cell? If not, how do they have an effect on the target cell?
8. Do all neurotransmitters cause the target cell to fire?
9. Why do a lot of nocturnal species have only rods in their retinas?
10. Are your rods or cones more important for reading a book?
11. In some people, the middle ear bones stiffen with age. This can result in deafness. Why?
12. How are the senses of smell and taste similar? How are they different?
13. Provide three examples of hormones that help maintain homeostasis in the body.
14. Suppose you know that the receptor for a hormone you are studying lies in the cytoplasm of cells. Can you tell whether the hormone is a protein or a steroid hormone?
15. Osteoporosis is a disease that primarily affects postmenopausal women, causing decreased bone density and brittle bones vulnerable to fracture. The hormone calcitonin is sometimes used to treat osteoporosis. Why? (*Hint:* It might be helpful to start by considering what parathyroid hormone does and how it has its effect.)
16. On a brilliant, sunny day, you take a long hike through open country. You sweat a lot, losing a lot of water. What hormone does this cause you to produce? Why?
17. How does meiosis in women differ from meiosis in men?
18. Vasectomy is a form of male sterilization in which a section of each vas deferens is removed. How does this cause sterility?
19. Tubal ligation is a form of female sterilization in which the oviduct is cut and the tubes tied. How does this cause sterility?
20. Does a fertilized human egg make anything other than the embryo?
21. Each time myosin heads pull on actin, the sarcomere contracts only 10 nanometers or so. Given that, how are we able to produce large motions?
22. Can muscle contraction occur without the presence of calcium ions? Why or why not? Where are the calcium ions in a muscle cell stored, and what causes their release?
23. Both the arrow poison curare and the nerve gas sarin affect the nerve-to-muscle connection. Do they work the same way? If not, how do they differ?
24. Write a letter to Grandma explaining how the new generation of antidepressants, selective serotonin reuptake inhibitors (SSRIs) such as Prozac and Zoloft, work. Explain to her how neurons signal each other and what SSRIs do to influence this process.

Problems

1. Action potentials travel at speeds anywhere between 0.5 and 120 meters/second, depending on factors such as temperature, the size of the axon, and whether the axon is myelinated. We have two different types of neurons that conduct pain signals to the central nervous system. The slower type conducts signals at 0.5 meters/second. The faster type conducts signals at 25 meters/second. Let's say that the distance from your hand to your central nervous system is about 1 meter. Now: You touch a hot stove. Show that you become aware of the first type of pain in 0.04 seconds, but only become aware of the second type of pain after 2 seconds. (You may have noticed that when you do something like touch a hot stove, you feel a flash of sharp pain first, followed by a slow throbbing pain).
2. The human retina has an area of about 1000 mm^2. If we have a total of 125 million rods and 6.5 million cones in each eye, show that we have about 131,500 sensory cells per square millimeter in the retina.
3. We have about 1000 different kinds of smell receptors. Each of these is a distinct protein coded for by a specific gene. In Chapter 16, we learned the Human Genome Project revealed that humans have a total of about 30,000 genes. Show that about 3.3 percent of our genes are dedicated to helping us smell.
4. The egg is a large cell and contributes almost all the nutrients to the zygote (fertilized egg) created at fertilization. The sperm contributes little more than its set of chromosomes. Just how much bigger is a human egg than a human sperm? The human egg is about 100 micrometers in diameter. The head of a human sperm is about 4 micrometers in diameter. (And, in case you're curious, human sperm are about 50 micrometers long.) Show that the volume of a human egg is 15,625 times larger than the volume of a human sperm. Recall that the volume of a sphere is $\frac{4}{3}\pi r^3$.

Conceptual Science Place

CHAPTER 19 ONLINE RESOURCES

Tutorials Neuron Structure, Nerve Signals, Neuron Communication, Overview of Cell Signaling, Nonsteroid Hormone Action, Steroid Hormone Action, Muscle Contraction • ***Videos*** Ultrasound of Human Fetus 1, Ultrasound of Human Fetus 2 • ***Quiz*** • ***Flashcards*** • ***Links***

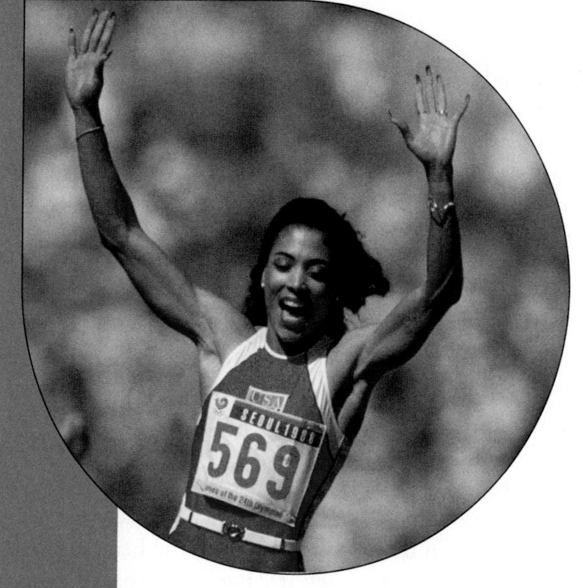

CHAPTER 20

Human Biology II—Care and Maintenance

Track and field star Florence Griffith Joyner crosses the finish line, winning a gold medal at the 1988 Olympics in Seoul, Korea.

CHAPTER OUTLINE

Whether you are engaged in a remarkable display of athleticism and endurance—or merely resting and reading a newspaper—your body relies on the coordinated efforts of multiple body systems to support its activities. In this chapter, we focus on the care and maintenance of the human body, examining the circulatory, respiratory, digestive, excretory, and immune systems. As we explore these systems, we'll learn many things: What makes the "lub-dubb" sound of the heartbeat? Is it possible to forget to breathe? What is a hiccup? Is swallowing a voluntary or involuntary action? If you cut or burn your finger, why does the tissue become red and swollen? How does the body defend itself against bacteria, viruses, and other pathogens?

20.1 Integration of Body Systems

As we learned in Chapter 19, body systems rarely act alone. Most of the body's primary functions require the efforts of two or more organ systems. For example, supplying the body with oxygen is a job that's split by two systems—the respiratory system brings oxygen into the body, and the circulatory system distributes it to the tissues. Similarly, getting rid of cellular wastes requires the coordinated efforts of the circulatory, respiratory, and excretory systems—the circulatory system collects wastes from the tissues, and the respiratory and excretory systems remove them from the body via exhalation and urine production, respectively. In this chapter, we will look at how five different organ systems work together to maintain the body. Among their many tasks are two of the most essential—acquiring energy for the body's activities and protecting the body from disease.

20.2 The Circulatory System

Path of Blood Flow in Mammals

Imagine if everything you needed came right to your door. Running low on milk? Here come a couple of quarts now, floating down the street and in the door. Need a steak for the barbecue? Wait a bit, there should be a T-bone along any moment. Fuji apple? This is your lucky day. No need to take garbage out either, just toss it over your shoulder and it'll float its way out. Is this a couch potato's dream come true? Perhaps, but it's also just a day in the life of our cells. The

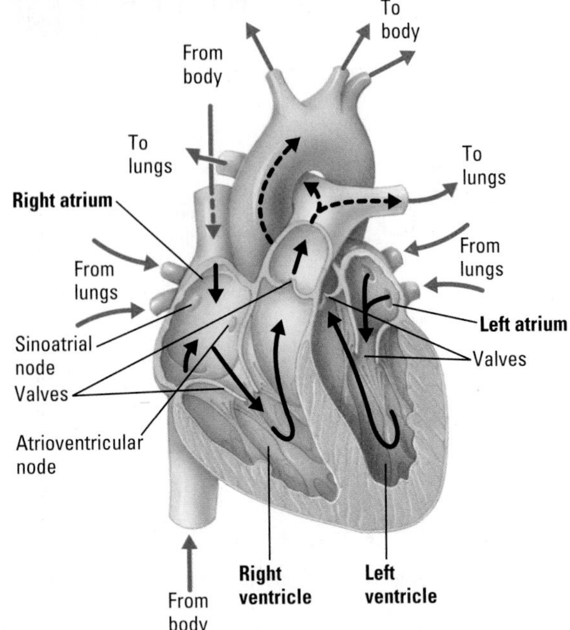

Figure 20.1 The heart has four chambers, a right atrium, right ventricle, left atrium, and left ventricle. The right side of the heart pumps blood to the lungs. The left side pumps blood to the body. Each heartbeat begins in the sinoatrial node or "pacemaker." The signal sweeps across the right and left atria, which contract simultaneously. It also passes to the atrioventricular node and from there to the two ventricles. Valves prevent blood from flowing backwards. (Courtesy of Campbell, Reece, Simon, *Essential Biology with Physiology,* © 2004.)

circulatory system takes all of the body's cells everything they need and removes all the stuff they need to dispose of. It is the body's system for moving things around, a sort of food, garbage, and mail service, rolled into one. The circulatory system delivers oxygen and glucose, takes away the wastes that cells produce, and carries special items like hormones or immune cells to specific target cells. It consists of three components, the heart, blood vessels, and blood.

The Heart

The heart, shown in Figure 20.1, is the pump that drives blood around the body. It is about the size of a clenched fist and has four separate chambers, a right atrium, right ventricle, left atrium, and left ventricle. The right side of the heart is responsible for pumping blood to the lungs, where it picks up oxygen, and the left side is responsible for pumping blood to the body.

Heart muscle does not require nervous system stimulation in order to contract—it contracts all on its own. (In fact, in developing embryos, the heart begins to beat very early, long before developing nerves reach it). Each heartbeat begins in a part of the right atrium called the *sinoatrial node*, or pacemaker. The pacemaker initiates an action potential that sweeps quickly through the right and left atria, which contract simultaneously. The pacemaker's signal also passes to the *atrioventricular node*, and from there to the two ventricles, which then also contract simultaneously. The atrioventricular node conducts action potentials slowly, producing a delay between the contraction of the atria and the contraction of the ventricles. This is why each heartbeat consists of two separate sounds, as you will notice if you listen closely. This miraculous "lub-dubb… lub-dubb… lub-dubb…" is perhaps the phenomenon we associate most closely with life. Note, however, that the "lub-dubb" sounds themselves come not from the contractions of heart muscle (after all, we don't make thudding noises when we contract muscles in our arms or legs!), but from *valves* that snap shut after each contraction. Heart valves lie between the atria and ventricles and between the ventricles and outgoing blood vessels. These keep blood from flowing backwards. The heart beats about 70 times a minute. This adds up quickly—to over 100,000 heartbeats a day.

Blood Vessels

Blood makes its way around the body in tubes called *blood vessels*. All blood vessels, except the thin-walled capillaries, are surrounded by a layer of smooth muscle and a layer of elastic tissue. **Arteries** carry blood away from the heart. Arteries stretch when blood is pumped, then recoil—you can feel this stretch and recoil by touching the pulse at your wrist or temple. From the arteries, blood moves into smaller vessels called *arterioles*. The smooth muscle around an arteriole controls its diameter, either dilating or constricting the vessel to control the amount of blood flow. How much blood a tissue receives depends largely on its activity level—for example, lots of blood goes to our digestive organs when we've just eaten a meal and are ready to digest it. From the arterioles, blood flows into tiny vessels called **capillaries** (Figure 20.2). It is from the capillaries that materials are exchanged between blood and tissues. To allow for this exchange, the walls of capillaries are very thin, sometimes no more than a single cell thick. Furthermore, no cell in the body is more than 130 micrometers away from a capillary, a distance small enough to allow for efficient exchange. Many molecules, including oxygen and carbon dioxide, diffuse freely between the capillaries and tissues. Others are moved in and out of the body's cells through facilitated diffusion, active transport, endocytosis, or exocytosis (Chapter 15). From the capillaries, blood flows back towards the heart in small *venules* and then larger **veins**. Blood flow in the veins depends partly on our voluntary motions. When we move, our muscle contractions squeeze blood along the veins. This is why long periods of sitting or standing cause our legs and ankles to swell. Valves in the veins help make sure the blood doesn't flow backwards (Figure 20.3).

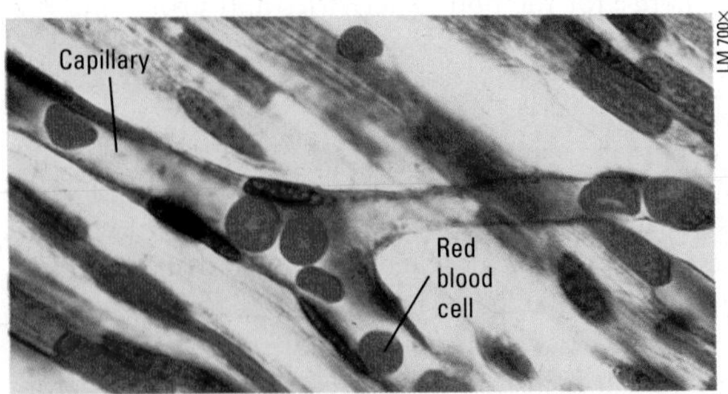

Figure 20.2 Capillaries are the tiny, thin-walled blood vessels from which materials are exchanged with body tissues. This photo shows a capillary with red blood cells.

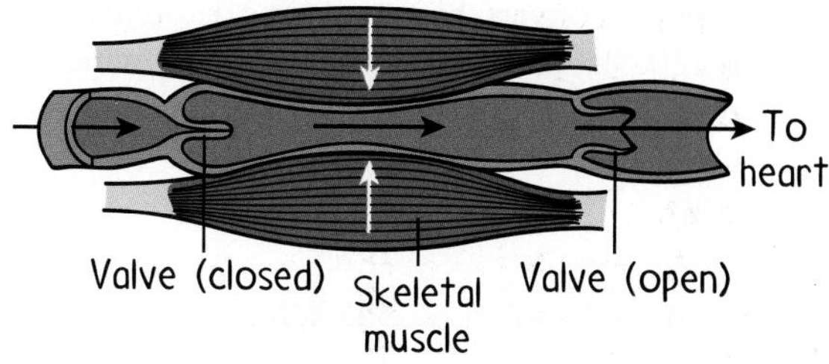

Figure 20.3 The contractions of our voluntary muscles squeeze blood along the veins. Valves make sure the blood doesn't flow backwards.

Like all good food, garbage, and mail services, blood does not move haphazardly around the body, but follows a set route, shown in Figure 20.4. The path of blood flow enables the circulatory system to efficiently carry out one of its primary tasks, delivering oxygen to tissues. How does blood flow? Let's begin with blood returning from the body to the heart. Blood returning from the body is deoxygenated, that is, it contains low levels of oxygen. This deoxygenated blood flows from veins into the right atrium of the heart. The right atrium pumps it to the right ventricle. The right ventricle pumps it out arteries that go to the lungs. There, the blood picks up oxygen and drops off carbon dioxide—the blood is now oxygenated. This oxygenated blood flows back to the heart, along veins that lead to the left atrium. The left atrium pumps it to the left ventricle which then pumps it out arteries that go to tissues all over the body. After carrying oxygen to the tissues, the blood becomes deoxygenated again and returns to the heart via veins. The entire circuit takes about a minute.

Blood

Now that we know how blood is circulated, let's look more closely at blood itself. We each have about 11 pints of blood in our bodies, making up about 8 percent of our body weight. A little more than half of this is *plasma*. Plasma is mostly water, but it also contains important molecules such as salts, proteins, hormones, glucose, other nutrients, and wastes. The rest of our blood is made up of cells—red blood cells, white blood cells, and platelets. **Red blood cells** carry oxygen. Unlike other cells in the body, red blood cells have no nucleus or mitochondria. This makes them smaller in size and easier to transport in the blood, but it also means they are unable to make the proteins necessary to maintain themselves. As a result, red blood cells have a life span of only about 4 months. Every second, some 2 million red blood cells die and 2 million new cells, made in the bone marrow, are called into action. The oxygen in red blood cells is carried by a large protein called **hemoglobin**. Each molecule of hemoglobin carries up to four oxygen molecules, and there are as many as 300 million molecules of hemoglobin in a single red blood cell. **White blood cells** are part of the immune system and help our bodies defend against disease. **Platelets** are involved in blood clotting. When body tissues are damaged, platelets attach to damaged blood vessels

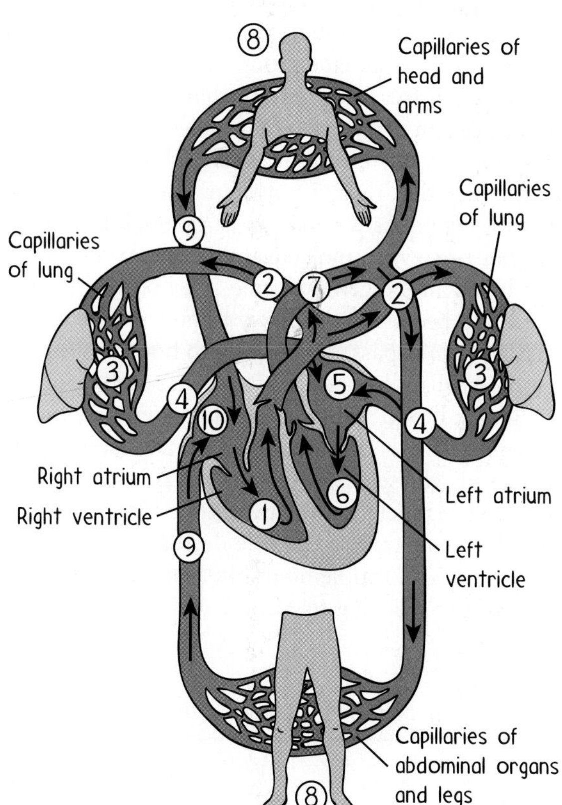

Figure 20.4 Blood flows in set pathways around the body. Blue represents deoxygenated blood, and red represents oxygenated blood. (Courtesy of Campbell, Reece, Simon, *Essential Biology with Physiology*, © 2004.)

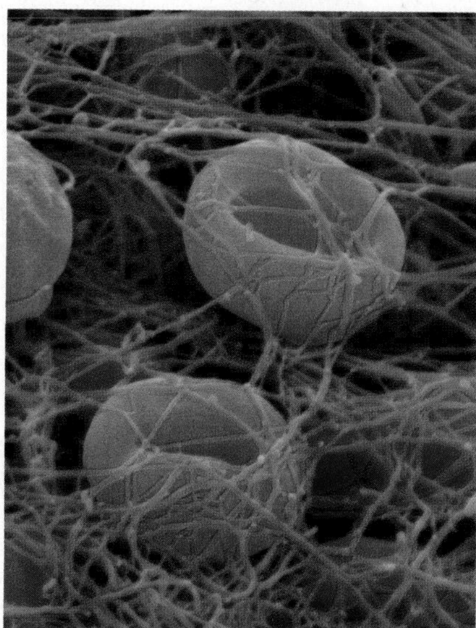

Figure 20.5 Clotting prevents blood from leaking. Here, we see red blood cells enmeshed in threads of fibrin.

INSIGHT During pregnancy, maternal and fetal blood never come into direct contact in the placenta. This means maternal red blood cells aren't passed directly to the fetus. How does the fetus get oxygen? It turns out that fetuses have a special type of hemoglobin in their red blood cells. Fetal hemoglobin, which is present from a few weeks after conception to a few months after birth, has a higher affinity for oxygen than adult hemoglobin. As a result, oxygen molecules move readily across the placenta from maternal red blood cells to fetal red blood cells.

and release special clotting factors, which convert blood plasma proteins to a sticky substance called fibrin (Figure 20.5). The disease hemophilia, which is associated with a deficiency in blood clotting, is the result of genetic mutations that affect clotting factors.

CHECK YOURSELF

1. Which chambers of the heart contain oxygenated blood? Which chambers contain deoxygenated blood?
2. Most arteries carry oxygenated blood, and most veins carry deoxygenated blood. Are there any exceptions?

CHECK YOUR ANSWERS

1. The left atrium and left ventricle contain oxygenated blood that has just returned from the lungs. The right atrium and right ventricle contain deoxygenated blood that has just returned from the body.
2. Recall that arteries are blood vessels carrying blood away from the heart, and veins are vessels carrying blood to the heart. The arteries that carry blood from the heart to the lungs carry deoxygenated blood. The veins that carry blood from the lungs back to the heart carry oxygenated blood.

INTEGRATED SCIENCE CHEMISTRY

Hemoglobin

There are 300 million molecules of hemoglobin, the oxygen-carrying protein, in every human red blood cell. A molecule of hemoglobin consists of four subunits, each of which contains a component known as a heme group that includes an iron atom at its center (Figure 20.6). It is this iron atom that binds oxygen. Consequently, each hemoglobin molecule can carry up to four molecules of oxygen.

When oxygen binds to one of the four heme groups in a hemoglobin molecule, the other three are altered in such a way that their affinity for oxygen increases; that is, they are more likely to bind oxygen. As a result, most hemoglobin molecules will carry the maximum number of oxygen molecules away from the lungs. Similarly, when one heme group unloads an oxygen molecule at a body tissue, the other three heme groups are altered and become more likely to give up their oxygen molecules as well, assuring that oxygen is passed efficiently to body tissues.

The oxygen affinity of a hemoglobin molecule also varies depending on its local environment. For example, lower blood pH (a more acidic environment) decreases hemoglobin's oxygen affinity. Why is this adaptive? An active, working tissue makes and uses more ATP and so releases more carbon dioxide during cellular respiration. Because carbon dioxide reacts with water in the blood to form carbonic acid, the presence of high levels of carbon dioxide decreases blood pH. This acidity decreases the oxygen affinity of local hemoglobin molecules, making it easier for them to unload oxygen to the working tissue.

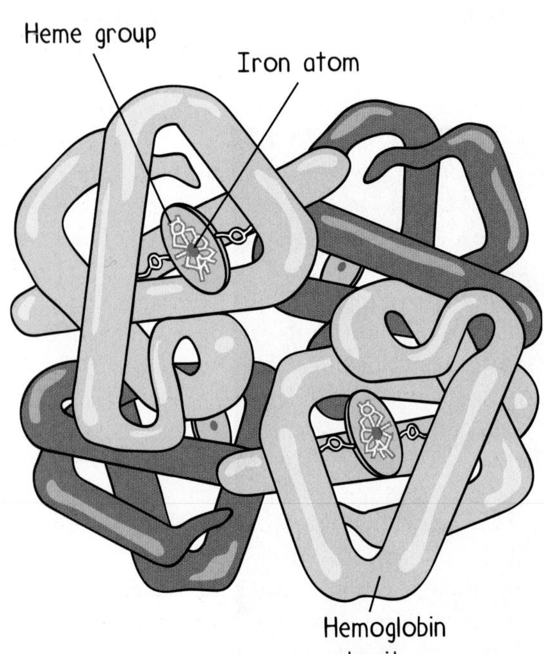

Figure 20.6 Hemoglobin consists of four subunits. Each subunit is associated with a heme group that includes an iron atom at its center.

CHECK YOURSELF

Why is iron an essential component of our diets? What symptoms are associated with insufficient iron intake?

CHECK YOUR ANSWER

Iron is found in hemoglobin, which carries oxygen to body tissues. Insufficient iron intake can result in a low hemoglobin count and cause anemia, a deficiency in the oxygen-carrying capacity of blood. Because tissues do not get enough oxygen to support their normal activity levels, anemia is associated with weakness and fatigue.

TECHNOLOGY

Transplanting Bone Marrow in the Fight Against Cancer

Chemotherapy and radiation therapy are two of the most common treatments for cancer. How do they work? In a way that may seem counterintuitive—by damaging DNA. This is effective, however, because extensive DNA damage can kill tumor cells (Chapter 16). Aren't the body's healthy cells affected too? Yes—but cells are most vulnerable to DNA damage when they divide frequently, and cancer cells, by the very nature of the disease, divide very frequently. For this reason, chemotherapy and radiation therapy target tumor cells particularly.

But some healthy cells in the body also divide very frequently, and these are harmed by chemotherapy and radiation therapy as well. The *blood-forming stem cells* of the bone marrow, which produce red blood cells, white blood cells, and platelets, are one such group. Fear of irreparably damaging a patient's blood-forming stem cells has, until recently, always limited the level of chemotherapy and radiation therapy administered. The development of successful techniques for transplanting bone marrow has changed that, and allowed for much more aggressive cancer treatments.

Bone marrow transplants are performed after cancer therapy destroys a patient's blood-forming stem cells. Patients receive either their own stem cells, harvested before cancer treatment, or cells from another individual. Blood-forming stem cells are collected from the bone marrow or blood of the donor. To collect them from bone marrow, a needle is inserted into the hip bone and the marrow harvested. To collect cells from the bloodstream, drugs are given to trigger the release of more stem cells into the blood, and blood is then processed to remove stem cells. If a cancer patient is donating to himself or herself, the cells must be treated before transplantation to remove cancer cells. The recipient receives stem cells directly into the bloodstream by IV. The cells migrate to the bone marrow and begin to function a few weeks after the transplant. It can be some time before the new stem cell population is fully functional, however, and the patient's immune function usually does not return to normal for several years.

20.3 Respiration

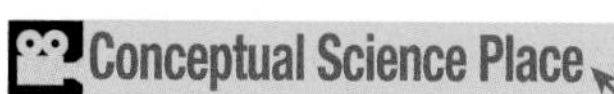

The Human Respiratory System; Transport of Respiratory Gases

Many of us only think about breathing when we're jogging or running up and down a soccer field—and suddenly huffing and puffing. In fact, of course, it's a constant feature of our existence—we take some 17,000 breaths over the course of a single day! Breathing is the body's way of taking in oxygen, a key ingredient for cellular respiration, the process cells use to obtain energy in the form of ATP (Chapter 15). After about 4 minutes without oxygen, severe brain damage sets in; after about 6, death. Through breathing, the respiratory system moves oxygen from the air to the circulatory system, which then delivers it to all our tissues. In the process, the respiratory system also gets rid of carbon dioxide, a side product of making ATP.

INSIGHT What's going on when you have the hiccups? Hiccups come from sudden spasmic contractions of the diaphragm. Air is sucked into the respiratory tract, and the vocal cords snap shut, creating that "hic" noise. Hiccups may be caused by any irritation to the diaphragm, including eating too much or too quickly. Hiccups normally go away by themselves after a few minutes. According to *Guinness Book of World Records*, however, Mr. Charles Osborne of Anthon, Iowa, a hog farmer, hiccupped nonstop from 1922 to 1990—for a total of about 430 million spasms!

The respiratory system is shown in Figure 20.7. As we inhale, air comes in through the nose (or, sometimes, the mouth—a useful backup, since otherwise we would all die of stuffy noses!). Hairs in our nostrils trap dust and any other particles we might breathe in. Air continues up the nasal passages, where it is moistened by mucus and warmed by a dense network of capillaries. At this time, odor molecules present in the air are also captured and detected by the cells involved in our sense of smell (Chapter 19). From the nasal passages, air passes through the *pharynx*, the part of the throat above the esophagus. Then it proceeds through the **larynx**, or voice box, and down the *trachea*, or windpipe. The trachea is a short tube stiffened by a series of cartilaginous rings. These rings keep the breathing passage open when swallowed food causes the esophagus, directly behind the trachea, to bulge. The trachea branches into two tubes called *bronchi* that lead to the right and left lungs and then into smaller and smaller tubules that finally dead-end at small sacs called **alveoli**. It is in the alveoli that gas exchange occurs.

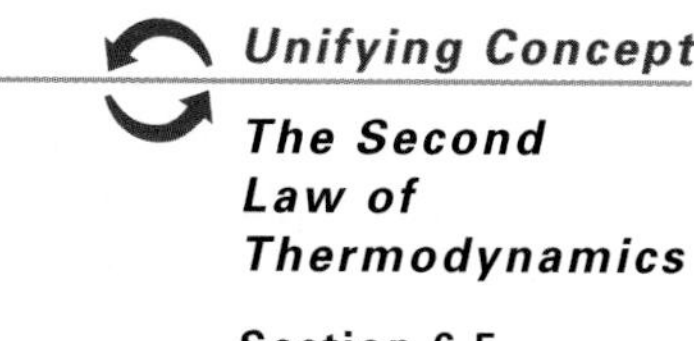

We have a total of about 300 million alveoli in our lungs. Each alveolus is enmeshed in a net of capillaries. Both the alveolus and the surrounding capillaries have extremely thin walls, consisting of only a single flattened cell. Gas exchange occurs through diffusion (Chapter 15)—that is, gas molecules move from an area of higher concentration to an area of lower concentration. Because there is more oxygen in the alveolus than in the deoxygenated blood of the capillaries, oxygen diffuses from the alveolus into the blood. Similarly, there is more carbon dioxide

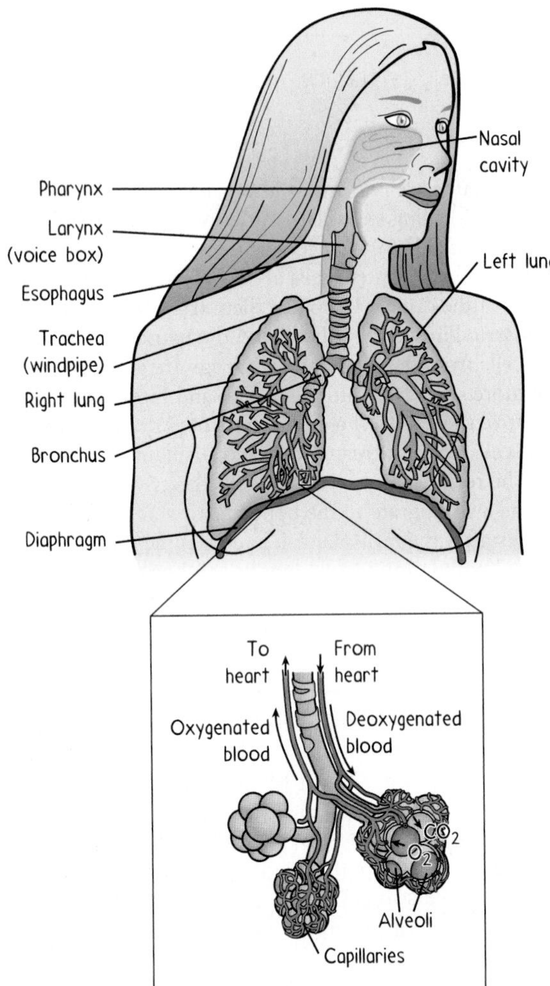

Figure 20.7 The respiratory system is responsible for obtaining oxygen for the body and getting rid of carbon dioxide. Gas exchange occurs in the lungs, in tiny sacs called alveoli. Oxygen diffuses from the alveoli into surrounding capillaries, and carbon dioxide diffuses from the capillaries into the alveoli. (Courtesy of Campbell, Reece, Simon, *Essential Biology with Physiology*, © 2004.)

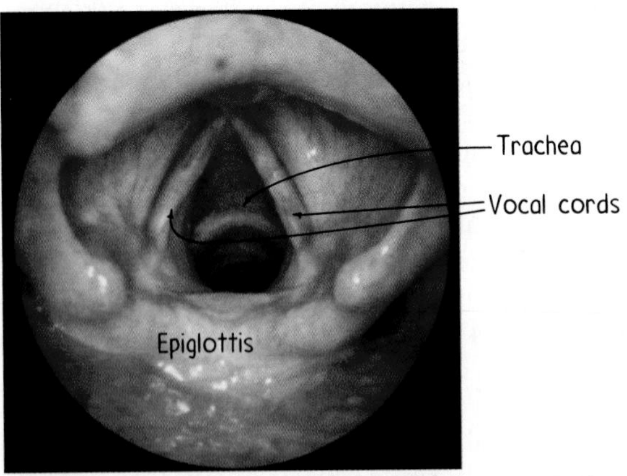

Figure 20.8 Air moves past the vocal cords as we exhale, causing them to vibrate and produce the sounds of human speech. We control the sounds we make by loosening or tightening the cords.

in deoxygenated blood than in the alveolus, so carbon dioxide diffuses from the blood to the alveolus. When we exhale, air reverses its path, going from the alveoli up through tubules to the bronchi, trachea, and out the nose or mouth. As air passes the larynx, it may vibrate a set of *vocal cords* that generate sound waves (Figure 20.8). We control muscles that stretch the vocal cords to produce the different sounds of speech.

How do we move air in and out of the lungs? The lungs sit inside the ribcage in an air-filled pocket called the thoracic cavity. Forming the bottom portion of the thoracic cavity is a sheet of muscle called the *diaphragm* (Figure 20.9). When the diaphragm is relaxed, it is dome-shaped. When we inhale, the diaphragm contracts. This causes it to flatten, increasing the volume of the thoracic cavity. Muscles between our ribs also contract, pulling the ribcage up and out from the chest and further increasing the volume of the thoracic cavity. So, the volume of the thoracic cavity increases, while the amount of air inside it remains constant. What happens? The air pressure in the thoracic cavity drops. Air is sucked into the lungs and fills the alveoli. This is not that different from the way a squashed rubber ducky (one of the ones with a hole in the bottom) sucks in air as it unsquashes and increases in volume or the way a bicycle pump sucks in air when you pull back its plunger. When we exhale, the diaphragm and rib muscles relax, decreasing the volume of the thoracic cavity. This increases the air pressure in the thoracic cavity and pushes air out of the lungs. Only about 10 percent of the air in the lungs is exchanged with air outside during each breath, but that's enough to keep our body tissues supplied with oxygen.

We breathe about 12 times per minute. In contrast to our heartbeats, we have some control over how often and how deeply we breathe. However, we don't have to worry about "forgetting" to breathe—respiration is controlled, along with other unconscious activities, by the brainstem (Chapter 19).

CHECK YOURSELF

Why is it better to breathe through your nose than through your mouth?

CHECK YOUR ANSWER

The nasal passages have hairs that trap dust and other particles, mucus to moisten air, and capillary beds to warm it. If you breathe through your mouth, none of this happens, and you are more likely to irritate your respiratory tract.

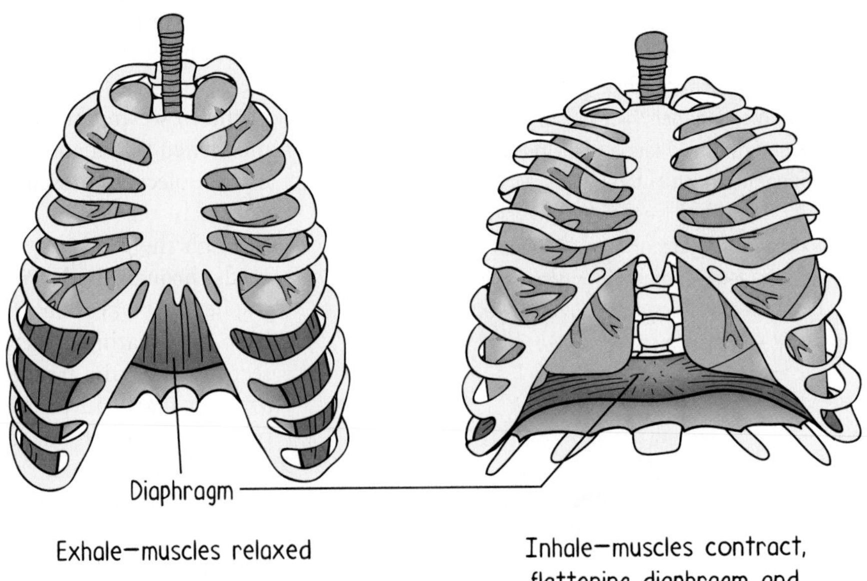

Figure 20.9 When we inhale, the diaphragm contracts and flattens and the ribcage expands, increasing the volume of the thoracic cavity and causing air to flow into the lungs.

20.4 Digestion

Conceptual Science Place

Digestive System Function

Food plays a central role in our lives. Even the busiest of us take the time to grab a sandwich, and for many of us, meals—between shopping, cooking, eating, and doing dishes—fill up a significant part of the day. We even devote an entire room—the kitchen—to storing, preparing, and eating food.

Food supplies the body with organic molecules that can be broken down for energy in the form of ATP. In addition, like other endotherms (Chapter 18), we rely on heat released during the breakdown of organic molecules to maintain a high, stable body temperature. Finally, the body obtains certain essential molecules, ones it can't produce on its own, from food. However, our cells wouldn't know what to do with a sliver of papaya or a slice of Swiss cheese. Food has to be broken down into organic molecules that can be absorbed and used by the body. This occurs during the process of **digestion**, which takes place in the organs of the digestive system (Figure 20.10).

Digestion begins in our mouths. First, we chew food, breaking it into little pieces that can more easily be attacked by enzymes. Second, we mix food with *saliva*, which contains an enzyme that breaks down starches. Saliva also has several other functions—it lubricates food, which can then be moved easily around the mouth, and it allows us to taste our food. This is because our taste buds can only detect chemicals dissolved in liquid. Eating, or just the anticipation of food, triggers the release of saliva (Figure 20.11).

After food is chewed and mixed with saliva, we swallow it. The tongue shapes the food into a compact mass and then pushes it to the back of the throat. Food descends into the pharynx and then enters the *esophagus*. Because there are actually two openings in the pharynx, one to the esophagus and one to the trachea, the body has to make sure swallowed food enters the right tube or we will choke. The *epiglottis*, a small flap of cartilage at the back of the tongue, covers the trachea during swallowing so that food can't get into it (Figure 20.12). Once in the esophagus, food is pushed down by a moving wave of muscular contractions known as **peristalsis**. Peristalsis squeezes food down the esophagus by constricting behind it and pushing it along, much the way you might use your fingers to squeeze a LifeSavers® candy out of its paper-tube wrapping (Figure 20.13). It is these muscular contractions, not gravity, that move food down the esophagus—for example, astronauts in the zero gravity of space have no trouble swallowing at all. Peristalsis is in fact so effective we can swallow

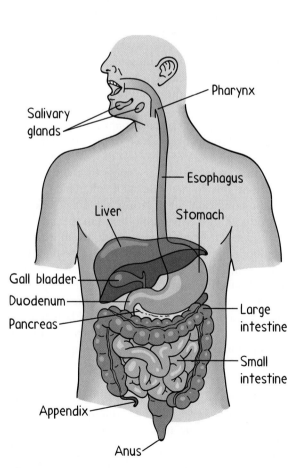

Figure 20.10 The digestive system breaks food down into organic molecules that can be used by the body.

Figure 20.11 The anticipation of food, or even the mere thought of it, can cause the release of saliva, which contains important digestive enzymes and also lubricates our food.

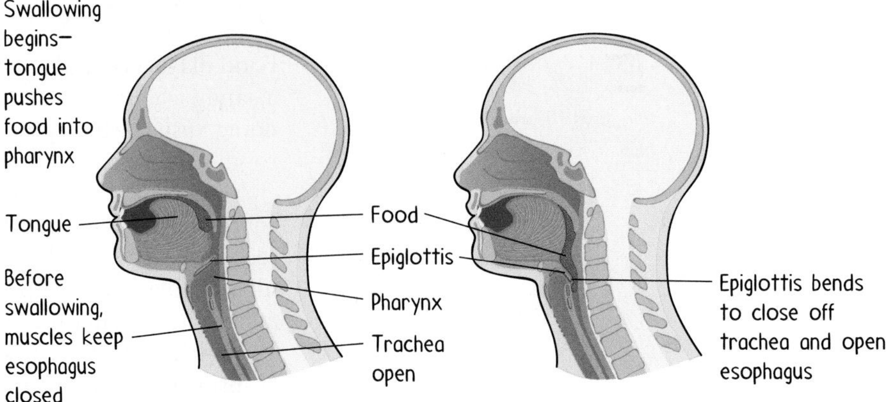

Figure 20.12 During swallowing, a flap of cartilage at the back of the tongue—the epiglottis—covers the trachea so that food goes down the esophagus rather than into the respiratory tract.

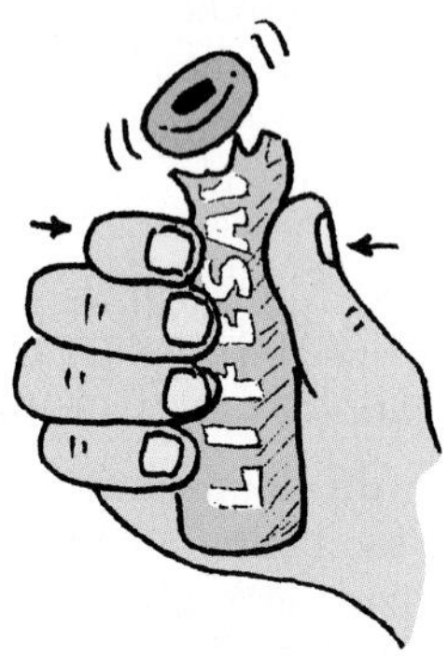

Figure 20.13 Peristalsis, the wave of muscle contractions that moves food down the esophagus, works in a way that is similar to how you squeeze a LifeSavers® candy out of a candy roll.

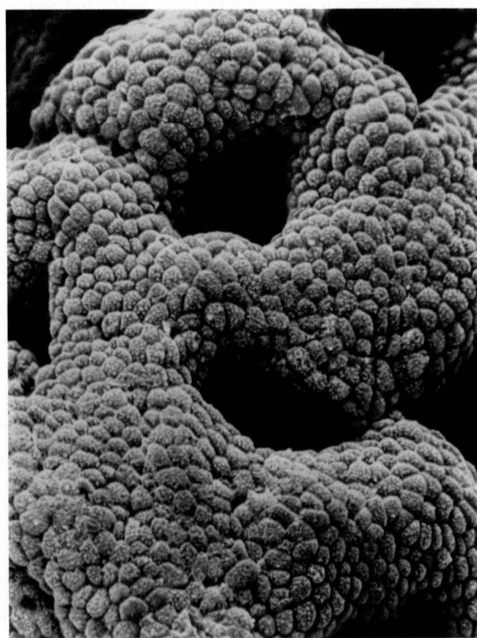

Figure 20.14 The inside surface of the stomach includes pits that house the gastric-juice-releasing glands.

even while upside down! Swallowing begins as a voluntary activity—the muscles at the top of the esophagus are voluntary muscles. However, at a certain point, swallowing becomes involuntary. The lower part of the esophagus is made of involuntary smooth muscle like that found in the rest of the digestive tract.

At the bottom of the esophagus, chewed food moves through a sphincter, or ring-shaped muscle, into the stomach. Glands in the stomach walls, shown in Figure 20.14, release *gastric juice*, a highly acidic mix of hydrochloric acid, digestive enzymes, and a protective mucus that prevents the stomach from digesting its own tissues. When we vomit, the acidic nature of our stomach contents becomes immediately apparent both from the taste and from the burning sensation in our throats. The purpose of this acidity is to kill any bacteria we swallow with our food. In the stomach, digestive enzymes and a muscular churning action combine to reduce our food to a thick liquid called *chyme*. Chyme exits the stomach through a second sphincter and enters the small intestine. Typically, it takes the stomach about 4 hours to process a meal.

The *small intestine* is about 20 feet long. In the *duodenum*, the first foot of the small intestine, digestion continues with the breakdown of proteins, fats, carbohydrates, and nucleic acids. Some of the digestive enzymes at work in the duodenum are made by the small intestine itself. Others are made by the pancreas. Pancreatic enzymes play an important role in neutralizing food, which arrives from the stomach in a highly acidic condition. In addition, the small intestine receives *bile*, a substance that is produced in the liver and stored in the gall bladder. Bile is an emulsifier—it breaks fats into tiny droplets that are more easily attacked by enzymes. Beyond the duodenum, the rest of the small intestine functions primarily in absorbing nutrients into the body. In order to be able to do this efficiently—that is, rapidly—the small intestine has a huge surface area. It is covered with numerous fingerlike projections called *villi*, each of which is in turn covered with tiny little projections called *microvilli* (Figure 20.15). Flattened, the small intestine would fill the area of a tennis court! Digested nutrients are absorbed across the surface of the small intestine into capillaries found inside each villus. Absorption of most types of molecules occurs through facilitated diffusion or active transport (Chapter 15).

After nutrient absorption, what's left of our food moves into the *large intestine*. In the large intestine, water and minerals such as sodium are absorbed into the body. The large intestine is also home to large numbers of *Escherichia coli* and other bacteria, which feed off our undigested materials. Some of these bacteria are useful to us because they synthesize vitamins, notably vitamin K and some of the B vitamins, which we absorb and use. From the large intestine, feces are eliminated from the body through the anus. Feces are composed primarily of living and dead bacteria and undigestible materials such as plant cellulose.

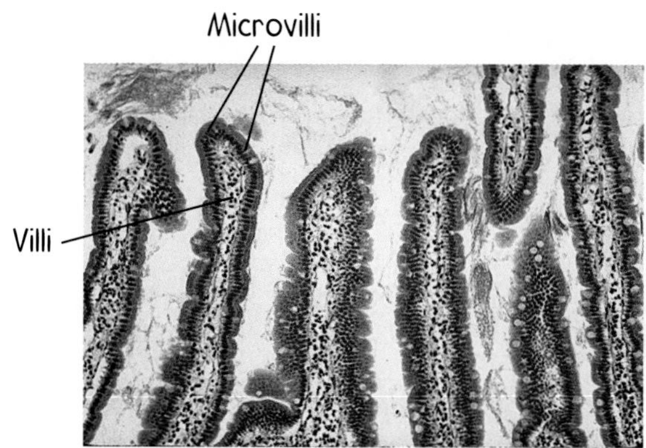

Figure 20.15 In order to efficiently absorb nutrients, the small intestine is covered with numerous fingerlike villi, which in turn are covered with tiny microvilli.

CHECK YOURSELF

Having the openings to both the trachea and the esophagus in the pharynx is problematic in that it can lead to choking. Are there any advantages to this arrangement?

CHECK YOUR ANSWER

Yes, it allows us to breathe through our mouths—air can flow into our mouths and down the trachea.

20.5 Nutrition, Exercise, and Health

We now know how the body obtains the oxygen and nutrients it needs to fuel its activities. These demanding functions are performed day in and day out by the circulatory, respiratory, and digestive systems. How does our lifestyle contribute to the functioning of these and other body systems? What can we do to maintain a healthy life?

Let's begin with nutrition and consider what makes for a healthy diet. First, we need a certain number of food Calories to support our activities—this is the fuel that our cells combine with oxygen to make ATP. What is a Calorie? You'll recall that a calorie (small "c") is the amount of heat required to raise the temperature of 1 gram of water 1°C (Chapter 6). Food Calories (big "C") are actually kilocalories—1000 of the calories just described. Depending on body size and activity level, most people need between 1500 and 3000 Calories a day.

Just getting enough Calories is not sufficient, however. We also need to make sure we take in all the essential nutrients the body can't make on its own. For example, humans are unable to make 8 of the 20 amino acids needed to build proteins. This is why it's important for us to eat a "complete protein"—one containing all the amino acids—regularly. Meat provides a complete protein, as does the combination of beans and rice, a staple of many vegetarian diets. Humans also require a number of vitamins and minerals in small amounts. *Vitamins* are essential nonprotein components of enzymes. Although the body needs many different vitamins, we'll look at just a few examples. Vitamin C, found in citrus fruits, dark green leafy vegetables, and certain other foods, helps the body resist infections and repair wounds. Vitamin A, which we obtain from carrots, eggs, fortified milk, and other foods, is important for proper eye function, allowing the eyes to adjust to dim light—insufficient vitamin A results in night blindness and other problems. Vitamin K, found in green leafy vegetables and synthesized by bacteria in our intestines, is essential for blood clotting. Insufficient amounts of vitamin K can result in hemorrhaging. *Minerals* are required as components of various body tissues. Minerals include the calcium found in bones and teeth, the phosphorus found in ATP, and the iron found in hemoglobin.

Beyond Calories and nutrients, does it matter what we eat? Yes! And unfortunately, trusting our instincts on this doesn't always work well. Some of the things we are inclined to eat—like sweets and fats—turn out not to be particularly good for us. This is probably because our taste for these foods evolved under a different set of circumstances, when they were less readily available. What to eat, then? The "Healthy Eating Pyramid" (Figure 20.16) provides guidelines developed by nutritionists at the Harvard School of Public Health.

In addition to a healthy diet, exercise (Figure 20.17) is an important part of any healthy lifestyle. For most of human history, people got plenty of physical activity from the things they had to do just to eat and survive. This was true when most people belonged to hunting and gathering societies and true when most people farmed. It is still true for a large fraction of the world's population, but not all of it. Desk jobs, cars, the Internet and television, and labor-saving devices of every sort at home and at work have brought many of us too close to a

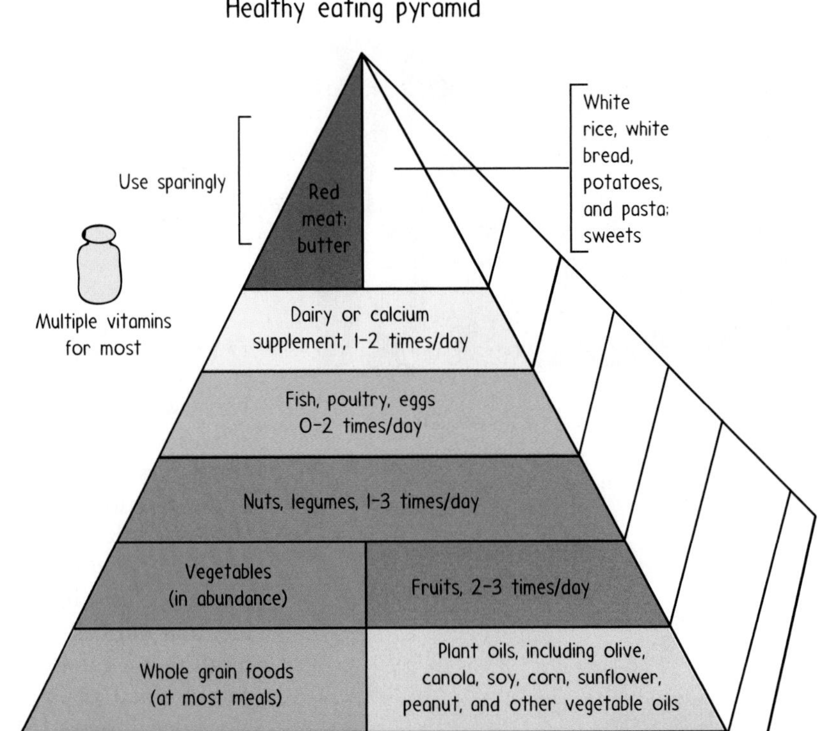

Figure 20.16 A healthy diet includes whole grains, plant oils, and fruits and vegetables in abundance. The "Healthy Eating Pyramid" is from *Eat, Drink, and Be Healthy: The Harvard Medical School Guide to Healthy Eating*, Simon and Schuster, 2001.

Figure 20.17 Even those of us with busy lives should make time for exercise. Here, Nils Gilman, busy father of two, shoots baskets in the Panhandle, Golden Gate Park, San Francisco.

sedentary, couch-potato lifestyle. By now, many of us have to go out of our way to get the physical activity we need to stay healthy.

What are the benefits of exercise? Regular exercise reduces the risk of many health problems, including heart disease, high blood pressure, colon and breast cancer, osteoporosis, diabetes, and obesity. Exercise improves the functioning of the heart and lungs by increasing oxygen consumption, the amount of blood the heart pumps, and lung capacity while decreasing heart rate and blood pressure. Exercise increases the amount of good cholesterol in the body, reducing the chance of heart disease. It reduces insulin resistance in the tissues, decreasing the chance of developing diabetes. It also maintains our joints, tendons, and ligaments and increases muscle mass and bone density. This not only improves strength and balance but helps us stay flexible and mobile as we age. By increasing our metabolism, regular exercise also helps us maintain a healthy body weight. Studies show that exercise even contributes to our mental well-being, reducing levels of depression, stress, and anxiety. And it helps us sleep better. Exercise may not cure all our ills, but there's no doubt it's a good thing.

INTEGRATED SCIENCE PHYSICS AND CHEMISTRY

Low-Carb Versus Low-Cal Diets

A whopping 64 percent—nearly 2 out of 3—Americans are overweight or obese, according to a Centers for Disease Control Study conducted in 2000. And, with healthy body weight becoming more and more obviously an important component of good health, many more Americans are trying to lose weight. But losing weight is a famously difficult task, and it seems that as long as there have been dieters, there have been fad diets.

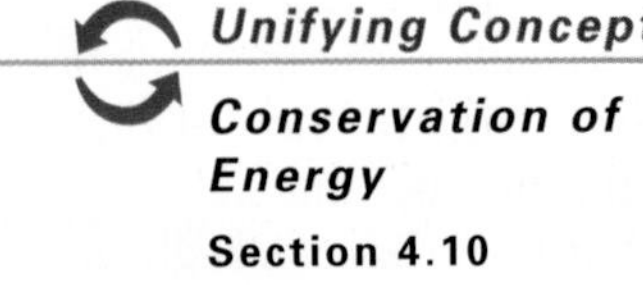

Unifying Concept
Conservation of Energy
Section 4.10

How do you lose weight? The answer is simple—use up more Calories than you take in. This forces your body to use stored energy—such as fat—to support its activities. A low-Calorie diet therefore makes perfect sense. The role of

SCIENCE AND SOCIETY

What Are the Odds? Current Major Health Risks

According to the latest data, life expectancy at birth in the United States is 80.1 years for females and 74.8 years for males. The overall life expectancy of 77.6 years is the highest ever in this country. What are the current major health risks?

Heart disease is the leading cause of death in the United States, followed by cancer and strokes. Cancer has been catching up to heart disease, however, and is already the leading cause of death in people under 85. Cancer is expected to overtake heart disease in the general population in 2018. This switch is due largely to a decrease in smoking—the number of smokers in the United States dropped from 42 percent of the population in 1965 to 22 percent in 2000. (Although smoking is an important risk factor for certain cancers, especially lung cancer, it also increases the risk of heart disease.) The 15 leading causes of death in the United States are shown here, along with worldwide data for comparison.

Leading Causes of Death in the United States (Data are for 2003, from the Centers for Disease Control.)

	Total Deaths (thousands)	% of Total
1. Heart disease	684	28.0
2. Cancer	555	22.7
3. Strokes	158	6.5
4. Chronic lung diseases (emphysema, chronic bronchitis, etc.)	126	5.2
5. Accidents (unintentional injuries)	106	4.3
6. Diabetes	74	3.0
7. Influenza and pneumonia	65	2.7
8. Alzheimer's disease	63	2.6
9. Kidney disease	43	1.7
10. Septicemia (blood poisoning)	34	1.4
11. Intentional self-harm (suicide)	31	1.3
12. Chronic liver disease and cirrhosis	27	1.1
13. High blood pressure	22	0.9
14. Parkinson's disease	18	0.7
15. Pneumonia resulting from sucking materials into the lungs	17	0.7

Leading Causes of Death Worldwide (Data are for 2002, from *The World Health Report, 2003*, published by the World Health Organization.)

	Total Deaths (thousands)	% of Total
1. Heart disease	7,181	12.6
2. Strokes	5,509	9.7
3. Lung infections	3,884	6.8
4. HIV/AIDS	2,777	4.9
5. Chronic lung disease	2,748	4.8
6. Diarrheal diseases	1,798	3.2
7. Tuberculosis	1,566	2.7
8. Malaria	1,272	2.2
9. Lung cancer	1,243	2.2
10. Road traffic accidents	1,192	2.1
11. Childhood diseases	1,124	2.0
12. Other unintentional injuries	923	1.6
13. Hypertensive heart disease	911	1.6
14. Self-inflicted (suicide)	873	1.5
15. Stomach cancer	850	1.5
16. Cirrhosis of the liver	786	1.4
17. Kidney disease	677	1.2
18. Colorectal cancer	622	1.1
19. Liver cancer	618	1.1
20. Measles	611	1.1

exercise is also clear—exercise uses up lots of Calories directly, and it also increases your metabolic rate, making your body consume Calories more quickly even when you're not exercising.

Now what about low-carb diets, such as the Atkins diet or the South Beach diet? Low-carb diets have become extremely popular, and it's easy to see why. Studies have confirmed that, for many people, low-carb diets do produce weight loss more quickly and more consistently than low-calorie diets. This appears to be because many people find low-carb diets easier to stick to because of their permissive attitude towards fats. People on low-carb diets lose weight for the same reason that people on low-Calorie diets lose weight—they consume fewer total calories. In addition, low-carb diets cause you to retain less water in the body—this water is used during excretion to flush out the extra proteins consumed.

Are there any problems with low-carb diets? Yes—they tend to be high in saturated fats and cholesterol, which are associated with heart disease, and they tend to be short on whole grains and fruit, which are known to protect against many diseases. Finally, not a whole lot is known about potential long-term effects of low-carb diets. For example, water loss and the processing of large amounts of proteins may be hard on organs such as the liver and kidneys.

CHECK YOURSELF

What are two ways in which exercise helps you lose weight?

CHECK YOUR ANSWER

Exercise uses up Calories directly. It also increases your metabolic rate, making your body consume Calories more quickly when you're not exercising.

20.6 Excretion and Water Balance

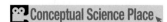

Conceptual Science Place

Structure of the Human Excretory System; Nephron Function; Control of Water Reabsorption

As cells go about their activities, they are constantly generating wastes. We already know what cells do with these wastes—let them diffuse into the bloodstream. But where do they go from there? How does the body get rid of them? This is the job of the excretory system (Figure 20.18). The excretory system filters our blood, removing waste materials while leaving the good stuff behind. At the same time, the excretory system controls the retention or excretion of important ions such as sodium, potassium, and calcium, as well as that most fundamental of substances, water. Our entire blood supply moves through the kidneys about 16 times a day. The end result? About 6 cups (1.4 liters) of urine.

Many different types of wastes are produced by the body, but some of the most prevalent come from making ATP. There's carbon dioxide, of course, which is removed from the body during respiration. In addition, there are all the unusable portions of the fuel molecules cells receive. For example, the amino groups (NH_2) are stripped from amino acids before they enter the ATP-making machinery. These amino groups are then combined with hydrogen ions to form the waste product ammonia (NH_3). But because ammonia is toxic even in small quantities, the liver quickly converts it to a less toxic waste substance, *urea*, that it releases back into the bloodstream for excretion.

Excretion begins in the kidneys. The functional unit of a kidney is called a **nephron**, shown in Figure 20.19. Our kidneys contain about a million nephrons apiece. From the circulatory system, fluid enters the nephron through a cup-shaped structure known as Bowman's capsule. It then flows through the rest of the nephron, moving through the proximal convoluted tubule, loop of Henle, distal convoluted tubule, and collecting duct. What enters the nephron is more or less blood plasma, and what comes out is urine. Let's look at how this happens.

Each nephron in our kidneys is associated with a cluster of capillaries called the *glomerulus*. The glomerulus is surrounded by a cup-shaped structure called *Bowman's capsule*. Blood pressure in the glomerulus pushes fluid out of the capillaries and into Bowman's capsule. This fluid is called the *filtrate* and is pretty similar to blood plasma. From Bowman's capsule, the filtrate flows into the *proximal convoluted tubule*. The proximal convoluted tubule acts like a sorting machine. In it, some "good" molecules in the filtrate—including many ions, as well as glucose, vitamins, and amino acids—are transported back into the blood to be kept by the body. Also, additional waste molecules are transported from the blood into the filtrate. This movement of molecules in and out of the filtrate occurs through active transport (Chapter 15), one of the reasons why excretion consumes a considerable amount of energy.

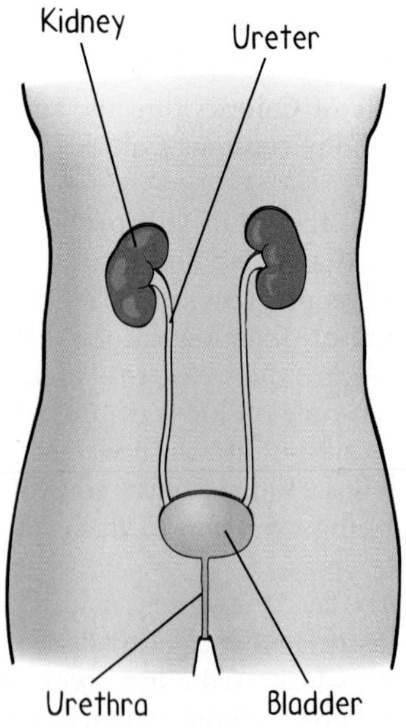

Figure 20.18 The excretory system consists of the kidneys, ureters, bladder, and urethra.

After moving through the proximal convoluted tubule, the filtrate continues into the *loop of Henle*, a hairpin-shaped loop. The loop of Henle functions in reabsorbing water from the filtrate. It does this by resting in a solute concentration gradient from one end of the loop to the other. Specifically, there is low solute concentration near the "ends" at the top of the hairpin and high solute concentration near the loop at the bottom of the hairpin (Figure 20.20). As the filtrate flows down the descending branch of the loop of Henle, it encounters the zone of high solute concentration. Water moves from the filtrate to the surrounding tissues via osmosis (Chapter 15), making the filtrate much more concentrated than before. Then the filtrate moves up the ascending branch of the loop of Henle.

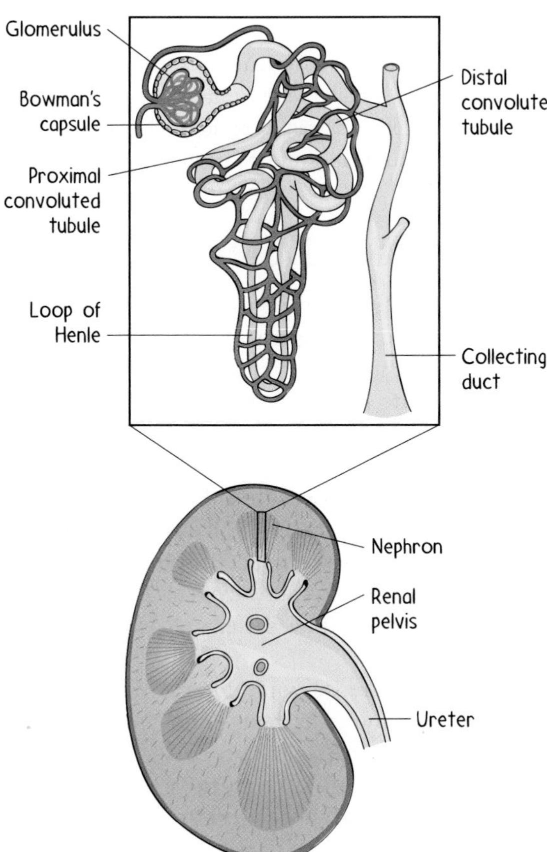

Figure 20.19 Each of our kidneys is made up of about a million functional units called nephrons.

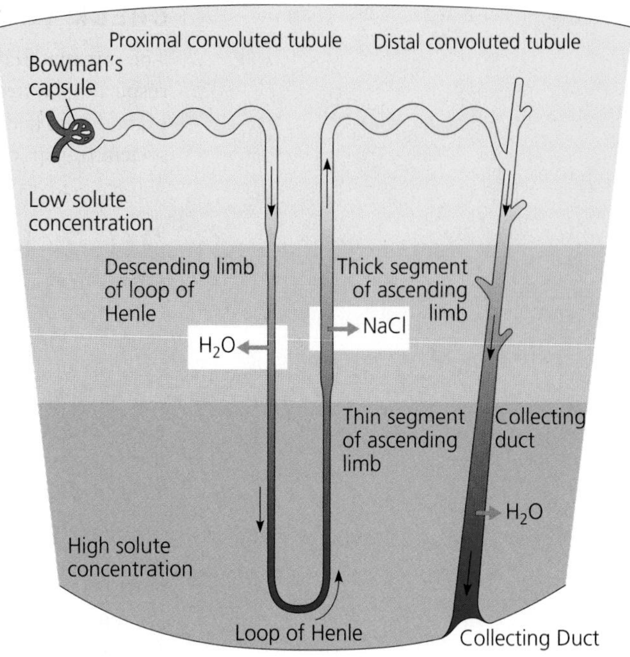

Figure 20.20 The loop of Henle functions in reabsorbing water from the filtrate. A solute concentration gradient from the top of the loop to the bottom causes water to move out of the filtrate as it moves down the descending branch. The ascending branch is impermeable to water. (Courtesy of Campbell, Reece, *Biology 7e*, © 2005.)

Water does not flow back into the filtrate as it moves through the low-solute part of the loop of Henle because the walls of the ascending branch simply aren't permeable to water. However, as the filtrate moves up the ascending branch, more ions are pumped out of it through active transport, so that the filtrate is once again less concentrated than the high solute concentration zone. We will see why this is important in a moment.

After ascending the loop of Henle, the filtrate moves into the *distal convoluted tubule*, where additional wastes are transported into it. Finally, the filtrate moves down the *collecting duct* where more water may be reabsorbed. Whether this happens or not depends on the presence or absence of antidiuretic hormone, which we first encountered in Chapter 19. Antidiuretic hormone causes the walls of the collecting duct to be permeable to water, allowing water to be reabsorbed. (Note that because additional ions were pumped out of the filtrate as it moved up the ascending branch of the loop of Henle, the filtrate is once again less concentrated than the high solute concentration zone of the kidney. This means water should move from the filtrate back into body tissues through osmosis—but only if the walls of the collecting duct are permeable to water.) If antidiuretic hormone is not present, the walls of the collecting duct are not permeable to water and no more water is reabsorbed. The amount of reabsorption that occurs in the collecting duct determines how concentrated the urine becomes.

Urine flows from the collecting duct into the *renal pelvis*, which is essentially a large funnel that catches the drippings of a million nephrons. From the renal pelvis, urine flows down the *ureter* to the *bladder*, a stretchy sac where it is temporarily stored. When the bladder is emptied, urine flows down the *urethra* and out the body.

INSIGHT Once upon a time, diabetes was diagnosed by ants. If ants were attracted to a person's urine, it meant there was sugar in it, and that meant diabetes. In fact, the full name of the disease is diabetes mellitus, *mellitus* being Latin for "sweet as honey." Why is diabetes associated with sugary urine? Diabetics are either unable to make the hormone insulin or unable to respond to it, causing large amounts of glucose to build up in the blood. Glucose is one of those "good" molecules the kidneys normally reabsorb. However, when blood glucose levels are very high, the glucose can't all be reabsorbed, and some is excreted in urine.

CHECK YOURSELF

We have now looked at all the organ systems that play a direct role in helping the body acquire energy in the form of ATP. What organ systems provide the raw materials necessary for making ATP (that is, for cellular respiration)? What systems are involved in getting rid of the waste products?

CHECK YOUR ANSWER

The raw materials are organic molecules from the digestive system and oxygen from the respiratory system. The wastes are carbon dioxide, which the respiratory system gets rid of, and urea and other substances, which the excretory system gets rid of. The circulatory system transports both raw materials and wastes to their destinations in the body.

20.7 Keeping the Body Safe: Defense Systems

Conceptual Science Place
Immune Responses

Without our body's defenses, we would quickly fall prey to the countless bacteria, viruses, and other **pathogens**—disease-causing organisms—that surround us. What protects us is our immune system, which can be divided into two parts, innate immunity and acquired immunity. **Innate immunity** is nonspecific—that is, its defenses work against a wide variety of potential pathogens. The innate immune system includes anatomical barriers that block the entry of pathogens, such as skin or the hairs in our nostrils, as well as a number of different types of immune cells. **Acquired immunity** is highly specific—the cells of this system recognize very specific features of specific pathogens and take action only when these features are encountered. Interestingly, the innate immune system appears to be present in all animals, whereas the acquired immune system is probably only found in vertebrates.

Some of the organs of the immune system are shown in Figure 20.21. All the immune cells of the innate and acquired immune systems—our "white blood cells"—are made in the bone marrow. Some mature there, whereas others mature in the thymus. Mature immune cells are found primarily in the circulatory system, lymphatic system, and spleen, where they prowl constantly for pathogens. The *lymphatic system* includes a series of lymph vessels that, like blood vessels, travel all over the body. Lymphatic vessels carry a clear fluid called **lymph** that contains numerous immune cells. *Lymph nodes* are small bean-shaped structures in the lymph vessels where large numbers of immune cells are concentrated. Lymph nodes are found in many parts of the body, including the throat (where the familiar tonsils are lymph nodes), armpits, and groin. Swollen lymph nodes are a sure sign that the body is fighting an infection.

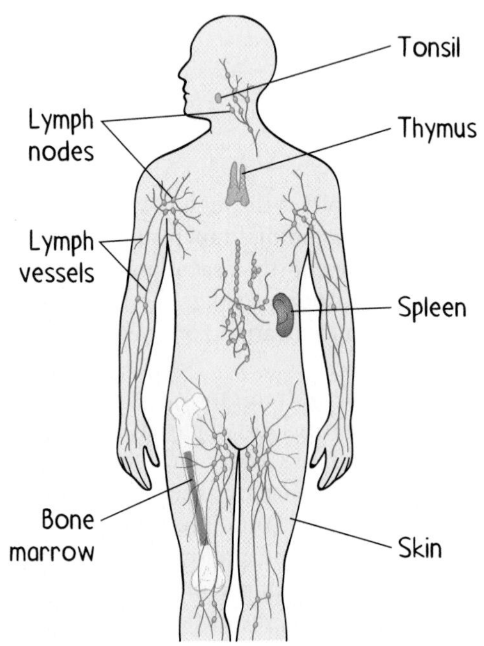

Figure 20.21 The immune system includes the skin, bone marrow, thymus, spleen, and lymph nodes and vessels.

Innate Immunity

The first line of defense against potential pathogens involves structures that try to keep them from entering the body in the first place. The skin is a crucial barrier, forming a tough outer layer that is nearly impenetrable when intact. The frequent shedding of skin cells makes it harder for potential pathogens to establish a foothold. In addition, hair follicles in the skin produce acidic secretions that help to kill bacteria and fungi. Many other body secretions contain bacteria-killing enzymes as well, including saliva, tears, sweat, and milk.

The mucous membranes that line the inside of the body—including the inner surfaces of the nose, mouth, eyelids, respiratory, digestive, urinary, and reproductive tracts—present a less formidable barrier than skin. However, all mucous membranes are covered by a layer of mucus that helps trap pathogens. In the respiratory tract, cilia sweep mucus up to the pharynx so that it can be swallowed. Stomach acid dispatches these pathogens as well as any that enter the body with food. Many mucous membranes are constantly flushed with fluids, again to help prevent pathogens from gaining a foothold—examples include tears, saliva, and urine.

What happens when pathogens do enter the body? Cells of the innate immune system launch an immediate attack. All innate immune cells respond to a wide variety of foreign pathogens. This is because their receptors recognize carbohydrates, proteins, or nucleic acids that characterize many different pathogens. It is estimated that the innate immune system includes several hundred different types of receptors. The innate immune system retains no memory of pathogens it has dealt with in the past, so its response to a given pathogen is similar each time.

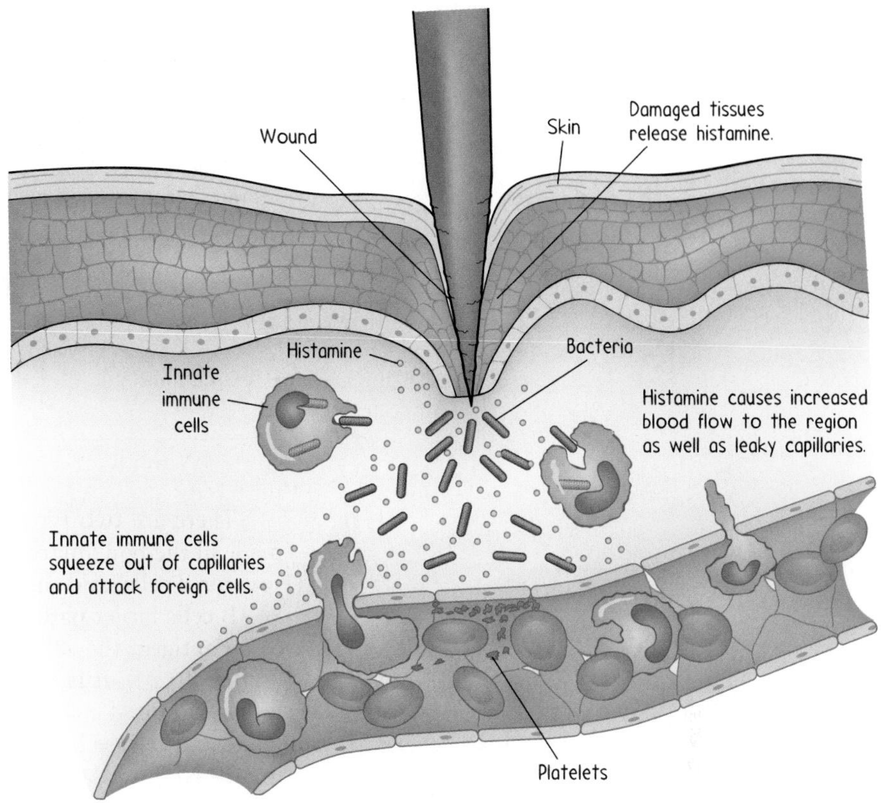

Wound
Skin
Damaged tissues release histamine.
Histamine
Bacteria
Innate immune cells
Histamine causes increased blood flow to the region as well as leaky capillaries.
Innate immune cells squeeze out of capillaries and attack foreign cells.
Platelets

Figure 20.22 The inflammatory response begins when damaged tissues release histamine, which causes increased blood flow to the wound. Local capillaries become leaky, producing swelling, and innate immune cells squeeze out of the capillaries and attack bacteria or other microorganisms that enter the body.

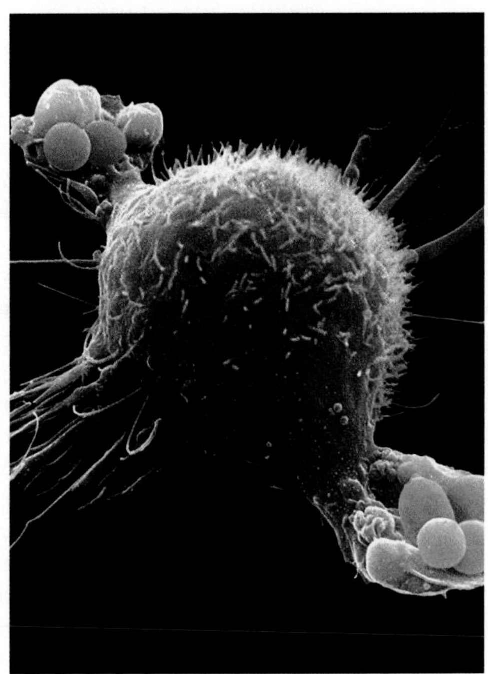

Figure 20.23 A cell of the innate immune system consumes bacteria. The stringlike structures are extensions of the immune cell's cytoplasm and are used by the cell to search its environment.

One of the most important functions of the innate immune system is to initiate the **inflammatory response**, shown in Figure 20.22. We have all experienced the inflammatory response many times—it is what makes your tissues swell and turn red when you cut your finger or scrape your knee. How does the inflammatory response happen? When tissues are damaged, they release chemicals called *histamines*. Histamines increase blood flow to the site of the injury and also cause local capillaries to leak fluid. This fluid causes swelling, which helps to isolate the injury from other body tissues. Histamines also attract innate immune system cells, such as the one shown in Figure 20.23, to the site of injury. Innate immune cells squeeze out of the capillaries, migrate to the site of the wound, and attack any microorganisms that have entered the body. Sometimes, the battle between innate immune cells and pathogens produces a whitish substance called pus. Pus is made up of dead bacteria, dead tissue, and dead and living innate immune system cells.

Acquired Immunity

The acquired immune system is very specific in its response to pathogens and other foreign substances. Each cell of the acquired immune system has receptors that respond to a single **antigen**—a molecule or part of a molecule belonging to a foreign pathogen. Most often, antigens are parts of foreign proteins. The acquired immune response is much slower than that of the innate immune system, usually taking between 3 and 5 days to reach full force. In addition, the acquired immune system retains a "memory" of pathogens it has encountered in the past, so that subsequent responses to the same pathogen can be faster and more aggressive. The acquired immune system includes a tremendous number of different receptors—on the order of 10 million. This makes it very likely that any foreign pathogen, whether it is a bacterium, a virus, or a worm, will trigger a response by one or more acquired immune cells. A comparison of innate and acquired immunity is provided in Table 20.1.

Table 20.1 A Comparison of Innate and Acquired Immunity

Innate Immunity	Acquired Immunity
• Not very specific, responds to molecules that characterize a wide variety of pathogens, including types of carbohydrates, proteins, and nucleic acids	• Very specific, responds to antigens, most often portions of foreign proteins
• Response is immediate	• Response is delayed, reaching a maximum 3 to 5 days after exposure
• No memory is retained of past encounters with pathogens	• Memory is retained of past encounters with pathogens—subsequent responses are faster and more aggressive
• Several hundred different receptors	• Huge number of receptors—somewhere on the order of 10 million

There are two types of acquired immune cells, B cells and T cells. Both are made in the bone marrow, but each matures in a different organ—the bone marrow for B cells, and the thymus for T cells.

B cells target pathogens in bodily fluids. When the receptor of a B cell binds to an antigen, the cell begins to divide, making many *clones*, or copies, of itself (Figure 20.24). It is the time required to make these clones that causes the delay in the acquired immune system's response. Once the clones are mature, they make and release large numbers of molecules called antibodies. **Antibodies** (Figure 20.25a) are large Y-shaped proteins that include two antigen-binding sites. They bind to their antigens in a very specific way, with a fit like a lock and key. A single B cell can make a huge number of antibodies—as many as 2000 molecules per second! These antibodies then travel around the body binding to antigens. What does this accomplish? In some cases, the binding of antibody to antigen interferes directly with a pathogen's ability to function. For example, a virus bound to a clunky antibody may be unable to enter and infect new cells. In other cases, the binding of antibodies causes pathogens to clump together, as shown in Figure 20.25b. These clumped masses become an easy target for other immune cells.

In any acquired immune reaction, a few of the clones produced are *memory cells* that remain in the body for a long time, years or even a lifetime. If the same antigen is encountered again, the memory cells initiate an immune response that is much faster and much more aggressive. In fact, these secondary infections are often obliterated before symptoms even develop.

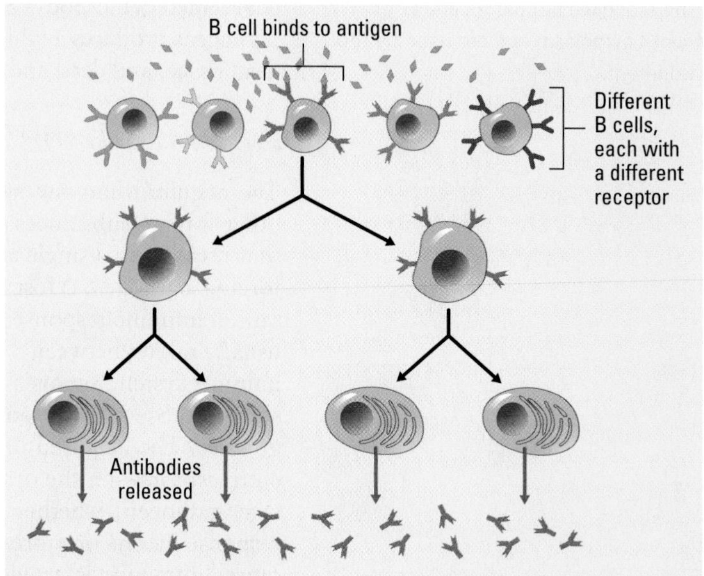

Figure 20.24 When a B cell binds to an antigen, it begins to divide, making many clones of itself. Mature clones make and release large numbers of antibodies. A few memory cells remain in the body a long time, ready to deal with subsequent infections by the same pathogen. (Courtesy of Campbell, Reece, Simon, *Essential Biology with Physiology*, © 2004.)

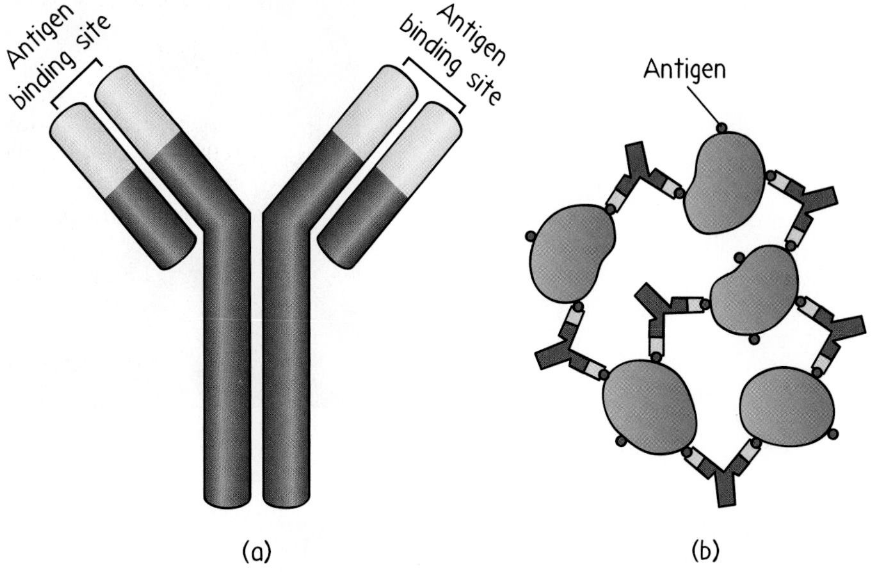

Figure 20.25 (a) Antibodies are large Y-shaped proteins that include two antigen-binding sites. The antigen-binding sites are very specific so that antibody binds to antigen with a lock-and-key fit. (b) Because antibodies each have two binding sites, they can cause pathogens to clump together.

T cells target pathogens that are inside the body's cells. When a cell becomes infected, it displays some of the pathogen's proteins on its surface, much the way you might wave a flag from a building to signal for help. This display of foreign proteins serves as an S.O.S. to T cells. A T cell called a *helper T cell* binds to an antigen on the pathogen protein and begins to grow and divide. Helper T cell clones then initiate a wide range of immune activities. They stimulate B cells and another group of T cells, *killer T cells*, to divide and produce clones. The B cells make and release antibodies against the pathogen, as we saw above. Killer T cells bind to displayed proteins on infected cells and then, as their name suggests, kill the infected cells. Besides stimulating B cells and killer T cells, helper T cells also produce memory cells and stimulate cells of the innate immune system. Finally, helper T cells make the *suppressor T cells* that inhibit the action of killer T cells and B cells once they are no longer needed.

The acquired immune system's "memory" for pathogens it has encountered in the past is what allows *vaccines* to work. Vaccines expose the body to antigens of a particular pathogen, but—this is the key—without introducing the pathogen itself! Most vaccines contain dead or very weak strains of the pathogen or use only a part of the pathogen—in the case of a bacterium, perhaps a flagellum or a part of the cell wall. The acquired immune system responds to antigens in the vaccine much as it would to the real pathogen—by making antibodies and memory cells. The next time the pathogen is encountered, the acquired immune system is ready.

Diseases of the Immune System

The immune system normally has no problem recognizing the body's own cells. In certain diseases, however, a malfunction causes it to identify certain body cells as foreign and attack them. These are called *autoimmune diseases*. Type I diabetes occurs when the immune system destroys the insulin-producing cells of the pancreas. Multiple sclerosis occurs when immune cells destroy the myelin sheaths surrounding neurons. Lupus is caused by immune attacks on a wide range of healthy tissues.

Acquired immunodeficiency syndrome, or AIDS, is a disease caused by the human immunodeficiency virus, HIV. (The "acquired" in the name of the disease refers to the fact that immunodeficiency is acquired during a person's lifetime rather than being inborn.) HIV attacks immune cells, particularly helper T cells (Figure 20.26). With these crucial immune cells compromised, infections and cancers that are usually easily fought off have a chance to overwhelm the body.

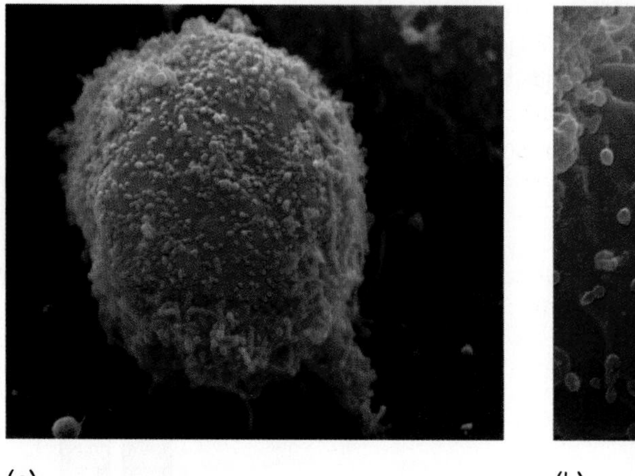

(a)

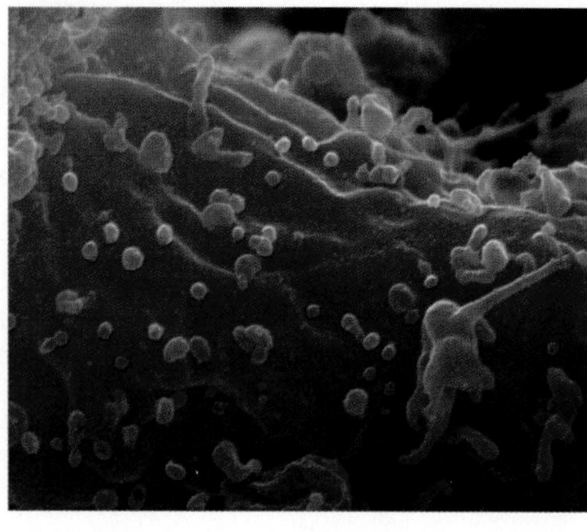

(b)

Figure 20.26 (a) HIV emerges on the surface of an infected helper T cell. (b) This is a close-up of the photo in (a).

SCIENCE AND SOCIETY

The Placebo Effect

A patient with knee pain goes in for surgery. He receives anesthesia, and cuts are made around his knee where the surgical instruments will be inserted. Afterwards, the surgery appears to be successful—both pain and swelling are greatly diminished. What's unusual about this story? Nothing, except that the operation was a sham. Cuts were made around the patient's knee, but nothing happened after that. Why does the patient feel so much better? Because of a phenomenon known as the placebo effect.*

Placebo is Latin for "I shall please." It refers to the once regular practice in which doctors prescribed sugar pills to patients whom they otherwise couldn't help. Although this is now considered unethical, it doesn't change the fact that sugar pills did often help. The *placebo effect* is defined as the improvement patients experience when they are given a treatment with no relevance to their medical problem.

Placebos appear to work for a wide variety of conditions and are usually far better than no treatment at all. In a study of patients with Parkinson's disease, for example, a placebo worked just as well as medication in inducing the release of dopamine by the brain. Placebos have also been found to work as well as modern antidepressants in the treatment of depression. The placebo effect is certainly real, though placebos work better for some maladies than others. Placebos appear to be particularly effective for conditions related to the nervous system, including pain, depression, anxiety, headaches, fatigue, and gastrointestinal symptoms. For most of these conditions, placebos have about 30–50 percent effectiveness—nearly as good as "real" treatments in some cases. Placebos are also believed to account for the "success" of certain alternative remedies with no medical basis.

The placebo effect is one of the oddest phenomena in medicine. What causes it? This question has not been fully answered and is the object of continued study. However, several possible mechanisms have been suggested. One idea is that the placebo effect operates through the release of endorphins (Chapter 19). Release of the body's natural opiates would explain why placebos are so good at treating pain. Further evidence comes from the fact that placebos become much less effective as painkillers when patients are given a drug that blocks the opiate receptors. However, because the placebo effect works for many symptoms other than pain, this can't be the whole story. Another idea is that receiving a placebo reduces stress, allowing the immune system to function more effectively. Numerous studies have shown that stress reduces the immune system's capabilities—consequently, stress relief would be expected to improve its function. Still, there must be more to it than this because the placebo effect is very specific—it doesn't help with *all* your ailments, only the one you think you're being treated for. There is no getting around the fact that a person's expectations somehow lie at the crux of the placebo effect. Some scientists have argued that the placebo effect is a conditioned reflex. The patient has, through numerous experiences with doctors, pills, injections, and so on, been conditioned to expect a positive effect after medical treatment. And somehow, the nervous system has become wired to comply.

Interestingly, studies have also demonstrated a "nocebo effect," sometimes called the placebo effect's "evil twin." Expectations of negative effects are realized too. For example, people on placebos often develop negative "side effects" from their treatments. Side effects of real medications may sometimes be caused by the nocebo effect as well. For example, studies have repeatedly shown that patients who were warned of specific side effects tend to experience them much more often than patients who weren't warned. The nocebo effect can be even more serious. One study showed that women who believed they were vulnerable to heart disease were four times as likely to die of it as women who didn't believe they were vulnerable, but had similar risk factors. The nocebo effect may also account for the effectiveness of voodoo death curses. Never underestimate the power of the mind.

*Patients enrolled in these types of clinical studies—which are usually designed to examine the effectiveness of a surgery, drug, or other treatment—are aware that they are participating in a clinical study and that they might be assigned to a "sham" treatment group. If the surgery or other treatment performs no better than the placebo (as occurred in this case), the conclusion is that the treatment is not effective. Testing treatments against placebos is important precisely because the placebo effect can be very powerful. Of course, if there is already a proven treatment for a particular condition, new treatments can be tested against the established one.

SCIENCE AND SOCIETY

Acquired Immunodeficiency Syndrome

Since the start of the AIDS epidemic in the 1980s, the disease has claimed over 20 million lives worldwide. Today, the number of people with HIV continues to increase in every part of the world. The World Health Organization estimates that about 39.4 million people were living with HIV at the end of 2004. Of this number, about 4.9 million—one of eight—had contracted the virus within the last year. Globally, 3.1 million people died of AIDS in 2004. Although sub-Saharan Africa remains the area hardest hit by HIV and AIDS, the virus is now spreading most rapidly in East Asia, Eastern Europe, and Central Asia.

In the past few years, increased international funding has helped to address some aspects of the global HIV epidemic. Education, testing, and counseling are now more widely available in some parts of the world, as are anti-HIV medications. Anti-HIV drugs are crucial both for decreasing the incidence of mother-to-child HIV transmission during childbirth and for allowing those with HIV to live longer, more productive lives. Nonetheless, international funding for AIDS still falls far short of what is required. Overall, it is estimated that prevention programs reach less than 20 percent of the people most at risk and that only 7 percent of individuals who need anti-HIV drugs have access to them. The "3 by 5" AIDS initiative, begun in 2003 by the World Health Organization (WHO) and Joint United Nations Programme on HIV/AIDS (UNAIDS), was an ambitious venture with the goal of providing the infrastructure and drugs necessary to treat 3 million people in developing countries with life-prolonging anti-HIV therapy by the end of 2005. Although this goal ultimately was not met, there were many local success stories. In addition, the initiative succeeded in bringing global attention to the need to expand HIV treatment access and helped to mobilize support. WHO and UNAIDS continue to work towards their goal of universal access to treatment by 2010.

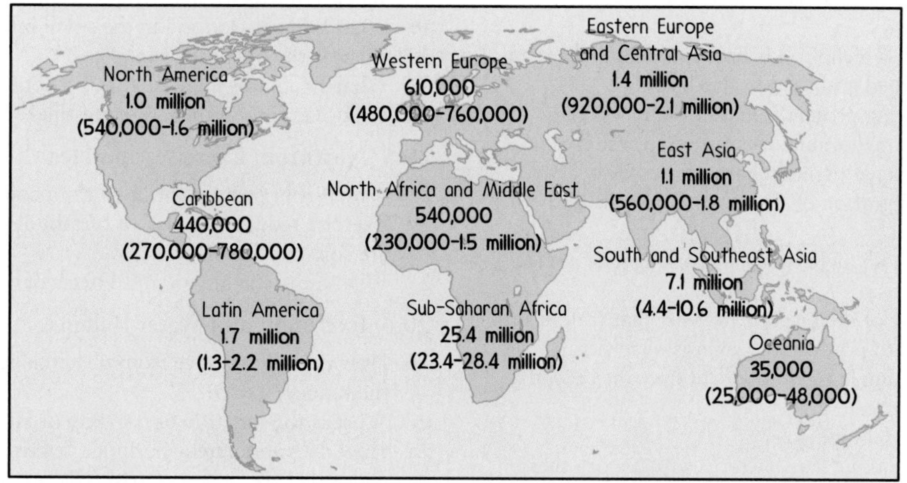

This map shows the number of adults and children estimated to be living with HIV in different parts of the world, as of December 2004. It is adapted from the World Health Organization's "AIDS Epidemic Update 2004."

In many countries, the impact of AIDS goes far beyond the death toll. AIDS has always hit young adults particularly hard—about half the new adult HIV infections occur in people between the ages of 15 and 24, with young women becoming infected more often than young men. The loss of large numbers of people in the prime of life means that fewer adults are available to take care of children and the elderly. About 15 million children have already been orphaned by AIDS, most of them in sub-Saharan Africa. The loss of working-age adults also affects the workforce, with important economic consequences in countries where HIV and AIDS are most prevalent.

Recent surveys have also found that women are increasingly affected by HIV and AIDS. Women now make up almost half the total cases of HIV and AIDS worldwide and represent well over half the cases in some of the hardest-hit areas, such as sub-Saharan Africa. Gender inequality is one underlying cause. In many places, women have less access than men to education, testing, counseling, and treatment. A UNICEF study showed that as many as half the women in countries where HIV and AIDS present the greatest danger lack basic information about the disease, such as how it is transmitted and what they can do to protect themselves. In addition, many women contract HIV not through their own high-risk behavior, but through the high-risk behavior of their partners.

In the United States, testing HIV-positive is not the death sentence it was two decades ago, largely because of the anti-HIV drugs now available. Although this is certainly good news, it has unfortunately also led to complacency and the return of high-risk behaviors in some quarters. No current anti-HIV therapy is a cure—treatments can only delay the onset of illness. Inevitably, drug-resistant strains of HIV have evolved in some people receiving treatment (see discussion of drug resistance in Chapter 17). Drug-resistant strains can then be transmitted to other individuals. In 2005, public health officials were alarmed to find in a New York City patient an aggressive strain of HIV that was resistant to 19 of the 20 licensed anti-HIV medications. It is almost inevitable that eventually, a transmissible strain of HIV will appear that can't be treated with current drugs.

CHAPTER 20 Review

Summary of Terms

Acquired immunity A system of highly specific body defenses that recognize and react to very specific features of pathogens.
Alveoli Tiny sacs of air found in the lungs; the site of gas exchange.
Antibody A large Y-shaped protein that binds to an antigen with a lock-and-key fit and that helps to disable a pathogen.
Antigen A part of a molecule from a pathogen or other foreign body that is recognized by cells of the acquired immune system.
Arteries Blood vessels that carry blood away from the heart.
B cells Cells of the acquired immune system that target pathogens in bodily fluids.
Capillaries Tiny thin-walled blood vessels from which materials are exchanged with body tissues.
Digestion The process through which food is broken down into organic molecules that can be absorbed and used by the body.
Hemoglobin The oxygen-carrying protein found in red blood cells.
Inflammatory response An innate immune response triggered by the release of histamine by damaged tissues; characterized by swelling, redness, and the migration of innate immune cells to the site of injury.
Innate immunity Nonspecific body defenses that work against a wide variety of potential pathogens.
Larynx The voice box; the part of the respiratory tract that contains the vocal cords, which are used to produce the sounds of speech.
Lymph The clear fluid, containing large numbers of immune cells, that flows through the lymphatic vessels.
Nephron The functional unit of a kidney.
Pathogens Disease-causing agents such as bacteria, viruses, or other organisms.
Peristalsis A moving wave of muscular contractions that moves food down the esophagus.
Platelets Cells found in blood that function in blood clotting.
Red blood cells The oxygen-carrying cells found in blood.
T cells Cells of the acquired immune system that target pathogens inside the body's cells.
Veins Blood vessels that carry blood towards the heart.
White blood cells Immune cells found in blood.

Review Questions

20.1 Integration of Body Systems

1. Which two body systems are involved in the task of supplying the body with oxygen?
2. How does getting rid of waste products require the integrated efforts of multiple body systems?

20.2 The Circulatory System

3. What stimulates the heart to beat?
4. Why do the two atria contract before the two ventricles?
5. What makes the "lub-dubb" sound of the heartbeat?
6. Which blood vessels are responsible for nutrient and waste exchange with tissues?
7. What function do the valves in veins serve?
8. Trace the path of blood through the body, beginning with blood returning from the tissues to the heart. Be sure to name each of the chambers of the heart.
9. What are the three types of blood cells, and what is the function of each?

20.3 Respiration

10. What path does air take through the respiratory tract?
11. How does gas exchange in the alveoli occur? What features of the alveoli and their surrounding capillaries facilitate gas exchange?
12. What causes air to flow into the lungs?
13. What happens when we exhale? That is, what causes air to flow out of the lungs?

20.4 Digestion

14. What does digestion accomplish?
15. Why is saliva essential for us to taste our food?
16. What prevents food from going into the trachea after it is swallowed?
17. Is swallowing a voluntary action?
18. What happens to food while it is in the stomach?
19. What happens to food in the small intestine?
20. What is the purpose of bile?
21. What structures increase the absorptive area of the small intestine?
22. What happens in the large intestine?

20.5 Nutrition, Exercise, and Health

23. Why is it important for us to eat a complete protein?
24. Exercise reduces the risk of certain diseases and conditions. What are some of these?
25. What are some measurable effects of exercise on the heart and lungs?

20.6 Excretion and Water Balance

26. How does fluid move from the circulatory system into nephrons in the kidney?
27. What is the function of the loop of Henle?
28. How does antidiuretic hormone determine whether additional water is reabsorbed as the filtrate passes down the collecting duct of a nephron?

20.7 Keeping the Body Safe: Defense Systems

29. What features of skin help keep pathogens out of the body?
30. What kinds of molecules are recognized by receptors of the innate immune system?
31. What happens during the inflammatory response?
32. How do acquired immune cells recognize pathogens?
33. What happens when a B cell first binds an antigen?
34. What is the function of a memory cell?

Multiple Choice Questions

Choose the BEST answer to the following.

1. Blood in the right ventricle of the heart will be pumped to
   (a) the right atrium.
   (b) arteries going to the lungs.
   (c) arteries going to body tissues.
   (d) the left atrium.
2. The blood cells that carry oxygen to body tissues are the
   (a) red blood cells.
   (b) white blood cells.
   (c) platelets.
   (d) hemoglobin.
3. Oxygen moves from the alveoli in the lungs into the bloodstream through the process of
   (a) active transport.
   (b) endocytosis.
   (c) diffusion.
   (d) exocytosis.

4. Air moves into the lungs when
   (a) the diaphragm contracts and flattens and the ribcage moves down and into the chest.
   (b) the diaphragm contracts and flattens and the ribcage moves up and out from the chest.
   (c) the diaphragm relaxes and the ribcage moves down and into the chest.
   (d) the diaphragm relaxes and the ribcage moves up and out from the chest.
5. All of the following occur in the small intestine except
   (a) bile from the gall bladder helps break fat into small droplets.
   (b) nutrients are absorbed into the body.
   (c) proteins are broken down.
   (d) a highly acidic mix of enzymes helps kill bacteria that were swallowed with food.
6. Aside from getting enough Calories, we also need to obtain the following through our diet:
   (a) a complete protein.
   (b) a complete carbohydrate.
   (c) vitamins and minerals.
   (d) a and c.
7. Which part of a nephron functions in reabsorbing water from the filtrate?
   (a) glomerulus
   (b) proximate convoluted tubule
   (c) loop of Henle
   (d) distal convoluted tubule
8. The stretchy sac where urine is temporarily stored is the
   (a) renal pelvis.
   (b) bladder.
   (c) ureter.
   (d) urethra.
9. All of the following are associated with the innate immune system except
   (a) skin.
   (b) tears.
   (c) the inflammatory response.
   (d) antibodies.
10. A difference between the innate immune system and the acquired immune system is
   (a) the innate immune system includes many more different types of receptors than the acquired immune system.
   (b) the acquired immune system response is immediate, whereas the innate immune response is delayed, peaking about 3 to 5 days after exposure.
   (c) the acquired immune system retains a memory for past exposure, with subsequent responses being faster and more aggressive; the innate immune system does not.
   (d) each acquired immune system cell responds to more different pathogens than each innate immune system cell.

INTEGRATED SCIENCE CONCEPTS

Chemistry—Hemoglobin

1. What is the structure of hemoglobin? Where does oxygen bind?
2. How is hemoglobin's oxygen affinity affected by blood pH?
3. How does the effect of pH on hemoglobin's oxygen affinity help oxygen transport function effectively?

Physics and Chemistry—Low-Carb Versus Low-Cal Diets

4. What is the key to losing weight?
5. Do low-carb diets work? How do they help people lose weight?
6. What are some potential problems with low-carb diets?

Active Explorations

1. Your heart rate is the number of times your heart beats in a minute. It changes over the course of the day as your activity level increases or decreases. Begin by measuring your resting heart rate. Ideally, you should do this when you have just woken up in the morning. Now measure your heart rate during various types of activities. What is it when you're reading quietly? When you're walking or doing chores around the house? What is it after you've eaten a big meal? Finally, measure your heart rate while you are exercising. Are you hitting your target zone for effective cardiovascular exercise? You can determine your target zone using the following calculations:

   First calculate your maximal heart rate:

   $$\text{maximal heart rate} = 220 - \text{your age}$$

   Then calculate your heart rate reserve:

   $$\text{heart rate reserve} = \text{maximal heart rate} - \text{resting heart rate}$$

   You are in your target zone if you are using between 60 percent and 80 percent of your heart rate reserve:

   Low end of target zone: 60 percent (heart rate reserve) + resting heart rate
   High end of target zone: 80 percent (heart rate reserve) + resting heart rate

2. How healthy is your diet? Write down everything you eat and drink for a day. Note the number of Calories when you can. About how many Calories do you consume per day? How does your diet compare to that recommended by the Healthy Eating Pyramid? What are you eating too much of? Not enough of?
3. The body mass index (BMI) adjusts your weight for your height. You can calculate your BMI as follows:

   $$\text{BMI} = \text{your mass in kilograms/(your height in meters)}^2 \text{ or}$$
   $$= \text{your weight in pounds/(your height in inches)}^2 \times 703$$

   What does BMI mean? BMI tells you if you are underweight or overweight.

BMI	
Below 18.5	Underweight
18.5–24.9	Normal
25.0–29.9	Overweight
30.0 and above	Obese

Note, however, that BMI does not measure body fat, as the figure below should convince you.

6'3"	Height	6'3"
220 lb	Weight	220 lb
27.5	BMI	27.5

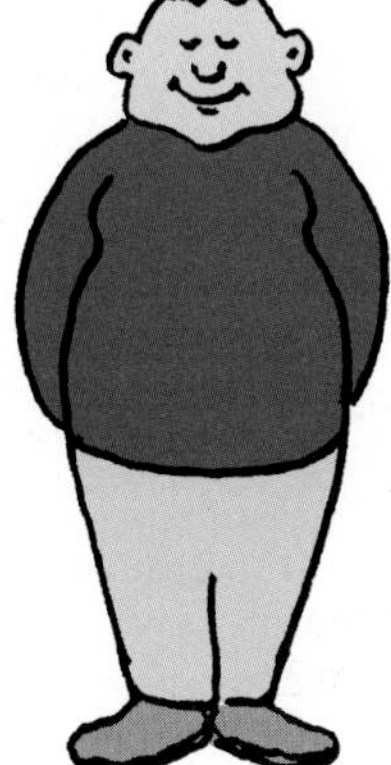

Exercises

1. Several of our senses provide examples of how multiple body systems work together to accomplish important tasks. What body systems are involved in hearing? In smelling? In tasting?
2. How does reproduction require the integrated action of multiple organ systems?
3. The pumping of the heart does the bulk of the work that is required to move blood around the body. What else contributes?
4. Why do you think the atria of the heart are less muscular than the ventricles? Why is the left ventricle more muscular than the right ventricle?
5. Where in the body is blood most oxygenated?
6. How does the body control the amount of blood that different tissues receive?
7. What bodily activities is the respiratory system involved in besides acquiring oxygen for the body?
8. Why shouldn't you talk with your mouth full (not just because it's impolite)?
9. What waste materials are produced in the process of making ATP, and what body systems are responsible for removing them from the body?
10. The liver is an organ that plays important roles in multiple organ systems. What role does the liver have in digestion? In excretion?
11. How does the endocrine system interact with the excretory system? Give examples.
12. Kangaroo rats, which live in dry desert habitats, have very long loops of Henle. Why might this be?
13. What is the difference between elimination (feces) and excretion (urine)? What is the body getting rid of in each case?
14. Why might eating a high-protein diet be particularly hard on the liver and kidneys?
15. Why is the innate immune system described as "nonspecific"? Why is the acquired immune system described as "specific"?
16. What are some differences between the innate immune system and the acquired immune system?
17. Allergies occur when the immune system is abnormally sensitive to particular substances. Why do people sometimes take antihistamines for their allergies?
18. How are the antigens B cells respond to different from the ones T cells respond to?
19. A mother kisses her child's "owie." Do you think this might result in the child feeling less pain? Why or why not?
20. Write a letter to Grandpa telling him about trendy low-carb diets. Tell him why many people have been able to lose weight on low-carb diets and explain some potential dangers of the diets as well.

Problems

1. Show that the blood can carry as many as 3×10^{22} molecules of oxygen. Here's some information you may find useful: You have 25 trillion red blood cells. Each red blood cell contains 300 million molecules of hemoglobin. Each molecule of hemoglobin can carry four molecules of oxygen.
2. Because red blood cells have no nuclei and are therefore unable to make the proteins necessary to maintain themselves, they have a relatively short life span of about 120 days. Given that we have about 25 trillion red blood cells in all, show that more than 208 billion red blood cells die and are replaced each day. Also show that, in the 20 seconds it took you to read this problem, about 48 million red blood cells died and were replaced.
3. A typical person has a heart rate of 70 beats per minute and takes 12 breaths in a minute. Show that her heart beats about 4200 times an hour, 100,800 times a day, and 36.8 million times a year. Show also that she takes about 720 breaths per hour, 17,280 breaths per day, and 6.3 million breaths per year.

Conceptual Science Place

CHAPTER 20 ONLINE RESOURCES

Tutorials Immune Responses, Path of Blood Flow in Mammals, The Human Respiratory System, Transport of Respiratory Gases, Digestive System Function, Structure of the Human Excretory System, Nephron Function, Control of Water Reabsorption • ***Flashcards*** • ***Quiz*** • ***Links***

CHAPTER 21

Ecosystems and Environment

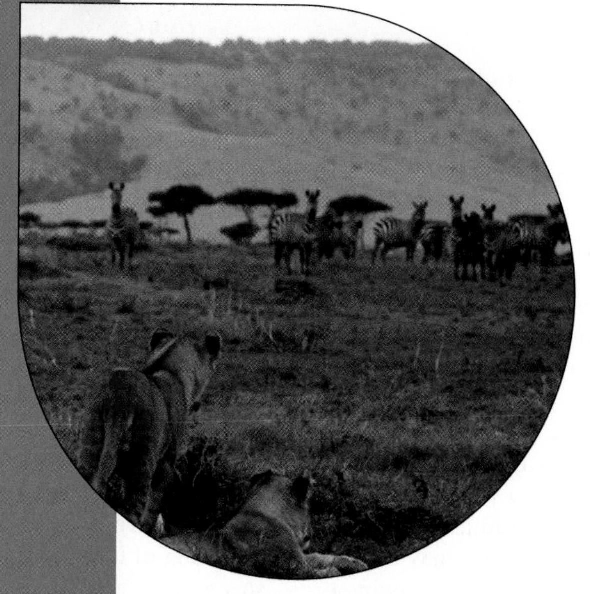

Two lionesses watch a herd of nervous zebras in the Masai Mara Reserve, Kenya.

CHAPTER OUTLINE

No living creature exists in a vacuum. In fact, all organisms interact with their environments and with other organisms. How do these interactions determine the features of ecosystems? For example, why is there more grass than zebras on the African savanna, and why are there more zebras than lions? Are ecosystems with more species more stable? Why does disturbing a habitat sometimes contribute to its biodiversity? Why do some animals, like the razor clam, make more than 100 million eggs at a time, whereas others, like the elephant, make only a single offspring every few years? How has human population grown over time? Will it continue to grow, or is it likely to crash one day? And finally, could planting a tree be the key to world peace?

21.1 Organisms and Their Environment

Ecology is the study of how organisms interact with their environments. An organism's environment includes nonliving, or **abiotic**, features, such as temperature, sunlight, precipitation, rocks, ponds, and so forth. It also includes **biotic** features—that is, other living organisms.

Ecology can be studied at many different levels, including that of the individual, population, community, and ecosystem. Studies that focus on individuals often ask how an organism's anatomy, physiology, and behavior help it to function in its environment. For example, why do certain plants invest more energy in building extensive root systems, whereas others invest more in building stems and leaves (anatomy)? Or, does the varying nutrient content of different types of seeds explain why seed-eating finches prefer one type of seed to another (physiology)? Or, how do gulls divide their time between four different strategies for obtaining food—soliciting handouts from humans, delving into garbage cans, foraging along the shoreline, and foraging at sea (behavior)? Because individual-level studies in ecology typically focus on adaptation (Chapter 17), there is often considerable overlap with evolutionary biology.

A **population** is a group of individuals of a single species that occupies a given area. The raccoons in Austin, Texas represent a population. So do the snow leopards of Central Asia and the orange-flanked bush robins of Langtang National Park, Nepal (Figure 21.1). Ecological studies at the population level frequently focus on the size of a population and on how population size changes over time. In addition, population ecologists may study how a specific population

Figure 21.1 Biologist Pamela Yeh holds an orange-flanked bush robin in Langtang National Park, Nepal. The bird was captured using mistnets hung between poles and will be measured, weighed, and given a unique identifying leg band before being released.

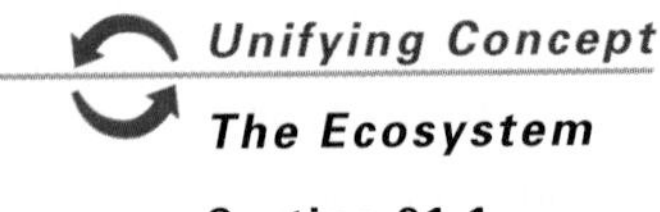

Unifying Concept

The Ecosystem

Section 21.1

uses resources, including what kinds of foods are eaten and what types of habitats are used.

A **community** consists of all the organisms that live within a given area. For example, the ecological community in a sagebrush habitat in Idaho might include sagebrush, prickly pear cactus, insects, squirrels, pronghorns, badgers, various lizards, mountain lions, coyotes, and, of course, numerous types of bacteria and other microorganisms. Community-level studies typically focus on interactions between species. For example, community ecologists may be interested in figuring out who eats whom in a community, or in how multiple species compete for resources.

An **ecosystem** consists of all the organisms that live within a given area *and* all the abiotic features of their environment. Ecosystem-level studies frequently focus on links between the biotic and abiotic worlds. For example, some ecosystem ecologists trace energy flow through ecosystems, determining how much energy is captured from sunlight by plants in the ecosystem, how much energy herbivores obtain from plants, how much energy carnivores obtain from herbivores, and so forth. Ecosystem ecologists may also study the cycling of important resources such as water and carbon between living organisms and Earth's atmosphere and crust. Additional examples of ecological studies at the different levels of interest are shown in Figure 21.2.

CHECK YOURSELF

1. An ecologist is interested in studying how hyenas and cheetahs compete for food. What type of ecological study is this?
2. An ecologist is interested in studying how the number of mountain lions in a certain park has changed over the past decade. What type of ecological study is this?

CHECK YOUR ANSWERS

1. This study relates to an interaction between two different species that occupy the same habitat, so it is a community-level study.
2. This study relates to a group of individuals belonging to a single species that occupies a given area, so it is a population-level study.

(a)

(b)

(c)

(d)

Figure 21.2 Ecologists explore questions at many different levels of interest. (a) An individual-level study in ecology: How do fireflies use light signals to attract mates? (b) A population-level study: How many giant pandas are there in the bamboo forests of central China? Are their numbers increasing or decreasing? (c) A community-level study: Who eats whom in a Pennsylvania woodland? Shown here is a green frog preying on a grasshopper. (d) An ecosystem-level study: How do important resources like water and carbon move between the biotic and abiotic components of an ecosystem? This is an elephant drinking from a water hole in Chobe National Park, Botswana.

21.2 Species Interactions

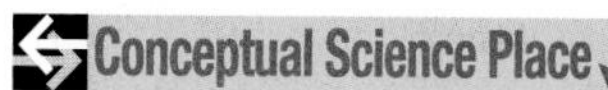

Interspecific Interactions; Food Webs

Mountain lions hunt mule deer, fleas suck the blood of dogs, and cleaner shrimp clean the mouths of moray eels. Within communities, species interact in a variety of ways.

Food Chains and Food Webs

One important set of species interactions is who eats what. These relationships are often diagrammed in a *food chain* that contains multiple feeding levels known as **trophic levels**. A food chain begins with **producers**, species that live by making organic molecules out of inorganic materials and energy. Most producers photosynthesize but some may chemosynthesize (recall the archaea described in Chapter 18). **Consumers** obtain food by eating other organisms. The primary consumers of a community are the species that eat the producers. For example, herbivores (plant-eaters) are primary consumers (Figure 21.3). Primary consumers are in turn eaten by secondary consumers (which include meat-eating carnivores), which may then be eaten by tertiary consumers, quaternary consumers, and so on. The species at the top of a food chain is sometimes called the top predator. Finally, at all levels, dead organic matter is consumed by decomposers such as bacteria and fungi. Organisms at different trophic levels within a woodland community are shown in Figure 21.4. In reality, most ecological communities are complicated enough that food "chains" are more accurately described as *food webs* (Figure 21.5).

Figure 21.3 Humans eat at varying levels of the food chain.

What position do humans occupy in our food chain? Like bears and raccoons, most humans are *omnivores* that eat at many levels of the food chain. We eat plants, animals, and even decomposers, in the form of mushrooms and truffles. Of course, some humans have a more restricted diet. In addition, because humans are not preyed upon by other species, we are top predators.

Competition

Another important species interaction is competition. It is easy to spot competition when you see a lion and a pack of hyenas fighting over a kill. However, direct confrontation isn't a necessary component of competition. In fact, any

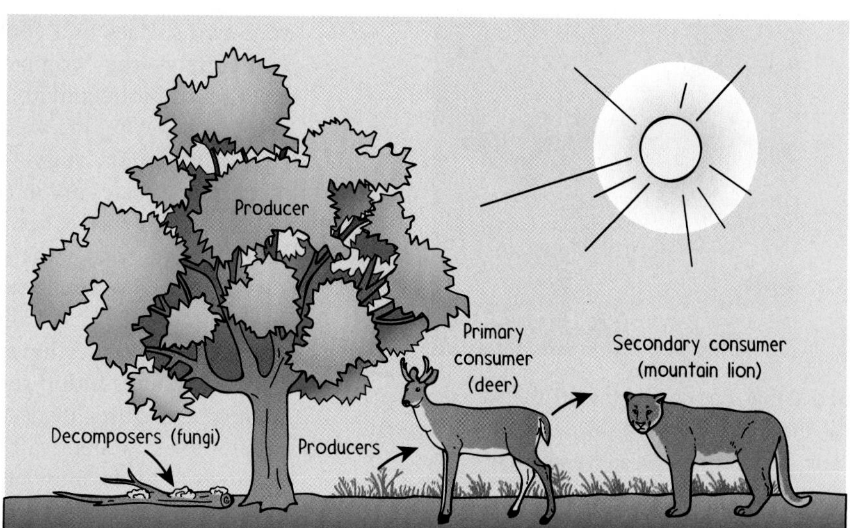

Figure 21.4 Plants such as trees and grass, which make their own organic molecules through photosynthesis, are producers. A plant-eating deer is a primary consumer. A deer-eating mountain lion is a secondary consumer. Decomposers, including many fungi and bacteria, consume dead organic matter at all trophic levels.

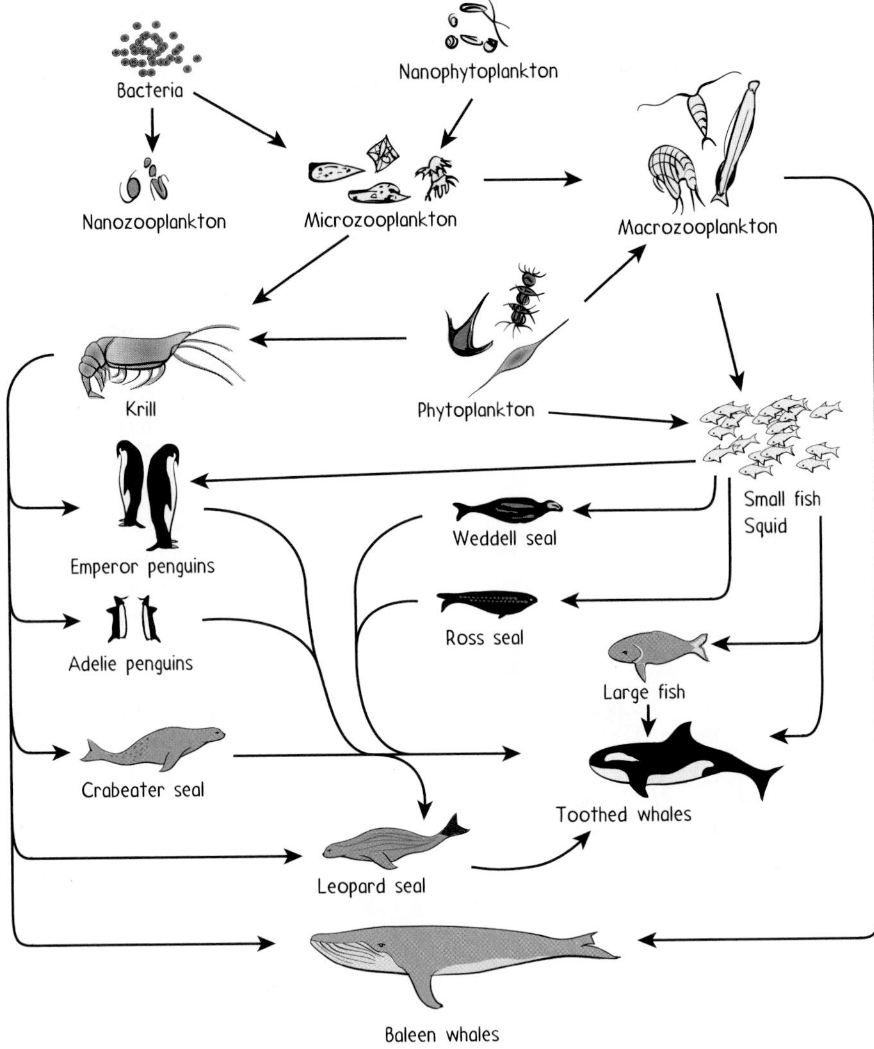

Figure 21.5 This diagram shows the Antarctic aquatic food web.

Figure 21.6 Species compete for limited resources such as the shelter of a tree cavity. Not all can excavate their own, like this red-bellied woodpecker.

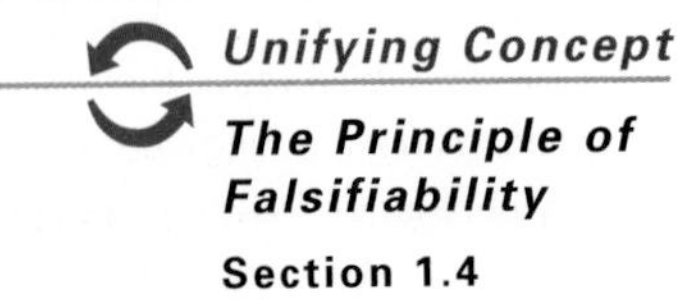

time two species in a community use the same resource—one that exists in limited supply—they compete (see Figure 21.6). A species' **niche** is defined as the total set of biotic and abiotic resources it uses within a community. This includes food eaten, water drunk, space occupied, and so forth. A species' niche defines its role in the community—you can think of a niche as a job that a species fills. Then, two species are in competition any time their niches overlap.

Although the niches of species frequently overlap, no two species in a community have *exactly* the same niche. Otherwise, the species that is better at capturing and using resources eventually outcompetes the other species and drives it to extinction—this is called the *competitive exclusion principle.* Is it really true that no two species have the same niche? Are there any exceptions? Ecologists have sometimes found species that appear to have identical niches within a community. On closer inspection, however, it has always turned out that the niches are not exactly identical. One of the most famous examples of this occurred in a group of birds now known as MacArthur's warblers. Ecologist Robert MacArthur studied five warbler species in the coniferous forests of the Northeast United States. All five lived in the same trees, and all five ate insects. Did they share the same niche? By watching the warblers closely, MacArthur discovered that there were crucial differences in the way they used resources—each species used a slightly different part of the tree (Figure 21.7), and each had a different way of hunting for insects.

Figure 21.7 The five species of MacArthur's warblers appear to occupy the same niche—they all live in the same coniferous forests, and they all eat insects. However, closer inspection reveals that their use of resources differs in significant ways. For example, each species occupies a different part of trees, as shown here (space occupied is shown in green). Each species also has a different way of hunting for insects.

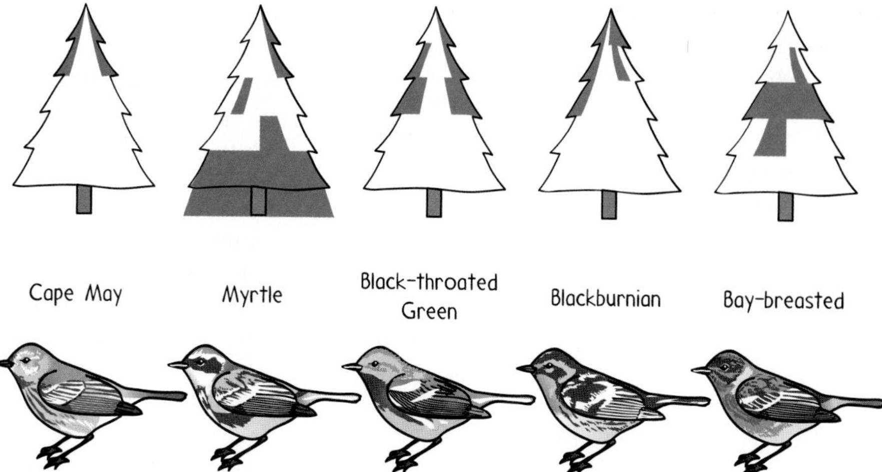

Symbiosis

Symbiosis occurs when individuals of two species live in close association with one another. We have already seen one instance of symbiosis that was crucial to the history of life as we know it—the close association between early eukaryotes and prokaryotes that resulted in the origin of eukaryotic organelles such as chloroplasts and mitochondria (Chapter 17).

Symbiotic relationships are sometimes divided into three categories based on the effects they have on participants. *Parasitism* benefits one member of the interaction and harms the other. Familiar examples of parasitism include fleas, tapeworms, and other organisms that live on or in their hosts and obtain nutrients from them. Pathogens such as bacteria or viruses are parasites as well. A strikingly different form of parasitism is brood parasitism, in which the female of one species lays eggs in the nest of another species, causing individuals of the other species to raise her young. Cuckoos are well-known brood parasites (Figure 21.8a). *Commensalism* is a form of symbiosis that benefits

(a)

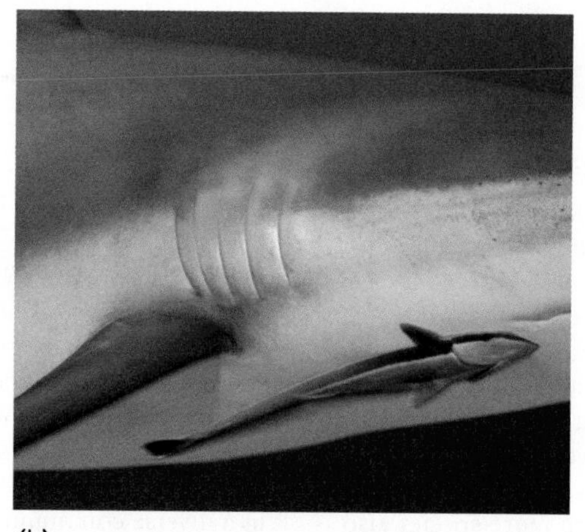
(b)

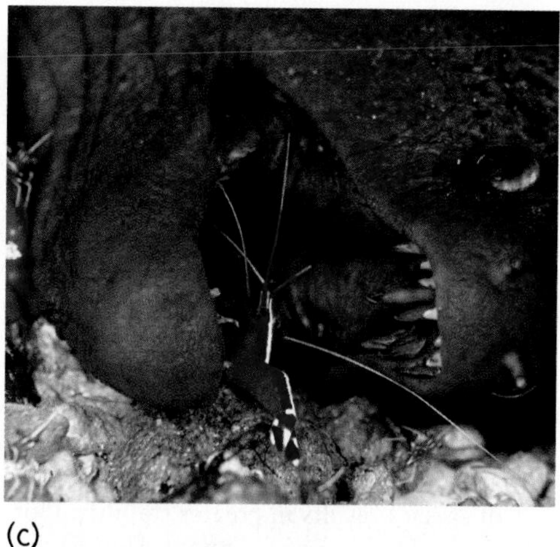
(c)

Figure 21.8 The three forms of symbiosis are parasitism, commensalism, and mutualism. (a) Parasitism: Is that my baby?! A reed warbler feeds a cuckoo chick. Cuckoos lay their eggs in the nests of other species, a form of parasitism known as brood parasitism. (b) Commensalism: A remora hitches a ride on a shark. The remora obtains protection from its host and feeds on leftover scraps from the shark's meals. (c) Mutualism: A cleaner shrimp removes parasites from the mouth of a moray eel. The shrimp obtains food, and the eel is freed from harmful parasites.

INSIGHT Everyone knows that cows eat grass, but did you know that they can digest it only with the help of microorganisms? Cows—and many other hoofed mammals—have a mutualistic relationship with single-celled archaea. The archaea are able to break down the cellulose in plant cell walls, something no mammal can do!

one species of the interaction while having no effect on the other. The remora (Figure 21.8b) is a small fish that attaches itself to a shark using a suction cup. In addition to obtaining protection from its host, the remora feeds on small scraps of food produced when the shark eats. *Mutualism* is a form of symbiosis that benefits both species (Figure 21.8c). Mycorrhizae, the close association between fungi and plant roots, represent a mutualistic relationship. The fungus receives nutrients from the plant while helping the roots to absorb water and minerals (Chapter 18). Another mutualism occurs between stinging ants and acacia trees. Ants live in hollowed-out thorns on the acacia tree and obtain special protein- and carbohydrate-rich nutrients from it. In return, the ants defend the tree from insects, mammals, and other herbivores and also trim away nearby plants that compete for sunlight and nutrients. This is why, in rainforests otherwise dense with life, acacias sometimes stand alone.

CHECK YOURSELF

1. Can an organism be both a producer and a consumer?
2. Oxygen is a resource many species use. Is there competition for oxygen in most communities?
3. In some places, dwarf mongooses and hornbills forage for food in close proximity to one another. Both look out for potential predators, but hornbills are better at detecting predatory birds overhead, and mongooses, with their exceptional sense of smell, are better at detecting terrestrial predators. What type of symbiosis does this represent?

CHECK YOUR ANSWERS

1. Yes. Certain parasitic plants both photosynthesize and obtain nutrients from their hosts—other plant species. These parasitic plants are both producers and primary consumers. There are even a number of predatory plants that eat higher up the food chain by trapping insects. The Venus flytrap, which snaps specialized leaves together to catch unsuspecting insects, is probably the most famous.
2. No—oxygen is a shared resource, but it is generally not in limited supply.
3. Mutualism—both species benefit.

SCIENCE AND SOCIETY

Biodiversity, Nature's Insurance Policy

Are ecological communities with more species more stable? Ecologist David Tilman tested this idea by studying several hundred equally sized plots of habitat in the Minnesota grasslands. For 11 years, Tilman and his students counted the number of plant species living on each plot and determined whether species number affected community stability. The result? Plots with more species were less affected by drought, showed less year-to-year variation in total vegetation produced, and were less vulnerable to invasion by new species—all of which suggests that a greater number of species results in greater stability. Plots with more species also showed greater productivity. That is, they had a greater total aboveground mass of all plants combined. Both stability and productivity are considered good indicators of how well an ecosystem is functioning.

Why might a diverse community be more stable and productive? More species means greater redundancy, with multiple species sharing a given role within a community. Because different species respond differently to changing conditions, one species may do all right when another species is doing poorly. For example, a community with only one producer is in dire straits if that producer has a bad year, but a community with three or four producers is much less likely to have all of them do poorly simultaneously. Diverse communities might be more productive because different species use resources in different ways—as a result, the species in a diverse community use up available resources more completely. The more complete use of ecosystem resources could also explain why it is harder for new species to successfully invade and become established in a diverse community.

Tilman's results have generated considerable controversy in recent years. Certain ecologists have attacked the productivity result in particular, arguing that the more species you have in a plot, the more likely it is that one of those species will happen to be a highly productive one. In addition, most studies addressing the effect of diversity on ecosystems have looked at grasslands, and it is not clear whether the same results will apply in other habitats.

SCIENCE AND SOCIETY

Invasive Species

Conceptual Science Place

Fire Ants as an Exotic Species

When the brown tree snake arrived in Guam from its native New Guinea, it found a snake's paradise—plentiful food in the form of forest birds and their eggs and not a single natural enemy in sight. The snake quickly spread through the small island and now occurs in startling numbers—in favorable habitats, there can be as many as 5,000 individuals per square kilometer. Even as the brown tree snake thrives, however, it devastates Guam's unique bird fauna. A dozen species are now extinct, and others are endangered.

Invasive species are species that have been introduced from their native habitat into a new, nonnative habitat, where they proceed to thrive. Although humans purposely introduced many species in the past (including the Australian rabbits discussed in Chapter 17), the introduction of nonnative species today is most often accidental, occurring when organisms hitch rides in far-ranging ships or airplanes. Most introduced species have only a limited impact on their new environments, if they are even able to survive and become established. Sometimes, however, a species will find its new home extremely conducive to survival and reproduction and will quickly increase its numbers, damaging native species by competing with them for food or by preying on them. This is when an introduced species becomes an invasive species.

A shopping cart left in zebra mussel-infested waters for several months now has mussels covering every available surface.

Invasive species are responsible for the decline of countless native species worldwide. It is estimated that over a third of the species listed as endangered under the Endangered Species Act are threatened wholly or partly because of an invasive species. Some invasive species also have significant economic impact. The zebra mussel, an invasive species that has spread across the eastern United States since its accidental introduction in 1988, clogs water pipes at power plants and water treatment facilities, causing an estimated $5 billion dollars of damage annually. The zebra mussel also threatens many native freshwater bivalves, which it competes with for food. Many of these native species are now listed as endangered.

21.3 Energy Flow in Ecosystems

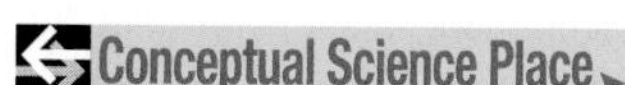

Energy Flow and Chemical Cycling; Energy Pyramids

All organisms need energy in order to grow, reproduce, and perform the activities necessary for survival. In most ecosystems, energy comes ultimately from the Sun. Earth receives a lot of sunlight energy—about 10^{19} kilocalories reach Earth's surface every day. Only a small fraction of this enters the biotic world, however, when plants and other organisms use it to build organic molecules during the process of photosynthesis (Chapter 15). Photosynthesizing organisms convert about 1 percent of the sunlight energy that strikes them into organic matter. Although this may sound inefficient, it is still enough, globally, to build 170 billion tons of organic material a year.

The amount of organic matter in an ecosystem is sometimes referred to as its **biomass**. And the rate at which an ecosystem's producers build biomass is the ecosystem's *primary productivity*. Ecosystems vary a great deal in their primary productivity, as Figure 21.9 shows. Rainforests, swamps, marshes, and coral reefs are among the most productive ecosystems on Earth. Desert and tundra are among the least productive.

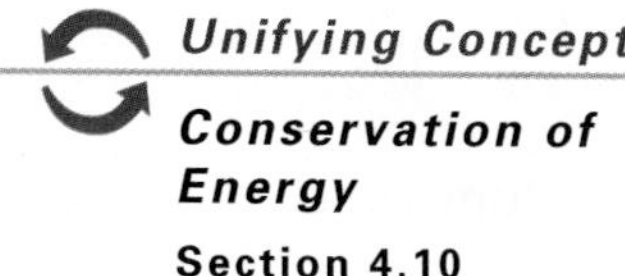

Unifying Concept

Conservation of Energy

Section 4.10

Once energy enters an ecosystem as producer biomass, it goes up the food chain, creating biomass in primary consumers, secondary consumers, and so on. Does all the energy taken in by plants ultimately go into growing rabbits, or eagles? Far from it—in all ecosystems, a huge amount of energy is lost as you go up the food chain. Typically, only about 10 percent of the energy at one trophic level becomes available to the next level. What happens to the other 90 percent? First of all, not every organism at one trophic level is exploited by the next level—for example, not every plant gets eaten by an herbivore. Second, when a consumer eats, the energy it receives from food goes into things other than building biomass—feces and maintenance, to be specific. Feces contain organic materials that the consumer is unable to digest. Maintenance is the energy the consumer requires to live—the energy it takes to find and eat food, run, mate, breathe, and so on. During these activities, a lot of energy is also lost to the environment as heat. So, by the time feces and maintenance

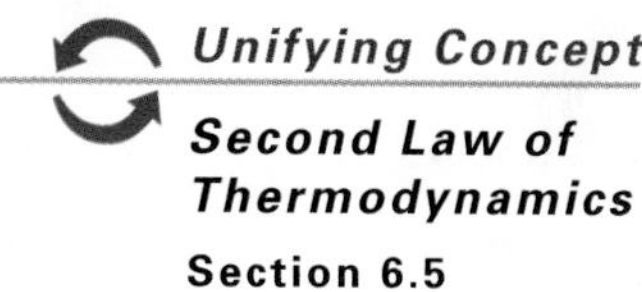

Unifying Concept

Second Law of Thermodynamics

Section 6.5

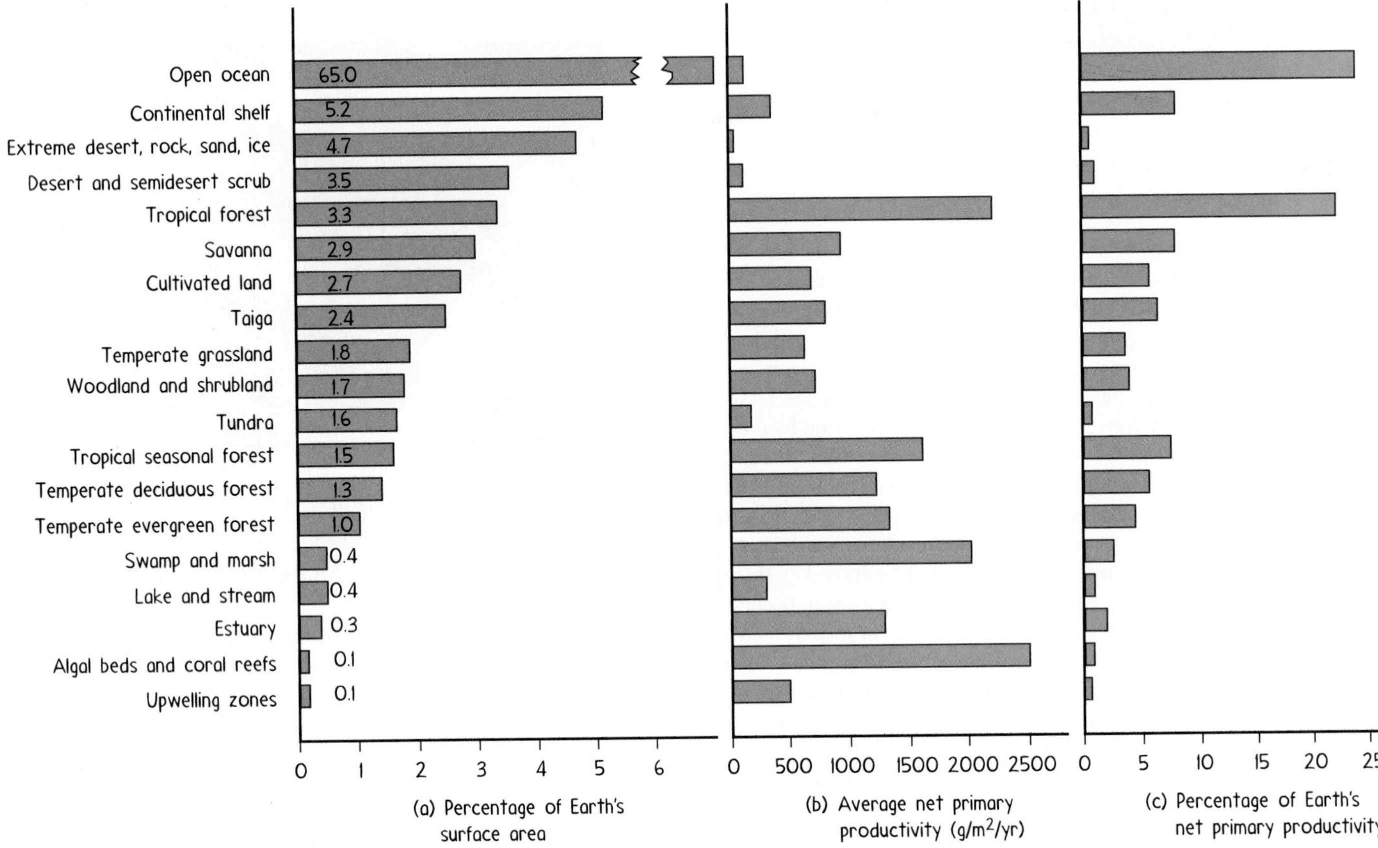

Figure 21.9 Primary productivity varies a great deal from one ecosystem to another. The most productive ecosystems include tropical forests, swamps and marshes, and algal beds and coral reefs. Deserts and tundra are among the least productive ecosystems. Although open ocean is not particularly productive on a per square meter basis, there is so much of it that it contributes a significant proportion of Earth's total primary productivity. (Courtesy of Campbell, Reece, Simon, *Essential Biology with Physiology*, © 2004.)

take their share, only about 10 percent is left for growth and reproduction—for building new biomass (Figure 21.10). This is why successive trophic levels of an ecosystem have less and less biomass—why there is more grass than zebras, and why there are more zebras than lions. This is also why most ecosystems are limited in the number of trophic levels they can support. The flow of energy through trophic levels is sometimes diagrammed in an *energy pyramid*. The shape of the pyramid emphasizes the loss of energy as you go up trophic levels (Figure 21.11).

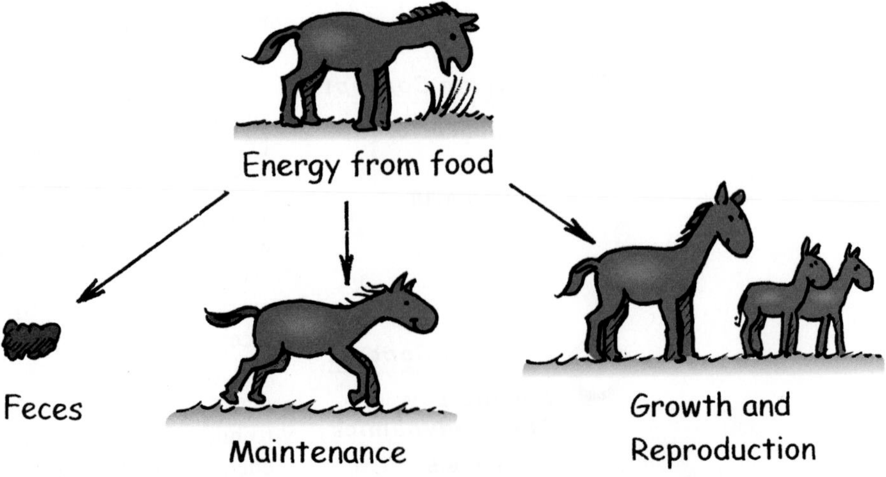

Figure 21.10 Energy taken in from food goes to three things: feces, maintenance, and growth and reproduction.

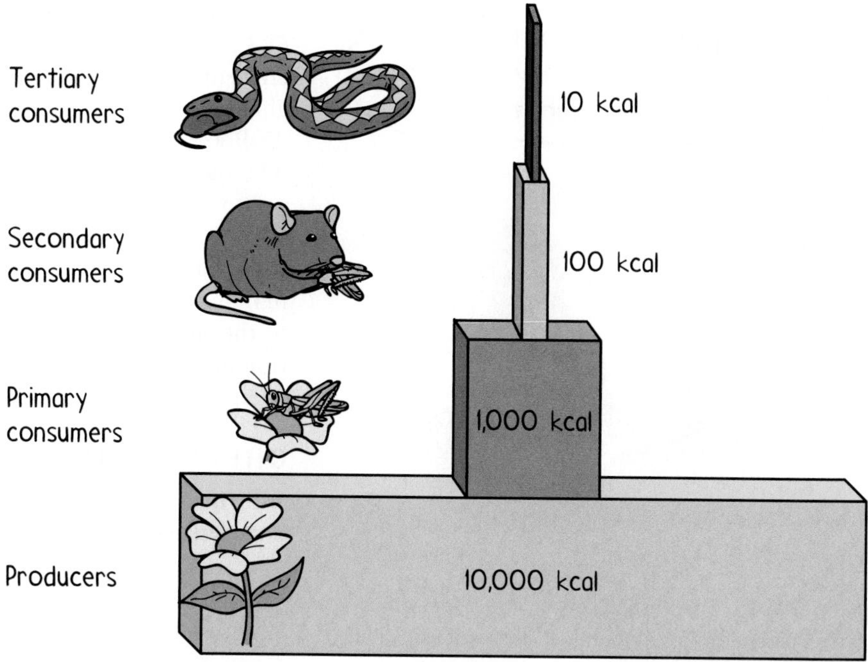

Figure 21.11 This idealized energy pyramid shows that the amount of energy at each successive trophic level decreases. This is because not every organism at one trophic level is eaten by an organism at the next level and because some energy is lost to feces and maintenance. (Adapted from Campbell, Reece, Simon, *Essential Biology with Physiology*, © 2004.)

INSIGHT Consumers vary a great deal in how efficiently they convert food to biomass. Of the energy they absorb (that is, energy not lost in feces), insects use 10 to 40 percent to build biomass. The rest goes to maintenance. Mammals and birds are strikingly inefficient energy converters—only 1 to 3 percent of their absorbed energy goes to biomass. Why? Mammals and birds are endotherms (Chapter 18). Endotherms burn a lot of calories just maintaining their high, stable body temperatures.

CHECK YOURSELF

A typical hectare of coniferous forest has a primary productivity of 30 metric tons per year. Assuming that 10 percent of the energy available at each trophic level moves to the next trophic level, what biomass of primary consumers is produced in a year? Of secondary consumers? Of tertiary consumers?

CHECK YOUR ANSWER

Biomass of primary consumers = 10% (30 metric tons) = 3 metric tons.

Biomass of secondary consumers = 10% (3 metric tons) = 0.3 metric tons.

Biomass of tertiary consumers = 10% (0.3 metric tons) = 0.03 metric tons.

INTEGRATED SCIENCE PHYSICS

Energy Leaks Where Trophic Levels Meet

Unifying Concept

The Second Law of Thermodynamics

Section 6.5

Why are energy pyramids always pyramids—that is, why is the amount of energy available to each trophic level always less than the amount of energy at the trophic level below it? We have already mentioned some of the reasons—because not all the organisms at each trophic level get eaten and because some of the energy in food that is eaten is lost in feces or goes into maintenance. We have also mentioned that significant energy is lost as heat. Why? Because of the Second Law of Thermodynamics (Chapter 6). The Second Law states that natural systems tend to move from organized energy states to disorganized energy states—that is, useful energy dissipates to unusable energy. Specifically, any time energy is converted from one form to another—including in any chemical reaction—some energy is lost to the environment as heat. And in fact, moving energy from one trophic level to another—such as by breaking down plant matter in the digestive tract of a rabbit and then using the molecules to build more rabbit muscle—involves nothing more than a long series of chemical reactions, one after another.

Let's consider one example—what happens when organisms burn glucose to make ATP. This is the reaction that defines cellular respiration, the process most organisms use to obtain energy (Chapter 15). Recall that the chemical reaction for burning glucose is

$$C_6H_{12}O_6 + 6\ O_2 \rightarrow 6\ CO_2 + 6\ H_2O + 673\ \text{kcal/mole}$$

Figure 21.12 The Second Law of Thermodynamics dictates that energy is lost to the environment as heat in every chemical reaction.

That is, glucose and oxygen react to form carbon dioxide and water, releasing 673 kilocalories of energy per mole in the process. If this reaction were perfectly efficient in organisms, the entire 673 kilocalories per mole released from burning glucose would be captured as ATP. Is it? We know that about 38 molecules of ADP are converted to ATP as the result of burning a single glucose molecule. ATP then provides 7 kilocalories per mole when it is broken down into ADP and phosphate during cellular processes. But, $38 \times 7 = 266$, much less than 673. Clearly, a lot of energy is missing! What happened to it? It was lost to the environment as heat. (In mammals and birds, heat lost this way contributes to the maintenance of stable, warm body temperatures, but this heat is eventually shed to the environment as well.) Once we consider the fact that *every* chemical reaction involves some energy loss to the environment, it becomes clear why so much energy leaks from one trophic level to the next (Figure 21.12).

CHECK YOURSELF

Why does the Second Law of Thermodynamics imply that each trophic level will have less energy available to it than the trophic level below?

CHECK YOUR ANSWER

Whenever energy is converted from one form to another, some energy is lost to the environment as heat. Moving energy from one trophic level to another is nothing more than a long series of chemical reactions, one after another.

21.4 Kinds of Ecosystems

A nature documentary opens with a familiar scene—yellowy grasslands, with scattered dull-green trees. The Sun is high and bright, and you can almost feel the heat, which causes the entire image to ripple. Immediately you have a sense of what to expect—and what not to expect. There isn't going to be a patch of red-leafed maple trees in the middle of this grassland. There aren't going to be pine trees, or soaring redwoods, or vine-covered rainforest trees with large buttressing roots. There aren't going to be moose, or polar bears, or tigers wandering this grassland. There surely aren't going to be dolphins or seahorses. On the other hand, you are not at all surprised when the camera pans across the field and slowly brings into focus—elephants, lions, leopards, giraffes. Earth includes many different types of ecosystems, each of which is occupied by characteristic suites of organisms. We can divide these ecosystems into two broad categories, terrestrial and aquatic.

Terrestrial Biomes

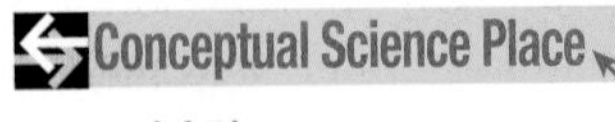

Terrestrial Biomes

The land area of Earth is divided into eight major types of ecosystems, known as **biomes**. Their distribution is shown on the map in Figure 21.13. Each biome is characterized by specific types of biological communities and, particularly, by specific types of plant life. The type of biome found in a habitat is determined primarily by climatic variables such as temperature, precipitation, and the severity of seasonal variation. As a result, latitude and altitude are major influences on the distribution of biomes on Earth. The same biome is often found in completely different parts of the world. For example, there are tropical forests—with their tall trees, dense vegetation, and astounding biodiversity—in the Amazon region of South America, in Southeast Asia, and in western Africa.

Tropical forests (Figure 21.14), sometimes called rainforests, are found close to the equator. Temperatures are warm and fairly constant throughout the year. Tropical forests receive between 200 and 400 centimeters of rain a year, distributed between a wet season and a dry season. Tropical forests are renowned for

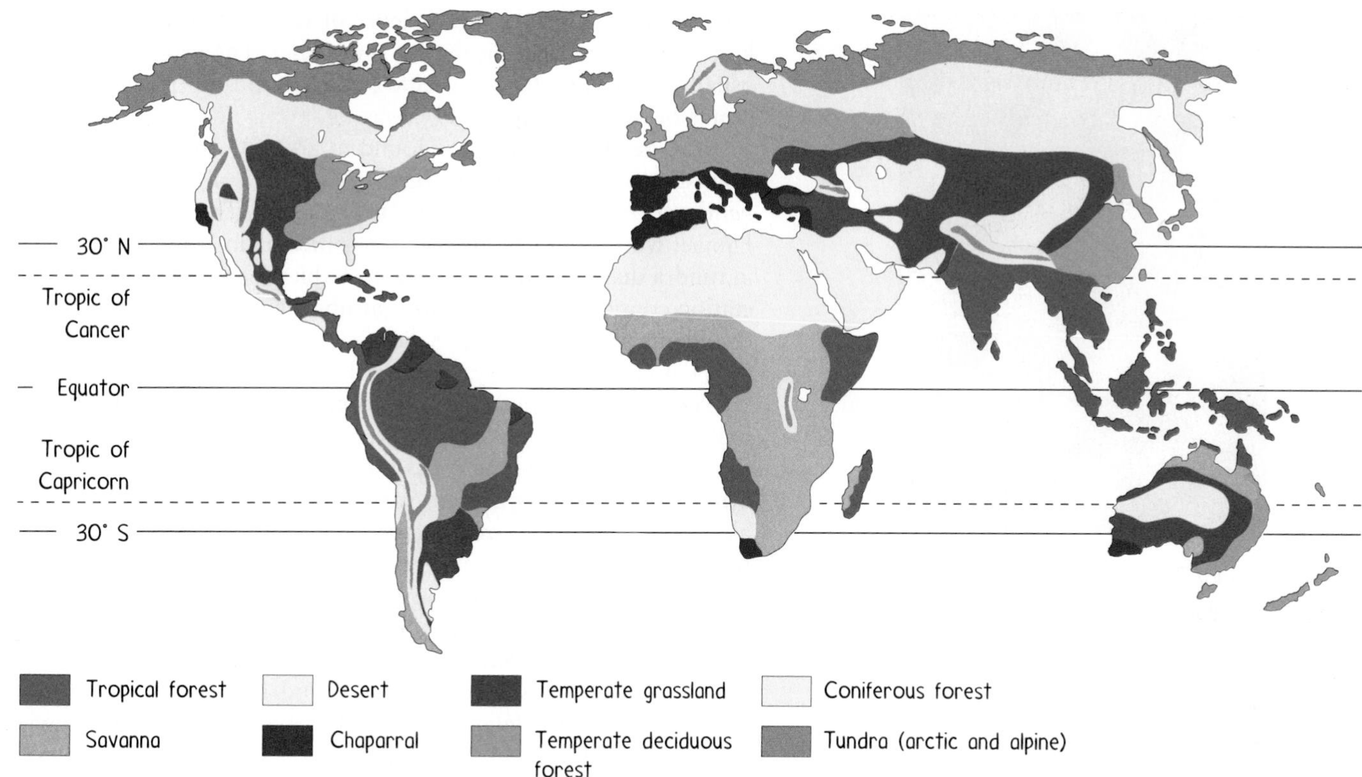

Figure 21.13 This map shows how terrestrial biomes are distributed across the world. Each biome is characterized by specific types of plant life. As you can see, latitude is one important influence on the type of biome that occurs in a particular area.

Figure 21.14 This tropical forest in Thailand shows the tremendous density and diversity of vegetation found in all tropical forests.

INSIGHT Tropical forest plants make much bigger seeds, on average, than plants in temperate regions—think of avocados versus apples, mangos versus raspberries. Why is this? Different explanations have been proposed, but the most likely appears to be that tropical seedlings grow in a shaded environment. Little sunlight means limited photosynthesis. As a result, having a big store of energy helps a seedling in its first phase of life.

their biodiversity—more species are found in this biome than in all other biomes combined. A single square kilometer of tropical forest may contain as many as 100 different species of trees. These form a dense canopy, shading the ground. Plant life in tropical forests also includes numerous epiphytes—plants that grow on top of other plants—and often, an understory of ferns. Tropical forests have little leaf litter on the ground because organic material is quickly decomposed by living organisms. Because of the tremendous density and diversity of life, most of the nutrients present in tropical forests are locked up in one or another living organism—as a result, the soil tends to be poor. Tropical forests are rapidly being destroyed worldwide for timber and agriculture. Unfortunately, the poor quality of the soil means that forest that has been cleared for farmland can only support crops for a few years.

Temperate forests (Figure 21.15) are found in areas with four distinct seasons, including a warm growing season and a cold winter. Temperatures vary considerably over the course of the year. Temperate forests receive between 75 and 150 centimeters of rainfall per year. There are usually three to four different species of trees per square kilometer. The leaves of temperate forest trees are unable to survive the freezing temperatures of winter. Consequently, these trees, which include elm, oak, beech, and maple, are *deciduous*—they drop their leaves in the autumn. Trees recover nutrients from their leaves before they are shed. The soil of temperate forests is fertile, and these forests make good farmland. In fact, many temperate forests have been cut down to make way for agriculture.

Coniferous forests (Figure 21.16), sometimes called evergreen forests, are found in areas with long, cold winters and short summers. They are relatively dry, receiving about 40 to 100 centimeters of precipitation per year, mainly in the form of snow. As their name suggests, coniferous forests are dominated by conifers such as pine, spruce, fir, and hemlock. The leaves (needles) of evergreen

Figure 21.15 This is a temperate forest in Great Smoky Mountains National Park, Tennessee. The season is autumn—the deciduous trees are preparing to shed their leaves.

Figure 21.16 These are evergreen trees in a coniferous forest in Denali National Park, Alaska.

trees are covered with a thick layer of wax and contain special substances that keep them from freezing during the winter. The ground in coniferous forests is usually covered with shed needles, and the soil is poor in nutrients. Many coniferous forests are threatened by logging.

Tundra (Figure 21.17) is found in areas of extreme cold and little precipitation. One of the defining features of tundra is a layer of *permafrost*, or permanently frozen subsoil, beneath the topsoil. The word *tundra* comes from a Finnish word, *tunturia*, meaning "treeless plain," and indeed, trees cannot grow in tundra due to permafrost and the habitat's short growing season. Instead, vegetation consists primarily of low shrubs, lichens, mosses, grasses, and some flowers. Plants are often found in close clumps as a defense against cold. The tundra biome is characterized by relatively low biodiversity.

Savannas (Figure 21.18) are tropical grasslands with a warm climate and a long dry season. They are covered with grass and occasional scattered trees. Savanna plants have long roots for dealing with drought, and the trees also have a thick bark that helps them survive periodic fires. During the dry season, the aboveground portions of the grasses die—only the roots survive. Dry season fires help maintain savannas by preventing tree growth. Living organisms also help maintain savanna habitats, in some cases preventing them from growing into tropical forest. For example, elephants eat and kill trees, and humans burn forests for cropland. Savannas receive a moderate amount of rainfall during the course of the year, usually between 75 and 100 centimeters.

Temperate grasslands (Figure 21.19) are found in areas with four distinct seasons, including a hot summer and cold winter. Temperate grasslands have more widely varying temperatures and less rainfall—about 50 to 90 centimeters per year—than savannas. The lower level of precipitation coupled with seasonal drought, fire, and grazing prevents trees from taking over the habitat and turning it into temperate forest. The soil of temperate grasslands is extremely fertile, making for ideal farmland.

Chaparral (Figure 21.20) is found in places with mild, rainy winters and hot, dry summers characterized by drought and fire. Chaparral occurs in parts of California, the Mediterranean, South Africa, and western Australia. Chaparrals are dominated by small trees and shrubs, many with small waxy leaves that are good at retaining moisture. Many chaparral plants also have extensive root systems that help them survive the summer drought.

Deserts (Figure 21.21) are habitats that receive very little precipitation, usually less than 50 centimeters per year. Although deserts are usually associated with hot climates, cold deserts, in which precipitation generally falls as snow, also exist. For example, Antarctica, the coldest and driest of the continents, is a desert. Many of the world's largest hot deserts, including the Sahara and the Australian desert, are found in a band around 30 degrees north latitude and 30 degrees south latitude, where high atmospheric pressure results in low levels of precipitation (Chapter 25).

Figure 21.17 This is a caribou in tundra habitat in Northwest Territories, Canada. Note the absence of trees.

Figure 21.18 This is a savanna in Masai Mara Reserve, Kenya.

Figure 21.19 This is a temperate grassland in Grasslands National Park, Saskatchewan, Canada.

Desert plants typically have special adaptations for living in dry conditions, including extensive root systems and the ability to store water when it is available. Desert soils are usually abundant in nutrients. This, combined with the long growing season and plentiful sunshine that characterizes many desert habitats, makes deserts into fertile farmland if water can be brought in through irrigation.

Aquatic Biomes

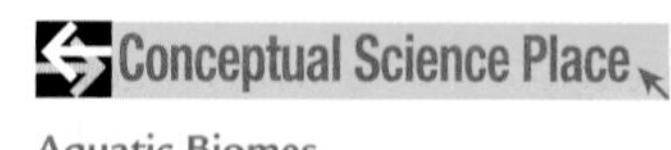

Aquatic Biomes

Life originated in the oceans, and many species continue to make use of aquatic habitats in both fresh water and saltwater.

Freshwater Habitats

Freshwater habitats include the still waters of lakes and ponds and the flowing waters of rivers and streams. Lakes and ponds vary tremendously in size and in biodiversity. Nonetheless, all can be divided into three major zones, illustrated in Figure 21.22. Habitats close to the water surface and to shore are part of the *littoral zone*. The littoral zone is relatively warm because of its exposure to sunlight. Organisms living in the littoral zone include photosynthetic plants and algae, insects, mollusks, crustaceans, fishes, amphibians, ducks, and turtles. The *limnetic zone* includes habitats that are close to the water surface but far from shore. The limnetic zone is occupied largely by *plankton*, organisms that float in the water rather than swimming actively through it. Phytoplankton are photosynthesizing plankton, whereas zooplankton are heterotrophic plankton. Fish may also be found in the limnetic zone. The *profundal zone* describes deep water habitats in ponds and lakes. Most organisms in the profundal zone consume organic debris that drifts down from above.

Figure 21.20 This is chaparral in Los Padres National Forest, California.

Species that live in the flowing waters of rivers and streams usually have adaptations that allow them to avoid being washed away. Many have hooks or suckers or are strong swimmers. Algae often occupy the base of the food chain in river and stream habitats.

Estuaries are habitats where freshwater rivers join oceans. Estuarine plants, such as certain seaweeds, marsh grasses, and mangroves (Figure 21.23), have adaptations that allow them to deal with changing salinity conditions. Estuaries are also home to many species of fishes, invertebrates, and birds and are essential nursery habitats for many fishes and invertebrates.

Figure 21.21 This is desert in Baja California, Mexico.

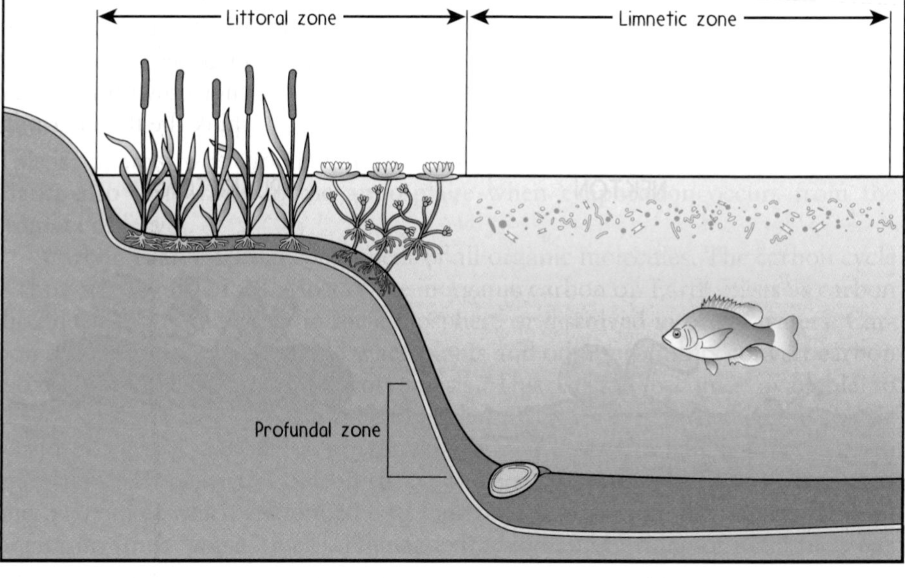

Figure 21.22 Lakes and ponds are divided into three zones, the littoral zone, the limnetic zone, and the profundal zone. Each is home to a distinct set of living organisms.

INSIGHT Knowledge of Earth's inner structure is very new. Ford was making cars and a World Series in baseball was being played before Earth scientists knew that Earth has a core.

text, *Principles of Geology*, and with that, laid out convincing arguments that Earth is indeed very old.

Around the turn of the twentieth century, the field of Earth science was gaining popular attention. A subject of parlor-room speculation was the question, "What would you encounter if you could dig a hole to the center of the Earth?" But geologists of the day could do little to answer that question. In the early 1900s, almost nothing was known about Earth's interior. But progress soon came when geologists began to use seismological data—earthquake measurements—to investigate the Earth's interior.

22.2 Using Seismology to Look Inside the Earth

Earthquakes, besides being fearsome and destructive events, are a key to understanding Earth's structure. An **earthquake** is the shaking or trembling of the ground that happens when rock under Earth's surface moves or breaks. Earthquakes release vast amounts of energy that radiates outward from the disturbance in the form of *seismic* waves. The study and measurement of seismic waves, called **seismology**, has provided most of what we know today about Earth's interior. Like any kind of wave, seismic waves reflect from surfaces and refract through others. (See Chapter 8 for more on waves.) Exactly how seismic waves reflect and refract, as well as variations in their speed and wavelength, reveals much about the medium in which these waves travel.

Unifying Concept
Waves
Section 8.1

Figure 22.1 Block diagrams show the effects of seismic waves. The yellow portion of the left side of each diagram represents the undisturbed area.
(a) Primary body waves alternately compress and expand Earth's crust, as shown by the different spacing between the vertical lines, similar to the action of a spring.
(b) Secondary body waves cause the crust to oscillate up-and-down and side-to-side. (c) Love surface waves whip back and forth in a horizontal motion. (d) Rayleigh surface waves operate much like secondary body waves, but they affect only the surface of the Earth.

Seismic waves come in two main varieties: *body waves*, which travel through the Earth's interior, and *surface waves*, which travel on the Earth's surface like ripples on water (Figure 22.1).

Body waves can be further classified either as *primary waves* (P-waves) or as *secondary waves* (S-waves). Primary waves are longitudinal—they compress and expand the material through which they move. P-waves are the fastest seismic waves, traveling at speeds between 1.5 and 8 km per second through any type of material—solid rock, magma, water, and air. S-waves, on the other hand, are transverse—they vibrate the particles of their medium up-and-down and side-to-side, and they travel more slowly than P-waves. S-waves can only travel through solid materials.

Surface waves come in two types: *Rayleigh* waves and *Love* waves (each named after its discoverer). Rayleigh waves roll over and over in a backward tumbling motion similar to ocean waves, except that ocean waves tumble in a forward direction. Love waves move just like S-waves, except the shaking motion is horizontally side-to-side. Since Love waves shake things side-to-side, they are particularly damaging to tall buildings.

Near the turn of the twentieth century, Irish geologist Richard Oldham was examining records of a massive earthquake in India when he discovered that its S-waves traveled some distance through the Earth and then stopped. He also observed that the P-waves traveled as far as the S-waves into the Earth but then refracted at an angle and lost speed. Since S-waves cannot travel through liquid but P-waves can, but at a reduced speed, Oldham inferred that the earthquake waves had hit an internal boundary—a place where the solid Earth becomes liquid. In other words, he discovered that the Earth has a distinct core, as shown in Figure 22.2. The year was 1906.

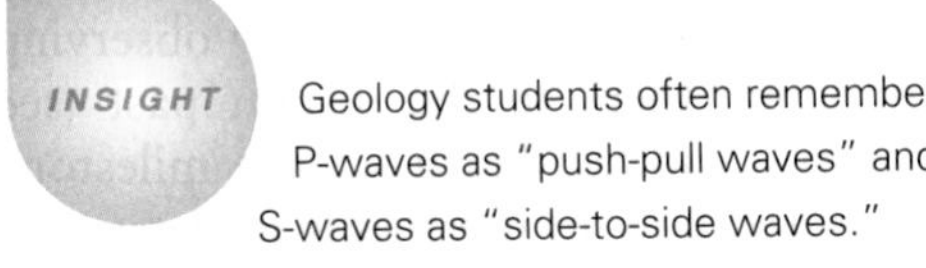

INSIGHT Geology students often remember P-waves as "push-pull waves" and S-waves as "side-to-side waves."

Three years later, seismologist Andrija Mohorovičić (pronounced mohorovu-chick) analyzed seismic readings from an earthquake near his town of Zagreb, Croatia. He detected a sharp increase in the speed of seismic waves at another boundary, one that lay shallower within the Earth. Mohorovičić figured that the wave speed increased because the wave was passing from a low-density solid to a high-density solid. Thus, Mohorovičić discovered that the Earth is composed of a thin, outer **crust** that sits upon a layer of denser material, the **mantle**. The dividing

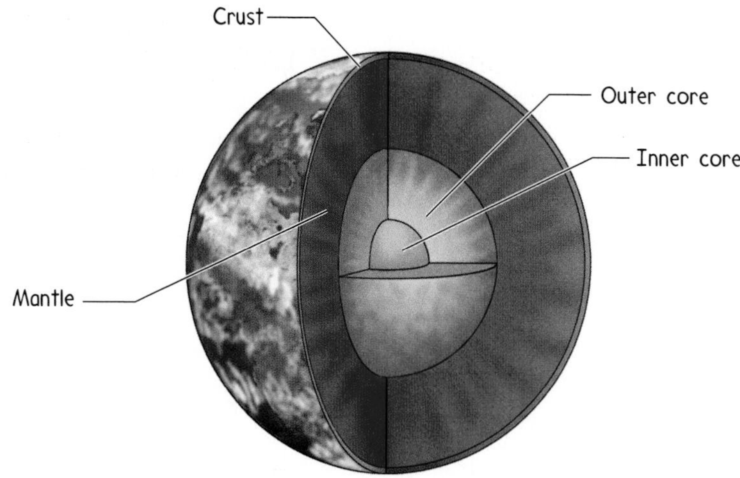

Figure 22.2 Earth has a layered internal structure. The layers—the crust, mantle, and core—differ in composition. The core consists of two distinct phases—a solid inner core and a liquid outer core.

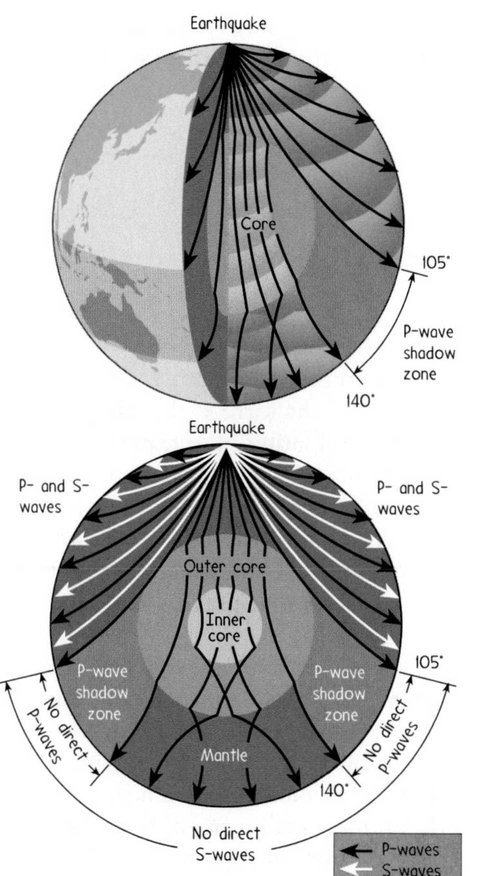

Figure 22.3 Cutaway- and cross-sectional diagrams showing the change in wave paths at the major internal boundaries and the P-wave and S-wave shadows. The P-wave shadow between 105° and 140° from an earthquake is caused by the refraction of waves at the core–mantle boundary. The S-wave shadow is even more extensive. Any location more than 105° from an earthquake does not receive S-waves since the liquid outer core does not transmit them.

line between the Earth's crust and mantle has been called the *Mohorovičić discontinuity*, or "Moho," ever since.

In 1913, Beno Gutenberg reinforced Oldham's earlier findings by showing that the mantle–core boundary is very distinct and is located at a depth of 2900 km. He observed that, when P-waves reach this depth, they are refracted so strongly that the boundary actually casts a P-wave shadow over part of the earth (Figure 22.3). The shadow is a region where no waves are detected. Further, the sharp boundary between the mantle and core casts an S-wave shadow that is even more extensive than the P-wave shadow, indicating that S-waves are unable to pass through the core. And since S-waves are transverse and unable to pass through liquids, Gutenberg realized that the core, or part of it, must be liquid.

Taken together, the discoveries of Oldham, Mohorovičić, and Gutenberg show that Earth consists of the three layers of materials of different composition: *crust*, *mantle*, and *core*. Each layer is a concentric sphere, so that, overall, the Earth's structure resembles that of a boiled egg.

This simple picture of Earth's layers was refined in 1936 by Ingre Lehmen, a Danish seismologist. Her research showed that P-waves refract not only at the core–mantle boundary but again at a certain depth within the core, where they gain speed. The way in which P-waves refract within Earth's core suggested that the core actually has two parts, a liquid outer core and a solid inner core. Adding Lehmen's work to the others brings us up to the current picture of Earth's internal layered structure.

22.3 More About Earth's Layers

The Crust

The **crust** is the Earth's surface layer. Like an eggshell, it is thin, brittle, and can crack. The crust has two distinct regions: *oceanic crust* and *continental crust.* These are made up of different types of rock.* The oceanic crust is mainly a dark, dense, and fine-grained rock named *basalt.* The continental crust, on the other hand, is composed mostly of *granitic* rocks that are lighter-colored, less dense, and

* As you will learn in the next chapter, a *rock* is defined as an aggregate of minerals that are joined together as a cohesive solid. And what's a mineral? A mineral is a naturally occurring solid element or compound with a definite composition and crystalline structure (its atoms are arranged in a regular, repeating pattern).

Table 22.1 Characteristics of Earth's Layers (average values)

Layer	Thickness	Density	Composition
Crust	5–70 km (19–43 mi)	2.7–3.0 g/cm^3	rock
Continental crust	10–70 km (12–37 mi)	2.7 g/cm^3	granitic rock
Oceanic crust	5–10 km (3–6 mi)	3.0 g/cm^3	basaltic rock
Mantle	2885 km (1779 mi)	3.4–4.4 g/cm^3	rocks including periodite, dunnite
Core (radius)	3486 km (2166 mi)	9.9–12.8 g/cm^3	iron + oxygen, sulfur, nickel
Outer core	2270 km (1407 mi)	9.9 g/cm^3	liquid iron + oxygen, sulfur, nickel
Inner core (radius)	1216 km (754 mi)	12.8 g/cm^3	solid iron + oxygen, sulfur, nickel

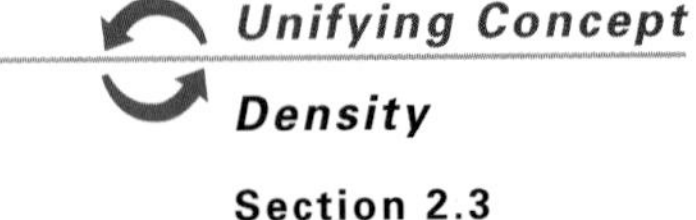
Unifying Concept
Density
Section 2.3

coarser-grained than basalt, as Table 22.1 indicates. Both portions of the crust are much less dense than the Earth's mantle, which consists of other kinds of rock. So, like ice floating on water, the low-density crust of Earth floats atop the denser mantle.

INTEGRATED SCIENCE PHYSICS

Isostasy

We say the low-density crust "floats" atop the denser mantle. Objects float because a *buoyant force* acts on them (the buoyant force is discussed in detail in Appendix E). The buoyant force acts on any floating object, pushing it upward to oppose the downward pull of gravity. In the case of the "floating" crust, gravity pulls the crust down, while the buoyant force exerted by the mantle pushes the crust upward. When the buoyant force and gravitational force are equal in magnitude, they balance and the crust maintains a stable vertical position. This concept—the vertical positioning of the crust so that gravitational and buoyant forces balance—is called *isostasy*.

The concept of isostasy can be made clear with an analogy. Imagine that the Earth's crust is a cargo ship and that the mantle is the ocean. The ship will establish its vertical position in the water when the net force on it is zero. This happens when the gravitational force pulling the ship downward (its weight) equals the buoyant force pushing it upward (Figure 22.4). Note that the heavily loaded and more dense ship will float lower, with more of it submerged, than the same ship when empty. Likewise, the crust's vertical position in the mantle rises and falls according to variations in density. Denser oceanic crust sits lower in the mantle than less dense continental crust.

Figure 22.4 Isostasy: The vertical position of the crust is stable when the gravitational and buoyant forces balance. Denser oceanic crust therefore has a lower vertical position than the less-dense continental crust, just as the loaded ship sits lower in the water than the unloaded ship. Note that these principles of flotation govern the position of the crust, even though the mantle is solid. The reason for this is that, over geologic time, the mantle actually flows like a fluid.

The continental crust is, on average, about 35 km thick. It's as much as twice this thickness at large mountain ranges, such as the Himalayas, Alps, and Sierra Nevada. Continental crust is not just taller where there are mountains; it's deeper below ground as well. Mountains are supported from below by "roots" that extend deep into the mantle as Figure 22.5 shows.

Taller mountains have deeper roots that maintain their isostatic balance. Why? As erosion carries the top of a mountain away over time, like an iceberg with its top shaved off, the lightened mountain is buoyed higher. So the submerged portion of continental crust also raises to shallower depths. This process of establishing a new vertical position is called *isostatic adjustment*.

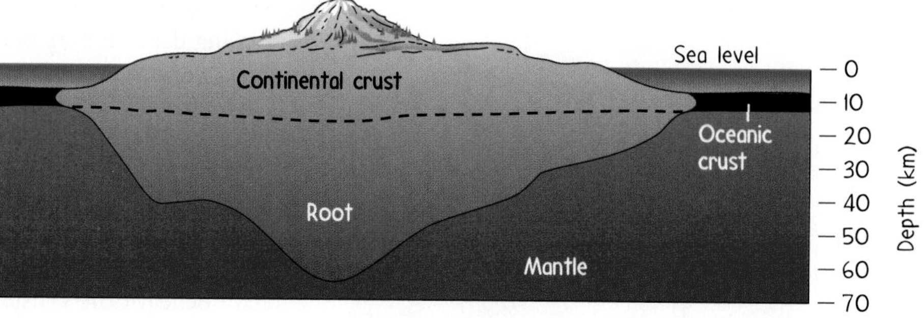

Figure 22.5 Mountains have "roots," blocks of thickened continental crust that support them. Tall mountains have large roots that maintain isostatic balance.

Unifying Concept

The Gravitational Force

Section 5.3

CHECK YOURSELF

1. How does the Earth's crust behave like a ship floating in water?
2. Why is the Earth's crust thicker beneath a mountain?

CHECK YOUR ANSWERS

1. The crust's vertical position, as well as the ship's, is determined by the balance of the buoyant and gravitational forces acting upon it.
2. Just as most of an iceberg is below sea level, likewise for mountains. Mountains sink until the upward buoyant force balances the downward gravitational force. Since mountains are so heavy, much of a mountain must be "sunk" deep into the mantle to support its massive above-ground portion.

The oceanic crust, with a relatively uniform thickness of between 5 and 10 km, is much thinner than the continental crust. On average, the thickness of the Earth's crust is akin to the thickness of an apple peel relative to the volume of an apple. The crust constitutes only about 0.3% of the planet's total volume.

The Mantle

Conceptual Science Place

The Earth's Mantle Activity

No one has thus far been able to drill a hole deep enough to penetrate Earth's crust. In the 1960s, United States investigators attempted to drill into the mantle. Dubbed *Project Mohole* (after *Moho*, the Mohorovičić discontinuity), the effort was unsuccessful and was therefore dubbed *Project No-hole.* The deepest hole ever drilled is in northern Russia on the Kola Peninsula. In 1989, after 19 years of drilling, the Kola well reached more than 12 km into the crust. But then drilling was abandoned, owing to mounting costs and equipment failures. It is unfortunate that no one has thus far drilled to the mantle to examine it, because most of the Earth—82% of its volume and 65% of its mass—*is* mantle. Also, the mantle is the Earth's thickest layer—2,900 km top to bottom.

The mantle, like the crust, is rocky. Although mantle rock is rich in silicon and oxygen, just as rocks of the crust are, mantle rock contains proportionately more of the heavier elements, such as iron, magnesium, and calcium. Hence the mantle is denser than the crust. The mantle also gains density due to the weight of the overlying crust, which bears down on the mantle and compresses it. Weight increases pressure and minerals are squeezed into rocks of denser structure.

The mantle is warm compared with the crust for several reasons. One reason for the mantle's heat is the persistent high pressure there. But the principal source of heat is the decay of radioactive elements in the mantle. In the lowermost mantle, the flow of heat from the core also plays a big role in elevating mantle temperature. Near Earth's surface, the temperature beneath increases by about

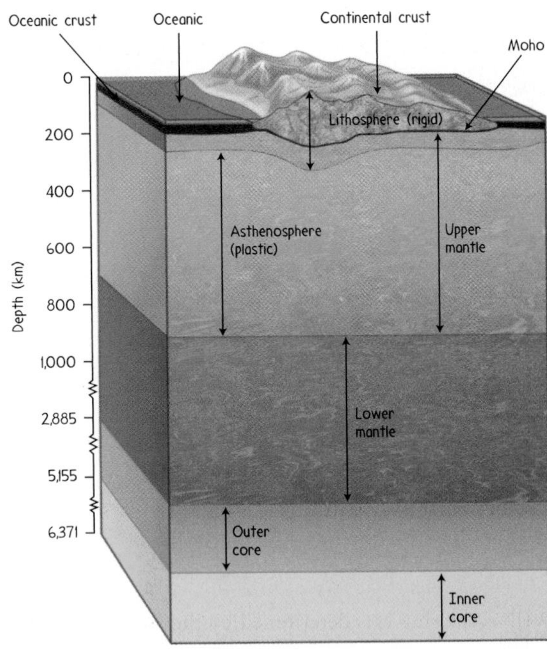

Figure 22.6 This diagram shows the relationship of the crust and mantle to the lithosphere and asthenosphere.

INSIGHT Though they are stiffer than the asthenosphere, rocks in the crust and lower mantle have some plasticity. In other words, they do not always break under stress. Under the right conditions, these rocks can bend and flow. For example, have you ever seen rock layers bent into graceful arches along cliffsides or road cuts? This lithospheric rock has deformed plastically—flowed and bent slowly without breaking.

30°C for every kilometer of depth. The rate of increase in temperature with increasing depth tapers off to as little as 1°C per kilometer deeper within the mantle, however.

Although the mantle has a fairly uniform material composition, it is divided into two different regions, based on their physical properties. The top region, the upper mantle, extends from a 660-kilometer depth to the crust–mantle boundary; and the second region, the lower mantle, extends from the base of the upper mantle to a depth of 2,900 kilometers at the mantle–core boundary.

The upper mantle itself is divided into two zones. The uppermost zone, directly beneath the crustal surface, is relatively cool and rigid. In many ways, it behaves like the stiff and breakable crust. In fact, because the crust and the upper region of the upper mantle are so similar, they act as a single layer of relatively rigid rock. This zone is called the **lithosphere** (Figure 22.6).

Beneath the lithosphere lies a layer of the mantle called the **asthenosphere**. Although the asthenosphere is solid, it actually flows over long periods of geologic time.* It behaves in a *plastic* manner. This is similar to Silly Putty or taffy; which acts like a solid under sudden stress (it can break) but behaves like a fluid when stress is applied to it slowly (it can flow). Hence, the rigid lithosphere rides like a raft on the slowly flowing asthenosphere. Beneath the asthenosphere, into the remaining 2,200 km of lower mantle, the rock becomes more rigid again. Although the lower mantle retains the ability to flow, it isn't as plastic as the overlying asthenosphere.

The Core

At the center of the Earth lies the **core**. The core has a radius of 3400 to 3500 km and consists of two layers. As a whole, the core occupies up to 15% of the Earth's volume and 30% of its mass. The core is nearly twice as dense as the mantle because it's made up largely of metallic iron. Iron is much heavier than the abundant silicon and oxygen of the crust and mantle.

The inner core is blazing hot. Estimates of its temperature range from 3,900°C to 7,200°C, which is about as hot as the surface of the Sun. There are three main sources of heat in Earth's core: (1) radioactivity—decaying radioactive elements produce energy that converts to heat, as discussed in Chapter 10; (2) heat generated during "The Great Bombardment," the time in Earth's early history when chunks of space debris frequently crashed into Earth and the energy of these impacts was largely converted to heat; and (3) the transformation of gravitational potential energy to heat as denser core material sank to the center of the planet early in its history.

You may be wondering why Earth's core is a solid, even though it's so hot. The answer is that intense pressure from the weight of the overlying earth prevents inner-core melting, in much the same way that increased pressure in a pressure cooker prevents high-temperature water from boiling. Pressure packs atoms together in the inner core too tightly to allow them to flow as liquid.

On the other hand, less weight is exerted on the outer core. Pressure there is less, and the iron and nickel there can flow in the liquid phase. As Earth rotates, this liquid outer core spins. Convection currents in the outer core are stirred up. These motions in the core have effects far outside the Earth's surface. The moving iron and other metals produce a flowing electric charge—a current (as described in Chapter 7). It is this electric current that creates Earth's magnetic field (Figure 22.7).

* When we say the athenosphere flows slowly, we mean *very slowly*. The hour hand on a clock moves about 10,000 times faster than the "flowing" asthenosphere. Its flow rate is so slow that even under close examination, rock of the asthenosphere would appear rigid.

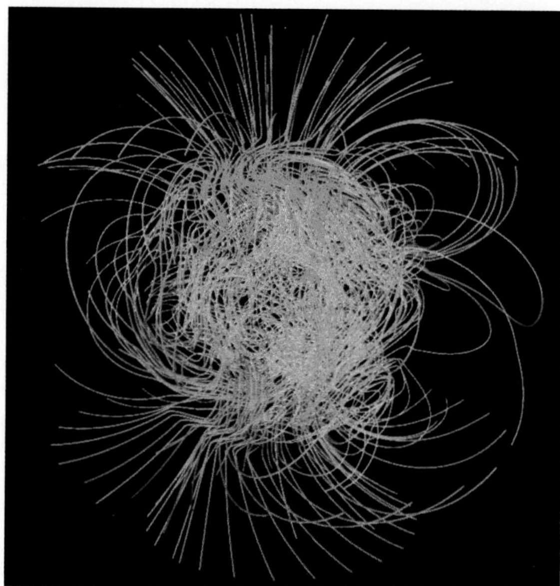

Figure 22.7 This computer simulation shows the magnetic field generated by the motion of Earth's fluid outer core. Magnetic field lines are blue where the field is directed inward and yellow where it is directed outward. You can see that the field beyond the mantle–core boundary has a smooth overall structure, while the field of the core is highly complicated.

Unifying Concept

The Principle of Falsifiability

Section 1.4

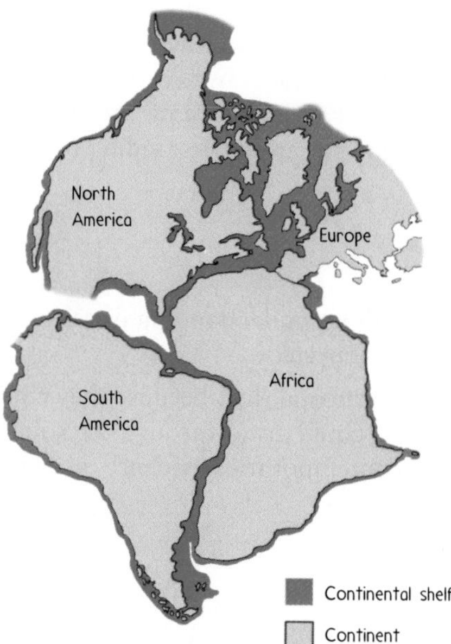

Figure 22.8 When you align the shorelines of South America and Africa, the continents fit together like pieces of a jigsaw puzzle. When you align the continents at their continental shelves, the fit is even better.

CHECK YOURSELF

1. Since it's not possible to drill a hole to Earth's inner core for direct observation, how do we know the inner core is solid?
2. Iron's normal melting point is 1535°C, yet the Earth's inner core temperature is greater than 4000°C. Why does the inner core remain solid?

CHECK YOUR ANSWERS

1. The change in velocities of P-waves at the boundary of the inner and outer core shows that the inner core is solid.
2. Melting is prevented by the intense pressure from the weight of the earth above. Atoms are crushed together so tightly that even high temperatures cannot budge them. Hence the core remains solid, despite the high temperature.

22.4 Continental Drift—An Idea Before Its Time

Have you ever noticed on a map that the eastern shoreline of Africa and the western shoreline of South America fit together like pieces of a jigsaw puzzle (Figure 22.8)? The first person with a detailed hypothesis to explain this was German Earth scientist Alfred Wegener (1880–1930), shown in Figure 22.9. Wegener's hypothesis, known as **continental drift**, states that the world's continents are in motion. They slowly move about the Earth's surface, coalescing and drifting apart to form different configurations over geologic time.

Wegener hypothesized that, at one time, the continents were joined together in a single landmass, a supercontinent that he named *Pangaea* ("universal land"). He proposed that the geological boundary of each continent lay not at its shoreline but at the edge of its *continental shelf*. (The continental shelf is the gently sloping platform between the shoreline and the steep slope that leads to the deep-ocean floor, as described in Chapter 24.) When Wegener tried to match South America and Africa together along their continental shelves, the fit was even better than along their coastlines. Wegener believed that organisms spread or moved about and mingled across the supercontinent of Pangaea, but, at some time later,

Figure 22.9 German naturalist Alfred Wegener (1880–1930). As a young man, Wegener yearned to explore Greenland. While training for this adventure, he established a record for the longest balloon flight (52 hours). Wegener's next passion was for astronomy, and he earned a doctoral degree in the subject. Later, he became a university professor of meteorology and geophysics. He proposed the hypothesis of continental drift, which led to the discipline of plate tectonics—a productive life indeed. In 1930, Wegener died while crossing an ice sheet on an expedition to Greenland.

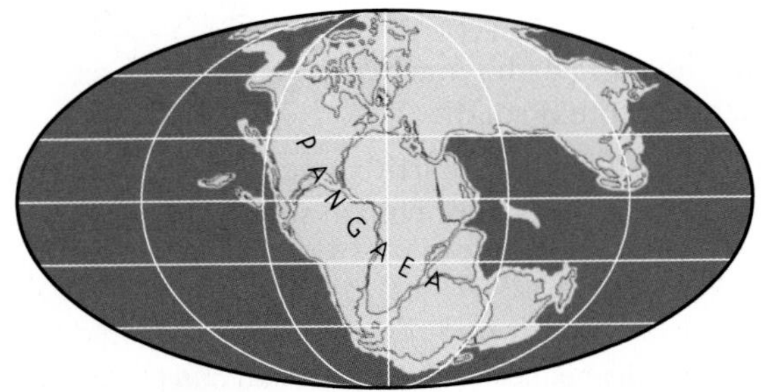

Figure 22.10 This diagram is of Pangaea, the "supercontinent" as it appeared 200 million years ago. Pangaea, like today's configuration of the continents, was but a temporary phase in the ongoing rearranging of the continental crust. The continents have been positioned in many different arrangements and have possessed many different shapes over geologic time; for example, they were joined together as the supercontinent Rodinia before Pangaea existed. Scientists predict that the continents will reconfigure as one supercontinent again in the future.

Figure 22.11 Matching fossils of *Glossopteris* flora from India, Australia, South America, Africa, and Antarctica were cited as evidence for continental drift. The identical plants could not have resulted from parallel evolution on such widely separated regions.

individual continents split from Pangaea and drifted away from one another. (Figure 22.10).

Wegener supported his hypothesis of continental drift with impressive biological, geological, and climatological evidence, which he put forth in his book *The Origins of Continents and Oceans*, published in 1912. Wegener cited fossil evidence—that nearly identical land-dwelling animals and plants had once lived in (what are today) South America and Africa and nowhere else. This was a strange finding because, today, animals and plants of these regions are notable for their striking differences. Also, fossils of the plants *Glossopteris flora*, shown in Figure 22.11, were found in rocks in India, Australia, South America, Africa, and Antarctica. Seeds from the plants were too large to have been distributed by air. How could these plants be the result of parallel evolution (Chapter 17) in such widely separated regions?

To explain the fossils, geologists of Wegener's time proposed that prehistoric land bridges once connected the continents. Yet there was no physical evidence that such land bridges ever existed. Continental drift provided a neat explanation for the fossil findings. The continents had once been a single landmass, with a single community of animals and plants. When the continents later split along the present day Mid-Atlantic Ocean, species were allowed to evolve independently in separate environments.

Wegener also found matching rock types on both sides of the Atlantic. For example, diamond-bearing formations occur in corresponding locations in Africa and South America on opposite sides of the Atlantic Ocean. Further, in many cases, the orientation of folds in mountain chains of the same age are continuous across the oceans—both in North America and Europe and in South America and Africa (Figure 22.12).

Besides the matching of fossils, rock types, and mountain folds, still more evidence supported Wegener's hypothesis. For example, continental drift explained puzzling evidence of ancient ice sheets on regions now located near the equator. When these regions were covered by ice sheets, they were located near the South Pole. Also, the location of young mountains along the edges of the continents was explained by the crumpling of Earth's crust when landmasses collided.

Yet Wegener's hypothesis was generally dismissed in scientific circles. No one, including Wegener, could provide a driving force. How could massive continents of solid rock "drift" across the Earth's surface? What force could be large enough to drive them? How could granitic continents plow through the much harder basaltic ocean floor? These questions put an understanding of continental drift on hold for decades.

CHECK YOURSELF

1. What evidence might lead someone with no understanding of science to suspect that the continents were once connected?
2. Scientists of Wegener's day rejected continental drift because they couldn't imagine how massive, rocky continents could grind through the solid rock of the ocean floor. What did Wegener's contemporaries evidently not know about the mantle?

CHECK YOUR ANSWERS

1. The most obvious evidence is the matching of the edges of the African and South American continents, which can be seen on any world map or globe.
2. Wegener's contemporaries assumed that, if the continents were to move, they would have to push through solid rock. They did not know that the mantle has a plastic layer, the asthenosphere, over which "floating" continents can slide.

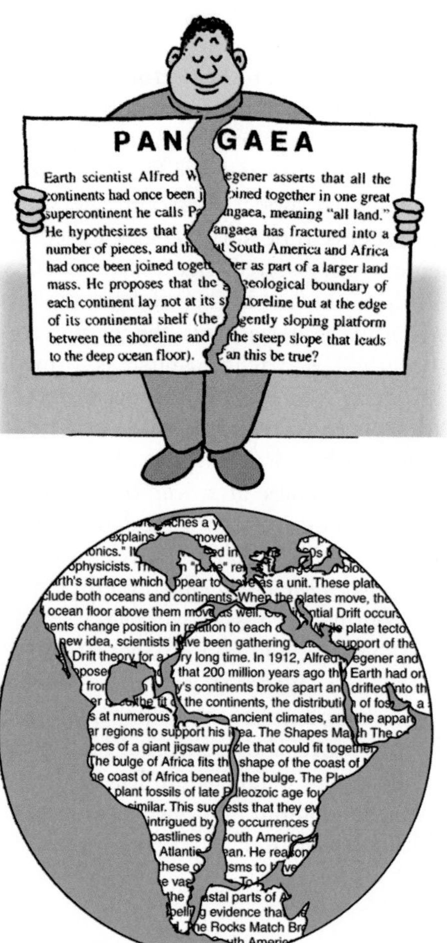

Figure 22.12 Wegener likened the fossil and rock matches to finding two pieces of torn newspaper with matching contours and lines of type. If the edges and the lines of type fit together, he said, the two pieces of newspaper must have originally been one.

22.5 A Mechanism for Continental Drift

Eventually, a mechanism for continental drift was discovered. It came in the 1960s, as a result of seafloor exploration.

During World War II, a mineralogist from Princeton University named Harry Hess was in charge of an attack transport ship. On board, Hess had a *fathometer*, an innovative depth sounder that could map underwater topography. The main purpose of the device was to help a ship maneuver near shore during beach landings. But Hess, a scientist as well as a sailor, continued using the fathometer in the open sea to collect data about the deep ocean bottom. Through his wartime scientific surveying, Hess collected profiles of the ocean floor across the North Pacific Ocean. The depth sounder enabled him to make surveys at greater depths than had been previously accomplished. His findings expanded upon and substantiated other recent discoveries and emerging ideas. The ocean bottom, it turned out, was not flat, with a thick layer of gooey, accumulated sediment, as had previously been thought. Instead, it was etched with deep canyons, trenches, and crevasses. Underwater mountains thrust upward from the seafloor, and there were many curious, flat-topped, submarine volcanoes, which Hess discovered and termed "guyots." Such features could not have formed if the sea bottom were the calm, stable place geologists had previously envisioned. Scientists began to wonder what kinds of geological activity could explain the seafloor's striking features (Figure 22.13).

With the improved technology of the 1950s, oceanographers were able to map the ocean floor in more detail. They made another extraordinary discovery—the Mid-Atlantic Ridge, the longest and tallest mountain range in the world (Figure 22.14). The Mid-Atlantic Ridge, which is 19,312 km long, winds through the center of the Atlantic basin and parallels the American, European, and African coastlines. Its highest peaks emerge above sea level to form oceanic islands, such as Iceland and the Azores. In the center of the ridge and all along its length, there is a valley or canyon—a *rift*. The Mid-Atlantic Ridge was subsequently found to be only one of many mid-ocean mountain ranges. In fact, a global mid-ocean ridge system winds all around the Earth like the seam on a baseball.

Oceanographers also discovered deep **ocean trenches**, which are long, deep, steep troughs in the seafloor. Ocean trenches were found near continental landmasses, particularly around the edges of the Pacific. It turned out that the deepest parts of the ocean are actually near some of the continents at the deep ocean trenches. And some of the shallowest waters are in the middle of the oceans, around underwater mountains. Oceanographers also discovered volcanic activity and high temperatures in the vicinity of the underwater mountain ranges.

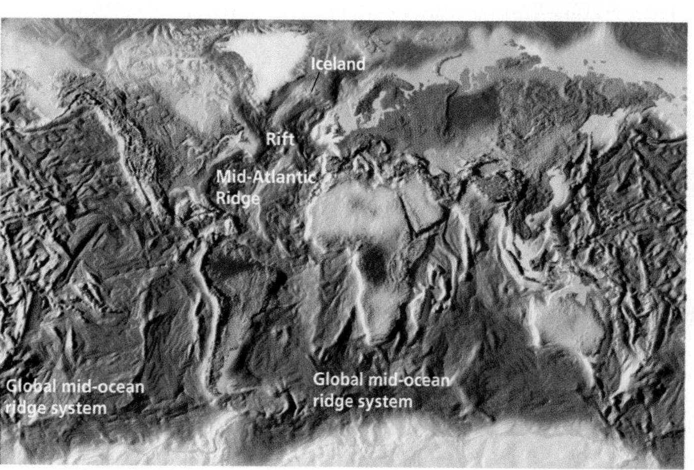

Figure 22.13 The ocean floor is an active geological region, and features canyons, trenches, and mountain chains. It is not the flat, featureless depository of eroded land sediments, as scientists had once believed.

Figure 22.14 The Mid-Atlantic Ridge runs down the center of the Atlantic Ocean. Its highest peaks emerge above the water in several places, creating oceanic islands (such as Iceland, shown here). In this photo, you can see the rift valley.

Seafloor Spreading

After returning from the war, Harry Hess returned to Princeton, put all the new information together, and developed a hypothesis that tied the topography of the ocean floor to Wegener's hypothesis of continental drift (Figure 22.15). As Hess formulated his hypothesis, Robert Dietz, a scientist with the U.S. Coast and Geodetic Survey, independently produced a similar model. Hess and Dietz proposed that the seafloor spreads apart in a process (not surprisingly) called **seafloor spreading.** The seafloor is not permanent, they stated, but instead is constantly being renewed. Mid-ocean ridges form over breaks in the lithosphere where hot material rises from Earth's interior. As rising material from the mantle oozes upward, cools, and collects at the mid-ocean ridge, new lithosphere is formed there. The old lithosphere is simultaneously destroyed in the deep ocean trenches. It works like a conveyor belt. New lithosphere forms at a mid-ocean ridge (also known as a *spreading center*) and older lithosphere is pushed from the ridge crest to eventually be recycled back into the mantle at a deep ocean trench (Figure 22.16). Can you see that seafloor spreading suggests a mechanism for continental drift? As new lithosphere forms and old lithosphere moves away from the rift, the seafloor widens, and therefore pushes the continents in a direction away from the ridge. If the seafloors spread, continents must move.

Hess published his ideas about seafloor spreading in a paper titled "History of Ocean Basins" in 1962. But Hess, like Wegener, encountered resistance because his ideas were elegant but too new and unproven to be widely accepted. But, the following year, scientists from Oxford University tied Hess's ideas together with "magnetic stripes"—evidence from magnetic studies of the seafloor. Their arguments convinced the scientific community once and for all that seafloor spreading and continental drift must occur.

Unifying Concept

The Principle of Falsifiability

Section 1.4

INSIGHT *Guyots*, the curious, flat-topped, submarine volcanoes first discovered by Harry Hess, fit into the seafloor-spreading hypothesis. Guyots were once active volcanoes that rose above the water's surface. They were worn flat on top by erosion while they were exposed to the atmosphere. As the seafloor spread it carried the guyots with it, moving them into deeper and deeper water.

CHECK YOURSELF

1. If the seafloor is spreading and the continents are moving, why don't people notice it?
2. Okay, you say, seafloor spreading explains why the continents move. But why does the seafloor spread? Haven't we just replaced one mystery (continental drift) with another mystery (seafloor spreading)?

CHECK YOUR ANSWERS

1. The motion occurs so slowly in comparison with human lifetimes that it is not perceived. The oceanic crust is moving away from the Mid-Atlantic Ridge at an average

Figure 22.15 This photo, taken in 1960, shows Harry Hess, who hypothesized that the seafloor spreads away from great rifts in the seafloor as new lithosphere is created there. His hypothesis helped establish a mechanism for continental drift.

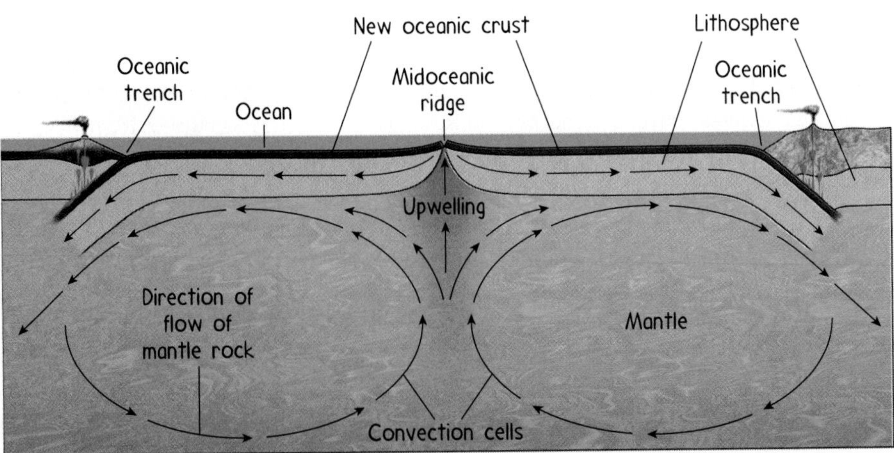

Figure 22.16 In conveyor-belt fashion, new lithosphere is formed at mid-ocean ridges (also called "spreading centers"), as old lithosphere is recycled back into the asthenosphere at a deep ocean trench.

speed of 5 cm/year. This means that, in 50 years, each side of the seafloor will have moved about (5 cm/year × 50 years) = 2.5 meters, or about 8 feet. This rate, about as fast as your fingernails grow, is too slow to produce effects that anyone might notice in their daily life.

2. Good observation—yes! Seafloor spreading doesn't explain the fundamental reason for continental drift. Fundamentally, the seafloor spreads because of the heat of the Earth's interior. In the Integrated Science feature on page 516, "What Forces Drive the Plates," you will learn *why* the seafloor spreads and thus *why* the continents drift.

INSIGHT *Paleo* means old or ancient. As a prefix, *paleo* is used to describe things that occurred in the past. For example, *paleoclimate* refers to ancient climates. *Paleomagnetism* is the study of ancient magnetic data, such as the history of Earth's magnetic reversals.

Magnetic Stripes

Recall, from Chapter 7, that Earth behaves as if it were a huge magnet with magnetic north and south poles near its geographic poles. Earth's magnetic poles are not stationary but wander slowly around the geographic poles. Once in a great while, the magnetic poles flip—the north and south poles exchange positions. This is called a *magnetic reversal.* Earth's magnetic field has been oriented in the same way for the past 700,000 years, but evidence shows that it's due for a reversal. There have been more than 300 magnetic reversals during the past 200 million years. (No one knows for sure how the magnetic poles reverse, when the process occurs, how long it takes, and what the effects on Earth's surface might be. For these reasons, magnetic-pole reversal has been called "the greatest unanswered question in the geological sciences.")

The lava that erupts at mid-ocean ridges is rich in iron (Fe). It contains small crystals of the mineral magnetite (Fe_3O_4). Magnetite crystals, as the name suggests, are magnetic—they can orient themselves with respect to an external magnetic field. When they are contained within fluid lava, these crystals are free to turn and align with the Earth's magnetic field. Like tiny compass needles, magnetite crystals move so that they point north and south, parallel to the field set up by Earth's magnetic poles. But, when the lava cools and solidifies, the magnetite crystals freeze in place. Thus, the alignment of the crystals becomes "locked in" when the lava solidifies as basaltic oceanic crust. So the basaltic rocks of the seafloor hold a record of Earth's magnetic field at the time they cooled, showing both the direction from the rock's location to the poles and the field's polarity.

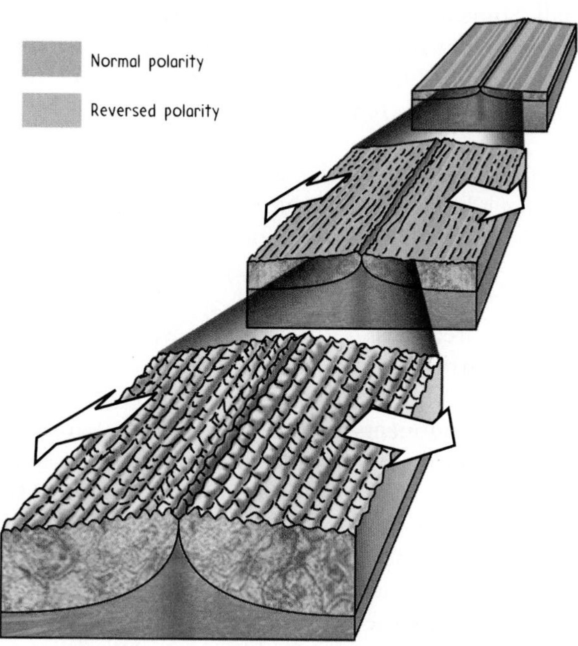

Figure 22.17 As new material is extruded at an oceanic ridge (spreading center), it is oriented according to the existing magnetic field. Magnetic surveys show alternating strips of normal and reversed polarity paralleling both sides of the rift area. Like a very slow magnetic-tape recording, the magnetic history of the Earth is thus recorded in the spreading ocean floors.

As scientists began to use magnetic instruments to survey the seafloor in the 1950s, they began to recognize odd patterns in its magnetic character. When magnetic polarity is mapped over a wide area, the ocean floor appears striped in a zebralike pattern. Alternating bands of rock are laid out symmetrically on either side of mid-ocean ridges. One stripe has "normal" (present-day) polarity; the band next to it has "reversed" (past-era) polarity. The youngest rocks at the ridge center are always observed to have normal polarity. The overall banded pattern of normal and reversed polarities have come to be known as **magnetic striping** (Figure 22.17).

Since the dates of pole reversal can be determined through radiometric dating (Chapter 10), the magnetic pattern of the spreading seafloor documents both the age of the spreading seafloor and the rate at which it spreads.

22.6 Plate Tectonics

Plate tectonics is the unifying theory that explains the dramatic, changing surface features of the Earth. The theory of plate tectonics has done for geology what the theory of evolution did for biology. As geology writer John McPhee once said, "The whole world suddenly makes sense."

Plate tectonics states that the Earth's rigid outer shell, the lithosphere, is divided into eight large pieces, plus a number of smaller ones. Each piece of lithosphere is called a **tectonic plate** or, simply, a **plate** (Figure 22.18). Plates are up to 100 km thick. Being slabs of lithosphere, tectonic plates consist of the uppermost

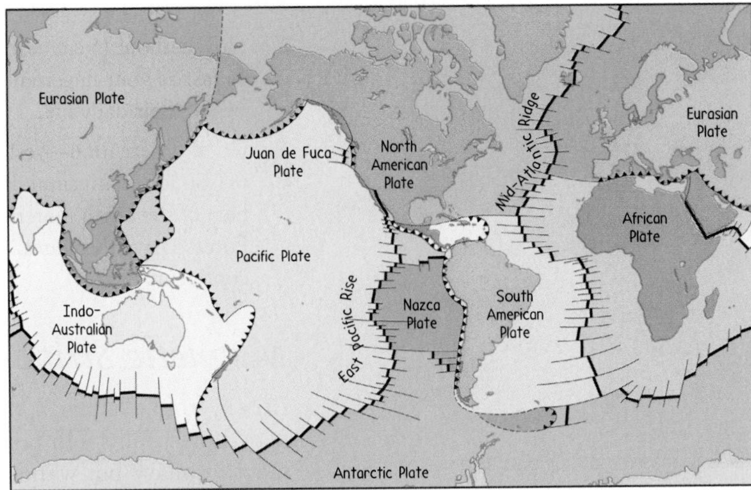

Figure 22.18 Earth's rigid outer shell, the lithosphere, is a mosaic of eight major plates plus a number of smaller ones. The major plates are shown here.

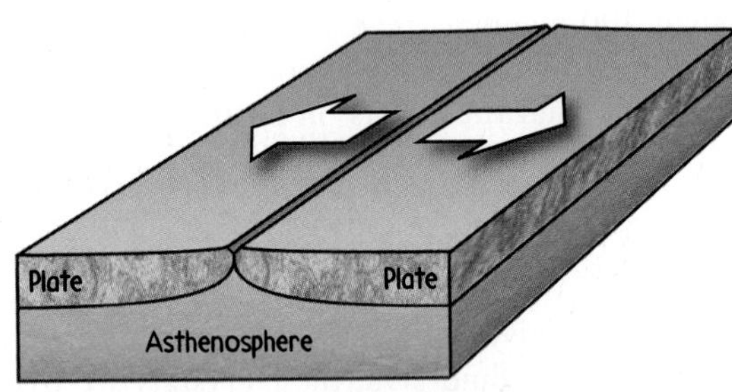

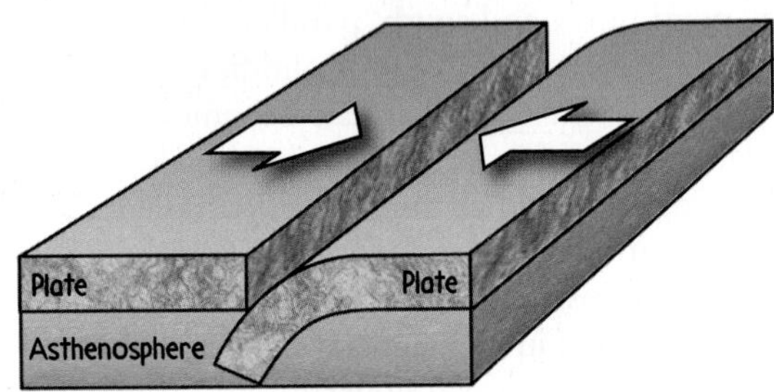

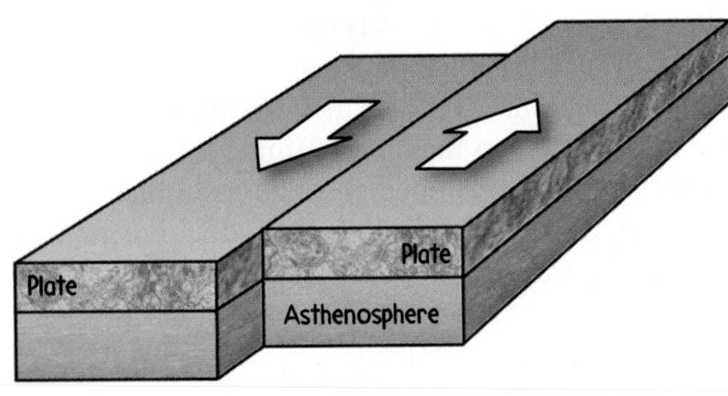

Figure 22.19 Plates interact at their boundaries, which results in intense geologic activity in those areas.

mantle plus the crust. The plates ride atop the plastic asthenosphere below. Some plates are large, others are small. Some are relatively immobile, while others are more active. Note that it is the plates that shuffle about Earth's surface. The continents only move because they are embedded in the plates. So we can see that Wegener's hypothesis was a bit incomplete, for whole plates are drifting—not merely the continents.

The continents don't have the same boundaries as the plates. The North American Plate, for example, is much larger than the North American continent. The western side of the North American Plate approximately traces the continent's West Coast, but the plate extends eastward halfway across the Atlantic Ocean to the Mid-Atlantic Ridge.

Earth's plates move in different directions at different speeds, ranging from 2 cm to about 15 cm per year. Plates carrying continents, such as the North American Plate, are slow movers, while oceanic plates, such as the Pacific Plate, tend to move much faster. This is because the continents act like heavy barges, with subsurface material—"roots"—extending farther down into the mantle than oceanic plates. Just as a barge dragging its bottom on a rocky river bottom encounters resistance, so do continental plates when they drag their extended bottoms through the asthenosphere. Plates pull apart, crash, merge, and separate from one another over geologic time. Because of all these interactions between plates, the edges of plates, *plate boundaries*, are regions of intense geologic activity (Figure 22.19). While interiors of plates are relatively quiet, most earthquakes, volcanic eruptions, and mountain building occur where plates meet. There are three types of tectonic plate boundaries:

Divergent Boundaries: where plates move away from each other
Convergent Boundaries: where plates move toward each other
Transform-fault Boundaries: where plates slide past each other

INTEGRATED SCIENCE PHYSICS

What Forces Drive the Plates?

The Earth's interior contains a great amount of heat energy, as you know. The Second Law of Thermodynamics tells us that heat naturally moves from a

Unifying Concept

The Second Law of Thermodynamics
Section 6.5

Convection
Section 6.9

The Gravitational Force
Section 5.3

warmer region to a cooler one (Chapter 6). Earth, obliged to follow the laws of physics, transfers heat from its core and mantle to its cooler outer surface. This heat transfer, it is thought, produces plate motion. *Plate tectonics arises because of heat transferring away from the Earth's interior.*

Heat transfers away from the Earth's interior mostly by *convection.* Convection is evident in a pot of boiling water. When water boils, the heated water at the bottom of the pot becomes less dense as it expands. The warmed water buoyantly rises in the pot, then cools as it spreads across the surface. Cooled, dense water sinks to complete the cycle. Such a region of rising and falling fluid currents is called a *convection cell* (Figure 22.20).

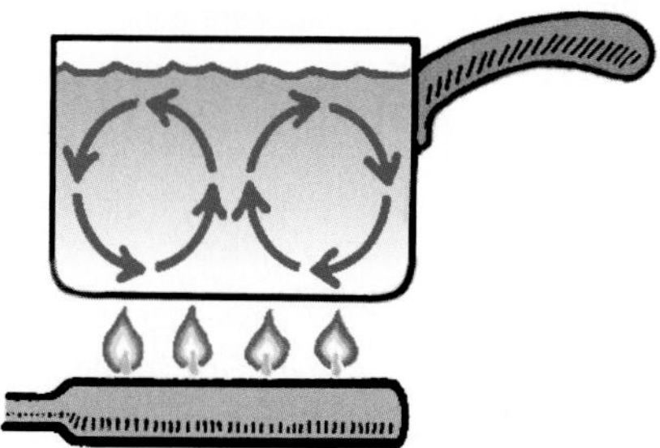

Figure 22.20 Convection cells in a pot of boiling water.

Rocks in the mantle are warm enough to convect, although very slowly. Earth scientists believe that convection cells within the mantle complete one cycle about every 200 million years. This is the time it takes hot rock from the lower mantle to rise to the upper mantle, cool, and then sink. The earth behaves something like a huge spherical stove with burners at its center and mantle rock convecting heat outward. It is not known if convection cells in the lower mantle interact with those in the upper mantle—and, if so, how much—or if there is one set of convection cells that traverses both the lower and upper mantle.

In any case, when the warmed, convecting mantle rocks of the asthenosphere encounter the thin, brittle rocks of the lithosphere, the lithosphere can crack. Under the rift zone of a mid-ocean ridge, lava flows through cracks and the seafloor spreads. The plates move laterally atop the convection cells. Friction acts as the glue, holding lithosphere to asthenosphere. So, summing up what we have said up to this point, thermal convection in plastically yielding asthenosphere rock carries along the lithosphere and is the driving force in plate motion.

Recently, however, some geophysicists have criticized this model. They say gravity plays a primary role in plate motion. The plates may slide, some scientists say, down and outward from elevated mid-ocean ridges like cookies sliding on a tilted cookie sheet. This is the "ridge-push" hypothesis. Another model is the "slab-pull" mechanism, in which gravity pulls the oldest (and therefore coolest and densest) edge of a plate into the Earth at an oceanic trench, where it sinks. The rest of the plate gets pulled along with the leading, sinking edge. The plate is pulled into the asthenosphere like a tablecloth being pulled slowly off a table (Figure 22.21).

The mantle convection model is the oldest and most widely recognized explanation of plate motion. But the ridge-push and slab-pull hypotheses are gaining adherents. Likely all three mechanisms contribute to plate motion. Geophysicists are currently testing these ideas to determine which is the dominant force that drives the plates.

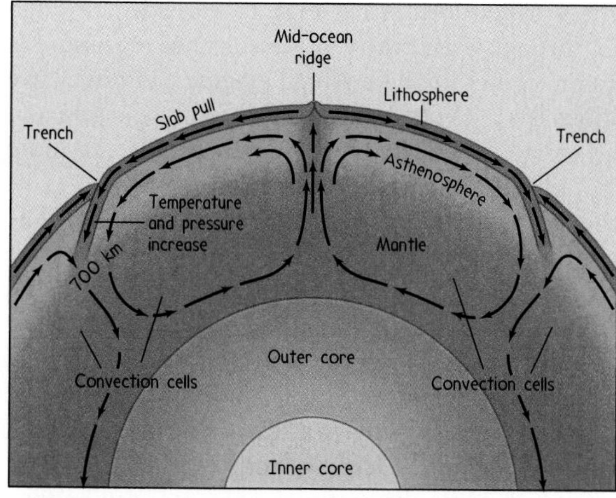

Figure 22.21 Simplified view of convection cells within the mantle. The descending part of the plate—the "slab"—heats up as temperature and pressure rise deeper in the mantle. At a depth of about 700 km, the slab softens and flows, merging with other warm rock of the asthenosphere.

CHECK YOURSELF

1. Why does the asthenosphere flow?
2. Which is a more correct summary of forces that drive tectonic plates:
   a. Plates move as gravity pulls dense slabs of lithosphere into the Earth.
   b. Plates move because they gravitationally slide away from mid-ocean ridges.
   c. Plates move because they are carried along with the convecting mantle.
   d. All of the above.

CHECK YOUR ANSWERS

1. The asthenosphere flows because it is convecting heat from the Earth's interior toward the surface.
2. Statement *a* describes the slab-pull model of plate motion; statement *b* summarizes the ridge-push model; statement *c* is true but incomplete, since it leaves out the contribution of gravity to plate motion. Statement *d* gets it right.

TECHNOLOGY

Direct Measurement of Continental Drift

Today, continental drift is not just deduced from evidence—it can be directly measured. The *Very Long Baseline Interferometry System* (VLBI), for example, was the first system to directly measure the relative motion of Earth's tectonic plates and continents. The VLBI used radio telescopes to detect and record radio signals emitted from quasars. Quasars are so far from Earth (billions of light-years away) that they are virtually pointlike. Their radio emissions, therefore, can be used like a surveyor's beam from a stationary source. The same signal from a quasar arrives at slightly different times at different measuring sites. So, when the VLBI tracked changes in the arrival times of radio signals over a period of years, it showed the rate of movement of the sites relative to each other.

The Global Positioning System (GPS) is currently used to measure the relative motion of different points on Earth. Because GPS results agree with the VLBI results, they provide a cross-check. The GPS system consists of twenty or so satellites that orbit the Earth at an altitude of 20,000 km. These satellites transmit signals back to Earth continuously. Scientists at ground stations around the world use the signals to pinpoint their position in terms of latitude, longitude, and altitude. Scientists repeatedly measure locations of ground stations, monitor change in their relative positions, and thus track continental movement.

The different measurements of continental drift agree with one another and with theoretical predictions. For example, results show that Hawaii is moving in a northwesterly direction toward Japan at a rate of 8.3 centimeters per year. Maryland is moving away from England at a rate of 1.7 centimeters per year.

GPS satellites are used to measure continental drift directly

22.7 Divergent Plate Boundaries

The regions where neighboring plates are moving away from one another are **divergent plate boundaries**. Here, new crust is created as lava fills in the widening gap between diverging plates. Plates move apart and volcanic mountains form as lava erupts, then cools and accumulates.

Lava erupting at a divergent boundary comes from rock in the asthenosphere that has partially melted. Recall that the asthenosphere consists of plastic but solid rock. However, when plates separate, the overlying weight on the asthenosphere decreases, so pressure there is reduced. Rock in the asthenosphere then partially melts, becoming magma. **Magma** is liquefied, molten rock. When magma emerges at the Earth's surface, it is called *lava*. Lava may be extruded through fissures in the Earth's surface or through a central vent, a volcano. (A fissure is simply a crack extending far into the planet through which magma can travel. We are better acquainted with eruptions from a volcano because they are exciting and dangerous, but the outpourings of magma from fissures are actually far more common—the rift at a mid-ocean ridge is a fissure, for example.

Divergent boundaries in the ocean floor produce seafloor spreading. The Mid-Atlantic Ridge is a divergent boundary, dividing the North American Plate and the Eurasian Plate in the North Atlantic and the South American Plate and the African Plate in the South Atlantic. The rate of spreading at the Mid-Atlantic Ridge ranges between 1 and 6 cm per year. However slow this spreading may seem, over geologic time, the effect is tremendous. Over the past 140 million years, seafloor spreading has transformed a tiny waterway through Africa, Europe, and the Americas into the vast Atlantic Ocean of today.

Divergent boundaries don't occur only on the seafloor. They can also happen in the middle of a continent, and, when they do, the continent tears apart. The East African Rift Zone is an example. If spreading continues there, the rift valley will lengthen and deepen and eventually extend out to the edge of the present-day African continent. The African continent will separate, and the rift will become a narrow sea as the Indian Ocean floods the area. The easternmost corner of Africa will then become a large island. Indeed, Earth is dynamic!

MATH CONNECTION

Calculate the Age of the Atlantic Ocean

It's easy to calculate the age of the Atlantic Ocean, if you can reasonably estimate the rate of seafloor spreading and the present width of the ocean. The Atlantic Ocean is currently about 7,000 km or 7×10^8 cm wide. We assume that the average rate at which the plates diverge in the Atlantic Ocean is 5 cm per year, and we make the assumption that this rate has been constant over geologic time. We then apply the familiar equation that relates speed, time, and distance:

$$\text{Time} = \frac{\text{distance}}{\text{speed}} = \frac{7 \times 10^8 \text{ cm}}{5 \text{ cm/year}} = 1.4 \times 10^8 \text{ years} = 140 \text{ million years}$$

Based on these estimates, the Atlantic Ocean is about 140 million years old.

Problem The Red Sea is presently a narrow body of water located over a divergent plate boundary. Based on current studies of the rate of seafloor spreading, the Red Sea will be as wide as the Atlantic Ocean in 200 million years. In order for this to be true, how fast is the seafloor spreading away at this divergent boundary?

Solution If we take the current width of the Red Sea as negligible, the seafloor will widen by a distance of 7×10^8 cm in 200 million years. The speed at which the seafloor spreads is then:

$$\text{Speed} = \frac{\text{distance}}{\text{time}} = \frac{7 \times 10^8 \text{ cm}}{2 \times 10^8 \text{ years}} = 3.5 \text{ cm/yr}$$

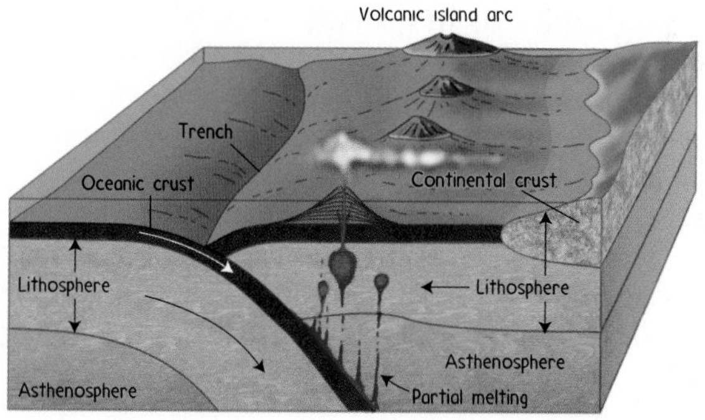

Figure 22.22 Oceanic–oceanic convergence occurs when plates that do not carry continental crust at their leading boundaries meet. A trench forms in the subduction zone, and an island arc of volcanoes grows from magma that is generated in the collision.

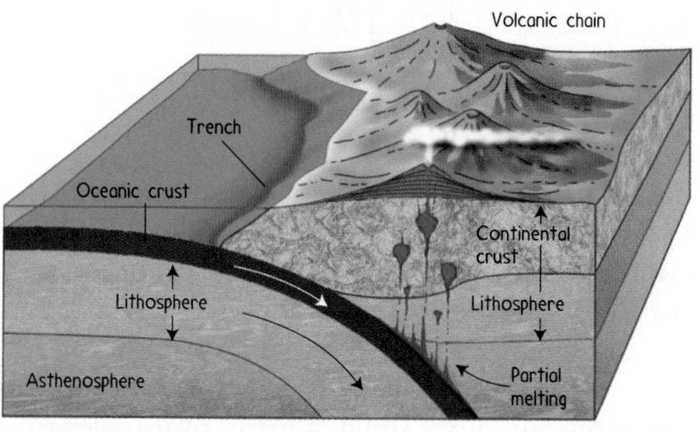

Figure 22.23 Oceanic–continental convergence occurs when a plate with oceanic crust at its leading edge is subducted beneath a plate with continental crust along its leading edge. A deep ocean trench and a coastal mountain range form as a result.

22.8 Convergent Plate Boundaries

Earth has remained more or less its present size since it formed about 4.5 billion years ago. This means that lithosphere must be getting destroyed about as fast as it is created. The destruction of Earth's lithosphere takes place at **convergent boundaries**. Here, plates come together in slow-motion collisions. Usually, one plate *subducts*, or descends, below another. The area around a subducting plate is called a **subduction zone.**

Plates converge in three different ways, depending on the kind of lithosphere that is involved. The three kinds of convergent plate boundaries are:

- Oceanic–oceanic convergence. (Figure 22.22)
- Oceanic–continental convergence. (Figure 22.23)
- Continental–continental convergence. (Figure 22.24)

In *oceanic–oceanic convergence*, both plates have an oceanic leading edge. When these plates meet, the older (and therefore cooler and denser) oceanic plate slides beneath the younger and less dense oceanic plate. At the convergent boundary, a deep ocean trench forms. So, ocean trenches mark active subduction zones (where one plate descends beneath the other). Ocean trenches run parallel to the edges of convergent boundaries. The Marianas Trench, for example, shows where the Pacific Plate hits the slower-moving Phillipine Plate. Trenches are the deepest places on Earth, inhabited by strange organisms called *extremophiles* that can survive at extreme high pressure, low temperature, and no light. Reaching 11,000 m (7 mi) below sea level, the Marianas Trench is the deepest. If the world's tallest mountain, Mt. Everest, were sunk to the bottom of the Marianas Trench, there would still be more than a mile of ocean above it!

Investigation of Figure 22.22 shows that mantle rock in the subduction zone partially melts to form magma. It buoyantly rises and erupts at the surface as lava. Over millions of years, the erupted lava and volcanic debris accumulate on the ocean floor until they grow tall enough to poke above sea level and form a volcanic island.

Volcanoes of this kind are grouped together in *island arcs* that parallel the trenches, such as the Aleutian Islands off the Alaskan Peninsula. Strong-to-moderate earthquakes are common along such boundaries as the descending plate sticks and slips by the overriding plate.

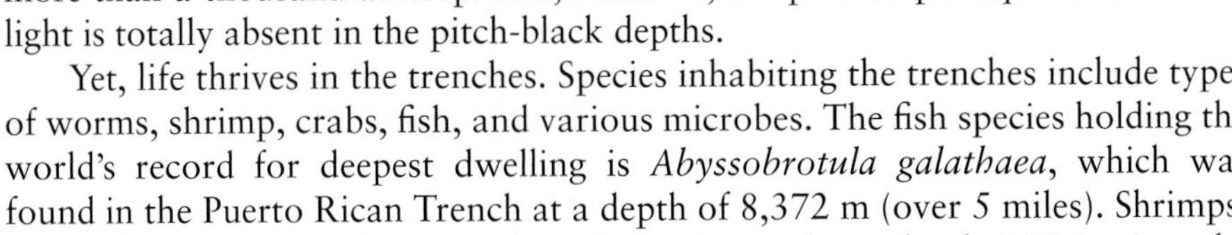

INTEGRATED SCIENCE BIOLOGY

Life in the Trenches

Conditions in the trenches are extreme. Water is nearly frozen. Pressure reaches more than a thousand atmospheres, some 16,000 pounds per square inch. Sunlight is totally absent in the pitch-black depths.

Yet, life thrives in the trenches. Species inhabiting the trenches include types of worms, shrimp, crabs, fish, and various microbes. The fish species holding the world's record for deepest dwelling is *Abyssobrotula galathaea*, which was found in the Puerto Rican Trench at a depth of 8,372 m (over 5 miles). Shrimps, scale worms, and sea cucumbers have been observed at a depth 3,326 m into the Marianas Trench.

How is it possible that life thrives in the trenches? The mechanisms are not well known, but several survival strategies have been uncovered that could explain how trench dwellers adapt to the pressure extremes. For example, trench dwellers' cell membranes tend to have more fat molecules (polyunsaturated fatty acids and phospholipids), providing greater tolerance to high pressures.

The problems of extreme cold and darkness are mitigated by the occurrence of *hydrothermal vents*. Hydrothermal vents are fissures in the seafloor. Most often, they are found in the newly formed oceanic crust in active rift zones of the global midocean ridge. However, some hydrothermal vents, which form as plates grind past one another, are located in the ocean trenches. Hydrothermal vents release heat, as well as hydrogen sulfide and other chemicals, from the Earth's interior. *Barophilic* (pressure-loving) microbes consume these chemicals and produce oxygen as a waste product in a process known as *chemosynthesis* (Chapter 15). Other organisms use the waste oxygen for respiration and consume the microbes as food. The organisms that consume the oxygen-producing microbes are, in turn, eaten by other organisms. Thus, a food chain exists in the trenches despite the lack of light for photosynthesis (Figure 22.25). (See Chapter 15 for more on photosynthesis.)

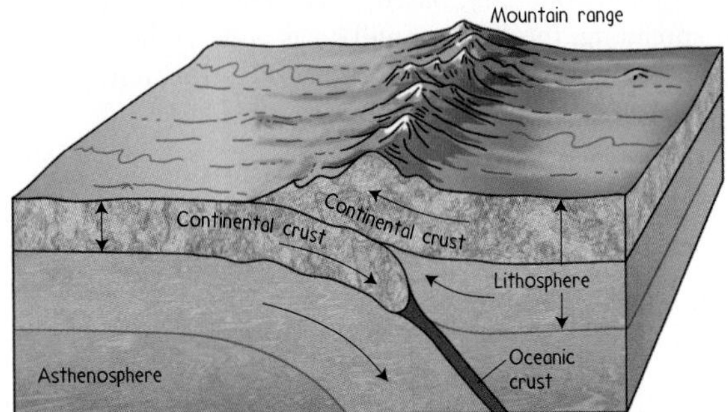

Figure 22.24 Continental–continental convergence occurs when continental crust caps the leading edge of each colliding plate. Mountains form where crust wrinkles and pushes upward.

Figure 22.25 The black anglerfish has a huge jaw and teeth that are well suited for ultraeffective predation, since food is scarce in the trenches. This fish also has a bioluminescent "lure" attached to its head to attract prey to within easy reach. Microbes that glow in the dark produce the light in the lure.

CHECK YOURSELF

1. What physical conditions prevail in the trenches that seem to make life impossible there?
2. Where does the heat that escapes from hydrothermal vents originate?

CHECK YOUR ANSWERS

1. Extreme cold, high pressure, and lack of light.
2. The heat comes from the convecting mantle; and much of this heat comes from the core.

A second kind of plate collision is *oceanic–continental convergence* (Figure 22.23). In this case, a plate with a continental leading edge slowly collides with a plate with an oceanic leading edge. Being denser, the basaltic oceanic plate subducts beneath the less dense, granitic continental plate. A deep ocean trench forms offshore where the converging plates meet. As mantle rock partially melts, magma forms in the subduction zone and rises up and erupts at the surface as lava. Lava erupts, cools, and accumulates many times over time and gives rise to a volcanic mountain chain. As more magma comes up from below, the volcanoes grow.

But why does rock partially melt in the subduction zone to form magma? It's not because the subducting plate is hot. Subducting lithosphere, remember, is cool and dense—it's relatively cold rock that sinks down into the asthenosphere.

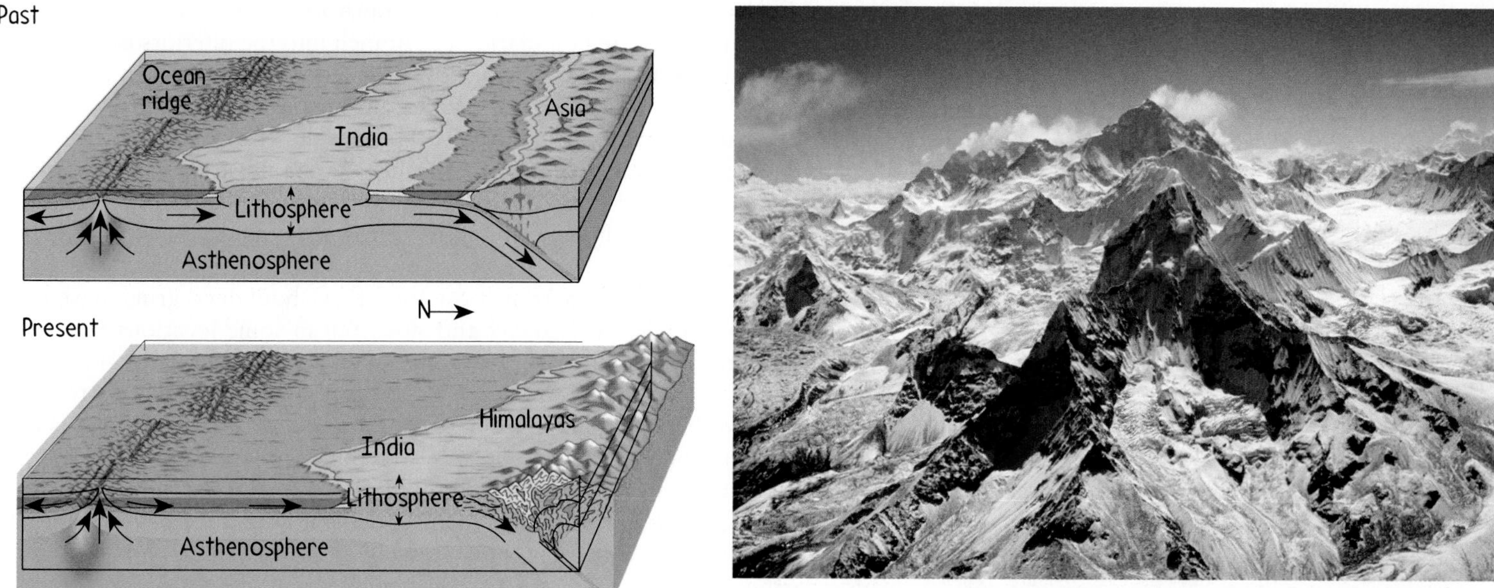

Figure 22.26 The continent-to-continent collision of India with Asia produced—and is still producing—the Himalayas.

When a descending plate reaches a depth of 100–150 km, the heat and pressure of the surrounding environment drive water trapped in the subducting plate into the overlying mantle. The water acts like salt on ice or like flux at a foundry—it lowers the freezing point of the mantle rock. So, with the injection of water, the overlying mantle partially melts without changing temperature. Magma is generated and buoyantly rises. Often, it pools beneath the continental crust, where it may partially melt some of the neighboring crustal rocks. Eventually, this molten rock migrates to the surface where it erupts—either calmly or explosively. The residents of the Pacific Northwest should appreciate this process, for it has created the beautiful mountains of the Cascade Range.

Finally, when plates that are converging head-on have continental crust along their leading edges, they display *continental–continental convergence* (Figure 22.24). In the case of continental–continental convergence, the blocks of colliding crust consist of the same type of buoyant granitic rock. Because the blocks of rock have the same density, neither sinks below the other when they collide—there is no subduction. Instead, the continents push one another upward, like crumpled cloth, and towering, jagged mountain chains result. For example, the Himalayas are the highest mountains in the world, rising to a majestic 8,854 m (5.5 miles) above sea level. The Himalayas formed when the subcontinent of India rammed into Asia about 50 million years ago. By welding together along the plate boundary, India and Asia merged as one. The Himalayas are still growing at a rate of about 1 cm per year, as the Eurasian and Indian plates continue to converge (Figure 22.26).

22.9 Transform Plate Boundaries

A **transform plate boundary**, or simply *transform boundary*, is a region where two tectonic plates meet, but rather than converging or separating, they slide past each other. Lithosphere is neither created nor consumed at these boundaries, since plates simply rub along each other as they move in opposite directions.

Transform boundaries are large *faults*. A fault, as you will learn in Chapter 24, is a crack that divides two blocks of rock that have moved relative to one

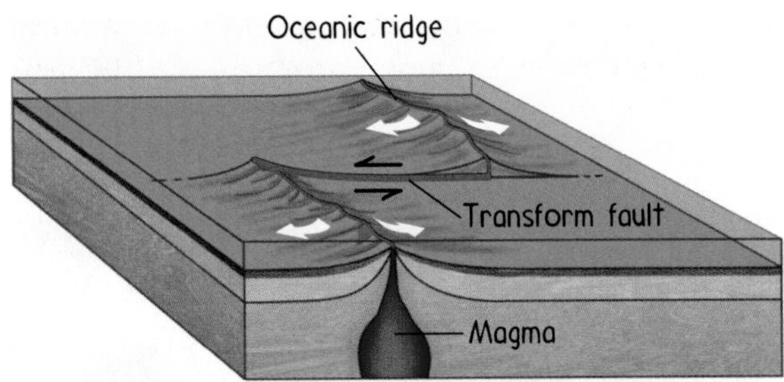

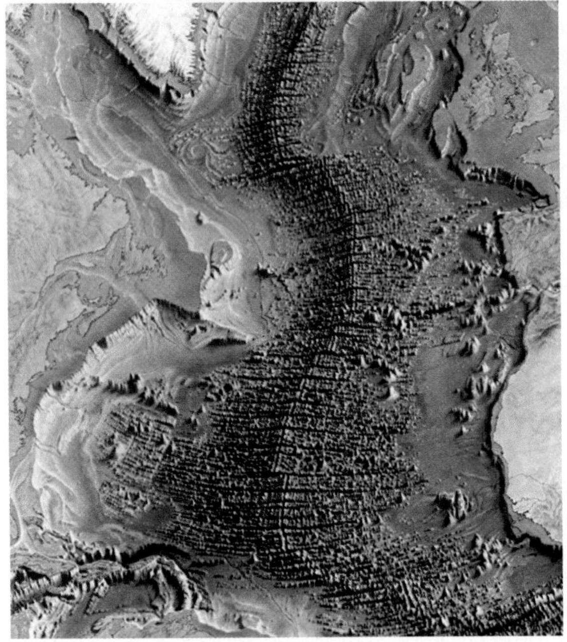

Figure 22.27 Most transform boundaries occur in ocean basins where they offset oceanic ridges. Transform boundaries offset segments of an oceanic ridge as shown, allowing plates to move laterally along Earth's surface. The plates stick and slip as they move along the boundary, causing earthquakes. Transform boundaries offset segments of the Mid-Atlantic Ridge as shown.

another. Faults can be much smaller than transform boundaries; networks of them often form near plate boundaries and branch into the interiors of plates.

Usually, transform boundaries join two segments of a mid-ocean ridge. For example, the Mid-Atlantic Ridge is broken up into segments, which are connected by transform faults (Figure 22.27). These faults "transform" (or transfer) the motion from one ridge segment to another. As the illustration shows, lithosphere at one ridge moves in a direction opposite that of the lithosphere of another ridge. In this way, slippage along the transform boundaries allows the tectonic plates to move.

As the rocky plates at a transform (or other) boundary grind past each other, their motion is usually steady and slow. But in some locations friction is so great that entire sections of rock become stuck against each other. Plate motion continues, and the blocks of "stuck" rock become compressed or stretched. (Recall from earlier in this chapter that, even though rock *seems* brittle, it is not. It is capable of compressing as well as stretching and storing energy like a spring). When the compressional or tensional forces pushing or pulling the rock exceed friction, the stressed rock suddenly breaks loose and slips, releasing its stored energy in a sudden jerk. This sticking-and-slipping of blocks of rock causes earthquakes.

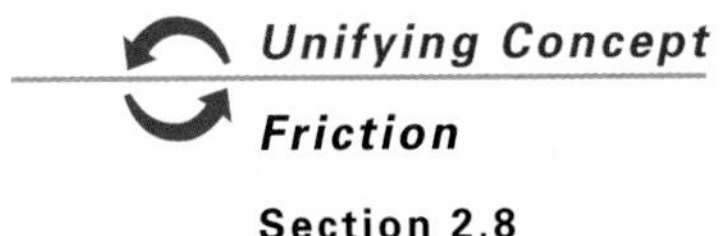

Although most transform faults are located within the ocean basins, a few are found on continental plates. One continental transform fault that gets a lot of attention is the San Andreas Fault, which is shown in the drawing and photo in Figure 22.28. The San Andreas Fault is the thickest thread in a tangle of faults that collectively accommodate the motion between the North American Plate and the Pacific Plate in California.

The slice of California to the west of the fault is slowly moving northwest while the rest of California is moving southeast. Contrary to popular opinion, you can see from the directions of the plates that Los Angeles is not going to split off and fall into the ocean. Instead it will steadily advance northwesterly toward

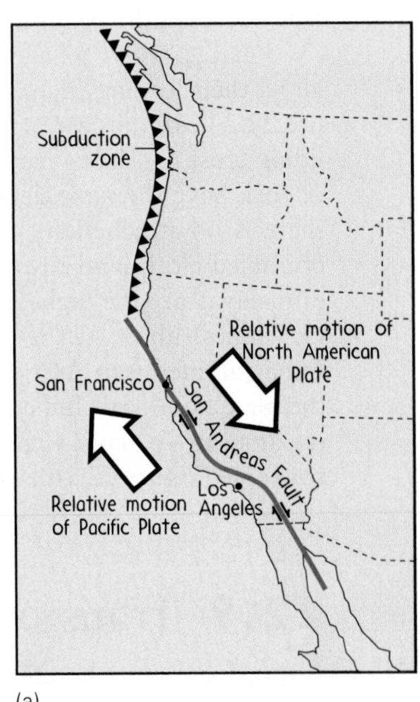

(a)

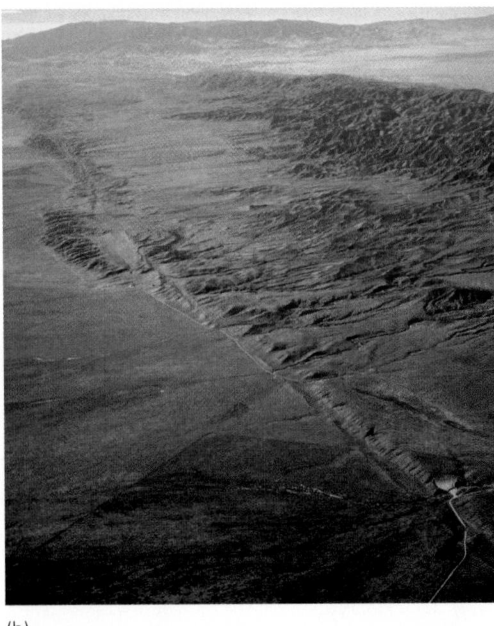

(b)

Figure 22.28 (a) The San Andreas Fault is a transform-fault plate boundary infamous for its earthquakes. The slice of California moving northwesterly lies on the Pacific Plate, while the rest of California sits on the North American Plate. (b) In this photo of the San Andreas Fault, notice the long valley created by many years of rock grinding along the fault.

San Francisco while San Francisco moves in the opposite direction. In about 16 million years the two cities will be side by side. In the meantime, California residents can expect plenty of earthquakes.

22.10 Earthquakes

Earthquakes are as devastating as they are common. More than 700 earthquakes strong enough to cause injury, death, or property damage occur annually. As briefly described, earthquakes are caused by stress (force) applied to rock. The stress causes strain (compression or tension). Rock is deformed until it can no longer bend without breaking. Then it suddenly breaks loose and slips into a new position, thereby releasing stored elastic energy. The released energy travels through the rock as seismic waves. The actual site of the initial slipping rock that generates seismic waves is called the *focus* of an earthquake (Figure 22.29). Seismic waves radiate out in all directions from the focus like sound from a ringing bell. The point at the Earth's surface directly above the focus is called the *epicenter*.

Earthquakes occur all over the world. Most occur along plate boundaries, although *intraplate* earthquakes (which are not well understood) do occur away from plate boundaries. An infamous series of three such quakes occurred in New Madrid, Missouri in the winter of 1811–1812. The first of them occurred just after midnight on December 16. People emerging from their shaking homes that night observed the land rolling in waves up to 1 m high. These intense surface waves opened deep cracks in the ground. The subsequent two quakes occurred within several weeks and were of similar intensity. The New Madrid earthquakes had a major effect on topography—new lakes were formed, the course of the Mississippi River was changed, more than 150,000 acres of forest were destroyed, and the contours of the land were reshaped over a wide area. Fortunately, because the region was sparsely settled at the time, there was little loss to human life and property.

Intraplate quakes aside, most of the world's earthquakes, about 90%, are associated with movement along the boundaries of tectonic plates. In general, mild,

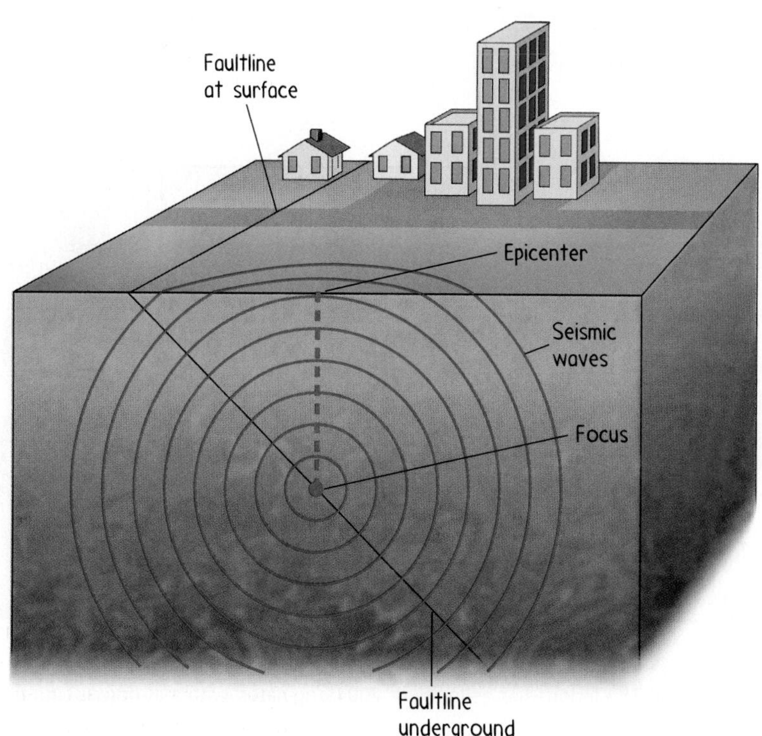

Figure 22.29 When rock suddenly moves, it is usually along a pre-existing fault, though such slippage of rock may actually create a fault. The movement of rock generates seismic waves that move outward from the focus. The point at ground level directly above the focus is called the *epicenter*.

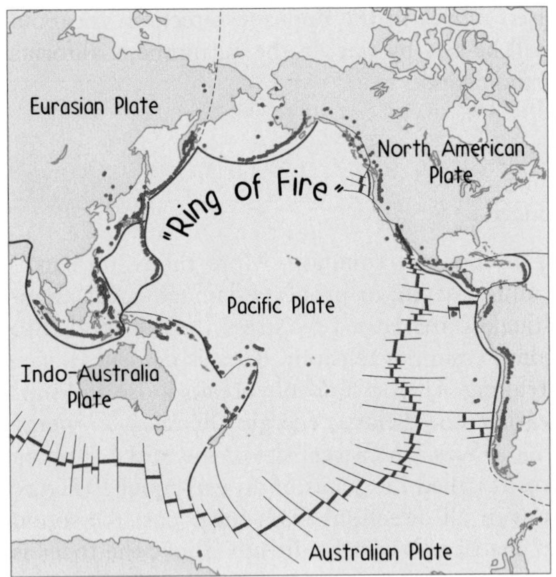

Figure 22.30 The Ring of Fire is a region of intense geologic activity, because plates meet at convergent boundaries all along it. Subduction earthquakes are common. Many volcanoes are found along the Ring of Fire, and they help to define its contours. Volcanoes are shown as red dots on the map.

shallow temblors occur at divergent plate boundaries. Earthquakes along transform plate boundaries are generally mild or moderate. The strongest jolts mostly occur along convergent plate boundaries, where subduction occurs.

Subducting plates grind, stick, and jostle beneath overlying plates. This occurs in the Ring of Fire, a region that encircles much of the Pacific Ocean (Figure 22.30). About 80% of the world's big earthquakes occur in the Ring of Fire. The extent of subduction in the Ring of Fire also produces great volcanic activity—about 75% of the world's volcanoes are located in the Ring of Fire. In Chapter 24, you will learn more about volcanoes, including how they form.

A powerful example of a subduction quake occurred most recently on December 26, 2004. The quake originated beneath the Indian Ocean, off the western coast of Sumatra. This location marks the boundary between the subducting India Plate and the overriding Burma Plate. Because of its size, the Sumatran quake is termed a *megathrust* earthquake. The energy released by this quake is comparable to the amount of energy released by a bomb made up of 100 billion tons of TNT (if it were possible to make such a bomb). Although the quake itself caused severe damage and casualties, the quake's aftermath was even more devastating. With its epicenter beneath the Indian Ocean, the quake generated huge seismic sea waves—a **tsunami**. Coastal areas throughout the Indian Ocean basin were destroyed. This Indonesian tsunami killed more than 184,000 people in 14 countries (Figure 22.31).

INTEGRATED SCIENCE PHYSICS

Anatomy of a Tsunami

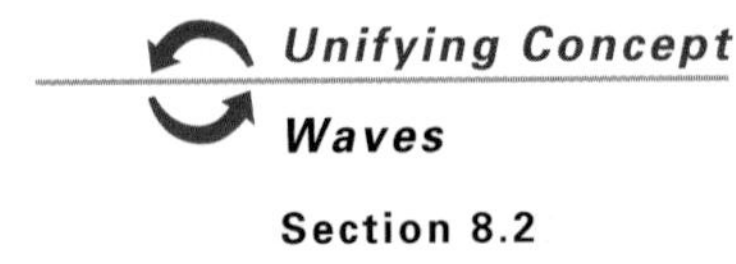

A tsunami is a tremendous seismic sea wave. It has properties typical of other ocean waves, but on a grander scale. Table 22.2 compares the characteristics of a tsunami with those of a typical, wind-driven ocean wave. Note that a tsunami is very fast very long, and is created by a very large disturbance—usually an earthquake.

Figure 22.32 shows how an earthquake in a subduction zone can cause a tsunami—this is the mechanism that caused the devastating Indonesian tsunami of 2004. The part of the seafloor that was connected to the overlying plate got bent as the sinking plate descended. When the stress reached the point where the

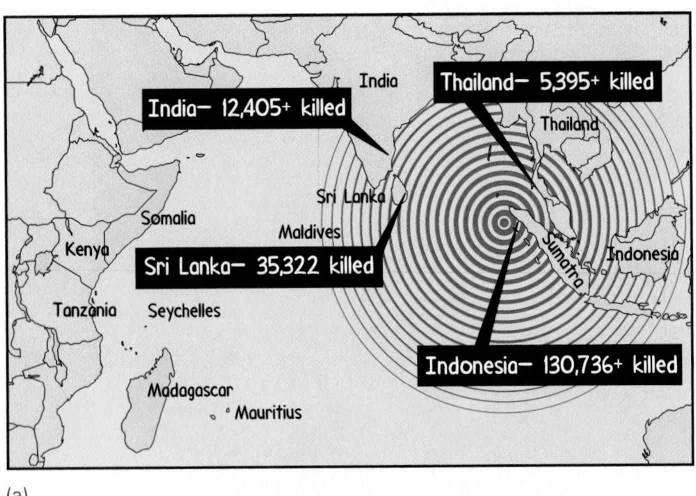

(a)

(b)

Figure 22.31 (a) The Indonesian tsunami of 2004 originated with a strong subduction earthquake off the coast of Sumatra. (b) The tsunami strikes a coastline in Sri Lanka.

Table 22.2 Characteristics of a Tsunami vs. a Typical Ocean Wave

Wave Feature	Ocean Wave	Tsunami
Wave speed	8–10 km/h (5–60 mph)	800–1000 km/h (500–600 m/hr)
Wave period	5–20 s	10–120 min
Wavelength	100–200 m (300–600 ft)	100–500 km (60–300 mi)
Initial Disturbance	Wind	Earthquake, explosion, volcanic eruption, or meteor impact

INSIGHT Modern electronic seismometers operate very much like the earliest devices. In place of a pen, they confine a small suspended mass by electrical forces. As the Earth moves, electronic feedback prevents the mass from moving. The amount of force needed to keep the mass steady is then recorded.

overlying plate could no longer bend without breaking, it snapped and jerked upward. Kilotons of rock thus shot upward like an enormous piston with huge force. The energy of this force was transferred to the water, pushing a column of water above normal sea level—and thus a tsunami was created.

When gravity pulls the initial wave of a tsunami back down to sea level, it and subsequent ripples propagate outward from the disturbance, traveling mostly underwater. A tsunami is usually less than 1 meter high until it gets close to shore. As it nears the shore, shallower water and the shape of the shore compress the tsunami. Like an accordion folding inward, the tsunami's wavelength decreases, but its height above sea level grows considerably. Tsunami wave heights can reach up to 30 m above sea level. When it finally reaches shore, the tsunami is a fast-moving wall of water capable of inflicting great damage. Though people tend to think of water as soft, it is also very heavy. A tsunami on land is like an enormous sledge hammer delivering an enormous blow.

CHECK YOURSELF

1. What is the source of a tsunami's huge energy?
2. How fast does a tsunami move compared with a typical ocean wave?
3. How does the overlying plate in a subduction zone create a tsunami?

CHECK YOUR ANSWERS

1. A massive underwater disturbance, such as an earthquake, a volcanic eruption, or an explosion.
2. A tsunami is much faster—about 10 times as fast.
3. When the overlying plate bends downward and then snaps upward, it acts like a piston, pushing a column of water vertically so that it emerges above sea level as a wave.

Can earthquakes be predicted? No—at least not very well. Scientists can measure the build-up of stress in rocks and so predict that an earthquake is likely within a certain time frame, perhaps a few decades. And history shows that faults

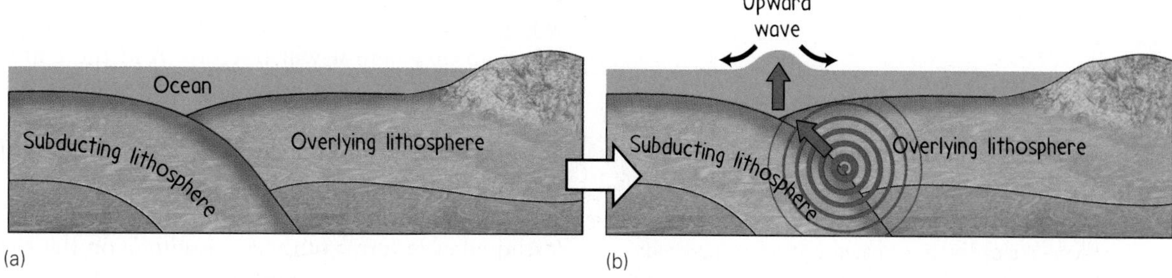

Figure 22.32 Typically, tsunamis are generated by earthquakes in subduction zones. This tsunami is generated as the stressed overlying lithosphere, which is bent downward as it is pulled by the subducting plate (a), suddenly releases, jerks upward, and pushes a column of water toward the surface (b).

MATH CONNECTION

The Richter Scale

Invented in 1935 by Charles F, Richter of the California Institute of Technology, the Richter Scale is *logarithmic* to accommodate the wide variation in earthquake magnitudes. Each one-point increase on the Richter scale indicates a ten-fold increase in an earthquake's ability to shake the ground. A magnitude 7.0 earthquake shakes the Earth 10 times more severely than a magnitude 6.0 earthquake but 100 times more severely than a magnitude 5.0 earthquake.

Note that the Richter scale measures shaking, not the total energy of the vibrations. There is a simple relationship, however, between shaking and energy. A one-point increase on the Richter Scale corresponds approximately to a 30-fold increase in the earthquake's energy. Thus, a 6.3 magnitude earthquake releases 30 times more energy than a 5.3 earthquake. The graph shows the energy released by various earthquakes and relates that released energy to Richter-scale magnitudes.

Problems

1. How many times more violent is the shaking of a magnitude 8.0 earthquake than that of a magnitude 5.0 earthquake?
2. How much more energy does a magnitude 7.5 earthquake release than a magnitude 6.5 earthquake? Than a magnitude 5.5 earthquake?

Solutions

1. 1000 times $(10 \times 10 \times 10)$
2. 30 times; 900 times (30×30)

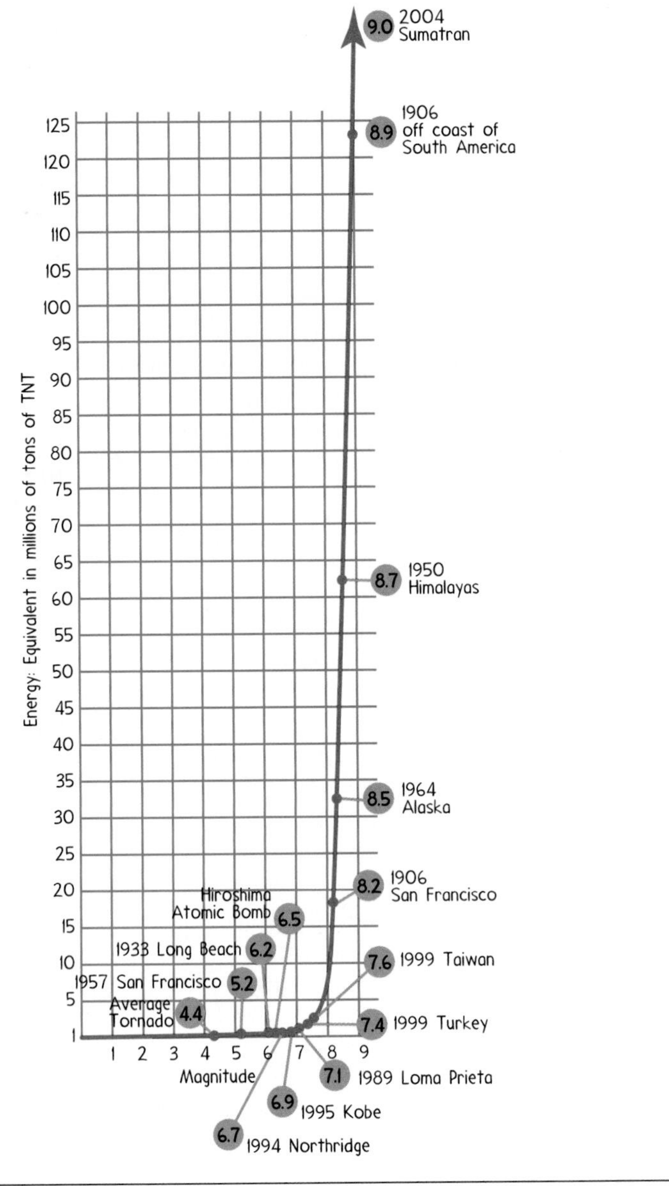

This graph shows the energy released by some earthquakes over the past century in terms of the amount of dynamite that would produce an explosion of the same size. Note that, since every point increase on the Richter scale corresponds to a 30-fold increase in the earthquake's energy, the energy released by an earthquake rises steeply as its Richter magnitude increases.

along plate boundaries are the most vulnerable places. But this is much different than being able to say specifically when and where an earthquake will occur. It's a bit like bending a twig slowly until it breaks. As you increase the stress on the twig you increase the likelihood of breakage, but you don't know exactly when it will snap or where it will fracture. And there are even more variables involved when it comes to trying to predict an earthquake—many more than predicting a twig quake!

There are several ways to assess the strength of an earthquake. For example, the Modified Mercalli Intensity Scale is used to quantify the damage from a quake at a specific location. However, the media generally reports the sizes of earthquakes in terms of their *magnitude* on the *Richter Scale*. Richter magnitude is a measurement of how much the ground shakes during a quake.

The instrument used to measure magnitude is called a *seismometer* (Figure 22.33). The traditional seismometer is basically just a needle or a pen suspended over a moving sheet of paper. It steadily traces out a straight line. But when the earth shakes, the paper jiggles relative to the pen, so the line on the paper

INSIGHT The magnitude of an earthquake generally relates to how big the area of ruptured rock is. If the patch of rock that breaks is a few square miles in area, a magnitude 5.0 earthquake may be the cause. If the patch is, say, two hundred square miles in area, the earthquake would likely register a 7.0 on the Richter scale. And if the patch that breaks has an area of a few thousand square miles, a devastating earthquake of magnitude 8 or higher would likely result.

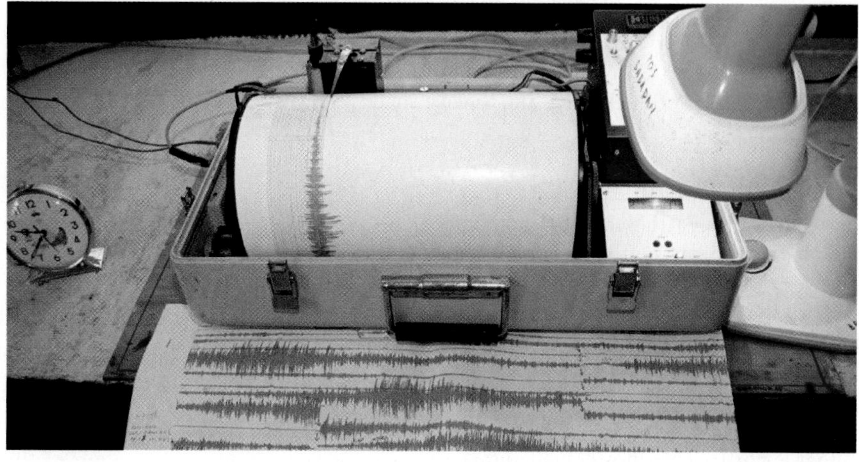

Figure 22.33 The standard seismograph correlates the shaking of the ground with the magnitude of an earthquake.

becomes wavelike. The stronger the earthquake, the harder the shaking and the larger the amplitude of the wave drawn on the seismometer.

Earthquakes vary greatly in magnitude. Weak earthquakes are very common, while large-magnitude earthquakes are fortunately rare. An earthquake with a magnitude of 3.0 on the Richter Scale goes usually unnoticed by people near the epicenter. The earthquake that destroyed much of San Francisco in 1906 measured 8.3, and there have been few earthquakes in recorded history with magnitudes reaching 9.0.

CHAPTER 22 Review

Summary of Terms

Asthenosphere The most plastic layer of the mantle, which flows slowly over time.

Continental drift The hypothesis that continents are in motion and that they travel over the face of the Earth rather than remaining fixed in one location.

Convergent plate boundary A place where neighboring plates move toward each other; old lithosphere is destroyed here.

Core The central, metallic, spherical layer of the Earth, consisting of a solid inner part and liquid outer part.

Crust The thin, rocky surface layer of the Earth.

Divergent plate boundary A place where neighboring plates move away from each other; new lithosphere is created here.

Earthquake An earthquake is the shaking or trembling of the ground that results when rock under Earth's surface moves or breaks.

Lithosphere A layer of rigid rock comprising the crust and upper mantle of the Earth.

Magma Subsurface, molten rock found mainly in the upper mantle or lower crust.

Magnetic striping The symmetrical pattern of alternating normal and reversed magnetic polarities found in the seafloor on either side of a spreading center.

Mantle The intermediate rocky layer of the Earth, which occupies most of the Earth's volume and carries most of its mass.

Ocean trench A long, narrow, and deep depression in the seafloor.

Plate A moving section of lithosphere; a "tectonic plate."

Plate tectonics The comprehensive geological theory stating that Earth's outer layer is composed of lithospheric plates that float on, and move along with, the relatively plastic asthenosphere.

Seafloor spreading The process by which new lithosphere forms at a mid-ocean ridge.

Seismology The study and measurement of seismic waves, called seismology, has provided most of what is known today about Earth's interior.

Subduction zone The area around a convergent plate boundary where one lithospheric plate descends beneath another, returning to the interior of the Earth.

Tectonic plate Each piece of lithosphere is called a tectonic plate or, simply, a plate.

Transform plate boundary A plate boundary at which the neighboring plates slide past each other, neither converging or diverging; lithosphere is neither created or destroyed.

Tsunami With its epicenter beneath the Indian Ocean, the quake generated huge seismic sea waves—a tsunami.

Review Questions

22.1 Earth Science Before the Twentieth Century

1. True or false: Much was known about the Earth's internal structure before the twentieth century. Justify your answer.
2. What contribution did Isaac Newton make to Earth science? Leonardo da Vinci? James Hutton?

22.2 Using Seismology to Look Inside the Earth

3. What are the major compositional layers of the Earth? Name these layers.

4. Cite the seismic evidence that the Earth has a liquid outer core.
5. What is the "Moho"?
6. In what way is Earth's structure like a boiled egg?
7. What did studies of the P-wave and S-wave shadows reveal about Earth's interior?

22.3 More About Earth's Layers

8. In what way is Earth's crust like an eggshell?
9. Name the two parts of Earth's crust. How do geologists distinguish between them?
10. What kind of rock is most abundant in Earth's continental crust?
11. What type of rock is oceanic crust made of?
12. What layer takes up most of Earth's mass and volume?
13. Describe the lithosphere.
14. Describe the asthenosphere.
15. What is Earth's core mostly made of?
16. How hot is Earth's core?
17. What is the evidence that Earth's core is solid?
18. Copy the diagram (which is not to scale) of the Earth's interior. Add the following labels: inner core; outer core; core; mantle; oceanic crust; continental crust; ocean.

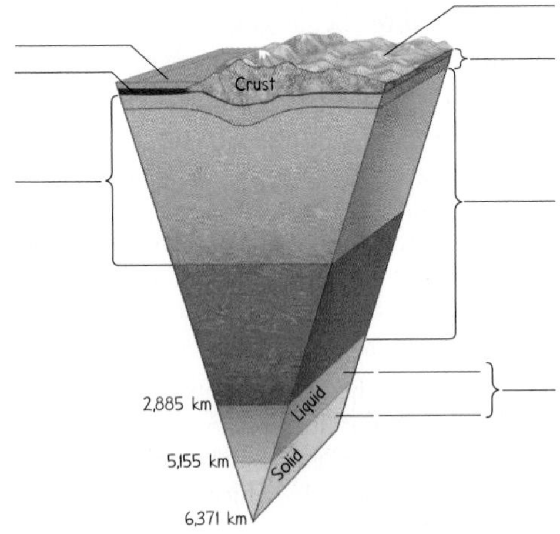

22.4 Continental Drift—An Idea Before Its Time

19. Explain the hypothesis of continental drift.
20. Explain how the contours of the African and South American continents suggest the idea that the continents are mobile.
21. Why was Wegener's hypothesis of continental drift rejected by the scientific community?
22. Why do people not notice that the continents drift?

22.5 A Mechanism for Continental Drift

23. Why was the ocean floor surveying conducted by Harry Hess of major geological importance?
24. Describe the process of seafloor spreading.
25. What are the deepest places on Earth, and what happens there?
26. How did the evidence of seafloor spreading advance the theory of continental drift?
27. How does the age of rocks in the seafloor provide evidence of seafloor spreading?
28. How does magnetic striping provide evidence of seafloor spreading?
29. The graph shows how the seafloor of the Atlantic Ocean varies with depth. Why was this picture of the seafloor surprising to scientists before Harry Hess? How does the picture support the hypothesis of continental drift?

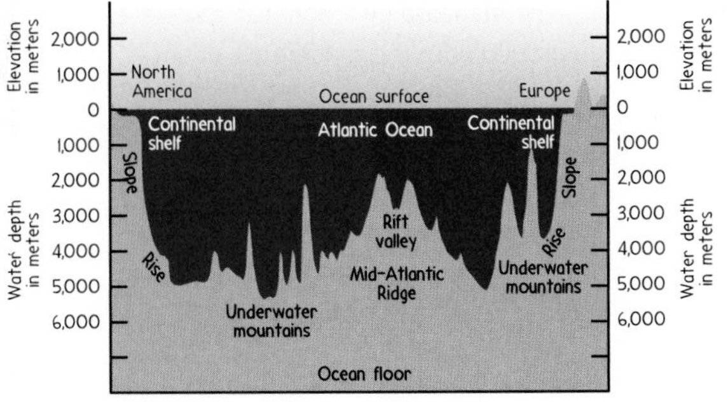

22.6 Plate Tectonics

30. What is a tectonic plate?
31. How many tectonic plates are there?
32. Are the boundaries of plates the same as the boundaries of continents? Explain.
33. In the context of plate tectonics, what is a rift?
34. Name the three major kinds of plate boundaries.
35. How fast do tectonic plates move?

22.7 Divergent Plate Boundaries

36. What happens at divergent plate boundaries at the seafloor?
37. What happens at divergent plate boundaries in continental crust?
38. What is created at divergent plate boundaries?

22.8 Convergent Plate Boundaries

39. Name the three types of convergent plate boundaries.
40. Briefly describe how lithosphere gets recycled at a subduction zone.
41. What striking surface features are created at continental—continental convergent boundaries?

22.9 Transform Plate Boundaries

42. Describe what happens at a transform boundary.
43. What type of surface feature might mark a transform boundary? Why does such a feature form?
44. Why are earthquakes common along the San Andreas Fault in California?

22.10 Earthquakes

45. Where do most earthquake occur?
46. What kinds of earthquakes don't occur at plate boundaries?
47. What is a tsunami?
48. Is the Richter scale linear or logarithmic? Why?
49. How much more does the ground shake during a 6.6 magnitude earthquake than it does during a 5.6 magnitude earthquake? How much less does the ground shake during a 6.6 magnitude earthquake than it does during a 7.6 magnitude earthquake?
50. Relate the energy released during an earthquake to its magnitude on the Richter scale.

Multiple Choice Questions

Choose the BEST answer to the following.

1. The addition of water to rock
   (a) decreases the melting point of the rock.
   (b) increases the viscosity of the rock.
   (c) increases the melting point of the rock.
   (d) changes the silica content of the rock.

2. For each increase of one on the Richter scale,
   (a) ground shaking increases 30 times.
   (b) the energy released increases 30 times.
   (c) the amplitude of the seismic wave doubles.
   (d) the energy released increases 10 times.
3. Which of the following statements is false?
   (a) The mantle includes part of the crust.
   (b) The lithosphere includes the entire crust.
   (c) The mantle includes part of the lithosphere.
   (d) The mantle includes the entire asthenosphere.
4. The first approximately accurate measurements of Earth's size were made
   (a) centuries before the time of Galileo.
   (b) during the time of Galileo.
   (c) more than a century after the time of Galileo.
5. Which of the following is not evidence of continental drift?
   (a) seafloor spreading.
   (b) paleomagnetism.
   (c) isostasy.
   (d) patterns of ancient glaciation.
6. Which part of Earth's crust is most dense?
   (a) the oceanic crust
   (b) the continental crust
   (c) the asthenosphere
   (d) the lithosphere
7. The source of a tsunami is usually
   (a) diverging tectonic plates.
   (b) an earthquake.
   (c) converging tectonic plates.
   (d) volcanism.
8. With a core as hot as the surface of the Sun, why doesn't Earth melt?
   (a) Actually, many parts of Earth's crust do melt as a result of heat escaping from the core.
   (b) The heat of the core is retained by the core.
   (c) Earth's mantle is made out of rock, an effective thermal insulator.
9. What physical change in metamorphic rock signals the end of metamorphism?
   (a) freezing
   (b) melting
   (c) crystallization
   (e) evaporation

INTEGRATED SCIENCE CONCEPTS

Physics—Isostasy

1. Why does continental crust "float" higher in the mantle than oceanic crust?
2. Why is the continental crust thickest under mountains?

Physics—What Forces Drive the Plates?

1. What law of physics is a major underlying cause of plate tectonics? Why do you think so?
2. What do convection cells have to do with the movement of tectonic plates?
3. In what way is the Earth like a pot of water on a huge spherical stove?
4. Distinguish between the slab-pull and ridge-push models of plate movement.

Biology—Life in the Trenches

1. What conditions make life in the trenches so difficult?
2. How can life exist in the trenches when there is no light for photosynthesis and thus no oxygen?
3. What are hydrothermal vents, and how do they relate to life in the trenches?

Physics—Anatomy of a Tsunami

1. How is a tsunami similar to a ripple in a pond? How does it differ?
2. What causes a tsunami?
3. Describe how a tsunami can be generated by subduction at a convergent boundary.

Active Exploration

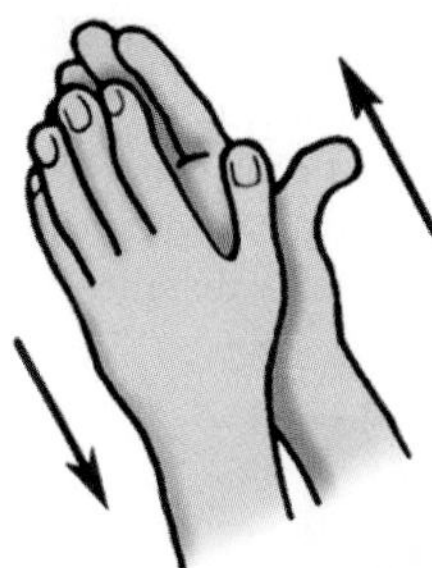

Simulate motion along a transform fault with your hands. Press your palms together and move your hands in opposite directions, as shown in the figure. Notice that friction keeps them "locked" together until opposing forces no longer balance, friction no longer holds, and your palms slip suddenly past each other. This nicely models the sticking and slipping that occurs along a transform fault.

Exercises

1. Is the Earth's inner core solid and the outer core liquid because the inner core is cooler than the outer core? Explain your answer.
2. What is the principal difference between the theory of plate tectonics and the continental drift hypothesis?
3. How is magma generated at divergent plate boundaries?
4. What would happen if new lithosphere were created faster than it is destroyed?
5. What kinds of plate boundaries feature subduction zones?
6. Why are earthquakes common in subduction zones?
7. Briefly describe how an island arc forms.
8. Why are there so many expressions such as "solid earth," "old as the hills," and "terra firma" that suggest that the Earth is unchanging?
9. How did seismic waves contribute to the discovery of Earth's deep internal layers?
10. Why does crust "float" on the mantle?
11. Briefly describe the fossil evidence that Wegener cited to support continental drift.
12. Why are volcanic mountain ranges often found along oceanic–continental convergent boundaries?
13. What is the evidence that tectonic plates move?
14. Why are most earthquakes generated near plate boundaries?
15. Magnetic stripes that were laid down on the Pacific Ocean seafloor are wider than the magnetic stripes laid down over the same time period on the Atlantic Ocean seafloor. What does this tell you about the rate of seafloor spreading of the Pacific Ocean compared with that of the Atlantic Ocean?
16. Are continents a permanent feature of our planet? Discuss why or why not. Estimates are that up to 80% of Earth's surface is covered with volcanic rocks. Use what you know about plate tectonics to explain why this would be true.
17. Could Los Angeles fall into the ocean, as is popularly thought? Why or why not?
18. Is plate tectonics a theory based on integrated science? Why or why not?
19. Evidence of ancient ice sheets has been found in areas near the equator. Give two possible explanations for this.
20. What is meant by magnetic pole reversals? What useful information do they provide about Earth's history?
21. What is a very likely cause for the Earth's magnetic field?

22. Relate the generation of magma to pressure changes at divergent plate boundaries.
23. Cite one piece of evidence that suggests subduction once occurred off the West Coast of the United States.
24. Where is the oldest oceanic crust? Where is the youngest?
25. Your friend who lives in Los Angeles says that he is relieved when there are tiny earthquakes because that releases strain and prevents "the big one." Do you share his sense of relief? Why or why not?

Problems

1. The San Andreas Fault separates the northwest-moving Pacific Plate, on which Los Angeles sits, from the North American Plate, on which San Francisco sits. If the plates slide past each other at an approximate rate of 3.5 cm per year, show that the two cities will become neighbors in about 17 million years. (The distance from San Francisco to Los Angeles is 600 km.

2. Suppose that a fence is built across the San Andreas Fault in the year 2005. Show that the two sections of the broken fence will be 70 cm apart in 2025. Assume that the average rate of movement along the fault is 3.5 cm (approximately 1 in) per year.
3. The Nazca and Pacific plates are spreading relatively fast, at the rate of 14.2 cm (5.6 in. per year). The East Pacific Rise is a mid-ocean ridge system that divides these plates. If the central segment of the rise is 200 km wide, show that it is about 1.4 million years old.
4. What is the rate of seafloor spreading at the divergent plate boundary shown in this diagram of magnetic striping?

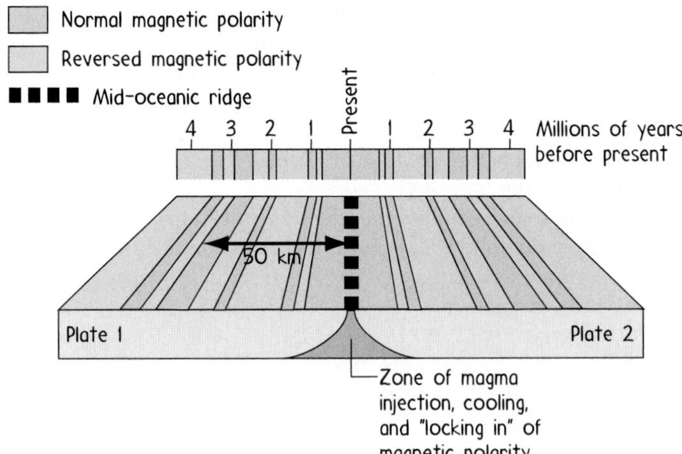

CHAPTER 22 **ONLINE RESOURCES**

Tutorials The Earth's Mantle Activity • ***Quiz*** • ***Flashcards*** • ***Links***

Rocks and Minerals

This is the Carajas iron mine in the Amazon Rainforest of Brazil—one of the largest iron mines in the world. The forest has been clear-cut to access the large supply of iron ore here. About 45 million tons of iron ore are extracted from this mine each year, to be used in products ranging from automobiles to pharmaceuticals.

CHAPTER OUTLINE

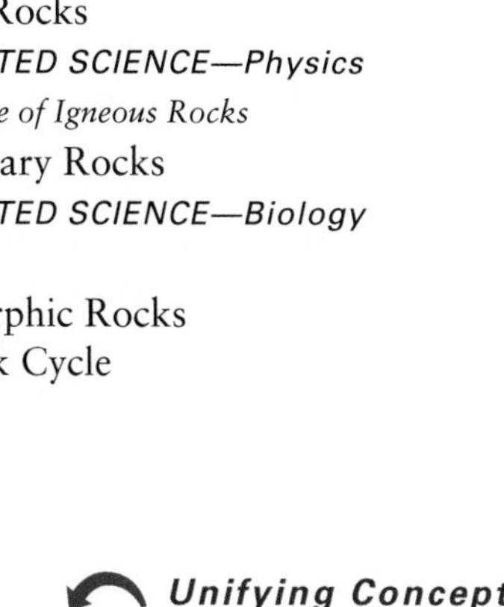

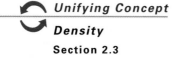

Unifying Concept
Density
Section 2.3

Almost all manufactured products contain minerals or substances derived from minerals. From the aluminum in cans to the graphite in pencils to the halite on your French fries, minerals are practically everywhere. What exactly is a mineral? Where do all the minerals we consume come from? Minerals join together to create rocks, which are the dominant material of the Earth. Rocks, like minerals, are economically important. To name just a few examples, coal is a rock burned for fuel, and marble and granite are popular ornamental stones. Rock is important scientifically too—what do rocks tell us about Earth's history? How do the properties of rock determine geological processes such as erosion and earthquakes? In what sense does rock recycle? As "rock hounds" know, rocks are fascinating once you learn their story.

23.1 Materials of the Earth

The Earth, the Moon, and many other planetary bodies are composed chiefly of rock. And rock is composed of what all matter in the Universe is made of—the chemical elements. There are 112 known naturally occurring elements, as discussed in Chapter 11 and as represented in the Periodic Table on page 229. Most chemical elements are rare on Earth. In fact, only eight elements make up about 98% of the Earth's entire mass, as Figure 23.1 shows. All of the other elements combined make up the remaining 2%.

Earth's elements are not distributed evenly. The heavier ones are concentrated near the center of the Earth while the lighter elements are more abundant near the surface. The explanation for this has to do with Earth's early history. Earth formed 4.5 billion years ago from debris orbiting the newly forming Sun (see Chapter 27). Some of this debris collided and coalesced into a rocky mass, forming a rudimentary Earth. Earth grew as its gravity attracted more debris to it. Each time material from space collided with and attached to Earth, the kinetic energy of the impacting bodies transformed to heat. Earth became *molten*—melted and able to flow. Under the influence of gravity, dense and heavy iron-rich material then sank to Earth's center to become the core. Less-dense, silicon- and oxygen-rich material rose toward the surface by **differentiation**, the process by which gravity separates materials of different densities. (A mixture of oil and water undergoes differentiation, for example, when the denser water sinks to the bottom and the less-dense oil

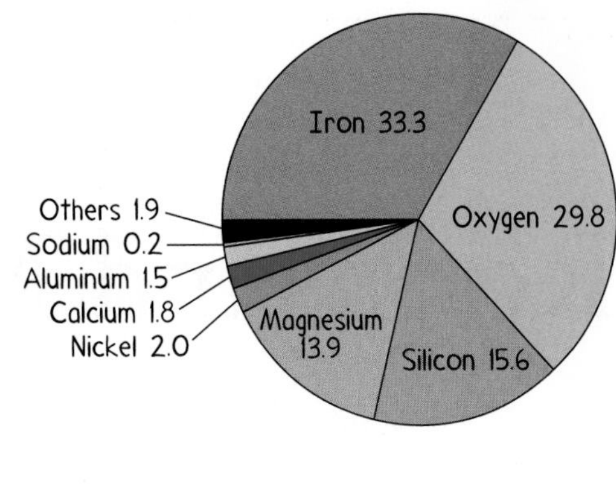

Figure 23.1 Only eight of the chemical elements are found in abundance on Earth.

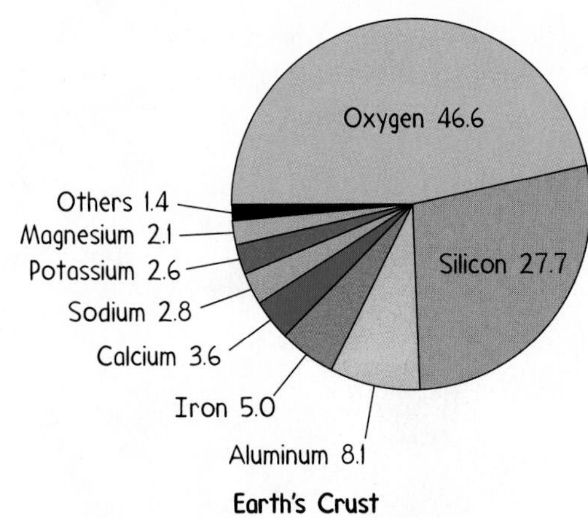

Figure 23.2 Percentage of elements in Earth's crust, by mass. Oxygen and silicon make up more than 75% of Earth's crust.

INSIGHT Density, as discussed in Chapter 2, is a measure of the mass of a material divided by its volume. Density relates to sinking and floating: If the density of an object is greater than the fluid in which it is immersed, it will sink.

rises to the top.) Similarly, when Earth was a hot fluid mass, it formed layers through the process of differentiation.

Eventually, the frequency of impacts slowed and Earth cooled. Lighter materials near the surface solidified to form minerals that clustered together to make rocks. Figure 23.2 shows the current composition of Earth's crust. Note that approximately 50% of the crust is oxygen (O) and 28% is silicon (Si) so that these two relatively lightweight elements make up about three-fourths of Earth's solid surface.

23.2 What Is a Mineral?

INSIGHT The dominance of oxygen in Earth's crust becomes even more apparent when you consider its abundance, not by mass, but in terms of numbers of atoms: 63 out of every 100 atoms making up Earth's crust are oxygen atoms. Oxygen isn't just important as a constituent of air—in the solid state, it makes up most of Earth's crust.

In everyday language, a *mineral* is a part of your diet ("vitamins and minerals"). In economics, *mineral resources* are materials including rock, sand, and coal extracted from the Earth for industrial purposes. But, in Earth science, a *mineral* is something else. A **mineral** is a material that has the following five characteristics.

- It is naturally occurring (formed naturally rather than manufactured.)
- It is a solid.
- It has a definite chemical composition.*
- It is *generally* inorganic—it's not alive and is not derived from living things.†
- It has a characteristic crystalline structure. (The particles in it—atoms, ions, or molecules—are positioned in a specific, orderly arrangement.)

Unifying Concept
The Atomic Nature of Matter
Section 7.1

All of these criteria are straightforward, except perhaps the last one. What exactly is a *crystalline* structure? A **crystalline structure**, is a repeating, orderly arrangement of atoms or molecules, as shown in Figure 23.3.

Usually, when a fluid solidifies, its atoms or molecules lose the random motion associated with the fluid state and they "freeze" into a crystalline structure (or *lattice*). The process of forming a crystalline structure is called **crystallization.** While solidification usually produces a crystalline material, under certain circumstances the atoms or molecules of a fluid may solidify in a random, disorganized manner to form an *amorphous* solid. In an amorphous material, constituent

*Although the chemical composition of a mineral is definite, it may vary within certain specified limits. Variations in mineral composition can produce differences in color, density, and other properties. These differences define different "varieties" of a particular mineral. For example, amethyst is a variety of quartz that is purple while rose quartz is a pink variety of quartz.

†There are some exceptions to the statement that all minerals are inorganic, since a few minerals can be produced organically as well as inorganically. Aragonite, a form of calcium carbonate, is an example. Oysters and pearls produce aragonite to build their shells. Because this organically produced compound is identical to inorganically produced aragonite, aragonite from either source is classified as a mineral.

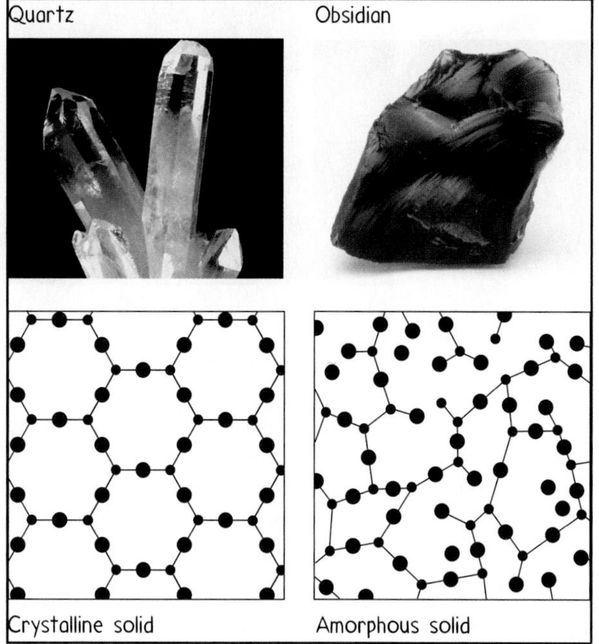

Figure 23.3 A two-dimensional drawing of the internal atomic arrangement of quartz shows its orderly, crystalline structure. Obsidian is a natural volcanic glass and an amorphous solid. Its atoms are not arranged in a repeating, geometric structure; therefore, obsidian is not a mineral.

particles have no long-range order. Glasses are amorphous because they cool so quickly from the liquid state that their atoms cannot travel to their lattice sites before they lose mobility (Figure 23.4). Plastics are amorphous because their molecules are too long to fit into a regular, geometric lattice.

Figure 23.4 The "crystal" in fine glassware is not really crystal at all. Fine crystal is glass, an amorphous solid. On an atomic level, the atoms and molecules in this fine crystal goblet are a disorderly jumble.

Sometimes, two or more minerals contain the same elements in the same proportions, but their atoms are arranged differently. As a result, they have different crystal structures and therefore different properties. Such minerals are called **polymorphs** of one another. Diamond and graphite are polymorphs because they both consist entirely of the same element, carbon. But the carbon atoms are arranged differently and so you would never mistake diamond for graphite. (Figure 23.5).

CHECK YOURSELF

1. Diamonds contain only carbon. Are diamonds minerals?
2. Obsidian is a kind of glass formed in volcanoes. Is it a mineral?

CHECK YOUR ANSWERS

1. Yes. Minerals can consist of one or more elements.
2. No. Because obsidian is a type of glass, it is amorphous and does not have a crystalline structure.

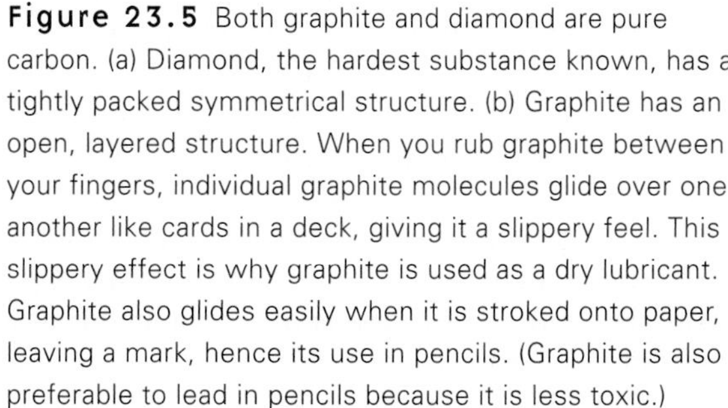
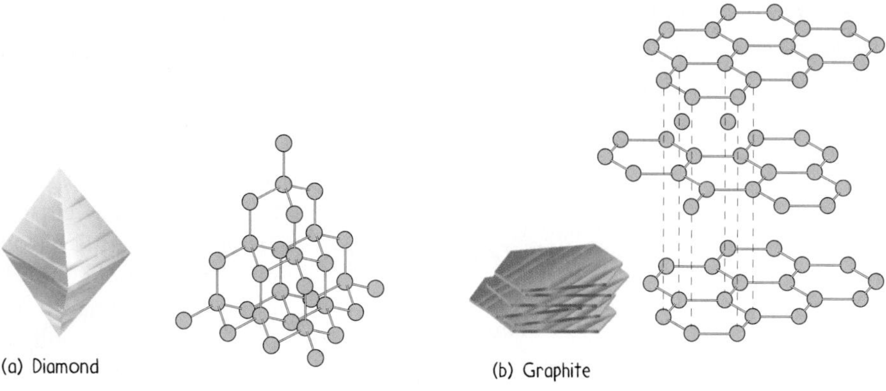

Figure 23.5 Both graphite and diamond are pure carbon. (a) Diamond, the hardest substance known, has a tightly packed symmetrical structure. (b) Graphite has an open, layered structure. When you rub graphite between your fingers, individual graphite molecules glide over one another like cards in a deck, giving it a slippery feel. This slippery effect is why graphite is used as a dry lubricant. Graphite also glides easily when it is stroked onto paper, leaving a mark, hence its use in pencils. (Graphite is also preferable to lead in pencils because it is less toxic.)

Figure 23.6 A few of the many minerals possessing a striking physical appearance. (a) The mineral hematite, with its grape-cluster shape. (b) Amethyst, the purple variety of quartz, which grows in hexagonal crystals with pointed ends. (c) Pyrite, or "Fool's Gold," typically forms cubic crystals marked with parallel lines called "striations." (d) Rosasite has fibrous, bluish-green crystals. (e) Rhodochrosite (whose name means "rose-colored") displaying its rhombohedral crystals. (f) Asbestos has strong, heat resistant fibers.

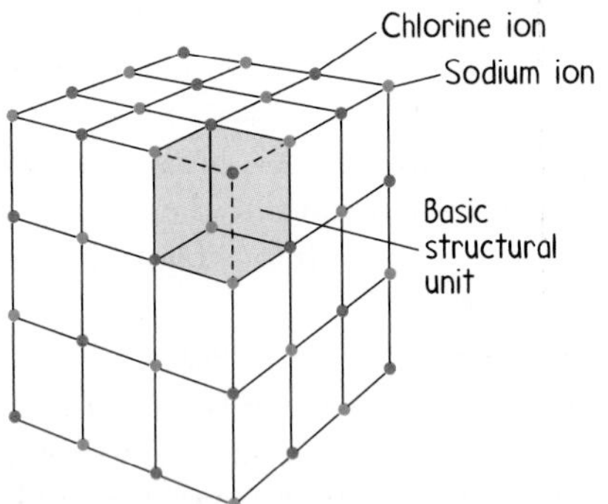

(a) Crystalline structure of halite

(b) Grains of the mineral halite (table salt)

Figure 23.7 The basic structural unit of halite (table salt) is cubic. This unit is repeated over and over in three dimensions. The internal order of halite crystals is reflected in its macroscopic mineral grains.

23.3 Mineral Properties

Mineral collectors and rock hounds know that minerals display a wide range of interesting and often useful physical properties. Some minerals are radioactive, others are fluorescent, and some are toxic. Minerals may possess a metallic *luster* (or shine) or, instead, have a pearly, waxy, greasy, or dull appearance. Though the majority of Earth's 4000 or so minerals look like nondescript rocks, there are also a great many minerals of vibrant color or intriguing form (Figure 23.6).

Certain mineral properties, such as those discussed here, are useful in helping to identify minerals. You can practice using these for mineral identification in the laboratory portion of this course. As you get to know the physical properties of minerals, keep in mind that they are an expression of the chemical composition and crystal structure of the mineral.

Crystal Form

The mineral specimens in Figure 23.6 each display characteristic *crystal form*, or shape. A mineral's *crystal form* mirrors the orderly, microscopic arrangement of its atoms. Figure 23.7 shows that halite (NaCl, or table salt) has a *cubic* crystal form. You can see this just as easily by examining some salt crystals with a hand lens. The sodium and chloride ions in halite are arranged in a cubic crystalline structure. (Ions are atoms that have gained or lost electrons to acquire an electric charge, as discussed in Chapter 11.)

Well-formed crystals are not the norm in nature. Most of the time, numerous crystals grow together in cramped locations. As they develop, each tiny crystal has to compete with neighboring crystals for space. The crystals that develop are so small and intergrown that the mineral's crystal form, if present, can't be observed without a microscope.

Hardness

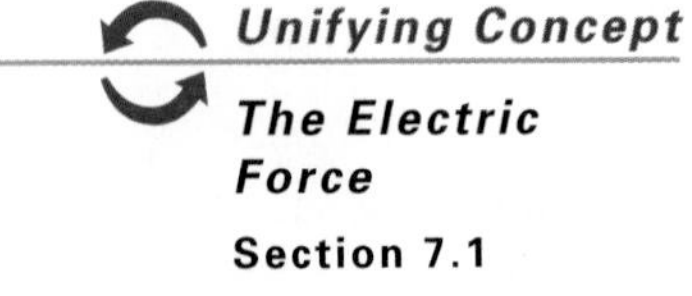

Unifying Concept

The Electric Force

Section 7.1

A mineral's *hardness* is its resistance to scratching. For example, a quartz crystal can scratch a feldspar crystal because quartz is harder than feldspar. Why are some minerals harder than others? Hardness depends on the strength of a mineral's chemical bonds—the stronger its bonds, the harder the mineral. The factors that influence bond strength are ionic charge, atom (or ion) size, and packing. Strong bonds are generally found between highly charged ions—the greater the attraction, the stronger the bond. Size affects bond strength as well, because small atoms and ions can generally pack closer together than

Table 23.1 Mohs' Scale of Hardness

Mineral	Hardness	Object of Similar Hardness
Talc	1	
Gypsum	2.5	Fingernail
Calcite	3	Copper wire or coin
Fluorite	4	
Apatite	5.5	Steel knife blade
Feldspar	6	Window glass
Quartz	7	Steel file
Topaz	8	
Corundum	9	
Diamond	10	

(a)

(b)

Figure 23.8 A mineral's cleavage is very useful in identification. (a) Muscovite, a mineral of the mica group, has perfect cleavage in one direction. (b) Calcite (calcium carbonate) has perfect cleavage in three directions.

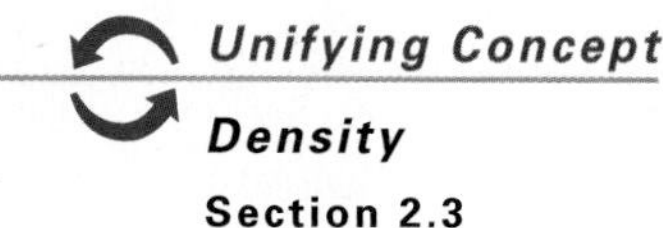

Figure 23.9 This rock, which is composed of quartz, does not exhibit cleavage. When it breaks, it develops a conchoidal fracture—a smooth, curved surface resembling that of broken glass.

large atoms and ions. Closely packed atoms and ions have a smaller distance between one another, and thus they form stronger bonds because they attract one another with more force.

Diamond is the hardest mineral known. Diamonds are hard because they are made up of small carbon atoms, which fit neatly together in a tightly packed structure, as shown in Figure 23.5. *Mohs' scale of hardness* (Table 23.1) compares the relative hardness of different minerals on a scale of 1 to 10. Diamond rates a 10.

Cleavage and Fracture

Another physical property determined by crystal structure and chemical bond strength is cleavage. When a mineral exhibits cleavage, it breaks along planes of weakness. For example, the crystal structure of the mineral mica consists of atoms arranged in sheets. The atoms within individual sheets are connected by strong covalent bonds, but the bonds between sheets are weak intermolecular bonds known as Van de Waals bonds. (These are discussed in Chapter 13.) So, if you hit a piece of mica with a hammer, the mica cleaves along the planes of its weak bonds and breaks into sheets (Figure 23.8). You can even peel mica off in thin layers. Shiny flakes of mica are used in glittering body paints and they add shimmer to auto body paints as well.

When a mineral with no cleavage breaks, the break is called a *fracture*. A fracture that is smooth and curved, so that it resembles broken glass, is called *conchoidal*. Quartz and olivine display smooth conchoidal fractures (Figure 23.9).

Color

Although color is an obvious feature of a mineral, it is not a reliable means of identification. Some minerals—copper and turquoise are two examples—have a distinctive color. But most minerals either occur in a variety of colors or can be colorless.

Chemical impurities affect color. For example, the common mineral quartz, SiO_2, can be clear and colorless if it has no impurities. Or it may be milky white from tiny fluid inclusions. Rose-colored quartz results from small amounts of titanium, and purple quartz (amethyst) contains small amounts of iron. The color of the mineral corundum, Al_2O_3, is commonly white or grayish. But impurities in corundum give us rubies and sapphires.

Specific Gravity

Density is a property of all matter, minerals included. In practical terms, the density of a mineral tells us how heavy a mineral feels for its volume. Mineralogists often use a related property—*specific gravity*—to identify minerals. Specific gravity is the ratio of the weight of a substance to the weight of an equal volume of water. For example, if 1 cubic centimeter of a mineral weighs three times as much as 1 cubic centimeter of water, its specific gravity is 3. Specific gravities range from 1–20.

Gold's particularly high specific gravity of 19.3 is nicely taken advantage of by miners panning for gold. Fine gold pieces hidden in a mixture of mud and sand settle to the bottom of the pan when the mixture is swirled in water. Water and less dense materials spill out of the pan when the mixture is swirled. After a succession of douses and swirls, only the substance with the highest specific gravity remains in the pan—gold!

CHECK YOURSELF

1. Why is identifying a mineral by its crystal form usually difficult?
2. The minerals muscovite and calcite both display very distinct cleavage. But if you take a hammer to the mineral quartz, it fractures. How does this relate to their crystal structure?

CHECK YOUR ANSWERS

1. Well-shaped crystals occur rarely in nature because minerals typically grow in cramped spaces.
2. Calcite and muscovite have planar directions of weakness in their structures. The bonds between these planes are weaker than the other bonds within the structure. Therefore, these minerals will cleave along such planes. Quartz has a structure with no planes of weakness. Therefore, quartz fractures.

SCIENCE AND SOCIETY

Mining

People use minerals in an almost endless variety of ways. Just look around. Iron, copper, aluminum and other metals are used in cars, buildings, refrigerators, and CD players. Building materials, such as concrete, sheet rock, cement, and brick are made from minerals. So are paint, plastics, films, drugs, explosives, money, lubricants, nuclear reactor cores, steel, food additives . . . the list goes on and on.

Currently, in the United States, several billion tons of mineral resources are needed each year for people to pursue their everyday activities. Each man, woman, and child in the U.S. uses thousands of kilograms of new mineral resources per year, and the average American home contains slightly more than a quarter million pounds (more than 113,000 kg) of minerals and metals (Figure 23.10).

Where do all these minerals come from? They are mined from the Earth's crust. As you might imagine, extracting such a huge volume of minerals has the potential to do environmental damage. Have you ever seen a mining site (Figure 23.11)? At typical mining sites, huge holes have been dug into the ground (*open pit mines*). Or, large strips of Earth's surface have been removed by bulldozers (*strip mines*). Since the land is disrupted, ecosystems can be destroyed. The large-scale removal of vegetation can cause erosion and landslides. Also, fossil fuels are consumed on a grand scale to extract and process the minerals. And pollutants, such as carbon dioxide and sulfur dioxide, and toxic waste, such as lead and chromium, are generated in processing some minerals and may be left behind. In the United States, mines produce more waste than all American cities and towns combined. How are these environmental issues being addressed?

Damage is reduced when mining companies practice *reclamation*—restoration of land to its condition before the mining was done. Current environmental legislation mandates that mining companies reclaim mining sites on public lands. Some states require that reclamation be practiced on private land as well. And, as with other natural resources, such conservation measures as reducing consumption and recycling can reduce environmental damage.

Each U.S. citizen will use . . .
700 kg copper
53 kg gold
14,200 kg halite
260,000 kg coal
10,200 kg clays
2800 aluminum
15,900 kg iron ore
385 kg lead
385 kg zinc
in his or her lifetime.

Figure 23.10 On average, each United States citizen will use about 1,633,000 kilograms of minerals and other mined resources during his or her lifetime. These mined materials provide each individual's food, clothing, and shelter, as well as his or her share of transportation, manufacturing, and recreational facilities.

Figure 23.11 The Kennecott Copper mine in Utah is the biggest hole on Earth. It is 1.2 km deep and 4 km wide. Billions of tons of rock have been removed to make this mine. From it, a vast supply of copper and other minerals has been extracted.

23.4 The Formation of Minerals

Crystallization

Minerals form by the process of *crystallization*. Crystallization starts as single microscopic crystals form whose boundaries are flat, planar surfaces. As more and more atoms bond to the microscopic crystal, the crystal grows from the outside.

Minerals crystallize from two primary sources: magma and water solutions. Consider first minerals that crystallize out of magma. Magma (as you learned in

Chapter 22) is molten rock found underground. There are a few major types of magma with differing compositions, but, in general, magmas are hot fluids with a texture like thick oatmeal. They contain some solids and gases, but magmas consist mainly of freely flowing atoms from the silicate group of minerals—silicon, and oxygen, plus aluminum, potassium, sodium, calcium, and iron. When magma starts to cool, the hot liquid atoms lose kinetic energy. Then the attractive forces among them pull the atoms into orderly crystal structures. Minerals crystallize systematically based upon their melting points. The first minerals to crystallize from magma have the highest melting points and, the last minerals to crystallize have the lowest melting points. For example, feldspar has a higher melting point than quartz, so feldspar crystallizes out of cooling magma before quartz does.

Minerals crystallize out of water solutions as well. For example, *evaporites* are minerals and rocks that crystallize when a restricted body of seawater, or the water of a salty lake, evaporates. Examples are gypsum, anhydrite, and halite. Evaporites crystallize in a way that is very similar to the crystallization of minerals from magma. The difference, though, is that solubility rather than melting point determines which minerals form first. Evaporite minerals with the lowest solubility—such as gypsum—precipitate first from water solutions. The low-solubility minerals are followed by minerals that dissolve more easily, such as anhydrite and then halite.

Stability

Not every collection of atoms can crystallize to form a mineral. A mineral will be *stable* only if combining atoms can attract one another sufficiently given prevailing environmental conditions. Temperature and pressure exert a great influence on mineral stability. For example, a mineral that is stable at the Earth's cool surface cannot exist in its warmer interior if the added thermal energy disrupts the mineral's chemical bonds. Similarly, a mineral that is stable at surface pressure may not be stable underground where pressure is higher. The increased pressure inside the Earth squeezes crystals, and even the individual atoms or ions within them. Negative ions are larger than positive ions because of their extra electrons orbiting the nucleus; in a sense, they are "fluffier" than positive ions. Hence, as pressure increases, negative ions are compressed more than positive ions are, and a mineral that was once tightly packed and stable at lower pressure can lose its crystal structure. Its constituent atoms would then migrate to form new minerals. (This occurs in metamorphic rock, as you'll see later in this chapter.)

Likewise, minerals that are stable at high pressures may be unstable at lower pressures. Diamonds grow at depths of at least 150 kilometers underground. Intense pressure there pushes carbon atoms together into tight networks. If that pressure is removed, the diamonds become unstable. So, at Earth's surface, a diamond is not really stable, and it will spontaneously rearrange its crystal structure to become the mineral graphite (Figure 23.12). Fortunately, the process takes about a billion years and diamond rings are safe over the span of human lifetimes.

Figure 23.12 Diamonds aren't forever after all!

TECHNOLOGY

Synthetic Diamonds

In the laboratory, it is possible to subject graphite to pressure high enough to turn it into diamond. Until recently, achieving the required pressure was so costly that synthetic gem-quality diamonds were actually more expensive than mined diamonds. However, in 2004, two companies introduced gem-quality diamonds to the market at a cost roughly 30% below that of mined diamonds. These new synthetic diamonds are visually indistinguishable from the real thing.

23.5 Classifying Minerals

There are more than 4000 named minerals, with new ones being identified each year. As Figure 23.13 indicates, this vast array of minerals can be classified on the basis of their chemical composition. Doing this produces two main divisions: the *silicates* and the *nonsilicates*. The **silicates** contain both silicon (Si) and oxygen (O). Most silicates contain other elements in their crystal structure as well. Oxygen is the most abundant element in Earth's crust; silicon is the second most abundant. Together these two elements constitute about 75% of the Earth's crust, as shown in Figure 23.2. Silicon has a great affinity for oxygen. In fact, silicon has such a strong tendency to bond with oxygen that silicon is never found in nature as a pure element; it is *always* chemically combined with oxygen. For these reasons, the silicates are the most common mineral group, constituting 92% of the Earth's crust.

The silicates are divided into two groups—ferromagnesian silicates and nonferromagnesian silicates. As their name suggests, ferromagnesian silicates contain iron (Fe), or magnesium (Mg), or both of these elements plus the standard oxygen and silicon atoms. Garnet, for example, is a ferromagnesian silicate. Because they contain the metals magnesium and iron, ferromagnesian silicates tend to be dark, dense, and have a glassy luster. Nonferromagnesian silicates are very common; they usually have relatively low densities and are

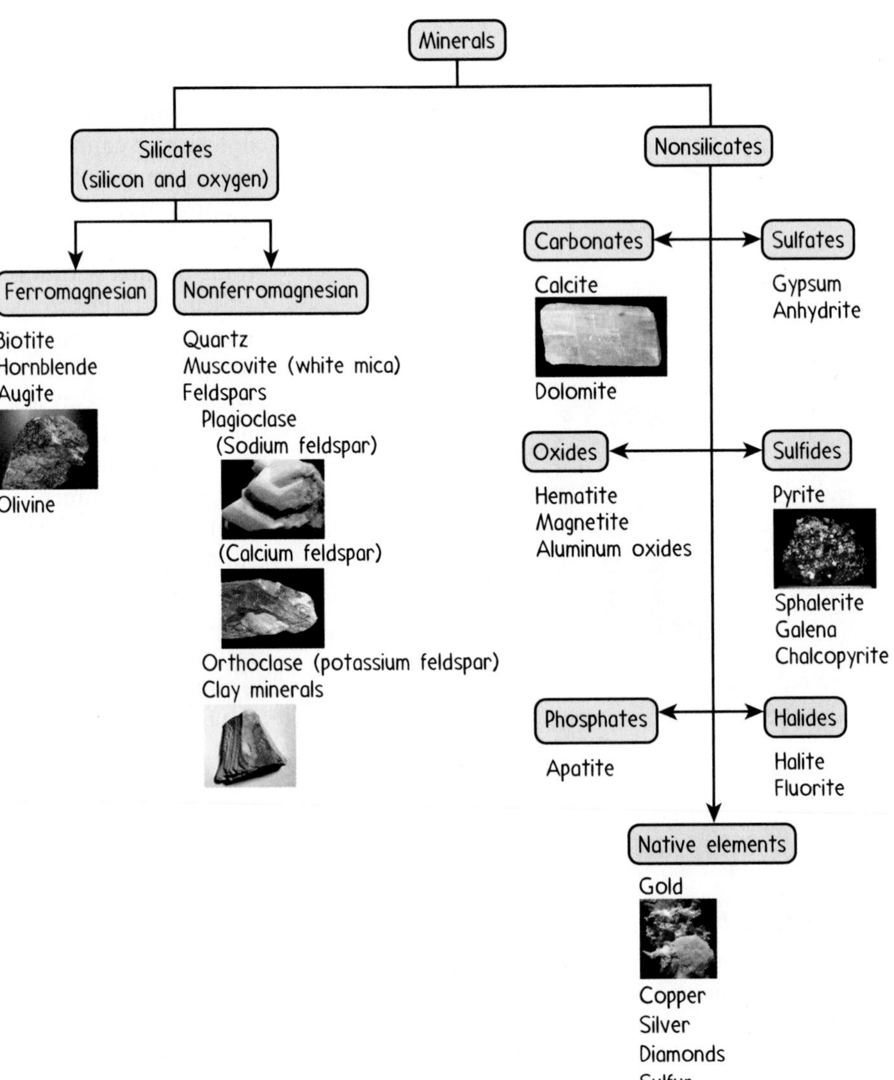

Figure 23.13 Classification of common minerals.

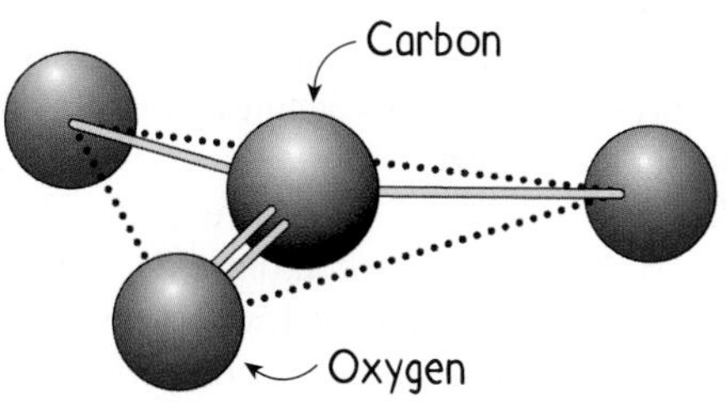

(a)

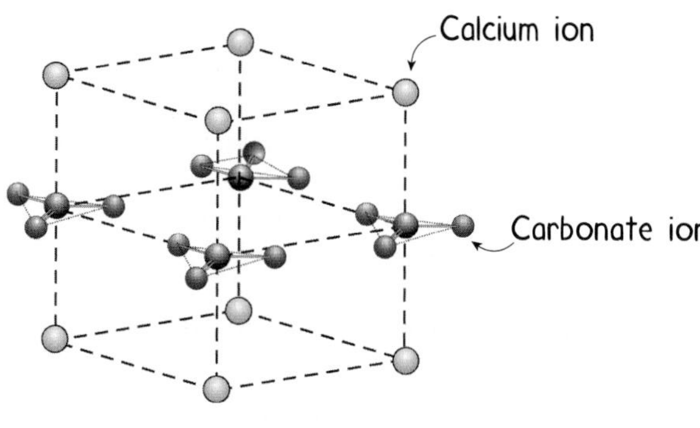

(b)

(c)

(d)

Figure 23.14 (a) The carbonate ion, CO_3^{2-}, has a triangular structure that features a central carbon atom bonded to three oxygen atoms. (b) Because carbonate minerals, such as calcite and dolomite, have a layered, sheetlike structure, they display the property of cleavage. (c) Dolomite rock (also called dolostone) is composed mainly of the mineral dolomite. (d) Cliffs consisting of dolomite rock along a stream in the Ozark Mountains of Missouri.

light in color. The most abundant mineral in the crust is feldspar, a nonferromagnesian silicate that contains aluminum, sodium, potassium, and/or calcium, plus silicon and oxygen. Feldspar makes up more than 50% of the crust. Quartz, the second most common mineral in the Earth's crust, is composed only of oxygen and silicon (SiO_2). You've probably seen the white, 6-sided quartz crystals called milky quartz rock crystal, which is a staple of mineral collections.

Nonsilicate minerals make up just 8% of the Earth's crust by mass. The nonsilicates include the carbonates, oxides, and such native elements as gold and silver, and a few others. The carbonates are the most abundant nonsilicate minerals. The carbonate minerals have a much simpler chemical structure than the silicates. Carbonate structure is triangular, with a central carbon atom bonded to three oxygen atoms, CO_3^{2-}. Groups of carbonate ions are arranged in sheets. Two common carbonate minerals are calcite and dolomite. Calcite consists of the chemical compound calcium carbonate, $CaCO_3$. Dolomite is a mixture of calcium carbonate and magnesium carbonate, $CaMg(CO_3)_2$. Calcite and dolomite are the main minerals found in the group of rocks called *limestone* (Figure 23.14).

In the oxide group, oxygen is combined with one or more metals. These metals include iron, chromium, manganese, tin, and uranium. The oxide mineral group is important economically because it contains many useful *ores*. An **ore** is a mineral deposit that is rich in valuable metals that can be extracted for a profit. For example, iron is obtained from iron oxide ore, as shown in the chapter opening photograph of the Carajas iron mine. Likewise, copper is obtained from copper oxide ore deposits occurring in such mines as the Kennecott Copper mine shown in Figure 23.11. The principle ore of the valuable industrial metal chromium is the mineral chromite—iron chromium oxide.

INTEGRATED SCIENCE CHEMISTRY

The Silicate Tetrahedron

All of the silicates are based on the same structural unit—the *silicate tetrahedron*. The silicate tetrahedron is an ion with four oxygen atoms joined to one silicon atom $(SiO_4)^{4-}$. It has the shape of a squat pyramid, with oxygen atoms at the four vertices and the smaller silicon atom tucked inside, as illustrated in Figure 23.15. All silicates feature this basic structural unit, either by itself or linked together with others.

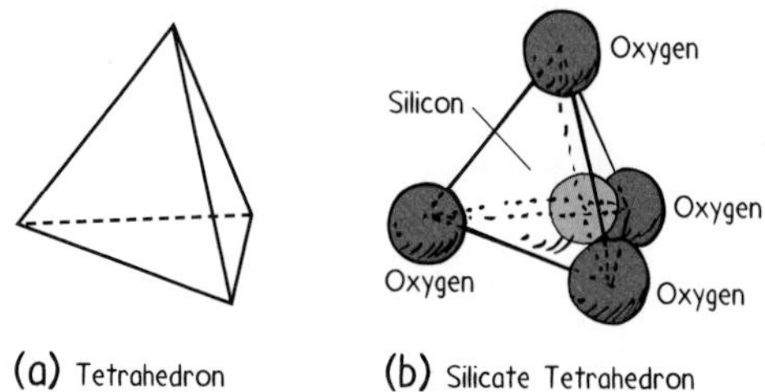

Figure 23.15 The silicate tetrahedron is the structural unit common to all silicate minerals. Four oxygen atoms are arranged in the shape of a tetrahedron with silicon in the center.

INSIGHT Metallic ores and nonmetal mineral deposits are mined on every continent. They are mined wherever they are sufficiently concentrated to be economically extracted.

There are many, many different kinds of silicates, because the silicate tetrahedron can exist in isolation or be linked with others in numerous different ways. Just as carbon is the basis for the vast array of organic molecules, the silicate tetrahedron is the basis for a vast array of silicate minerals.

What about the SiO_4^{4-} structure makes it so versatile? As Figure 23.16 indicates, each of the oxygen atoms on the tetrahedron has an unpaired electron available for bonding. These unattached electrons can be shared with individual metal ions, such as magnesium, iron, and so forth. Single tetrahedra with metal to ions bonded to them tend to pack closely and bond tightly, so they are hard and durable. Many gemstones, including garnets, peridot, and topaz, have this crystal structure.

On the other hand, a silicate tetrahedron can bond to others of its own kind by sharing oxygen atoms. Thus one SiO_4^{4-} tetrahedron can link to another and another. Single chains, double chains, sheets, and a variety of complicated structures can result (Figure 23.16). Mica and the variety of asbestos called chrysotile are sheet silicates.

CHECK YOURSELF

1. Quartz contains only silicon and oxygen. What can you infer about its structure?
2. Why is the silicate tetrahedron a constituent of so many different minerals?

CHECK YOUR ANSWERS

1. It is a silicate and therefore has the silicate tetrahedron as its basic structure. The fact that it consists only of silicon and oxygen suggests that the tetrahedra must be connected to one another.
2. The silicate tetrahedron can form many different minerals because its oxygen atoms can link together in so many different ways. Also, the silicate tetrahedron can exist in isolation bonded to metallic ions.

Silcate Mineral		Typical Formula	Example	Silicate Structure
Olivine		$(Mg, Fe)_2SiO_4$	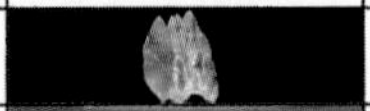	Single tetrahedron
Pyroxene		$(Mg, Fe)SiO_3$		Chains 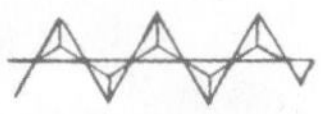
Amphibole		$(Ca_2Mg_5)Si_8O_{22}(OH)_2$	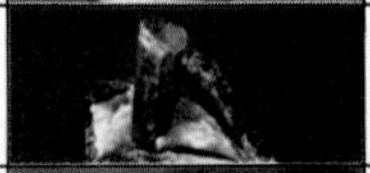	Double chains 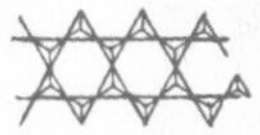
Micas	Muscovite	$KAl_3Si_3O_{10}(OH)_2$		Sheets 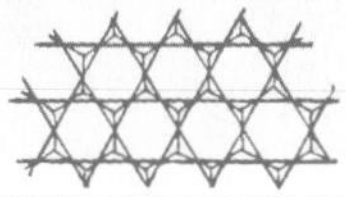
	Biotite	$K(Mg, Fe)_3Si_3O_{10}(OH)_2$		
Feldspars	Orthoclase	$KAlSi_3O_8$		Three-dimensional networks 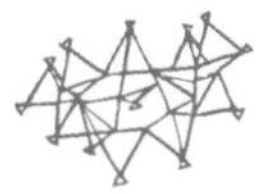
	Plagioclose	$(Ca, Na) AlSi_3O_8$		
Quartz		SiO_2		

Figure 23.16 As silicate tetrahedra link to one another, they polymerize to form chains, sheets, and various network patterns. The complexity of the silicate structure increases down the chart.

Figure 23.17 A rock is an aggregate of one or more minerals. Rocks are physical mixtures rather than chemical compounds.

23.6 Rocks

A **rock** can be defined simply as an aggregate of minerals.* While minerals are *chemical compounds*, rocks are *physical mixtures*. Sometimes, mineral grains in rock are "cemented" together by natural materials that act like glue; in other rocks, the grains tightly interlock.

Looking at a rock, you can sometimes see the mineral crystals in it. Granite, the most common type of rock in the Earth's continental crust, is shown in Figure 23.17. It contains visible crystals of the minerals feldspar, quartz, and hornblende, which make it an attractive ornamental stone. On the other hand, such fine-grained rocks as slate look homogeneous because the mineral crystals they contain are too small to see with the unaided eye.

When you go outside, you may not see much rock, even though that's what the Earth's crust is made of. That's because most surface rocks are covered with *soil*, or what many people call "dirt." Soil is a mixture of small bits of crushed rock plus *organic matter*—the decayed remains of plants, animals, and other organisms. But if you dig under the soil, you'll find rock. Rocks are divided into the following three categories (Figure 23.18), based on how they were formed:

Igneous rocks form by the cooling and crystallization of magma or lava. Recall, from Chapter 22, that magma is molten rock material that forms inside the Earth. Lava is simply magma that has erupted at the Earth's surface through a volcano or a crack in the crust. *Igneous* literally means "formed by fire." Basalt and granite are common igneous rocks.

Figure 23.18 The three main types of rock. (a) Basalt and granite are igneous rocks. (b) Sandstone and limestone are sedimentary rocks. (c) Marble and slate are metamorphic rocks.

*You will find that petrology, the study of rocks, consists of general principles that often have exceptions. For example, the statement "rocks are made of minerals" is true enough to be a good rule of thumb, but there are exceptions. Not all rocks are made only of minerals. Coal is considered a rock, for example, although it is made of plant remains that are not minerals. Don't let the occasional inconsistencies and exceptions leave you bewildered—learn the general principles first, and then try to save some room in your memory for interesting special cases.

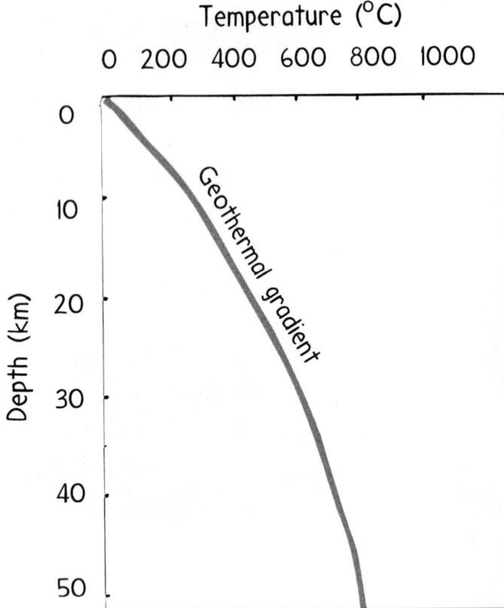

Figure 23.19 The temperature inside the Earth increases about 30°C for each kilometer of depth from the surface. This increase of temperature with depth is known as the geothermal gradient.

Sedimentary rocks generally form from pieces of preexisting rock, or from dissolved compounds worn away from other rocks, that are transported and deposited by water, wind, or ice. Sandstone, shale, and limestone are common sedimentary rocks.

Metamorphic rocks form from preexisting rocks (igneous, sedimentary, or metamorphic) that, without melting, are transformed by high temperature, high pressure, or both. The word metamorphic means "changed in form." Marble and slate are common metamorphic rocks. Marble is metamorphosed limestone, and slate is metamorphosed shale.

CHECK YOURSELF

Which of the following do not belong in a rock collection? Why?
a. quartz b. diamond c. granite d. soil e. magma

CHECK YOUR ANSWER

The only rock is granite, c. Quartz and diamond are minerals. Soil is a mixture of organic matter and small pieces of rock. Magma is liquefied rock, but rocks are defined as solids. Besides, magma, which typically has temperatures of 1000°C or more, would be too hot to handle for even the most avid rock collector!

23.7 Igneous Rocks

Igneous rocks are formed when **magma** (or erupted magma—lava) cools and solidifies. Before we discuss how this crystallization process occurs, let's investigate where magma comes from in the first place. Magma is produced when the rocky material of Earth's solid interior begins to partially melt; three factors can cause this melting to occur: (1) high temperature, (2) localized decreases in pressure, and (3) the addition of water.

First consider the role of heat. As you know, the Earth's temperature increases with depth (as described in Chapter 22). The **geothermal gradient** describes this change (Figure 23.19). It shows that, on average, the Earth's temperature rises about 30°C for each kilometer of depth near Earth's surface. Blocks of rock can also be heated enough to melt when bodies of magma move nearby and transfer their heat to the rock.

But temperature alone is not enough to make buried rocks melt; pressure plays an important role. Rock deep inside the Earth is kept solid by the enormous weight of overlying rock. Very high pressure in mantle rock prevents the atoms within the minerals from breaking chemical bonds and moving freely to form a liquid magma. So most rocks in the mantle don't melt, even though their temperature may be higher than what would be required to melt rock at the surface of the Earth. However, over geologic time scales, because of the plasticity of rock, even solid rock can flow. Ever so gradually, rock at depth can move closer to the surface of the Earth, encounter sufficiently reduced pressure there, and melt.

Finally, the addition of water can generate magma by partially melting rock. Water introduced into rock lowers the melting point of the rock, just as salt on an icy road dissolves into liquid water in the ice and lowers its freezing point.

INSIGHT During the early stages of the Earth's development, when the planet was molten, there were no rocks. Later, as the Earth's surface cooled, minerals crystallized, aggregated, and consolidated to form rocks. The first rocks to appear were *igneous rocks,* because these form directly from the crystallization of magma or lava.

From Magma to Rock

The mineral makeup of an igneous rock depends on the chemical composition of the magma from which it crystallizes. There are a large variety of igneous rocks; does this imply that there is an equally large variety of magmas?

Not really. There are just three principal kinds of magma: *basaltic, andesitic,* and *granitic.* These magmas differ in their chemical composition and properties, and they are found in different tectonic settings. From each of these magmas comes a large array of igneous rocks. The reason? As

INSIGHT Igneous rocks are the most common of the three types of rocks. Yet, these are not often observed because they are largely buried by sedimentary rocks, which form at Earth's surface.

you learned earlier in this chapter (Section 23.4), minerals crystallize in a systematic fashion from magma according to their melting points. The first minerals to crystallize have the highest melting points. They are followed sequentially by minerals with lower melting points. Because of this, the same magma can produce rocks with different mineral compositions. Also, this progressive crystallization can produce many different igneous rocks with mineral compositions quite different from that of the original or *parent* magma.*

Intrusive Versus Extrusive Igneous Rocks

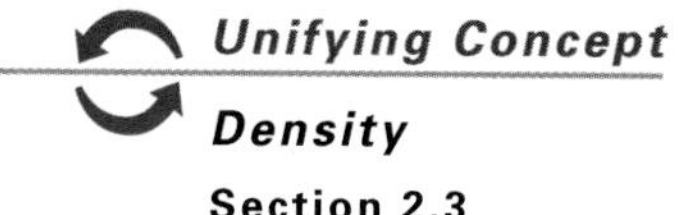

Magma tends to rise toward the Earth's surface because it is less dense than surrounding rock. As it rises, it cools. Magma that cools and solidifies underground is known as **intrusive igneous rock**. Intrusive igneous rock is also called *plutonic* rock, after Pluto, the mythological god of the underworld. Intrusive igneous rocks, such as granite, have characteristically large mineral crystals because they cool slowly at warm subsurface temperatures.

INSIGHT Intrusive igneous rocks form deep within the Earth, where they intrude into existing rock. We would never observe them at the surface if erosion didn't strip away the overlying rock.

The word *intrusive* means "pushed into." Intrusive igneous rock is so named because of its ability to intrude, or push into, areas of preexisting rock. All intrusive igneous rock bodies are called *plutons*. Plutons occur in a wide variety of forms, from slabs to wide, shapeless blobs. *Dikes* are plutons that form by the intrusion of magma into fractures that cut across the layers of existing rock. Dikes are old channels for rising magma, and they often occur near volcanic vents. A spectacular example of this can be seen in Figure 23.20, which is a photo of the radiating dikes that formed around the exposed volcanic neck at Shiprock, New Mexico. *Batholiths* are the largest plutons. They are defined as having more than 100 square kilometers of exposed surface. Batholiths form the cores of many major mountain systems of the world, including the Sierra Nevada.

Figure 23.20 Radiating dikes surround the eroded remains of a volcanic vent in Shiprock, New Mexico.

On the other hand, hot and mobile magmas can sometimes make it all the way to Earth's surface without solidifying. Magma may erupt as lava in a spectacular volcanic blast. More commonly, lava erupts as seepage from fissures in the seafloor or on land. Rock that forms from lava is called **extrusive igneous rock**. Extrusive igneous rock also goes by the name *volcanic rock* (Figure 23.21).

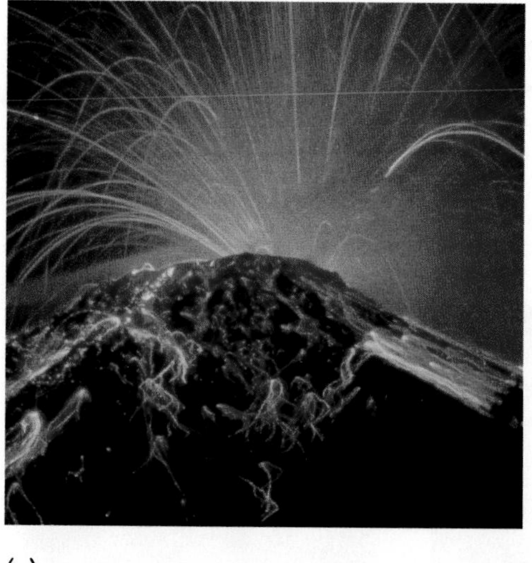

(a)

(b)

(c)

Figure 23.21 (a) Lava from volcanic eruptions cools to create extrusive igneous rock, which is also called volcanic rock. (b) Different kinds of lava produce different kinds of extrusive rocks. (c) This basaltic lava cools to form the rock known as basalt.

* The sequence of crystallization of minerals is described by Bowen's Reaction Series—a good topic for further research, if you're interested.

Extrusive igneous rock, such as basalt, cools rapidly at surface temperatures and therefore consists of small crystals.

INTEGRATED SCIENCE PHYSICS

The Texture of Igneous Rocks

Igneous rocks are classified by their *texture*—the size of their crystals. There are four categories of igneous textures:

a. *coarse-grained*: crystals that are visible without a microscope
b. *fine-grained*: microscopic crystals
c. *glassy*: no crystals
d. *porphyritic*: crystals of two different sizes

The size of a crystal in an igneous rock is determined by how fast the magma it derives from cools—slow cooling produces large crystals; rapid cooling produces small crystals; nearly instantaneous cooling produces glassy-textured, amorphous rocks, such as obsidian, that have no crystals at all. Why is this so?

Crystallization occurs in two steps: *nucleation* and *crystal growth*. During *nucleation*, small crystals begin to form. Nucleation may occur at a "seed crystal"—a small group of atoms precipitated out of the cooling magma, or some other imperfection in the magma. During crystal growth, atoms attach to the crystal surface, so that the crystal grows in an orderly and systematic way from the outside. When the growing crystals become large enough for their edges to touch, crystal growth stops because there is no more room—the entire magma body is solidified as a patchwork of interlocking crystals.

When a magma cools slowly over many, many years, nucleation occurs much more slowly than crystal growth. Crystals have time to get quite large before they have to compete with their neighbors for available atoms. Thus, when you see a coarse-grained rock, such as granite, or the gabbro shown in Figure 23.22a, with its big, chunky crystals, you know that it cooled slowly from magma, over a span of thousands or millions of years.

On the other hand, cooling can be rapid, on the order of days, weeks, or months. When cooling is fast, the rate of nucleation is high. Legions of tiny crystals pop up at the same time. Each little crystal doesn't have time to grow very large before its neighbors crowd in upon it. Rapid cooling thus produces a *fine-grained* rock made up of small crystals, such as the rhyolite shown in Figure 23.22b.

Sometimes, erupted magma—lava—cools almost instantly. This can happen, for example, to drops of lava spat out of a violent volcano. In this case, the ions

(a)

(b)

(c)

(d)

Figure 23.22 Igneous rocks classified by texture. (a) Gabbro, a coarse-grained rock. (b) Rhyolite, showing a fine-grained texture. (c) Pumice sample with an abrasive texture. (d) Andesite porphyry—a porphyritic rock.

lose energy before they can arrange themselves in *any* crystalline structure—they just freeze in place, and an abrasive texture results. Pumice is an example, shown in Figure 23.22c.

Porphyritic rocks, such as the andesite porphyry shown in Figure 23.22d, have a complex history. Because magma is a mixture of many minerals, it is possible for a cooling magma to contain some crystals that have grown very large while others have not even started to form. If magma that contains some large crystals rises toward the surface suddenly or erupts, the molten portion of the magma or lava cools quickly, and the large crystals will be embedded in a matrix of smaller crystals.

CHECK YOURSELF

1. Obsidian is a volcanic glass that looks like manufactured glass. What is its texture? Is it intrusive or extrusive—and how do you know? How is obsidian unlike most rocks?
2. What is the texture of granite? How quickly does it crystallize from magma?
3. What kind of texture will an igneous rock have if the rate of nucleation was high?

CHECK YOUR ANSWERS

1. Its texture is glassy. Obsidian is extrusive, because the instantaneous cooling that produces such rocks is a surface phenomenon. Obsidian is amorphous, so, unlike true rocks, it is not composed of minerals.
2. Granite has a coarse-grained structure, so it crystallizes slowly from magma over a period of thousands to millions of years.
3. The rock will have a fine-grained texture.

Figure 23.23 The sedimentary rock has formed strata (visible on the lower right side of this photo) at Golden Cap in Dorset, England. These strata were laid down more than 100 million years ago during the Jurassic period. Lias rock is a rich source of fossil ammonites, extinct mollusks with distinctive coiled shells.

23.8 Sedimentary Rocks

Only about 5% of Earth's crust is composed of sedimentary rock. Nevertheless, the rocks you are most likely to encounter are sedimentary rocks, because they are spread out in a thin, discontinuous layer, capping 75% of the continental crust. Some common sedimentary rocks are chalk, limestone, sandstone, and shale. The most characteristic feature of sedimentary rock is that it is built up of layers of material, or **strata**, that have been deposited and fused together over thousands or millions of years (Figure 23.23). Strata, when visible, give sedimentary rock formations a layered appearance. Sedimentary strata are built up of **sediments**—unconsolidated particles created by the weathering and erosion of preexisting rock; by chemical precipitation from water solutions; or by the secretions of organisms.

Layers of sedimentary rock provide a record of the past. The chemical composition and physical form of sedimentary strata reveal much about Earth's history and its past surface environments (Figure 23.24). Also, because sedimentary rocks are formed at Earth's surface,* they often contain the remains of life forms—*fossils*. Fossils tell the story of life on Earth, and they provide clues about the geologic past—for example, by helping to determine the ages of rock. Find out more about fossils in Chapter 26.

Figure 23.24 Hundreds of dinosaur fossils and skeletons have been found in the sandstone at Dinosaur Provincial Park in Canada. Sandstone is a common sedimentary rock.

The Formation of Sedimentary Rock

Sedimentary rock forms in a long process with four stages: *weathering*, *erosion*, *deposition*, and *sedimentation*. Weathering is the first step. **Weathering** is the disintegration or decomposition of rock at or near Earth's surface. Such

*Another reason fossils are found in sedimentary rock, rather than in igneous or metamorphic rock, is that the temperatures that are reached in igneous and metamorphic rock are too high for fossils to remain intact.

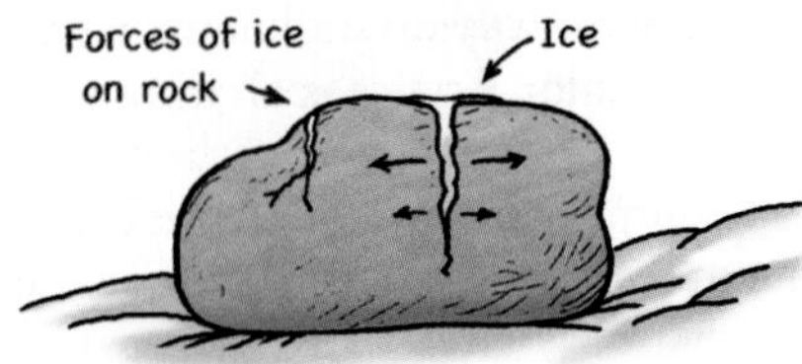

Figure 23.25 Do you remember, from Chapter 6, that water expands upon freezing? In ice wedging, a form of mechanical weathering, water flows into a crack, freezes, and expands, which widens the crack and ultimately splits the rock into smaller pieces.

Unifying Concept
The Gravitational Force
Section 5.3

INSIGHT As rocks containing the elements sodium and potassium are weathered and eroded, these elements tend to form soluble compounds that can be transported in solution far away from the rocks from which they were weathered. They are ultimately carried all the way to the sea, and this is why ocean water is salty.

agents as water, wind, ice, and reactive chemicals weather the rock—breaking it into smaller chunks, cracking its surface, rounding and smoothing its edges and corners, and sometimes transforming its chemical composition. There are two types of weathering—*mechanical* and *chemical.* Mechanical weathering, usually caused by water, physically breaks rock down into smaller pieces. *Ice wedging*, shown in Figures 23.25 and 23.26, is an example. In this process, water seeps into rock, freezes and expands, and eventually melts and contracts. This physically pushes sections of rock apart. Biological agents, such as trees, may produce mechanical weathering, too. For example, you have probably seen roots wedged into cracks in a rock, enlarging the cracks as they grow. Although not as forceful as water, wind mechanically weathers rocks through abrasion.

Chemical weathering is actually the main producer of sediment. In chemical weathering, the compounds in rock decompose into substances that are more stable in the surface environment. Again, water is the main agent. When rain falls, it reacts with carbon dioxide in the air and soil to produce carbonic acid, which makes rainwater slightly acidic. When it seeps downward into rock, acidified rainwater can partially dissolve the rock and alter the minerals it contains (Figure 23.27).

Erosion is the next step in the formation of sedimentary rocks. Erosion is the process by which weathered particles are removed from rocks and transported by streams, glaciers, wind, or other mobile agents. Because of gravity, the ultimate direction in which sediment erodes is always downhill toward the ocean bottom.

Figure 23.26 The rocks on this mountain peak in Wales have been split apart by ice wedging, a kind of mechanical weathering. In this cold environment, water freezing in the cracks of the rocks has frozen and expanded and thereby caused the rocks to crack and break apart.

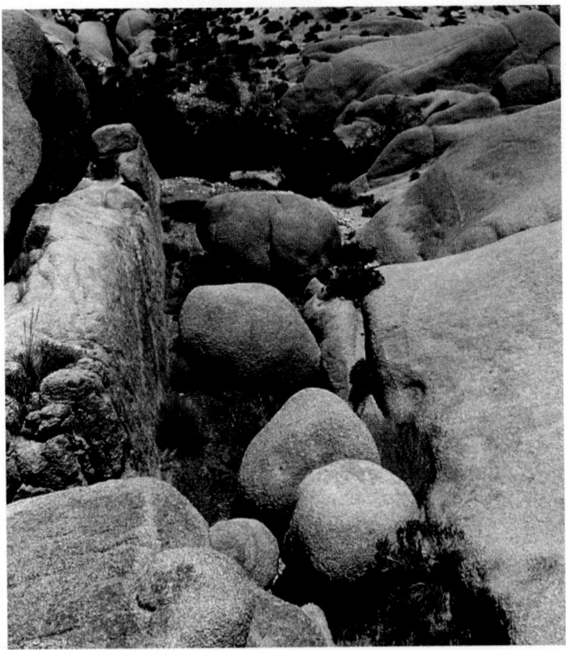

Figure 23.27 This picture shows granite rocks in the Mojave Desert of California that have been worn into rounded shapes through spheroidal weathering. In this process, rainwater chemically weathers the rock by reacting with its outer layers (hydrolysis). This makes the rock easier to erode. Rainwater erodes the rock by physically washing away its weakened outer layers, and leaving rounded boulders behind.

(a) Well-sorted sediments

(b) Poorly sorted sediments

Figure 23.28 A deposit that contains particles of similar sizes is called a well-sorted deposit. A poorly sorted deposit contains sediments of many different sizes. In general, poorly sorted sediments traveled a short distance before they were deposited, and well-sorted sediments traveled a long distance before being deposited.

As sediments are transported, they continue to weather. Pieces of mechanically weathered rock are normally quite angular and jagged when they are first produced. During transportation, especially by water, the chunks, pieces, and particles run into one another and break. This decreases their size and rounds off sharp edges. When transportation stops, deposition and sedimentation begin.

Deposition is the stage in which eroded particles come to rest. Sediments are deposited in horizontal layers, with each successive layer younger than the one beneath it.* The larger a sediment particle, the stronger a water current must be to carry it. Usually, a water current gets weaker as it gets farther from its source. As the water slows down, the larger sediments are the first to be deposited, while smaller ones are able to remain with the flow. In this way, sediments tend to be sorted according to size as they are deposited (Figure 23.28). You will find boulders along the base of a mountain waterfall, gravel along a stream, and sand and silt farther downstream near the river's mouth.

In the process of **sedimentation**, sediment particles *lithify*—literally "turn to rock." Lithification occurs in two steps—*compaction* and *cementation.* Compaction is the first step. As the weight of overlying sediment presses down upon deeper layers, sediment particles are compacted or squeezed together. Compaction squeezes much of the water out of the pores, or spaces, between the sediment particles. The released "pore water" often contains dissolved compounds, such as silica, calcium carbonate, and iron oxide. These compounds can precipitate from solution and partially fill the pore spaces with mineral matter. The mineral matter binds the particles together and acts as a cementing agent in the process of *cementation.* Silica cement, the most durable, produces some of the hardest and most resistant sedimentary rocks. When iron oxide acts as a cementing agent, it produces the red or orange stain often seen in sedimentary rocks. The rock colors of Bryce Canyon National Park in Utah, shown in Figure 23.29, provide a beautiful example of iron oxide stain.

Figure 23.29 The red and orange colors in the sedimentary rocks at Bryce Canyon in Utah are caused by the presence of iron oxides.

*This is formally stated as Steno's Law of Superposition: "In an undisturbed sequence of strata, the oldest strata lie at the bottom and successively higher strata are progressively younger." Danish scientist Nicholas Steno formulated this law in the 1600s.

(b) Silicon and oxygen are less dense than iron.
(c) Both of the above.
(d) Neither of the above.

8. Which minerals crystallize first from magma?
   (a) minerals with the lowest melting point
   (b) minerals with the highest melting point
   (c) minerals with the highest solubility
   (d) minerals with the lowest solubility

INTEGRATED SCIENCE CONCEPTS

Chemistry—The Silicate Tetrahedron

1. The silicate tetrahedron is a grouping of what elements?
2. Why do silicon and oxygen atoms form strong bonds?
3. Why does the silicate tetrahedron form so many different kinds of structures?

Physics—The Texture of Igneous Rock

1. A geologist finds a rock with large crystals embedded in a matrix of smaller crystals. What can he conclude about the rock? What is the texture of this rock?
2. Why are rocks made form slowly cooling magmas coarse-grained? Name an example of this kind of rock.
3. Why are rocks made from rapidly cooling magmas fine-grained? Name an example of this kind of rock.
4. Why are rocks made form instantly cooling magmas glassy? Name an example of this kind of rock.
5. During crystallization of magma, if the rate of nucleation is slow compared with the rate of crystal growth, will crystals be small or large?

Biology—Coal

1. By want biological process do fossil fuels such as coal obtain their stored energy?
2. Why do the plant remains that form coal not decay after they die?
3. By what process is the energy stored in coal released and made available to do useful work?
4. Why does coal become harder and have a higher energy content as it ages from lignite to bituminous coal to anthracite?
5. Why is coal unusual among rocks?
6. Coal formation began in the most recent 10% of Earth's history. Why could coal formation not occur before this?

Active Explorations

1. Look at some halite crystals under a magnifying glass or microscope and observe their general cubic shapes. There's no machine at the salt factory specifically designed to give salt these cubic shapes, as opposed to round or triangular ones. The cubic shape occurs naturally and is reflection of how the atoms of salt are organized—cubically. Smash a few of these salt cubes and then look at them carefully again. What you'll see are smaller salt cubes. Explain why salt breaks into cubes in terms of the property of cleavage.
2. Model the chemical weathering of rock. Take a piece of steel wool that does not contain soap, moisten it with water, and put it on a saucer. After 3 days, pick it up, observe the rust, and notice how the "rock" crumbles when you rub it between your fingers. Steel wool contains iron, which forms iron oxide or rust when water is present. Rocks that have streaks of red, yellow, or reddish-brown usually contain iron. Like the steel wool, they are chemically weathered when exposed to water and air.

Exercises

1. What does roundness tell us about sediment particles?
2. What can we say about a rock that is composed of various sizes of sediments in a disorganized fashion?
3. Describe the process of crystallization.
4. How does the atomic structure of glass differ from the atomic structure of the mineral calcite?
5. Calcite is a nonsilicate mineral. Is it therefore very rare?
6. The chemical formula for quartz is SiO_2. What is the chemical formula of coesite, a polymorph of quartz?
7. An impression—a type of fossil—is made by an organism that is buried quickly, before it can decompose. Is this impression of a fish contained in an igneous, sedimentary, or metamorphic rock? Why do you think so?

8. Your friend makes the following remark: "Minerals in Earth's crust generally do not contain oxygen because oxygen exists in the gaseous state at surface temperatures." Is he right or wrong? Defend your answer.
9. Retrograde metamorphosis is the process of a metamorphic rock returning to its original unmetamorphosed state. (a) What surface conditions encourage retrograde metamorphism to occur? (b) Why don't rocks undergo retrograde metamorphism as easily as they undergo metamorphism?
10. What is more plentiful on Earth—the group of minerals known as feldspars or the group of minerals known as silicates?
11. Does a mineral's stability depend on temperature and pressure? Explain.
12. How does recrystallization produce a metamorphic rock?
13. Name a common intrusive igneous rock often found on the surface of mountains. Explain how this intrusive igneous rock, formed underground, is exposed at the surface.
14. Describe the different conditions that produce the four different kinds of igneous rocks.
15. Why are metamorphic rocks created underground?
16. Cycles in nature, such as the rock cycle, can be viewed as consisting of materials and processes. What are the processes of change that take place in the rock cycle?
17. Is rock conserved in the same sense that energy is conserved? Why or why not?
18. Why does rock melt to form magma?
19. How can one parent magma produce a variety of igneous rocks?
20. Why are coarse-grained igneous rocks generally intrusive igneous rocks? Why are extrusive igneous rocks usually fine-grained?
21. Would you expect to find any fossils in limestone? Why or why not?
22. What makes gold so soft while quartz and diamond are so much harder?
23. The Earth's mineral resources are used in many ways. Many of these resources are plentiful, yet they are also nonrenewable. Once extracted and used, they do not grow back over the span of human lifetimes. What are some possible problems associated with the extraction of minerals?

24. Many minerals can be identified by their physical properties, such as hardness, crystal form, cleavage, color, luster, and density (specific gravity). Why is identifying a mineral by its crystal form usually difficult?
25. Can dikes and plutons be observed above ground? Can they form above ground? Explain.
26. Why is the ocean salty?
27. Why is asbestos in drinking water much less harmful than asbestos in air?
28. Is cleavage the same thing as crystal form? Why or why not?

Problem

Refer to the geothermal gradient. Does temperature change faster with increasing depth between 0 km and 10 km or between 40 km and 50 km? Can you offer an explanation?

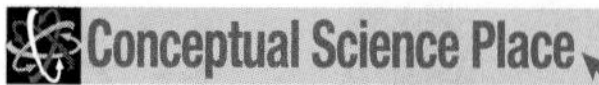

CHAPTER 23 ONLINE RESOURCES

Tutorials The Rock Cycle Activity • ***Quiz*** • ***Flashcards*** • ***Links***

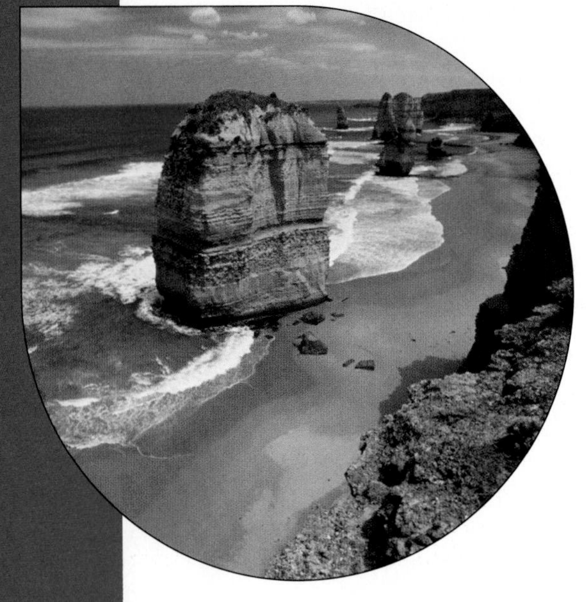

CHAPTER 24

Earth's Surface—Land and Water

The prominent rocky features shown offshore at the Port Campbell National Park in Australia are limestone sea stacks. Sea stacks form as a result of differential rates of erosion along a coastline. As ocean waves wear away a coast, softer rocks and rocks weakened by faults and joints erode first. Rocky formations such as these sea stacks are left behind. Sea stacks do not last long in geological terms due to the constant erosive action of ocean waves.

CHAPTER OUTLINE

Imagine viewing the Earth from a satellite and surveying its entire surface. What would you see? The first thing you'd likely notice is that most of Earth—71%—is covered by ocean. The remaining 29% is taken up by seven continental landmasses. You'd see thousands of volcanoes, and you'd likely wonder: Why are most of them found along the rim of the Pacific Ocean? Earth's crust, in some places, is wrinkled by tight folds; in other places, it's etched by cracks and faults. What titanic forces could have cracked and warped the crust on such a massive scale? Jagged peaks form mountain chains—but why are so many mountain chains located near coastlines? From your aerial perspective, you'd see other interesting surface features, such as smoldering craters, snaking rivers, stark white glaciers, and fan-shaped deltas where rivers meet the sea. Taken as a whole, Earth's surface would appear quite varied—as if it had been broken and bent, twisted and pulled, built up and worn down. And, of course, over geologic time, it has! In this chapter, you'll explore Earth's surface and the processes that build up, tear down, and reshape its ever-changing form.

24.1 A Survey of the Earth

Few of us will ever view the Earth from space, but we can discover much about this awesome planet from satellite transmissions and from field scientists and explorers. We know that Earth consists of seven continents: Africa, Antarctica, Asia, Australia, Europe, North America, and South America. On average, the continents lie 840 meters above sea level. Earth's continental elevation varies between the extremes of its highest point, Mt. Everest (8848 m, or 29,028 ft, above sea level), and its lowest point, on the shores of the Dead Sea (400 m, or 1,312 ft, below sea level).

The most prominent topographic features of the continents are the linear mountain belts—for example the Circum-Pacific Belt and the Alpine-Himalayan Belt. The continents also feature expansive flat areas called *plains,* typically found at low elevations. You would also notice plateaus—uplifted areas composed of horizontal layers of rocks. Like plains, plateaus are flat areas, but they are found at higher elevations (Figure 24.1).

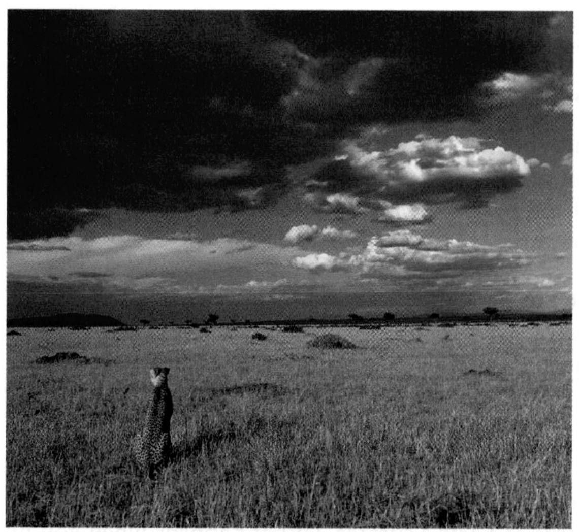

Figure 24.1 Plains in Kenya's Masai Mara reserve.

The three major oceans on Earth are the Pacific Ocean (Earth's largest, deepest, and oldest); the Atlantic Ocean (Earth's coldest and saltiest); and the Indian Ocean (the smallest). However, these oceans are connected; so, in truth, there is only one vast, global ocean. Beneath the oceans, paralleling some coastlines, can be found long, narrow ocean trenches. These are the Earth's deepest places. The lowest point on Earth is in the Marianas Trench, with a depth of 11,033 m (36,198 ft)—that's deeper than any surface mountain is tall. Also under the ocean lies the global mid-ocean ridge system, a continuous mountain belt 65,000 km (40,000 mi) long. Between the ocean trenches and ridges are deep undersea plains.

The topography of the Earth (the shape of its surface) is defined by its *landforms*, or surface features. There is a great diversity of landforms in addition to the plains, plateaus, trenches, and mountains so far mentioned (Figure 24.2). Some of these—faults and folds, for example—were created largely through tectonic processes. Other landforms, such as valleys, canyons, deltas and dunes result mainly from the action of water, wind, or ice as they wear down and redistribute rock.

TECHNOLOGY

Remote Sensing

Today, *remote sensing* is the primary tool for mapping landforms and observing rapid or slow changes on Earth's surface. As the term is generally applied, remote sensing consists of imaging Earth's surface with satellite-based cameras, radio receivers, scanners, thermal sensors, and other instruments. Such tools are used to create digital images, maps, and graphs of Earth's surface features. Most often, remote-sensing devices gather data in the form of electromagnetic waves—visible light, infrared radiation, microwaves, etc. If an image, map, or graph is a "satellite image," you know it was obtained with remote-sensing technology.

There are many remote-sensing applications. Satellite images are used for mapmaking and surveying, to detect temperature variations over Earth's surface, and to track weather changes. Satellite images provide clues that point to subsurface deposits of mineral ores, oil, gas, and groundwater, and they help scientists and others to manage the land, the ocean, and the atmosphere. Satellite pictures also allow us to compare landscapes before and after such natural events as floods, earthquakes, and fires, and to assess the damage done. In 2005, satellite images helped rescuers, homeowners, and the public identify the damage to New Orleans and the Gulf Coast after Hurricane Katrina. Satellite images, posted to the National Oceanographic and Atmospheric Administration's Web site, were the first images of many parts of the devastated Gulf Coast that were made available to the public.

(a)

(b)

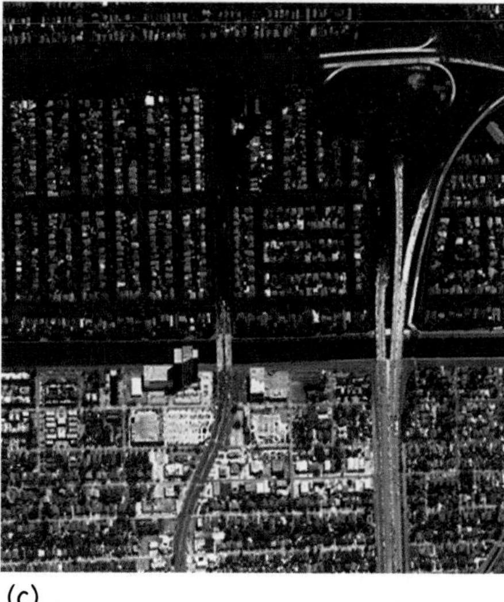

(c)

(a) This satellite image shows New Orleans, Louisiana on March 9, 2004. At the time, New Orleans was a city with 1.3 million people and extensively developed buildings and roads. Because New Orleans lies below sea level, a system of levees was built to prevent flooding by surrounding waters. (b) New Orleans after Hurricane Katrina struck on August 31, 2005. The storm surge and rains led to the breaching of levees. Flooding, as shown here, occurred in more than 80% of the city. (c) An enlarged view of a portion of part (b), showing submerged roads and buildings.

MATH CONNECTION

How Long Can a Mountain Exist?

Due to their impressive bulk, mountains are often considered to be symbols of permanence. But how long can a mountain really exist, given the steady erosion of its rock and soil? We can figure this out with a little math and a few approximations. The overall problem-solving strategy is to estimate the size of a typical mountain, to estimate how much material erodes from it per unit time, and then to divide the size of the mountain by its rate of erosion.

To estimate the volume of a mountain, approximate its shape as a rectangular box. Mountains come in an array of sizes and shapes, but a box 3 km long, 2 km wide, and 5 km high would give us the approximate volume of an average mountain:

$$\begin{aligned}\text{Volume} &= \text{length} \times \text{width} \times \text{height}\\ &= 3\,\text{km} \times 2\,\text{km} \times 5\,\text{km} = 30\,\text{km}^3.\end{aligned}$$

Expressing this in standard units of meters, we have

$$\begin{aligned}\text{Volume} &= (3000\,\text{m}) \times (2000\,\text{m}) \times (5000\,\text{m})\\ &= 3.0 \times 10^{10}\,\text{m}^3.\end{aligned}$$

Now estimate how much rock, sand, and gravel are eroded in a typical day. This would vary according to the kinds of rock that are present, the amount of precipitation, and other factors. But an average mountain could have four principal streams, and each one could easily carry a tenth of a cubic meter of material off the mountain per day. This is a conservative estimate—a tenth of a cubic meter is a box about a foot and a half on each side, about the size of a kitchen sink. So the estimated rate of erosion per day would be:

$$\begin{aligned}\text{Rate of erosion per day} &= (0.1\,\text{m}^3/\text{stream}\cdot\text{day}) \times (4\text{ streams})\\ &= 0.4\,\text{m}^3/\text{day}\end{aligned}$$

From this, we can calculate the rate of erosion per year:

$$\begin{aligned}\text{Rate of erosion per year} &= (0.4\,\text{m}^3/\text{day}) \times (365\text{ days/year})\\ &= 146\,\text{m}^3/\text{year}\end{aligned}$$

Divide the volume of the mountain by the rate at which it is worn away in a year to find out how long it can last:

$$\begin{aligned}\text{Duration of a mountain} &= (3.0 \times 10^{10}\,\text{m}^3)/(146\,\text{m}^3/\text{year})\\ &= 2.05 \times 10^8\text{ years} = 205\text{ million years.}\end{aligned}$$

Thus, a typical mountain would exist only about two hundred million years, even given a conservative estimate of the rate of erosion. Compared with the age of the Earth, which is 4.5 billion years, mountains are young features with short lifetimes.

24.4 Earth's Waters

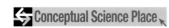

Conceptual Science Place
Hydrologic Cycle Activity

Water—it is vital to life as we know. Water seems plentiful—most of the Earth's surface is covered with it. However, the vast majority of Earth's water, 97.6 %, resides in the oceans as salt water. About 2% of fresh water is frozen in polar ice caps and glaciers. The remainder of Earth's water, less than 1%, consists of fresh water in the form of water vapor in the atmosphere, water in the ground, and water in rivers, streams, and lakes (Figure 24.16).

Water on Earth is constantly circulating, driven by the heat of the Sun and the force of gravity. As the Sun's energy evaporates water, a cycle begins (Figure 24.17). Evaporation moves water molecules from Earth's surface to the *atmosphere*—the thin envelope of gases surrounding the Earth. The resulting moist air may be transported great distances by the wind. Some of the water molecules condense to form clouds and then precipitate as rain, sleet, or snow. The total amount of water

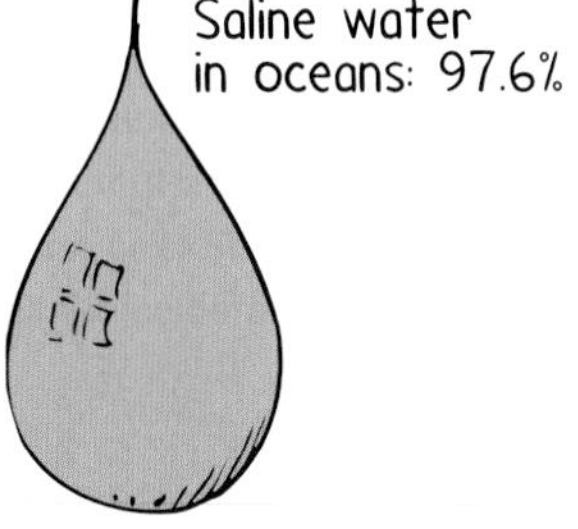

Figure 24.16 Distribution of Earth's water supply.

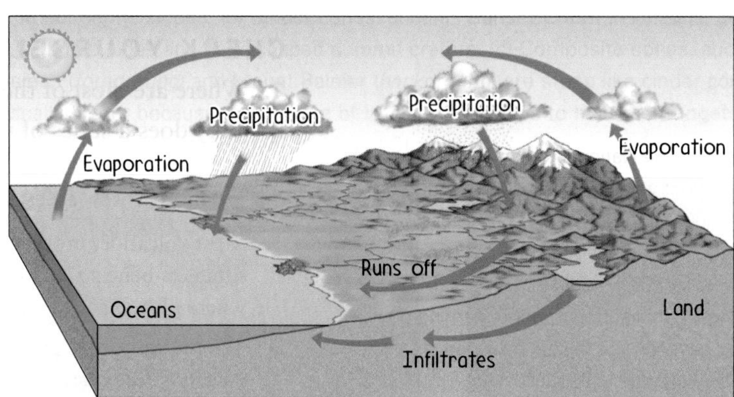

Figure 24.17 The hydrologic cycle. Water evaporated at the Earth's surface enters the atmosphere as water vapor, condenses into clouds, precipitates as rain or snow, and falls back to the surface, only to evaporate again and go through the cycle yet another time.

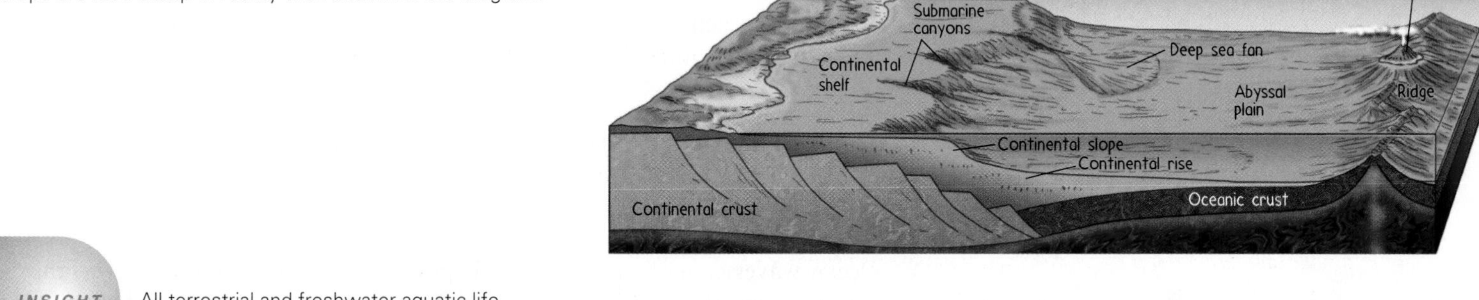

Figure 24.18 The continental margin is the transition zone between the coast and ocean basin. The vertical drops are less steep in reality than shown in the diagram.

INSIGHT All terrestrial and freshwater aquatic life depends on the tiny portion—just 1% of Earth's total water—that is not saline.

vapor in the atmosphere remains relatively constant, because evaporation and condensation balance each other.

If precipitation falls on the ocean, the cycle is complete—water goes from the ocean back to the ocean. A longer cycle occurs when precipitation falls on land, for water may drain to streams, then rivers, and then journey back into the ocean. Or it may soak deep into the ground, or evaporate back into the atmosphere before reaching the ocean. Also, water falling on land may become part of a snow pack or a glacier. Although snow or ice may lock up water for many years, it eventually melts or evaporates, so the water once again can move through the cycle. This natural circulation of water is known as the **hydrologic cycle** or, simply, the *water cycle.*

One part of the water cycle has particular relevance to the shaping of Earth's landforms, and that is the journey fresh water takes from the time it falls as rain, makes its way across Earth's surface, and eventually returns to the ocean. You will learn much more about this in Section 24.6 and Section 24.7.

24.5 The Ocean Floor

The ocean floor covers most of Earth's surface. What is the topography of this vast, hidden area? As Figure 24.18 shows, it is a varied and active geologic region. It consists of *continental margins, deep ocean basins,* and *mid-ocean ridges.* A **continental margin** is a transition zone between dry land and the ocean bottom.

The shallowest portion of the margin, the part closest to the continent, is the *continental shelf.* The shelf, a nearly flat underwater surface extending from the shoreline toward the ocean basin, is an underwater extension of the continent. The area of the continental shelf changes over geologic time as sea level rises and falls and the ocean advances and retreats. It also varies widely from place to place. It is practically nonexistent in certain locations, but it can extend seaward to a width of up to1,500 km in others.

The *continental slope* is the sloping outer edge of a continent; it marks the boundary between oceanic crust and continental crust. The continental slope is steeper along mountainous coastlines than along coastal plains. If you were driving across the continental slope away from coastal mountains, your elevation would decrease about as abruptly as if you were driving down a hillside. On the other hand, if you were driving outward from a coastal plain, the continental slope would barely be a dip in the road.

A *continental rise* is a wedge of sediment that has accumulated at the base of the continental slope. Turbulent currents and gravity transport sediment to the continental rise from the shelf and slope. The falling sediment carves submarine canyons, and these become havens for diverse ocean organisms. When sediment collects at the base of a canyon, it initially forms a fan-shaped deposit called a *deep-sea fan.* Over time, deep-sea fans can grow and merge to form a continental rise.

The *deep ocean basins* cover about 30% of Earth's surface. Deep ocean basins are generally 3 to 5 km deep and are characterized by abyssal plains, ocean trenches, and seamounts. **Abyssal plains** are the flattest and deepest part of the ocean floor. Abyssal plains are mostly flat because thick accumulations of fine-grained sediment bury the underlying uneven oceanic crust.

Figure 24.19 There is almost no light, little food, and extreme high pressure (5,880–8,820 pounds per square inch) in the abyssal plains. Nevertheless, life exists there. Residents of the abyssal plains include the giant sea squid, the largest of all invertebrates. It's more than 150 feet long and it weighs more than a ton. Vicious anglerfish use light lures to attract their prey. Sea stars have luminescence on their arms that can be left behind to trick predators.

Ocean trenches, discussed in detail in Chapter 22, are the deepest places in the oceans, sometimes exceeding 10,000 meters (more than 6 miles) in depth—much deeper than the height of most of Earth's surface mountains. Trenches are places of great tectonic activity. They occur at subduction zones, where lithosphere is forced down into the mantle. This action causes earthquakes and creates volcanoes.

INTEGRATED SCIENCE PHYSICS

Ocean Waves

Ocean waves come in a variety of shapes and sizes, from tiny ripples to the gigantic waves powered by hurricanes. Water waves, like all other waves, begin with some kind of disturbance. The most common disturbance that causes ocean waves is wind. If you blow on a bowl filled with water, you'll see a succession of small ripples moving across the water's surface. The generation of waves in the ocean is similar. As wind speed increases, the ripples grow to full-sized waves; as stronger winds blow, larger waves are created. As waves travel away from their origin, they develop into regular patterns of smooth, rounded waves called *swells*—the mature undulations of the open ocean.

Unifying Concept
Waves
Section 8.1

Wave motion can be understood in terms of a sine curve, as shown in Figure 24.20. Recall, from Chapter 8, that it is the disturbance that is carried by a wave, not the medium the wave moves through. The waveform travels across the ocean while the material making up the wave, for the most part, remains in one place.

However, because ocean waves have both transverse and circular components, ocean waves are more complicated than the simple transverse waves described in Chapter 8. As water passes a given point, the water particles at that point move in circular paths. This circular motion can be seen by observing the behavior of a piece of floating wood on the ocean's surface. The wood sways to and fro while bobbing up and down, actually tracing a circle during each wave cycle. This circular motion occurs near the water's surface, and it decreases gradually with depth (Figure 24.21). At depths greater than half a wavelength, circular motion is negligible. For this reason we can say, with reasonable accuracy, that water waves occur mainly at the surface.

Figure 24.20 Ocean waves (a) have characteristics of simple sine waves (b).

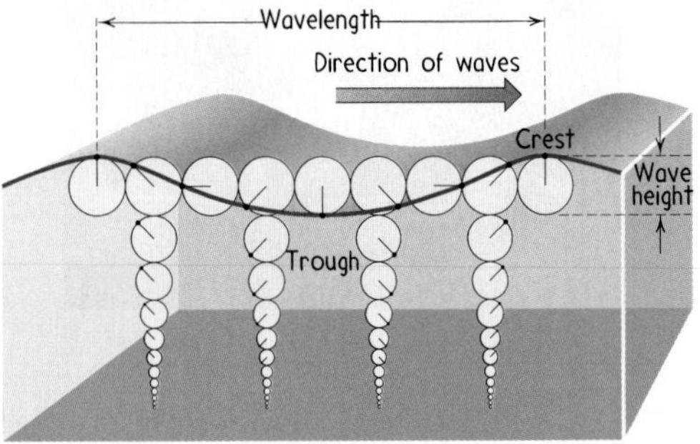

Figure 24.21 Movement of water particles with the passage of a wave. The particles move in a circular orbit. Orbital motion is greatest at the surface and gradually decreases with depth. At depths greater than half a wavelength, orbital motion is negligible.

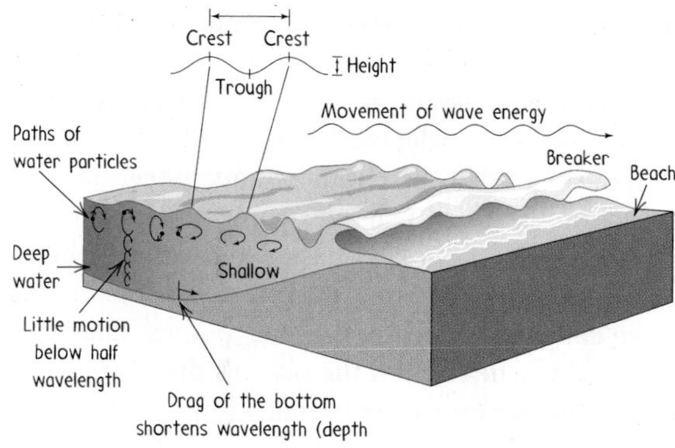

Figure 24.22 Waves change form as they travel from deep water through shallow water to shore. In deep water, orbital motion is circular. In shallow water, orbital motion becomes elliptical as a result of contact of the wave with the bottom. When waves reach a critical height, they break and crash shoreward into the surf zone.

When a wave approaches the shore, where water depth decreases, the circular-motion is interrupted by the ocean bottom. As the water's depth gets shallower and approaches half the wave's wavelength, the bottom of the circular path flattens, slowing the wave. The wave period remains unchanged because the swells from deeper waters continue to advance. As a result, incoming waves gain on leading waves, and the distance between waves decreases. This bunching up of waves in a narrower zone produces higher, steeper waves. When wave height steepens to the point at which the water can no longer support itself, the wave overturns, breaking shoreward with a crash. This breaking water is called *surf*—the area of wave activity between the line of breakers and the shore (Figure 24.22).

CHECK YOURSELF

1. Why does a surfer stay at the same spot in an area of ocean swells unless she paddles to the shore or further out to sea?
2. When does a wave slow down as it nears the shore?

CHECK YOUR ANSWERS

1. The water under the surfer is moving in a circular path, moving backward as much as forward.
2. A wave slows when the bottom part of its circular path flattens.

INTEGRATED SCIENCE CHEMISTRY

Ocean Water

Ocean water is a complex solution of mineral salts, dissolved gases, and decomposed biological material. More than 70 chemical elements are found in it. Indeed, seawater has been described as a weak solution of almost everything! Despite this, the composition of seawater is surprisingly simple, because only a few elements and compounds are present in abundance. Seawater contains about 35 g of dissolved *salts* (see Chapter 11) for every 1000 g of solution. Sodium chloride (NaCl) plus four other abundant salts make up more than 99% of the salts in the sea, as Table 24.1 shows.

Table 24.1 Abundant Salts of the Sea

Salt of seawater	Weight per 1000 grams
Sodium chloride (NaCl)	23.48 g
Magnesium chloride ($MgCl_2$)	4.98 g
Sodium sulfate (Na_2SO_4)	3.92 g
Calcium chloride ($CaCl_2$)	1.10 g
Sodium fluoride (NaF)	0.66 g
Total:	34.8 g

Salinity is defined as the proportion of dissolved salts to pure water. Thus the salinity of seawater is about 35 grams per 1000 grams of seawater, or 35 parts per thousand. Scientists note this with the symbol ‰. So the salinity of the ocean is written: 35‰. How salty is this in everyday terms? One cubic foot of seawater yields 2.2 pounds of salts.

Water, "the universal solvent," has a strong ability to dissolve salts, so salts exist as ions in the sea. In terms of the concentrations of ions, seawater is about 55% chloride (Cl^-), 31% sodium (Na^+), and 8% sulfate (SO_4^{2-}). Each of the other ions make up 1% or less of the weight of a given sample of seawater.

The salinity of seawater varies in a narrow range. A particular ocean's salinity decreases when freshwater is added to it. For example, heavy rains, river runoff, and the melting of ice decrease the saltiness of an ocean on a seasonal basis. Conversely, formation of sea ice and the evaporation of water make an ocean saltier. Evaporation increases salinity because only pure water vapor leaves the seawater solution; the salts are left behind. And when sea ice forms, only the water molecules freeze. Salts once again are left behind in solution. There are slight regional variations in salinity as well. The salinity of oceans in the dry subtropics where evaporation is high can reach 37‰. Heavy precipitation dilutes the oceans in equatorial locations to as low as 33‰.

It is a remarkable fact that the composition of seawater has remained virtually constant over millions—perhaps even a billion—years. After all, there is a great deal of activity in the oceans. Salts and other compounds are continually added to the oceans by eroded continental rock, volcanic vents, dead marine organisms, and other sources.

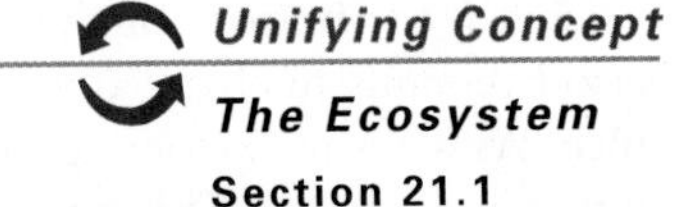

To keep the composition of seawater constant, salts and other compounds must be removed from the sea as fast as they are deposited. Much salt is removed from seawater by chemical precipitation. Also, precipitates and other material are taken into the mantle by subduction. Marine life has a strong influence on the composition of seawater by removing salts, dissolved gases, and other solutes. Multitudes of tiny *foraminifers* (marine protozoans) and various *crustaceans* (such as crabs and shrimp) remove calcium salts to build their bodies. Diatoms, which are microscopic marine algae, draw heavily on the ocean's dissolved silica to form their shells. Some animals concentrate elements that are present in seawater in minute, almost undetectable amounts. Lobsters extract copper and cobalt; certain seaweeds concentrate iodine; and (how's this for an esoteric fact?) sea cucumbers extract vanadium!

CHECK YOURSELF

1. If you want to mix up a batch of saltwater with the same salinity as seawater, how many grams of salt would it take to make 1 kilogram of seawater? How many grams of water?
2. Name three sources of the salts in seawater.
3. Why is the composition of seawater so stable?

CHECK YOUR ANSWERS

1. The salinity of saltwater equals 35‰. So measure out 35 grams of salt to the beaker then pour in 965 grams of water to make 1 kilogram of seawater.
2. Weathered rock; bodies of dead ocean-dwelling organisms; volcanic vents.
3. Salt and other materials are removed from saltwater as fast as they are deposited.

24.6 Fresh Water

You know that only about 2% of Earth's water is nonsaline. Most of this fresh water—85%—is frozen in ice sheets and glaciers and so is unavailable for human use (Table 24.2). The largest supply of fresh water for human use is groundwater, which is water that has soaked underground. Surface water—liquid water in lakes, ponds, rivers, streams, springs, and puddles—is far less plentiful than groundwater, but it has a huge impact on life and surface geology. Rivers and streams, for example, contain only one one-thousandth of one percent (0.001%) of Earth's total water at any given time. Yet, running water moves through them quickly, so rivers and streams play a major role in moving rock and sculpting Earth's surface. The ultimate source of Earth's fresh water is precipitation.

Table 24.2 Distribution of Earth's Fresh Water

Parts of the Hydrosphere	Volume of Fresh Water (km^3)	Percentage of Total Volume of Fresh Water
Ice sheets and glaciers	24,000,000	84.945
Groundwater	4,000,000	14.158
Lakes and reservoirs	155,000	0.549
Soil moisture	83,000	0.294
Water vapor in the atmosphere	14,000	0.049
River water	1,200	0.004
Total	28,253,200	100.0

Surface Water

When rain falls on land, most of it goes back up into the atmosphere through evaporation. Most of what doesn't evaporate soaks into the ground. This absorption of water by the ground is called **infiltration.** Whatever the ground can't absorb becomes **runoff,** water that moves over the Earth's surface. The proportion of rainfall that becomes runoff depends on the type of soil and on how wet or dry it is, as well as the steepness of the slope, the presence of plant life, and the rate at which the rain falls—whether it is a sudden downpour, or a sustained, gentle rain.

INSIGHT You see a lot of runoff on city streets because pavement reduces the ground's ability to soak up water.

During and after a rainstorm, runoff collects in sheets that move downhill and merge to form *streams.* A stream is any body of flowing surface water, from the tiniest woodland creek to the mightiest river (a river is just a large stream.) As streams move downhill, they merge with other streams. In this process, some streams can become quite large. Whatever the size, most streams eventually discharge into the sea.

The area of land that drains into a stream is called the stream's *drainage basin* or **watershed.** Watersheds can be large or small, and every stream, tributary, or river has one. Large rivers that gather many streams also claim their watersheds. In other words, the watershed of a large river is a patchwork of the many smaller watersheds that service the smaller streams.

Watersheds are separated from one another by **divides,** lines that trace the highest ground between streams. Under most circumstances, the separation is complete—rain that falls on one side of a divide cannot flow into an adjacent basin. A divide can be hundreds of kilometers long if it separates two large watersheds, or it can just be a short mountain ridge separating two small gullies. The Continental Divide, a continuous line running north to south down the length of North America, separates the Pacific basin on the west from the Atlantic basin on the east. Water to the west of the Divide eventually flows to the Pacific Ocean, and water to the east of it flows to the Atlantic Ocean (Figure 24.23).

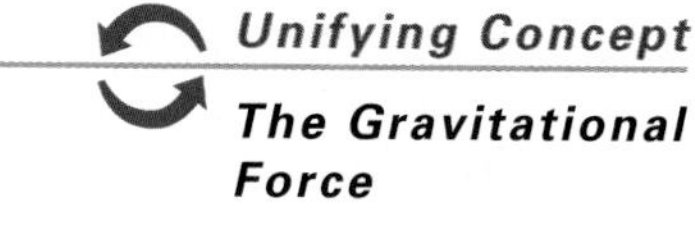

Unifying Concept
The Gravitational Force
Section 5.3

Surface runoff pauses on its way to the sea when it flows into a lake. A lake is formed by surface and subsurface water flowing into a depression in the Earth's surface. For a lake with a stable water level, evaporation plus drainage out of the lake balances the water flowing into it.

Groundwater

Rain and snowmelt absorbed by the ground may be retained by the soil. Some of this soil moisture is taken in by plants, and some of that water is transpired into the atmosphere. Water that isn't captured by soil percolates downward, moving between rocks and sediments and into the narrow joints, faults, and fractures in rock. Percolating water continues to move lower until it reaches the *saturated zone.* In the saturation zone, all the open spaces between sediments and rocks—

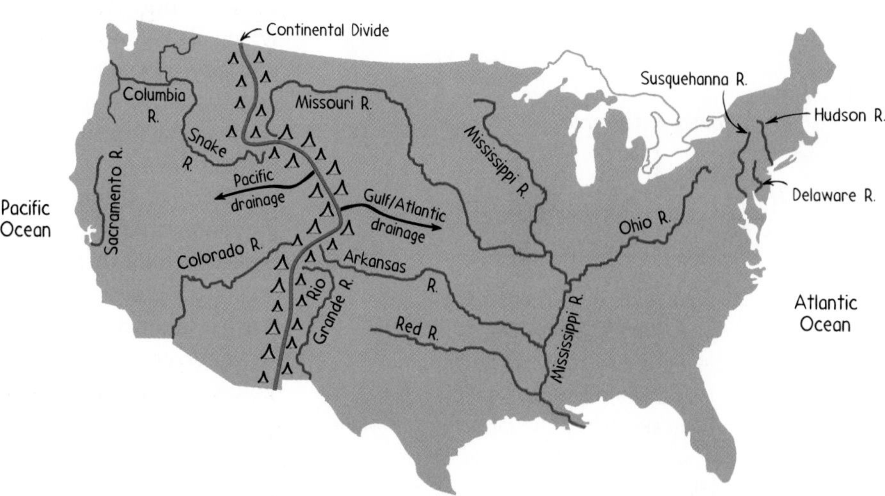

Figure 24.23 The Continental Divide in North America separates the Pacific basin on the west from the Atlantic basin on the east.

SCIENCE AND SOCIETY

Whose Water Is It?

If you have ever stood on the shore of one of the Great Lakes, or experienced the power of a waterfall, or been caught in a major downpour, it may seem to you that the supply of freshwater on Earth is inexhaustible. From the perspective of one resident of the United States, water may be quite abundant. The world population, however, has grown to more than 6 billion people. If we were to spread ourselves evenly throughout all habitable land, there would be about 50 of us in every square kilometer. Thus it should come as no surprise that, collectively, we have a big impact on Earth's limited resources, including fresh water.

We are reminded that fresh water is a limited resource in the United States when farmers fight for the privilege to irrigate, or when our water utility bills rise, or when the water supply of downstream municipalities is endangered as upstream municipalities release sewage into the water. Globally, there are many nations in which the primary supply of fresh water is rivers that originate in some other nation. As the upstream nation diverts fresh water for its own expanding population, political tensions rise. Over the next decade, for example, it is projected that agricultural development in Ethiopia and Sudan will reduce the flow of the Nile River into Egypt by 15%. Similarly, Turkey is currently escalating its damming and irrigation projects along the headwaters of the Euphrates River. Once fully implemented, these projects could result in a 40% reduction of the river's flow into Syria and an 80% reduction of the river's flow into Iraq. Not surprisingly, this issue has been a major source of political tension among these nations.

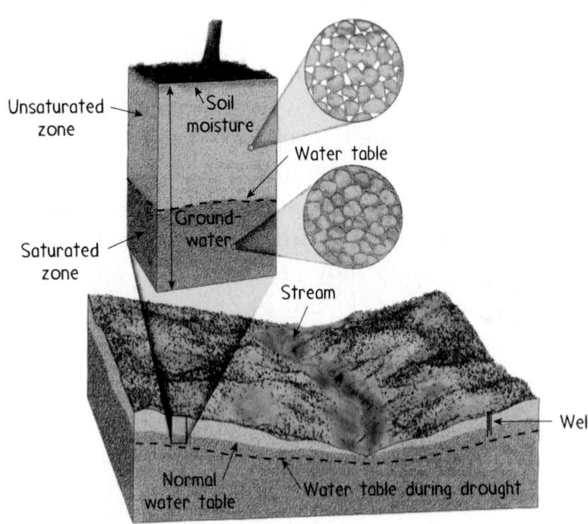

Figure 24.24 The unsaturated zone is above the saturated zone. Water in the unsaturated zone does not completely fill the open pore spaces. This is soil moisture. Water in the saturated zone completely fills all open pore spaces. This is groundwater.

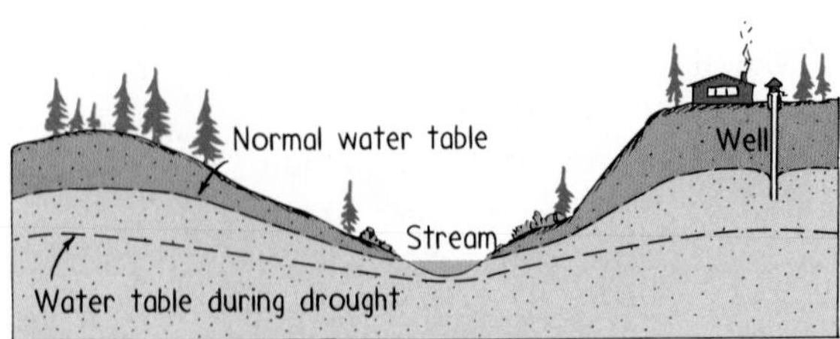

Figure 24.25 The water table roughly parallels the surface of the ground. If a well is drilled, the water level in the well is at the water table. In time of drought, the water table falls, reducing stream flow and drying up wells. The water table also falls if the rate at which water is pumped out of a well exceeds the rate at which groundwater is replaced.

and even the spaces between mineral grains *inside* rocks—are filled with water. Water residing in the saturated zone is called **groundwater.** Groundwater continues to flow both downward and laterally in the saturated zone, eventually finding a stream or other outlet often at the land surface (Figure 24.24).

Groundwater supplies streams, lakes, swamps, springs, and other surface waters. Most surface water is not obtained directly from runoff; instead it flows into surface reservoirs from underground. This is because a large percentage of precipitation does not become runoff—it infiltrates into the ground and slowly moves until it empties into stream channels. During dry periods, when rain does not fall, it is groundwater that feeds surface waters. So groundwater can provide needed water to the surface in time of drought.

Wells pump groundwater to the surface, where it is used for drinking, agriculture, and industry. In some regions, however, overuse of this resource has led to environmental problems, including land subsidence, groundwater contamination, and streamflow depletion.

The upper boundary of the saturated zone is the **water table.** The level of the water table beneath Earth's surface varies with precipitation and climate. The water table is not flat like a kitchen table; rather, it tends to rise and fall with surface topography, as shown in Figure 24.25. Where the water table intersects the surface of the land, we find marshes, swamps, and springs. At lakes and perennial streams (streams that run all year), the water table is above the land surface.

You may have noticed how, during a rainstorm, sandy ground soaks up rain like a sponge. Sandy soils soak up water very easily; other soils, such as clay soils, do not. Rocky surfaces with little or no soil are the poorest absorbers of water. The amount of groundwater that a material can store depends on its **porosity.** Porosity is the percentage of the total volume of rock or sediment that consists of *pore spaces* (open spaces.) Pore spaces are usually found between sediments, but fractures in rock, spaces between mineral grains within a rock, and pockets formed in soluble rock also contribute to porosity.

Rock or sediment may be very porous, but, if the pore spaces are small and not interconnected, groundwater cannot freely move through it. **Permeability** is the ease with which fluids flow through pore spaces within a rock or between rocks or sediments. Sand and gravel are highly permeable because they are composed of rounded particles that do not fit together tightly. Their pore spaces are large and connected, so water can flow easily from one pore space to the next. On the other hand, if pore spaces are too small or poorly connected, water cannot flow through them at all. Clay sediments, for example, are composed of flat particles that fit tightly together. This is why clay, which can be quite porous, is practically impermeable (Figure 24.26).

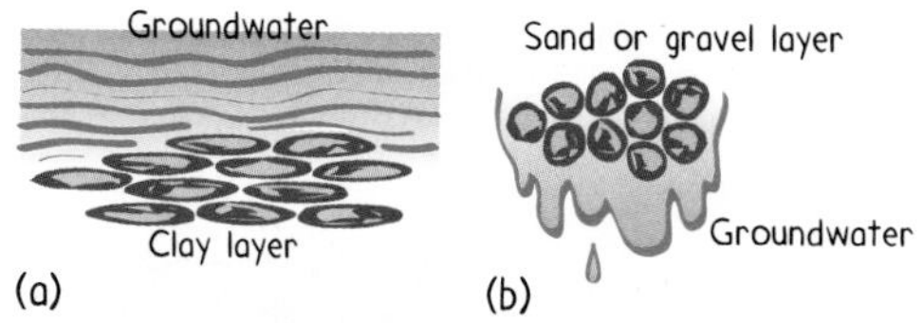

Figure 24.26 (a) The sediment particles in clay are flat and tightly packed, so pore spaces are poorly connected. Thus, they do not transmit water, so clay is said to be impermeable. (b) sediment particles in sand and gravel are relatively uniform in size and shape, with large and well-connected pore spaces. This allows water to flow freely; sand and gravel are permeable materials.

An **aquifer** is a zone of water-bearing rock through which groundwater can flow. Aquifers generally have high porosity and high permeability. These reservoirs, which underlie the land surface in many places, contain an enormous amount of groundwater. They are important because wells can be drilled into them and water can be removed. More than half of the land surface in the United States is underlain by aquifers. One such is the vast Ogallala aquifer, which stretches from South Dakota to Texas and from Colorado to Arkansas.

Groundwater continues to collect as precipitation percolates down from the surface. However, the process of restoring lost groundwater, called *recharging*, is slow. An aquifer constantly gains water from its recharge zone (the area of land from which the groundwater originates), but only a small amount of water reaches it each year. Completely recharging a depleted aquifer may take thousands or even millions of years. Thus, groundwater is considered a nonrenewable resource.

INSIGHT Groundwater, the largest supply of fresh water available for human use, has great economic importance. In the United States, more than 40% of the water used for everything except hydroelectric power production and power-plant cooling is groundwater.

CHECK YOURSELF

1. What's the difference between a saturation zone and an aquifer?
2. Wells are drilled into aquifers. Why? Can a well run dry?
3. At a certain depth, percolating subsurface water encounters impermeable rock. What happens to this water?

CHECK YOUR ANSWERS

1. A saturation zone may consist of water-bearing but impermeable rock or sediments, such as clay. Aquifers have high porosity *and* high permeability.
2. Aquifers readily transmit water to wells. Withdrawn water is replaced eventually by precipitation, though the process can be very slow. Pumping too much water too fast depletes the water in the aquifer. Eventually, this causes the well to yield less water and even to run dry. Pumping your own well too fast can make your neighbor's well run dry, too.
3. It can no longer move downward, so it moves laterally—sideways—and eventually finds a surface outlet.

INSIGHT Want to see the water table? Most ponds and lakes are simply places where the land surface is below the water table. Swamps and marshes are places where the water table is level with the ground surface.

INTEGRATED SCIENCE CHEMISTRY

Groundwater Contamination

Because groundwater is used in irrigation and industry, and because it supplies surface rivers and lakes, groundwater pollution is a problem for everyone—not just for those who get their drinking water from wells. Sewage is a major cause of groundwater pollution. Sewer water contains bacteria that, if untreated, can cause such waterborne diseases as typhoid, cholera, and infectious hepatitis. Agricultural areas where nitrate fertilizers are used extensively also contribute to groundwater pollution. Nitrate levels in groundwater must be closely monitored because these compounds, in concentrations as small as 15 parts per million, are toxic to humans. Pesticides and other agricultural chemicals are also primary pollutants that seep into the groundwater. Leaky chemical storage tanks and industrial wastewater lagoons further contribute to the problem. Figure 24.27 illustrates how these pollutants get into the groundwater.

A groundwater contaminant that has received much attention is the gasoline additive known as MTBE (methyl *tert*-butyl ester). This chemical is widely used to boost the octane rating of gasoline and as an oxygenate, meaning it reduces the amount of unburned hydrocarbons and carbon monoxide in auto exhaust. However, MTBE is a suspected carcinogen. Gas stations generally store gasoline in

Figure 24.31 Large sinkhole in the state of Virginia.

The Work of Groundwater

Flowing groundwater changes surface landscapes. For example, by dissolving and transporting soluble rock underground, groundwater allows surface depressions known as sinkholes to form (Figure 24.31). Sinkholes are funnel-shaped cavities in the ground that are open to the sky.

Land subsidence is another example. Land subsidence is a lowering of the land-surface elevation caused by changes underground, including changes caused by the overpumping of aquifers. When groundwater is removed from pore spaces, water pressure is reduced, so sediment layers are compressed. A lowering of the land surface is the observable result. While land subsidence has occurred in nearly every state of the United States, it is especially prevalent in the West and Southwest, as a result of the large-scale development of groundwater resources that began there a half-century ago (Table 24.3, Figure 24.32).

Land subsidence is permanent. Even if an aquifer were to be recharged so that groundwater levels returned to what they were prior to subsidence, the land-surface elevation would not appreciably recover. Measures for reducing land subsidence include switching from groundwater to surface water supplies. If this is not possible, water usage may need to be reduced or alternate well sites that minimize subsidence must be determined.

The dissolving action of groundwater carves out magnificent caves and caverns (caverns are simply large caves). Typically, the process occurs in limestone where groundwater has been acidified by its reactions with the calcium carbonate in the limestone (Figure 24.33).

Table 24.3 Subsidence Levels in Selected Areas of the Western United States

State	City	Subsidence (m)
Arizona	West of Phoenix	5.5
	Tucson	<0.3
California	Davis	1.2
	Southwest of Mendota	8.8
	Santa Clara Valley	3.7
Nevada	Las Vegas	1.8
Texas	El Paso	<0.03
	Houston	2.7

Figure 24.32 The land surface in California's San Joaquin Valley subsided by about 9 meters (30 feet) over a 50-year period because of groundwater withdrawal and the resulting compaction of sediments.

The Work of Wind

Water is the dominant agent of change to the natural landscape, but wind plays a role too. It plays a lesser role in shaping the land than water and ice because it cannot erode or carry sediments as effectively. Further, wind is intermittent, and it cannot chemically weather sediment. Nevertheless, if you've ever been in a windstorm or at the beach on a windy day, you can appreciate wind's sandblasting effect. Once in the air, particles of sediment can be carried great distances by wind. Red dust from the Sahara of North Africa is found on glaciers on the Swiss Alps and on islands in the Caribbean Sea. Fine grains of quartz from central Asia are blown onto the beaches of the Hawaiian Islands.

In the desert, winds move over surfaces of dry sand, picking up the small, easily transported particles but leaving the large, harder-to-move particles behind. The small particles bounce across the desert floor, knocking more particles into the air, to form ripple marks, which are actually tiny sand dunes (Figure 24.34).

Figure 24.33 The various limestone formations inside the Blanchard Springs Caverns in Arkansas were created by calcium carbonate dissolved in groundwater dripping through the caves. The calcium carbonate was deposited slowly over thousands of years to create large pillars of limestone, such as those in this photograph.

Figure 24.34 Generated by blowing winds, ripple marks are narrow ridges of sand separated by wider troughs. They are small, elongated sand dunes. Large sand dunes are visible in the background of the photograph.

Even in the desert, though, water does more work on the land than the wind does. The occasional intense rains in the desert produce surface runoff that moves more sediment than the much more frequent action of the wind.

The Work of Glaciers

A **glacier** is a mass of dense ice that forms from unmelted snow that has accumulated for several thousand years. A glacier moves under its own weight due to the pull of gravity, so, in the simplest terms, you can think of a glacier as a slowly moving river of ice. Glaciers are powerful agents of erosion and deposition. Glaciation has created the beautiful landscapes of Tibet, Nepal, and Bhutan in Asia; the Alps of Switzerland; the fjords of Norway; and Yosemite Valley and the Great Lakes in North America. As it moves across the Earth's surface, a glacier loosens and lifts up blocks of rock, incorporating them into the ice. In many ways, a glacier is like a plow as it scrapes and plucks up rock and sediment (erosion). But it is also like a sled as it carries its heavy load to distant places (deposition). The large rock fragments carried at the bottom of a glacier scrape the underlying bedrock and leave long, parallel scratches (like sled tracks) aligned in the direction of ice flow (Figure 24.35). These are called *striations*.

Figure 24.35 Striations mark the location of a former glacier.

There are two main types of glaciers, *alpine* and *continental*. Alpine glaciers develop in mountainous areas and are often confined to individual valleys, while continental glaciers cover much larger areas. Alpine glaciers occur in most high mountain chains in the world, such as the Cascade Range, the Rocky Mountains, the Andes and the Himalayas. They can transform V-shaped stream valleys into rounded, U-shaped valleys by eroding the sides and bottom of an angular V-shaped valley. Continental glaciers cover broader areas; they can cover entire continents and reach a thickness of 1 km or more. Huge continental glaciers, called *ice sheets*, are currently found on Greenland and Antarctica. They spread out over the land surface, smoothing and rounding the underlying topography as they move.

Glaciers advance across the land from the poles and retreat according to fluctuations in Earth's climate. In an *ice age*, Earth's temperature takes a sustained downturn and ice sheets extend into the Northern and Southern hemispheres. We are currently in the middle of an ice age. Specifically, we are in an *interglacial period*, a time between two major advances of glaciers (Figure 24.36).

INSIGHT Twenty thousand years ago, continental glaciers covered much of North America. These glaciers retreated to Greenland 10,000 years ago.

As a glacier advances across the land, it acquires and transports great amounts of debris. When the glacier retreats, this debris is left behind as it is melted out of the ice. Because a glacier abrades and picks up everything in its path, glacial deposits are characteristically composed of unsorted rock fragments

(a)

(b)

Figure 24.36 (a) Glacier des Bossons in the French Alps is an Alpine glacier. At approximately 7 km in length, it is the longest glacier in Europe. It is also one of the steepest glaciers, with an average inclination of 45 degrees. (b) An aerial view of the Greenland ice sheet, a continental glacier.

(a)

(b)

(c)

Figure 25.1 Tropical rain forests, such as this one in Trinidad, are biomes found in the tropical climate zone. Tropical climates promote the world's largest diversity of plant life. (b) Temperate grasslands, such as the Oklahoma prairie shown here, are biomes found in the temperate climate zone. (c) The tundra is a biome located in the polar climate zone. Here, within the Arctic Circle, grasses and tough shrubs grow in the frozen soil.

below 10°C (50°F) even in the warmest months. In the polar climate zone, at the peak of summer, the Sun never sets; and in the deep winter, it never rises.

25.2 Solar Radiation

Seasons

Temperature—whether it's hot or cold—is the most fundamental element of weather and climate. Temperature depends in large part on **solar radiation**—electromagnetic energy emitted by the Sun.

The Sun heats up the atmosphere indirectly—it warms the ground, and then the ground warms the air. This is how it works: The ground is warmed as its molecules absorb incoming high-energy electromagnetic waves from the Sun and transform some of this energy to increased molecular motion. This increased molecular energy is measurable as a rise in temperature, as you may recall from Chapter 6. But most of the energy absorbed by molecules at Earth's surface is reradiated. Radiation reemitted from Earth's surface is called *terrestrial radiation* (Figure 25.3). Whereas incident solar radiation is in the visible, short-wavelength part of the electromagnetic spectrum, terrestrial radiation is in the longer-wavelength infrared range. (The reason for this, as explained in Chapter 8, is that an emitting body radiates electromagnetic waves with a peak frequency that is directly proportional to its own temperature.) Gases of the lower atmosphere, while largely transparent to the incoming short-

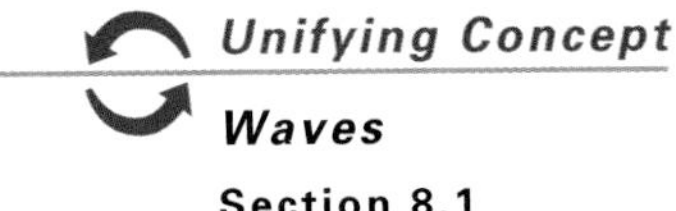

Waves
Section 8.1

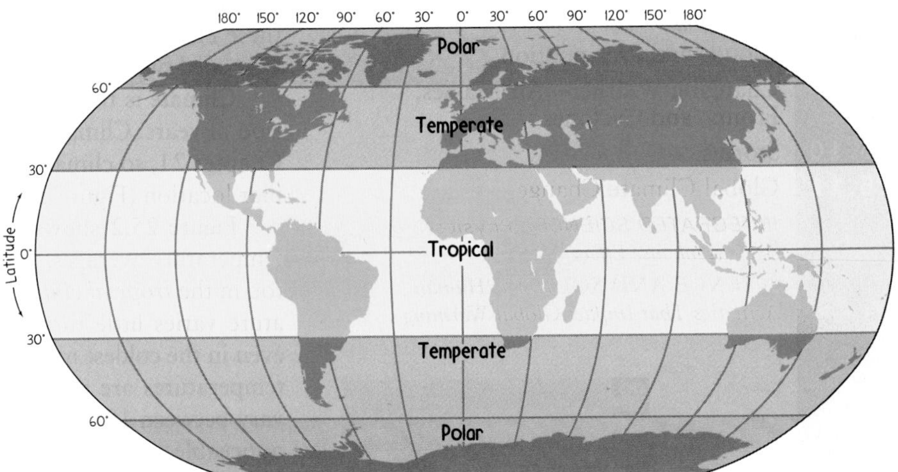

Figure 25.2 Yearly average temperatures vary predictably in Earth's principal climate zones, shown here.

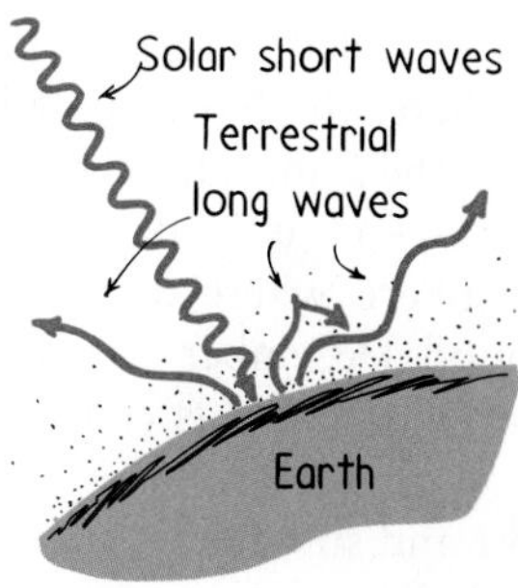

Figure 25.3 The hot Sun emits short-wavelength electromagnetic radiation, and the cooler earth reemits long wavelengths. Radiation emitted from the Earth is called *terrestrial radiation.*

INSIGHT The surface of the Sun has a high temperature and therefore emits electromagnetic waves at a high frequency—mostly in the visible portion of the spectrum. The surface of the Earth, by comparison, is relatively cool, and so the radiation it emits has a frequency lower than that of visible light. The radiation Earth emits is in the infrared range. You can't see infrared—it's below our threshold of sight—but you can feel it. Infrared waves absorbed by our skin produce the sensation of heat. When you absorb infrared radiation from the atmosphere, the atmosphere feels warm to you.

wavelength solar radiation, readily absorb the longer-wavelength terrestrial radiation. As atmospheric gases absorb and transform terrestrial radiation to increased molecular motion, they are warmed.*

But Earth's temperature is not constant. It varies with latitude as shown in Figure 25.2. Further, temperature varies seasonally at nonequatorial latitudes. Why? The basic cause of geographic and seasonal temperature variation is uneven heating of Earth's surface. Consider geographic temperature variation first.

The surface is generally warmer where *solar intensity*, the amount of solar radiation per area, is highest. Equatorial regions are Earth's warmest places because they experience maximum solar intensity. The Sun's rays strike these regions most directly, as shown in Figure 25.4. Polar regions, on the other hand, are Earth's coolest places because solar intensity is at a minimum there. The Sun's rays strike the high latitudes most obliquely, and they therefore receive less solar radiation per area compared with equatorial regions.

Variation in solar intensity with latitude also explains the seasons. This is where the tilting of Earth's axis of rotation comes into play. For example, the northern United States and Canada have distinct summer and winter seasons because they experience different solar intensities at these times of the year, due to Earth's tilting axis and the corresponding variations in the angle that the Sun's rays make with Earth's surface. To see this, carefully examine Figure 25.5. It shows how the tilt of Earth's axis produces the differences in solar intensity in the yearly cycle of seasons. At any spot on Earth, when the Sun's rays are the closest to perpendicular as they will ever get at that place, that region experiences summer. Six months later, the rays fall upon the same region more obliquely, and there is winter. In between are fall and spring.

Another effect of the angle of the Sun's rays with respect to Earth's surface is the length of daylight each day. Can you see, in Figure 25.5, that a location in summer has more daylight per daily rotation of the Earth than the same location in winter when the Earth is on the opposite side of the Sun? If you have trouble visualizing this, take a look at the high latitudes at the poles. Consider the special latitude in the Northern Hemisphere where daylight lasts nearly 24 hours during the summer solstice (around June 21) and night lasts about 24 hours at the winter

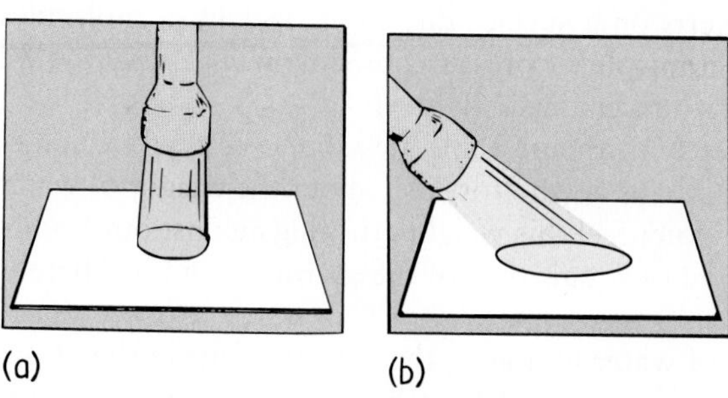

Figure 25.4 This simple demonstration shows why temperature depends on the angle of incident solar radiation. (a) When the flashlight is held directly above at a right angle to the surface, the beam of light produces a bright circle. (b) When the light is shone at an angle, the light beam elongates into ellipses, spreading the same amount of light energy over more area and therefore decreasing the intensity of the light striking the surface. The same is true for sunlight on Earth's surface. High noon in equatorial regions is like a vertically held flashlight; high noon at higher latitudes is like the flashlight held at an angle.

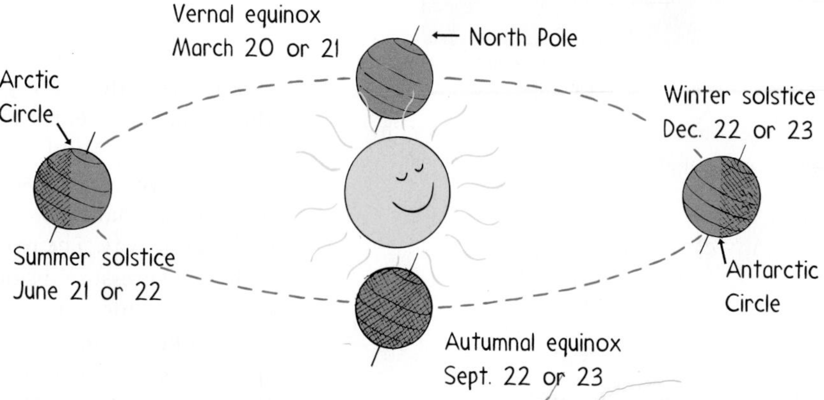

Figure 25.5 The tilt of the Earth and the corresponding differences in the intensity of solar radiation produce the yearly cycle of the seasons. It is interesting to note that Earth follows an elliptical path around the Sun, and that the Earth is actually farthest from the Sun when the Northern Hemisphere experiences summer. So the angle of the Sun's rays, not the distance from the Sun, is most responsible for Earth's surface temperatures.

* Only about half of the solar radiation striking the top of the atmosphere reaches Earth's surface and is absorbed. The rest is either (1) absorbed by the clouds and atmosphere or (2) reflected skyward by clouds, the atmosphere, or Earth's surface. This portion of solar radiation that is reflected by a surface is called its *albedo*. Albedo varies; a snowy hillside, for example, reflects much of the solar energy incident upon it, so it would have a high albedo.

INSIGHT The thickness of our atmosphere is determined by two competing factors: the kinetic energy of its molecules (temperature), which enables molecules to move off in different directions; and gravity, which tends to hold them to the ground. If Earth's gravity were somehow shut off, atmospheric molecules would disappear into space. Or, if gravity acted but the molecules moved too slowly to form a gas (as might occur on a remote, cold planet), our atmospheric gases would become a liquid or solid layer—just so much more matter lying on the ground. The balance between gravity and thermal energy on Earth is delicate, and it is just right to support life in the biosphere.

solstice (around December 21). This latitude is called the Arctic Circle. During the summer solstice in the Northern Hemisphere, the North Pole leans toward the Sun and the South Pole leans away from the Sun. Summer and winter are reversed, of course, in the two hemispheres, and so are the dates of the solstices. The special latitude in the Southern Hemisphere where daylight lasts nearly 24 hours during the summer solstice (around December 21) and night lasts about 24 hours at the winter solstice (around June 21) is called the Antarctic Circle.

Halfway between the peaks of the winter and summer solstice, around mid-September and mid-March, the hours of daylight and night are of equal length. These are called the equinoxes (Latin for "equal nights"). The equal hours of day and night during the equinoxes are not restricted to high latitudes but occur all over the world. The length of daylight plays a role in Earth's temperature; it's not just the amount of solar radiation striking the Earth's surface that matters, the length of time that the solar radiation is being absorbed matters as well.

CHECK YOURSELF

1. Your friend says, "The tilt of the Earth's axis, not Earth's distance from the Sun, is the cause of Earth's seasons." Do you agree or disagree? Explain.
2. Why are daylight hours fewer in winter months?

CHECK YOUR ANSWERS

1. Agree. The tilt of Earth's axis affects the angle at which solar radiation strikes a given location. When a location experiences winter, the angle at which the Sun's rays strike the Earth's surface is most acute. In summer, the rays strike most directly.
2. The Earth is tilted on its axis, like a top leaning in one direction all the time. As the Earth revolves around the Sun, the Northern Hemisphere is tilted toward the Sun in summer and away from the Sun during the winter. When the Earth is tilted away from the Sun, the Sun is lower on the horizon. Therefore, the Sun rises later and sets earlier, resulting in shorter days. If it weren't for the 23.5° tilt of the Earth's axis, there would be no seasons.

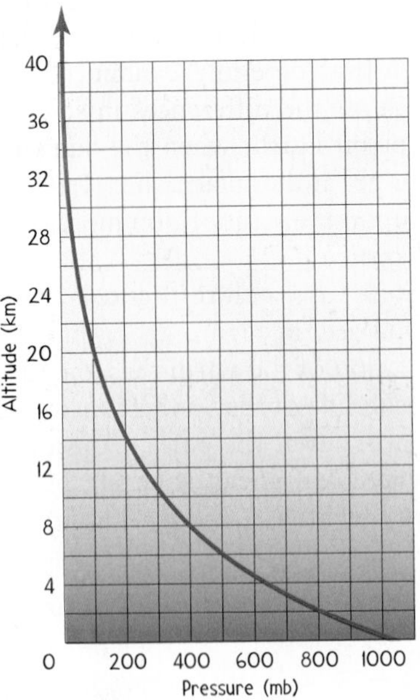

Figure 25.6 Atmospheric pressure declines with altitude as density of atmospheric gas diminishes.

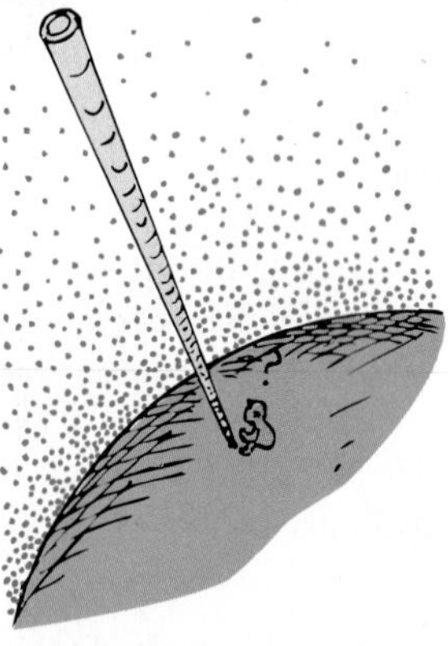

Figure 25.7 The mass of air that would occupy a bamboo pole that extends to the "top" of the atmosphere is about 1 kg. This air has a weight of 10 N.

25.3 Atmospheric Pressure

We live at the bottom of the atmosphere. Since the atmosphere has weight, it pushes against Earth's surface with a certain force per unit area—a pressure. The pressure that the atmosphere exerts on a surface due to the weight of molecules above that surface is known as **atmospheric pressure**, or simply *air pressure*. (We use the terms *air* and *atmosphere* interchangeably here.)

We have adapted so completely to the invisible air around us that we sometimes forget that it has weight. Perhaps a fish "forgets" about the weight of water in the same way. The reason we don't feel this weight crushing against our bodies is that the pressure inside our bodies equals that of the surrounding air. There is no net force for us to sense.

Unlike the uniform density of water in a lake, the *density* of air in the atmosphere (its mass per volume, as discussed in Chapter 2) decreases with altitude (Figure 25.6). This is the well-known phenomenon of the thinning of the air with altitude, which you have noticed if you have ever gone mountain climbing. The higher you go, the fewer gas molecules there are in a given volume of air, and the harder it is to breathe. At sea level, 1 cubic meter of air at 20°C has a mass of about 1.2 kilograms; at a height of 10 kilometers, 1 cubic meter of air has a mass of about 0.4 kilograms (Figure 25.7).

What is the value of atmospheric pressure at sea level? A column of air with a cross-sectional area of 1 square meter extending up through the atmosphere has a mass of about 10,000 kilograms. The weight of this air is about 100,000 newtons (10^5 N). This weight produces a pressure of 100,000 newtons per square meter, so we estimate atmospheric pressure to be 10^5 N/m^2 (Figure 25.8).

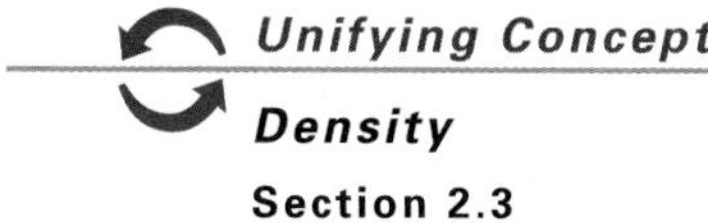

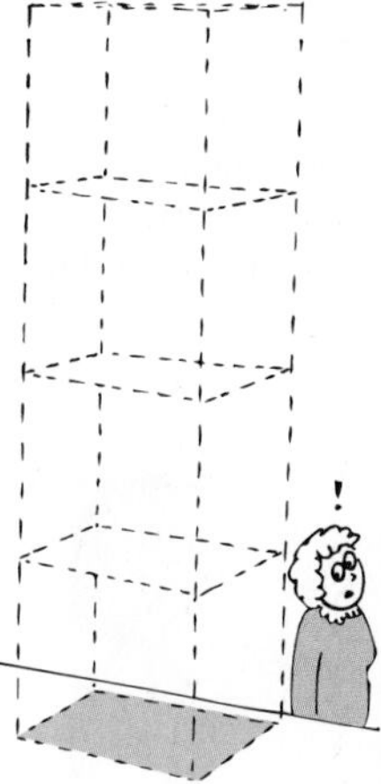

Figure 25.8 The weight of air that bears down on a one-square-meter surface at sea level is about 100,000 N. So atmospheric pressure is about 10^5 N/m^2 or 100 kPa.

The SI unit of pressure is the pascal (Pa), and 1 Pa equals 1 N/m^2. Our estimate of atmospheric pressure thus could be stated as 100 kilopascals. Actually, the average atmospheric pressure at sea level is precisely 101.3 kilopascals (101.3 kPa). This is often expressed as 1 atmosphere (1 atm) of pressure. There are other popular units for atmospheric pressure. News programs report atmospheric pressure in *inches of mercury* or *millimeters of mercury*. This unit refers to the mercury barometer, the instrument traditionally used to measure air pressure. Meteorologists often express air pressure in *bars* or *millibars* (mb). Conversion factors for air pressure are listed in Table 25.1.

Table 25.1 Equivalent Measurements for Atmospheric Pressure

1 standard atmosphere (1 atm)
101,325 pascals
1.013 bars
14.7 pounds per square inch (14.7 psi or 14.7 lb/in^2.)
760 torr
760 millimeters of mercury (760 mm Hg)

CHECK YOURSELF

1. Why doesn't the pressure of the atmosphere break windows?
2. The density of atmospheric gas molecules diminishes with altitude. How does this affect atmospheric pressure?

CHECK YOUR ANSWERS

1. Atmospheric pressure is exerted on both sides of a window, so no net force is exerted on the window. If, for some reason, the pressure is reduced on one side only, as in a strong wind, watch out!
2. Atmospheric pressure decreases due to the decreasing density of atmospheric gas molecules.

25.4 Structure and Composition of Earth's Atmosphere

Earth's present-day atmosphere is a mixture of nitrogen and oxygen, with small percentages of water vapor, argon, and carbon dioxide, and trace amounts of other elements and compounds (Table 25.2). (See Chapter 26 for an interesting account of how Earth's atmosphere has changed over time.)

Our atmosphere is divided into layers, each with different characteristics. As Figure 25.9 shows, the lowest layer of the atmosphere is the **troposphere**. The troposphere is the atmosphere's thinnest layer, extending from the ground to a height of 16 kilometers over the equatorial region and to a height of 8 kilometers over the poles.

Because of gravity's pull, most of Earth's atmospheric gas molecules are held close to Earth in the troposphere. Though this layer is relatively thin, the troposphere contains 90% of the atmosphere's mass and almost all of Earth's water vapor and clouds. This makes the troposphere the densest atmospheric layer.

Vertical Structure of the Atmosphere

Table 25.2 Composition of the Atmosphere

Gas	Symbol	Percentage by Volume	Gas	Symbol	Percentage by Volume
Permanent Gases			*Variable Gases and Particulates*		
Nitrogen	N_2	78	Water vapor	H_2O	0 to 4
Oxygen	O_2	21	Carbon dioxide	CO_2	0.038
Argon	Ar	0.9	Ozone	O_3	0.000004*
Neon	Ne	0.0018	Carbon monoxide	CO	0.00002*
Helium	He	0.0005	Sulfur dioxide	SO_2	0.000001*
Methane	CH_4	0.0001	Nitrogen dioxide	NO_2	0.000001*
Hydrogen	H_2	0.00005	Particles (dust, pollen)		0.00001*

*Average value in polluted air

TECHNOLOGY

The Barometer

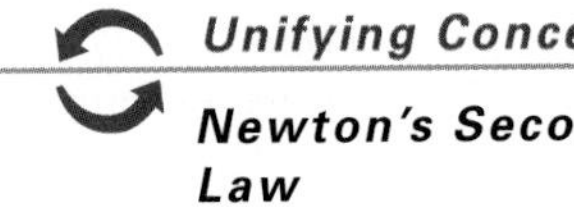
Unifying Concept
Newton's Second Law
Section 3.2

The *mercury barometer* was invented by Evangelista Torricelli, a student of Galileo, in 1643. People have been using this simple but clever bit of technology ever since. The term *barometric pressure* is synonymous with atmospheric pressure and air pressure, reflecting the historical importance of the barometer.

760 mm

A simple barometer consists of a glass tube longer than 76 centimeters (760 millimeters). The tube is closed at one end and filled with mercury. At standard atmospheric pressure, the mercury in the tube runs out of the submerged open bottom until the level in the tube is 76 centimeters above the level in the dish. The empty space trapped above, except for some mercury vapor, is a vacuum. The vertical height of the mercury column remains constant even when the tube is tilted—unless the tube is shorter than 76 centimeters, in which case the mercury completely fills the tube.

Why does mercury behave in this way? The explanation is similar to the reason a simple seesaw will balance when the weights of the people at the two ends are equal. The barometer "balances" when the weight of liquid in the tube exerts the same pressure as the atmosphere outside. Whatever the width of the tube, a 76-centimeter column of mercury weighs the same as the air that would fill a tube of the same width reaching up to the "top" of the atmosphere. If the atmospheric pressure increases, then the atmosphere pushes down harder on the mercury and the column is pushed higher than 76 centimeters. The mercury is literally pushed up into the tube of a barometer by the weight of the atmosphere pushing down on the surface of mercury in the bowl.

An aneroid barometer, which is small and portable, works differently. It consists of a small metal box that is partly exhausted of air and has a flexible lid that bends in or out with changes in atmospheric pressure. The motion of the lid is passed along by a mechanical spring-and-lever system to a dial that shows the air pressure. Modify an aneroid barometer slightly, and you have a cylindrical chamber called a *bellows*, which is squeezed inward when air pressure is high and expands outward when air pressure is lower. Connect the bellows to a device that records air pressure, and you have an *aneroid barograph*. An aneroid barograph, shown below, provides a record of air pressure changes over time.

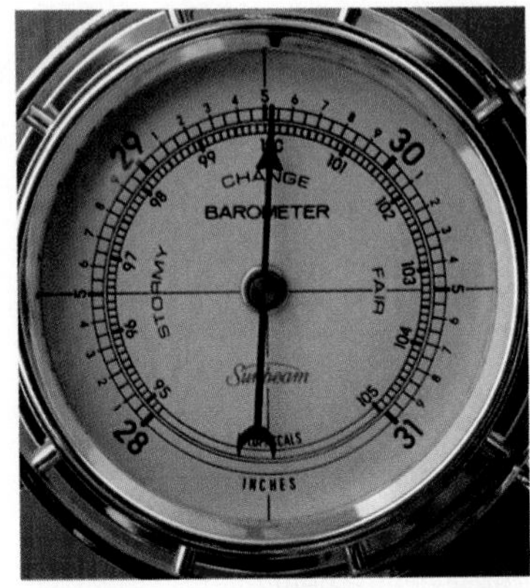
CHANGE
BAROMETER
STORMY
FAIR
INCHES

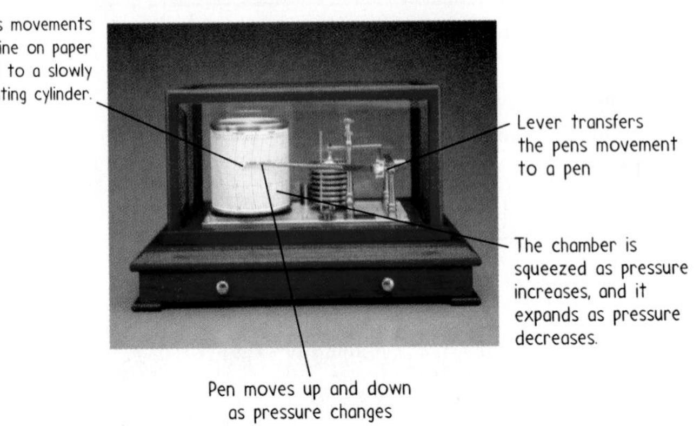
The pen's movements draw a line on paper attached to a slowly rotating cylinder.
Lever transfers the pens movement to a pen
The chamber is squeezed as pressure increases, and it expands as pressure decreases.
Pen moves up and down as pressure changes

Temperature in the troposphere decreases steadily (at 6°C per kilometer) with increasing altitude. At the top of the troposphere, temperature averages a frigid −50°C.

Weather occurs primarily in the troposphere. Commercial jets, for example, generally fly just above the troposphere in order to avoid turbulence caused by weather disturbances. Weather is restricted to the troposphere because uneven heating of Earth's surface drives the many terrestrial radiation-absorbing molecules there to move and mix. This creates weather patterns.

Above the troposphere is the **stratosphere**, which reaches a height of 50 km above the ground. Ozone molecules form in the stratosphere and absorb ultraviolet

INSIGHT Generally, higher elevations are cooler than those near sea level. On an average basis, temperature decreases 6°C (11°F) for every 1000 meters of elevation. The reason? Atmospheric gas molecules exist in a dense, compressed layer near Earth's surface. When they emit infrared radiation, there are many neighboring molecules to absorb it. In this way, heat is retained. At higher elevations, there are fewer neighboring molecules to capture the infrared radiation and so the warming infrared radiation increasingly escapes to outer space.

radiation (UV) from the Sun. As the molecules absorb the UV radiation, they transform some of its electromagnetic energy to kinetic energy, and so they become warmer. (Do you remember from Chapter 6 that temperature is a measurement of the kinetic energy of molecules?) Due to the absorption of UV radiation by ozone, the temperature rises from about −50°C at the bottom of the stratosphere to about 0°C at the top.

The **mesosphere** extends upward from the top of the stratosphere to an altitude of about 80 kilometers. The gases that make up the mesosphere absorb little solar radiation. As a result, temperature decreases from 0°C at the bottom of the mesosphere to about −90°C at the top.

Temperature varies in the opposite way in the atmosphere's next layer, the **thermosphere**. Here, temperature generally increases with altitude. There is little air in this layer, but what air there is readily absorbs solar radiation. For this reason, temperatures are high in the thermosphere, ranging from 500°C to 1500°C depending on solar activity. Because of the very low density of air molecules in the thermosphere, however, this extreme temperature has little significance. Very little heat would be transferred to a slowly moving body in this region. You would be quite chilly if you could visit the thermosphere.

The **ionosphere** is an ion-rich region within the thermosphere and upper mesosphere. The ions are produced by the interaction of high-frequency solar radiation with atoms of atmospheric gases. Incoming solar rays strip electrons from nitrogen and oxygen atoms, producing a large concentration of free electrons and positively charged ions. Ions in the ionosphere cast a faint glow that prevents moonless nights from being stark black. Near Earth's magnetic poles, fiery light displays called *auroras* occur as the solar wind (high-speed charged particles ejected by the Sun) strikes and excites molecules of atmospheric gases. Auroras are most spectacular during times of *solar flares*—storms or eruptions of hot gases on the Sun.

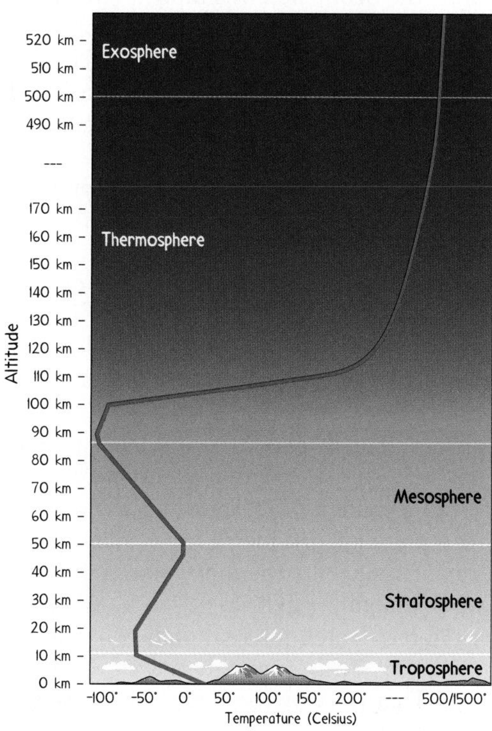

Figure 25.9 The average temperature of Earth's atmosphere varies in a zig-zag pattern with altitude.

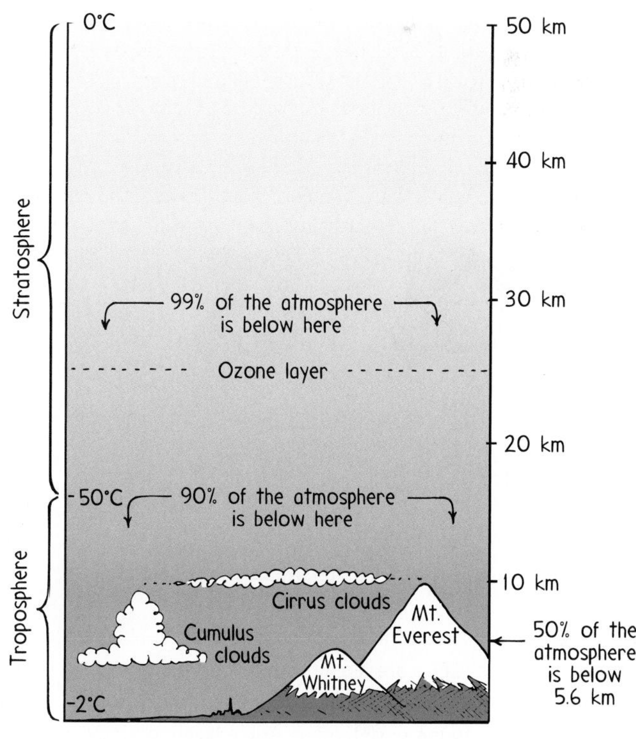

Figure 25.10 The two lowest atmospheric layers, the troposphere and the stratosphere.

Finally, above 500 kilometers, is the **exosphere**. The exosphere is the sparse, outermost layer of atmosphere that gradually thins until it yields to the radiation belts and magnetic fields of interplanetary space.

INTEGRATED SCIENCE CHEMISTRY

The Atmospheric "Ozone Hole"

INSIGHT You learned in Chapter 22 that the Earth is a sphere composed of concentric shell-like layers: crust, mantle, and core. You're learning in this chapter that the atmosphere, too, consists of concentric shell-like layers with different characteristics: troposphere, stratosphere, and so forth. But Earth has yet another shell-like layer of great importance—the *biosphere*. The biosphere is the delicate layer in which life is found. It extends from below the seafloor to high in the atmosphere, and it envelops the entire globe. Bacteria, for example, live in the pore spaces of rock beneath us, up on the highest mountain peaks, and in the thin air of great altitudes.

INSIGHT The atmosphere rises up from Earth's surface, but where does the atmosphere end and outer space begin? There is no distinct boundary. The atmosphere thins rapidly with distance from the Earth, then dissipates gradually, and finally terminates where there are too few molecules to detect, at an altitude of around 500 km.

In its most stable form, oxygen exists as molecular oxygen (O_2). Oxygen also forms the less stable molecule **ozone**, which consists of three oxygen atoms (O_3). Ozone is a pale blue gas with a sharp odor that is characteristic of the air after a thunderstorm or that of the air near an old electric motor. It is very reactive, and it is toxic even at low concentrations, particularly to the lungs. Ozone is a component of smog, and it is carefully monitored when it is found near ground level. These are the reasons why tropospheric ozone has been called "bad ozone."

But ozone isn't always bad. "Good ozone" is O_3 that is found in the stratosphere. There, ozone the pollutant becomes ozone the protector. It is created when high-energy ultraviolet radiation breaks diatomic oxygen down to atomic oxygen, which then reacts with additional O_2 to form ozone:

$$O_2 + \text{UV radiation} \longrightarrow 2O$$
$$2O + 2O_2 \longrightarrow 2O_3$$
$$\text{net reaction} \quad 3O_2 \longrightarrow 2O_3$$

This synthesis of ozone is essential to life on Earth because it absorbs harmful ultraviolet (UV) radiation. Ultraviolet radiation can kill bacteria, as well as cause genetic damage, cancer, and eye and skin injury.

Scientists discovered an "ozone hole" over Antarctica in the 1970s; since the early 1980s, they have been measuring its size. It is not a literal "hole" in an "ozone shield." Rather, the **ozone hole** is simply a large area of the stratosphere with extremely low levels of ozone. Atmospheric ozone has been depleted (beyond its natural variation) as a result of the presence of *chlorofluorocarbons* (CFCs) in the atmosphere. CFC's are inert gas molecules that were once commonly used in air conditioners and aerosol propellants. Two of the most frequently used CFCs are shown in Figure 25.11.

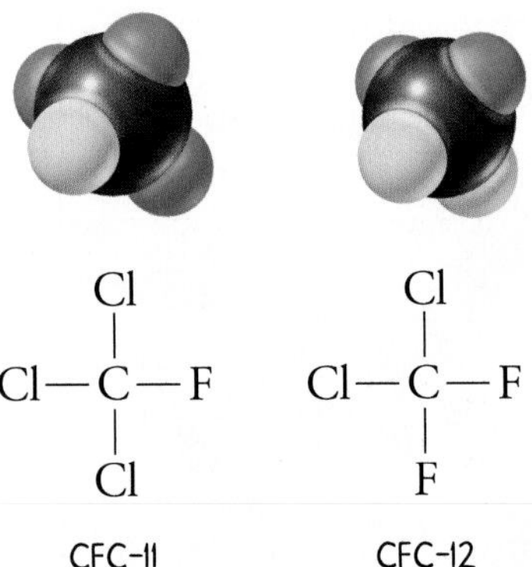

Figure 25.11 Two of the most common CFCs, also known as freons. CFCs are highly stable, inert gases. Because of their inertness, CFCs were once thought to pose little threat to the environment. At the height of CFC production in 1988, some 1.13 million tons were produced worldwide.

CFCs are so stable that they remain in the environment for 80 to 120 years, and so they are now thoroughly spread throughout the atmosphere. However, CFC molecules don't remain inert when they reach the stratosphere. There, the strong UV radiation fragments them, liberating their chlorine atoms. The chlorine atoms react with ozone and catalytically destroy it. Estimates are that one chlorine atom can destroy at least 100,000 ozone molecules in one or two years before it forms a hydrogen chloride molecule (HCl) and is carried away by atmospheric moisture.*

Ozone depletion is seasonal, and it occurs most markedly over Antarctica and the North Pole because the frigid winters there favor the formation of stratospheric ice crystals (Figure 25.12). Chlorine-containing compounds cling to the crystals and react with them, producing molecular chlorine Cl_2. In the spring, when sunlight returns, it fragments the molecular chlorine into ozone-depleting atomic chlorine: $Cl_2 + \text{sunlight} \rightarrow 2Cl^-$.

There has been unprecedented international cooperation in banning ozone-destroying chemicals. In 1987, an international agreement called the Montreal Protocol on Substances That Destroy the Ozone Layer phased out the production

* A catalyst, as explained in Chapter 13, is a substance that increases the rate of a chemical reaction without being consumed by the reaction.

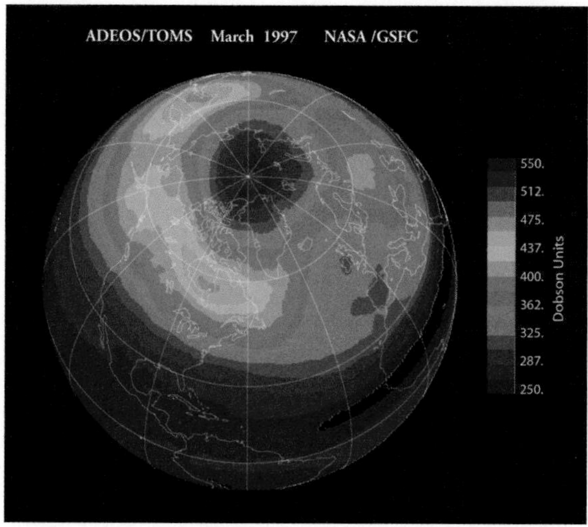

Figure 25.12 False-color image of ozone levels over the Northern Hemisphere, recorded by the Total Ozone Mapping Spectrometer (TOMS) of the National Aeronautics and Space Administration (NASA). Purple and blue areas are areas of ozone depletion; green through red areas are areas of higher-than-normal ozone levels.

of CFCs by 2000. The good news is that satellite measurements show that the rate of ozone depletion in the upper stratosphere is slowing; the bad news is that the total amount of ozone up there is still declining. Year to year, monitoring of the ozone hole shows that its size varies from encouragingly small to disappointingly large. But now that CFC production has been curtailed, we should see the ozone hole trending smaller until eventually the Antarctic ozone layer recovers.

CHECK YOURSELF

1. Is there any chemical difference between stratospheric ozone and ozone found in air pollution?
2. Where do the chlorine atoms that destroy ozone in the stratosphere come from?

CHECK YOUR ANSWERS

1. No, there is no chemical difference. Ozone, no matter where it is found, and whether it is "good" or "bad," is a molecule consisting of three oxygen atoms (O_3).
2. There are natural sources for some of the chlorine that ends up in the stratosphere (such as volcanoes), but the chlorine atoms generated by human activity are freed from chlorofluorocarbon (CFC) molecules.

25.5 Circulation of the Atmosphere—Wind

Air, or any gas, moves from an area of higher pressure to one of lower pressure. If your car tire gets punctured, for example, you hear a hissing sound as air rushes out of it. Similarly, **wind** is air flowing horizontally from a region of high-pressure to one of lower pressure.

Every day, thousands of weather stations around the world at sea level record air pressure in units of millibars. When the data are mapped, **isobars** are used to connect places with the same average pressure (Figure 25.13). The space between isobars represents the change in pressure over a given distance; closely

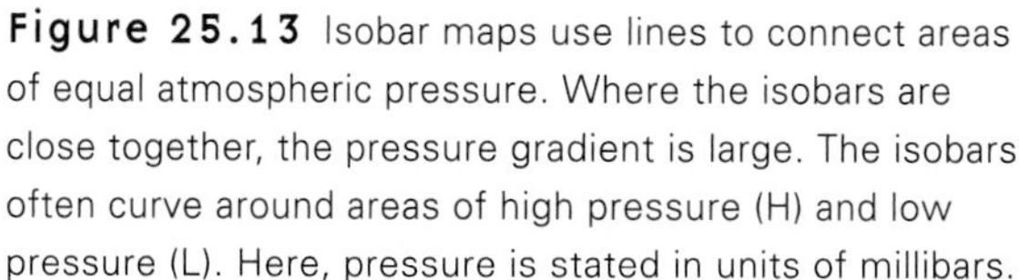

Figure 25.13 Isobar maps use lines to connect areas of equal atmospheric pressure. Where the isobars are close together, the pressure gradient is large. The isobars often curve around areas of high pressure (H) and low pressure (L). Here, pressure is stated in units of millibars.

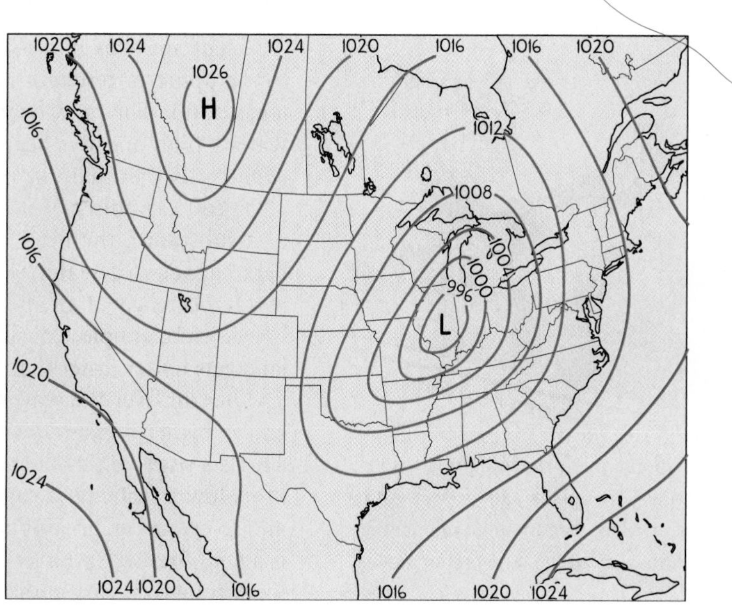

Figure 25.14 Air moves along a pressure gradient from higher pressure to lower pressure.

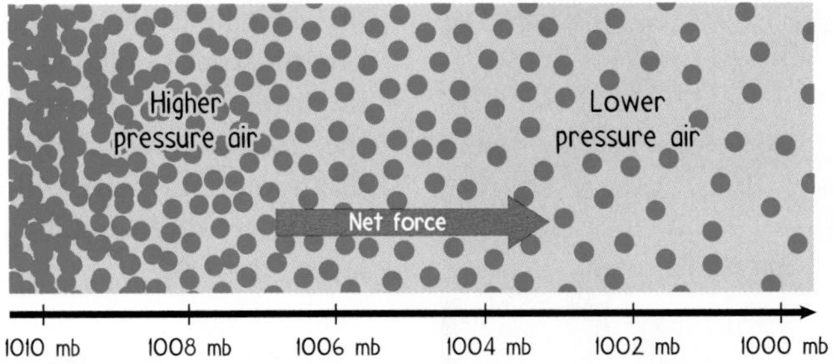

INSIGHT A downhill skier's speed depends on the steepness or *gradient* of the slope. Similarly, wind speed is a function of the pressure gradient.

INSIGHT Can you see that the Sun is the ultimate cause of wind?

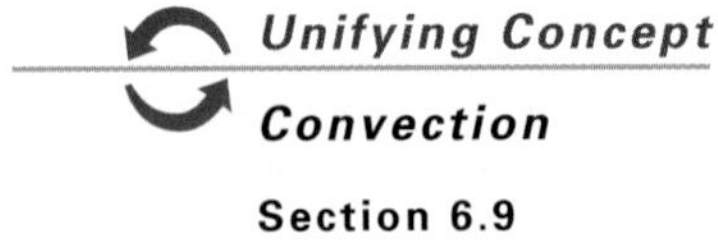

spaced isobars show an abrupt pressure change, and widely spaced isobars signal a more gradual change in pressure. The spacing of isobars is called the **pressure gradient.**

The higher the pressure gradient, the stronger the resulting winds. The strength of the wind is measured in terms of its speed: a gentle breeze is air moving at between 12 and 19 km per hour; a wind strong enough to rattle power lines is moving between 39 and 51 km per hour; and hurricane winds move at more than 120 km per hour. Winds are named according to the direction from which they blow, so a "30 km/hr northwesterly wind" blows from the northwest at a speed of 30 km/hr.

Uneven heating of Earth's surface underlies the differences in air pressure that produce winds. Because warm air expands and cool air contracts, warm air is characterized by low density and low air pressure, while cool air is characterized by high density and high pressure. As air heated at Earth's surface becomes less dense and rises, it cools, moves laterally, and then sinks, only to be heated by the surface and to rise again. This upward, horizontal, and downward movement of air is called a *convection cell*, which you first learned about in Chapter 6, and which is illustrated in Figure 25.15. The vertically rising air in a convection cell is called a *convection current*. The average horizontal motion is called wind.

Local differences in surface heating give rise to small-scale convection cells and pressure gradients, and these create small-scale *local winds*. Planet-scale temperature differences that occur because equatorial regions experience greater solar intensity produce much larger convection cells and pressure gradients. These give rise to global wind patterns called *prevailing winds*.

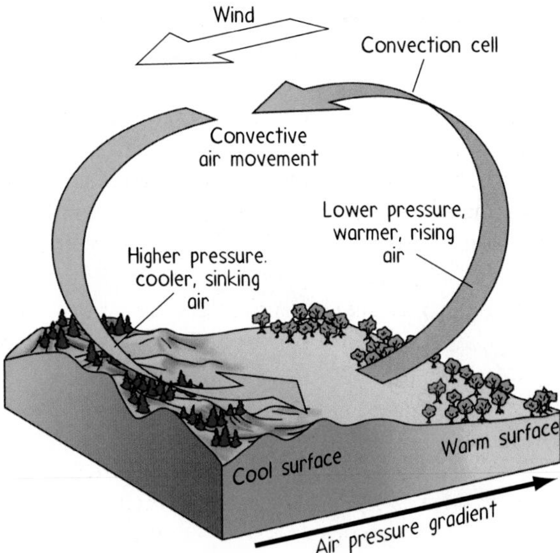

Figure 25.15 This model of air movement in a convection cell shows that warm air rises as it is warmed by the surface. As it rises, it cools, becomes denser, and sinks. The horizontal motion of air in a convection cell is wind.

Local Winds

Adjacent surfaces may have different temperatures because of compositional or topographic differences between them. A notable example occurs where water meets land. The land heats and cools more rapidly than water, primarily due to water's high specific heat capacity. (As you learned in Chapter 6, *specific heat capacity* is the quantity of heat required to change one gram of a substance by 1°C. You can think of it as *thermal inertia*—the higher the specific heat capacity of a substance, the more resistant it is to changing its temperature.) The specific heat capacity of water is 5 times the specific heat of soil; hence, water absorbs much more heat than soil before it warms by the same amount.

As an example, consider land and sea breezes (Figure 25.16). During the day, land gets hotter faster than the ocean because land has a lower specific heat capacity. The hot air over the warmed land rises, creating an area of lower air pressure. The cooler, higher-pressure air from over the ocean then blows from the sea to the land. This is a *sea breeze*. But at night, the land cools off faster than the sea, again because of its low specific heat capacity relative to water. Cooler air descends over the land and so creates an area of higher pressure. Wind blows from the land to the sea. This is a *land breeze*. You may have experienced a refreshing sea breeze if you have spent time along the coast in the summer, or even near the shores of the Great Lakes.

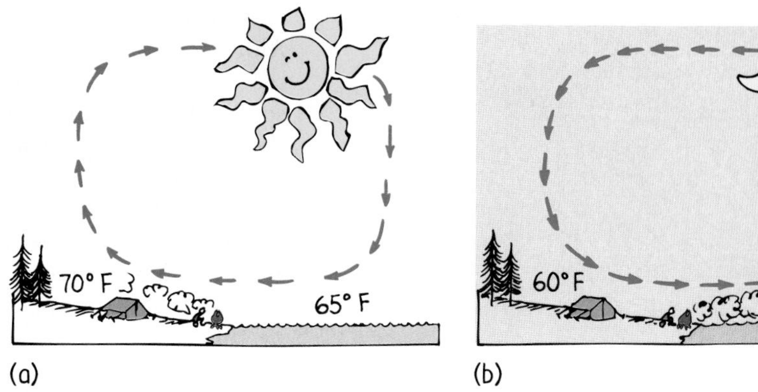

Figure 25.16 Convection currents produced by unequal heating of land and water. (a) During the day, warm air above land rises, and cooler air over the water moves in to replace it. (b) At night, the direction of airflow is reversed, because then the water is warmer than the land.

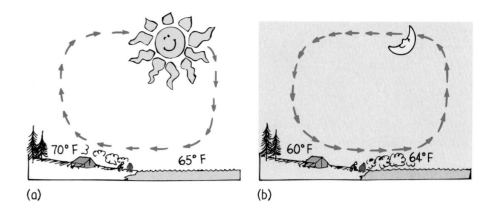

Besides great bodies of water, other irregularities in the Earth's surface also influence wind behavior. Mountains, valleys, deserts, and forests, all play a part in determining which way the wind blows.

CHECK YOURSELF

1. What is the ultimate cause of winds on Earth?
2. How are land and sea breezes alike? How are they different?

CHECK YOUR ANSWERS

1. The Sun. Uneven heating of the Earth's surface sets up the convection cells that produce the wind.
2. Both land breezes and sea breezes are local winds that occur where land is adjacent to a large body of water. They are different in that a sea breeze blows from the sea to the land during the day, and a land breeze blows from the land to the sea at night.

Prevailing Winds

Local winds and convection currents keep the air mixed on a small scale, but long-range horizontal mixing in the atmosphere comes from **prevailing winds.** Local winds sometimes obscure the effects of these giant air circulation patterns, but the prevailing winds are nevertheless present.

Like local winds, the underlying cause of the prevailing winds is the unequal heating of Earth's surface. The high solar intensity at equatorial latitudes causes intense heating there, which produces powerful convection currents. As the warm, moist air rises, it creates a zone of lowered pressure marked by clouds and precipitation along the equator. When the rising air eventually reaches the troposphere, it can rise no higher and it spreads laterally toward the poles. As the now dry air spreads, it cools and becomes denser, and so it sinks back down to Earth's surface at latitudes of about 30° north and south of the equator. The sinking, dense air produces areas of higher pressure with drier conditions. Many of the world's deserts are located in these high-pressure zones, where sinking, dry air "piles up." Examples? Consider the Sahara desert of North Africa, the Great Victoria Desert of Australia, and the Sonoran Desert of Arizona and Mexico.

INSIGHT The winds between the equator and 30°N latitude are called the *northeast trade winds* because European sailing ships of two centuries ago used them in foreign trade. The ships took the trade winds west toward the New World, sailed up the coast of the British Colonies, then caught the prevailing winds blowing out of the west, the *"westerlies,"* back to Europe. Note that winds are named according to the direction from which they blow.

Some of the sinking air moves from the high-pressure zones back toward the equator. As it flows, it produces the prevailing winds between 0° and 30° latitude known as the *trade winds*, so named for their role in propelling trading ships centuries ago. Near the equator, where the trade winds die, there is a zone of still air. Seamen of long ago cursed the equatorial seas as their ships floated listlessly for lack of wind, and they named the area the *doldrums*.

The pair of convection cells between 0° and 30° north and south which produce the prevailing winds are called *Hadley cells*. There are two other pairs, making a total of six Hadley cells on Earth's surface, as shown in Figure 25.17. Although most of the air that sinks at 30° north and south latitude returns to the equator, some of it moves poleward. Somewhere around 60° north and south, this low-altitude air flowing from the equator meets cold air coming from the poles. The air moving in from the lower latitudes is usually warmer, and so it is buoyed upward by the cold, polar air. It then moves back to the equator, cooling and finally sinking at about 30° north and south. This sinking air contributes to the high-pressure air found at 30° north and south. According to legend, sailors were frequently stalled at these latitudes, both north and south. As food and water supplies dwindled, horses on board were either eaten or cast overboard to conserve fresh water. As a result, these regions became known as the *horse latitudes*.

Not all of the air that rises at 60° north and south moves back to the equator. Some continues to move to the poles, then rises and moves back toward 60° north and south. As it sinks, it completes the polar Hadley convection cells.

If Earth didn't rotate, the prevailing winds—the air flowing parallel to Earth's surface within the convection cells—would flow in a north–south direction. High-altitude winds would blow from the equator to the poles, while low-altitude winds would blow from the poles back to the equator.

But this is not what happens. Earth *does* rotate, and so the *Coriolis effect* comes into play, turning the moving air. The Coriolis effect is the tendency for moving bodies not attached to the Earth (such as air molecules) to move to the right in the Northern Hemisphere and to the left in the Southern Hemisphere. Due to the Coriolis effect, air in the planet's six major convection cells turns in the directions indicated by the arrows in Figure 25.18. Prevailing winds in these cells are named the *polar easterlies*, the *westerlies*, and the *trade winds*.

INSIGHT Why does Earth's atmosphere break up into six convection cells, instead of four, or nine, or 101? The number depends on how fast a planet revolves. Jupiter, for example, revolves much faster than Earth, and so its atmosphere is broken up into many bands of convection cells. Venus rotates slowly and only has two cells in each of its hemispheres. If the Earth were to spin faster, we would experience more air-circulation cells and faster mixing of our atmosphere.

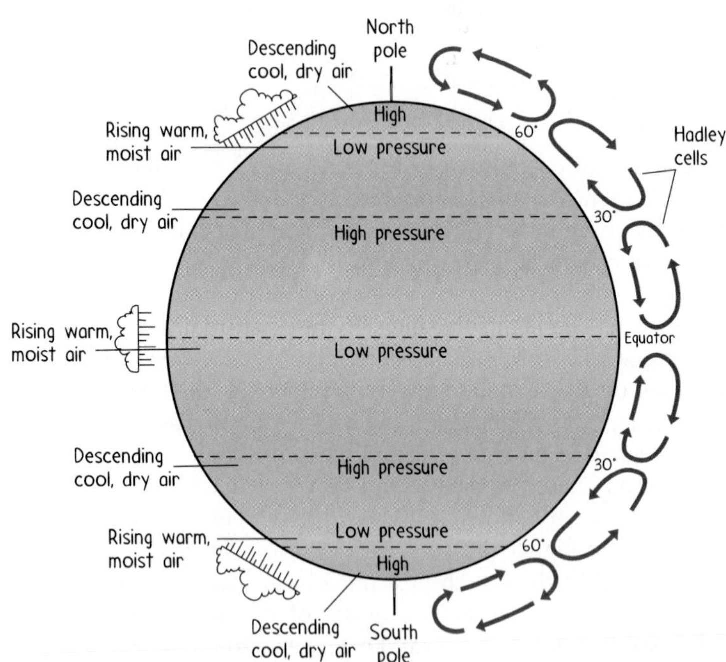

Figure 25.17 Hadley cells describe the convective air movement that creates the persistent, global-scale prevailing winds.

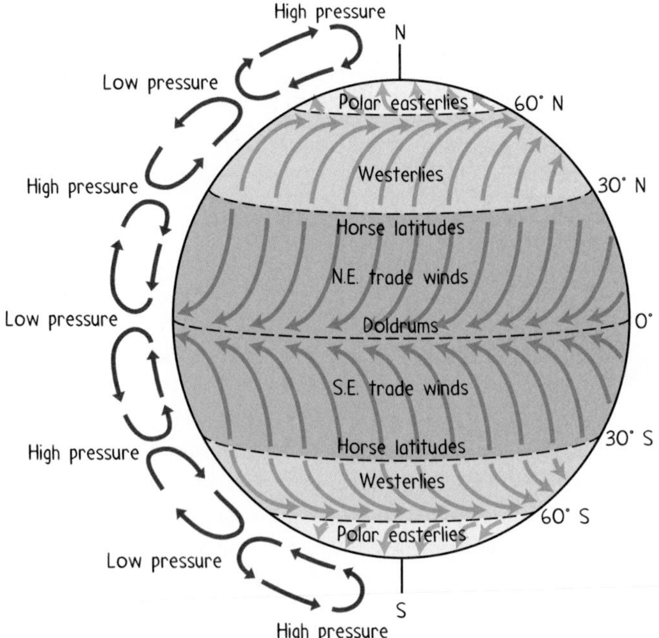

Figure 25.18 Global circulation of the atmosphere results from a combination of two main factors: uneven heating of Earth's surface (which sets up convection cells) and the Earth's rotation. Note that there are six cell-like circulation patterns; prevailing winds blow in the directions indicated by the arrows. Also note the locations of the major prevailing winds—the westerlies, easterlies, and trade winds.

Figure 25.19 (a) On the nonrotating merry-go-round, a thrown ball travels in a straight line. (b) On the counterclockwise-rotating merry-go-round, the ball moves in a straight line. However, because the merry-go-round is rotating, the ball appears to deflect to the right of the intended path.

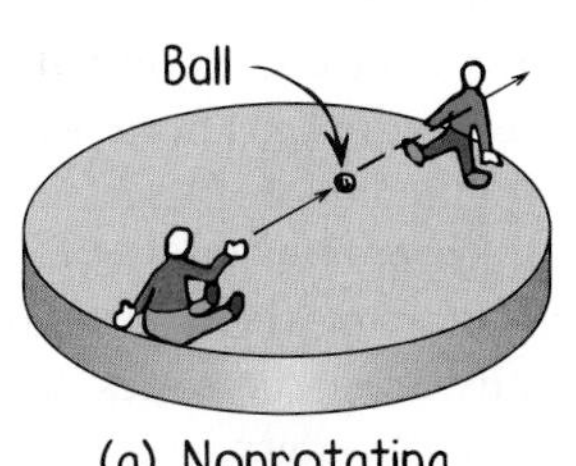

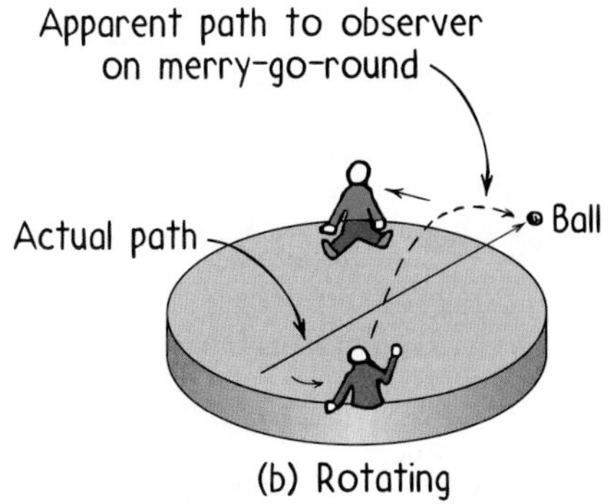

INTEGRATED SCIENCE PHYSICS

The Coriolis Effect

Unifying Concept
Newton's First Law
Section 3.1

The rotation of the Earth greatly affects the path of moving air. To see why, consider an analogy. Think of the Earth as a large merry-go-round rotating in a counterclockwise direction (in the same direction the Earth spins, as viewed from the North Pole). Pretend that you and a friend are playing catch on this merry-go-round. When you throw the ball to your friend, the circular movement of the merry-go-round affects the direction the ball appears to travel. Although the ball really travels in a straight-line path, it appears to curve to the right, as shown in Figure 25.19. (The ball travels straight, but your friend never catches it, because the movement of the merry-go-round causes his position to change. This apparent curving is similar to what happens on the Earth. As the Earth spins, all freely moving objects—air and water, aircraft and ballistic missiles, and even snowballs to a small extent—appear to deviate from their straight-line paths as the Earth rotates beneath them.) This apparent deflection due to the rotation of the Earth is the Coriolis effect.

A significant impact of the Coriolis effect is the apparent deflection of the winds toward the right in the Northern Hemisphere and toward the left in the Southern Hemisphere. The impact of the Coriolis effect varies according to the speed of the wind. The faster the wind, the greater the deflection. Latitude also influences the degree of deflection. Deflection is greatest at the poles and decreases to zero at the equator (Figure 25.20).

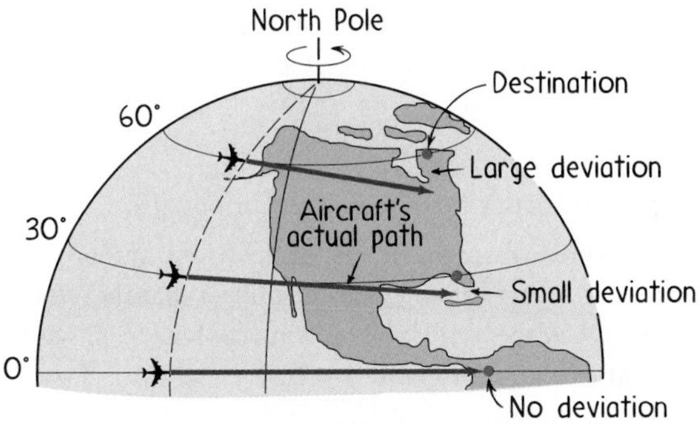

Figure 25.20 Latitude influences the apparent deflection resulting from the Coriolis effect. A free-moving object, such as an airplane heading east or west, appears to deviate from its straight-line path as the Earth rotates beneath it. Deflection is greatest at the poles and decreases to zero at the equator.

CHECK YOURSELF

1. Why are prevailing winds subject to the Coriolis effect?
2. If Earth rotated clockwise, as viewed from the North Pole, in what direction would Northern Hemisphere winds be deflected? How would they be deflected in the Southern Hemisphere?

CHECK YOUR ANSWERS

1. Winds are composed of moving air molecules, which are not attached to the Earth. Because the Earth rotates beneath them, their motion, relative to Earth's surface, is deflected.
2. In this reversed reference frame, the prevailing winds would be turned to the left in the Northern Hemisphere and toward the right in the Southern Hemisphere.

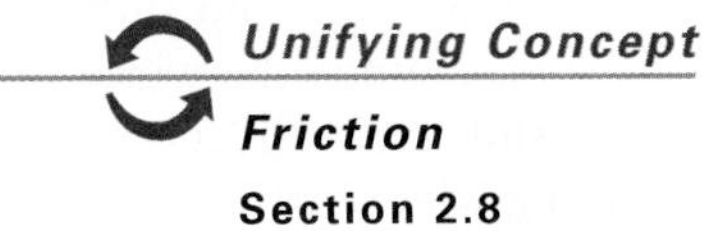

Air moving close to Earth's surface encounters a frictional force. The rougher the surface, the greater the friction, and so the greater the drag. Because surface friction reduces wind speed, it reduces the Coriolis effect. This causes winds in the Northern Hemisphere to spiral out clockwise from a high-pressure

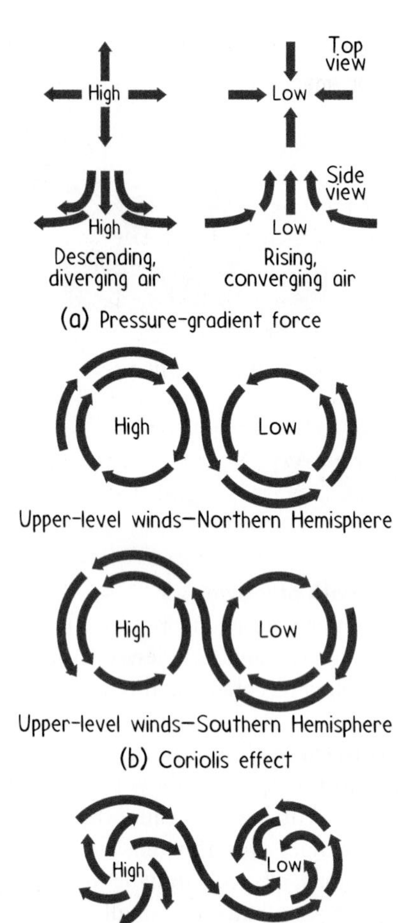

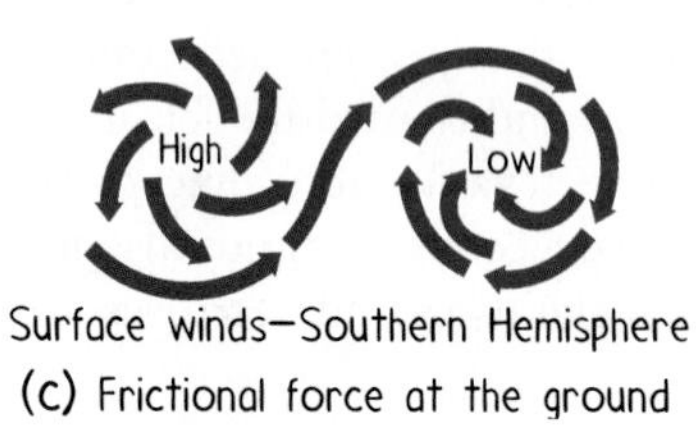

Figure 25.21 Three major factors affect the direction of prevailing winds. (a) Air moves along a pressure gradient from high pressure to low pressure. (b) Once the air is moving, it appears deflected as the Earth rotates beneath it. This is the Coriolis effect. (c) Air moving close to the ground is slowed by the frictional force, which reduces the Coriolis effect.

region and spiral counterclockwise into a low-pressure region (Figure 25.21). In the Southern Hemisphere, these circulation patterns are reversed.

In the upper troposphere, where friction is not significant, "rivers" of rapidly moving air meander around the Earth at altitudes of 9 to 14 km. These high-speed winds are called the *jet streams*. With wind speeds averaging between 95 and 190 km per hour, they play an essential role in the global transfer of energy from the equator to the poles.

Prevailing winds, like land and sea breezes, affect temperature patterns by transporting warm or cool air over land and water. For example, the prevailing winds in the latitudes of North America are westerly—they blow west to east. On the western coast of the continent, therefore, air moves from the Pacific Ocean to the land. Because of water's high specific heat capacity, ocean temperatures do not vary much from summer to winter. In winter, the water warms the air, which is then blown eastward over the coastal regions. In summer, the water cools the air and the coastal regions are cooled. On the eastern coast of the continent, the temperature-moderating effects of the Atlantic Ocean are significant, but because the winds blow from the west toward the east, they are blowing in the wrong direction to help moderate the inland temperature. So temperature variation in the east is much greater than in the west. San Francisco, for example, is warmer in winter and cooler in summer than Washington D.C., which is at about the same latitude.

CHECK YOURSELF

1. Why doesn't air moving along the surface of the Earth in a Hadley cell flow in a straight north–south path?
2. The East coast of the United States is bordered by water, just as the West Coast is. Why, then, does the East Coast have wider seasonal temperature variation than the West Coast?
3. Referring back to Figure 25.13, do prevailing winds blow perpendicular to adjacent isobars, from areas of greater pressure to lesser pressure?

CHECK YOUR ANSWERS

1. The flow of air is turned, with respect to Earth's surface, by the Coriolis effect.
2. Because the prevailing winds blow warmed water to the east in the winter rather than toward the shore, the East Coast does not get the same warming effect that the West Coast does. Similarly, in the summer, when the winds are blowing cool air from west to east, they carry this air offshore rather than onto land.
3. No, the Coriolis effect and friction alter the direction of prevailing winds.

25.6 Oceanic Circulation—Currents

Like the gas molecules of the atmosphere, ocean water is not static but mixes and circulates. The oceans contain *currents*, streams of water that move relative to the larger ocean. **Surface currents** are usually created by wind pushing across the water's surface, moving the water along with it. These steady, global surface currents are relatively permanent features of the ocean and are set up by the prevailing winds. Compare Figure 25.18 closely with Figure 25.22. Notice that the surface currents match up quite well with the prevailing winds if you account for the interference of land masses. You can also appreciate the close link between atmospheric circulation and oceanic circulation when you consider currents in the northern Indian Ocean. There are seasonal wind shifts in this region, which are called summer and winter *monsoons*. When the winds reverse direction during these monsoons, surface currents shift their direction accordingly.

Surface currents, like the prevailing winds, are turned and twisted along their path by Earth's rotation. Because of the Coriolis effect and other factors, surface

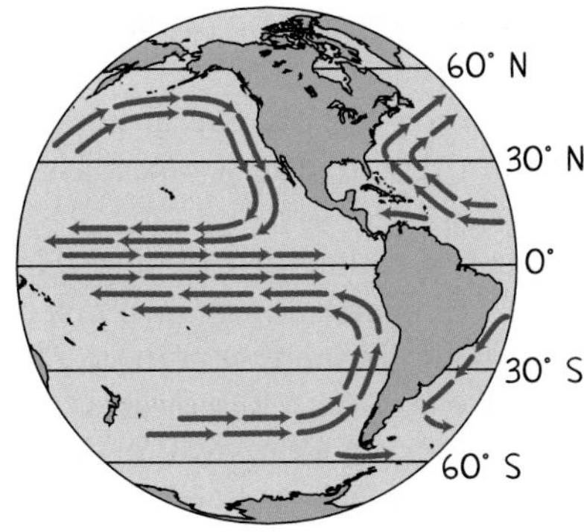

Figure 25.22 The ocean's major surface currents, shown by the arrows, closely match the pattern of prevailing winds illustrated in Figure 25.18, except in regions where land masses disrupt the water's flow.

water currents tend to form giant circular flow patterns called *gyres*. Five gyres exist in the world's oceans. There are two gyres each in the Pacific Ocean and the Atlantic Ocean and one in the Indian Ocean. In the Northern Hemisphere, gyres rotate clockwise (turning to the right as seen from the North Pole). In the Southern Hemisphere, they rotate counterclockwise (turning to the left) as Figure 25.23 shows.

Surface currents play a vital role in redistributing heat. The uneven heating of Earth's surface warms equatorial waters and keeps polar regions chilly. But as currents transport waters far from their source, they transport heat as well. For example, heat is transported in the North Atlantic Ocean as warm equatorial water flows westward into and around the Gulf of Mexico then northward along the eastern coast of the United States. This warm-water current is called the *Gulf Stream*. As the Gulf Stream flows northward along the eastern coast of North America, the prevailing westerlies steer the warm current eastward toward Europe. Great Britain and Norway benefit from the warm waters of the Gulf Stream, because warm waters from the Gulf of Mexico liberate heat to the atmosphere in their vicinity. As the warm current encounters Europe, it is turned southward toward the equator, where it once again is picked up by the trade winds to move westward into the Gulf of Mexico.

CHECK YOURSELF

In what two ways does the proximity of Great Britain to the Atlantic Ocean affect its climate?

CHECK YOUR ANSWER

Great Britain is affected by its proximity to the Atlantic Ocean because of (1) the moderating influence that water has on land temperatures due to the high specific heat capacity of water and (2) the warming effect of the Gulf Stream.

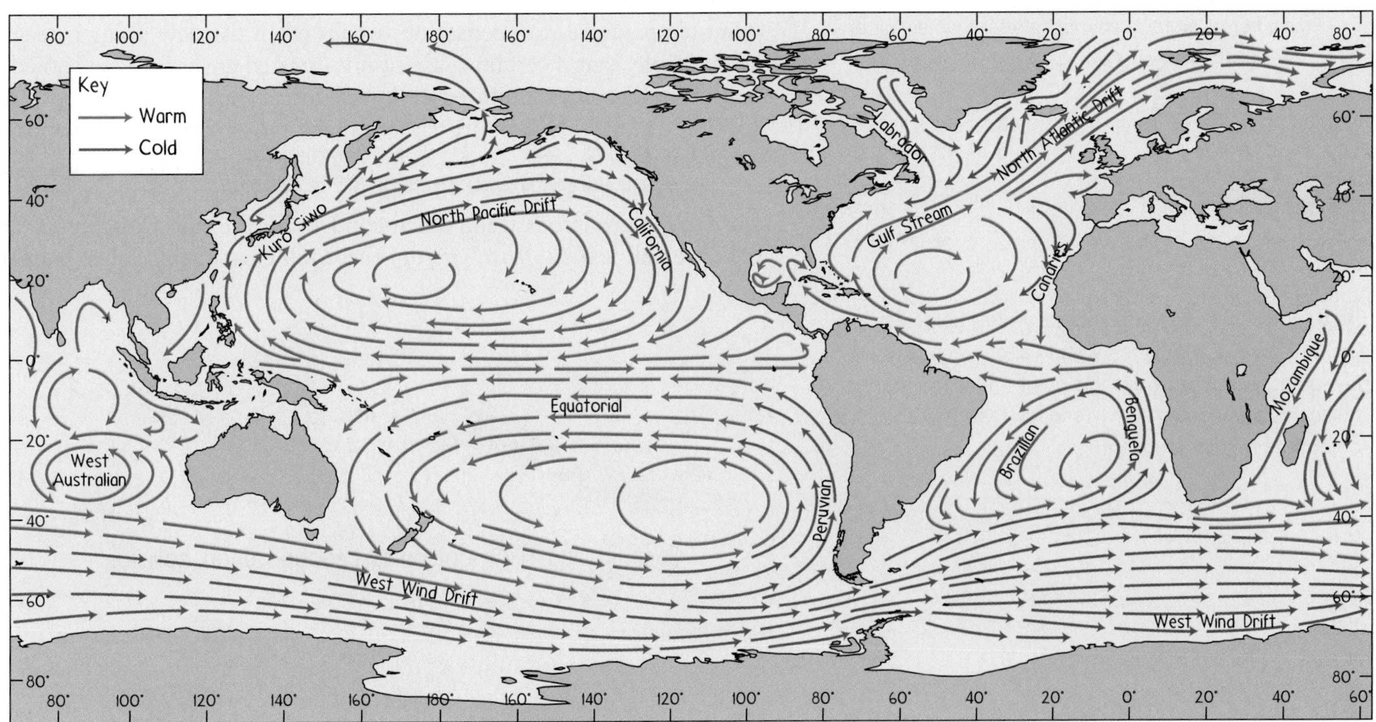

Figure 25.23 The world's major surface currents are indicated here. Notice the clockwise rotation of gyres in the Northern Hemisphere and counterclockwise rotation in the Southern Hemisphere.

Table 25.3 Maximum Amount of Water Vapor That a 1-kg Mass of Air Can Hold at Different Temperatures

Temperature (°C)	Grams of Water Vapor per kg of Air
−30	0.3
−20	0.75
−10	2
7	3.5
20	14
30	26.5
40	47

25.7 Humidity

All air, even the driest air, contains some water vapor. The amount of water vapor in air is its **humidity**. More specifically, humidity is the mass of water vapor a given volume of air contains.

There is a limit to the amount of water vapor that air can hold. How much water vapor it can hold depends on the air's temperature: as air temperature increases, the volume of water vapor that it can hold increases. Table 25.3 shows the amount of water vapor that air can hold at various temperatures.

Weather reports describe humidity in terms of relative humidity. **Relative humidity** is the ratio of the amount of water vapor currently in the air compared with the largest amount of water vapor that it is possible for the air to hold at that temperature. Stated as an equation, the relationship is:

$$\text{Relative humidity} = \left(\frac{\text{water-vapor content}}{\text{water-vapor capacity}}\right) \times 100\%$$

Thus, if air contains 7 grams of water on a 25°C day, the relative humidity is expressed as 7/14, or 50%, because the water content of the air is half the amount that it can hold at a temperature of 25°C.

Relative humidity can change if the amount of water vapor in the air changes, and a change in temperature also changes relative humidity. For instance, if the amount of water vapor in the air stays constant as the temperature rises, the relative humidity drops because warm air has a higher water-vapor capacity. So the relative humidity has decreased even though the actual quantity of water vapor in the air is unchanged.

INSIGHT When perspiration evaporates, it takes the energy it needs to change state from liquid to gas from your skin. This energy, water's *latent heat of vaporization*, thus leaves your body, and you become cooler as a result. In humid weather, perspiration doesn't cool you as well as it does in dry air because evaporation is slowed. This is why humid air feels so much hotter than dry air of the same temperature.

When air contains as much water as it can possibly hold, the air is *saturated*. This is equivalent to saying that saturated air at a given temperature has 100% relative humidity. As an air mass cools, it can hold less and less water vapor before becoming saturated. If it cools down enough, the air mass reaches a point at which the water vapor present is the amount required to saturate the air at the lower temperature. This temperature, the temperature at which saturation occurs, is the **dew point** temperature. Condensation occurs when the dew point is reached. Suppose for example, that a certain mass of unsaturated air at 30°C is cooled to 15°C and that the air becomes saturated at that temperature. The dew point of this air is then 15°C; if the air is cooled further, its capacity for holding water vapor would be exceeded and the excess vapor would condense (Figure 25.24).

Water vapor condenses high in the atmosphere, forming clouds. It condenses close to the ground as well. When condensation in the air occurs near Earth's surface, we call it *dew*, *frost*, or *fog*. On cool, clear nights, objects near the ground

WHAP!
Fast-moving H_2O molecules rebound upon collision

Figure 25.24 Condensation of water molecules. Condensation occurs when water vapor reaches its dew-point temperature. At this point, the water molecules are moving slowly enough that they coalesce, rather than rebound, upon impact.

Figure 25.25 Water vapor condenses from a mass of air when the air's dew point is exceeded. This results in the formation of (a) fog, (b) dew, (c) frost, or (d) clouds.

INSIGHT In rainy weather, when your car windshield fogs up, turn on the air conditioner rather than the defroster. What causes window fogging is the humidity in the car caused by rain, by wet clothes, and by the breath of passengers. Since the air from the air conditioner is very dry, it clears a foggy windshield very nicely in a short time.

cool down more rapidly than the surrounding air. As the air cools to its dew point, it cannot hold as much water vapor as it could when it was warmer. Water from the now-saturated air condenses on any available surface. This may be a twig, a blade of grass, or the windshield of a car. We often call this type of condensation early-morning dew. When the dew point is at or below freezing, we have frost. When a large mass of air cools and reaches its dew point, the relative humidity approaches 100%. And this produces a cloud near the ground—*fog* (Figure 25.25).

CHECK YOURSELF

1. What is the major difference between fog and a cloud?
2. What is the humidity of an air mass that holds 0.13 kg of water vapor and occupies a volume of 7.2 m^3?
3. At −10°C, 1 kg of air contains 1.25 g of water vapor. What is its relative humidity?

CHECK YOUR ANSWERS

1. Fog forms near the surface of the Earth while a cloud forms some distance above its surface.
2. Humidity = mass of water/volume of air = 0.13 kg/7.2 m^3 = 0.018 kg/m^3.
3. $$\text{Relative humidity} = \left(\frac{\text{water-vapor content}}{\text{water-vapor capacity}}\right)$$
$$= 1.25\ \text{g/kg} \times 100\% = 62.5\%$$
$$= 2\ \text{g/kg}$$

25.8 Clouds and Precipitation

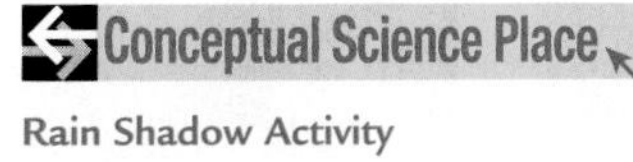

Rain Shadow Activity

As air rises, it expands and cools. As the air cools, water molecules move more slowly and condensation occurs. If there are larger and slower-moving particles or ions present in the air, water vapor condenses on these particles, and this creates a **cloud**—a visible aggregate of minute water droplets or tiny ice crystals. Usually (except in the case of fog), moist air becomes cold enough to form a cloud because something has pushed it up to an altitude at which it reaches saturation. One of the mechanisms by which air is lifted is *orographic*

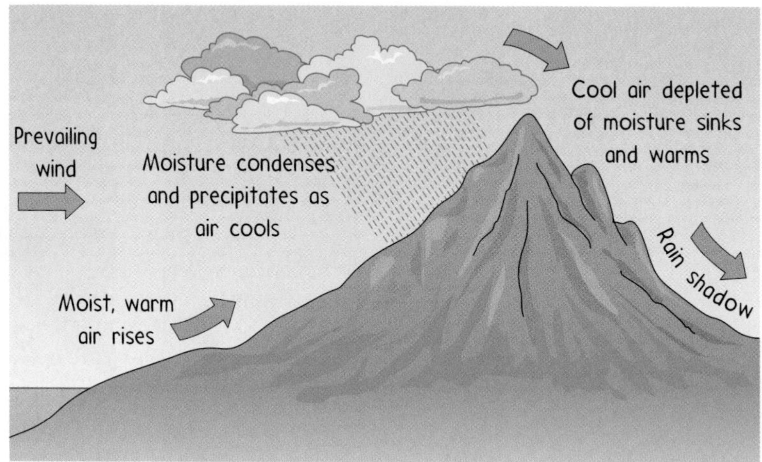

Figure 25.26 Orographic lifting. As warm, moist air rises on the windward (upslope) side of a mountain, the air cools, water vapor condenses, and a cloud forms. Precipitation may develop. By the time the air parcel reaches the leeward (downsloping) side of the mountain, the air is depleted of moisture. As a result, the leeward slope is dry. It lies in a "rain shadow."

lifting. In this case, an air parcel (a patch of air of uniform temperature and moisture) is pushed upward over an obstacle, such as a mountain range, as shown in Figure 25.26. Moving air rises because it can't go forward when it hits the mountain. The rising air cools. If the air is humid, clouds form as moisture condenses.

In 1803, British weather observer Luke Howard was the first to classify clouds according to their shapes. He recognized three cloud forms: *cirrus* (Latin for "curl"); *cumulus* (L., "piled up"); and *stratus* (L., "spread out") (Figure 25.27). If you have searched for these basic shapes in the sky, however, you know that clouds do not usually come in these simple forms; instead, they usually occur as composites of these forms. For this reason, Howard's simple classification has been modified so that clouds are generally classified by their form as well as their altitude. This results in ten basic cloud types, each of which belongs to one of the four major cloud groups (Table 25.4).

High clouds form at altitudes above 6,000 meters and are denoted by the prefix *cirro-*. The air at this elevation is quite cold and dry, so clouds this high are made almost entirely of ice crystals. The most common high clouds are *cirrus* clouds, which are blown by strong high-altitude winds into their classic wispy shapes, such as the "mare's tail" and "artist's brush." *Cirrocumulus* clouds are arrays of rounded white puffs that rarely cover more than a small patch of the sky. Small ripples and a wavy appearance make the cirrocumulus clouds look like the scaled body of a mackerel. Hence, cirrocumulus clouds make up what is often referred to as a "mackerel sky" (Figure 25.28a). Cirrus clouds usually indicate fair weather.

(a)

(b)

(c)

Figure 25.27 Clouds classified by shape: (a) "wispy" cirrus clouds, (b) "piled-up" cumulus clouds, and (c) "spread-out" stratus clouds.

Table 25.4 The Four Major Cloud Groups

1. High Clouds (above 6000 m)	2. Middle Clouds (2000–6000 m)	3. Low Clouds (below 2000 m)	4. Clouds of Vertical Development
Cirrus	Altostratus	Stratus	Cumulus
Cirrostratus	Altocumulus	Stratocumulus	Cumulonimbus
Cirrocumulus	Nimbostratus		

Middle clouds are denoted by the prefix *alto-*. They are made up of water droplets and, when temperature allows, ice crystals. *Altostratus* clouds (Figure 25.28b) are gray to blue-gray, and they often cover the sky for hundreds of square kilometers. Altostratus clouds are often so thick that they diffuse incoming sunlight to the extent that objects on the ground don't produce shadows. Altostratus clouds often form before a storm. So if you can't see your shadow when you're going on a picnic, cancel!

Low clouds are most often composed of water droplets, but they can contain ice crystals in colder climates. *Stratus* clouds tend to be the lowest of the low clouds. They are uniformly gray and they cover the whole sky, often resembling a high fog. Stratus clouds are not associated with falling precipitation, but they sometimes generate a light drizzle or mist. *Nimbostratus* clouds are dark and foreboding (Figure 25.28c). They are a wet-looking cloud layer associated with rain and snow.

Clouds of vertical development do not fit into any of the three height categories. These clouds typically have their bases at low altitudes, but they reach up into the middle or high altitudes. Though cumulus clouds are fair-weather clouds, they are called "clouds of vertical development" because they can grow dramatically. Vertically moving air currents in them can produce a towering cloud with an anvil head—a *cumulonimbus* cloud (Figure 25.28d). Cumulonimbus clouds, popularly called thunderheads, may produce heavy rain showers, thunder and lightning, and hail.

Precipitation is water in the liquid or solid state that returns to Earth's surface from the atmosphere. Types of precipitation include mist, drizzle, and rain (in the liquid form) and hail, snow, and sleet (in the form of solid ice crystals). Mist and drizzle generally fall from stratus clouds. Rain falls from nimbostratus and cumulonimbus clouds.

(a)

(b)

(c)

(d)

Figure 25.28 Clouds from each of the major cloud groups: (a) cirrocumulus clouds, (b) altostratus clouds, (c) nimbostratus clouds, and (d) cumulonimbus clouds.

Since clouds are denser than the surrounding air, why don't they fall from the sky? Ah, they do! They fall at least as fast as the air below rises. So without updrafts, we'd have no clouds.

Figure 25.29 Explanation of updrafts.

Here's a challenge question: If clouds are made of water droplets and ice crystals, which are denser than air, why don't we see them sinking to the ground? The gravitational force pulling a droplet down *is* enough to make it fall. So why don't all droplets in clouds fall to the ground? The answer has to do with updrafts—rising air currents. A typical cumulus cloud has an updraft speed of at least 1 meter per second, which is faster than the droplet can fall. So the droplets are supported by the upward-rising air. Without updrafts, the droplets drift so slowly out of the bottom of the cloud and evaporate so quickly that they have no chance to reach the ground. They are replaced by new droplets forming above.

Raindrops, on the other hand, are huge compared with typical cloud droplets. A drop of rain big enough to reach the ground contains about a million times more water than a cloud droplet. Raindrops fall faster than most updrafts can push upward and faster than they evaporate.

25.9 Changing Weather—Air Masses, Fronts, and Cyclones

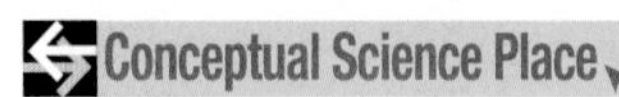

Cold Front Activity; Warm Front Activity

With the changing seasons, solar radiation, pressure belts, prevailing winds, and surface currents shift along Earth's latitudes. These general seasonal patterns influence three atmospheric phenomena that affect day-to-day weather: (1) large bodies of air called *air masses*; (2) *weather fronts*; and (3) the local pressure systems called *cyclones* and *anticyclones* that are associated with air masses and fronts.

Air Masses

An **air mass** is a huge volume of air, usually 1600 km or more across and several km thick, that has characteristic temperature and humidity throughout and that tends to remain intact as it travels. An air mass develops when it can remain stagnant over an expansive source region of land or water for a long time. The air mass acquires the temperature characteristics of the source region through the heat-transfer processes of conduction, convection, and radiation, which you studied in Chapter 6. It acquires the moisture characteristics of its source region through evaporation and condensation, also discussed in Chapter 6. An air mass tends to stay intact and to maintain its characteristics even as it travels away from its source region.

Various distinct air masses cover large portions of the Earth's surface, each with its own characteristics. An air mass formed over water in the tropics is different from one formed over land in a polar region. Air masses are divided into six general categories according to what types of land or water over which they form and the latitude at which their formation occurs (Table 25.5 and Figure 25.30). The type of surface over which an air mass forms is designated by a lower-case letter ("m" for maritime or "c" for continental). The source region in which an air mass forms is designated by a capital letter ("A" for arctic, "P" for polar, or "T" for tropical.)

As an air mass moves away from its source region, it brings its weather conditions to the areas over which it travels. At the same time, its own properties are

Table 25.5 Classification of Air Masses and Their Characteristics

Typical Source Region	Classification	Symbol	Characteristics
Arctic	maritime arctic	mA	cool, moist, unstable
Greenland	continental arctic	cA	cool, dry, stable
North Atlantic, Pacific Ocean	maritime polar	mP	cool, moist, unstable
Alaska, Canada	continental polar	cP	cold, dry, stable
Caribbean Sea, Gulf of Mexico	maritime tropical	mT	warm, moist, usually unstable
Mexico, Southwestern U.S.	continental tropical	cT	hot, dry, stable aloft; unstable at surface

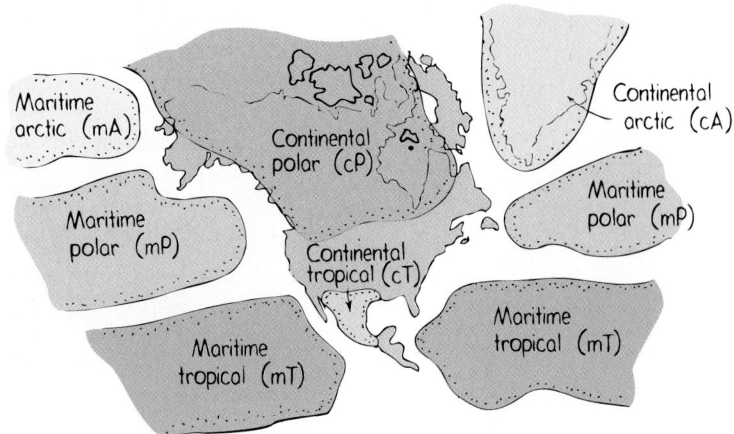

Figure 25.30 Typical source regions of air masses for North America.

slowly modified as it interacts with new environments. Continental polar (cP) and continental arctic (cA) air masses generally produce very cold, dry weather in winter and cool, pleasant weather in summer. Maritime polar (mP) and maritime arctic (mA) air masses, picking up moisture as they travel across the oceans, generally bring cool, moist weather to a region. Continental tropical (cT) air masses are generally responsible for the hot, dry weather of summer, and warm, humid conditions are due to maritime tropical (mT) air masses.

Fronts

Air masses meet at a boundary called a *weather front* or **front**. Fronts are associated with rapid changes in weather. When air masses collide, the less dense, warmer air mass flows upward over the more dense, cooler air. This process is called *frontal lifting*. Although the warmer air always rises vertically above the cooler air, the horizontal movements of the air masses vary. Sometimes, it's the colder, denser air mass that advances into and displaces a stationary warm air mass. In this case, the contact zone between the air masses is called a **cold front**. But, if it's the warm air that moves into territory that had been occupied by a cold air mass, the zone of contact is called a **warm front**. If neither of the air masses is moving, the contact zone is called a **stationary front**. At the front—a region perhaps a few kilometers wide—the air masses mix and often produce clouds, rain, winds, and storms.

You can generally predict that a cold front is moving in if you observe high cirrus clouds, a shift in wind direction, a drop in temperature, and a drop in air pressure. As cold air moves into and displaces a warm air mass at a cold front, the warm air is forced upward, rises, and cools, and may condense to form a series of cumulonimbus clouds (Figure 25.31). If so, the advancing wall of clouds

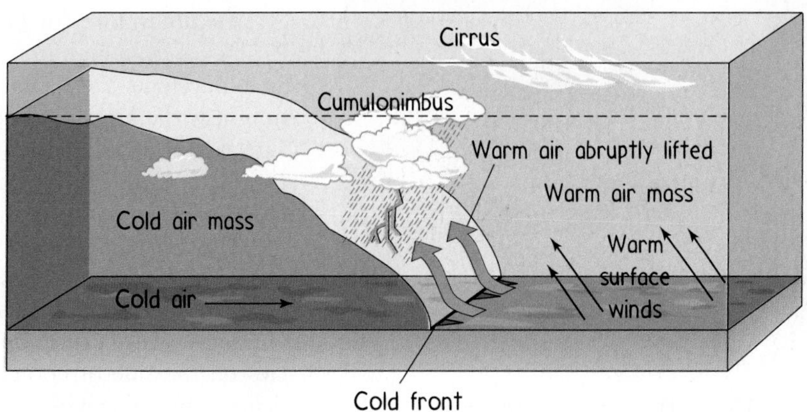

Figure 25.31 A cold front forms when a cold air mass moves into a warm air mass. The cold air forces the warm air upward, where it condenses to form clouds. If the warmer air is moist and unstable, heavy rainfall and gusty winds develop.

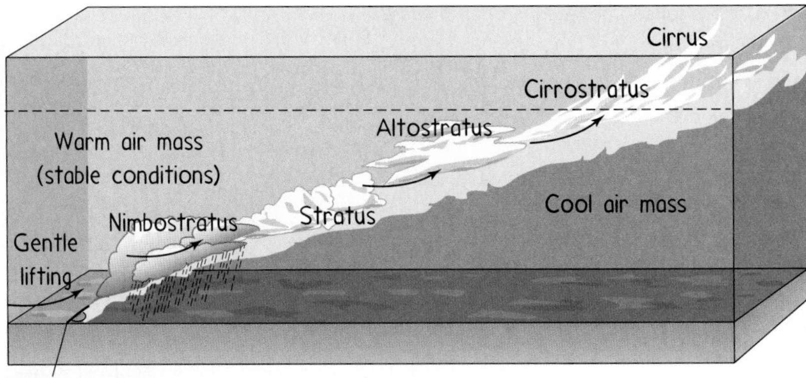

Figure 25.32 A warm front forms when a warm air mass moves into and displaces a cold air mass. The less-dense, warmer air rides up and over the colder, denser air, resulting in widespread cloudiness and light-to-moderate precipitation that can cover great distances.

at the front may well produce thunderstorms with heavy showers and gusty winds. After the cold front passes, the lifted warm air cools and sinks, pressure rises, and rain ceases. Except for a few fair-weather cumulus clouds, the skies clear and there is the calm that typically comes after a storm.

A warm front approaches more gradually than a cold front. The arrival of a warm front is also indicated by cirrus clouds. A warm front forms when warm air moves into an area formerly occupied by a cooler air mass. At the warm front, the less dense warm air gradually rises up and over the colder, denser air (Figure 25.32). Ahead of the front, the cirrus clouds thicken into altocumulus and altostratus clouds that turn the sky an overcast gray. Closer to the front, light to moderate rain or snow can develop, and winds may become brisk. At the front, air gradually warms, and the rain or snow turns to drizzle. Behind the front, the air is warm and the clouds scatter.

Cyclones and Anticyclones

Cyclones are associated with the onset of rough weather (Figure 25.33). In popular speech, "cyclone" can refer to various kinds of severe storms, from tornadoes to hurricanes. In meteorology, however, a **cyclone** (or a *low-pressure center* or simply a *low*) is an area of low pressure around which winds flow. Due to the Coriolis effect, the winds in a cyclone move counterclockwise in the Northern Hemisphere and clockwise in the Southern Hemisphere. Since the cyclone's center is the region of lowest pressure, air converges into the center, but it is then forced to rise upward. Rising air in a cyclone can produce clouds and precipitation, ranging from rain and thunderstorms in the summer and fall to rain, thunderstorms, and possibly snow in the winter.

Figure 25.33 Satellite image of a cyclone.

An **anticyclone** (or *high-pressure center* or *high*) is an area of high pressure. Air moves from high pressure to low, so air moves downward and outward from an anticyclone. This sinking motion leads to generally fair skies and no precipitation near the high. The Coriolis effect turns the moving air around a high-pressure center so that anticyclonic winds blow clockwise around a high in the Northern Hemisphere and counterclockwise around a high in the Southern Hemisphere.

The positions of fronts and pressure systems are plotted on weather maps because of their role in predicting the weather (Figure 25.34). On a weather map, a high-pressure system is denoted with an *H*, and a low-pressure system is denoted with an *L*. The surface position of a warm front is shown as a line with semicircles on the side of cooler air. Likewise, the surface position of a cold front is denoted by a line with triangles extending into the region of warmer air.

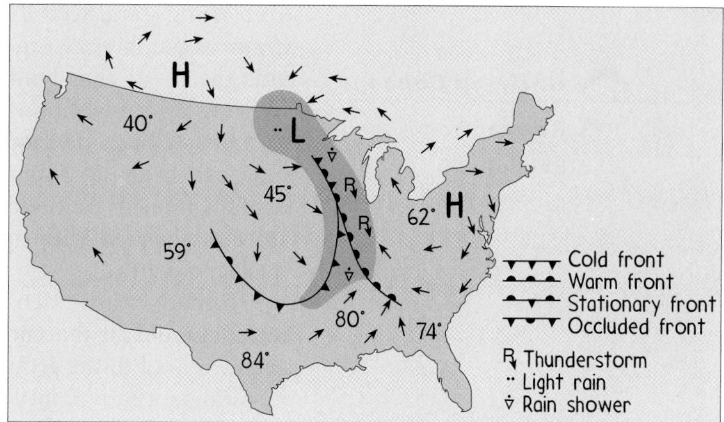

Figure 25.34 Weather maps show pressure systems and fronts. In this weather map, a cold front moves into an area that spans from the southwestern to the southeastern United States. Note the direction of air movement away from high-pressure centers and toward centers of low pressure. Because sinking air does not usually produce clouds, we find clear skies and fair weather over the high-pressure zones. Conversely, rising air out of a low-pressure zone can cause cloud formation and precipitation.

25.10 Storms

INSIGHT Lightning can—and often does—strike the same place twice. Tall, metallic objects are the most likely sites for lightning to strike.

Storms, which are typically brought in by weather fronts, are defined as violent and rapid changes in weather. The three major types of severe storms are thunderstorms, tornadoes, and hurricanes (Figure 25.35).

A thunderstorm begins with a cumulus cloud. A cumulus cloud can grow into the kind of towering cloud associated with thunderstorms—a cumulonimbus cloud—if it is fed by moist, uplifted, *unstable* air. Unstable air is air that continues to rise because it is warmer than the air around it. Unstable air can

(a)

(b)

(c)

Figure 25.35 (a) The mature stage of a thundercloud. Such a cloud typically has a base of several kilometers in diameter and can tower to altitudes as high as 12 kilometers. At these high altitudes, horizontal winds typically stretch the flattened thunderhead into a characteristic anvil shape. (b) The winds of a tornado travel up to 800 kilometers per hour in a counterclockwise direction in the Northern Hemisphere and in a clockwise direction in the Southern Hemisphere. A tornado is a vortex that causes damage through suction as well as through the force of its winds. (c) Hurricanes form in the tropics when winds spiral inward and rotate around a central low-pressure area. Hurricanes may have wind speeds of 300 km/hr or more. They are fueled by the latent heat liberated when vast amounts of water vapor condense.

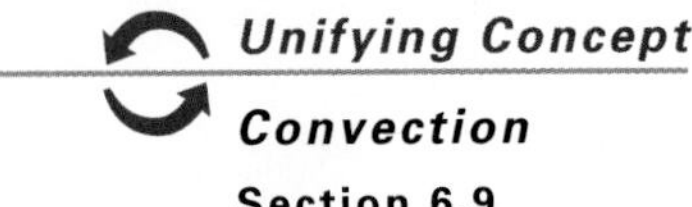

Section 6.9

keep rising—and feed a thundercloud—until it becomes stable, at which point it moves out laterally and gives the cloud a flattened top. Many things can provide the uplift that builds a cumulonimbus cloud—from convection, to wind blowing up a mountain slope, or winds accompanying a weather front. Once icy cirrus clouds develop along the flattened top of the towering cloud, lightning can begin to occur in it. The thunderstorm has begun. Thunderstorms contain immense amounts of energy. A supercell, which is a strong thunderstorm associated with tornadoes, can contain 20 to 100 times the energy of a nuclear warhead!

Tornadoes are extensions of thunderstorms. A tornado is defined as a rotating column of air that moves around a low-pressure core and that reaches from a thundercloud to the ground. It is the Coriolis effect that turns the air, and makes it spiral as it moves into a low-pressure area. Tornados extend from a thunderstorm all the way to the ground—a vortex that doesn't reach the ground is called a funnel cloud. As a tornado moves across the land, it follows a path controlled by its parent thundercloud, and it may appear to bounce and skip. A tornado acts like a gigantic vacuum cleaner, picking up objects in its path. It wreaks havoc not only by suction but also because of the battering power of its winds. Tornadoes are classified on the Fujita scale. On this scale, a rating of F0 indicates a weak tornado, which has wind speeds between 64 km/h and 100 km/h. The most severe tornados are rated F5 and have wind speeds between 420 km/h and 512 km/h. The Fujita scale indicates how much damage is done by a tornado. An F0 tornado can rip branches from trees or overturn a car; an F5 tornado will destroy almost anything in its path.

INSIGHT Lightning develops when water drops and soft hail in a cloud bump into and rub against one another, knocking off electrons. The electrons gather, usually along the bottom of the cloud. This induces a positive electrical charge at the ground surface. The air between the charged areas acts as an insulator. When sufficient charge builds up, the insulating ability of the air is exceeded and an electric spark—lightning—arcs across the sky.

In the tropics, the transfer of heat to the atmosphere by evaporation and conduction is so thorough that air and water temperatures are about equal. The high humidity of this part of the world favors the development of cumulus clouds and afternoon thunderstorms. Most of the individual storms are not severe. However, if the moisture content and temperature of the air increase and a large low-pressure center develops, rising warm, moist air can spiral around the low to produce a more violent storm—a hurricane—with winds speeds up to nearly 300 kilometers per hour.

A hurricane gains energy from the latent heat (Chapter 11) released when huge quantities of water condense. This energy warms the air, which buoys the hurricane. As the hurricane lifts, pressure near its surface is reduced, and this draws in more moist air. Increasing winds rotate around the central low-pressure area, the area known as the *eye* of the hurricane. But a hurricane needs a continual supply of warm, moist air to keep going. Once it moves over land, a hurricane weakens because its source of moisture and latent heat is cut off.

25.11 Global Climate Change

Earth's climate has changed dramatically during its history. Great ice ages have come and gone, as have periods of worldwide warmth. Climate transitions usually occur over hundreds of thousands of years, or even longer. Today, Earth's climate is transitioning again—it is warming up, and at an unprecedented rate. There is general consensus in the scientific community that this accelerated warming is at least in part due to human activities.

Specifically, scientists believe that the increased amounts of carbon dioxide and certain other gases exhausted into the air are enhancing the greenhouse effect. The **greenhouse effect** is the warming of the atmosphere that results from terrestrial radiation being trapped by these "greenhouse gases." Most of the scientists who do climate modeling predict that the average temperature of Earth will increase by at least 2°C (about 4°F) by 2050. This predicted rise in global temperature is what is known as **global warming**.

Short-wavelength visible light
from the sun is transmitted
through the glass.
Long-wavelength infrared
radiation is not transmitted
out through the glass and
is trapped inside.

Figure 25.36 Glass acts as a one-way valve, letting visible light in and preventing infrared energy from exiting.

INTEGRATED SCIENCE PHYSICS

The Greenhouse Effect

Park your car with the windows closed in the bright sunlight, and your car's interior soon becomes quite toasty. The inside of a greenhouse is similarly toasty. This happens because glass is transparent to visible light but not infrared, as illustrated in Figure 25.36. As you may recall from Chapter 8, wavelengths of visible light are shorter than wavelengths of infrared. Short-wavelength visible light from the Sun enters your car or a greenhouse and is absorbed by various objects—car seats, plants, soil, whatever. The warmed objects then emit infrared energy, which cannot escape through the glass, and so the infrared energy builds up inside, increasing the temperature.

A similar effect occurs in Earth's atmosphere, which, like glass, is transparent to visible light emitted by the Sun. As you learned earlier in this chapter, the ground absorbs this energy but radiates infrared waves. Atmospheric carbon dioxide, water vapor, and certain other gases absorb and reemit much of this infrared energy back to the ground, as Figure 25.37 illustrates. This process is called "the greenhouse effect." The greenhouse effect is quite desirable because the Earth's average temperature would be a frigid −18°C otherwise. A human-caused greenhouse effect is not a desirable thing, however, because it heats up the Earth too much and too quickly.

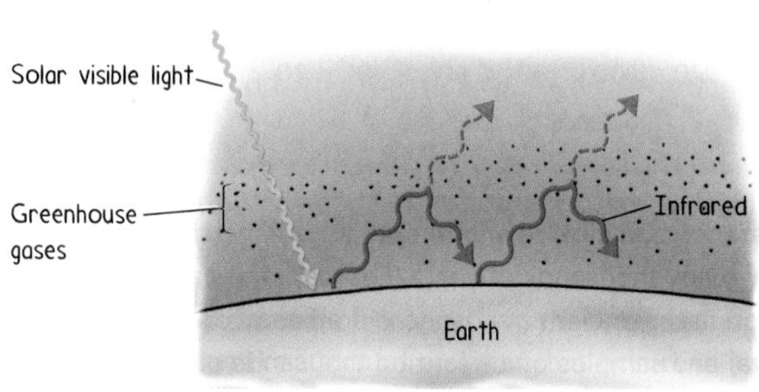
Solar visible light
Greenhouse gases
Infrared
Earth

Figure 25.37 The greenhouse effect in the Earth's atmosphere. Visible light from the Sun is absorbed by the ground, which then emits infrared radiation. Carbon dioxide, water vapor, and other greenhouse gases in the atmosphere absorb and reemit heat that would otherwise be radiated from the Earth into space.

CHECK YOURSELF

1. Explain how a one-way valve makes a good analogy for the greenhouse effect.
2. How does increased carbon dioxide in the atmosphere enhance the greenhouse effect?

CHECK YOUR ANSWERS

1. The Earth's atmosphere and glass both allow incoming visible light waves to pass but not outgoing infrared waves. As a result, radiant energy is trapped.
2. Carbon dioxide traps infrared radiation.

Scientists agree that the planet is warming, but no one knows specifically how fast or what the effects of the warming will be. Climatologists cannot specify whether sea levels will rise and, if so, by how much; what the climate in Kansas or Japan will be 50 years from now; or exactly what the average world temperature will be in 2075. This uncertainty is due to the large number of variables that contribute

Unifying Concept

Exponential Growth and Decay

Appendix E

date sedimentary rock, principles of relative dating are combined with absolute dates from certain radiometric dating techniques.

CHECK YOURSELF

1. A mineral sample contains $\frac{1}{4}$ of its original uranium-235 atoms. How old is the mineral?
2. Radiometric dating is used to determine the age of a potassium-40-bearing mineral. During the lifetime of that mineral, however, some of the gaseous daughter product, argon-40, has diffused out of the crystalline structure of the mineral. Does this affect the accuracy of the test?

CHECK YOUR ANSWERS

1. Looking at the graph of half-life and Table 26.1, we can see that the mineral must be as old as two half-lives of uranium-235—that's 2 × 704 million years = 1408 million years.
2. Yes, the diffusion of decay products diminishes the accuracy of the test. Radiometric dating is based on the assumption that there is no "leakage" of parent or daughter products into or out of the mineral. But this assumption is not always true, as this example suggests.

26.3 Geologic Time

INSIGHT Currently, the record for the world's oldest rock belongs to rock found at Porpoise Cove in Canada. This granitic rock is over 3.8 billion years old and narrows the gap between the origin of Earth and the origin of the rock record to only about 700 million years.

To chart the history of Earth, we use the geologic time scale (Figure 26.5). This scale was developed through the use of relative dating, and specific dates have been applied to it with radiometric dating. The geologic time scale subdivides Earth's 4.5-billion-year history into time units of different sizes. As you can see by looking at the chart, *eons* are the largest units of geologic time. The eon we are living in began about 540 million years ago. It is called the Phanerozoic,

Eon	Era	Period	Sub-period	Epoch	Ma
Phanerozoic	Cenozoic	Quaternary		Holocene	0.01
				Pleistocene	1.8
		Tertiary		Pliocene	5.3
				Miocene	23.8
				Oligocene	33.7
				Eocene	54.8
				Paleocene	65
	Mesozoic	Cretaceous			145.5
		Jurassic			199.6
		Triassic			248
	Paleozoic	Permian			290
		Carboniferous	Pennsylvanian		318
			Mississippian		354
		Devonian			417
		Silurian			443
		Ordovician			490
		Cambrian			542
Precambrian Time	Proterozoic				2500
	Archean				3800
	Hadean				4500

Figure 26.5 The geologic time scale divides Earth's history into time units of different size. Units of time on this scale are Ma, which stands for "millions of years ago." The Paleozoic era began about 542 Ma, or 542 million years ago, for example.

which means *visible life*. This is a fitting name because rocks and deposits of the Phanerozoic eon are rich in fossils illustrating the evolution of life.

Look at the geologic time chart again and you can see that the Phanerozoic eon is subdivided into three eras: the Paleozoic, Mesozoic, and Cenozoic. Paleozoic literally means "time of ancient life," Mesozoic means "time of middle life," and Cenozoic means "time of recent life." As the names suggest, these eras are characterized by profound differences in the nature of organisms living at the time. For example, the Cenozoic is the "age of the mammals." Eras are divided into smaller units of time called *periods*. Periods, like eras, are characterized by differences among life forms. Finally, periods are divided into the smallest chunk of geologic time—the epoch. Epochs are generally defined in terms of geological distinctions rather than differences among life forms.

Note that the vast majority of Earth's history occurred prior to the Paleozoic era. The vast span of time that preceded the Paleozoic, over 4 billion years, is divided into three eons: the *Hadean*, the *Archaen*, and the *Proterozoic*. Together, the three ancient eons are called **Precambrian** time.

CHECK YOURSELF

1. Describe the present time in Earth's history in terms of all units of the geologic time scale from eons to epochs.
2. The time units on the geologic scale are Ma. What does 1 Ma stand for?

CHECK YOUR ANSWERS

1. We are living in the Phanerozoic eon, in the Cenozoic era, in the Quaternary period, and in the Holocene epoch.
2. One Ma means "one million years ago."

26.4 Precambrian Time (4500 to 542 Million Years Ago)

Precambrian time ranges from about 4.5 billion years ago, when Earth formed, to about 542 million years ago, when abundant macroscopic life appeared. The Precambrian—the time about which we know the least—comprises almost 90 percent of Earth's history. Most of the rocks that formed in this early part of Earth's history have been eroded away, metamorphosed, or recycled into Earth's interior. Although organisms ranging from simple bacteria to more complex marine invertebrates such as marine worms and jellyfish lived during Precambrian time, there are few fossils in Precambrian rocks. Organisms of that time did not have easily fossilized hard body parts, which evolved only later in the history of life.

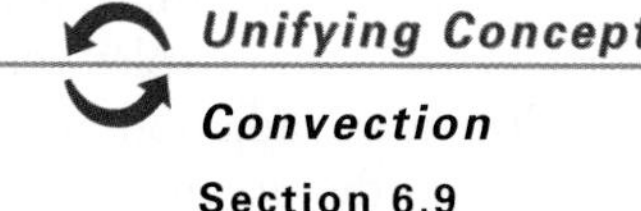

The beginning of the Precambrian was likely a time of considerable volcanic activity and frequent meteorite impacts, as described in Chapter 22. During the Hadean eon, between 4500 and 3900 million years ago, large and small chunks of interplanetary debris left over from the formation of the solar system continually smashed into Earth. Earth was an oceanless planet covered with volcanoes erupting gases and steam from its scorching interior. Huge holes and gashes left by falling debris scarred its surface. Intense convection in the mantle, and tremendous heat escaping from the interior, left Earth's early crust in turmoil. The early atmosphere consisted mostly of gases that had erupted from the many volcanoes. Carbon dioxide may have comprised 80 percent or more of this atmosphere. Water vapor made up most of the balance, with molecular nitrogen, ammonia, sulfur dioxide, and nitric oxide as minor constituents. There was no free oxygen.

Slowly, Earth's surface cooled. This set the stage for the formation of the oceans in the Archean Eon. When Earth's surface cooled below 100°C, water vapor in the atmosphere condensed to form clouds. The clouds poured torrents of rain onto the Earth and the rains filled basins to form oceans. Remarkably, the amount of water at Earth's surface has remained constant since that time, billions of years ago.

Also in the Archaean, Earth's permanent crust formed. Evidence from folded and faulted rocks and radiometric dating indicates that the first significant continental crust movements occurred about 2.5 billion years ago. Lithospheric plates developed. Continents then began to form as small landmasses came together.

The early atmosphere was rich in carbon dioxide and water vapor but very poor in free oxygen. We know from fossils that life has existed on Earth a very long time—at least 3.5 billion years. The success of the earliest organisms depended on their ability to survive in the primitive oxygen-poor environment. As you recall from Chapter 17, cyanobacteria, which lived in abundance some 2.7 billion years ago in shallow seas, produced oxygen as a by-product of photosynthesis. Because of these and other early photosynthesizing organisms, Earth's atmosphere gradually evolved to support more complex land-dwelling organisms, as detailed in the following Integrated Science section (Figure 26.6).

INTEGRATED SCIENCE BIOLOGY AND CHEMISTRY

The Great Transformation of Earth's Atmosphere

Today, the atmosphere contains less than one-tenth of 1 percent carbon dioxide. Where did all the carbon dioxide of Earth's early atmosphere go? Some dissolved in the oceans, reacted with other substances, and formed limestone deposits on the ocean floor. However, most of the carbon dioxide of the early atmosphere was absorbed by one-celled life forms such as *cyanobacteria*. As you read about in Chapter 17; the evolutionary impact of cyanobacteria is hard to overstate. Like green plants, cyanobacteria use photosynthesis to convert carbon dioxide and water to a carbohydrate plus free oxygen:

$$CO_2 + H_2O \text{ (light)} \rightarrow CH_2O + O_2$$

Over a period spanning a billion years, the cumulative effect of these early one-celled, photosynthesizing organisms was to transform Earth's atmosphere by removing huge amounts of carbon dioxide from the atmosphere and generating free oxygen (O_2). How did this process work? When photosynthesizing organisms died, they sank to the bottom of the oceans and were buried. The organic molecules they produced, and which composed their tissues, were incorporated in the sediments, and hence in the solid Earth. In this way, carbon was pumped from the air, through the simple organisms living in the ocean, to Earth's rocky layers. Oxygen, meanwhile, was pumped into the atmosphere as a product of photosynthesis.

When sufficient free oxygen became available, an ozone (O_3) layer formed in the upper atmosphere. Many people understand that O_2 was necessary for the development of complex organisms that breathe it in. But few people appreciate the significance of this other form of oxygen—ozone—to the development of land-dwelling organisms. Ozone in the upper atmosphere forms a protective shield, which screens land dwellers from ultraviolet radiation (Chapter 25). Among its other, harmful effects, ultraviolet radiation disrupts DNA. So without ozone, life would have been confined to the oceans and complex land-dwelling organisms with their intricate genetic material would never have evolved.

Figure 26.6 This artwork of Precambrian Earth shows characteristic features of the time, including space debris, such as meteorites and comets, crashing into Earth's surface. The bright green material on the edge of the water is algae, while cyanobacteria thrive in the darker green, round structures called *stromatolites* in the water.

CHECK YOURSELF

1. In what two ways was the development of free oxygen essential to the development of terrestrial animals?
2. Where did the carbon dioxide that characterized the early atmosphere go?

CHECK YOUR ANSWERS

1. Free oxygen provided oxygen for respiration and was necessary for the formation of atmospheric ozone.
2. Some carbon dioxide was dissolved in the oceans and later formed limestone deposits; the rest was consumed by one-celled photosynthesizing organisms and became incorporated into rock.

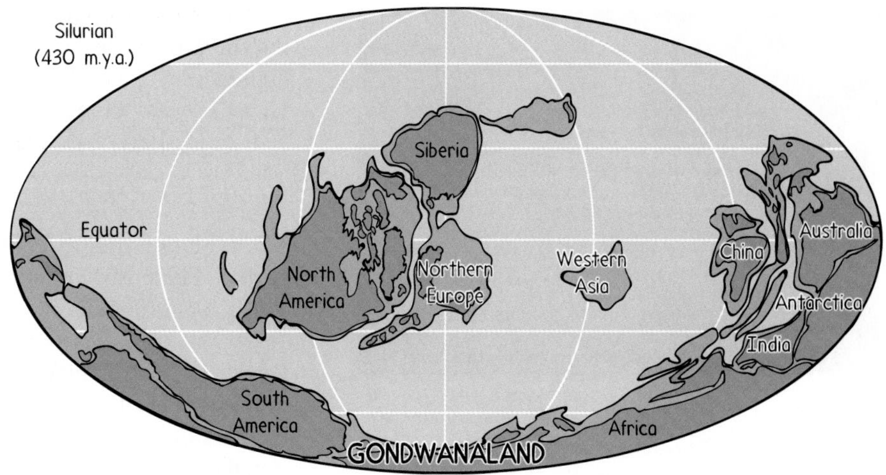

Figure 26.10 This map represents the ancestral continents as they may have been positioned during the Silurian period.

casually known as the "age of the fishes." Some groups, such as the sharks and bony fishes, are still present today. Among the bony fishes, the lobe-finned fishes are of particular interest because they gave rise to land-living, terrestrial vertebrates. Some lobe-finned fishes evolved internal nostrils, which enabled them to breathe air. In addition, the fins of these fishes were lobed and muscular, enabling the animals to support their bodies and "walk." Today, the lungfishes and the coelacanth (pronounced SEE-la-kanth), a "living fossil," are the only lobe-finned fishes still in existence. The first truly terrestrial vertebrates, descendants of lobe-finned fishes, made their appearance during the late Devonian. These vertebrates shared many features with the amphibians of today. For example, like existing amphibians, they lay unshelled eggs and could only live in moist environments (Figure 26.11).

Also during the Devonian, Europe and North America merged and lay near the equator. Africa, South America, India, Antarctica, and Australia were collected as Gondwanaland, which remained in the Southern hemisphere.

Figure 26.11 This artwork of a late Devonian forest shows an Acanthostega tetrapod, an amphibian, climbing over a rock while a dragonfly flies above. Tetrapods flourished in the Devonian, having evolved from fishes.

INSIGHT The coelacanth was thought to have become extinct after the Mesozoic era. However, in 1938, the first living specimen was caught off the coast of East Africa. Since then other specimens have been discovered. The coelacanth is considered a "living fossil." A "living fossil" is any living species that closely resembles other species known only from fossils.

The Carboniferous Period (354 to 290 Million Years Ago)

The Carboniferous period includes both the Mississippian and the Pennsylvanian subperiods. Warm, moist climatic conditions contributed to lush vegetation, forests, and swamps. Dense swamps covered large portions of what are now North America, Europe, and Siberia. As plants and trees died, their remains settled to the bottoms of these stagnant swamps and decayed anaerobically to produce coal (Chapter 23). Most of the coal used today derives from these Carboniferous coal swamps. In fact the name "Carboniferous" or "carbon-bearing" is a reference to the coal swamps that characterized this part of Earth's history.

In the Carboniferous period, insects underwent rapid changes that led to diverse forms, including giant cockroaches and dragonflies with wingspans of 80 centimeters. The first amniotes, the vertebrate group that includes today's reptiles and mammals, also appeared during the Carboniferous. Amniotes are characterized by the presence of a shelled, or amniote, egg. The amniote egg provides a completely self-contained environment for an embryo. Most crucially, it protects embryos from drying out, allowing amniotes to live in a diverse array of terrestrial habitats (Figure 26.12).

Figure 26.12 Warm, moist climatic conditions contributed to the lush vegetation and swampy forests of the Carboniferous period. These forests produced most of the coal deposits around the world.

The Appalachian mountain belt of eastern North America was formed during the Carboniferous when Laurussia (the landmass that is now Europe and the United States) and Gondwanaland collided.

The Permian Period (290 to 248 Million Years Ago)

The evolution and diversification of amniote vertebrates continued in the Permian period. These early amniotes, which had scales like the reptiles of today, ruled the Earth for 200 million years. (By comparison, modern humans evolved 195,000 years ago. See Section 17.9.) Two major amniote groups appeared during the Permian. One of these groups, which includes ancestors of the earliest mammals, dominated the Permian. The other amniote group appearing in the Permian, which includes the reptiles, eventually gave rise to the dinosaurs early in the Mesozoic era. Yet, the flourishing of life was greatly interrupted in the Permian. The largest extinction in Earth's history occurred in the Paleozoic, at the end of the Permian period. What caused the Permian Extinction, the biggest mass extinction in Earth's history?

INTEGRATED SCIENCE BIOLOGY

The Permian Extinction

Extinction is the fate of every species. A species can go extinct for any of a number of reasons. Its habitat may be destroyed, or environmental conditions may change in such a way that populations are unable to adapt. Through most of Earth's history, then, species have gone extinct at a regular "background" rate. Sometimes, however, a huge number of species suddenly go extinct over a very short time period—this is known as a *mass extinction*. Mass extinctions typically occur very rapidly, over a few million years or less, and affect many different kinds of species all over the world

The Permian extinction caused the demise of about 90 percent of species living at the time. Interestingly, the extinction coincided with the largest known volcanic eruption in Earth's history, which occurred in what is now western Siberia. This eruption lasted for about a million years and may have produced two bouts of rapid climate change. First, sulfate aerosols from the eruption could have rapidly cooled the climate by blocking solar radiation. Then, as the eruption trailed off, Earth could have quickly warmed again. These two periods of rapid climate change, combined with other effects of the eruption such as a disrupted ozone layer, acid rain, and the release of large amounts of carbon dioxide into the atmosphere, are hypothesized to have contributed to the huge die-offs. Photosynthesizing species may have been most drastically affected first, with effects reverberating up the entire food chain.

In addition to effects of the volcanic eruption, another geological factor likely triggered the Permian extinction. By the end of the Paleozoic, all the continents had fused into the one global supercontinent Pangaea (Figure 26.13). With

Figure 26.13 With the collision of landmasses, the supercontinent Pangaea was formed in the late Paleozoic.

this redistribution of land and water, and changes in the elevations of landmasses, the world's climate became very seasonal. Although the cause of the Permian extinction is uncertain, the climate changes associated with the formation of Pangaea, including changes in temperature, and precipitation, and the lowering of sea level, surely stressed many species.

CHECK YOURSELF

1. Is the occurrence of a major cataclysm near the time of a mass extinction—such as a huge volcanic eruption—sufficient to link the extinction to the event?
2. Why would the consolidation of landmasses into the supercontinent Pangaea increase the seasonal variation they experience?

CHECK YOUR ANSWERS

1. No. It is certainly suggestive when major cataclysms closely coincide with major extinctions. However, there have also been huge volcanic explosions and meteor strikes that were not associated with mass extinctions, so one does not lead inevitably to the other.
2. The high specific heat capacity of water has a moderating effect on temperature (as explained in Chapter 25). Therefore, with less surrounding water, consolidated landmasses would tend to experience more seasonal variation.

26.6 The Mesozoic Era (248 to 65 Million Years Ago)

The Mesozoic era is informally known as the "age of the reptiles" because of its famous reptilian inhabitants.

The Mesozoic era, known informally as "the age of reptiles," consisted of three periods: Triassic, Jurassic, and Cretaceous. Reptiles that survived the Permian extinction at the end of the Paleozoic era evolved to become the dominant species of the world. The Mesozoic era was the time of the rise of the dinosaurs. Mammals evolved early in the Mesozoic, but they were relatively small and insignificant, compared with the dinosaurs.

Land plants greatly diversified during the Mesozoic era. True pines and redwoods appeared, and they rapidly spread throughout the land. Flowering plants arose in the Cretaceous period, and they diversified so quickly that, by the end of the period, they were the dominant plants. The emergence of the flowering plants also accelerated the evolution and specialization of insects because of the close association between flowers and their insect pollinators.

The end of the Cretaceous period, 65 million years ago, brought another mass extinction. Many dinosaurs, flying reptiles, and marine reptiles were wiped out, as were other organisms, both on land and in the seas.

INTEGRATED SCIENCE BIOLOGY AND ASTRONOMY

The Cretaceous Extinction

The mass extinction that ended the Cretaceous killed 60 percent of Earth's species over a period of about 2 million years. Most famously, the dinosaurs were devastated, with only a few lineages surviving. (One surviving lineage was birds; for this reason, technically speaking, birds are dinosaurs). In addition, many marine groups were hard hit, as were terrestrial plants, with the striking exception of ferns. Mammals generally survived well, as did most reptiles other than the ill-fated dinosaurs. What caused the Cretaceous extinction?

Scientists generally agree that climate change was a key factor, with mild warm Mesozoic climates suddenly turning cool in the Cenozoic. This cooling was followed by a period in which sunlight was scarce and a longer period of greenhouse warming. In the rock record, geologists have found a thin layer of clay containing high levels of iridium. The position of this layer in the rock record corresponds to the time of the Cretaceous extinction. Scientists are di-

Figure 26.14 Artwork of the Chicxulub crater on the Yucatan Peninsula of Mexico, soon after its formation. The crater is about 180 km wide. The impact that produced the crater may have caused the mass extinction that ended the reign of the dinosaurs, 65 million years ago at the end of the Cretaceous.

vided as to what caused this iridium deposit as well as the crucial climatic changes—some favor volcanism on Earth, others the impact of a large meteorite. There is evidence of both phenomena near the end of the Cretaceous.

Those who believe that a meteorite caused the mass extinction point out that the concentration of iridium in a meteorite is much higher than the average concentration of iridium in Earth's crust. This suggests, they say, that the iridium found in the rock record was probably spread worldwide by the impact event. The timing is right—the iridium was deposited about 65 million years ago—the time of the great Cretaceous extinction. A large impactor almost certainly did hit Earth at the time, leaving "shocked" quartz, spheres of rock that melted from the energy of the impact but then cooled and solidified, and layers of soot from widespread fires. A buried impact crater, called Chicxulub (Figure 26.14), has also been found on the Yucatan Peninsula in Mexico. However, the precise role of the impact in the Cretaceous extinction is still far from clear.

The possibility that the Cretaceous extinction was caused by a meteorite is the starting point for the Nemesis theory. The Nemesis theory argues that Earth experiences a mass extinction approximately once every 26 million years and that this periodicity is due to a hypothetical star named Nemesis, binary to the Sun. Every 26 million years, proponents argue, Nemesis comes close enough to the solar system to disturb comets, which proceed to bombard the inner solar system. Variations of the Nemesis idea suggest a rogue planet in the solar system rather than a binary star or attribute periodic extinctions to regular crossings of the solar system through the Oort cloud. (The Oort Cloud, as you will learn in Chapter 27, is a comet-filled region at the outer edge of the solar system.)

However, there is no evidence for the existence of Nemesis or a rogue planet, or even of any periodicity in mass extinctions on Earth. In addition, most mass extinctions appear to have a purely terrestrial cause—aside from the Cretaceous extinction, only one other is potentially linked to a meteor impact.

CHECK YOURSELF

1. What are the two major hypotheses that would explain the extinction at the end of the Cretaceous? What key piece of evidence do these both try to explain?
2. If a meteorite caused the Cretaceous extinction, does this firmly establish the Nemesis theory?
3. In terms of the number of species destroyed, which was the bigger extinction—the Permian or Cretaceous?

CHECK YOUR ANSWERS

1. One hypothesis is that the extinction was triggered by climate changes following a massive volcanic eruption; the other hypothesis holds that a large meteorite impact triggered the climate change that caused the eruption. Both of these hypotheses explain the high concentration of iridium in the clay layer at the boundary of the Cretaceous.
2. No; the Nemesis theory is an elaboration of the meteorite impact theory.
3. The Permian was the bigger mass extinction, having killed off 50 percent more species than the Cretaceous.

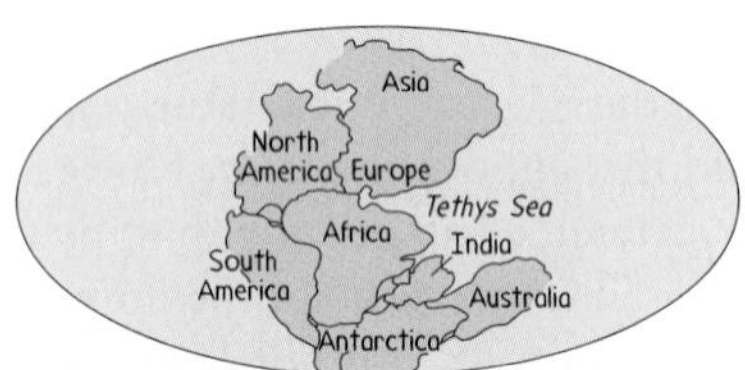

200 million years ago
Mesozoic Era

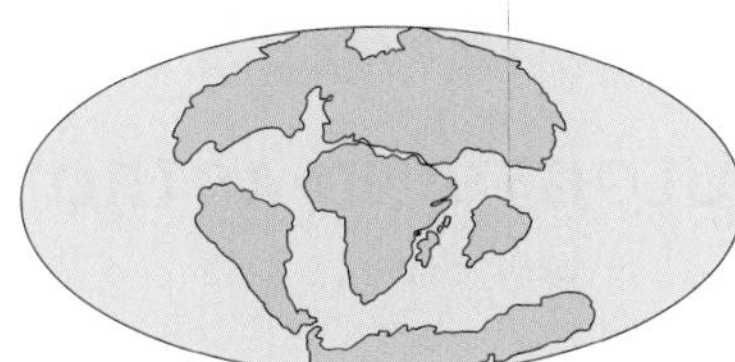

65 million years ago
Cenozoic Era

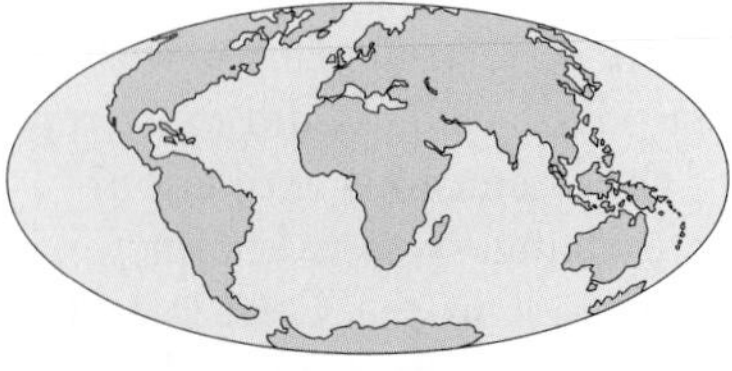

Present

Figure 26.15 Stages during the breakup of Pangaea.

The major geological event of the Mesozoic was the breakup of Pangaea. The breakup continued for 200 million years through the Triassic, Jurassic, and Cretaceous periods and extended into the Cenozoic era (Figure 26.15). The first step in the breakup, which occurred in the Triassic, was the development of a rift between what is now the eastern United States and Africa. As the rift developed, the Atlantic Ocean began to form. The rifting of Pangaea continued during the Jurassic, and by the mid-Cretaceous it had split into several smaller continents. India shifted northward, while ancestral South America and Africa separated from ancestral Australia and Antarctica. The South Atlantic Ocean formed after South America and Africa split apart, carried by diverging tectonic plates. Of all the former landmasses that were joined in the Paleozoic time, only the merger of Europe and Asia has survived to the present time.

The break-up of Pangaea impacted biological evolution, just as drastic environmental changes always do. The breaking apart of Pangaea created large-scale geographic isolation. Organisms were separated from one another and exposed to different environmental conditions, which caused a divergence in the evolution of land-based life.

26.7 The Cenozoic Era (65 Million Years to the Present)

The Cenozoic era is made up of two periods—the Tertiary and the Quaternary. Most of the Cenozoic belongs to the Tertiary period. The Quaternary makes up only the most recent 1.8 million years, as you can see by looking back at the Geologic Time Scale (Figure 26.5).

Enormous tectonic disturbances occurred rapidly throughout the world during the Tertiary. For example, there was a spreading center off the western margin of North America, with the Pacific Plate on the west and the Farallon Plate on the east (Figure 26.16). As the Farallon Plate subducted beneath North America, the spreading center approached the North American continental margin. The collision between the westward moving North American Plate and the Pacific ridge system occurred about 30 million years ago, creating the San Andreas fault.

Considerable mountain building occurred in the Tertiary period. For example, the landmass of Africa-Arabia collided with Europe to produce the Alps.

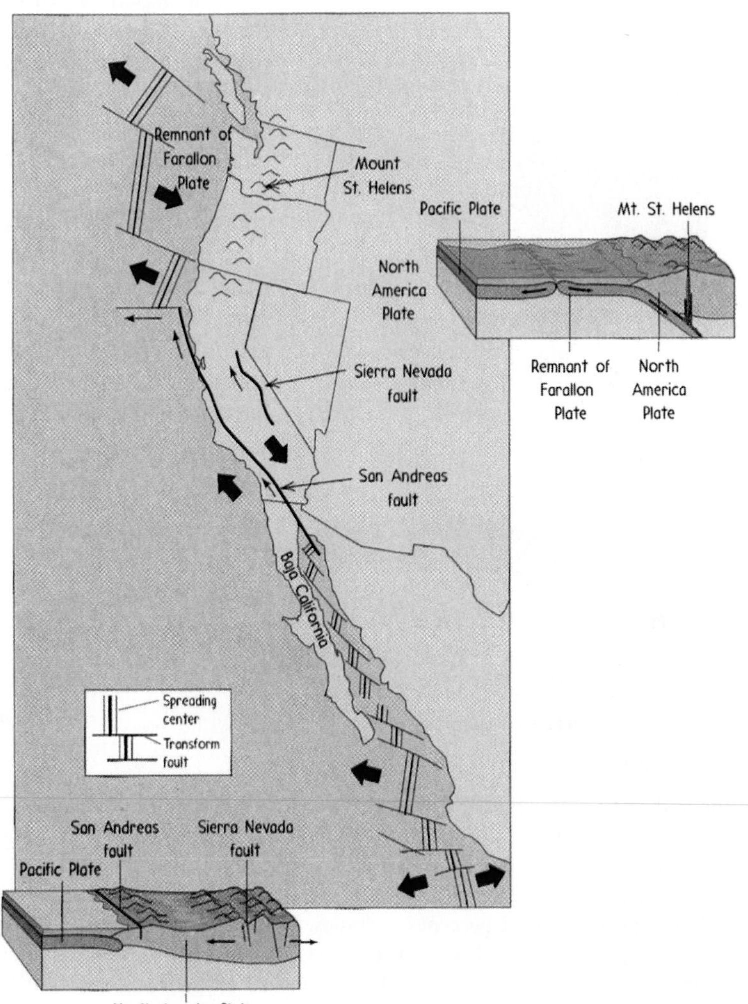

Figure 26.16 The San Andreas fault is the result of an encounter between the North American Plate and the Pacific Plate. As the fault grew longer, the area of Baja California was torn from the continental margin.

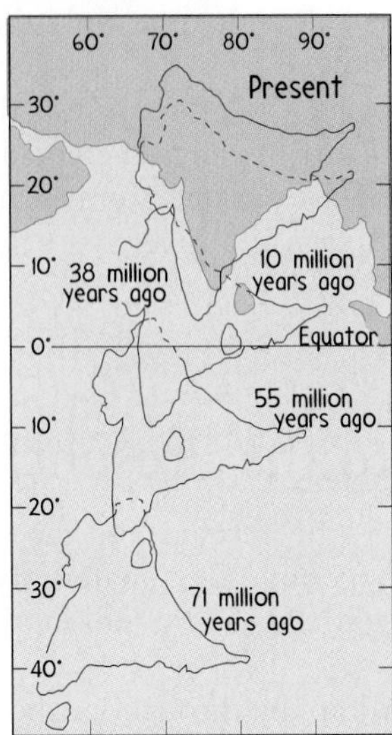

Figure 26.17 The formation of the Himalayas was a result of the collision of India with Asia. Because this was a continent-to-continent collision, the Himalayas have an unusually thick accumulation of continental lithosphere. Like icebergs, the mountains run deeper below the surface than they are high.

India collided with Asia to produce the Himalayas. As shown in Figure 26.17, the leading edge of the Indian Plate was forced partially under Asia, producing a thick accumulation of continental lithosphere. Due to isostasy, (discussed in Chapter 22) the thick lithosphere provided additional uplift to the Himalayas.

Climates cooled during much of the Cenozoic. It was during the Pleistocene that the most recent episodes of global cooling, or *ice ages*, took place. Much of the world's temperate zones were alternately covered by glaciers during cool periods and uncovered during the warmer interglacial periods when the glaciers retreated. During glacial episodes, as much as one-third of the land was covered by great thicknesses of ice.

After the mass extinctions at the end of the Mesozoic era, many environmental niches were left vacant. These openings allowed the relatively rapid evolution of mammals in habitats formerly occupied by their extinct predecessors. Bats, some large land mammals, and marine animals such as whales and dolphins evolved to occupy niches left vacant by the extinction of many of the Mesozoic reptiles.

Humans evolved during the Cenozoic era in the Quaternary period and Pleistocene epoch. During the Pleistocene epoch, global temperatures dipped as an ice age ensued. The cool temperatures had a profound impact on life. Mammoths, rhinos, bison, reindeer, and musk oxen all evolved warm, woolly coats for protection from the frigid cold. Extensive glaciation caused sea level to drop as water became bound up in glaciers. Even though the distribution of landmasses was essentially the same as it is today, the lowered sea level resulted in "land bridge" connections between landmasses that are now separated by water. One of these land bridges existed across the present-day Bering Strait, and it provided the route for the human migration from Asia to North America.

During the Pleistocene, Earth's large mammals evolved wooly coats for protection against the frigid temperatures characteristic of an ice age.

The expansion of humans, not only into North America but also throughout the world, coincided with a period of extinction that occurred during the Pleistocene. Pleistocene extinctions primarily involved large terrestrial mammals, leaving marine animals unaffected. In North America, many large mammals became extinct after humans arrived, and in Africa, mammalian extinctions can be related to the appearance of the Stone Age hunters.

The cause of the Pleistocene extinction is a much-debated issue. The extreme climatic variation that existed at the time could have been partly responsible. However, though large-scale glaciation was occurring in some regions, the climate in many areas was relatively mild. This leads some scientists to believe that harsh climate likely played only a small role in the Pleistocene extinctions and that humans essentially hunted and ate the large mammals to extinction. However, this theory is controversial as it is unclear whether *Homo sapiens* was present in sufficient numbers and had sufficient technological know-how to kill off entire species.

SCIENCE AND SOCIETY

Mass Extinctions and the Modern World

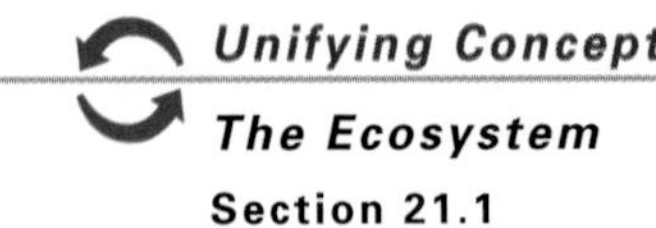

Section 21.1

Are we in the midst of another mass extinction? Human population continues to grow, and humans continue to destroy natural habitats at an unprecedented rate. As of 2004, the World Conservation Union includes in its "Red List" of threatened species 23 percent of all mammals, 12 percent of birds, 61 percent of reptiles, 31 percent of amphibians, 46 percent of fishes, 73 percent of insects, 45 percent of mollusks, 86 percent of mosses, 67 percent of ferns, and 73 percent of flowering plants. And this is *now*—it doesn't even take into account what global warming will do. Recently, a group of scientists estimated the number of extinctions that global warming is expected to produce. Using a midrange prediction of temperature rise, they calculated that the combination of climate change and the scarcity of new habitats for species to migrate to will cause between 15 and 37 percent of all species to go extinct by 2050. That's massive. One of the most important questions to ask ourselves is: Will the human activities that adversely affect our planetary life-support system—from the large-scale burning of fossil fuels, to overpopulation, to the destruction of the Amazon rainforest—ultimately lead to our own extinction?

Following the Pleistocene epoch is the Holocene epoch. To observe the Holocene environment, just look around. It is the most recent 10,000 years of Earth's history or so, including the present time. There has been some climatic variation, for example a "Little Ice Age" occurred between about 1200 and 1700 A.D. However, in general, the Holocene has been a relatively warm interglacial period.

The Holocene is sometimes called the "Age of Man." This is somewhat inaccurate because *Homo sapiens* had evolved and occupied many regions of the globe well before the start of the Holocene. Yet the Holocene has witnessed all of humanity's recorded history and the rise and fall of all its civilizations. Humanity has had a great impact upon the Holocene environment. All organisms influence their environments to some degree, but few have ever changed Earth as much, or as fast, as we are doing in modern times.

CHAPTER 26 Review

Summary of Terms

Cenozoic era The time of "recent life," from 65 million years ago to the present.

Cross-cutting A relative dating principle stating that, where an igneous intrusion or fault cuts through other rocks, the intrusion or fault is younger than the rock it cuts.

Faunal succession A relative dating principle stating that fossil organisms succeed one another in a definite, irreversible, determinable order.

Geologic time scale The time scale that subdivides Earth's 4.5-billion-year history into time units of different sizes, from eons to epochs.

Inclusions A relative dating principle stating that any inclusion (a piece of one rock type contained within another) is older than the rock containing it.

Mass extinction The rapid extinction of numerous species; the rate of extinction is much greater than the background rate of species extinction.

Mesozoic era The time of "middle life," from 245 million years ago to about 65 million years ago.

Original horizontality A relative dating principle stating that layers of sediment are deposited evenly, with each new layer laid down almost horizontally over the older sediment.

Paleozoic era The time of "ancient life," from 542 million years ago to 245 million years ago.

Precambrian time The time of hidden life, which began about 4.5 billion years ago when Earth formed, lasted until about 544 million years ago (the beginning of the Paleozoic era), and makes up 85 percent of Earth's history.

Principle of Uniformitarianism The present is the key to the past.

Radiometric dating A method for calculating the age of geologic materials based on the nuclear decay of naturally occurring radioactive isotopes.

Relative dating the ordering of rocks in sequence by their comparative ages.

Superposition A relative dating principle stating that in an undeformed sequence of sedimentary rocks, each bed or layer is older than the one above and younger than the one below.

Unconformity A break or gap in the geologic record, caused by an interruption in the sequence of deposition or by erosion of preexisting rock.

Review Questions

26.1 A Model of Earth's History

1. If all of history were collapsed into a period of 1 year, when would bacterial life appear? When would *Homo sapiens* appear? When would recorded history commence?
2. What is the principle of Uniformitarianism? What does it tell us about the natural laws of physics, chemistry, and biology?

26.2 The Rock Record

3. What five principles are used in relative dating? Describe each one.
4. What is the "rock record"?
5. What is a gap in the rock record called?

26.3 Geologic Time

6. Which of the geologic time units spans the greatest length of time?
7. How old is Earth?

26.4 Precambrian Time

8. What percentage of Earth's history is Precambrian time?
9. Cite one Earth-shaping event from the Hadean eon and another from the Archaean eon.

26.5 The Paleozoic Era

10. Name the periods of the Paleozoic era.
11. For what is the Silurian period best known?
12. During what time period were most coal deposits laid down?

26.6 The Mesozoic Era

13. By what informal name is the Mesozoic era known?
14. What effect did the breakup of Pangaea have on sea level?
15. What Pangaean landmass survives to this day?

26.7 The Cenozoic Era

16. Which epochs make up the Tertiary period? The Quaternary period?
17. Stated in terms of the units of the Geologic Time Scale, when did humans evolve?
18. What adaptation did mammals of the Pleistocene epoch make to the prevailing climate?
19. Describe the climate of the Holocene.

Multiple Choice Questions

Choose the BEST answer to the following.

1. Uniformitarianism is the idea that
   (a) ideas across all branches of Earth science are consistent.
   (b) certain sedimentary deposits tend to be well sorted.
   (c) present-day processes also occurred in the geologic past.
   (d) younger sedimentary rocks are always deposited on top of metamorphic or intrusive igneous rocks.

2. The earliest division of geologic time for which abundant fossils have been found is
   (a) the Precambrian.
   (b) the Silurian.
   (c) the Pliocene.
   (d) the Cambrian.
3. Before radiometric techniques were available to determine the ages of rock, scientists
   (a) had no way to determine the ages of rock.
   (b) used relative dating to determine the relative ages of rocks.
   (c) used radiometric techniques to determine the composition of rock.
4. Gaps in the rock record are called
   (a) unconformities.
   (b) fossils.
   (c) inclusions.
   (d) faults.
5. While Paleozoic literally means "time of ancient life," Cenozoic literally means
   (a) time of no life.
   (b) time of middle life.
   (c) time of recent life.
6. Why do we know so much more about the Paleozoic era compared to the Precambrian?
   (a) There was no life on Earth during the Precambrian.
   (b) Shelled organisms developed during the Paleozoic.
   (c) The Paleozoic era began a mere 50,000 years ago.
7. How did the breakup of Pangaea during the Mesozoic affect the evolution of life?
   (a) It had no significant effect.
   (b) It led to a mass extinction of land-based life.
   (c) It triggered the divergence of land-based life.

INTEGRATED SCIENCE CONCEPTS

Physics—Radiometric Dating of Rock

1. What are the half-lives of uranium-238, potassium-40, and carbon-14?
2. What isotope is preferred in dating very old rocks?
3. What isotope is commonly used for dating sediments or organic material from the Pleistocene period?

Biology and Chemistry—The Great Transformation of Earth's Atmosphere

1. (a) How does the concentration of carbon dioxide in Earth's early atmosphere compare to the concentration of this gas today? (b) How does the concentration of free oxygen compare then and now?
2. How did the transformation of Earth's atmosphere affect the evolution of life?
3. By what mechanism, did the transformation of Earth's atmosphere occur?

Biology—The Permian Extinction

1. What percentage of Earth's species was killed in the Permian extinction?
2. What two factors may have been the root cause of the Permian extinction?

Biology and Astronomy—The Cretaceous Extinction

1. What important clue lies in the rock record pertaining to the Cretaceous extinction?
2. Describe the Nemesis theory and how this is different from the idea that a meteorite triggered the Cretaceous extinction.

Exercises

1. Suppose a certain type of sediment is deposited in all modern streams. On a geologic expedition into unknown territory, we find the same type of deposit in ancient rocks. What can we say about the ancient rocks? What assumption are we making?
2. Why don't all rock formations show a continuous sequence from the beginning of time to the present?
3. How are fossils used in determining geologic time?
4. In a sequence of sedimentary rock layers, the oldest layer is on the bottom and the youngest layer is at the top. What relative dating principle applies here?
5. What role did tectonic activity play in the formation of the San Andreas fault?
6. What kinds of mammals evolved to occupy niches left vacant by the extinction of Mesozoic reptiles?
7. Did life exist on Earth in the Precambrian? If so, cite two or three examples.
8. If a sedimentary rock contains inclusions of metamorphic rock, which rock is older? Defend your answer.
9. Which isotopes are most appropriate for dating rocks from the following ages? (a) Early Precambrian time. (b) The Mesozoic era. (c) The late Pleistocene epoch.
10. Geologists often refer to the early Paleozoic era as the "Cambrian Explosion." What is meant by this phrase? What conditions prevailed at the time, allowing the "explosion"?
11. Suppose that, in an undeformed sequence of rocks, you find a trilobite embedded in shale layers at the bottom of the formation and fossil leaves embedded in shale at the top of the formation. From your observation, what can you say about the ages of the formation?
12. What key developments in life occurred during Precambrian time?
13. Coal beds are formed from the accumulation of plant material that has become trapped in swamp floors. Yet coal deposits are found on the continent of Antarctica, where no swamps or vegetation exists. What is your explanation?
14. In what ways can sea level be lowered? What effect might the lowering of sea level have on existing life forms?
15. What can cause a rise in sea level? Is this likely to happen in the future? Why or why not?
16. What general principle is used to make sense of what must be the processes that occurred throughout Earth's history?

Problems

1. If fine muds were laid down at the rate of 1 cm per 1000 years, show that it would take 100 million years to accumulate a sequence 1 km thick.
2. How long has *Homo sapiens* existed? State your answer in terms of a percentage of Earth's history.
3. Chart the data given in "Science and Society: Extinction in the Modern Times" in pie graphs; one for each group of organisms: mammals, birds, reptiles, amphibians, fishes, insects, mollusks, mosses, ferns, and flowering plants.

CHAPTER 26 ONLINE RESOURCES

Tutorials Formation of an Angular Unconformity Activity • ***Quiz*** • ***Exercises*** • ***Flashcards*** • ***Links***

PART FIVE

ASTRONOMY

Starlight, star bright . . . the stars look really beautiful tonight! They're even more beautiful knowing their faint color indicates their temperature. And stars, including the Sun, are hot because they convert part of their mass to energy, as described by $E = mc^2$. Energy produced by the Sun radiates into space, and a small quantity is intercepted by Earth, with some trapped by plants, then consumed by nonphotosynthesizing organisms like you and me—providing the energy we need to live. Stars are hot . . . but they are also really cool!

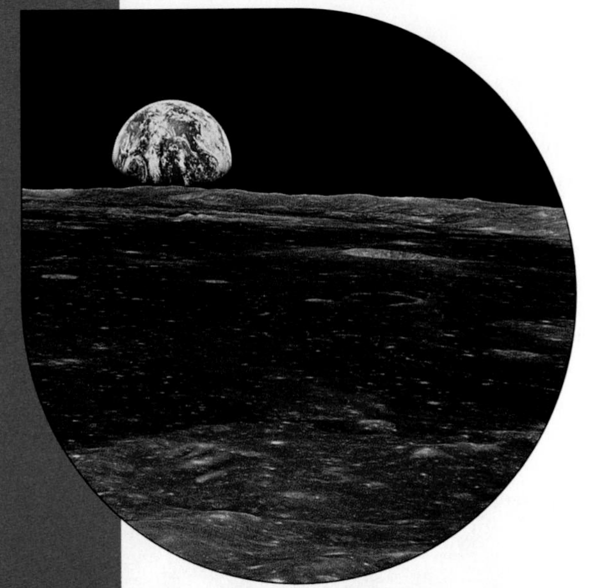

The Solar System

This photograph was taken by Apollo 13. *It shows Earth rising over the surface of the Moon.*

CHAPTER OUTLINE

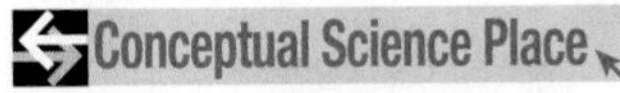

A Tour of the Solar System; Formation of the Solar System

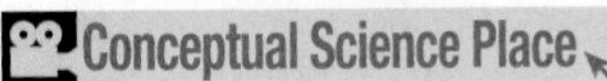

Orbits in the Solar System; History of the Solar System

Five billion years ago, there was no Sun. Instead, the region of the galaxy that would become our solar system was dark, cold, and almost empty. A diffuse cloud of gas and dust gently swirled in the blackness. But step-by-step, that cloud became the Sun and planets. How did mere gas and dust become the Sun and its planets, including the highly complex and organized Planet Earth? By applying principles you learned earlier in this book, you can understand how the solar system formed and why it has the structures and features it has. You'll also find answers to questions such as: How does the Sun produce energy for life on Earth? How do the planets differ from each other and why? What are solar and lunar eclipses and why are they rare? What are meteors, asteroids, and comets? What planetary conditions are necessary to support life? Does life exist elsewhere besides on Earth?

Let's begin with an overall tour of the solar system for context. Then, we'll describe how the solar system developed and explore some of its fascinating details.

27.1 Overview of the Solar System

The solar system is the collection of objects gravitationally bound to our Sun. In addition to the Sun itself, the solar system contains at least nine planets, their approximately 150 moons, a large number of asteroids (small, rocky bodies), and comets (small, icy bodies). These objects exist in the interplanetary medium, a sparse blend of dust and gas particles.

The Sun is at the center of the solar system and contains most of its mass—a whopping 99.86 percent. Moving outward from the Sun are, in order, the **inner planets:** Mercury, Venus, Earth, and Mars. Next is the main asteroid belt, which lies between the orbits of Mars and Jupiter. Then, there are the **outer planets:** Jupiter, Saturn, Uranus, Neptune, and, as we shall see, controversial Pluto. Beyond Neptune, containing Pluto, lies the disk-shaped **Kuiper Belt** (pronounced "koy-per") of comets and assorted objects. Far beyond the Kuiper Belt is the **Oort Cloud** (rhymes with *short*), a giant cometary sphere completely surrounding the solar system.

The planets exhibit a high degree of orderliness in their motions and in their positioning. For example, all the planets travel in elliptical orbits around the Sun.

INSIGHT The words *revolve* and *rotate* are often mistaken to mean the same thing. To *revolve* means to travel in a nearly circular path around a central point. To rotate, however, is to turn about an internal axis—to spin. Earth rotates on its axis as it revolves around the Sun.

And except for Pluto, all the planets and their larger moons follow orbits that lie roughly in the same plane. This plane, called the **ecliptic**, is defined as the plane of Earth's orbit. Further, all the planets and almost all of their moons orbit in the same direction—counterclockwise (when viewed from the Sun's north pole). This is also the direction in which the Sun and almost all of the planets and their moons spin or *rotate*. Also, the planets are neatly divided such that the inner planets are small, solid, and rocky while the outer planets are large and gaseous.

The solar system, like the interior of an atom, is mostly empty space. Figure 27.1 gives a hint of this but no illustration can be to scale because the planets are so tiny in comparison to the distances between them. To appreciate the true relative sizes of objects in the solar system, try this mental exercise. Imagine reducing

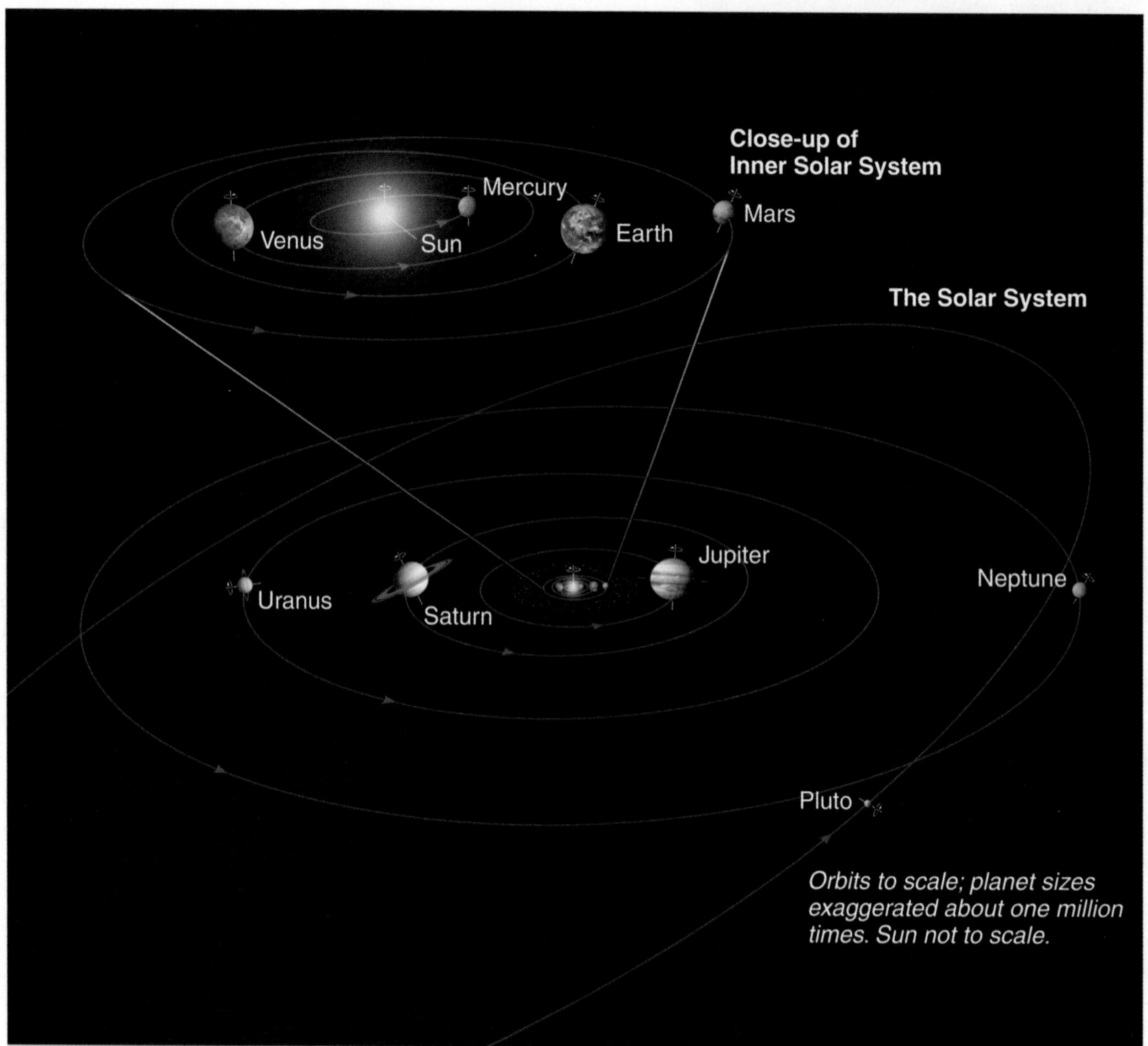

Figure 27.1 Interactive Figure

The layout of our solar system as it would appear from beyond recently demoted Pluto if we could magnify the sizes of the planets by about a million times. Notice that all the planets orbit the Sun in the same direction. The tilt of each planet's rotation axis is also shown, with a circling arrow to designate the direction of rotation. This illustration shows the main asteroid belt between Mars and Jupiter but not the remote regions where comets are concentrated—the Kuiper Belt and Oort Cloud. (Courtesy of Bennett, Donahue, Schneider, Voit, *The Essential Cosmic Perspective,* 3e, © 2005.)

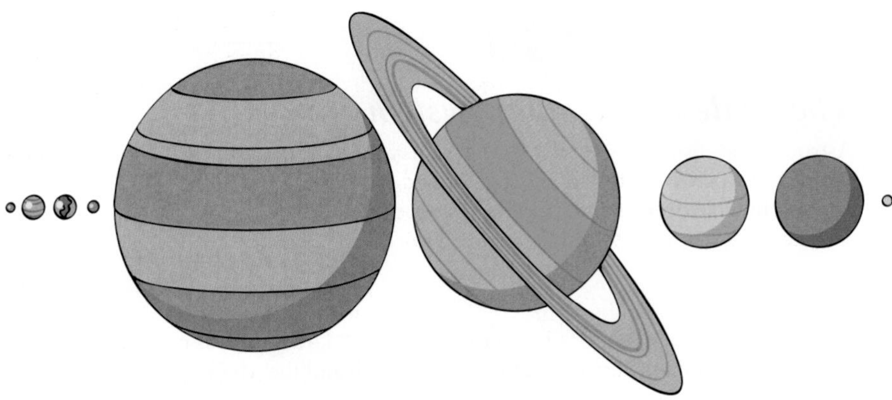

Figure 27.2 This illustration shows the order and relative sizes of planets. Moving away from the Sun (not shown to the left), we have in order: Mercury, Venus, Earth, Mars, Jupiter, Saturn, Uranus, Neptune, and recently demoted Pluto. The planets range greatly in size, but the Sun dwarfs them all—containing over 99 percent of the mass in the solar system.

the size of everything by a factor of a billion (Figure 27.2). Now Earth is 1.3 centimeters (cm) in diameter (the size of a grape). The Moon, the size of a pea, is a distance of 30 cm—about a foot—from Earth. The Sun is about 1.5 meters in diameter, the size of a Sumo wrestler. The distance between the Sun and Earth is 150 meters—the length of one-and-a-half football fields. Jupiter is 15 cm in diameter, the size of a large grapefruit, and at a distance of about 12 football fields away from the Sun. Saturn, the size of an apple, is 15 football fields away from the Sun. Tiny outermost "almost-a-planet" Pluto is about the size of an apple seed and is located at a distance of about 60 football fields from the Sun.

Because distances in the solar system are so great, astronomers use the *astronomical unit* to measure them. One astronomical unit (AU) is about 1.5×10^8 kilometers (about 9.3×10^7 mi) or the distance from Earth to the Sun. Table 27.1 gives the distances of planets from the Sun in kilometers (km) as well as in AU. The data in Table 27.1 also shows the division of the planets into two groups with similar properties. The inner planets—Mercury, Venus, Earth, and Mars—are solid and relatively small and dense. For this reason they are often called the "terrestrial planets." The outer planets are large, have many rings and satellites, and are composed primarily of hydrogen and helium gas. The outer planets are often referred to as the "Jovian planets" because they resemble Jupiter in terms of their large sizes and gaseous compositions.

Table 27.1 Planetary Data

	Mean Distance from Sun (Earth-distances, AU)	Orbital Period (years)	Diameter (km)	Average Mass (Earth = 1)	Density (kg)	(Earth = 1)	(g/cm^3)	Inclination to Ecliptic
Sun			1,392,000	109.1	1.99×10^{30}	3.3×10^5	1.41	
Mercury	0.39	0.24	4,880	0.38	3.3×10^{23}	0.06	5.4	7.0°
Venus	0.72	0.62	12,100	0.95	4.9×10^{24}	0.81	5.2	3.4°
Earth	1.00	1.00	12,760	1.00	6.0×10^{24}	1.00	5.5	0.0°
Mars	1.52	1.88	6,800	0.53	6.4×10^{23}	0.11	3.9	1.9°
Jupiter	5.20	11.86	142,800	11.19	1.90×10^{27}	317.73	1.3	1.3°
Saturn	9.54	29.46	120,700	9.44	5.7×10^{26}	95.15	0.7	2.5°
Uranus	19.18	84.0	50,800	3.98	8.7×10^{25}	14.65	1.3	0.8°
Neptune	30.06	164.79	49,600	3.81	1.0×10^{26}	17.23	1.7	1.8°
Pluto	39.44	247.70	2,300		0.18×10^{22}	0.002	1.9	7.2°

MATH CONNECTION

The Scale of the Solar System

Astronomical distances are mind-boggling. Try the following problems to better appreciate the sizes of bodies in the solar system and distances between them. (Use the distance formula: distance = speed × time and data from Table 27.1.)

Problems

1. The distance between Earth and the Moon is 384,401 km. How many Earth-diameters would fit between Earth and the Moon?
2. How long would it take to drive from Earth to the Moon if you drive at 55 miles per hour? State your answer in hours and in years.
3. If you could fly to the Sun on a jet that moves at 1000 km/h, how long would it take? State your answer in years.
4. The diameter of the Sun is 1,390,000 km. What is its diameter stated in units of AU? How many times greater is the mean distance between Earth and the Sun compared to the diameter of the Sun?

Solutions

1. $\dfrac{38{,}401 \text{ km}}{12{,}760 \text{ km}} \approx 30$

   About 30 Earth-sized planets would fit in the distance between Earth and its Moon.
2. $d = rt \rightarrow t = d/r$

   Convert all units to metric. (55 mi/h)(1 km/0.62 mi) ≈ 89 km/h. Then: t = 384,410 km/89 km/h = 4319 h (1 day/24 h) (1 y/365 days) ≈ 0.5 y

   It would take about one-half year to drive to the moon traveling freeway speeds.
3. $t = d/r = 1.5 \times 10^8$ km/1×10^3 km/h $= 1.5 \times 10^5$ h

   $(1.5 \times 10^5$ h)(1 day/24 h)(1 y/365 days) ≈ 170 y

   If you could fly to the Sun in a jet without becoming vaporized, it would take about 170 years.
4. 1.39×10^6 km ≈ 0.01 AU

   1.5×10^8 km

   $\dfrac{1 \text{ AU}}{1.01 \text{ AU}} = 100$

   The diameter of the Sun is approximately equal to 0.01AU, so the distance between Earth and the Sun is about 100 times the diameter of the Sun. Big as the Sun is, the solar system is mostly empty space.

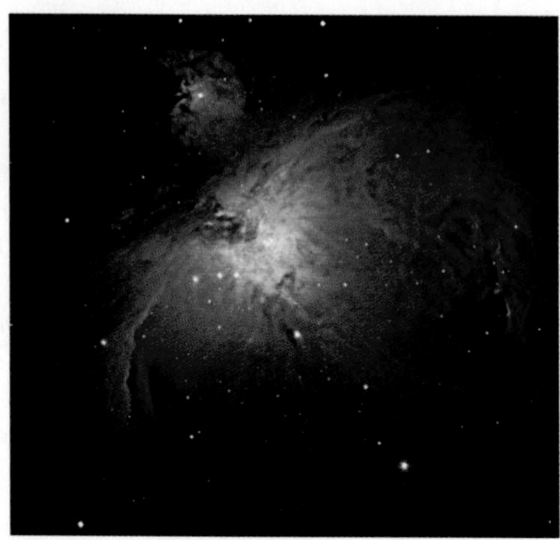

Figure 27.3 This photograph, provided by the *Hubble Telescope*, shows the Orion Nebula. The Orion Nebula, like the nebula from which our solar system formed, is an interstellar cloud of gas and dust and the birthplace of stars.

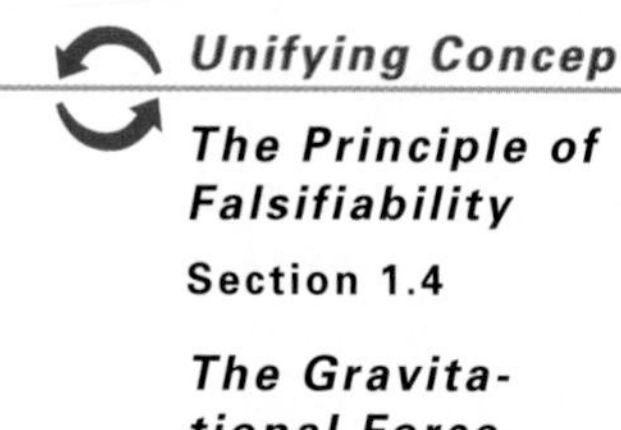

Unifying Concept

The Principle of Falsifiability
Section 1.4

The Gravitational Force
Section 5.3

27.2 The Nebular Theory

The solar system is not a random distribution of celestial objects, but as we have seen, it is a collection of bodies exhibiting features and trends that would be unlikely to arise from coincidence. The patterns that we observe in the solar system are clues as to how it formed. Thus, any theory of solar system formation must be able to explain these two major regularities: (1) *The orderly motions among large bodies of the solar system* and (2) *the neat division of planets into two main types*—terrestrial *and* Jovian. Further, a viable theory of solar system formation must explain other known features of the solar system, including the existence of asteroids, comets, and moons, and the chemical compositions of the Sun and planets.

The modern scientific theory that satisfies these requirements is called the **nebular theory** (Figure 27.3). The nebular theory holds that the Sun and planets formed together from a cloud of gas and dust, a *nebula* (Latin for "cloud"). According to the theory, the solar system began to condense from the cloud of gas and dust about 5 billion years ago. The cloud would have been very diffuse and large, with a diameter thousands of times larger than Pluto's orbit. The nebula would have had to have at least a slight net rotation, possibly due to the rotation of the galaxy itself.

Perhaps the nebula arose from a random density fluctuation so that it was just slightly more dense than the surrounding interstellar medium. In any event, it must have been gravitationally unstable, so that the force of gravity pulling particles together exceeded gas pressure—the tendency of a gas to expand to fill all available space. Once the gravitational collapse of the cloud started, gravity assured that it would continue. The Universal Law of Gravitation (Chapter 5) is an inverse-square law: The strength of the gravitational force weakens with the square of the distance between attracting masses. The cloud maintained a constant mass as it shrank, so that the gravitational force grew ever stronger and the cloud took on a spherical shape.

27.3 The Sun

Conceptual Science Place
The Sun

Ancients who worshipped the Sun seem to have realize life on Earth. We are able to see, hear, touch, feel, and lion tons of mass in the Sun convert to radiant energy tion of this energy reaches Earth and is converted by isms to chemical energy stored in large molecules. T are the primary energy source for almost all of the or Sun, Earth's nearest star, is the solar system's power su

Solar energy is generated deep within the core o comprises about 10 percent of the Sun's total volu 15,000,000 degrees Celsius. The core is also very den density of solid lead. Pressure in the core is 340 billion pressure! Despite this, the hydrogen, helium, and min ments exist in the plasma state. (Plasma, recall, is a s gas except that it consists of ions and electrons rather energies have stripped atoms of their electrons.)

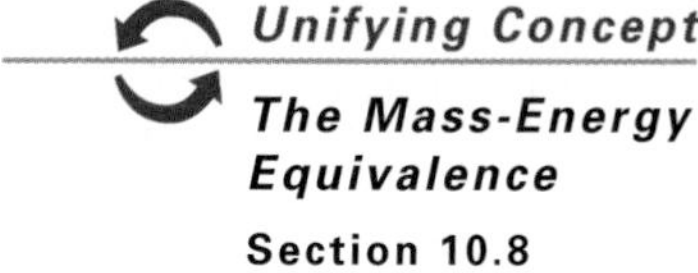

INSIGHT The conversion of hydrogen to helium in the Sun has been going on since it formed nearly 5 billion years ago, and it is expected to continue at this rate for another 5 billion years. At the end of its life, the Sun will start to fuse helium into heavier elements and it will grow to become a red giant star—large enough to swallow Earth. It will ultimately collapse as a white dwarf, and after a trillion years, cool off completely. (More about the life cycle of stars in Chapter 28.)

The most important thing about the core is that th ated through *nuclear fusion*. Recall from Chapter 10 th of nuclear reaction in which lighter atomic nuclei con clei. Fusion brought about by high temperatures, as o *thermonuclear fusion*. All fusion reactions release ener of reactants is greater than the total mass of the produc reaction is converted to energy in accordance with Ei the mass-energy equivalence: $E = mc^2$. Each thermon the Sun causes four hydrogen nuclei to fuse together to The resulting helium has 99.3 percent of the original h ence in mass is converted to energy, which transfers a form of X rays and gamma rays. At the surface, much bit of which nicely reaches Planet Earth.

Unifying Concept
The Mass-Energy Equivalence
Section 10.8

CHECK YOURSELF

1. Was the Sun more massive 1000 years ago than it is to
2. In what sense is the Sun the giver of life on Earth?

CHECK YOUR ANSWERS

1. Yes, the Sun loses mass as hydrogen nuclei combine to ma Sun continues this "nuclear burning" it grows lighter.
2. The Sun transfers energy to the Earth in the form of electr Photsynthesizing organisms convert this radiant energy to which provides most organisms on Earth the energy neede

Figure 27.6 shows that energy travels through two t way from the Sun's core to its surface. In the radiation

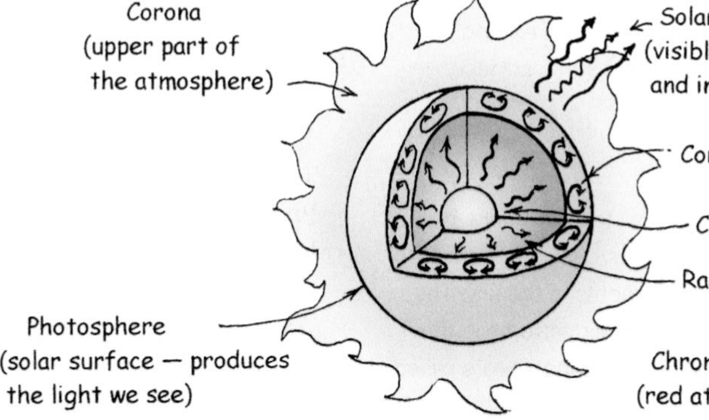

Figure 27.6 The structure of the Sun.

INTEGRATED SCIENCE PHYSICS

The Solar Nebula Heats Up, Spins Faster, and Flattens

INSIGHT Recall Newton's Law of Universal Gravitation in Chapter 5: $F = \frac{G(m_1 m_2)}{d^2}$, where m_1 and m_2 are the masses of attracting bodies, d is the distance between them, and G is a constant. As distance between attracting particles in the nebula decreased, the force of gravity increased, shrinking the cloud ever faster.

Over the millions of years during which the solar nebula collapsed, it heated up, spun faster, and flattened into a disk shape. As a result, the nebula transformed from a large, diffuse, cloud to a much smaller spinning disk with a hot center (Figure 27.4). Why did these changes occur?

As the nebula shrank under the influence of the gravitational force, it became hotter. Why? In accordance with the law of conservation of energy, the gravitational potential energy of attracting particles transformed to heat as they grew closer together.

Also, as the nebula shrank, it spun faster and faster. The reason for this is the *conservation of angular momentum*. Any rotating object, whether a cloud of gas, a bicycle wheel, or an acrobat doing a somersault, tends to keep rotating until a force makes it stop. We say that rotating objects have "inertia of rotation." Recall from Chapter 4 that all moving objects have inertia of motion or *linear momentum*. Similarly, rotating objects have inertia of rotation or *angular momentum*. More can be learned about angular momentum in Appendix B, but regarding the nebular theory the important thing to know about angular momentum is that it is *conserved*. The conservation of angular momentum tells us that angular momentum remains constant so that when a rotating body contracts, it spins faster. A familiar example is an ice skater who pulls in her arms going into a spin, resulting in increased spin. So the cloud of dust and gas that initially rotated very slowly sped up as it shrank, gaining rotational speed.

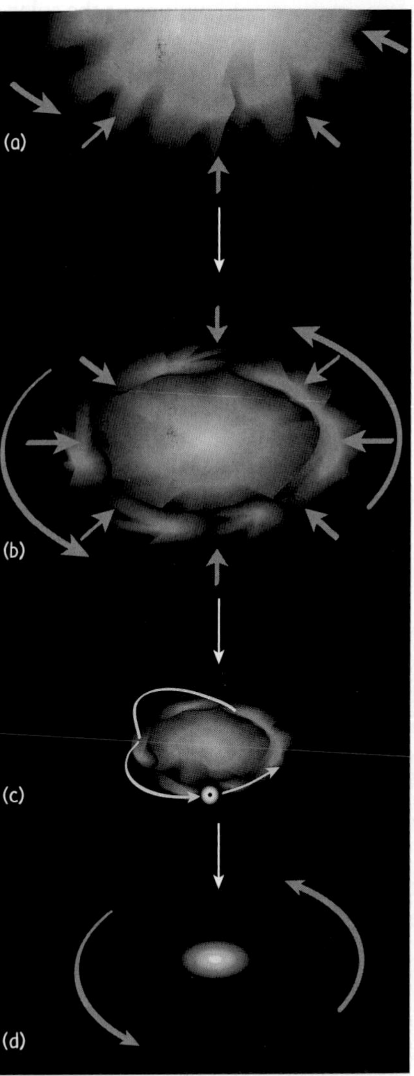

Figure 27.4 (a) The nebula from which the solar system formed was originally a large, diffuse cloud that rotated imperceptibly slowly. The cloud began to collapse under the influence of gravity. (b) As the cloud collapsed, it heated up as gravitational potential energy converted to heat. It spun faster by the conservation of angular momentum. (c) The cloud flattened into a disk as a result of its fast rotation. (d) A spinning, flattened disk was produced whose mass was concentrated at its hot center. (Courtesy of Bennett, Donahue, Schneider, Voit, *The Essential Cosmic Perspective, 3e*, © 2005.)

What happens to the shape of a sphere as it spins faster and faster? The answer is, it flattens. You have seen this if you have ever watched a chef turn a ball of pizza dough into a disk by spinning it on his hands. Even Planet Earth is a slightly "flattened" sphere due to its daily spin. Jupiter, with a greater spin, noticeably departs from a purely spherical shape. When a ball of gas increases its spin rate, flattening is very evident—witness the flattening of our galaxy, the Milky Way. So the initially spherical ball of gas progressed to a spinning disk, the center of which became the *protosun*.

The formation of the spinning disk explains the orderly motions of our solar system today. The planets all orbit the Sun in nearly the same plane because they formed from that same flat nebular disk. The direction in which the disk was spinning became the direction of the Sun's rotation and the orbits of planets. It was also the preferred direction of rotation for planets—which is why most planets rotate in the same way today—though the small sizes of planets compared to the entire disk allowed some exceptions to occur.

CHECK YOURSELF

As a nebula contracts, its rate of spin increases. What rule of nature is at play here?

CHECK YOUR ANSWER

The rule of nature that applies to all spinning bodies is the *conservation of angular momentum*. It and the conservation of energy are dominant players in the universe.

The hot, central portion of the solar nebula, the protosun, was a clump of gas and dust that would become the Sun once thermonuclear fusion ignited within it. The surrounding disk, on the other hand, was the source of material that would become the planets. In the spinning disk, matter collected in some regions more densely than in others. Perhaps small particles of gas and dust stuck together via gravity or electrostatic attraction. Because of their extra mass, these clumps exerted a stronger gravitational force on one another than on neighboring regions of the disk, and so they pulled in even more material to them. This accretion of matter

Unifying Concept
Conservation of Energy
Section 4.10
Conservation of Angular Momentum
Section 4.4

led to the breakup of the nebular disk into small objec which ranged in size from boulders to objects several kilor etesimals grew larger through countless collisions until th nated surrounding matter and finally became full-blown p

The planetesimals were growing into planets at about protosun was commencing thermonuclear fusion (We'll retu sion in the Sun in Section 27.3.) Once thermonuclear fusic began to radiate energy, the nebular disk warmed, with the higher temperatures than the outer portions. As a result, the developed differently. The inner planets formed from mate at high temperatures; hence, the inner planets are rocky. Th trast, consist mainly of hydrogen and helium gas that coal of the solar system far away from the Sun. In these cold reg the gravitational forces among gas particles overtook the ga disperse them. Thus, we can see that the nebular theory acc of the planets and the neat division of them into two group

INTEGRATED SCIENCE CHEMISTRY

The Chemical Composition of the Solar System

According to the evidence as we interpret it today, the uni Big Bang (Chapter 28) about 14 billion years ago. Hyd the only chemical elements present when the universe heavier elements up through iron were forged by fusion i large stars (Chapter 10). Elements heavier than iron we sions of giant stars—supernovae. Dying stars spill much into space. So the heavy elements along with hydrogen a into space from dying stars and can be recycled into n (Figure 27.5).

The process of fusing heavy elements (meaning all el lium) and recycling them has probably gone on throughou verse. Interestingly, only a small fraction of the original h been converted into heavier elements. When our solar sy billion years ago, about 2 percent (by mass) of the origin in the galaxy had been converted to heavier elements. solar nebula was about the same as the composition of the So the solar nebula was about 98 percent hydrogen and h cent were heavy elements produced in the recycling of ele large stars. The Sun and outer planets retain this composit restrial planets contain a higher proportion of heavier eler

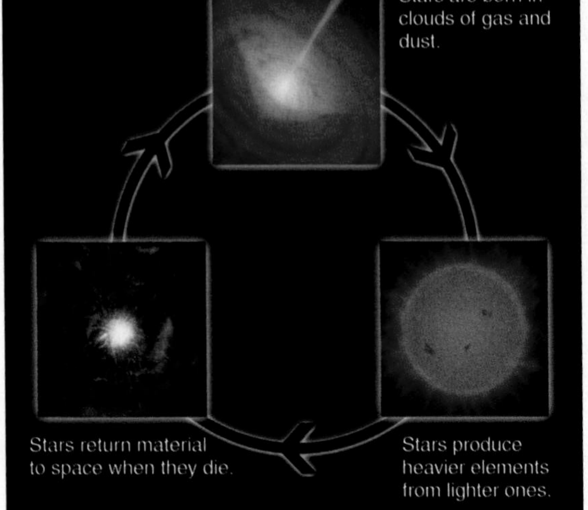
Stars are born in clouds of gas and dust.
Stars return material to space when they die.
Stars produce heavier elements from lighter ones.

Figure 27.5 The heavy elements, forged deep in the interiors of large stars, are released into space when stars die and are taken up by the next generation of stars. Here, elements heavier than helium are the "heavy elements." (Courtesy of Bennett, Donahue, Schneider, Voit, *The Essential Cosmic Perspective, 3e*, © 2005.)

CHECK YOURSELF

1. Elements up through iron are manufactured by the fu atoms. How are the elements heavier than iron manuf
2. Our Sun contains cobalt, which is heavier than iron. D our Sun formed at the beginning of our universe, or w

CHECK YOUR ANSWERS

1. Elements heavier than iron were manufactured in violent ex Such events are called supernovae (plural for supernova).
2. The fact there is cobalt in our Sun tells us that it is compose supernovae that preceded the birth of the Sun. The Sun is sc a universe some 14 billion years old. So our Sun is a relative neighborhood.

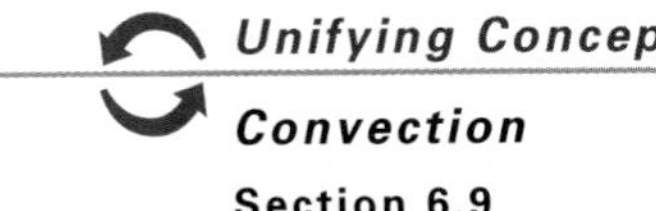

Unifying Concept
Convection
Section 6.9

INSIGHT Convection cells are plentiful in nature—occurring in diverse environments such as Earth's mantle (where they result in the motions of tectonic plates), to our atmosphere (where they produce wind), to the Sun (where they move heat to the Sun's surface).

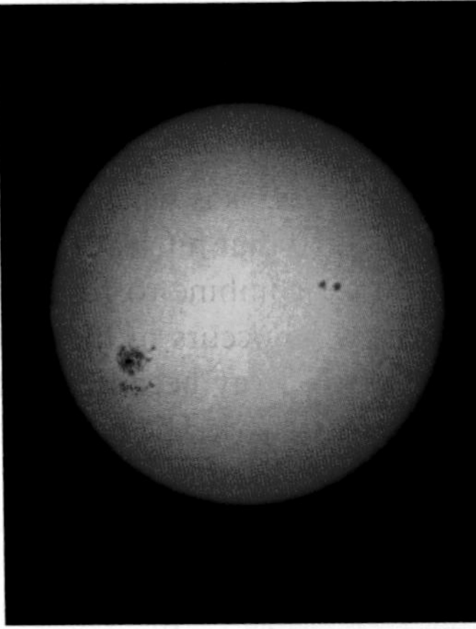

Figure 27.7 Sunspots on the solar surface are relatively cool regions. We say relatively cool because they are hotter than 4000 K. They look dark only in contrast with their 5800-K surroundings.

Figure 27.8 The chromosphere is not ordinarily visible because the intense light from the photosphere overwhelms it. However, during an eclipse, the Moon blocks the photosphere and the chromosphere becomes visible as a thin reddish band of light. The corona is the pearly white, irregularly shaped halo of gases that extends several million kilometers beyond the Sun's surface.

INSIGHT The ancients could tell the difference between planets and stars because of the differences in their movements in the sky. The stars remain relatively fixed in their patterns in the sky, but the planets wander. The planets were called the *wanderers*.

reradiate electromagnetic energy generated in the core, moving it toward the Sun's surface. The process is slow, taking perhaps a million years, because the X rays and gamma rays from the core undergo countless collisions with atoms as they trace an indirect route through the radiation zone. The convection zone is a turbulent layer consisting of low-density gases that are stirred by convection, a mode of heat transfer described in detail in Chapter 6. At the bottom of the convection zone, atoms of gas are heated by radiation from the radiation zone. As the gases warm and become less dense, they rise to the Sun's surface. The gases emit energy into space from the surface in the form of visible light, ultraviolet light, and infrared radiation. The atoms of gas in the convection zone, having lost some of their energy as radiation, lose volume, become more dense, and sink back to the radiation zone. There, they become heated again as they absorb radiation from the Sun's core. The heated gas atoms rise again, carrying energy from the bottom to the top of the convection zone, then losing it at the surface by radiation, and sinking again. This movement of gas atoms in the solar convection zone is the familiar pattern of a convection cell, which occurs in the Earth's mantle, Earth's atmosphere, a pot of boiling water, and many other physical systems.

The visible regions of the Sun are its surface and its atmosphere. The Sun's surface, called the photosphere (sphere of light) is a glowing, 6000-K plasma, probably about 100 kilometers thick. It emits most of the light we see. The photosphere features relatively cool regions that appear as spots when viewed from Earth—sunspots (Figure 27.7). Sunspots are cooler and darker than the rest of the photosphere and are caused by magnetic fields that impede hot gases from rising to the surface. Sunspots can be seen by the unaided eye through protective filters or when the Sun is low enough on the horizon not to damage the eyes. Sunspots are typically twice the size of Earth. They move around due to the sun's rotation, and they last about a week or so.

The layer of the Sun's atmosphere just above the photosphere is a transparent, 10,000-kilometer-thick shell of plasma called the chromosphere (sphere of color), seen during an eclipse as a pinkish glow surrounding the eclipsed sun (Figure 27.8). Beyond the chromosphere are streamers and filaments of outward-moving, high-temperature plasmas curved by the Sun's magnetic field. This outermost region of the Sun's atmosphere is the corona, extending out several million kilometers to where it merges into a whirl of high-speed protons and electrons—the solar wind. It is the solar wind that powers the aurora borealis on Earth and produces the tails of comets. Also in the corona are solar prominences, dense clouds of plasma (Figure 27.9). The clouds of plasma are pulled into looped and twisted shapes by the Sun's magnetic field. Large prominences can cause electrical blackouts if they are directed toward Earth.

The Sun spins slowly on its axis. Since the Sun is a fluid rather than a solid, different latitudes of the Sun spin at different rates. Equatorial regions spin once in 25 days, but higher latitudes take up to 36 days to make a complete rotation. This differential spin means the surface near the equator pulls ahead of the surface farther north or south. The Sun's differential spin wraps and distorts the solar magnetic field, producing sunspots and prominences. The form of the solar magnetic field is not constant. A reversal of magnetic poles occurs every 11 years, and the number of sunspots also reaches a maximum every 11 years (currently).

27.4 The Inner Planets

Today we know that **planets** are relatively large and cool bodies that orbit a star. Planets emit no visible light of their own; like the moon, they simply reflect sunlight. They are massive enough for their gravity to make them spherical, but small enough to avoid thermonuclear fusion in their cores.

Compared with the outer planets, the four planets nearest the Sun are close together. These are Mercury, Venus, Earth, and Mars. These small and dense inner planets all have atmospheres (though Mercury barely has one). They are

INTEGRATED SCIENCE PHYSICS

The Solar Nebula Heats Up, Spins Faster, and Flattens

INSIGHT Recall Newton's Law of Universal Gravitation in Chapter 5: $F = \frac{G(m_1 m_2)}{d^2}$, where m_1 and m_2 are the masses of attracting bodies, d is the distance between them, and G is a constant. As distance between attracting particles in the nebula decreased, the force of gravity increased, shrinking the cloud ever faster.

Over the millions of years during which the solar nebula collapsed, it heated up, spun faster, and flattened into a disk shape. As a result, the nebula transformed from a large, diffuse, cloud to a much smaller spinning disk with a hot center (Figure 27.4). Why did these changes occur?

As the nebula shrank under the influence of the gravitational force, it became hotter. Why? In accordance with the law of conservation of energy, the gravitational potential energy of attracting particles transformed to heat as they grew closer together.

Also, as the nebula shrank, it spun faster and faster. The reason for this is the *conservation of angular momentum*. Any rotating object, whether a cloud of gas, a bicycle wheel, or an acrobat doing a somersault, tends to keep rotating until a force makes it stop. We say that rotating objects have "inertia of rotation." Recall from Chapter 4 that all moving objects have inertia of motion or *linear momentum*. Similarly, rotating objects have inertia of rotation or *angular momentum*. More can be learned about angular momentum in Appendix B, but regarding the nebular theory the important thing to know about angular momentum is that it is *conserved*. The conservation of angular momentum tells us that angular momentum remains constant so that when a rotating body contracts, it spins faster. A familiar example is an ice skater who pulls in her arms going into a spin, resulting in increased spin. So the cloud of dust and gas that initially rotated very slowly sped up as it shrank, gaining rotational speed.

What happens to the shape of a sphere as it spins faster and faster? The answer is, it flattens. You have seen this if you have ever watched a chef turn a ball of pizza dough into a disk by spinning it on his hands. Even Planet Earth is a slightly "flattened" sphere due to its daily spin. Jupiter, with a greater spin, noticeably departs from a purely spherical shape. When a ball of gas increases its spin rate, flattening is very evident—witness the flattening of our galaxy, the Milky Way. So the initially spherical ball of gas progressed to a spinning disk, the center of which became the *protosun*.

The formation of the spinning disk explains the orderly motions of our solar system today. The planets all orbit the Sun in nearly the same plane because they formed from that same flat nebular disk. The direction in which the disk was spinning became the direction of the Sun's rotation and the orbits of planets. It was also the preferred direction of rotation for planets—which is why most planets rotate in the same way today—though the small sizes of planets compared to the entire disk allowed some exceptions to occur.

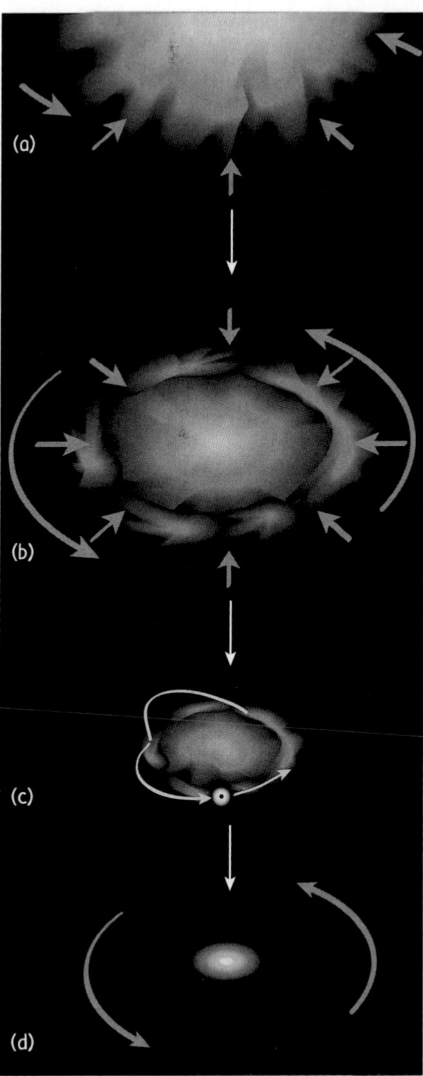

Figure 27.4 (a) The nebula from which the solar system formed was originally a large, diffuse cloud that rotated imperceptibly slowly. The cloud began to collapse under the influence of gravity. (b) As the cloud collapsed, it heated up as gravitational potential energy converted to heat. It spun faster by the conservation of angular momentum. (c) The cloud flattened into a disk as a result of its fast rotation. (d) A spinning, flattened disk was produced whose mass was concentrated at its hot center. (Courtesy of Bennett, Donahue, Schneider, Voit, *The Essential Cosmic Perspective, 3e*, © 2005.)

CHECK YOURSELF

As a nebula contracts, its rate of spin increases. What rule of nature is at play here?

CHECK YOUR ANSWER

The rule of nature that applies to all spinning bodies is the *conservation of angular momentum*. It and the conservation of energy are dominant players in the universe.

The hot, central portion of the solar nebula, the protosun, was a clump of gas and dust that would become the Sun once thermonuclear fusion ignited within it. The surrounding disk, on the other hand, was the source of material that would become the planets. In the spinning disk, matter collected in some regions more densely than in others. Perhaps small particles of gas and dust stuck together via gravity or electrostatic attraction. Because of their extra mass, these clumps exerted a stronger gravitational force on one another than on neighboring regions of the disk, and so they pulled in even more material to them. This accretion of matter

Unifying Concept

Conservation of Energy
Section 4.10

Conservation of Angular Momentum
Section 4.4

led to the breakup of the nebular disk into small objects called *planetesimals*, which ranged in size from boulders to objects several kilometers in diameter. Planetesimals grew larger through countless collisions until they gravitationally dominated surrounding matter and finally became full-blown planets.

The planetesimals were growing into planets at about the same time that the protosun was commencing thermonuclear fusion (We'll return to thermonuclear fusion in the Sun in Section 27.3.) Once thermonuclear fusion occurred and the Sun began to radiate energy, the nebular disk warmed, with the inner portions reaching higher temperatures than the outer portions. As a result, the inner and outer planets developed differently. The inner planets formed from materials that remained solid at high temperatures; hence, the inner planets are rocky. The outer planets, by contrast, consist mainly of hydrogen and helium gas that coalesced in the cold regions of the solar system far away from the Sun. In these cold regions of the solar system, the gravitational forces among gas particles overtook the gas pressure that tended to disperse them. Thus, we can see that the nebular theory accounts for the formation of the planets and the neat division of them into two groups.

INTEGRATED SCIENCE CHEMISTRY

The Chemical Composition of the Solar System

According to the evidence as we interpret it today, the universe was created in the Big Bang (Chapter 28) about 14 billion years ago. Hydrogen and helium were the only chemical elements present when the universe was young. Eventually, heavier elements up through iron were forged by fusion reactions in the cores of large stars (Chapter 10). Elements heavier than iron were produced by explosions of giant stars—supernovae. Dying stars spill much of their contents back into space. So the heavy elements along with hydrogen and helium are released into space from dying stars and can be recycled into new generations of stars (Figure 27.5).

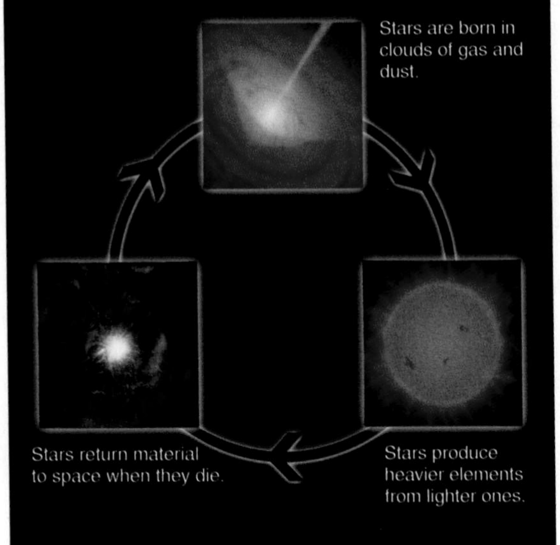
Stars are born in clouds of gas and dust.
Stars produce heavier elements from lighter ones.
Stars return material to space when they die.

Figure 27.5 The heavy elements, forged deep in the interiors of large stars, are released into space when stars die and are taken up by the next generation of stars. Here, elements heavier than helium are the "heavy elements." (Courtesy of Bennett, Donahue, Schneider, Voit, *The Essential Cosmic Perspective, 3e*, © 2005.)

The process of fusing heavy elements (meaning all elements heavier than helium) and recycling them has probably gone on throughout the history of the universe. Interestingly, only a small fraction of the original hydrogen and helium has been converted into heavier elements. When our solar system formed about 4.5 billion years ago, about 2 percent (by mass) of the original hydrogen and helium in the galaxy had been converted to heavier elements. The composition of the solar nebula was about the same as the composition of the galaxy when it formed. So the solar nebula was about 98 percent hydrogen and helium. The other 2 percent were heavy elements produced in the recycling of elements in generations of large stars. The Sun and outer planets retain this composition today, while the terrestrial planets contain a higher proportion of heavier elements.

CHECK YOURSELF

1. Elements up through iron are manufactured by the fusion of hydrogen atoms. How are the elements heavier than iron manufactured?
2. Our Sun contains cobalt, which is heavier than iron. Does this tell us whether our Sun formed at the beginning of our universe, or was it a late comer?

CHECK YOUR ANSWERS

1. Elements heavier than iron were manufactured in violent explosions of giant stars. Such events are called supernovae (plural for supernova).
2. The fact there is cobalt in our Sun tells us that it is composed of the remnants of supernovae that preceded the birth of the Sun. The Sun is some 5 billion years old, in a universe some 14 billion years old. So our Sun is a relative newcomer to the cosmic neighborhood.

27.3 The Sun

Conceptual Science Place
The Sun

Ancients who worshipped the Sun seem to have realized that it is the source of all life on Earth. We are able to see, hear, touch, feel, and love only because 4.5 million tons of mass in the Sun convert to radiant energy every second. A tiny fraction of this energy reaches Earth and is converted by photosynthesizing organisms to chemical energy stored in large molecules. These energy-rich molecules are the primary energy source for almost all of the organisms of this planet. The Sun, Earth's nearest star, is the solar system's power supply.

Solar energy is generated deep within the core of the Sun. The solar core comprises about 10 percent of the Sun's total volume. It is very hot—over 15,000,000 degrees Celsius. The core is also very dense, with over 12 times the density of solid lead. Pressure in the core is 340 billion times Earth's atmospheric pressure! Despite this, the hydrogen, helium, and minute quantities of other elements exist in the plasma state. (Plasma, recall, is a state of matter similar to a gas except that it consists of ions and electrons rather than atoms, because high energies have stripped atoms of their electrons.)

INSIGHT The conversion of hydrogen to helium in the Sun has been going on since it formed nearly 5 billion years ago, and it is expected to continue at this rate for another 5 billion years. At the end of its life, the Sun will start to fuse helium into heavier elements and it will grow to become a red giant star—large enough to swallow Earth. It will ultimately collapse as a white dwarf, and after a trillion years, cool off completely. (More about the life cycle of stars in Chapter 28.)

The most important thing about the core is that this is where energy is liberated through *nuclear fusion*. Recall from Chapter 10 that nuclear fusion is a kind of nuclear reaction in which lighter atomic nuclei combine to form heavier nuclei. Fusion brought about by high temperatures, as occurs in the Sun, is called *thermonuclear fusion*. All fusion reactions release energy because the total mass of reactants is greater than the total mass of the products. The mass "lost" in the reaction is converted to energy in accordance with Einstein's famous equation, the mass-energy equivalence: $E = mc^2$. Each thermonuclear fusion reaction in the Sun causes four hydrogen nuclei to fuse together to form one helium nucleus. The resulting helium has 99.3 percent of the original hydrogen mass. The difference in mass is converted to energy, which transfers away from the core in the form of X rays and gamma rays. At the surface, much is emitted as light, a tiny bit of which nicely reaches Planet Earth.

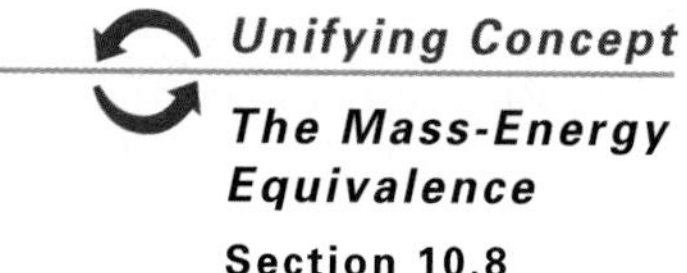

Unifying Concept
The Mass-Energy Equivalence
Section 10.8

CHECK YOURSELF

1. Was the Sun more massive 1000 years ago than it is today? Defend your answer.
2. In what sense is the Sun the giver of life on Earth?

CHECK YOUR ANSWERS

1. Yes, the Sun loses mass as hydrogen nuclei combine to make helium nuclei, so as the Sun continues this "nuclear burning" it grows lighter.
2. The Sun transfers energy to the Earth in the form of electromagnetic waves—sunshine. Photsynthesizing organisms convert this radiant energy to chemical energy—food—which provides most organisms on Earth the energy needed to conduct life processes.

Figure 27.6 shows that energy travels through two thick internal layers on its way from the Sun's core to its surface. In the radiation zone, atoms absorb and

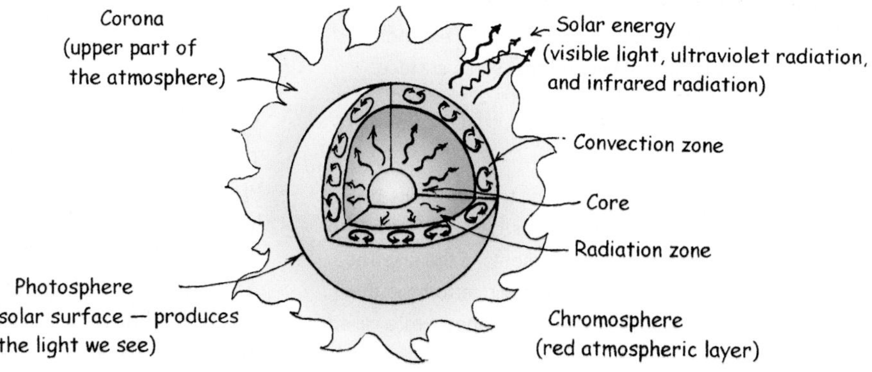

Figure 27.6 The structure of the Sun.

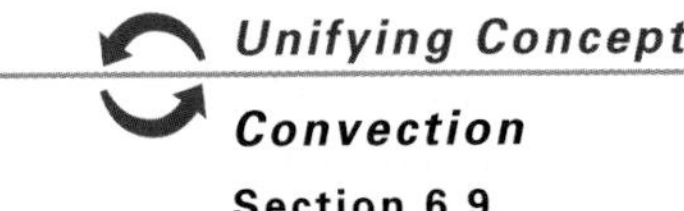

Unifying Concept
Convection
Section 6.9

INSIGHT Convection cells are plentiful in nature—occurring in diverse environments such as Earth's mantle (where they result in the motions of tectonic plates), to our atmosphere (where they produce wind), to the Sun (where they move heat to the Sun's surface).

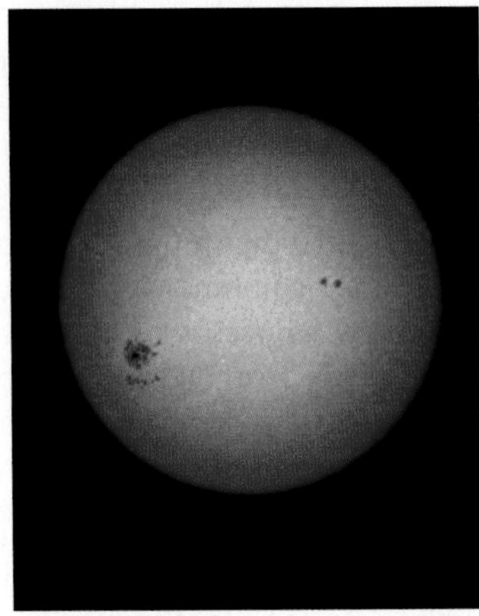

Figure 27.7 Sunspots on the solar surface are relatively cool regions. We say relatively cool because they are hotter than 4000 K. They look dark only in contrast with their 5800-K surroundings.

Figure 27.8 The chromosphere is not ordinarily visible because the intense light from the photosphere overwhelms it. However, during an eclipse, the Moon blocks the photosphere and the chromosphere becomes visible as a thin reddish band of light. The corona is the pearly white, irregularly shaped halo of gases that extends several million kilometers beyond the Sun's surface.

INSIGHT The ancients could tell the difference between planets and stars because of the differences in their movements in the sky. The stars remain relatively fixed in their patterns in the sky, but the planets wander. The planets were called the *wanderers*.

reradiate electromagnetic energy generated in the core, moving it toward the Sun's surface. The process is slow, taking perhaps a million years, because the X rays and gamma rays from the core undergo countless collisions with atoms as they trace an indirect route through the radiation zone. The convection zone is a turbulent layer consisting of low-density gases that are stirred by convection, a mode of heat transfer described in detail in Chapter 6. At the bottom of the convection zone, atoms of gas are heated by radiation from the radiation zone. As the gases warm and become less dense, they rise to the Sun's surface. The gases emit energy into space from the surface in the form of visible light, ultraviolet light, and infrared radiation. The atoms of gas in the convection zone, having lost some of their energy as radiation, lose volume, become more dense, and sink back to the radiation zone. There, they become heated again as they absorb radiation from the Sun's core. The heated gas atoms rise again, carrying energy from the bottom to the top of the convection zone, then losing it at the surface by radiation, and sinking again. This movement of gas atoms in the solar convection zone is the familiar pattern of a convection cell, which occurs in the Earth's mantle, Earth's atmosphere, a pot of boiling water, and many other physical systems.

The visible regions of the Sun are its surface and its atmosphere. The Sun's surface, called the photosphere (sphere of light) is a glowing, 6000-K plasma, probably about 100 kilometers thick. It emits most of the light we see. The photosphere features relatively cool regions that appear as spots when viewed from Earth—sunspots (Figure 27.7). Sunspots are cooler and darker than the rest of the photosphere and are caused by magnetic fields that impede hot gases from rising to the surface. Sunspots can be seen by the unaided eye through protective filters or when the Sun is low enough on the horizon not to damage the eyes. Sunspots are typically twice the size of Earth. They move around due to the sun's rotation, and they last about a week or so.

The layer of the Sun's atmosphere just above the photosphere is a transparent, 10,000-kilometer-thick shell of plasma called the chromosphere (sphere of color), seen during an eclipse as a pinkish glow surrounding the eclipsed sun (Figure 27.8). Beyond the chromosphere are streamers and filaments of outward-moving, high-temperature plasmas curved by the Sun's magnetic field. This outermost region of the Sun's atmosphere is the corona, extending out several million kilometers to where it merges into a whirl of high-speed protons and electrons—the solar wind. It is the solar wind that powers the aurora borealis on Earth and produces the tails of comets. Also in the corona are solar prominences, dense clouds of plasma (Figure 27.9). The clouds of plasma are pulled into looped and twisted shapes by the Sun's magnetic field. Large prominences can cause electrical blackouts if they are directed toward Earth.

The Sun spins slowly on its axis. Since the Sun is a fluid rather than a solid, different latitudes of the Sun spin at different rates. Equatorial regions spin once in 25 days, but higher latitudes take up to 36 days to make a complete rotation. This differential spin means the surface near the equator pulls ahead of the surface farther north or south. The Sun's differential spin wraps and distorts the solar magnetic field, producing sunspots and prominences. The form of the solar magnetic field is not constant. A reversal of magnetic poles occurs every 11 years, and the number of sunspots also reaches a maximum every 11 years (currently).

27.4 The Inner Planets

Today we know that **planets** are relatively large and cool bodies that orbit a star. Planets emit no visible light of their own; like the moon, they simply reflect sunlight. They are massive enough for their gravity to make them spherical, but small enough to avoid thermonuclear fusion in their cores.

Compared with the outer planets, the four planets nearest the Sun are close together. These are Mercury, Venus, Earth, and Mars. These small and dense inner planets all have atmospheres (though Mercury barely has one). They are

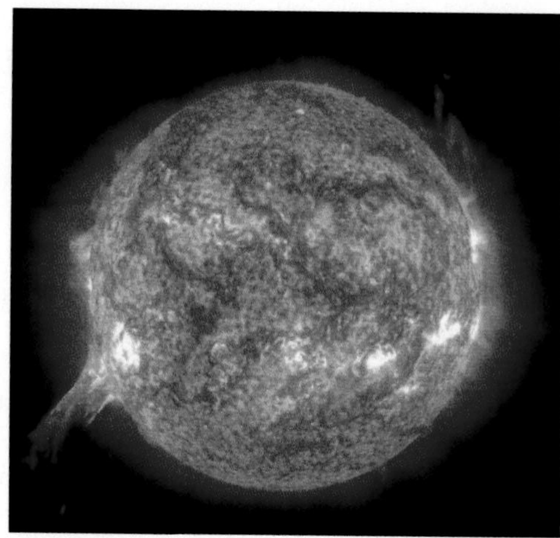

Figure 27.9 This image of a solar prominence was obtained with the *SOHO* (*Solar and Heliospheric Observatory*) satellite. The Sun's surface has a granular appearance because of turbulence in its gases.

also rocky planets, each with a solid, mineral-containing crust and an earthlike composition. This is why they are sometimes called the *terrestrial planets.*

Mercury

Mercury (Figure 27.10) is somewhat larger than the Moon and similar in appearance, and it is the closest planet to the Sun. Because of its closeness to the Sun, it is the fastest planet, circling the Sun in only 88 Earth days. Thus, one "year" on Mercury lasts for only 88 Earth days. Mercury spins about its axis only three times for each two revolutions about the Sun. This makes its "daytime" very long and very hot, with temperatures as high as 430 degrees Celsius.

Because of Mercury's smallness and weak gravitational field, it holds very little atmosphere, which is about a trillionth as dense as Earth's atmosphere (a better vacuum than laboratories on Earth can produce). So, without a blanket of atmosphere, and because there are no winds to transfer heat from one region to another, nighttime on Mercury is very cold, about −170 degrees Celsius. Mercury is a fairly bright object in the nighttime sky and is best seen as an evening "star" during March and April or as a morning star during September and October. It is seen near the Sun at sunup or sunset.

Venus

Venus, the second planet from the Sun, is frequently the first starlike object to appear after the Sun sets, so it is often called the evening "star" (Figure 27.11). Compared with the other planets, Venus most closely resembles Earth. It is similar in size, density, and distance from the Sun. However, Venus has a very dense atmosphere, opaque cloud cover, and high average temperature (460 degrees Celsius)—too hot for oceans. The atmosphere of Venus is about 96% CO_2. Remember from Chapter 25 that carbon dioxide is a "greenhouse gas." By this we mean that CO_2 blocks the escape of infrared radiation from Earth's surface and contributes to the warming of our planet. The thick blanket of CO_2 surrounding Venus effectively traps heat near the Venutian surface. This, as well as the proximity of Venus to the Sun make Venus the hottest planet in the solar system.

INSIGHT Satellites circling Earth are continually falling due to Earth's gravity (Chapter 5). Because of their tangential speeds, they fall around (rather than into) Earth. The same is true of the planets. Because they continually circle the Sun with appropriate tangential speeds, they avoid crashing into the Sun.

Another difference between Venus and Earth is in how the two planets spin about their axes. Venus takes 243 Earth days to make one full spin, and only 225 Earth days to make one revolution around the Sun. This means that a day on Venus lasts longer than a year! Venus spins in a direction opposite to the direction of Earth's spin. A space-traveling observer hovering about the solar system who sees Earth spinning counterclockwise sees Venus spinning clockwise.

In recent years, 17 probes have landed on the surface of Venus. There have been 18 flyby spacecraft (notably *Pioneer Venus* in 1978 and *Magellan* in 1993).

Figure 27.10 This is a mosaic image of Mercury compiled from data obtained by the *Mariner 10* spacecraft on flyby mission during 1974–75. Areas for which data are missing are blank. Mercury is heavily cratered from the impacts of many meteorites. Mercury is a small planet, with only about 5 percent of the volume and mass of Earth.

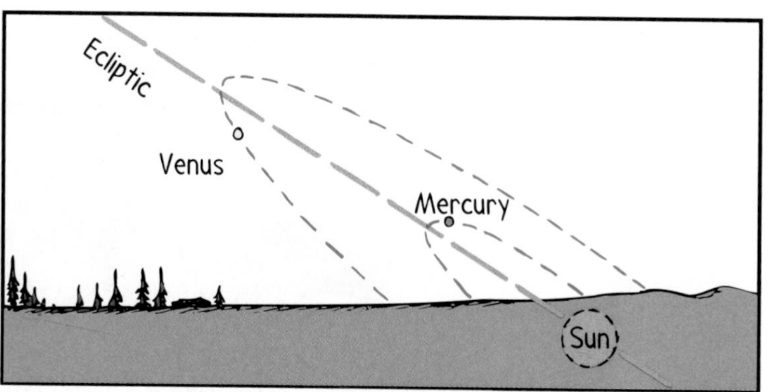

Figure 27.11 Because the orbits of Mercury and Venus lie inside the orbit of Earth, they are always near the Sun. Near sunset (or sunrise) they are visible as "evening stars" or "morning stars."

Figure 27.12 This computer-generated picture of Venus shows the planet's rocky surface partially illuminated by the Sun.

Figure 27.13 Earth, the blue planet.

From spacecraft data, we know that Venus has been very active volcanically and is an extremely harsh place.

Earth

The Earth science chapters of this book are dedicated to a description of the Planet Earth, so our description of it here will be brief. Our home planet Earth is the blue planet, with more water surface than land (Figure 27.13). Liquid water nurtures life and facilitates the surface processes that shape the land.

Earth has an atmosphere that contains just enough water vapor and carbon dioxide to maintain a moderate greenhouse effect to keep temperatures favorable for life. Our distance from the Sun is also just right to maintain an average surface temperature delicately balanced between that of freezing and boiling water. Our relatively high daily spin rate allows only a brief and small lowering of temperature on the nighttime side of Earth. So temperature extremes of day and night are favorable for life on Earth.

Mars

Mars captures our fancy as another world, perhaps even as a world with life. This is because of the similarities between Earth and Mars: Mars is a little more than half Earth's size, its mass is about one-ninth that of Earth, and it has a core, mantle, crust, and a thin, nearly cloudless atmosphere. It has polar ice caps and seasons that are nearly twice as long as Earth's because Mars takes nearly 2 Earth years to orbit the Sun. When Mars is closest to Earth, a situation that occurs once every 15 to 17 years, its bright, ruddy color outshines the brightest stars.

The Martian atmosphere is about 95 percent carbon dioxide, with only about 0.15 percent oxygen. Yet, because the Martian atmosphere is relatively thin, it does not trap heat through the greenhouse effect as much as Earth's and Venus's atmospheres do. So the temperatures on Mars are generally colder than on Earth, ranging from about 30 degrees Celsius in the day at the equator to a frigid −130 degrees Celsius at night. Night is only slightly longer than on Earth, for the Martian day lasts 24 hours and 37.4 minutes. If you visit Mars, never mind your raincoat, for there is far too little water vapor in the atmosphere for rain. Even the ice at the planet's poles consists primarily of carbon dioxide. And don't give a second thought to waterproof footwear, for the very low atmospheric pressure won't permit the existence of any puddles or lakes.

(a)

(b)

Figure 27.14 (a) NASA's Mars Exploration Rover, *Spirit*, with cameras mounted on the white mast, took panoramic photos of the Martian surface. (b) *Spirit* took photographs in June 2004 for this composite, true-color image of the region named Columbia Hills on Mars. The vehicle later traveled to the hills to analyze their composition.

INSIGHT Why are planets round? All parts of a forming planet pull close together by mutual gravitation. No "corners" form because they're simply pulled in. So gravity is the cause of the spherical shapes of planets and other celestial bodies.

The presence on Mars of features that appear to be dry ocean beds has been taken as an indication that water was once abundant in the Martian past. Channels on the Martian surface that appear to have been carved by water are seen by visiting spacecraft. Although these could not be seen through the telescopes of early investigators, some surface features on Mars were imagined by those early investigators to be canals, reinforcing the notion then of a Martian civilization. Although there are questionable traces of life in Martian meteorites found in Antarctica (see Chapter 17), landings on Mars show no evidence of current life at the surface and no canals. The 1997 *Pathfinder* mission showed it to be a very dry and windy place. Since the Martian atmosphere has a very low density, unequal heating produces Martian winds that are about ten times faster than the winds on Earth.

Mars has two small moons—Phobos, the inner one, and Deimos, the outer. Both are potato-shaped and have cratered surfaces. Phobos orbits in the same easterly direction in which Mars spins (like our Moon), at a distance of almost 6000 kilometers in a period of 7.5 hours. From Mars it appears about half the size of our Moon. Deimos is about half the size of Phobos, and it orbits Mars in 30.3 hours at a distance of 20,000 kilometers from the Martian surface.

27.5 The Outer Planets

The more widely spaced outer planets beyond Mars are much different from the inner planets. They're different in size, in composition, and in the way they were formed. The outer planets, Jupiter, Saturn, Uranus, and Neptune, are gigantic, gaseous, low-density worlds. Because Saturn, Uranus, and Neptune are similar to Jupiter, the nearest and largest of the outer planets, all four of them are called *Jovian* planets. All have ring systems, Saturn's being the most prominent. Beyond these giants is outermost Pluto, much dissimilar to the other planets. We will consider the outer planets in the order of their distance from the Sun.

Jupiter

Jupiter is the largest of all the planets. Its yellow light in the night sky outshines the stars. In prespacecraft years, Jupiter was thought of as a failed star, because its composition is closer to that of the Sun than to that of the terrestrial planets. Jupiter is more liquid than gaseous or solid. It spins rapidly about its axis in about 10 hours, a speed that flattens it so that its equatorial diameter is about 6 percent greater than its polar diameter. As with the Sun, all parts do not rotate in unison. Equatorial regions complete a full revolution several minutes before nearby regions in higher and lower latitudes. The atmospheric pressure at Jupiter's rocky surface is more than a million times the atmospheric pressure of Earth. Jupiter's atmosphere is about 82 percent hydrogen, 17 percent helium, and 1 percent methane, ammonia, and other molecules.

Figure 27.15 This artist's rendering shows aurorae (pink) in the upper atmosphere of Jupiter. Thunderclouds are seen below the aurorae, and the nearest major moon, Io, is seen at center left. Aurorae, like the northern lights on Earth, are caused by charged particles from the solar wind exciting gas molecules in the upper atmosphere. The gas molecules emit light as they return to an unexcited energy state.

The average diameter of Jupiter is about 11 times greater than Earth's, which means Jupiter's volume is more than 1000 times Earth's. Jupiter's mass is greater than the combined masses of all the other planets. Due to its low density, however—about one-fourth of Earth's—Jupiter's mass is barely more than 300 times Earth's. Jupiter's core is a solid sphere about 15 times as massive as the entire Earth, and it is composed of iron, nickel, and other minerals.

More than half of Jupiter's volume is an ocean of liquid hydrogen. Beneath the hydrogen ocean lies an inner layer of hydrogen compressed into a sort of liquid metallic state. In it are abundant conduction electrons that flow to produce Jupiter's enormous magnetic field. The strong magnetic field about the planet captures high-energy particles and produces radiation belts 400 million times as energetic as Earth's Van Allen radiation belts. Radiation levels surrounding Jupiter are the highest ever recorded in space.

Figure 27.16 Jupiter, with its moons Io (orange dot over planet) and Europa (white dot to right of planet), as seen from the *Voyager I* spacecraft in February 1979. The great red spot (lower left), larger than Earth, is a cyclonic weather pattern of high winds and turbulence.

Figure 27.17 Saturn surrounded by its famous rings, which are believed to be composed of rocks and ice.

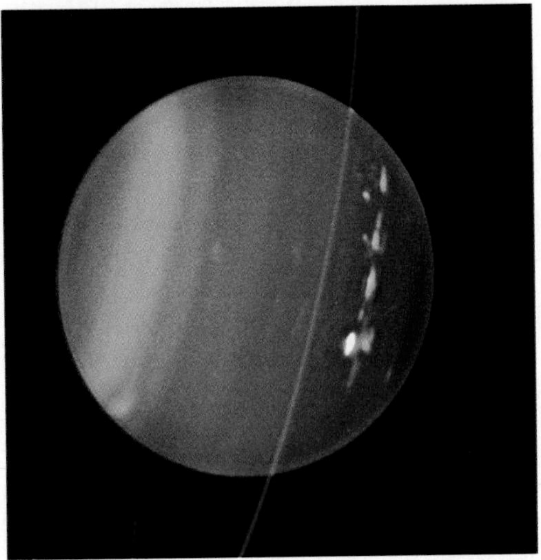

Figure 27.18 This infrared image of Uranus was gathered by the 10-meter *Keck Telescope*, Hawaii, on July 11–12, 2004. It is thought that Uranus tilts on its rotational axis as a result of a collision it had with a large body early in the solar system's history. The bright white and blue spots in the Southern hemisphere of Uranus are clouds. Methane in the upper atmosphere absorbs red light, giving Uranus its blue-green color.

Surface temperatures are about the same day and night. Jupiter radiates about twice as much heat as it receives from the Sun. The excess heat likely comes from internal heat generated long ago by gravitational contraction at the time the planet formed. Recall that when forming planets contract, gravitational potential energy is converted to thermal energy (Figure 27.16).

If you're planning to visit Jupiter, choose one of its moons instead. There are at least 28 of them, in addition to a faint ring orbiting the planet. Among the four largest moons (which were discovered by Galileo in 1610), Io and Europa are about the size of our Moon, and Ganymede and Callisto are about as large as Mercury. The most intriguing of Jupiter's moons seems to be Io, which has more volcanic activity than any other body in the solar system.

Saturn

Saturn is one of the most remarkable objects in the sky, mainly because its rings are clearly visible with a small telescope. It is brighter than all but two stars, and it is second among the planets in mass and size. Saturn is twice as far from Earth as Jupiter. Its diameter, not counting its ring system, is nearly 10 times that of Earth, and its mass is nearly 100 times greater. It is composed primarily of hydrogen and helium, and it has the lowest density of any planet, only 0.7 times the density of water. These characteristics mean that Saturn would easily float in a bathtub, if the bathtub were large enough. Its low density and its 10.2-hour rapid spin produce more polar flattening than can be seen in any other planet. Like Jupiter, Saturn radiates about twice as much heat energy as it receives from the Sun.

Saturn's rings, likely only a few kilometers thick, lie in a plane coincident with Saturn's equator. Four concentric rings have been known for many years, and spacecraft missions have detected others. The rings are composed of chunks of frozen water and rocks, believed to be the material of a moon that never formed or the remnants of a moon torn apart by tidal forces. All the rocks and bits of matter that make up the rings pursue independent orbits about Saturn. The inner parts of the ring travel faster than the outer parts, just as any satellite near a planet travels faster than a more distant satellite (Figure 27.17).

Saturn has some 24 moons beyond its rings. The largest is Titan, 1.6 times larger than our Moon and even larger than Mercury. It spins once every 16 days and has a methane atmosphere with atmospheric pressure that is likely greater than Earth's. Its surface temperature is cold—roughly −170 degrees Celsius. So bring a heavy coat and breathing gear if you plan to visit Titan. If that doesn't work out, try another of Saturn's large moons, Iapetus. One side of Iapetus is very bright and the other very dark. Try the region between these two extremes.

Uranus

Uranus is twice as far from Earth as Saturn is, and it can barely be seen with the naked eye. Uranus was unknown to ancient astronomers. It has a diameter four times larger than Earth's and a density slightly greater than that of water. So, if you were able to place Uranus in a giant bathtub, it would sink. The most unusual feature of Uranus is its tilt. Its axis is tilted 98 degrees to the perpendicular of its orbital plane, so it lies on its side (Figure 27.18). Unlike Jupiter and Saturn, it appears to have no internal source of heat. Uranus is a cold place.

Uranus has at least 21 moons, in addition to a complicated faint ring system. Recall, from Chapter 4, that perturbations in the planet Uranus led to the discovery in 1846 of a farther planet, Neptune.

Neptune

Neptune has a diameter about 3.9 times greater than Earth's, its mass is 17 times greater, and its mean density is about a third that of Earth's. Its

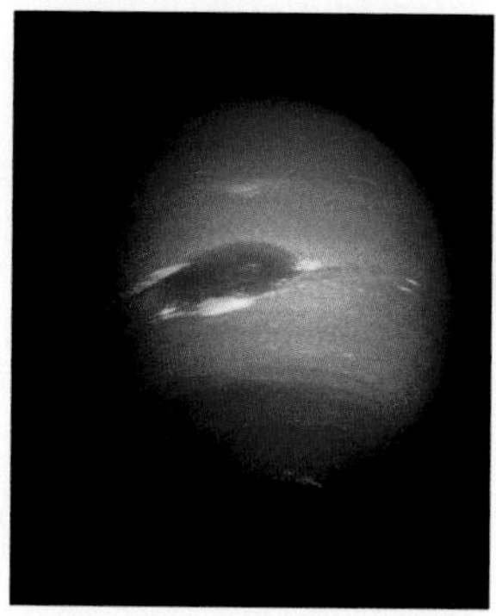

Figure 27.19 Cyclonic disturbances on Neptune in 1989 produced a great dark spot, which was even larger than Earth and similar to Jupiter's great red spot. The spot has now disappeared.

atmosphere is mainly hydrogen and helium, with some methane and ammonia. Like Jupiter and Saturn, it emits about 2.5 times as much heat energy as it receives from the Sun.

Neptune has at least eight moons in addition to a ring system. Recent findings suggest a total of eleven moons. The largest moon is Triton, which orbits Neptune in 5.9 days in a direction opposite to the planet's eastward spin. Triton's diameter is three-quarters the size of our Moon's diameter, and Triton has twice as much mass as our Moon. It has bright polar caps and geysers of liquid nitrogen. A smaller moon, Nereid, takes nearly a year to orbit Neptune in a highly elongated elliptical path.

Pluto

Pluto is neither an inner planet nor an outer planet. From 1930 to 2006 it has been classified along with Mercury, Venus, Earth, Mars, Jupiter, Saturn, Uranus, and Neptune as a major planet, mainly for historical reasons. It is smaller than the planets and smaller than Earth's moon. It is composed primarily of rock and nitrogen ice, unlike the planets. Also, Pluto has a very eccentric (elongated) and highly inclined (tilted) orbit. On August 25, 2006, Pluto was demoted by the International Astronomical Union (IAU), which is the scientific organization charged with classifying celestial objects. pluto is no longer classified as a planet.

Pluto's misplaced identity began early in the twentieth century when investigators, excited about the discovery of Neptune, searched the skies for yet another planet. It is well known that we often see what we expect or hope to see, rather than what is there. Finding further perturbations of Uranus prompted astronomers to search a certain region of the sky, and, in 1930, there it was—an object beyond Neptune. The object was named Pluto, and it joined the list of planets.

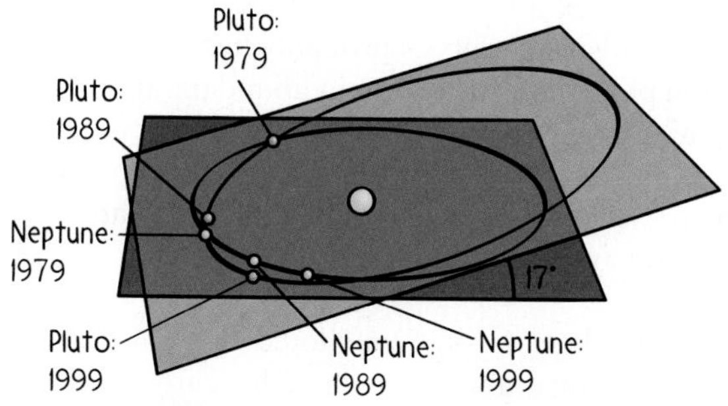

Figure 27.20 Pluto's orbit is significantly more elliptical and inclined at a steeper angle with respect to the ecliptic compared to the other planets. It comes closer to the Sun than Neptune for 20 years in its 248-year orbit, as occurred from 1979 to 1999.

So Pluto isn't a planet, than what is it? Consider three clues:

- Pluto spends most of its time well beyond Neptune, in the region where the comets of the Kuiper Belt are located. Though Pluto is most often located in the Kuiper Belt region, its very elliptical orbit brings it closer to the Sun than Neptune at rare intervals.
- Pluto's orbit is highly elliptical, like the orbits of comets, and is steeply inclined (17 degrees) to the planetary plane, like the brim of a hat tipped to one side (Figure 27.20).
- Pluto's composition does not match any of the other planets but is a near-perfect match to the many Kuiper Belt objects.

So, Pluto is now regarded as one of the many Kuiper Belt objects—much like a comet. Data on the orbits of Kuiper Belt comets support this idea. Over a dozen Kuiper Belt comets are known to have the same orbital period and average distance from the Sun as Pluto itself. These comets are nicknamed "Plutinos." Like Pluto, several Plutinos have moons.

However, while Pluto is tiny for a planet, it is huge for a comet. So Pluto cannot be considered a comet in the usual sense. Nor is Pluto an asteroid, since it is larger, more icy, and possesses a more elliptical orbit than typical asteroids.

Interestingly, other Pluto-size and larger objects have recently been found in the Kuiper Belt. Examples are the objects nick-named Buffy, Xena, and Sedna. Xena is about 5 percent larger than Pluto and Sedna (named after the Inuit goddess of the ocean) is roughly the size of Pluto. Sedna appears to be a mixture of rock and ice like Pluto. Sedna is three times further away from the Sun than Pluto. Beyond Sedna, an as yet un-named object, 2005 UB313, is larger than Pluto. It is not yet known if it is more massive than Pluto. Despite their remote locations, many astronomers have felt that if Pluto retained its classification as a planet, then others ought to be classified as planets—perhaps distinguished as "dwarf planets." Support for this comes

INSIGHT Recent studies of Galileo's notebooks show that Galileo saw Neptune in December 1612 and again in January 1613. He was interested in Jupiter at the time, and so he merely plotted Neptune as a background star.

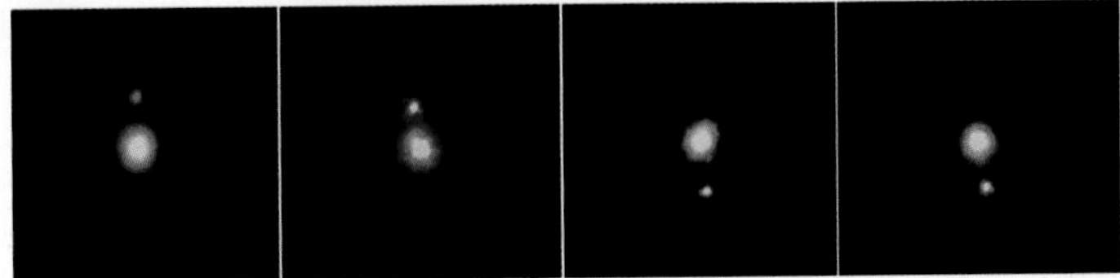

Figure 27.21 These infrared images of Pluto being orbited by its moon Charon are fuzzy because of the small size of these bodies and their great distance from Earth.

from the fact that Pluto and objects of its kind often have moons (Figure 27.21).

So, our solar system presently consists of a sun and eight planets. For the latest on planetary science, please consult the websites recommended at the end of the book.

INTEGRATED SCIENCE BIOLOGY

What Makes a Planet Suitable for Life?

What does a planet need in order to support life? Astrobiologists, scientists who study the origin and distribution of life in the universe, have a number of ideas about planetary characteristics that are conducive to the evolution of life. First, most familiar life on Earth is carbon-based, and there are good reasons to expect life elsewhere to be carbon-based as well. Carbon has the unusual ability to bind to as many as four other atoms at a time, allowing it to serve as the basis for a wide variety of complex molecules. This versatility is probably necessary for life. So, the first attribute of a habitable planet is that it contain abundant amounts of carbon. Second, living things require energy. This energy can come from sunlight or from chemical reactions that take place on or inside planets. Third, the evolution of life probably requires the presence of a liquid medium such as water. Liquids allow molecules to move around and react with one another, which is probably essential for any life form.

Given these constraints, where in the solar system should we look for life? Interestingly, the first two points mentioned above, the need for carbon and the need for an energy source, are satisfied by most planetary bodies in the solar system. The presence of water is, on the other hand, extremely rare. However, there is abundant geological evidence that the planet Mars once had liquid water. Parts of the Martian surface appear to have been produced by flowing water, resembling floodplains or dry riverbeds, and Mars's poles are still covered in water ice. The one-time presence of liquid water also implies that Mars used to be significantly warmer. Mars almost certainly was habitable, then, in the past. But was it inhabited? The discovery of bacteria-like "fossils" in a Martian meteorite (Chapter 17) fueled speculation that Mars not only was inhabited, it was the source of life on Earth. However, these supposed bacteria are much tinier than any organisms found on Earth, and perhaps too tiny to contain all the cellular structures a bacterium needs to function. The question of whether life existed—or even exists—on Mars is still open, and further exploration is needed to provide definitive answers.

CHECK YOURSELF

Many scientists believe that life is most likely to evolve on planets with liquid in their environments. One of the perceived advantages of water over other potential liquids (such as ammonia, methane, or ethane) is that frozen water—ice—floats. Why might this be important for life?

Figure 27.22 The Earth and Moon as photographed in 1977 from the *Voyager 1* spacecraft on its way to Jupiter and Saturn (NASA).

CHECK YOUR ANSWER

On Earth, a layer of floating ice insulates the underlying water, allowing it to remain liquid. If ice sank, it would expose more liquid water to the environment, which would then also freeze and sink. In a cold period, lakes and oceans would eventually freeze solid—not good for the organisms living in them.

27.6 Earth's Moon

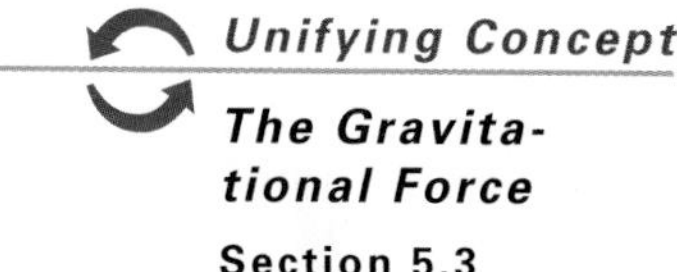

Unifying Concept
The Gravitational Force
Section 5.3

During the early history of the solar system, when Earth was still a very young planet, a great amount of unaccreted rocky debris orbited the Sun and frequently collided with Earth. Most of the rocky bodies hitting Earth ranged from sand-grain to boulder size, but some were larger, tens of kilometers or more in diameter. Many scientists believe that a huge rocky mass the size of the planet Mars collided with Earth about 20 million years after Earth had formed. Material blasted out from Earth as well as fragments from the impactor formed a ring-shaped cloud around Earth. The particles in the cloud later aggregated through mutual gravitational attraction to form a spherical body, the Moon, which captured by Earth's gravity, continues to orbit our planet today. As described in Chapter 11, this idea is known as the *giant impact theory* of the origin of Earth's Moon.

We know more about the Moon than any other celestial body. The Moon is small, with a diameter of about the distance from San Francisco to New York City. It once had a molten surface, but it cooled too rapidly for the establishment of moving crustal plates, like those of Earth. In its early history, it was intensely bombarded by meteoroids. A little more than 3 billion years ago, meteoroid bombardment and volcanic activity filled basins with lava to produce its present surface. It has undergone very little change since then. Its igneous crust is thicker than Earth's. The Moon is too small with too little gravitational pull to have an atmosphere, and so, without weather, the only eroding agents have been meteoroid impacts.

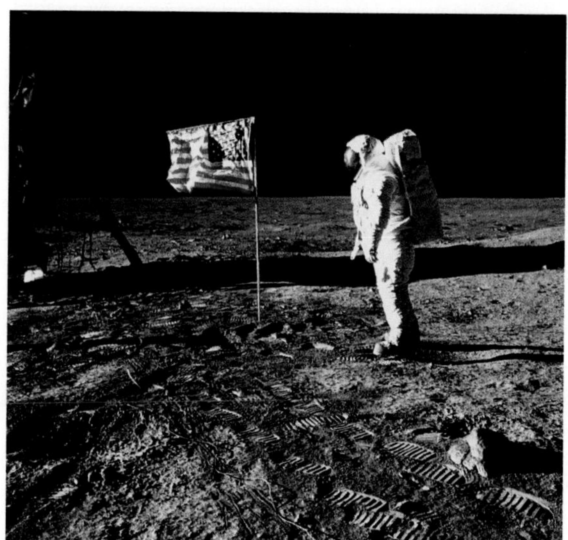

Figure 27.23 Edwin E. Aldrin, Jr., one of the three *Apollo 11* astronauts, stands on the dusty lunar surface. To date, 12 people have stood on the Moon.

The Phases of the Moon

Sunshine illuminates one-half of the Moon's surface. The Moon shows different amounts of its sunlit half as it circles around Earth each month. These changes are the **Moon's phases** (Figure 27.24). The Moon cycle begins with the new Moon. In this phase, its dark side faces us and we see darkness. This occurs when the Moon is between Earth and the Sun (position 1 in Figure 27.25).

During the next seven days, we see more and more of the Moon's sunlit side (position 2 in Figure 27.25). The Moon is going though its waxing crescent phase ("waxing" means increasing). At the first quarter, the angle between the Sun, Moon, and Earth is 90 degrees. At this time, we see half the sunlit part of the Moon (position 3 in Figure 27.25).

During the next week, we see more and more of the sunlit part. The Moon is going through its waxing gibbous phase (position 4 in Figure 27.25). ("Gibbous" means more than half.) We see a full Moon when the sunlit side of the Moon

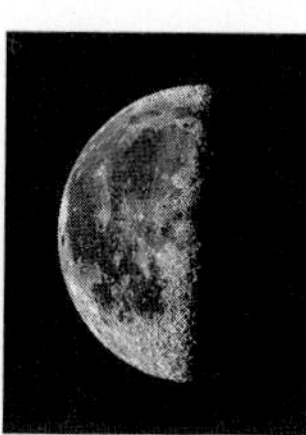

Figure 27.24 The Moon in its various phases.

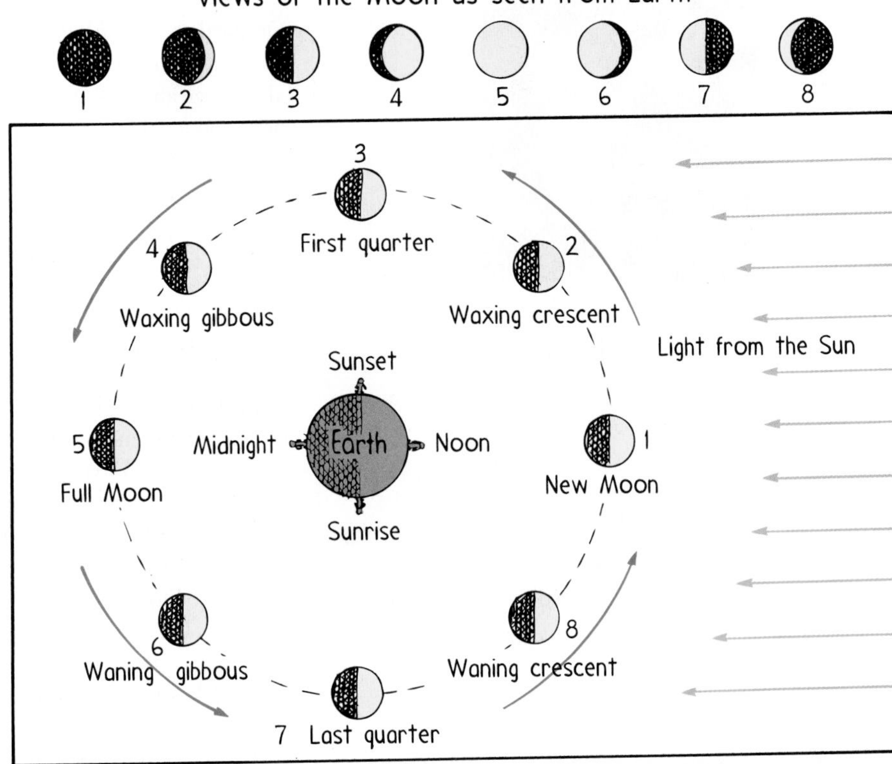

Figure 27.25 Sunlight illuminates only one-half of the Moon. As the Moon orbits Earth, we see varying amounts of its sunlit side. One lunar phase cycle takes 29.5 days.

faces us squarely (position 5). At this time, the Sun, Earth, and Moon are lined up, with Earth in between.

The cycle reverses during the following two weeks, as we see less and less of the sunlit side while the Moon continues in its orbit. This movement produces the waning gibbous, last quarter, and waning crescent phases. ("Waning" means shrinking.) The time for one complete cycle is about 29.5 days.*

INSIGHT If someone shone a flashlight on a ball in a dark room, you could tell where the flashlight was by looking at the illumination on the ball. The same is true of how the Moon is lit by the Sun.

CHECK YOURSELF

1. Can a full Moon be seen at noon? Can a new Moon be seen at midnight?
2. Astronomers prefer to view the stars when the Moon is absent from the night sky. When, and how often, is the Moon absent from the night sky?

CHECK YOUR ANSWERS

1. Inspection of Figure 27.25 shows that, at noontime, you would be on the wrong side of Earth to view the full Moon. Likewise, at midnight, the new Moon would be absent. The new Moon is in the sky in the daytime, not at night.
2. At the time of the new Moon and during the week on either side of the new Moon, the night sky does not show the Moon. Unless an astronomer wishes to study the Moon, these dark nights are the best time for viewing other objects. Astronomers usually view the night skies during two-week periods every two weeks.

INTEGRATED SCIENCE PHYSICS

Why One Side of the Moon Always Faces Us

The first images of the back side of the Moon were taken by the unmanned Russian spacecraft *Lunik 3* in 1959. The first human witnesses of the Moon's back were Apollo 8 astronauts, who orbited the Moon in 1968. From Earth, we see only a single lunar side. We can see this is true even with naked-eye observations—the

* The Moon actually orbits Earth once every 27.3 days relative to the stars. The 29.5-day cycle is relative to the Sun and is due to the motion of the Earth–Moon system as it revolves about the Sun.

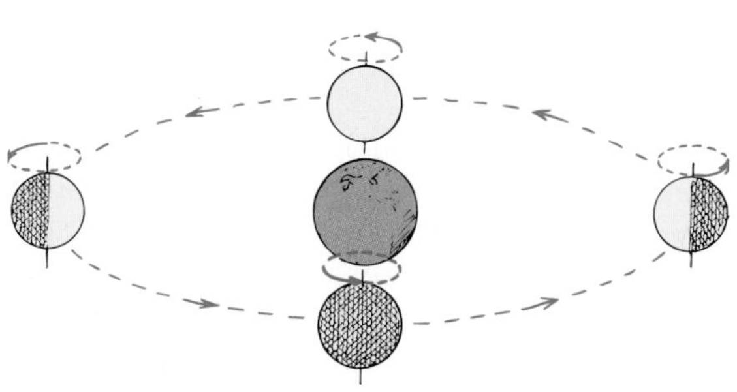

Figure 27.26 The Moon spins about its own polar axis just as often as it circles Earth. So, as the Moon circles Earth, it spins so that the same side (shown in yellow) always faces Earth. In each of the four successive positions shown here, the Moon has spun $\frac{1}{4}$ of a turn.

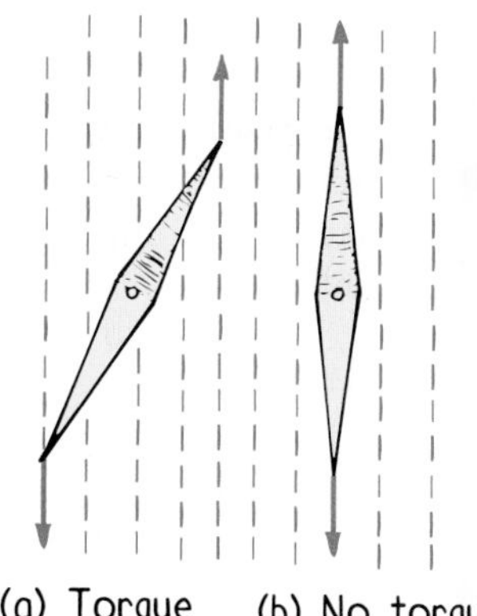

Figure 27.27 (a) When the compass needle is not aligned with the magnetic field (dashed lines), the forces represented by the blue arrows at either end produce a pair of torques that rotate the needle. (b) When the needle is aligned with the magnetic field, the forces no longer produce torques.

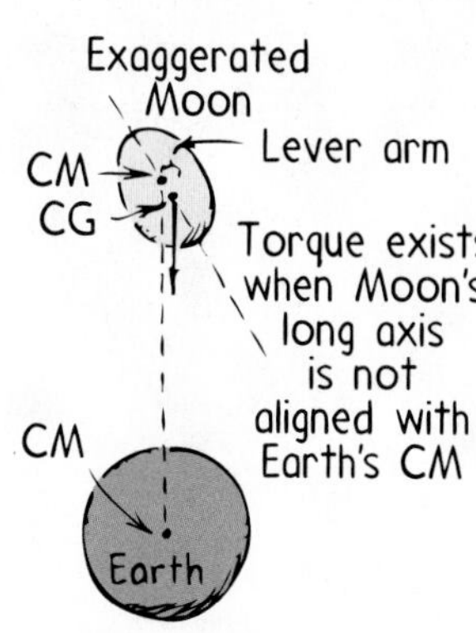

Figure 27.28 When the long axis of the Moon is not aligned with Earth's gravitational field, Earth exerts a torque that rotates the Moon into alignment.

familiar facial features of the "man in the Moon" are always turned toward us on Earth. Does this mean that the Moon doesn't spin about its axis like the Earth does daily? No, but, relative to the stars, the Moon in fact does spin, although quite slowly—about once every 27 days. This monthly rate of spin matches the rate at which the Moon revolves about Earth. This explains why the same side of the Moon always faces Earth (Figure 27.26). This matching of monthly spin rate and orbital revolution rate is not a coincidence. Let's see why.

Think of a compass needle that lines up with a magnetic field. This lineup is caused by a *torque*—a "turning force with leverage" (like that produced by the weight of a child at the end of a seesaw). The compass needle on the left in Figure 27.27 rotates because of a pair of torques. The needle rotates counterclockwise until it aligns with the magnetic field. In a similar manner, the Moon aligns with Earth's gravitational field.

We know from the law of gravitation that gravity weakens with the inverse square of distance, so the side of the Moon nearer to Earth is gravitationally pulled more than the farther side. This stretches the moon out to a football shape. (The moon does the same to Earth and gives us tides). If its long axis doesn't line up with Earth's gravitational field, a torque acts upon it (Figure 27.28). Like a compass in a magnetic field, it turns into alignment. So the Moon lines up with Earth in its monthly orbit. One hemisphere always faces us.

It's interesting to note that for many moons orbiting other planets, a single hemisphere faces the planet. We say these moons are "tidally locked." The long-range fate of Earth is to be tidally locked to the Sun.

CHECK YOURSELF

A friend says that the Moon does not spin about its axis, and evidence for a nonspinning Moon is the fact that its same side always faces Earth. What do you say?

CHECK YOUR ANSWER

Place a quarter and a penny on a table. Pretend the quarter is Earth and the penny the Moon. Keeping the quarter fixed, revolve the penny around it in such a way that Lincoln's head is always pointed to the center of the quarter. Ask your friend to count how many rotations the penny makes in one revolution (orbit) around the quarter. He'll see that it rotates once with each revolution. The key concept is that the Moon takes the same amount of time to complete one rotation as it does to revolve around Earth.

INSIGHT The corona of the Sun is about as bright as the full Moon to those in the path of totality.

Figure 27.29 A solar eclipse occurs when the Moon passes in front of the Sun as seen from Earth. The Moon's shadow has two portions: a dark, central umbra surrounded by the lighter penumbra. A total eclipse is seen from within the umbra and may last several minutes.

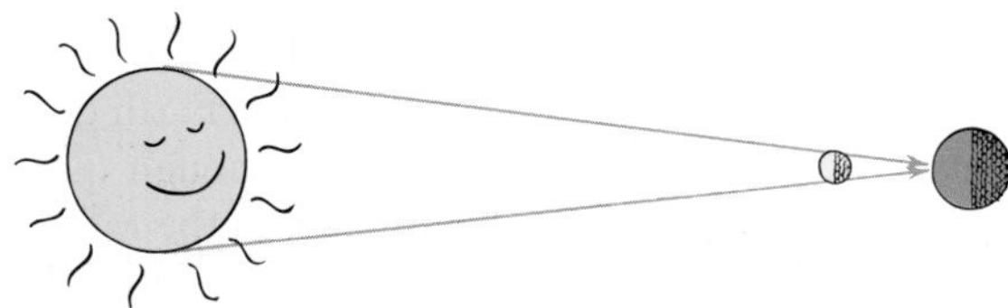

Figure 27.30 Geometry of a solar eclipse. During a solar eclipse, the Moon is directly between the Sun and Earth and the Moon's shadow is cast on Earth. Because of the small size of the Moon and tapering of the solar rays, a solar eclipse occurs only on a small area of Earth.

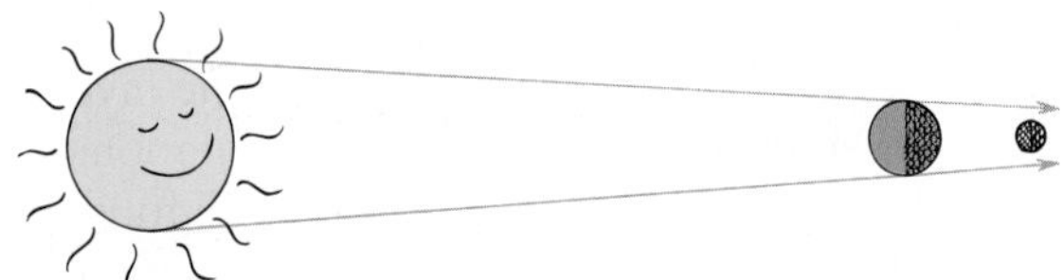

Figure 27.31 A lunar eclipse occurs when Earth is directly between the Moon and the Sun and Earth's shadow is cast on the Moon.

Figure 27.32 A fully eclipsed Moon is not completely dark in the shadow of Earth but is quite visible. This is because Earth's atmosphere acts as a lens and refracts light into the shadow region—sufficient light to faintly illuminate the Moon.

Eclipses

Although the Sun is 400 times larger in diameter than the Moon, it is also 400 times farther away. So, from Earth, both the Sun and Moon subtend the same angle (0.5 degrees) and appear to be the same size in the sky. It is this coincidence that allows us to see solar eclipses.

Both Earth and the Moon cast shadows when sunlight shines upon them. When the path of either of these bodies crosses into the shadow cast by the other, an eclipse occurs. A **solar eclipse** occurs when the Moon's shadow falls on Earth. Because of the large size of the Sun, the rays taper to provide an umbra and a surrounding penumbra (Figures 27.29 and 27.30). An observer in the umbra part of the shadow experiences darkness during the day—a total eclipse, *totality*. Totality begins when the Sun disappears behind the Moon, and ends when the Sun reappears on the other edge of the Moon. The average time of totality is about 2 or 3 minutes, with a maximum no longer than 7.5 minutes. The eclipse time in any location is brief because of the Moon's motion. An observer in the penumbra experiences a partial eclipse, and can still see part of the Sun.* The darkness of totality is not complete, however, because of the bright corona that surrounds the sun.

The alignment of Earth, Moon, and Sun also produces a **lunar eclipse** when the Moon passes into the shadow of Earth (Figure 27.31). Usually a lunar eclipse precedes or follows a solar eclipse by 2 weeks. Just as all solar eclipses involve a new Moon, all lunar eclipses involve a full Moon. They may be partial or total. All observers on the dark side of Earth see a lunar eclipse at the same time. Interestingly enough, when the Moon is fully eclipsed, it is still visible (Figure 27.32).

CHECK YOURSELF

1. Does a solar eclipse occur at the time of a full Moon or a new Moon?
2. Does a lunar eclipse occur at the time of a full Moon or a new Moon?

CHECK YOUR ANSWERS

1. A solar eclipse occurs at the time of a new Moon, when the Moon is directly in front of the Sun. Then the shadow of the Moon falls on part of Earth.
2. A lunar eclipse occurs at the time of a full Moon, when the Moon and Sun are on opposite sides of Earth. Then the shadow of Earth falls on the full Moon.

27.7 Asteroids, Comets, and Meteoroids

Like scraps of material left over at a construction site, asteroids and comets are thought to be material that was unsuccessful in accreting to become a planet during the formation of the solar system. Asteroids are the rocky leftover planetesimals of the inner solar system, while comets are the icy leftover planetesimals of the outer solar system. Evidence that asteroids and comets are leftover planetesimals comes from analysis of meteorites, spacecraft visits to comets and asteroids, and computer simulations of solar system formation. In fact, the nebular theory allowed scientists to make predictions about the locations of comets that weren't yet verified until decades later, when we discovered large comets orbiting in the vicinity of Neptune and Pluto.

Asteroids are small, rocky bodies that orbit the Sun. Most are located in the main asteroid belt, which lies in the plane of the ecliptic between Mars and Jupiter. Tens of thousands of asteroids reside here. The smallest asteroids are

* People are cautioned not to look at the Sun at the time of a solar eclipse because the brightness and the ultraviolet light of direct sunlight are damaging to the eyes. This good advice is often misunderstood by those who then think that sunlight is more damaging at this special time. However, staring at the Sun when it is high in the sky is harmful whether or not an eclipse occurs. In fact, staring at the bare Sun is more harmful than when part of the Moon blocks it. The reason for special caution at the time of an eclipse is simply that more people are interested in looking at the Sun during this time.

(a)

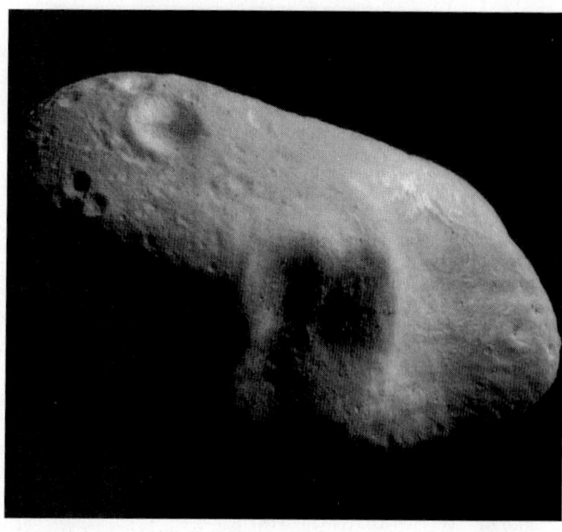

(b)

Figure 27.33 (a) Most, though not all, asteroids in the solar system occupy the main asteroid belt. Earth has been struck by asteroids in the past, resulting in tremendous explosions. It is a statistical certainty that there will be other large asteroid impacts in the future. (b) The asteroid Eros (photographed from the *NEAR* spacecraft). Eros is about 40 kilometers long and, like other small objects in the solar system, it is not spherical.

irregular in shape, like boulders, and the larger ones are roughly spherical (Figure 27.33). They vary in size from less than a kilometer across to hundreds of kilometers in diameter. This means that even the largest asteroids are dwarfed in comparison to Earth's moon.

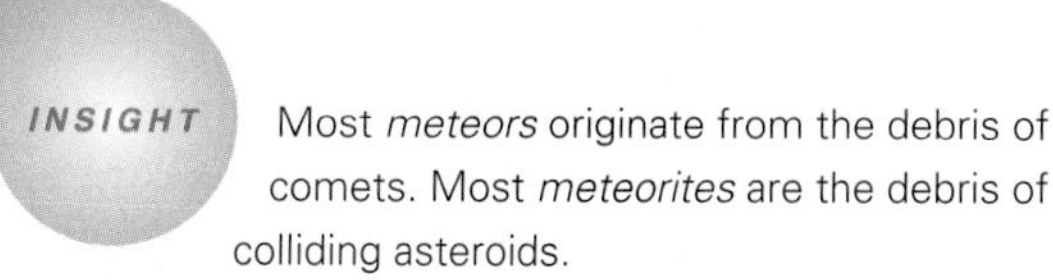

Most *meteors* originate from the debris of comets. Most *meteorites* are the debris of colliding asteroids.

Why are asteroids located in the main asteroid belt between Mars and Jupiter, rather than spread out throughout the solar system? The answer is that the asteroid belt is the only place where asteroids could survive for billions of years. During the birth of the solar system, planetesimals formed throughout the solar system. However, most of those within Mars' orbit ultimately accreted into one of the four inner planets. The relatively few asteroids that orbit in the inner solar system today are almost certainly "impacts waiting to happen." Some of these asteroids do pass near Earth's orbit and may pose a potential threat to our planet. (More on this in Chapter 26.) By contrast, asteroids in the asteroid belt stay clear of any planet and can therefore survive on their own for billions of years.

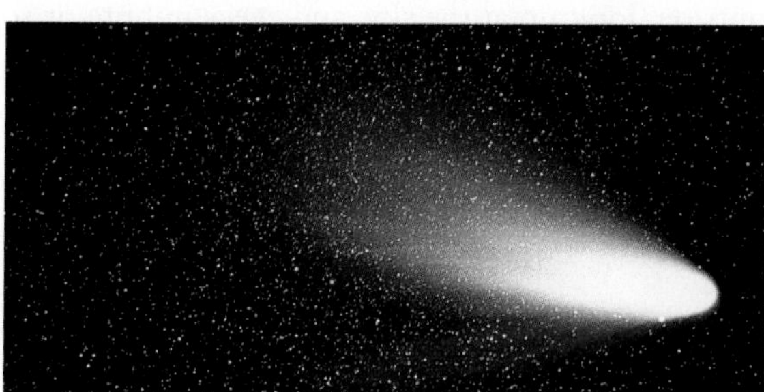

Figure 27.34 Comet Hale-Bopp in 1997.

Like asteroids, **comets** orbit the Sun (Figure 27.34). But comets differ from asteroids in chemical composition. Rather than being rocky as asteroids are, comets are "dirty snowballs"—masses of water, methane, or other ice in which chunks of rock, metal, and dust are embedded. There are two major groupings of comets in the solar system: One group we mentioned in Section 27.1, the Kuiper Belt orbits the Sun in a wide track beyond Pluto's orbit. Kuiper Belt comets have relatively short periods, less than 200 years. They orbit in the same direction and nearly in the same plane as the planets. The other group of comets is a vast realm of drifting icy bodies at the outermost fringes of the solar system. This is the Oort Cloud. Comets of the Oort cloud have long-period orbits (over 200 years). They do not orbit in a simple pattern as the Kuiper Belt comets do.

Whereas asteroids travel between the planets in elliptical orbits with low eccentricity, the orbits of comets are highly elliptical. As a comet approaches the Sun, solar heat vaporizes its ices. Escaping vapors glow to produce a fuzzy, luminous ball called a *coma*. Within the bright coma is the frozen solid part of the comet, the *nucleus*. Occasionally, a comet will be disturbed in its orbit and fall toward the Sun. Then, the solar wind and radiation push particles from the coma, creating the long "tail" of the comet. This tail can extend over 100 million kilometers. The density of the material in a comet's tail is quite low—less than that of the best laboratory vacuums on Earth. Nevertheless, particles of the tail reflect light and shine visibly on Earth. Comets move slowly and gracefully

Figure 27.35 A meteor is produced when a meteoroid enters Earth's atmosphere, usually about 80 km high. Most are sand-sized grains, which are seen as "falling" or "shooting" stars.

Table 27.2 Meteor Shower Data

Shower Name	Radiant*	Dates	Peak Dates	Meteors per Hour
Quadrantids	Pegasus	Jan 1–6	Jan 3	60
Eta Aquarids	Aquarius	May 1–10	May 6	35
Perseids	Perseus	Jul. 23–Aug. 20	Aug. 12	75
Orionids	Orion	Oct. 16–27	Oct. 22	25
Geminids	Gemini	Dec. 7–15	Dec. 13	75

* Meteors appear to radiate from a certain region of the sky, appropriately called a *radiant*. Radiants refer to constellations. See Chapter 28 for more on where the various constellations are located in the night sky.

Figure 27.36 The Barringer Crater in Arizona, made 25,000 years ago by an iron meteorite having a diameter of about 50 meters. The crater extends 1.2 km across and reaches 200 m deep.

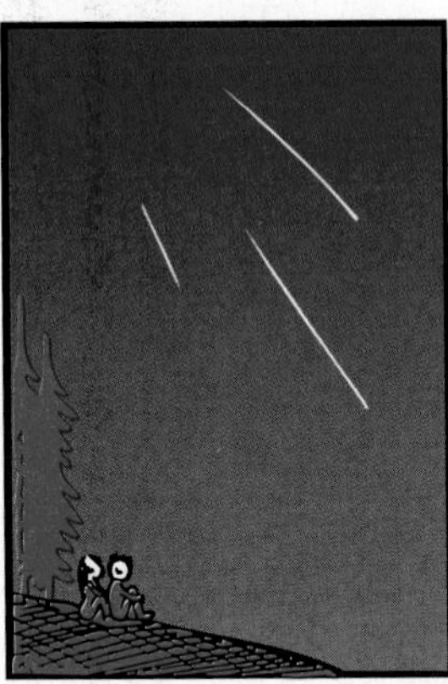

Figure 27.37 When Earth crosses the orbit of a comet, we see a meteor shower.

across the sky for weeks or months to display one of nature's most beautiful astronomical spectacles.

A **meteoroid** is a relatively small (sand-grain to boulder size) piece of debris chipped off from an asteroid or comet. A **meteor** is a meteoroid that strikes Earth's atmosphere, usually at an altitude of about 80 kilometers, heated white-hot by friction with the atmosphere and seen from Earth as a flash of light—a "falling star." Most of the meteors we see are little, about the size of a grain of sand. They usually disintegrate in the atmosphere. But any meteor that survives its fiery descent through the atmosphere and reaches the ground is called a **meteorite** (Figure 27.35).

Most meteorites are small and strike Earth with no more energy than a falling hailstone. Some are big, however, and evidence of their impact is seen as craters. If Earth were without weather and erosion, its surface would be as cratered as the Moon's. Most impact craters on Earth were eroded or covered by geologic processes long ago. More recent impacts, however, leave telltale marks (Figure 27.36). The most dramatic impact on record, though not the largest,* was near the Yucatan Peninsula in Mexico 65 million years ago. The effects of that impact likely led to the extinction of dinosaurs and half of the other species living at the end of the Cretaceous period.

Meteor showers are fun and inspiring to watch. They generally result when Earth passes through a stream of particles left behind by an orbiting comet. These events occur at certain times on a regular basis, so plan on attending the next one (Table 27.2). Just go outside, look up at the sky, and every minute or so you will see a shooting star. Each streak is a tiny chip of a comet, once so very far away, that has fallen into Earth's neighborhood (Figure 27.37).

* The Vredefort crater in South Africa and the Sudbury crater in Canada are both larger than the crater that was found at Yucatan.

CHAPTER 27 Review

Summary of Terms

Asteroid A small, rocky, planetlike fragment that orbits the Sun. Tens of thousands of these objects make up an asteroid belt between the orbits of Mars and Jupiter.

Comet A body composed of ice and dust that orbits the Sun, usually in a very eccentric orbit, and that casts a luminous tail when it is close to the Sun.

Ecliptic The plane of Earth's orbit.

Full Moon The phase of the Moon when its sunlit side is the side facing Earth.

Inner planets The planets close to the Sun: Mercury, Venus, Earth, and Mars.

Kuiper Belt The disk-shaped region of the sky beyond Neptune that is populated by many icy bodies and is a source of short-period comets.

Lunar eclipse The phenomenon whereby the shadow of the Earth falls upon the Moon, producing the relative darkness of the full Moon.

Meteor The streak of light produced by a meteoroid burning in Earth's atmosphere; a "shooting star."

Meteorite A meteoroid, or a part of a meteoroid, that has survived passage through Earth's atmosphere to reach the ground.

Meteoroid A small piece of debris from a comet or asteroid in interplanetary space.

Moon phases The cycles of change of the "face" of the Moon, changing from new, to waxing, to full, to waning, and back to new.

Nebular theory The idea that the Sun and planets formed together from a cloud of gas and dust, a *nebula*.

New Moon The phase of the Moon when darkness covers the side facing Earth.

Oort Cloud The region beyond the Kuiper Belt that is populated by trillions of icy bodies and is a source of long-period comets.

Outer planets Jupiter, Saturn, Uranus, Neptune, and Pluto.

Planets The major bodies orbiting the Sun or other star, which are massive enough for their gravity to make them spherical but small enough to avoid having nuclear fusion in their cores.

Solar eclipse The phenomenon whereby the shadow of the Moon falls upon Earth, producing a region of darkness in the daytime.

Review Questions

27.1 Overview of the Solar System

1. If Earth were the size of a grape, how big would the Moon be? How far away would the Moon be from Earth? How large would the Sun be? How far would it be from the grape-sized Earth?
2. Describe the orbital motion of the planets in term of the direction of their rotation, revolution, and the plane in which they move.
3. What are the five classes of objects that orbit the Sun?

27.2 The Nebular Theory

4. What two major regularities must any successful theory of the origin of the solar system explain?
5. How long ago did the solar system begin to form?
6. What force pulled the particles of gas and dust together, making the solar nebula shrink? What force tended to push the particles apart?

27.3 The Sun

7. In what way does the Sun's mass supply the energy used by most of Earth's organisms?
8. Describe each of the layers of the Sun: *core*, *radiation zone*, *convection zone*, *photosphere*, *chromosphere*, *corona*.
9. What is the solar wind? What are sunspots? Solar prominences?

27.4 The Inner Planets

10. Name the inner planets.
11. Why are the inner planets also called "terrestrial planets"?
12. Why are days on Mercury very hot and the nights very cold?
13. What two planets are evening or morning "stars"?

27.5 The Outer Planets

14. How are the outer planets different than the inner planets?
15. Name the outer planets.
16. Why are the outer planets also called the "Jovian planets"?
17. How is Pluto unlike the planets?

27.6 Earth's Moon

18. Why is it that the only agents of erosion on the Moon are impacting meteorites?
19. Where is the Sun when you view a full Moon?
20. Where are the Sun and the Moon at the time of a new Moon?
21. What is the cause of a solar eclipse? A lunar eclipse?

27.7 Asteroids, Comets, and Meteoroids

22. What are the two main groupings of comets in the solar system?
23. What are two differences between comets and asteroids?
24. How is the coma of a comet produced? What creates a comet's tail?
25. What produces a meteor shower?

Multiple Choice Questions

Choose the BEST answer to the following.

1. With each pass of a comet around the Sun, its mass
   (a) is greatly reduced.
   (b) remains virtually unchanged.
   (c) actually increases.
2. Asteroids orbit about the
   (a) Sun.
   (b) Moon.
   (c) Earth.
   (d) none of these.
3. In a museum collection you can likely see a
   (a) meteoroid but not a meteorite.
   (b) meteor but not a meteoroid.
   (c) meteorite but not a meteoroid.
4. Most meteors seen as shooting stars are about the size of
   (a) grains of sand.
   (b) baseballs.
   (c) small buildings.
   (d) very large buildings.
   (e) small continents.
5. The predominant gas in the atmosphere of Venus is
   (a) oxygen.
   (b) nitrogen.
   (c) water vapor.
   (d) carbon dioxide.
   (e) methane.
6. The Sun tends to bloat outward by nuclear fusion, and contract due to
   (a) gravitation.
   (b) nuclear fission.
   (c) mass decrease.
   (d) its relatively slow spin.
   (e) reduced gas pressure.
7. No greenhouse effect occurs on Mercury because it lacks
   (a) daily spin.
   (b) atmosphere.
   (c) relatively cool regions.
   (d) plant life.
   (e) terrestrial radiation.
8. The reason that the Moon takes on a crescent shape each month has to do with the
   (a) Earth's shadow.
   (b) Sun's position.
   (c) both of these.
   (d) neither of these.

INTEGRATED SCIENCE CONCEPTS

Physics—The Solar Nebula Heats Up, Spins Faster, and Flattens

1. What three changes occurred as the solar nebula was collapsing that transformed it into a star?
2. How was the solar nebula like a spinning ice skater?

Chemistry—The Chemical Composition of the Solar System

3. How are heavy chemical elements formed?
4. How is the chemical composition of the terrestrial planets different from that of the Jovian planets? What is the explanation for this?

Biology—What Makes a Planet Suitable for Life?

5. Why do scientists believe that most, perhaps all, life in the universe is based on carbon?

6. Why does the evolution of life probably require the presence of liquid on a planet?

Physics—Why One Side of the Moon Always Faces Us

7. How does the Moon's rate of rotation about its own axis compare with its rate of revolution around the Earth?
8. What does the Moon have in common with a compass needle?
9. Is the fact that we see only one side of the Moon evidence that the Moon spins or that it doesn't rotate? Defend your answer.

Active Explorations

1. Look at Table 27.2 to find the next major meteor shower. Watch it and enjoy! While you are doing so, review in your mind the distinctions among meteoroids, meteors, meteorites, and comets.
2. Simulate the lunar phases. Insert a pencil into a Styrofoam ball. This will be the Moon. Position a lamp (representing the Sun) in another room near the doorway. Hold the ball in front and slightly above yourself. Slowly turn yourself around keeping the ball in front of you as you move. Observe the patterns of light and shadow on the ball. Relate this to the phases of the Moon.

Exercises

1. Copy the diagram and label the planets.

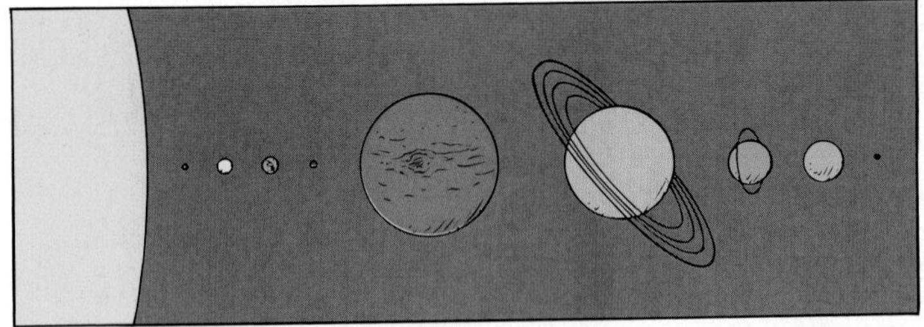

2. Why aren't eclipses more common events?
3. What causes comet tails to point away from the Sun?
4. Cite as many of the features of the solar system as you can that are explained by the nebular theory.
5. Is there evidence that Mars was at one time wetter than it presently is?
6. We know that the Sun is much larger than the Moon, but both appear the same size in the sky. What is your explanation?
7. The giant impact theory is believed to explain the origin of the Moon. Does this mean that the nebular theory does not explain the Moon's formation? Explain.
8. Why are there many more craters on the surface of the Moon than on the surface of Earth?
9. Do star astronomers make stellar observations during the full Moon part of the month or during the new Moon part of the month? Does it make a difference?
10. Nearly everyone has witnessed a lunar eclipse, but relatively few people have seen a solar eclipse. Why?
11. Because of Earth's shadow, a partially eclipsed Moon looks like a cookie with a bite taken out of it. Explain, with a sketch, how the curvature of the bite indicates the size of Earth relative to the size of the Moon. How does the tapering of the Sun's rays affect the curvature of the bite?
12. Why is it not totally dark in the location where a total solar eclipse occurs?
13. What energy processes make the Sun shine? In what sense can it be said that gravity is the prime source of solar energy?
14. A TV screen is normally light gray when it is not illuminated. How is the blackness of sunspots similar to the blackness in images on a TV screen?
15. When a contracting ball of hot gas spins into a disk shape, it cools. Why?
16. The greenhouse effect is very pronounced on Venus, but it doesn't exist on Mercury. Why?
17. Where are the elements heavier than hydrogen and helium formed?
18. What is the cause of winds on Mars (and on almost every other planet, too)?
19. What is the major difference between the terrestrial and Jovian planets?
20. What does Jupiter have in common with the Sun that the terrestrial planets don't? What differentiates Jupiter from a star?
21. Using the following mnemonic to state the order of the planets: *My very excellent mother just served us nine pizzas*, state the order of the planets.
22. Why are the seasons on Uranus different from the seasons on any other planet?
23. Why are meteorites so much more easily found on Antarctica than on the other continents?
24. Draw a comic strip illustrating the formation of the Sun.
25. A meteor is visible only once, but a comet may be visible at regular intervals throughout its lifetime. Why?
26. Explain the connection between meteor showers and comets.
27. Why are the inner planets rocky while the outer planets are gaseous?
28. Chances are about 50–50 that, in any night sky, there is at least one visible comet that has not been discovered. This keeps amateur astronomers busy looking, night after night, because the discoverer of a comet receives the honor of having it named for him or her. With this high probability of comets being visible in the sky, why aren't more of them found?

Problems

1. How many Earth diameters fit between Earth and the Moon?
2. What is the diameter of the Earth–Moon system?
3. To send a radio signal from Earth to Saturn, show that it would take about 71 minutes to arrive there. Then show that it would take about 321 minutes for a radio signal from Earth to reach Pluto.
4. Show that it takes about 8.3 minutes for visible light from the Sun to reach Earth. How long does it take infrared radiation to travel this distance? Ultraviolet radiation?

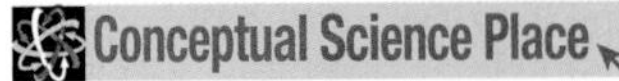

CHAPTER 27 ONLINE RESOURCES

Interactive Figures 27.1 • ***Tutorials*** A Tour of the Solar System, Formation of the Solar System, The Sun • ***Videos*** Orbits in the Solar System, History of the Solar System • ***Quiz*** • ***Flashcards*** • ***Links***

CHAPTER 28

The Universe

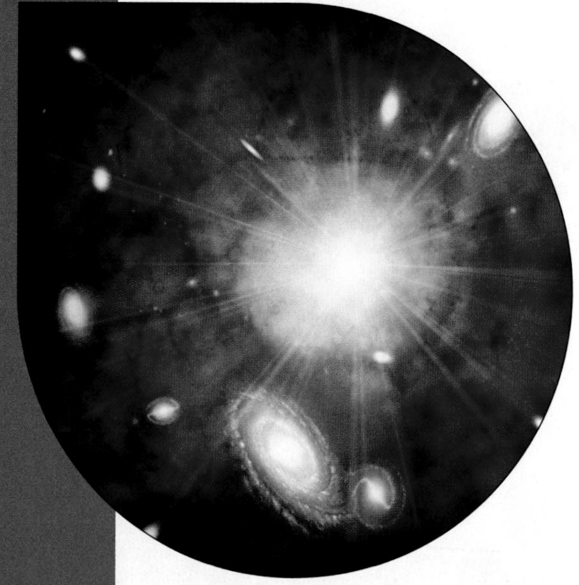

The universe has been expanding since the time of the Big Bang some 13.7 billion years ago. Matter coalesced into galaxies (around the edge of the picture), which are observed to be moving away from one another.

CHAPTER OUTLINE

The roots of astronomy reach back to prehistoric times when humans began to observe star patterns in the night sky. Though ancient people developed ways to measure the position and periodic movement of the stars in the night sky, they knew nothing about the stars themselves. Today we know that Earth orbits a star—our Sun—with its night side always facing away from the Sun. We understand why the background of stars varies in the nighttime sky throughout the year. And we know a great deal about how stars form and die. In this chapter, we will explore the night sky and the universe at large, which extends well beyond what we can observe with the naked eye. Where do stars come from? What is a black hole, a galaxy, a pulsar, and a quasar? What do we know about the origin of the universe, the cosmological events of the past, and the future?

28.1 Observing the Night Sky

Early astronomers divided the night sky into groups of stars, called *constellations*, such as the group of seven stars we now call the Big Dipper. The names of the constellations today carry over mainly from the names assigned to them by early Greek, Babylonian, and Egyptian astronomers. The Greeks, for example, included the stars of the Big Dipper into a larger group of stars that outlined a bear. The large constellation, Ursa Major (the Great Bear) is illustrated in Figure 28.1. The grouping of stars and the significance given to them has varied from culture to culture. To some cultures, the constellations stimulated storytelling and the making of great myths; to other cultures, the constellations honored great heroes, such as Hercules and Orion; to others, they served as navigational aids for travelers and sailors. To many cultures, including the African Bushmen and Masai, the constellations provided a guide for the planting and harvesting of crops because they were seen to move in the sky in concert with the seasons. Charts of this periodic movement became some of the first calendars. We can see, in Figure 28.2, why the background of stars varies throughout the year.

The stars are at different distances from Earth. However, because all the stars are so far away, they appear equally remote. This illusion led the Ancient

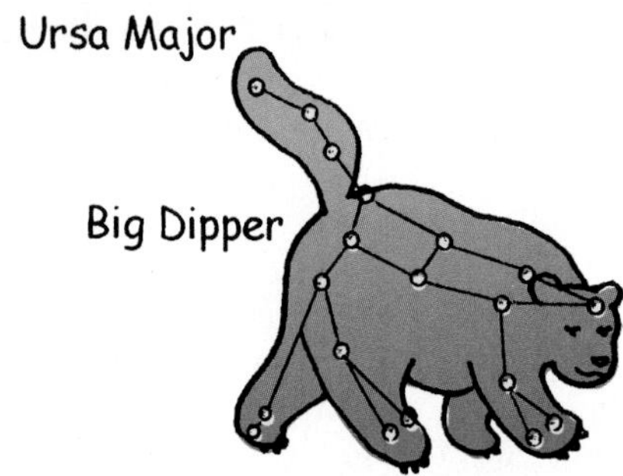

Figure 28.1 The constellation Ursa Major, the Great Bear. The seven stars in the tail and back of Ursa Major form the Big Dipper.

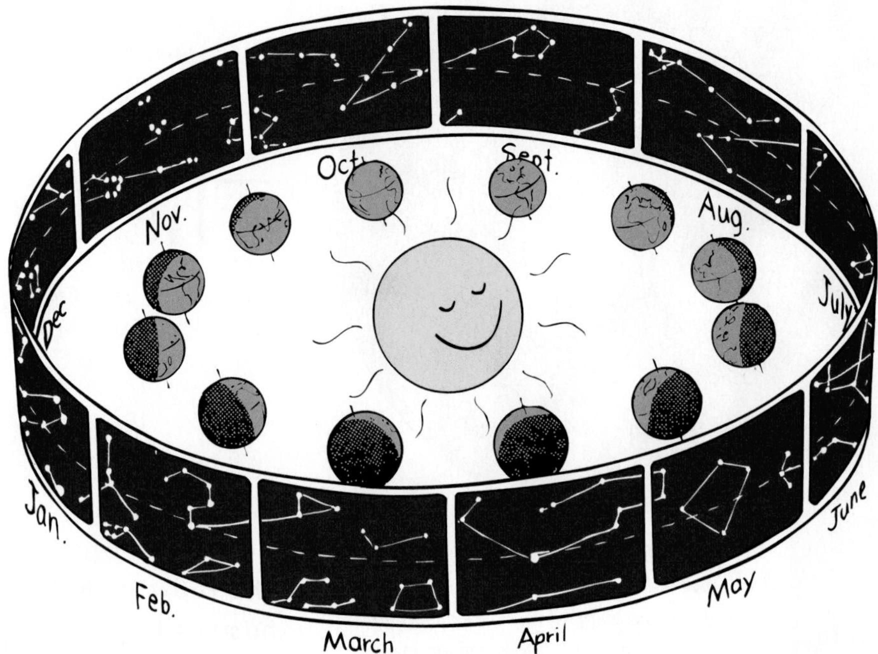

Figure 28.2 The night side of Earth always faces away from the Sun. As Earth circles the Sun, different parts of the universe are seen in the nighttime sky. Here the circle, representing 1 year, is divided into 12 parts—the monthly constellations. The stars in the nighttime sky change in a yearly cycle.

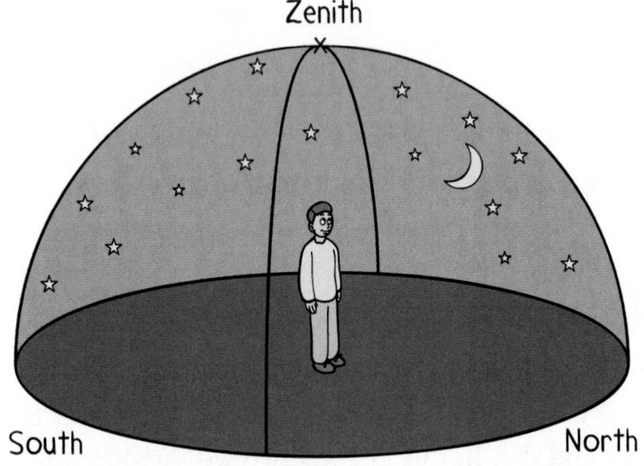

Figure 28.3 The celestial sphere is an imaginary sphere to which the stars are attached. We see no more than half of the celestial sphere at any given time. The point directly over out heads at any time is called the *zenith*.

Greeks and others to conceive of the stars as being attached to a gigantic sphere surrounding Earth, called the **celestial sphere** (Figure 28.3). Though we know it is imaginary, the celestial sphere is still a useful construction for visualizing the motions of the stars.

The stars appear to turn around an imaginary north-south axis once every 24 hours. This is the *diurnal motion* of the stars. Diurnal motion is easy to visualize as a rotation of the celestial sphere from east to west. This motion is a consequence of the daily rotation of Earth on its axis. When we speak of the diurnal motion of the stars, we are referring to the motions of celestial objects as a whole; this motion does not change the relative positions of objects. Figure 28.4 shows the diurnal motion of the stars making up the Big Dipper. Long-time exposure photographs show that the Big Dipper appears to move in circles around the North Star (Figure 28.5). The North Star appears stationary as the celestial sphere rotates because it lies very close to the projection of Earth's rotational axis.

In addition to the diurnal motion of the sky, there is *intrinsic* motion of certain bodies that change their positions with respect to the stars. The Sun, Moon, and planets, called "wanderers" by ancient astronomers, appear to migrate across the fixed backdrop of the celestial sphere.

INSIGHT The grouping of stars into constellations tells us about the thinking of astronomers of earlier times, but it tells us nothing about the stars themselves.

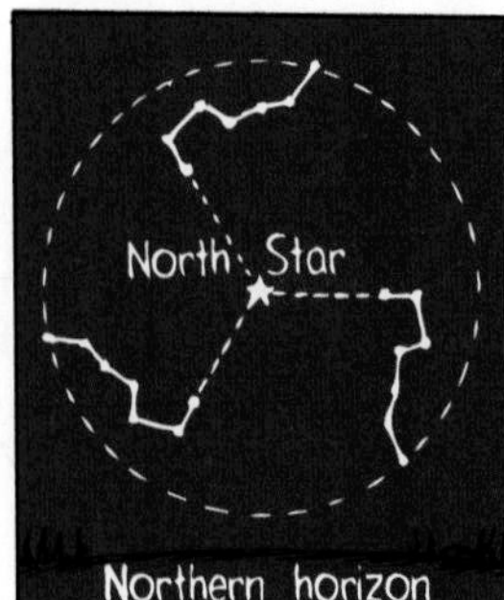

Figure 28.4 The pair of stars in the end of the Big Dipper's bowl point to the North Star. Earth rotates about its axis and therefore about the North Star, so, over a 24-hour period, the Big Dipper (and other surrounding star groups) makes a complete revolution.

CONCEPT CHECK

1. Which celestial bodies appear fixed relative to one another, and which celestial bodies appear to move relative to the others?
2. What are two types of observed motions of the stars in the sky?

CHECK YOUR ANSWERS

1. The stars appear fixed as they move across the sky. The Sun, Moon, and planets move relative to one another as they move across the backdrop of the stars.

Figure 28.5 A time exposure of the northern night sky.

INSIGHT The motions of bodies in the nighttime sky can be separated into three types: (1) the changing positions of constellations that correlates with seasonal variation, (2) diurnal motion—the apparent rotation of the celestial sphere every 24 hours, and (3) intrinsic motion of Sun, Moon, and planets with respect to the backdrop of the stars. There is a fourth kind of motion to mention as well: Due to the different speeds and directions that the stars are moving, the stars actually *do* move with respect to one another. But this motion is not noticeable over the span of human lifetimes.

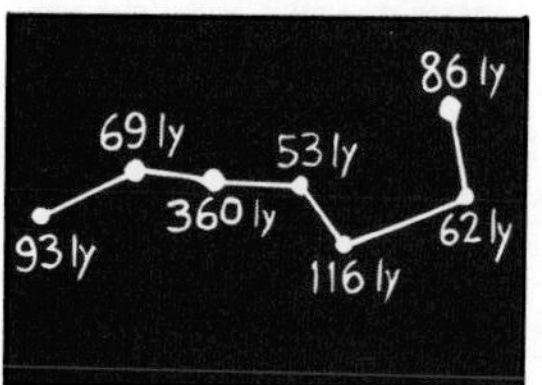

Figure 28.6 Interestingly, the seven stars of the Big Dipper are at varying distances from Earth. Note their varying distances in light-years.

2. One type of motion of the stars is their nightly rotation as if they were painted on a rotating celestial sphere; this is due to the Earth's rotation on its own axis. Stars also appear to undergo a yearly cycle around the Sun because of the Earth's revolution about the Sun.

Though the stars' great distance from Earth makes them appear as though they were all positioned on a celestial sphere, some stars are actually much farther away than others from Earth. Astronomers measure the vast distances between Earth and the stars using **light-years**. One light-year is the distance that light travels in 1 year, nearly 10 trillion kilometers. Figure 28.6 shows the distances to the seven stars making up the Big Dipper in light-years.

The speed of light (as we know from Chapter 8) is 3×10^8 meters/second. Although this is very fast, it nevertheless takes light appreciable time to travel large distances. And so when you see the light emitted by a very distant object, you are actually seeing the light it emitted long ago—you are looking back in time. Consider the example of Supernova 1987a (a *supernova* is the explosion of a star, as you will learn more about in Section 28.3.) This supernova occurred in a galaxy 190,000 light-years from Earth. Although we witnessed the supernova in 1987, the light from this explosion took 190,000 years to reach our planet, so the explosion actually occurred 190,000 years earlier. "News" of the supernova took 190,000 years to reach Earth!

28.2 The Brightness and Color of Stars

All stars have much in common with the Sun. All are born from clouds of interstellar dust having roughly the same chemical composition as the Sun (Chapter 27). About three-fourths of the interstellar material from which any star forms is hydrogen; one-fourth is helium; and no more than 2 percent of the material from which a star forms consists of heavier chemical elements. Stars shine brilliantly for millions or billions of years because of the nuclear fusion reactions that occur in their cores. And all stars, the Sun included, ultimately exhaust their nuclear fuel and die.

Yet, not all stars are the same. If you look into the night sky, you will see that stars differ in two very visible ways: brightness and color. Brightness relates to how much energy a star is producing, while its color indicates its surface temperature.

MATH CONNECTION

The Light-Year

Problems

1. If 1 light-year is the distance light travels in one year, how many kilometers is 1 light-year equal to?
2. The nearest star to Earth beyond the Sun is Proxima Centauri, which is located 4.2 light-years away. (a) When you look at this star, how old is the light you are seeing? (b) When will you see the star as it is right now?

Solutions

1. The distance light travels in 1 year, like any other distance, can be computed from the distance formula that was presented in Chapter 2:

$$\text{Distance} = \text{speed} \times \text{time. So we have:}$$
$$\text{distance} = (\text{speed of light}) \times (1\ \text{year}) = (3 \times 10^8\ \text{m/s}) \times (1\ \text{year})$$

To find an answer in kilometers, we convert meters to kilometers and a year to seconds:

$$\text{distance} = \frac{3 \times 10^8\ \text{m}}{1\text{s}} \times \frac{1\ \text{km}}{10^3\ \text{m}} \times 1\ \text{year} \times \frac{365\ \text{days}}{1\ \text{year}} \times \frac{24\ \text{hour}}{1\ \text{day}} \times \frac{3600\ \text{s}}{1\ \text{hour}} = 9{,}460{,}000{,}000{,}000\ \text{km}$$

So one light-year is another name for the huge distance 9.46 trillion kilometers.

2. (a) The light is 4.2 years old. (b) You'll have to wait 4.2 years to see Proxima Centauri as it is right now.

SCIENCE AND SOCIETY

Astrology

There is more than one way to view the cosmos and its processes—astronomy is one and astrology is another. Astrology is a belief system that began more than 2000 years ago in Babylonia. Astrology has survived nearly unchanged since the second century A.D., when some revisions were made by Egyptians and Greeks who believed that their gods moved heavenly bodies to influence the lives of people on Earth. Astrology today holds that the position of Earth in its orbit around the Sun at the time of birth, combined with the relative positions of the planets, has some influence over one's personal life. The stars and planets are said to affect such personal things as one's character, marriage, friendships, wealth, and death.

Could the force of gravity exerted by these celestial bodies be a legitimate factor in human affairs? After all, the ocean tides are the result of the Moon's and Sun's positions, and the gravitational pulls between the planets perturb one another's orbits. Since slight variations in gravity produce these effects, might not slight variations in the planetary positions at the time of birth affect a newborn? If the influence of stars and planets is gravitational, then credit must also be given to the effect of the gravitational pull between the newborn and Earth itself. This pull is enormously greater than the combined pull of all the planets, even when lined up in a row (as occasionally happens). The gravitational influence of the hospital building on the newborn would surely exceed that of the distant planets. So planetary gravitation cannot be an underlying agent for astrology.

Astrology is not a science, for it doesn't change with new information as science does, nor are its predictions borne out by fact. So the realm of astrology may be superstition. Or it may be a primitive psychology where the stars serve as a point of departure for musings about personality and personal decisions. Or astrology may be in the realm of numerology or phrenology—a pseudoscience, a body of untestable beliefs purporting to utilize scientific methods. Astrology means different things to different people, but in any case, it's far outside the realm of science.

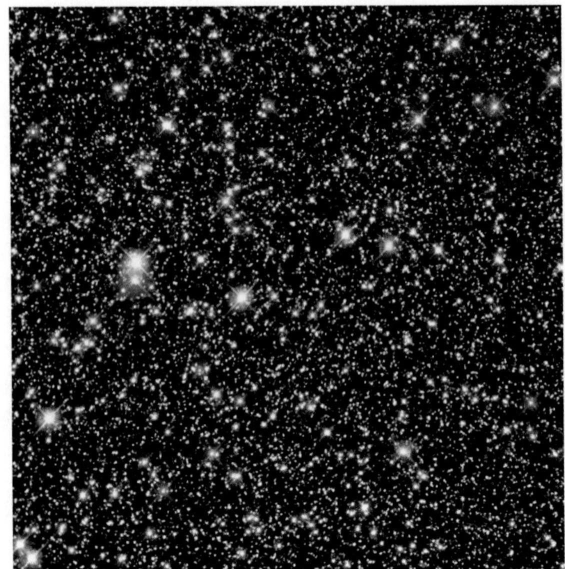

Figure 28.7 Most of the stars in this photograph are approximately the same distance—2000 light-years—from the center of the Milky Way galaxy. A star's color indicates its surface temperature—a blue star is hotter than a yellow star, and a yellow star is hotter than a red star. This photo was taken by the *Hubble Telescope*.

However, note that although a star's brightness is related to its energy output, its brightness also depends on how far away it is from Earth.* For example, the stars Betelgeuse and Procyon appear equally bright even though Betelgeuse emits about 5000 times as much light as Procyon. The reason? Procyon is much closer to Earth than Betelgeuse.

To avoid confusing brightness with energy output, astronomers clearly distinguish between *apparent brightness* and the more important property *luminosity*. **Apparent brightness** is the brightness of a star as it appears to our eyes. **Luminosity**, on the other hand, is the total amount of light energy that a star emits into space. Luminosity is usually expressed relative to the Sun's luminosity, which is noted L_{Sun}. For example, the luminosity of Betelgeuse is $38{,}000L_{Sun}$. This indicates that Betelgeuse is a very luminous star emitting about 38,000 times as much light into space as the Sun. On the other hand, Proxima Centauri is quite dim, with a luminosity of $0.00006L_{Sun}$.

Astronomers have measured the luminosity of many stars and found that stars vary greatly in this respect. The Sun is somewhere in the middle of the luminosity range. The most luminous stars are about a million times as luminous as the Sun, while the dimmest stars produce about 10,000 times less light than the Sun.

Besides apparent brightness, a star's color is another property that varies widely among stars. Figure 28.7, a photograph of stars taken with the *Hubble Telescope*, shows this–stars come in every color of the rainbow. A star's color directly tells you about its surface temperature—a blue star is hotter than a yellow star, and a yellow star is hotter than a red star, for example. In fact, astronomers use color to measure the temperatures of stars. Why is it that a star's color corresponds to its temperature?

INTEGRATED SCIENCE PHYSICS

Radiation Curves of Stars

As you learned in Chapter 6, all objects with a temperature above absolute zero emit energy in the form of electromagnetic radiation. The peak frequency f of the radiation is directly proportional to the absolute temperature T of the emitter:

$$f \sim T$$

* Recall that light follows the inverse-square law: the intensity of light diminishes as the reciprocal of the square of the distance from the source.

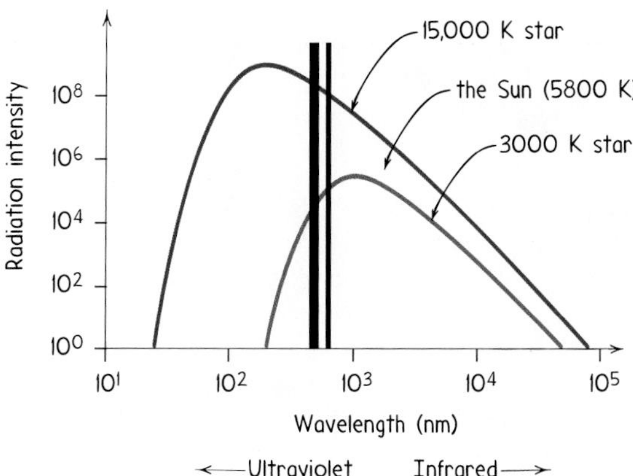

Figure 28.8 These idealized radiation curves for stars of different surface temperatures show two important facts: (1) Hotter stars emit radiation with higher average frequency than cooler stars, and (2) hotter stars emit more radiation per unit surface area at every frequency than cooler stars.

Stars have different colors simply because they are emitters of electromagnetic waves in the visible range and because our eyes sense different frequencies of visible radiation as different colors. Figure 28.8 shows the radiation curves, which are graphs of the frequency of emitted radiation versus temperature of the emitter, for three stars with different temperatures. The radiation curves show that the hotter a star is, the shorter the wavelength of its peak frequency and the bluer it looks. So the blue stars in the night sky have higher temperatures than the red ones. The Sun, for example, with its approximately 5800-K temperature, emits most strongly in the middle of the visible spectrum and so it would appear yellow from outer space. Betelgeuse, on the other hand, appears red because its cooler surface temperature (about 3400 K). Betelgeuse emits more red light than blue light.

Notice also from Figure 28.8 that the hotter a star is, the more radiant energy it emits. Thus we see that hot blue stars are more luminous than cooler red stars of the same size.

CHECK YOURSELF

The temperature of Sirius is about 9400 K. What color is this star—and why?

CHECK YOUR ANSWER

Sirius has a slightly blue color. It emits more blue light than red light owing to its high surface temperature.

28.3 The Hertzsprung–Russell Diagram

When you compare the luminosity of stars to their temperature, interesting patterns emerge. Early in the twentieth century, Danish astronomer Ejnar Hertzsprung and American astronomer Henry Norris Russell did just this. They produced a diagram known as the Hertzsprung–Russell diagram, or H–R diagram, which is of key importance in astronomy (Figure 28.9). The **H–R diagram** is a plot of the luminosity versus surface temperature of stars. Luminous stars are near the top of the diagram, and dim stars toward the bottom. Hot bluish stars are toward the left side of the diagram and cool reddish stars are toward the right side.

The H–R diagram shows several distinct regions of stars. Most stars are plotted on the band that stretches diagonally across the diagram. This band is called the **main sequence**. Stars on the main sequence, including our Sun, generate energy by fusing hydrogen to helium. As we would expect, the hottest main-sequence stars are the brightest and bluest stars and the coolest main-sequence stars are the most dim and red stars. Take a moment to locate the Sun on the H–R diagram. Can you see that the Sun is a roughly average main-sequence star in terms of its luminosity and temperature?

Toward the upper right of the diagram is a distinct group of stars—the **red giants**. These stars clearly do not follow the pattern of the hydrogen-burning main-sequence stars. Because these stars are red, we know they must have low surface temperatures. If they were main-sequence stars, the red giants would be dim. Yet, notice how high the red giants are on the luminosity scale—they are very bright. The fact that the red giants are both much cooler and much brighter than the Sun tells us that these stars must also be much larger than the Sun. (Hence the name "giant.") Above the red giants on the H–R diagram are a few rare stars, the supergiants. The supergiants are even larger and brighter than the red giants. As you will see in the next section, the giants and supergiants are stars nearing the end of their lives because the hydrogen fuel in their cores is running out.

Toward the lower left are some stars that are so dim they cannot be seen with the unaided eye. The surfaces of these stars can be hotter than the Sun,

INSIGHT Because the red giants and supergiants are so luminous, they are easy to see in the night sky even if they are not close to Earth. You can often identify them by their reddish color.

Figure 28.9 The H–R diagram shows a star's surface temperature on the horizontal axis and its luminosity on the vertical axis. Plotted here are some of the most visible stars in the sky, along with a few that are notable because they are close to Earth. The stars shown are not drawn to scale.

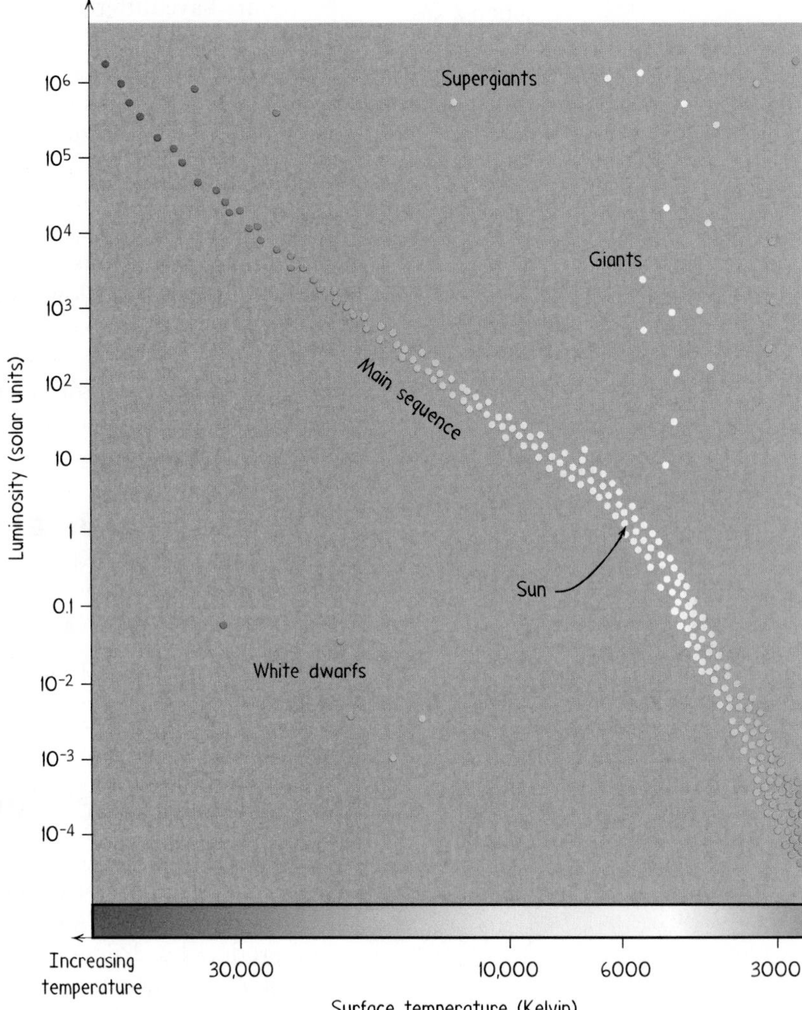

which makes them blue or white in color. Yet their luminosities are quite low—on the order of $0.1L_{sun}$ to $0.0001L_{Sun}$. To be so hot and radiate so little light, these stars must be very small—they are called the **white dwarfs**. White dwarfs are typically the size of Planet Earth or even smaller, yet they have mass comparable to the Sun. The density (or mass per volume) of a white dwarf is thus extremely high—higher than anything found on Earth. As you will learn in the next section, white dwarfs are dead stars, the remnants of stars that have exhausted their nuclear fuel.

INSIGHT The H–R diagram is to astrophysicists what the periodic table is to chemists—an extremely important tool. A star's location on the H–R diagram can reveal its age. The age of our galaxy can be estimated by looking at the locations of our oldest stars and their white-dwarf remnants.

CHECK YOURSELF

1. What characteristic do all main-sequence stars share?
2. Red giants have cool surface temperatures yet are highly luminous. Does this mean that the frequency of light emitted by a red giant does not depend on its surface temperature as described by Figure 28.8?

CHECK YOUR ANSWERS

1. All main-sequence stars generate energy by the nuclear fusion of hydrogen to helium.
2. No, radiation curves hold for a red giant star as for any other radiating body. Red giants do have a relatively low energy output per unit surface area; they are highly luminous only because they are very large.

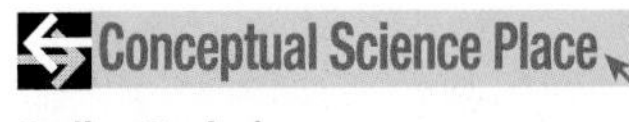

Stellar Evolution

Conceptual Science Place

Lives of Stars

28.4 The Life Cycles of Stars

In Chapter 27, we discussed the nebular theory, which explains how the Sun formed from an expansive, low-density cloud of gas and dust called a *nebula* (Figure 28.10). Other stars form in essentially the same way. That is, over time, a nebula flattens, heats, and spins more rapidly as it gravitationally contracts. The center of the nebula becomes dense enough to trap infrared radiation so that this energy is no longer radiated away. The hot central bulge of a nebula is called a *protostar*.

Mutual gravitation between the gaseous particles in a protostar results in an overall contraction of this huge ball of gas, and its density increases still further as matter is crunched together, with an accompanying rise in pressure and temperature. When the central temperature reaches about 10 million K, hydrogen nuclei begin fusing to form helium nuclei. This thermonuclear reaction, converting hydrogen to helium, releases an enormous amount of radiant and thermal energy, as discussed in Chapter 27. The ignition of nuclear fuel marks the change from protostar to star. Outward-moving radiant energy and the gas accompanying it exert an outward pressure called *thermal pressure* on the contracting matter. When nuclear fusion occurs fast enough, thermal pressure becomes strong enough to halt the gravitational contraction. At this point, outward thermal pressure balances inward gravitational pressure, and the star's size and mass stabilize.

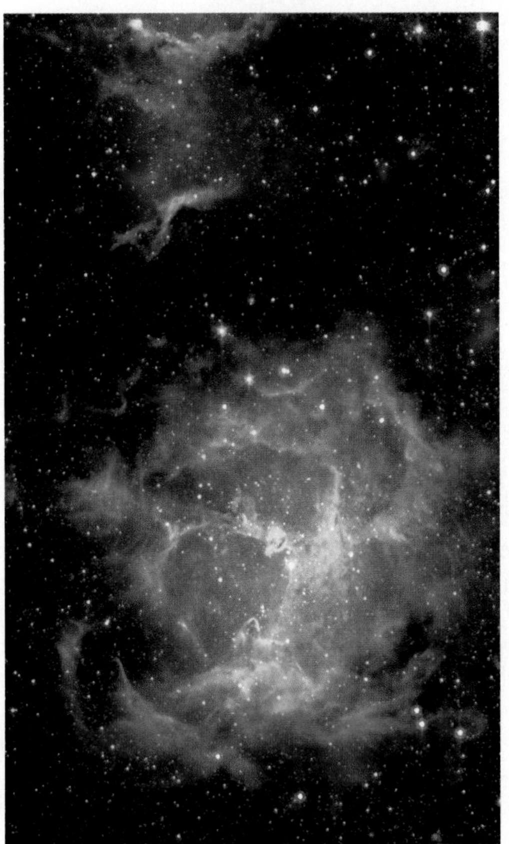

Figure 28.10 This image of the Trefid Nebula was obtained by the *Spitzer Space Telescope*. This nebula is located 5400 light-years from Earth in the constellation Sagittarius. Within each of the four red dust clouds, there are developing stars.

CHECK YOURSELF

What do the processes of thermonuclear fusion and gravitational contraction have to do with the physical size of a star?

CHECK YOUR ANSWER

The size of a star is the result of these two continually occurring processes. Energy from thermonuclear fusion tends to blow the star outward like an ongoing hydrogen bomb explosion, and gravitation tends to contract its matter in an ongoing implosion. The outward thermonuclear expansion and inward gravitational contraction produce an equilibrium that accounts for the star's size.

Though all stars are born in the same way from contracting nebulae, they do not all progress through their lives in the same way. *A star's mass is what determines the stages a star will go through from birth to death.* There are limits on the mass that a star can attain. A star with a mass less than 0.08 times the mass of the Sun ($0.08M_{Sun}$) would never reach the 10 million K threshold needed for sustained fusion of hydrogen. On the other hand, stars with masses above 100_{Sun} would undergo fusion at such a furious rate that gravity could not resist thermal pressure and the star would explode. So stars exist within the limits of about a tenth of the mass of the Sun and 100 times the solar mass.*

The majority of stars have masses a few times smaller or larger than the Sun. Such stars inhabit a central place on the main sequence of the H–R diagram. If you plot the life-cycle stages of average stars on a H–R diagram, they will trace a curve similar to the one for our Sun, which is shown in Figure 28.11. The Sun was born about 4.5 billion years ago at position 1, when the fusion of hydrogen ignited. The Sun will spend the most of its lifetime—some 10 billion years—on the main sequence, with thermal pressure keeping gravity at bay. Speaking more generally, a star's hydrogen-burning lifetime lasts for a period of a few million to 50 billion years, depending on its mass. More massive stars have *shorter* lives than less massive stars.

INSIGHT In addition to the multitudes of particles in interstellar space, a montage of electromagnetic waves engulfs the universe. There are also virtual particles, popping in and out of existence. Then there's the dark matter and "vacuum energy," about which we know nothing as yet. Outer space certainly is not empty.

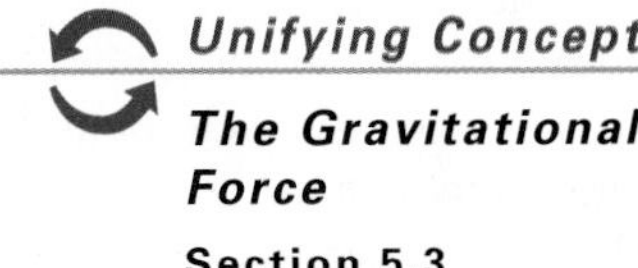

Unifying Concept

The Gravitational Force

Section 5.3

* One solar mass, $1M_{Sun,}$, is a unit of mass equivalent to the Sun: 2×10^{33} kilograms.

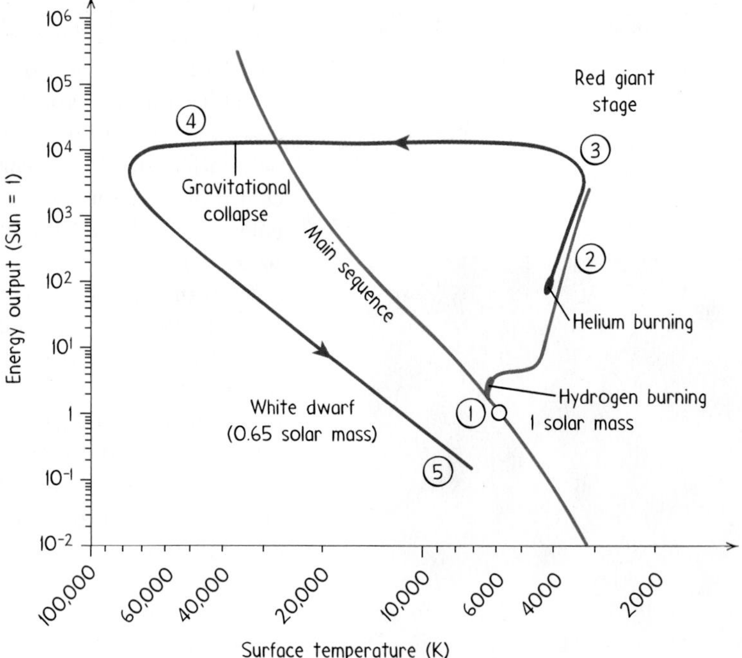

Figure 28.11 The stages of the Sun's life cycle are plotted on this H–R diagram.

INSIGHT A star's life cycle depends on its mass. The lowest-mass stars are brown dwarfs, dim but long-lived stars. Medium-mass stars progress from main-sequence stars to red giants or supergiants then to white dwarfs. Very massive stars have short lives and die in massive explosions called supernovae.

INSIGHT The net nuclear reaction for helium burning is

$$^4He + ^4He + ^4He \rightarrow ^{12}C$$

Helium undergoes nuclear fusion to create carbon late in the life of an average-mass star such as our Sun.

Why should this be so? High-mass stars are more luminous than low-mass stars, meaning that they burn their hydrogen fusion fuel at a faster rate. Massive stars *must* be more luminous than small-mass stars so that the outward pressure of their nuclear fusion can offset the greater gravitational force of their contraction. Massive stars start out with more hydrogen fuel than small-mass stars, but they consume their fuel so much faster that they die billions of years younger than smaller stars.

No star lasts forever. In the old age of an average-mass star like our Sun, the supply of hydrogen fuel is diminished so gravity overwhelms thermal pressure and the star pulls inward. As the burned-out hydrogen core contracts due to gravity, its temperature rises. At a certain point, the temperature becomes high enough in the core to launch *helium burning*—the fusion of helium to carbon. The star then has a structure consisting of concentric shells. Helium fuses to carbon at the star's center while hydrogen fuses to helium in a surrounding shell. Energy output soars, moving the star off the main sequence.

With such intensified nuclear fusion within a star, the outward force of thermal pressure wins out over the inward force of gravity. The star balloons to become a red giant (position 2). When our Sun reaches the red giant stage about 5 billion years from now, its swelling and increased energy output will escalate Earth's temperatures. Earth will be stripped of its atmosphere and the oceans boiled dry. Ouch!

As fusion continues, carbon will continue to accumulate in the Sun's core, but temperatures will never become hot enough to allow the carbon to undergo fusion. Instead, carbon "ash" accumulates inside the star and fusion gradually tapers off. Now gravity takes over, and the star contracts, which boosts its temperature. The Sun continues to emit vast amounts of energy through an ever-shrinking surface. Its color changes to blue and its position shifts to the left in the H–R diagram.

At a certain point, a strange force called *degeneracy pressure* comes into play, and the Sun can shrink no more. Degeneracy pressure is explained by quantum mechanics (Chapter 10). Though the details of degeneracy pressure are beyond the scope of this book, the essential idea is that electrons cannot be pushed ever closer together. As a star collapses, its electrons are pushed closer and closer together but at a certain point, degeneracy pressure limits further contraction. Degeneracy pressure thus counteracts gravity and stabilizes the size of a star in much the same way that thermal pressure does, though degeneracy pressure is permanent and independent of

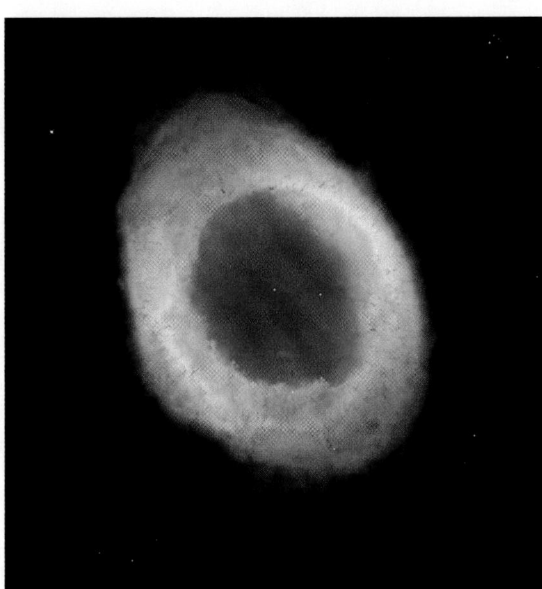

Figure 28.12 The planetary Ring Nebula in the constellation Lyra, which can be seen through a modest telescope.

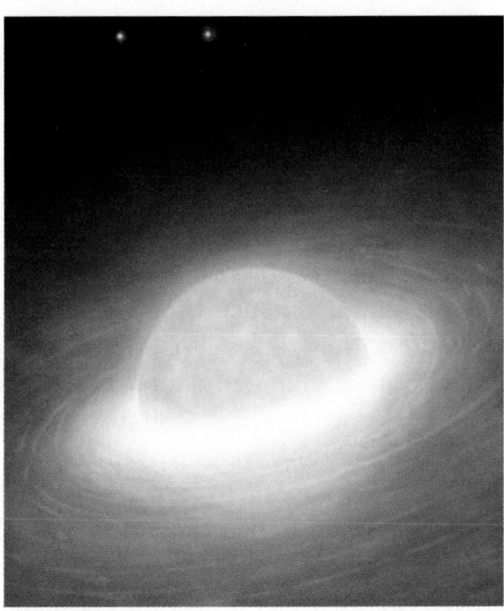

Figure 28.13 A white dwarf, shown here, is the final stage in the evolution of low- and medium-mass stars. After a star has used all its nuclear fuel, its outer layers escape into space, leaving the dense core behind as a white dwarf. The strong gravitational field of a white dwarf causes it to attract matter from surrounding space to form an accretion disk around the rotating star. The disk is heated by friction where it meets the star, causing it to glow brightly.

temperature. When the Sun gets so small that degeneracy pressure halts its contraction, its nuclear fuel will have been spent and it will no longer be producing energy.

When the Sun dies, it will do what other low- and medium-mass stars do. In a display that would be beautiful to watch (from a safe distance), solar winds and other processes will carry the Sun's outer layers into space. A huge shell of expanding gas will move, smoke-ring-like, away from the ash-filled core. Such an expanding shell is a *planetary nebula** (Figure 28.12). The nebula will disperse within a million years, leaving the Sun's cooling carbon core behind as a white dwarf. White dwarfs have the mass of a star but the volume of a planet, and are thus far more dense than anything on Earth. Because the nuclear fires of a white dwarf have burned out, it is not actually a star anymore. It's more accurate to call it a stellar remnant. In any case, a white dwarf cools for eons in space until it becomes too cold to radiate visible light (Figure 28.13).

There is another possible fate for a white dwarf, if it is part of a binary. A **binary star** is a double star—a system of two stars that revolve about a common center, just as Earth and the Moon revolve about each other. If a white dwarf is a binary and if its partner is close enough, the white dwarf may gravitationally pull hydrogen from its companion star. It then deposits this material on its own surface as a very dense hydrogen layer. Continued compacting increases the temperature of this layer, which ignites to embroil the white dwarf's surface in a thermonuclear blast that we see as a **nova.** A nova is an event, not a stellar object. After a while, a nova subsides until enough matter again accumulates to repeat the event. A given nova flares up at irregular intervals that may range from decades to hundreds of thousands of years.

While low- and medium-mass stars become white dwarfs, the fate of stars more than about $10M_{\text{Sun}}$ is quite different. When such a massive star contracts following its red giant or supergiant phase, more heat is generated than in the contraction of a small star. Such a star does not shrink to become a white dwarf. Instead, carbon nuclei in its core fuse and liberate energy while synthesizing heavier elements, such as neon and magnesium. Thermal pressure halts further gravitational contraction until all the carbon is fused. Then the core of the star contracts again to produce even greater temperatures and a new fusion series produces even heavier elements. The fusion cycles repeat until the element iron is formed.

The fusion of elements with larger atomic numbers than iron consumes energy rather than liberates energy. (The reason for this, as you may recall from Chapter 10, is that the average mass per nucleon is lower for iron than any other element.) With no possibility of energy coming from fusion in an iron core, the center of the star collapses without rekindling. The entire star begins its final collapse.

The collapse is catastrophic. When the core density is so great that all the nuclei are compressed against one another, the collapse momentarily comes to a halt. Then it explodes violently, hurling into space the elements previously manufactured over billions of years. The entire episode can last a few minutes. It is during this brief time that the heavy elements beyond iron are synthesized, as protons and neutrons mash into other nuclei to produce such elements as silver, gold, and uranium. Because the time available for synthesizing these heavy elements is so brief, they are far less abundant than iron and the lighter elements.

Such a stellar explosion is a **supernova,** one of nature's most spectacular events. A supernova flares up to millions of times its former brightness. In 1054 A.D., Chinese astronomers recorded their observation of a star so bright that it could be seen by day as well as by night. This was a supernova, its glowing plasma remnants now making up the spectacular Crab Nebula (Figure 28.14). A less spectacular but more recent supernova was witnessed in 1987. This gave astronomers an exciting firsthand look at one of these seldom-seen events. Supernovae are fiery furnaces that generate the elements essential to life, for all the elements beyond iron that make up our bodies originated in far-off, long-ago supernovae.

* Despite its name, a planetary nebula has nothing to do with planets. The name is derived from the fact that planetary nebula look like planets when observed through small, Earth-bound telescopes.

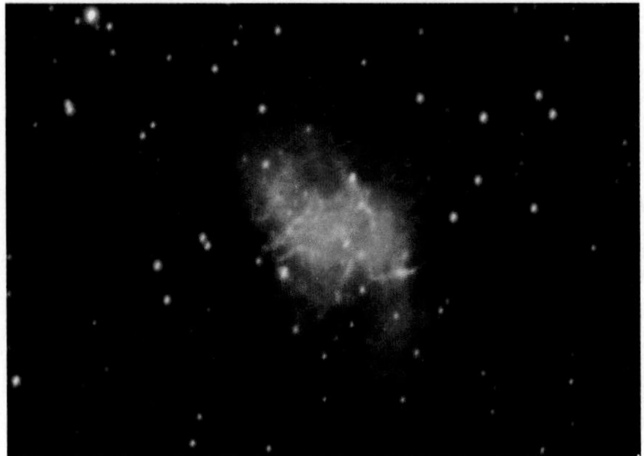

Figure 28.14 The Crab Nebula, the remnant of a supernova explosion that was seen from Earth in 1054 A.D.

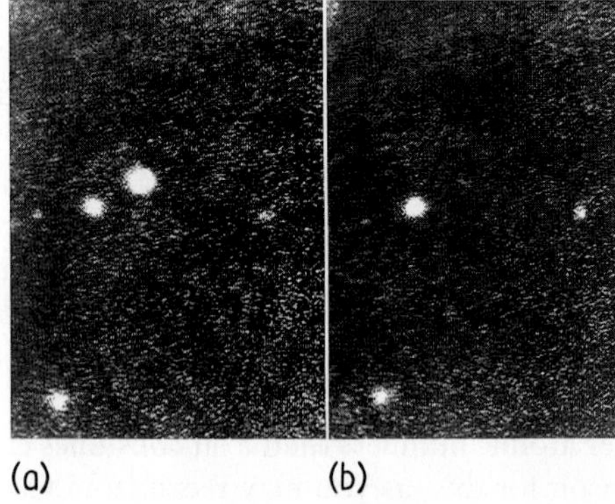

(a) (b)

Figure 28.15 The pulsar in the Crab Nebula rotates like a searchlight, beaming light and X rays toward Earth about 30 times per second, blinking on and off: (a) pulsar on, (b) pulsar off.

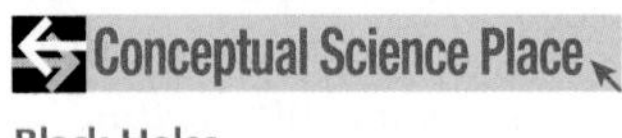

Black Holes

INSIGHT The loss of stellar material through stellar winds and pulsations greatly affect a star's evolution. For low-mass stars, the planetary nebula is what we see after a star's "final hiccup." For high-mass stars, the amount of mass lost before running out of nuclear fuel determines whether they become neutron stars or black holes.

The inner part of a supernova star implodes to form a core compressed to neutron density. Incredibly, protons and electrons compress together to form a core of neutrons just a few kilometers wide. This superdense, central remnant of a supernova survives as a *neutron star*. In accord with the law of conservation of angular momentum,* these tiny bodies, with densities hundreds of millions times greater than those of white dwarfs, spin at fantastic speeds. Neutron stars provide an explanation for the existence of pulsars. *Pulsars,* which are in fact neutron stars, are rapidly varying sources of low-frequency radio emissions. As a pulsar spins, the beam of radiation it emits sweeps across the sky. If the beam sweeps over Earth, we detect its pulses. Of the approximately 300 known pulsars, only a few have been found emitting X rays or visible light. One is in the center of the Crab Nebula (Figure 28.15). It has one of the highest rotational speeds of any pulsar studied, rotating more than 30 times per second. This is a relatively young pulsar, for it is theorized that X radiation and optical radiation are emitted only during a pulsar's early history.

Dying stars with coresmuch greater than the mass of our Sun collapse so violently that no physical forces are strong enough to halt continued contraction. The bigger they are, the harder they fall. The enormous gravitational field about the imploding concentration of mass makes explosion impossible. Collapse continues and the star disappears from the observable universe. What is left is a black hole.

28.5 Black Holes

A **black hole** is the remains of a supergiant star that has collapsed into itself. It is so dense and has such an intense gravitational field that light cannot escape from it. We can see why gravity is so great in the vicinity of a black hole by considering the change in the gravitational field at the surface of any star that collapses. In accord with Newton's law of gravity, any mass at the surface of a star, whether object or particle, has weight that depends both on its mass and on the mass of the star. But, more importantly, weight also depends on the distance between the object and the center of the star. So, if a star collapses, the distance to its center decreases. Weight increases, without a change in total mass. By how much? That depends on the amount of collapse. If a star collapses to half its size, then, in accord with the inverse-square law, the weight of an object at its surface quadruples (Figure 28.16). If a star collapses to a tenth its size, the weight at the surface is 100 times as much. Along with the increase in gravitational field, the escape velocity from the surface of the collapsing star also increases. If a star such as our Sun were to collapse to a radius of 3 kilometers, the escape velocity from its surface would exceed the speed of light, and nothing—not even light—could escape![†] The Sun would be invisible. It would be a black hole.

The Sun, in fact, has too little mass to experience such a collapse, but when some stars with core masses many times greater than the mass of the Sun reach the end of their nuclear resources, they undergo collapse; and, unless their rate of rotation is high enough, their collapse continues until the stars reach infinite densities. Gravitation near the surfaces of these shrunken stars is so enormous that light cannot escape from them. They have crushed themselves out of visible existence.

* Whereas linear momentum, as studied in Chapter 4, is inertia × velocity, angular momentum is rotational intertia × angular velocity. The law of the conservation of angular momentum states that angular momentum is conserved during internal processes. So, just as a spinning figure skater spins faster when her arms are drawn in, a rotating ball of gas spins faster when it contracts (a decreased rotational inertia is compensated for by an increased rotational velocity). And just as the spinning skater slows when her arms are extended, a star such as our Sun slows as its planets form. (See Appendix B.)

† Light, just like massive things, is affected by gravity. Just as we fail to see the curvature of a high-speed bullet when viewed along short segments, we most often fail to see the curvature by gravity of even higher-speed light. If you take a follow-up physics course and study relativity, you will understand how scientists have come to know that light does curve in a gravitational field.

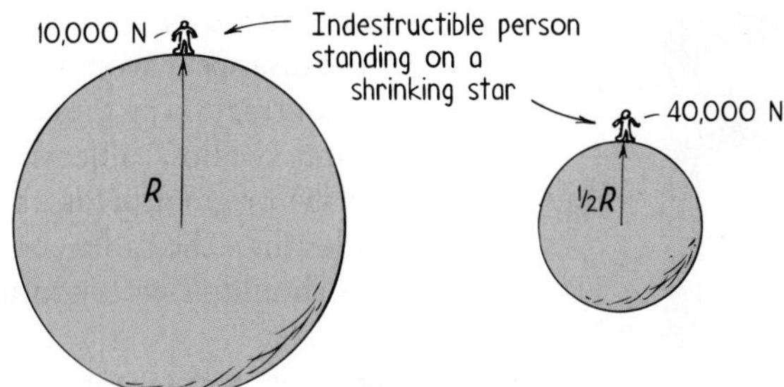

Figure 28.16 If a star collapses to half its radius with no mass change, gravitation at its surface increases by 4-fold (in accordance with the inverse-square law). If the star collapses to one-tenth its radius, gravitation at its surface increases 100-fold.

A black hole has the same amount of mass after its collapse as before its collapse, so the gravitational field in regions at and beyond the original star's radius is no different in either case. But closer distances near the vicinity of a black hole are nothing less than the collapse of space itself, with a surrounding warp into which anything that passes too close—light, dust, or a spaceship—is drawn (Figure 28.17). Astronauts in a powerful spaceship could enter the fringes of this

TECHNOLOGY

Telescopes

Galileo was the first to person to build a telescope and he was the first to direct it to the nighttime sky. Early telescopes, such as Galileo's, could only collect visible light. Still today, the telescopes familiar to backyard astronomers are light telescopes—they collect and focus electromagnetic radiation in the range of visible light. Such devices gather light from a distant object and reflect it from a mirror onto a lens that focuses the image for viewing. Modern light-gathering telescopes, such as the Keck Telescope in Hawaii, extend this idea. Instead of using a single mirror, the Keck Telescope uses more than a dozen mirrors. All other factors being equal, the larger the reflective surface of a telescope, the more light it can collect and the more useful it will be for viewing faint objects.

Radio telescopes, first built in the 1930s and now common throughout the world, collect radio waves rather than visible light. Typically, a radio telescope is a large parabolic antenna ("dish") or an array of them. Many heavenly bodies, such as pulsars and quasars, emit strongly in the radio-frequency range so radio telescopes are essential tools for studying them. Also, radio telescopes are sometimes used to search for signals transmitted by hypothetical intelligent extraterrestrial life (the SETI project) and for tracking spacecraft. The Very Large Array (VLA) in Soccorro, New Mexico is one of the famous radio telescopes in the United States, another is "The Big Ear" at Ohio State University. The largest single radio telescope in use today is the RATAN-600 in Russia, with a circular antenna 576-m in diameter.

Visible light and radio waves can be intercepted by ground-based telescopes, however Earth's atmosphere is quite opaque to other frequencies of electromagnetic radiation. The atmosphere absorbs the X rays, gamma rays, most infrared and ultraviolet radiation, and long-wavelength microwaves emitted by distant astronomical objects, so orbiting observatories are positioned far above Earth's atmosphere to capture it. There are a variety of orbiting observatories and they sample the entire range of electromagnetic radiation from the cosmos.

The most celebrated of these instruments is the *Hubble Space Telescope* (*HST*). Launched in 1990, the *HST* is a reflecting telescope with a 2.4-meter mirror that gathers data in the visible and ultraviolet portions of the electromagnetic spectrum. The *HST*, despite intermittent problems, has proven to be an instrument of supreme scientific value. Among its many accomplishments, the *Hubble* has detected the most distant supernovas and quasars and improved the accuracy of the Hubble constant.

This photo shows a row of several dish antennae which make up the *Very Large Array* (*VLA*) radio telescope near Socorro, New Mexico. At this writing, the *VLA* is the world's largest radio telescope array, consisting of 27 dish antennae, each 25 meters in diameter. The data from all the dishes can be combined to produce a single radio image, the 27 antennae effectively forming one giant 27-kilometer radio dish.

The *Hubble Telescope* is the most sensitive optical telescope yet constructed. It is seen here with its protective front door open and with antennae and solar panels outstretched.

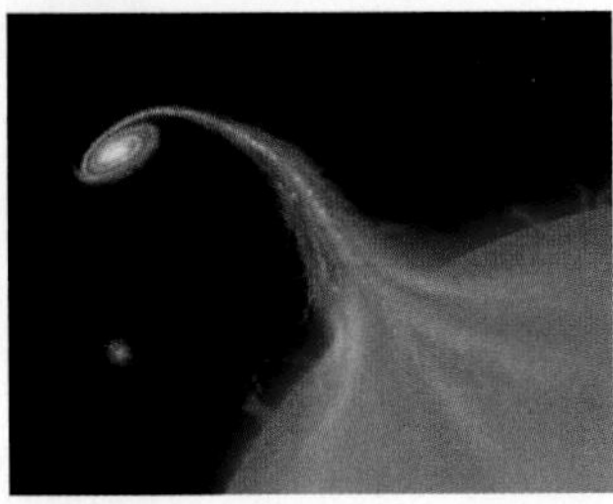

Figure 28.17 A rendering of a black hole stealing matter from a companion star.

INSIGHT An invisible concentration of material at the Milky Way's core is some 4 million times as heavy as the sun and squeezed into a region no larger than the distance between the sun and Earth. The nucleus of our galaxy is a black hole.

Figure 28.18 A wide-angle photograph of the Milky Way, from the constellation Cassiopeia on the left to the constellation Sagittarius on the right. The dark lanes and blotches are interstellar gas and dust obscuring the light of background stars.

Unifying Concept

Conservation of Momentum

Section 4.4

INSIGHT Planets in our solar system don't crash into the Sun because their tangential velocities are sufficient for orbit; likewise for stars in galaxies. Stars with sufficient tangential velocities orbit about the galactic center. But slower stars are pulled into, and gobbled up by, the galactic nucleus, which, if massive enough is usually a black hole.

warp and still escape. Below a certain distance, however, they could not, and they would disappear from the observable universe.

Contrary to stories about black holes, they're nonaggressive and don't reach out and swallow objects at a distance. Their gravitational fields are no stronger than the original fields about the stars before their collapse—except at distances smaller than the radius of the original star. Except when they are too close, black holes shouldn't worry future astronauts.

CHECK YOURSELF

What determines whether a star becomes a white dwarf, a neutron star, or a black hole?

CHECK YOUR ANSWER

The mass of a star is the principal factor that determines its fate. Stars that are about as massive as the Sun, and those that are less massive, evolve to become white dwarfs; stars with masses $10M_{Sun}$ or greater evolve to become neutron stars; the most massive stars ultimately become black holes.

28.6 Galaxies

A **galaxy** is a large assemblage of stars, interstellar gas, and dust. Galaxies are the breeding grounds of stars. Our own star, the Sun, is an ordinary star among some roughly 200 billion others in an ordinary galaxy known as the **Milky Way** (Figure 28.18). With unaided eyes, we see the Milky Way as a faint band of light that stretches across the sky. The early Greeks called it the "milky circle" and the Romans called it the "milky road" or "milky way." The latter name has stuck.

Most astronomers believe that, 10 to 15 billion years ago, galaxies formed from huge clouds of primordial gas pulled together by gravity, similar to our description of the solar system's formation in the previous chapter. Formation begins with gravitational attraction between distant particles. Then contraction is accompanied by an increased rotational rate (like a skater who spins faster when her arms are drawn in). In most cases, rotation causes a galaxy to flatten into a disk. This is what happened to the Milky Way. A most striking feature of our galaxy is the spiral arms that wind outward through the disk. These arms are swarms of hot, blue stars and clusters of young stars, amidst clouds of dust and gas.

The masses of galaxies range from about a millionth the mass of our galaxy to some 50 times more. Galaxies are calculated to have much more mass than has been detected. This undetected mass is known as *dark matter*. The nature of this dark matter is still in question. The millions of galaxies visible on long-exposure photographs can be separated into three main classes—elliptical, spiral, and irregular.

Elliptical galaxies are the most common galaxies in the universe. Most contain little gas and dust and cannot make new stars. An exception is the giant elliptical galaxy M87 (Figure 28.19). The largest ellipticals are about 5 times larger than our galaxy, and the smallest are 100 times smaller. Stars in elliptical galaxies are more crowded toward the center, and the outer parts of some larger ellipticals are occupied by hundreds of globular clusters that contain up to a million old stars.

Irregular galaxies are normally small and faint and are difficult to detect. They don't have spiral arms or dense centers. They contain large clouds of gas and dust mixed with both young and old stars. The irregular galaxy first described by the navigator on Magellan's voyage around the world in 1521 is our nearest neighboring galaxy—the Magellanic Clouds. This galaxy consists of two "clouds," called the Large Magellanic Cloud (LMC) and the Small

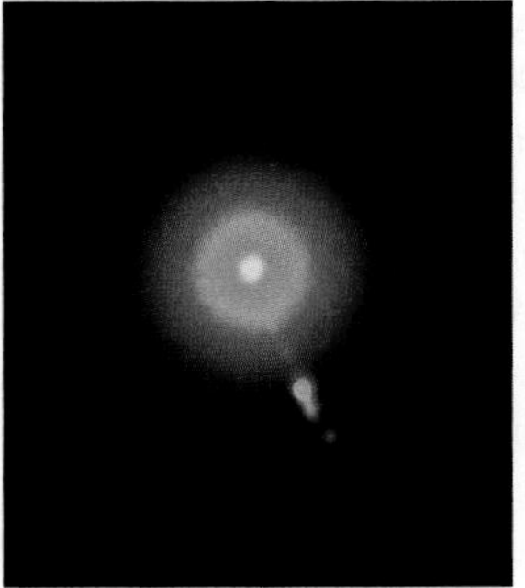

Figure 28.19 The giant elliptical galaxy M87, one of the most luminous galaxies in the sky, is located near the center of the Virgo cluster, some 50 million light-years from Earth. It is about 40 times more massive than our own galaxy, the Milky Way.

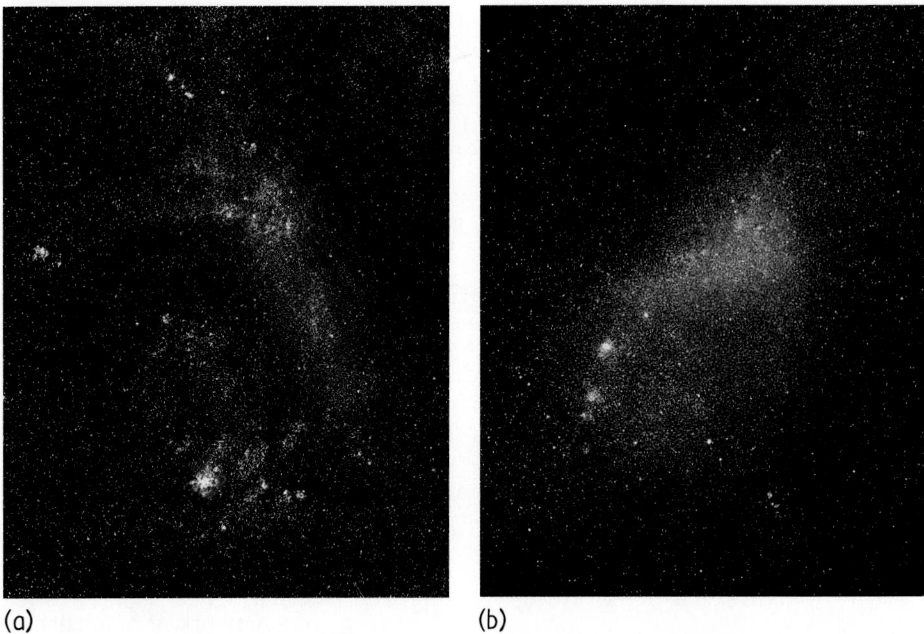

(a) (b)

Figure 28.20 (a) The Large Magellanic Cloud and (b) the neighboring Small Magellanic Cloud are a pair of irregular galaxies. The Magellanic Clouds are our closest galactic neighbors, about 150,000 light-years distant. They likely orbit the Milky Way.

Magellanic Cloud (SMC). The LMC is dotted with hot young stars having a combined mass of some 20 billion solar masses, and the SMC contains stars having a combined mass of about 2 billion solar masses (Figure 28.20). The combined mass is small for a galaxy. Irregular galaxies are probably as common as spiral galaxies.

Spiral galaxies are perhaps the most beautiful arrangements of stars in the heavens (Figures 28.21, 28.22, and 28.23). Spiral galaxies have a thin disk that extends outward from a dense, central nucleus. These galaxies are bright with the light of newly formed stars. The brightness of most spiral galaxies makes them easy to see at great distances. About two-thirds of all known galaxies are spirals, although they probably make up only about 15 to 20 percent of all galaxies. We do not see the greater number of fainter elliptical galaxies that are thought to exist.

Figure 28.21 Spiral galaxy M83 in the southern constellation Centaurus, about 12 million light-years from Earth.

Figure 28.22 An edge-on view of a spiral galaxy, much like our own Milky Way, which makes four rotations every billion years.

Figure 28.23 The great spiral nebula in Andromeda, a spiral galaxy about 2.3 million light-years from Earth.

Our Milky Way is a typical spiral galaxy. When we look at the Milky Way that crosses the night sky, we are looking through the galaxy's disk. Interstellar dust obscures our view of most of the visible light that lies along the plane of this disk. Most of our knowledge about our galaxy is via infrared and radio telescope observations, which reveal many details of the 25,000 light-year-diameter nucleus, but astronomers are still puzzled by the processes occurring there. The nucleus seems to be crowded with stars and hot dust, and at the very center is thought to be a massive black hole (of about a million solar masses) that generates energy by swallowing surrounding matter.

Galaxies collide with one another. Although the stars in a galaxy are normally so far apart that collisions of individual stars are unlikely, interstellar gases and dust do collide violently, so that matter is stripped from one galaxy and deposited in another. Low-speed collisions can result in the merger of two galaxies. There is evidence that the Milky Way may be presently consuming the Magellanic Clouds. At high velocities, colliding galaxies can distort each other and create tails and bridges. The collisions of spiral galaxies are thought to form huge elliptical galaxies.

Galaxies are vast yet they are not the largest things in the universe. Galaxies come in clusters. And galaxy clusters appear to be part of even larger clusters, the galaxy superclusters. It doesn't stop there; superclusters, in turn, seem to be part of a network of filaments surrounding empty voids.

INTEGRATED SCIENCE BIOLOGY

The Search for Extraterrestrial Life

Search for Extraterrestrial Life

The desire to contact extraterrestrial life—if it exists—has endured through the ages. For several decades, there has been an organized effort to do just that. The SETI (Search for Extraterrestrial Intelligence) program is an effort to locate evidence of past or present communicative civilizations in the universe, particularly within our own galaxy. SETI was established by NASA in 1984. Though Congress cut funding in the 1990s, the effort continues today through private support.

SETI methodology dates back to 1959 when physicists from Cornell suggested that radio waves could be used for interstellar communication. The idea is that since human civilization leaks radio messages into space, the inhabitants of other planets might be doing the same thing. Large radio telescopes on Earth could detect such radio leaks from civilizations in nearby star systems, as well as stronger signals dispatched from planets thousands of light-years away. The young astronomer Frank Drake (who later became president of the SETI Institute) was the first to conduct a radio search throughout the galaxy. Though SETI telescopes have been analyzing signals for decades, the antenna have not yet picked up any unique frequencies that could signal distant civilizations, but there is plenty of sky yet to be searched . . .

Is it logical to search for extraterrestrial civilizations? What are the odds that such civilizations could exist, and if they do, what are the chances that we could contact them? Frank Drake's answer to this question is the Drake equation. Since its introduction in 1961, this tool has framed the debate. The Drake equation estimates the odds of our ever making contact with another intelligent civilization by multiplying seven quantities related to the prevalence of life. The Drake equation is

$$N = R \times F_p \times N_e \times F_l \times F_i \times F_c \times L$$

where

R = the number of suitable stars—stars like the Sun—that form in our galaxy per year.

F_p = The fraction of these stars that have planets.

N_e = The number of Earth-like planets—meaning planets that have liquid water—within each planetary system.

F_l = The fraction of Earth-like planets where life develops.

F_i = The fraction of life sites where intelligent life develops.

F_c = The fraction of intelligent life sites where communication develops.

L = The "lifetime" (in years) of a communicative civilization.

N = The number of civilizations with which we could possibly communicate.

A problem with Drake's equation, however, is that we don't know the value of any of the terms! We can make rough estimates of some of them—astronomers believe that the product of $R \times F_p \times N_e$, which would give us the number of habitable planets in the Milky Way, maybe in the neighborhood of 100 billion. The technology for detecting extrasolar planets—planets that orbit stars beyond our solar system—is rapidly advancing, so we may be able to improve this estimate within the next couple decades.*

But there is no way presently to know how many of these 100 billion hypothetical planets have produced life. Life arose quickly on Earth (Chapter 17), which suggests that life develops fairly easily under the right conditions, but this is not definitive—F_l could be anywhere between 1 (if Earth is the only planet in the Milky Way with life) and 100 billion (if they all have life.) The fact that life flourished on Earth some 4 billion years before humans developed suggests that the development of intelligent life is much more tenuous than the rise of microbial life. But how much more difficult? As to F_c, the fraction of *intelligent* life sites, there could be intelligent beings who haven't yet invented radio telescopes. (We humans belonged to this category until the twentieth century). Or there could be others who have the means to communicate but don't—perhaps because they are not interested or fear that they might endanger themselves by advertising where they live. And L, the lifetime of an intelligent, communicative civilization . . . humans have been communicative for less than 100 years. Do you think we, or the average intelligent species, remains willing and able to communicate with other star systems for a century? A thousand years? A million? If organizations of intelligent beings—civilizations—last long enough, the Milky Way may now be brimming with communicative, advanced beings. One day, we just may make contact with them.

28.7 The Big Bang

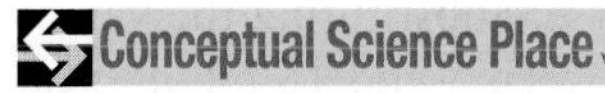

Hubble's Law; Fate of the Universe

The study of the universe in its totality, and especially its origin and structure, is called **cosmology**. The cornerstone of modern cosmology is the theory of the Big Bang, the idea that the physical universe began in a primordial explosion 13.7 billion years ago.† The Big Bang marks the beginning of both space and time for our universe. Since its violent beginning, cosmologists argue, the universe has developed in a steady progression of physical processes governed by physical laws.

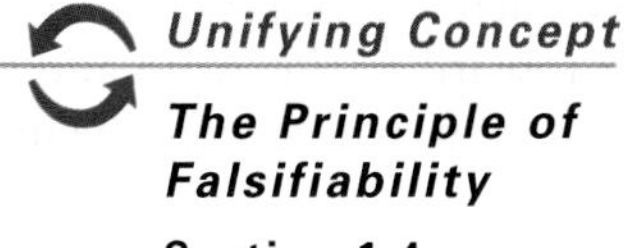

Evidence that the universe began as a primordial explosion—the Big Bang—is the *expansion of the universe*. It is thought that, at the start, all space existed as one point that contained all the matter that exists today. Since the Big Bang, the Universe has continued to expand outward from this point. The present expansion of the universe is evident in a Doppler red shift in the light that we

* Scientists and science fiction writers have long hypothesized the existence of extrasolar planets, but none were discovered until the 1990s. Today, over 100 extrasolar planets have been found and new ones are discovered each year. Extrasolar planets are too distant and too faint compared to the stars they orbit to be detected by telescopes. However, there are two powerful means of detecting them. The *transit method* detects a planet's shadow when it transits (moves in front of) its host star. Alternatively, planet hunters employ the Doppler effect. As a planet orbits a star, the star experiences a slight gravitational pull toward the planet, causing the star to wobble as the planet moves. The tell-tale wobble is revealed by alternating red-shifts and blue-shifts in a host star's spectral emissions.

† Immediately following the Big Bang, the universe likely began a period of exaggerated outward expansion, with matter flying outward faster than the current speed of light. This is the *inflation theory*, widely accepted in the astrophysics community.

Figure 28.27 The universe was created during the Big Bang (*upper right*). Galaxies of stars (upper center) formed from the simplest elements, hydrogen and helium. These fused into heavier elements, which were ejected into the interstellar medium when the star died. Earth (*upper left*) and the other planets incorporated these heavier elements along with hydrogen and helium. The earliest single-celled organisms on Earth (*left center*) evolved into more complex plants and animals—and eventually, humans. The development of an individual is depicted in the last three stages: human sperm and egg (*lower left*), an embryo, and an adult human—a self-conscious, curious, intelligent explorer of the universe!

universe may be incorrect, but it is most likely less wrong than the views of others before us. Our present view of the universe began with the findings of Copernicus, Galileo, and Newton. What they found was very much opposed by others at the time, mainly because established order was based on Aristotle's teachings. These "new" ideas were thought to diminish the role of humans in the universe, to undermine our importance. It was believed that people are important because we are higher than nature—apart from nature. We have expanded our vision since then by enormous effort, painstaking observation, and an ongoing desire to comprehend our surroundings. Seen from today's understanding of the universe, we find our importance not in being apart from nature but in being very much a part of it. We are the part of nature that is becoming more and more conscious of itself (Figure 28.27).

CHAPTER 28 Review

Summary of Terms

Apparent brightness The brightness of a star as it appears to our eyes.

Big Bang The primordial explosion of space at the beginning of time.

Binary star A pair of stars that orbit about a common center.

Black hole The remains of a giant star that has collapsed upon itself, so dense, and with a gravitational field so intense, that light itself cannot escape from it.

Celestial sphere An imaginary sphere surrounding Earth to which the stars are attached.

Cosmology The study of the origin and structure of the physical universe.

Elliptical galaxy A galaxy that is round or elliptical in outline. It has little gas and dust, no disk or spiral arms, and few hot, bright stars.

Galaxy A large assemblage of stars, interstellar gas, and dust.

H–R diagram (Hertzsprung–Russell diagram) A plot of intrinsic brightness versus surface temperature for stars. When so plotted, stars' positions take the form of a main sequence for average stars, with exotic stars above or below the main sequence.

Hubble's Law The farther away a galaxy is from Earth, the more rapidly it is moving away from us; $V = H \times d$.

Irregular galaxy A galaxy with a chaotic appearance and with large clouds of gas and dust, but without spiral arms.

Light-year The distance light travels in one year.

Luminosity The total amount of light energy that a star emits into space.

Main sequence The diagonal band of stars on an H–R diagram; such stars generate energy by fusing hydrogen to helium.

Milky Way The name of the galaxy to which we belong—our cosmic home.

Nova An event wherein a white dwarf suddenly brightens and appears as a "new" star.

Quasar A quasi-stellar object: a small but powerful source of energy believed to be the active core of a very distant galaxy.

Red giant Cool giant stars above main-sequence stars on the H–R diagram.

Spiral galaxy A disk-shaped galaxy with hot, bright stars and spiral arms. Our Milky Way is a spiral galaxy.

Supernova The explosion of a massive star caused by gravitational collapse with the emission of enormous quantities of matter.

White dwarf A dying star that has collapsed to the size of Earth and is slowly cooling off; located at the lower left on the H–R diagram.

Review Questions

28.1 Observing the Night Sky

1. What are constellations?
2. Why do the stars appear to turn on an imaginary north-south axis once every 24 hours?
3. Is a light-year a measurement of time or distance?

28.2 The Brightness and Color of Stars

4. What does the color of a star tell you about the star?
5. Distinguish between luminosity and apparent brightness for a star.

28.3 The Hertzsprung-Russell Diagram

6. What is an H-R diagram?
7. Where are the great majority of stars plotted on an H-R diagram?
8. Where does our Sun reside on an H-R diagram?

28.4 The Life Cycles of Stars

9. What process changes a protostar to a full-fledged star?
10. (a) What are the outward forces that act on a star? (b) What are the inward forces that act on a star?
11. Which have the greater lifetimes, high-mass or low-mass stars?
12. What is expected to happen to the Sun in its old age?
13. When a high-mass star collapes, it doesn't rekindle. Why?

28.5 Black Holes

14. What is the relationship between a supergiant star and a black hole?
15. Why is it that light cannot escape a black hole?

28.6 Galaxies

16. What type of galaxy is the Milky Way?
17. What is the evidence for galaxies not being the largest bodies in the universe?

28.7 The Big Bang

18. What significant event is marked by the Big Bang?
19. How does the Doppler shift support the idea of an expanding universe?
20. State Hubble's Law mathematically. Now in words, state what Hubble's Law tells us about the universe.
21. How are conditions that existed immediately after the Big Bang duplicated in today's laboratories?

28.8 Quasars

22. What does the enormous redshift of quasars tell us about their velocity?
23. What likely powers a quasar?

Multiple Choice Questions

Choose the BEST answer to the following.

1. Summer and winter constellations are different because
   (a) of the spin of the Earth about its polar axis.
   (b) the night sky faces in opposite directions in summer and winter.
   (c) of the tilt of Earth's polar axis.
   (d) the universe is symmetric and harmonious.
2. Polaris is always directly over
   (a) the north pole.
   (b) any location north of the equator.
   (c) the equator.
3. The star nearest Earth is
   (a) Alpha Centuri.
   (b) Polaris.
   (c) Mercury.
   (d) the Sun.
4. The longest-living stars are those of
   (a) low mass.
   (b) high mass.
   (c) intermediate mass.
5. We do not see stars in the daytime because
   (a) the Sun blocks them.
   (b) they simply don't exist in the daytime part of the sky.
   (c) skylight overwhelms starlight.
   (d) of the lack of contrast with moonlight.
   (e) the solar wind obscures them from view.
6. Metals are relatively more abundant in
   (a) old stars.
   (b) new stars.
   (c) neither in particular.
7. After our Sun burns its supply of hydrogen, it will become a
   (a) white dwarf.
   (b) black dwarf.
   (c) black hole.
   (d) red giant.
   (e) blue giant.
8. A black hole is
   (a) an empty region of space with a huge gravitational field.
   (b) a small region that has the mass of many galaxies.
   (c) the remains of a giant collapsed star.

INTEGRATED SCIENCE CONCEPTS

Physics—Radiation Curves of Stars

1. How is the radiation curve of a high-temperature star different from the radiation curve of a low-temperature star?
2. Why are stars different colors?

Biology—The Search for Extraterrestrial Life

3. What is SETI? Was it a mistake for Congress to cut its funding? Why or why not?
4. Suppose life is found on Mars. How would this affect the values we can substitute into Drake's equation?
5. The total number of stars in the universe is greater than the number of grains of sand on all the beaches on Earth. Given this, do you think it's possible that humans are alone in the universe?

Active Exploration

To observe the diurnal motion of the stars, go star watching tonight. Pick a star or constellation that lines up with a stationary landmark such as a tree or house. Then come back in an hour or so and you will see that the star has moved west of the landmark but remains in place relative to the other stars.

Exercises

1. Thomas Carlyle wrote, "Why did not somebody teach me the constellations and make me at home in the starry heavens, which are always overhead and which I don't half know to this day?" What besides the names of the constellations did Thomas Carlyle not know?

2. Why do we not see stars in the daytime?
3. Which figure in the chapter best shows that a constellation seen in the background of a solar eclipse is one that will be seen 6 months later in the night sky?
4. We see the constellations as distinct groups of stars. Discuss why they would look entirely different from some other location in the universe, far distant from Earth.
5. In what sense are we all made of stardust?
6. How is the gold in your mother's ring evidence of the existence of ancient stars that ran through their life cycles long before the solar system came into being?
7. Would you expect metals to be more abundant in old stars or in new stars? Defend your answer.
8. Why is there a lower limit on the mass of a star? Why is there an upper limit to the mass of a star?
9. What keeps a main-sequence star from collapsing?
10. How does the energy of a protostar differ from the energy that powers a star?
11. Why do nuclear fusion reactions not occur on the outer layers of stars?
12. Why are massive stars generally shorter-lived than low-mass stars?
13. With respect to stellar evolution, what is meant by the statement, "The bigger they are, the harder they fall"?
14. Why will the Sun not be able to fuse carbon nuclei in its core?
15. Some stars contain fewer heavy elements than our Sun contains. What does this indicate about the age of such stars relative to the age of our Sun?
16. Which has the highest surface temperature: a red star, a white star, or a blue star?
17. A black hole is no more massive than the star from which it collapsed. Why, then, is gravitation so intense near a black hole?
18. If the nucleus of our galaxy undergoes a gigantic explosion at this very moment, should we be concerned about its possible effects on us during our lifetime? Defend your answer.
19. Are there galaxies other than the Milky Way that can be seen with the unaided eye? Discuss.
20. Quasars are the most distinct objects we know of in the universe. Why do we therefore say their existence goes back to the earliest times in the universe?
21. What is meant by saying that the universe does not exist in space? Change two words around to make the statement agree with the standard model of the universe.
22. Why are the long-wavelength microwaves that permeate the universe considered to be evidence of the Big Bang?
23. In your own opinion, do you have to be at the center of your class to be special? Does Earth have to be at the center of the universe to be special?
24. How is the universe like a lump of rising raisin-bread dough?
25. Why does the Big Dipper change its position in the night sky over the course of the evening but Polaris remains relatively fixed in its position?
26. Explain why, in terms of the life cycle of the Sun, Earth cannot last forever.
27. Elements heavier than iron are created in stars—are they formed in the same way as elements lighter than iron? Explain.
28. Will the Sun become a supernova? A black hole? Defend your answer.
29. Why do the colors of stars vary? Why does their brightness vary?
30. What property of a star relates to the amount of energy it is producing?
31. Why did ancient cultures study the constellations?
32. Write a letter to Grandma telling her why we think the universe is 13.7 billion years old.

Problems

1. Suppose Star A is four times as luminous as Star B. If these stars are both 500 light-years away from Earth, how will their apparent brightness compare? How will the apparent brightness of these stars compare if Star A is twice as far away as Star B?
2. The brightest star in the sky, Sirius, is about 8 light-years from Earth. Show that if you could somehow travel there at jet-plane speed, 2000 kilometers per hour, the trip would take about 4.3 million years.

Conceptual Science Place

CHAPTER 28 **ONLINE RESOURCES**

• ***Tutorials*** Stellar Evolution, Black Holes, Hubble's Law, Fate of the Universe • ***Videos*** Lives of Stars, Search for Extraterrestrial Life, From the Big Bang to Galaxies • ***Quiz*** • ***Flashcards*** • ***Links***

Appendix A
On Measurement and Unit Conversion

Two major systems of measurement prevail in the world today: the United States Customary System (USCS, formerly called the British system of units), used in the United States of America, Burma/Myanmar, and Liberia, and the Système International (SI) (known also as the international system and as the metric system), used everywhere else. Each system has its own standards of length, mass, and time. The units of length, mass, and time are sometimes called the fundamental units because, once they are selected, other quantities can be measured in terms of them.

United States Customary System

Based on the British Imperial System, the USCS is familiar to everyone in the United States. It uses the foot as the unit of length, the pound as the unit of weight or force, and the second as the unit of time. The USCS is presently being replaced by the international system—rapidly in science and technology (all 1988 Department of Defense contracts) and some sports (track and swimming), but so slowly in other areas and in some specialties it seems the change may never come. For example, we will continue to buy seats on the 50-yard line. Camera film is in millimeters but computer disks are in inches.

For measuring time, there is no difference between the two systems except that in pure SI the only unit is the second (s, not sec) with prefixes; but in general, minute, hour, day, year, and so on, with two or more lettered abbreviations (h, not hr), are accepted in the USCS.

Système International

During the 1960 International Conference on Weights and Measures held in Paris, the SI units were defined and given status. Table A.1 shows SI units and their symbols. SI is based on the metric system, originated by French scientists after the French Revolution in 1791. The orderliness of this system makes it useful for scientific work, and it is used by scientists all over the world. The metric system branches into two systems of units. In one of these the unit of length is the meter, the unit of mass is the kilogram, and the unit of time is the second. This is called the *meter-kilogram-second* (mks) system and is preferred in physics. The other branch is the *centimeter-gram-second* (cgs) system, which, because of its smaller values, is favored in chemistry. The cgs and mks units are related to each other as follows: 100 centimeters equal 1 meter; 1000 grams equal 1 kilogram. Table A.2 shows how several units of length are related to each other.

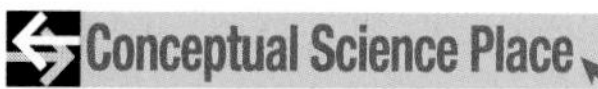

The Metric System

One major advantage of the metric system is that it uses the decimal system, where all units are related to smaller or larger units by dividing or multiplying by 10. The prefixes shown in Table A.3 are commonly used to show the relationship among units.

Table A.1 SI Units

Quantity	Unit	Symbol
Length	meter	m
Mass	kilogram	kg
Time	second	s
Force	newton	N
Energy	joule	J
Current	ampere	A
Temperature	kelvin	K

Table A.2 Conversions Between Different Units of Length

Unit of Length	Kilometer	Meter	Centimeter	Inch	Foot	Mile
1 kilometer =	1	1000	100,000	39,370	3280.84	0.62140
1 meter =	0.00100	1	100	39.370	3.28084	6.21×10^{24}
1 centimeter =	1.0×10^{25}	0.0100	1	0.39370	0.032808	6.21×10^{26}
1 inch =	2.54×10^{25}	0.02540	2.5400	1	0.08333	1.58 × 1025
1 foot =	3.05×10^{24}	0.30480	30.480	12	1	1.89 × 1024
1 mile =	1.60934	1609.34	160,934	63,360	5280	1

Table A.3 Some Prefixes

Prefix	Definition
micro-	One-millionth: a microsecond is one-millionth of a second
milli-	One-thousandth: a milligram is one-thousandth of a gram
centi-	One-hundredth: a centimeter is one-hundredth of a meter
kilo-	One thousand: a kilogram is 1000 grams
mega-	One million: a megahertz is 1 million hertz

Meter

The standard of length of the metric system originally was defined in terms of the distance from the North Pole to the equator. This distance was thought at the time to be close to 10,000 kilometers. One ten-millionth of this, the meter, was carefully determined and marked off by means of scratches on a bar of platinum-iridium alloy. This bar is kept at the International Bureau of Weights and Measures in France. The standard meter in France has since been calibrated in terms of the wavelength of light—it is 1,650,763.73 times the wavelength of orange light emitted by the atoms of the gas krypton-86. The meter is now defined as being the length of the path traveled by light in a vacuum during a time interval of 1/299,792,458 of a second.

Kilogram

The standard unit of mass, the kilogram, is a block of platinum, also preserved at the International Bureau of Weights and Measures located in France (Figure A.1). The kilogram equals 1000 grams. A gram is the mass of 1 cubic centimeter (cc) of water at a temperature of 4°C. (The standard pound is defined in terms of the standard kilogram; the mass of an object that weighs 1 pound is equal to 0.4536 kilogram.)

Second

The official unit of time for both the USCS and the SI is the second. Until 1956, it was defined in terms of the mean solar day, which was divided into 24 hours. Each hour was divided into 60 minutes and each minute into 60 seconds. Thus, there were 86,400 seconds per day, and the second was defined as 1/86,400 of the mean solar day. This proved unsatisfactory because the rate of rotation of Earth is gradually becoming slower. In 1956, the mean solar day of the year 1900 was chosen as the standard on which to base the second. In 1964, the second was officially defined as the time taken by a cesium-133 atom to make 9,192,631,770 vibrations.

Figure A.1 The standard kilogram.

Newton

One newton is the force required to accelerate 1 kilogram at 1 meter per second per second. This unit is named after Sir Isaac Newton.

Joule

One joule is equal to the amount of work done by a force of 1 newton acting over a distance of 1 meter. In 1948, the joule was adopted as the unit of energy by the International Conference on Weights and Measures. Therefore, the specific heat of water at 15°C is now given as 4185.5 joules per kilogram Celsius degree. This figure is always associated with the mechanical equivalent of heat—4.1855 joules per calorie.

Ampere

The ampere is defined as the intensity of the constant electric current that, when maintained in two parallel conductors of infinite length and negligible cross section and placed 1 meter apart in a vacuum, would produce between them a force equal to 2×10^{27} newton per meter length. In our treatment of electric current in this text, we have used the not-so-official but easier-to-comprehend definition of the ampere as being the rate of flow of 1 coulomb of charge per second, where 1 coulomb is the charge of 6.25×10^{18} electrons.

Kelvin

The fundamental unit of temperature is named after the scientist William Thomson, Lord Kelvin. The kelvin is defined to be 1/273.15 the thermodynamic temperature of the triple point of water (the fixed point at which ice, liquid water, and water vapor coexist in equilibrium). This definition was adopted in 1968 when it was decided to change the name *degree Kelvin* (°K) to *kelvin* (K). The temperature of melting ice at atmospheric pressure is 273.15 K. The temperature at which the vapor pressure of pure water is equal to standard atmospheric pressure is 373.15 K (the temperature of boiling water at standard atmospheric pressure).

Area

The unit of area is a square that has a standard unit of length as a side. In the USCS, it is a square with sides that are each 1 foot in length, called 1 square foot and written 1 ft^2. In the international system, it is a square with sides that are 1 meter in length, which makes a unit of area of 1 m^2. In the cgs system it is 1 cm^2. The area of a given surface is specified by the number of square feet, square meters, or square centimeters that would fit into it. The area of a rectangle equals the base times the height. The area of a

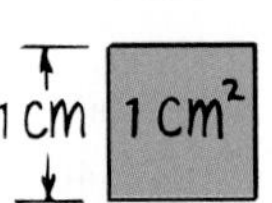

Unit square.

circle is equal to πr^2, where $\pi \approx 3.14$ and r is the radius of the circle. Formulas for the surface areas of other objects can be found in geometry textbooks.

Volume

The volume of an object refers to the space it occupies. The unit of volume is the space taken up by a cube that has a standard unit of length for its edge. In the USCS, one unit of volume is the space occupied by a cube 1 foot on an edge and is called 1 cubic foot, written 1 ft^3. In the metric system it is the space occupied by a cube with sides of 1 meter (SI) or 1 centimeter (cgs). It is written 1 m^3 or 1 cm^3 (or cc). The volume of a given space is specified by the number of cubic feet, cubic meters, or cubic centimeters that will fill it.

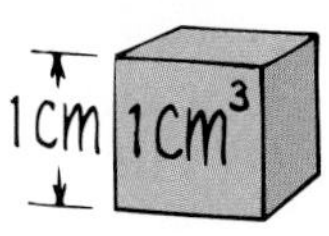

Unit volume.

In the USCS, volumes can also be measured in quarts, gallons, and cubic inches as well as in cubic feet. There are 1728 (12 × 12 × 12) cubic inches in 1 ft^3. A U.S. gallon is a volume of 231 in^3. Four quarts equal 1 gallon. In the SI volumes are also measured in liters. A liter is equal to 1000 cm^3.

Unit Conversion

Often in science, and especially in a laboratory setting, it is necessary to convert from one unit to another. To do so, you need only multiply the given quantity by the appropriate *conversion factor*.

All conversion factors can be written as ratios in which the numerator and denominator represent the equivalent quantity expressed in different units. Because any quantity divided by itself is equal to 1, all conversion factors are equal to 1. For example, the following two conversion factors are both derived from the relationship 100 centimeters = 1 meter:

$$\frac{100 \text{ centimeters}}{1 \text{ meter}} = 1 \qquad \frac{1 \text{ meter}}{100 \text{ centimeters}} = 1$$

Because all conversion factors are equal to 1, multiplying a quantity by a conversion factor does not change the value of the quantity. What does change are the units. Suppose you measured an item to be 60 centimeters in length. You can convert this measurement to meters by multiplying it by the conversion factor that allows you to cancel centimeters.

CHECK YOURSELF

Convert 60 centimeters to meters.

CHECK YOUR ANSWER

(60 centimeters)(1 meter)/(100 centimeters) = 0.6 meter

↓ quantity in centimeters ↓ conversion factor ↓ quantity in meters

To derive a conversion factor, consult a table that presents unit equalities, such as Table A.2. Then multiply the given quantity by the conversion factor, and voilà, the units are converted. Always be careful to write down your units. They are your ultimate guide, telling you what numbers go where and whether you are setting up the equation properly.

CHECK YOURSELF

Multiply each physical quantity by the appropriate conversion factor to find its numerical value in the new unit indicated. You will need paper, pencil, a calculator, and a table of unit equalities.

a. 7320 grams to kilograms
b. 235 kilograms to pounds
c. 2.61 miles to kilometers
d. 100 calories to kilocalories

CHECK YOUR ANSWERS

a. 7.32 kg
b. 518 lb
c. 4.20 km
d. 0.1 kcal

Conceptual Science Place

APPENDIX A ONLINE RESOURCES

Tutorials The Metric System

Appendix B
Linear and Rotational Motion

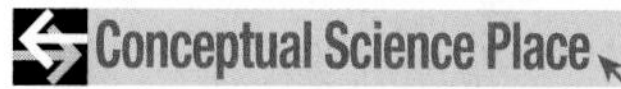

Rotational Motion

Advanced Concepts of Motion

To take the concepts of motion developed in the physics part of this text to the next level of complexity, we must be mindful of certain subtleties. In some instances, the idea of the *reference frame* is important.

When we describe the motion of something, we say how it moves relative to something else (Chapter 2). That "something" we measure motion with respect to is a *reference frame.* A reference frame can be defined mathematically by an origin and axes; in everyday terms, a reference frame is like a "background" we measure motion against. We are free to choose this frame's location and to have it moving relative to another frame. When our frame of motion has zero acceleration, it is called an *inertial frame.* In an inertial frame, force causes an object to accelerate in accord with Newton's Second Law. But when our frame of reference is accelerating, we observe fictitious forces and motions. Observations from a carousel, for example, are different when it is rotating and when it is at rest. Our description of motion and force depends on our "point of view."

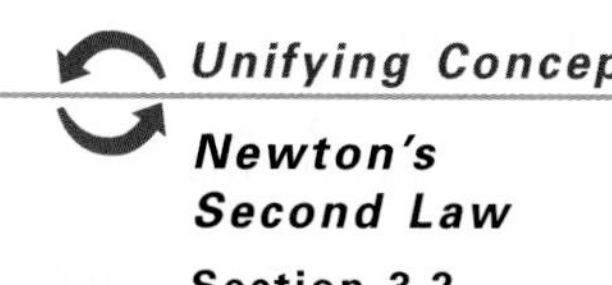

We distinguish between speed and velocity (Chapter 2). Speed is how fast something moves, or the time rate of change of position (excluding direction): a scalar quantity. Velocity includes direction of motion: a vector quantity whose magnitude is speed. Objects moving at constant velocity move the same distance in the same time in the same direction.

But here's the subtlety: There is a distinction between speed and velocity, which has to do with the difference between distance and net distance, or *displacement.* Speed is distance per duration while velocity is displacement per duration. Displacement differs from distance. For example, a commuter who travels 10 kilometers to work and back travels 20 kilometers, but has "gone" nowhere. The distance traveled is 20 kilometers and the displacement is zero. Although the instantaneous speed and instantaneous velocity have the same value at the same instant, the average speed and average velocity can be very different. The average speed of this commuter's round-trip is 20 kilometers divided by the total commute time—a value greater than zero. But the average velocity is zero. In science, displacement is often more important than distance. (To avoid information overload, we have not treated this distinction in the text.) As you learned in Chapter 2, acceleration is the rate at which velocity changes. This can be a change in speed only, a change in direction only, or both. Negative acceleration is often called *deceleration.*

Finally, in Newtonian space and time, space has three dimensions—length, width, and height—each with two directions. We can go, stop, and return in any of them. Time has one dimension, with two directions—past and future. We cannot stop or return, only go. In Einsteinian space-time, these four dimensions merge (but this *very* advanced topic awaits you in a follow-up course.)

Computing Velocity and Distance Traveled on an Inclined Plane

A staple of any physics course is the study of motion on an inclined plane. We develop this topic here to help you sharpen your analytical concepts of motion. Recall from Chapter 2 Galileo's experiments with inclined planes. We considered a plane tilted such that the speed of a rolling ball increases at the rate of 2 meters per second each second—an acceleration of 2 m/s^2. So at the instant it starts moving its velocity is zero, and 1 second later it is rolling at 2 m/s, at the end of the next second 4 m/s, the end of the next second 6 m/s, and so on. The velocity of the ball at any instant is simply velocity = acceleration × time. Or, in shorthand notation $v = at$. (It is customary to omit the multiplication sign, ×, when expressing relationships in mathematical form. When two symbols are written together, such as the *at* in this case, it is understood that they are multiplied.)

How fast the ball rolls is one thing; how far it rolls is another. To understand the relationship between acceleration and distance traveled, we must first investigate the relationship between *instantaneous* velocity (velocity at a particular point in time) and *average* velocity (which is computed over some extended time interval.) If the ball shown in Figure B.1 starts

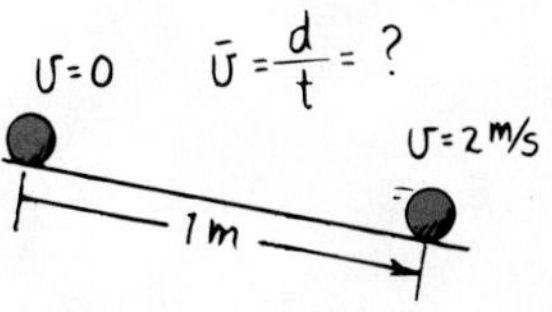

Figure B.1 The ball rolls 1 meter down the incline in 1 s and reaches a speed of 2 m/s. Its average speed, however, is 1 m/s. Do you see why?

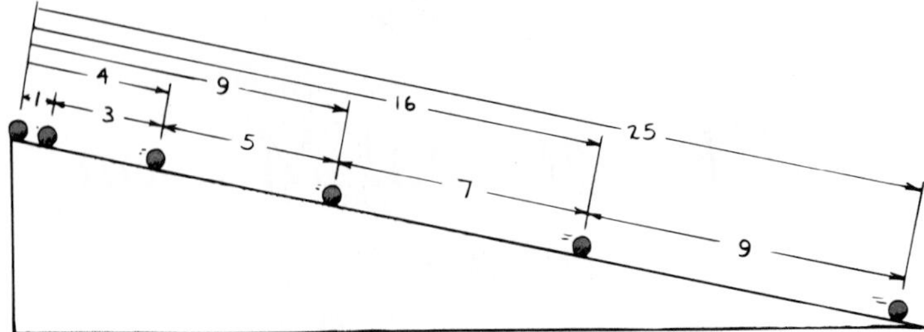

Figure B.2 If the ball covers 1 m during the first second, then it will cover the odd-numbered sequence of 3, 5, 7, 9 m, in each successive second. Note that distance increases as the square of time.

from rest, it will roll a distance of 1 meter in the first second. What will be its average speed? The answer is 1 m/s (it covered 1 meter in the interval of 1 second). But we have seen that the instantaneous velocity at the end of the first second is 2 m/s. Since the acceleration is uniform, the average in any time interval is found the same way we usually find the average of any two numbers: add them and divide by 2. (Be careful not to do this when acceleration is not uniform!) So if we add the initial speed (zero in this case) and the final speed of 2 m/s and then divide by 2, we get 1 m/s for the average velocity.

In each succeeding second we see the ball roll a longer distance down the same slope in Figure B.2. Note the distance covered in the second time interval is 3 meters. This is because the average speed of the ball in this interval is 3 m/s. In the next 1-second interval the average speed is 5 m/s, so the distance covered is 5 meters. It is interesting to see that successive increments of distance increase as a sequence of odd numbers. Nature clearly follows mathematical rules!

Investigate Figure B.2 carefully and note the total distance covered as the ball accelerates down the plane. The distances go from zero to 1 meter in 1 second, zero to 4 meters in 2 seconds, zero to 9 meters in 3 seconds, zero to 16 meters in 4 seconds, and so on in succeeding seconds. The sequence for total distances covered is of the squares of the time. We'll investigate the relationship between distance traveled and the square of the time for constant acceleration more closely in the case of free fall.

Computing Distance When Acceleration Is Constant

How far will an object released from rest fall in a given time? To answer this question, let us consider the case in which it falls freely for 3 seconds, starting at rest. Neglecting air resistance, the object will have a constant acceleration of about 10 meters per second each second (actually more like 9.8 m/s^2, but we want to make the numbers easier to follow).

$$\begin{aligned}\text{velocity at the } \textit{beginning} &= 0 \text{ m/s}\\ \text{velocity at the } \textit{end} \text{ of 3 seconds} &= (10 \times 3) \text{ m/s}\\ \text{average velocity} &= \frac{1}{2} \text{ the sum of these two speeds}\\ &= \frac{1}{2} \times (0 + 10 \times 3) \text{ m/s}\\ &= \frac{1}{2} \times 10 \times 3 = 15 \text{ m/s}\\ \text{distance traveled} &= \text{average velocity} \times \text{time}\\ &= \left(\frac{1}{2} \times 10 \times 3\right) \times 3\\ &= \frac{1}{2} \times 10 \times 32 = 45 \text{ m}\end{aligned}$$

We can see from the meanings of these numbers that

$$\text{distance traveled} = \frac{1}{2} \times \text{acceleration} \times \text{square of time}$$

This equation is true for an object falling not only for 3 seconds but for any length of time, as long as the acceleration is constant. If we let d stand for the distance traveled, a for the acceleration, and t for the time, the rule may be written, in shorthand notation,

$$d = \frac{1}{2} at^2$$

This relationship was first deduced by Galileo. He reasoned that if an object falls for, say, twice the time, it will fall with twice the average speed. Since it falls for twice the time at twice the average speed, it will fall four times as far. Similarly, if an object falls for three times the time, it will have an average speed three times as great and will fall nine times as far. Galileo reasoned that the total distance fallen should be proportional to the square of the time.

In the case of objects in free fall, it is customary to use the letter g to represent the acceleration instead of the letter a (g because acceleration is due to gravity). Although the value of g varies slightly in different parts of the world, it is approximately equal to 9.8 m/s^2 (32 ft/s^2). If we use g for the acceleration of a freely falling object (negligible air resistance), the equations for falling objects starting from a rest position become

$$v = gt$$

$$d = \frac{1}{2} gt^2$$

When Chelcie Liu releases both balls simultaneously, he asks, "Which will reach the end of the equal-length tracks first?" (*Hint*: On which track is the average speed of the ball greater?)

Much of the difficulty in learning physics, like learning any discipline, has to do with learning the language—the many terms and definitions. Speed is somewhat different from velocity, and acceleration is vastly different from speed or velocity.

CHECK YOURSELF

1. An auto starting from rest has a constant acceleration of 4 m/s^2. How far will it go in 5 s?
2. How far will an object released from rest fall in 1 s? In this case the acceleration is $g = 9.8$ m/s^2.
3. If it takes 4 s for an object to freely fall to the water when released from the Golden Gate Bridge, how high is the bridge?

CHECK YOUR ANSWERS

1. distance $= \frac{1}{2} \times 4 \times 5^2 = 50$ m
2. distance $= \frac{1}{2} \times 9.8 \times 1^2 = 4.9$ m
3. distance $= \frac{1}{2} \times 9.8 \times 4^2 = 78.4$ m

Notice that the units of measurement when multiplied give the proper unit of meters for distance:

$$d = \frac{1}{2} \times 9.8 \times 16 = 78.4 \text{ m}$$

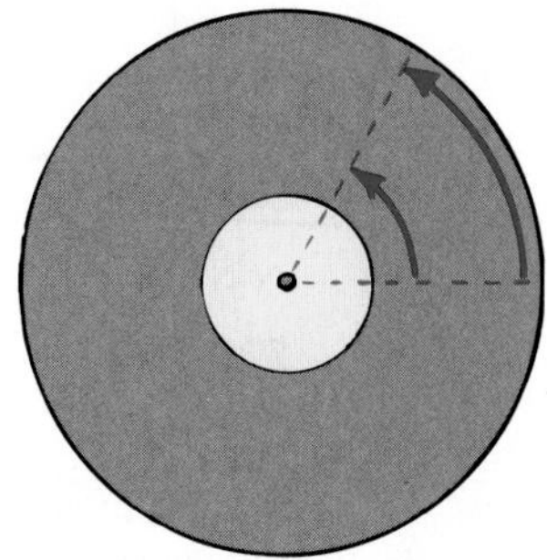

Figure B.3 Interactive Figure

When a phonograph record turns, a ladybug farther from the center travels a longer path in the same time and has a greater tangential speed.

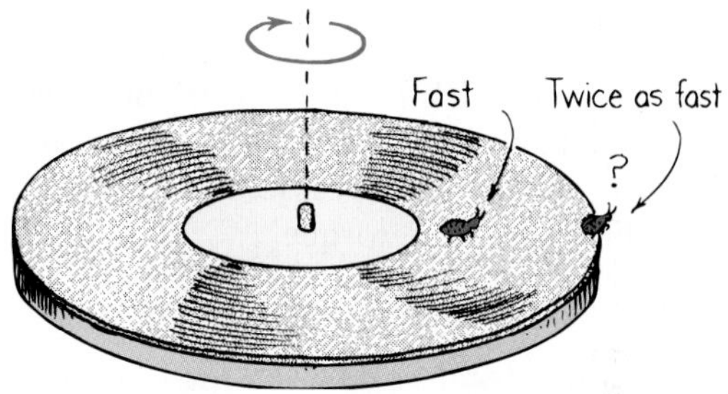

Figure B.4 The entire disk rotates at the same rotational speed, but ladybugs at different distances from the center travel at different tangential speeds. A ladybug twice as far from the center moves twice as fast.

Circular Motion

Linear speed is what we have been calling simply *speed*—the distance traveled in meters or kilometers per unit of time. A point on the perimeter of a merry-go-round or turntable moves a greater distance in one complete rotation than a point nearer the center. Moving a greater distance in the same time means a greater speed. The speed of something moving along a circular path is **tangential speed,** because the direction of motion is tangent to the circle.

Rotational speed (sometimes called angular speed) refers to the number of rotations or revolutions per unit of time. All parts of the rigid merry-go-round turn about the axis of rotation *in the same amount of time.* All parts share the same rate of rotation, or *number of rotations or revolutions per unit of time.* It is common to express rotational rates in revolutions per minute (rpm).* Phonograph records that were common some years ago rotate at 33 1/3 rpm. A ladybug sitting anywhere on the surface of the record revolves at 33 1/3 rpm (Figure B.3).

Tangential speed is *directly proportional* to rotational speed (at a fixed radial distance). Unlike rotational speed, tangential speed depends on the distance from the axis (Figure B.4). Something at the center of a rotating platform has no tangential speed at all and merely rotates. But, approaching the edge of the platform, tangential speed increases. Tangential speed is directly proportional to the distance from the axis (for a given rotational speed). Twice as far from the rotational axis, the speed is twice as great. Three times as far from the rotational axis, there is three times as much tangential speed. When a line of people locked arm in arm at the skating rink makes a turn, the motion of "tail-end Charlie" is evidence of this greater speed. So tangential speed is directly proportional both to rotational speed and to radial distance.†

CHECK YOURSELF

On a rotating platform similar to the disk shown in Figure B.4, if you sit halfway between the rotating axis and the outer edge and have a rotational speed of 20 rpm and a tangential speed of 2 m/s, what will be the rotational and tangential speeds of your friend who sits at the outer edge?

CHECK YOUR ANSWER

Since the rotating platform is rigid, all parts have the same rotational speed, so your friend also rotates at 20 rpm. Tangential speed is a different story; since she is twice as far from the axis of rotation, she moves twice as fast—4 m/s.

Torque

Whereas force causes changes in speed, *torque* causes changes in rotation. To understand torque (rhymes with *dork*), hold the

* Physics types usually describe rotational speed in terms of the number of "radians" turned in a unit of time, for which they use the symbol ω (the Greek letter omega). There's a little more than 6 radians in a full rotation (2π radians, to be exact).

† When customary units are used for tangential speed v, rotational speed ω, and radial distance r, the direct proportion of v to both r and ω becomes the exact equation $v = r\omega$. So the tangential speed will be directly proportional to r when all parts of a system simultaneously have the same ω, as for a wheel, disk, or rigid wand. (The direct proportionality of v to r is not valid for the planets because planets don't all have the same ω.)

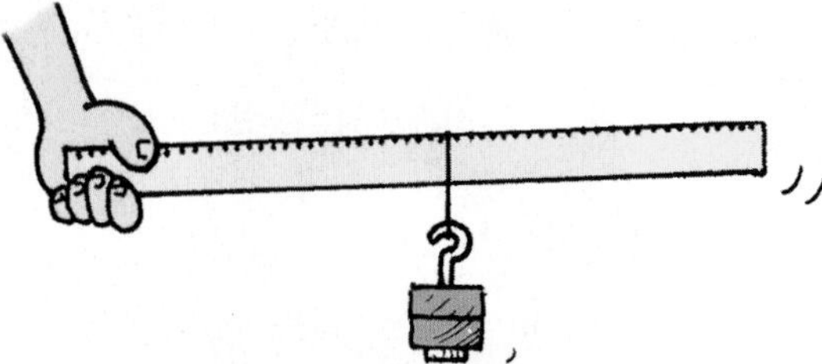

Figure B.5 If you move the weight away from your hand, you will feel the difference between force and torque.

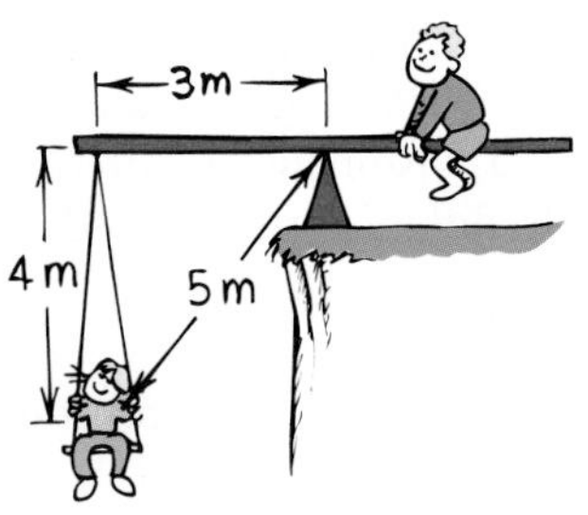

Figure B.7 The lever arm is still 3 m.

end of a meter stick horizontally with your hand (Figure B.5). If you dangle a weight from the meter stick near your hand, you can feel the meter stick twist. Now if you slide the weight farther from your hand, the twist you feel is greater, although the weight is the same. The force acting on your hand is the same. What's different is the torque.

$$\text{torque} = \text{lever arm} \times \text{force}$$

Lever arm is the distance between the point of application of the force and the axis of rotation—the axis about which the body turns around. The lever arm is the shortest distance between the applied force and the rotational axis. Torques are intuitively familiar to youngsters playing on a seesaw. Kids can balance a seesaw even when their weights are unequal. Weight alone doesn't produce rotation. Torque does, and children soon learn that the distance they sit from the pivot point is every bit as important as weight (Figure B.6). When the torques are equal, making the net torque zero, no rotation is produced.

Recall the equilibrium rule in Chapter 2—that the sum of the forces acting on a body or any system must equal zero for mechanical equilibrium. That is, $\sum F = 0$. We now see an additional condition. The net torque on a body or on a system must also be zero for mechanical equilibrium. Anything in mechanical equilibrium doesn't accelerate—neither linearly nor rotationally.

Suppose that the seesaw is arranged so that the half-as-heavy girl is suspended from a 4-meter rope hanging from her end of the seesaw (Figure B.7). She is now 5 meters from the fulcrum, and the seesaw is still balanced. We see that the lever-arm distance is 3 meters, not 5 meters. The lever arm about any axis of rotation is the perpendicular distance from the axis to the line along which the force acts. This will always be the shortest distance between the axis of rotation and the line along which the force acts.

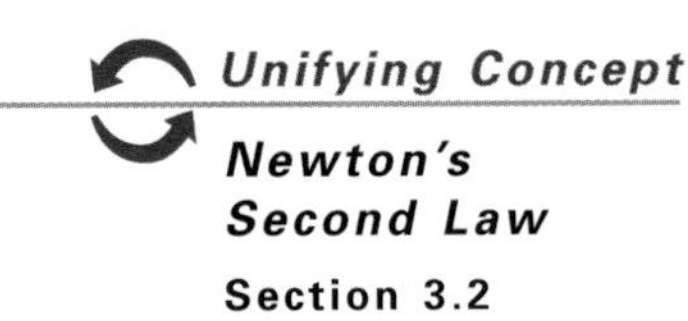

This is why the stubborn bolt shown in Figure B.8 is turned more easily when the applied force is perpendicular to the handle, rather than at an oblique angle, as shown in the first figure. In the first figure, the lever arm is shown by the dashed line and is less than the length of the wrench handle. In the second figure, the lever arm is equal to the length of the wrench handle. In the third figure, the lever arm is extended with a pipe to provide more leverage and a greater torque.

Angular Momentum

Things that rotate—whether a cylinder rolling down an incline, an acrobat doing a somersault, or the cloud of gas and dust that would become the solar system (Chapter 27)—keep on rotating until something stops them. A rotating object has an "inertia of rotation." Recall, from Chapter 5, that all moving objects have "inertia of motion" or momentum—the product of mass and velocity. This kind of momentum is **linear momentum**. Similarly, the "inertia of rotation" of rotating objects is called **angular momentum**.

Any object that rotates turns about its *axis of rotation*. For the case of an object that is small compared with the radial distance to its axis of rotation, like a tetherball swinging from a long string or a planet orbiting around the Sun, the angular momentum can be expressed as the magnitude of its linear

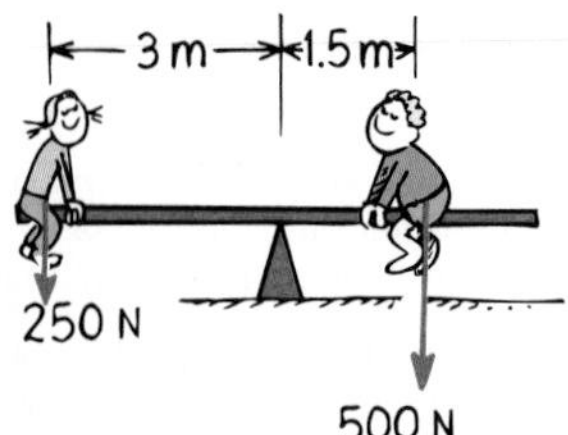

Figure B.6 No rotation is produced when the torques balance each other.

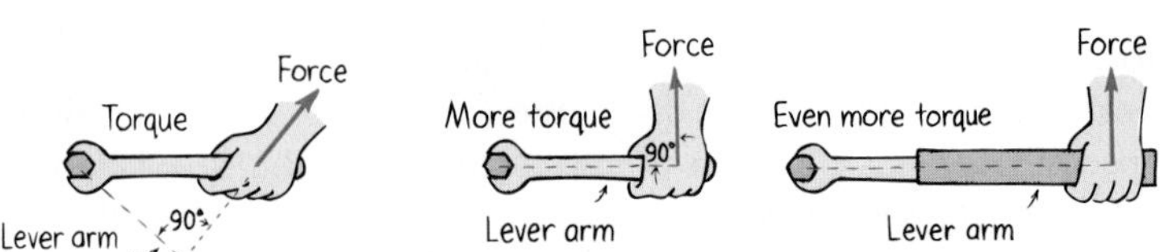

Figure B.8 Although the magnitudes of the force in each case are the same, the torques are different.

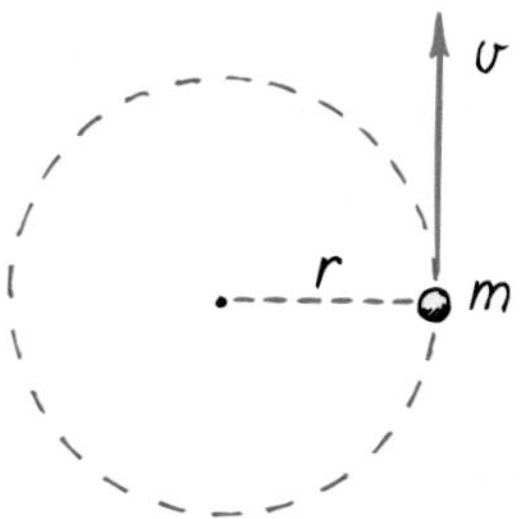

Figure B.9 A small object of mass *m* whirling in a circular path of radius *r* with a speed *v* has angular momentum *mvr*.

Figure B.10 Angular momentum keeps the wheel axle nearly horizontal when a torque supplied by Earth's gravity acts on it. Instead of causing the wheel to topple, the torque causes the wheel's axis to turn slowly around the circle of students. This is called precession.

momentum, *mv*, multiplied by the radial distance, *r* (Figure B.9).* In shorthand notation, angular momentum = *mvr*. Like linear momentum, angular momentum is a vector quantity and has direction as well as magnitude. In this appendix, we won't treat the vector nature of angular momentum (or even of torque, which also is a vector).

Just as an external net force is required to change the linear momentum of an object, an external net torque is required to change the angular momentum of an object. We can state a rotational version of Newton's First Law (the Law of Inertia):

> **An object or system of objects will maintain its angular momentum unless acted upon by an unbalanced external torque.**

We see application of this rule when we look at a spinning top. If friction is low and torque also is low, the top tends to remain spinning. Earth and the planets spin in torque-free regions, and once they are spinning, they remain so.

Angular momentum is a vector. However, in this book we will not treat the vector nature of momentum except to acknowledge the amazing action of the gyroscope. The rotating bicycle wheel in Figure B.10 shows what happens when a torque supplied by Earth's gravity acts to change its angular momentum (which is along the wheel's axle.) The pull of gravity that normally acts to topple the wheel over and change its rotational axis causes it instead to *precess* about a vertical axis, like a gyroscope. The best way to appreciate this is to hold a spinning bicycle wheel while standing on a platform that is free to rotate—the wheel turns you around. Precession is another advanced motion topic you may learn more about in a follow-up physics course.

Conservation of Angular Momentum

Just as the linear momentum of any system is conserved if no net forces are acting on the system, angular momentum is conserved if no net torque acts on the system. In the absence of an unbalanced external torque, the angular momentum of that system is constant. This means that its angular momentum at any one time will be the same as at any other time.

Conservation of angular momentum is shown in Figure B.11. The man stands on a low-friction turntable with weights extended. To simplify, consider only the weights in his hands. When he is slowly turning with his arms extended, much of the angular momentum is due to the distance between the weights and the rotational axis. When he pulls the weights inward, the distance is considerably reduced. What is the

Figure B.11 Conservation of angular momentum. When the man pulls his arms and the whirling weights inward, he decreases the radial distance between the weights and the axis of rotation, and the rotational speed increases correspondingly.

* For rotating bodies that are large compared with radial distance—for example, a planet rotating about its own axis—the concept of rotational inertia must be introduced. Then angular momentum is rotational inertia × rotational speed. See any of Paul Hewitt's *Conceptual Physics* textbooks for more information.

result? His rotational speed increases!* This example is best appreciated by the turning person, who feels changes in rotational speed that seem to be mysterious. But it's straight physics! This procedure is used by a figure skater who starts to whirl with her arms and perhaps a leg extended and then draws her arms and leg in to obtain a greater rotational speed. Whenever a rotating body contracts, its rotational speed increases.

* When a direction is assigned to rotational speed, we call it rotational velocity (often called *angular velocity*). By convention, the rotational velocity vector and the angular momentum vector have the same direction and lie along the axis of rotation.

The Law of Angular Momentum Conservation is seen in the motions of the planets and the shape of the galaxies. When a slowly rotating ball of gas in space gravitationally contracts, the result is an increase in its rate of rotation. The conservation of angular momentum is far-reaching.

Unifying Concept
The Law of Conservation of Momentum
Section 4.4

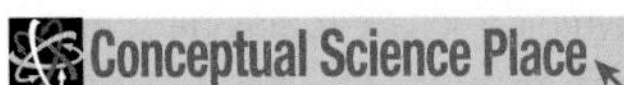

APPENDIX B ONLINE RESOURCES

Interactive Figures B.3 • ***Tutorial*** Rotational Motion

Appendix C
Working with Vector Components

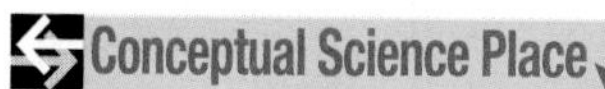

Vectors

A vector quantity is a directed quantity—one that must be specified not only by magnitude (size) but by direction as well. Recall from Chapter 2 that velocity is a vector quantity. Other examples are force, acceleration, and momentum. In contrast, a scalar quantity can be specified by magnitude alone. Some examples of scalar quantities are speed, time, temperature, and energy.

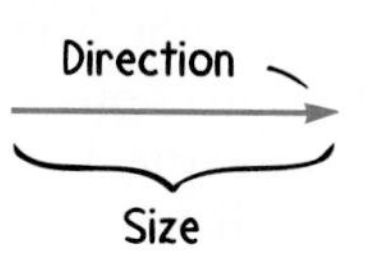

Figure C.1

Vector quantities may be represented by arrows (Figure C.1). The length of the arrow tells you the magnitude of the vector quantity, and the arrowhead tells you the direction of the vector quantity. Such an arrow drawn to scale and pointing appropriately is called a vector.

Adding Vectors

Vectors that add together are called **component vectors**. The sum of component vectors is called a **resultant**. To add two vectors, make a parallelogram with two component vectors acting as two of the adjacent sides (Figure C.2). (Here our parallelogram is a rectangle.) Then draw a diagonal from the origin of the vector pair; this is the resultant (Figure C.3).

Caution: Do not try to mix vectors! We cannot add apples and oranges, so velocity vectors combine only with velocity vectors, force vectors combine only with force vectors, and acceleration vectors combine only with acceleration vectors—each on its own vector diagram. If you ever show different kinds of vectors on the same diagram, use different colors or some other method of distinguishing the different kinds of vectors.

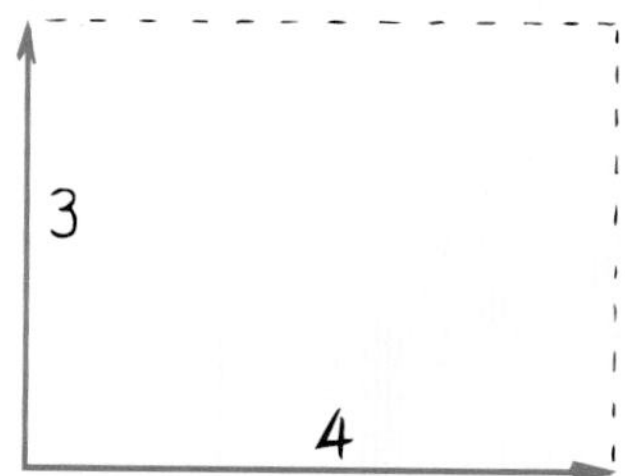

Figure C.2

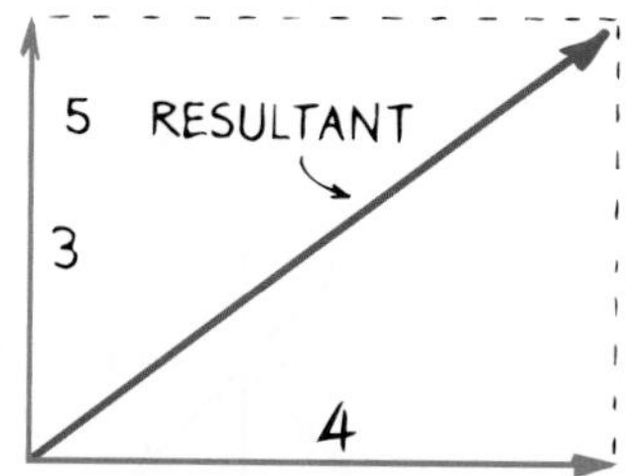

Figure C.3

Finding Components of Vectors

Recall from Chapter 3 that to find a pair of perpendicular components for a vector, first draw a dashed line through the tail of the vector (in the direction of one of the desired components as in Figure C.4). Second, draw another dashed line through the tail end of the vector at right angles to the first dashed line (Figure C.5). Third, make a rectangle whose diagonal is the given vector (Figure C.6). Draw in the two components. Here we let **F** stand for "total force," **U** stand for "upward force," and **S** stand for "sideways force."

Examples

1. Ernie Brown pushes a lawnmower and applies a force that pushes it forward and also against the ground. In Figure C.7, **F** represents the force applied by the man. We can separate this force into two components. The vector **D** represents the downward component, and **S** is the sideways component, the force that moves the lawnmower forward. If we know the magnitude and direction of the vector **F**, we can estimate the magnitude of the components from the vector diagram.
2. Would it be easier to push or pull a wheelbarrow over a step? Figure C.8 shows a vector diagram for each case.

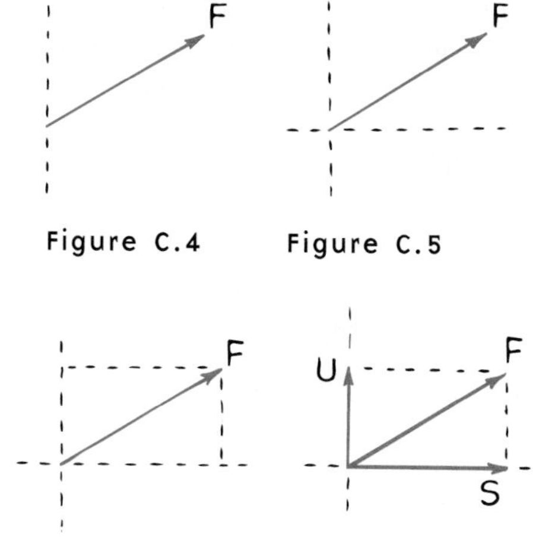

Figure C.4

Figure C.5

Figure C.6

Figure C.7

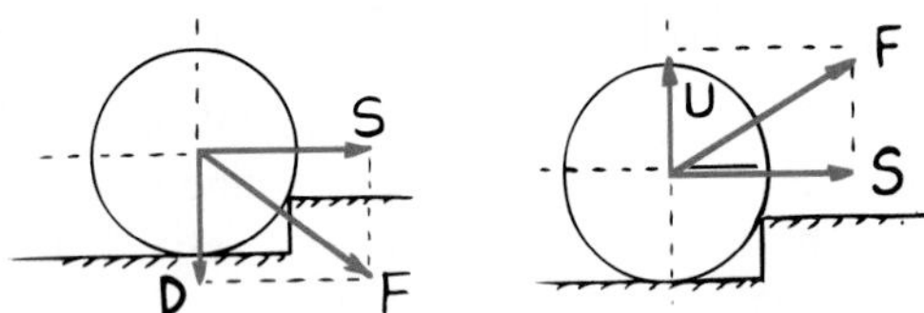

Figure C.8

When you push a wheelbarrow, part of the force is directed downward, which makes it harder to get over the step. When you pull, however, part of the pulling force is directed upward, which helps to lift the wheel over the step. Note that the vector diagram suggests that pushing the wheelbarrow may not get it over the step at all. Do you see that the height of the step, the radius of the wheel, and the angle of the applied force determine whether the wheelbarrow can be pushed over the step? We see how vectors help us analyze a situation so that we can see just what the problem is!

The Polarization of Light—An Application of Vector Components

Recall from Chapter 8 that light travels as a transverse electromagnetic wave. You can create a transverse wave in a rope by shaking it. If you shake the rope vertically up and down, the wave vibrates in a vertical plane. If you shake the rope horizontally side-to-side, the wave vibrates in a plane that is horizontal, as the wave moves forward (Figure C.9). The vibrations in each case are to and fro in one direction. For this reason we say the wave is **polarized**—it vibrates in a single plane.

Figure C.9 A vertically plane-polarized plane wave and a horizontally plane-polarized plane wave.

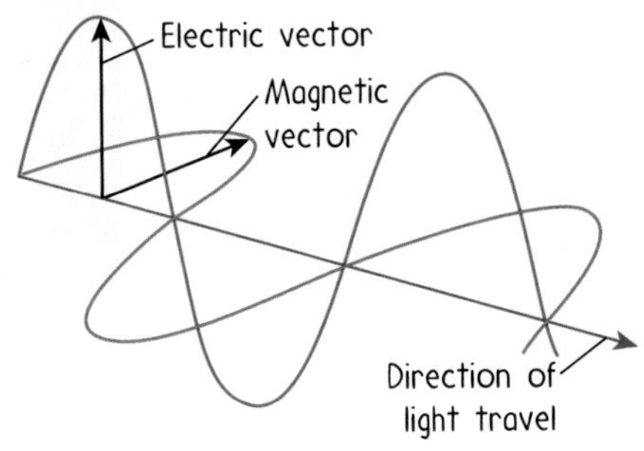

Figure C.10

A light wave is made up of an oscillating electric field vector and an oscillating magnetic field vector (Figure C.10). It is the orientation of the electric field vector that defines the direction of the polarization of light. A single vibrating electron emits an electromagnetic wave that, like the rope shaken in one direction, is polarized. Its electric field vibrates in a single plane.

However, the electric field vectors in waves of light from the Sun or from a lamp vibrate in all conceivable directions as they move. That is, they are *unpolarized* because the light waves are emitted from so many electrons vibrating in so many directions.

Now, light can be polarized, or filtered so that it consists only of electromagnetic waves with electric vectors vibrating all in the same plane. One type of polarizing filter is the material known as Polaroid, invented in the 1930s and now so widely used in sunglasses. Regular light incident upon a polarizing filter, such as the lenses in Polaroid sunglasses, emerges as polarized light. Think of a beam of unpolarized light coming straight toward you. Consider the electric field vector in that beam. Some of the possible directions of the vibrations are shown in Figure C.11. There are as many vectors in the horizontal direction as there are in the vertical direction since the light is unpolarized. The center sketch shows the light falling on a polarizing filter with its polarization axis vertically oriented. Only vertical components of the light pass through the filter, and the light that emerges is vertically polarized, as shown on the right.

Figure C.12 shows that no light can pass through a pair of Polaroid sunglasses when their axes are at right angles to one another, but some light does pass through when their axes are at a nonright angle. This fact can be understood with vectors and vector components.

Recall from Chapter 3 that any vector can be thought of as the sum of two components at right angles to each other. The

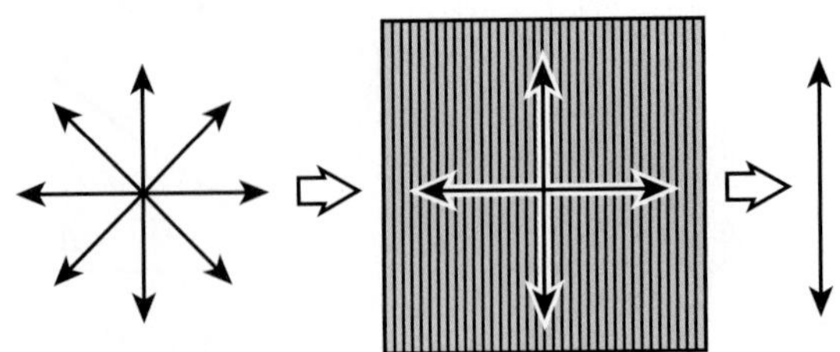

Figure C.11

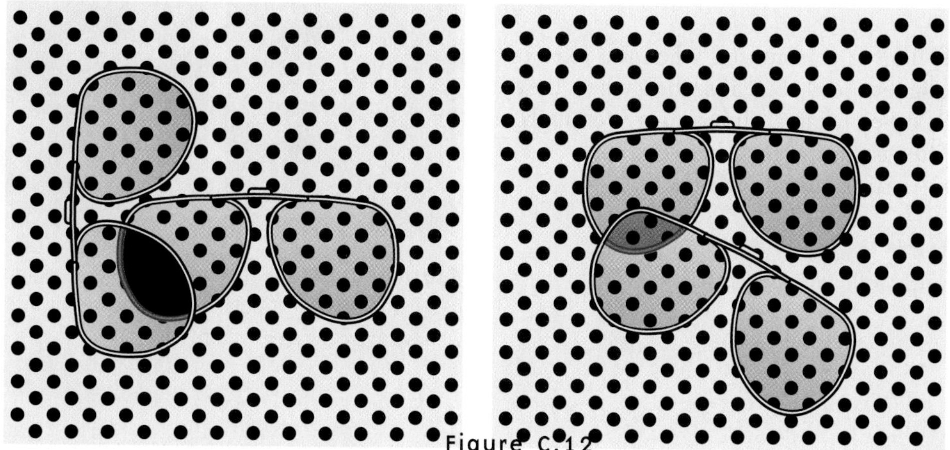

Figure C.12

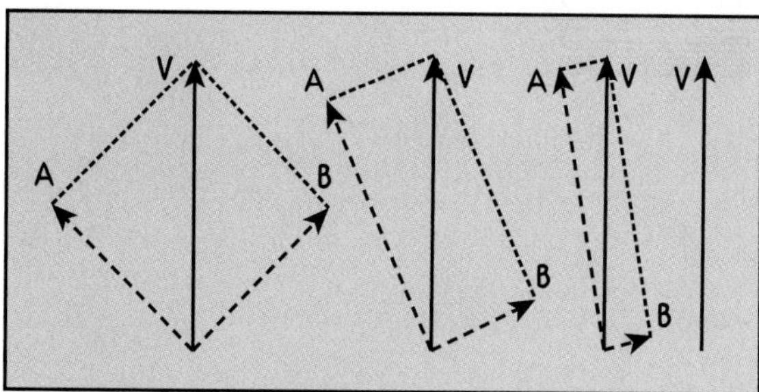

Figure C.13

two components are often chosen to be in the horizontal and vertical directions, but they can be in any two perpendicular directions. In fact, the number of sets of perpendicular components possible for any vector is infinite. A few of them are shown for the vector V in Figure C.13. In every case, components A and B make up the sides of a rectangle that has V as its diagonal.

You can see this somewhat differently by thinking of component **A** as always being vertical and **B** as being horizontal and picturing vector **V** as rotating instead (Figure C.14). This time the different orientations of **V** are superimposed on a polarizing filter with its polarization axis vertical. In the first sketch on the left, all of **V** gets through. As **V** rotates, only the vertical component **A** passes through, and it gets shorter and shorter until it is zero when **V** is completely horizontal.

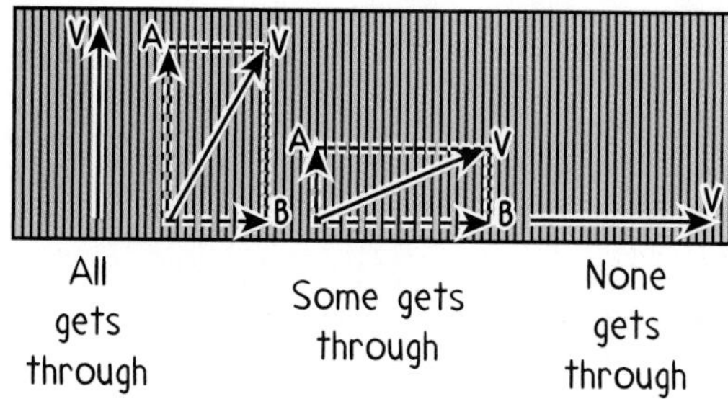

Figure C.14

Can you now understand how light gets through the second pair of sunglasses in Figure C.12? Look at Figure C.15, where for clarity, the two crossed lenses of Figure C.12, which are one atop the other, are instead shown side by side. The vector **V** that emerges from the first lens is vertical. However, it has a component **A** in the direction of the polarization axis of the second lens. Component **A** passes through the second lens, while component **B** is absorbed.

To really appreciate this, you must play around with a couple of polarizing filters, which you can do in a lab exercise. Rotate one above the other and see how you can regulate the amount of light that gets through. Can you think of practical uses for such a system?

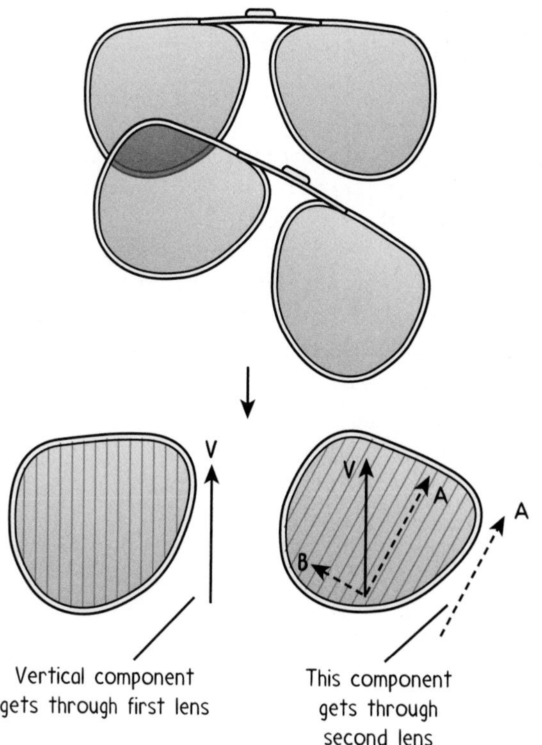

Figure C.15

CHECK YOURSELF

As shown in Figure C.16, light is transmitted when the axes of the polarizing filters are aligned (left), but absorbed when they are at right angles to each other (center). Interestingly enough, when a third polarizing filter is sandwiched between the crossed filters (right), light is transmitted. Why?

Figure C.16 Two polarizing filters at right angles transmit light when a third filter with a polarization axis at an intermediate angle is sandwiched between them. Why?

CHECK YOUR ANSWER

The explanation is shown in the following diagram.

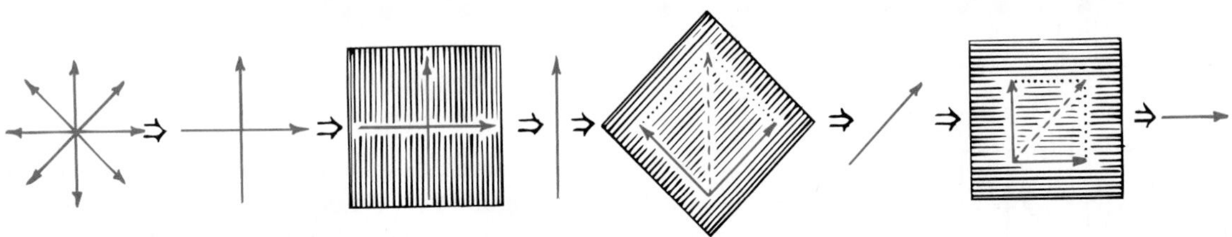

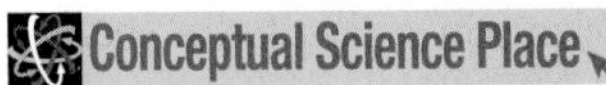

APPENDIX C ONLINE RESOURCES

- ***Tutorials*** Vectors

Appendix D
Exponential Growth and Doubling Time*

Try to fold a piece of paper in half and then fold it again upon itself successively for 9 times. You won't be able to do it. It gets too thick for folding. Even if you could fold a fine piece of tissue paper upon itself 50 times, it would be more than 20 kilometers thick! The continual doubling of a quantity builds up exponentially. Double one penny 30 times, so that you can begin with one penny, then have two pennies, then four, and so on, and you'll accumulate a total of $10,737,418.23! One of the most important things we seem unable to perceive is the process of exponential growth. If we could, we could "see" why compound interest works the way it does, prices of goods rise the way they do, and populations and pollution proliferate out of control.

When a quantity such as money in the bank, population, or the rate of consumption of a resource steadily grows at a fixed percent per year, we say the growth is *exponential*. Money in the bank may grow at 4 percent per year; electric power generating capacity in the United States grew at about 7 percent per year for the first three-quarters of the twentieth century. The important thing about exponential growth is that the time required for the growing quantity to double in size (increase by 100 percent) is also constant. For example, if the population of a growing city takes 12 years to double from 10,000 to 20,000 inhabitants and its growth remains steady, in the next 12 years the population will double to 40,000 and in the next 10 years to 80,000 and so on.

When a quantity is *decreasing* at a rate that is proportional to its value, that quantity is undergoing *exponential decay*. As discussed in Chapters 10 and 26, radioactive elements are subject to radioactive decay, meaning that "parent" isotopes are transformed by nuclear processes to "daughter" isotopes (see Figure 10.15). The time required for an exponentially decaying quantity to be reduced to half of its initial value is called its half-life. Just as doubling time is constant for an exponentially growing quantity, half-life is constant for an exponentially decaying quantity.

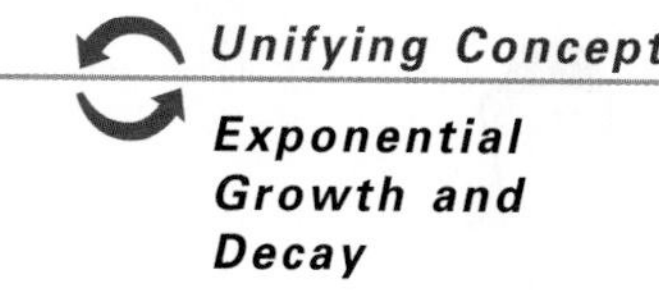

There is an important relationship between the percent growth rate and its *doubling time*, the time it takes to double a quantity:

$$\text{doubling time} = \frac{69.3}{\text{percent growth per unit time}} \quad \frac{\sim 70}{\%}$$

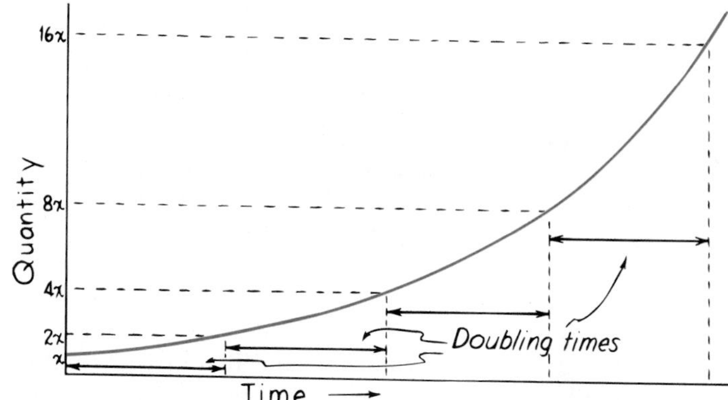

Figure D.1 An exponential curve. Notice that each of the successive equal time intervals noted on the horizontal scale corresponds to a doubling of the quantity indicated on the vertical scale. Such an interval is called the *doubling time*.

So to estimate the doubling time for a steadily growing quantity, we simply divide the number 70 by the percentage growth rate. For example, the 7 percent growth rate of electric power generating capacity in the United States means that in the past the capacity has doubled every 10 years ($\frac{70\%}{7\%/\text{year}} = 10$ years) A 2 percent growth rate for world population means the population of the world doubles every 35 years ($\frac{70\%}{2\%/\text{year}} = 35$ years). A city planning commission that accepts what seems like a modest 3.5 percent growth rate may not realize that this means that doubling will occur in 70/3.5 or 20 years; that's double capacity for such things as water supply, sewage-treatment plants, and other municipal services every 20 years (Figure D.1).

What happens when you put steady growth in a finite environment? Consider the growth of bacteria that grow by division, so that one bacterium becomes two, the two divide to become four, the four divide to become eight, and so on. Suppose the division time for a certain strain of bacteria is 1 minute. This is then steady growth—the number of bacteria grows exponentially with a doubling time of 1 minute. Further, suppose that one bac-

* This appendix is drawn from material by University of Colorado physics professor Albert A. Bartlett, who strongly asserts, "The greatest shortcoming of the human race is man's inability to understand the exponential function."

Table D.1 The Last Minutes in the Bottle

Time Empty	Part Full (%)	Part
11:54 A.M.	$\frac{1}{64}$ (1.5%)	$\frac{63}{64}$
11:55 A.M.	$\frac{1}{32}$ (3%)	$\frac{31}{32}$
11:56 A.M.	$\frac{1}{16}$ (6%)	$\frac{15}{16}$
11:57 A.M.	$\frac{1}{8}$ (12%)	$\frac{7}{8}$
11:58 A.M.	$\frac{1}{4}$ (25%)	$\frac{3}{4}$
11:59 A.M.	$\frac{1}{2}$ (50%)	$\frac{1}{2}$
12:00 noon	full (100%)	none

terium is put in a bottle at 11:00 A.M. and that growth continues steadily until the bottle becomes full of bacteria at 12 noon. Consider seriously the following question.

CHECK YOURSELF

When was the bottle half full?

CHECK YOUR ANSWER

11:59 A.M.; the bacteria will double in number every minute!

It is startling to note that at 2 minutes before noon the bottle was only $\frac{1}{4}$ full. Table D.1 summarizes the amount of space left in the bottle in the last few minutes before noon. If you were an average bacterium in the bottle, at which time would you first realize that you were running out of space? For example, would you sense there was a serious problem at 11:55 A.M., when the bottle was only 3 percent filled, ($\frac{1}{32}$) and had 97 percent of open space (just yearning for development)? The point here is that there isn't much time between the moment that the effects of growth become noticeable and the time when they become overwhelming.

Suppose that at 11:58 A.M. some farsighted bacteria see that they are running out of space and launch a full-scale search for new bottles. Luckily, at 11.59 A.M. they discover three new empty bottles, three times as much space as they had ever known. This quadruples the total resource space ever known to the bacteria, for they now have a total of four bottles, whereas before the discovery they had only one. Further suppose that, thanks to their technological proficiency, they are able to migrate to their new habitats without difficulty. Surely, it seems to most of the bacteria that their problem is solved—and just in time.

Table D.2 Effects of the Discovery of Three New Bottles

Time	Effect
11:58 A.M.	Bottle 1 is $\frac{1}{4}$ full
11:59 A.M.	Bottle 1 is $\frac{1}{2}$ full
12:00 noon	Bottle 1 is full
12:01 P.M.	Bottles 1 and 2 are both full
12:02 P.M.	Bottles 1, 2, 3, and 4 are all full

CHECK YOURSELF

If the bacteria growth continues at the unchanged rate, what time will it be when the three bottles are filled to capacity?

CHECK YOUR ANSWER

12:02 P.M.!

We see from Table D.2 that quadrupling the resource extends the life of the resource by only two doubling times. In our example the resource is space—but it could as well be coal, oil, uranium, or any nonrenewable resource.

Continued growth and continued doubling lead to enormous numbers. In two doubling times, a quantity will double twice ($2^2 = 4$; quadruple) in size; in three doubling times, its size will increase eightfold ($2^3 = 8$); in four doubling times, it will increase sixteenfold ($2^4 = 16$); and so on. This is best illustrated by the story of the court mathematician in India who years ago invented the game of chess for his king. The king was so pleased with the game that he offered to repay the mathematician, whose request seemed modest enough. The mathematician requested a single grain of wheat on the first square of the chessboard, two grains on the second square, four on the third square, and so on, doubling the number of grains on each succeeding square until all squares had been used (Figure D.2). At this rate there would be 2^{63} grains of wheat on the 64th square. The king soon saw that he could not fill this "modest" request, which amounted to more wheat than had been harvested in the entire history of Earth!

Figure D.2 A single grain of wheat placed on the first square of the chessboard is doubled on the second square, this number is doubled on the third, and so on, presumably for all 64 squares. Note that each square contains one more grain than all the preceding squares combined. Does enough wheat exist in the world to fill all 64 squares in this manner?

Table D.3 Filling the Squares on the Chessboard

Square Number	Grains on Square	Total Grains Thus Far
1	1	1
2	2	3
3	4	7
4	8	15
5	16	31
6	32	63
7	64	127
.	.	.
.	.	.
.	.	.
64	2^{63}	$2^{64} - 1$

It is interesting and important to note that the number of grains on any square is 1 grain more than the total of all grains on the preceding squares. This is true anywhere on the board. Note from Table D.3 that when 8 grains are placed on the fourth square, the eight is 1 more than the total of 7 grains that were already on the board. Or the 32 grains placed on the sixth square is one more than the total of 31 grains that were already on the board. We see that in one doubling time we use more than all that had been used in all the preceding growth!

So if we speak of doubling energy consumption in the next however many years, bear in mind that this means in these years we will consume more energy than has heretofore been consumed during the entire preceding period of steady growth. And if power generation continues to use predominantly fossil fuels, then except for some improvements in efficiency, we would burn up in the next doubling time a greater amount of coal, oil, and natural gas than has already been consumed by previous power generation, and except for improvements in pollution control, we can expect to discharge even more toxic wastes into the environment than the millions upon millions of tons already discharged over all the previous years of industrial civilization. We would also expect more human-made calories of heat to be absorbed by Earth's ecosystem than have been absorbed in the entire past! At the previous 7 percent annual growth rate in energy production, all this would occur in one doubling time of a single decade. If over the coming years the annual growth rate remains at half this value, 3.5 percent, then all this would take place in a doubling time of two decades. Clearly this cannot continue!

The consumption of a nonrenewable resource cannot grow exponentially for an indefinite period, because the resource is finite and its supply finally expires. The most drastic way this could happen is shown in Figure D.3a, where the rate of consumption, such as barrels of oil per year, is plotted against time, say in years. In such a graph the area under the curve represents the supply of the resource. We see that when the supply is exhausted, the consumption ceases altogether. This sudden change is rarely the case, for the rate of extracting the supply falls as it becomes more scarce. This is shown in Figure D.3b. Note that the area under the curve is equal to the area under the curve in (a). Why? Because the total supply is the same in both cases. The principal difference is the time taken to finally extinguish the supply. History shows that the rate of production of a nonrenewable resource rises and falls in a nearly symmetric manner, as shown in Figure D.3c. The time during which production rates rise is approximately equal to the time during which these rates fall to zero or near zero.

Production rates for all nonrenewable resources decrease sooner or later. Only production rates for renewable resources, such as agriculture or forest products, can be maintained at steady levels for long periods of time (Figure D.4), provided such production does not depend on waning nonrenewable resources such as petroleum. Much of today's agriculture is so petroleum-dependent that it can be said that modern agriculture is simply the process whereby land is used to convert petroleum into food. The implications of petroleum scarcity go far beyond rationing of gasoline for cars or fuel oil for home heating.

The consequences of unchecked exponential growth are staggering. It is important to ask: Is growth really good? In answering this question, bear in mind that human growth is an

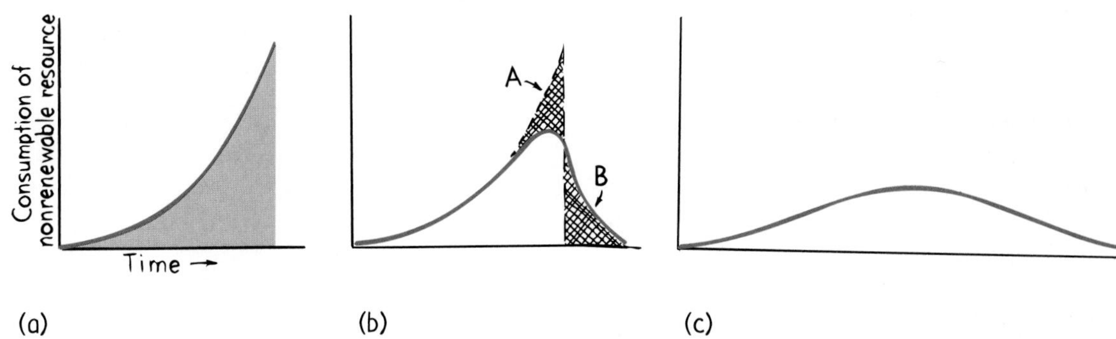

Figure D.3 If the exponential rate of consumption for a nonrenewable resource continues until it is depleted, consumption falls abruptly to zero. The shaded area under this curve represents the total supply of the resource. (b) In practice, the rate of consumption levels off and then falls less abruptly to zero. Note that the crosshatched area A is equal to the crosshatched area B. Why? (c) At lower consumption rates, the same resource lasts a longer time.

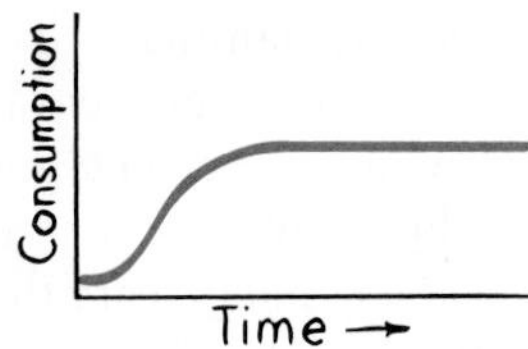

Figure D.4 A curve showing the rate of consumption of a renewable resource such as agricultural or forest products, where a steady rate of production and consumption can be maintained for a long period, provided this production is not dependent upon the use of a nonrenewable resource that is waning in supply.

early phase of life that continues normally through adolescence. Physical growth stops when physical maturity is reached. What do we say of growth that continues in the period of physical maturity? We say that such growth is obesity—or worse, cancer.

PROBLEMS

1 According to a French riddle, a lily pond starts with a single leaf. Each day the number of leaves doubles, until the pond is completely covered by leaves on the 30th day. On what day was the pond half-covered? One-quarter covered?

2 In an economy that has a steady inflation rate of 7 percent per year, in how many years does a dollar lose half its value?

3 At a steady inflation rate of 7 percent, what will be the price every 10 years for the next 50 years for a theater ticket that now costs $30? For a coat that now costs $300? For a car that now costs $30,000? For a home that now costs $300,000?

4 If the sewage treatment plant of a city is just adequate for the city's current population, how many sewage treatment plants will be necessary 42 years later if the city grows steadily at 5 percent annually?

5 If world population doubles in 40 years and world food production also doubles in 40 years, how many people then will be starving each year compared to now?

6 Suppose you get a prospective employer to agree to hire your services for wages of a single penny for the first day, 2 pennies for the second day, and double each day thereafter, providing the employer keeps to the agreement for a month. What will be your total wages for the month?

In the preceding exercise, how will your wages for only the 30th day compare to your total wages for the previous 29 days?

Appendix E
Physics of Fluids

Liquids and gases have the ability to flow; hence they are called *fluids*. In order to discuss the physics of fluids properly, we first need to understand two concepts—*density* and *pressure*.

Density

As introduced in Chapter 2, a basic property of materials—whether in the solid, liquid, or gaseous phases—is the measure of compactness: **density**.

$$\text{density} = \frac{\text{mass}}{\text{volume}}$$

A loaf of bread has a certain mass, volume, and density. When squeezed, the volume decreases and its density increases—but its mass remains the same. Mass is measured in either grams or kilograms and volume in either cubic centimeters (cm^3) or cubic meters (m^3). Another unit of volume is the liter, which is 1000 cm^3. One gram per cubic centimeter = 1 kg per liter (Figure E.1). The densities of a few materials are given in Table E.1.

A quantity known as *weight density* is commonly used when discussing liquid pressure.

$$\text{weight density} = \frac{\text{weight}}{\text{volume}}$$

Weight density is common to British units, in which 1 cubic foot of fresh water (almost 7.5 gallons) weighs 62.4 pounds. So fresh water has a weight density of 62.4 lb/ft^3. Saltwater is a bit denser, 64 lb/ft^3.

Figure E.1 A liter of water occupies a volume of 1000 cm^3, has a mass of 1 kg, and weighs 9.8 N. Its density may therefore be expressed as 1 kg/L and its weight density as 9.8 N/L. (Seawater is slightly denser, about 10 N/L.)

Pressure

Pressure is defined as the force exerted over a unit of area, such as a square meter or square foot:

$$\text{pressure} = \frac{\text{force}}{\text{area}}$$

We see in Figure E.2 that force and pressure are different from each other. In the figure we see pressure due to the weight of a solid. Pressure occurs in fluids as well.

Table E.1 Densities of Some Materials

Material	Grams per Cubic Centimeter	Kilograms per Cubic Meter
Liquids		
Mercury	13.60	1,360
Glycerin	1.26	1,260
Seawater	1.03	1,025
Water at 4°C	1.00	1,000
Benzene	0.90	899
Ethyl alcohol	0.81	806
Solids		
Osmium	22.5	22,480
Platinum	21.5	21,450
Gold	19.3	19,320
Uranium	19.0	19,050
Lead	11.3	11,344
Silver	10.5	10,500
Copper	8.9	8.920
Brass	8.6	8,560
Iron	7.8	7,800
Tin	7.3	7,280
Aluminum	2.7	2,702
Ice	0.92	917
Gases (atm, pressure, at sea level)		
Dry air		
0°C	0.00129	1.29
10°C	0.00125	1.25
20°C	0.00121	1.21
30°C	0.00116	1.16
Hydrogen at 0°C	0.00090	0.090
Helium at 0°C	0.00178	0.178

Figure E.2 Although the weight of both books is the same, the upright book exerts greater pressure against the table.

Figure E.3 The pressure exerted by a liquid is the same at any given depth below the surface, regardless of the shape of the containing vessel.

$$\text{liquid pressure} = \text{weight density} \times \text{depth}$$

It is important to note that pressure does not depend on the amount of liquid—but on its depth. You feel the same pressure a meter deep in a small pool as you do a meter deep in the middle of the ocean. The pressure is the same at the bottom of each of the connected vases in Figure E.3, for example. Depth, not volume, is the key to liquid pressure.

Pressure in a liquid at any point is exerted in equal amounts in all directions. For example, if you are submerged in water, no matter which way you tilt your head, you feel the same amount of water pressure on your ears. When liquid presses against a surface, there is a net force directed perpendicular to the surface (Figure E.4). If there is a hole in the surface, the liquid spurts at right angles to the surface before curving downward due to gravity (Figure E.5). At greater depths, the pressure is greater and the speed of the escaping liquid is greater.

Figure E.4 The forces that produce pressure against a surface add up to a net force that is perpendicular to the surface.

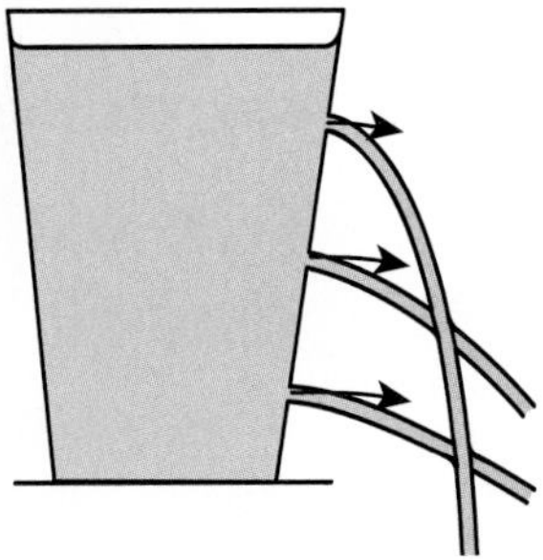

Figure E.5 Water pressure pushes perpendicularly against the sides of a container and increases with increasing depth.

Buoyancy in a Liquid

When an object is submerged in water, the greater pressure on the bottom of the object results in an upward force called the **buoyant force.** We see why in Figure E.6. The arrows represent the forces at different places due to water pressure. Forces produce pressures against opposite sides of different places due to water pressure. Forces that produce pressures against opposite sides cancel one another because they are at the same depth. Pressure is greater against the bottom of the object because the bottom is deeper (more pressure). Because the upward forces against the top are less, the forces do not cancel, and there is a net force upward. This net force is the buoyant force.

If the weight of the submerged object is greater than the buoyant force, the object sinks. If the weight is equal to the buoyant force, the object remains at any level, like a fish. If the buoyant force is greater than the weight of the completely submerged object, the object rises to the surface and floats.

Understanding buoyancy requires understanding the meaning of the expression "volume of water displaced." If a stone is placed in a container that is brimful of water, some water will overflow (Figure E.7). Water is *displaced* by the stone. A little thought tells you that the *volume of the stone*—that is, the amount of space (cubic centimeters) it takes up—is equal to the volume of water displaced. Place any object in a container partially filled with water, and the level of the surface rises (Figure E.8). By how much? By exactly the same amount

Figure E.6 The pressure against the bottom of a submerged object produces an upward buoyant force.

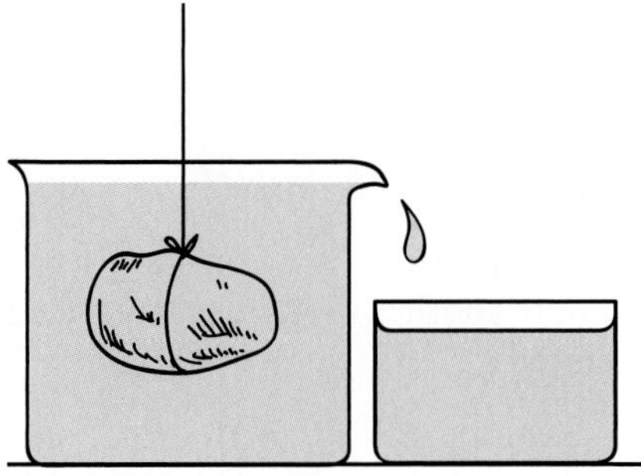

Figure E.7 When a stone is submerged, it displaces a volume of water equal to the volume of the stone.

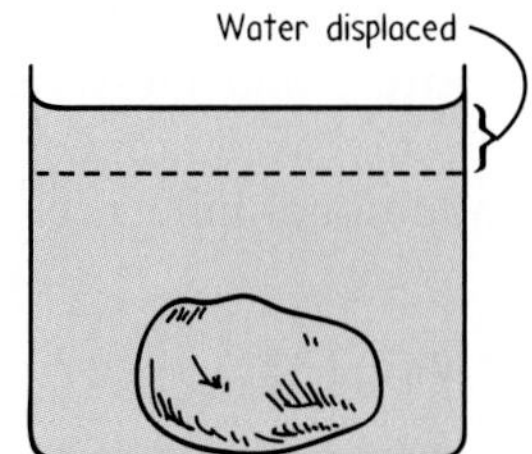

Figure E.8 The increase in water level is the same as that which would occur if, instead of placing the stone in the container, we had poured in a volume of water equal to the stone's volume.

as if we had added a volume of water equal to the volume of the immersed object. This is a good method for determining the volume of irregularly shaped objects: *A completely submerged object always displaces a volume of liquid equal to its own volume.*

Archimedes' Principle

The relationship between buoyancy and displaced liquid was first discovered in the third century B.C. by the Greek scientist Archimedes. It is stated as follows:

> **An immersed body is buoyed up by a force equal to the weight of the fluid it displaces.**

This relationship is called **Archimedes' principle** and is true of all fluids, both liquids and gases (Figure E.9). By *immersed*, we mean either *completely* or *partially submerged*. If we immerse a sealed 1-liter container halfway into a tub of water, it displaces a half-liter of water and is buoyed up by a force equal to the weight of a half-liter of water. If we immerse it completely (submerge it), a force equal to the weight of a full liter or 1 kilogram of water (which is 9.8 newtons) buoys it up. Unless the container is compressed, the buoyant force is 9.8 newtons at *any* depth, as long as the container is completely submerged. A gas is a fluid and is also subject to Archimedes' principle. Any object less dense than air, a gas-filled balloon for example, rises in air.

Flotation

Iron is nearly eight times as dense as water and therefore sinks in water, but an iron ship floats. Why? Consider a solid 1-ton block of iron. When submerged it doesn't displace 1 ton of water

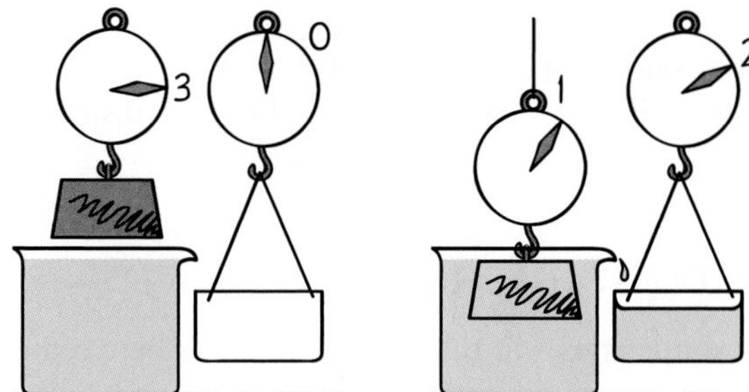

Figure E.9 Objects weigh more in air than in water. When submerged, this 3-N block appears to weigh only 1 N. The "missing" weight is equal to the weight of water displaced, 2 N, which equals the buoyant force.

because it's eight times more compact than water. It displaces only $\frac{1}{8}$ ton of water—certainly not enough to make it float. Suppose we reshape the same iron block into a bowl. It still weighs 1 ton, but when placed in water, it displaces a greater volume of water than before. Its larger volume displaces more water. The deeper it is immersed, the more water it displaces and the greater the buoyant force acting on it. When the buoyant force equals 1 ton, it sinks no farther. It floats. When any object displaces a weight of water equal to its own weight, it floats (Figure E.10). This is the **principle of flotation:**

> **A floating object displaces a weight of fluid equal to its own weight.**

A 500-N friend floating in a swimming pool displaces 500 N of water. To accomplish this, your friend must be slightly less dense than water (which may or may not involve the use of a life preserver). Any floating object is less dense than the fluid it floats upon. The density of ice, for example, is 0.9 that of water, so icebergs float in water. Interestingly, mountains are less dense than the semimolten mantle beneath them. So they float. Just as most of an iceberg is below the water surface (90 percent), most of a mountain (about 85 percent) extends into the mantle. If you could shave off the top of an iceberg, the iceberg would be lighter and be buoyed up to nearly its original height before its top was shaved. Similarly, when mountains erode and wear away, they become lighter and are pushed up from below to float to nearly their original heights. So when a kilometer of mountain erodes away, about 0.85 kilometer of mountain pops up from below. That's why it takes so long for mountains to weather away. Floating objects, whether mountains or ships, must displace a weight of fluid equal to their own weight. Thus, a 10,000-ton ship must be built wide enough to displace 10,000 tons of water before it sinks too deep in the water. The same

Figure E.10 A floating object displaces a weight of fluid equal to its own weight.

holds true for vessels in the air. A dirigible or huge balloon that weighs 100 tons displaces at least 100 tons of air. If it displaces more, it rises; if it displaces less, it falls. If it displaces exactly its weight, it hovers at constant altitude.

Pressure in a Gas

There are similarities and there are differences between gases and liquids. The primary difference between a gas and a liquid is the distance between molecules. In a gas, the molecules are far apart and free from the cohesive forces that dominate their motions when in the liquid and solid phases. The motions of gas molecules are less restricted. A gas expands, fills all the space available to it, and exerts a pressure against its container. Only when the quantity of gas is very large, such as the Earth's atmosphere or a star, do gravitational forces limit the size or determine the shape of the mass of gas.

Dalton's Law

The pressure in a gas depends only on the number of gas particles in a given volume and their kinetic energy. The type of atom or molecule is unimportant. Many gases—for example air—are mixtures. If we know the pressure that each type of gas in a mixture exerts, we can add the individual pressures together to get the total pressure. The pressure exerted by each type of gas is called the *partial pressure* exerted by that gas. Stated mathematically, this is Dalton's Law:

$$P_{\text{Total}} = P_1 + P_2 + P_3$$

Dalton's Law tells us that, at constant volume and temperature, the total pressure exerted by a mixture of gases is equal to the sum of the partial pressures.

Boyle's Law

Pressure and volume in a confined gas are nicely related. "Pressure × volume" for a quantity of gas at any specified time is equal to any "different pressure × different volume" at any other time. In shorthand notation,

$$P_1V_1 = P_2V_2$$

where P_1 and V_1 represent the original pressure and volume, respectively, and P_2 and V_2 the second pressure and volume. This relationship is called **Boyle's Law**. Boyle's law applies to ideal gases. An *ideal* gas is one in which the disturbing effects of the forces between molecules and the finite size of the individual molecules can be neglected. Air and other gases under normal pressure approach ideal-gas conditions.

The Ideal Gas Law

There is another important gas law, called the **Ideal Gas Law**. The Ideal Gas Law relates all four variables that are used to measure changes in a gas—pressure (P), volume (V), temperature (T), and number of moles n. The law is stated as

$$PV = nRT$$

where R is a proportionality constant, whose value depends on the units you are using for pressure and volume.

The Combined Gas Law

If the number of gas particles is a constant, for example, for an enclosed gas, there is another useful gas law called the **Combined Gas Law**. It is stated

$$\frac{P_1V_1}{T_1} = \frac{P_2V_2}{T_2}$$

where T_1 and T_2 represent the initial and final *absolute* temperatures, measured in kelvins (Chapter 6). The Combined Gas Law is often useful, for example, when determining the volume of a gas at standard temperature and pressure (STP).* Thus far we have treated pressure only as it applies to stationary fluids. Motion produces an additional influence.

Bernoulli's Principle

When a fluid flows through a narrow constriction, its speed increases. This is easily noticed by the increased speed of water that spurts from a garden hose when you narrow the opening of the nozzle. The fluid must speed up in the constricted region if the flow is to be continuous.

The Swiss scientist Daniel Bernoulli experimented with fluids in the eighteenth century. He wondered how the fluid gained this extra speed, and reasoned that it is acquired at the expense of a lowered internal pressure. His discovery, now called **Bernoulli's Principle**, states

> **When the speed of a fluid increases, pressure in the fluid decreases.**

Bernoulli's Principle is a consequence of the conservation of energy. In a steady flow of fluid, there are three kinds of energy: kinetic energy due to motion, gravitational potential energy due to elevation, and work done by pressure forces. In a steady fluid flow where no energy is added or removed, the sum of these forms of energy remains constant. If the elevation of the flowing fluid doesn't change, then an increase in speed means a decrease in pressure, and vice versa. Bernoulli's Principle is accurate only for steady flow. If the speed is too great, the flow may become turbulent and follow a changing, curling path known as an *eddy*. This type of flow exerts friction on the fluid and causes some of its energy to be transformed to heat. Then Bernoulli's Principle does not hold. Hold a sheet of paper in front of your mouth. When you blow across the top surface, the paper rises. This is because the pressure of the moving air against the top of the paper is less than the pressure of the air at rest against the lower surface.

If we imagine the rising paper as an airplane wing, we can better understand the lifting force that supports a heavy airliner. A blend of Bernoulli's Principle and Newton's laws account for the air flight we see today. Quite awesome!

* To study temperature and pressure changes in a gas, it is useful to specify a set of standard conditions that can be used for comparison. These conditions are known as standard temperature and pressure and they have a value of 0°C and 1000 atm.

Appendix F
Chemical Equilibrium

The arrow of a chemical equation indicates the direction of the reaction. In the equation for the formation of water, for example, the arrow indicates that hydrogen and oxygen combine to form water:

$$2\,H_2 + O_2 \rightarrow 2\,H_2O$$

The reverse reaction is also possible; that is, oxygen and hydrogen can be formed from water:

$$2\,H_2 + O_2 \leftarrow 2\,H_2O$$

Most chemical reactions are *reversible*, but the extent of the reverse reaction depends very much on conditions. Under usual conditions, for example, water does not readily convert to oxygen and hydrogen. When electricity passes through water, however, the water does decompose to O_2 and H_2 (Figure F.1).

Nitrogen dioxide, NO_2, a brown gas and one of the toxic components of air pollution, provides a good example of a reaction that is fairly easy to reverse. At room temperature nitrogen dioxide molecules pair off to become colorless dinitrogen tetroxide, N_2O_4:

$$NO_2 + NO_2 \rightarrow N_2O_4$$

Once dinitrogen tetroxide molecules are formed, however, they break apart to re-form nitrogen dioxide:

$$N_2O_4 \rightarrow NO_2 + NO_2$$

Of course, once the nitrogen dioxide molecules are re-formed, they can get together to re-form the dinitrogen tetroxide. The net result is two competing reactions, which we depict using double arrows:

$$NO_2 + NO_2 \leftrightarrow N_2O_4$$

Initially, we may start with pure nitrogen dioxide. In time, however, the percentage of nitrogen dioxide decreases and the percentage of dinitrogen tetroxide increases. This shift in amounts present continues to a point where the two reactions balance each other and the percentages of NO_2 and N_2O_4 settle to constant values (at room temperature, about 31 percent NO_2 and 69 percent N_2O_4). At this point, the forward and reverse reactions have achieved what is called **chemical equilibrium**, where the rate at which products are formed is equal to the rate at which they are converted back into reactants (Figure F.2).

Chemists do not usually express the balance of reactants and products at equilibrium in terms of percentages. Rather, they use equilibrium constants. An equilibrium constant is a number that relates the amounts of reactants to products at equilibrium. For example, consider the reaction in which *a* moles of reactant A and *b* moles of reactant B react to give *c* moles of product C and *d* moles of product D at equilibrium.

$$aA + bB \leftrightarrow cC + dD$$

Figure F.1 In principle, applying a strong electric current to the ocean for an incredibly long time would cause all the water to be converted into gaseous hydrogen and oxygen. The first strike of a match, however, would initiate a massive explosion as the two gases exothermically convert back into water.

Figure F.2 The NO_2 and N_2O_4 in this flask are in equilibrium at 16 percent NO_2 and 84 percent N_2O_4. Although the nonchanging brown color seems to indicate no activity in the flask, individual molecules are continuously converting back and forth with a balance between forward and reverse reactions.

Suggested Websites

Chapter 1. About Science

http://www.howstuffworks.com
An intriguing website of technological devices.

http://www.csicop.org
Home page for the Committee for the Scientific Investigation of Claims of the Paranormal.

http://www.nsf.gov
Home page of the National Science Foundation, one of the leading sponsors of scientific research and education.

http://www.sciencenews.org/
Archives of *Science News*, a widely read weekly magazine covering current developments in science

Part 1: Physics

Chapter 2. Describing Motion

http://www.awphysicalscience.com
Many interactive and animated tutorials that focus on physical science concepts, plus much more.

http://www.fearofphysics.com/
A comprehensive site for beginning physics covering a wide range of mechanics topics plus others. Includes animations and quizzes.

http://www.merlot.org
A rich smorgasbord of teaching and learning materials for the instructor with links to good material in science (and in other fields). Well organized and searchable.

http://honolulu.hawaii.edu/distance/sci122/Programs/p12/p12.html
This University of Hawaii site provides a detailed overview of Aristotle's views on motion and related modern concepts of motion.

http://www.aw.com
Information on other *Conceptual* textbooks.

http://www.conceptualphysics.com
Paul Hewitt's personal website.

Chapter 3. Newton's Laws of Motion

http://www.explorescience.com,
Collisions and air-track physics, as well as many other topics are offered in this rich Exploratorium website.

http://www.sciencemaster.com/jump/physical/newton_law.php
ScienceMaster.com presents an overview of Newton's Laws of Motion, supplemented with related web links and animations.

http://www.pilotsweb.com/principle.htm
Explore the applications of Newton's laws in aviation on Pilot's Web: The Aviator's Journal online.

http://www.learner.org/exhibits/parkphysics/
How do the laws of physics affect amusement park ride design? In this exhibit, you'll have a chance to find out by designing your own roller coaster. You can also experiment with bumper car collisions and other amusement part rides.

http://www.braeunig.us/space/
Visit this rocket and space technology site to learn more about how Newton's laws are used in modern space flight technology. Includes tutorials and additional testing resources.

Chapter 4. Energy and Momentum

http://www.energyquest.ca.gov/index.html
A fascinating and entertaining look at energy—all levels.

http://www.physicsclassroom.com/Class/momentum/momtoc.html
This site introduces impulse-momentum change theorem and the law of conservation of momentum, and explains and applies them to the analysis of colliding objects.

Chapter 5. Gravity

http://www.explorescience.com
Projectiles, independence of horizontal and vertical motion, and many other topics are offered in this rich Exploratorium website.

http://www.fearofphysics.com/
Set tangential speed and see if a projectile becomes a satellite. This and much more. Includes animations and quizzes.

http://www.thetech.org/exhibits/online/satellite/
Satellites check weather, transmit TV signals, carry orbiting observatories, and perform other fascinating operations. A construction set helps you assemble three kinds of fully operational satellites! The Satellite Site!

http://liftoff.msfc.nasa.gov/RealTime/JTrack/
Spot the Hubble, the MIR, or another satellite with J-Pass. Sign up, specify your location, and the satellite you wish to track. Satellite prediction reports will be emailed to you telling when and where the satellite is coming, its brightness, and more.

Chapter 6. Heat

http://eo.ucar.edu/skymath/tmp2.html
What is temperature? What is a thermometer? What is heat? Find out the answers to these fundamental questions in this tutorial written by Beverly Lynds from Project Skymath.

http://www.entropylaw.com/
All about entropy, the law of thermodynamics, and order from disorder.

http://library.thinkquest.org/3659/thermodyn/first_law.html
When the Law of Energy Conservation is applied to thermal systems, we call it the First Law of Thermodynamics. This page from the CHEMystery tutorial explains the First Law.

http://library.thinkquest.org/3659/thermodyn/second_law.html
The Second Law of Thermodynamics explains why your room tends to get messier and not cleaner. Read more about the Second Law on this page from CHEMystery.

Atomic symbol An abbreviation for an element or atom.

ATP Adenosine triphosphate, the basic unit of energy used in cellular processes.

Autotrophs Living organisms that make their own food and organic materials.

Axon The part of a neuron that transmits information to other cells, either effector cells or other neurons.

B cell A cell of the acquired immune system that targets pathogens in bodily fluids.

Bacteria One of the three domains of life, consisting of a wide range of generally single-celled prokaryotic organisms.

Base A substance that accepts hydrogen ions.

Basic research Research dedicated to the discovery of the fundamental workings of nature.

Basic solution A solution in which the hydroxide ion concentration is higher than the hydronium ion concentration.

Beta particle An electron (or positron) emitted during the radioactive decay of certain nuclei.

Big Bang The primordial explosion of space at the beginning of time.

Binary star A pair of stars that orbit about a common center.

Biogeochemical cycles The movement of substances such as water, carbon, and nitrogen between the tissues of living organisms and the abiotic world.

Biomass The amount of organic matter present in an ecosystem.

Biomes Major types of terrestrial ecosystems, as classified by their plant life, including tropical forest, temperate forest, coniferous forest, tundra, savanna, temperate grassland, chaparral, and desert.

Biotic Pertaining to living organisms.

Black hole The remains of a giant star that has collapsed upon itself, so dense, and with a gravitational field so intense, that light itself cannot escape from it.

Boiling Evaporation in which bubbles form beneath the liquid surface.

Calorie The amount of heat needed to change the temperature of 1 gram of water by 1 degree Celsius.

Capillary A small blood vessel from which materials are exchanged with body tissues.

Carbohydrates Sugars, starches, and other organic molecules composed of carbon, hydrogen, and oxygen atoms.

Carboxylic acid An organic molecule containing a carbonyl group in which the carbon is bonded to a hydroxyl group.

Carrier proteins Proteins that help molecules normally unable to cross the cell membrane to do so, either with or without energy input.

Carrying capacity A maximum number of individuals or maximum population density that a habitat can support.

Catalyst A substance that increases the rate of a chemical reaction by lowering its activation energy.

Celestial sphere An imaginary sphere surrounding Earth to which the stars are attached.

Cell cycle The series of steps cells go through when they divide.

Cell membrane The membrane that separates the inside of the cell from the outside.

Cells The units that make up all living organisms (except viruses).

Cellular respiration Following glycolysis, the aerobic breakdown of glucose that results in production of ATP.

Cenozoic era The time of "recent life," from 65 million years ago to the present.

Central nervous system The brain and spinal cord.

Centripetal force Any force that is directed at right angles to the path of a moving object and that tends to produce circular motion.

Chain reaction A self-sustaining reaction in which the products of one reaction event stimulate further reaction events.

Chemical bond The attraction between two atoms that holds them together in a compound.

Chemical change During this kind of change, atoms in a substance are rearranged to give a new substance having a new chemical identity.

Chemical equation A representation of a chemical reaction.

Chemical formula A notation used to indicate the composition of a compound, consisting of the atomic symbols for the different elements of the compound and numerical subscripts indicating the ratio in which the atoms combine.

Chemical property A property that relates to how a substance changes its chemical identity.

Chemical reaction A rearrangement of atoms so that one or more new compounds are formed from preexisting compounds or elements.

Chemistry The study of matter and the transformations it can undergo.

Chemoautotrophs Autotrophs that use energy from inorganic chemicals to make their food and organic materials.

Chemoreception A form of sensing in which chemicals bind to receptors on chemosensory cells, causing ion channels to open and action potentials to happen.

Chloroplast The organelle in plant cells where photosynthesis occurs.

Chromosomes Linear structures found in the nucleus of eukaryotic cells that contain the cell's DNA.

Clade A group of species that includes an ancestor and all its descendants.

Cladogram A diagram that shows the history of speciation events among a group of species.

Climate The general pattern of weather that occurs in a region over a period of years.

Cloud A visible aggregate of minute water droplets or tiny ice crystals.

Codominance A situation in which the combination of two alleles in a heterozygote results in both traits being expressed.

Codon A sequence of three nucleotides in an mRNA molecule that codes for a single amino acid.

Coefficient A number used in a chemical equation to indicate either the number of atoms/molecules or the number of moles of a reactant or product.

Cold front A front along which a cold air mass moves under and displaces a warm air mass.

Combustion An exothermic oxidation–reduction reaction between a nonmetallic material and molecular oxygen.

Comet A body composed of ice and dust that orbits the Sun, usually in a very eccentric orbit, and that casts a luminous tail when it is close to the Sun.

Community All the organisms that live within a given area.

Compound A material in which atoms of different elements are bonded to one another.

Concentration A quantitative measure of the amount of solute in a solution.

Conceptual model A representation of a system that helps in making predictions about how the system behaves.

Condensation A transformation from a gas to a liquid.

Conduction The transfer of thermal energy by molecular and electronic collisions within a substance (especially within a solid).

Conductor Any material having free charged particles that easily flow through it when an electric force acts on them.

Conformation One of the possible spatial orientations of a molecule.

Conservation of energy In the absence of external work input or output, the energy of a system remains unchanged. Energy cannot be created or destroyed.

Conservation of momentum In the absence of an external force, the momentum of a system remains unchanged. Hence, the momentum before an event involving only internal forces is equal to the momentum after the event:

$$mv\text{(before event)} = mv\text{(after event)}$$

Consumer An organism that obtains food by eating other organisms.

Continental drift The hypothesis that continents are in motion and that they travel over the face of the Earth rather than remaining fixed in one location.

Continental margin A transition zone between dry land and the ocean bottom.

Control A test that excludes the variable being investigated in a scientific experiment.

Convection The transfer of thermal energy in a gas or liquid by means of currents in the heated fluid. The fluid flows, carrying energy with it.

Convection current Air that rises upward carrying thermal energy.

Convergent plate boundary A place where neighboring plates move toward each other; old lithosphere is destroyed here.

Core The central, metallic, spherical layer of Earth, consisting of a solid inner part and liquid outer part.

Coriolis effect The tendency for moving bodies not attached to Earth (such as air molecules) to turn in their path relative to Earth's surface.

Corrosion The deterioration of a metal, typically caused by atmospheric oxygen.

Cosmology The study of the origin and structure of the physical universe.

Coulomb The SI unit of electrical charge. One coulomb (symbol C) is equal in magnitude to the total charge of 6.25×10^{18} electrons.

Coulomb's Law The relationship among force, charge, and distance:

$$F = k\frac{q_1 q_2}{d^2}.$$

If the charges are alike in sign, the force is repelling; if the charges are unlike, the force is attractive.

Covalent bond A chemical bond in which atoms are held together by their mutual attraction for two electrons they share.

Critical mass The minimum mass of fissionable material in a reactor or nuclear bomb that will sustain a chain reaction.

Cross-cutting A relative dating principle stating that where an igneous intrusion or fault cuts through other rocks, the intrusion or fault is younger than the rock it cuts.

Crust The thin, rocky surface layer of the Earth.

Crystalline structure A group of atoms arranged in an orderly, repeating geometric structure.

Crystallization The process of forming a crystalline structure during solidification.

Current Value given by voltage/resistance.

Cyclone An area of low pressure around which winds flow.

Cytoplasm The portion of the cell outside the nucleus.

Delta An accumulation of sediment where a stream enters a lake or ocean.

Dendrites The parts of a neuron that receive information from other cells, either sensory cells or other neurons.

Density A measure of mass per volume for a substance.

Deposition The process of eroded particles coming to rest at a particular location.

Dew point The temperature at which saturation is reached and condensation occurs.

Differentiation The process by which gravity separates materials of different densities.

Diffraction Any bending of light by means other than reflection and refraction.

Diffusion The random movement of molecules resulting in the transport of molecules from an area of high concentration to an area of low concentration.

Digestion The process through which food is broken down into organic molecules that can be absorbed and used by the body.

Diploid Cells containing two of each kind of chromosome.

Dipole A separation of charge that occurs in a chemical bond because of differences in the electronegativities of the bonded atoms.

Direct current (DC) An electric current flowing in one direction only.

Dispersion The separation of light into colors arranged by frequency.

Dissolving The process of mixing a solute in a solvent.

Divergent plate boundary A place where neighboring plates move away from each other; new lithosphere is created here.

Divide Line that traces the highest ground between streams.

DNA Deoxyribonucleic acid, the cell's genetic material, a double-stranded molecule consisting of sugar-phosphate backbones attached by pairs of matched nitrogenous bases; in the form of a double helix.

Dominant The allele that is expressed in a heterozygote.

Doppler effect The change in frequency of a wave due to the motion of the source (or due to the motion of the receiver).

Earthquake The shaking or trembling of the ground that results when rock under Earth's surface moves or breaks.

Ecliptic The plane of Earth's orbit.

Ecological succession Changes in the species composition of an ecosystem following a disturbance.

Ecology The study of how organisms interact with their environments.

Ecosystem All the organisms that live within a given area and all the abiotic features of their environment.

Ectotherms Organisms that regulate their body temperature behaviorally, by seeking either warm or cool areas.

Efficiency The percentage of the work put into a machine that is converted into useful work output. (More generally, efficiency is useful energy output divided by total energy input.)

Elastic collision A collision in which colliding objects rebound without lasting deformation or the generation of heat.

Electric current The flow of electric charge that transports energy from one place to another. It is measured in amperes, where 1 A is the flow of 6.25×10^{18} electrons per second, or 1 coulomb per second.

Electric field Defined as force per unit charge, it can be considered as an energetic "aura" surrounding charged objects. About a charged point, the field decreases with distance according to the inverse-square law, like a gravitational field. Between oppositely charged parallel plates, the electric field is uniform.

Electric potential The electric potential energy per amount of charge, measured in volts, and often called *voltage*: Voltage = electric energy/amount of charge.

Electric potential energy The energy a charge possesses by virtue of its location in a magnetic field.

Electric power The rate of energy transfer, or rate of doing work; the amount of energy per unit time, which can be measured by the product of current and voltage.

Electric resistance The property of a material that resists the flow of electric current through it. It is measured in ohms (Ω).

Electrically polarized Term applied to an atom or molecule in which the charges are aligned so that one side has a slight excess of negative charge.

Electromagnet A magnet whose field is produced by an electric current. It is usually in the form of a wire coil with a piece of iron inside the coil.

Electromagnetic induction The induction of voltage when a magnetic field changes with time.

Electromagnetic spectrum The range of electromagnetic waves that extends in frequency from radio waves to gamma rays.

Electromagnetic wave An energy-carrying wave produced when an electric charge accelerates.

Electron A negatively charged particle in an atom.

Electron-dot structure A shorthand notation of the shell model of the atom in which valence electrons are shown around an atomic symbol.

Electronegativity The ability of an atom to attract a bonding pair of electrons to itself when bonded to another atom.

Electrostatics The study of electric charge at rest (not *in motion*, as in electric currents)

Element Any material that is made up of only one type of atom.

Elemental formula A notation that uses the atomic symbol and (sometimes) a numerical subscript to denote how atoms are bonded in an element.

Ellipse The sum of the distances from any point on the path to two points called foci is a constant; also the oval path followed by a satellite.

Elliptical galaxy A galaxy that is round or elliptical in outline. It has little gas and dust, no disk or spiral arms, and few hot, bright stars.

Endocytosis A process in which materials are moved into a cell through the pinching off of a vesicle from the cell membrane.

Endoskeleton An internal skeleton, such as that found in echinoderms and chordates.

Endosymbiotic theory The theory that the mitochondria and chloroplasts of eukaryotic cells evolved from prokaryotes living inside the earliest eukaryotic cells.

Endotherms Organisms that rely on food metabolism to maintain body temperature.

Energy The property of a system that enables it to do work.

Entropy The measure of the energy dispersal of a system. Whenever energy freely transforms from one form to another, the direction of transformation is toward a state of greater disorder and, therefore, toward one of greater entropy.

Enzyme A protein that catalyzes a chemical reaction in a living organism.

Equilibrium rule The vector sum of forces acting on a nonaccelerating object equals zero: $\Sigma F = 0$.

Erosion The process by which weathered particles are removed and transported by a stream, a glacier, the wind, or some other mobile agent.

Escape speed The speed that a projectile, space probe, or similar object must reach in order to escape the gravitational influence of Earth or of another celestial body to which it is attracted.

Ester An organic molecule containing a carbonyl group in which the carbon is bonded to one carbon atom and one oxygen atom is bonded to another carbon atom.

Ether An organic molecule containing an oxygen atom bonded to two carbon atoms.

Eukaryotes Organisms whose cells have a nucleus and organelles, including protists, animals, plants, and fungi.

Evaporation A transformation from a liquid to a gas.

Evolution Heritable changes in living organisms over time—or, as Darwin put it, "descent with modification."

Exocytosis A process in which materials are moved out of a cell through the fusion of a vesicle with the cell membrane.

Exons Portions of the mRNA transcript that contribute to building the protein.

Exoskeleton An external skeleton, such as that found in arthropods.

Exponential growth A model of population growth that characterizes populations with unlimited resources.

Extrusive igneous rock Igneous rock that forms at the Earth's surface.

Facilitated diffusion Movement of molecules across the cell membrane by a carrier protein without energy input.

Fact A phenomenon about which competent observers can agree.

Faraday's Law An electric field is induced in any region of space in which a magnetic field is changing with time. The magnitude of the induced electric field is proportional to the rate at which the field changes. The direction of the induced magnetic field is at right angles to the changing electric field.

Fault A break in a rock along which movement has occurred.

Fault-block mountain Mountains that form from tension and that have at least one side bounded by a normal fault.

Faunal succession A relative dating principle stating that fossil organisms succeed one another in a definite, irreversible, determinable order.

Fermentation Following glycolysis, the anaerobic breakdown of glucose that results in the production of ethanol and carbon dioxide gas.

First Law of Thermodynamics A restatement of the Law of Energy Conservation, usually as it applies to systems involving changes in temperature: Whenever heat flows into or out of a system, the gain or loss of thermal energy equals the amount of heat transferred.

Fitness The number of offspring an organism produces in its lifetime compared to other organisms in the population.

Flower The reproductive structure of flowering plants, which may include stamens (male reproductive structures) and/or carpels (female reproductive structures), as well as petals.

Fold A mountain that is built up of folding on a grand scale.

Folded mountain Mountains that feature extensive folding of rock layers.

Force Simply stated, a push or a pull.

Force vector An arrow drawn to scale so that its length represents the magnitude of a force and its direction represents the direction of the force.

Forced vibration The setting up of vibrations in an object by a vibrating source.

Free fall Motion under the influence of gravitational pull only.

Freezing A transformation from a liquid to a solid.

Frequency The number of to-and-fro vibrations an oscillator makes in a given time, or the number of times a particular point on a wave (for example the crest) passes a given point in a given time.

Friction The resistive force that opposes the motion or attempted motion of an object through a fluid or past another object with which it is in contact.

Front The boundary zone between two air masses.

Frontal lifting The rising of warm, less dense air over colder, dense air at a weather front.

Full moon The phase of the Moon when its sunlit side is the side facing Earth.

Functional group A specific combination of atoms that behave as a unit in an organic molecule.

Fungi A group of heterotrophic eukaryotes that obtain food by secreting digestive enzymes over organic matter and then absorbing the nutrients.

Galaxy A large assemblage of stars, interstellar gas, and dust.

Gamma ray High-frequency electromagnetic radiation emitted by the nuclei of radioactive atoms.

Gas Matter that has neither a definite volume nor a definite shape, always filling any space available to it.

Gene A section of DNA that contains the instructions for making a protein.

Genetic mutations Changes in the DNA nucleotide sequence of an organism.

Genome The total genetic material of an organism.

Genotype The genetic makeup of an organism.

Geologic time scale The time scale that subdivides Earth's 4.5-billion-year history into time units of different sizes, from eons to epochs.

Geothermal gradient The increase of temperature with depth within the Earth.

Glacier A mass of dense ice that forms when snow on land is subjected to pressure from overlying snow, so that it is compacted and recrystallized.

Global climate change A trend of gradual, increasing global temperature thought to be caused by the greenhouse effect and responsible for changes in global climate patterns.

Glycolysis The first step cells use to break down glucose to obtain energy.

Golgi apparatus A eukaryotic organelle that processes and packages products received from the endoplasmic reticulum.

Gravitational force The attractive force between objects due to mass.

Greenhouse effect The warming of the atmosphere that occurs because of the ability of certain atmospheric gases to absorb and reemit infrared radiation.

Groundwater The water that resides in a saturation zone.

Group A vertical column in the periodic table, also known as a family of elements.

Half-life The time required for half the atoms in a sample of a radioactive isotope to decay.

Half-reaction One portion of an oxidation–reduction reaction, represented by an equation showing electrons as either reactants or products.

Haploid Cells containing one of each kind of chromosome.

Heat The thermal energy that flows from a substance of higher temperature to a substance of lower temperature, commonly measured in calories or joules.

Heat of fusion The amount of energy needed to change any substance from solid to liquid (and vice versa).

Heat of vaporization The amount of energy needed to change any substance from liquid to gas (and vice versa).

Hemoglobin An iron-containing protein found in red blood cells that carries oxygen.

Heritable Traits that are passed from parents to offspring because they are at least partially determined by genes.

Heteroatom Any atom other than carbon or hydrogen in an organic molecule.

Heterotrophs Living organisms that obtain their energy and organic materials from other living organisms or other outside sources.

Heterozygotes Organisms with two different alleles for a given gene.

Homeostasis The maintenance of a stable internal environment.

Homozygotes Organisms with two identical alleles for a given gene.

Hormones Chemical messengers produced in one place in the body, released into the bloodstream, and received by target cells elsewhere in the body.

Hot spot A stationary, exceptionally hot region deep in Earth's interior usually near the mantle-core boundary.

H-R (Hertzsprung–Russell) diagram A plot of intrinsic brightness versus surface temperature for stars. When so plotted, stars' positions take the form of a main sequence for average stars, with exotic stars above or below the main sequence.

Hubble's Law The farther away a galaxy is from Earth, the more rapidly it is moving away from us; $v = H \times d$.

Hydrocarbon A chemical compound containing only carbon and hydrogen atoms.

Hydrogen bond A strong dipole–dipole attraction between a slightly positive hydrogen atom on one molecule and a pair of nonbonding electrons on another molecule.

Hydrologic cycle The natural circulation of water, or simply, the *water cycle*.

Hydronium ion A water molecule after accepting a hydrogen ion.

Hydrophilic A substance that is soluble in water.

Hydrophobic A substance that is not soluble in water.

Hydroxide ion A water molecule after losing a hydrogen ion.

Hypothesis An educated guess or a reasonable explanation. When the hypothesis can be tested by experiment, it qualifies as a *scientific hypothesis*.

Igneous rock Rock formed by the crystallization of magma or lava.

Impulse The product of the force acting on an object and the time during which it acts.

Inclusions A relative dating principle stating that any inclusion (a piece of one rock type contained within another) is older than the rock containing it.

Incomplete dominance A situation in which the combination of two alleles in a heterozygote produces an intermediate trait.

Induced dipole A dipole temporarily created in an otherwise nonpolar molecule, induced by a neighboring charge.

Inelastic collision A collision in which the colliding objects become distorted, generate heat, and possibly stick together.

Inertia The property of things to resist changes in motion.

Infiltration Absorption of water by the ground.

Inflammatory response An innate immune defense response triggered by the release of histamine by damaged tissues; characterized by swelling, redness, and the movement of innate immune cells to the site of injury.

Innate immunity Nonspecific body defenses that work against a wide variety of potential pathogens.

Inner planets The planets close to the Sun: Mercury, Venus, Earth, and Mars.

Insulator Any material without free charged particles and through which current does not easily flow.

Interaction Mutual action between objects in which each one exerts an equal and opposite force on the other.

Interference The combined effect of two or more waves overlapping.

Interneuron A neuron that connects one neuron to another neuron.

Introns Portions of the mRNA transcript that are removed because they do not contribute to building the protein.

Intrusive igneous rock Igneous rock that forms underground.

Inverse-square law Law relating the intensity of an effect to the inverse square of the distance from the cause: intensity ~ 1/distance^2.

Ion An electrically charged particle created when an atom either loses or gains one or more electrons.

Ionic bond A chemical bond in which an attractive electric force holds ions of opposite charge together.

Ionic compound Any chemical compound containing ions.

Irregular galaxy A galaxy with a chaotic appearance and with large clouds of gas and dust, but without spiral arms.

Isotopes Different forms of an element whose atoms contain the same number of protons but different numbers of neutrons.

Joule The SI unit of energy and work, equivalent to a newton-meter.

Ketone An organic molecule containing a carbonyl group in which the carbon is bonded to two carbon atoms.

Kilogram The unit of mass. One kilogram (symbol kg) is the mass of 1 liter (symbol L) of water at 4 degrees Celsius.

Kinetic energy Energy of motion, described by the relationship

$$\text{Kinetic energy} = \Delta m v^2$$

Kinetic Theory of Matter Matter is made up of tiny particles—atoms or molecules. The particles are always moving, and they move in a number of ways. They rotate, vibrate and move in straight lines between collisions.

Kuiper Belt The disk-shaped region of the sky beyond Neptune that is populated by many icy bodies and is a source of short-period comets.

Larva A stage in the growth and development of animals that is distinct from the adult in form and ecology.

Larynx The voice box; the part of the respiratory tract that contains the vocal cords, which are used to produce the sounds of speech.

Latent heat Energy that is released or absorbed in a change of phase.

Law A general hypothesis or statement about the relationship of natural quantities that has been tested over and over again and has not been contradicted; also known as a *principle*.

Law of Mass Conservation Matter is neither created nor destroyed during a chemical reaction.

Law of Reflection The angle of reflection equals the angle of incidence.

Law of Universal Gravitation Every body in the universe attracts every other body with a mutually attracting force. For two bodies, this force is directly proportional to the product of their masses and inversely proportional to the square of the distance separating them: $F = G(m_1 \times m_2)/d^2$.

Life history strategy The position a population of organisms occupies on the continuum between producing a large number of "inexpensive" offspring and a small number of "expensive" offspring.

Light-year The distance light travels in one year.

Linked genes Genes that are often inherited together because they are located near each other on the same chromosome.

Lipids Hydrophobic organic compounds, many of which include fatty acids as a primary component.

Liquid Matter that has a definite volume but no definite shape, assuming the shape of its container.

Lithosphere A layer of rigid rock comprising the crust and upper mantle of Earth.

Lock-and-key model A conceptual model that explains how drugs interact with receptor sites.

Logistic growth A model of population growth in which growth slows as the carrying capacity of the habitat is approached.

Longitudinal wave A wave in which the medium vibrates in a direction parallel (longitudinal) with the direction in which the wave travels.

Luminosity The total amount of light energy that a star emits into space.

Lunar eclipse The phenomenon whereby the shadow of Earth falls upon the Moon, producing relative darkness of the full Moon.

Lymph The clear fluid, containing large numbers of immune cells, that flows through the lymphatic vessels.

Lysosome A eukaryotic organelle that breaks down organic materials.

Magma Subsurface, molten rock found mainly in the upper mantle or lower crust.

Magnetic domains Clustered regions of aligned magnetic atoms. When these regions themselves are aligned with one another, the substance containing them is a magnet.

Magnetic field The region of magnetic influence around a magnetic pole or a moving charged particle.

Magnetic force Between magnets, it is the attraction of unlike magnetic poles for each other and the repulsion between like magnetic poles. Between a magnetic field and a moving charge, it is a deflecting force due to the motion of the charge; the deflecting force is perpendicular to the velocity of the charge and perpendicular to the magnetic field lines.

Magnetic striping The symmetrical pattern of alternating normal and reversed magnetic polarities found in the seafloor on either side of a spreading center.

Main sequence The diagonal band of stars on the H-R diagram; such stars generate energy by fusing hydrogen to helium.

Mantle The intermediate rocky layer of Earth, which occupies most of Earth's volume and carries most of its mass.

Mass The quantity of matter in an object. More specifically, it is a measure of the inertia or sluggishness that an object exhibits in response to any effort made to start it, stop it, deflect it, or change its state of motion in any way.

Mass extinction The rapid extinction of numerous species; the rate of extinction is much greater than the background rate of species extinction.

Mass movement The down-slope movement of Earth's materials acting under the influence of gravity alone.

Mass number The total number of nucleons in an atomic nucleus.

Matter Anything that occupies space.

Maxwell's counterpart to Faraday's Law A magnetic field is induced in any region of space in which an electric field is changing with time. The magnitude of the electric field is proportional to the rate at which the electric field changes. The direction of the induced magnetic field is at right angles to the changing electric field.

Meiosis A form of cell division in which one diploid cell divides into four haploid daughter cells.

Melting A transformation from a solid to a liquid.

Membrane protein Proteins found in the cell membrane.

Mesozoic era The time of "middle life," from 245 million years ago to about 65 million years ago.

Messenger RNA (mRNA) A form of RNA made during transcription that is used to carry genetic information from DNA to the ribosomes.

Metal An element that is shiny, opaque, and able to conduct electricity and heat.

Metallic bond A chemical bond in which the metal ions in a piece of solid metal are held together by their attraction to a "fluid" of electrons in the metal.

Metalloid An element that exhibits some properties of metals and some properties of nonmetals.

Metamorphic rock Rock made of preexisting rock that has been altered by high pressure, high temperature, or hot chemical solutions to become more stable under new conditions.

Meteor The streak of light produced by a meteoroid burning in Earth's atmosphere; a "shooting star."

Meteorite A meteoroid, or a part of a meteoroid, that has survived passage through Earth's atmosphere to reach the ground.

Meteoroid A small piece of debris from a comet or asteroid in interplanetary space.

Milky Way The name of the galaxy to which we belong—our cosmic home.

Mineral A naturally occurring inorganic material that has both a definite chemical composition and a crystalline structure.

Mitochondria A eukaryotic organelle that breaks down organic molecules to obtain ATP.

Mixture A combination of two or more substances in which each substance retains its properties.

Molarity A unit of concentration equal to the number of moles of a solute per liter of solution.

Mole 6.02×10^{23} of anything.

Molecule A group of atoms held tightly together by covalent bonds.

Momentum The product of the mass of an object and its velocity.

Monomer The small molecular unit from which a polymer is formed.

Moon phases The cycles of change of the "face" of the Moon, changing from new, to waxing, to full, to waning, and back to new.

Motor neuron A neuron that carries messages from the central nervous system to effector cells.

Mycorrhizae Close associations between fungi and the roots of plants allowing the fungi to obtain nutrients from the plant while helping the roots absorb water and minerals from the soil.

Natural frequency A frequency at which an elastic object naturally tends to vibrate.

Natural selection Organisms with heritable, advantageous traits leave more offspring than organisms with other traits, causing advantageous traits to become more common in populations over time.

Nebular theory The idea that the Sun and planets formed together from a cloud of gas and dust; a *nebula*.

Nephron The functional unit of a kidney.

Net force The combination of all forces that act on an object.

Neuron A cell that receives and transmits information from one part of the body to another.

Neutral solution A solution in which the hydronium ion concentration is equal to the hydroxide ion concentration.

Neutralization A reaction in which an acid and base combine to form a salt.

Neutron An electrically neutral subatomic particle in an atomic nucleus.

Neutron star A small, extremely dense star composed of tightly packed neutrons formed by the welding of protons and electrons.

New Moon The phase of the Moon when darkness covers the side facing Earth.

Newton The scientific unit of force.

Newton's First Law of Motion Every object continues in a state of rest, or in a state of motion in a straight line at a constant speed, unless it is compelled to change that state by forces exerted on it.

Newton's Second Law of Motion The acceleration produced by a net force on an object is directly proportional to the net force, is in the same direction as the net force, and is inversely proportional to the mass of the object.

Newton's Third Law of Motion Whenever one object exerts a force on a second object, the second object exerts an equal and opposite force on the first object.

Niche The total set of biotic and abiotic resources a species uses within a community.

Nonmetal An element located toward the upper right of the periodic table and that is neither a metal nor a metalloid.

Nonpolar bond A chemical bond that has no dipole.

Nova An event wherein a white dwarf suddenly brightens and appears as a "new" star.

Nuclear fission The splitting of the nucleus of a heavy atom, such as uranium-235, into two main parts, accompanied by the release of much energy.

Nuclear fusion The combining of nuclei of light atoms to form heavier nuclei, with the release of much energy.

Nucleic acids Organic molecules composed of a nucleotide base attached to a sugar and phosphate backbone.

Nucleon A nuclear particle; a proton or a neutron in an atomic nucleus.

Nucleus A structure in eukaryotic cells that is surrounded by a double membrane and contains the cell's genetic material.

Ocean trench A long, narrow, and deep depression in the seafloor.

Octet rule A rule stating that atoms gain or lose electrons to acquire the outer shell electron configuration of a noble gas, usually neon or argon, which each have eight electrons in their outermost shell.

Ohm's Law The statement that the current in a circuit varies in direct proportion to the potential difference or voltage and inversely with the resistance.

Oort Cloud The region beyond the Kuiper Belt that is populated by trillions of icy bodies and is a source of long-period comets.

Opaque The term applied to materials that absorb light without reemission.

Ore A mineral deposit that is rich in valuable metals that can be extracted for a profit.

Organ A structure in the body that has a certain function.

Organ system Multiple organs that work together to perform a certain bodily function.

Organelle A structure in the cytoplasm of eukaryotic cells that is bound by a membrane and performs a specific function for the cell.

Organic chemistry The study of carbon-containing compounds.

Original horizontality A relative dating principle stating that layers of sediment are deposited evenly, with each new layer laid down almost horizontally over the older sediment.

Outer planets Jupiter, Saturn, Uranus, Neptune, and Pluto.

Ovulation The release of a mature egg cell that occurs once during each menstrual cycle.

Oxidation The process whereby a reactant loses one or more electrons.

Oxidation–reduction reaction A reaction involving the transfer of electrons from one reactant to another.

Ozone hole A large area of the stratosphere with extremely low levels of ozone.

Paleozoic era The time of "ancient life," from 544 million years ago to 245 million years ago.

Parabola The curved path followed by a projectile near the Earth under the influence of gravity only.

Parallel circuit An electric circuit with two or more devices connected in such a way that the same voltage acts across each one, and any single one completes the circuit independently of all the others.

Parts per million A unit of concentration.

Pathogen A disease-causing agent, such as a bacterium, virus, or other organism.

Period The time required for a vibration or a wave to make a complete cycle; a horizontal row in the periodic table.

Periodic table A chart in which all known elements are listed in order of atomic number.

Periodic trend The gradual change of any property in the elements across a period.

Peripheral nervous system All the nerves in your body that are not part of the central nervous system.

Peristalsis A moving wave of muscular contractions that moves food down the esophagus.

Permeability The ease with which fluids flow through interconnected pore spaces.

pH A measure of the acidity of a solution, equal to the negative of the base-10 logarithm of the hydronium ion concentration.

Phenol An organic molecule in which a hydroxyl group is bonded to a benzene ring.

Phenotype The traits of an organism.

Phospholipids Organic molecules with hydrophilic heads and hydrophobic tails that represent a primary component of cell membranes.

Photon A particle of light.

Photosynthesis The process in plants and some other organisms in which light energy from the Sun is converted to energy in organic molecules.

Physical change A change in which a substance changes its physical properties without changing its chemical identity.

Physical model A representation of an object on some convenient scale.

Physical property Any physical attribute of a substance, such as color, density, or hardness.

Placenta The organ that provides for nutrient and waste exchange between the mother and a developing embryo. It includes both embryonic and maternal tissues.

Planetary nebula An expanding shell of gas ejected from a low-mass star during the latter stages of its evolution.

Planets The major bodies orbiting the Sun, which are massive enough for their gravity to make them spherical but small enough to avoid having nuclear fusion in their cores.

Plants Autotrophic, multicellular, terrestrial eukaryotes that obtain energy through photosynthesis.

Plate A moving section of lithosphere; a "tectonic plate."

Plate tectonics The comprehensive geological theory stating that Earth's outer layer is composed of lithospheric plates that float on, and move along with, the relatively plastic asthenosphere.

Platelets Cells found in blood that are involved in blood clotting.

Pleiotropy A situation where a single gene affects more than one trait.

Polar bond A chemical bond that has a dipole.

Pollen In seed plants, an immature male gametophyte wrapped in a protective coating.

Polyatomic ion Molecules that carry a net electric charge.

Polygenic traits Traits determined by more than one gene.

Polymer A long organic molecule made of many repeating units.

Polymorph Two or more minerals that contain the same elements in the same proportions but have different crystal structures.

Population A group of individuals of a single species that occupies a given area.

Porosity The amount of groundwater that a material can store.

Postzygotic reproductive barriers A barrier that causes hybrids resulting from the breeding of two different species that either don't survive or are sterile.

Potential difference The difference in potential between two points, measured in volts, and often called voltage difference.

Potential energy The stored energy that a body possesses because of its position.

Power The time rate of work: power = work/time, or alternatively, power = current × voltage. (More generally, power is the rate at which energy is expended.)

Precambrian time The time of hidden life, which began about 4.5 billion years ago when Earth formed, lasted until about 544 million years ago (the beginning of the Paleozoic era), and makes up 85 percent of Earth's history.

Precipitate A solute that has come out of solution.

Precipitation Water in the liquid or solid state that returns to Earth's surface from the atmosphere.

Pressure gradient Change in pressure over a given distance.

Prevailing wind A wind that consistently blows from one direction.

Prezygotic reproductive barriers A barrier that prevents members of different species from mating in the first place, or that keeps fertilization from occurring if they do mate.

Principal quantum number, *n* An integer that specifies the quantized energy level of an atomic orbital.

Principle of Falsifiability For a hypothesis to be considered scientific it must be testable—it must, in principle, be capable of being proven wrong.

Principle of Uniformitarianism The present is the key to the past.

Probability cloud The pattern of electron positions plotted over time to show the likelihood of an electron's being at a given position at a given time.

Producer An organism that lives by making organic molecules from inorganic materials and energy.

Product A new material formed in a chemical reaction, appearing after the arrow in a chemical equation.

Projectile Any object that moves through the air or through space under the influence of gravity.

Prokaryotes Organisms whose cells lack a nucleus and organelles, including bacteria and archaea.

Proteins Organic molecules composed of strings of carefully folded amino acids.

Protists Eukaryotic organisms that are not plants, animals, or fungi.

Proton A positively charged particle in an atomic nucleus.

Pseudoscience A theory or practice that is considered to be without scientific foundation but purports to use the methods of science.

Pulsar A celestial object (most likely a neutron star) that spins rapidly, sending out short, precisely timed bursts of electromagnetic radiation.

Punctuated equilibrium The theory that species do not change very much over long periods of time and then change a lot suddenly, during speciation.

Quantum A small, discrete packet of light energy.

Quantum hypothesis The idea that light energy is contained in discrete packets called quanta.

Quasar A quasi-stellar object; a small but powerful source of energy believed to be the active core of a very distant galaxy.

Rad A unit of absorbed energy (*r*adiation *a*bsorbed *d*ose).

Radiation The transfer of energy by means of electromagnetic waves.

Radioactivity The process whereby unstable atomic nuclei break down and emit radiation.

Radiometric dating A method for calculating the age of geologic materials based on the nuclear decay of naturally occurring radioactive isotopes.

Reactant A starting material in a chemical reaction, appearing before the arrow in a chemical equation.

Receptor A protein that binds to and "receives" a chemical messenger molecule.

Recessive The allele that is not expressed in a heterozygote.

Recombination The production of new combinations of genes different from those found in the parental chromosomes as a result of crossing over during meiosis.

Recrystallization The growth of new mineral crystals at the expense of old ones.

Red blood cells The oxygen-carrying cells found in blood.

Red giant Cool giant stars above main-sequence stars on the H–R diagram.

Reduction The process whereby a reactant gains one or more electrons.

Reflection The returning of a wave to the medium from which it came when it hits a barrier.

Refraction The bending of waves due to a change in the medium.

Relationship of impulse and momentum Impulse is equal to the change in the momentum of the object upon which the impulse acts. In symbolic notation,

$$Ft = \Delta mv$$

Relative dating The ordering of rocks in sequence by their comparative ages.

Relative humidity The ratio of the amount of water vapor currently in the air compared to the largest amount of water vapor that the air can hold at that temperature.

Rem The unit of measure for radiation dosage based on potential damage (roentgen equivalent man).

Replication The process through which a DNA molecule is copied.

Resonance A dramatic increase in the amplitude of a wave that results when the frequency of forced vibrations matches an object's natural frequency.

Resultant The net result of a combination of two or more vectors.

Retina The portion of the eye that contains the light-sensitive cells for sight, the rods and cones.

Ribosomes An organelle in which proteins are assembled.

RNA Ribonucleic acid, a single-stranded molecule consisting of a sugar-phosphate backbone attached to a series of nitrogenous bases.

Rock A consolidated mixture of two or more minerals.

Rock cycle A model that summarizes the formation, breakdown, and re-formation of rock as a result of igneous, sedimentary, and metamorphic processes.

Rough endoplasmic reticulum A eukaryotic organelle studded with ribosomes that assembles proteins for the cell membrane and for export out of the cell.

Runoff Precipitation not absorbed by the ground that moves over Earth's surface.

Salt An ionic compound formed from the reaction between an acid and a base.

Sarcomere The contractile unit of muscle.

Satellite A projectile or small body that orbits a larger body.

Saturated A term describing a solution containing the maximum amount of solute that will dissolve.

Saturated hydrocarbon A hydrocarbon containing no multiple covalent bonds, with each carbon atom bonded to four other atoms.

Scaler quantity A quantity, such as mass, volume, speed, and time, that can be completely specified by its magnitude.

Science The collective findings of humans about nature and a process of gathering and organizing knowledge about nature.

Scientific method An orderly method for gaining, organizing, and applying new knowledge.

Seafloor spreading The process by which new lithosphere forms at a midocean ridge.

Seamount Undersea volcanic peaks.

Second Law of Thermodynamics Heat never spontaneously flows from a low-temperature substance to a high-temperature substance. Also, all systems tend to become more and more disordered as time goes by.

Sedimentary rock Rocks formed from the accumulation of weathered material (sediments) carried by water, wind, or ice.

Sedimentation The process of layering and lithification of sediments that produces sedimentary rock.

Sediments Unconsolidated particles obtained from preexisting rock, chemical precipitation, or by the secretions of organisms.

Seed In seed plants, a structure consisting of a sporophyte plant embryo, a food supply, and a tough outer coating.

Seismology The study and measurement of seismic waves.

Semiconductor A material that can be made to sometimes behave as an insulator and sometimes as a conductor.

Sensory neuron A neuron that carries messages from sense receptors to the central nervous system.

Series circuit An electric circuit with devices connected in such a way that the same electric current flows through each of them.

Sex-linked traits Traits determined by genes found on sex chromosomes.

Shell A set of overlapping atomic orbitals of similar energy levels; in other words, a region of space in which electrons of similar energy levels in an atom have a 90 percent chance of being located.

Silicate A mineral that contains both silicon and oxygen and perhaps other elements.

Single nucleotide polymorphism (SNP) A location in the human genome where the base-pair sequence differs among human beings.

Smooth endoplasmic reticulum A eukaryotic organelle that assembles membranes and other lipids, in addition to performing other functions.

Solar eclipse The phenomenon whereby the shadow of the Moon falls upon Earth, producing a region of darkness in the daytime.

Solar radiation Electromagnetic energy emitted by the Sun.

Solid Matter that has a definite volume and a definite shape.

Solubility The ability of a solute to dissolve in a given solvent.

Soluble Capable of dissolving to an appreciable extent in a given solvent.

Solute Any component in a solution that is not the solvent.

Solution A homogeneous mixture in which all components are dissolved in the same phase.

Solvent The component in a solution present in the largest amount.

Speciation The formation of new species.

Species A group of organisms whose members can interbreed among themselves, but not with members of other species.

Specific heat capacity The quantity of heat per unit mass required to raise the temperature of a substance by 1 degree Celsius.

Spectroscope A device that uses a prism or a diffraction grating to separate light into its component colors.

Speed The distance traveled per unit of time.

Spiral galaxy A disk-shaped galaxy with hot, bright stars and spiral arms. Our Milky Way is a spiral galaxy.

Stationary front A front along which neither of the meeting air masses is advancing.

Storm Violent and rapid change in weather.

Strata Layers of sediment that have been deposited and fused over geologic time.

Strong force The powerful force that attracts nucleons to one another over a short distance.

Structural isomers Molecules that have the same molecular formula but different chemical structures.

Subduction zone The area around a convergent plate boundary where one lithospheric plate descends beneath another, returning to the interior of the Earth.

Sublimation The change of phase of a solid directly to a gas.

Submicroscopic The realm of atoms and molecules, where objects are smaller than can be detected by optical microscopes.

Supernova The explosion of a massive star caused by gravitational collapse with the emission of enormous quantities of matter.

Superposition A relative dating principle stating that in an undeformed sequence of sedimentary rocks, each bed or layer is older than the one above and younger than the one below.

Support force The force that supports an object against gravity, often called the normal force.

Surface current Wind-driven shallow ocean currents.

Symbiosis A situation in which individuals of two species live in close association with one another.

Sympatric speciation Speciation that occurs without the introduction of a geographic barrier.

Synapse A connection between a neuron and its target cell.

T cell A cell of the acquired immune system that targets pathogens inside the body's cells.

Tangential velocity Velocity that is parallel to (tangent to) a curved path.

Technology The means of solving practical problems by applying the findings of science.

Tectonic plate A piece of the Earth's rigid outer shell, the lithosphere.

Temperature A measure of the hotness or coldness of substances, related to the average translational kinetic energy per molecule in a substance; measured in degrees Celsius, or in degrees Fahrenheit, or in kelvins.

Terminal speed The speed at which the acceleration of a falling object terminates when air resistance balances weight.

Theory A synthesis of a large body of information that encompasses well-tested hypotheses about certain aspects of the natural world.

Thermal (internal) energy The total energy (kinetic plus potential) of the submicroscopic particles that make up a substance.

Thermodynamics The study of heat and its transformation to different forms of energy.

Thermonuclear fusion Nuclear fusion produced by high temperature.

Third Law of Thermodynamics No system can reach absolute zero.

Tissue A group of similar cells that performs a certain function.

Transcription The creation of a molecule of RNA from a DNA template, an intermediate step in building a protein from DNA.

Transfer RNA (tRNA) A form of RNA that transfers the appropriate amino acid to a protein being built.

Transform boundary A plate boundary at which the neighboring plates slide past each other, neither converging nor diverging; lithosphere is neither created nor destroyed.

Translation The assembly of a protein based on directions from an RNA transcript.

Transmutation The conversion of an atomic nucleus of one element into an atomic nucleus of another element through a loss or gain in the number of protons.

Transparent The term applied to materials through which light can pass in straight lines.

Transportation The movement of eroded material.

Transverse wave A wave in which the medium vibrates in a direction perpendicular (transverse) to the direction in which the wave travels.

Trophic level One of the different feeding levels in a food chain, including producers, primary consumers, secondary consumers, tertiary consumers, and so forth.

Troposphere The layer of atmosphere closest to the Earth's surface.

Tsunami A seismic sea wave caused by an earthquake.

Unconformity A break or gap in the geologic record, caused by an interruption in the sequence of deposition or by erosion of preexisting rock.

Universal constant of gravitation, *G* The proportionality constant in Newton's law of gravitation.

Unsaturated Describes a substance in which more solute can dissolve.

Unsaturated hydrocarbon A hydrocarbon containing at least one multiple covalent bond.

Upwarped mountain A dome-shaped mountain produced by a broad arching of Earth's crust.

Valence electron An electron that is located in the outermost occupied shell in an atom and can participate in chemical bonding.

Valence shell The outermost occupied shell of an atom.

Variation Differences in a trait from one individual to another.

Vector An arrow whose length represents the magnitude of a quantity, and whose direction represents the direction of the quantity.

Vector components Parts into which a vector can be separated and that act in different directions from the vector.

Vector quantity A quantity that specifies direction as well as magnitude.

Vein A blood vessel that carries blood to the heart.

Velocity The speed of an object along with specification of its direction of motion.

Volcano A hill or mountain formed by the extrusion of lava, ash, and rock fragments.

Warm front A front along which a warm air mass overrides a retreating mass of cooler, denser air.

Water table The upper boundary of the saturated zone.

Watershed The area of land that drains into a stream.

Wave A disturbance that travels from one place to another transporting energy, but not necessarily matter, along with it.

Wavelength The distance from the top of one crest to the top of the next one or, equivalently, the distance between successive identical parts of the wave.

Weather The state of the atmosphere at a particular time and place.

Weathering Slow changes by mechanical or chemical agents at or near Earth's surface that disintegrate or decompose rock.

Weight Simply stated, the force of gravity on an object; more specifically, the gravitational force with which a body presses against a supporting surface.

Weightlessness A condition encountered in free fall wherein a support force is lacking.

White blood cells Immune cells found in blood.

White dwarf A dying star that has collapsed to the size of Earth and is slowly cooling off; located at the lower left on the H–R diagram.

Wind Horizontally flowing air in the atmosphere from a region of high pressure to low pressure.

Work The product of the force and the distance through which the force moves:

$$W = Fd$$

Work–energy theorem The work done on an object equals the change in kinetic energy of the object:

$$\text{work} = \Delta KE$$

Credits

Art Credits

Chapter 16
Figure 16.23 Adapted figure from the EXACT Sciences Corporation, © 2003.

Chapter 17
Figure 17.32 Adapted figure reprinted with permission from the UC Museum of Paleontology's *Understanding Evolution* (http://evolution.berkeley.edu).
Figure 17.37 Adapted figure reprinted with permission from Dr. Donald Levin, University of Texas, Austin.
Chapter 19, Figure A, *Technology: Visual Prostheses for the Blind.* Adapted figure reprinted with permission of USC Biomimetic MicroElectronic Systems.

Chapter 20
Figure 20.22 Adapted figure reprinted with permission from Jo Oreo, BiologyMad.com.

Chapter 21
Figure 21.4 Adapted figure from *Environmental Effects of Ozone Depletion: 1991 Update,* © 1991 by the United Nations Environment Programme.
Figure 21.22 Reprinted with permission from Encyclopaedia Britannica, © 1996 by Enclopaedia Britannica, Inc.
Chapter 21, Problem 22: *Survivorship, India vs. Great Britain (1999) vs. 17th-C. England.* Reprinted with permission from Keith Montgomery, University of Wisconsin, Marathon County.

Chapter 22
Figure 22.8 Reprinted with permission from the U.S. Geological Survey.
Chapter 22, Exercise 28: Reprinted with permission from the U.S. Geological Survey.
Chapter 22, Problem 4: Reprinted with permission from the U.S. Geological Survey.

Chapter 23
Figure 23.11 Adapted figure reprinted with permission from the Mineral Information Institute.

Photo Credits

Part Openers: Part 1 p 15 Leslie Hewitt; **Part 2 p 215** Suzanne Lyons; **Part 3 p 317** Darren Robert Yee; **Part 4 p 503** Paul G. Hewitt; **Part 5 p 625** Suzanne Lyons

Chapter 1: Opening Photo NASA; **1.1** CORBIS; **1.2 (right)** University of Alabama, **(left)** Jim McClintock; **1.3 (a,b)** Jim Mastro

Chapter 2: Opening Photo Nature Picture Library, Doug Perrine; **p 18** Bettmann/CORBIS; **History of Science p 19** Alinari/Art Resource, Artist: Justus Sustermans; **2.10** Paul G. Hewitt; **2.16** Omni-Photo Communications, Inc., Tom Stillo; **2.17** Terry Murphy/Animals Animals/Earth Scenes; **2.18** Alan Schein Photography/CORBIS; **p 28** Suzanne Lyons, **p 32** NBAE/Getty Images

Chapter 3: Opening Photo Larry Seargeant/Comstock Images; **History of Science p 38** Erich Lessing/Art Resource. Artist: German School; **3.11** John Dalton/Photo Researchers, Inc.; **3.12** Fundamental Photographs; **3.26** Paul G. Hewitt; **History of Science p 53** Giraudon/Art resource. Artist: Godfrey Kneller; **p 55** Steve Hewitt

Chapter 4: Opening Photo Peter Arnold, Inc., Tom Murphy; **4.3** Dorling Kindersley; **4.7** Paul G. Hewitt; **Math Connection p 63** Paul G. Hewitt; **4.13** CORBIS; **4.15** Charles Krupa/AP; **4.16** Paul G. Hewitt; **4.21** Paul G. Hewitt; **4.22 (a, b)** Michael Vollmer

Chapter 5: Opening Photo NASA; **5.16** NASA; **5.37** Pearson Benjamin Cummings; **5.40** NASA, Goddard Space Flight Center

Chapter 6: Opening Photo Brad Lewis/Omjalla Images/G. Brad Lewis; **6.3** Kasia Werel, Collection of Paul G. Hewitt; **6.10** Suzanne Lyons; **6.11** Getty Images; **6.16** Paul G. Hewitt; **6.17** Paul G. Hewitt, Meidor Hu; **6.19** Paul G. Hewitt; **6.25** Tracy Suchocki; **6.26** Don Hynek, Wisconsin Division of Energy; **6.28** Paul G. Hewitt; **6.29** Paul G. Hewitt

Chapter 7: Opening Photo Steven Hunt/Getty Images; **7.13** Zig Leszczynski/Animals Animals; **7.14** Pearson Addison Wesley; **7.18** Paul G. Hewitt; **7.21** Pearson Education; 7.22 Pearson Education; **7.23** Pearson Education; **7.26** George Haling/Photo Researchers, Inc.; **7.27 (a, b)** Richard Megna/Fundamental Photographs; **7.33 (a, b, c)** Richard Megna/Fundamental Photographs; **7.35** John Downer/naturepl.com; **7.41 (left, right)** Pearson Addison Wesley, **p 141** Paul G. Hewitt

Chapter 8: Opening Photo Getty Images, S. Cazenave/Vandy; **8.2** The Image Bank/Getty Images; **8.8 (a)** AP/World Wide Photos; **8.8 (b)** CORBIS; **(c)** AP/World Wide Photos; **8.15** Paul G. Hewitt; **8.17** Roger Ressmeyer/CORBIS; **8.18** Institute of Paper Science and Technology; **8.24** Suzanne Lyons; **8.26** Suzanne Lyons; **8.42 (a, b, c)** Education Development Center Inc.; **8.44** Ken Kay/Fundamental Photographs; **8.46 (top, bottom)** Richard Megna/Fundamental Photographs; **p 165** Getty Images

Chapter 9: Opening Photo Driscoll, Youngquist & Baldeschwieler, CALTECH/Photo Researchers, Inc.; **9.3 (a)** Charles D. Winters/Photo Researchers, Inc., **(b)** Charles D. Winters/Photo Researchers, Inc., **(c)** Andrew Lambert Photography/Photo Researchers, Inc., **(d)** Andrew Lambert Photography/Photo Researchers, Inc.; **9.4** John Suchocki/Lilian Lee; **Technology p 170 (left)** National Institute of Standards and Technology; **(center)** IBM/Peter Arnold, Inc.; **(right)** IBM Research, Almaden Research Center; **9.6** Suzanne Lyons; **9.9 (a)** Pearson Benjamin Cummings, **(b)** UK Meteorological Office/Photo Researchers, Inc; **9.10 (left)** Phototake NYC; **9.10 (right)** John Suchocki; **9.11** John Suchocki; **9.12** (flame test: top of four), (flame test: second of four) Tom Bochsler/Prentice Hall, Inc., (flame test: third of four) Andrew Lambert Photography/Photo Researchers, Inc., (flame test: fourth of four) Dorling Kindersley, (spectra: all) Alan J. Jircitano; **9.13** Science Photo Lab/Photo Researchers Inc.; **9.21 (a: left, center, right)** John Suchocki; **9.25** John Suchocki; **9.26 (a, b)** John Suchocki; **p 188 (left)** John Suchocki; **p 188 (right)** Paul G. Hewitt

Chapter 28: Opening Photo Mark Garlick/Photo Researchers, Inc.; **28.5** Roger Ressmeyer/CORBIS; **28.7** NASA; **28.10** NASA; **28.12** NASA; **28.13** Mark Garlick/Photo Researchers, Inc.; **28.14** Ronald Royer/Photo Researchers, Inc.; **28.15 (a, b)** Lick Observatory; **Technology p 659 (left)** David Nunuk/Photo Researchers, Inc., **(right)** Chris Butler/Photo Researchers, Inc.; **28.17** Julian Baum/New Scientist/JPL/Photo Researchers, Inc.; **28.18** NASA; **28.19** Dr. Steve Gull/SPL/Photo Researchers, Inc.; **28.20 (a)** National Optical Astronomy Observatories, **(b)** Royal Observatory, Edinburgh/SPL/Photo Researchers, Inc.; **28.21** NRAO/SPL/Photo Researchers, Inc.; **28.22** NASA; **28.23** Hale Observatory/Photo Researchers, Inc.; **28.25** Emilio Segre Visual Archives/American Institute of Physics/Photo Researchers, Inc.; **28.26** Royal Greenwich Observatory/SPL/Photo Researchers, Inc.; **28.27** Christian Jegou, Publiphoto Diffusion/ Photo Researchers, Inc.

Index

Page numbers followed by "n" indicate a footnote reference.

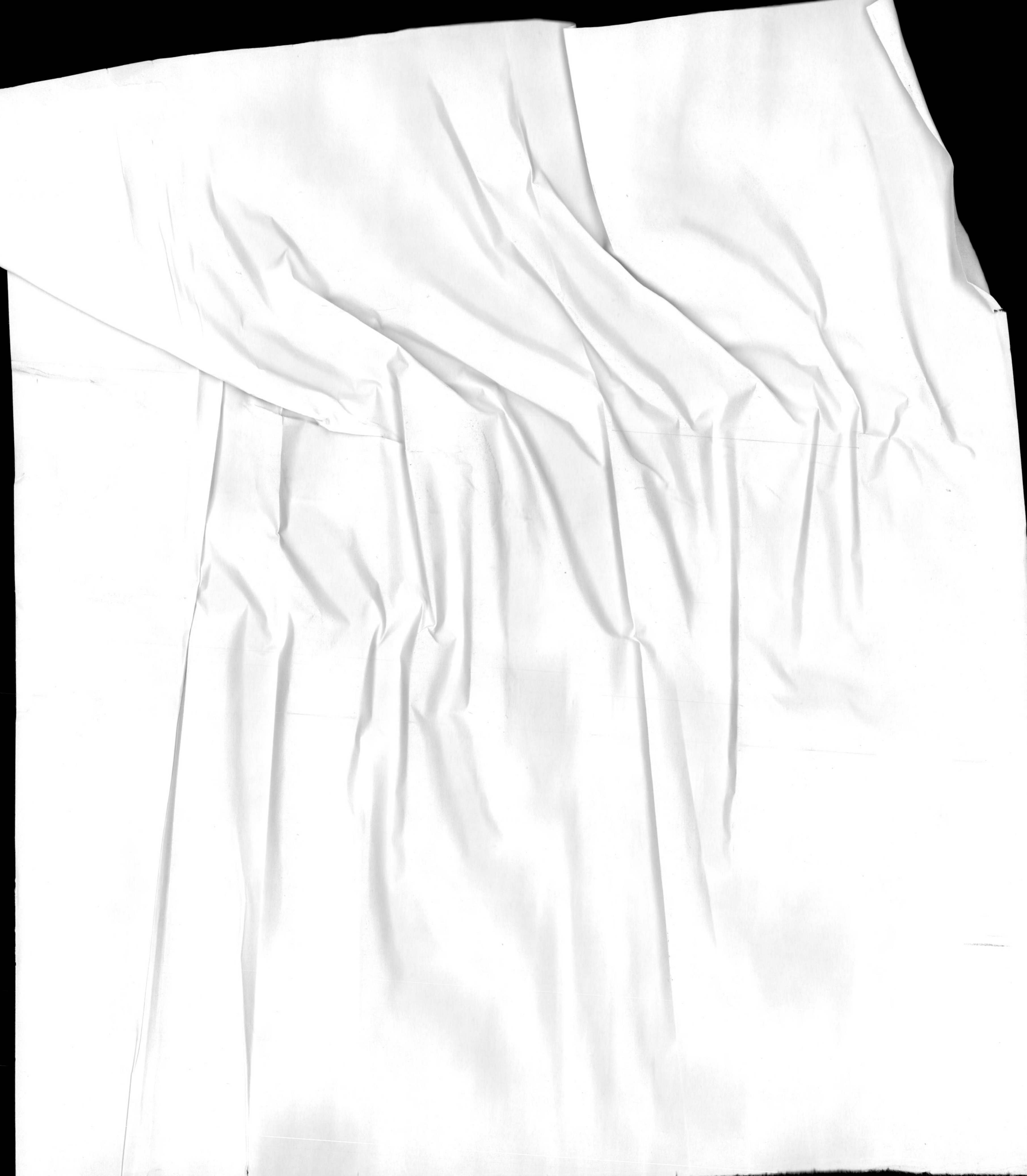